Automotive Chassis Systems

Online Services

Delmar Online
To access a wide variety of Delmar products and services on the World Wide Web, point your browser to:
http://www.delmar.com
or email: info@delmar.com

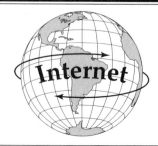

Automotive Chassis Systems

Thomas W. Birch

Delmar
Thomson Learning™

Africa • Australia • Canada • Denmark • Japan • Mexico
New Zealand • Philippines • Puerto Rico • Singapore
Spain • United Kingdom • United States

NOTICE TO THE READER

Publisher does not warrant or guarantee any of the products described herein or perform any independent analysis in connection with any of the product information contained herein. Publisher does not assume, and expressly disclaims, any obligation to obtain and include information other than that provided to it by the manufacturer.

The reader is expressly warned to consider and adopt all safety precautions that might be indicated by the activities herein and to avoid all potential hazards. By following the instructions contained herein, the reader willingly assumes all risks in connection with such instructions.

The publisher makes no representation or warranties of any kind, including but not limited to, the warranties of fitness for particular purpose or merchantability, nor are any such representations implied with respect to the material set forth herein, and the publisher takes no responsibility with respect to such material. The publisher shall not be liable for any special, consequential, or exemplary damages resulting, in whole or part, from the readers' use of, or reliance upon, this material.

Delmar Staff:
Business Unit Director: Alar Elken
Executive Editor: Sandy Clark
Acquisitions Editor: Vernon R. Anthony
Developmental Editor: Denise Denisoff
Editorial Assistant: Bridget Morrison
Executive Marketing Manager: Maura Theriault
Channel Manager: Mona Caron

Marketing Coordinator: Kasey Young
Executive Production Manager: Mary Ellen Black
Production Coordinator: Karen Smith
Project Editor: Barbara Diaz
Art/Design Coordinator: Cheri Plasse

COPYRIGHT © 2000
Delmar is a division of Thomson Learning. The Thomson Learning logo is a registered trademark used herein under license.

Printed in the United States of America
4 5 6 7 8 9 10 XXX 08 07 06 05 04

For more information, contact Delmar Publishers at 3 Columbia Circle, PO Box 15015, Albany, New York 12212-5015; or find us on the World Wide Web at http://www.delmar.com

Asia
Thomson Learning
60 Albert Street, #15-01
Albert Complex
Singapore 189969

Australia/New Zealand
Nelson/Thomson Learning
102 Dodds Street
South Melbourne, Victoria 3205
Australia

Canada
Nelson/Thomson Learning
1120 Birchmont Road
Scarborough, Ontario
Canada M1K 5G4

International Headquarters
Thomson Learning
Scarborough, Ontario
International Division
290 Harbor Drive, 2nd Floor
Stamford, CT 06902-7477
USA

Japan
Thomson Learning
Palaceside Building 5F
1-1-1 Hitotsubashi, Chiyoda-ku
Tokyo 100 0003
Japan

Latin America
Thomson Learning
Seneca, 53
Colonia Polanco
11560 Mexico D. F. Mexico

South Africa
Thomson Learning
Zonnebloem Building
Constantia Square
526 Sixteenth Road
P.O. Box 2459
Halfway House, 1685
South Africa

Spain
Thomson Learning
Calle Magallanes, 25
28015-Madrid
Espana

UK/Europe/Middle East
Thomson Learning
Berkshire House
168-173 High Holborn
London
WC1V 7AA United Kingdom

Thomas Nelson & Sons Ltd.
Nelson House
Mayfield Road
Walton-on-Thames
KT 12 5PL United Kingdom

All rights reserved Thomson Learning © 2000. The text of this publication, or any part thereof, may not be reproduced or transmitted in any form or by any means, electronic or mechanical, including photocopying, recording, storage in an information retrieval system, or otherwise, without prior permission of the publisher.

You can request permission to use material from this text through the following phone and fax numbers.
Phone: 1-800-730-2214; Fax: 1-800-730-2215; or visit our Web site at http://www.thomsonrights.com

Library of Congress Cataloging-in-Publication Data

Birch, Thomas W. (Thomas Wesley), 1933–
    Automotive chassis systems / Thomas W. Birch.
      p. cm.
    Includes index.
    ISBN 0-7668-0001-6 (alk. paper)
    1. Automobiles—Chassis. I. Title.
TL255.B54 1999
629.2'4—dc21   99-26156
                    CIP

Contents

Preface xiii

Chapter 1
Introduction to the Chassis 1

Objectives 1
1.1 Introduction 1
1.2 Chassis Components—Their General Purpose 1
1.3 Brake Systems 1
1.4 Brake System Components 2
1.5 Brake System Repair 9
1.6 Practice Diagnosis 10
1.7 Suspension and Steering Systems 10
1.8 Chassis Components 11
1.9 Nonindependent Suspensions 12
1.10 Independent Suspensions................ 12
1.11 Swing Axle Suspension 12
1.12 Short-Long Arm Suspension/S-L A 13
1.13 MacPherson Strut Suspension/Strut Suspension 14
1.14 Multilink Suspension.................... 15
1.15 Trailing Arm Suspension 16
1.16 Control Arms 16
1.17 Suspension Bushings.................... 17
1.18 Ball Joints 18
1.19 Kingpins............................... 18
1.20 Springs and Shock Absorbers 19
1.21 Spring Types 19
1.22 Steering System........................ 22
1.23 Steering Gears 22
1.24 Steering Linkage 22
1.25 Practice Diagnosis 23
Terms to Know 23
Review Questions 24

Chapter 2
Principles of Braking 27

Objectives 27
2.1 Introduction and Federal Requirements 27
2.2 Deceleration........................... 28
2.3 Energy of Motion....................... 31
2.4 Friction and Heat Energy................ 31
2.5 Heat Energy 32
2.6 Brake Lining Friction 33
2.7 Tire Friction........................... 37
2.8 Brake Power 38
2.9 Weight Transfer 41
2.10 Skids 43
2.11 Stopping Sequence 44
2.12 Antidive Suspensions 44
2.13 Practice Diagnosis 44
Terms to Know 46
Review Questions 46

Chapter 3
Drum Brake Theory 48

Objectives 48
3.1 Introduction 48
3.2 Shoe Energization and Servo Action 49
3.3 Brake Shoes 51
3.4 Lining Attachment...................... 52
3.5 Backing Plate.......................... 52
3.6 Shoe Anchors 53
3.7 Brake Springs 54
3.8 Shoe Adjusters 56
3.9 Brake Drums 58
3.10 Practice Diagnosis 61
Terms to Know 61
Review Questions 61

Chapter 4
Disc Brake Theory — 63

Objectives 63
4.1 Introduction 63
4.2 Calipers 64
4.3 Brake Pads 69
4.4 Caliper and Pad Mounting Hardware 70
4.5 Rotors 72
4.6 Rear-Wheel Disc Brakes 74
4.7 Practice Diagnosis 75
Terms to Know 75
Review Questions 75

Chapter 5
Parking Brake Theory — 77

Objectives 77
5.1 Introduction 77
5.2 Cables and Equalizer 77
5.3 Lever and Warning Light 79
5.4 Integral Drum Parking Brake 80
5.5 Auxiliary Drum Parking Brake 81
5.6 Disc Parking Brake 81
5.7 Practice Diagnosis 83
Terms to Know 83
Review Questions 83

Chapter 6
Hydraulic System Theory — 85

Objectives 85
6.1 Introduction 85
6.2 Hydraulic Principles 85
6.3 Master Cylinder Basics and Components ... 89
6.4 Master Cylinder Basic Operation 89
6.5 Master Cylinder Construction 90
6.6 Residual Pressure 92
6.7 Dual Master Cylinder Construction 92
6.8 Tandem Master Cylinder Operation 94
6.9 Diagonally Split Hydraulic System 96
6.10 Quick-Take-Up Master Cylinder 97
6.11 Miscellaneous Master Cylinder Designs 98
6.12 Wheel Cylinder Basics 99
6.13 Wheel Cylinder Construction 99
6.14 Caliper Basics 101
6.15 Caliper Pistons and Sealing Rings 101
6.16 Caliper Boot 103
6.17 Caliper Body and Bleeder Screw 105
6.18 Tubing and Hoses 105
6.19 Hydraulic Control Valves and Switches ... 108
6.20 Brake Fluid 115
6.21 Practice Diagnosis 119
Terms to Know 119
Review Questions 119

Chapter 7
Power Brake Theory — 122

Objectives 122
7.1 Introduction 122
7.2 Vacuum Booster Theory 123
7.3 Vacuum Booster Construction 125
7.4 Vacuum Booster Operation 128
7.5 Hydraulic Booster Theory 129
7.6 Hydraulic Booster Construction 130
7.7 Hydraulic Booster Operation 130
7.8 Electrohydraulic Booster Theory 131
7.9 Electrohydraulic Booster Construction 133
7.10 Electrohydraulic Booster Operation 133
7.11 Practice Diagnosis 133
Terms to Know 135
Review Questions 135

Chapter 8
Antilock Braking System Theory — 137

Objectives 137
8.1 Introduction 137
8.2 Electronic Wheel Speed Sensors 140
8.3 Electronic Controller 142
8.4 Electronic Modulator 144
8.5 Recirculation Pump 146
8.6 Antilock Braking Systems 148
8.7 ABS Brake Fluid 154
8.8 Automatic Traction Control 154
8.9 Practice Diagnosis 155
Terms to Know 155
Review Questions 156

Contents

Chapter 9
Tire Theory   157

Objectives 157
9.1 Introduction 157
9.2 Tire Construction 158
9.3 Tire Sidewall Information 163
9.4 Tire Sizing 163
9.5 Tire Load Ratings 165
9.6 Uniform Tire Quality Grading Labeling 166
9.7 Speed Ratings 167
9.8 Tubes and Tubeless Tires 168
9.9 Replacement Tire Selection 170
9.10 Spare Tires 170
9.11 No-Flat, Run-Flat Tires 171
9.12 Retreads 173
9.13 Practice Diagnosis 174
Terms to Know 174
Review Questions 175

Chapter 10
Wheel Theory   177

Objectives 177
10.1 Introduction 177
10.2 Construction 177
10.3 Wheel Sizes 179
10.4 Offset and Backspacing 180
10.5 Aftermarket Wheels 181
10.6 Wheel Attachment 182
10.7 Practice Diagnosis 185
Terms to Know 185
Review Questions 185

Chapter 11
Wheel Bearing Theory   187

Objectives 187
11.1 Introduction 187
11.2 Bearings and Bearing Parts 187
11.3 Bearing End Play and Preload 189
11.4 Seals 190
11.5 Serviceable Nondrive Axle Wheel Bearings . . 191
11.6 Nonserviceable Nondrive Axle
      Wheel Bearings 191
11.7 Drive Axle Wheel Bearing, Solid Axle 191
11.8 Drive Axle Wheel Bearing, Independent
      Suspension 193
11.9 Nonserviceable Drive Axle Wheel Bearings . . 194
11.10 Practice Diagnosis 195
Terms to Know 195
Review Questions 195

Chapter 12
Front-Wheel-Drive Driveshaft Theory   197

Objectives 197
12.1 Introduction 197
12.2 Driveshaft Construction 197
12.3 Practice Diagnosis 199
Terms to Know 201
Review Questions 201

Chapter 13
Front Suspension Types   202

Objectives 202
13.1 Introduction 202
13.2 Short-Long Arm Suspensions 203
13.3 Multilink Suspensions 206
13.4 Strut Suspensions 206
13.5 Solid Axles 210
13.6 Swing Axles, Twin I-Beam Axles 210
13.7 Miscellaneous Suspension Types 212
13.8 Front-Wheel Drive Axles 212
13.9 Practice Diagnosis 213
Terms to Know 213
Review Questions 214

Chapter 14
Rear Suspension Types   216

Objectives 216
14.1 Introduction 216
14.2 Solid Axle, RWD Suspension 216
14.3 Independent Rear Suspension 219
14.4 Miscellaneous RWD Axle and Suspension
      Types 222
14.5 FWD Rear Axles 223
14.6 Rear-Wheel Steering 224
14.7 Practice Diagnosis 226
Terms to Know 227
Review Questions 227

Chapter 15
Springs 229

Objectives...229
- 15.1 Introduction.....................................229
- 15.2 Sprung and Unsprung Weight..............230
- 15.3 Spring Rate and Frequency................230
- 15.4 Wheel Rate and Frequency................232
- 15.5 Spring Materials...............................233
- 15.6 Leaf Springs....................................235
- 15.7 Coil Springs....................................236
- 15.8 Torsion Bars...................................239
- 15.9 Electronically Controlled Air Suspensions...240
- 15.10 Active Hydraulic Suspensions.............243
- 15.11 Aftermarket Air Suspensions..............246
- 15.12 Overload Springs.............................246
- 15.13 Stabilizer or Antiroll Bar....................246
- 15.14 Practice Diagnosis............................248

Terms to Know..249
Review Questions....................................250

Chapter 16
Shock Absorbers 252

Objectives...252
- 16.1 Introduction.....................................252
- 16.2 Shock Absorber Operating Principles.....253
- 16.3 Shock Absorber Damping Ratios..........254
- 16.4 Shock Absorber Damping Force...........255
- 16.5 Double-Tube Shock Absorbers.............256
- 16.6 Monotube Shock Absorber Operation....259
- 16.7 Strut Shock Absorbers......................261
- 16.8 Computer-Controlled Shock Absorbers..261
- 16.9 Load-Carrying Shock Absorbers...........264
- 16.10 Shock Absorber Quality....................264
- 16.11 Shock Absorber Failures..................265
- 16.12 Practice Diagnosis..........................265

Terms to Know..266
Review Questions....................................266

Chapter 17
Steering Systems 269

Objectives...269
- 17.1 Introduction.....................................269
- 17.2 Airbag or Supplemental Restraint System....271
- 17.3 Steering Columns.............................273
- 17.4 Standard Steering Gears....................276
- 17.5 Rack-and-Pinion Steering Gears..........278
- 17.6 Power Steering................................280
- 17.7 Steering Linkage.............................291
- 17.8 Drag Links......................................297
- 17.9 Four-Wheel Steering........................298
- 17.10 Practice Diagnosis..........................298

Terms to Know..300
Review Questions....................................300

Chapter 18
Chassis Electrical and Electronic Systems 303

Objectives...303
- 18.1 Introduction.....................................303
- 18.2 Basic Electricity...............................303
- 18.3 Basic Electronics..............................305
- 18.4 Electrical Circuit Problems..................307
- 18.5 Measuring Electrical Values................308
- 18.6 Interpreting Readings........................311
- 18.7 Electronically Controlled System Diagnosis...313
- 18.8 Practice Diagnosis............................315

Terms to Know..319
Review Questions....................................319

Chapter 19
Wheel Alignment: Front and Rear 321

Objectives...321
- 19.1 Introduction.....................................321
- 19.2 Measuring Angles.............................322
- 19.3 Camber..322
- 19.4 Camber and Scrub Radius..................324
- 19.5 Camber-Caused Tire Wear.................325
- 19.6 Camber Spread and Road Crown.........325
- 19.7 Camber Change...............................326
- 19.8 Toe-In..326
- 19.9 Factors That Affect Toe.....................327
- 19.10 Toe-Caused Tire Wear.....................327
- 19.11 Toe Change..................................328
- 19.12 Caster..328

| Contents | ix |

19.13 Caster Effects.................329
19.14 Double Ball Joint Steering Axis.........330
19.15 Factors That Affect Caster............330
19.16 Caster and Road Crown..............331
19.17 Effects of Too Little or Too Much Caster....331
19.18 Steering Axis Inclination.............332
19.19 Included Angle...................333
19.20 Toe-Out on Turns..................334
19.21 Toe-Out on Turns: Problems...........336
19.22 Setback........................336
19.23 Rear-Wheel Alignment..............337
19.24 Track or Thrust Line................337
19.25 Practice Diagnosis.................338
Terms to Know......................338
Review Questions...................338

Chapter 20
Brake System Service   341

Objectives.........................341
20.1 Introduction.....................341
20.2 Brake Inspection.................343
20.3 Troubleshooting Brake Problems......354
20.4 Brake Repair Recommendations.......359
20.5 Special Notes on Surface Finish......359
20.6 Fastener Security................361
20.7 Practice Diagnosis...............363
Terms to Know....................363
Review Questions.................363

Chapter 21
Wheel Bearing Service   365

Objectives.........................365
21.1 Introduction.....................365
21.2 Repacking Serviceable Wheel Bearings.....365
21.3 Repairing Nonserviceable Wheel Bearings...374
21.4 Repairing Drive Axle Bearings, Solid Axles...................375
21.5 Repairing Serviceable FWD Front Wheel Bearings.....................381
21.6 Repairing Nonserviceable FWD Front Wheel Bearings................382
21.7 Practice Diagnosis...............384

Terms to Know....................384
Review Questions.................385

Chapter 22
Drum Brake Service   387

Objectives.........................387
22.1 Introduction.....................387
22.2 Drum Removal....................388
22.3 Drum Inspection..................393
22.4 Drum Machining..................397
22.5 Shoe Assembly: Predisassembly Cleanup.....................403
22.6 Brake Spring Removal and Replacement....404
22.7 Brake Shoe Removal...............410
22.8 Component Cleaning and Inspection.....411
22.9 Component Lubrication............414
22.10 Brake Shoe Preinstallation Checks.....415
22.11 Regrinding Brake Shoes...........416
22.12 Brake Shoe Installation..........418
22.13 Completion.....................420
22.14 Practice Diagnosis..............421
Terms to Know....................421
Review Questions.................421

Chapter 23
Disc Brake Service   424

Objectives.........................424
23.1 Introduction....................424
23.2 Removing a Caliper...............425
23.3 Rotor Inspection................431
23.4 Rotor Refinishing...............436
23.5 Off-Car Machining of a Rotor......440
23.6 On-Car Machining of a Rotor.......446
23.7 Resurfacing a Rotor..............447
23.8 Rotor Replacement...............449
23.9 Caliper Service.................450
23.10 Removing and Replacing Pads......451
23.11 Caliper Installation............457
23.12 Completion.....................459
23.13 Practice Diagnosis..............459
Terms to Know....................460
Review Questions.................460

Chapter 24
Hydraulic System Service 462

Objectives 462
24.1 Introduction 462
24.2 Working with Tubing 463
24.3 Hydraulic Cylinder Service 466
24.4 Master Cylinder Service 468
24.5 Wheel Cylinder Service 477
24.6 Caliper Service 483
24.7 Bleeding Brakes 495
24.8 Diagnosing Hydraulic System Problems ... 506
24.9 Testing a Warning Light 511
24.10 Brake Light Switch Adjustment 511
24.11 Practice Diagnosis 514
Terms to Know 514
Review Questions 514

Chapter 25
Power Booster Service 517

Objectives 517
25.1 Introduction 517
25.2 Vacuum Supply Tests 518
25.3 Booster Replacement 520
25.4 Booster Service 523
25.5 Practice Diagnosis 529
Terms to Know 530
Review Questions 530

Chapter 26
Parking Brake Service 531

Objectives 531
26.1 Introduction 531
26.2 Cable Adjustment 532
26.3 Cable Replacement 534
26.4 Parking Brake Warning Light Service 538
26.5 Practice Diagnosis 539
Terms to Know 539
Review Questions 539

Chapter 27
Antilock Braking System Service 541

Objectives 541
27.1 Introduction 541
27.2 ABS Warning Light Operation 544
27.3 Warning Light Sequence Test 544
27.4 ABS Problem Codes and Self-Diagnosis ... 546
27.5 Electrical Tests 550
27.6 Hydraulic Pressure Checks 556
27.7 Repair Operations 557
27.8 Completion 560
27.9 Practice Diagnosis 560
Terms to Know 561
Review Questions 561

Chapter 28
Suspension and Steering System Diagnosis and Inspection 563

Objectives 563
28.1 Introduction 564
28.2 NVH 564
28.3 NVH Diagnostic Procedure 569
28.4 Tire Wear Inspection 571
28.5 Tire-Related Problem Diagnosis 577
28.6 Tire Pull 577
28.7 Radial Tire Waddle 578
28.8 Vibrations 578
28.9 Tire and Wheel Runout 578
28.10 Tire Balance 582
28.11 Wheel Bearing Problems: Diagnostic Procedure 583
28.12 Suspension and Steering System Inspection 585
28.13 Spring and Shock Absorber Inspection 586
28.14 Ball Joint Checks 592
28.15 Control Arm Bushing Checks 601
28.16 Strut Rod Bushing Checks 602
28.17 Strut Checks 602
28.18 Strut Damper or Insulator Checks 603
28.19 Steering Linkage Checks 603
28.20 Steering Gear Checks 607
28.21 Power Steering Checks 608
28.22 Completion 611
28.23 Practice Diagnosis 611
Terms to Know 613
Review Questions 613

Contents

Chapter 29
Tire and Wheel Service — 618

Objectives 618
- 29.1 Introduction 618
- 29.2 Removing and Replacing a Tire and Wheel .. 618
- 29.3 Lug Bolt and Stud Replacement 620
- 29.4 Tire Rotation 623
- 29.5 Removing and Replacing a Tire on a Wheel 624
- 29.6 Removing and Replacing a Tubeless Tire Valve 628
- 29.7 Repairing a Tire Leak 629
- 29.8 Tire Truing 632
- 29.9 Tire Siping 633
- 29.10 Balancing Operations 633
- 29.11 Practice Diagnosis 641

Terms to Know 641
Review Questions 642

Chapter 30
Front-Wheel-Drive Driveshaft Service — 644

Objectives 644
- 30.1 Introduction 644
- 30.2 CV Joint Problem Diagnosis 644
- 30.3 Split CV Joint Boots 644
- 30.4 Driveshaft Removal 645
- 30.5 Driveshaft Disassembly 647
- 30.6 Servicing CV Joints 649
- 30.7 Installing CV Joint Boots 654
- 30.8 R & R ABS Tone Ring 655
- 30.9 Installing a FWD Driveshaft 656
- 30.10 Practice Diagnosis 658

Terms to Know 658
Review Questions 658

Chapter 31
Spring and Shock Absorber Service — 660

Objectives 660
- 31.1 Introduction 660
- 31.2 Removing and Replacing Shock Absorbers .. 660
- 31.3 Removing and Replacing Coil Springs 663
- 31.4 Leaf Spring Service 668
- 31.5 Torsion Bar Service 670
- 31.6 Strut Service 671
- 31.7 Stabilizer Bar Service 679
- 31.8 Completion 682
- 31.9 Practice Diagnosis 682

Terms to Know 682
Review Questions 682

Chapter 32
Suspension Component Service — 685

Objectives 685
- 32.1 Introduction 685
- 32.2 Taper Breaking 685
- 32.3 Ball Joint Replacement 690
- 32.4 Removing a Control Arm 690
- 32.5 Removing and Replacing a Ball Joint 692
- 32.6 Control Arm Bushing Replacement 696
- 32.7 Installing a Control Arm 701
- 32.8 Removing and Replacing Strut Rod Bushings 703
- 32.9 Removing and Replacing Kingpins 704
- 32.10 Rear Suspension Service 706
- 32.11 Completion 707
- 32.12 Practice Diagnosis 707

Terms to Know 708
Review Questions 708

Chapter 33
Steering System Service — 711

Objectives 711
- 33.1 Introduction 711
- 33.2 Steering Linkage Replacement 712
- 33.3 Steering Gear Service 725
- 33.4 Steering Wheel and Column Service ... 735
- 33.5 Power Steering Service Operations 739
- 33.6 Completion 752
- 33.7 Practice Diagnosis 752

Terms to Know 752
Review Questions 752

Chapter 34
Wheel Alignment: Measuring and Adjusting **755**

Objectives................................... 755
  34.1 Introduction 755
  34.2 Measuring Alignment Angles 756
  34.3 Wheel Alignment Sequence.............. 761
  34.4 Measuring Camber 762
  34.5 Camber Specifications.................. 766
  34.6 Adjusting Camber 767
  34.7 Measuring Caster...................... 777
  34.8 Adjusting Caster 781
  34.9 Measuring SAI........................ 784
  34.10 Adjusting SAI 786
  34.11 Measuring Toe-Out on Turns............ 786
  34.12 Adjusting Toe-Out on Turns 786
  34.13 Measuring Toe....................... 787
  34.14 Toe and Steering Wheel Position.......... 791
  34.15 Toe Specifications.................... 792
  34.16 Adjusting Toe........................ 792
  34.17 Rear-Wheel Alignment 797
  34.18 Measuring Rear-Wheel Camber 797
  34.19 Measuring Rear-Wheel Toe............. 798
  34.20 Measuring Rear-Wheel Track or Thrust 799
  34.21 Adjusting Rear-Wheel Camber and Toe 800
  34.22 Frame and Body Alignment.............. 801
  34.23 Road Testing and Troubleshooting 802
  34.24 Practice Diagnosis 803
Terms to Know 803
Review Questions 803

Appendix A
ASE Certification **807**

Appendix B
English–Metric Conversion **812**

Appendix C
Distance and Angular Equivalents **814**

Appendix D
Bolt Torque Tightening Chart **815**

Appendix E
Torque Tightening Chart for Line Connections and Bleeder Screws **816**

Appendix F
Shoe Size Chart **817**

Glossary...................................... 819
Index 831

Preface

Automotive Chassis Systems is designed for technicians who will be repairing the chassis or under-car systems not only of today's complex automobiles but of the even more complex cars of tomorrow. Systems covered in this book include the braking, suspension, and steering systems.

The evolution of automotive chassis systems in the past 50 years has included the development of many automotive features that are now seen in the repair shops: self-adjusting drum brakes, disc brakes, tandem hydraulic systems, diagonally split hydraulic systems, vacuum-operated boosters, hydraulically operated boosters, electrically powered boosters, antilock braking systems, self-adjusting parking brakes, ball joints, air springs, MacPherson struts, front-wheel drive, four-wheel alignment, computer-controlled air suspension, computer-controlled shock absorbers, composite spring materials, active suspensions, multilink suspensions, multi–ball joint steering axis, rack-and-pinion steering, four-wheel steering, computer-controlled power steering, driver-side airbags, and electronic steering systems. The near future promises even further evolution that will probably include electric drive-by-wire brakes, radar-operated collision-avoidance systems, and electronic drive-by-wire steering systems.

As automotive systems have become more and more complex and precise, mechanics have needed to become technicians by learning repair methods that are more exacting; this approach requires far more knowledge and skill with specialized tools and equipment than did the methods of the past. Repair shops are also changing. New-car dealerships play much the same role they always have: they employ factory-trained technicians who have the specialized knowledge, tools, and equipment to repair the most complicated features of each new model. Many dealership computers are linked to those of the vehicle manufacturer to share the latest information. But many of the service stations and independent garages of the 1950s are gone, and in their place are shops that specialize in particular repair areas, such as air-conditioning or tune-up, or in particular makes of domestic or imported cars. The repair shop of yesterday, with a mechanic who could repair any car driven through the door, is gone. The evolution of the automobile has made that shop and mechanic extinct.

Automotive Chassis Systems presents enough information about the systems of present-day cars to provide a basis for understanding, diagnosing, and repairing them. Some people will place much of this information in the category of being interesting but unnecessary; however, technicians should realize that a complete understanding of the why and how of a car's operation aids greatly in understanding why the car does not operate properly and how to go about repairing it. The highly desirable ability to solve problems is a direct result of a thorough understanding of basic operating principles. Many of our new developments are based on physical principles that have been understood for some time, and problem-solving techniques that are needed to diagnose the various types of automotive failures require a broad base of knowledge and understanding of how all these systems work. Except for computer controls, not much is really new in automotive systems.

This book describes the various styles of braking, suspension, and steering systems and their components. Changes in system design have made some repair operations faster and easier but more complex. All the National Institute for Automotive Service Excellence (ASE) braking, suspension, and steering tasks are explained, and review questions are presented in the style of ASE certification tests. The ASE task list for Suspension and Steering (Test A4) and Brakes (Test A5) are included in this text as Appendix A. The most common service and repair methods are described in a general manner.

One style of ASE question includes Technician A and Technician B discussing a vehicle service problem. Because these two technicians normally discuss service-related problems only, questions with Statement A and Statement B format are used in this text for similar questions that are based on theory.

Specialization is a definite advantage for technicians and repair shops. Rarely can any one person know everything about cars anymore; there is simply too much to know. It is also financially prohibitive for one shop to own all the special tools and equipment needed to make every possible car repair. A general repair shop, such as a dealership, might pay rent of several thousand dollars a month for a desirable location and might have a tool and equipment inventory worth well over a hundred thousand dollars. That kind of shop must do a large volume of work to stay in business. It must have technicians who can use their specialized skills and knowledge to perform complex tasks and ensure that customers' cars are operating safely and efficiently. Also, these repairs must be made correctly and completely the first time the car is worked on; motorists do not want to waste time or money to have an inadequate repair done over.

A textbook of this type is not meant to show you how to do repair operations or adjustments for specific vehicles or how to use specific pieces of equipment. These operations are covered in the printed and software tools of the trade (vehicle service manuals and equipment instruction manuals). Instead, textbooks teach the general methods of making a repair or doing a wheel alignment so you, as a future technician, will be able to understand and fully utilize the service manuals.

The most thorough and reliable sources of information about a particular car are its manufacturer's service manual and training centers. However, these sources cover only one make and sometimes only one model. Manuals are often followed up with specific service bulletins when a need for more information or a change in a new-car part or system makes them necessary. Broader coverage is available in technician service manuals issued by such publishers as Alldata, Chilton, Mitchell, and Motor or by aftermarket manufacturers, including Ammco, Bendix, CR Industries, Dana Corporation, Federal-Mogul, Hunter, Moog, TRW, Raybestos, SKF, and Wagner. These manuals have varying amounts of information for particular cars; some of it is very thorough and specific to particular cars or systems, and some of it is very general. *Automotive Chassis Systems* supplements the manuals by presenting the theoretical and practical knowledge needed for a complete understanding of the operation of these systems.

Technicians in different parts of the United States use different terms for the same item or task. Terms commonly used on the West Coast are slightly different from those used in the East. To the greatest extent possible, the terms used in this book are those used by automobile and aftermarket manufacturers and automotive technicians. Many of these terms are defined in the glossary at the end of the book.

In describing repair procedures, I have assumed that readers have a basic working knowledge of hand tools, fasteners, and general automotive repair procedures and of the safety precautions that should be exercised during general service and repair operations. Space does not permit including them in a book of this nature, which concentrates on a few specialized systems. Wise technicians know that improper use of a wrench or other hand tool, or of a bolt or nut, often leads to a more difficult job and that violation of a commonsense safety rule can cause injury to the technician, the vehicle operator, or an innocent bystander.

To the Instructor: A complete *Instructor's Guide* to accompany this text is available from Delmar Publishers. Answers to the end-of-chapter review questions can be found in the *Instructor's Guide*.

Acknowledgments

Accu Industries, Ashland, VA

Alston Engineering, Sacramento, CA

American Honda Motor Co., Inc

Amermac, Inc., Ellaville, GA

Ammco Tools, Inc., North Chicago, IL

Armstrong Tire Comp., New Haven, CT

Arn-Wood Co., Englewood, CO

Arvin Replacement Products Group—Gabriel Ride Control, Brentwood, TN

ASE, the National Institute for Automotive Service Excellence, Herndon, VA

ATE/USA, Annapolis, MD

Dave Becker, Marysville, CA

Bee Line Company, Bettendorf, IA

Bendix Brake, Allied Aftermarket Division, Jackson, TN

B.F. Goodrich, Akron, OH

Bilstein Corporation of America, San Diego, CA

Kerry Birch, Yuba City, CA

Branick Industries, Fargo, ND

Bridgestone/Firestone, Inc., Nashville, TN

Carlson Quality Brake Division, International Brake Industries, Lima, OH

Carrera Shocks, Atlanta, GA

Century Wheel Division of Precision Switching, Spring Grove, IL

Chrysler Corporation

CR Industries/Chicago Rawhide, Elgin, IL

Cosmos International, Elbow Lake, MN

Dana Corporation, Toledo, OH

Delco Moraine Division, General Motors, Dayton, OH

John M. Demko, Burmah-Castrol, Inc., Edison, NJ

Dorman Products, Cincinnati, OH

Easco/K-D Tools, Lancaster, PA

EIS Brake Parts, Berlin, CT

Preface

Everco Industries, St. Louis, MO

Forrest E. Folck, Motor Vehicle Forensic Service, San Diego, CA

Federal-Mogul Corporation, Detroit, MI

Fluke Corporation, Everett, WA

Ford Motor Company

FMC Corporation, Automotive Aftermarket Division, Conway, AR

Ken Gaal, Yuba City, CA

Gabriel Ride Control Products, Carol Stream, IL

Gates Rubber Company, Denver, CO

General Motors Corporation, Service Technology Group

Goodyear Tire & Rubber Company, Akron, OH

Gramlich Tools, Ontario, Canada

Grigg Automotive Manufacturing, Huntington Park, CA

Lee Grimes, Koni, Hebron, KY

Guldstrand Engineering, Culver City, CA

Hennessy Industries, Inc., LaVergne, TN

Hickok Incorporated, Cleveland, OH

Hunter Engineering Company, Bridgeton, MO

JFZ Engineered Products, Chatsworth, CA

Bruce D. Kirk, Smartsville, CA

Stephen P. Klien, Yuba City, CA

Koni America, Culpepper, VA

Kwik-Way, Marion, OH

Lee Manufacturing, Cleveland, OH

Lisle Corporation, Clarinda, IA

Loctite Corporation, Cleveland, OH

Lucas Girling Limited, Englewood, NJ

Anthony Lux, Allied Signal/Bendix Brake, Rumford, RI

Mazda Motors of America

McQuay Norris, Mishawaka, IN

Michelin North America, Greenville, SC

Mitchell Information Services, San Diego, CA

Moog Automotive Inc., St. Louis, MO

Garland Morehead, Colusa, CA

Neapco, Inc., Pottstown, PA

Nilfisk of America, Malvern, PA

Nissan Motor Corporation

Nissan North America, Inc.

Nuturn Corporation, Smithville, TN

John Nissen, Colusa, CA

Patch Rubber Company, Roanoke Rapids, NC

Phoenix Systems, Tucson, AZ

Pro-Cut International, West Lebanon, NH

Quickor Suspension Company, Portland, OR

Raybestos (Brake Systems, Inc.), Franklin, IL

Rubber Manufacturers' Association, Washington, DC

Raymond Savage, Robert H. Wager Company, Chatham, NJ

SPX Corporation, Aftermarket Tool and Equipment Group, Kalamazoo, MI

SPX Kent-Moore, Warren, MI

SPX/OTC, Owatonna, MN

Saturn Corporation

Schrader Bridgeport, Altavista, VA

Sisuner International, Ann Arbor, MI

Snap-on Diagnostics, Crystal Lake, IL

Snap-on Tools Company, Kenosha, WI

Specialty Products, Longmont, CO

Stainless Steel Brakes Corporation, Clarence, NY

Storm-Vulcan, Dallas, TX

Superior Pneumatic, Cleveland, OH

Tapley Instrumentation Ltd., Chatham, NJ

Tenneco Automotive, Monroe, MI

McLane Tilton, Buellton, CA

Tilton Engineering, Buellton, CA

The Timken Company, Canton, OH

Toyota Motor Corporation

Uniroyal Tire Company, Troy, MI

Vacula Automotive Products, Williamsville, NY

Vette Products, St. Petersburg, FL

Volkswagon of America

Wagner Division, Cooper Industries, Inc., Parsippany, NJ

Warner Electric Brake and Clutch Company, South Beloit, IL

Western Wheel Corporation, La Mirada, CA

Woody Woodward, Bendix Aftermarket Brake Division, Jackson, TN

Thomas W. Birch

Introduction to the Chassis

Objectives

Upon completion and review of this chapter, you should be able to:

- ❏ Identify the major components of the automotive braking system.
- ❏ Identify the major components of the suspension and steering systems.
- ❏ Understand the general purpose of these components, how they relate to each other, and the role they play in the operation of the car.

1.1 Introduction

This chapter is an overview of the vehicle chassis. It includes the chassis varieties used in cars, pickups, utility vehicles, vans, and light trucks. You should learn the purpose of the two major portions—the suspension system and the steering system—and the variations of these systems that you will encounter. In later chapters, you should gain a working knowledge of the components, how to diagnose problems that might occur, how to replace faulty parts, and how to make any necessary adjustments and alignments.

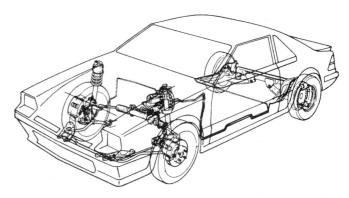

Figure 1–1 A car's chassis includes the braking, suspension, and steering systems. *(Courtesy of Ford Motor Company)*

1.2 Chassis Components— Their General Purpose

We all realize that a car can move because the tires and wheels can roll along a road. Tires and wheels make a vehicle's motion possible. The car's **chassis** connects the axles of these tires and wheels to the body of the car, where the driver and passengers sit. The chassis consists of the frame or body, suspension members (springs and shock absorbers), steering gear and linkage, axles or spindles or both, brake assemblies, and tires and wheels (Figure 1–1).

When these units work together, we have a car that approaches the ideal; that is, one that will:

- ■ Travel smoothly on the road so the driver and passengers feel no bumps or vibrations
- ■ Travel in a straight line until the driver wishes to turn and then will respond quickly and easily to the turning motions of the steering wheel
- ■ Stop smoothly and quickly as the brakes are applied
- ■ Enable the tires to roll down the road with a minimum amount of drag, thereby yielding maximum tire life and fuel mileage

1.3 Brake Systems

The braking system provides the means to stop a car. Obviously, to be in full control of a machine, we have to be able to start it and stop it. There must be a stop for every start. To control an automobile, we need to be able to start it moving, make it turn, accelerate and decelerate, and—of major importance—stop it. A car with a braking system that is not working properly is a candidate for the wrecking yard and may be a cause of injury to the driver and passengers as well as to others. Many drivers think about their braking systems only at the time of a panic stop—when it is too late to do anything but hope.

The braking system is considered by many people to be the most important system involved in the operation of a vehicle. More than a few federal and state requirements govern stopping ability, and many states make periodic inspections of vehicles' (especially buses') brake systems and stopping ability.

We do not normally use brakes at their maximum capability, but we want them to work flawlessly in emergencies, much like a parachute. Most people find hard braking unpleasant and avoid hard stops. Also, as we will see later, brakes waste energy. A vehicle is most efficient when the brakes are not used. People who normally do not stop hard, that is, most of us, seldom get a chance to really test their brakes. The braking systems of most cars are rarely tested to ensure that they are working at maximum efficiency. (I wonder how many pilots would jump out of an airplane using a parachute that had been assembled and packed 5 years ago and had deteriorated from being out in the weather.) Another problem related to brake operation is that most people are poorly trained in using their brakes in emergency situations. Many drivers are not sure how quickly a particular car can stop or how to make that car stop in the shortest distance.

A controversy occurs when it is time to repair braking systems. In the interest of saving money, many people believe they can fix brakes and replace only the items that are actually worn out—the brake shoes and sometimes a drum or rotor. This is the do-it-yourself or home-mechanic approach. At the other end of the spectrum is the technician who replaces or repairs whatever items are necessary to achieve "new-car" braking performance. This second approach ensures that the "parachute" will work properly when needed.

The ideal braking system is one that allows the driver to bring a vehicle to a stop in the shortest possible distance. To be able to do this, it should have enough power to lock up and skid all four tires while stopping on clean, dry pavement. The tire lockups should occur in a controllable fashion so that stops can be made without brake lockup. Also, stopping should occur with a moderate amount of pedal pressure so that even weaker drivers can achieve tire lockup. Brake lockup is not desirable because it results in a longer stopping distance, probable loss of vehicle control, and excessive tire wear. Brake lockup indicates that one or more brake assemblies are stronger than necessary. On the other hand, a brake that cannot achieve lockup is possibly not as strong as it should be. Another requirement for good brakes is a stable stop without pulling or darting to the side. An antilock braking system (ABS) is used to prevent brake lockup so the driver can safely use stronger brakes.

I intend to present enough information in this book so that you can rebuild brake systems to achieve this ideal braking performance. In today's rapidly paced world, with liability lawsuits a common occurrence, it is best to always aim for new-car performance when making a repair. *Brake job* is an undefined term. No legislation specifies exactly what to do, and there are no commonly used test devices or instruments to assess exactly how good or bad a brake system is. Many customers do not know how brakes work, exactly what parts are involved in stopping a car, or how these parts can deteriorate and cause failure. Through this book, you should develop an understanding of the principles involved in stopping a car, how these principles can make stopping difficult, the interrelationship among braking system components, the things that can occur to reduce their performance, and the various methods you can use to check their operation and repair them as needed. Understanding these things makes diagnosing and correcting brake problems easy.

1.4 Brake System Components

Many of us are somewhat familiar with the parts of a braking system (Figure 1–2). To give us a common ground for discussion of the principles and to help us understand their interrelationship, I briefly describe them here. Each part is related to the others, and proper operation of each part is necessary for correct operation of the whole system (Figure 1–3). The major parts of the **base** or **foundation brakes** plus ABS are discussed in detail two more times: first in chapters on the theory of their operation (Chapters 3 and 4) and again in chapters on service and repair (Chapters 22 and 23).

1.4.1 Brake Shoes and Friction Materials

Brakes are heat machines. They provide stopping power by generating heat from the rubbing of a **friction material,** the **brake lining,** against a rotating **drum** or **rotor.** The car slows down as friction produced by this rubbing action converts the energy of the moving car into heat.

The brake lining is attached to the **brake shoes,** often called **pads** when disc brake linings are discussed (Figure 1–4). The lining must be able to rub against the drum

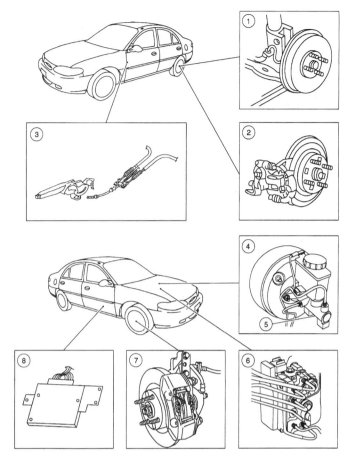

Figure 1–2 A typical automotive brake system consists of rear drum (1) or disc (2) brake units, a parking brake system (3), a power booster (4), a master cylinder (5), and front disc brake units (7). Cars with ABS also have a hydraulic controller (6) and an electronic brake control module (8). *(Courtesy of Ford Motor Company)*

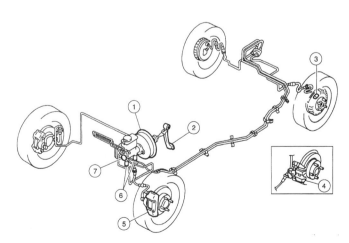

Figure 1–3 The basic brake system uses tubing to transmit hydraulic brake pressure from the master cylinder (7) to the front disc brake units (5) and rear drum (3) or disc (4) brake units. *(Courtesy of Ford Motor Company)*

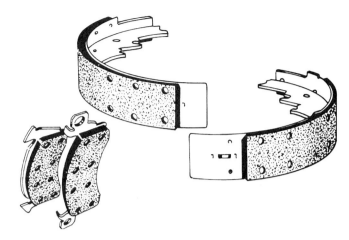

Figure 1–4 Brake lining is attached to the shoes in drum brakes (top) and to the shoes or pads in disc brakes (lower left). *(Reprinted from Mitchell Anti-Lock Brake Systems, with permission of Mitchell Repair Information, LLC)*

or rotor without causing an excessive amount of wear to the drum or rotor surface. It also must be able to operate at very high temperatures (several hundred degrees) without failing.

1.4.2 Drum Brakes

The drum brake is the traditional type of brake on older cars, and it was used on all four wheels of most cars before the 1970s. Drum brakes are currently used on the rear wheels of many cars (Figure 1–5).

The pan-shaped drum is attached to the axle or hub flange, just inside the wheel, and it rotates directly with the wheel. The brake shoes are positioned just inside the drum and are mounted on the **backing plate.** The shoes are anchored to the backing plate so they can pivot into and out of contact with the drum but cannot rotate with it. The **anchors** can be arranged so an opening of the shoe is placed over a round anchor or so that the smooth end of the shoe butts up against a flat anchor block. Braking forces are transmitted from the shoes to the anchors, to the backing plate, and then to the suspension members.

To operate, the brake shoes are pushed outward so the lining is forced against the drum. This force can come from the hydraulic wheel cylinder or from the mechanical linkage of the parking brake. As the lining touches the drum, the rotation of the drum tends to either pull the shoe along with it or push the shoe away, depending on the position of the shoe anchor relative to the direction of rotation. These actions are commonly referred to as **energizing** and **deenergizing** the shoes (Figure 1–6). The energized shoe is the **forward** or **leading shoe,** and the deenergized shoe is the **trailing shoe.**

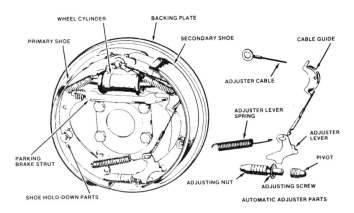

Figure 1-5 A drum brake assembly of the duoservo type; this particular unit is a rear-wheel assembly with a parking brake and uses an automatic adjuster or self-adjuster. (*Courtesy of Bendix Brakes, by AlliedSignal*)

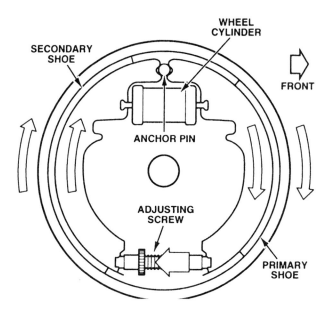

Figure 1-7 A servo brake. The primary shoe is energized by contact with the rotating drum and, by pushing through the adjuster screw, applies pressure on the secondary shoe. (*Courtesy of General Motors Corporation, Service Technology Group*)

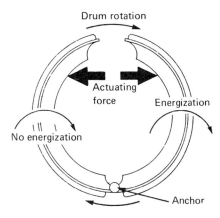

Figure 1-6 Depending on how the brake shoe is mounted, it can be energized or deenergized by the rotation of the drum. (*Reprinted from Mitchell Anti-Lock Brake Systems, with permission of Mitchell Repair Information, LLC*)

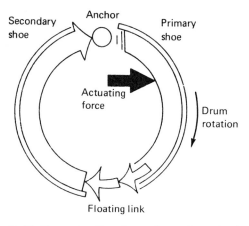

Figure 1-8 The actuating force from the wheel cylinder pushes the primary shoe against the drum. In turn, the primary shoe pushes the secondary shoe against the drum through servo action. (*Reprinted from Mitchell Anti-Lock Brake Systems, with permission of Mitchell Repair Information, LLC*)

If the two shoes are connected so that one shoe can transmit pressure to the other, and both shoes are using a single anchor, the leading shoe will apply pressure on the trailing shoe. This is called **servo action** (Figure 1-7 and Figure 1-8). In **servo brakes,** the leading shoe is called the **primary shoe,** and the trailing shoe is called the **secondary shoe.** Most servo brakes are designed so that servo action can occur during both forward and reverse stops. This style of brake is called a **duoservo brake.**

In **nonservo** designs, the brake shoes are mounted individually, so there is no interaction between them. There are several different styles of nonservo brakes, depending on how the shoes are positioned on the backing plate. Nonservo brakes are usually found in a leading–trailing shoe relationship. They can also be mounted in a two–leading shoe arrangement, with both shoes in an energized position, or in a two–trailing shoe arrangement, with both shoes in a deenergized position (Figure 1-9).

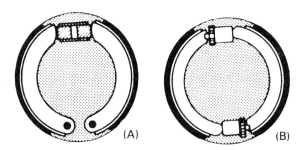

Figure 1-9 A leading–trailing shoe (A) and a two–leading shoe brake (with counterclockwise rotation) (B). Both are nonservo brake designs. (*Courtesy of ITT Automotive*)

Chapter 1 ■ Introduction to the Chassis

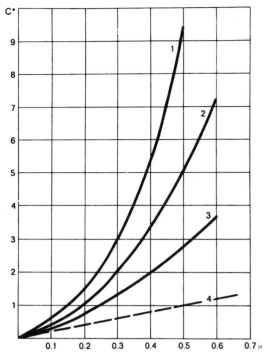

1. Servo brake (duo)
2. Two-leading shoe brake (duplex)
3. Leading and trailing shoe brake (simplex)
4. Disc brake

Figure 1-10 This graph compares the power of four brake designs. The vertical height represents the relative amount of stopping power, and the horizontal represents the amount of pedal pressure. (*Courtesy of ITT Automotive*)

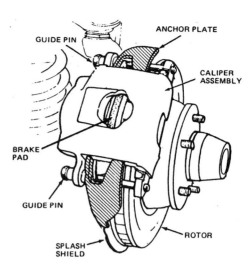

Figure 1-11 A sliding caliper disc brake assembly used on many front wheels. (*Courtesy of Wagner Brake*)

Duoservo and nonservo brakes are discussed more completely in Chapter 3.

The reason for the different brake shoe arrangements is to create a brake that operates more easily with less pedal pressure, to achieve a front-to-rear balance with a pair of brakes of different designs, or to create a more stable design. The duoservo design generates the most stopping power, but it is also the most sensitive to pressure and frictional drag between the primary shoe and the drum. Severe pull can occur if there is a slight difference in power between the two primary shoes on one axle. A two-leading shoe brake is almost as powerful, but it is very sensitive to direction of rotation. There is much greater strength in the forward direction (Figure 1-10).

1.4.3 Disc Brakes

A disc brake uses a flat, round disc, or rotor, attached to the wheel hub instead of a drum. The brake shoes, also called pads, are positioned on opposite sides of the rotor and are mounted in the brake **caliper.** The caliper contains the hydraulic piston(s) used to apply the shoes and to transmit the braking forces from the shoes to the suspension members. Disc brakes are found on the front wheels of most cars manufactured after 1970 and the rear wheels of many of the more expensive or higher-performance cars (Figure 1-11). More and more vehicles are being equipped with four-wheel disc brakes.

Most early disc brake designs used **fixed calipers.** This caliper is bolted solidly to the steering knuckle. Fixed calipers use a pair of pistons, with the inboard (of the rotor) piston applying the inboard shoe and the outboard piston applying the outboard shoe. The fixed calipers used on heavier vehicles have two pistons at each side to apply more pressure evenly over a longer shoe. Most recently designed calipers are of the **floating** or **sliding** style. A single piston applies the inboard shoe, and hydraulic reaction causes the caliper to move in an inward direction and apply the outboard shoe (Figure 1-12). Floating- and sliding-caliper designs use a slightly different type of attachment to the mounting bracket. Floating calipers use a caliper support, which allows the slight sideways motion required for brake operation and release, while holding the caliper in position and transmitting the braking forces from the caliper to the suspension members.

All disc brakes are nonenergized, nonservo brakes; lining pressure is directly proportional to brake pedal pressure. Consequently, disc brakes tend to be more directionally stable than drum brakes. Hard stops can be made with fewer brake-pull problems. A brake rotor also stays cleaner—more free of water, dust, or dirt—than a drum brake. Centrifugal force throws contaminants off a rotor, whereas a drum's inner friction surface tends to collect them. A disc brake also operates more coolly than a drum brake because of the increased area exposed to the air flowing past it. Still another advantage is that the clamping action of disc brake pads causes no distortion of the rotor, whereas the spreading action of drum brake

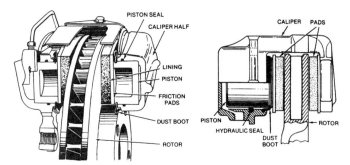

Figure 1–12 A fixed caliper (left) is mounted solidly on the steering knuckle and has pistons on each side of the rotor. A floating caliper (right) has a piston on the inboard side, and the caliper moves to apply the outboard pad. (*Reprinted from Mitchell Anti-Lock Brake Systems, with permission of Mitchell Repair Information, LLC*)

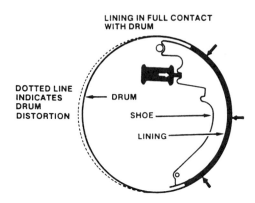

Figure 1–13 When a brake shoe is forced into the drum with sufficient force, the drum distorts to match the shape and curvature of the shoe. (*Courtesy of Bendix Brakes, by AlliedSignal*)

shoes causes the drum to elongate into an elliptical or oval shape. This distortion of the drum causes a lowering of the brake pedal and possible pinching of the ends of the shoes (Figure 1–13).

1.4.4 Brake Hydraulic Systems

All modern automotive brake systems use a hydraulic system to transmit the application forces from the brake pedal to the brake shoes. Hydraulics can be used to greatly increase force. A hydraulic system also has the benefit of transmitting an equal force to two or more places in the system at the same time, an important feature in brake systems. Both brake units on an axle must be applied with the same force at the same time during a stop. If they are not, a brake pull will be the likely result.

The brake's hydraulic system begins at the **master cylinder.** The master cylinder is basically a piston-type hydraulic pump operated by the brake pedal. Steel tubing and reinforced rubber hoses connect the master cylinder to the pistons in the disc brake caliper and the drum brake wheel cylinders. As the brake pedal is pushed, **brake fluid** is pumped through this tubing to the caliper or wheel cylinder piston. This fluid pushes on the pistons, which in turn push the brake shoes against the rotor or drum. Brake engineers vary the diameters of the master cylinder, caliper, or wheel cylinder pistons and utilize a proportioning valve to provide a system that is balanced from front to rear.

In the past, a single master cylinder piston was connected to all four of the caliper or wheel cylinder pistons on the car. This construction provided a nice, simple system with a frightening possibility for failure. A fluid leak in any part of the hydraulic system could stop the operation of the whole system, and there would be a total loss of braking power if the fluid pressure escaped.

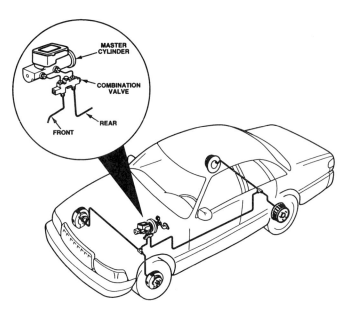

Figure 1–14 A tandem split hydraulic system. Note the position of the combination valve and that the two front brakes connect to one end of the master cylinder and the rear brakes connect to the other end. This system is used on most RWD vehicles. (*Courtesy of General Motors Corporation, Service Technology Group*)

In the late 1960s, **tandem hydraulic systems** were introduced (Figure 1–14). The system was split into two parts: front and rear. A fluid leak in either part can take pressure only from that part of the system. Tandem master cylinders use two pistons, one in front of the other. One piston operates the front brakes, and the other operates the rear brakes. With a tandem system, one part of the system will still function if a leak develops in a wheel cylinder, caliper, or brake line. **Front-wheel drive (FWD)** cars can use a **diagonal split system** (Figure 1–15). One piston of the master cylinder operates the right front and left rear

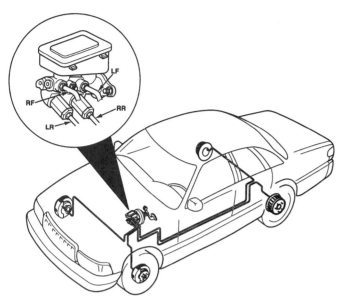

Figure 1–15 A diagonal split hydraulic system. Note that the brakes on the diagonal corners of the car are connected to one end of the master cylinder, and the other two corners are connected to the other end. This system is used on most FWD vehicles. (*Courtesy of General Motors Corporation, Service Technology Group*)

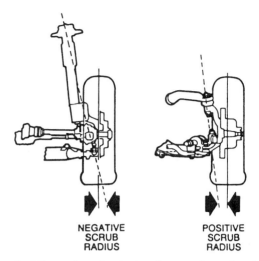

Figure 1–16 Scrub radius is the distance from the tire center to the steering axis at ground level. A FWD vehicle (left) has a negative scrub radius, and a RWD vehicle (right) has a positive scrub radius. (*Courtesy of Chrysler Corporation*)

brakes, and the other piston operates the left front and right rear brakes. A diagonal split system provides 50 percent of the braking power if a hydraulic failure occurs. A failure of the front part of a tandem split system leaves substantially less than 50 percent of the braking power. Diagonal split systems can be used only on cars with a negative scrub radius (the tire centerline is inboard of the steering axis) on the front wheel, as on most FWD cars (Figure 1–16).

Automotive brake hydraulic systems often include one or more of the following valves (Figure 1–17):

- A **pressure differential valve and switch.** This unit turns on a **brake warning light** to inform the driver that a hydraulic failure has occurred in one part of a split system.
- A **proportioning valve.** This unit decreases the hydraulic pressure to rear drum brakes to reduce the possibility of rear-wheel lockup during hard braking.
- A **metering valve.** This unit delays the application of the front disc brakes under light pedal operation so the front and rear brakes apply at the same time.

These valves can be used individually or in a group of two or three. They are often combined into a single assembly called a **combination valve.**

The entire hydraulic system is described in more detail in Chapter 6.

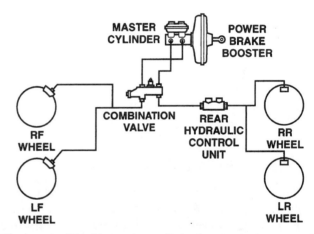

Figure 1–17 This tandem split brake system has a combination valve that includes the metering and proportioning valves and the brake failure warning light switch. This pickup also includes a rear hydraulic control unit for ABS. (*Courtesy of Chrysler Corporation*)

1.4.5 Power Brakes

Power brakes use a booster to assist the driver in pushing on the master cylinder. Normally, a **vacuum booster,** which fits into the linkage between the brake pedal and the master cylinder, is used. The rest of the system differs very little from an ordinary, nonpower brake system. The booster can multiply the force of the driver's foot many times. It uses the pressure differential between atmospheric pressure and the intake manifold vacuum against a large diaphragm to obtain this boost (Figure 1–18).

Some cars use a **hydro-boost** unit, a brake booster that uses power steering system hydraulic pressure for

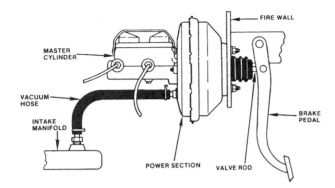

Figure 1-18 This power section is a vacuum power brake booster. The vacuum hose connects the booster to the intake manifold. (*Courtesy of Bendix Brakes, by AlliedSignal*)

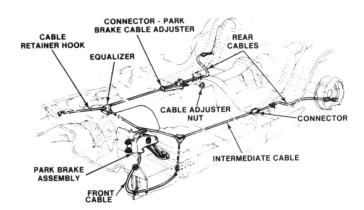

Figure 1-19 A parking brake system utilizes the brake shoes and drum or rotor of the rear brakes and uses cables and a pedal or lever to mechanically apply these brakes. (*Courtesy of Chrysler Corporation*)

the increased force. Hydro-boost units are mounted in the same manner as vacuum boosters, but they are usually smaller and more powerful.

Some newer cars use an electric motor–operated hydraulic booster. This unit operates much like a hydro-boost unit, but it develops its own hydraulic pressure using a pump driven by the electric motor. This compact unit operates only when needed to provide an assist. With some ABS, the booster is integrated with the master cylinder and control valve assembly. All three styles of power boosters are described in Chapter 7.

1.4.6 Parking Brakes

A mechanically operated **parking brake** is required on all motor vehicles with four or more wheels. This brake system normally uses the rear drums or rotors and shoes of the service brake system. It consists of the familiar pedal or lever in the passenger compartment, a means of holding the lever in the applied position, the cables that connect the pedal or lever to the linkage on the shoes, and the levers and struts that attach the cables to the shoes. This system is designed to lock two wheels of the vehicle—usually the rear ones—so that they will not rotate, holding the vehicle stationary while parked (Figure 1–19).

The parking brake system is often called an **emergency brake**. But anyone who has tried to use the parking brake for a high-speed stop knows that it is not designed for true emergencies. Some drivers might be a little careless in maintaining their brake systems if they believe they have an effective system for use under emergency conditions. It is best to think of this system as a parking brake and to maintain the service brake so there will be no need to use an emergency brake.

1.4.7 Antilock Braking Systems

As we discuss in the next chapter, it is quite difficult to make a very hard stop without locking up one or more of the brakes. Brake lockup causes tire skid, which in turn can cause loss of vehicle control. Loss of control can be especially severe when stopping on wet or icy roads or roads with loose material on the surface. An experienced driver can sense or feel brake lockup beginning to occur and can adjust the brake pedal pressure to apply the brakes just short of lockup. Doing so maintains vehicle control, but the reduction in pedal pressure reduces the power of all four brakes, not just the one that is beginning to lock up.

An **ABS** consists of electronic sensors for all four tires to monitor wheel speed, a control module to determine whether lockup is beginning and to operate the valves, and a series of valves to control the brakes. If the control module senses a wheel decelerating or stopping too quickly, it activates the valve assembly for that wheel, which releases that particular brake or pair of brakes. When the wheel is revolving again at the correct speed, the brake is reapplied by the system. The electronic control module is a computer device that quickly determines the speed of each wheel; it then determines whether the speed of any wheel is too slow and, if so, quickly releases and reapplies one or more of the wheel brakes. The brake can be cycled off and on as many times per second as necessary to keep the wheel rotating, so that vehicle control can be maintained. All the driver needs to do in an emergency situation is push on the brake pedal and steer the car (Figure 1–20).

At this time, ABS is standard equipment on most luxury, high-performance cars, pickups, and some vans and utility vehicles and is an extra-cost option on most other cars. ABS is an effective and expensive addition to the brake system that is becoming quite common.

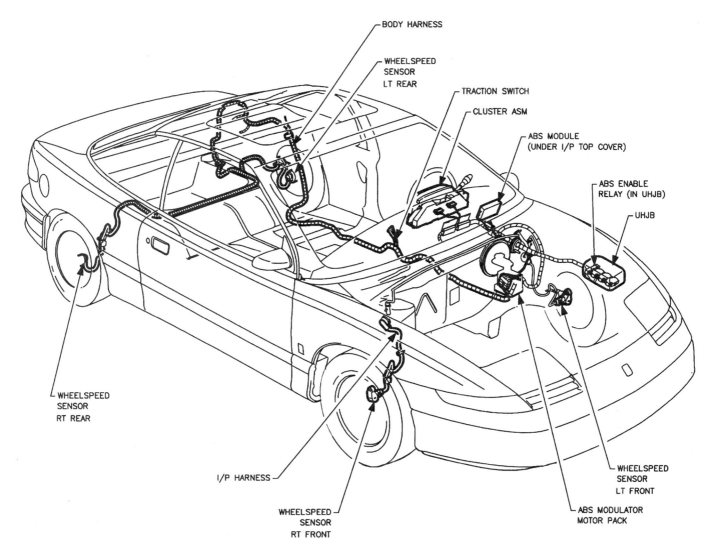

Figure 1-20 An ABS adds wheel speed sensors and electronic controls to the basic brake system. (©Saturn Corporation, used with permission)

1.5 Brake System Repair

The brake lining, like the tires, wears out. Each time the brakes are used, a small amount of lining wears off during the rubbing action. Brake lining wear is to be expected. The rate of lining wear varies depending on the weight of the car, its speed, traffic conditions, and, probably most importantly, the aggressiveness of the driver. We often see cars going through traffic with their brake lights on. The highly aggressive "left-foot braker," with his or her right foot on the gas pedal and left foot on the brake, pays for this habit in both low fuel mileage and short brake lining life.

Brake lining wear is usually fairly slow and can be easily measured by pulling a brake drum or looking at the disc brake pads. Wear of the rotor or drum surfaces can be easily measured using a standard outside micrometer caliper or a brake drum micrometer. The size of these components can be compared with specifications indicated on the drum or rotor. At the same time the friction surfaces are wearing, the hydraulic system is also deteriorating. The rubber seals wear slightly as the pistons stroke in their bores and also become harder and less flexible as heat and time act on them. Also, most brake fluids absorb water, which causes rusting and etching of the cylinder bores as well as a lowering of the fluid's boiling point. A total loss of braking power occurs if braking heat causes the fluid to boil.

Worn-out brake lining is noisy and usually quite audible. The metallic grinding, grating noise is easily noticed by most drivers. At this point, the stopping power of the brake is greatly reduced, and repair will probably be expensive because one or more of the drums or rotors will need to be replaced. A faulty wheel cylinder either sticks and fails to apply, fails to release, or leaks. A leaky wheel cylinder allows fluid to leak onto the brake lining, which ruins the lining. Leaks can also cause a loss in the effectiveness of that portion of the brake system or, in some cases, a loss of all braking power.

Wise technicians check the entire braking system when the brake lining is replaced. Both the friction surfaces and the hydraulic portions wear, and a failure in the hydraulic system can have catastrophic results. The goal of the brake technician, like that of the tune-up technician, engine overhaul technician, or any other technician, is to try to achieve new-car operation.

1.6 Practice Diagnosis

You are working in a general repair shop as a technician who specializes in brake repair and encounter the following problems:

CASE 1: The customer brought in a 1987 Ford Crown Victoria with a complaint that the left rear wheel locks up during a medium-to-hard stop. He says that the problem is probably caused by a faulty valve in the hydraulic system. Is his diagnosis correct? What should you do to confirm his diagnosis? What should you do to find the cause of the problem?

CASE 2: The customer has brought in her 1992 Saturn with a complaint that the emergency (parking) brake does not work. On checking it, you find the parking brake lever moves through its full travel with very little resistance. What is probably wrong? What should you do to find the cause of this problem?

CASE 3: The customer has brought in a recently purchased, 2-year-old Camry. He complains that the rear tires chirp when he applies the brakes hard and he feels a slight shudder in the brake pedal. Could this be normal for an ABS-equipped vehicle? How can you determine whether this car is equipped with ABS?

1.7 Suspension and Steering Systems

The **suspension system** includes the springs, shock absorbers, struts, control arms, spindle or axle, and the bushings that allow the necessary motions. These parts should hold the tire and wheel in correct alignment with the car and the road. They also allow the tires and wheels to move up and down relative to the body over bumps and holes in the road surface. The tires can thus follow the road surface and maintain traction without transmitting the roughness of the road to the driver and passengers (Figure 1–21). Any of you who have ridden in a vehicle with a solid suspension (e.g., a farm tractor, bicycle, or forklift) realized the value of a well-adjusted suspension system when you went over the first bad bump. The solid suspension probably transferred most of the bump from the road to you.

In relation to a car's chassis, the side-to-side distance between the centerlines of the tires on an axle is called the **track.** The distance between the centers of the front tires and the centers of the rear tires is called the **wheelbase** (Figure 1–22).

To be in correct alignment, a tire should roll on a path that is parallel to the centerline of the vehicle. This is called zero **toe-in** or **toe-out.** The tire should also be straight up and down, or at a right angle (90 degrees) to the road surface. This is called **zero camber.** If one or more of the tires is not positioned this way, it is out of

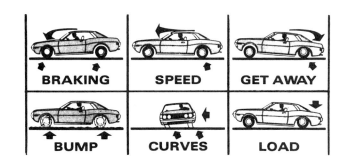

Figure 1–21 The car's suspension system allows the tires to maintain contact with the road when the car is operated under various conditions. (*Reprinted by permission of Hunter Engineering Company*)

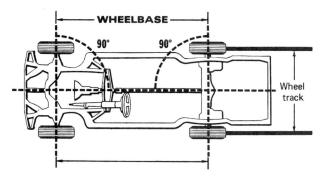

Figure 1–22 The suspension keeps the car's tires parallel to the centerline of the car when it is going straight ahead. The lengthwise distance between the axle is called the wheelbase; the crosswise distance between two tires on an axle is called track. (*Reprinted by permission of Hunter Engineering Company*)

alignment and will scuff or scrub sideways as it rolls down the road. This movement causes tire wear and a loss in fuel economy because a scuffing tire does not roll as freely as one that is aligned correctly. In actual practice, the tires are often aligned to have a slight amount of toe or camber to compensate for suspension travel and road forces (Figure 1–23).

The steering system allows the driver to control the direction the car travels. Turning the steering wheel causes the front wheels to point in the direction the driver wants to go. The steering system consists of the steering wheel, steering gear, and tie-rods (Figure 1–24).

1.8 Chassis Components

At one time, the **frame** was the mounting point onto which the rest of the car was attached. The steering, suspension, engine, drivetrain, and body were all bolted to the frame. The frame became the skeleton and backbone of the car and was generally made from strong steel channel or square, tube-shaped members (Figure 1–25).

Most new cars do not use a traditional frame; they use a **unibody** type of construction. The sheet metal body of the car has reinforced sections that are strong enough for the suspension, steering, and drivetrain components to be connected to them. The reinforced body has made the separate frame unnecessary. The center and rear portions of a body are quite strong. The roof and floor pan of the body, reinforced by the door frames, make a fairly strong and rigid box. The area in front of the windshield is not nearly as strong because the only permanent body sections in that area are the inner fender panels. The fenders themselves are usually removable to allow easy replacement. In most cases, a unibody car has subframe sections bolted or welded into it for the engine and front suspension. This subframe is often called an engine cradle (Figure 1–26).

Any severe accident that bends or twists the frame or suspension-carrying portions of a unibody car can cause the suspension mounts to change position relative to each other or the body. This movement of the suspension mounting points changes the alignment of the tires and wheels, which usually results in increased tire wear, reduced fuel mileage, or a car that pulls to one side or wanders back and forth down the road.

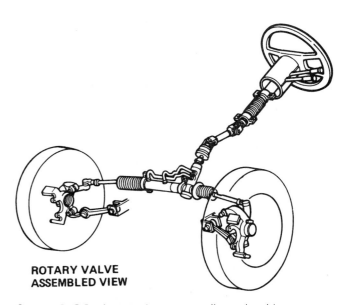

Figure 1–24 The steering system allows the driver to steer the car. (*Courtesy of Ford Motor Company*)

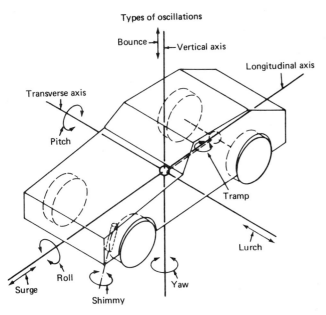

Figure 1–23 The body and suspension members of a car can move in several different directions, as shown on this engineer's model.

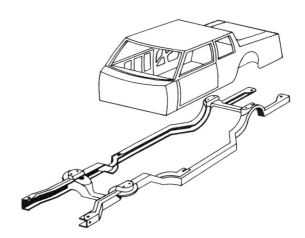

Figure 1–25 On many cars and all pickups, the body and frame are two separate units that are attached to each other by bolts passing through rubber mounts.

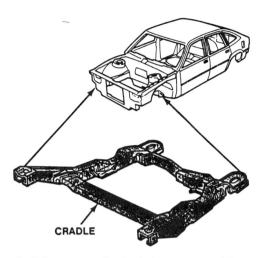

Figure 1-26 Some unibody designs use a subframe, often called a cradle, to support the engine, powertrain, and front suspension. (*Courtesy of General Motors Corporation, Service Technology Group*)

1.9 Nonindependent Suspensions

The suspension arms or axles connect the wheels and tires to the frame and allow the tires and wheels to move up and down relative to the frame and body. The different styles of suspensions fall into two general classifications, independent and nonindependent. These suspension types are briefly described in this chapter and again, in much more detail, in Chapters 13 and 14.

In nonindependent suspensions, the front wheels are mounted on the same axle, and the rear wheels are mounted on another, single axle. A **solid axle,** also called a **beam axle,** is used. (See Figure 13–20.) In the early days of automobile development, every car had two solid axles, one in front and one in back. Many trucks and pickups still use a solid front axle. Nearly all **rear-wheel-drive (RWD),** front-engine cars, trucks, and pickups use a solid rear axle. Solid axles are strong and relatively inexpensive. They are also quite simple and trouble free, especially on drive axles. Solid axles have several drawbacks, however. First of all, they are heavy, and because of their weight, they increase the vehicle's unsprung weight (i.e., the vehicle weight that is not on the springs). This increase in unsprung weight requires more spring and shock absorber control to keep the tires in contact with the road. Also, when one tire of a solid axle goes over a bump, the other tire on that axle has to change position and may lose traction (Figure 1–27). A car with solid axles usually has a harsher ride than a car with an independent suspension.

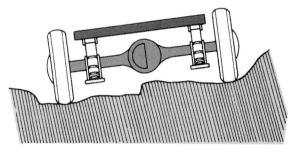

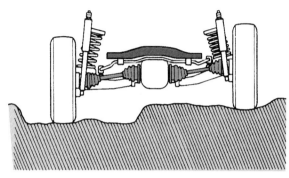

Figure 1-27 When one end of a solid or rigid axle goes over a bump, the position of the other tire is affected. An independently suspended tire does not affect any of the other tires when it encounters a bump.

1.10 Independent Suspensions

In independent suspension systems, each of the front wheels, and sometimes the rear wheels, are mounted on separate spindles and control arms. Mounting the wheels separately allows them to travel up or down independently of the other wheel on the same axle. This usually reduces the unsprung weight, gives a softer ride, and allows for greater control of the tire and wheel position. However, an independent suspension is a more complicated system with more parts and movable bushings. There are more parts that may wear out, possibly leading to misalignment of the tires. All passenger cars, many pickups, and a few trucks have independent front suspensions. A few RWD and some FWD cars have independent rear suspensions (Figure 1–28).

1.11 Swing Axle Suspension

The simplest type of independent suspension is the **swing axle.** This is essentially a portion of a solid axle

Chapter 1 ■ Introduction to the Chassis

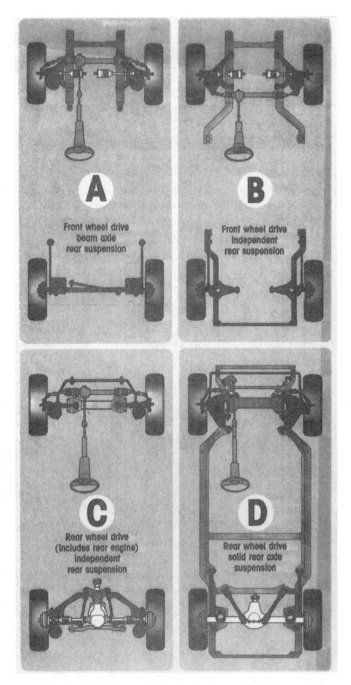

Figure 1–28 The four most common combinations of front and rear suspensions are shown here.

in that as the tire moves up and down, it also moves in and out. The outer end travels in an arc centered at the inner pivot. This movement causes a track (tire-to-tire width on an axle) change and a camber (vertical tire position) change, which in turn cause tire wear or vehicle steering problems. Swing axles have been used at the rear of some cars and on the front end of Ford pickups, **four-wheel drives (4WDs),** and trucks.

1.12 Short-Long Arm Suspension

Most RWD cars use a front suspension of the **short-long arm (S-L A)** type that is also called an **unequal arm suspension.** S-L A suspensions use two arms of unequal length, a short upper arm and a longer lower arm. The inner ends of each control arm are attached to the frame with bushings that allow the control arm to pivot. The outer ends of each control arm are attached to the steering knuckle or spindle support with either bushings or ball joints. These pivots allow steering to take place (on front suspensions) as well as vertical suspension motion (Figure 1–29).

The primary purpose of the lower control arm is to control the track width of the tires. Consequently, the lower control arm should be horizontal or nearly so. If it were mounted at an angle, vertical wheel travel would cause a severe track change. The upper control arm has the primary purpose of controlling the camber angle of the tire. This arm is seldom horizontal; it is angled either downward or upward at the other end (Figure 1–30).

Using control arms of two different lengths that are mounted in a nonparallel position causes the tire to go through a camber change as it travels up and down. This does not sound good, but the alternative is a track change, which is definitely bad. If the control arms are of equal length and parallel, the tire travels in a sideways arc during vertical travel. This causes a sideways motion and track change of the tire while the tire is going over a bump, which is called **scrub.** Scrubbing the tire sideways under this load causes tire drag and wear (Figure 1–31). S-L A suspensions are designed so the track changes as little as possible while the tire goes over bumps. The only way to keep the road-contact track width the same while the steering knuckle moves up and down in an arc is to change the camber. The tread, at the road surface, has negligible side scrubbing, and the camber change does not cause much tire wear during this short period. Improper wheel alignment can also cause tire scrub.

that pivots from the frame at the inner end. The inner pivot of a nondriving, swing axle is a pivot bushing. (See Figure 13–21.) The inner pivot of a driving, swing axle is usually the universal joint. The axle's outer end, which is attached to the tire and wheel, swings up and down much like a door swings on its hinge. The only difference is that a door is vertical, and the axle is horizontal. (See Figure 13–23.) A **radius rod** or **control arm** is usually used to keep the outer end from moving forward or backward. Swing axles have a disadvantage

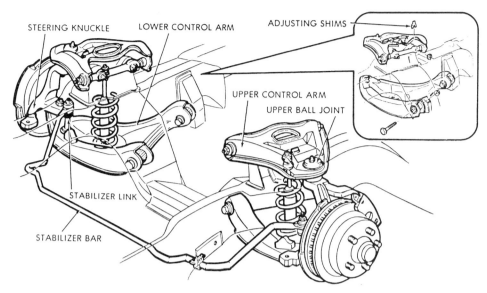

Figure 1–29 An unequal or short-long arm suspension showing the relationship of the various parts. (*Courtesy of General Motors Corporation, Service Technology Group*)

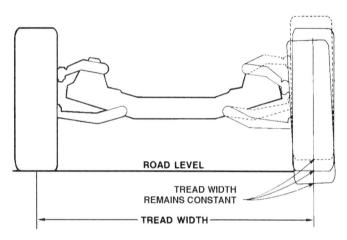

Figure 1–30 The geometry of S-L A control arms keeps tread width constant during bump travel. (*Courtesy of Moog Automotive Inc.*)

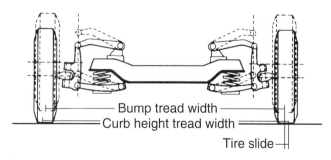

Figure 1–31 If equal-length control arms were used, a track change and tire scrub would occur, which would cause tire wear. (*Courtesy of Ammco Tools, Inc.*)

1.13 MacPherson Strut Suspension/Strut Suspension

Another major independent suspension design is called the **MacPherson strut** or merely **strut suspension** (Figure 1–32). This suspension was named for the designer. Mr. MacPherson was an engineer for Ford of Britain during the introduction of strut suspensions on the 1951 Ford Consul and Zephyr. Strut suspensions are currently used on the front end of most FWD cars; on many smaller, front engine, RWD cars; and on the rear end of some RWD and FWD cars. It is sometimes called a Chapman strut when used on rear suspensions.

The strut consists of a spindle and steering knuckle that are built onto an oversized, telescoping shock absorber. The spring is usually mounted around this shock absorber (Figure 1–33). The upper end of the strut is attached to the car's inner fender housing with a bearing assembly that allows the steering knuckle to pivot for turns. This bearing is built into a rubber mount assembly that allows a slight angle change of the strut during suspension travel as well as a dampening of road vibrations. The lower end of the strut is attached to the car with a lower control arm, much like an S-L A suspension. Like the S-L A suspension, the lower arm controls the track width. Some cars use a modified strut suspension with the spring mounted between the lower control arm and the frame, much like most S-L A suspensions. (See Figure 13–18.)

Chapter 1 ■ Introduction to the Chassis

1.14 Multilink Suspension

Multilink suspensions are also called double-wishbone, wishbone-strut, and long-spindle S-L A (Figure 1–34). Some see them as a modified strut, whereas others see them as a variation of an S-L A suspension. This assembly uses an extended steering knuckle so the upper ball joint is close to the top of the tire. The upper control arm on some cars is mounted at a rather severe angle so as not to interfere with the engine compartment. The spring is mounted around the shock absorber and to the lower control arm much like a strut. This lower attachment can be forklike to allow room for the front driveshaft to pass through it on FWD cars.

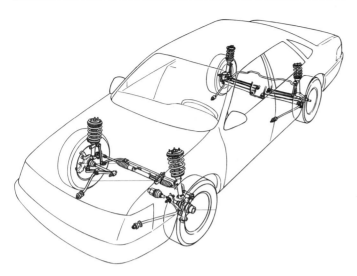

Figure 1–32 A car with MacPherson strut front and rear suspension. (*Courtesy of Moog Automotive Inc.*)

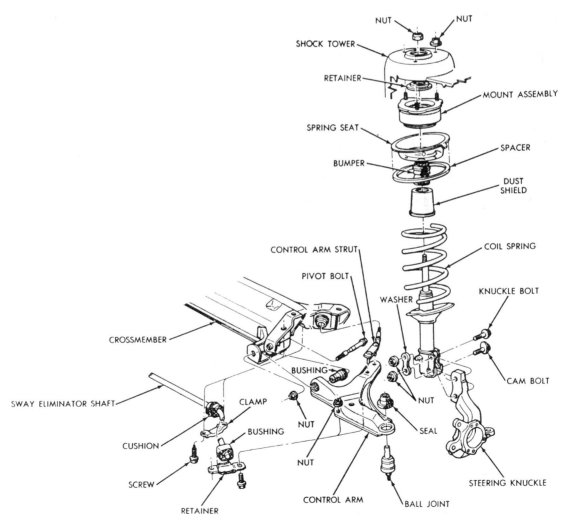

Figure 1–33 An exploded strut assembly showing the various parts. This unit is for a FWD car front suspension. (*Courtesy of Chrysler Corporation*)

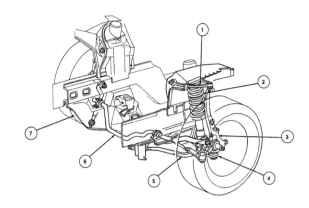

Item	Part Number	Description
1	3085	Upper Front Suspension Arm Bushing Joint
2	18124	Front Shock Absorber
3	3105	Front Wheel Spindle

Item	Part Number	Description
4	3079	Front Suspension Lower Arm
5	3468	Front Suspension Lower Arm Strut (2 Req'd)
6	5482	Front Stabilizer Bar
7	5C145	Front Sub-Frame Assy

Figure 1–34 A multilink suspension, which is basically an S-L A suspension with the spring mounted on the strutlike shock absorber. (*Courtesy of Ford Motor Company*)

1.15 Trailing Arm Suspension

Another style of independent suspension involves mounting the spindle and wheel assembly onto one or a pair of **trailing arms.** A trailing arm pivots from the frame at the front (leading) end, and the spindle is at the rear (trailing) end. A leading arm is arranged just the opposite, with the pivot at the rear. This is a relatively simple suspension that has been used for front suspensions on only a few car models. On modern vehicles, it is more popular as a rear suspension. The arms, which pivot on metal or rubber bushings at the front to allow suspension travel, must be strong enough to withstand side loads so the tire will be kept in proper alignment. (See Figure 13–24.)

If the trailing arm pivot is at a right angle (90 degrees) to the car centerline, there is no camber or track change during suspension travel, but vehicle lean causes an undesirable and equal change in camber angle. A more desirable camber change can be obtained by changing the position of the arm's pivot bushing. When the pivot bushings are mounted at an angle other than 90 degrees, the suspension is called a **semitrailing arm** (Figure 1–35).

1.16 Control Arms

Control arms for S-L A and strut suspensions can use either two inner bushings on a wide, triangular-shaped A-arm or one inner bushing on a simple, narrow control arm with a **strut rod** (Figure 1–36). A-arms are also

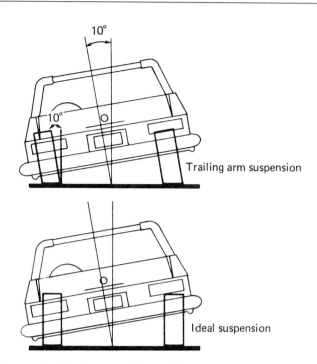

Figure 1–35 If a car with a trailing arm suspension leans 10 degrees outward on a turn, the tires will also lean 10 degrees (top). A better situation is to allow the body to lean but keep the tires vertical.

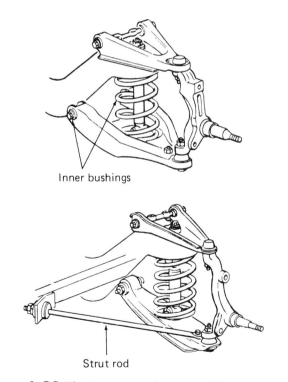

Figure 1–36 The top suspension uses a lower A-arm; the control arm has two inner pivot bushings. The lower control arm (bottom) has a single pivot bushing plus a strut rod. (*Courtesy of McQuay Norris*)

Chapter 1 ■ Introduction to the Chassis

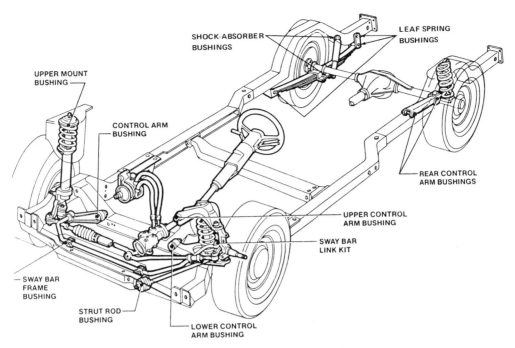

Figure 1–37 Rubber bushings are used at many locations in a car's suspension system. (*Courtesy of Moog Automotive Inc.*)

called wishbones. These arms are usually made from steel stampings, but some are cast from aluminum or iron. The outer end of a control arm must be able to move up and down but must not move forward or backward. Having two inner pivot points, an A-arm forms a triangle, which is one of the strongest structural shapes. The point of a triangle cannot move sideways unless the sides change length or the base moves. The two inner pivots form the triangle's base on an A-arm, and the inner pivot and the strut rod bushing form the base on a control arm and strut rod triangle. All of these control arm pivots must be designed to allow the necessary movement required for steering and bump control but cannot allow free play, slacks, or slop, which could cause unwanted, uncontrolled tire and wheel motion.

1.17 Suspension Bushings

Most modern cars use a rubber, **torsilastic** bushing (Figure 1–37). Rubber is a favored bushing material because it requires no lubrication, does not transfer minor road vibrations, is usually silent, and offers a relatively large degree of compliance (i.e., allows for changes in position). Bushing compliance permits control arm motion without binding. Harder plastic or urethane bushings are used where less bushing compliance is desired. Torsilastic refers to the natural elastic nature of rubber to allow movement of the bushing in a rotating or twisting

Figure 1–38 The two inner pivot bushings are called torsilastic bushings because the rubber must twist in a torsional manner. Note the notches at the ends of the inner sleeve of the bushing (inset). (*Courtesy of McQuay Norris*)

plane. This bushing is not a bearing in the common sense. Motion is allowed by the twisting of the rubber, not by one part rotating or sliding into another (Figure 1–38 and Figure 1–39).

The inner sleeve of the bushing is usually locked to the control arm pivot shaft and is bolted solidly to the frame. The outer sleeve is pressed into the control arm, and the rubber must twist to allow the control arm to pivot. This movement also causes the control arm to have a memory; that is, the rubber will want to return to the

Figure 1–39 A complete torsilastic bushing is shown at the upper left. The bushing below it has had a section removed to show the inner sleeve and the rubber squeezed tightly between the two sleeves. The bushing to the right has been pressed apart (three pieces).

position from which it started. This action increases the resistance to movement of the control arms. Some sources state that 10 percent of the resistance to body roll comes from the rubber bushings (Figure 1–40).

Rubber bushings are also used to allow a swinging motion necessary at the end of a strut rod, stabilizer bar end link, or shock absorber. The rubber bushings provide the needed compliance and dampen road vibrations (Figure 1–41).

Metal bushings are used for some inner control arm pivots. This bushing usually resembles a bolt thread on each end of the pivot shaft and large nuts that are secured into the control arm. These nuts are threaded onto the pivot shaft threads. As the control arm pivots up and down, the nutlike bushings rotate on the threads of the shaft. Metal bushings require lubrication, and they also need seals to keep the lubricant in and dirt and water out. They can become quite noisy if they lose their lubricant and become dry, rusty, or dirty (Figure 1–42).

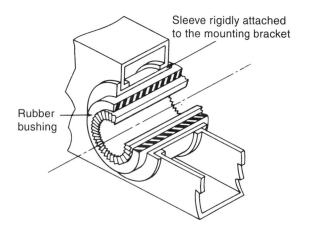

Figure 1–40 The outer sleeve of a torsilastic bushing is pressed into the control arm, and the inner sleeve is locked to the mounting bracket. The rubber twists to allow control arm motion.

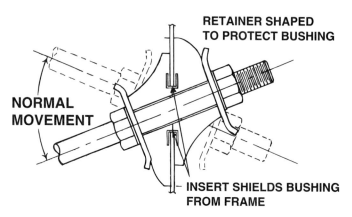

Figure 1–41 The rubber strut rod bushing allows enough strut rod motion for suspension travel. (*Courtesy of Moog Automotive Inc.*)

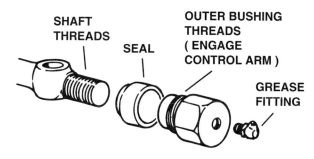

Figure 1–42 A metal bushing locks into the control arm at the outer threads; bushing motion occurs at the inner threads. (*Courtesy of Moog Automotive Inc.*)

1.18 Ball Joints

A ball-and-socket joint is used to provide both a swinging or pivoting motion and a rotating motion. These movements are necessary at the top and bottom of the steering knuckle of an S-L A suspension, the bottom of a strut suspension's steering knuckle, and where portions of the steering linkage connect (Figure 1–43). These joints must be loose enough to allow free movement but not so loose that sloppy, uncontrolled motions result. When a ball joint carries a large amount of load, such as the vehicle load–carrying joint on S-L A suspension, the ball joint is arranged so vehicle load is used to keep play and clearance out. When this joint is used in a normally unloaded position such as a tie-rod end, preload inside the joint, often provided by spring tension, keeps the joint tight yet flexible.

1.19 Kingpins

At one time, all vehicles used a **kingpin** or **king bolt** to connect the spindle to the steering knuckle or axle. The

Chapter 1 ■ Introduction to the Chassis

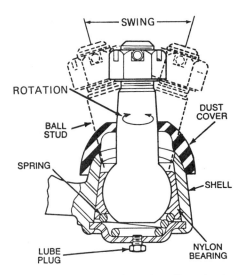

Figure 1–43 A ball-and-socket joint allows both rotary and swinging motions of the ball stud; they are used for ball joints and tie-rod ends. (*Courtesy of Chrysler Corporation*)

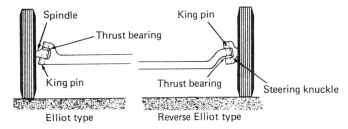

Figure 1–44 An Elliott (left) and a reverse Elliott (right) kingpin. The reverse Elliott type is much more common. (*Courtesy of Ammco Tools, Inc.*)

kingpin provided the pivot or steering axis for steering the front wheels. Most solid axles and swing axles still use kingpins. An **Elliott** style of kingpin locks the steel kingpin into the spindle, and the kingpin rotates in a pair of bushings in the end of the axle. A **reversed Elliott** style locks the steel kingpin in the end of the axle, and there are a pair of bushings in the spindle. The reversed Elliott is the most common type (Figure 1–44).

1.20 Springs and Shock Absorbers

The springs make the load-carrying connection between the suspension members and the frame. Springs have the ability to bend or twist and absorb energy when they are compressed to shorter lengths. When a tire meets an obstruction, it is forced upward, and the energy of this upward motion is absorbed by the spring rather than transmitted to the frame and body of the vehicle. The spring, however, only absorbs this energy for a brief period; as soon as possible, it releases the energy by extending back to its original length. This extension either pushes the tire back down on the road or lifts the car if the obstacle is still under the tire.

When a spring releases its stored energy, it does so with such quickness and momentum that the end of the spring usually extends too far. The spring goes through a series of oscillations, contractions, and extensions until all of the energy in the spring is used up or released. The speed of these oscillations depends on the natural frequency of the spring and suspension. A car with undampened springs (i.e., no shock absorbers) tends to bounce up and down in time to these oscillations. In most cases, this bounce frequency is disturbing to the driver or passengers. A stronger spring oscillates at a faster frequency than a softer one. Many manufacturers purposely mismatch the spring frequency at the front and the rear to obtain a flatter, more acceptable ride.

Spring oscillations are normally dampened or reduced by **shock absorbers.** Shock absorbers are very poorly named. They do not absorb shock; the springs do. The shock absorbers stop excessive spring oscillations. The shock absorber absorbs some of the energy that was put into the spring by the bump, and it converts that energy into heat that is dissipated into the air. A shock absorber is usually mounted inside or next to each of the four springs on a car (Figure 1–45). Shock absorbers are discussed in more detail in Chapter 16.

1.21 Spring Types

At one time, the most commonly used spring was the **leaf spring.** This is usually a group of long strips of flattened spring steel. The longest, or main leaf, has a pivot bushing at each end. One end is attached to the frame through a bushing, the center of the spring is attached to the axle, and the second end of the main leaf is attached to the frame through a shackle. The shackle allows the spring to change length as the spring bends or flattens (Figure 1–46). The frame end of the spring is usually used as a control arm; it locates the front-to-rear position of the axle. Additional spring leaves are added to the main or master leaf. These additional leaves are usually of different lengths to obtain a variable spring rate. Leaf springs can be used in pairs, with one spring at each end of the axle, or can be mounted **transversely,** across the car, with each end of the spring at each wheel (Figure 1–47). Spring rates are discussed in more detail in Chapter 15.

Coil springs are now the most commonly used type of spring. They require the use of control arms to locate the wheel and axle, but they provide a good ride without

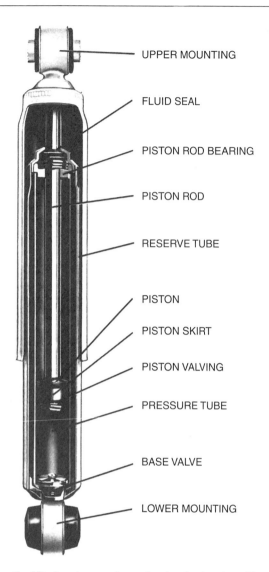

Figure 1-45 A cutaway view of a shock absorber. The pressure tube is filled with oil. Suspension travel forces the piston through this oil, which generates resistance in the shock absorber. (*Courtesy of Monroe Auto Equipment*)

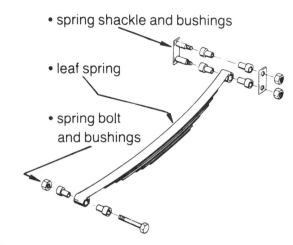

Figure 1-46 A semielliptic leaf spring. The axle is clamped near the center, bushings are used at each end, and a shackle is used at one end so the spring can change length. (*Courtesy of Dana Corporation*)

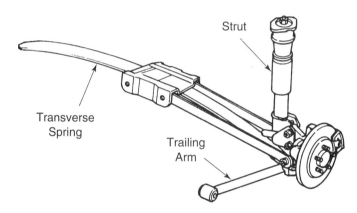

Figure 1-47 A transverse leaf spring is used on this FWD car rear suspension. The center is clamped to the car body, and the ends of the spring are attached to each rear suspension.

wear or interleaf friction (Figure 1-48). Also, the coil shape often fits easily into most installations. A coil spring is simply a spring steel wire or rod that is wound into a helical, coiled shape.

Torsion bars are often used where coil springs would get in the way or take up too much vertical room. Some FWD cars and 4WD vehicles with S-L A suspensions use torsion bars for this reason. The torsion bar is anchored to the frame at one end and attached to a lever arm (often a suspension control arm) at the other end. The control arm is free to pivot with the suspension, and it twists the torsion bar as the wheel moves up and down. An adjustment is usually built into the attachment at one end of the torsion bar to allow the bar to be twisted tighter or looser. This adjustment compensates for sag. Sag is deterioration that causes a spring to shorten and lower the car as it gets older.

Metal springs can be replaced by air-filled rubber bags or chambers. **Air suspension** offers a real advantage in that it allows for a change in spring rate and length by increasing or decreasing the air pressure inside the air chamber. A disadvantage is that the system is rather complex. It requires an air chamber at each wheel, a height control sensor and valve at each wheel, an air compressor, and tubing to connect all of these parts together (Figure 1-49).

Modern air suspension can be controlled and adjusted by a computer as the car moves. The computer adjusts the car's height so it stays correct regardless of the weight that might be added or removed, such as passengers or luggage.

Chapter 1 ■ Introduction to the Chassis

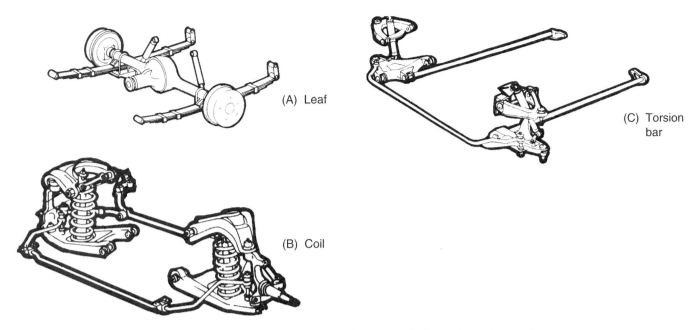

Figure 1-48 A rear axle with a leaf spring (A). The shackle at the rear end allows the spring to change length as it flattens and bends. An S-L A suspension with a coil spring mounted between the frame and lower control arm (B). This is the most common RWD front spring suspension arrangement. An S-L A front suspension with a torsion bar for a spring (C). The lower control arm connects to one end of the torsion bar to become a lever. (*Courtesy of Tenneco Automotive*)

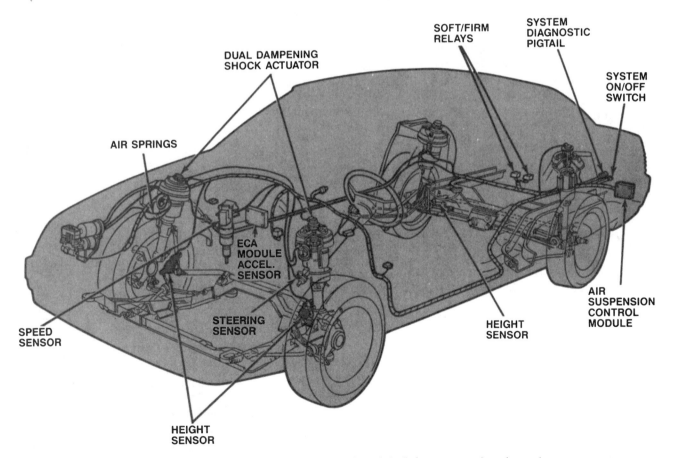

Figure 1-49 Most air suspension systems use an electronic control module (left rear corner) to determine the correct air pressure of the springs to give the proper height or ride. (*Courtesy of Ford Motor Company*)

On some cars, adjustments can also be made in the vehicle height or spring rate as the car speeds up or even as the road surface changes. Both changes improve high-speed handling, and lowering the car can improve gas mileage.

1.22 Steering System

Each front wheel is mounted onto a spindle that is attached to a steering knuckle or strut assembly. The steering knuckle pivots on ball joints, a kingpin with bushings, or an upper bearing and ball joint. These pivots are called the **steering axis.** The steering knuckle has a steering arm attached to it that is connected to the tie-rods and steering gear. Movement of the end of the steering arm toward or away from the center of the car causes the steering knuckles to pivot on the steering axis and the front wheels to make a right or left turn. (See Figure 17–1 and Figure 17–2.)

1.23 Steering Gears

Two types of steering gears are used to transmit the steering motions from the steering wheel to the tie-rods: the **conventional standard gear** and the **rack-and-pinion gear.** Most conventional steering gears are of the recirculating ball nut type. Both types change the rotary motion of the steering wheel into the back-and-forth motion of the tie-rod. They also provide a gear ratio to increase the driver's effort and slow down the steering speed. The conventional steering gear has traditionally been used on larger cars; it offers the advantage of smooth, easy steering with very little road shock. Rack-and-pinion steering is more compact and lighter weight and often has a faster steering ratio. A faster steering ratio provides faster steering, but a slower ratio provides more precise steering with less effort.

Two main shafts are used in conventional steering gears: the **steering shaft,** which connects the worm shaft in the steering gear to the steering wheel, and the **Pitman shaft,** which with the **Pitman arm** is connected to the steering linkage. Internal gearing provides gear reduction and a 90-degree change in motion.

Rack-and-pinion gears have a steering shaft that connects the pinion gear to the steering wheel and a rack that is connected directly to the tie-rods. The contact between the pinion gear mounted on the steering shaft and the rack provides gear reduction and a change in motion. Operating theory and service procedures of both styles of steering gears are discussed in more detail in Chapters 17 and 33.

1.24 Steering Linkage

The steering linkage consists of the tie-rods and sometimes other rods or links that connect the steering gear to the steering arms. Several types of steering linkages are used with conventional steering gears. With rack-and-pinion steering gears, the two tie-rods connect to the ends or center of the rack. In all cases, the steering linkage is designed to transfer steering motions to the wheels in such a way that the vertical motion of the wheels over a bump will not steer the wheels; that is called bump steer.

Most cars with conventional steering gears use a **parallelogram steering linkage.** A center link or relay rod attaches to the Pitman arm and is supported at the other end by an idler arm. The Pitman arm and idler arm are mounted in the same relative position in the car, so the center link moves back and forth, straight across the car, during steering maneuvers. Two tie-rods connect the center link to each of the steering arms (Figure 1–50). The simplest steering linkage is used with rack-and-pinion steering gears; a tie-rod connects each end of the rack to a steering arm.

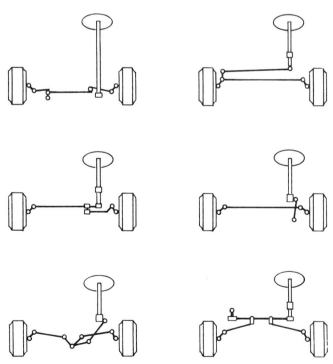

Above are shown various types of tie-rod assemblies

Figure 1–50 Some of the different steering linkage arrangements. The parallelogram linkage (upper left and lower right) is the most common. (*Reprinted by permission of Hunter Engineering Company*)

Pickups, trucks, and cars with solid axles use a single tie-rod that connects one steering arm to the other. A **drag link** connects the steering gear to a third steering arm, which is connected to one of the steering knuckles. Sometimes, instead of using a third arm, an extension of one of the steering arms is connected to the drag link. The drag link can run either across the car to the right wheel or lengthwise to the left wheel.

Connections at each end of the tie-rods and other links are made through a ball-and-socket joint commonly called a tie-rod end. This ball-and-socket joint allows the suspension to move up and down and the linkage to swing back and forth while transmitting motion. It also eliminates free play that would cause sloppy and dangerous steering.

1.25 Practice Diagnosis

While driving a car, you notice that if you let go of the steering wheel momentarily, it turns to the right. When you stop the car and look at the tires, all seem to be fully inflated. What kinds of things can cause this problem? What is the first thing you would do to locate the cause?

Terms to Know

- air suspension
- anchors
- antilock braking system (ABS)
- backing plate
- base brake
- beam axle
- brake fluid
- brake lining
- brake shoes
- brake warning light
- caliper
- chassis
- coil springs
- combination valve
- control arm
- conventional standard gear
- deenergizing
- diagonal split system
- drag link
- drum
- duoservo brake
- Elliott
- emergency brake
- energizing
- fixed caliper
- floating caliper
- forward shoe
- foundation brake
- four-wheel drive (4WD)
- frame
- front-wheel drive (FWD)
- friction material
- hydro-boost
- king bolt
- kingpin
- leading shoe
- leaf spring
- MacPherson strut
- master cylinder
- metering valve
- nonservo
- pads
- parallelogram steering linkage
- parking brake
- Pitman arm
- Pitman shaft
- pressure differential switch
- pressure differential valve
- primary shoe
- proportioning valve
- rack-and-pinion gear
- radius rod
- rear-wheel drive (RWD)
- reverse Elliott
- rotor
- scrub
- secondary shoe
- semitrailing arm
- servo action
- servo brake
- shock absorbers
- shoe
- short-long arm (S-L A)
- sliding caliper
- solid axle
- steering axis
- steering shaft
- strut rod
- strut suspension
- suspension system
- swing axle
- tandem hydraulic system
- toe-in
- toe-out
- torsilastic
- torsion bars
- track
- trailing arms
- trailing shoe
- transversely
- unequal arm suspension
- unibody
- vacuum booster
- wheelbase
- zero camber

Review Questions

1. The chassis consists of the car's frame and the:
   a. Suspension system
   b. Steering system
   c. Braking system
   d. All of the above

2. The ideal brake is one that:
   a. Can lock up all four wheels
   b. Gives the driver a means of easily controlling brake lockup
   c. Will stop the car without pulling or darting to the side
   d. All of the above

3. Brake lining:
   a. Is called friction material
   b. Is attached to the brake shoes
   c. Rubs against the brake drum or rotor during a stop
   d. All of the above

4. Statement A: Most modern cars use drum brakes.
   Statement B: Most modern cars use disc brakes.
   Which statement is correct?
   a. A only
   b. B only
   c. Both A and B
   d. Neither A nor B

5. In a duoservo brake:
   a. The leading shoe increases the pressure on the trailing shoe
   b. The leading shoe energizes the trailing shoe
   c. The primary shoe applies the secondary shoe
   d. Neither the primary nor the secondary shoe is energized by drum rotation

6. Which of the following brake designs will provide the most stopping power?
   a. Two–leading shoe
   b. Leading–trailing shoe
   c. Two–trailing shoe
   d. Duoservo

7. In a disc brake, the piston used to apply the brakes is in the:
   a. Wheel cylinder   c. Backing mount
   b. Caliper   d. Either a or b

8. Statement A: Floating calipers use a single piston, and sliding calipers use two pistons.
   Statement B: Fixed calipers use either two or four pistons.
   Which statement is correct?
   a. A only   c. Both A and B
   b. B only   d. Neither A nor B

9. Statement A: Disc brake designs can be classed as nonservo.
   Statement B: Disc brakes can be classed as duoservo.
   Which statement is correct?
   a. A only   c. Both A and B
   b. B only   d. Neither A nor B

10. The major reason for using hydraulic brakes on modern cars is that hydraulic operation:
    a. Produces an increase in force between the brake pedal and the shoes
    b. Allows the driver to move the shoes in a slow, controlled manner
    c. Produces an equal amount of pressure at each wheel cylinder
    d. Keeps the brakes cooler

11. Statement A: A tandem split brake system has two hydraulic circuits, one in front and the other in back.
    Statement B: A diagonal split brake system has one circuit with the right front and left rear brakes and one circuit with the other two brakes.
    Which statement is correct?
    a. A only   c. Both A and B
    b. B only   d. Neither A nor B

12. A pressure differential valve is used to:
    a. Delay the application of the front brakes
    b. Reduce pressure to the rear brakes
    c. Warn the driver of a hydraulic failure
    d. Ensure even application pressure at all the brake assemblies

13. Front-wheel drive cars use a diagonal split hydraulic system because:
    A. The different front-wheel geometry allows steering control while applying a single front brake
    B. These cars would have very poor stopping ability with only the rear brakes
    Which option best completes the statement?
    a. A only   c. Both A and B
    b. B only   d. Neither A nor B

14. Power brakes:
    A. Are essentially a standard brake system with the addition of a booster between the brake pedal and the master cylinder
    B. Significantly reduce stopping distances when very high pedal pressures are used
    Which option best completes the statement?
    a. A only   c. Both A and B
    b. B only   d. Neither A nor B

Chapter 1 ■ Introduction to the Chassis

15. Power boosters provide a pedal assist through the use of:
    a. A vacuum
    b. Hydraulic pressure
    c. An electric motor
    d. Any of these

16. Statement A: Cars use a hydraulically applied service brake system for normal stops.
    Statement B: Cars use a mechanically applied parking brake system to hold them stationary while parked.
    Which statement is correct?
    a. A only   c. Both A and B
    b. B only   d. Neither A nor B

17. A parking brake usually uses the rear brake shoes and drum or rotor and applies them through the use of a:
    a. Mechanical linkage
    b. Separate hydraulic system
    c. Series of vacuum lines and cylinders
    d. None of the above

18. An antilock braking system is designed to:
    a. Deliver more stopping power
    b. Operate with less pedal pressure
    c. Provide complete driver control over the car while stopping on wet or icy roads
    d. All of the above

19. An antilock braking system consists of:
    a. A set of wheel speed sensors
    b. Pressure control valves
    c. An electronic control module
    d. All of the above

20. Through time and use, as the car is driven, the:
    a. Brake lining wears out
    b. Brake fluid absorbs moisture and deteriorates
    c. Hydraulic seals harden and wear
    d. All of the above

21. When a brake job is done, the major criterion for the repair should be:
    a. Low cost
    b. Future driver and passenger comfort
    c. New-car braking performance
    d. Both a and b

22. A front suspension system consists of the:
    a. Springs and shock absorbers
    b. Control arms
    c. Steering knuckle
    d. All of the above

23. Statement A: The distance between the tire centers, measured across the car, is called *track*.
    Statement B: *Wheelbase* refers to the distance between the rear of the front tires and the front of the rear tires.
    Which statement is correct?
    a. A only   c. Both A and B
    b. B only   d. Neither A nor B

24. Statement A: Wheels are aligned to reduce tire scrub and wear.
    Statement B: When a tire leans inward at the top, it is toed in.
    Which statement is correct?
    a. A only   c. Both A and B
    b. B only   d. Neither A nor B

25. Most modern cars are of a _____ type construction.
    A. Frame plus body
    B. Unibody
    Which option best completes the statement?
    a. A only   c. Both A and B
    b. B only   d. Neither A nor B

26. A vehicle with nonindependent suspension uses:
    a. A swing axle
    b. A solid axle
    c. A 4WD axle
    d. None of the above

27. Which of the following is not a type of independent suspension?
    a. Swing axle
    b. Short-long arm
    c. Solid axle
    d. Trailing arm

28. Statement A: An S-L A type of suspension causes a camber change as the wheel moves up and down.
    Statement B: The upper arm is shorter than the lower arm.
    Which statement is correct?
    a. A only   c. Both A and B
    b. B only   d. Neither A nor B

29. A strut suspension uses _____ to control the position of the tire.
    a. Two equal-length control arms
    b. Two unequal-length control arms
    c. A lower control arm and an oversized, telescoping shock absorber
    d. A pair of trailing arms

30. The assembly that connects the top of the strut to the fender well provides:
    a. A bearing to allow steering
    b. A flexible mounting to allow strut-angle changes.
    c. A dampener to isolate road vibrations
    d. All of the above

31. Technician A says that an A-arm has two inner bushings.
    Technician B says that a strut rod is used with control arms that have a single inner bushing.
    Who is correct?
    a. A only   c. Both A and B
    b. B only   d. Neither A nor B

32. A rubber control arm bushing does not:
    a. Reduce the transmission of road vibrations
    b. Operate without lubrication
    c. Cause resistance to suspension movement
    d. Allow the control arm to swing freely

33. A ball joint or tie-rod end allows a:
    A. Swinging or pivoting motion
    B. Rotating motion
    Which option best completes the statement?
    a. A only   c. Both A and B
    b. B only   d. Neither A nor B

34. Statement A: Springs absorb energy when they are compressed.
    Statement B: The shock absorbers help the spring support the car's weight.
    Which statement is correct?
    a. A only   c. Both A and B
    b. B only   d. Neither A nor B

35. A car's springs can be of the _____ type.
    a. Leaf spring   c. Torsion bar
    b. Coil spring   d. All of the above

36. A(n) _____ spring system has the ability to automatically maintain a constant car height.
    a. Air   c. Coil
    b. Leaf   d. Torsion bar

37. When a car's tires meet a bump in the road:
    A. The springs and shock absorbers compress to allow upward tire travel
    B. The springs and shock absorbers extend to allow downward tire travel
    Which option best completes the statement?
    a. A only   c. Both A and B
    b. B only   d. Neither A nor B

38. The steering gear:
    A. Changes the rotating steering wheel into a back-and-forth motion
    B. Provides a gear ratio to make steering easier
    Which option best completes the statement?
    a. A only   c. Both A and B
    b. B only   d. Neither A nor B

39. Statement A: A parallelogram type of steering linkage is usually used with rack-and-pinion steering gears.
    Statement B: Tie-rods are used to connect the steering arms to the steering gear or steering linkage.
    Which statement is correct?
    a. A only   c. Both A and B
    b. B only   d. Neither A nor B

40. The steering linkage:
    A. Transmits steering motions from the steering gear to the steering knuckle
    B. Helps hold the front wheels in a straight-ahead position or turning direction
    Which option best completes the statement?
    a. A only   c. Both A and B
    b. B only   d. Neither A nor B

2 Principles of Braking

Objectives

Upon completion and review of this chapter, you should be able to:

- ❑ Understand the physical forces that act on a vehicle to produce a stop and the limitations that these forces place on stopping ability.
- ❑ Identify the various methods used to measure braking power.
- ❑ Understand the purpose of friction material and the requirements placed on it.
- ❑ Understand the safety hazards that result from working around asbestos-based lining materials.
- ❑ Understand the effect of wheel lockup during a stop.

2.1 Introduction and Federal Requirements

Many aspects of slowing and stopping a car are controlled by simple physics. These are natural laws dealing with the deceleration of a body in motion that cannot be violated. An understanding of these basic principles aids greatly in thoroughly understanding brakes and the limitations placed on them by the laws of physics.

The federal (U.S.) laws pertaining to brakes apply to new car or vehicle braking systems. Federal Motor Vehicle Safety Standard (FMVSS) number 105 (developed by the Society of Automotive Engineers [SAE] and the National Highway Traffic Safety Administration [NHTSA]) requires that a passenger vehicle with a gross vehicle weight (GVW) of less than 8,000 lbs (3,630 kg) be able to stop from a speed of 60 mph (100 km/h) in less than 216 ft (65.9 m) and stay within a 12-ft-wide (3.7-m-wide) lane (Figure 2–1). With a loss of power assist, it must stop in less than 456 ft (139 m). This standard also requires that the parking brake be able to hold the vehicle stationary for at least 5 minutes in both a forward and reverse direction while parked on a 30 percent grade. Federal standards also place requirements and a timetable for eliminating asbestos from original equipment manufacturer (OEM) brake lining. Other FMVSS standards that pertain to motor vehicle brakes are 106, brake hoses; 108, lighting; 116, motor vehicle brake fluid; 121, air brake systems; 122, motorcycle brake systems; and 211, wheel nuts, disc, and hub caps.

Much of the discussion in this chapter refers to maximum stopping ability as encountered in panic stops.

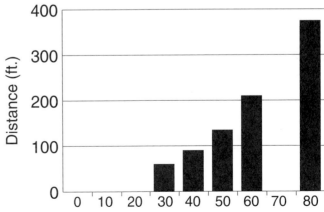

*The Post-Burnish Test is conducted to measure the stopping ability of the "seated in" brake pads/shoes.
Source - Abex Friction Products

Figure 2–1 The stopping requirements of FMVSS number 105 for a new vehicle with burnished brake lining. (*Courtesy of Wagner Brake*)

These are unusual conditions, but each brake system must be prepared for such emergency situations.

2.2 Deceleration

The brakes stop a car by decelerating it until its speed is zero. Deceleration is the opposite of acceleration, which is the action of increasing speed or velocity. The rate of deceleration or acceleration is commonly measured using **g** (an abbreviation for gravity) as the standard unit.

The force of gravity causes an object to fall when dropped. While falling, the object accelerates to a speed of 32.2 ft per second (fps) in 1 second; this speed is equal to 9.8 m per second (mps), 21.95 miles per hour (mph), and 35.3 km per hour (km/h). The object continues to accelerate at this rate for each second it falls until it either reaches the ground or aerodynamic drag prevents any further speed increase. For example, a falling steel ball reaches a speed of about 64.4 fps (19.6 mps) in 2 seconds and 322 fps (98 mps) in 10 seconds (Figure 2–2).

A reduction in speed or deceleration of 32.2 fps in 1 second (fps per second, fps/s, or fps^2) is referred to as a stopping rate of 1 **g**. The average passenger car can obtain a best stop of about 0.6 to 0.8 **g** (19.32 to 25.76 fps/s, 5.88 to 7/85 mps/s), with 0.7 **g** (22.54 fps/s, 6.87 mps/s) about the average. An Indianapolis-type or Formula 1 race car can stop at rates greater than 1.25 to 1.5 **g** (40.25 to 48.3 fps/s, 12.27 to 14.72 mps/s).

The typical driver rarely encounters these stopping rates. The average stop made with a passenger car is less than 0.2 **g** (6.44 fps, 1.96 mps); higher stopping rates are considered unpleasant by most drivers. A 100-lb object experiences a forward pressure of 20 lb at a 0.2-**g** stopping rate, 50 lb at 0.5 **g**, 100 lb at 1 **g**, and 150 lb at 1.5 **g** (Figure 2–3). A 1-**g** stop creates a rather heavy load on the seat belt, and anything that is not fastened down will fly forward—relative to the car—during such a stop (Figure 2–4).

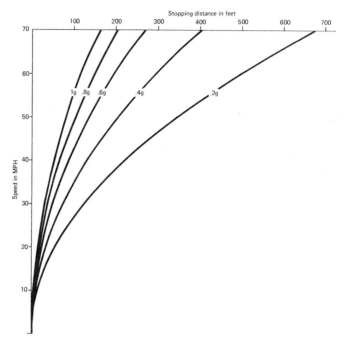

Figure 2–3 Relative stopping distances at five different deceleration rates. Note how the stopping distance increases as the speed increases.

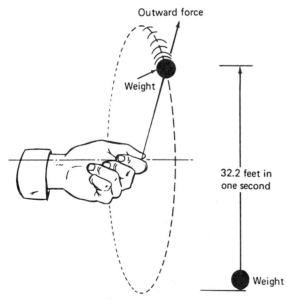

Figure 2–2 The force of gravity causes an object to accelerate at a rate of 32.2 fps during each second that it falls (assuming no aerodynamic drag). If an object (left) is spun fast enough, it can generate enough centrifugal force to keep the string tight even in a vertical position. This force is 1 **g** or greater.

Figure 2–4 During a hard stop, inertia places a forward-acting force on everything in the car. The driver pushes on the steering wheel and floorboards to resist this forward motion. The seat belt also resists inertia forces.

If a stop is made at the rate of 0.5 **g** from 55 mph (88.5 km/h, 80.6 fps, 24.6 mps), the speed is reduced at the rate of 16.1 fps for each second of the stop. The speed of the vehicle at succeeding 1-second intervals is 64.5, 48.4, 32.3, 16.2, and 0.1 fps (almost zero). This stop takes slightly more than 5 seconds, and the car will travel about 161 ft (49.1 m) (Figure 2–5). A 0.2-**g** (6.44-fps, 1.96-mps) stop from the same speed would decelerate the car at 1-second intervals to speeds of 74.16, 67.72, 61.28, 54.18, 48.4, 41.96, 35.52, 29.08, 22.64, 16.2, 9.76, and 3.32 fps. This stop takes more than 12 seconds, and the car will travel about 504 ft (153.72 m). It is easy to see that it takes less time and distance to stop a car at a higher stopping rate (Figure 2–6).

The term **braking efficiency,** often used when testing brakes, refers to the brake's ability to stop a vehicle. The amount of braking force generated by the brakes is divided by the weight of the vehicle to determine braking efficiency. To achieve 100 percent efficiency, the braking force must equal the vehicle weight, which will produce a deceleration rate of 1 **g.** If the braking force were 60 percent of the weight of the vehicle, the braking efficiency would be 60 percent and the stopping rate would be 0.6 **g.** Like the deceleration rate, braking efficiency is affected by the condition of the road surface, the tires, and the brakes.

2.2.1 Measuring Deceleration Rates

Automotive brake system repair technicians and engineers use several methods to measure deceleration rates in order to determine the quality of a braking system's overall operation. A device called a **decelerometer** can be mounted in the car. Some versions are attached to the windshield; others sit on the floor or seat. A decelerometer gives a readout of the stopping rate in **g** rate, feet per second^2, or meters per second per second; the readout is on a meter or a liquid column (Figure 2–7).

A rather crude but fairly accurate method of determining deceleration rates is to time the stop (i.e., measure the

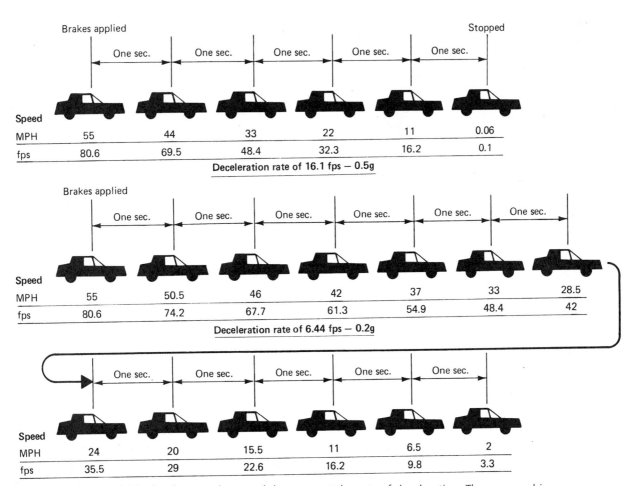

Figure 2–5 When a vehicle decelerates, the speed decreases at the rate of deceleration. The upper vehicle is decelerating at a rate of 0.5 **g**—16.1 fps/s. The lower vehicle is decelerating at a rate of 0.2 **g**—6.44 fps/s.

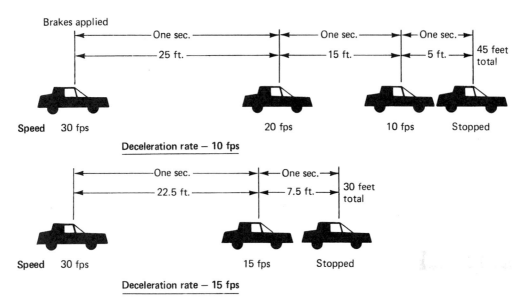

Figure 2-6 When a vehicle slows at a deceleration rate of 10 fps/s—about 0.3 **g**—it stops from a speed of 30 fps in 3 seconds and travels 45 ft. When stopping at a rate of 15 fps/s—almost 0.5 **g**—the lower vehicle stops in 2 seconds and travels 30 ft.

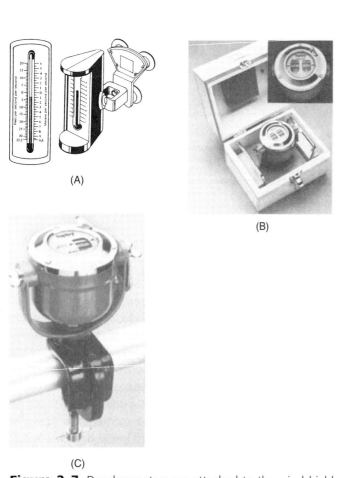

Figure 2-7 Decelerometers are attached to the windshield (A), set on the floor or seat (B), or clamped onto some part of the vehicle (C). As the vehicle makes a stop, the meter displays the deceleration rate in stopping distance, deceleration rate, or braking efficiency. *(Courtesy of Ammco Tools, Inc.)*

distance of the stop from the point where the brakes were applied to the point of complete stop). The deceleration rate is then calculated by using the appropriate formula:

$$\mathbf{DRg} = \frac{\mathbf{S}}{22\,\mathbf{T}} \qquad \mathbf{DRg} = \frac{\mathbf{S^2}}{29.9\,\mathbf{D}}$$

where **DRg** = deceleration rate (in **g**)
  S = speed at the beginning of the stop (in mph)
  T = time (in seconds)
  D = distance traveled during stop (in feet)

or

$$\mathbf{DRFPS^2} = \frac{1.476\,\mathbf{S^2}}{\mathbf{T}} \qquad \mathbf{DRFPS^2} = \frac{1.076\,\mathbf{S^2}}{\mathbf{D}}$$

where **DRFPS²** = deceleration rate (in feet per second per second)

A high number such as 0.7 **g**, 22.5 fps, or greater means the brakes are working quite well, and a low number indicates that they are not as effective.

A brake technician seldom computes deceleration rates, because the numbers are not actually useful when repairing brake systems. Also, the time involved in meeting the requirements of a proper road surface with no traffic makes deceleration measurements difficult. Deceleration rates are measured from fairly high speeds and represent the maximum values a vehicle can achieve. This requires good tires, a stretch of clean, dry pavement with no traffic, and usually a passenger or observer to take the readings or measurements. These requirements, plus the time involved, make actual deceleration tests practical for the engineer developing a brake system for a particular car but impractical for the repair technician. The brake technician usually makes an educated guess about stopping rates during a road test.

Most states require that a passenger car be able to stop in a distance of 25 ft (7.6 m) from a speed of 20 mph (32 km/h). This is a stopping rate of 0.53 **g**, or 17.3 fps.

2.3 Energy of Motion

It takes energy to move a car. **Energy** is the ability or capacity to do work. **Work,** in this sense, is the process of moving something or changing the speed or direction of its travel. According to the laws of physics, energy cannot be created or destroyed. It can, however, be changed from one form to another. Energy is available in many forms. Heat, light, various fuels, and electricity are common examples. In an automobile, the potential energy of gasoline is converted to heat and then to mechanical motion within the engine. When the car is put into motion, this energy becomes **kinetic energy:** energy of motion.

The amount of energy in a moving object, such as a car, can be calculated by using the following formula:

$$\text{kinetic energy} = \frac{W = S^2}{29.9}$$

where **W** = weight of car (in pounds)
S = speed (in miles per hour)

Unless you are an engineer, you will have little use for this formula other than to realize that a heavy car in motion has more kinetic energy than a light car and, even more important, that the amount of energy increases at a rate equal to the square of the speed. For example, a 1,000-lb (454.5-kg) car going 30 mph (48.27 km/h) represents 30,100.3 ft-lb (40,816 N-m) of kinetic energy. A 2,000-lb (909-kg) car going the same speed represents 60,200.6 ft-lb (81,632 N-m) of energy (twice as much). At 60 mph (96.54 km/h) the lighter car would represent 120,401.3 ft-lb (163,264.16 N-m) of kinetic energy and the heavier car 240,802.6 ft-lb (326,528.3 N-m) of kinetic energy. Note that as the weight of the car is doubled, the amount of energy is also doubled, but as the speed is doubled, the amount of energy is increased about four times (Figure 2–8).

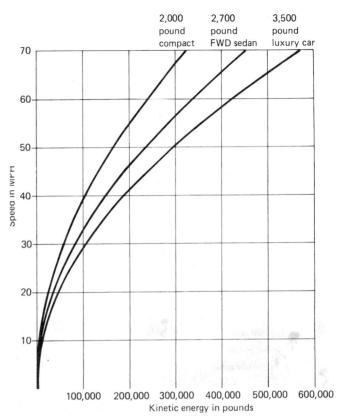

Figure 2–8 The amounts of kinetic energy of three different vehicles at various speeds. Note how the amount of energy increases with an increase in weight and speed.

Every moving object contains kinetic energy, and energy must be converted from some form and put into that object to start it moving. Because energy cannot be created or destroyed, the energy in an object must be removed in order to stop it. The kinetic energy needs to be converted to a different form of energy (usually heat), and as the speed increases, the amount of energy needed for this conversion greatly increases.

2.4 Friction and Heat Energy

The simplest way to stop a car is to convert the kinetic energy to heat. Heat is one of the most common, versatile forms of energy. This conversion occurs naturally when a car coasts to a stop. The rolling friction of the axle shafts and wheel bearings, the flexing friction of the sidewalls and treads of the tires, and the molecular friction of the air drag over the body and undercarriage all create small amounts of heat. As the kinetic energy becomes heat energy, the car decelerates and comes to a stop. A car with very efficient bearings, tires, and aerodynamic design coasts for a long distance before using up its kinetic energy.

When we lack the time or distance needed to make coasting stops, we use the brakes. Brakes are essentially heat machines. They generate heat from friction by rubbing the lining against the rotating rotors or drums. Friction can generate large amounts of heat. Prove this to yourself by vigorously rubbing your hands together or by running and then sliding on a rug or carpet. Be careful when you do this—anyone who has experienced "rug burn" knows why. Another good example of heat generated by friction is the smoke pouring off a tire that is locked up and skidding. This tire is generating enough heat to burn up the tread rubber.

2.5 Heat Energy

Heat is measured in two forms: **intensity** and **quantity.** Heat intensity is what we feel—how hot it is—and is measured in degrees. The two commonly used scales are **Fahrenheit (F)** and **Celsius (C).** These two scales do the same thing, but they have different lengths and starting points. Easily remembered places on these scales are the melting point of ice, 32°F or 0°C, and the boiling point of water, 212°F or 100°C.

Heat quantity is the amount of heat in an object or area, or the amount of heat that an object can absorb or release. Heat quantity also has two measurement systems: **British thermal units (BTUs)** and **calories (c).** One BTU is the amount of heat it takes to increase the temperature of 1 lb of water by 1°F. One calorie is the amount of heat it takes to increase the temperature of 1 g of water by 1°C. A BTU is a much larger quantity: 1 BTU = 252 c.

It is important for the brake technician to understand heat energy. One BTU is equal to 778 ft-lb of energy. Conversely, 1 ft-lb of energy is equal to 0.0013 BTUs. The energy of a 2,000-lb car going 30 mph is equal to 77.38 BTUs and that of the same car going 60 mph is equal to 309.5 BTUs. If we were to remove that much heat energy from these two moving cars, they would stop.

Heat can be transferred from one object to any other object that is cooler. The object can be in solid, liquid, or vapor form, but it must be cooler. Heat always flows from a warmer object to a cooler object. As the cooler object absorbs heat, its temperature increases, and it becomes hotter. As the hotter object loses its heat, it becomes cooler.

Heat generated at the brake lining and rotor or drum surfaces flows to the lining and into the rotor or drum and then to the cooler air passing over them (Figure 2–9). The temperature of the brake components increases during

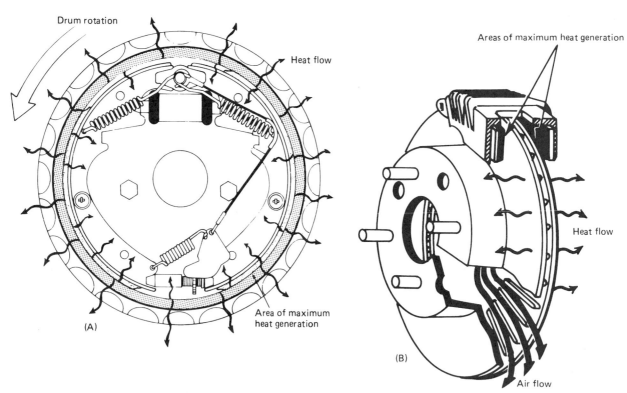

Figure 2–9 As the brakes are used, heat is generated where the lining rubs on the drum or rotor. This heat is either absorbed by the brake components or dissipated into the surrounding air.

each stop. The amount of increase is determined by the vehicle speed and weight, the rate of the stop, and the mass of the brake components, especially that of the drums and rotors. If the stop is gradual, from a slow speed, and short in duration, the heat will probably flow away as quickly as it is generated, and the components will not get much hotter than the **ambient** temperature (the temperature of the surrounding air). If the stop is rapid, from a high speed, if the car is heavy, or if there are several stops in a short period of time, the brakes will generate more heat than can be easily dissipated. Then the brake components will get very hot. Carefully feel the rotors or drums after a few hard stops; they can be very hot and could easily boil water. Under extreme conditions, temperatures can be high enough to cause brake lining fade or even high enough to boil the brake fluid in the wheel cylinders.

In the past, passenger-car brake temperatures stayed relatively low—about 250°F (120°C) for average usage and 350°F (175°C) for heavy usage. In today's lighter cars the brake components have become substantially smaller, so there is less metal to absorb the heat and act as a heat sink during stops. An average brake temperature for a modern FWD car is about 350°F (177°C), with heavy usage increasing it to as high as 500°F to 800°F (260°C to 425°C) (Figure 2–10). The brake components of modern systems, especially the lining and fluid, have to withstand considerably higher temperatures than those of the past.

2.6 Brake Lining Friction

Friction is the resistance as one surface slides over another. Friction produces heat. The brake lining is designed to produce heat from friction as it rubs against the rotor or drum and is commonly called **friction material.** The amount of heat produced is determined by the coefficient of friction between the lining and the rotor or drum, the amount of pressure pushing them together, and the relative speed of each.

The term **coefficient of friction** refers to the amount of resistance preventing one item from sliding across another. It is calculated by sliding one object across the surface of another, measuring the amount of force required, and then dividing that force by the weight of the moving object. For example, if it takes a 2-lb (0.9-kg) force to slide a 100-lb (45.45-kg) block of ice across a certain surface, the coefficient of friction is 2 divided by 100, or a very low 0.02. If it takes 70 lb (31.81 kg) of force to slide a 100-lb block of rubber across the same surface, the coefficient of friction is 70 divided by 100, or a rather high 0.70 (Figure 2–11).

Friction material, normally called brake lining, has to provide the proper amount of drag or friction as it rubs against the drum or rotor under various heat conditions from extremely cold—at the beginning of a stop in the winter—to extremely hot—at the end of a high-speed stop in the summer. At the same time, there must be a minimum amount of wear on the drum or rotor friction surfaces.

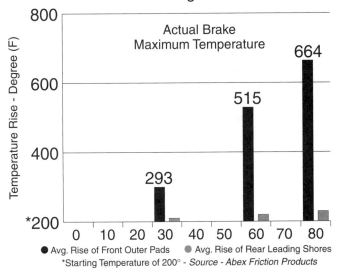

Figure 2–10 This graph shows the temperature increase of the front and rear brakes during stops at different speeds. *(Courtesy of Wagner Brake)*

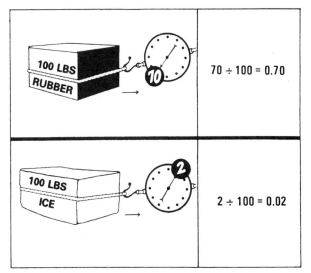

Figure 2–11 The coefficient of friction between two different materials is determined by dividing the amount of drag needed to move one of the materials by the amount of force pushing them together. *(Courtesy of Ford Motor Company)*

The coefficient of friction of most passenger-car brake linings is about 0.3. This represents a usable and controllable amount of friction. If the coefficient of friction is too low, the shoes will not produce enough friction and heat to stop effectively. The result will be a "hard" brake pedal with poor stopping power. If the coefficient of friction is too high, the brakes will be too "grabby" and will be very hard to control. Wheel lockup and skids will occur too easily.

The coefficient of friction of some lining blends changes as the lining heats up. An inferior lining often fades—undergoes a loss or lowering of the coefficient of friction—as it gets hotter. The stopping power drops as the brake pedal gets harder. Some linings experience an increase in the coefficient of friction and become more grabby as they get hotter. Quality lining materials undergo a very minor change, if any, in their coefficient of friction as they heat and cool (Figure 2–12).

2.6.1 Brake Lining

Brake lining is formed using different recipes and may contain the primary ingredient in several different forms. The major ingredient in the lining is the base frictional material. It may consist of a powder, short fibers, or longer fibers that are woven into cords or mats, and it provides the friction and the ability to resist heat. Other important ingredients combined in a lining are the following:

- **Binders:** usually a resin that holds all the other ingredients together
- **Friction modifiers:** materials that change or adjust the coefficient of friction
- **Fillers:** materials that can improve the strength and the ability to transfer heat or that provide quiet operation
- **Curing agents:** materials that provide the correct chemical reaction during manufacture and curing

Some of the methods for producing brake lining material involve an expensive manufacturing or curing process. The exact recipe for a lining type is usually a carefully guarded secret (Figure 2–13).

At one time, brake lining material was placed in one of the following two categories: **organic** and **inorganic**. Most so-called organic linings had **asbestos** as the base friction material, and most inorganic linings were **semimetallic**. These categories identified the primary friction material for the linings. This classification system seemed somewhat odd because both compounds—asbestos and metal—are inorganic. With the reduction in the use of asbestos (it has not been used on new domestic cars since 1993 but is still used in replacement linings), the major classifications for newer lining types are **nonasbestos organic (NAO)** and semimetallic. Most NAO linings use fiberglass or aramid fibers as a base. Asbestos is known to cause health hazards. Do not breathe the dust from any lining because the other materials can also cause problems (Figure 2–14).

Semimetallic lining has become the standard **original equipment (OE)** lining for most new FWD cars. The brownish-black dust on many front wheels is evidence of semimetallic lining. The brown color comes from rust of the metal dust, and the black is from graphite, a friction modifier. Typical front rotor temperatures have risen from about 350°F (177°C) in the early 1970s to about 550°F (288°C) in the mid-1990s. Metallic-based linings have superior high-temperature wear characteristics. Most of the other lining types wear quite rapidly at the higher temperatures (Figure 2–15). It is important that the replacement lining be of the same type as the OE lining so it will have the same wear and friction characteristics (Figure 2–16).

Brake lining quality is difficult to judge because there is no government or industry standard for aftermarket lining. Well-known lining manufacturers with good reputations spend a good deal of time and money to ensure the quality of their product. Their reputation is usually a sufficient basis for selecting good lining. Most manufacturers

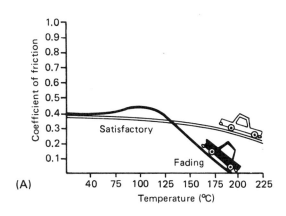

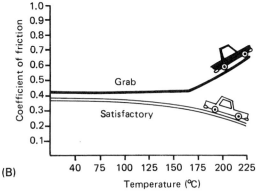

Figure 2–12 The dark brake lining curve (A) is showing severe fade as it gets hot. The dark curve in (B) shows a grabby condition as it gets hot. *(Courtesy of Ford Motor Company)*

Friction Material

Type	Ingredients	Advantages	Disadvantages
Organic	Early: wood and leather Recent: asbestos and other ingredients	Quiet, cheap, low abrasiveness, good cold friction	Asbestos content, brake fade when hot
Metallic	Powdered metal	Fade resistant	Poor cold friction, high pedal pressure, abrasive, noisy
Semimetallic	Combination	Fade resistant, long wear life	Expensive, brittle, poor cold friction
Synthetic	Fiberglass	Good lining life, quiet, nonabrasive	Expensive, not good for very high temperatures
	Aramid	Very good lining life, quiet, nonabrasive	Poor cold performance

Figure 2–13 Some of the advantages and disadvantages of different lining types.

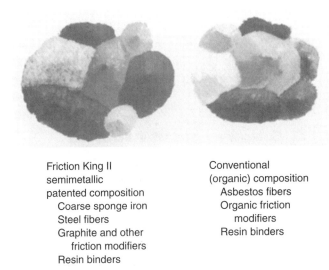

Friction King II semimetallic patented composition
  Coarse sponge iron
  Steel fibers
  Graphite and other friction modifiers
  Resin binders

Conventional (organic) composition
  Asbestos fibers
  Organic friction modifiers
  Resin binders

Figure 2–14 A brake lining is formed when the various ingredients are mixed and molded into shape. *(Courtesy of Bendix Brakes, by AlliedSignal)*

produce several types of lining that differ in quality, from an **economy** or **competitively priced** form that is equal to the OE lining to a **premium** or **heavy-duty** quality; these types are usually available in both an NAO and semimet (short for semimetallic) material. Some of the criteria used to judge the quality of a lining are as follows:

- High resistance to fade
- Quick, full recovery if fade occurs
- Long wear life
- Little or no rotor or drum wear
- Quiet operation
- Good friction characteristics when damp or wet (Figure 2–17)

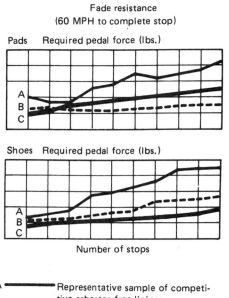

A ———— Representative sample of competitive asbestos-free linings

B ------- Wagner standard asbestos grade linings (PD pads and PAB shoes)

C ———— Wagner asbestos-free linings (NA pads and WNA shoes) and Wagner premium asbestos linings (WD pads and WEB shoes)

Figure 2–15 A comparison of the amount of pedal pressure required to stop a car for three different linings for a series of ten stops. Note the differing amounts of pedal pressure required to bring the car to a complete stop as the brakes heat up. *(Courtesy of Wagner Brake)*

The coefficient of friction of a lining can be easily determined by reading the **edge code** or **edge brand** printed on the edge of the lining (Figure 2–18). This code was established by the **SAE**. It is composed of three groups of letters and numbers. The first group is a series of letters that identify the manufacturer of the lining. The second group is a series of numbers, letters, or both that identify the lining compound or formula. The third group is two letters that identify the coefficient of friction; the first letter is determined from cold tests and the second from hot tests. The meaning of these code letters is provided in Figure 2–19. The friction code determines only the relative amount of drag that will be generated in that brake. Most passenger cars have a code of EE or EF. A low coefficient of friction, less than 0.15, will probably produce a hard brake pedal with poor braking power in most cars. A high coefficient of friction, more than 0.55, will probably produce a grabby brake. Other factors that determine lining quality such as fade resistance, wear on drums or rotors, wear life of the lining, moisture resistance, erratic behavior, quiet operation, and structural integrity are not included in this coding system.

The terms *hard* and *soft* are often used to refer to the coefficient of friction and fade characteristics of a lining. A soft lining is one with a fairly high coefficient of friction. It tends to break in fairly quickly and easily, have quiet operating characteristics, and fade at relatively low temperatures. A hard lining has a lower coefficient of friction. It tends to take longer to break in, has a greater tendency to squeal, and has better high-temperature characteristics.

2.6.2 Concerns about Asbestos

It has been determined that asbestos can cause health problems. If the fibers are inhaled, the human body is not able to expel them, and these fibers can cause asbestosis or form the base for a cancerous growth.

Asbestos is a mineral that easily separates into long, flexible fibers. It makes a near-perfect friction material and heat insulator because these fibers are noncombustible, do not conduct electricity, and do not transmit

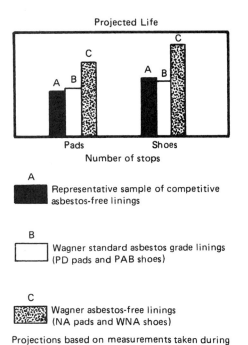

Figure 2–16 The projected life of three different linings calculated under severe operating conditions. *(Courtesy of Wagner Brake)*

	Brake Lining Characteristics[a]						
	Organic			Metallic			
Lining Material	DM-5470	DM-5490	DM-8010	DM-8011	DM-8015	DM-8015A	DM-8032
Duty	All-purpose moderate	All-purpose moderate	All-purpose moderate	All-purpose moderate	All-purpose heavy duty	All-purpose heavy duty	All-purpose heavy duty
Noise characteristic[b]	Superior	Superior	Excellent	Good	Good	Excellent	Good
Life characteristic	Superior	Superior	Excellent	Excellent	Superior	Superior	Superior
Displacement characteristic	Good	Excellent	Excellent	Excellent	Superior	Superior	Superior
MVSS[c] characteristic	Good	Good	Excellent	Excellent	Superior	Superior	Superior

[a] Based on comparable Delco Moraine product standards.
[b] Noise performance is affected by rotor design and internal rotor dampening. Ratings shown are based on moderately dampened systems.
[c] Motor Vehicle Safety Standard.

Figure 2–17 This chart shows the operating characteristics for three organic and four metallic lining types. *(Courtesy of General Motors Corporation, Service Technology Group)*

heat. Most health problems are related to the amphibole type of asbestos, which is used mostly for insulation. Brake and clutch friction material comes from the chrysotile asbestos family.

There are several forms of asbestos. The amphibole group contains silica as part of its chemical makeup; the outer layer of this fiber is totally resistant to human bodily fluids. It will not dissolve, so it can be expelled from the body. The outer layer of a chrysotile fiber is magnesia, which is readily attacked by bodily fluids and expelled from the body. Fewer health problems come from friction materials than from insulation. This fact should ease the mind of the technician performing brake repairs, but some degree of concern for personal safety should still be exercised. The cases of health problems traced to brake repair involve a lag time of about 15 to 20 years between exposure and the time that symptoms develop. In one case involving the child of a technician, the problem was traced to brake dust brought home on the technician's clothing.

As brakes are used and heat up, the resins that bind the fibers together break down. The asbestos fibers that have worn off the shoe are released inside the drum. These fiber particles are very small—usually too small to be seen by the unaided eye. Breathing them along with other dust particles may cause injury. Many sources recommend that a face mask be worn while doing brake work and that dust from worn brakes be carefully collected and then disposed of in the proper, approved manner.

The **Occupational Safety and Health Administration (OSHA)** has set standards concerning the acceptable level of asbestos fibers that can be present in a workplace or area where brakes are being repaired. OSHA has also determined safe methods that should be used to reduce exposure to these fibers. It has been determined that asbestos fibers cling together and mat easily with water and do not become airborne when wet. These fibers can be captured using a vacuum cleaner equipped with a high-efficiency particulate air (HEPA) filter, which provides a relatively easy and safe method of cleanup. These procedures are described in Chapter 22.

2.7 Tire Friction

Tire-to-road friction—commonly called **traction**—allows the driver to start and stop a car and control where it goes. A rolling tire follows a path in the direction it is pointed. A car, of course, goes where the wheels are pointed. The amount of traction that a tire can produce is determined by the load on the tire and the coefficient of friction between the tire and the road. A large amount of traction allows a car to accelerate, brake, and turn corners quickly. Poor traction produces tire spinning on acceleration, lockup on braking, and skidding on cornering and braking maneuvers.

Interestingly, a tire produces its highest traction when it is slipping slightly on the road surface, about 15 to 25 percent (Figure 2–20). This increased traction is a result of the elastic tread surface creeping into maximum contact with the road surface. This slippage occurs when an accelerating tire is turning slightly faster than the vehicle's speed or a decelerating tire is being braked to a speed slightly slower than that of the car. For example, maximum braking traction will occur for a car that is traveling 55 mph (88.5 km/h) if the tires are slowed to a speed of about 45 mph (70 km/h). At this slippage rate, about 20 percent, the tire usually produces a very light skid mark.

A tire that is slipping at a rate greater than 30 percent begins to suffer a loss of traction and a loss of directional control. A skidding tire skids sideways as easily as forward or backward.

Figure 2–18 The lettering on the edge of this lining is called an edge code. This code—DELCO 235FE—indicates that the lining was made by Delco Moraine and the friction class is FE. The manufacturer's code to identify the lining type and composition is 235.

Code Letter	Coefficient of Friction
C	Not over 0.15
D	Over 0.15 but not over 0.25
E	Over 0.25 but not over 0.35
F	Over 0.35 but not over 0.45
G	Over 0.45 but not over 0.55
H	Over 0.55
Z	Not classified

Figure 2–19 The last two letters of the edge code indicate the coefficient of friction of the lining. The coding for the letters is shown here.

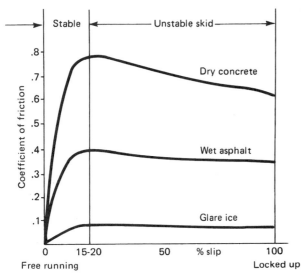

Figure 2–20 The road-to-tire coefficient of friction also varies with tire-to-road slip rate. The greatest amount of traction occurs at about 15 to 20 percent slip.

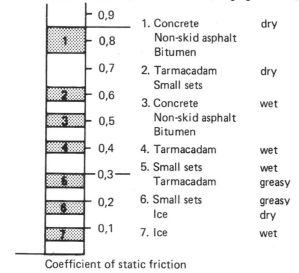

Figure 2–21 Tire traction depends on the tire-to-road coefficient of friction, which depends greatly on the type and condition of the road surface. *(Courtesy of ITT Automotive)*

If braking force exceeds the traction of the tires, lockup and a skid will result. A skidding tire slows a car down because of the heat generated by the tire-to-road friction, but this is usually much less than the heat that can be generated by the shoes and the rotor or drum while the tire rotates. The highest deceleration rate that a car can attain is determined by the traction of the tires. The highest value of the tire-to-road coefficient of friction is about 0.8 on dry pavement. Wet or icy pavement greatly reduces this coefficient of friction and the amount of traction and thus greatly reduces stopping ability. The coefficient of friction between a tire and wet concrete pavement is about 0.5 and for ice it is less than 0.1 (Figure 2–21).

2.8 Brake Power

The amount of power that a brake assembly can generate varies with the radius of the tire, the radius of the rotor or drum, the coefficient of friction of the lining, the application force at the lining, the weight on the tire, and the coefficient of friction between the tire and the road. Brake power is often called **brake torque** because it resists the turning of the rotor or drum (Figure 2–22). The tire-to-road factor enters into the amount of brake torque that can be generated because lockup will result if the amount of brake torque exceeds the amount of traction. The brake shoes develop a rotating retarding force (torque) between the tires and the suspension, measured in foot-pounds or newton-meters.

The actual amount of braking force for a given vehicle is calculated using a formula similar to that in Figure 2–23. The most accurate way to determine the amount of braking force is to work from the deceleration rate using the following formula:

$$F = DRg = 2{,}000$$

where F = force, braking force per ton of vehicle weight (in pounds)
DRg = deceleration rate (in g)

As the brakes are used, the linings become hot. The amount of heat (the number of BTUs or calories) is determined by the velocity change and the weight of the car. The amount of temperature increase in the lining and shoes is determined by the amount of lining contact area, the amount of rotor or drum surface area, the ambient temperature, and the amount of airflow present. The lining contact area—usually measured in square inches or square centimeters—is important because it influences the rate of heat absorption. If we were to reduce the lining contact area by one-half, we would double the rate of temperature rise in the lining that remains. Temperatures of more than a few hundred degrees can have a drastic effect on the lining's coefficient of friction and wear rate. Of concern to brake technicians, the lining contact area on newly replaced shoes should be as complete as possible to prevent the possibility of overheating. Good contact between the disc brake lining and the rotor is fairly easy to obtain because the surfaces are flat. Good drum-to-shoe contact is more difficult because both surfaces

Chapter 2 ■ Principles of Braking

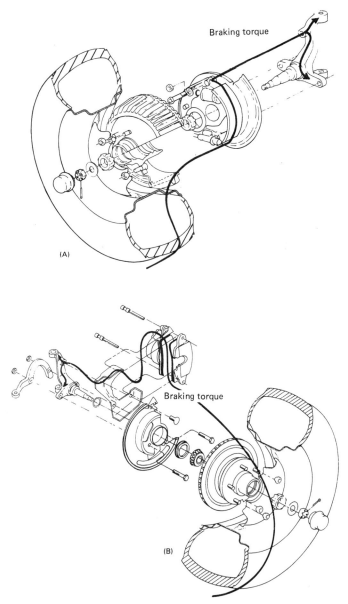

Figure 2-22 As the brakes are applied, a retarding force—between the tire's road contact and the car's suspension—is created. Because it is a rotating force, it is often called brake torque.

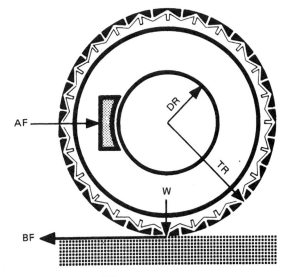

$$BF = W \times \frac{TR}{DR} \times \frac{RCF}{LCF}$$

BF = Brake force
W = Weight
TR = Tire radius
DR = Brake drum radius
RCF = Coefficient of friction, road
LCF = Coefficient of friction, lining

Figure 2-23 The amount of braking force that can be generated by a tire and brake assembly can be determined using this formula. *(Courtesy of ITT Automotive)*

---

are curved, and the curvature of the shoe must match that of the drum. Poor contact occurs if the curves do not match. Hard stops with a newly replaced, improperly fitted lining can cause excessive temperatures in those areas of the lining that are making contact. Excessive temperatures can prematurely and permanently damage the lining. Methods for checking and correcting the lining contact are described in Chapter 22.

An engineering term commonly used to describe potential brake power is **swept area.** This is the area of the rotors or drums that is swept or rubbed by the shoes. To calculate the swept area for a rotor (illustrated in Figure 2-24), the following formula can be used:

$$SA = Pr2 = \pi - [(Pr\ 2\ Pw)2 \times \pi]$$

where **SA** = swept area
Pr = pad radius to outer edge
Pw = pad width

To calculate the swept area of a drum (illustrated in Figure 2-25), use the following formula:

$$SA = Dd = \pi = Sw$$

where **Dd** = drum diameter
Sw = shoe width

Most passenger cars use a swept area of about 200 square inches (sq. in.) of lining per ton of vehicle weight, or about 10 lb of weight per square inch (4.45 kg per 6.452 cm^2) of lining. Road racing cars that place a severe load on their brakes use about twice as much swept area—about 400 sq. in. per ton of vehicle weight. Recent changes in passenger cars to improve fuel economy have caused a reduction in the size of the brake components to help reduce the weight of the car. Reductions in the size of the rotors

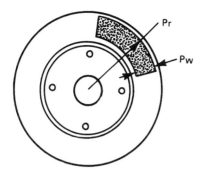

Figure 2-24 The amount of swept area of a disc brake can be determined by using these dimensions.

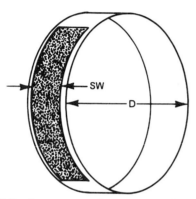

Figure 2-25 The amount of swept area of a drum brake assembly can be determined by using these dimensions.

and drums have decreased the amount of swept area by as much as 30 percent; the effect has been an increase in the operating temperature of the shoes and rotors or drums.

The swept area of both drum and disc brakes is limited by the inside diameter of the wheel. The brake drum must fit inside the wheel, and there must be room between the rotor and the wheel for the caliper. One German car manufacturer developed a disc brake in which the caliper mounted through the rotor, thus allowing a larger-diameter rotor, but the unit proved to be too expensive.

Clearly related to the swept area is the area of the lining surface. Some manufacturers provide this specification, which simply refers to the lining area that is in contact with the drums and rotors. At one time, most manufacturers produced cars with about 200 lb (91 kg) of vehicle weight for each square inch (6.452 cm^2) of lining. Today some downsized cars have more than 350 lb (159 kg) of weight per square inch of lining.

2.8.1 Measuring Brake Power

Braking power can be measured on a **brake dynamometer** (Figure 2-26). This equipment has two pairs of rollers,

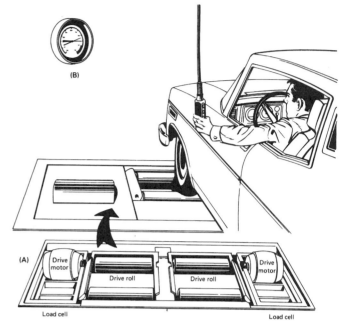

Figure 2-26 A brake dynamometer has two pairs of rollers that are driven by electric motors (A). The car is positioned with the front or rear tires on the rollers, and then the operator starts the drive motors, applies the brakes, and reads the amount of each wheel on the meter (B).

one pair for each of the tires on an axle. The rollers are driven by a pair of electric motors. The car is placed over the rollers so that the front or rear tires are on the rollers, and the motors are started to rotate the tires. The brakes are then applied, and the effective power of each brake is measured by the amount of resistance it offers the driving motor. The amount of power is indicated on a meter. The power and application rate of the right brake can be compared easily with that of the left brake, and problems such as weak return springs or sticky calipers or wheel cylinders will show up. The overall brake power is determined by applying the brakes completely. Some dynamometers are powerful enough to stress the brakes to the point of fade so that this aspect, as well as overall power, can be observed. After measuring the braking power on one axle, the car is moved to allow a similar measurement of braking power on the other axle.

Another style of dynamic brake tester is the **computerized plate tester.** This unit consists of four plates, one for each wheel of the car. The car is driven onto the plates at a low speed, 4 to 8 mph, and the brakes are applied while the wheels are on the plates. Braking force is measured, and the force generated by each wheel is displayed as a separate plot on a computer screen. Each plot shows the amount and the profile of the brake application, which allows the technician to diagnose many potential problems (Figure 2-27).

Chapter 2 ■ Principles of Braking

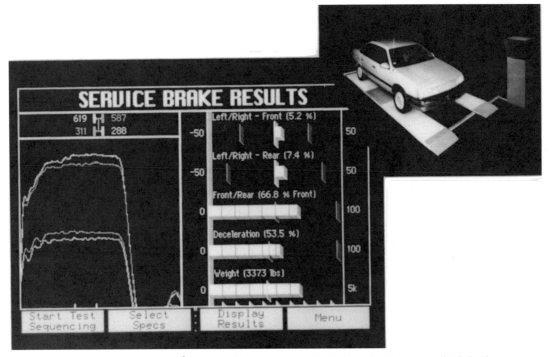

Figure 2–27 As the car is driven over the plates of the computerized plate brake tester (right), the brakes are applied. The computer screen then displays the amount of braking power of each wheel. Other tests display the deceleration rate, vehicle weight, parking brake effectiveness, and any sideslip at the front or rear tires. *(Reprinted by permission of Hunter Engineering Company)*

Brake dynamometers are too expensive to be used in most repair shops. Their use is generally limited to engineering and test laboratories and vehicle inspection or diagnostic lanes. The computerized plate tester is used in larger service shops.

2.9 Weight Transfer

Weight transfer is one of the most difficult forces to deal with in connection with brakes. It refers to the weight that moves from the rear tires to the front tires as the car decelerates (Figure 2–28). **Inertia** resists changes in motion. This physical law states that a body at rest tends to remain at rest and a body in motion tends to remain in motion. Inertial forces act on the whole car, but to make things easier to understand and calculate, we consider that inertia acts on the **center of gravity (CG)** of the car.

A car's CG is its balance point—the point on which the car would balance if it were picked up. In a front-engine, RWD car the CG is about 55 to 60 percent of the wheelbase distance in front of the rear axle, at the left-to-right (side-to-side) center of the car, and about as high as the upper one-third of the engine. In a modern FWD car, the CG is closer to the front axle than it is in a RWD car.

As the car is stopping, inertia creates a force equal to the deceleration force on the CG. This force acts against

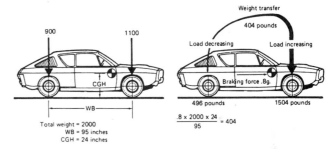

Figure 2–28 As a car stops, inertia causes weight to transfer from the rear tires to the front tires.

the traction force of the tires. The effect of the interaction of these two forces is to increase the load on the front tires, while the load on the rear tires is reduced by an equal amount. This is weight transfer.

Weight transfer produces two noticeable results: a lowering of the front end of the car along with a raising of the rear end (called **dive**) and a change in the relative amounts of traction. The transfer of weight increases the available traction of the front tires, but it also reduces the traction of the rear tires. As the loading and traction of the rear tires are reduced, the possibility of a rear-wheel lockup increases. The harder the stop, the greater the weight transfer and the greater the probability of a rear-wheel skid (Figure 2–29).

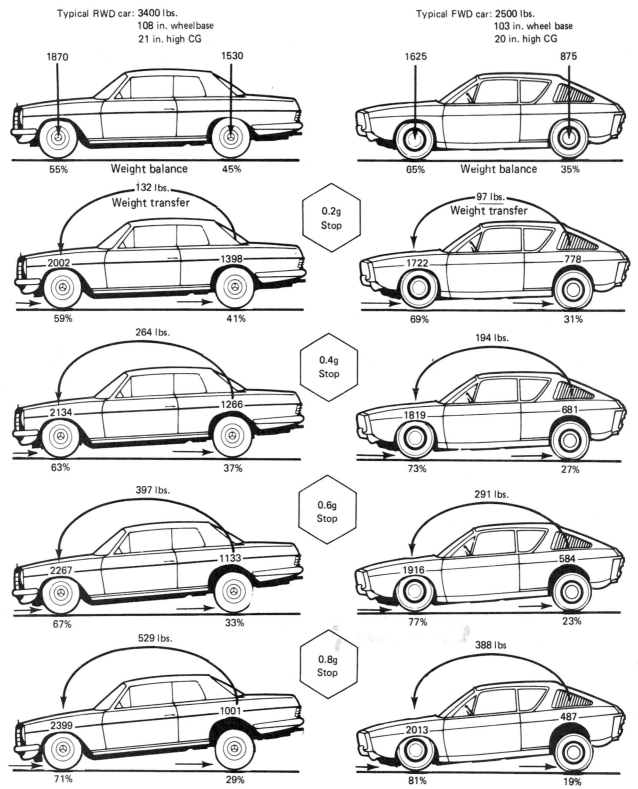

Figure 2–29 As the stopping rate increases, the amount of weight transfer also increases. The loading for typical RWD and FWD cars is shown here.

The formula used to calculate weight transfer is as follows:

$$WT = \frac{DR \times W \times CGh}{WB}$$

where **WT** = lengthwise weight transfer (in pounds)
DR = deceleration rate (in **g** value)
W = car weight (in pounds)
CGh = height of the CG (in inches)
WB = wheelbase (in inches)

There are three ways to reduce the amount of weight transfer: reduce the deceleration rate (certainly not desirable in an emergency), lengthen the wheelbase (rather impractical), or reduce the height of the CG. Reducing the CG height—lowering the car—is a common change made on race cars but is often impractical on street-driven passenger cars. The point to remember is that raising a car's CG increases weight transfer, and an increase in weight transfer increases the probability of rear-wheel lockup during a hard stop. If a vehicle is raised up, its stopping ability should be carefully checked after the modification has been made.

2.10 Skids

When a tire locks up during braking, a **skid** will result. A skidding tire produces a loss of traction and stopping power and also a loss of directional control. A two-wheel, rear-end lockup is a fairly common occurrence. Front-to-rear balance on a brake system is a compromise between too little rear braking power on wet-pavement, low-rate stops and too much power on dry pavement with harder stops.

If rear-wheel brake lockup does occur, the rear end of the car will slide out, either to the right or to the left, as road forces or road slope dictates (Figure 2–30). There is a natural tendency for the car to swing around and change ends when inertia is pushing forward on the CG and being retarded mainly by the front tires. Locking of the rear wheels reduces their ability to resist sideways motion. This effect is utilized in a "bootlegger's turn," a rapid U-turn made at a high speed, which uses a quick application of the parking brake to cause a controlled rear-wheel lockup and skid as the steering wheel is flicked to make the turn. This maneuver requires quite a bit of skill and can be very dangerous.

A car is more stable if the front tires lock up, but the ability to steer is then lost. A skidding front tire has no turning power, so the car will travel in a direction dictated by inertia or the slope of the road.

If all four wheels lock up, the car will soon be in an extremely unstable skid. During the stop, it will probably spin sideways—completely at the mercy of inertia, the road surface, and any objects that might be encountered.

If a trailer is being towed and the brakes of the trailer lock up, the trailer will skid sideways—much like the rear end of a car with the rear brakes locked. If it is not stopped, the trailer will swing around far enough to run into the tow vehicle; this is called **jackknifing.**

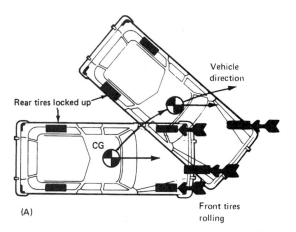
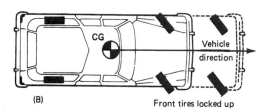

Figure 2–30 If the rear tires lock up (A) during a hard stop, the car will probably skid sideways because it is in an unstable condition—inertia is pushing forward on the CG and is being resisted primarily by the front tires. Front-tire lockup (B) causes the front tires to lose turning power, and inertia causes the car to travel in the original direction with no steering ability.

When brake systems are designed, the engineers consider the following factors in balancing front and rear braking power: the type of rear drum brake used, the relative diameter of the front and rear brake rotors or drums, the width of the rear shoes and drums, the size of the wheel cylinder and caliper pistons, and the hydraulic proportioning valve. Normally, the correct balance produces a slight tendency to lock the rear tires on dry pavement while carrying a normal load.

Antilock braking systems are close to the ideal solution for wheel lockup. An ABS allows an increase in rear-wheel braking power to provide for fully loaded stops, and because of the electronic controls that automatically reduce or pulse the pressure in the wheel cylinders or calipers, wheel lockup and skids are prevented.

2.11 Stopping Sequence

Braking a car to a stop normally follows a sequence of events that occur in a slow, relaxed manner during normal driving and in the same order but with shorter time and distance intervals during panic stops.

A stop begins when the driver recognizes a danger or a reason to stop, makes a decision, and then takes action to move his or her foot from the throttle to the brake pedal. The time elapsed is referred to as **reaction time.** The average reaction time is slightly more than 1 second. The braking operation starts as the pedal is depressed. Pressure is built up in the hydraulic system, and the lining of the brake shoes is moved into contact with the rotor or drum surface (Figure 2–31). As braking force is generated at the brake assemblies, the car will begin undergoing weight transfer, and brake dive will occur as the car sets up for the actual stopping process. Then actual braking occurs and continues until the car comes to a stop or slows down sufficiently. The driver often varies the pedal pressure and stopping rate to suit road and traffic conditions or to correct for a skid during a rapid stop.

2.12 Antidive Suspensions

Many cars have a front suspension that is designed to resist the brake dive that results from weight transfer. Dive is not only annoying to many drivers as the front end drops but can also cause severe load changes if it occurs so rapidly that it bottoms the suspension. An **antidive suspension** places the control arms of the front suspension in

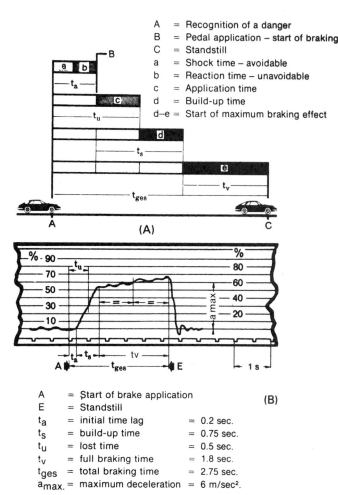

Figure 2–31 The normal sequence of events during a stop (A). The chart (B) shows the recorded time for these events during a fairly hard (60 to 65 percent) stop. *(Courtesy of ITT Automotive)*

a position so that a lever arm of the suspension tends to lift the front of the car during braking. Braking force is thus used to counteract dive (Figure 2–32).

Production cars are designed to have about 50 percent antidive. A certain amount of dive is desirable as feedback to the driver, because the amount of dive tells the driver how effective the braking is. Too much antidive tends to bind suspension action during braking and also causes excessive caster change during normal front-end action.

2.13 Practice Diagnosis

You are working in a tire shop that also specializes in brake repair and encounter the following problems:

Chapter 2 ■ Principles of Braking

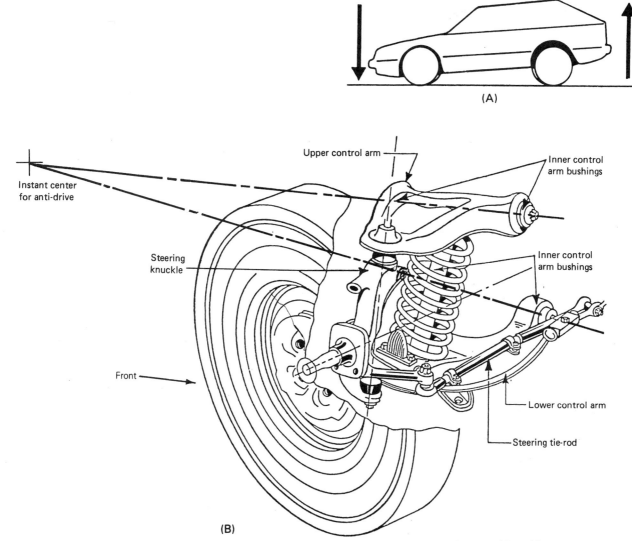

Figure 2–32 Brake dive causes the front of the car to lower (A) during braking. Antidive is achieved by placing the mounting points of the front control arms at an angle, creating a lever arm that creates a lifting force at the instant center (B).

CASE 1: The customer has brought in his 3/4-ton pickup that has an extended camper on it. He complains that when he goes downhill, the rig does not stop very well. The brakes get very hot (they smell bad) if he has to make several stops. Is this a typical problem? What should you do to help correct the problem? What should you tell this customer?

CASE 2: The customer has brought in her 5-year-old compact pickup, V6 with five-speed transmission, with a complaint that the rear tires lock up and skid during hard stops, especially on rainy days. On checking it, you notice that the inside of the bed is in like-new condition. What should you do to help correct the problem? What should you tell the customer?

Terms to Know

ambient	economy	Occupational Safety and Health
antidive suspension	edge brand	Administration (OSHA)
asbestos	edge code	organic
brake dynamometer	energy	original equipment (OE)
brake torque	Fahrenheit (F)	premium
braking efficiency	friction	quantity
British thermal unit (BTU)	friction material	reaction time
calorie (c)	heavy duty	semimetallic
Celsius (C)	inertia	skid
center of gravity	inorganic	Society of Automotive Engineers (SAE)
coefficient of friction	intensity	swept area
competitively priced	jackknifing	traction
computerized plate tester	kinetic energy	work
decelerometer	nonasbestos	
dive	nonasbestos organic (NAO)	

Review Questions

1. Deceleration rate is normally measured on a scale calibrated in:
   a. **g** values
   b. Feet per second per second
   c. Braking efficiency
   d. Any of the above

2. Statement A: Stopping rates are affected by the strength of the brake assemblies.
   Statement B: Stopping rates are affected by the tire-to-road coefficient of friction.
   Which statement is correct?
   a. A only
   b. B only
   c. Both A and B
   d. Neither A nor B

3. Deceleration rates can be measured using:
   a. A decelerometer
   b. A stopwatch
   c. Specially sized rectangular blocks
   d. Any of the above

4. The energy of a moving car is referred to as:
   a. Inertia
   b. Kinetic energy
   c. Horsepower
   d. All of the above

5. Statement A: As the weight of a vehicle increases, the amount of energy of motion increases at the same rate.
   Statement B: As the speed of a vehicle increases, the amount of energy of motion increases at the same rate.
   Which statement is correct?
   a. A only
   b. B only
   c. Both A and B
   d. Neither A nor B

6. Friction can be used to convert energy of motion into:
   a. Heat energy
   b. BTUs
   c. Calories
   d. Any of the above

7. Statement A: Heat always travels from a warmer object to a cooler object.
   Statement B: Heat intensity is measured using a Fahrenheit or Celsius scale.
   Which statement is correct?
   a. A only
   b. B only
   c. Both A and B
   d. Neither A nor B

8. Statement A: A stop from 55 mph produces higher brake temperatures than a stop from 40 mph.
   Statement B: During a stop, the temperature of the brake lining is affected by the weight of the rotor.
   Which statement is correct?
   a. A only
   b. B only
   c. Both A and B
   d. Neither A nor B

9. If it takes 30 lb of force to slide a 100-lb block of material, the coefficient of friction is:
   a. 0.30
   b. 3.0
   c. 0.7
   d. None of the above

10. Statement A: If the lining coefficient of friction increases during a stop, this is called fade.
    Statement B: Fading lining requires that pedal pressure be decreased to prevent wheel lockup.
    Which statement is correct?
    a. A only
    b. B only
    c. Both A and B
    d. Neither A nor B

11. Statement A: A tire has the greatest traction when it is slipping slightly relative to the road.
    Statement B: A skidding front wheel does not have the ability to steer the car.
    Which statement is correct?
    a. A only
    b. B only
    c. Both A and B
    d. Neither A nor B

12. Swept area refers to the size of the:
    a. Contact area between the tire and the road
    b. Rotor and drum area rubbed by the brake lining
    c. Bores in the master cylinder, calipers, and wheel cylinders
    d. All of the above

13. During a stop, weight transfer:
    a. Reduces the weight on the rear tires
    b. Reduces the weight on the front tires
    c. Increases the traction of the rear tires
    d. All of the above

14. Statement A: Rear-wheel lockup can cause the rear end of the car to slide sideways.
    Statement B: Front-wheel lockup can cause the front end of the car to slide sideways.
    Which statement is correct?
    a. A only
    b. B only
    c. Both A and B
    d. Neither A nor B

15. Antidive is caused by the:
    a. Location of the center of gravity
    b. Location of the rear suspension arms
    c. Position of the front suspension arms
    d. All of the above

3 Drum Brake Theory

Objectives

Upon completion and review of this chapter, you should be able to:

❑ Identify the different styles of drum brake units.

❑ Comprehend the terms commonly used with drum brakes.

❑ Understand how brake shoe-to-drum pressure is increased or decreased because of self-energizing or deenergizing actions.

❑ Understand the operation of nonservo, duoservo, and uniservo brakes and how they differ.

❑ Understand the purpose of each of the components used in a drum brake assembly.

3.1 Introduction

At one time, drum brakes were the type of brake used on most vehicles. The very earliest cars used **brake drums** attached to the rear wheels with a lined band wrapped around the drum. When the driver wanted to stop, the band was tightened onto the drum. This type of brake was called an *external contracting brake* because the lining was on the outside and contracted—was made smaller—to apply braking pressure (Figure 3–1). This brake design was adversely affected by road dust, dirt, and water, which could easily find their way between the lining and drum and severely reduce braking power or cause erratic brake operation along with severe wear.

A major advance in brake designs was the change from an external band to internal shoes. Being on the inside requires that the shoes expand to apply pressure, so this design is called an *internal expanding brake* (Figure 3–2). The shoes are mounted on the backing plate, also called a platform, and the edge of the backing plate is often fitted so it intermeshes with a groove in the brake drum. This fit is fairly effective at keeping water, dust, and dirt out of the brake assembly (Figure 3–3).

The first drum braking systems were applied through a mechanical linkage. Metal rods or cables, along with levers, transmitted pressure from the brake pedal or lever to the shoes. Very early cars had brakes only on the rear

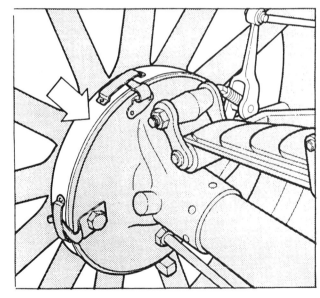

Figure 3–1 An early brake design using an external contracting brake. The brake band (arrow) was wrapped around the drum and was pulled tighter to apply the brake. (*Courtesy of Ford Motor Company*)

wheels until a rather complicated linkage was developed to transmit motion to the steerable front wheels. All modern cars use hydraulics to easily transmit force equally to all four wheels. Hydraulic application is described in Chapter 6.

Chapter 3 ■ Drum Brake Theory

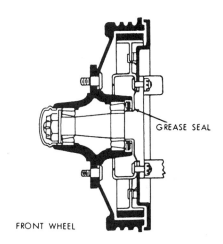

Figure 3–2 In most drum brake units, the backing plate and drum are arranged to form a labyrinth seal to help prevent the entrance of dirt and water. (*Courtesy of John Bean Company*)

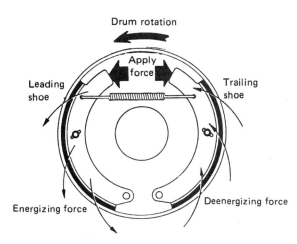

Figure 3–4 Because this drum is rotating in a counterclockwise direction, the leading shoe (left) is energized and pulled tighter against the drum. The trailing shoe (right) will be pushed away or deenergized.

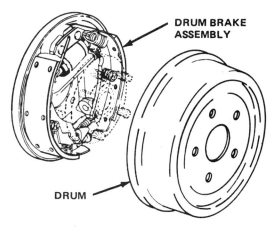

Figure 3–3 A drum brake assembly. The drum fits over the shoes and axle flange. (*Courtesy of General Motors Corporation, Service Technology Group*)

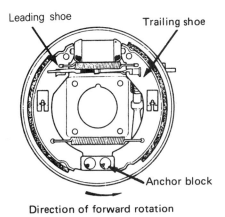

Figure 3–5 A leading–trailing shoe, nonservo brake. Note that the anchor block is positioned between the two shoes. (*Courtesy of LucasVarity Automotive*)

3.2 Shoe Energization and Servo Action

As mentioned earlier, brake shoes can be mounted onto the backing plate in different ways. In duoservo designs, rotation of the drum can be used to help apply the pressure of the lining against the drum. During application, the frictional drag tends to rotate the shoe around its anchor. Depending on the position of the anchor—at the leading or trailing end of the shoe relative to the direction of drum rotation—this rotational force will either increase or decrease the lining-to-drum pressure.

If the shoe is applied at the leading end and anchored at the trailing end—relative to drum rotation—the shoe will be pulled tighter into the drum. This occurs because the shoe attempts to rotate with the drum. It is called an **energized, forward,** or **leading shoe.** If the shoe is anchored at the leading end and applied at the trailing end, drum rotation will push the shoe away, reducing the application pressure. This is called a **deenergized, reverse,** or **trailing shoe** (Figure 3–4). A brake that uses a leading and a trailing shoe is often called a leading–trailing shoe or a nonservo brake (Figure 3–5).

Servo brakes use one anchor for both shoes; the shoes are arranged so that one can apply pressure on the other. With this arrangement, the leading shoe will apply pressure on the trailing shoe; this is called servo action. In servo brakes, the leading shoe is called the primary shoe and the trailing shoe is called the secondary shoe. The primary shoe is normally positioned toward the front of the car, and the secondary toward the rear. When the

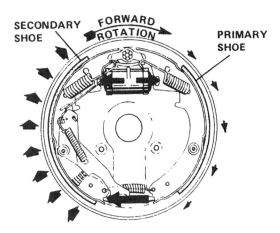

Figure 3–6 A duoservo brake is designed so the energized primary shoe increases the application pressure on the secondary shoe. (*Courtesy of General Motors Corporation, Service Technology Group*)

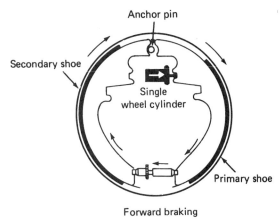

Figure 3–7 If there is only one wheel cylinder piston (uniservo brake), servo action can only occur during forward-direction stopping.

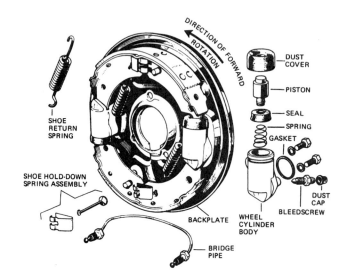

Figure 3–8 A two–leading shoe, nonservo brake. Note the two single-piston wheel cylinders and the relative positions of the shoes and anchors at the wheel cylinder bodies. (*Courtesy of LucasVarity Automotive*)

brakes are applied, the rotation of the drum energizes the primary shoe and it attempts to rotate with the drum. The motion of the primary shoe in turn applies the secondary shoe (Figure 3–6). About two-thirds of the stopping force from this brake comes from the secondary shoe, and two-thirds of the force applying the secondary shoe comes from servo action. Because it does more work, the secondary shoe usually wears faster than the primary shoe. Servo action produces a powerful brake that requires a relatively small amount of application force. When this brake is used with a wheel cylinder that has two pistons, servo action can occur in both forward and rearward directions, and it is called a duoservo brake. Occasionally this brake is used with a single-piston wheel cylinder, which can apply pressure only on the primary shoe. It is then called a **uniservo** brake because servo action occurs in only a forward direction (Figure 3–7).

The brake shoes on a nonservo brake are mounted independently, each with its own anchor and wheel cylinder piston; there is no interaction between them and, as mentioned earlier, usually one shoe is a leading shoe and the other a trailing shoe. This type of brake is often referred to as a **leading–trailing shoe** or *simplex* brake. Because it exerts a greater pressure against the drum, the leading shoe does more of the braking and usually wears out faster than the trailing shoe. The shoes can also be arranged so that they are both leading shoes—called a **two–leading shoe** or *duplex* brake—or both trailing shoes—called a **two–trailing shoe** brake. Two–leading shoe and two–trailing shoe designs normally use two separate single-piston wheel cylinders (Figure 3–8).

When duoservo drum brakes are used on the rear wheels and disc brakes on the front wheels, there is a definite tendency for rear-wheel lockup during heavy braking. At the moment of braking, weight transfer is changing the traction balance, increasing the traction of the front tires and reducing the rear traction, while servo action is increasing the relative rear-brake strength. When the combination of rear drum brakes and front disc brakes is used with a tandem front–rear split of the hydraulic system, a fairly good front-to-rear balance of the two styles of brakes can be achieved by using a proportioning valve in the rear system. The proportioning valve is used to reduce the power of the rear brakes during a hard stop. On cars with a diagonal split hydraulic system, two proportioning valves are required, one between each of the master cylinder sections and a rear brake. Many FWD cars with diagonal split braking systems use

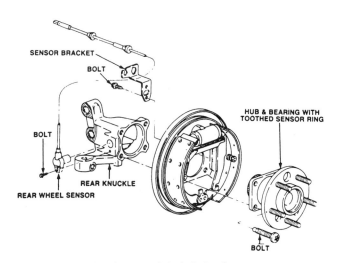

Figure 3–9 This late-model, full-sized passenger car uses leading–trailing brakes at the rear to help reduce the possibility of rear-wheel lockup. Note that it has a sensor and a toothed sensor ring for an ABS and also uses a unitized hub and wheel bearing assembly. (*Courtesy of General Motors Corporation, Service Technology Group*)

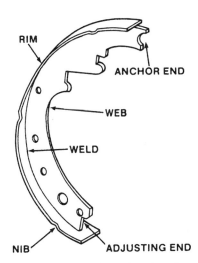

Figure 3–10 A steel brake shoe showing the various parts. (*Courtesy of Bendix Brakes, by AlliedSignal*)

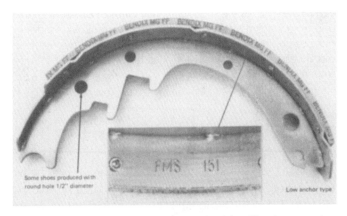

Figure 3–11 This shoe has an FMSI identification number of 151. It is a 10 × 1¾-in. shoe with the dimensions and shape as shown. Note that it has an edge code indicating Bendix lining MG with an FF friction code.

nonservo rear brakes. The stopping force characteristics of a nonservo brake are quite similar to those of a disc brake, which reduces the tendency of rear-wheel lockup and reduces or eliminates the need for proportioning valves (Figure 3–9).

3.3 Brake Shoes

Most passenger-car brake shoes are fabricated from two pieces of stamped steel. The *shoe rim* is curved to match the curvature of the drum and is slightly narrower than the width of the drum's inner surface; the rim provides a surface for lining attachment. The *shoe web* is welded to the rim, reinforcing it and providing a place for the anchor, application force, hold-down and return springs, parking brake attachment, and adjusting mechanisms. The shoe rim usually has a series of nibs—bent areas—on the edges where the shoe rests against the backing plate. These nibs improve the bearing contact where the shoe slides on the backing plate during application and release (Figure 3–10).

Brake shoes are also made from aluminum castings. These shoes tend to be lighter in weight and conduct heat away from the lining better than steel shoes. They also tend to be weaker, especially when hot.

The application end of the shoe is often referred to as the **toe**, and the end at the anchor is called the **heel**. The heel and toe of a nonservo brake shoe are easy to determine. The heel is at the anchor, and the toe is at the wheel cylinder. With duoservo brakes, the heels and toes switch depending on the direction of drum rotation and the action of the shoe.

Brake shoes come in many different shapes and sizes (curvatures and widths) with webs of different shapes and different placement of the holes. The various types of shoes are identified by a shoe number assigned by the *Friction Materials Standards Institute (FMSI)* (Figure 3–11). Normally, a shoe set can be ordered by using the make, model, and year for a particular car with a very good chance of getting an exact replacement. Occasionally it is necessary to specify the diameter or the width of the drum or both. Some manufacturers use drums of one diameter on their standard model and larger brakes on their sportier version, larger-size version, or station wagon. When replacing the lining, it is always a good practice to compare the new and old shoes to ensure a correct replacement (Figure 3–12).

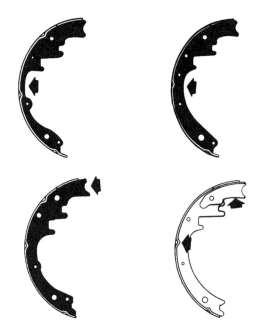

Figure 3–12 Brake shoes can have webs of different shapes and hole placement and reinforcement or rims with differing nib positioning. Always compare replacement shoes with the old shoes to ensure proper replacement. (*Courtesy of Wagner Brake*)

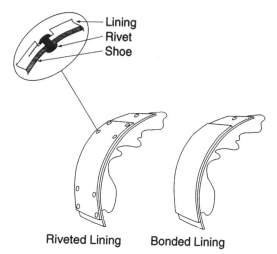

Figure 3–13 Passenger-car and light-truck brake lining is normally riveted or bonded onto the shoe rim. (*Courtesy of John Bean Company*)

3.4 Lining Attachment

On passenger cars and light trucks, the lining is attached to the shoe by one of two methods: bonding or riveting (Figure 3–13). Occasionally, the lining is bolted to the shoe on heavy trucks. Large truck shoes are expensive, and bolted linings, available in predrilled blocks, allow easy lining replacement in the field.

A **bonded lining** is secured to the shoe by a high-temperature adhesive. The lining is clamped onto the clean brake shoe rim with a layer of adhesive between the lining and the shoe. Then the shoe is placed in a high-temperature oven to set the adhesive. Some disc brake linings are *mold bonded* or *integral molded*. The pad backing is made with a series of holes into which the lining material is forced during the molding process. As the lining is molded, it is also bonded to the pad.

A **riveted lining** is secured by a series of brass or aluminum rivets. They pass through holes that are drilled and countersunk in the lining. The rivets are upset or flattened inside the shoe rim to hold the lining tightly in place.

Bonded linings are often preferred because there is more usable lining on the shoe. When a riveted lining wears out, the rivets can contact the drum or rotor surface and cause severe scoring or grooving. When a bonded lining wears out, the shoe rim can also contact the rotor or drum and cause severe scoring, but bonding does allow the use of nearly all of the lining's thickness. In some linings—usually those of inferior grade—cracks can begin at the rivet holes and cause breakup and separation of the lining from the shoe. Cracks are rare and usually a result of high-temperature conditions. Some technicians prefer riveted linings because they tend to be quieter. The bonding process laminates the lining to the shoe and creates a more rigid assembly that has a greater tendency to vibrate. In some circumstances, this more rigid shoe produces vibrations, which in turn can cause squealing noises. A premium or heavy-duty lining is often riveted for cooler operation. Bonding adhesive tends to insulate the lining from the shoe, which results in poorer heat transmission than when rivets are used.

Normally, the lining is centered on the shoe rim, and it can be full length or shorter. Lining length is designed to obtain even wear characteristics in a pair of shoes, whether primary and secondary or leading and trailing. A secondary shoe always has longer lining than a primary shoe. Sometimes a leading shoe has longer lining than a trailing shoe. Occasionally, a lining shorter than full length is placed in a high or low position on the shoe to change the self-energization or servo characteristics of the shoe (Figure 3–14).

3.5 Backing Plate

The **backing plate** is the *foundation* on which the brake assembly is mounted. The backing plate is bolted securely to the rear axle or front steering knuckle. It locates the shoe anchor and is usually used to transmit the brake torque from the anchor to the axle or steering knuckle. An exception is the *direct torque* design used on late-model

Chapter 3 ■ Drum Brake Theory

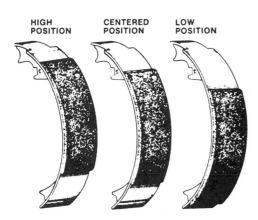

Figure 3–14 Brake lining can be attached to the shoe in different positions depending on the desired stopping characteristics. (*Courtesy of Bendix Brakes, by AlliedSignal*)

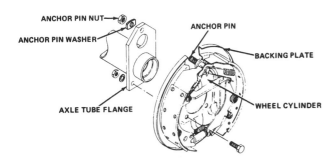

Figure 3–15 This direct-torque drum brake assembly attaches the anchor pin and wheel cylinder, along with the backing plate, to the axle tube flange. (*Courtesy of General Motors Corporation, Service Technology Group*)

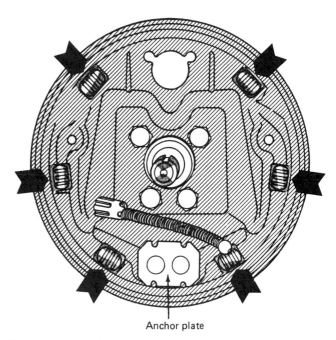

Figure 3–16 This shoe support–backing plate has six shoe platforms (arrows) and the anchor plate. (*Courtesy of Chrysler Corporation*)

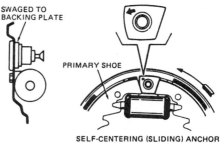

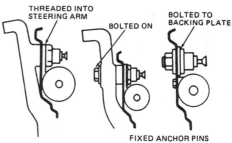

Figure 3–17 The anchor pin can be attached to the backing plate or steering knuckle by one of these methods. Most anchors are swaged, welded, or riveted in place. (*Courtesy of Wagner Brake*)

General Motors cars, in which a flange on the axle contains mounting points for the wheel cylinder and anchor (Figure 3–15).

Most backing plates have *ledges* or *platforms* on which the nibs of the brake shoes ride. These ledges support the shoes in a square position relative to the drum surface. In conjunction with the shoe anchor, they ensure proper positioning between the shoe and the drum. Also contained in the backing plate are the holes and bosses for attaching the wheel cylinder, the shoe hold-down springs, and the parking brake cable (Figure 3–16).

3.6 Shoe Anchors

Traditionally, domestic cars (those from U.S. manufacturers) use a round shoe anchor pin, and the shoes have a semicircular opening that butts against the anchor. The primary purpose of the **anchor,** sometimes called a *support,* is to absorb braking torque from the shoes and transmit that force to the backing plate and the car's suspension (Figure 3–17).

Round anchor pins position the shoes vertically on the backing plate and prevent shoe rotation with the drum. The **anchor pin** is a steel pin welded or riveted solidly to the backing plate or threaded into the steering knuckle through the backing plate. These are called **fixed anchors.** At one time, some anchor pins were adjustable; they could be moved up or down to center the shoes in

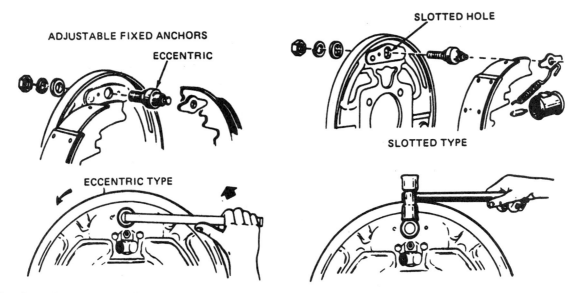

Figure 3–18 In the past, some brake designs used adjustable anchors so the shoe curvature could be aligned to the brake drum. (*Courtesy of Wagner Brake*)

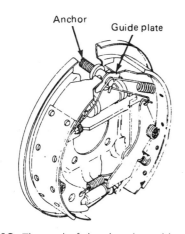

Figure 3–19 The end of the shoe is positioned vertically by the anchor pin and horizontally by the guide plate and backing plates. (*Reprinted from Mitchell Anti-Lock Brake Systems, with permission of Mitchell Repair Information, LLC*)

the drum (Figure 3–18). This was called a **major brake adjustment.** No car manufactured after the mid-1950s uses **adjustable anchors** that require this adjustment.

A shoe guide plate and a step on the anchor are often used to position the anchor end of the shoe laterally on the backing plate. This positions the anchor end of the shoe the proper distance from the backing plate (Figure 3–19).

Many cars currently use a flattened, grooved anchor along with a shoe that is flat or slightly curved at the anchor or heel end. This design allows the shoe to slide up or down and center itself in the drum. The groove at the anchor, like the guide plate and steps on the anchor pin, serves to position the shoe laterally (Figure 3–20).

3.7 Brake Springs

Drum brake assemblies commonly use two sets of springs—one set returns the shoes to the released position, and the other set holds the shoes against the backing plate platforms. Additional springs are often used to operate the self-adjuster mechanism, to hold the ends of the shoes in a particular position, or to prevent looseness and rattling in the parking brake mechanism (Figure 3–21).

The shoe **return springs** have a critical job, especially on servo brakes. During brake release, they pull the shoes back and push the wheel cylinder pistons inward as they return the shoes to the released position. A weak return spring can cause slow release or shoe drag during release. A weak return spring can also allow an earlier application on one wheel of an axle; this early application can cause a pull or even a grab or brake lockup. Servo action multiplies the frictional drag between the primary shoe and the drum to apply the secondary shoe. A weak return spring allows increased primary-shoe application pressure. Like other springs, return springs are designed for a particular installation. There are many shapes and sizes. Return springs are also called *pullback springs* or *retracting springs* (Figure 3–22).

Several different shapes of shoe **hold-down springs** are used. These spring assemblies are used to provide a slight pressure between the shoe nibs and the backing plate ledges (Figure 3–23). They ensure that the shoes stay against the backing plate platforms so the lining is kept straight with the drum.

Checking and servicing of brake springs are described in Chapter 22.

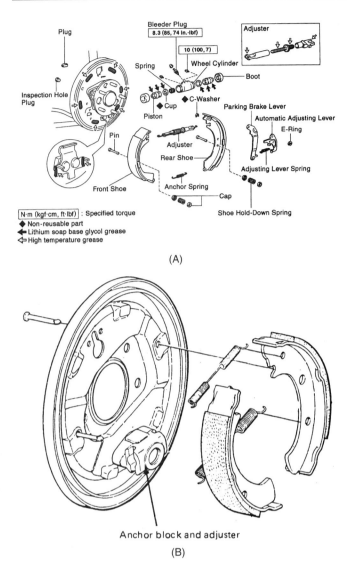

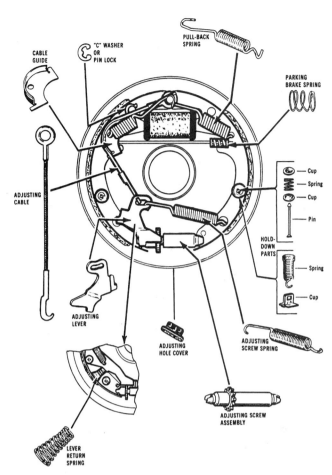

Figure 3–21 Some varieties of springs used in drum brakes. (*Reprinted from Mitchell Anti-Lock Brake Systems, with permission of Mitchell Repair Information, LLC*)

Figure 3–20 This anchor (A) allows the shoe to center itself with the drum by sliding up or down. Some sliding anchors include the adjuster mechanism (B). (*B is reprinted from Mitchell Anti-Lock Brake Systems, with permission of Mitchell Repair Information, LLC*)

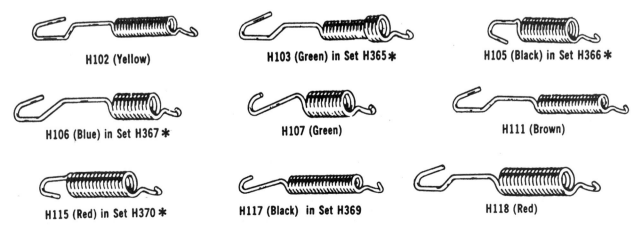

Figure 3–22 Brake shoe return springs come in many shapes and sizes. Compare any replacement so you are sure to use the correct spring. This supplier's part numbers and color coding are shown. (*Reprinted from Mitchell Anti-Lock Brake Systems, with permission of Mitchell Repair Information, LLC*)

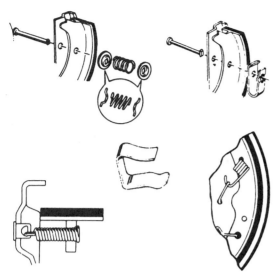

Figure 3–23 Five different styles of brake shoe hold-down springs. The upper two are the most common. *(Courtesy of Wagner Brake)*

Figure 3–24 A typical starwheel-type adjuster commonly used to adjust the lining clearance on duoservo brakes. *(Courtesy of Wagner Brake)*

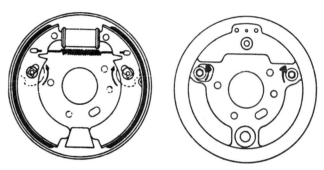

Figure 3–25 Nonservo brake designs without self-adjustment commonly use a pair of eccentric cams (which act as shoe stops) to adjust the lining clearance. *(Courtesy of Wagner Brake)*

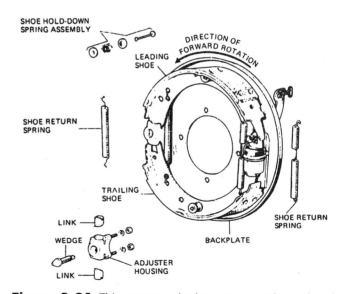

Figure 3–26 This nonservo brake uses a wedge-style adjuster. Threading the wedge inward causes the links to position the lining closer to the drum. *(Courtesy of LucasVarity Automotive)*

3.8 Shoe Adjusters

Brake shoes require periodic adjustment to keep the lining fairly close to the drum surface. The shoes need to be moved closer to the drum as the lining wears. Too much lining clearance requires more brake pedal movement as the shoes are applied. This situation can cause a "low brake pedal" or even a pedal that goes to the floor without applying the brakes.

At one time, brake shoe clearance was adjusted manually. The car was lifted off the ground, and a special tool, called a **brake spoon,** was used to adjust the shoe position. On servo brakes, the adjuster was a threaded adjuster assembly with a starwheel positioned between the lower ends of the two shoes. Turning the starwheel moved both shoes closer to the drum (Figure 3–24). For nonservo brakes, the most common adjuster was an eccentric cam positioned in back of each shoe in the backing plate; turning the cam moved the released position of each shoe closer to the drum (Figure 3–25). Other adjuster styles were a tapered wedge at the heels of the shoes, threaded shoe supports, a threaded parking brake strut, and threaded wheel cylinder caps (Figure 3–26).

Today, most cars use **self-adjusting** brakes. There are several different types depending on the shoe design and manufacturer. Most duoservo brakes use either *cable-style self-adjusters* or *lever-style self-adjusters* (Figure 3–27).

General Motors cars traditionally use the lever style. The lever is attached to the secondary shoe by a bushing at the shoe hold-down spring. This lever is connected to the anchor pin by a wire link. During a stop in reverse, the

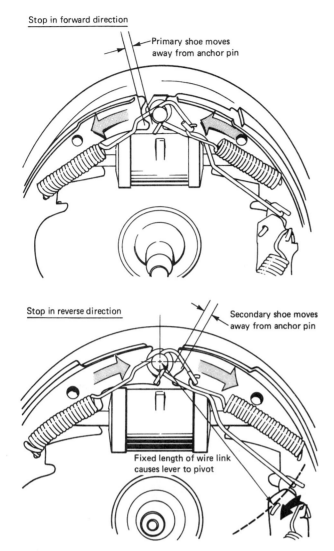

Figure 3-27 During a forward stop, servo action forces the secondary shoe against the anchor; during a stop in the reverse direction, servo action moves the secondary shoe away from the anchor pin. This movement is used to operate the self-adjuster mechanism. (*Courtesy of General Motors Corporation, Service Technology Group*)

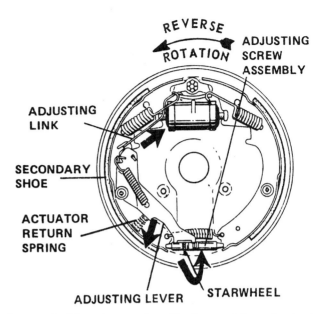

Figure 3-28 A lever-style self-adjuster. When the brakes are applied while backing up, the shoe movement after a certain point causes the adjuster to operate. (*Courtesy of General Motors Corporation, Service Technology Group*)

secondary shoe forces the primary shoe against the anchor pin as a result of servo action. The secondary shoe and lever rotate slightly with the drum, depending on the amount of clearance. More shoe clearance causes more shoe rotation and movement during a stop. When this motion is far enough, it forces the adjuster lever downward far enough to turn the starwheel, which in turn reduces the shoe clearance (Figure 3-28).

Most other domestic cars with duoservo brakes use a cable-type self-adjuster. The cable connects the adjuster lever, mounted on the secondary shoe, to the anchor and passes over a guide that is also attached to the secondary shoe. During a stop in reverse, as the secondary shoe forces the primary shoe against the anchor pin, the rotation of the secondary shoe causes the cable to lift the adjuster lever. As the brakes are released, the adjuster lever spring pulls the lever downward against the adjuster starwheel. If the travel is sufficient, the starwheel is rotated, which in turn reduces the shoe clearance (Figure 3-29). Some cable-type self-adjusters position the lever below the adjuster screw to provide a more positive adjustment. This type usually includes an override or overload spring at the cable-to-lever connection.

Several styles of self-adjusters are used with nonservo brakes. The most common leading-trailing brake self-adjuster is a *lever and starwheel adjuster* at the parking brake strut (Figure 3-30). Parking brake application pulls the lever away from the trailing shoe; if the movement is far enough, it causes the starwheel to adjust, making the parking brake strut longer and reducing the lining clearance. The other common styles include a pair of *self-operating cams* (Figure 3-31), a *rotating ratchetlike adjuster* at the parking brake strut (Figure 3-32), and a *telescoping rod and strut assembly* (Figure 3-33). The first two styles make their adjustments whenever the shoe travels farther than the gap between a slot in the shoe and an adjuster pin. When this occurs, the adjustment ratchet or cam is moved or turned, moving the pin outward. The operation of the other self-adjuster style is based on the movement of the parking brake strut during application of the parking brake.

Each self-adjuster is designed to make an adjustment only when necessary to decrease excessive shoe clearance.

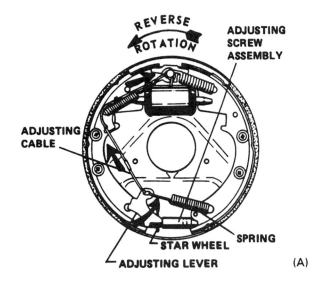

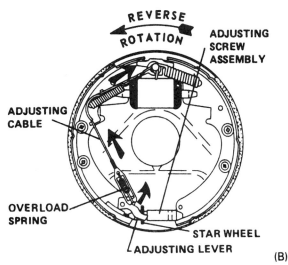

Figure 3-29 A cable-type self-adjuster operates as the brakes are applied while backing up (A). Some cable-type self-adjusters position the adjuster lever under the starwheel (B). (*Courtesy of General Motors Corporation, Service Technology Group*)

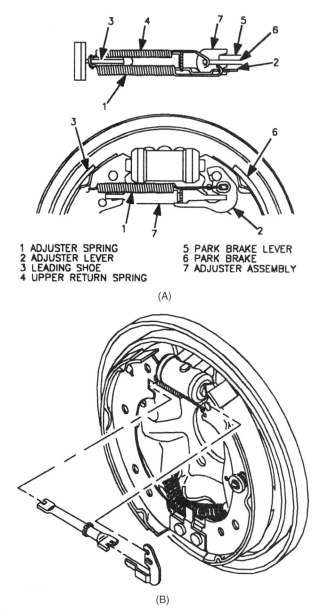

Figure 3-30 This nonservo brake design uses a self-adjuster mechanism that also serves as the parking brake strut. Applying the parking brake causes the adjuster mechanism to lengthen if there is sufficient travel. (© *Saturn Corporation, used with permission*)

An overadjustment that might cause shoe drag is very rare. Underadjustment, which causes a low brake pedal, is much more common and can be a fault of the adjuster mechanism or of the driver. The self-adjuster styles for duoservo brakes operate when the brakes are applied while backing up. Some drivers of cars with automatic transmissions seldom apply the brakes after backing; they simply shift into drive and go forward. Consequently, the self-adjusters never have a chance to operate. A similar problem applies in the case of brakes that operate from the parking brake. Some drivers use an automatic transmission's park gear or a manual transmission's first or reverse gear for parking and seldom use the parking brake. Diagnosing and correcting poor self-adjuster operation are discussed in Chapters 20 and 22.

3.9 Brake Drums

The **brake drums** have a rather simple job: They provide the rotating surface for the shoes to rub against. While doing so, they should present a hard, wear-resistant surface, have the physical strength to prevent excess deflection or distortion, and act as a heat sink.

The gray cast iron that most drums are made of is hard and quite wear resistant, partially because of the high amount of carbon it contains. As far as wear resistance and strength are concerned, cast iron is an ideal drum material. However, cast iron does have some disadvantages. It is

Chapter 3 ■ Drum Brake Theory

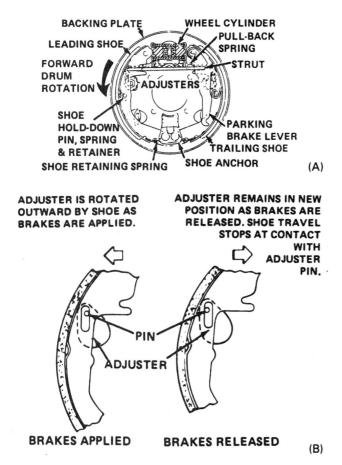

Figure 3-31 This nonservo brake uses an automatic cam adjuster (A). Movement of the shoe during brake application can reposition the adjuster (B). (*Courtesy of General Motors Corporation, Service Technology Group*)

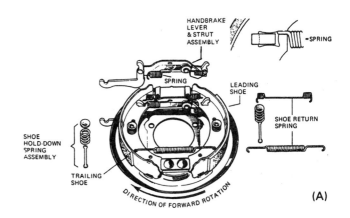

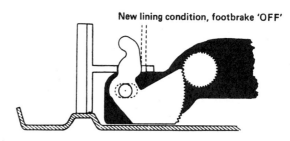

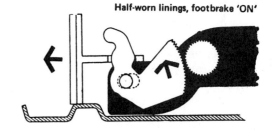

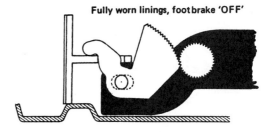

Figure 3-32 The nonservo, self-adjuster design is built into the parking brake lever and strut (A). As the lining wears, the quadrant lever rotates and positions the shoes closer to the drum (B). (*Courtesy of LucasVarity Automotive*)

heavy and can crack or break fairly easily. Consequently, many drums are made of a composite with a stamped-steel center section and a cast-iron rim and friction surface. The cast iron provides the strength to resist distortion and bell mouthing (in which the size of the drum opening increases), especially if it is reinforced with ribs or fins (Figure 3-34).

A **heat sink** is an area that can absorb heat. When the brakes are applied, heat is generated at the rubbing surface of the shoes and drum. This heat increases the temperature of these two surfaces or is transmitted to cooler areas by conduction. Any heat that is not conducted away causes a rise in temperature. Cast iron can absorb heat at a rate of about 10°F per BTU per pound (5.5°C per 252 c per 0.45 kg); in other words, the temperature will increase about 10°F if 1 lb of cast iron absorbs 1 BTU of heat. A stop that generates 20 BTUs of heat will increase the temperature of a drum that weighs 5 lb about 40°F. This same stop with a drum that weighs 1 lb will cause the temperature of the drum to rise 200°F. A heavier drum is a much better heat sink; its temperature will not

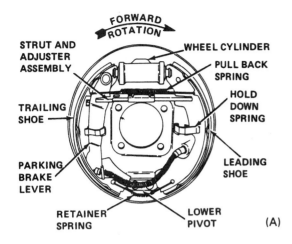

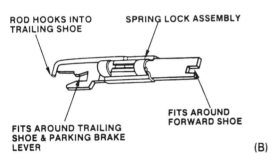

Figure 3-33 When the parking brake is applied on this design, the rod in the strut and adjuster assembly can pull the spring lock assembly to a longer position if the shoes need adjustment. (*A is courtesy of General Motors Corporation, Service Technology Group; B is courtesy of Bendix Brakes, by AlliedSignal*)

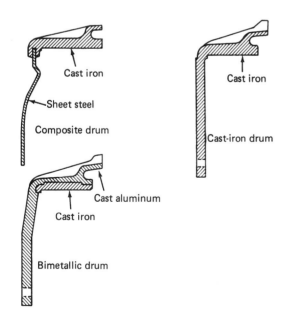

Figure 3-34 Most automotive brake drums are composites of stamped steel and cast iron. Some are all cast iron, and a few are cast aluminum with a cast-iron ring for the friction surface.

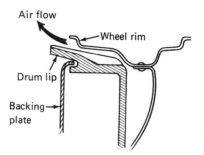

Figure 3-35 Some wheel and drum designs work together to cause airflow around the drum to help cool it.

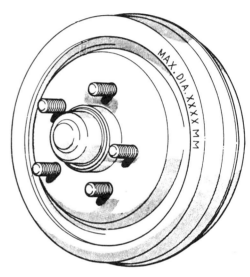

Figure 3-36 The maximum allowable diameter, also called the discard dimension, is indicated on all modern brake drums. (*Courtesy of Chrysler Corporation*)

increase as rapidly as it absorbs heat. During a stop, some of this heat passes on to the air flowing past the drum, but heat transmission from metal to air is rather slow. Some of the features designed to improve brake drum cooling include finned drums to increase airflow, bimetallic—aluminum and cast-iron—drums to improve heat conductivity, and wheels with finned center sections to improve the airflow past the drum (Figure 3-35).

As the brakes are used, the inner friction surface of the drum normally wears very slightly, but dirt or grit on the lining or contact with a rivet or shoe rim causes very rapid drum wear. During lining replacement, this surface should be checked and, ideally, reconditioned to make sure it is true and has the right surface texture. All modern drums have the discard dimension indicated on one of the surfaces (Figure 3-36). Any drum with a diameter larger than its discard diameter should be thrown away. Using a drum that is too large can result in higher brake temperatures because of the heat sink loss or a spongy brake pedal because of the increased deflection that results from the reduced physical strength. Another potential drum problem is out-of-roundness, which can cause a pulsation of the brake pedal because the shoes move in and out as they follow the fluctuating friction surface.

Brake drums are always checked during brake service for wear and cracks. These checks, along with their service procedures, are described in Chapter 22.

3.10 Practice Diagnosis

You are working in a brake and front-end shop as a service technician and encounter the following problems:

CASE 1: A customer has brought in her 1993 Buick Roadmaster (large RWD car) with a complaint of a low brake pedal. On checking, you note that this car has a V8 engine and automatic transmission. What is probably wrong? What should you do to confirm this problem? What should you tell this customer?

CASE 2: The customer has brought in his 1985 Ford Thunderbird with a complaint that the car swerves when he applies the brakes. He does not blame the lining because he and his neighbor installed new shoes a couple of months ago. Your road test confirms the swerving problem. When you get the car back in the shop, you inspect the lining and find that the right rear has lining that is the full length of both shoes but the lining is shorter on both left shoes. What is wrong? What should you do to correct this problem? What should you tell the customer?

Terms to Know

adjustable anchors	fixed anchors	reverse shoe
anchor	forward shoe	riveted lining
anchor pin	heat sink	self-adjusting
backing plate	heel	toe
bonded lining	hold-down springs	trailing shoe
brake drums	leading shoe	two–leading shoe
brake spoon	leading–trailing shoe	two–trailing shoe
deenergized	major brake adjustment	uniservo
energized	return springs	

Review Questions

1. If the anchor of a brake shoe is at the trailing end and the shoe is applied at the leading end, when the lining rubs against the drum, the shoe will be:
   a. Energized
   b. Deenergized
   c. Servoed
   d. All of the above

2. The brake shoe in Question 1 is called a:
   a. Leading shoe
   b. Trailing shoe
   c. Primary shoe
   d. Both a and c

3. Servo brakes use:
   a. An anchor for each brake shoe
   b. Two brake shoes with a single anchor
   c. One brake shoe with two anchors
   d. None of the above

4. Statement A: A leading–trailing shoe brake is also called a simplex brake.
   Statement B: The trailing shoe of a leading–trailing shoe brake provides more stopping power than the leading shoe.
   Which statement is correct?
   a. A only
   b. B only
   c. Both A and B
   d. Neither A nor B

5. The major advantage of a duoservo brake is that it:
   a. Operates well while backing up
   b. Offers a great deal of driver control
   c. Provides a large amount of stopping power for the amount of pedal pressure
   d. All of the above

6. Statement A: The brake lining is attached to the shoe web.
   Statement B: The shoe rim is held against the backing plate by springs.
   Which statement is correct?
   a. A only
   b. B only
   c. Both A and B
   d. Neither A nor B

7. During the 1960s and 1970s, the most common ingredient in brake linings was:
   a. Asbestos
   b. Metal
   c. Fiberglass
   d. Kevlar

8. Statement A: The lining on a secondary shoe is usually longer than that on a primary shoe.
   Statement B: The lining on a trailing shoe is usually longer than that on a leading shoe.
   Which statement is correct?
   a. A only
   b. B only
   c. Both A and B
   d. Neither A nor B

9. Brake service operations should be performed in a careful, planned manner because:
   a. The dust released can cause health problems
   b. Vehicle accidents can result from an improper or incomplete repair
   c. Personal injury can result from improper use of tools and equipment
   d. All of the above

10. Statement A: Bonded lining is secured to the brake shoe by an adhesive that is cured at high temperatures.
    Statement B: Riveted lining is secured to the shoe by a group of hardened steel rivets.
    Which statement is correct?
    a. A only
    b. B only
    c. Both A and B
    d. Neither A nor B

11. Statement A: The backing plate is bolted securely to the rear axle.
    Statement B: The brake shoe anchor is usually welded or riveted solidly onto the backing plate.
    Which statement is correct?
    a. A only
    b. B only
    c. Both A and B
    d. Neither A nor B

12. The platforms on the backing plate are used to support the:
    a. Wheel cylinders
    b. Shoe rim
    c. Self-adjuster lever
    d. Hold-down springs

13. Statement A: A weak return spring can cause a brake pull because that brake can apply early.
    Statement B: A weak return spring can cause shoe drag and lining wear.
    Which statement is correct?
    a. A only
    b. B only
    c. Both A and B
    d. Neither A nor B

14. During brake release, brake shoe return springs are used to:
    a. Pull the brake shoes back against the anchor(s) or shoe stops
    b. Push the wheel cylinder pistons back in the bore
    c. Return brake fluid to the master cylinder reservoir
    d. All of the above

15. Statement A: The self-adjuster mechanism for most duoservo brakes operates as the parking brake is applied.
    Statement B: The self-adjuster mechanism for most nonservo brakes operates as the brakes are applied while backing up.
    Which statement is correct?
    a. A only
    b. B only
    c. Both A and B
    d. Neither A nor B

16. A brake drum:
    a. Provides a smooth surface for the lining to rub against
    b. Provides a heat sink to absorb braking heat during a stop
    c. Should not be used if the inside diameter is larger than the number indicated on the drum
    d. All of the above

4 Disc Brake Theory

Objectives

Upon completion and review of this chapter, you should be able to:

❑ Identify the different styles of disc brake units.

❑ Comprehend the terms commonly used with disc brakes.

❑ Understand how a disc brake operates and its advantages as compared with drum brakes.

❑ Understand the purpose of each of the components used in a disc brake assembly.

❑ Understand the different styles of lining wear indicators.

4.1 Introduction

The operation of disc brakes is simpler than that of drum brakes. The brake shoes or pads are squeezed against the disc or rotor during braking. For release, the pads merely relax their pressure. Though the terms are interchangeable, *pad* is used more often than *shoe,* and *rotor* is used more often than *disc,* when discussing disc brakes. Disc brakes have no self-energizing or servo action. Pad pressure increases in direct proportion to the brake application force.

Anyone familiar with the caliper brakes on a modern bicycle understands disc brakes in their simplest form (Figure 4–1). Two pads are tightened against the wheel rim by a simple, hinged, mechanically operated caliper. Most important, though, these caliper brakes are light in weight and very effective. They can usually lock up the wheel under very hard application.

Disk brakes have several advantages over drum brakes. The two pads press on each side of the rotor in opposition to each other. Consequently, there is no distortion such as elongation of a drum, which can change shoe-to-drum contact or cause a low brake pedal (Figure 4–2). A major portion of the rotor's friction surface is exposed directly to air. This surface stays much cooler than a drum's friction surface. During rotation, centrifugal force throws any contaminants off of a rotor's friction surface (Figure 4–3), but in the case of a drum, the contaminants are forced onto the

Figure 4–1 The simple mechanical caliper used on a bicycle is an example of a disc brake. The pads push directly on the wheel rim as the cable is pulled to apply the brake.

friction surface. Also, the pads of a disc brake release to a position right next to the rotor's surface. This placement creates a wiping action that keeps dust, dirt, and water from entering between the lining and the rotor.

Next, and possibly most important, disc brakes have no self-energization or servo action, so both brakes on an

63

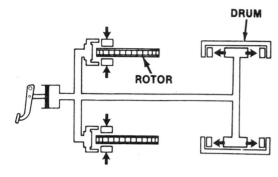

HYDRAULIC PRESSURE IS TRANSFORMED INTO MECHANICAL MOVEMENT BY CALIPERS OR WHEEL CYLINDERS

Figure 4–2 There is no distortion of a rotor during braking because the same pressure is applied on each side of a noncompressible surface by the two pads. A drum can expand slightly from the pressure of the shoes. (*Courtesy of Brake Parts, Inc.*)

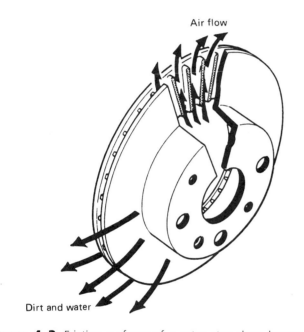

Figure 4–3 Friction surfaces of a rotor stay clean because centrifugal force throws off water and dirt. Centrifugal force also causes airflow through the internal ventilating fins to help cool the rotor. (*Courtesy of American Honda Motor Co., Inc.*)

axle usually generate the same braking power. With servo action, slight variations—common to high-mileage drum brakes—can change the frictional drag between the primary shoe and the drum, changing the amount of servo action, which in turn can cause a large difference in braking power. These variations can result from a weak return spring, sticky wheel cylinder, improperly fitting shoe, contaminated lining, faulty adjustment, and so on. Brake pull and uneven stops were fairly common when drum brakes were used on front wheels. Today, with front disc brakes, straight, even stops are easily achieved. An additional advantage of passenger-car disc brakes is that they are naturally self-adjusting.

Disc brakes do have a couple of disadvantages, however. Without servo action, a disc brake cannot develop the same braking power as a drum brake with the same hydraulic pressure. Much more force is necessary. Power boosters are required on most disc brake applications; they are an absolute necessity on mid-sized and larger cars. Also, it is difficult to include a mechanically operated parking brake in a caliper design. Several disc brake parking brake designs have been developed, but they tend to be expensive, complicated, heavy, weak, and prone to stick. The problems associated with the parking brake are probably the primary reasons why four-wheel disc brakes are not very common on lower-priced cars. These parking brake designs are described in Chapter 5.

Another characteristic of disc brakes, and also of drum brakes to a lesser degree, is squeal. Squeal is a highly annoying, high-frequency noise caused by vibration of a pad on the rotor or of a shoe on the drum. The large, flat rotor surface emits sound better than a drum's surface, making squeal a more common problem with disc brakes. The intensity of squeal is affected by the hardness of the lining, the rigidity of the shoe or caliper, the rigidity and speed of the rotor, and the amount of caliper pressure. Harder linings and thinner rotors increase the tendency to squeal, whereas firmer pad-to-caliper attachments and thicker pads, with dampening material on the backing, reduce vibrations and the tendency to squeal.

4.2 Calipers

The **caliper** is the casting that is mounted over the rotor. It contains the brake pads and the hydraulic piston(s) that apply the pads. It must be strong enough to transmit the high clamping forces needed and also to transfer the braking torque from the pads to the steering knuckle (Figure 4–4).

The pressure between the brake pads and each side of the rotor should be equal to prevent flexing and bind at the wheel bearings and flexing or distortion of the rotor or caliper. Two major types of caliper design are found on both front and rear brakes: fixed calipers and floating calipers.

4.2.1 Fixed Calipers

The first calipers used on passenger cars were of the **fixed-caliper** design. The caliper is fixed or fastened securely onto the steering knuckle. This caliper does not move relative to the steering knuckle. Fixed-caliper pistons are arranged in pairs, with a piston on each side

Chapter 4 ■ Disc Brake Theory

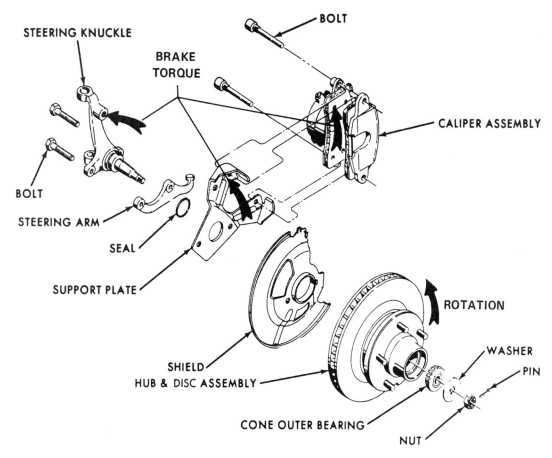

Figure 4–4 When the brakes are applied, the brake torque is transferred from the pads to the support plate and then onto the steering knuckle and front suspension. (*Courtesy of General Motors Corporation, Service Technology Group*)

of the rotor. The inboard piston applies the inboard pad, and the outboard piston applies the outboard pad (Figure 4–5).

The fixed calipers on heavier vehicles use four pistons (two pairs) to generate sufficient stopping force (Figure 4–6). Lighter-weight cars commonly use two pistons (one pair) because the lighter weight of the car does not require the same stopping power (Figure 4–7). Fixed calipers are no longer used on new domestic cars. They tend to be more expensive to manufacture than floating calipers because of their complexity and number of parts.

The brake pads of many fixed calipers can be slid in or out after removing a single retainer. The caliper does not usually have to be removed for pad replacement. The ends of the pads push against machined abutment surfaces in the caliper during braking. A slight clearance at the abutments allows pad movement during application and release.

4.2.2 Floating Calipers

Floating calipers are a simpler design. Normally only one piston is used. Some passenger cars and light trucks use a two-piston floating caliper. Floating calipers are used with a **caliper mount,** also called an **adapter** or *anchor plate,*

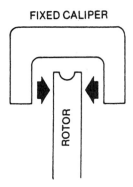

Figure 4–5 In a fixed-caliper brake, the caliper is stationary, and a piston on each side of the rotor puts pressure on the lining. (*Reprinted from Mitchell Anti-Lock Brake Systems, with permission of Mitchell Repair Information, LLC*)

bolted solidly to the steering knuckle. This mount transfers braking forces from the shoes in the caliper to the steering knuckle. In some designs, the inboard shoe is fitted directly into the caliper mount. The caliper is fitted into the mount so that it can move sideways relative to the rotor and steering knuckle (Figure 4–8).

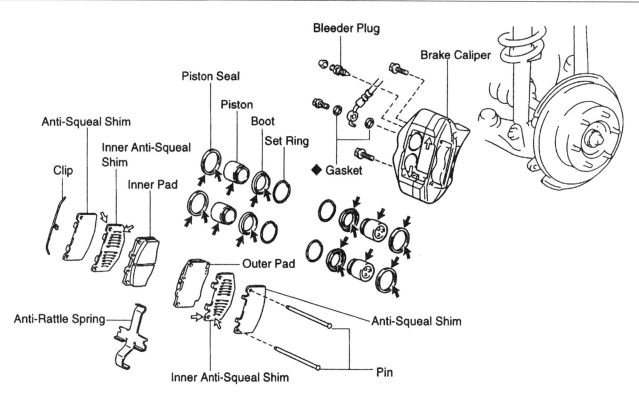

◆ Non-reusable part
← Lithium soap base glycol grease
⇐ Disc brake grease

Figure 4–6 A four-piston, fixed-caliper brake is bolted securely to the steering knuckle and has two pairs of pistons.

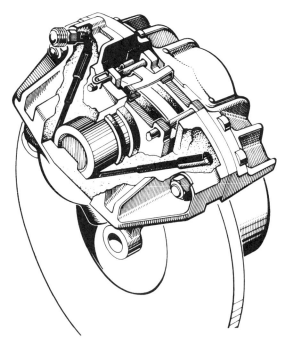

Figure 4–7 This fixed-caliper brake uses two pistons; note the internal passage to carry fluid to the outboard piston and air to the bleeder screw. (*Courtesy of ITT Automotive*)

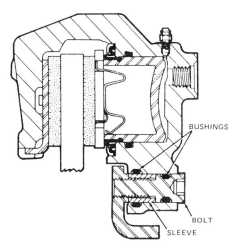

Figure 4–8 This floating caliper is designed to move sideways on its bushings, sleeve, and mounting pin. (*Courtesy of General Motors Corporation, Service Technology Group*)

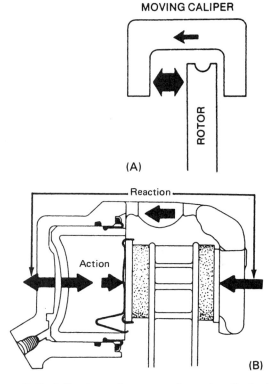

Figure 4–9 A floating caliper moves inward as the brakes are applied (A). The action of the piston puts pressure on the inboard pad while the caliper reaction puts pressure on the outboard pad. (*A is reprinted from Mitchell Anti-Lock Brake Systems, with permission of Mitchell Repair Information, LLC; B is courtesy of General Motors Corporation, Service Technology Group*)

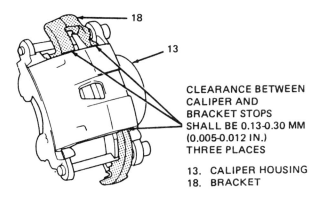

Figure 4–10 To allow caliper movement, a slight clearance must exist between the caliper and bracket stops. (*Courtesy of General Motors Corporation, Service Technology Group*)

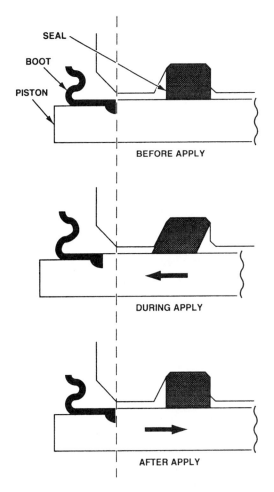

Figure 4–11 As the caliper piston moves to apply, the square seal ring becomes distorted. During release, the O-ring can then relax and pull the piston back slightly. (*Courtesy of General Motors Corporation, Service Technology Group*)

The caliper piston applies the inboard pad, and the caliper applies the outboard pad. An important law of physics is that for every action, there is an equal and opposite reaction. As hydraulic pressure pushes the piston, it also pushes on the end of the piston bore in the opposite direction. The *action* is the pressure on the piston in an outward direction, and the *reaction* is the pressure on the caliper in an inward direction. The piston slides outward to apply the inboard pad, and the caliper slides inward to apply the outboard pad. The outboard pad is usually secured to the caliper body (Figure 4–9).

Many calipers float on one or two **guide pins** or **bolts.** These pins are threaded into either the mount or the caliper body, and they pass through the other. Most designs use rubber or Teflon **sleeves,** also called *bushings* or *insulators,* around the guide pins for a better bearing surface. The guide pins allow the required sideways motion of the caliper body. Braking torque is transferred as the caliper body is forced against the abutment of the caliper mount. Normally, there is a slight clearance at the caliper-to-mount abutments to allow the necessary sideways motion but not enough to cause a slap or knock on brake application (Figure 4–10).

Most calipers are designed so that a certain amount of pad release does occur. As explained more completely in Chapter 6, the piston-sealing O-ring produces a slight pullback of the piston when the hydraulic pressure drops off (Figure 4–11). Because of its relatively light weight, the

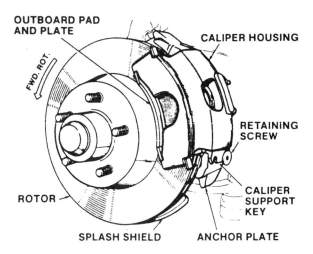

Figure 4–12 A sliding caliper. Note how the caliper fits in its mount. The braking action is like that of a floating caliper. (*Courtesy of Bendix Brakes, by AlliedSignal*)

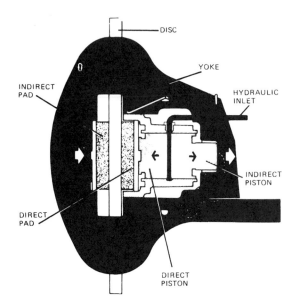

Figure 4–13 A caliper that uses a yoke straddling the rotor to transmit the application force from the indirect piston to the indirect (outboard) pad. (*Courtesy of LucasVarity Automotive*)

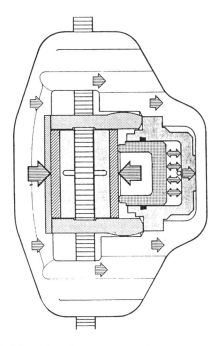

Figure 4–14 This caliper uses a yoke to transmit the force from caliper reaction to the outboard pad; the caliper body must float. (*Courtesy of ITT Automotive*)

piston is fairly easy to release, much easier than the caliper. The release of the outboard pad is very slight. It occurs mainly from the relaxing of the caliper and the slight wobble or flexing of the rotor and is aided by the sleeves and bushings around the guide pins. Proper floating of the caliper body is very important. Because of the piston size, a caliper can generate close to 10,000 lb (4,545 kg) of clamping force. This much force will definitely press the pads against the rotor, even if the piston or caliper is sticky. Drag will probably occur if the piston or caliper is sticky, because there will not be nearly this amount of force to push the piston or caliper back to release the pads.

There are several variations of floating calipers. Some early domestic designs were called **sliding calipers,** or mechanical guide calipers. The body of a sliding caliper is fitted between two V-shaped, machined grooves or ways in the caliper mount (Figure 4–12). These guides allow only a slight sideways motion, whereas the guide pins of a floating caliper also allow a slight twisting or flexing motion. Sliding calipers have more drag between the caliper body and the mount, and they tend to stick or drag unless the V-ways are properly lubricated.

Several important designs utilize an additional part called a **yoke.** One manufacturer calls this yoke a caliper and the piston body a cylinder. The yoke straddles the caliper or piston body and both the outboard and inboard pads. In one design, the caliper body is bolted to the steering knuckle, and one piston applies the inboard pad while a second piston pushes the yoke and the outboard pad inward (Figure 4–13). In a similar design, both the yoke and the piston body float on a retainer that is bolted onto the steering knuckle. Only one piston is used, because cylinder reaction pushes inward on the yoke to apply the outboard pad (Figure 4–14).

Another design uses a caliper plate that attaches to the mounting bracket on a pivot pin so that it can swing on the pivot during application and release. As in a floating caliper, the piston applies the inboard pad, and reaction

Chapter 4 ■ Disc Brake Theory

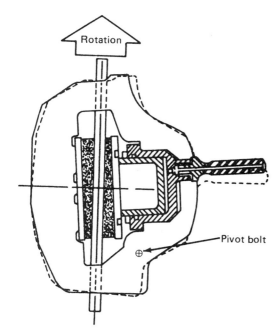

Figure 4–15 A swing or pivoting caliper. Note that the caliper pivots to apply the outboard pad. Also note the normal taper of the lining. (*Courtesy of Wagner Brake*)

force from the cylinder moves the caliper to apply the outboard pad. This caliper does not stay parallel to the rotor, and the lining is normally tapered or wedge shaped (Figure 4–15).

The pads are fitted into most floating calipers from the open, inner portion of the caliper. Typically, the caliper must be removed from the mounting bracket to allow pad replacement. In most cases, the inboard pad floats in the caliper or mounting bracket while the outboard pad is secured tightly to the caliper. Pad replacement is described in Chapter 23.

4.3 Brake Pads

Most disc brakes use a pad with a flat metal backing. Most pad backings include mounting ears, clips, or projections; some have only a mounting hole. The pads of a fixed-caliper brake and the inboard pad of a floating-caliper brake are normally designed to drop into place between two abutments with just enough clearance for application and release movement. Braking pressure is transferred to the abutments in the caliper or caliper mounting bracket. As mentioned earlier, most floating-caliper outboard pads are secured solidly to the caliper. Any motion between the outboard pad and the caliper might cause vibration and brake squeal (Figure 4–16).

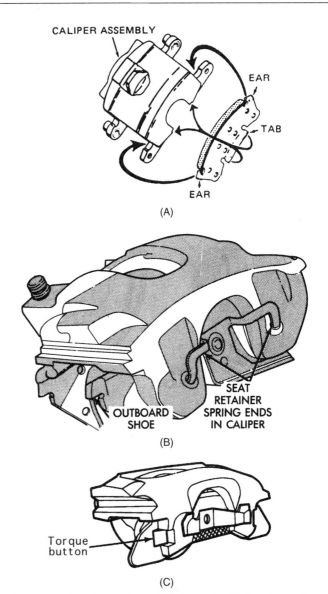

Figure 4–16 The outboard pads used with floating calipers are normally attached to the caliper by locking tabs and ears (A) or spring pressure (B). Torque is transmitted from the pad to the caliper by the mounting tabs and ears (A) or torque buttons (C). (*A is courtesy of Wagner Brake; B is courtesy of Chrysler Corporation; C is courtesy of Ford Motor Company*)

On most fixed-caliper designs, the inboard and outboard pads are identical and interchangeable. Most floating-caliper designs have an inboard pad that is noticeably different from the outboard pad. A few vehicles have pads that fit in only one particular location, such as inboard left side or inboard right side (Figure 4–17). Some use *torque buttons* to transfer braking torque between the pad and the caliper and also to keep the pad properly positioned in the caliper.

Disc brake lining is essentially the same as that used on drum brakes. With many newer FWD cars, semimetallic linings are fairly common because of their ability to

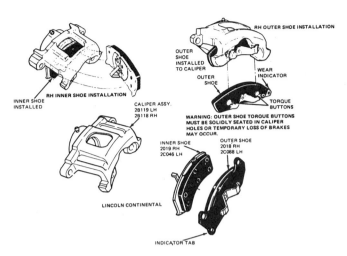

Figure 4–17 This brake pad set has four different shoes, each of which must be fitted in the correct location. Note the warning concerning proper installation. (*Courtesy of Ford Motor Company*)

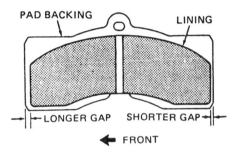

Figure 4–18 A few brake pads have offset lining; this places slightly more lining at the trailing edge to produce more even wear. (*Courtesy of Wagner Brake*)

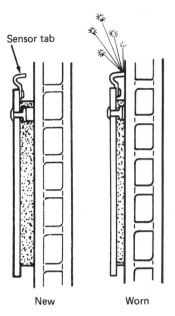

Figure 4–19 This pad wear sensor, also called an audible wear indicator, is attached to the trailing edge of the outboard pad. It will produce a high-pitched squeal if the lining wears to the point where the sensor tab contacts the rotor. (*Courtesy of Wagner Brake*)

withstand higher operating temperatures. The lining is riveted, bonded, or mold bonded to the pad backing. In all cases the lining surface is flat and, except for swing-caliper designs, the lining surface is parallel to the pad backing. Pads used with swing calipers are tapered, or thicker at one end than the other. In a few cases, the lining is offset sideways on the shoe—closer to one end than the other—and this feature produces more even lining wear. The trailing end of a shoe is always hotter than the leading end and therefore wears more rapidly. Excessive taper wear can lead to excessive caliper flex and a low brake pedal (Figure 4–18).

Many pads include **pad wear indicators.** The most common is a tab secured to the trailing edge of the outboard pad, which is called an *audible sensor*. When the lining is worn to the point where replacement is necessary, the wear indicator tab rubs against the outer edge of the rotor and produces a noticeable high-pitched squeal (Figure 4–19). Some cars use *visual sensors*. A warning light on the dash lights up when the pads are worn. In this system, each pad includes an electrical sensor that is connected to the warning light by an electrical wire. A few cars use a *tactile sensor*. Here a projection on the rotor strikes the pad backing place when the lining is worn out and creates a pulsating action of the brake pedal during braking.

4.4 Caliper and Pad Mounting Hardware

Every caliper design uses a few supplementary parts, commonly referred to as **mounting hardware.** Hardware, also called small parts, comes in various forms and shapes to suit each of the different calipers. Most hardware can be placed in one of four categories: *antisqueal, pad retention, pad antirattle,* and *pad* or *caliper positioning* (Figure 4–20).

As mentioned previously, the pads on most fixed calipers drop into place. They are usually retained by a steel pin that prevents them from working out or bouncing out. This pin passes through a hole in the pads and caliper body and is retained by a spring clip or cotter pin (Figure 4–21).

In many cases, the pads are loose and free to move between the rotor and caliper while the brakes are released. These loose pads can rattle and make jingling noises as the

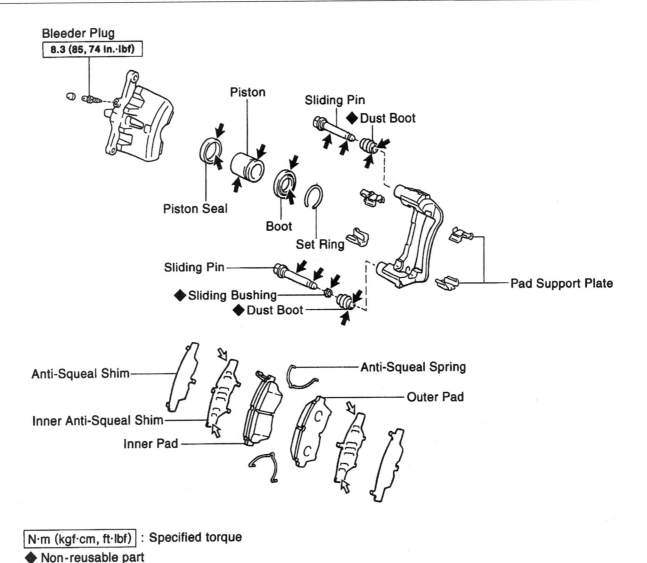

Figure 4–20 This caliper uses several types of hardware: the antisqueal shims and springs and the pad support plates ensure proper pad operation, and the sliding bushings and boots ensure proper caliper operation.

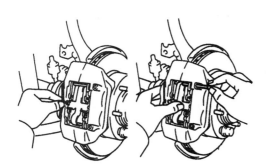

Figure 4–21 The pads on most fixed-caliper brakes are held in the caliper by retaining pins. Removal of the pins allows the pads to slide out.

vehicle is driven. In some cases, a spring is used to apply a slight pressure to keep them quiet. The **antirattle spring** or **clip** is sometimes also used to move the pad away from the rotor, reducing drag or vibration that can cause squeal (Figure 4–22).

Almost every floating-caliper design uses bushings or insulators along with the guide pins or bolts. These parts help locate the caliper and pads, return the caliper to a released position, cushion the caliper movement, and prevent metal-to-metal contact between the caliper and the mounting pins. Improper or worn caliper mounting hardware can cause excessive pad drag and wear or brake squeal (Figure 4–23).

In all cases, caliper and pad hardware should be serviced or replaced whenever the brake pads are replaced. The high heat produced by the brakes and contamination from dirt, sand, and salt from the road, as well as water and ice, have a drastic effect on these parts. Cleaning, replacement, and correct lubrication of the parts are mandatory to achieve a properly operating disc brake. These procedures are described in Chapter 23.

4.5 Rotors

A rotor or disc, like a drum, provides the friction surface for the lining to rub against. Also, like a drum, it is made from gray cast iron for all of the same reasons. In some cases the rotor and the front hub are cast as one piece. Typically, they are two pieces so either the rotor or the hub can be replaced separately. Some rotors are a composite with cast-iron braking surfaces and a much thinner, stamped-steel center section (Figure 4–24). A composite rotor is lighter than a similar cast-iron rotor.

Two styles of rotors are used on passenger cars: solid and vented. **Solid rotors** are often used on smaller cars and are smaller in width, lighter, and less expensive (Figure 4–25). Heavier cars, which produce more braking heat, use **vented rotors.** These rotors are cast with cooling fins between the two friction surfaces. As the rotor turns, these fins pump air from the inner eye (the center of the inboard side) through the rotor to the outer edge. Because of this airflow, a

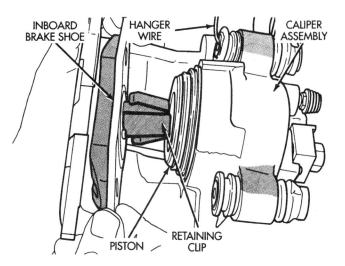

Figure 4–22 This inboard pad is connected to the piston by the retaining clip; this helps pull the lining away from the rotor and reduces rattles while released. (*Courtesy of Chrysler Corporation*)

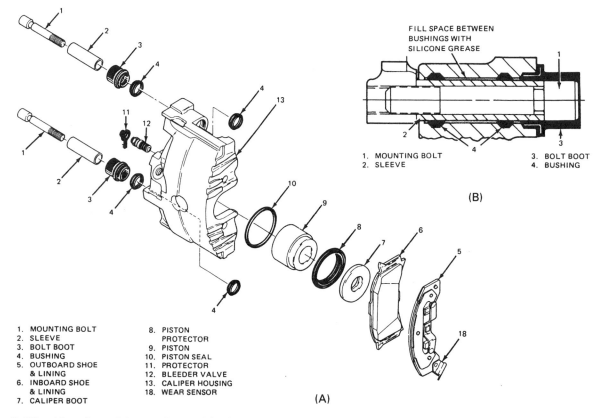

Figure 4–23 This caliper slides on sleeves (2); silicone grease at the sleeve and bushings ensures good caliper action for a long time. (*Courtesy of General Motors Corporation, Service Technology Group*)

Chapter 4 ■ Disc Brake Theory

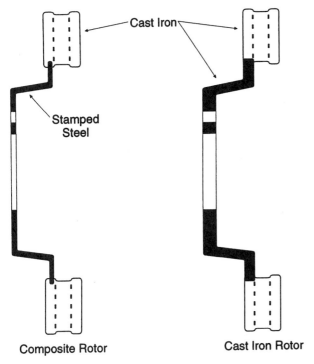

Figure 4–24 Many newer rotors are composites; the cast-iron braking surfaces are combined with a stamped-steel center section. Rotors may also be all cast iron.

vented rotor operates substantially more coolly than a solid rotor. Vented rotors are thicker, heavier, and more expensive than solid ones.

The fins in most vented rotors are straight and point straight toward the center of the rotor. Some rotors have angled or curved fins; such rotors pump more air and therefore run cooler. Angled or curved fins are directional. They must be mounted on the proper side of the car and are used in pairs. Mounting them on the wrong side of the car greatly reduces airflow and can cause overheating of the caliper and lining. A rule of thumb is that the curved fin should point toward the front of the car at the top of the rotor (Figure 4–26).

Like brake drums, modern rotors have the discard dimension indicated on one of the surfaces (Figure 4–27). Any rotor with a thickness less than the discard dimension should be thrown away. Using a rotor that is too thin will result in higher operating temperatures because of the heat sink loss. Higher operating temperatures can cause higher lining temperatures, which in turn can cause faster lining wear, possible fade, and possible brake fluid boiling. Another problem with rotors that are too thin is that the pad, caliper, and/or piston will move to an improper position. Two other potential rotor problems are rotor runout and thickness variation, also called parallelism. **Lateral runout** is a condition in which the rotor surface wobbles sideways. It can cause pad knock back and a low brake pedal, or pedal pulsation and possible grab (Figure 4–28). Thickness variation is a condition in which some parts of the rotor wear faster than others, creating thick and

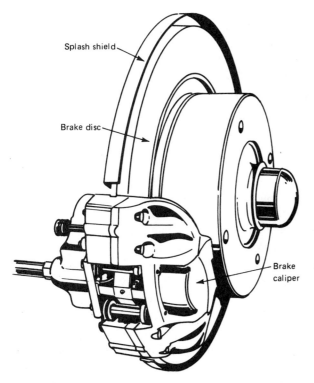

Figure 4–25 Many smaller cars use a solid, nonvented rotor. (*Courtesy of Volkswagen of America, Inc.*)

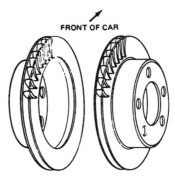

Figure 4–26 These rear rotors use curved internal fins; they must be mounted on the proper side of the car to achieve correct airflow. (*Courtesy of Bendix Brakes, by AlliedSignal*)

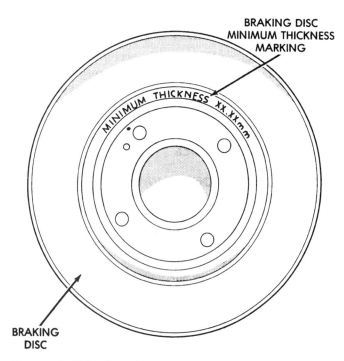

Figure 4–27 All modern rotors are marked with the minimum thickness dimension. (*Courtesy of Chrysler Corporation*)

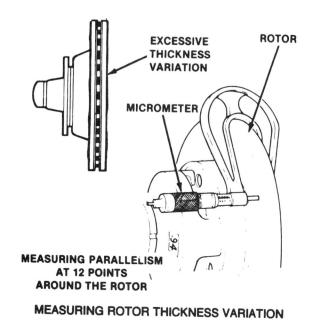

Figure 4–29 A rotor with excessive thickness variation has different measurements at different points on its surface. (*Courtesy of Brake Parts, Inc.*)

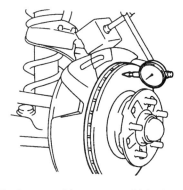

Figure 4–28 A rotor with runout wobbles back and forth laterally. The amount of runout is measured using a dial indicator.

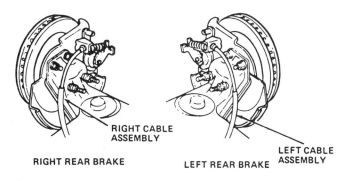

Figure 4–30 These rear-wheel disc brake assemblies use calipers with a mechanical parking brake mechanism. (*Courtesy of General Motors Corporation, Service Technology Group*)

thin areas, that cause pedal pulsations. Thickness variation is probably caused by overtightening or uneven tightening of the wheel lug nuts (Figure 4–29).

4.6 Rear-Wheel Disc Brakes

As mentioned earlier, it is difficult and expensive to incorporate a parking brake into a disc brake assembly. Because the need for equal stopping power is not as great on the rear brakes, most manufacturers choose to install drum brakes at the rear as a cost-saving compromise. A car with rear- or four-wheel disc brakes is more expensive to build. Where cost is not a factor, four-wheel disc brakes have be- come standard or optional equipment. The Corvette has used them since 1965. Four-wheel disc brakes have been standard equipment on many domestic luxury cars and performance-oriented cars since the late 1970s.

Rear-wheel disc brakes are of both fixed- and floating-caliper design (Figure 4–30). They normally use the same style of caliper as on the front. Some rear-wheel calipers are identical to the front calipers except for the piston diameter. Rear caliper pistons are usually smaller in diameter. Fixed-caliper units use a small drum brake assembly for a parking brake. Most modern rear-wheel floating calipers include a means of applying the pads mechanically for the parking brake. The mechanical operation of these calipers is described in Chapter 5.

4.7 Practice Diagnosis

You are working in a brake and front-end shop and encounter the following problems:

CASE 1: The customer has brought in her 1994 Honda Accord with a complaint of a very harsh noise at the right front when she applies the brakes. The car has 76,000 miles on it, and your road test confirms the noise. What is probably wrong? What should you do to confirm your suspicions?

CASE 2: A 10-year-old Mazda has a loud, unpleasant noise at the right front when the brakes are applied. When you remove the right front wheel, you find the outside of the rotor severely worn. This car has a single-piston floating caliper. When you remove the caliper, you find that the outboard pad is wearing on the metal backing and the inside pad still has about 3/32 in. of usable lining. What probably caused this wear problem? What will you need to do to repair this car?

CASE 3: The customer has brought in his 1990 Dodge Dakota pickup with a complaint of a jerky brake pedal when he makes a hard stop. Your road test confirms a pulsating pedal problem that causes the pedal to move up and down during the stop. What is probably wrong? What should you do to confirm your suspicions? What will you probably have to do to fix this problem?

Terms to Know

adapter	guide bolts	sleeves
antirattle clip	guide pins	sliding calipers
antirattle spring	lateral runout	solid rotor
caliper mount	mounting hardware	vented rotor
fixed caliper	pad retention	yoke
floating caliper	pad wear indicators	

Review Questions

1. An advantage with disc brakes over drum brakes is that disc brakes:
   a. Cool better
   b. Have friction surfaces that stay cleaner
   c. Provide better stopping control
   d. All of the above

2. Statement A: A disc brake has the same stopping power as a drum brake with less pedal pressure.
   Statement B: Disc brakes have fewer noise and squeal problems than drum brakes.
   Which statement is correct?
   a. A only   c. Both A and B
   b. B only   d. Neither A nor B

3. Statement A: Fixed calipers are bolted solidly to the steering knuckle.
   Statement B: Most fixed-caliper designs use one or two pistons.
   Which statement is correct?
   a. A only   c. Both A and B
   b. B only   d. Neither A nor B

4. In a floating-caliper design, the:
   a. Caliper has to move sideways during brake application and release
   b. Piston applies the inboard pad
   c. Outboard pad is applied by caliper reaction to hydraulic pressure
   d. All of the above

5. A floating caliper and a sliding caliper operate in the same manner except for the way the:
   a. Outboard pad is applied
   b. Caliper is located in the mounting bracket
   c. Piston and inboard pad are connected
   d. Outboard pad is secured in the caliper

6. The lining clearance of most disc brake units is adjusted by a:
   a. Starwheel self-adjuster
   b. Special cam
   c. Rubber O-ring
   d. None of the above

7. Statement A: Springs are used to release the brake shoes on floating-caliper designs.
   Statement B: The disc brake piston is moved to a released position by the rolling action of a rubber O-ring.
   Which statement is correct?
   a. A only   c. Both A and B
   b. B only   d. Neither A nor B

8. Statement A: Lining wear indicators produce a squealing noise when they rub on the rotor.
   Statement B: Some pad wear indicators turn on a warning light when the lining is worn.
   Which statement is correct?
   a. A only   c. Both A and B
   b. B only   d. Neither A nor B

9. Small parts are used with disc brake calipers and pads to:
   a. Keep the pads from rattling
   b. Retain the pads in the caliper
   c. Help position the caliper
   d. All of the above

10. Statement A: On most floating-caliper designs, the inboard pad is secured tightly to the piston.
    Statement B: In most floating-caliper designs, the outboard pad floats when the caliper is released.
    Which statement is correct?
    a. A only   c. Both A and B
    b. B only   d. Neither A nor B

11. On most floating-caliper brakes, the caliper mounting bracket is bolted securely to the:
    a. Rear axle   c. Lower control arm
    b. Steering knuckle   d. Caliper

12. Statement A: A vented rotor runs cooler than a solid rotor.
    Statement B: Solid rotors are easier to replace but cost more than vented rotors.
    Which statement is correct?
    a. A only   c. Both A and B
    b. B only   d. Neither A nor B

13. Potential rotor problems include:
    a. Lateral runout
    b. A variation in rotor thickness
    c. Rotors that are worn too thin
    d. All of the above

14. Statement A: Cars with four-wheel disc brakes use a special caliper that applies the pads hydraulically for parking brake usage.
    Statement B: Some cars with four-wheel disc brakes use a brake drum and shoes for a parking brake.
    Which statement is correct?
    a. A only   c. Both A and B
    b. B only   d. Neither A nor B

15. Excessive lateral runout of the rotor surface can cause:
    a. Pad knock back   c. Brake grab
    b. A low brake pedal d. All of the above

5 Parking Brake Theory

Objectives

Upon completion and review of this chapter, you should be able to:

- ❑ Understand the purpose for which parking brakes are designed.
- ❑ Understand how parking brakes operate.
- ❑ Comprehend the terms commonly used with parking brakes.
- ❑ Identify the different styles of parking brake control units and how they produce parking brake operation.
- ❑ Identify the different styles of parking brakes used with disc brake assemblies.

5.1 Introduction

The **parking brake** is a mechanically operated brake designed to hold the vehicle stationary when parked. The parking brake must be able to hold the vehicle while parked on any grade. In cases where there is poor traction, the parking brake should be able to lock the braked wheels of the parked vehicle. In passenger cars, the parking brake is normally applied by the muscular efforts of the driver using a hand or foot lever. The lever must include a latch or ratchet to hold it in the applied position. Parking brakes may share brake shoes and rotors or drums with the service brakes, but they must use a separate method of application. Also, the parking brake linkage must not interfere with the operation of the service brake.

A parking brake is often called an emergency brake, but this is a misnomer. There is no legislative or manufacturing design requirement concerning the ability of a parking brake to *stop* a car, only a requirement to hold the car after it has been stopped. As mentioned earlier, anyone who has tried to stop a car using only the parking brake realizes how inadequate it is for this purpose.

Potential problems can occur from drag of the parking brake if it is not released completely while driving. Driving with the parking brake on causes premature wear and glazing of the lining, but, even worse, the heat can cause fluid boiling. This, in turn, can cause pedal fade, and in a diagonal split system, the brake pedal can go to the floor, with a total brake loss.

5.2 Cables and Equalizer

All passenger-car parking brakes operate by a pulling force on a metal cable or rod. This pulling force originates at a hand-operated lever or a foot-operated pedal. The front cable, often called a **control cable,** is attached through an equalizer to a second cable or pair of cables, sometimes called *application cables*. The cable ends are attached to a lever at each brake assembly (Figure 5–1).

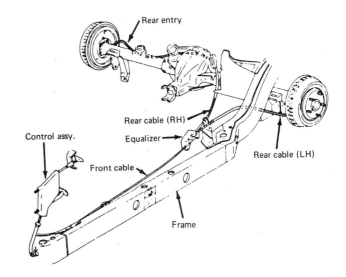

Figure 5–1 A typical RWD parking brake system consists of the control assembly and the cables for transmitting the application force to the rear brake shoes. (*Courtesy of General Motors Corporation, Service Technology Group*)

The **equalizer** allows equal force and motion to be exerted at each brake. It splits, or equalizes, the operating force so each wheel has equal brake application.

The cable runs through a flexible *housing* or *conduit* at the axle end. It is usually exposed between the equalizer and the housing end. Metal guides are often used to route the cable around obstacles such as the driveshaft or an exhaust pipe. A cable functions by pulling; when in operation, it is tight and tries to pull in a straight line. The cable housing permits the cable to operate with enough slack to allow vertical axle movement. In some cases, a single cable is used at the back. This cable starts at one brake, runs through the equalizer, and ends at the other brake. Many cars use a separate cable for each brake, with the two cables connected at the equalizer (Figure 5–2).

The equalizer assembly is normally mounted under the floor of the passenger compartment, near the center of the car, under or slightly to the rear of the driver's seat (Figure 5–3). When the parking brake lever is mounted between the front seats, a metal rod is often used to connect it to the equalizer. Occasionally the equalizer is built into the lever assembly. When the lever is mounted under the instrument panel, a control cable and housing are normally used to attach it to the equalizer (Figure 5–4).

In some cars, the equalizer is at the rear axle. The front inner cable runs back to the axle and is connected to the cable from one of the brakes, and the cable housing is connected to the cable from the other brake. The action of pulling the cable produces an equal and opposite reaction in the cable housing (Figure 5–5).

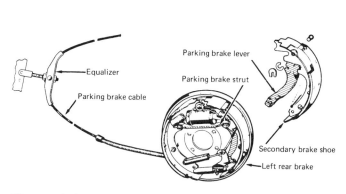

Figure 5–2 Most parking brake cables include an equalizer mechanism that ensures that the same application force is going to each of the rear brakes. (*Courtesy of Bendix Brakes, by AlliedSignal*)

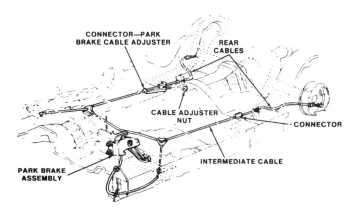

Figure 5–4 This parking brake system uses a foot-operated pedal to pull the front cable, which in turn pulls the intermediate and then the rear cables. (*Courtesy of Chrysler Corporation*)

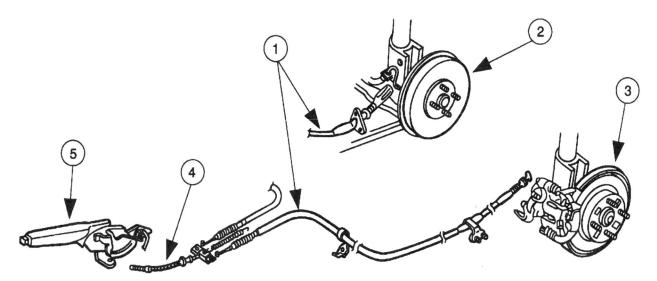

Figure 5–3 This parking brake system uses a hand lever or control (5). It pulls on the front cable (4), equalizer, and rear cables (1) to apply either the rear drum (2) or disc (3) brakes. (*Courtesy of Ford Motor Company*)

Chapter 5 ■ Parking Brake Theory

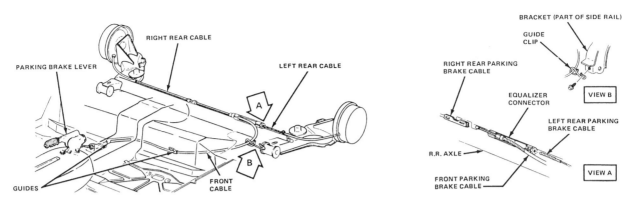

Figure 5-5 This parking brake uses an equalizer at the rear of the car. The inner cable of the front parking brake cable connects to the right rear brake cable, and the housing connects to the left rear brake cable. (*Courtesy of General Motors Corporation, Service Technology Group*)

5.3 Lever and Warning Light

Every driver should be familiar with the **parking brake lever.** This lever is designed to pull on the parking brake cable when the lever is depressed. As its name implies, the lever provides a leverage or a mechanical advantage that multiplies the driver's force substantially. The lever can be hand or foot operated (Figure 5–6). A latch is included in all levers to automatically hold them in the applied position; the latch must be released before the lever will release. In some luxury cars, this latch is automatically released when the car is started and shifted into drive or reverse.

Some modern parking brake systems include an **automatic adjuster** at the parking brake lever to remove any slack from the cables. When the lever is in the released position, a pawl is lifted from the ratchet mechanism, and this allows a clock spring to wind up, creating a 19-lb pull on the cables. Pulling the parking brake lever causes the pawl to engage the ratchet so normal action can occur (Figure 5–7).

Most cars include a **warning light** to remind the driver that the parking brake is applied. This light helps prevent an individual from driving a car with the parking brake partially applied, which causes added wear, possible heat damage to the shoes and drums, and possible brake failure. The light can be a separate unit, but it usually shares the same bulb as the brake failure warning light. It is activated by a switch at the parking brake

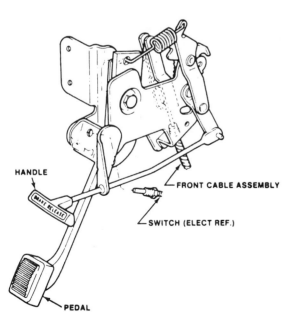

Figure 5-6 Pulling on the handle of this pedal-operated control assembly releases the latch and allows the pedal and parking brakes to release. (*Courtesy of Chrysler Corporation*)

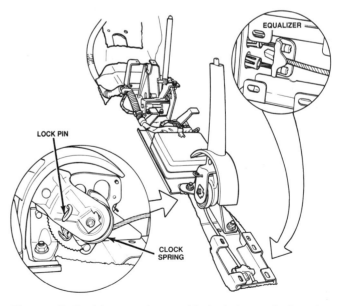

Figure 5-7 This control assembly includes a clock spring and ratchet that automatically take up any cable slack when the parking brake is released. The lock pin is installed to keep the clock spring from unwinding if a cable or control is disassembled. (*Courtesy of Chrysler Corporation*)

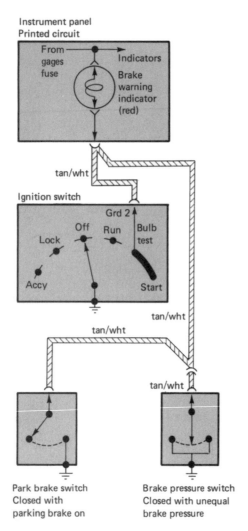

Figure 5–8 In most cars, the parking brake switch is open when the brake is released and is in a parallel circuit with the brake warning light switch. (*Courtesy of General Motors Corporation, Service Technology Group*)

lever. The switch is opened or turned off when the lever is released and closed or turned on when the lever is applied (Figure 5–8). Most switches have a simple and easy adjustment for resetting if necessary (Figure 5–9).

5.4 Integral Drum Parking Brake

Most of us are familiar with the common **integral parking brake** mechanism that is added to rear-wheel drum brakes. It consists of a lever that pivots on the web of the rear shoe and a strut that transmits force from the lever to the front shoe (Figure 5–10). Application of the parking brake produces a forward pull from the parking

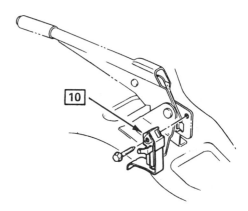

Figure 5–9 The parking brake switch (10) is usually mounted next to the lever or pedal so that movement of the lever closes the switch and turns on the warning light. (*Courtesy of General Motors Corporation, Service Technology Group*)

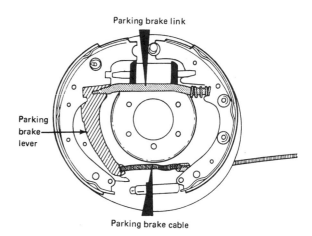

Figure 5–10 When the parking brake is applied, the cable pulls the parking brake lever forward, which pushes the parking brake link and strut to the right and the secondary shoe to the left. Note the spring at the right end of the link to keep the link from rattling while released. (*Courtesy of Ford Motor Company*)

brake cable to the lever. This causes a rearward force at the pivot between the lever and the secondary shoe and a forward force at the connection between the strut and the primary shoe.

When the brakes are released, the shoe return springs return the shoes to the anchors or stops, and the spring at the end of the parking brake cable returns the parking brake lever against the rim of the secondary shoe. At this point, there should be clearance at the ends of the parking brake strut. The coil spring normally positioned between the strut and the primary shoe prevents rattles (Figure 5–11). In many cases, a spring washer is also included at the parking brake lever pivot to eliminate rattles.

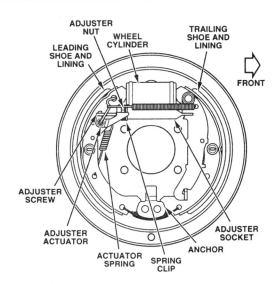

Figure 5-11 When this leading–trailing brake is applied, shoe travel can cause the adjuster actuator to turn the adjuster screw and adjust the brake clearance. The shoes are right against this adjuster screw and socket while they are released. (*Courtesy of General Motors Corporation, Service Technology Group*)

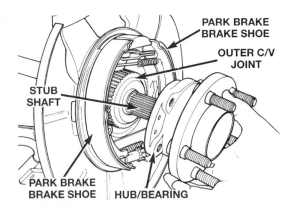

Figure 5-12 This small drum brake assembly is used only for a parking brake on this four-wheel disc brake vehicle. The drum is in the center of the rear rotors. (*Courtesy of Chrysler Corporation*)

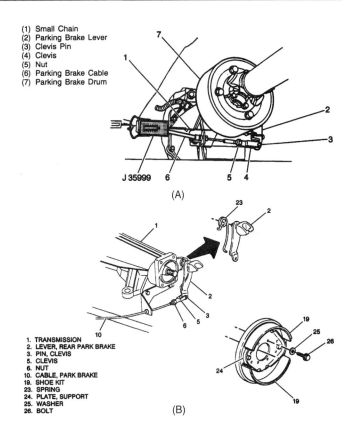

Figure 5-13 Some trucks and older cars use an auxiliary parking brake mounted at the rear of the transmission (A); tool J35999 is being used for an adjustment. (B) shows the cable and inner parts. (*Courtesy of General Motors Corporation, Service Technology Group*)

5.5 Auxiliary Drum Parking Brake

Some cars use separate drum brake assemblies whose only function is as a parking brake (Figure 5–12). This arrangement is found on some vehicles with four-wheel disc brakes. On these cars, the drum for the parking brake is cast as the inner part of the rotor. The drum is usually rather small (only $6^1/_2$ in. [165 mm] on the Corvette), but its only purpose is to hold the wheel stationary. In the past, some cars used a single auxiliary drum assembly mounted at the rear of the transmission. Auxiliary parking brakes are still used on many trucks (Figure 5–13).

Most auxiliary drum brakes are simply small versions of a duoservo drum brake without the wheel cylinder. The wheel cylinder is replaced by an actuating lever or mechanical cam. Shoe adjustment is accomplished by a threaded starwheel, as in the mechanically adjusted duoservo brakes of the past. Because the shoes are normally applied while the drum is stopped, there is very little rubbing of the lining. Since there is negligible lining wear, there is little need for a periodic shoe adjustment.

5.6 Disc Parking Brake

Compared with the integral drum parking brake, disc parking brake systems are rather complicated. There are essentially three different styles of parking brakes that operate through the brake caliper. Domestic cars use one of two different rotary action or lead screw mechanisms, and some import cars use a cam and lever type of arrangement (Figure 5–14). Caliper-type parking brakes must be self-adjusting so that both hydraulic and mechanical operation can occur in a normal manner and not be affected by lining wear. The Delco Moraine Division

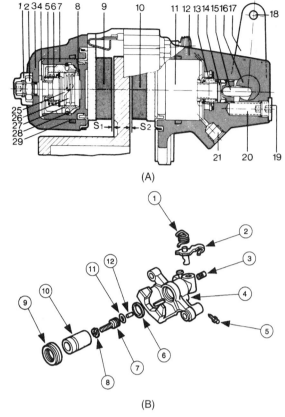

Figure 5-14 A rear fixed caliper (A) and floating caliper (B) showing the mechanical parking brake mechanism. (*A is courtesy of ITT Automotive; B is courtesy of Ford Motor Company*)

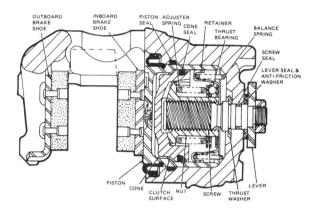

Figure 5-15 This cutaway view of a Delco Moraine rear caliper shows the relationship of the internal parts for both hydraulic service brake and mechanical parking brake operation. (*Courtesy of General Motors Corporation, Service Technology Group*)

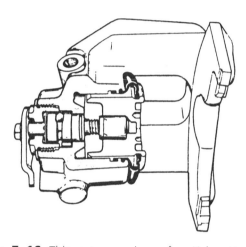

Figure 5-16 This cutaway view of a Kelsey-Hayes rear caliper shows the relationship of the internal parts for both hydraulic service brake and mechanical parking brake operation. (*Courtesy of Bendix Brakes, by AlliedSignal*)

of General Motors Corporation and Kelsey-Hayes, two major domestic manufacturers, use different approaches in their designs.

Delco Moraine calipers operate and self-adjust hydraulically in the same manner as front, no-parking-brake calipers (Figure 5-15). This operation is described in the next chapter. When the parking brake lever is applied, the cable movement rotates a lever that is attached to a high lead screw, which in turn passes through the inboard side of the caliper cylinder. A high lead screw is threaded with a very high pitch. It threads into an adjusting nut in the caliper piston assembly. Rotation of the high lead screw causes a sideways movement of the caliper piston and application of the brake pads.

Mechanical self-adjustment occurs during hydraulic application of the piston whenever the piston travels farther than the design allows. When adjustment is necessary, the piston pulls on the high lead screw and adjusting nut. If it pulls hard enough, a gap will develop between the adjusting nut and the cone, which will allow the nut to rotate on the screw threads, changing the position of the nut and thereby making an adjustment. During mechanical application, the high pressure between the nut, the cone, and the piston prevents nut rotation.

Kelsey-Hayes parking brake calipers also use normal operation for hydraulic application and self-adjustment (Figure 5-16). However, the mechanical operation is different from that of the Delco Moraine design. Movement of the parking brake lever at the back of the caliper causes a rotation of the operating shaft. The operating shaft is like a cam in that it has three ramp-shaped detent pockets at its inner face; there are also three pockets at the outer face of the thrust screw. A ball is placed in each of the three pockets between the thrust screw and the operating shaft. The thrust screw is held from rotating by an antirotation pin. When the operating shaft is rotated by the operating lever, the detents in the face of the operating shaft and the thrust screw, along with the three balls, generate an inward motion of the thrust screw. The thrust screw then pushes inward on the piston and applies the brake shoes.

Mechanical self-adjustment of this caliper also occurs during hydraulic operation. Whenever the piston travels farther than the preset amount, the pulling force between the piston and the thrust screw causes the adjuster assembly in the piston to rotate, thereby making the adjustment.

5.7 Practice Diagnosis

You are working in a tire shop that also does brake and front-end repair and encounter the following problems:

CASE 1: While doing a brake inspection on a 4-year-old pickup, you find that the parking brake pedal goes almost to the floor before the brakes are applied. The rest of the brakes are normal. What is probably wrong? What will you need to do to correct it?

CASE 2: The customer brought in her Nissan Stanza with a complaint that the brakes completely failed; she was able to stop without an accident and noticed a burned smell at the time. After a while the brakes seemed okay so she drove the car to the shop. Your checks show the brake pedal and front brakes to be normal, but the rear brake shoes are badly glazed and show some cracking. The shoes and drums also show a bluish-gold color and heat damage. What probably caused this problem? What will you need to do to fix it? What should you check out before the car leaves the shop?

CASE 3: A 1994 Camaro has four-wheel disc brakes, a V8 engine, an automatic transmission, and a complaint of a useless parking brake. Your check confirms that the parking brake pedal goes all the way to the floor. What probably caused this problem? What should you do to correct it?

Terms to Know

automatic adjuster
control cable
equalizer

integral parking brake
parking brake

parking brake lever
warning light

Review Questions

1. Statement A: Every car sold in the United States must be equipped with a braking system that will hold a car stationary while parked.
   Statement B: Every car sold in the United States must be equipped with a mechanically applied braking system that will stop a car within the distances established by the vehicle code.
   Which statement is correct?
   a. A only
   b. B only
   c. Both A and B
   d. Neither A nor B

2. A parking brake is normally applied by the driver's hand pressure or foot pressure on a pedal or lever, and the amount of effort is multiplied through:
   a. A power booster
   b. One or more levers
   c. A series of cables
   d. Electrical magnetism

3. As the parking brake is applied, the equal application of force at the two brake assemblies is ensured by the:
   a. Equalizer
   b. Cable guides
   c. Lever arrangements
   d. All of the above

4. Statement A: On most modern cars, the parking brake warning light circuit consists of a separate circuit containing a fuse, a warning light, and a switch.
   Statement B: The parking brake warning light switch is usually mounted close to the pedal or lever.
   Which statement is correct?
   a. A only
   b. B only
   c. Both A and B
   d. Neither A nor B

5. In a drum parking brake assembly, the spring at the end of the strut is used to:
   a. Help apply the brakes
   b. Ensure a complete release
   c. Prevent rattles
   d. None of the above

6. When the parking brake is released, the parking brake mechanism is returned to a released position by a spring:
   a. Around the cable and between the backing plate and the parking brake lever
   b. Between the parking brake lever and the secondary shoe
   c. At the end of the parking brake strut
   d. All of the above

7. Statement A: Some cars with four-wheel disc brakes use small drums and shoes that function only as parking brakes.
   Statement B: Some trucks and older cars use a drum-type parking brake mounted on the output shaft of the transmission.
   Which statement is correct?
   a. A only
   b. B only
   c. Both A and B
   d. Neither A nor B

8. On most domestic cars with four-wheel disc brakes, the parking brake:
   a. Is applied by a mechanical series of cables and levers
   b. Is an integral part of the rear calipers
   c. Has a self-adjusting mechanism built into the pistons
   d. All of the above

6 Hydraulic System Theory

Objectives

Upon completion and review of this chapter, you should be able to:

- ❑ Understand the purpose of the brake hydraulic system and how it operates.
- ❑ Understand the components used in a brake hydraulic systems.
- ❑ Comprehend the terms commonly used with brake hydraulic systems.
- ❑ Understand how a hydraulic system can be used to transmit force from a brake pedal to the brake lining.
- ❑ Understand the reason for split hydraulic systems and how the two styles of split systems differ.
- ❑ Identify the different styles of master cylinders used on automobiles.
- ❑ Understand how a master cylinder, wheel cylinder, and brake caliper operate.
- ❑ Understand the purpose and operation of the various valves and switches that are used in a hydraulic brake system.
- ❑ Identify the different types of brake fluid and their different characteristics.

6.1 Introduction

The **hydraulic brake system** is used to apply the brakes. It is designed to do three basic things: transmit motion from the driver's foot to the brake shoes, transmit force along with motion, and multiply that force by varying amounts to the different wheel assemblies (Figure 6–1). These operations could also be performed by a mechanical system of rods, cables, and levers, but hydraulic operation has the definite advantage in that an exactly equal force is applied to both brake assemblies on an axle (Figure 6–2). This equality of hydraulic pressure helps ensure even, straight stops.

The stopping power of the right and left sides of the vehicle is equalized by the hydraulics and the similarity of the brake units. The stopping power of the front and rear of the vehicle is balanced by the relative size of the caliper and wheel cylinder pistons, the type and size of the brake assemblies, and, sometimes, the hydraulic valves.

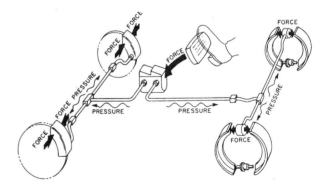

Figure 6–1 In hydraulic brakes, fluid pressure is used to transmit the force from the driver's foot to the brake shoes. (*Reprinted from Mitchell Anti-Lock Brake Systems, with permission of Mitchell Repair Information, LLC*)

6.2 Hydraulic Principles

Hydraulics, often called *fluid power,* is a method of transmitting motion or force. Hydraulics is based on the fact that liquids can flow easily through complicated paths yet cannot be compressed (squeezed into a smaller volume) (Figure 6–3). Another important feature is that when liquids transmit pressure, that pressure is transmitted equally in all directions. This is a simplified version of **Pascal's law** (Figure 6–4).

If we were to fill a strong container with liquid, we would find it impossible to add more liquid, even by force. The only way for more liquid to enter would be for the container to rupture and leak. Once the container is full, any added force on the fluid becomes fluid pressure. Pressure is defined as the amount of force pushing on a certain area. In the United States, pressure is measured in **pounds per square inch (psi);** 10 psi means a force of 10 lb acting on an area of 1 sq. in. A smaller area would have a smaller force on it, and a larger area would have a larger force on it. Traditionally, in Europe and other parts of the world that use the metric system, pressure was measured using bar or kilograms per square centimeter (kg/cm^2). Today pressure is also measured in kilopascals (kPa); 1 psi is equal to 6.895 kPa or 0.07 kg/cm^2.

Pressure can enter a hydraulic system in several ways. It is easier to describe and understand this concept if we use a piston as the pressure input and one or more pistons for the output. This is similar to a brake hydraulic system. The amount of pressure in a system is a product of three factors: the ability of the system to contain the pressure, the size or area of the input piston, and the amount of force on the piston. The strength of the system is important, because if the pressure gets too high, the system will rupture and release the pressure. Imagine a brake system with the brake drums removed. We could not develop much fluid pressure because the wheel cylinders could move too far and pop out of their bores. When force is exerted on the piston of a closed system, that force becomes fluid pressure. The amount of pressure is equal to the force divided by the area of the piston. A 200-lb (90.9-kg) force on a piston that is 1 sq. in. (6.45 m^2) in area generates a force of 200 psi (13.8 bar or 14 kg/cm^2). This pressure can be converted to 1,379 kPa by multiplying 200 by 6.895 (Figure 6–5). The amount of pressure is determined by dividing the force by the area of the piston. The same 200-lb force acting on an area of 0.5 sq. in. (3.2 cm^2) would generate a pressure of 200 ÷ 0.5, or 400 psi (27.6 bar, 28 kg/cm^2, or 2,758 kPa).

When discussing hydraulic pistons and computing fluid pressures and forces, it is important to use the area

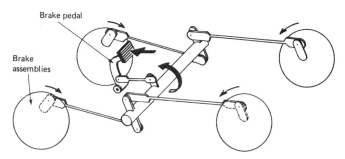

Figure 6–2 Early automobiles used mechanical brakes. A series of metal rods or cables, levers, and a cross shaft connected each of the wheel assemblies to the brake pedal. If the rods were of the wrong length, a brake would apply too early or too late.

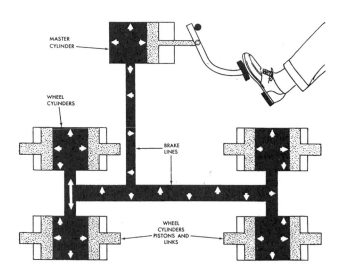

Figure 6–3 Fluid is forced out of the master cylinder as the driver applies the brake. The fluid transmits the pressure equally throughout the system. (*Courtesy of General Motors Corporation, Service Technology Group*)

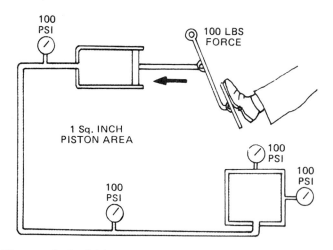

Figure 6–4 Fluid transmits pressure equally throughout a hydraulic circuit and acts with an equal force on each surface of the same size. This is an example of Pascal's law. (*Courtesy of Wagner Brake*)

rather than the diameter. The area of a piston or a circle can be easily determined using either of the following formulas:

Area $= \pi r^2$ or **Area** $= 0.785 d^2$

where π (pi) $= 3.1416$

r = radius, or one-half diameter
d = diameter

The pressure in a hydraulic system becomes a force to produce work and make things move. The amount of force can be determined by multiplying the area of the output piston by the pressure. A pressure of 200 psi pushing on a piston with an area of 1 sq. in. produces a force of 200 lb. The same pressure on a piston with an area of 4 sq. in. (26.17 cm^2) produces a force of 800 lb (363.6 kg) (200 × 4 = 800). The application force is multiplied whenever the output piston is larger than the input piston. Force is divided or made smaller if the input piston is larger (Figure 6–6).

The output piston motion is also related to the input piston. Remember that a hydraulic system can only transmit the energy, force, and motion that are put into it; it cannot create energy. What goes in at one place is all that will come out at another. It is possible, however, to change force to motion and vice versa. Most brake systems multiply and increase force with a loss in motion, but some systems increase motion. As the input piston moves, it displaces or pushes fluid through the tubing to the system. The amount of fluid displaced is equal to the piston area times the length of the piston stroke. A piston with a diameter of 1 in. (2.54 cm) has an area of 0.785 sq. in. (5.067 cm^2). If the piston strokes 2 in. (5.08 cm), 1.57 cubic inches (cu. in.) (25.7 cm^3) of fluid is displaced (0.785 sq. in. × 2) (5.067 cm^2 × 5.08). This much fluid can move a 1-in.-diameter piston 2 in. A piston with a diameter of 3 in. (7.62 cm) has an area of 7.07 sq. in. (45.6 cm^2). It will travel a distance of 0.22 in.—1.57 cu. in. ÷ 7.07 sq. in. = 0.22 in. (0.558 cm—25.7 cm^3 ÷ 45.6 cm^2 = 0.558 cm) (Figure 6–7).

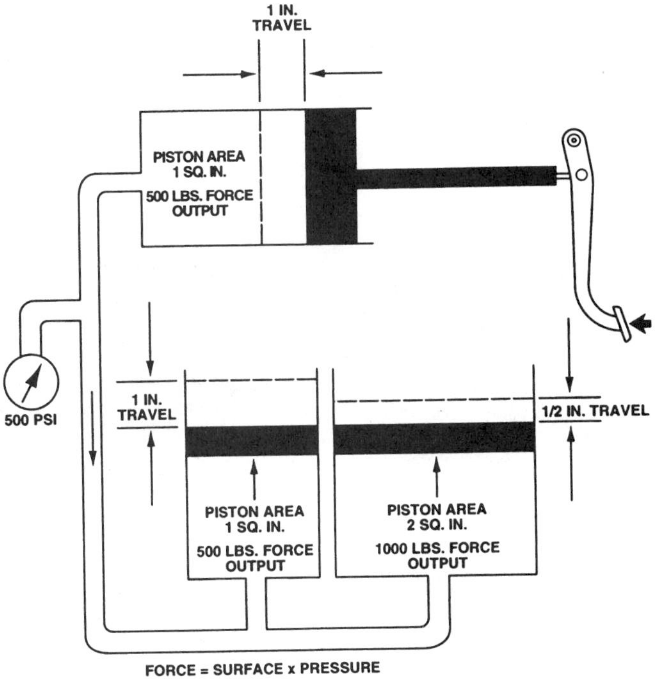

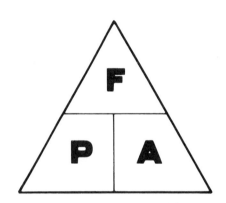

Figure 6–6 A memory triangle for working with hydraulic forces and pressures. Cover up the unknown quantity—F for force, P for pressure, or A for area—and the remainder of the triangle will indicate the solution. For example, P = F ÷ A.

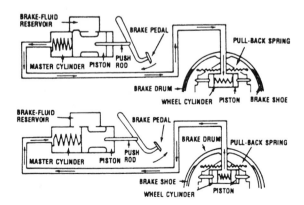

Figure 6–7 When the brakes are applied (top), fluid is displaced out of the master cylinder and into the wheel cylinders and calipers. When the brakes are released (bottom), the brake shoe pullback, return springs force the fluid back to the master cylinder. (*Courtesy of Chrysler Corporation*)

Figure 6–5 Dividing the application force by the area of the master cylinder piston tells us the fluid pressure. Multiplying fluid pressure by the output piston area tells us the output force. In this example, 500 lb of force at the master cylinder piston will produce 500 psi (500 ÷ 1 = 500). The smaller output piston will exert 500 lb of force (500 × 1 = 500), and the larger piston will exert 1,000 lb of force (500 × 2 = 1,000). (*Courtesy of General Motors Corporation, Service Technology Group*)

A typical hydraulic brake system on a domestic passenger car has a master cylinder piston with a diameter of about 1 in. (2.54 cm), two disc brake caliper pistons with diameters of about 2.5 in. (6.35 cm) each, and two rear-wheel cylinders with diameters of about 7/8 or 0.875 in. (2.22 cm) each. The actual diameters depend on the weight of the car, whether power brakes are used, and the weight balance of the car. Heavier cars need more braking power, which often means a smaller master cylinder piston. But the smaller amount of fluid displacement gives only a small amount of brake shoe movement. Smaller cars can use a larger-diameter piston in the master cylinder, which will displace more fluid and give more wheel cylinder and caliper piston movement. A typical car with a master cylinder diameter of 1 in. (2.54 cm) (0.78 sq. in.) (5 cm^2) and a 6:1 brake pedal ratio develops a fluid pressure of 382 psi (2,634 kPa) under a pedal pressure of 50 lb (22.7 kg). If the master cylinder bore is increased to $1\frac{1}{8}$ in. (2.8 cm), the area will increase to 0.99 sq. in. (6.4 cm^2), and this area increase will drop the fluid pressure to 303 psi (2,089 kPa) at the same pedal pressure.

Cars with power boosters often use larger-diameter master cylinders. Larger wheel cylinder and caliper pistons can also be used to increase braking power, whereas smaller pistons increase piston and brake shoe travel. When it is necessary to increase the rear-wheel braking power relative to the front, larger-diameter wheel cylinders are used. Calipers with larger-diameter pistons are used when it is necessary to increase the relative power of the front brakes. The relative power between the front and rear brakes is called balance.

If the size of both the front and the rear pistons is increased to gain more application force, the brake engineer must remember that the increase will require more fluid to move the pistons. If the wheel cylinder and caliper pistons require more fluid than can be displaced by the master cylinder, the brake pedal will go to the floor without applying the shoes. The farther the shoes must travel or the larger the wheel cylinder and caliper pistons, the farther the master cylinder must travel or the larger the master cylinder bore must be (Figure 6–8).

The greatest potential problems for a hydraulic system are leaks and air. A leak allows fluid to escape when it leaves the master cylinder. Thus there is less fluid to apply the wheel cylinder and caliper pistons, and the pedal sinks to the floor. Air is compressible; its volume decreases under pressure. This volume reduction takes place at the expense of fluid displacement and is experienced by the driver as a "spongy" brake pedal. The expansion and contraction of air give a springiness to the brake pedal (Figure 6–9).

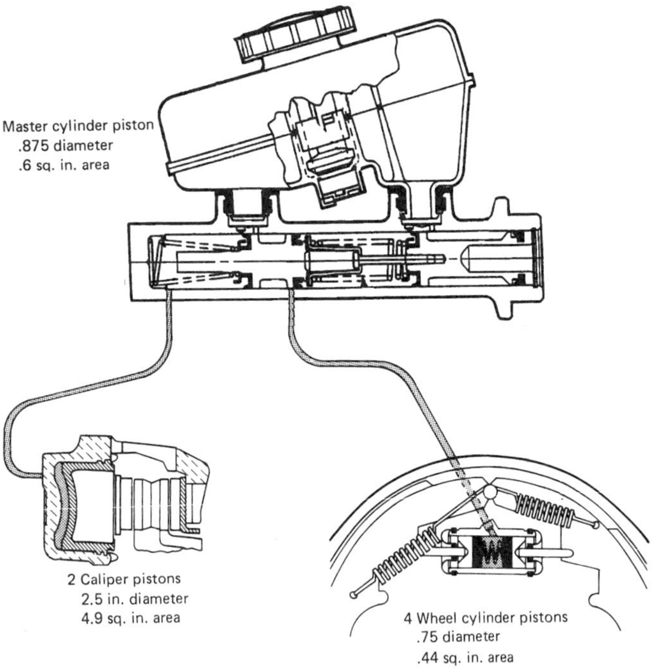

Figure 6–8 A typical manual brake master cylinder on a RWD, full-sized car has a bore diameter of 0.875 in. In 1 in. of travel, the pistons will displace 2 × 0.6 cu. in. of fluid. The two caliper pistons and four wheel cylinder pistons have a combined area of 10.68 sq. in. If they all moved an equal distance, 1.2 cu. in. of fluid would move each of them 0.10 in.

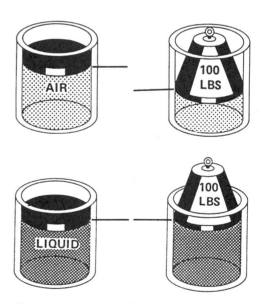

Figure 6–9 Air is compressible; its volume becomes smaller under pressure, whereas the volume of a fluid remains the same because it is noncompressible. (*Courtesy of General Motors Corporation, Service Technology Group*)

6.3 Master Cylinder Basics and Components

The **master cylinder** is the input piston for the car's brake system. It is connected to the brake pedal so that movement of the brake pedal is transmitted to the master cylinder piston by a pushrod. The **brake pedal** is a simple lever. The ratio is usually about 6:1 or 7:1. This means that the application force from the foot is multiplied six or seven times, but the motion of the foot is reduced by the same amount. With a 6:1 ratio, 50 lb (22.7 kg) of pedal pressure produces 300 lb (136.4 kg) of pressure on the master cylinder piston, but it takes 6 in. (15.24 cm) of pedal motion to produce 1 in. (2.54 cm) of motion at the piston.

The first master cylinders used a single piston and cylinder bore with one outlet at the end of the bore (Figure 6–10). From this outlet, tubing branched out to each of the wheel cylinders. The master cylinder body has two major areas, the piston and cylinder bore and the reservoir. These two areas are connected by two passages, a small **compensating port** and a larger **bypass port.** The bypass port is also called an *intake* or *replenishing port.* The cylinder bore is a smooth, precision-sized bore in which the piston and seals slide. The piston moves inward when the brake pedal is pushed. A pushrod connects the brake pedal to the piston. When there is no pedal pressure, the piston return spring pushes the piston against a retaining ring or clip at the end of the cylinder bore. The piston uses two seals called cups: a **primary cup** at the inner face of the piston and a **secondary cup** at the outer end. The primary cup pumps fluid, and the secondary cup keeps fluid from leaking out of the end of the cylinder. The two passages between the cylinder bore and the reservoir are located close to the primary cup. The compensating port is just in front of it, and the bypass port is behind it (Figure 6–11).

6.4 Master Cylinder Basic Operation

In a normal brake system, the whole system is interconnected and filled with fluid up to the upper part of the master cylinder reservoir. The reservoir is vented to the atmosphere so fluid can expand and contract and still stay at atmospheric pressure. When the fluid in the wheel cylinders and calipers gets hot it can expand and move through the open compensating port to the reservoir. When it cools and contracts, it can move the other way, keeping the system filled with fluid. Modern master cylinders use a rubber diaphragm over the reservoirs to separate the fluid from air and to help reduce fluid contamination (Figure 6–12).

When sufficient force is placed on the brake pedal to overcome the piston return spring, the master cylinder piston moves inward. Fluid from the bore is displaced to the reservoir until the lip of the primary cup moves past and closes the compensating port. From this point, further force and movement of the brake pedal will displace fluid to the system and move the brake shoes into contact with the rotors or drums (Figure 6–13). Lining contact stops further movement of the shoes and wheel cylinder or caliper pistons. From this point, system pressure will increase. Any increase in pedal force causes a corresponding increase in wheel cylinder and lining pressure.

When the brake pedal is released, the piston return spring moves the piston very rapidly back to its stop at the retaining ring—much faster than the fluid returns from the wheel cylinder or calipers. During this motion, the primary cup collapses or relaxes its wall pressure slightly

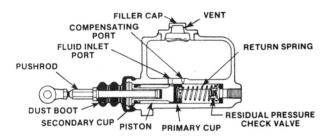

Figure 6–10 A cutaway view of a single master cylinder in the released position. Note the position of the primary cup relative to the compensating port. (*Courtesy of Bendix Brakes, by AlliedSignal*)

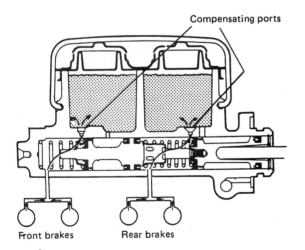

Figure 6–11 If the fluid in the calipers and wheel cylinders heats up and expands, the increased volume will flow through the compensating ports to the reservoir. (*Courtesy of General Motors Corporation, Service Technology Group*)

and moves through the fluid. The fluid flows past the edges of the primary cup (Figure 6–14). Some pistons have a series of holes in the primary cup face to improve this flow. The flow past the primary cup allows more fluid to pump into the system if the brake pedal is pumped or cycled rapidly. It also prevents low fluid pressure or a vacuum during brake release. A vacuum might cause air to enter the system past the secondary cup of the master cylinder piston or past one of the wheel cylinder cups.

6.5 Master Cylinder Construction

In the past, master cylinder bodies were made of cast iron. This strong material has a relatively low cost, can be cast into complex shapes, and is easily machined. Today, because of the need to reduce vehicle weight to improve fuel mileage, most master cylinders are made from cast aluminum and plastics. These materials can also be cast into complex shapes and are easily machined.

The cylinder bore of a master cylinder has the following critical requirements: It must be strong enough to contain the pressure; it must be of an exact size, no larger than

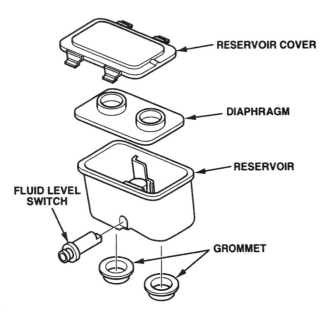

Figure 6–12 The rubber diaphragm under the reservoir cover helps keep the fluid clean and free from water while allowing atmospheric pressure to act on the fluid. (*Courtesy of General Motors Corporation, Service Technology Group*)

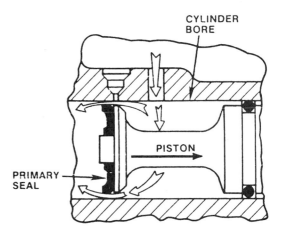

Figure 6–14 During master cylinder release, fluid moves from the reservoir through the bypass port and past the primary seal. (*Courtesy of Bendix Brakes, by AlliedSignal*)

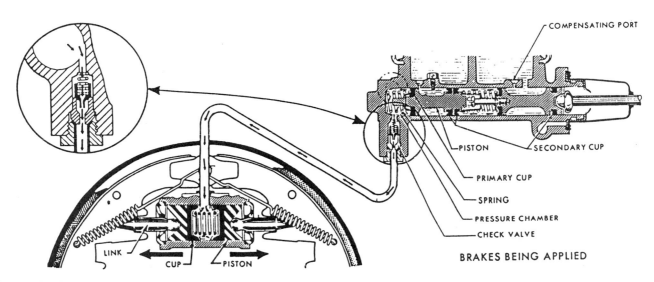

Figure 6–13 As soon as the primary cup moves past the compensating port during brake application, fluid is forced out of the master cylinder. If a residual valve is used, fluid will flow through the valve and out to the wheel cylinder and caliper pistons. (*Courtesy of General Motors Corporation, Service Technology Group*)

Chapter 6 ■ Hydraulic System Theory

0.006 in. (0.15 mm) greater than the piston diameter; it must be round and straight; and it must be smooth. In production, cylinder bores are given their final size by **roller burnishing.** A hardened-steel, precision-sized roller is forced through the bore that has been machined to a slightly undersized diameter. As the burnishing tool passes through the bore, the machine scratches are smoothed out as the bore is expanded to the correct size. This action also produces a harder bore because of the work hardening of the metal (Figure 6–15). Aluminum bores are **anodized** to harden the bore after machining. Anodizing is an electrochemical process that helps the normally soft aluminum resist corrosion, galling, and wear. The hard anodized layer is extremely thin, however, and when this layer wears through, the soft aluminum underneath is exposed to rapid corrosion and wear.

The **reservoir** can be cast into the master cylinder body or attached to it by other means. Many modern units use a plastic or nylon reservoir that is clamped onto the body or connected to it by a pair of rubber grommets (Figure 6–16). In congested engine compartments, the reservoir can be mounted in a remote location and connected to the cylinder body by rubber or metal tubing (Figure 6–17).

Master cylinder pistons are cast from aluminum. They are often anodized to resist corrosion. They usually have one flat face for the primary cup and a groove for the ring-shaped secondary cup. The secondary end of the piston has a provision for the pushrod (Figure 6–18).

At one time, **cups** were made from natural rubber; now synthetic rubber is used. A rubber cup is a somewhat amazing sealing device. One expert has determined that the average American driver uses the brakes about

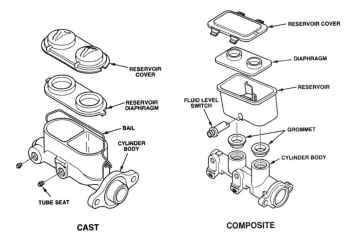

Figure 6–16 Most modern master cylinders (above right) use a cast-aluminum body with a plastic or nylon reservoir. (*Courtesy of General Motors Corporation, Service Technology Group*)

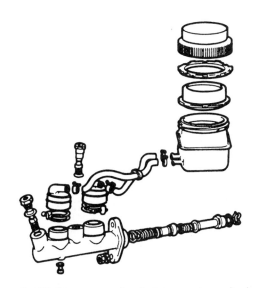

Figure 6–17 The reservoir of this master cylinder is remotely mounted and connected to the cylinder body by a pair of hoses. (*Courtesy of Bendix Brakes, by AlliedSignal*)

Figure 6–15 A cast-iron master or wheel cylinder bore is given a very exacting finish by being "bearingized." The final manufacturing step is to force a hardened-steel ball through the bore to smooth out any imperfections. (*Courtesy of Brake Parts, Inc.*)

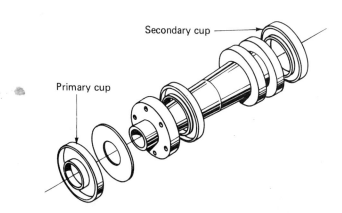

Figure 6–18 Some master cylinder piston faces are drilled so fluid can bypass the primary cup during release. Ring-type secondary cups are normally used.

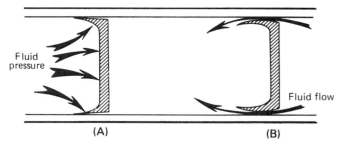

Figure 6–19 Rubber cups seal tighter when fluid pressure is exerted on the front of the cup (A) because pressure forces the lips against the bore. Pressure at the rear of the cup (B) moves the lips away from the bore.

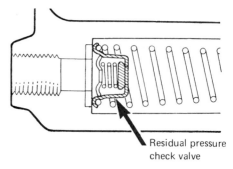

Figure 6–20 Most single master cylinders use a residual check valve at the outlet end of the cylinder bore. (*Reprinted from Mitchell Anti-Lock Brake Systems, with permission of Mitchell Repair Information, LLC*)

75,000 times a year. Each application requires that the master cylinder cups slide down the bore, and while this is occurring, no fluid or fluid pressure should escape past the cup. Normally, some wear of the cup occurs. This wear causes a blackish coloration of the fluid and residue at the bottom of the master cylinder reservoir. A cup is especially good at sealing pressure because of its shape. Pressure pushes the lips of the cups tightly against the bore, which improves the seal. The greater the pressure, the tighter the seal. The primary cup has to slide down the bore as the system's pressure increases. There is very little pressure on the cup lips when the brakes are released. A cup is a single-direction seal, somewhat like a check valve. Pressure from the front seals the lips tightly to the bore, whereas pressure from the back causes flow past the cup. This can be a desirable part of the design, as with a primary cup during release. It can be a problem if air enters past a cup at a wheel cylinder (Figure 6–19).

6.6 Residual Pressure

At one time, each master cylinder contained one or two **residual pressure check valves.** A single master cylinder uses one valve placed at the outlet end of the cylinder bore (Figure 6–20). A **tandem master cylinder** has a valve under one or both outlet tube seats (Figure 6–21).

These valves are designed to allow a free flow of fluid out of the master cylinder but to stop some of the fluid from returning. They shut off the return flow when the system pressure drops down to about 5 to 25 psi (34.5 to 172 kPa). This pressure stays in the tubing and wheel cylinders of the system while the brakes are released.

At one time, residual pressure was thought to be necessary to keep air from entering the system past the wheel cylinder cups. This pressure also kept the wheel cylinder cups against the pistons, the piston next to the pushrods, and the pushrods next to the brake shoes, so that there would be no lag or lost motion during brake application.

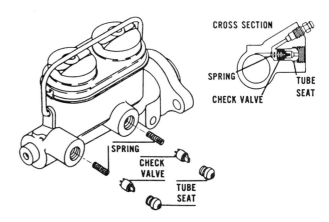

Figure 6–21 Residual check valves for a tandem master cylinder are mounted under the outlet port tube seats. Fluid flows through the center of the check valve on the way out and around the check valve on the way back in. (*Courtesy of General Motors Corporation, Service Technology Group*)

Today the **cup expanders** and the springs in the wheel cylinders take care of this duty, and residual pressure valves are no longer commonly used. Residual pressure is never used with disc brakes. It would provide enough pressure for partial application and cause lining drag.

6.7 Dual Master Cylinder Construction

One major drawback with a brake hydraulic system is that a sudden failure can result from a major fluid leak. If a tube, hose, wheel cylinder, or other component ruptures, there is no braking power. One solution for this problem is to split the hydraulic system into two parts, a requirement for all cars sold in the United States since 1967. The hydraulic system and master cylinder were divided so that the front brakes operate from one piston of the master cylinder and the rear brakes operate from the other (Figure 6–22).

Chapter 6 ■ Hydraulic System Theory

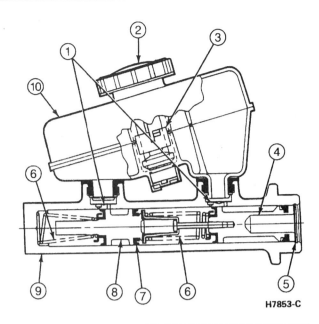

Item	Part Number	Description
1	—	Compensating Ports (Part of 2140)
2	2162	Brake Master Cylinder Filler Cap
3	—	Float Magnet Assembly (Part of 2K478)
4	—	Primary Piston (Part of 2140)
5	—	Bore End Seal (Part of 2140)
6	—	Spring (Part of 2140)
7	—	Seal (Part of 2140)
8	—	Secondary Piston (Part of 2140)
9	2140	Brake Master Cylinder
10	2K478	Brake Master Cylinder Reservoir

Figure 6–22 A cutaway view of a dual master cylinder showing the relative position of the internal parts. (*Courtesy of Ford Motor Company*)

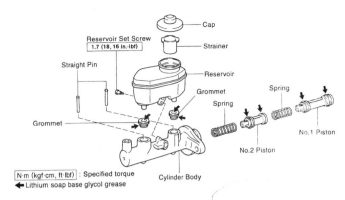

Figure 6–23 An exploded view of a dual master cylinder.

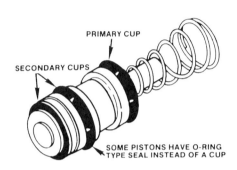

Figure 6–24 The secondary piston assembly from a dual master cylinder. Note that the lip of the rear cup (left) faces to the left and the other two face to the right, toward the fluid to be sealed. (*Courtesy of Bendix Brakes, by AlliedSignal*)

A dual **tandem master cylinder** has two pistons in tandem—one in front of the other—that operate in the same cylinder bore. The **primary piston** operates mechanically, directly from the pushrod, and the **secondary piston** operates hydraulically, from the hydraulic pressure in the primary section. Normally, the hydraulic pressure is the same on each side, front and rear, of the secondary piston. The primary piston has two rubber cups, a primary and a secondary, just like the piston in a single master cylinder. The secondary piston usually has three cups, a primary cup and two secondary cups (Figure 6–23). Some designs use a secondary sealing ring in place of one of the cups. The primary cup of the secondary piston, like the other primary cups, is a pumping cup. The secondary cups on the secondary piston are slightly different. One prevents a fluid leak from the secondary system into the primary, and the other prevents a high-pressure fluid leak from the primary system into the secondary reservoir. The lip of the latter cup faces the primary piston, and all of the other cups face the other direction (Figure 6–24). Some new master cylinders use slightly different secondary cup arrangements on the secondary piston.

A tandem master cylinder bore has a compensating port and a bypass port for each piston. The reservoir is divided into two parts so that a failure in one of the systems will not drain the fluid out of the other (Figure 6–25). In a master cylinder with four-wheel drum brakes, both reservoir sections are the same size. A master cylinder for a car with a disc-front and drum-rear combination has a larger section for the disc brakes. Disc brake pistons do not return but gradually creep outward as the lining wears. The reservoir fluid level drops in relation to the lining wear (Figure 6–26). This master cylinder also has two fluid outlets. They usually exit through the side of the cylinder body. These ports are normally drilled so that they intersect the cylinder bore at the top and at the end of each chamber. This feature helps when bleeding the air

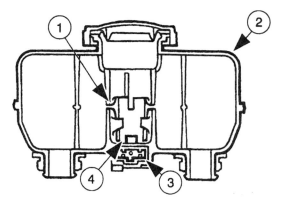

Figure 6–25 This reservoir is divided into two fluid cavities on each side of the well for the float (1), magnet (4), and fluid level switch (3). (*Courtesy of Ford Motor Company*)

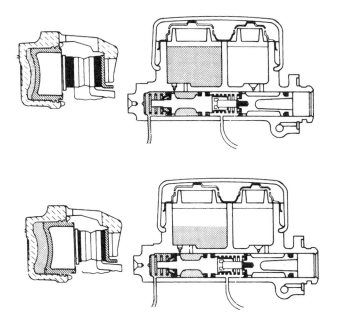

Figure 6–26 As disc brake lining wears, the caliper and piston move closer to the rotor, and the piston moves outward in the bore. The master cylinder fluid level decreases as this occurs.

out of the cylinder and lines. As mentioned earlier, some master cylinders incorporate a residual check valve under one or both of these line fittings.

6.8 Tandem Master Cylinder Operation

The operation of the primary section of a tandem master cylinder is the same as that of a single master cylinder. The secondary piston operates in a similar manner except that there is no pushrod. Pressure buildup in the primary section moves the secondary piston so that it builds up

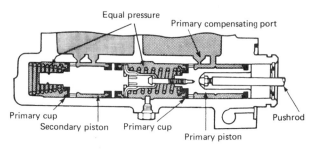

Figure 6–27 Normally, the secondary piston is operated by the hydraulic pressure in front of the primary piston. The pressure in front of the two pistons is equal. (*Courtesy of Bendix Brakes, by AlliedSignal*)

the same pressure in the secondary section. The secondary piston floats according to the pressure in the primary and secondary sections (Figure 6–27).

When released, the primary piston returns to the retaining ring at the end of the bore. It is pushed there by both primary and secondary piston return springs. The released position of the secondary piston is determined by the length and strength of the primary and secondary return springs. The values of these two springs are adjusted during manufacture to ensure correct secondary piston positioning. The primary cup should be returned just past its compensating port. If the port is not uncovered during release, brake drag will occur because the fluid will not be able to flow back from the wheel cylinders or calipers. If the piston returns too far past the port, a low brake pedal can result from loss of the pump stroke. Pumping will not begin until the primary cup moves past the compensating port. In the past, a stop screw was threaded into the cylinder bore to ensure proper secondary piston positioning. The secondary piston stopped against this screw during piston return. This stop screw was threaded into the bore from the side, the bottom, or the top or reservoir side (Figure 6–28).

Some newer master cylinders used with ABS use two central valves instead of compensating ports. These valves, at the centers of the primary and secondary pistons, are open while the brakes are released to allow flow between the reservoirs and the area in front of the pistons, like a compensating port. When the brakes are applied, these valves are forced shut, blocking any flow to the reservoirs, so the fluid in front of the pistons can be forced to the calipers and wheel cylinders (Figure 6–29). During ABS activation in some systems, the master cylinder pistons shift back and forth rapidly, with the lip of the primary cups sliding past the compensating ports and pressure in front of the cups; this action can cause wear or etching of the lips of the cups and possible failure.

In case of hydraulic failure in one of the systems, two features are built into this unit to ensure operation of the

Chapter 6 ■ Hydraulic System Theory

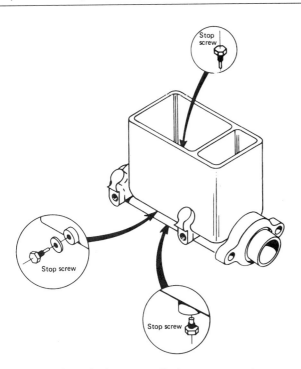

Figure 6-28 A dual master cylinder can use a stop screw or pin to locate the released position of the secondary piston. A stop screw or pin can enter the cylinder bore from the top, bottom, or side.

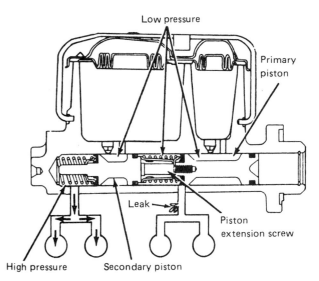

Figure 6-30 If a hydraulic failure occurs in the primary hydraulic section, the secondary piston is operated mechanically by the piston extension screw. (*Courtesy of General Motors Corporation, Service Technology Group*)

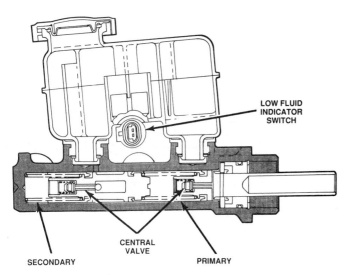

Figure 6-29 This ABS master cylinder uses a central valve in place of the compensating ports; the central valves are opened when the brakes are released. (*Courtesy of Chrysler Corporation*)

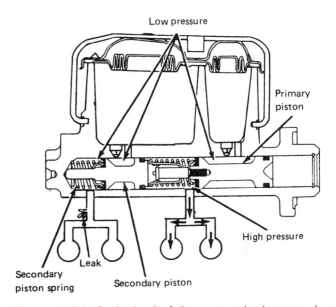

Figure 6-31 If a hydraulic failure occurs in the secondary hydraulic section, the secondary piston moves until its extension stops at the end of the bore. The primary section then operates in the normal manner. (*Courtesy of General Motors Corporation, Service Technology Group*)

other half. The primary piston has an extension screw threaded into the forward, primary cup end. If the primary section experiences loss of pressure, this extension will push on and apply the secondary piston mechanically. There will be some lost motion until the extension meets the secondary piston. The driver will notice a lower brake pedal and a loss in braking power. The brake failure warning light should also come on to indicate a hydraulic failure (Figure 6-30). There is also an extension on the forward end of the secondary piston. If a pressure loss occurs in the secondary system, the secondary piston will move easily until this projection strikes the end of the bore. From this point, the primary section can begin to operate in a normal manner. There will also be an obviously lower brake pedal and a loss in braking power (Figure 6-31).

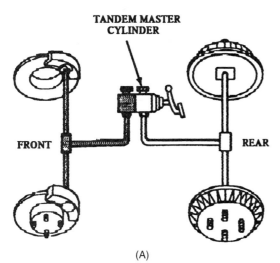
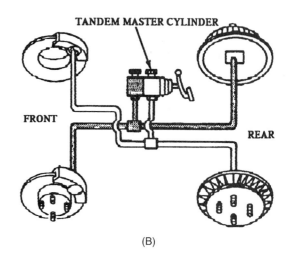

Figure 6-32 A tandem split system pairs the front and rear brakes at the primary and secondary ends of the master cylinder (A). A diagonal system (B) pairs the right front with the left rear and the left front with the right rear. (*Reprinted by permission of Hunter Engineering Company*)

6.9 Diagonally Split Hydraulic System

The first split hydraulic systems used a **front–rear split**, also called a *tandem split*. Because of the unequal front and rear axle weights and weight transfer, a front–rear split does not mean a fifty-fifty split. In actual practice, a rear-system failure leaves 60 to 70 percent of the braking power, whereas a front-system failure leaves only 30 to 40 percent. The only way to produce a true fifty-fifty split would be to divide the system in a right-side–left-side manner or diagonally. A car with only right-side or left-side brakes would be extremely hard to control during a stop. A RWD car with a diagonal split system would be very difficult to control while stopping with one front brake and the rear brake at the opposite corner.

FWD cars have a different wheel geometry; they use a negative scrub radius, whereas RWD cars use a positive scrub radius. The *scrub radius* is the distance between the center of the tire and the steering axis at the road surface. The steering axis is the pivot point for the front wheels when they steer or turn a corner. With a *positive scrub radius*, the tire is outboard of the steering axis, and with a negative scrub radius, the tire is inboard. With a positive scrub radius, a tire under braking tries to turn or steer outward. During stops, this tendency is balanced out between the two front tires. With a *negative scrub radius*, a braked tire tries to turn or steer inward, but this movement is offset by the braking drag on the tire, which tries to pull the car in the opposite direction. With a negative scrub radius, a controlled stop can be made with only one front brake. A **diagonal split**, sometimes called a *crisscross system*—right front with left rear and left front with

Figure 6-33 Five different styles of split or dual braking systems are shown here. Most vehicles use version 1 (tandem split) for RWD vehicles and version 2 for FWD vehicles. Versions 3, 4, and 5 provide better stopping control if failure occurs, but they are complex and expensive. (*Reprinted with permission of Robert Bosch Corporation*)

right rear—provides 50 percent braking if either system fails (Figure 6-32). Other versions of split hydraulic systems are used in some of the more expensive vehicles (Figure 6-33).

The master cylinder used with diagonal split systems is usually a seemingly ordinary tandem master cylinder. Some of them have four outlet ports, with a front and rear outlet for both the primary and secondary sections. In some of these master cylinders, a proportioning valve is installed in each of the two rear-wheel outlets. These valves are discussed later in this chapter. Also, in some designs, the pressure differential valve for the brake failure warning light switch is included in the master cylinder body.

A potentially serious problem with diagonal split systems is that heat from parking brake drag can cause total brake loss. If the car is driven with partially applied parking brakes, the heat generated can cause the fluid to boil in the rear wheel cylinders. Because the boiling fluid is in both brake circuits, it can cause total brake failure. After the brakes cool down, the brakes usually return to normal operation.

6.10 Quick-Take-Up Master Cylinder

A **quick-take-up master cylinder,** also called a *dual-diameter bore, step bore,* or *fast-fill* master cylinder, is used with newer, low-drag disc brake calipers (Figure 6–34). These calipers have a redesigned piston-sealing, square-cut O-ring and groove, which have the ability to pull the piston back a greater distance. This design provides a slight clearance between the front pads and the rotor, reducing brake drag and improving fuel mileage, but it can also cause a very low brake pedal. More pad clearance requires more movement of the caliper piston during application. More movement of these two rather large pistons requires more fluid displacement from the master cylinder. A large-bore master cylinder is needed, but it usually provides for lower hydraulic pressures. Remember that hydraulic pressure equals application force divided by piston area.

A quick-take-up master cylinder has a step bore in the primary section and uses a primary piston that has a primary cup of normal diameter with a larger, oversized secondary cup. This secondary cup is a low-pressure pumping cup. It pumps the extra fluid needed to take up the clearance at the calipers. The stepped-down area between the primary and secondary cups is now called the low-pressure section.

One manufacturer uses a master cylinder that has a 7/8- or 0.875-inch (22.2-mm) bore that steps to a $1\frac{1}{4}$- or 1.25-in. (31.75-mm) bore. A bulge at the outside of the cylinder body in this section usually identifies this master cylinder type. The larger bore has twice the area, so it displaces twice as much fluid as the smaller bore.

During brake application, fluid is pumped out of the low-pressure chamber past the lips of the primary cup. This fluid goes into the high-pressure chamber of the primary system to move the caliper pistons and also to move the secondary piston in the master cylinder. As soon as the brake lining clearance is taken up, the hydraulic pressure begins to increase. This could produce

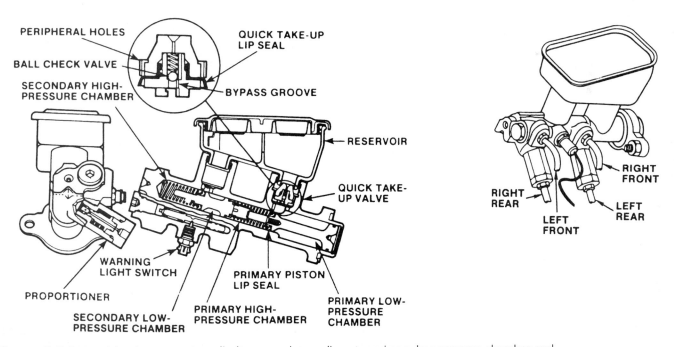

Figure 6–34 A quick-take-up master cylinder uses a larger-diameter primary low-pressure chamber and a quick-take-up valve. Also note that this particular master cylinder has four outlet ports, proportioning valves at the rear ports, and a hydraulic failure warning light switch. (*Courtesy of Bendix Brakes, by AlliedSignal*)

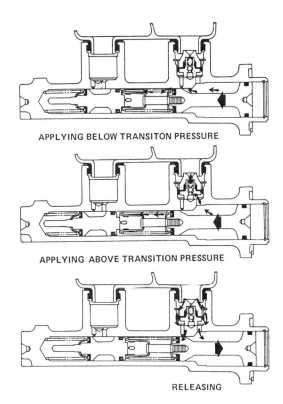

Figure 6–35 During brake application, fluid flows from the primary low-pressure chamber around the primary cup to move the lining into contact with the rotor. As pressure is generated (above transition), fluid goes from this section through the quick-take-up valve and back to the reservoir. (*Courtesy of General Motors Corporation, Service Technology Group*)

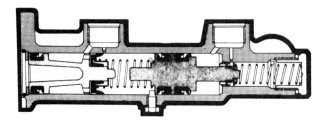

Figure 6–36 A step bore tandem master cylinder uses a smaller-diameter primary cup on the secondary piston. This design produces higher pressure in the secondary hydraulic circuit. (*Courtesy of ITT Automotive*)

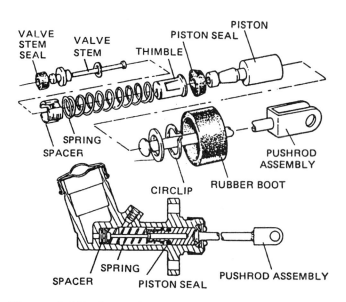

Figure 6–37 This single master cylinder design was used on many British-made cars. Note that the outlet port is above the middle of the cylinder bore. (*Courtesy of Wagner Brake*)

enough backpressure on the larger-displacement piston to cause a hard brake pedal, but before this point is reached, the quick-take-up valve between the low-pressure section and the reservoir opens. This valve opens at a preset pressure to allow the extra fluid, which is no longer needed, to flow into the reservoir with little effort. From this point, the primary cups of the pistons produce braking pressure in a normal manner. Fluid flows during release are also very similar to those in a conventional master cylinder (Figure 6–35).

6.11 Miscellaneous Master Cylinder Designs

Other master cylinder designs have very limited usage.

The **dual master cylinder** is essentially two complete master cylinders that are side by side in the same body casting. One of them is used for the brakes, and the other operates the clutch slave cylinder. The two bores and two reservoirs are completely separate from each other so that possible failure of one does not affect the operation of the other. This design has been used on some domestic pickups.

The *step bore master cylinder* is a tandem master cylinder with a bore that has two different diameters. The secondary piston and bore section are smaller. This design provides two different hydraulic pressures in the primary and secondary sections, which produces automatic brake proportioning (Figure 6–36).

Lucas-Girling master cylinders are used on many British-made cars. Older, single-piston designs have a single cup on the piston. The compensating port to the reservoir is at the end of the master cylinder bore, and the flow through this port is controlled by a valve. When the brakes are released, this valve is open to allow compensation. It is closed as soon as the piston is moved for brake application (Figure 6–37).

6.12 Wheel Cylinder Basics

Wheel cylinders are *output pistons* for the hydraulic system. Pressure generated in the master cylinder pushes against the cups in the wheel cylinders to develop the force to apply the brake shoes (Figure 6–38). The amount of force can be determined by multiplying the amount of hydraulic pressure by the area of the wheel cylinder cups or pistons. The diameter of many wheel cylinders is indicated by a number on the cylinder body, inner face of the cup, or piston.

Most wheel cylinders use a straight bore with a cup and piston at each end. The piston and cup apply the same force to each brake shoe. Two–leading shoe, two–trailing shoe, and some uniservo designs use a single wheel cylinder. One piston and one cup are used in a cylinder bore that is closed at one end (Figure 6–39). One nonservo design uses a sliding single wheel cylinder. It is mounted to the backing plate in such a way as to allow the cylinder bore to slide from hydraulic reaction and apply one shoe while the piston is applying the other one. A few brake designs use a *step bore wheel cylinder*. The cylinder bore has two diameters with two different-sized pistons and cups. The smaller cup and piston in the smaller bore will exert less pressure on one shoe than the larger piston at the other end applies on the other shoe (Figure 6–40).

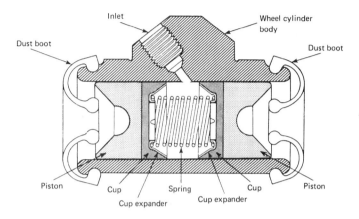

Figure 6–38 A cutaway of a typical wheel cylinder showing the internal parts. (*Courtesy of Bendix Brakes, by AlliedSignal*)

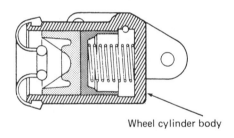

Figure 6–39 This single-piston wheel cylinder can apply pressure to only one brake shoe. It is used in a two–leading shoe or a uniservo brake. (*Courtesy of Bendix Brakes, by AlliedSignal*)

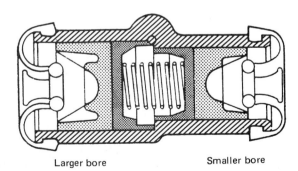

Figure 6–40 This step bore wheel cylinder applies greater force on the shoe to the left than on the shoe to the right. (*Courtesy of Bendix Brakes, by AlliedSignal*)

6.13 Wheel Cylinder Construction

A common double-piston, straight-bore wheel cylinder consists of the cylinder body, two pistons, two piston cups, two boots, a center spring (often with expanders), and a bleeder screw. The cylinder body is made from cast iron or aluminum. As in a master cylinder, the bore must be straight, smooth, and of a precise size. Aluminum bores are anodized to resist corrosion and wear (Figure 6–41).

The cylinder body is normally manufactured with a mounting projection that extends through the backing plate. During brake application, there is a lot of sideways or shear force between the wheel cylinder and the backing plate. This projection ensures that the wheel cylinder will not move relative to the backing plate. Most wheel cylinders are secured to the backing plate by a pair of machine screws or bolts. A recent Delco Moraine design uses a special clip (Figure 6–42).

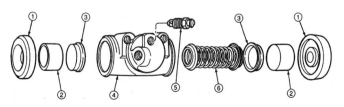

Figure 6–41 A disassembled wheel cylinder showing the two boots (1), pistons (2), cups (3), body (4), bleeder screw (5), and spring (6). (*Courtesy of Ford Motor Company*)

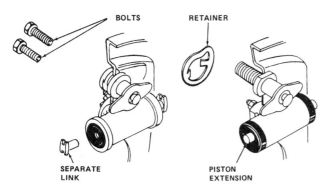

Figure 6–42 A wheel cylinder can be attached to the backing plate by a pair of bolts or a retainer. Note that the left unit uses an internal boot and a separate link to contact the brake shoe; the other unit uses a piston extension and external boot. (*Courtesy of General Motors Corporation, Service Technology Group*)

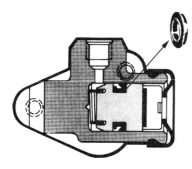

Figure 6–43 Some wheel cylinders use a ring-type seal. (*Courtesy of ITT Automotive*)

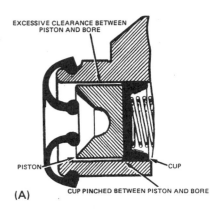

Figure 6–44 If there is excessive piston-to-bore clearance, the cup can be forced into this gap by very high fluid pressure (A). This will cause heel drag and nibbling of the cup edges (B). (*A is courtesy of Wagner Brake; B is courtesy of ITT Automotive*)

Wheel cylinder pistons are made from anodized cast aluminum, sintered iron, or plastic and are shaped to accept the end of the brake shoe link or pushrod or a projection of the brake shoe. The inner side of the piston is flat and smooth for the cup to rest on and push against. Some wheel cylinders use a piston that is grooved to accept a ring-shaped cup similar to a master cylinder secondary cup (Figure 6–43).

A *wheel cylinder cup* operates just like a master cylinder cup. Internal pressure forces the lips of the cup against the cylinder, ensuring a pressure-tight seal. A wheel cylinder cup has a definite wear advantage over a master cylinder cup in that pressure moves the cup and piston away from the lip of the cup and not toward it. Also, when hydraulic pressure gets high enough to really pin the lips of the cup against the bore, the cup does not need to travel much farther. Most cups have tapered lips; they are much thicker at the inner section. When the hydraulic pressure gets very high, the thicker section can stretch to allow more cup movement without sliding the lip of the cup. Seals of the synthetic rubber cup or lip type can be easily damaged if the lip of the cup is cut or torn by faulty installation or rough bore surfaces. During hard stops, the piston-to-bore clearance becomes very critical. Excessive clearance provides enough room for hydraulic pressure to force the cup material between the piston and the bore. If this happens, brake drag can occur because the piston will not return or the edges of the cup will wear or be nibbled away. This condition is called **heel drag** (Figure 6–44).

The spring has a simple job. It pushes outward on the cups so everything between them and the brake shoes stays in contact. Many springs include a pair of **expander washers** to maintain a slight pressure between the lips of the cups and the cylinder bores. This pressure helps prevent air from entering into the system when there is no hydraulic pressure on the cups.

Each end of the cylinder bore is enclosed in a rubber **dust boot.** This boot prevents contaminants from entering the bore and becoming wedged between the piston and bore or causing corrosion of the bore or piston. Either condition can cause the piston to stick. Most boots are *external*. They fit over a raised lip at the outer end of the cylinder body and are held in place by the elastic nature of the rubber material. Some boots are *internal*. They fit into a recess at the end of the bore and are held tightly

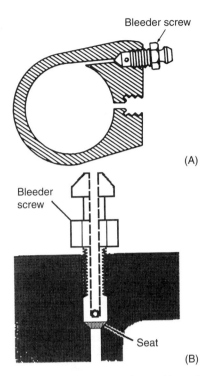

Figure 6-45 The bleeder screw is positioned so that the passage from the valve seat connects to the very top of the cylinder bore (A). The hardened end of the screw seals against the tapered seat in the cylinder (B).

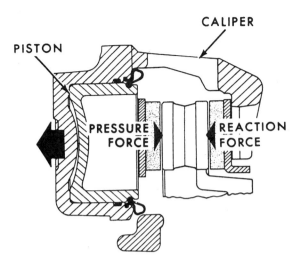

Figure 6-46 In a floating caliper, hydraulic pressure causes the piston to push on the inboard pad. The same hydraulic pressure at the end of the cylinder bore develops a reaction force to move the caliper and apply the outboard pad. *(Courtesy of Ford Motor Company)*

in place by an internal metal ring. The center of either style of boot fits snugly over the brake shoe link or over a projection of the piston to make a seal.

The **bleeder screw** is fitted into a passage that extends from the top center of the cylinder bore. It is actually a valve; the hardened, tapered end of the screw fits tightly against a tapered seat to close off this passage. When the screw is loosened, air and fluid can flow past the seat and through a passage in the screw, allowing air to be bled from the cylinder. The bleeder screw is normally closed tightly to keep fluid from escaping. Many include a rubber or plastic cup to keep dirt, water, and salt from entering the outer opening and plugging it (Figure 6–45).

6.14 Caliper Basics

As previously mentioned, two major styles of calipers are used on passenger cars: *fixed* and *floating*. The primary hydraulic differences include the number of pistons used, the diameter of the pistons, the type of piston seal used, and the fact that many fixed calipers need to be split for service. The caliper pistons, like wheel cylinder pistons, are output pistons; hydraulic pressure pushing on them applies the brake pads. The amount of force that they exert on the pads is determined by multiplying the area of the piston or pistons by the hydraulic pressure. With floating calipers, this force is doubled because of the cylinder reaction (Figure 6–46).

Most calipers use a cup-shaped bore in a cast-iron caliper body. Each bore includes a piston with a seal and boot and a bleeder screw. A few pistons include an insulator to block heat flow from the lining to the piston. Disc brakes experience higher brake fluid operating temperatures than drum brakes because the fluid in the caliper is fairly close to the friction material, and the caliper body and piston conduct quite a bit of heat to the fluid.

6.15 Caliper Pistons and Sealing Rings

Most calipers use a *square-cut* (also called a *lathe-cut*) *O-ring seal* that has a square cross section for a sealing ring. This O-ring fits into a groove in the cylinder bore. The piston that fits through this sealing ring has smooth, straight sides. The O-ring seal has a static (stationary) contact with the cylinder groove and both a static and a dynamic (moving) contact with the piston.

In the released position, the elastic nature of the synthetic rubber O-ring causes it to assume its normal square shape in the groove, with the piston held snugly inside it. During brake application, hydraulic pressure moves the

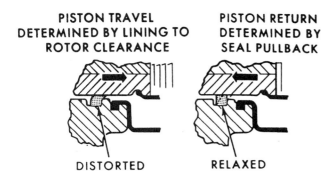

Figure 6–47 As the piston moves to apply the brake pads, the O-ring seal is pulled into a distorted position. It relaxes and pulls the piston back slightly when the hydraulic pressure is released. (*Courtesy of Ford Motor Company*)

Figure 6–48 The thick-walled, dark piston on the left is made of a phenolic material; the shiny, thin-walled piston on the right is a common chrome-plated steel piston.

piston outward in the bore as the fluid pushes on the face of the piston and on the inside of the O-ring seal. The O-ring seal deflects or twists enough so that the piston and caliper can move to apply the brake pads. Release of the hydraulic pressure allows the O-ring to relax and pull the piston back to the released position (Figure 6–47). The O-ring seal serves two major purposes: It seals the hydraulic pressure and returns the piston. Some newer calipers have a redesigned O-ring seal and groove to obtain more piston pullback, which provides slight pad clearance instead of drag and therefore allows better fuel mileage. As the brake lining wears, the piston slips outward through the O-ring seal whenever necessary to compensate for the amount of wear. All disc brake calipers are self-adjusting.

The piston used with O-ring seals is precision sized, with straight, smooth sides. Traditionally, disc brake pistons are made from steel that has been stamped into the cup shape. After stamping, steel pistons are machined to size and plated to protect them from corrosion. A few cars use cast-aluminum pistons as weight-saving measures. Many newer pistons are made from phenolic resin. These pistons are much thicker and have a brownish-gray color. Phenolic pistons have the advantage of being lighter in weight, corrosion free, and very good heat insulators. Good heat insulation reduces the fluid temperature in the caliper. Phenolic pistons earned a reputation of sticking or rocking in the bore when they were first introduced. After a slight revision of the bore clearance dimension, these pistons now have a very favorable reputation. It is not recommended that steel or aluminum pistons be used to replace damaged phenolic pistons because of the possibility of fluid overheating and boiling as a result of the increased heat transfer (Figure 6–48).

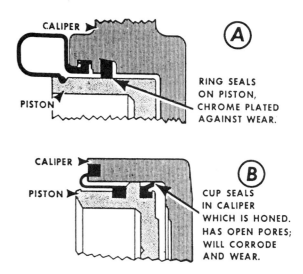

Figure 6–49 Modern calipers use a stationary, square-cut O-ring seal in the bore (A) that seals against the piston. Some older calipers use a stroking seal mounted on the piston (B) that seals against the bore. (*Reprinted from Mitchell Anti-Lock Brake Systems, with permission of Mitchell Repair Information, LLC*)

Some fixed-caliper designs use a cup or lip seal similar to those of wheel cylinders. A larger, ring-shaped cup is used. Like the cup in a wheel cylinder, it has a **dynamic** (or moving) **seal** with the bore and a **static** (or stationary) **seal** with the piston. This is often called a **stroking seal.** A caliper that uses stroking seals must have straight, smooth bores just like a master cylinder or wheel cylinder. A light spring is commonly used with these pistons because the seal does not have much wall drag, and vehicle vibrations can cause the piston to overrelease. Overrelease can cause too much pad clearance and a low brake pedal on the next application. No residual hydraulic pressure is used with disc brake calipers (Figure 6–49).

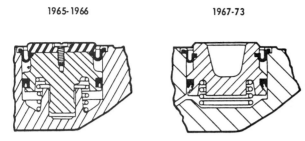

Figure 6–50 The calipers on 1965 and 1966 Corvettes used pintle-type pistons that had a guide in the caliper. From 1967 to 1973, Corvettes used the piston style shown on the right. Note that the early pistons used a heat insulator next to the brake pad. (*Courtesy of General Motors Corporation, Service Technology Group*)

Figure 6–51 This caliper was saved and improved by machining the bore and installing a stainless steel sleeve. (*Courtesy of Stainless Steel Brakes Corp.*)

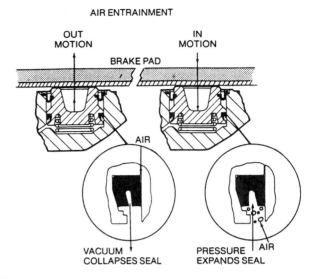

Figure 6–52 Excessive rotor runout can cause an in-and-out piston motion that can pump air past a stroking seal. (*Courtesy of Stainless Steel Brakes Corp.*)

Some of the early pistons used with cup seals were of the *pintle* type (Figure 6–50). The piston had a round pintle—a projection on its inner side that fit into a smaller bore in the cylinder. The pintle was used to hold the piston straight in the bore. Most calipers use a cup-shaped piston similar to the pistons used with O-ring seals but with a groove for the piston cup. Most of these pistons are made from cast aluminum.

Calipers that use cup seals have a definite disadvantage in that the cylinder bore must be smooth and free from corrosion. Any imperfections can damage the sealing lip of a cup. A rusty bore can easily ruin an expensive caliper body. Several companies are remanufacturing fixed-caliper bodies with stainless steel sleeves in the bore. These calipers will not corrode and are much less expensive than new calipers (Figure 6–51). The condition of the cylinder bore is not as critical when O-ring seals are used. A positive seal will occur as long as the piston is clean and smooth and can slide in the bore.

Another possible problem with calipers that use cup seals is entrance of air. While the brake is released, excessive rotor runout causes the pad to move sideways—in and out—on each revolution. The piston and seal move in and out along with the pad. As the piston seal moves outward without pressure behind it, a slight low pressure or vacuum is created inside the seal lip. This pressure drop can pull air inward past the seal lip. If enough air enters, the next brake application will produce a low, spongy pedal as this trapped air is compressed (Figure 6–52).

6.16 Caliper Boot

The piston boot in a caliper serves the same purpose as a wheel cylinder boot, but it is somewhat more critical because it is directly exposed to road elements. Also, its proximity to pads results in very high operating temperatures. A faulty boot can allow dirt, water, or road salt to enter the cylinder bore and cause corrosion (Figure 6–53). The buildup of contaminants or the corrosion that they produce can cause the piston to stick, resulting in pad drag and excess lining wear. On a stroking-seal-type caliper, the corrosion usually causes leakage to occur when pad wear allows the piston to move outward to the corroded area.

Boots are made from synthetic rubber and need to be securely attached to both the piston and the caliper body. Some designs experience problems during the heating

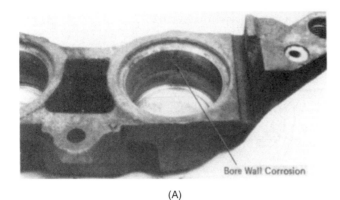

(A)

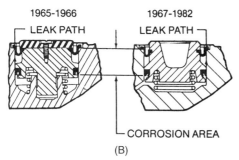

(B)

Figure 6–53 A caliper bore (A) can be ruined by moisture in the brake fluid or by moisture entering past the outer boot (B). (*Courtesy of Stainless Steel Brakes Corp.*)

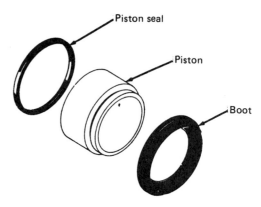

Figure 6–54 The O-ring seal prevents fluid pressure from leaking past the piston, and the boot keeps contaminants from entering the cylinder bore. (*Courtesy of General Motors Corporation, Service Technology Group*)

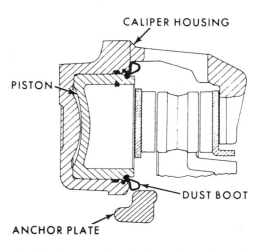

Figure 6–55 This dust boot is locked into the cylinder bore when the piston is installed. (*Courtesy of Ford Motor Company*)

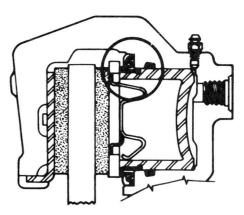

Figure 6–56 This boot is sealed to the cylinder bore by a soft metal ring that is part of the boot. (*Courtesy of General Motors Corporation, Service Technology Group*)

and cooling cycles of the caliper. At this time some calipers "breathe." The expansion and contraction of the air inside the boot can cause air to be expelled during the heat cycle and new air to be drawn back in during the cooling phase. This new air can bring in moisture, which in turn can cause corrosion buildup and damage to the cylinder bore (Figure 6–54).

In most cases, the boot is secured around the outer portion of the piston by the elastic nature of the rubber boot material. There are several different methods of securing the boot to the caliper. One manufacturer, Kelsey-Hayes, places the end of the boot into a groove at the outer end of the cylinder bore. It is locked in place when the piston is installed. This boot style requires some skill to install and forms a rather effective seal as long as the piston and cylinder groove are clean (Figure 6–55). Another manufacturer, Delco Moraine, locks the boot into a groove outside the cylinder bore with a metal ring that is positioned in the boot. This style is fairly easy to install if the correct tool is used. Installing this boot without the correct tool will probably result in a poor seal and future failure (Figure 6–56). Several import caliper designs use a boot that fits into a groove or over a projection and is secured in place with a metal retaining ring. Servicing of these different boots is described in Chapter 23. In each

Chapter 6 ■ Hydraulic System Theory

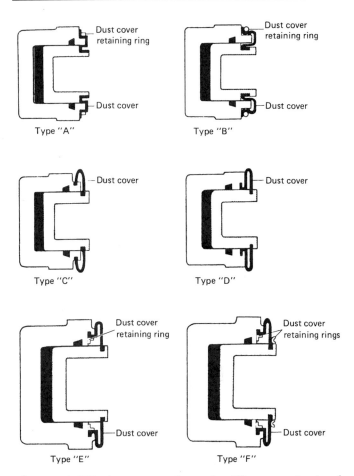

Figure 6–57 There are at least six different methods of sealing the piston to the caliper. (*Courtesy of LucasVarity Automotive*)

case, it is very important to obtain a clean, weather-tight seal to help prevent any future internal contamination (Figure 6–57).

6.17 Caliper Body and Bleeder Screw

As mentioned earlier, most caliper bodies are made from gray cast iron. A caliper must be strong enough to resist distortion during high brake pressures. Caliper bodies are manufactured in pairs and are almost interchangeable side for side. The major difference between the right and left calipers, in most cases, is the location of the bleeder screw (Figure 6–58). As in a wheel cylinder, the passage to the bleeder screw must be drilled into the uppermost location of the cylinder bore. Switching the right and left calipers will position the bleeder screw and passage incorrectly and make it impossible to bleed all of the air out of the caliper.

The caliper is usually machined for the cylinder bore, the groove for the boot and the O-ring are machined in

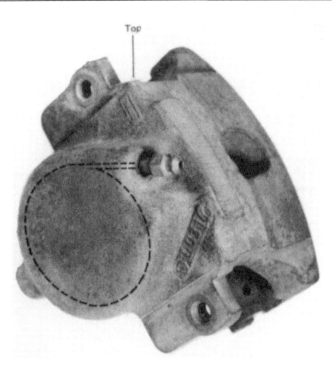

Figure 6–58 In many cars, the right and left calipers are identical, with the exception of the bleeder screw location. A caliper is always mounted so that the passage to the bleeder screw is at the top.

the bore, and the mounting pads for the outboard shoe and the abutments where the caliper body meets the caliper mount are machined on the outside. On fixed calipers, the two halves of the caliper body are machined where they join at the center of the caliper. Two or more **bridge bolts** secure the two halves together. The mating surfaces of the two parts usually include a drilling to each of the caliper bores. These drillings are sealed by O-rings where they meet at the center. They are used to bring fluid to the outboard cylinder or cylinders and to bring air to the bleeder screw (Figure 6–59). Some caliper designs use two bleeder screws, one for each caliper half. Some caliper designs use an external tube to deliver fluid to the outboard cylinder or cylinders.

6.18 Tubing and Hoses

Steel tubing is used for the lines that transfer fluid from the master cylinder to the wheel cylinders or calipers. Flexible hoses are used in those portions next to the front and rear suspensions where movement is required (Figure 6–60). Tubing is similar to pipe except it is sized by the outside diameter (OD) and does not use threads to join at the ends. The common sizes for automotive brake tubing are 3/16 in. (4.6 mm) for disc brakes and 1/4 in. (6.4 mm) for drum brakes.

Occasionally 5/16-in. (7.9-mm) tubing is used. It is usually made by wrapping a steel sheet into a tubular shape and welding or brazing the seam closed (Figure 6–61).

Tubing must be flared to make leak-free high-pressure connections. Two different styles of flares are used. Traditionally, a standard SAE 45-degree **double flare** was made at the end of the tube and locked onto the tube seat with an **inverted flare** nut (Figure 6–62). A double flare folds the end of the tube inward, toward the center of the tube. A single flare cannot be made of steel tubing because the tube will split as the end is stretched to make the flare. The **International Standards Organization (ISO) flare,** sometimes called a *bubble flare,* also stretches the tube a short distance from the end so splitting is not a

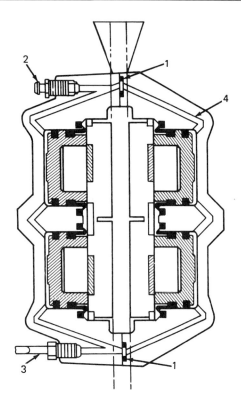

1. "O" ring seal (fluid crossover internal passage)
2. Bleeder valve
3. Brake fluid inlet hole
4. Internal fluid passage

Figure 6–59 This fixed caliper uses internal passages to transfer fluid pressure from the inlet (3) to each of the bores and to transfer the air up to the bleeder screw (2). Note the O-rings (1), which are used to seal the passages where the two caliper halves join. (*Courtesy of General Motors Corporation, Service Technology Group*)

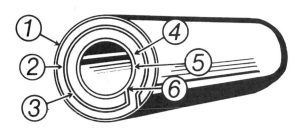

1. Tinned copper-steel alloy protects outer surface.
2. Long-wearing and vibration-resisting soft steel outer wall.
3. Fused copper-steel alloy center unites two steel walls.
4. Long-wearing and vibration-resisting soft steel inner wall.
5. Copper-steel alloy lining protects inner surface.
6. Bundyflex's beveled edges and single close-tolerance strip results in no inside bead at joint. The tubing is uniformly smooth, inside and out.

Figure 6–61 Quality brake tubing is made by wrapping a strip of steel into a two-layer tube, fusing it together, and applying corrosion protection. (*Courtesy of Wagner Brake*)

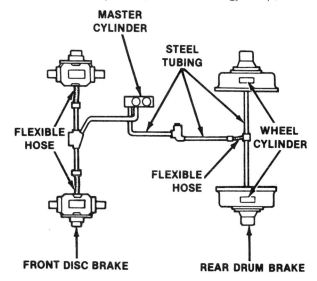

LOCATION OF FLEXIBLE HOSES AND STEEL TUBING IN A TYPICAL SYSTEM

Figure 6–60 Most vehicles use a flexible hose at each front caliper and one (tandem split) or two (diagonal split) hoses at the rear axle or wheel assemblies. Steel tubing is used for the remaining lines. (*Courtesy of Brake Parts, Inc.*)

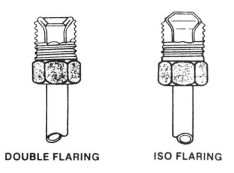

DOUBLE FLARING   ISO FLARING

Figure 6–62 Brake tubing uses either a double flare or a single, bubble flare. The tube nuts look alike, but the end of the tube has a different shape. (*Courtesy of Bendix Brakes, by AlliedSignal*)

Chapter 6 ■ Hydraulic System Theory

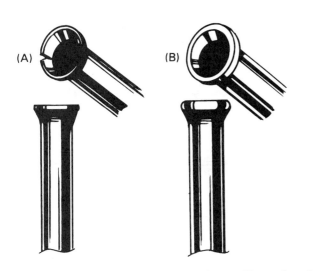

Figure 6–63 A single flare on steel tubing will usually split (A). A double flare (B) will not. (*Courtesy of General Motors Corporation, Service Technology Group*)

Figure 6–64 A standard flare nut (right) has female threads; an inverted flare nut (left) has male threads. (*Courtesy of ITT Automotive*)

problem (Figure 6–63). ISO flares were introduced on domestic cars beginning in the early 1980s.

Both the inverted flare nut and the ISO flare nut have male (or external) threads. Plain flare nuts are commonly used in household gas lines and have female (or internal) threads (Figure 6–64). The standard-sized nut used with 3/16-in. tubing has a 3/8-in.-by-24 thread; that is, the threads are 3/8 in. in diameter with 24 threads per inch. With 1/4-in. tubing, a 7/16-in.-by-24 thread is used. ISO flare nuts use metric threads. Manufacturers commonly use tubing with oversized nuts for certain locations to ensure that the components are assembled correctly on the assembly lines. Occasionally this creates a problem when replacing a tube or a master cylinder. Step-up or step-down fittings are available to allow connection of standard-sized tube nuts to nonstandard-sized tube seats or vice versa (Figure 6–65).

Tubing failure and damage to the tube nut are usually caused by collapse of the tubing from impact, rusting through of the tubing, twisting of the tubing because of improper wrench technique, or kinking of the tubing from trying to bend it too tight. Note that most of these problems are the result of poor work habits. Replacement tubing is available in different diameters and lengths. It is usually sold as a unit, assembled with flares and tube nuts. Replacement tubing is frequently plated to resist rust and corrosion and is often available with a spring-shaped guard around it. Guarded tubing is used where road hazards, abrasion, and salt corrosion are particularly problematic (Figure 6–66).

Copper tubing must never be used for brake lines. It might burst, because its working pressure is less than that

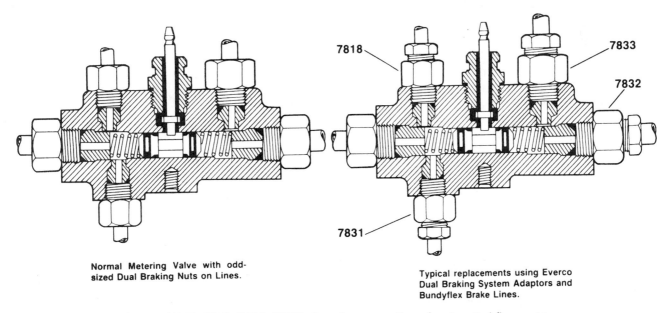

Figure 6–65 Fitting adapters (7818, 7833, 7832, 7831) allow the connection of an inverted flare nut to a port of a different size. Many master cylinders and valve assemblies use oversize tube nuts to prevent improper assembly (*Courtesy of Wagner Brake*)

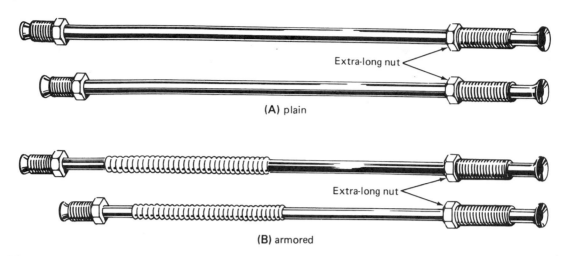

Figure 6-66 Replacement tubing is available in a variety of lengths for the more popular sizes and in a plain or an armored form. (*Courtesy of Wagner Brake*)

Figure 6-67 Flexible rubber hoses are usually of a two-ply construction. The fluid-carrying inner liner is reinforced by two layers of woven fabric. Note how the end of the hose is crimped tightly into the end fitting. (*Courtesy of Wagner Brake*)

of steel tubing. Also, copper tends to work harden from vibration, and this can cause cracks to form in the tubing.

Brake hoses are generally composed of rubber hose that is reinforced with woven fabric. The fabric strengthens the rubber to prevent expansion under pressure. The inner layer of rubber (sometimes plastic) contains and transfers the fluid. The middle layers or plies of fabric and rubber, usually two, reinforce the hose. The outer layer, called a jacket, protects the inner layers from weathering and abrasion. The outer layer is usually ribbed so that any twists are easily seen during installation. Hoses should not be twisted or kinked too tightly or internal damage will result (Figure 6-67).

Brake hoses are available in different lengths and with one of three different forms of end fittings: male threads, female threads, or banjo fittings (Figure 6-68). The male thread can be a straight SAE or metric thread that is sealed by a copper gasket or one that is sealed by a tapered seat at the end of the thread. The more common female threads include a seat for either an SAE or an ISO flare. A **banjo** fitting uses a ringlike nipple. A special hollow bolt is used to transfer fluid from the fitting to the unit where the hose is connected. A pair of copper gaskets, one on each side, is required to seal a banjo-type connection.

The most common causes for hose failure are weather and abrasion that breaks through the jacket and weakens the hose (which can cause the hose to burst). Other problems include a collapse of the inner liner or a torn flap in the inner liner. The first of these problems will cause constant fluid restriction, which usually results in brake drag. In the second case, the torn flap acts like a one-way valve, which usually causes brake drag also. Occasionally, a leak can form in the inner liner and cause bulges or bubbles to form in the jacket. Normal repair of a faulty hose involves replacing it with the same type and length of hose.

6.19 Hydraulic Control Valves and Switches

All modern brake systems include two or more valves and switches to help control the system pressure to the wheel cylinders or calipers; to warn the driver of a system failure, pad wear, or low brake fluid; and, in some cases, to automatically turn on the stoplights. All tandem or diagonal split systems include a pressure differential valve and switch, which turns on the brake failure warning light to alert the driver if a hydraulic failure occurs in one of the systems (Figure 6-69). In all cars, stoplights are activated when the brakes are applied to warn following drivers that

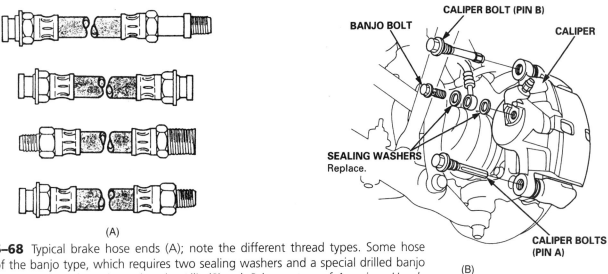

Figure 6–68 Typical brake hose ends (A); note the different thread types. Some hose ends are of the banjo type, which requires two sealing washers and a special drilled banjo bolt (B). (*A is courtesy of Bendix Brakes, by AlliedSignal; B is courtesy of American Honda Motor Company*)

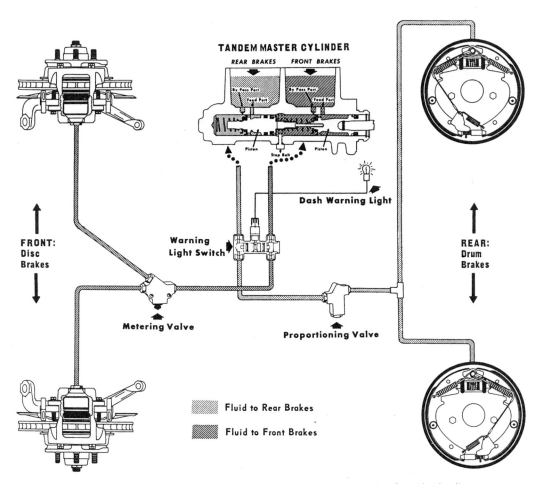

Figure 6–69 This tandem system uses three separate valves—a metering valve in the front brake line, a proportioning valve in the rear brake line, and a pressure differential valve to operate the warning light switch. (*Reprinted from Mitchell Anti-Lock Brake Systems, with permission of Mitchell Repair Information, LLC*)

the brakes are being applied. Most cars have a switch on the parking brake, as described in Chapter 5, to remind the driver that the parking brake is on.

Many cars that use a combination of disc and drum brakes include a **proportioning valve** to reduce the possibility of rear-wheel lockup during hard stops and/or a **metering valve** to ensure that both the disc and drum brakes apply at the same time. These two valves are often combined into a single unit with the differential pressure valve. This assembly is called a **combination valve.**

Many cars also include a warning light to inform the driver that the brake fluid is low or that the front pads are excessively worn. The lights are controlled by a switch at the master cylinder reservoir or by contacts in the brake pads, as described in Chapter 4.

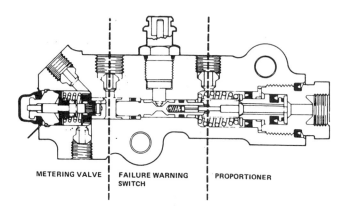

Figure 6–70 This combination valve contains the three valve sections. (*Courtesy of General Motors Corporation, Service Technology Group*)

6.19.1 Pressure Differential Valve and Switch

The *pressure differential valve* and *switch*, also called a *warning light switch*, a *dash lamp switch*, or a *system effectiveness switch*, is a combination of a valve and a switch; the valve operates the switch (Figure 6–70). The valve is connected hydraulically to the primary and secondary hydraulic sections. The valve piston floats between these two pressures or two springs of equal pressure. When the brakes are released, the valve is held out or moved to the center position by the two springs. When the brakes are applied, the valve is still centered by the equal hydraulic pressure in the two systems.

If there is a pressure loss in the primary or secondary system, the pressure remaining in the other section will move the piston off center. The movement of the piston will then make an electrical connection with the switch contact. This will provide a ground for the warning light and turn on the light. In some units, the piston movement operates a switch that also provides a ground for the warning light. When the brakes are released in most systems, the springs recenter the piston and turn off the light (Figure 6–71).

Some early valves, used mostly on Ford products in the late 1960s, had no centering springs. This type of valve would not automatically recenter (Figure 6–72). If a pressure loss occurred, the light would stay on until the valve was recentered and rearmed. To rearm the valve, a bleeder screw at the front or rear—whichever side did not fail—is opened while the brake pedal is slowly applied. At the instant that the light goes out (indicating a centered piston), the bleeder should be closed. Too much speed or pressure at the brake pedal will cause the valve to go past center. If this happens, the procedure must be repeated while opening a bleeder screw at the other end of the car.

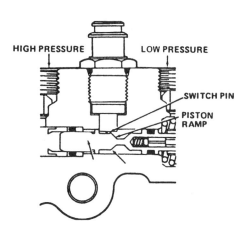

Figure 6–71 If there is a pressure loss—failure in one of the sections—the valve piston will move off center and turn on the switch to illuminate the brake warning light. (*Courtesy of General Motors Corporation, Service Technology Group*)

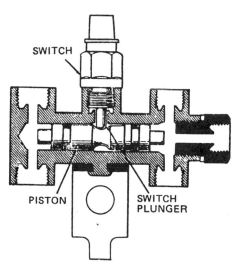

Figure 6–72 This pressure differential valve has no centering springs, so it can remain in a noncentered position. (*Courtesy of Wagner Brake*)

Chapter 6 ■ Hydraulic System Theory

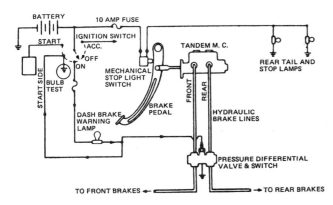

Figure 6–73 A basic brake warning light and a simple stoplight circuit. (*Courtesy of Wagner Brake*)

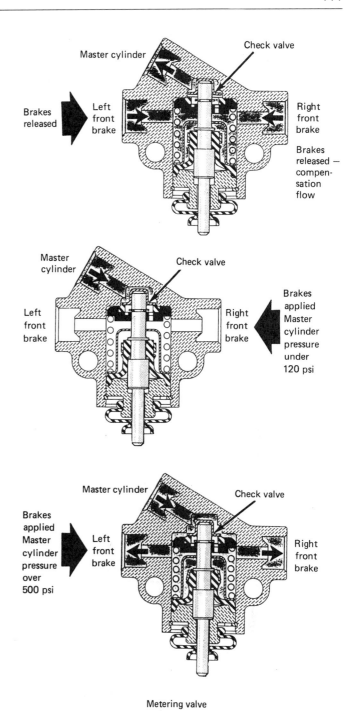

Figure 6–74 A metering valve. Note that the valve is partially open while the brakes are released, closed when there is a low pressure (less than 120 psi), and reopened at higher pressures. (*Courtesy of Ford Motor Company*)

The *failure warning light* uses a rather simple circuit (Figure 6–73). An electrical wire is connected to the ignition terminal of the ignition switch, then to the fuse block, and then to one side of the warning light. The second side of the warning light is connected to the differential switch and also to the lamp test or *proof* terminal of the ignition switch. The same grounded side is usually also connected to the switch at the parking brake. The warning light turns on when the ignition switch is on and the differential switch is grounded by an off-center piston, when the parking brake is applied (on most systems), and when the ignition switch is turned to the start position and the lamp test circuit is closed. This last circuit is used each time the car is started to inform the driver that the light works.

In a few cases, the differential valve is built into the body of the master cylinder, with a passage from each end of the valve to the primary and secondary systems. In most cases, the outlets of the master cylinder are connected to each end of the valve by steel brake tubing. Brake fluid passes the ends of the valve on its way to the wheel cylinders or calipers.

6.19.2 Metering Valve

The metering valve, also called the *hold-off* or *disc-balancing valve*, is used in some disc–drum combination systems. It is connected between the pressure differential valve and the front calipers.

Disc brakes can be applied at lower pressures than drum brakes because there are no brake shoe return springs. It takes a pressure of about 100 to 150 psi (690 to 1,034 kPa) in drum brake wheel cylinders to move the brake shoes into contact with the drum. The metering valve is designed to be closed at lower pressures and wide open at about 75 to 135 psi (517 to 931 kPa) so that disc brake application will be held off until drum brake application. A metering valve is also open at very low pressures, less than 5 to 15 psi (34 to 103 kPa), so that compensation can take place for fluid expansion or contraction in the calipers (Figure 6–74).

Metering valves have an external stem so that they can be held open during operations using a pressure bleeder. Brake bleeder pressure is usually high enough to close off the metering valve. The valve stem can be of either the *push type* or the *pull type*. A pull-type stem is pulled outward to open the valve, and a push-type stem is pushed inward. If you touch this stem while the brakes are being applied, you can usually feel the valve operate (Figure 6–75).

Metering valves are not used on all modern systems. In mild climates, the worst thing that will occur without one is early disc pad wear on cars with low pressure or easy stops or when drivers rest their left foot on the brake pedal while driving. In severe climates with icy roads, the metering valve holds back front brake application until the rear brakes can apply. This reduces the possibility of front-brake lockup under light, gentle pedal application.

The metering valve can be a separate valve connected to the outlet of the pressure differential valve or it can be built into the same assembly with the pressure differential valve and/or the proportioning valve. When used, a line to each front caliper is connected to the metering valve outlet.

6.19.3 Proportioning Valve

The proportioning valve, also called a *pressure-reducing valve*, a *pressure ratio valve*, a *pressure-regulating valve*, a *pressure control valve*, or an *apportioning valve*, is used to reduce the hydraulic pressure going to the rear wheel cylinders. During a hard stop, while weight transfer is moving weight off the rear end, the rear wheels tend to lock up and skid. This is especially true if the car has duoservo brakes at the rear in combination with front disc brakes (Figure 6–76).

The proportioning valve is located between the pressure differential valve and the rear wheel cylinders. At a preset pressure, usually somewhere between 400 and 600 psi (2,758 and 4,137 kPa), the valve will begin to reduce the rate of pressure increase at the rear. This is called the **changeover** or **split point.** The amount of pressure reduction is called the **slope.** After the changeover point, the rear pressure increases at a rate slower than that of the

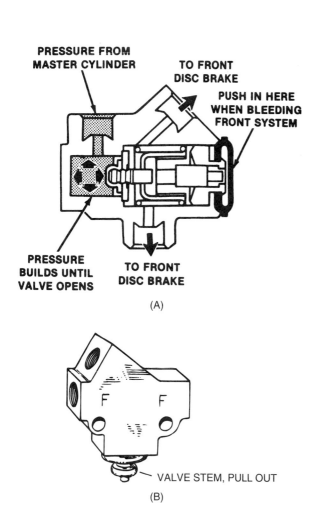

Figure 6–75 Push-type (A) and pull-type (B) metering valves. The name refers to the direction the valve stem must be moved to open it when using pressure brake bleeders. (*A is courtesy of Brake Parts, Inc.; B is courtesy of Wagner Brake*)

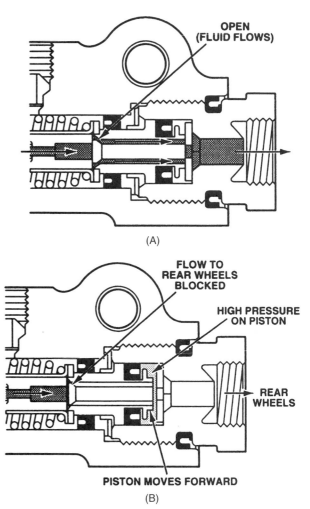

Figure 6–76 At lower pressures, the proportioning valve is open and does not change fluid pressures (A). Above the split point, the valve reduces the pressure to the rear brakes (B). (*Courtesy of General Motors Corporation, Service Technology Group*)

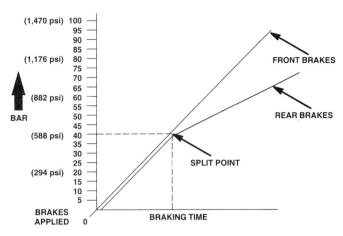

Figure 6–77 A typical brake pressure curve shows a split point at 40 bar and a slope of 50 percent; the rear brake pressure is increasing at 50 percent of the front pressure rate of increase. (*Courtesy of Chrysler Corporation*)

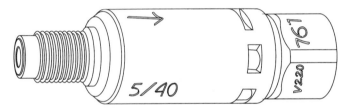

Figure 6–78 This proportioning valve is marked with the split point and slope calibration. (*Courtesy of Chrysler Corporation*)

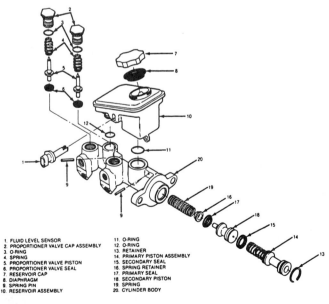

Figure 6–79 This diagonal split system master cylinder has a proportioning valve at each rear outlet port. (*Courtesy of General Motors Corporation, Service Technology Group*)

front brakes. A slope of 0.50 allows the rear-brake pressure to increase at a rate of 50 percent of that of the front. Most passenger-car valves let the rear pressure increase at somewhere between 25 and 60 percent of the front pressure increase. As mentioned earlier, the balance between the front and rear brakes during a hard stop is a compromise between too much rear brake and a skid or too little rear brake and a loss in braking power. Engineers try to design the changeover point, and the ratio of the pressure increase from that point, according to the size, weight, and type of car (Figure 6–77). The car should be balanced just short of a rear-wheel lockup under heaviest-load conditions on dry pavement. Some proportioning valves are marked with the split point and slope (Figure 6–78).

Many cars with diagonal split systems have a double proportioning valve arrangement. Some cars use two individual proportioning valves mounted directly on the master cylinder (Figure 6–79). Other cars use a pair of proportioning valves or a dual proportioning valve mounted remotely. The dual valve contains two separate valves mounted in the same housing (Figure 6–80). Two lines come from the master cylinder, and a line goes to each rear wheel cylinder.

Some cars, pickups, and light trucks are equipped with a **height-sensing proportioning valve.** A vehicle needs more rear braking power as weight is added to the back end. A height-sensing valve has an external lever that is attached by a link to the rear axle (Figure 6–81). Heavier loads, which lower the rear of the vehicle, cause the valve to move the changeover point to a higher setting. Most valves are somewhat adjustable for fine-tuning of the brake balance.

6.19.4 Stoplight Switches

The **stoplight switch,** also called the *brake light switch,* turns on the stoplights at the rear of the car when the brakes are applied. Modern cars use a mechanical switch that is activated by brake pedal movement. Older cars use a hydraulic pressure switch that is activated by system pressure. Some switches contain two circuits. The second circuit shuts off the cruise control when the brakes are applied.

Several styles of mechanical switches are used (Figure 6–82). They are all normally closed (completed circuit) switches that are opened (switched off) when the brake is released. Most of them are mounted on a bracket next to the pedal so that release of the pedal will open the switch and turn the brake lights off. Switch brackets and switches are often adjustable so that the point of switch operation can be changed. When adjusting a switch, it is

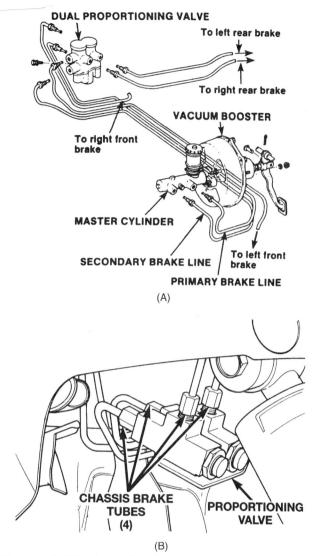

important to make sure that the brake pedal is not held in a partially applied position by the switch.

Several styles of hydraulic switches are also available; the major differences among them are the size and type of electrical connections. This switch is usually threaded into a tee connection in the brake tubing. Sometimes it threads directly into the master cylinder body. It is usually a normally open (no circuit) switch that is closed (turned on) by system pressure. As the brakes are applied, hydraulic pressure closes the switch and turns the stoplights on.

Stoplights use a fairly simple electrical circuit. It begins at a fuse or circuit breaker, and a wire carries battery positive (B+) power to one terminal of the switch. The second wire of the stoplight switch goes to the turn indicator switch, which splits the circuit to the left and right stoplights. In cars with three rear lights—tail, turn, and stop—on each side, the wire from the stoplight switch connects directly to the two or three stoplights (Figure 6–83).

Figure 6–80 Diagonal split systems with proportioning valves can use a dual proportioning valve (A) or a pair of proportioning valves (B). (*A is courtesy of Bendix Brakes, by AlliedSignal; B is courtesy of Chrysler Corporation*)

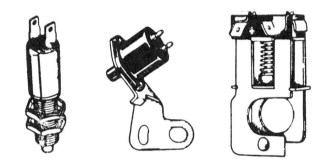

Figure 6–82 Three different mechanical stoplight switches. They are mounted so that brake pedal movement will close the switch and turn on the stoplights. (*Courtesy of Wagner Brake*)

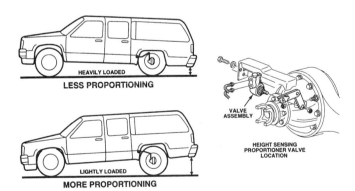

Figure 6–81 Many light trucks, vans, and station wagons use a height-sensing proportioning valve connected between the body and rear axle. This valve allows more rear braking while the vehicle is heavily loaded and less while it is lightly loaded to reduce rear-wheel lockup. (*Courtesy of General Motors Corporation, Service Technology Group*)

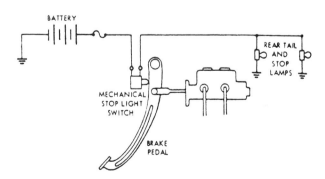

Figure 6–83 This simple stoplight circuit is for a system with separate turn indicators. (*Reprinted from Mitchell Anti-Lock Brake Systems, with permission of Mitchell Repair Information, LLC*)

Chapter 6 ■ Hydraulic System Theory

6.19.5 Fluid Level Switches

Low-brake-fluid warning lights are used on many new cars. This light is controlled by a switch in the master cylinder reservoir. The switch is operated by a float. When the brake fluid level drops, the float lowers and closes the switch contacts. When the contacts close, the light is turned on (Figure 6–84).

The low-brake-fluid warning light also has a fairly simple electrical circuit. A wire connects B+ voltage from the ignition switch through the light to one terminal of the fluid level switch at the master cylinder. Another wire connects the other switch terminal to ground. If the switch closes because of a low brake fluid level, current will flow through the completed path to ground and the light will turn on.

An odd thing related to this switch can occur on some vehicles. Electrical radiation from a spark plug wire can turn on the low-brake-fluid warning light. Spark plug wires should be kept at least 2 in. away from the switch area to keep this from happening.

6.20 Brake Fluid

No hydraulic system can operate without fluid, and the automotive hydraulic brake system has some critical requirements. Brake fluid quality is regulated in the United States by federal standards established by the Department of Transportation (DOT) and the National Highway Traffic Safety Administration (NHTSA) and also by various departments in many states. These standards were established by the SAE. Brake fluid must possess the following characteristics:

- **Viscosity.** It must be free flowing at all temperatures.
- **High boiling point.** It must remain in the liquid state at the highest temperatures that are normally encountered.
- **Noncorrosive.** It must not attack plastic, rubber, or metal parts.
- **Water tolerance.** It must be hygroscopic, or able to absorb and retain moisture that collects in the system.
- **Lubricating ability.** It must lubricate the pistons and cups to ensure free movement and reduce wear and internal friction.
- **Low freezing point.** It must stay fluid and flow at the lowest operating temperatures.
- **Compatibility.** It must be compatible with other brands of brake fluid as well as all components in the system.

In addition, there are specified minimum boiling points for the three classes of brake fluid, as shown in Figure 6–85.

DOT 3 and DOT 4 fluids must be amber to clear in color and are generally made from a *polyglycol* base. DOT 5 fluid must be purple in color and uses a *silicone*

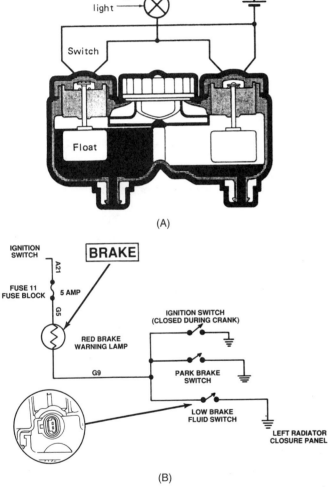

Figure 6–84 Low-brake-fluid warning circuits use a switch or switches in the master cylinder reservoir (A). The switch can be a parallel circuit to the parking brake switch (B). (*A is courtesy of ITT Automotive; B is courtesy of Chrysler Corporation*)

Boiling Point	DOT 3	DOT 4	DOT 5
Dry	401°F (205°C)	446°F (230°C)	500°F (260°C)
Wet	284°F (140°C)	311°F (155°C)	356°F (180°C)

Figure 6–85 The dry and wet boiling point requirements of DOT 3, DOT 4, and DOT 5 brake fluid.

Figure 6-86 Brake fluid containers are marked with the DOT rating and the wet and dry boiling points.

base. DOT 3 and DOT 4 fluids can be mixed with each other, but neither one should be mixed with DOT 5 fluid. Silicone fluid is lighter and will float on glycol fluid. A water film may form at the layer between them and cause corrosion at that level. Also, if these two fluids are mixed and then agitated, they will make a foam and encapsulate air bubbles.

The dry boiling points specified are for new fluid with no absorbed water (Figure 6-86). The wet boiling points are for fluid that has absorbed 2 percent water (Figure 6-87). The boiling points are often printed along with the term **equilibrium reflux boiling point (ERBP),** which refers to the method of measuring the boiling point.

There are a few other types of brake fluids. A few import cars—Rolls Royce and Citroen—have seal materials that are designed to be used with a special hydrocarbon brake fluid called *hydraulic system mineral oil (HSMO)*. Many older British cars required the use of a special brake fluid. One of the original brake fluids used a castor oil base. It is important to follow the fluid requirements of the manufacturer to ensure long life of the seal materials. Older glycol fluids had a boiling point much lower than that of today's fluids; at one time the brake fluid standards were SAE J1703 or 70R3—boiling point 374°F (190°C)—and SAE 70R1—boiling point 302°F (150°C). These older fluids are no longer used.

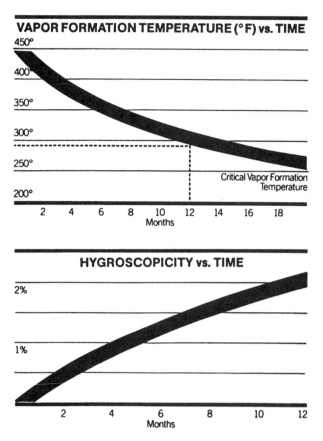

Figure 6-87 After about 12 months in humid areas, the average brake system will absorb about 2 percent water. Note that as water content increases, the fluid's boiling point decreases. (*Courtesy of Stainless Steel Brakes Corp.*)

It is important that industrial hydraulic oils not be mixed with automotive brake fluid. They are petroleum based and will cause excessive expansion and swelling of the seal materials.

Glycol fluids have two serious drawbacks. They are **hygroscopic,** which means they absorb water. Glycol fluids also attack paint if fluid is spilled on the car's finish; the paint will be discolored or even lifted. Spilled fluid should be immediately flushed with a large amount of water.

Water enters most brake systems by permeating the rubber hoses and diaphragm at the reservoir. Water molecules (vapor) can pass through the pores of the rubber. The rate of water absorption varies with the level of humidity of the environment. It has been found that the average automobile reaches 2 percent water in the brake fluid in about 18 months. In humid areas, brake fluid gains water at a rate of 2 percent or more per year. Water absorption by the fluid presents two serious problems: boiling point loss and corrosion. Water absorption to a

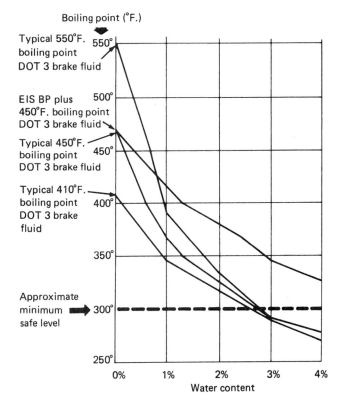

Figure 6-88 This chart shows the dry and wet boiling points for several brake fluids. (*Reprinted from Mitchell Anti-Lock Brake Systems, with permission of Mitchell Repair Information, LLC*)

level of 2 percent has a serious effect on the boiling point, lowering it to about three-fourths of the dry boiling point (Figure 6–88). Cars operating under severe braking conditions, with fluid that is several years old, can experience fluid vaporization, which causes a spongy pedal or even a total loss of braking power as the fluid heats up. During an overhaul of a system that is a few years old, the corrosive effects of brake fluid are seen as rust in the lower portions of cast-iron cylinders and corrosion in the lower portions of aluminum cylinder bores and on aluminum pistons. When the rust or corrosion is removed, the severe pitting that results still remains. It is this pitting that usually ruins these parts (Figure 6–89).

Silicone fluids do not share these problems. They are nonhygroscopic, so they do not absorb water, and they have no effect on painted surfaces. Silicone fluid, though, will not accept paint on top of it. It must be carefully cleaned from any surface to be painted. Because no water is absorbed, there will be very little or no corrosion in the system. Silicone fluid does have a few drawbacks. It is expensive, costing at least three to four times as much as glycol fluids; in some areas, it costs as much as ten times as much. Silicone fluid is also compressible when it gets hot. This compressibility causes a spongy brake pedal under severe operating conditions. Silicone fluid also has a greater tendency to foam when it is forced through small openings or experiences rapid vibrations. Because of this tendency to foam, silicone fluid should not be used in an ABS unless specified by the manufacturer of the system.

Silicone fluid can also cause some of the rubber seals to swell as much as 10 percent. In at least one case after a car was switched to silicone fluid, the primary master cylinder cup swelled enough to stick in the bore and not return completely; this caused a serious brake drag because the fluid could not return through the compensating port.

The following precautions can help minimize the amount of water and its resulting corrosive effects in the system:

- The master cylinder should be kept tightly closed, and it should be closed immediately after filling.
- Brake fluid should be purchased in the smallest practical amount. It should not be stored in partially filled containers for long periods of time.
- Brake fluid containers should be kept tightly closed and capped.
- Brake fluid containers should be kept in clean, dry areas.
- Only clean containers should be used for dispensing or transferring fluid.
- Old fluid should never be reused.
- If there is a possibility that fluid is contaminated, it should be thrown away.

Some sources recommend periodic brake fluid changes so that the deterioration of glycol fluids will have no adverse effects; however, in today's world, this is often not economically practical. The cost of the periodic changes would probably be greater than that of the parts ruined by corrosion. With the increased use of ABS, with its very expensive hydraulic modulator, periodic fluid changes make sense economically and are a wise safety practice. With vehicles that normally operate at very high fluid temperatures, it is probably worthwhile to change brake fluid because of the safety aspect. The cost of a fluid change is minor compared with that of an accident resulting from brake failure. Most owners of road racing cars change the brake fluid after every race. Most informed sources recommend that it be changed at the time of a major brake repair if not more often. At this time, unless

Figure 6-89 Contaminated fluid has caused corrosion and failure of these components. The aluminum master cylinder bore (A) has deep pits, the cast-iron master cylinder (B) and wheel cylinder bore (C) have mounds of rust, the caliper bores (D) have severe rust pits, and the aluminum pistons (E, F, G) show severe corrosion.

Chapter 6 ■ Hydraulic System Theory

the hydraulic system is disassembled and rebuilt, the system should be flushed with new fluid until all the old, dirty, contaminated fluid is removed.

When servicing brakes, the hydraulic system should never be exposed to petroleum-based chemicals, oils, or solvents. These substances attack the synthetic rubber compounds used in sealing cups and O-rings, hoses, and the diaphragm in the master cylinder reservoir. Petroleum products will soften and swell the rubber, thereby ruining the sealing ability, the ability to slide in a bore, or the ability to transfer fluid from one place to another. If a petroleum product such as motor oil, gasoline, or solvent enters the hydraulic system, the system should be drained and disassembled and every rubber part should be replaced (Figure 6–90).

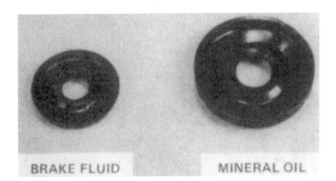

Figure 6–90 These two cups were the same size until the one on the right was placed in mineral (motor) oil. Note how much it has swollen. (*Courtesy of General Motors Corporation, Service Technology Group*)

6.21 Practice Diagnosis

You are working in a tire shop that also specializes in brake repair and encounter the following problems:

CASE 1: The customer has brought in his 1989 Ford Bronco with a complaint of a low brake pedal. He said that his neighbor told him the brakes need to be bled. On checking it, you find that the pedal goes almost to the floor fairly easily before becoming solid and firm. What is probably wrong with this vehicle? Is the neighbor's diagnosis correct? What should you do to find the exact cause of this problem?

CASE 2: The customer has brought in a 1983 Honda Prelude with a complaint of a dragging brake at the right front. On checking it you find that the wheel is very hot, and when you loosen the bleeder screw, a small amount of fluid squirts out and the brake becomes free. Is this a sign of a faulty master cylinder? A faulty metering or proportioning valve? What should you do to locate the exact cause of this problem?

CASE 3: The customer has brought in his 1991 Chevrolet C1500 pickup with a problem of a grabby left rear brake. Your road test confirms lockup of the tire with only a moderate amount of pedal pressure. What is probably wrong? Could this be a proportioning valve problem? A wheel cylinder problem? What should you do to find the exact cause of this problem?

Terms to Know

anodized	equilibrium reflux boiling point (ERBP)	primary piston
banjo	expander washers	proportioning valve
bleeder screw	front–rear split	quick-take-up master cylinder
brake pedal	heel drag	reservoir
bridge bolts	height-sensing proportioning valve	residual pressure check valves
bypass port	hydraulic brake system	roller burnishing
changeover	hydraulics	secondary cup
combination valve	hygroscopic	secondary piston
compensating port	International Standards Organization (ISO) flare	slope
cup expanders		split point
cups	inverted flare	static seal
diagonal split	master cylinder	stationary seal
double flare	metering valve	stoplight switch
dual master cylinder	Pascal's law	stroking seal
dust boot	pounds per square inch (psi)	tandem master cylinder
dynamic seal	primary cup	wheel cylinders

Review Questions

1. Hydraulic brake systems are designed to:
   a. Transmit motion from the driver's foot to the shoes
   b. Transmit force from the driver's foot to the shoes
   c. Multiply the amount of force being transmitted
   d. All of the above

2. Hydraulic brakes offer a major advantage in brake systems because they:
   a. Are self-lubricating
   b. Multiply force
   c. Transmit pressure equally to two or more places
   d. None of the above

3. A force of 100 lb acting on a piston with an area of 1/2 sq. in. will generate a pressure of:
   a. 400 psi
   b. 200 psi
   c. 100 psi
   d. 200 kPa

4. A fluid pressure of 400 psi acting on a piston with an area of 2 sq. in. will generate a force of:
   a. 400 lb
   b. 200 lb
   c. 600 lb
   d. 800 lb

5. Statement A: A hydraulic leak can keep a system from generating high pressures.
   Statement B: Because it is compressible, air in a hydraulic system will cause excessive pressures.
   Which statement is correct?
   a. A only
   b. B only
   c. Both A and B
   d. Neither A nor B

6. Statement A: The compensating port in a master cylinder allows a fluid flow between the wheel cylinders and the master cylinder reservoir while the brakes are released.
   Statement B: As the brakes are released, fluid will flow from the master cylinder reservoir to the cylinder bore through the bypass port.
   Which statement is correct?
   a. A only
   b. B only
   c. Both A and B
   d. Neither A nor B

7. Statement A: The master cylinder begins pumping fluid right after the secondary cup passes by the compensating port.
   Statement B: The main purpose of the primary cup is to keep fluid from leaking out of the end of the cylinder bore.
   Which statement is correct?
   a. A only
   b. B only
   c. Both A and B
   d. Neither A nor B

8. In some brake systems, pressure is kept in the system while the brakes are released by the:
   a. Wheel cylinder cups
   b. Residual check valve
   c. Replenishing port
   d. All of the above

9. In a tandem master cylinder, the primary piston is applied by the pedal pushrod, and the secondary piston is applied by:
   a. A second pushrod
   b. An extension of the primary piston
   c. Hydraulic pressure
   d. All of the above

10. In the case of hydraulic failure in the secondary circuit, the primary piston will operate normally after:
    a. An extension of the secondary piston meets the end of the bore
    b. An extension of the primary piston meets the secondary piston
    c. Both pistons bottom at the end of the bore
    d. The engine is started

11. The master cylinder for some diagonal split systems has:
    a. Four outlet ports
    b. A pair of proportioning valves
    c. An internal differential valve
    d. All of the above

12. A quick-take-up master cylinder is designed to:
    a. Displace more fluid at low pressure and develop the same high pressures normally used
    b. Provide higher pressures during light pedal operation
    c. Operate bigger caliper pistons with less pedal pressure
    d. All of the above

13. Statement A: Most wheel cylinders contain two pistons, one spring, and a rubber cup.
    Statement B: Some wheel cylinders use only one piston and cup.
    Which statement is correct?
    a. A only
    b. B only
    c. Both A and B
    d. Neither A nor B

14. The O-ring around a caliper piston is used to:
    a. Seal the piston to the bore
    b. Return the piston to a released position
    c. Adjust for lining wear
    d. All of the above

15. Statement A: A stroking seal used in a master cylinder moves along with the piston.
    Statement B: A static seal is stationary in the cylinder and allows the piston to move inside it.
    Which statement is correct?
    a. A only
    b. B only
    c. Both A and B
    d. Neither A nor B

16. All cars use steel brake tubing that has:
    A. A double SAE flare at the ends
    B. An ISO flare at the ends
    Which option best completes the statement?
    a. A only
    b. B only
    c. Both A and B
    d. Neither A nor B

17. Statement A: Many cars use a proportioning valve in the hydraulic circuit to reduce the tendency for rear-wheel lockup during hard stops.
    Statement B: Many cars use a metering valve to reduce the tendency of front-wheel lockup during hard stops.
    Which statement is correct?
    a. A only
    b. B only
    c. Both A and B
    d. Neither A nor B

18. Statement A: A hydraulic pressure failure in one of the hydraulic circuits will center the differential valve and turn on the brake warning light.
    Statement B: The differential valve is mounted so that pressure from each hydraulic circuit is at the ends of the valve.
    Which statement is correct?
    a. A only
    b. B only
    c. Both A and B
    d. Neither A nor B

19. Statement A: The stoplight switch is turned on by brake pedal movement as the brakes are applied.
    Statement B: Some cars use a stoplight switch that is turned on by hydraulic pressure.
    Which statement is correct?
    a. A only
    b. B only
    c. Both A and B
    d. Neither A nor B

20. Statement A: As brake fluid absorbs water, the boiling point increases.
    Statement B: Automotive brake fluid standards are controlled by the DOT.
    Which statement is correct?
    a. A only
    b. B only
    c. Both A and B
    d. Neither A nor B

Power Brake Theory

Objectives

Upon completion and review of this chapter, you should be able to:

- ❏ Understand the different types of power boosters and how they operate.
- ❏ Comprehend the terms commonly used with power brake units.
- ❏ Understand how each type of power booster operates.

7.1 Introduction

In many driving situations, the driver uses the brakes regularly, and the brake designs on some vehicles require a large amount of pedal effort. Imagine a four-wheel disc brake system on a full-sized domestic car that is equipped with a full range of accessories and equipment. Without power brakes, very few drivers would have the power to apply the brakes strongly enough to make a hard stop. Power brakes reduce the brake pedal pressure requirement while retaining much of the feel and sensitivity of a nonpower brake. Many vehicles are equipped with power brakes as standard equipment, and they are optional equipment on others.

A power brake system is really a standard brake system, sometimes with a different master cylinder but always with the addition of a **booster.** The booster is the source of the power. The booster is usually mounted between the brake pedal and the master cylinder, and it multiplies the force coming from the brake pedal (Figure 7–1). A common term is **power assist.** The booster assists the driver in pushing on the pedal.

In many cases, the power booster is used with a larger-bore master cylinder or a brake pedal with less leverage. The added assist from the booster makes the larger bore or faster leverage practical. Also, the increased displacement from the larger bore helps ensure that there is enough fluid to completely apply the brakes and have a high pedal. This also allows complete brake application with a short stroke of the brake pedal.

Traditionally, power boosters have used a vacuum assist. Engine intake manifold vacuum is readily available

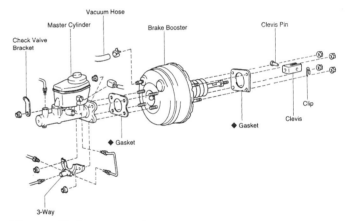

Figure 7–1 A power brake system uses a booster to increase the force from the pushrod to the master cylinder. This system uses a vacuum booster.

and almost free. In the early and mid-1970s, hydraulic boosters were introduced. These devices use hydraulic pressure from the power steering system for the assist (Figure 7–2). Hydraulic boosters are smaller in size, are more powerful, have a quicker response, and can be easily used in cars with diesel or turbo-charged gasoline engines. Diesel engines do not have a vacuum in the intake manifold; turbo-charged gasoline engines have pressure while under turbo boost. A vacuum pump is required for vacuum booster operation when these engines are used. In the mid-1980s, electrohydraulic boosters were introduced. These units operate similarly to hydraulic boosters and have their own electric motor–powered hydraulic pump (Figure 7–3). Electrohydraulic boosters run the pump only when necessary for an assist, so they are very efficient and offer improved fuel economy. They are also smaller in size and do not need the hoses between the booster and the power steering pump. Electrohydraulic boosters are integrated into the master cylinder and valve assembly of some ABS. Some trucks and buses use a dual power booster. This device is a combination of a vacuum and a hydraulic booster. Another truck booster design combines a hydraulic booster with a reserve electric motor pump. In this case, the electric pump provides hydraulic pressure for stops in case of pressure loss in the power steering system.

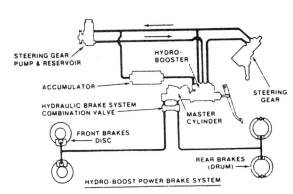

Figure 7–2 Some power booster systems use a hydro-boost booster. The hydraulic pressure is supplied by the power steering system. *(Courtesy of Chrysler Corporation)*

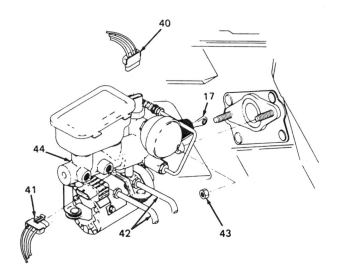

17. PUSHROD
40. ELECTRICAL CONNECTOR
41. ELECTRICAL CONNECTOR
42. BRAKE PIPE
43. NUT
44. POWERMASTER UNIT

Figure 7–3 A Powermaster brake booster combines the master cylinder with an electric motor and hydraulic pump. Note that this booster (44) is being removed for service. *(Courtesy of General Motors Corporation, Service Technology Group)*

7.2 Vacuum Booster Theory

When a gasoline engine operates at a speed less than wide-open throttle, the closing of the throttle plates will cause a **vacuum** in the intake manifold. The action of the pistons pulls air out of the manifold faster than **atmospheric pressure** can push it past the throttle plates.

The term *vacuum* refers to a pressure lower than atmospheric, which is the air pressure created by the weight of the atmosphere around our planet. This pressure is about 15 psi (100 kPa) (1 bar). Pressure less than atmospheric has traditionally been measured in **inches of mercury (in. Hg),** with atmospheric pressure being 0 in. Hg and a total vacuum being about 30 in. Hg. An absolute vacuum, with zero air pressure, is equal to 29.92 in. Hg (Figure 7–4). At idle, many gasoline engines have a manifold vacuum of about 15 to 18 in. Hg, at cruising speeds it is about 10 in. Hg, and while decelerating the vacuum measures about 20 to 21 in. Hg. The actual measurement for a particular engine varies depending on the throttle opening, the size of the engine relative to the car, the axle–gear ratio, the mechanical condition of the engine, and the type of emission-control devices. The design of many newer cars has greatly reduced the strength of the manifold vacuum.

In an engine with a manifold vacuum of 10 in. Hg, the pressure in the manifold is 10 psi (68.9 kPa) absolute. This pressure is 5 psi (34.5 kPa) less than atmospheric. Imagine a piston with a diameter of 10 in. (25.4 cm) and an area of 78.5 sq. in. (506.7 cm^2). With atmospheric pressure on one side and a pressure that is 5 psi lower (10 in. Hg) on the other, the piston will generate a force of 5 × 78.5 (392.5) lb (178.4 kg) toward the low-pressure side.

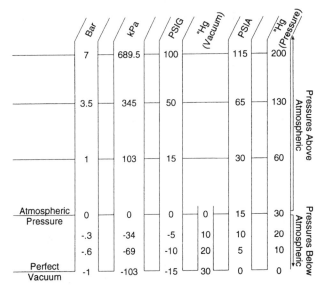

Figure 7–4 Different systems are used to measure air pressure. Pressures below atmospheric are commonly called vacuum.

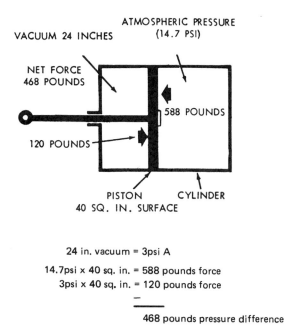

Figure 7-5 If there were a manifold vacuum of 24 in. Hg on one side of a 40-sq.-in. piston and atmospheric pressure on the other side, there would be a pressure difference of 468 lb to push the piston toward the left. *(Courtesy of John Bean Company)*

This is approximately the amount of power that power boosters generate. Considering that 50 lb (22.7 kg) of foot pressure on a brake pedal is fairly high, it is clear that this is a significant assist. The pedal ratio usually increases the pedal force about six times, to around 300 lb (136 kg). Booster force is greater with higher manifold vacuums, larger-diameter boosters, or tandem boosters (Figure 7-5).

Diesel engines do not use a throttle plate so they do not have vacuum in the intake manifold. Vehicles with diesel engines and vacuum boosters must use a pump to generate the vacuum for the booster. Vacuum pumps can be powered by an electric motor; run off the engine drive belt, camshaft like a fuel pump, or camshaft like a distributor; or mounted on the alternator (Figure 7-6).

Modern boosters are of the **vacuum suspended** type. Manifold vacuum is on both sides of the booster diaphragm when the booster is released. Early booster designs were **atmospheric suspended**. Both sides of the diaphragm were under atmospheric pressure during release. To apply this brake, a valve opened the master cylinder side to the manifold vacuum when the brake pedal was depressed. This system required a vacuum storage chamber to make certain that the necessary vacuum was there when needed. All brake boosters must be able to supply a brake assist if the engine suddenly stalls (Figure 7-7).

Most booster and master cylinder combinations are of the **tandem booster** or **integral booster** design, in which the booster is mounted directly behind the master cylinder. Two other booster arrangements are the *linkage booster*

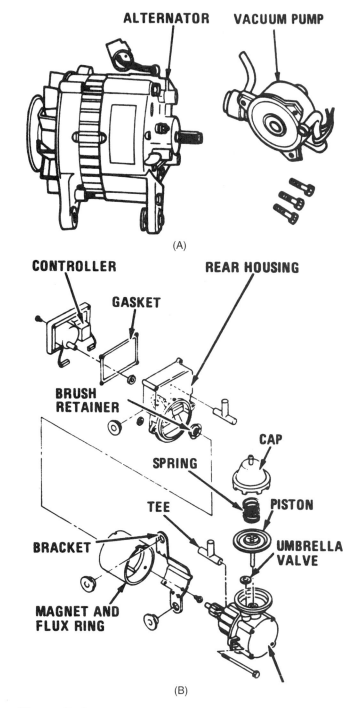

Figure 7-6 Vacuum pumps are used with diesel engine vehicles to supply vacuum for booster operation. They can be driven by the engine camshaft, drive belt, or alternator (A) or by an electric motor (B). *(Courtesy of Brake Parts, Inc.)*

style and a remotely mounted unit often called a *pressure multiplier*. The linkage booster unit was mounted above or below the master cylinder and used the pedal linkage to transfer the assist to the master cylinder. These units were used in the 1950s (Figure 7-8). Pressure multiplier units can be mounted anywhere in a vehicle, with the only necessary connections being the brake lines and the vacuum

Chapter 7 ■ Power Brake Theory

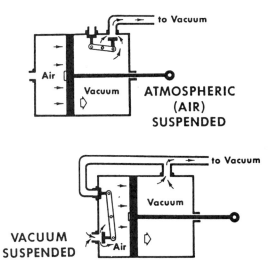

Figure 7-7 Some early boosters were air suspended; atmospheric pressure was on both sides of the piston while released. All modern boosters are vacuum suspended, with vacuum on each side of the piston while released. *(Courtesy of Wagner Brake)*

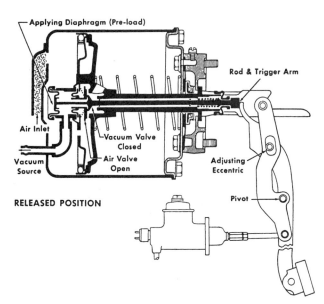

Figure 7-8 In the past, linkage booster or pedal assist boosters were used. Booster force was transmitted to the master cylinder through an extension of the brake pedal linkage. *(Courtesy of Wagner Brake)*

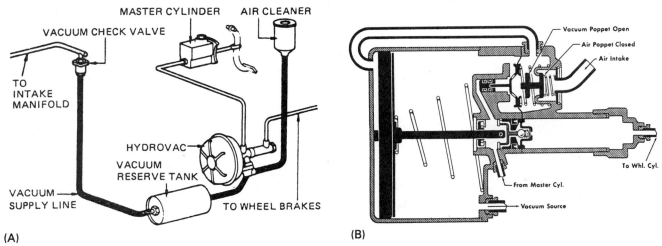

Figure 7-9 A pressure multiplier or Hydrovac booster can be mounted separately from the master cylinder (A). Its internal operation is similar to that of the other styles of vacuum boosters, with the addition of its own fluid piston. *(Courtesy of Wagner Brake)*

hose. Their input is the output pressure from the master cylinder. The booster output pressure is many times greater than the input pressure. The *Bendix Hydrovac* unit, used on many trucks, is a booster of this type (Figure 7-9). This booster style has been successfully retrofitted—added to vehicles that were not originally equipped with power brakes. Occasionally a pressure multiplier is mounted so that only the front-wheel disc brakes are assisted. On some light trucks and vans, two boosters are used, one for each hydraulic circuit.

7.3 Vacuum Booster Construction

Vacuum boosters are constructed with a rather large—6 to 11 in. (15 to 28 cm) in diameter—metal housing that is divided into two sealed chambers. The piston or diaphragm plate is usually sealed to the two chamber halves by a rolling or flexible rubber diaphragm. When the pressure is different on each side, the diaphragm plate

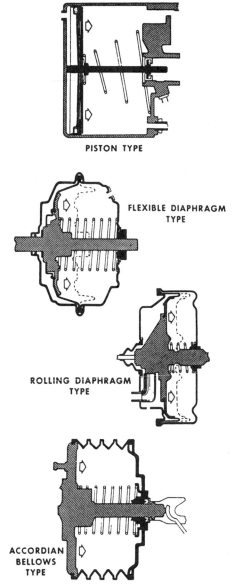

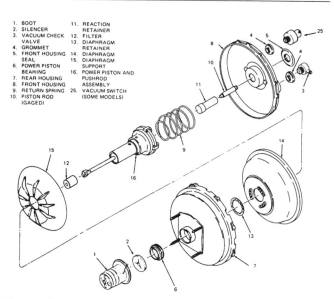

Figure 7-11 An exploded view of a vacuum booster. Note that the front and rear housing (7 and 8) interlock on the outer edge of the diaphragm (14) to form two airtight chambers. *(Courtesy of General Motors Corporation, Service Technology Group)*

Figure 7-10 Some early booster designs used a flexible bellows or piston as the power unit; modern units use either a rolling or flexible diaphragm. *(Courtesy of Wagner Brake)*

moves forward in the housing. In the past, some boosters used a piston with a sliding seal. Others used a booster chamber that was a collapsible bellows (Figure 7-10).

Most booster chambers are constructed from two interlocking stamped-steel housings. One housing half mounts on the car's engine compartment bulkhead or fire wall, and the other half has mounting provisions for the master cylinder (Figure 7-11).

The front or master cylinder section of the booster housing contains the vacuum connection. This connector is often also a check valve. The manifold vacuum pulls air out of the booster, but atmospheric pressure must not push air back into the booster if the engine is not running. The check valve allows air to flow in only one direction—toward the engine. A rubber hose leads from this connector to a fitting on the intake manifold. In some cars, the check valve is in the intake manifold fitting or in a separate unit in the hose (Figure 7-12). The rear or bulkhead side of the booster contains a support bushing and seal for the rear of the diaphragm support plate. An extension of the diaphragm support passes through this opening and slides inward during brake application.

The outer edge of the power piston diaphragm is usually locked between the two booster housing halves. A seal is made between each half as they are locked together. The diaphragm is attached to the diaphragm support plate or piston. This design allows it to move the plate and apply force through the plate to the hydraulic pushrod and the primary piston in the master cylinder. The rear portion of the diaphragm support contains a filter element and is the air intake for the booster. The filter element quiets the airflow and removes dirt particles. Two control valves are built into the center of the support plate, one for vacuum and one for atmospheric pressure or air. The valve plunger slides in the bore in the diaphragm plate. This type of valve is often called a **floating valve** because the valve bore travels with the piston (Figure 7-13).

Chapter 7 ■ Power Brake Theory

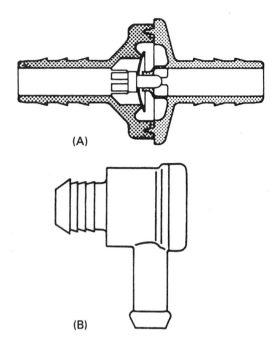

Figure 7-12 Vacuum check valves can be installed in the vacuum hose (A) or into the booster (B). Note that the upper valve is sectioned to show how air can flow from left to right but not the other direction. *(Courtesy of ITT Automotive)*

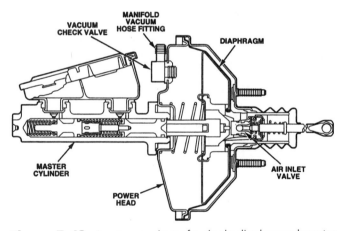

Figure 7-13 A cutaway view of a single diaphragm booster and master cylinder. *(Courtesy of General Motors Corporation, Service Technology Group)*

Some boosters use **tandem diaphragms.** Two separate diaphragms and support plates are tandem mounted so they move together to apply the same hydraulic pushrod. A dividing wall is placed between the two diaphragms so the housing will have two more pressure chambers. A tandem booster produces twice the assist of a single booster of the same diameter (Figure 7-14).

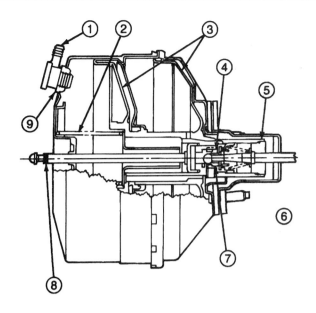

Item	Part Number	Description
1	2365	Power Brake Booster Check Valve
2	—	Return Spring (Part of 2005)
3	—	Tandem Power Diaphragms (Part of 2005)
4	—	Vacuum Port Closed — Brakes On (Part of 2005)
5	—	Filter (Air Inlet) (Part of 2005)
6	—	Brake Pedal Push Rod (Part of 2005)
7	—	Atmospheric Port Open — Brakes On (Part of 2005)
8	—	Master Cylinder Push Rod (Part of 2005)
9	2B176	Check Valve Grommet

Figure 7-14 A cutaway view of a tandem booster showing the two diaphragms (3). *(Courtesy of Ford Motor Company)*

Runout, which is the point at which a booster is producing its maximum assist, on an 8-in. (200-mm) single booster with a 1-in. (2.54-cm) bore master cylinder, produces a hydraulic pressure of 560 psi (3.861 kPa). A tandem booster with two 8-in. (200-mm) diaphragms and the same master cylinder produces 1,080 psi (7.447 kPa) at runout. Runout occurs when there is enough brake pedal pressure to completely compress the reaction disc. At this point, there is a full manifold vacuum on one side of the diaphragm(s) and atmospheric pressure on the other (Figure 7-15). A tandem booster is about 20 percent longer because of the added parts.

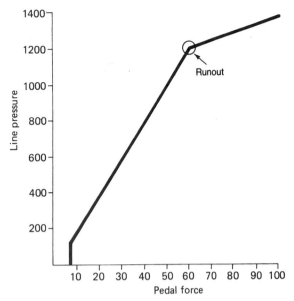

Figure 7–15 This illustration shows the output of a typical vacuum booster. Note that after runout, line pressure increases only from increased pedal pressure; the booster is giving the maximum amount to assist.

7.4 Vacuum Booster Operation

Most boosters have three operating positions—released, applying, and holding. These positions are determined by the amount of pressure on the valve pushrod. During release, there is no pressure on the valve rod. During application, there is enough pressure to compress the reaction disc, and during holding, the reaction disc is partially compressed. In the released position, the vacuum valve is open and the air valve is closed. During application, the vacuum valve is closed and the air valve is open. Both valves are closed while holding. These positions are determined by the amount of force on the brake pedal and whether the pedal is moving or stationary (Figure 7–16). The valve is in the application position when there is enough pedal force to compress the reaction disc. Movement of the disc and valves starts the boost, which in turn moves the diaphragm and support plate. As the support plate moves, it changes the position of the reaction disc and valves, and the unit position changes from applying to holding. The driver can increase or decrease the amount of braking from the holding position by increasing or decreasing pedal pressure. Runout or manual application occurs when there is enough pedal pressure to collapse the reaction disc completely. From this point, pedal pressure is transmitted through the reaction disc to the hydraulic pushrod along with the booster output force.

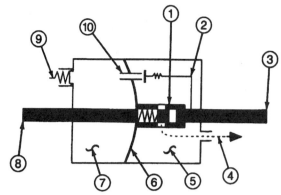

Figure 7–16 This vacuum booster schematic shows the vacuum valve (1), valve equalizer linkage (2), pedal pushrod (3), atmospheric air inlet (4), vacuum chamber B (5), diaphragm (6), vacuum chamber A (7), master cylinder pushrod (8), vacuum check valve (9), and equalizer–vacuum valve (10). During brake application, the pressure in vacuum chamber B increases. *(Courtesy of Ford Motor Company)*

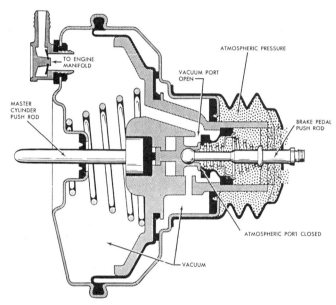

Figure 7–17 In the released position, the control valve opens the vacuum port so manifold vacuum is on both sides of the diaphragm; the atmospheric port is closed. *(Courtesy of General Motors Corporation, Service Technology Group)*

When the booster is released, the diaphragm return spring moves the diaphragm plate with the diaphragm and the control valve to the rear of the booster. The control valve spring then positions the control valve so the air valve is closed and the vacuum valve is open. The manifold vacuum moves equally to both sides of the diaphragm (Figure 7–17). This is the only time that air should flow through or into a booster. A booster should have no effect on engine performance other than causing a slight increase in engine speed for a moment as the brakes are released.

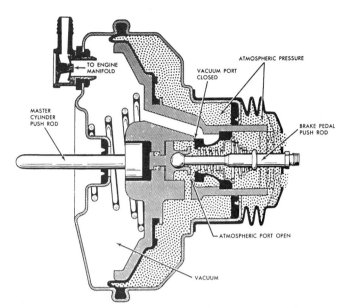

Figure 7–18 As the pedal pushrod moves inward, it moves the control valve so the vacuum port closes and the atmospheric port opens. This raises the pressure on the right side of the diaphragm to produce a power assist. *(Courtesy of General Motors Corporation, Service Technology Group)*

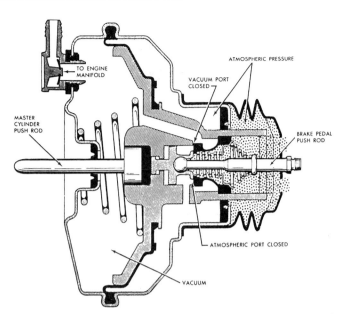

Figure 7–19 In the holding position, both the vacuum and atmospheric ports are closed so the pressure on the right side of the diaphragm will not change. This keeps the same pressure in the brake system. *(Courtesy of General Motors Corporation, Service Technology Group)*

When the brakes are applied, force from the brake pedal through the valve rod or pushrod pushes on the inner portion of the valve through the reaction disc to the diaphragm. Usually the reaction disc compresses and allows the valve plunger to change position in its bore to perform two separate operations. The vacuum valve is closed first. This shuts off any flow from one side of the diaphragm to the other. The second operation is to open the air valve so air can enter the rear booster chamber. This provides the pressure differential across the diaphragm for the power assist (Figure 7–18). The diaphragm then moves forward, applying the brakes, and because of the floating valve, it resets the operation to hold. While in a holding position, the pressure on each side of the diaphragm and the force on the master cylinder do not change (Figure 7–19). During application there is no airflow between the booster and the manifold, so there should be no change in engine operation. A certain amount of air does enter through the air valve into the booster. Sometimes you can hear this airflow, especially if the engine is shut off.

If the engine stalls, the booster will work normally for at least one cycle. It will give an assist for several cycles, but each application will allow more and more air to enter, which reduces the amount of vacuum. Each cycle will provide less and less assist.

7.5 Hydraulic Booster Theory

Hydraulic boosters are called **hydro-boost** by Bendix, the company that first developed them. The external force used to provide the assist with this booster is hydraulic pressure from the power steering pump. This pressure is much higher, about 1,000 to 1,500 psi (6,895 to 10,340 kPa), than the 5- to 10-psi (34- to 69-kPa) pressure differential that can be used with a vacuum booster. Therefore the piston in a hydraulic booster can be made much smaller and still produce a greater assist.

This booster is connected to the power steering pump by a steel pressure hose and a reinforced rubber return hose. It is also connected to the power steering gear by another pressure hose. Hydraulic flow begins at the pump, goes first to the brake booster, and then continues on to the steering gear (Figure 7–20).

Hydraulic boosters use an **accumulator** to store pressurized fluid, to ensure that the vehicle can stop even if the engine is off. Early accumulators operated by compressing a spring-loaded piston, but newer units use a nitrogen gas–charged accumulator. An accumulator uses a pressure chamber that is divided by a movable piston or flexible diaphragm. It has a static charge of nitrogen gas on the side opposite the oil. When the engine runs and the

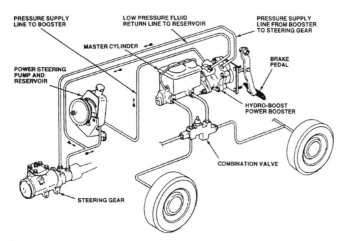

Figure 7-20 A hydro-boost system showing the arrangement of the hydraulic pressure and return lines. *(Reprinted by permission of Hunter Engineering Company)*

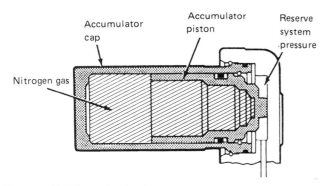

Figure 7-21 Hydraulic boosters store enough pressurized fluid to apply the booster in case the engine stops. The stored oil moves the accumulator piston to the left, compressing the nitrogen gas charge. *(Courtesy of Ford Motor Company)*

power steering pump develops pressure, pressurized fluid flows into the accumulator. Then the pressure of the fluid compresses the spring or gas charge in the accumulator (Figure 7–21). The volume stored in the accumulator is large enough to ensure pressurized fluid for at least one brake application. More than one application is possible, but each will be weaker as the accumulator charge bleeds down. A check valve between the accumulator and the pressure port of the booster prevents a pressure loss back to the power steering pump when the engine stops.

It is important to remember the accumulator when working on a hydraulic booster. The accidental and unexpected discharge of an accumulator that is full of oil under high pressure may cause injury and certainly will cause a mess. The accumulator should be discharged by pumping the brake pedal. Repeat this operation until the feel of the brake pedal indicates a full discharge. The pedal will get harder and harder on each application as the accumulator bleeds down.

7.6 Hydraulic Booster Construction

The hydraulic booster uses a cast-iron body or housing that is mounted between the master cylinder and the car's bulkhead. The body section contains the power piston, an open-center spool valve, and a lever assembly, as well as the input and output pushrods. A removable housing cover, sealed by a rubber seal ring, allows access to these parts and helps enclose the high-pressure fluid chamber. The power piston and input rod use lip seals to prevent fluid leaks at these openings (Figure 7–22).

A **spool valve** has precision-sized bands that fit the valve housing bore closely enough to prevent excessive fluid leaks along the bore, yet the spool valve is still able to slide in the bore. This valve is used to control the pressure in the booster pressure chamber at the end of the power piston. A spool valve has grooves that allow fluid to flow from one side port to another. Sliding the valve covers and uncovers these side ports. A hollow spool valve is hollow from one end down through the valve to a side passage, allowing another possible fluid passageway.

The position of the spool valve is controlled by a lever. This lever is unique in that the fulcrum can be at either pin A or pin B (Figure 7–23). Pin A is on the power piston, and pin B is on the input rod. In this way, movement of either the input rod or the power piston will change the position of the spool valve and therefore the pressure in the chamber.

7.7 Hydraulic Booster Operation

In a relaxed or released position, the power piston return spring positions the power piston rearward, toward the brake pedal. With the piston in this position and the brake pedal released, the spool valve is positioned so that the power cavity is hydraulically connected to the return port. Hydraulic pressure then returns to the power steering pump reservoir (Figure 7–24).

When the driver applies the brakes, the movement of the input rod moves the lever and repositions the spool valve. This action connects the pressure port to the pressure cavity so that power steering pump pressure reaches the cavity. This pressure at the end of the power piston causes it to stroke toward the master cylinder and apply the brakes (Figure 7–25). The amount of pressure in the cavity varies depending on how hard the brake pedal is applied and the pressure setting of the power steering pump. Because the fulcrum is traveling with the piston, if the driver holds the brake pedal steady, the power piston

Chapter 7 ■ Power Brake Theory

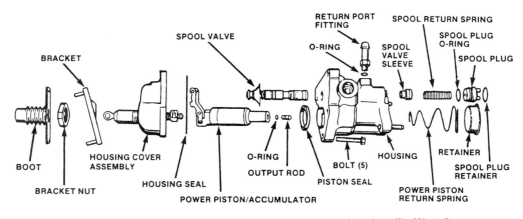

Figure 7-22 An exploded view of a hydro-boost II unit. *(Courtesy of Bendix Brakes, by AlliedSignal)*

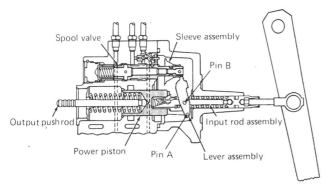

Figure 7-23 A cutaway view of a hydro-boost unit; the pressure chamber is at the right of the power piston. *(Courtesy of General Motors Corporation, Service Technology Group)*

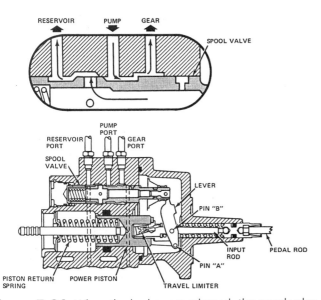

Figure 7-24 When the brakes are released, the spool valve is positioned so pump pressure goes to the steering gear and pressure chamber pressure is released to the pump reservoir. *(Courtesy of General Motors Corporation, Service Technology Group)*

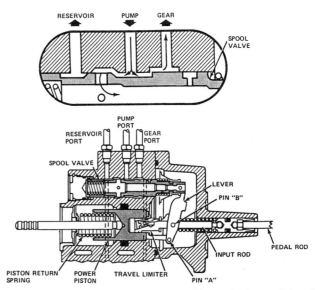

Figure 7-25 Brake pedal pressure (through the pedal rod) moves the spool valve to send pump pressure to the pressure chamber. *(Courtesy of General Motors Corporation, Service Technology Group)*

and lever will move the spool valve to the holding position. The pressure in the power chamber will then be held constant. The pressure in this chamber and the amount of braking will change as the driver varies the foot pressure (Figure 7-26).

7.8 Electrohydraulic Booster Theory

The electrohydraulic booster, called a *Powermaster*, has its own electric motor–driven pump to provide the boost pressure. This pump operates only when needed to provide this pressure. A hydraulic booster, as just described, places a light but constant drag on the power steering pump, which

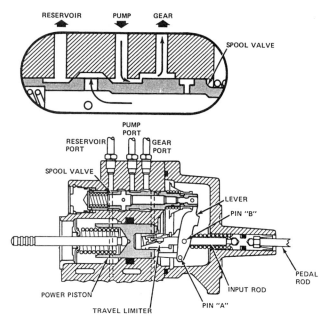

Figure 7–26 In the holding position, the spool valve shuts off the flow to or from the pressure chamber. This keeps the same pressure in the brake system. *(Courtesy of General Motors Corporation, Service Technology Group)*

causes a slight loss in fuel economy. Electrohydraulic boosters are more compact (Figure 7–27). They do not require hose connections to the engine, provide a large amount of boost, and can be used with any engine and accessory combination.

The small, 12-volt (V), electric motor–driven vane pump charges an accumulator, and the booster operates off this accumulator plus added pump output, if necessary. The operation of the pump is controlled by a pressure switch. When the accumulator pressure falls below 510 psi (3,516 kPa), the switch closes and the pump runs. When the pressure exceeds 675 psi (4,654 kPa), the switch opens and the pump stops. The pump normally runs in a short cycle, taking less than 20 seconds to charge the accumulator, and normally does not run again until the brakes are used (Figure 7–28).

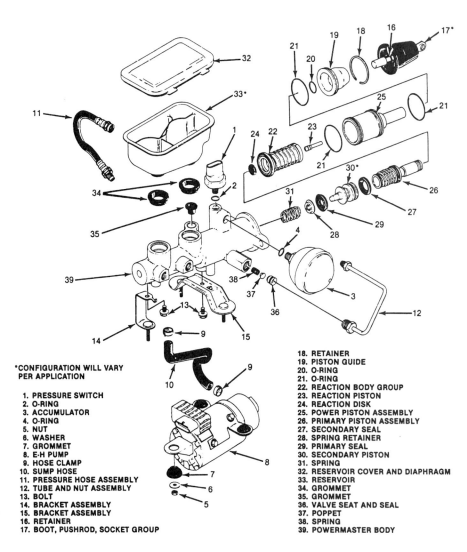

*CONFIGURATION WILL VARY PER APPLICATION

1. PRESSURE SWITCH
2. O-RING
3. ACCUMULATOR
4. O-RING
5. NUT
6. WASHER
7. GROMMET
8. E-H PUMP
9. HOSE CLAMP
10. SUMP HOSE
11. PRESSURE HOSE ASSEMBLY
12. TUBE AND NUT ASSEMBLY
13. BOLT
14. BRACKET ASSEMBLY
15. BRACKET ASSEMBLY
16. RETAINER
17. BOOT, PUSHROD, SOCKET GROUP
18. RETAINER
19. PISTON GUIDE
20. O-RING
21. O-RING
22. REACTION BODY GROUP
23. REACTION PISTON
24. REACTION DISK
25. POWER PISTON ASSEMBLY
26. PRIMARY PISTON ASSEMBLY
27. SECONDARY SEAL
28. SPRING RETAINER
29. PRIMARY SEAL
30. SECONDARY PISTON
31. SPRING
32. RESERVOIR COVER AND DIAPHRAGM
33. RESERVOIR
34. GROMMET
35. GROMMET
36. VALVE SEAT AND SEAL
37. POPPET
38. SPRING
39. POWERMASTER BODY

Figure 7–27 An exploded view of a Powermaster unit. *(Courtesy of General Motors Corporation, Service Technology Group)*

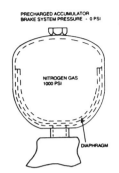
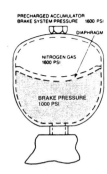

Figure 7–28 An empty accumulator (left) is charged with nitrogen gas at a fairly high pressure. A charged accumulator (right) is partially filled with brake fluid or hydraulic oil at an even higher pressure. *(Courtesy of Chrysler Corporation)*

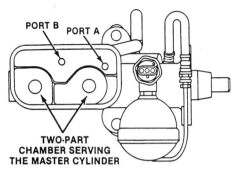

Figure 7–29 The reservoir of a Powermaster unit shows ports A and B for the booster pump and the two ports for the brake system. *(Courtesy of General Motors Corporation, Service Technology Group)*

7.9 Electrohydraulic Booster Construction

The electrohydraulic booster is combined with the master cylinder in the same cast-aluminum body section. The power piston and the reaction group operate directly against the primary master cylinder piston. The unit is only slightly larger than a nonpower master cylinder.

Brake fluid is used as booster fluid. A three-section reservoir stores this fluid in a section separate from the two master cylinder fluid sections. Two ports at the bottom of the booster section lead to the pump intake and the booster return ports. The fluid level in this reservoir section drops significantly when the accumulator is charged and should be checked only after discharging the accumulator (Figure 7–29). Otherwise, the reservoir may overfill and cause overflow when the accumulator is discharged.

The pump and motor form a single assembly that is secured to a bracket attached to the booster–master cylinder body. The pump intake is connected to the reservoir port by a rubber hose, and the pump outlet is connected to the booster by a reinforced high-pressure hose. Electric power for the pump originates at a 30-ampere (amp) fuse in the car's fuse block. An electrical connector brings this battery positive (B+) voltage to a relay mounted on the motor. The relay is basically a switch controlled by the pressure switch. When the accumulator pressure drops, the pressure switch closes, the relay closes, and the motor runs. When the accumulator pressure rises, the pressure switch opens, the relay opens, and the motor stops. The pressure switch has a second set of contacts that turn on the brake failure warning light if the accumulator pressure drops below 400 psi (2,758 kPa) (Figure 7–30).

The power piston fits into its bore in the body casting and is held in place by the piston guide. Rubber O-rings are used at several points to prevent fluid leaks. A group of parts, called the *reaction group,* fits inside the power piston. These parts include the inner and outer control valves, the reaction disc, and the reaction piston. The chamfered edges at the ends of the inner and outer control valves control the booster pressure in the power piston cavity.

7.10 Electrohydraulic Booster Operation

Even though a different style of valve is used, the electrohydraulic booster operates in essentially the same way as a hydraulic booster. A fluid passage leads from the high-pressure port at the accumulator to the end of the outer control valve, and a return passage leads from the inner control valve to the reservoir return port. The inner and outer valves are positioned by the pushrod, reaction disc, and power piston. They provide release (relaxation), application, and holding of the hydraulic pressure acting on the power piston. These operations are shown in Figure 7–31, Figure 7–32, and Figure 7–33.

7.11 Practice Diagnosis

You are working in a brake and front-end repair shop and encounter the following problems:

CASE 1: The customer has brought in her 15-year-old Chevrolet Caprice with a complaint that she has to push very hard on the brake pedal to stop the car. On checking

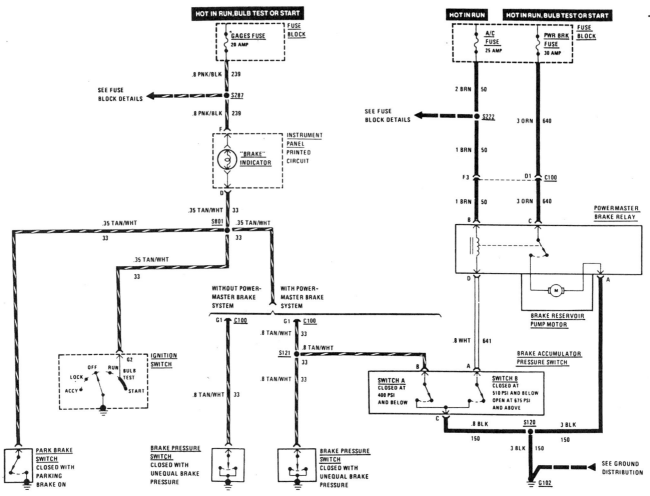

Figure 7–30 The electrical circuit for a Powermaster system shows the brake pressure switch on the accumulator that operates the pump; it can also turn on the brake warning light. *(Courtesy of General Motors Corporation, Service Technology Group)*

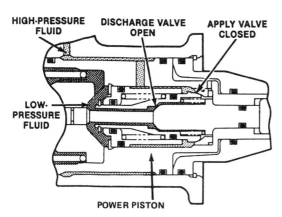

Figure 7–31 With the system at rest, the application valve is closed and the discharge valve is open so there will be no boost pressure on the power piston. *(Courtesy of General Motors Corporation, Service Technology Group)*

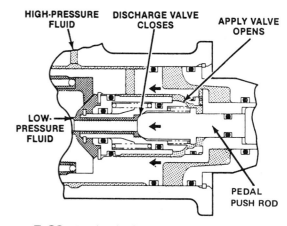

Figure 7–32 As the brakes are applied, the application valve opens to allow fluid pressure to enter the power piston chamber. *(Courtesy of General Motors Corporation, Service Technology Group)*

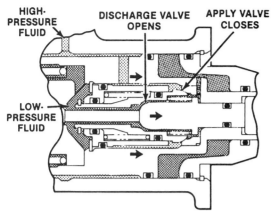

Figure 7-33 When the brakes are released, spring pressure moves the discharge valve to open and release pressure to the reservoir. *(Courtesy of General Motors Corporation, Service Technology Group)*

it you find a very hard brake pedal on a high-mileage car; it does not seem to make much difference in pedal effort when you start the engine, and the engine sounds a little rough. What could be wrong with this car? What should you do to check your suspicions?

CASE 2: While inspecting the brake system on a 1994 Buick, you notice that the car has an electric motor on the master cylinder instead of a vacuum booster. The fluid level in the master cylinder appears low. What should you do next? Is this a normal or abnormal condition?

Terms to Know

accumulator	hydro-boost	spool valve
atmospheric pressure	inches of mercury (in. Hg)	tandem booster
atmospheric suspended	integral booster	tandem diaphragms
booster	power assist	vacuum
floating valve	runout	vacuum suspended

Review Questions

1. Statement A: Power brakes use larger, more powerful wheel brake assemblies so they are substantially stronger than standard brakes. Statement B: Power brakes use a power booster between the brake pedal and master cylinder to produce higher hydraulic system pressures. Which statement is correct?
   a. A only
   b. B only
   c. Both A and B
   d. Neither A nor B

2. Power boosters produce their assist with the aid of:
   a. Engine intake manifold vacuum
   b. Power steering hydraulic pressure
   c. Hydraulic pressure from an electric motor–driven pump
   d. All of the above

3. In the released position, a vacuum-suspended booster has:
   a. Atmospheric pressure on both sides of the diaphragm
   b. A vacuum on both sides of the diaphragm
   c. A vacuum on the pedal side and atmospheric pressure on the master cylinder side of the diaphragm
   d. A vacuum on the master cylinder side and atmospheric pressure on the pedal side of the diaphragm.

4. In the applied position, a vacuum-suspended booster has:
   a. Atmospheric pressure on both sides of the diaphragm
   b. A vacuum on both sides of the diaphragm
   c. A vacuum on the pedal side and atmospheric pressure on the master cylinder side of the diaphragm
   d. A vacuum on the master cylinder side and atmospheric pressure on the pedal side of the diaphragm

5. To ensure a power-assisted stop if the engine stalls, vacuum boosters have:
   a. A vacuum accumulator in the vacuum supply system
   b. A check valve in the vacuum hose
   c. A vacuum pump attached to the booster
   d. All of the above

6. Statement A: A linkage booster power assist unit is mounted above the master cylinder.
   Statement B: A hydrovac is a pressure multiplier booster that is mounted separately from the master cylinder.
   Which statement is correct?
   a. A only   c. Both A and B
   b. B only   d. Neither A nor B

7. Statement A: A tandem master cylinder is always used with tandem diaphragm boosters.
   Statement B: Booster runout occurs when there is atmospheric pressure on the pedal side of the diaphragm and the booster is producing maximum assist.
   Which statement is correct?
   a. A only   c. Both A and B
   b. B only   d. Neither A nor B

8. A hydro-boost unit:
   a. Develops higher braking pressures than a vacuum booster
   b. Can be easily used with diesel engines
   c. Is always mounted integrally with the master cylinder
   d. All of the above

9. To ensure an assisted stop if the engine stalls, a hydro-boost unit uses:
   a. An accumulator in the booster
   b. A check valve in the hydraulic hose
   c. An electric motor–driven hydraulic pump
   d. All of the above

10. Statement A: Brake booster valves have three basic operating positions: applying, holding, and released.
    Statement B: Booster valve position is determined by the pressure of the driver's foot relative to the braking pressure.
    Which statement is correct?
    a. A only   c. Both A and B
    b. B only   d. Neither A nor B

11. A Powermaster unit:
    a. Is a self-contained booster and master cylinder combination
    b. Uses brake fluid for the booster and brake hydraulic systems
    c. Uses an accumulator to store pressurized fluid for an assist if the engine stalls
    d. All of the above

12. Statement A: The pump of a Powermaster unit is controlled by the booster control valve.
    Statement B: The pump of the Powermaster unit operates continuously when the engine runs.
    Which statement is correct?
    a. A only   c. Both A and B
    b. B only   d. Neither A nor B

Antilock Braking System Theory

Objectives

Upon completion and review of this chapter, you should be able to:

❑ Understand how an antilock braking system operates.

❑ Be familiar with the terms commonly used with antilock braking systems.

❑ Be familiar with the operating differences in the various types of antilock braking systems.

8.1 Introduction

Antilock braking systems (ABS), sometimes called *antiskid braking,* have been used to some degree on domestic vehicles since the late 1960s. The early systems were used primarily on luxury cars as an extra-cost option and had limited popularity. Since the early to mid-1980s, ABS has become more popular and is now standard equipment on many vehicle models. Most experts predict almost 100 percent usage of ABS on cars in the near future.

Wheel lockup during braking causes skidding, which in turn causes a loss of traction and vehicle control. A tire generates its greatest amount of traction when it is slipping at a rate of 15 to 25 percent. Wheel lockup will result in longer stopping distances and possible accidents. ABS is designed to prevent wheel lockup and the resulting skid, even under the worst driving conditions, by automatically compensating for changes of traction or tire loading. ABS does not necessarily produce shorter stops, but it greatly improves the driver's ability to control the vehicle when trying to stop quickly (Figure 8–1). On good, dry pavement, an ABS-equipped car usually stops at about the same rate as a non-ABS-equipped car (Figure 8–2). The stopping distances are about equal. Under poor traction conditions, such as on wet or icy pavement, ABS allows the car to stop in a significantly shorter distance with a controlled, steerable stop. On low-traction surfaces, non-ABS systems allow the wheels to lock up easily in most cases, which greatly reduces braking power (Figure 8–3). Remember that braking power is limited by tire traction and driver reaction time and skill

Figure 8–1 ABS helps a driver steer the car to a straight stop in all traction conditions (top). Wheel lockup causes skidding and loss of control (bottom). *(Courtesy of General Motors Corporation, Service Technology Group)*

(Figure 8–4). A four-wheel ABS cycles the braking power on the tire(s) with poor traction while retaining full stopping power in those tires with good traction. In a non-ABS system, if both front or both rear wheels lock up, the ability to steer and maneuver the car is lost.

ABS begins with a standard brake system and adds one, two, three, or four modulator or control valves, one to four speed sensors, and an electronic controller for the valves. The modulator valves are used to cycle the hydraulic pressure at the brake assemblies, the speed sensors

137

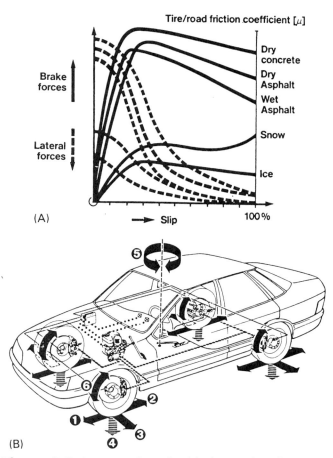

Figure 8-2 By preventing wheel lockup under all traction conditions (A), the tire has maximum traction for acceleration (1), braking (2), cornering (3), normal reaction forces (4), yaw motions (5), and inertia of the tire and wheel (6). *(Courtesy of ITT Automotive)*

ROAD CONDITION	TIRE CONDITION	RESULTANT COEFFICIENT OF FRICTION
Dry pavement	New tire	1.0 (highest)
Dirt road	New tire	0.9
Dry pavement	Old, worn tire	0.8
Dirt road	Old, worn tire	0.7
Gravel	New tire	0.6
Gravel	Old, worn tire	0.5
Wet road	New tire	0.4
Wet road	Old, worn tire	0.3
Ice	New tire	0.2
Ice	Old, worn tire	0.1 (lowest)

1.0 = Highest coefficient
0.1 = Lowest coefficient

(A)

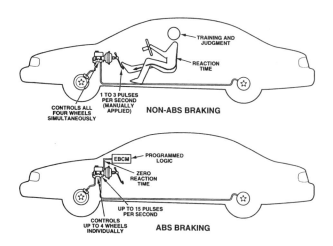

Figure 8-3 The coefficient of friction between the tire and road varies depending on tire and road conditions (A). If a heavily braked wheel exceeds the amount of friction it will lock up (B). *(A is courtesy of Chrysler Corporation; B is courtesy of General Motors Corporation, Service Technology Group)*

Figure 8-4 Without ABS, the driver must try to prevent wheel lockup in a panic stop by pulsing the brake pedal. With ABS, the system can pulse any or all of the wheels up to fifteen times per second. *(Courtesy of General Motors Corporation, Service Technology Group)*

determine the rotating speed of the wheel, and the electronic controller monitors the speed of the tires and operates the modulator valves to prevent wheel lockup.

An ABS may have either a tandem or a diagonal split hydraulic system, and pickups with ABS may have rear-wheel-only systems (Figure 8–5). Most of these are four-wheel systems using either three or four circuits or channels; a circuit is one or more wheels with modulator valves controlling the hydraulic pressure. Rear-wheel-only, single-circuit systems are standard equipment on most new pickups (Figure 8–6). One manufacturer names these systems RWAL (rear-wheel antilock) and uses the term 4WAL (four-wheel antilock) for some four-wheel systems. Most of the early domestic ABS units were two-wheel, rear-wheel-only systems.

As a braked wheel begins to lock up and skid, ABS cycles the brake off until the wheel is rotating at the correct speed. Then it cycles the brake back on. If the wheel locks up again, the cycle is repeated. This cycling is fast, occurring about five to fifteen times a second, and each braking circuit is cycled individually as needed (Figure 8–7).

Modern electronic systems use three stages of ABS operation: (1) normal operation (sometimes called pressure increase or buildup), in which normal braking occurs; (2) pressure holding, in which the pressure in the

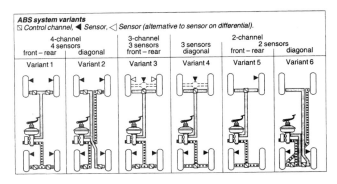

Figure 8–5 Antilock braking systems vary according to the number and location of sensors and the number of controlled channels or circuits. A rear-wheel-only system is similar to variant 4 or 5 without the front control channel. *(Courtesy of Robert Bosch Corp.)*

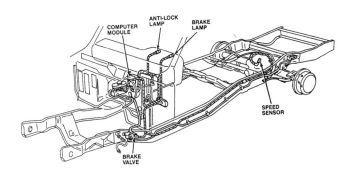

Figure 8–6 A two-wheel (rear-only) ABS is used on many pickups as standard equipment. Note the wheel speed sensor built into the rear axle. *(Courtesy of Ford Motor Company)*

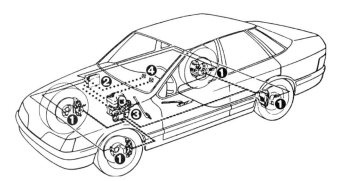

Figure 8–7 In many passenger cars, the ABS is a four-wheel system with sensors for each of the wheels. A three-circuit system that combines both rear brakes in the same circuit is shown. The wheel sensors (1) provide the EBCM (2) with the speed of each wheel; if a wheel locks up, the hydraulic control (3) cycles the brake unit. The warning light (4) informs the driver of a system malfunction. *(Courtesy of ITT Automotive)*

brake assembly is not increased or decreased if wheel lockup occurs; and (3) pressure reduction, decay, or reapplication, in which pressure is released from the brake assembly if necessary to get the wheel turning.

Most ABS units use **electronic wheel speed sensors** with an **electronic control module** and **electronically controlled modulator valves.** Most current systems use three or four wheel speed sensors—one for each front wheel, and either one for each rear wheel or one for the rear axle or driveshaft. A rear-wheel-only system uses just a rear-wheel speed sensor. Some systems fit the speed sensor to the speedometer cable or into the transmission or rear axle assembly. Rear-wheel-only systems use a single modulator valve to control both rear wheels at the same time. This is a one-circuit system. Most four-wheel systems use a modulator or control valve for each front wheel and one for both rear wheels. This arrangement is called a *three-circuit* brake. The three circuits are right front, left front, and both rear brake units. Some cars use a *four-circuit* system in which a control valve regulates the hydraulic pressure in each brake assembly (Figure 8–8). A vehicle tends to yaw or turn sideways when the brake is applied at one rear wheel and released at the other; modern four-circuit systems use more sophisticated controls to reduce this problem.

A *mechanical ABS,* called a *Stop Control System (SCS)* by its manufacturer, is used on smaller FWD cars.

An aftermarket system called Brake-guard ABS is marketed to give ABS benefits to all vehicles with hydraulic brakes. Brake-guard is not a true ABS; it simply connects into each hydraulic circuit to provide a small amount of hydraulic pressure accumulator action to reduce very high pressure spikes.

Most ABS designs revert to standard braking if the system fails, and normal, non-ABS braking becomes available. The electronic systems turn on an amber antilock warning light if failure occurs. Systems that use ABS booster pressure to operate the rear brakes lose the rear brakes and power assist if there is a failure in the boost section (Figure 8–9). The electronic portion of an ABS is designed to check itself each time the car is started. You can sometimes hear or feel the system go through its self-checks as the car is started or as it begins moving. On some cars, if you maintain pressure on the brake pedal during this time, you can feel the effect of the modulators exercising as part of the self-check. Presently, most ABS problems are the result of poor electrical connections and debris or foreign material in the modulator valve, which causes leakage. Be aware that any system as complex as ABS has many potential problem areas.

Antilock brake systems can also be classified as *integral* or *nonintegral.* Nonintegral is also called add-on.

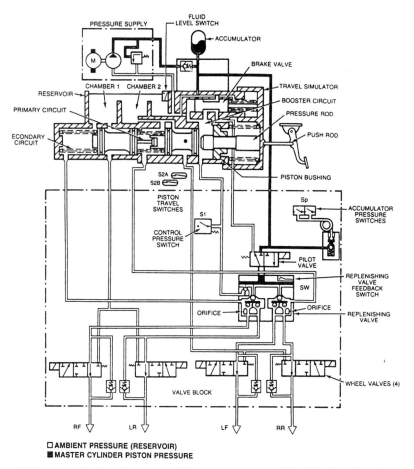

Figure 8-8 A four-circuit ABS. Four wheel valves control the pressure to each of the brake calipers to prevent wheel lockup. *(Courtesy of General Motors Corporation, Service Technology Group)*

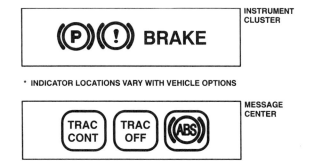

Figure 8-9 Cars equipped with ABS have an amber antilock brake light on the instrument panel in addition to the red brake warning light. This vehicle also has traction control. *(Courtesy of Chrysler Corporation)*

With some older systems, when ABS is standard equipment, an integral system is used; when ABS is an option, an add-on system is often used. Integral systems have the master cylinder and modulator valve assembly in one unit. Add-on systems mount the modulator between the master cylinder and an existing brake system. Also, some integral systems use a hydraulic pump and accumulator for power booster assist; one manufacturer calls this a *closed system* and the nonintegral ABS an *open system*. An add-on system has the same master cylinder and power booster as the standard, non-ABS brake, with the modulator valve assembly mounted separately. Some nonintegral systems use a pump for fluid recirculation; this prevents the pedal from moving to the floor as the modulator valves cycle (Figure 8-10).

8.2 Electronic Wheel Speed Sensors

Electronic *wheel speed sensors* produce an electronic signal as the wheel revolves. These sensors are the "eyes" of the electronic controller, allowing the controller to "see" the rate of deceleration or lockup of the wheel (Figure 8-11). Each sensor uses a gearlike, *toothed rotor*, also called a *sensor ring, toothed ring, exciter ring, tone wheel,* or *reluctor*—on the wheel hub or axle—that rotates with the wheel. Most sensor rings are replaceable and are often pressed onto the part (e.g., constant velocity [CV] joint, rotor, axle shaft) where they operate. When replacing a sensor ring, make sure that the replacement has the correct number of teeth and is positioned correctly. The wheel speed sensor can be mounted at the wheel hub or built into the pinion shaft or differential case of a rear axle assembly (Figure 8–12).

The **sensor** is a magnetic induction coil—a coil of wires with a magnet core—that is mounted right next to the sensor ring. The air gap between the toothed rotor and the sensor coil is a precise distance to ensure inductance without contact and wear (Figure 8–13). This same type of speed sensor is used in the distributors of some late-model cars and in the speedometer cable for some cruise control systems.

As the wheel rotates, an electrical signal is generated in the induction coil each time one of the teeth of the sensor ring passes the coil. The frequency of this signal is proportional to the speed of the wheel, and this frequency

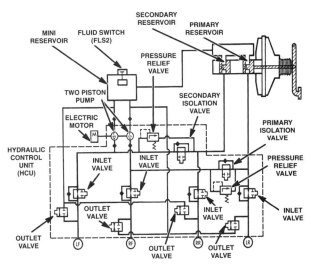

Figure 8–10 This ABS–traction control system has a two-piston pump that returns fluid from the outlet valve(s) to the inlet valve(s) and prevents the brake pedal from dropping during ABS action. This pump also provides pressure to apply a brake to prevent wheel spin and control traction. *(Courtesy of Chrysler Corporation)*

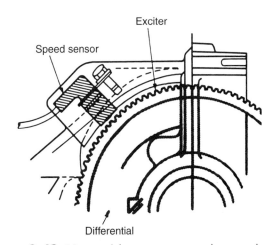

Figure 8–12 Many pickups use an axle speed sensor mounted in the rear axle. This exciter ring is mounted on the differential case, inside the rear axle housing. *(Courtesy of Ford Motor Company)*

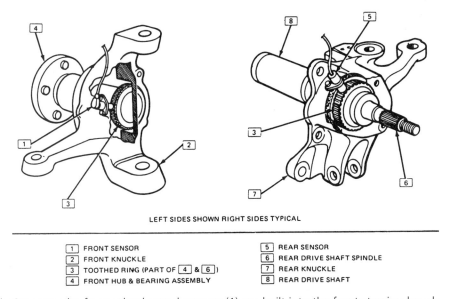

Figure 8–11 On this Corvette, the front-wheel speed sensors (1) are built into the front steering knuckles (2), and the rear-wheel speed sensors (5) are built into the rear-wheel support knuckles (7). *(Courtesy of General Motors Corporation, Service Technology Group)*

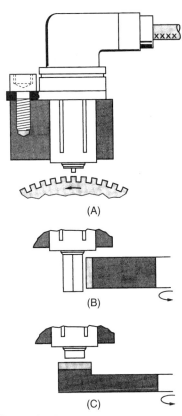

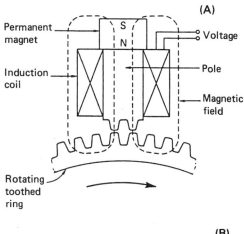

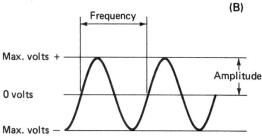

Figure 8–13 A wheel speed sensor can be mounted so the pole pin is in a radial position outside of the reluctor (A), in an axial position outside the reluctor (B), or alongside the reluctor (C). *(Courtesy of Robert Bosch Corp.)*

Figure 8–14 As the toothed reluctor ring moves past the induction coil, an AC electrical signal is generated (A). The frequency of the signal increases as the wheel turns faster (B). *(Courtesy of ITT Automotive)*

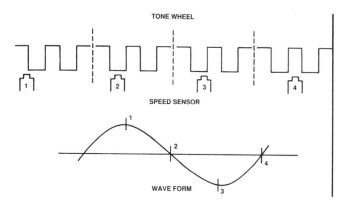

Figure 8–15 The waveform shows a positive voltage signal generated as the front of a tone wheel tooth passes the speed sensor (1). The signal voltage drops to zero as the speed sensor passes the center of the tooth (2) or between two teeth (4). A negative signal is generated as the back of the tooth passes the speed sensor (3). *(Courtesy of Chrysler Corporation)*

increases and decreases as the wheel speeds up and slows down (Figure 8–14). The electrical signal occurs because the metal teeth of the sensor ring pull the magnetic lines of flux over the coils of wire (Figure 8–15). If a wheel were to lock up under braking, the frequency of the sensor signal for that wheel would drop to zero. Each sensor is connected to the electronic controller by a pair of wires.

8.3 Electronic Controller

The *electronic controller*, also called an *electronic control module (ECM)*, *electronic brake control module (EBCM)*, or *controller antilock brake (CAB)*, is a microprocessor with about 8 kilobytes of memory (Figure 8–16). The EBCM receives the signal from each wheel speed sensor as input and actuates the modulator valves as output. It compares the speed of each wheel (the frequency of the sensor signal) with that of the others and to a deceleration profile programmed and stored in its memory. It determines if wheel lockup is beginning to occur, if the sensor signal frequency is slowing at too fast a rate when compared with the deceleration profile, or if the signal has a significantly lower frequency than that of the other wheels. The EBCM constantly monitors and compares the wheel speeds. It also checks itself and the system to ensure proper operation. Some systems use two identical controllers (inside the EBCM) to compare with each other also. If they do not agree with each other, they shut the system off and turn on a warning light.

When the EBCM determines that a wheel is slowing too quickly or that it has already stopped, it activates the brake modulator for that wheel (Figure 8–17). The

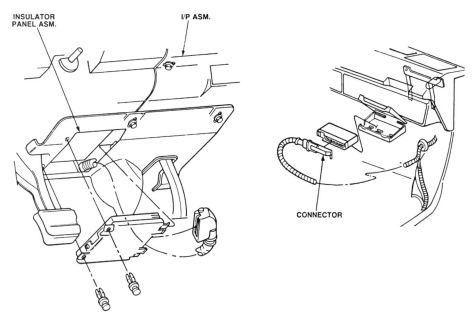

Figure 8–16 The EBCM is mounted in a relatively cool, clean, well-protected location. In this car, it is mounted under the left or right end of the bottom of the instrument panel. *(Courtesy of General Motors Corporation, Service Technology Group)*

modulator releases the brake enough to allow the wheel speed to increase. As the signal frequency from the wheel comes within the correct profile, the controller activates the modulator to reapply the brake. Because electrical signals travel at close to the speed of light, these actions can occur very rapidly.

The EBCM is somewhat fragile, in addition to being very expensive. The controller is normally mounted in one of the cleaner, drier, cooler locations where it is somewhat protected from impact and corrosion. In some cars it is mounted in the trunk area, behind the rear seat, or in one of the side panels (Figure 8–18). In other vehicles it is mounted in the bottom of the instrument panel, behind or above the glove compartment, in the trunk, or under a seat. It is mounted under the hood in the engine compartment in some cars and pickups.

The electrical circuit for an ABS is somewhat complex and varies depending on the design of the system and the particular car model. It always includes a power supply for the controller with the ability to power the controller output to the modulators. The circuit also includes connections to the individual wheel speed sensors and modulators and the brake failure warning light circuit. A typical circuit is shown in Figure 8–19. This circuit complexity requires advanced electronic troubleshooting skills, well beyond those of the typical brake technician of the 1970s and 1980s.

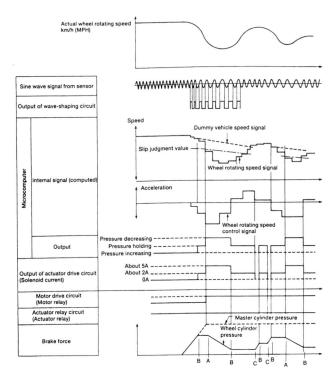

Figure 8–17 A time profile of an ABS-assisted stop. At points A, the EBCM noted the wheel speed was slowing too rapidly and cycled the control valve to hold the brake pressure steady. At points B the brake pressure was lowered, and at points C the pressure was allowed to increase. *(Courtesy of Nissan North America, Inc.)*

Normal outputs controlled by the EBCM are the solenoid valves in the modulator, the amber ABS warning light, and the pump motor. The normal inputs are the wheel speed sensors and, in some systems, pump pressure, fluid level, stoplight switch, and a brake pedal travel sensor (Figure 8–20).

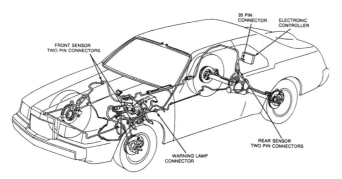

Figure 8–18 This car has the EBCM mounted behind the rear seat, and it is connected to the wiring harness by a 35-pin connector. *(Courtesy of Ford Motor Company)*

8.4 Electronic Modulator

The **modulator,** also called the *hydraulic control unit* or *actuator,* is the device that cycles or pumps the brakes. During normal stops, the modulator does not change normal brake operation. During very hard stops, the brake pressure increases in the brake assemblies in a normal manner. If a wheel begins to lock, the modulator must stop any further hydraulic pressure increase at that particular wheel cylinder or caliper. It can hold that pressure in the brake assemblies. If this action is not enough to get the wheel turning at the correct speed, the modulator must reduce the pressure. As soon as the wheel is turning at the correct speed, the modulator must let the pressure in the wheel cylinder or caliper increase again. If reapplication of the brake causes a lockup, the modulator must go through the cycle over again. Most systems can cycle the brake pressure about five to fifteen times a second.

The early domestic units used individual modulators that were mounted between the master cylinder and the wheel cylinder or caliper and connected to the controller

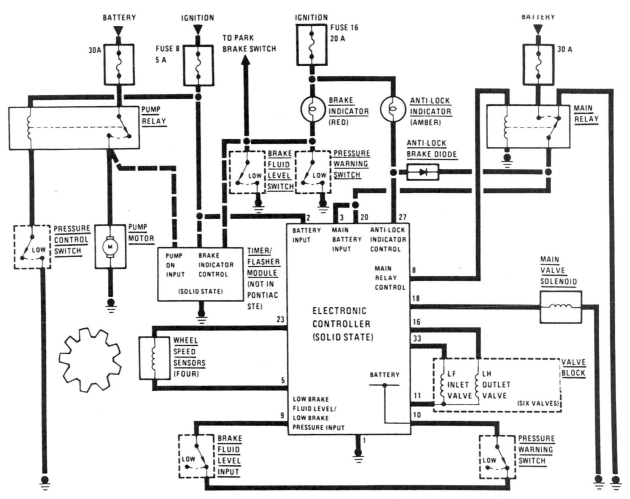

Figure 8–19 An electrical ABS schematic showing the EBCM and most of the connections to it. *(Courtesy of General Motors Corporation, Service Technology Group)*

Chapter 8 ■ Antilock Braking System Theory

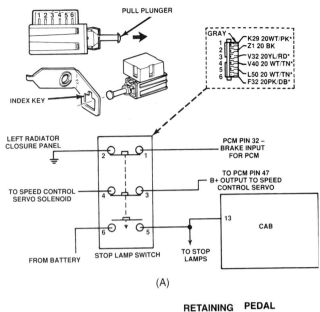

(A)

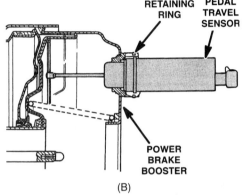

(B)

Figure 8–20 This stoplight switch also provides signals to the power control module (PCM), speed control servo, and controller antilock brake (CAB) to let them know the brakes are applied (A). The pedal travel sensor (PTS) provides a signal so the CAB knows how far the brake pedal has traveled (B). *(Courtesy of Chrysler Corporation)*

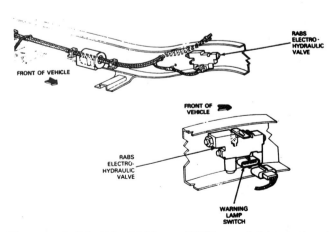

Figure 8–21 This RABS or RWAL modulator valve is mounted inside the frame rail. *(Courtesy of Ford Motor Company)*

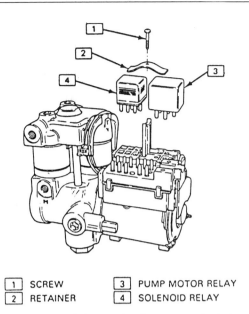

| 1 | SCREW | 3 | PUMP MOTOR RELAY |
| 2 | RETAINER | 4 | SOLENOID RELAY |

Figure 8–22 A modulator valve from a Bosch-designed system. It has three solenoids that control the three-circuit system. *(Courtesy of General Motors Corporation, Service Technology Group)*

and a source of power. Modern rear-wheel-only systems use a single control valve assembly that is operated electronically (Figure 8–21). Early Bosch-designed units have all the modulator valves mounted in a single hydraulic unit that is connected to the master cylinder by two hydraulic brake lines (Figure 8–22). Early Teves-designed systems combine all the valves into a valve block assembly that is mounted directly on the master cylinder. This unit combines a master cylinder, an electrohydraulic booster, and the valve block (Figure 8–23). The Delco ABS-VI system also combines the master cylinder and the hydraulic modulator; this unit uses a vacuum power booster (Figure 8–24).

The actual valves inside a modulator assembly vary greatly. Many valves are balls that are normally open or off the seat and can be closed or pushed onto the seat by a solenoid-operated pin. Others can be pushed onto a seat by a spring or fluid pressure and off the seat by the solenoid-operated pin. Some systems use sliding spool valves to open or close fluid passages. In some cases, a solenoid controls two ball valves, and this solenoid can have either two or three operating positions (Figure 8–25). In most systems, one solenoid controls pressure increase, inlet, or build, and another controls pressure decrease, outlet, or decay. The Bosch ABS 2S uses one three-position solenoid to allow pressure increase, hold, and decrease. The Delco ABS-VI system combines an isolation solenoid that can block pressure increase along with a motor-controlled piston to decrease pressure. Up to ten solenoids and valves are combined into a modulator assembly.

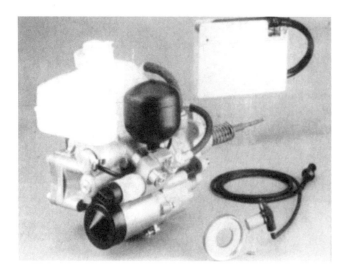

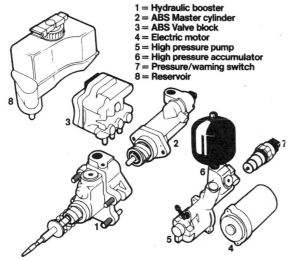

Figure 8–23 The master cylinder–power booster–control valve combination used in an early Teves, ATE-designed systems. *(Courtesy of ITT Automotive)*

1 = Hydraulic booster
2 = ABS Master cylinder
3 = ABS Valve block
4 = Electric motor
5 = High pressure pump
6 = High pressure accumulator
7 = Pressure/warning switch
8 = Reservoir

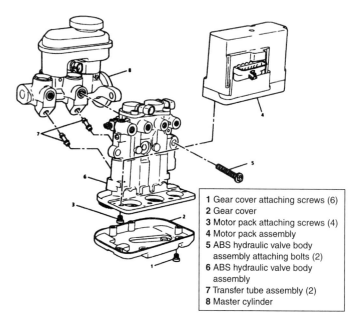

1 Gear cover attaching screws (6)
2 Gear cover
3 Motor pack attaching screws (4)
4 Motor pack assembly
5 ABS hydraulic valve body assembly attaching bolts (2)
6 ABS hydraulic valve body assembly
7 Transfer tube assembly (2)
8 Master cylinder

Figure 8–24 This Delco (now Delphi) ABS-VI hydraulic modulator assembly is secured to the master cylinder. Note that the gear cover can be removed to service the gear assemblies and that the motor pack can also be serviced. A special operation must be performed before beginning any service work. *(Courtesy of General Motors Corporation, Service Technology Group)*

Early modulator valves use a design in which a check valve is normally held open by the end of a piston rod. When closed, this valve allows flow in only one direction—from the wheel cylinder or caliper to the master cylinder. If lockup occurs, the piston rod moves back, and the check valve closes to limit any further increase in brake pressure. Further movement of the piston rod lowers the pressure in the modulator and the wheel cylinder or caliper. Movement of the piston rod, on these early systems, is accomplished by manifold vacuum, hydraulic pressure (power steering pump), or an electric solenoid. When vacuum or hydraulic pressure is used, the vacuum or hydraulic flow is controlled by an electric solenoid. In all cases, the control solenoids are actuated by the EBCM.

8.5 Recirculation Pump

In some systems during an ABS stop, the valve action to hold and release a brake takes fluid from the system, which causes the brake pedal to lower, possibly to the floor. This in turn can cause the circuit to run out of brake pressure. To prevent this problem from occurring, pumps are used to recirculate or return the fluid from the pressure-reducing or decay valves back into the fluid lines (Figure 8–26). The pump has to be able to develop pressures equal to that in the brake system. The driver should feel a bump or bumps from the brake pedal when this occurs. Some systems return this fluid to the master cylinder reservoir.

Most systems use a single- or double-piston pump that is controlled by the EBCM receiving a signal from the brake pedal travel switch or accumulator pressure switch. Some systems store the fluid from the brake valves in an accumulator before starting the pump (Figure 8–27).

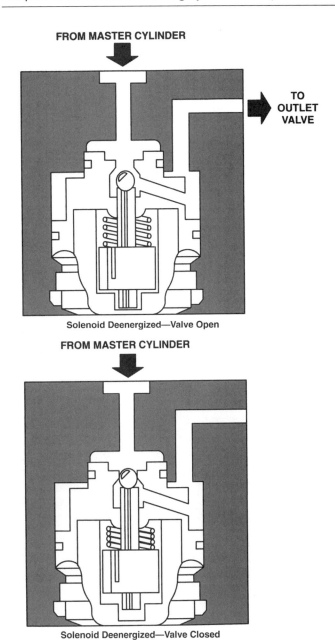

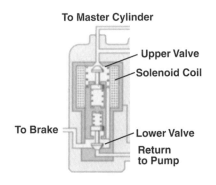

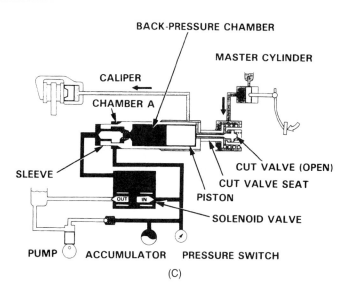

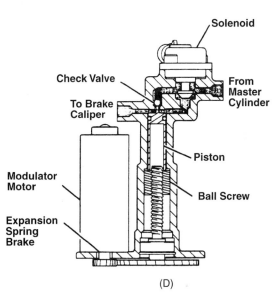

Figure 8–25 A variety of modulator valve types. The inlet–isolation valve (A) is normally open and is closed when the solenoid moves the ball onto the seat. The upper (inlet) valve is normally open and the lower (outlet) valve is normally closed (B); both are closed when the solenoid moves upward to midposition, and the outlet valve opens when the solenoid moves to the top position. When the inlet valve is open, the piston moves to the right to open the cut valve, allowing brake pressure to flow to the caliper (C); closing the IN valve and opening the OUT valve cause the piston to move to the left, closing the cut valve and causing a pressure drop in the caliper. The check and solenoid valves are normally open (D); activating the solenoid closes that valve, and lowering the piston closes the check valve and reduces the pressure in the caliper. (A is courtesy of Chrysler Corporation; B is courtesy of Robert Bosch Corp.; C is courtesy of American Honda Motor Co., Inc.; D is courtesy of General Motors Corporation, Service Technology Group)

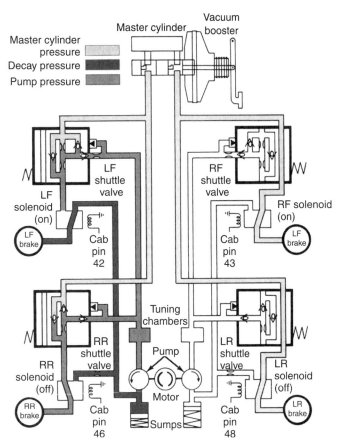

Figure 8–26 This ABS has activated the left front solenoid to move the shuttle valve to reduce (decay) the pressure in the caliper. It has also started the pump to return the fluid to the brake lines for left front and right rear brakes. *(Courtesy of Chrysler Corporation)*

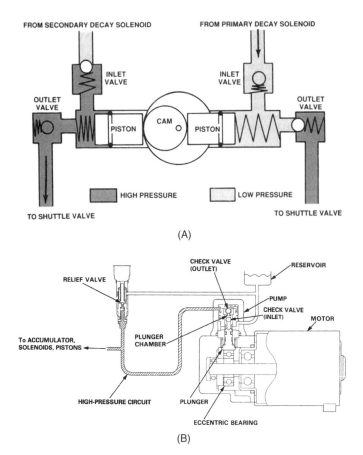

Figure 8–27 This twin-piston pump isolates the primary and secondary circuits (A); the two check valves control the flow in and out of the pump chambers. The pump plunger is operated by an eccentric bearing mounted on the motor shaft (B). *(A is courtesy of Chrysler Corporation; B is courtesy of American Honda Motor Co., Inc.)*

Recirculator pump pressure is also used in some traction control systems to apply the brake and slow down the spinning wheel.

8.6 Antilock Braking Systems

ABS modulator designs vary widely. These systems are manufactured by several domestic (U.S.), European, and Japanese companies, as well as some vehicle manufacturers. Common manufacturers include Bendix, Bosch, Delco Moraine (now Delphi), Kelsey-Hayes, and Teves (now ITT Automotive) (Chart 8–1).

Early Teves systems used electrohydraulic boosters in which the booster pressure also became the pressure source for the rear brakes and the modulator valves were integral with the master cylinder. Some versions of this design use a different valve arrangement and also use boost pressure to operate the rear brakes (Figure 8–28).

Many modern systems use a master cylinder with a vacuum booster, mount the modulator assembly separately, and connect these together with tubing. Other designs use a separate modulator valve assembly with three (three-circuit system) or four (four-circuit system) pairs of valves (Figure 8–29). The operation of these systems is similar to that just described. During normal braking, fluid flows unchanged through the modulator to the brake assemblies (Figure 8–30). If a wheel starts to lock up, the inlet (or isolation) modulator valve closes to prevent any further increase in brake pressure. If the wheel stays locked up, the outlet or decay modulator valve opens to release pressure. Some systems include an accumulator to store this released fluid and a pump to return the fluid to the master cylinder reservoir or back into the pressure lines.

Newer modulator designs use a variety of control methods. Several system designs use one or a pair of valves for each hydraulic circuit, and each valve is operated by an

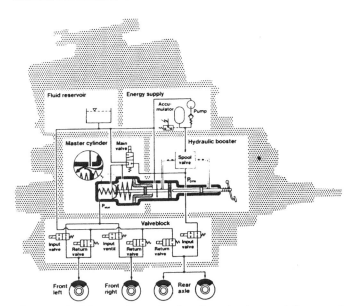

Figure 8-28 A schematic showing the internal flow and the arrangement of the valves in an ABS master cylinder–power booster–control valve assembly. *(Courtesy of ITT Automotive)*

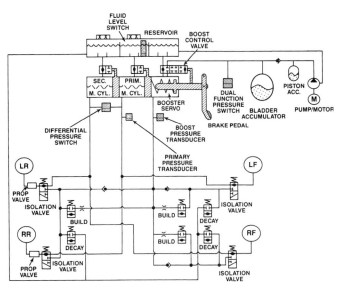

Figure 8-29 This three-circuit system is a closed system and uses a hydraulic pump and accumulator for boost pressure. There are ten valves in the modulator portion that are positioned for normal operation. *(Courtesy of Chrysler Corporation)*

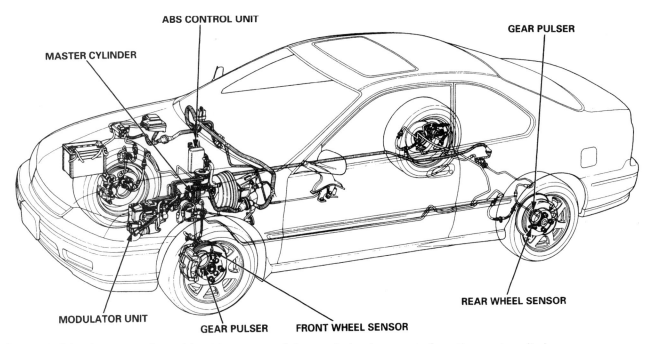

Figure 8-30 Like many others, this ABS uses a modulator unit that is separate from the master cylinder and vacuum booster. *(Courtesy of American Honda Motor Co., Inc.)*

electric solenoid. One valve, the inlet, is normally open and normally allows flow between the master cylinder and the wheel cylinder or caliper (Figure 8-31). The other valve, the outlet, is normally closed. When it opens, fluid can flow from the wheel cylinder or caliper to the master cylinder reservoir. If wheel lockup occurs, the EBCM closes the inlet valve to prevent any further increase in brake application pressure. If the wheel remains locked, the outlet valve is opened to reduce the braking pressure (Figure 8-32).

The RWAL system uses a single modulator, called the dual solenoid control valve, rear antilock braking system (RABS) valve, and isolation–dump valve, for the rear brakes. This valve is mounted next to the master cylinder or in a remote location closer to the rear.

Chart 8–1 Common Manufacturers of ABS.

ABS Manufacturer and System	Period of Usage	Integral or Nonintegral	Vehicle Make	Car, Light Truck (LT), or Utility Vehicle (UV)	Booster Type	Number of Channels	Fluid Return Pump	Diagnostic Trouble Codes	Traction Control Option
Bendix ABX-4	1995–	Nonintegral	Chrysler	Car	Vacuum	4	Yes	16	
Bendix 4	1993–1995	Integral	Jeep	Car	Electrohydraulic	3	Yes	14	
Bendix 6	1991–1993	Nonintegral	Chrysler	UV	Vacuum	4	Yes	16	
Bendix 9	1989–1991	Integral	Jeep	UV	Electrohydraulic	3	No	20	
Bendix 10	1990–1993	Integral	Chrysler	Car	Electrohydraulic	4	No	Yes	
Bendix Mecatronic II	1995–	Nonintegral	Ford	Car	Vacuum	4	Yes	Yes	Yes
Bosch 2	1981–1990	Nonintegral	SEE NEXT LINE		Vacuum	3	Yes	BC	Yes
(includes 2E/2S/2U)			Car models include Audi, BMW, Corvette, Ford, G.M., Lexus, Mazda, Mercedes, Mitsubishi, Nissan, Porsche, Rolls Royce, Sterling, Subaru, and Toyota; *utility vehicle models include Isuzu and Suzuki*						
Bosch III	1987–1992	Integral	Chrysler	Car	Electrohydraulic	4	No	16	
			G.M.	Car					
Bosch 5	1995–	Nonintegral	G.M.	Car	Vacuum	4	Yes	Yes	Yes
(includes ABS/ASR)			Ford	Car					
Bosch 5.3	1997–	Nonintegral	G.M.	Car	Vacuum	4	Yes	Yes	Yes
			Toyota	Car					
			Subaru	Car					
Bosch VDC	1996	Nonintegral	Mercedes	Car	Vacuum	4	Yes	Yes	Yes
Delco Moraine III	1989–1991	Integral	G.M.	Car	Electrohydraulic	3	No	63	
Delco Moraine VI	1991–	Nonintegral	G.M.	Car	Vaccum	3	No	79	Yes
(now Delphi)			Geo	Car					
			Saturn						
Delphi DBC7	1998–	Nonintegral	G.M.	Car	Vacuum	4	Yes	Yes	Yes
Honda	1988–	Nonintegral	Honda/Acura	Car	Vacuum	3	Yes	Yes	
Kelsey-Hayes RWAL	1987–	Nonintegral	SEE NEXT LINE		Vacuum	1	No	17	
(includes RABS and EBC2)			*Light truck and utility vehicle models include Dodge, Ford, Geo, G.M., Isuzu, Mazda, and Nissan*						
Kelsey-Hayes 4WAL	1988–	Nonintegral	Dodge	LT	Vacuum	3	Yes	41	
(includes versions EBC5, EBC10,			G.M.	LT/UV					
EBC325, and EBC430)			Isuzu	UV					
			Kia	UV					
Nippondenso	1990–	Nonintegral	Infiniti	Car	Vacuum	4	Yes	Yes	
			Lexus	Car					

ABS Manufacturer and System	Period of Usage	Integral or Nonintegral	Vehicle Make	Car, Light Truck, or Utility Vehicle	Booster Type	Number of Channels	Diagnostic Trouble Codes	Traction Control Option
Sumitono	1987–	Nonintegral	Ford Probe, Honda, Mazda	Car	Vacuum	3/4	Yes	Yes
Teves Mark II	1985–1991	Integral	Ford, G.M., Saab, Volkswagen	Car	Electrohydraulic	4	42	No
Teves Mark IV (now ITT Teves)	1990–	Nonintegral	Chrysler, Ford, G.M.	Car	Vacuum	4	29	Yes
ITT Teves Mark 20	1997–	Nonintegral	BMW, Chrysler, Ford, Honda	Car	Vacuum	3/4	Yes	Yes

Note: This chart is printed to give you an idea of which vehicles use the major types of ABS. When servicing a particular system, it is very important to use printed or computerized service information for that particular vehicle. A vehicle manufacturer often uses one ABS type for one vehicle model and other versions from the same system manufacturer or from a different manufacturer for different models. Most of the newest ABS types include a traction control system (TCS) as an extra-cost option; this option adds valves to the hydraulic modulator and possibly an additional channel. Also, some vehicle manufacturers use a system that they have designed and manufactured or a system designed by one of the major ABS manufacturers and manufactured under license of that company. When seeking service information, always use the vehicle make, model, and year and then the ABS type.

ABS Manufacturer and System: indicates the basic design of the system; many systems have one or more versions; Robert Bosch Corporation now owns the ABS manufacturing assets of Bendix

Period of Usage: not necessarily the same for each vehicle manufacturer

Integral or Nonintegral: self-explanatory

Vehicle Make: can include any vehicle made by the manufacturer

Car, Light Truck, or Utility Vehicle: self-explanatory

Booster Type: most systems use either a vacuum or electrohydraulic booster

Number of Channels: most systems have either three or four braking channels; one system has one

Fluid Return Pump: this pump returns fluid from the hydraulic modulator to the master cylinder

Diagnostic Trouble Codes: indicates if diagnostic trouble codes (DTC) are available and in most cases how many codes are used; BC indicates blink or flash codes at the malfunction indicator light (MIL)

Traction Control Option: indicates whether the system can include traction control

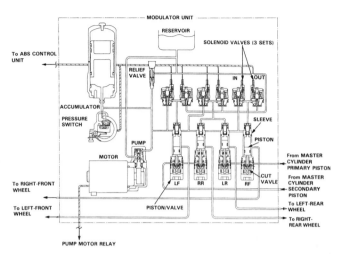

Figure 8-31 This three-circuit system uses a cut valve for each brake unit, but both rear cut valves are controlled by the same pair of inlet and outlet valves. These cut valves block pressure from the master cylinder to the brake circuits. *(Courtesy of American Honda Motor Co., Inc.)*

This assembly contains two solenoid valves and an accumulator. During an ABS-assisted stop, the isolation solenoid closes the valve and stops any further pressure increase (Figure 8-33). The other solenoid can open the dump valve if needed, allowing rear pressure to dump into the accumulator, thereby reducing rear-brake pressure. Any fluid that enters the accumulator returns to the master cylinder when the brakes are released. A potential problem with this design is that with some vehicles, the brake pedal lowers during each modulator cycle. Approximately eight cycles can bring the pedal to the floor, and then the pedal must be pumped to restore braking action.

In some Bosch units, a pump is used instead of an outlet valve. During the pressure-reduction phase, fluid is pumped from the brake cylinder side of the valve back to the master cylinder side of the valve or master cylinder body (Figure 8-34). When the wheel is turning again, the valves are returned to their normal positions.

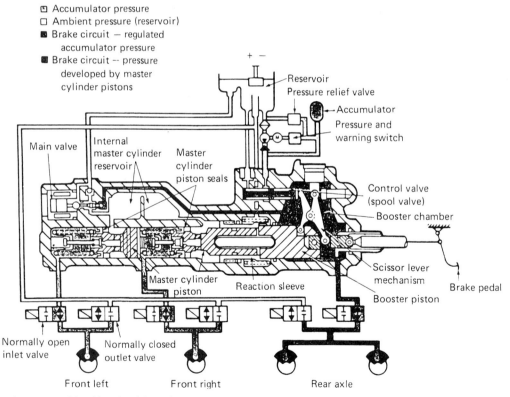

Figure 8-32 During normal braking in this early Teves system, the inlet valves are open and the outlet valves are closed. During the pressure-holding stage, the inlet valves close, and during the pressure-reducing stage, the outlet valves open. *(Courtesy of General Motors Corporation, Service Technology Group)*

Chapter 8 ■ Antilock Braking System Theory

The Delco/Delphi ABS-VI hydraulic modulator assembly contains two solenoids and three electric motor–driven pistons. Each front brake circuit uses a single piston and motor; the rear brake circuits use a pair of pistons driven by a single motor. Each piston drive mechanism contains a braking mechanism to hold the piston in a fixed position unless driven by the motor. During normal brake operation, the pistons are kept in their uppermost, "home" position, where an extension of

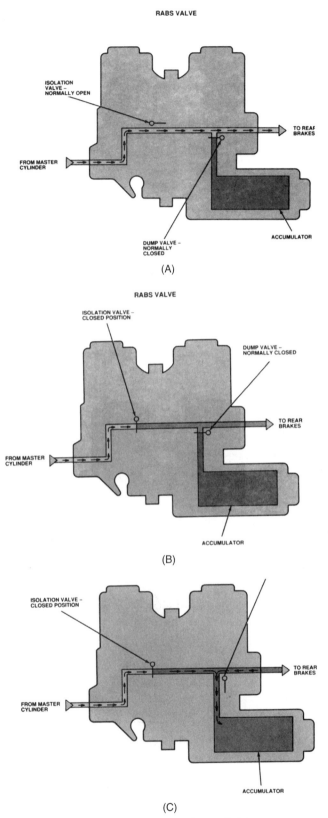

Figure 8–33 During normal braking of this RWAL system, the isolation valve is open and the dump valve is closed so there is a fluid flow between the master cylinder and the rear brakes (A). During the pressure-holding stage, the isolation valve closes (B). During the pressure-reducing stage, the dump valve opens to bleed pressure to the accumulator (C). *(Courtesy of Ford Motor Company)*

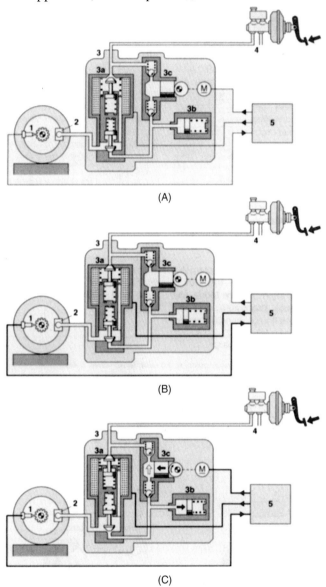

Figure 8–34 During normal braking in this Bosch system (A), braking pressure passes through the control valves. During the pressure-holding stage (B), the solenoid valve is stroked to midposition. During the pressure-reducing stage (C), the solenoid valve is stroked to full up, opening the outlet, and the pump is actuated to return the fluid. *(Courtesy of Robert Bosch Corp.)*

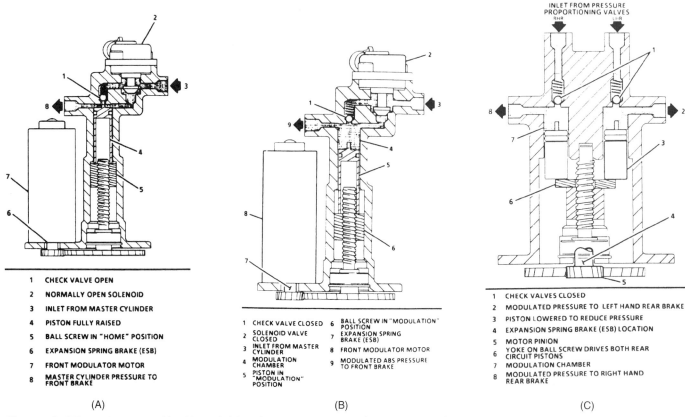

Figure 8–35 During normal braking of this Delco ABS-VI system, the piston is in the uppermost position and the solenoid valve is open (A). During the pressure-holding and pressure-reducing stages, the solenoid is actuated to isolate the circuit and the piston is lowered to reduce the pressure as needed (B). A piston pair is used to control both rear brakes (C). *(Courtesy of General Motors Corporation, Service Technology Group)*

the piston holds a check valve open (Figure 8–35). During an ABS-assisted stop, one or more of the electric motors is driven to move the piston(s) downward. If the circuit being controlled is a front brake, one or both of the solenoids are also closed. As a piston is driven downward, the check valve is closed to isolate the brake circuit from the master cylinder. From this point, brake circuit pressure is controlled by the piston(s); the lower the piston position, the lower the pressure. The motor(s) are operated by the EBCM to lower or raise the pressure and obtain the maximum braking pressure without wheel lockup. The solenoids in the front brake circuits provide a redundant circuit so that braking will still be possible if the ABS fails with the pistons lowered.

8.7 ABS Brake Fluid

ABS uses either DOT 3 or DOT 4 fluid; you should use the fluid recommended by the manufacturer. Silicone (DOT 5) fluid should not be used. Air does not move easily through silicone fluid, and when you shock the fluid, any air bubbles tend to become smaller bubbles. During an ABS-controlled stop, the rapid action of the solenoid valves shocks the fluid.

8.8 Automatic Traction Control

Wheel spin can occur while a driver is trying to accelerate, much like the wheel lockup that can occur during braking. Traction, like braking, can be adversely affected by road conditions and weather. Traction control, also called traction control system (TCS), antispin regulation, or automatic slip regulation (ASR), is a system that senses wheel spin and limits the amount of wheel spin that can occur (Figure 8–36).

Wheel spin occurs when the amount of torque at a drive wheel exceeds the available traction; if a wheel spins more than 20 percent over a normal speed, traction

Chapter 8 ■ Antilock Braking System Theory

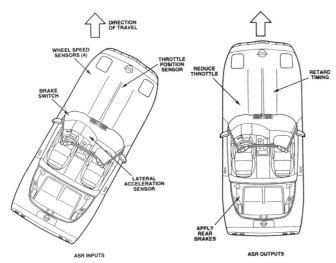

Figure 8–36 This acceleration slip regulation (ASR) uses several sensors to determine if wheel spin is occurring (left) and several methods to control wheel slip (right). *(Courtesy of General Motors Corporation, Service Technology Group)*

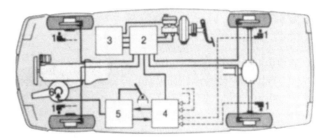

Figure 8–37 This TCS system shares the wheel speed sensors (1), hydraulic modulator (2), and control unit (4) with the ABS. Added are a hydraulic modulator that can apply the rear brakes (3), a control unit (5), and an engine throttle control (6). Some systems also include an engine timing or fuel-injector control. *(Courtesy of Robert Bosch Corp.)*

will drop to a lower amount. Because of the differential in the drive axle, the spinning wheel will reduce the amount of torque available to the other drive wheel that is not spinning. Most differentials split torque so an equal amount goes to each drive wheel. Wheel spin causes a slower acceleration because of the reduced traction and possible sideslip and spinout if it occurs while on a curve. Wheel spin also causes unnecessary and excessive tire and differential wear.

TCS shares the ABS wheel speed sensors and microprocessor. The microprocessor compares the speeds of the two drive wheels with each other and with the non–drive wheels. Excessive speed at a drive wheel indicates wheel spin. Depending on the particular system, wheel spin is controlled by one or more methods: applying the brakes on the drive wheel that is spinning (on a RWD car) or reducing engine torque by retarding the timing, closing the throttle, or shutting off one or more fuel injectors (Figure 8–37).

8.9 Practice Diagnosis

You are working in a new-car dealership and encounter the following problems:

CASE 1: You are prepping a new car for delivery to the customer, and you notice the car has ABS. The ABS warning light operates as you start the engine and goes out a few seconds after start-up. But as you start on your road test, the light comes back on. What is this light indicating? Is this a normal or abnormal condition? If abnormal, what is probably wrong with the car?

CASE 2: One of your top-of-the-line cars that was sold about 11 months ago has come back with a complaint of a low brake pedal. Along with most of the possible accessories, this car has a big engine, automatic transmission, four-wheel disc brakes, and ABS. Is this an ABS problem? What could be causing the low pedal? What should you do next?

Terms to Know

antilock braking system (ABS)
electronically controlled modulator valves
electronic control module
electronic wheel speed sensors
modulator
modulator valves
sensor
wheel speed sensor

Review Questions

1. Antilock braking systems are designed:
   a. With more powerful brakes to provide quicker stops
   b. To prevent wheel lockup during stops
   c. To provide a warning system to inform the driver that a skid is occurring
   d. All of the above

2. Statement A: Antilock brake systems measure wheel speed using electronic sensors.
   Statement B: Rear-wheel speed sensors can be mounted in the rear axle housing.
   Which statement is correct?
   a. A only
   b. B only
   c. Both A and B
   d. Neither A nor B

3. Statement A: If wheel lockup occurs, an ABS cycles the braking pressure in all the wheel brake assemblies.
   Statement B: A control valve is used to allow pressure buildup, stop pressure increase, or reduce pressure in the wheel brake assemblies.
   Which statement is correct?
   a. A only
   b. B only
   c. Both A and B
   d. Neither A nor B

4. An antilock braking system can use:
   a. One braking control circuit
   b. Two braking control circuits
   c. Three braking control circuits
   d. Any of the above

5. Statement A: The brake warning light for cars with ABS is changed from red to an amber color.
   Statement B: If an electrical failure occurs in the ABS controls, the system will revert to a standard brake system and turn on an amber warning light.
   Which statement is correct?
   a. A only
   b. B only
   c. Both A and B
   d. Neither A nor B

6. As a wheel rotates, the wheel speed sensor sends out an electrical signal with a:
   a. Frequency that varies with the speed of the wheel
   b. Voltage that decreases as the wheel speed changes
   c. Current that increases as the wheel speed changes
   d. All of the above

7. The electronic brake control module:
   a. Monitors the speed signals from the wheel speed sensors
   b. Operates the hydraulic control unit(s)
   c. Checks the electrical circuits constantly to make sure they are operating correctly
   d. All of the above

8. Statement A: If the speed of one wheel drops faster than that of the others, an ABS will operate to cycle that brake.
   Statement B: If the speed of a wheel drops faster than that in the profile programmed into the EBCM, an ABS will operate to cycle that brake.
   Which statement is correct?
   a. A only
   b. B only
   c. Both A and B
   d. Neither A nor B

9. Statement A: The first stage of ABS control is to stop an increase in the hydraulic pressure in a caliper or wheel cylinder.
   Statement B: The second stage of ABS control is to reduce the hydraulic pressure in the brake assembly.
   Which statement is correct?
   a. A only
   b. B only
   c. Both A and B
   d. Neither A nor B

10. Two technicians are discussing TCS.
    Technician A says that a TCS can reduce engine power if wheel spin occurs.
    Technician B says that you can tell if a TCS is working if the accelerator pedal pulses during wide-open throttle acceleration on wet roads.
    Who is correct?
    a. A only
    b. B only
    c. Both A and B
    d. Neither A nor B

Tire Theory

Objectives

Upon completion and review of this chapter, you should be able to:

- ❏ Comprehend the terms commonly used to describe tire construction, sizing, and operation.
- ❏ Have a basic understanding of how a tire is constructed.
- ❏ Understand the current tire size designations.
- ❏ Select a correct replacement tire for a car.
- ❏ Have a basic understanding of the various types of tires used for spares and the operating and service requirements of each tire type.
- ❏ Know what a recap tire is and its benefits.

9.1 Introduction

Almost everyone knows that tires are fitted on wheels, and the tires and wheels roll down the road. Most drivers also know that the tread of the tire and its grip on the road provide traction so the car can accelerate, turn corners, and stop. Some drivers have been made painfully aware that it is extremely difficult to control a car when there is not enough traction. The car can skid and cause an accident.

Most of the discussion in this chapter concentrates on passenger-car tires. Much of it also applies to truck, heavy-duty off-road, farm, and industrial tires, but each tire type differs slightly in its own way.

Traction refers to the amount of grip between the tire and the road. It can be affected by anything on the road surface including ice, water, sand, leaves, and so forth. Traction can also be affected by the following tire conditions: the depth of the tread, the pattern of the tread, the hardness or softness of the tread rubber, the inflation pressure, the width, the load on the tire, the alignment, and the temperature of the tire and the road surface. One vehicle-handling expert, who specializes in road racing cars, maintains that there are thirty-one factors that affect tire-to-road traction. Depending on the tire tread and road surface, a tire will reach maximum traction at a slip rate of about 10 to 25 percent of the vehicle speed. Greater than 25 percent slippage usually results in a skid and, with the skid, a severe loss in traction. Besides the traction loss, a skidding tire also loses its directional control. A rolling tire usually travels in the direction that it is pointed. A skidding or sliding tire slides sideways almost as easily as it moves in a forward direction. A tire that is spinning at a rate greater than 25 percent of the vehicle speed loses traction and directional control in the same way as a skidding tire.

The tires also serve as springs. Small bumps are absorbed as the tire sidewall flexes. We all realize that when a load is put on a tire, the tire and tread flatten at the road surface. This movement produces a bulge in the tire's sidewall. The ability to absorb a bump varies depending on tire construction and pressure. A nonbelted tire usually has a softer and smoother ride than a belted tire; a fabric-belted tire usually rides smoother than a steel-belted tire. A tire inflated to 28 psi (193 kPa) usually has a smoother ride than one that is inflated to 34 psi (234.5 kPa). When a car is designed, differences in tire ride quality can be compensated for by changing shock absorber settings. They can also be compensated for by changing the rubber isolation bushings between the suspension and the tires or between the suspension and the frame.

A tire designer has to make numerous conflicting choices. A belted tire with a rigid tread might produce better fuel mileage, but the ride quality and traction would suffer. A softer tread rubber compound might improve traction, but tire life and possibly gas mileage would suffer. A lower, wider tire might look better and offer better dry-pavement traction, but vehicle ground clearance would be reduced, axle revolutions per mile would increase, and the tires would tend to hydroplane in wet conditions. Tire design is definitely a compromise and has led to many special tires for specific purposes or conditions.

9.2 Tire Construction

Today's passenger-car tire begins as several layers of rubber of various mixtures or compounds, some cording, and two wire rings. A rubber compound can include from eight to fifteen different raw materials. A few of the reasons for using different rubber blends include to change the tensile strength; to increase the resistance to age, light, or different chemicals; to increase the resistance to abrasion; and to change the adhesion tendencies between the rubber and various cord materials (Figure 9–1).

The cording for the **plies** is formed from two or more strands of various fabrics. The cords can be twisted, braided, or woven together. The wire rings, called the **bead wires,** reinforce the bead portion. This part of the tire must stay firmly fixed in place on the wheel. The bead wires help transmit starting and stopping forces between the tire tread and the wheel and resist the centrifugal force that is trying to separate the tire from the wheel. The **sidewall plies,** also called **body** or **carcass plies,** reinforce the body of the tire by wrapping around and attaching to the bead wires (Figure 9–2). These plies form a strong envelope to hold the inner liner and the air inside of the tire. The inner edge of the bead, where the plies wrap around the bead wires, is angled slightly so as to wedge the bead tightly onto the angled bead seat of the wheel. This ensures that the tire fits tightly and is centered on the wheel when it is inflated (Figure 9–3).

9.2.1 Ply Materials

Sidewall plies are commonly made from several different materials. **Cotton,** a natural fabric, was the first commonly used ply material in the early days of tire building, prior to World War II. Cotton plies were relatively weak and tended to deteriorate easily when exposed to moisture. Today the most commonly used materials for sidewall plies are **polyester, nylon,** and **rayon.** Each of these synthetic, man-made materials offers different advantages.

Rayon is fairly strong, resists fatigue, does not flat spot, and has good fabric-to-rubber adhesion characteristics. Of the three synthetic materials, rayon has the least

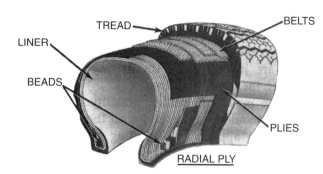

Figure 9–1 A sectioned view showing a tire's internal construction. Note that the body plies wrap around the bead bundle and that the belt plies reinforce the body plies in the tread area. *(Courtesy of Chrysler Corporation)*

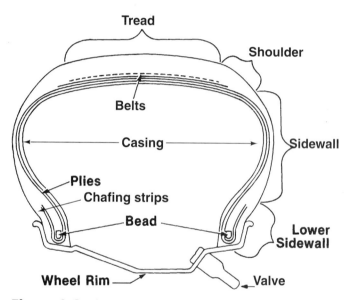

Figure 9–2 This cross section of a tire shows the major internal parts. Note how the sidewall plies wrap around the bead wires.

tendency to change length because of temperature change. Nylon is a strong and very durable fabric that is relatively inexpensive. Because of its strength and its ability to dissipate heat, nylon is the most commonly used ply cord material in truck tires. When warm, nylon cords are quite flexible and tend to shrink or shorten. Early nylon cord tires tended to flat spot as they cooled; that is, as the bottom portion of the cord cooled, it took the shape of the flat road surface. This tread portion tended to stay flat until it warmed up again. Polyester is a thermoplastic material that flexes easily, is strong and lightweight, and resists flat spotting. Polyester is the least expensive of the three synthetic fibers. Polyester cords tend to extend or grow longer with heat; they also tend to lose strength as they get hotter. A comparison of the various body ply and belt materials is shown in Table 9–1.

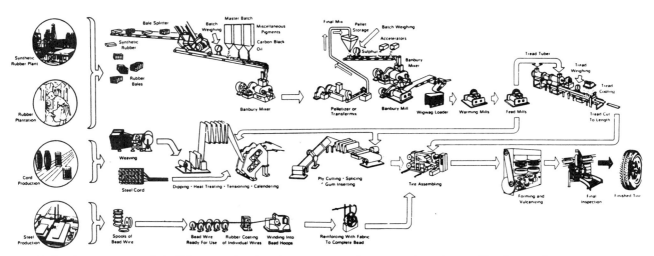

Figure 9–3 The process of building a tire involves many rather complex steps. *(Courtesy of Bridgestone/Firestone, Inc.)*

Table 9–1 Comparison of Different Cord Materials Used in the Sidewall and Belt Plies.

Cord Material	Relative Strength[a]	Where Used	Comments
Cotton	2.3	Not used	Weak; rots and mildews; poor rubber adhesion; variable quality; not used after World War II; flexible
Steel, stranded	3.4	Belt	Heavy, strong, and hard to cut; strength not affected by heat; lengthens when hot; very rigid
Rayon	4.6	Sidewall	Medium strong; resists fatigue
Polyester	8.8	Sidewall	Strong, light, and low in cost; resists flat spots; weakens when hot; flexible
Nylon	9.0	Sidewall	Very strong; shrinks when hot; tends to flat spot; flexible
Fiberglass	11.0	Belt	High strength-to-weight ratio; resistant to rot; temperature stable; fairly rigid
Aramid	17.0	Belt	Expensive; excellent physical properties; semirigid

[a]Relative strength shows the approximate strength in grams per denier. Denier is a measuring system for the fineness of cord or yarn.

The layers of sidewall plies are joined together by thin layers of rubber. A layer of rubber **latex,** sometimes called **gum,** is coated onto the ply cords. This latex bonds the cords together in a ply and also bonds the cord ply to the rubber layer right next to that ply. Each different type of ply cord requires a latex mixture to suit that particular cord material. There are usually several different types of bonding latex in a tire, one for each type of cord. Tight bonding of the materials in a tire is important to give the tire strength and to help dissipate heat. Poor ply bonding causes tire separation.

Another layer of rubber forms an airtight inner liner in tubeless tires. This inner liner is made from butyl rubber, which is airtight. Still another strip of rubber, called a filler, fills the area where the plies wrap around the bead wires. This strip of rubber can change the rate or stiffness of sidewall flexing.

The outside layer of rubber on the sidewall protects the plies from corrosion and weather damage, as well as abrasion and curb damage. Sidewall rubber must be highly resilient for good flexibility. Natural rubber is often used in radial tire sidewalls because of its excellent flexibility. The sidewall rubber also provides decoration in the form of whitewalls or special lettering. The white oxide rubber used for this purpose has a tendency to crack and bond weakly to other types of rubber. This is the major reason why white rubber is used very little on high-performance tires. The sidewall is also the location for any lettering that provides information about the tire.

The tread area is usually composed of two different rubber compounds. The combination of the inner and outer layers is often called **dual compounding.** The inner layer, next to the plies, is called a **cap base** or **under tread.** It is compounded for good heat dissipation. The

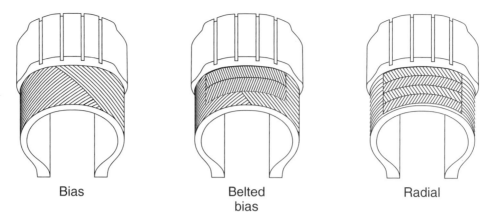

Figure 9-4 Three tire sections showing three different styles of body and belt ply construction. A bias or diagonal ply is at the left, a radial ply is in the center, and a bias-belted ply is at the right. Note the different angles of the body plies. *(Courtesy of General Motors Corporation, Service Technology Group)*

outer rubber layer is bonded to the cap base. The tread pattern is molded into this tough, wear-resistant layer of rubber.

9.2.2 Ply Arrangement

At one time, the sidewall plies were always used in pairs, with the two plies of a pair overlapping as they crossed the tire at different angles. This angle is the reason for calling these tires **bias ply.** The cords of one ply cross the cords of the second ply at an angle of about 30 to 40 degrees (Figure 9–4). Because of the overlapping of the cord plies, bias-ply construction is strong. However, it also forms a tire envelope that has trouble flexing in a direction that tires need to flex. The ply laminations form a tire body that cannot change its shape very easily. The bottom area of the tire must flatten to provide tread contact with the road. If it does not, there will be very limited traction and a very firm ride (Figure 9–5).

The cords in a **radial** ply run almost straight across the tire from one bead to the other. When the ply cords run in this direction, it is very easy for them to bend and allow sidewall flex. Tread-to-road contact is good, with little to no tread squirm. Squirm causes a scuffing of the tread rubber because the tread narrows and widens as it comes in contact with the road surface. Because the sidewall can flex easier with less internal friction, a radial tire rolls easier and delivers better fuel economy. The radial sidewall ply direction, however, is vulnerable to penetration and road damage. The sidewall must be protected by a belt under the tread. The belt helps hold the body plies in place, thus resisting the effects of centrifugal force. The belt also reinforces the body of the tire to provide good cornering and straightaway strength. Radial tires tend to be more expensive, partly because

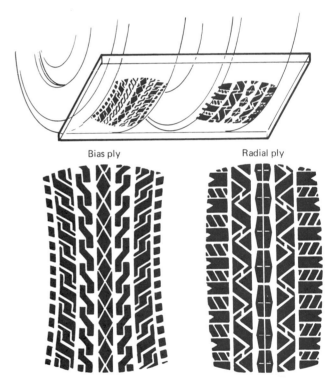

Figure 9-5 As a tire rolls, the tread area must flatten into the contact pattern, or tire footprint. The tread of a bias-ply tire closes up or pinches together at the center, which causes the tread to squirm across the road.

they are more difficult to build and partly because more expensive materials are used in them. Radial tires offer the advantages of better tire and fuel mileage, cooler and quieter operation, better dry- and wet-pavement handling and braking, and better resistance to sideslip caused by crosswinds. The only major drawback, besides cost, is a slightly rougher ride with more vibrations in the 20- to 30-mph (32- to 48-km/h) range.

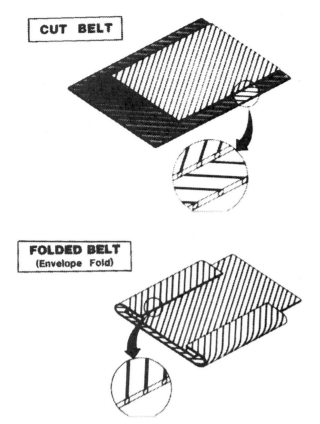

Figure 9–6 Belt plies can be cut or folded. This difference and the ways the belts are arranged in the tire allow the tire designer a variety of ways for changing the operating characteristics of a tire. *(Courtesy of Pirelli Tire North America)*

Figure 9–7 A tire mold or press has the shape and size of a tire on the inside surface. The "green," uncured tire is forced to take the shape of the mold as it cures or vulcanizes.

9.2.3 Belts

The **belt,** often called **tread plies,** is made from cording that is similar to that used in sidewall plies. The usual materials are **aramid** (Kevlar, Dupont's brand name, or Flexten, Goodyear's brand name), fiberglass, rayon, and steel. Several combinations of these materials are often used for different plies on the same tire. A tire's outside diameter can be made more stable by placing a belt of a material that tends to shrink—for example, nylon—over a belt that tends to expand—for example, steel.

The belt runs around the tire in the area just under the tread. For strength, the belt has to bond tightly to both the sidewall plies and the tread rubber. The belt can be cut to width or cut oversized with the edges folded over. The belt can be woven or braided (Figure 9–6). The layers of belt plies can be the same width, called **stacked construction,** or different widths, called **pyramid construction.** The various belt construction methods allow the tire engineer to design different handling, impact resistance, and heat buildup characteristics into the tire. Belts are always used in radial tires. When belts are used in bias tires, the tires are called **bias belted.**

9.2.4 Tread

The tread pattern is formed in the outermost layer of rubber, when the tire is vulcanized or cured in a mold. The inner surface of the mold is the same size and shape that the outside of the tire will become; that is, the tread pattern is determined by the shape of the mold surface. The "green," uncured tire is placed into the mold. Pressure from a bladder inside the tire forces the soft tread and sidewall rubber into the mold's shape while heat vulcanizes the rubber (Figure 9–7). Vulcanizing stiffens the rubber to develop the hardness and elasticity that we are all familiar with in tires. Before vulcanizing, the rubber is quite soft.

The rubber compound for the tread can be formulated to give the desired wear rate, ride quality, traction, weather resistance, or other characteristic desired for a particular tire. For many years, performance tires used a rather soft, good traction rubber blend for the tread; they had good traction but poor tire life. A silica-based compound now being used by tire manufacturers gives excellent traction characteristics along with a long tread life. A slight drawback with silica-based tread rubber is an increased tendency for the car to generate static electricity; more motorists are getting shocked when they touch the door handle or use a key to unlock the car door. The tread pattern is designed to provide the traction characteristics for the intended use of the tire.

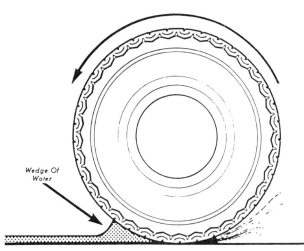

Figure 9–8 Hydroplaning occurs when a wedge of water builds up in front of a tire, causing the tire to ride up and over a film of water. *(Courtesy of B.F. Goodrich. © The B.F. Goodrich Company)*

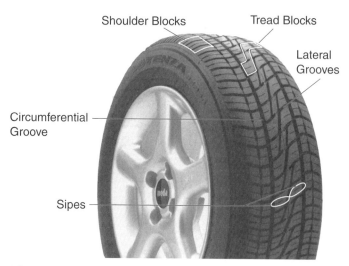

Figure 9–9 The circumferential and lateral grooves form channels to let water escape to resist hydroplaning. Tread blocks and sipes provide traction and cornering grip.

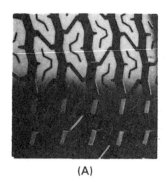
(A)

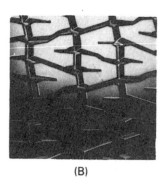
(B)

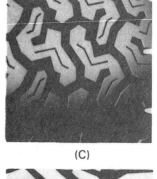
(C)

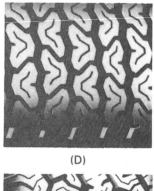
(D)

(E)

(F)

Figure 9–10 A variety of tread patterns are available for tires depending on how the tires are to be used. (A) and (B) are street performance designs with good wet and dry traction characteristics. (C) and (E) are off-road designs. (D) is a street performance, mild off-road design, and (F) is a passenger-car tire with a long life. *(Courtesy of B.F. Goodrich. © The B.F. Goodrich Company)*

The primary reason for the grooves in the tread is to let water run out from between the tire and the road, but tire stability, dry traction, wear rate, and noise are also considered when a particular tread pattern is designed. If water cannot get out from under the tread, the tire might **hydroplane,** plane, or climb up and over the layer of water much like a water-skier does. A hydroplaning tire does not have any traction at all because the tire actually leaves the road surface (Figure 9–8). Factors that increase the tire's tendency to hydroplane are small tread grooves, shallow or worn tread grooves, wider tires, lower tire pressures, and high vehicle speeds. An **all season** tire has wide tread grooves to provide good wet road traction. An all season tire has **M + S, M & S,** or **M–S** imprinted on the sidewall, indicating mud and snow traction ability. A few street tires and many racing tires have a **directional tread** because some rain grooves work better in one direction than the other. Directional tires have an arrow on the sidewall to indicate the correct direction of rotation. Many tractor tires use a directional tread pattern. More

Chapter 9 ■ Tire Theory

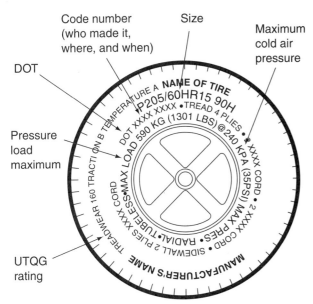

Figure 9–11 The sidewall of a tire contains a great deal of information about the tire. *(Courtesy of Rubber Manufacturers' Association)*

Figure 9–12 A Department of Transportation (DOT) number. In this case it is DOT: JFFH MN6 063. The JF is a code for the manufacturing plant; FH and MN6 tell the manufacturer about the construction and the materials used; and 063 indicates that the tire was built in the sixth week of 1973, 1983, or 1993.

aggressive tread designs or those with wider grooves and larger blocks are used for mud and snow traction. Some of these patterns become very noisy at road speeds (Figure 9–9). There is also a possibility of traction loss while cornering or braking because these rubber blocks can bend or squirm on the road surface when a traction load is put on them (Figure 9–10).

9.3 Tire Sidewall Information

The sidewall of the tire carries most of the information the motorist needs to know about the tire. Letters are molded into the sidewall to give:

- The tire size
- The number of sidewall plies and the material used
- The number of tread plies and the materials used
- The maximum air pressure to which the tire can be inflated and the maximum amount of load the tire can safely carry at that pressure
- The tire brand name
- The DOT number
- The speed rating
- Marking to indicate an all season type of tire
- Uniform tire quality grade labeling
- Information about whether the tire is directional and, if so, which direction it should run (Figure 9–11)

All passenger-car tires must be approved by the Department of Transportation and carry a **DOT number,** as well as the other information listed. The DOT number is coded to indicate the manufacturing plant that made the tire (the first two letters), pertinent information about the tire construction or sizing (the middle letters), and when the tire was made (the last three numbers). The last number is the final digit of the year in which the tire was made, and the two numbers preceding it indicate the week it was made (Figure 9–12). A DOT number ending in 306 indicates the tire was made during the 30th week of 1976, 1986, or 1996. A booklet published by *Tire Guide,* an independent trade publication, can be purchased in order to read the tire maker and plant location portion of the code.

9.4 Tire Sizing

Three different tire dimensions are important to the tire purchaser: wheel size, tire width, and aspect ratio. The wheel size in inches (sometimes millimeters) is indicated by the last two digits of a tire size. The most common passenger-car wheel sizes are 13, 14, and 15 in. (330.2, 355.6, and 381 mm). This size is the diameter of the tire measured between the beads across the tire and the diameter of the wheel measured between the bead flanges. A tire will only fit on a wheel of the same size and bead seat taper. Tires with a metric diameter will only fit metric-sized wheels and vice versa (Figure 9–13).

Numerous methods have been used for measuring tire widths; the systems seem to change periodically

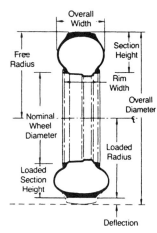

Figure 9–13 Tire and wheel size designations are based on several different measurements. *(Courtesy of Bridgestone/Firestone, Inc.)*

(Figure 9–14). Currently, the **P-metric system** is being used. The *P* designates a passenger-car tire; a *T* would indicate a truck tire. The letter is followed by three numbers that indicate the section width, in millimeters (Figure 9–15). Some of the older systems measured the section width in inches and some measured it in millimeters. The alphanumeric system was based on the load-carrying capability of the tire.

The **aspect ratio** is the height-to-width relationship of the tire's cross section (Figure 9–16). The number used is the ratio of the height to the width, expressed as a percentage. A 70 series tire is seven-tenths, or 70 percent, as high as it is wide; a 50 series tire is twice as wide as it is high. The two numbers following the slash in the P-metric size indicate the aspect ratio. The aspect ratio affects the tire's overall diameter, as well as the loaded or effective radius. This, in turn, affects the circumference and therefore the number of revolutions the tire must turn to cover a certain distance (Figure 9–17). A P205/70R14 tire from one particular manufacturer has an overall diameter of 25.35 in. (644 mm); this tire turns 824 revolutions to go 1 mi. (1.61 km) at 50 mph (80.5 km/h). A P205/60R14 size of the same tire has an overall diameter of 23.7 in. (602 mm); it takes 882 revolutions (58 more than the larger tire) to go the same distance. The tread width is slightly wider, 0.05 in. (1.3 mm), on the 60 series tire.

Next in the P-metric system of tire size marking is a letter to designate the type of sidewall construction: **R** for a radial ply tire, **B** for a bias-belted tire, or **D** for a diagonal–bias ply, nonbelted tire.

A P205/70R15 tire is a P-metric passenger-car tire with a section width of 205 mm (8.2 in.). It has a height that is 70 percent of the section width, or about 143.5 mm (5.7 in.). It is of radial construction and fits a 15-in. (38.1-mm) wheel.

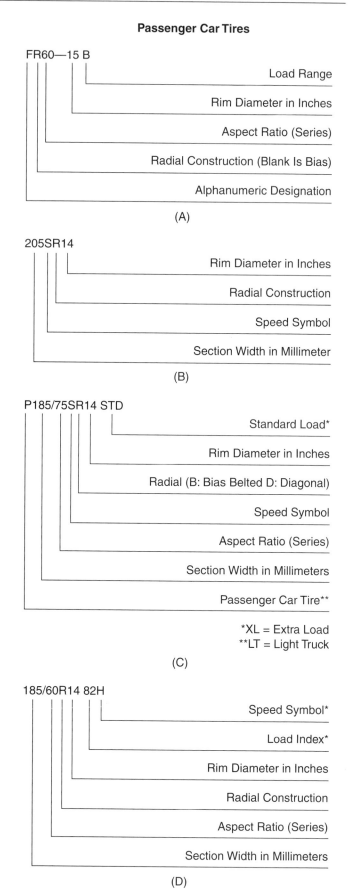

Figure 9–14 Four different tire measuring systems: (A), (B), and (C) are obsolete; (D) is in current use. *(Courtesy of Bridgestone/Firestone, Inc.)*

9.5 Tire Load Ratings

The tire does not support the car; it contains the air that supports the car. Those familiar with hydraulic power transmission know that the force from a hydraulic cylinder equals the hydraulic pressure multiplied by the piston area in the cylinder. Pneumatic power transmission is essentially the same thing except the air is used to transmit the force instead of oil, and that air can be compressed into a smaller space. If pneumatics are applied to the automobile tire, it can be determined that the weight of the car is supported by four pneumatic areas, or the road contact points of the tire. If we multiply these areas, in square inches (square millimeters), by the tire pressure, in pounds per square inch (kilograms per square millimeter), we would get a result that would be very close to the weight of the car in pounds (kilograms). We all know that a tire that is going flat will spread out its tread contact area as it loses pressure. Raising the tire pressure will make the tire more round, which will reduce the tread contact area.

If a tire is not strong enough to contain this air pressure, it will blow out if the failure occurs quickly or simply go flat if the leak is slow. In addition to the normal, stationary vehicle load, the tire has to withstand even greater shock loads. Each time a tire hits a bump or obstruction, the load on the tire increases substantially. Imagine a car traveling 55 mph (88.5 km/h); it will be traveling 80.6 ft (24.6 m) every second. Imagine a bump in the road that is 1 in. (25.4 mm) high. This bump will force the tire tread to rise as it passes over, but the car's weight will try to keep the wheel from lifting. So, for a brief time, the sidewall of the tire will flex as the tire compresses and the pressure in the tire will increase substantially, depending on the amount of compression. Tires are built with extra strength to provide a safety margin for cases much more severe than the example just given. Imagine the pressure in the tires of a car that has just landed on the ground after flying through the air in a movie chase scene. Again, if a tire is not strong enough, it

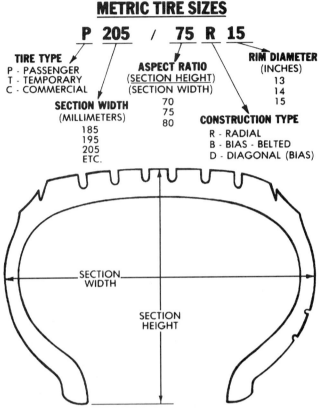

Figure 9–15 The P-metric system for measuring the widths and what each portion of the code represents.

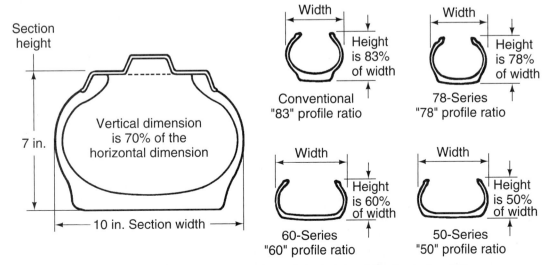

Figure 9–16 The profile or aspect ratio of a tire is a comparison of the height of the tire to its width.

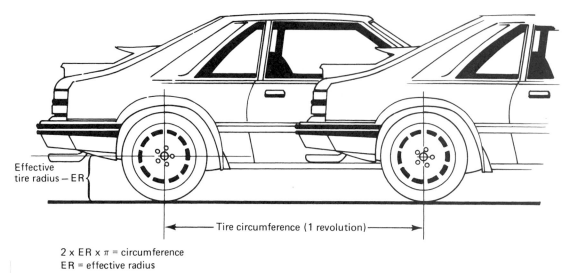

Figure 9-17 A tire will roll down the road a distance equal to its circumference times the number of revolutions. The circumference is equal to the effective or loaded radius times 2 times pi ($C = 2\pi r$).

will blow out. The strength of the tire is indicated by the tire's load rating, which is molded into the sidewall. In this country, it is illegal to drive a vehicle that is carrying a load greater than the load rating of its tires.

P-metric sized tires are available in the following two load ratings: standard load tires, which can contain a maximum of 35 psi (241 kPa), and extra load tires, which can be inflated to 41 psi (283 kPa). This system determines the maximum inflation pressure for the tire. The actual load a tire can carry is determined by the tire size and the load rating. Since we ride on air, a larger tire can carry more load than a smaller tire. Charts are available to help determine how much actual load a particular tire can carry.

At one time, a **load range** system was used to designate a tire's load-carrying ability. Load range was related to ply ratings. Load range A was essentially the same as a two-ply sidewall, load range B equaled a four-ply sidewall, C equaled a six-ply sidewall, and so on. Because plies can be made using different-sized cords, ply numbers and load ratings can vary. The load rating of the tire is still related to the strength of the sidewall plies, and tire pressure is related to tire strength. A cold tire can be inflated to the pressure imprinted on the sidewall; at that pressure, the tire will support the load imprinted on the sidewall. The pressure in a hot tire will increase above the cold pressure, but this increase will do no harm. A tire can be inflated to a lower pressure to reduce ride harshness but not below 80 percent of the rated pressure (e.g., 36 psi $\times$ 0.80 = 29 psi). Watch for tire wear at the lower pressure. The maximum pressure rating for load range A is 28 psi (196 kPa), for load range B it is 32 psi (221 kPa), for load range C it is 36 psi (248 kPa), and so on.

Figure 9-18 This tire has a Uniform Tire Quality Grade Label (UTQGL) of tread wear 220, traction B, and temperature B.

9.6 Uniform Tire Quality Grading Labeling

In the late 1970s, the **National Highway Safety Administration (NHSA)** established the standards for the **Uniform Tire Quality Grading Labeling (UTQGL)** of passenger-car tires. This act established standards in three areas of tire performance: tread wear, traction, and temperature resistance. Every new tire is tested and graded against these standards. The score (or grade) is molded into the sidewall of the tire (Figure 9–18).

The tread wear standard is in a state of flux, with some degree of conflict among the **Rubber Manufacturers' Association (RMA),** the **National Highway Traffic Safety Administration (NHTSA),** some tire manufacturers, and some consumers' groups. The base grade for

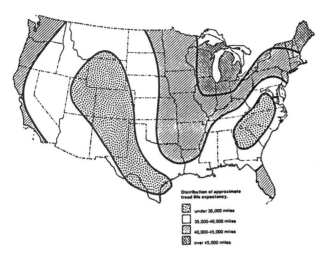

Figure 9-19 Because of different road construction materials and methods and different climatic conditions, motorists in different parts of our country can expect to get different tread life from their tires. *(Courtesy of Uniroyal)*

tread wear is 100. If a tire has twice the tread life as the base tire, it is graded 200 (the maximum grade). If it goes 50 percent farther than the base, the grade is 150; if it goes 50 percent less than the base, the grade is 50. There is some question as to the validity of this particular standard. Some tire manufacturers have been able to revise their scores and use this modified score in their advertising. Also, in the real world of automobile operation, tread wear depends on several other factors including inflation pressures, alignment, tire pressure, vehicle type, driving habits, road type, climate, and so forth. However, the tread wear grade does give an indication of probable tread life differences between tires (Figure 9-19).

Traction grades are A (highest), B (middle), and C (lowest). Traction tests are used to test the tire's ability to stop on wet pavement. A-graded tires have a wet-pavement coefficient of friction greater than 0.35 on concrete and 0.5 on asphalt, B-graded tires have a coefficient of friction greater than 0.26 on concrete and 0.38 on asphalt, and C-graded tires have a lower coefficient of friction. *Coefficient of friction* is a term that refers to the relative friction between two surfaces. A tire marked C probably has poor stopping ability on wet pavement. Traction tests give a poor indication of the tire's ability to corner; this ability is often quite different from the tire's ability to accelerate or stop. Also, a tire with good wet-pavement traction does not necessarily have good dry-pavement traction.

Temperature grades are A (highest), B (middle), and C (lowest). They indicate the tire's ability to resist heat generation, its ability to get rid of heat, or both. Heat is one of the tire's worst enemies. It can reduce the bonding strength of the rubber materials, which can cause separation of the cords and plies from the rubber layers or cause separation of the rubber layers from each other. Heat also causes the rubber in the tire to become harder and more brittle over time, as the added heat causes the rubber to continue vulcanizing. A grade of C corresponds to the minimum level of performance a tire must meet to pass a DOT tire safety test. A and B grades indicate a tire quality somewhat better than the minimum, with A being better than B.

9.7 Speed Ratings

Another tire rating related to speed is commonly used in Europe. In the United States, the maximum speed limits make this rating somewhat unnecessary. A letter (S, H, or V) is added to the tire size just after the aspect ratio; for example, a 195/70HR14 has an H speed rating. A speed rating of S indicates a tire that is speed rated to 112 mph (180 km/h), H indicates a speed rating to 130 mph (210 km/h), and V indicates a tire that can sustain speeds of over 130 mph (210 km/h). These are the most common speed ratings; there are at least nineteen different ratings. In Europe, even tractor tires have a speed rating. The ones that will probably be increasingly used on passenger cars are the following: P, to 93 mph (150 km/h), Q, to 99 mph (160 km/h), R, to 106 mph (170 km/h), T, to 118 mph (190 km/h), and U, to 124 mph (200 km/h), in addition to the S, H, and V ratings already discussed (Table 9-2). The speed rating indicates at what speed a tire can be safely operated without danger of failure. In Europe, it is illegal to drive faster than the speed rating of the tires. It is a good practice to replace tires with ones that have the same speed rating as the original tires.

For anyone operating at speeds near a tire's speed rating, a load index (available at most tire stores) should be consulted to determine the load that the tires can safely carry. No tire should be operated at speeds greater than the speed rating or carry loads greater than the load rating printed on the sidewall or in the load index.

The speed itself does not damage the tire, but the tire can fail because of heat generated by a tire phenomenon called the **standing wave.** As mentioned earlier, the road-contact portion of the tread and the sidewalls next to it are deformed by the weight of the vehicle. The natural elastic properties of the tire cause the tread and sidewalls to return to their normal shape as they roll away from the road surface. At high speeds, these parts of the spinning tire cannot return to their normal shape fast enough and can still be deformed when they return to the road surface. This will cause a noticeable distortion to develop in the tire. The standing wave creates internal friction that will cause the tire to become hotter, wear faster, and lose handling and traction capabilities. This can quickly destroy a tire (Figure 9–20).

Table 9–2 Speed Ratings.

Speed Symbol	Speed (mph)	Speed (km/h)
A1	3	5
A2	6	10
A3	9	15
A4	12	20
A5	16	25
A6	19	30
A7	22	35
A8	25	40
B	31	50
C	37	60
D	40	65
E	43	70
F	50	80
G	56	90
J	62	100
K	68	110
L	75	120
M	81	130
N	87	140
P	93	150
Q	99	160
R	106	170
S	112	180
T	118	190
U	124	200
H	130	210
V[a] (VR)	149	240
W[b] (ZR)	168	270
Y[b] (ZR)	186	300

[a]Current tire speed rating markings include the use of the service description to identify the tire's speed capability (P215/65R15 95V—maximum speed 149 mph). Previous customs included the speed symbol in the size designation only (P215/65VR15), and the speed capability was listed as above 130 mph.

[b]Any tire with a speed capability above 149 mph (240 km/h) can, at the tire manufacturer's option, include ZR in the size designation (P275/40ZR17). If a service description IS NOT included, the tire manufacturer must be consulted for the maximum speed capability (P275/40ZR17—speed capability is 149 mph). If a service description IS included with the size description, the speed capability is limited by the speed symbol in the service description (P275/40ZR17 93W = maximum speed 168 mph).

Source: Courtesy of Bridgestone/Firestone, Inc.

Figure 9–20 This high-speed photograph was taken of a cut steel-belted tire running on a test roll at 120 mph (193 km/h). Note the standing wave in the sidewall and tread areas. *(Courtesy of Pirelli Tire North America)*

9.8 Tubes and Tubeless Tires

Originally, all tires used tubes to seal the air inside of the tire. A tube is much like its name implies, a tube of rubber. It has no end, similar to a doughnut. A tube does have an opening at the valve stem to make it possible to put air into it or let air out of it. The valve core is threaded into the stem to provide the valve action and hold the air pressure; it is commonly called a Schrader valve. This valve is opened when a projection on the tire inflator presses inward on the stem of the valve. A cap is usually installed over the valve to protect it from dirt and other debris and to provide a secondary seal for holding the air in the tube (Figure 9–21).

There are two general types of tubes used with passenger-car tires: radial and nonradial. A nonradial tube should only be used in a bias or a bias-belted tire. A radial tube can be used in any type, but it is more expensive. Radial tubes have improved splices where the tube is joined during manufacture. In many cases, they are also

Chapter 9 ■ Tire Theory

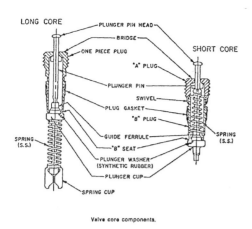

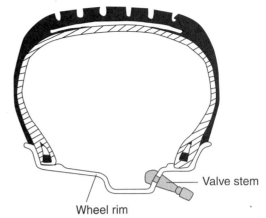

Figure 9–22 A tubeless tire's bead makes an airtight seal at the wheel's bead seat and flange, and the tire's body is airtight. The valve stem is also sealed at the wheel and allows the tire to be inflated.

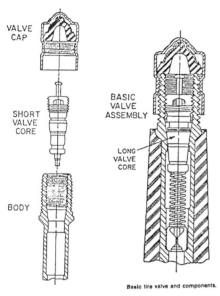

Figure 9–21 Tire valve cores and stems. Note the valve core sealing surfaces and valve core-to-stem sealing. *(Courtesy of Schrader-Bridgeport International, Inc.)*

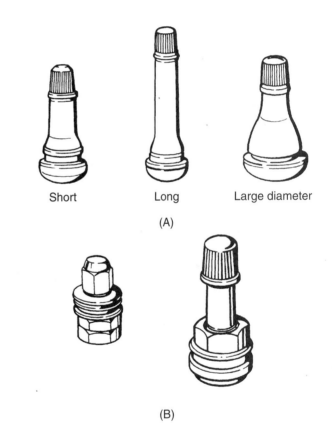

Figure 9–23 Snap-in (A) and metal clamp-in (B) tubeless tire valves are available in different diameters and lengths. *(Courtesy of Plews/Edlemann Division, Stant Corporation)*

more elastic to contend with the increased sidewall flex of the radial tire.

Today, most passenger-car tires are **tubeless**; that is, they do not use or need a tube. Tubeless tires have a definite safety advantage. A tube tends to deflate very rapidly or blow out when pierced. This can be compared with the action of a balloon that is stuck by a pin; the stretched-out rubber pulls away from the pinhole, which very rapidly gets larger. When pierced, the rubber in a tubeless tire tends to move toward the hole. This causes a much slower deflation and much greater car control while the tire is losing air pressure. Tubeless tires also run cooler without the increased rubber thickness and friction from the tube.

The wheel rim for a tubeless tire must be airtight, and the tire bead must make an airtight seal with the rim. The inside of the tire body is lined with an airtight layer of rubber called a butyl liner. The **valve stem** and **Schrader valve** are installed in the wheel rim (Figure 9–22). There are two general styles of tubeless tire valve stems used with passenger cars. The rubber, snap-in type is held in place by the compression of rubber; the metal, clamp-in type is locked in place by tightening a nut (Figure 9–23). Valve stems are available for the two common wheel hole sizes of 0.453 in. (11.5 mm) and 0.625 in. (15.87 mm). Each size is available in various lengths. A valve stem should be just long enough to extend through the wheel cover to allow for easy tire inflation. Wheels smaller than 14 in. usually use the smaller 0.453-in. stem. Wheel sizes of 15 in. and larger use the 0.625-in. stem.

A tubeless tire is normally used without a tube; use of a tube makes the tire run hotter and probably shortens its life. The increased heat is caused by tube-to-sidewall chafing and the increased rubber thickness of the sidewall. Use of a tube is sometimes necessary with a porous alloy or mag wheel, since porous rims let the air leak through minute holes in the metal. It is usually better to paint the inside of the rim or treat it with sealant rather than use a tube. Spoke wheels require a tube if the spokes extend into the wheel rim, which they usually do. A rubber or tape rim strip should also be used to prevent the tube from chafing against the spoke ends.

9.9 Replacement Tire Selection

Most car owners are quite satisfied with the **original equipment manufacturer (OEM)** tires that they have on their car, and their major concern is how to get the most tire life from their investment. They usually purchase a tire of the same type and size as the original for replacement. When choosing a slightly different replacement tire for these cars, it is wise to follow these recommendations:

- A replacement tire should be of the same size as or slightly larger than the original. The replacement tire should have a load-carrying capacity that is equal to or greater than the load to be carried.
- A tire that is one size larger (cross section) usually fits on the wheel and in the fender well with no interference problems. It costs slightly more but because of its larger size will deliver more miles before wearing out. During the life of this tire, the cost per mile is usually slightly lower.
- A replacement tire should have tire pressure characteristics that match those on the tire inflation sticker on the car.
- The speed rating of a replacement tire should be equal to that of the OEM tire.
- When the replacement tire is a different size than the original, the rim width should be checked to ensure its suitability. The fender clearance should also be checked after the tire and wheel are mounted to ensure that the tire can operate in all possible suspension travel and steering positions without interference.

Many people are not completely satisfied with their present tires and want to replace them to improve their car's handling or wet-weather traction, increase their fuel mileage, or improve the looks of the car. When selecting replacement tires that are different from the OEM tires, it is wise to follow these recommendations:

- Ideally, all four tires on a car will be of the same size and construction; it is very important that the two tires on an axle be of the same size, tread pattern, and construction. Mismatching tires on an axle can cause a pulling or self-steering to one side as the car goes down the road (Figure 9–24).
- If the rear tires are to be different from the front tires, radial-ply tires should never be put on the front with bias-ply tires on the rear. Also, a lower-profile tire should never be on the front. If the most responsive tires (i.e., those with the best handling characteristics) are put on the front of the car, a potentially dangerous oversteer condition can develop.
- On vehicles equipped with ABS, tire diameter is extremely critical. Tires of the wrong size will cause the brake control module to shut the system down and turn on the amber ABS brake warning light.

9.10 Spare Tires

From the beginning, motorists have worried about flat tires and the effect they have on vehicle operation. A **spare tire** that can be temporarily used to replace the flat tire is very common. At one time, the spare was a fifth tire and wheel of the same size and type as the four standard tires. On older cars, the spare is often the best tire from the last, worn-out set. Today, a new car might come with one of three types of temporary-use spares: **stowaway, compact,** or **temporary**. All three of these are of bias-ply construction and are usable for about 3,000 mi. (4,830 km) at speeds up to 50 mph (80.5 km/h), unless a slower speed is required. They are all marked "temporary" on the sidewall. They offer the advantages of being lighter, which makes them easier to handle and helps improve fuel mileage, and also smaller, which allows more storage space in the car (Figure 9–25). Vans and pickups use a full-size spare.

In all cases in which the spare is different from the other tires on the car, driving speed while using the spare should be reduced because of the change in traction and handling. Also, if the spare is being used on the drive axle and that axle is equipped with a limited-slip differential, the spare must have the same effective diameter as the other tire on that axle.

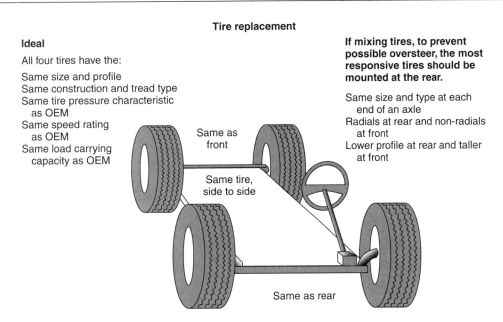

Figure 9–24 Ideally, all of the tires on a passenger car, light truck, or SUV will be of the same size, aspect ratio, and construction type. With care, some mixing of tire type and aspect ratio can be done while keeping safe handling characteristics.

Figure 9–25 Three different types of temporary-use spares. The stowaway spare is stored flat and inflated when it is used. *(Courtesy of General Motors Corporation, Service Technology Group)*

A folding, stowaway spare is not inflated until it is needed and is then inflated either with a pressurized canister that is carried in the car or with a standard air hose. During inflation, the tire should be either mounted on the car axle with the lugs slightly tightened or on a tire machine to prevent personal injury. After use, this tire can be deflated, refolded, and returned to its storage space. This tire is usually not serviceable; that is, the tire and wheel are replaced as an assembly if the tire becomes worn-out or damaged.

A compact spare is a narrow, lightweight, high-pressure tire. It is mounted on a wheel that is usually narrower in width and larger in diameter than the standard wheels on the car. A compact spare is fully serviceable, just like the standard tires.

A temporary spare is simply a full-size, lightweight tire of the same effective diameter as the standard tires. It is usually used on cars with limited-slip differentials. This tire is also serviceable using standard equipment.

9.11 No-Flat, Run-Flat Tires

The ideal spare tire is none at all. It would cost nothing, add no weight to the car, and take up no storage space. But to operate without a spare, we need tires that either will not go flat or that can be operated with no air pressure in them.

If a pneumatic tire is operated for very long without pressure, the excessive sidewall flexing will generate so much heat that the tire will start to disintegrate, or catch fire, or the bead will move into the well at the drop center of the wheel. If the latter happens, the bead can work its way off of the wheel.

No-flat or **self-sealing tires** are built with a soft, pliable inner lining that is gummy. If a puncturing object penetrates the tire, this sealant seals to the puncturing object; if the object is withdrawn, the sealant seals the hole. Self-sealing tires tend to be heavy and expensive.

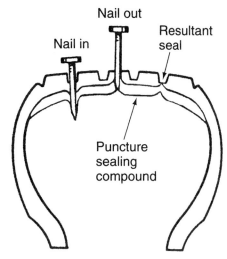

Figure 9-26 A self-sealing tire has a puncture sealing compound on the inside of the tread area. This compound seals to a projection or to a hole if the projection is removed and prevents a flat. *(Courtesy of General Motors Corporation, Service Technology Group)*

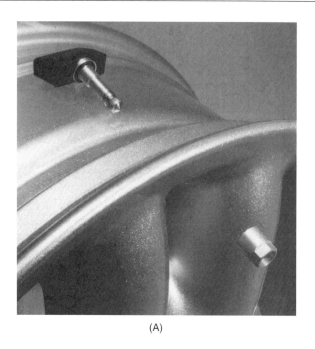

(A)

(B)

Figure 9-28 The Schrader Smart Valve System consists of a tire pressure transducer/transmitter mounted in place of the valve stem (A) and a receiver/display unit mounted in the vehicle (B). The display will give an audible and visual alarm if a tire has low pressure or is flat. *(Courtesy of Schrader-Bridgeport International, Inc.)*

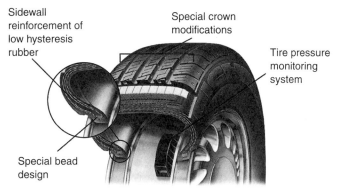

Figure 9-27 A cutaway view showing the features of a zero pressure tire and the tire pressure monitor. *(Courtesy of Michelin)*

They also generate slightly more heat than a standard tire (Figure 9-26).

Run-flat tires, also called *extended mobility tires* (*EMT*) or *zero pressure* (*ZP*) *tires,* must be strong enough to support the car, ensuring vehicle stability and mobility after being deflated. They can be driven up to 50 mi. (80 km) at speeds up to 55 mph (88 km/h) with no air pressure in them. They also must have enough internal strength to keep the beads seated or be used with a wheel design that will do the same.

To support the car, a run-flat tire uses thicker, reinforced sidewalls, which are about six times as strong as a standard tire. Run-flat tires tend to be heavy and expensive. They also require an air pressure monitoring system so the driver will be aware of any loss in air pressure (Figure 9-27).

Tire pressure monitor systems consist of a pressure sensor–transmitter module inside of each tire, attached to the wheel; a receiver; and a driver display module (Figure 9-28). Each sensor is powered by a special battery that has a life expectancy of about 10 years or 100,000 mi. The driver display will inform the driver of the actual pressure in each tire, warn the driver of any pressure drop below about 18 psi, and set off an alarm if the pressure in any tire drops below 10 psi.

Chapter 9 ■ Tire Theory

9.12 Retreads

When the tread of the tire becomes worn down to a depth of 2/32 in. (1.587 mm) from the bottom of the groove, a tire is considered to be worn-out. (For a long time, 1/32 in. was the measurement standard for the tire industry.) Many states not only cite motorists with worn-out tires but also remove the cars from the road. Many tires still have strong, usable bodies at this time, and it is a waste of energy and materials to discard these tires. **Retreading,** also called **recapping,** is the process of applying new tread rubber and a new tread pattern to the old body. Two slightly different processes, **hot-cap** and **cold-cap,** are used to secure the new tread. Cold-caps are now commonly called **precured** recaps. At this time, retreading is used mostly on truck tires.

At one time, tires were **regrooved** to appear less worn. Regrooving cuts the tread grooves deeper. It is illegal to regroove the tread on tires for passenger cars and two-wheel vehicles.

Both retread processes begin with a sound, undamaged tire body or casing. The casing is carefully inspected to ensure that there are no internal defects or structural flaws. Next, the remaining old tread is ground or rasped off of the casing. This provides a clean, properly roughened rubber surface for new tread rubber.

New tread rubber is then wrapped around the casing. In the hot-cap process, uncured rubber in one of two forms is used. In one form, a strip about 3/8 in. (9.5 mm) thick, as wide as the tread, and as long as the tread circumference is wrapped once around the tire. In the other form, a much longer strip of the same thickness, but about 1 in. (25.4 mm) wide, is wrapped around the tire the number of revolutions required to make up the tread width. The single-wrap process can leave either too much or too little rubber at the splice junction if the tread length is not exactly right for the tire circumference. This can cause tire runout or balance problems. Some recappers use the splice junction to their advantage. The tire balance is checked after the body is buffed, and as the tread rubber is wrapped, the splice is placed at the light portion of the body. The overlap is then adjusted so it is the correct weight to balance the tire. The continuous-wrap method does not have this junction problem.

After the tread has been applied, the tire is put into a mold to form the tread pattern, a process similar to the original tire-molding process. A thick, heavy-duty curing tube is placed in the tire and inflated to a pressure of over 100 psi (689.5 kPa) while in the mold. This internal pressure is used to force the tread rubber into the mold to ensure complete bonding and tread design formation. The mold is heated to about 300°F (149°C) to form the tread pattern, cure the tread rubber, and bond it to the body.

Figure 9–29 A precured strip of tread rubber is being applied to this tire over a strip of cushion gum. It will need to be stitched firmly in place and then heat-bonded to the carcass.

The secrets of getting a good recapped tire are preparing and conditioning the bonding surface on the tire body and using pure bonding cement, good bonding rubber, the correct air pressure, the correct mold temperature, and the correct curing time. After molding, the tire is ready to be cleaned and returned to service.

Precured recapping uses cured tread rubber that is premolded with the tread pattern and is the right width for the tire. This process was developed by the Bandag Corporation. An adhesive and an intermediate layer of bonding rubber, called a cushion strip, are applied to the body, and the new tread is wrapped around it. Pressure is applied to squeeze the tread onto the casing, and it is then heated to about 212°F (100°C) to cure the adhesive. Precured recaps are commonly used on truck tires. Because the tread is precured and does not include the cord materials of the body, it can be molded under a much higher pressure and temperature, which results in a more dense, longer-running tread. Precured recaps often deliver more mileage than the original tread. Also, the lower temperature used in this process does not shorten the life of the tire body as much as the hot process (Figure 9–29).

Although recaps are becoming less and less common in passenger cars, they remain quite popular in the trucking industry, where the price of a new tire is about $250 to $300. A worn carcass, worth about $75, is too expensive to throw away. Also, since many trucks average about 200,000 mi. (321,800 km) a year and have eighteen or more tires, tire costs can be significant. It is possible to save about one-fourth of the tire costs by using recaps. This can amount to a few thousand dollars a year in savings.

Retreads are less expensive than standard tires. Depending on the skill and care of the recapper and the quality of the tread rubber, retreads can be as strong and as durable as original tires. There is often a problem of obtaining a supply of sound carcasses for a particular size and type of tire. Some recappers recap a set of tires and return them to the owner, but this usually takes a certain amount of time. This service is referred to as custom capping.

9.13 Practice Diagnosis

You are working in a tire shop and encounter the following problems:

CASE 1: The customer bought a set of all season tires last year and has brought the car back in with a complaint of noisy tires. You drive the car on a road test and confirm the harsh noise from the tires. What should you do next? What might have kept the tires from becoming noisy?

CASE 2: The customer has come in to buy a new set of tires for his pickup truck. You notice that the truck has an extended over-cab camper on it. What replacement tires should you recommend? How can you tell if the current tires are adequate?

Terms to Know

- all season
- antilock braking systems (ABS)
- aramid
- aspect ratio
- bead wires
- belt
- bias belted
- bias ply
- body or carcass plies
- cap base
- cold-cap
- compact
- cotton
- Department of Transportation (DOT)
- directional tread
- DOT number
- dual compounding
- gum
- hot-cap
- hydroplane
- latex
- load range
- M & S
- M–S
- M + S
- National Highway Safety Administration (NHSA)
- National Highway Traffic Safety Administration (NHTSA)
- no-flat tires
- nylon
- original equipment manufacturer (OEM)
- plies
- P-metric
- polyester
- precured
- pyramid construction
- R, B, and D (P-metric system designations)
- radial
- rayon
- recapping
- regrooved
- retreading
- Rubber Manufacturers' Association (RMA)
- run-flat tires
- Schrader valve
- self-sealing tires
- sidewall plies
- spare tire
- stacked construction
- standing wave
- stowaway
- temporary
- traction
- tread plies
- tubeless
- under tread
- Uniform Tire Quality Grading Labeling (UTQGL)
- valve stem

Review Questions

1. Statement A: The amount of grip between the tire and the road is called traction.
   Statement B: A tire has its best grip when it is slipping slightly across the road surface.
   Which statement is correct?
   a. A only
   b. B only
   c. Both A and B
   d. Neither A nor B

2. The bead wires of the tire:
   A. Clamp onto the wheel rim
   B. Help transmit starting and stopping torque from the wheel to the tire
   Which option best completes the statement?
   a. A only
   b. B only
   c. Both A and B
   d. Neither A nor B

3. Which of the following is not a common sidewall ply material?
   a. Cotton
   b. Rayon
   c. Nylon
   d. Polyester

4. The cords in a tire's sidewall plies:
   A. Cross each other at about a 35-degree angle in a bias-ply tire
   B. Are used with a belt in a radial-ply tire
   Which option best completes the statement?
   a. A only
   b. B only
   c. Both A and B
   d. Neither A nor B

5. Statement A: The belt helps reinforce the tread area of the tire.
   Statement B: Each type of belt material requires a special rubber blend to bond the belt to the material next to it.
   Which statement is correct?
   a. A only
   b. B only
   c. Both A and B
   d. Neither A nor B

6. Which of the following materials is used in tire belts?
   a. Rayon
   b. Steel
   c. Glass fiber
   d. All of the above

7. The tire mold is used to:
   A. Form the tread pattern and sidewall markings
   B. Vulcanize the rubber
   Which option best completes the statement?
   a. A only
   b. B only
   c. Both A and B
   d. Neither A nor B

8. Statement A: The name of the manufacturer and the tire size must be molded into the tire sidewall.
   Statement B: The maximum air pressure a tire can safely hold and whether the tire has a directional tread must be molded into the sidewall.
   Which statement is correct?
   a. A only
   b. B only
   c. Both A and B
   d. Neither A nor B

9. The DOT number of a tire is coded to tell:
   a. What day of the year a tire was made
   b. The name and location of the plant that made the tire
   c. What the ply construction of the tire is
   d. All of the above

10. The DOT number of a tire that ends in 506 indicates that the tire was made:
    a. In December of 1976, 1986, or 1996
    b. On the 50th day of 1976
    c. On the 50th day of 1986
    d. Any of the above

11. A tire size of 185/70R14 indicates that:
    A. The tire has a tread width of 185 mm
    B. The distance between the beads across the tire diameter is 14 cm
    Which option best completes the statement?
    a. A only
    b. B only
    c. Both A and B
    d. Neither A nor B

12. A tire size of P185/70R14 indicates that:
    A. The width of the tire's cross section is 70 percent of the tread width
    B. The tire is of a radial construction
    Which option best completes the statement?
    a. A only
    b. B only
    c. Both A and B
    d. Neither A nor B

13. A P185/70R14 tire is designed for use on:
    A. Passenger cars
    B. Light trucks
    Which option best completes the statement?
    a. A only
    b. B only
    c. Both A and B
    d. Neither A nor B

14. Statement A: The tire does not support the car; air does.
    Statement B: A tire with more plies generally has a lower load rating.
    Which statement is correct?
    a. A only
    b. B only
    c. Both A and B
    d. Neither A nor B

15. Statement A: A tire with a tread wear grade of 75 should last a relatively long time and give good tire mileage.
    Statement B: A tire with an A traction grade should give good wet-condition stopping.
    Which statement is correct?
    a. A only
    b. B only
    c. Both A and B
    d. Neither A nor B

16. Statement A: Generally, the higher the letter in a tire's speed rating, the faster the tire can be operated.
    Statement B: Tire speed ratings are only enforced in Europe.
    Which statement is correct?
    a. A only
    b. B only
    c. Both A and B
    d. Neither A nor B

17. A tubeless tire must:
    a. Be used on an airtight wheel
    b. Make an airtight connection at the wheel flange
    c. Have an airtight inner liner
    d. All of the above

18. Statement A: Replacement tires must be the same size or slightly larger than the OEM tires.
    Statement B: Replacement tires can have a lower load rating than the OEM tires.
    Which statement is correct?
    a. A only
    b. B only
    c. Both A and B
    d. Neither A nor B

19. Statement A: A temporary spare tire can be used just like the other tires on the car.
    Statement B: A stowaway spare cannot be patched or repaired.
    Which statement is correct?
    a. A only
    b. B only
    c. Both A and B
    d. Neither A nor B

20. A retread tire has:
    A. A new tire tread cut into the original tire
    B. New tread rubber with a new tread formed into it
    Which option best completes the statement?
    a. A only
    b. B only
    c. Both A and B
    d. Neither A nor B

Math Questions

1. Your car has P225/70R15 tires that measure 26 in. from the ground to the top of the tire and 12½ in. from the ground to the center of the wheel. What is the circumference of these tires? How many revolutions must this tire make to go 1 mi. (5,280 ft)?

2. Your compact pickup has P165/80R13 tires on it, and its total weight is 2,800 lb, with 70 percent of the unloaded weight on the front tires. When it is unloaded, how much weight is on the front tires? How much weight is on the rear tires? The tire sidewall marking indicates that each tire can carry 1,025 lb. How much additional weight can you safely carry in the rear?

10 Wheel Theory

Objectives

Upon completion and review of this chapter, you should be able to:

❏ Comprehend the terms commonly used to describe wheel construction, sizing, and operation.

❏ Have a basic understanding of how wheels are constructed.

❏ Understand the current wheel size designations.

❏ Select a correct replacement wheel for a car.

10.1 Introduction

A wheel has a fairly easy job; all it has to do is secure the tire to the hub. But in doing this, the tire must be held exactly centered in two directions, radially and laterally. It also must be held centered under some rather severe loads. These loads are the vertical load of the vehicle's weight; the shock loads caused by bumps; the twisting, torsional loads of braking and accelerating; and the sideways load that occurs during cornering (Figure 10-1).

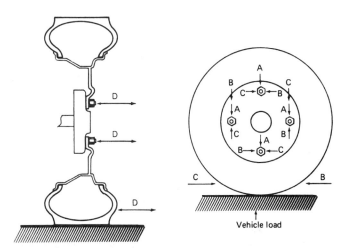

Figure 10-1 A wheel is subject to forces generated by the weight of the vehicle (A), braking (B), acceleration (C), and cornering (D).

10.2 Construction

A wheel is made of two main parts—the **center section,** also called a **spider** or **disc,** which attaches to the hub, and the **rim,** where the tire is attached.

The rim is usually made from rolled steel, but it can also be made from stamped, cast, or spun aluminum or magnesium alloy. The rim has two **bead flanges** that the tire beads push against. These flanges must be strong enough to keep the tire beads in place. Internal tire pressure exerts a constant, outward pressure that forces the tire beads against the wheel flanges. Just inside of the flanges are the **bead seats,** which are at a slight angle of about 15 degrees for passenger cars. The flanges control the tire position in a lateral, sideways direction, while the tire beads wedging onto the bead seats position the tire in a radial, vertical direction. Friction between the tire bead and the wheel bead seats and flanges transmits braking and acceleration forces from the tire to the wheel or vice versa.

Flanges are made with different heights and curvatures to match the tire bead configuration. Flange type is designated by a classification system using one or two letters. The most commonly used flange heights are J = 0.68 in. (17.3 mm), JJ = 0.69 in. (17.5 mm), JK = 0.71 in. (18 mm), K = 0.77 in. (19.5 mm), and L = 0.85 in. (21.6 mm). Passenger-car rims have a raised bead just inside of the bead seat called a **safety bead.** This safety bead helps keep the beads of a flat tire from moving into the well if the tire deflates. If a flat occurred without this safety bead, the bead of the flat tire might work its way

177

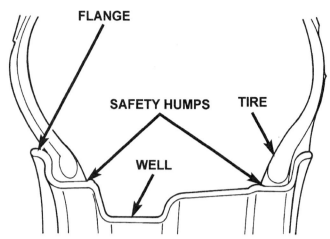

Figure 10–2 Passenger-car wheels have a pair of safety humps just inside of the tire beads; this wheel is often called a safety wheel. Note the dropped well of the rim and that it is off center, toward the left. *(Courtesy of Chrysler Corporation)*

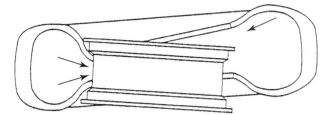

Figure 10–3 The dropped well in the wheel rim allows us to remove and replace the tire on the wheel. Note that when the tire bead is in the well (left arrows), the other side of the bead is outside of the wheel flange (right arrow). *(Courtesy of Pirelli Tire Corporation)*

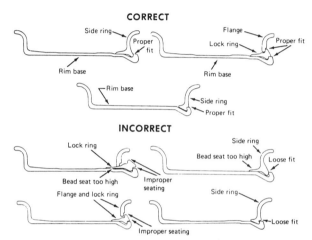

Figure 10–4 Three styles of two- and three-piece rims. Internal tire pressure pushing sideways on the beads holds the side and lock rings in place. Removal of the side ring or flange and lock ring allows the tire to be slid off the side of the rim. If the side ring or lock ring is not properly seated, it can fly off the wheel with a great deal of force.

off the wheel before the driver could stop the car. The safety bead is required for DOT approval on tubeless tire wheels (Figure 10–2).

All passenger cars use **drop center rims.** The rim is shaped with a lowered well section near the center. This drop center, or well, is necessary so the tire can be installed or removed. As we learned in the previous chapter, the tire bead is reinforced with steel wire; it will not stretch. The wheel well is deep enough so that the bead will pass over the flange on one edge of the rim when the other side of the bead is in the well (Figure 10–3).

Most truck and heavy equipment wheel rims are of a two- or three-piece configuration, commonly called **split rims.** One of the bead flanges is removable for tire removal or installation. Service work on this tire and wheel combination requires special skill and equipment and extreme care (Figure 10–4). Split rims have separated during tire inflation and even while running on the road. The flying rim flanges have killed or severely injured service personnel and bystanders. Space does not permit covering split rim service operations in this text. Instructions are published by various wheel manufacturers and the Rubber Manufacturers' Association.

The center section of a passenger-car wheel, like the rim, is usually made from stamped steel; it can also be made from cast, stamped, or spun aluminum or magnesium alloy. It is usually welded to the rim, but it can be riveted, bolted, or connected using wires or spokes.

A spindle hole is centered in this wheel part and surrounded by three to six lug bolt holes. The size and number of lug bolts are determined by the vehicle load and expected driving conditions. Heavy-duty vehicles commonly use six or eight lug bolts; most passenger cars use four or five. **Lug bolt circles,** also called **bolt patterns,** are of different diameters, depending on the design strength or the available space. Bolt pattern sizes are usually printed as 5-4½ in. or 4-100 mm; this would mean there are five holes that are evenly spaced around a 4½-in.-diameter circle or four holes evenly spaced around a 100-mm-diameter circle. It is easy to measure the bolt circle diameter on even-numbered, four-, six-, or eight-lug circles; just measure from the center of one lug bolt to the center of the lug bolt across the wheel. A three- or five-bolt circle diameter is difficult to measure. Most shops use charts that specify the pattern for a particular car or a template to check a particular wheel or hub (Figure 10–5).

Composite wheels made from plastics will probably be commonly used in the near future. Composite wheels are made from various plastic compounds and are usually reinforced with fiberglass or carbon fibers

Chapter 10 ■ Wheel Theory

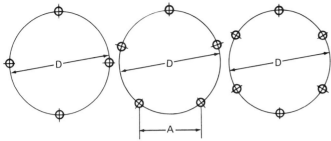

Figure 10–5 A wheel's bolt circle diameter is measured as shown here. For five-hole wheels, the bolt circle diameter can be determined by multiplying the center-to-center distance between two wheel studs, A, by 1.7013.

to produce adequate strength. It is expected that composite wheels will be stronger, lighter, less expensive, better looking, and free from corrosion.

10.2.1 AH2 Standard

The AH2 standard wheel design, sometimes called a *bead lock rim*, uses a redesigned flange and bead area to lock the tire onto the rim. The purpose is to prevent slippage of the tire around the wheel under conditions of hard acceleration or braking that lead to tire runout or imbalance. It also tends to keep the tire on the bead for a longer period to prevent an instant loss of air if a tire loses air pressure. Because tires are much more difficult to remove and replace from these wheels, only well-trained technicians using modern equipment should attempt this procedure.

10.3 Wheel Sizes

The two most commonly used wheel dimensions are diameter and width. These dimensions match the size of the tire. Wheel diameter is measured from bead seat to bead seat across the diameter of the wheel and must be exactly the same as the tire diameter. The flange height is not included in the wheel diameter (Figure 10–6).

Wheel width, measured across the rim at the inside of the bead flanges, is usually smaller than the tire width. Some tire manufacturers specify a particular wheel "design" width for their tires. Tables are published that show recommended rim widths for each tire size (Figure 10–7). If no other information is available, a rule of thumb is that the rim width should be about 75 to 80 percent of the tire cross-sectional width. A wider rim tends to reduce sidewall flexing, which results in increased steering response, a stiffer ride, and increased tread shoulder wear. A narrower rim tends to increase sidewall flexing, sidewall damage, and center tread wear and produce a softer ride (Figure 10–8).

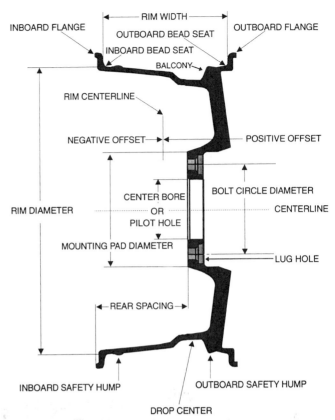

Figure 10–6 A diagram of a wheel showing the various parts. *(Reprinted by permission of Hunter Engineering Company)*

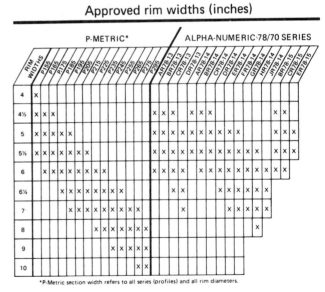

Figure 10–7 Charts are available to show the approved rim widths for different size tires; for example, 5½-, 6-, 6½-, and 7-in.-wide rims are correct for a P205 tire. *(Courtesy of Rubber Manufacturers' Association)*

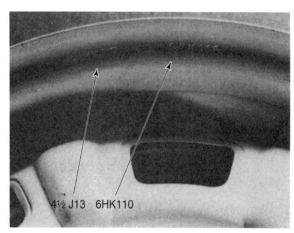

Figure 10–8 Many wheels have their size stamped into the rim. This wheel is 4½ in. wide, has a J flange, and is 13 in. in diameter; the other numbers are the manufacturer's part number.

10.4 Offset and Backspacing

Another important wheel dimension to consider when replacing a wheel is **offset.** Offset specifies where the wheel mounting face is relative to the rim and determines where the tire centerline will be relative to the hub. If the wheel mounting face is centered and directly in line with the wheel centerline, there is zero offset. In such a case, there is an equal amount of rim width on each side of the wheel mounting surface at the hub.

If the wheel mounting face and tire centerline are not centered or in line, the wheel has a certain amount of positive (+) or negative (−) offset, and the rim and tire will be offset from the hub. There is some controversy regarding to which direction positive or negative offset refers. Respected sources agree that negative offset moves the tire and rim outboard from the mounting flange and increases the track width. The mounting flange will be inboard from the wheel centerline. Positive offset (used on many FWD cars) does just the opposite. The tire is moved inward relative to the mounting flange (Figure 10–9).

Determining offset requires measuring the wheel width between the flanges and measuring the distance from the mounting flange to the back side of the wheel, often called **backspacing, backside spacing,** or **rear spacing.** The amount of offset can be determined by using the formula given in Figure 10–10. If the backspacing dimension is less than one-half of the rim width, the wheel has negative offset. If the backspacing dimension is greater, the offset is positive (Figure 10–10).

Offset is a little difficult to handle by some people and a little difficult to measure by all. Many wheel manufacturers simply use the backspacing dimension instead of offset. It is much easier to measure, and it is almost

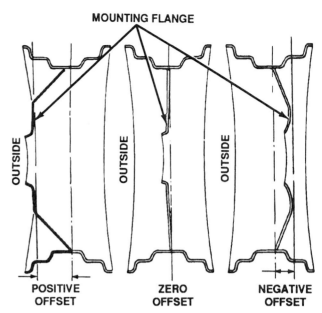

Figure 10–9 If the mounting flange is centered in the wheel, the wheel has zero offset. If it is outward from the wheel center, the wheel has positive offset; if it is inward, the wheel has negative offset.

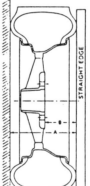

You can measure offset with or without the tire mounted on the wheel. Place wheel with the outboard side down and take dimensions "A" and "B". Offset is calculated with the formula shown. If "B" is larger than "A/2" the offset is positive. If tire is not on wheel, take measurements to edge of rim rather than tire.

$$\text{OFFSET} = \frac{A}{2} - B$$

Figure 10–10 Wheel offset can be measured using the procedure shown here. *(Courtesy of Bridgestone/Firestone, Inc.)*

the same as using offset dimensions (Figure 10–11). However, you should remember that a 6-in. (15.24-cm) wheel with a 2-in. (5.08-cm) backspacing will have a different offset, −1 in. (−2.54 cm), than a 5-in. (12.7-cm) wheel with the same backspacing, −1/2 in. (−1.27 cm). The tire centerline will move 1/2 in. (1.27 cm) and the track will change 1 in. (2.54 cm) if one pair of these wheels is replaced by the other.

Wheel offset has several effects on vehicle operation. Negative offset (i.e., the tire centerline is moved outward) increases the track width, which reduces the lateral weight transfer. This is a benefit, but it also reduces the tire-to-fender clearance, moves the wheel bearing load

Chapter 10 ■ Wheel Theory

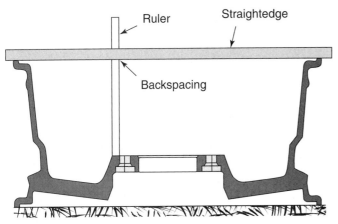

Figure 10–11 Wheel backspacing is measured by placing a straightedge across the wheel at the bead flanges and measuring the distance from the mounting flange to the straightedge.

outward, increases the load on the smaller wheel bearing, increases the wheel leverage on the springs and suspension, and changes the **scrub radius.** The scrub radius, also called the **kingpin offset** or **steering axis offset,** is the distance from the center of the tire to the steering axis measured at the road surface. Most RWD cars position the tire centerline slightly outboard of the steering axis, whereas most FWD cars place the tire centerline inward. This can be easily seen by comparing the wheel offset of the two styles of cars. Changing the scrub radius can have drastic effects on vehicle handling, especially when a front tire meets a bump or pothole. It is highly recommended that replacement wheels have the same or very close to the same offset as the OEM wheels in order to maintain the correct scrub radius (Figure 10–12).

10.5 Aftermarket Wheels

Building and selling aftermarket wheels has become a rather strong industry. The major reason for replacing the OEM wheels is probably cosmetic, but other common reasons include a desire to change the wheel width to accommodate wider-profile tires or to change the wheel diameter to allow plus 1, 2, or 3 changes (Figure 10–13).

An increasingly popular concept with car enthusiasts is the **plus 1, plus 2,** or **plus 3** conversion. If you desire a wider tire with a wider and shorter aspect ratio, you usually have to give up tire diameter. The plus 1 concept is to use a wheel that is 1 in. (2.54 cm) larger in diameter and a tire that is one step greater in aspect ratio—80 to 70, 70 to 60, 60 to 50, or even 50 to 40. Plus 2 uses a wheel that is 2 in. larger in diameter with a tire that has an aspect ratio that is two steps wider. Plus 3 (which is not very common) uses a wheel that is 3 in. larger in diameter with a

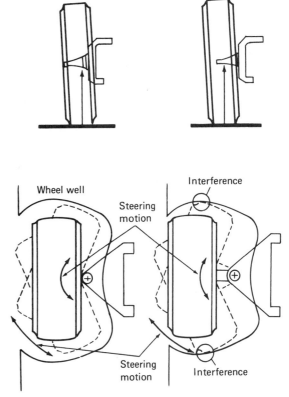

Figure 10–12 If an OEM wheel (top left) is replaced with a wheel that has a more negative offset (top right), the load on the spindle will move outward and place more load on the outer wheel bearing. The tire will also move outward in the fender well (lower right) and often cause interference during turning maneuvers.

tire that is three steps wider. Some manufacturers are offering this option on their sport and performance cars. Plus 1, 2, and 3 allow the motorist to end up with a wider, sportier tire that fits the car as well as the OEM tire recommendations. Whenever the tire width is increased, adequate side clearance for normal suspension and steering motions of the tire must be ensured.

Many people believe that aftermarket wheels look better than the stock OEM wheels and that a change in wheel style allows them to improve the appearance of their car, as well as customize it to their personal tastes. Many different styles are available to suit almost every preference (Figure 10–14).

When a wider replacement tire is desired, a change in wheel width is often mandatory. A stock wheel will often accommodate a tire one or two sizes larger than standard width, but a greater change will probably exceed the correct tire-to-rim width range. When wider wheels are desired, it is often possible to use an OEM wheel that was available as optional equipment. Many car manufacturers offer larger or sport or rally tires and wheels as an option.

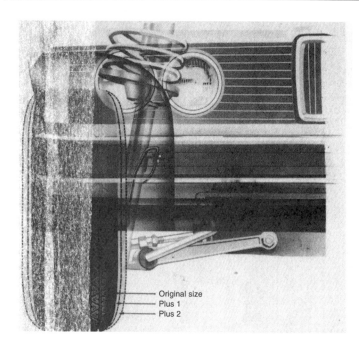

Figure 10-13 The plus 1, plus 2 concept replaces the original tire and wheel with a wider but shorter tire on a wider and larger diameter wheel. The end result is a wider tire with close to the original diameter. (Courtesy of B.F. Goodrich. © The B.F. Goodrich Company)

These wheels are usually wider than the standard wheel and offer the benefit of having the correct offset for a particular car. You are also assured that this wheel will have no interference problems with the brake assemblies and brake drum balance weights.

A change in wheel diameter usually cannot be done without a change in the tire's aspect ratio. Otherwise, fender clearance and gear ratios will increase with larger tires and wheels, ground clearance and gear ratio losses will increase with smaller tires and wheels, and speedometer error problems will occur.

Engine revolutions per minute (rpm), gear ratio, tire diameter, and car speed are all directly related. A tire will roll down the road a distance that is equal to its circumference on each revolution. You can use the formula $C = 2\pi R$ to compute the circumference, but be sure to use the distance from the center of the wheel to the ground as R, the tire's effective radius. This is important because car load causes the tire to bulge, which shortens this distance. Some manufacturers publish revolutions per mile specifications for their tires for easy comparison. You can use the following formula to determine engine rpm or car speed:

$$\text{Engine rpm} = \frac{\text{mph} \times \text{Gear Ratio} \times 336}{\text{Tire Diameter}}$$

$$\text{mph} = \frac{\text{rpm} \times \text{Tire Diameter}}{\text{Gear Ratio} \times 336}$$

When using these formulas, the gear ratio equals the final drive ratio (rear or front axle ratio) times the transmission ratio. Also, the tire diameter equals the effective tire radius times two.

On vehicles equipped with ABS tire diameter is critical. On these cars, wheel speed sensors monitor tire rpm to determine if wheel lockup is occurring during braking. If a different-sized replacement tire is used, the electronic controller will see the different wheel speed sensor signal, determine there is a problem, shut off the ABS, set a problem code, and turn on the amber brake warning light.

The quality of aftermarket wheels is ensured by the **Specialty Equipment Manufacturers' Association (SEMA)**, through the **SEMA Foundation, Inc. (SFI)**. SEMA and SFI are nongovernmental agencies that were formed by concerned aftermarket manufacturers. A wheel that carries SFI certification has been manufactured to rather stringent standards, which helps ensure its safe operation.

10.6 Wheel Attachment

Most of us are familiar with the normal **wheel lug bolt and nut.** This bolt has a serrated shank that is pressed through the hub to prevent bolt rotation, and it is often upset or riveted at the outer side to retain it securely in place. When the wheel is installed, the tapered, conical face of the lug nut enters the tapered holes in the wheel's nut bosses to center

Chapter 10 ■ Wheel Theory

Figure 10–14 Aftermarket wheel manufacturers offer a large selection of wheels in many styles and sizes. *(Courtesy of Hayes Lemmerz International, Inc.)*

the wheel to the hub. This allows the use of a larger hole in the wheel lug holes and bosses for easy wheel installation and removal, and the tapers ensure wheel centering on the hub. Some wheels have a center hole that closely fits the hub flange. These *hub centric wheels* are centered as the wheel's center hole fits over the hub (Figure 10–15).

The lug nuts must hold the wheel tightly against the hub. This tight fit reduces the bending loads on the bolts as well as the shearing loads during braking and acceleration. A properly tightened wheel transmits about 90 percent of its load through the friction between the wheel and the hub and about 10 percent of the load through the bolts. If the lug bolts get loose, their portion of the load will increase to as high as 100 percent, and they will probably bend or break. The lug bolts selected are large enough to hold these loads. The most common sizes are 7/16-20 (7/16-in. thread diameter and 20 threads per inch), 1/2-20, 3/8-24, M12×1.5 (12-mm thread diameter and 1.5-mm thread pitch, i.e., distance from one thread to the next), M12×1.25, M14×1.5, and M10×1.5.

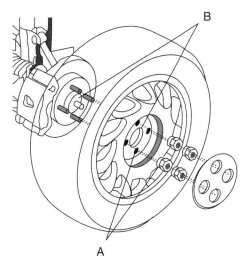

Figure 10–15 At one time, most wheels were centered to the hub by the tapered end of the lug nut entering the tapered bosses in the wheel (A). Now, many wheels are hub centric, with a snug fit between the center hole and the hub (B). (© Saturn Corporation, used with permission)

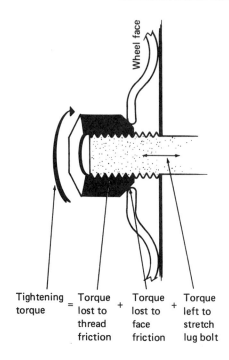

100% = 15-20% + 60-80% + 4-8%

Figure 10–16 When a lug nut is tightened, most of the tightening torque goes to friction at the threads and the interface between the nut and the wheel. Part of the torque is used to stretch the bolt and compress the nut boss in the wheel.

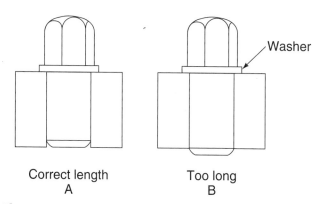

Figure 10–17 An alloy wheel often has a thick wheel section that requires a straight-shank nut; this nut should be slightly shorter than the thickness of the wheel and have a diameter slightly smaller than the holes. Steel wheels normally use a tapered lug nut.

A lug nut stays in place on the lug bolt because of the friction between the threads of the nut and bolt and between the face of the nut and the wheel. If the nut is tightened against a solid object, the total amount of these frictions will increase rapidly as soon as the nut meets the object. They will also decrease rapidly if the nut backs off slightly. When a bolt and nut are tightened, a stretching load is placed on the bolt as soon as the clearance is taken up. This stretching action maintains interthread and interface friction to ensure nut retention. If the bolt is overtightened, the bolt can be stretched to the fatigue or yield point, and it might break, especially if stress or road shock places a greater load on it. An overtightened lug nut can also cause distortion loads on the brake rotors, drums, or both. If the lug nut is undertightened, it will soon come loose and fall off because there will not be enough interthread friction to hold the nut in place (Figure 10–16).

The area immediately around the lug nut boss of most stamped wheels is usually raised slightly so this portion of the wheel center section will deflect as the lug nuts are tightened. This deflection has the effect of reducing bolt stretch as the lug nuts are tightened and also maintaining an outward pressure to maintain nut friction.

Cast wheels usually have thick center sections and normally use straight-shank lug nuts plus a washer. The washer helps reduce galling of the wheel as the nuts are tightened and loosened. Wheel centering is accomplished by a close fit between the nut shank and the wheel holes. Nuts are available with different shank lengths; the shank length should be slightly shorter than the thickness of the wheel boss. Some cast center section wheels use tapered nut faces much like stamped wheels. On special installations, you should make sure there is sufficient nut-to-bolt thread contact. Sometimes, with thick wheel bosses, it is necessary to replace the lug bolts with ones that have a longer thread length (Figure 10–17).

With either style of lug nut, the nut shank or the bolt length must be long enough to ensure adequate thread contact or thread stripping might result. The thread contact length should be equal to 0.8 times the bolt diameter for tapered seat nuts and 0.85 times the bolt diameter for straight-shank nuts—1 to 1.5 times the bolt diameter is preferred. For a 1/2-20 in. lug bolt (12.7-mm diameter with 20 threads per 25.4 mm), there should be a minimum of 0.8 × **D,** or 0.8 × 0.5, which equals 0.4 in. (10.16 mm) of thread contact. This would mean a contact of at least eight (0.4 × 20) threads. Most people prefer a thread length with a few more threads than this to provide a margin of safety and a hedge against bolt and nut wear or stripping during wheel installation and removal.

Knockoff or **quick-change wheels** use a large, single nut to secure the wheel to the hub. Driving and braking torque loads are transmitted by splines in the wheel center circle and hub or by a group of drive pins on the hub that fit into holes in the wheel center section. Knockoff wheels were commonly used on sports cars and also are currently in use on many racing cars.

10.7 Practice Diagnosis

You are working in a tire shop that also does front-end repair and wheel alignment. A customer brings in a late-model car she recently purchased. Her complaint is a rubbing noise and a binding condition when she makes tight steering maneuvers. You notice that the wheels are a set of aftermarket replacements, and your road test confirms the problem. When you lift the car on a rack, you find rub marks on both the inner fender panel and the inside of the tread at the shoulder. What should you do next to give the customer the best information?

Terms to Know

backside spacing
backspacing
bead flanges
bead seats
bolt patterns
composites
drop center rims
kingpin offset

knockoff wheels
lug bolt circles
offset
plus 1
plus 2
plus 3
quick-change wheels
rear spacing

safety bead
scrub radius
SEMA Foundation, Inc. (SFI)
Specialty Equipment Manufacturers' Association (SEMA)
split rims
steering axis offset
wheel lug bolt and nut

Review Questions

1. Statement A: The center portion of the wheel is sometimes called the disc.
   Statement B: The tire fits onto the rim of the wheel.
   Which statement is correct?
   a. A only
   b. B only
   c. Both A and B
   d. Neither A nor B

2. The tire bead seats of a wheel are:
   A. Formed at a slight angle
   B. About 0.060 in. smaller than the bead of the tire
   Which option best completes the statement?
   a. A only
   b. B only
   c. Both A and B
   d. Neither A nor B

3. The center portion of the rim is dropped downward to allow:
   A. A stronger attachment to the wheel center section
   B. The tire to be removed and replaced
   Which option best completes the statement?
   a. A only
   b. B only
   c. Both A and B
   d. Neither A nor B

4. A wheel with a bolt pattern of 4-100 mm has:
   A. Four mounting holes on a 4-in.-diameter circle
   B. Mounting holes in a square shape, 100 mm long, on each side of the square
   Which option best completes the statement?
   a. A only
   b. B only
   c. Both A and B
   d. Neither A nor B

5. Statement A: The width of the wheel should equal the width of the tire tread.
   Statement B: The diameter of the wheel should equal the inside diameter of the tire.
   Which statement is correct?
   a. A only
   b. B only
   c. Both A and B
   d. Neither A nor B

6. Statement A: The scrub radius will change if the wheel is replaced with one that has a different offset.
   Statement B: Changing the wheel offset will change the loading on the wheel bearings.
   Which statement is correct?
   a. A only
   b. B only
   c. Both A and B
   d. Neither A nor B

7. The backspacing on a 6-in.-wide wheel is 2 in., so the wheel has an offset of:
   a. Positive 2 in.
   b. 1 in.
   c. Negative 2 in.
   d. Negative 1 in.

8. A plus 1 tire changeover means that the tires and wheels are replaced with a:
   A. Tire that has the next wider profile
   B. Wheel that is 1 in. larger
   Which option best completes the statement?
   a. A only
   b. B only
   c. Both A and B
   d. Neither A nor B

9. Statement A: Most of the braking and accelerating torque between the wheel and the hub is transmitted by the lug bolts.
   Statement B: Improperly tightened lug bolts can cause brake rotor and drum distortion as well as lug bolt breakage.
   Which statement is correct?
   a. A only
   b. B only
   c. Both A and B
   d. Neither A nor B

10. The wheel is centered to the hub as the:
    A. Tapered lug nuts enter the tapered wheel bosses
    B. Center of the wheel is placed over the hub
    Which option best completes the statement?
    a. A only
    b. B only
    c. Both A and B
    d. Neither A nor B

Math Questions

1. You have just bought a really good-looking 2-year-old Camaro or Mustang (take your pick). It has a HP V8 engine with a five-speed transmission. The fifth-gear ratio is 0.75:1, and the rear axle ratio is 3.23:1. The current tires are P215/70R15 on the OEM wheels. You decide to replace the tires with a set of P255/60SR15. The effective radius of the OEM tires is 12½ in., and the effective radius of the new tires will be 11¾ in. What is the engine rpm at 65 mph now? What will it be after the tires are replaced? What percentage will the rpm change? What other noticeable change will occur?

2. You are measuring the offset on a wheel. It has a width of 6 in. and has a backspacing of 2½ in. How much offset is there? Is it positive or negative?

11 Wheel Bearing Theory

Objectives

Upon completion and review of this chapter, you should be able to:

- ❑ Comprehend the terms commonly used with wheel bearings.
- ❑ Understand the various types of wheel and axle bearing arrangements.

11.1 Introduction

The wheel bearing's job is to allow the hub, wheel, and tire to rotate freely while it holds them in alignment with the car's steering knuckle or axle. In doing this, the wheel bearing has to transfer the vertical load of the car, any side loads that result from cornering maneuvers, and other forces. In addition, on drive axles, the bearing allows the drive axle to rotate while transmitting driving torque to the wheel and tire.

A mechanic has five basic styles of wheel bearings to learn, and these are based on the type and function of the bearings. As mentioned earlier, the term **wheel bearing** is normally used for the bearings on all front wheels and nondrive axle rear wheels. The term **axle bearing** is commonly used for RWD drive axles. The five various bearing types are:

1. Serviceable nondrive axle bearings: front wheels on RWD vehicles and rear wheels on FWD cars
2. Nonserviceable nondrive axle bearings: rear wheels on some FWD vehicles
3. Drive axle bearings on a solid axle: rear wheels on most RWD cars
4. Drive axle bearings on independent suspension: rear wheels on independent rear suspension (IRS) and front wheels on FWD cars
5. Nonserviceable drive axle bearings: front wheels on some FWD cars and rear wheels on some IRS

The two nonserviceable styles are very similar to each other. They both support the weight of the car, with the drive axle style capable of transmitting torque from the axle to the wheel. Like their name implies, nonserviceable bearings are not serviced. There is no service required or possible for the bearing unit. If the bearing is faulty, the bearing assembly is removed and replaced with a new one; **R & R** is the common term for this procedure.

There is also a strong similarity between the serviceable nondrive axle bearings and the IRS drive axle bearings in that they are usually a pair of tapered roller bearings. They both normally require occasional cleaning, repacking with grease, and a clearance or preload adjustment.

11.2 Bearings and Bearing Parts

The term **bearing,** when used by most technicians and manufacturers, usually refers to either ball, roller, or tapered roller bearings. These bearing types are often referred to as frictionless bearings. They are called **frictionless** because the balls or rollers, fitted between the two races, roll easily whenever either of the two races rotates (Figure 11–1). A **bushing,** commonly used for crankshaft bearings, has a sliding action. Bushing friction is usually reduced by grease or a constant flow of oil.

Frictionless bearings are usually made up of three major parts: (1) a cone or inner race, (2) a cup or outer race, and (3) the balls or rollers (Figure 11–2). All of these parts are made from hardened-steel alloys and are precision ground to very close size and finish tolerances. The bearing balls or rollers are usually placed in a cage or separator so they will not rub against each other. The cage also ensures that the balls or rollers stay spread out in the correct spacing for proper load distribution. Sealed bearings position a seal between the inner and outer races on one or both sides of the bearing. The seal is used to

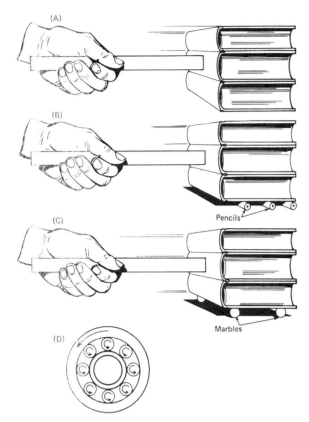

Figure 11-1 Try pushing a stack of books using a stick or ruler as shown in (A). You will note a certain amount of drag. Place a few pencils (B) and then some marbles (C) under the books and try pushing them again. You should notice a definite reduction in the amount of drag and friction. The rolling action of the pencils and marbles is similar to the action in a bearing (D).

keep lubrication in the bearing and dirt out. A frictionless bearing must be lubricated to reduce friction and prevent wear. Lubricants also carry heat away from the bearing and protect the metal surfaces from corrosion. Dirt and other abrasive materials must be kept out to prevent bearing damage or wear.

Ball bearings run in concave grooves that are ground into each of the races. A ball bearing usually has the ability to control radial movement and loads as well as thrust movement (Figure 11-3). For example, it can support a shaft and allow the shaft to rotate, while keeping the shaft from moving sideways. Because the balls only contact the races in a tiny spot, the amount of side or radial load and end or thrust load that a ball bearing can support is rather limited. Note that only a few of the balls are carrying the load at one time. The balls on the nonloaded side of the bearing are not really doing much (Figure 11-4).

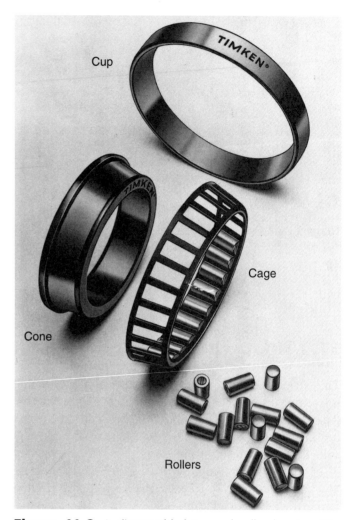

Figure 11-2 A disassembled tapered roller bearing. The cup is also called the outer race and the cone the inner race. *(Courtesy of The Timken Company)*

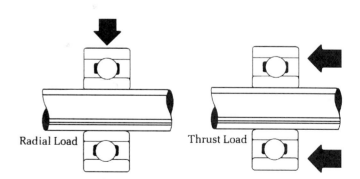

Figure 11-3 The two basic loads that a bearing is subject to are radial loads from a right angle to the bore and thrust loads parallel to the bore. *(Courtesy of Chicago Rawhide)*

Chapter 11 ■ Wheel Bearing Theory

Roller bearings can carry a greater side load than ball bearings because the rollers, being longer, have a much greater load-carrying surface area. A roller bearing cannot control end thrust. When mounted where side thrust control is important, thrust bearings are used along with the roller bearings. Very thin roller bearings are called **needle bearings.** Needle bearings can carry still more load because there are more of the thin needles and, therefore, more surface area to transfer loads.

Tapered roller bearings are used when side thrust as well as radial loads need to be controlled. Front wheel bearings are the perfect example of this. The wheel bearings carry the radial load of the car's weight as well as the side load that is trying to take the hub off of the spindle. Tapered roller bearings are used in pairs, with the tapers of each bearing facing in opposite directions to each other. It is these tapers that provide sideways control (Figure 11–5).

11.3 Bearing End Play and Preload

An important step when adjusting bearings is to set the correct bearing **end play.** Too much end play gives a loose, sloppy fit, which would let the wheel wobble and change alignment angles, in addition to reducing the cone-to-bearing contact. The bearing will get maximum life if it has a free running clearance (i.e., no preload) with no appreciable end play. Free play also provides room for expansion as the bearings, shafts, and housings heat up

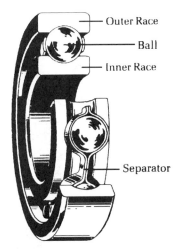

Figure 11–4 A cutaway view of a ball bearing. *(Courtesy of Chicago Rawhide)*

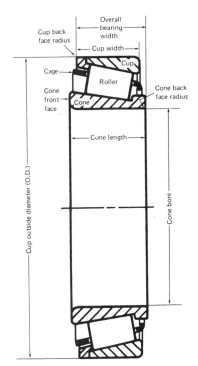

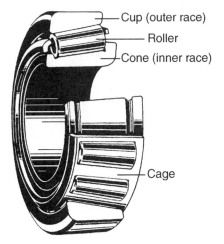

Figure 11–5 A section view of a tapered roller bearing that shows the normally used dimensions. *(Courtesy of Chicago Rawhide)*

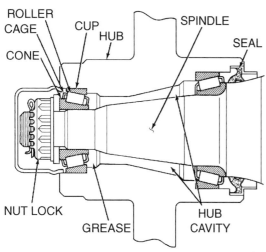

Figure 11-6 A hub, bearing, and spindle assembly as used on many nondrive wheels. Bearing end play or preload is adjusted with the nut just inside of the nut lock. (*Courtesy of Chrysler Corporation*)

during operation. A bearing will have end play if there is a slight clearance between the balls or rollers and the races. A bearing will have preload when there is a pressure between the balls or rollers and the races (Figure 11-6).

Preload is used on bearings when the shaft must not change position. The pinion gear and shaft in a rear axle transmit a large amount of torque, and because of gear pressure, the gears try to move lengthwise as well as sideways. The pinion bearings must be preloaded to keep the gear from changing position under these heavy loads. Bearing preload increases the friction of the bearing and consumes power. If possible, tapered roller bearings are adjusted to a slight end play to allow freer running. The exact amount of end play for a particular bearing set is usually specified by the vehicle manufacturer. A rule of thumb for the clearance of wheel bearings is 0.001 to 0.005 in. (0.03 to 0.13 mm) of end play. Ball bearings are normally adjusted to a slight preload to keep the balls properly positioned in the races.

11.4 Seals

Frictionless bearings are always used with some sort of seal at each side of the bearing. The seal is used to keep the lubricant in the bearing and dirt and other foreign material out of the bearing. Many seals are designed to do both operations at the same time. A lip seal is the most common type of wheel or axle bearing seal (Figure 11-7).

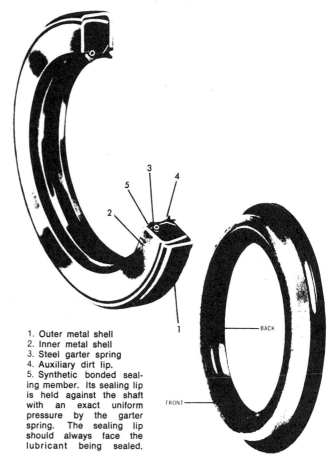

1. Outer metal shell
2. Inner metal shell
3. Steel garter spring
4. Auxiliary dirt lip.
5. Synthetic bonded sealing member. Its sealing lip is held against the shaft with an exact uniform pressure by the garter spring. The sealing lip should always face the lubricant being sealed.

Figure 11-7 A typical seal showing its various parts. (*Courtesy of Chicago Rawhide*)

The sealing lip of the seal, sometimes called a wiping lip, is made from a flexible material, usually neoprene rubber. Different materials can be used depending on the speed of the seal, the type of lubricant, and the operating conditions under which the seal is used. The flexible lip is usually molded into the seal's outer case. The seal's housing forms a **static (stationary)** seal to the hub or axle housing when it is pressed into position in its bore. The seal lip forms a **dynamic (moving)** seal by wiping oil or grease off of the shaft. The open side of the seal lip always faces toward the lubricant (inside) so that any internal pressure will increase, not decrease, the wiping pressure between the seal lip and the shaft. A garter spring is often placed around the seal lip to increase this wiping pressure. Most seal failures occur when this wiping lip fails. Some seals have a second lip that faces outward. This lip stops dirt and grit from working its way under the inner sealing lip.

11.5 Serviceable Nondrive Axle Wheel Bearings

Serviceable nondrive axle bearings are very common and are used on the front wheels of most RWD cars and the rear wheels of many FWD cars. A pair of tapered roller bearings are placed over the spindle with the cups pressed into the hub. The bearings are placed so the smaller diameters of the bearing tapers are toward each other to give the bearing a large amount of control over end play.

The larger of the two bearings normally carries most of the radial load. This bearing is positioned close to the center of the tire where most of the vehicle load is. It is also located over the larger and stronger portion of the spindle. The smaller bearing serves primarily to keep the hub from wobbling. With the large bearing, it controls end play.

End play of most front wheel bearings is adjusted by the tightness of the spindle nut. Turning the nut inward decreases the end play and, if turned far enough, preloads the bearings. After the bearing end play is adjusted, the nut is locked in place to hold this adjustment. The usual lock is a cotter pin placed through a hole in the end of the spindle and a slot in the castellated nut or castellated style of nut lock. Other methods of locking the nut in place include:

- Using a second nut tightened against the adjusting nut, called a **lock** or **jam nut**
- Bending or staking a portion of the adjusting nut into a groove in the spindle
- Tightening a clamp portion of the adjusting nut onto the spindle
- Bending a sheet metal washer over the adjusting nut and also over the jam nut; this washer is often kept from rotating by a tang of the washer that fits a groove in the spindle (truck and some 4WD)
- Tightening set screws or positioning pins in the lock nut against the adjusting nut; these set screws pass through slots in the washer that is tanged to the spindle (truck and 4WD) (Figure 11-8)

Occasionally, a bearing is adjusted by changing spacer shims or washers that are positioned between the two bearings. Thicker shims give more end play and less preload.

11.6 Nonserviceable Nondrive Axle Wheel Bearings

Nonserviceable nondrive axle wheel bearings are used on the rear wheels of some FWD cars. This bearing assembly is permanently lubricated, adjusted, and sealed during manufacture and requires no further service. In fact, it is not possible to service it internally. This bearing assembly is bolted to the car's suspension, and the tire and wheel and the brake drum or rotor are bolted to it. If this bearing becomes noisy, rough, or loose, the whole assembly is removed and replaced with a new unit (Figure 11-9).

11.7 Drive Axle Wheel Bearing, Solid Axle

A drive axle bearing is used on the rear end of most RWD cars. The bearing, commonly called an axle bearing, is at the outer end of the axle housing, and the axle, with the tire and wheel bolted to it, turns in this bearing.

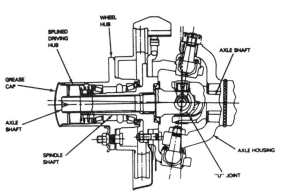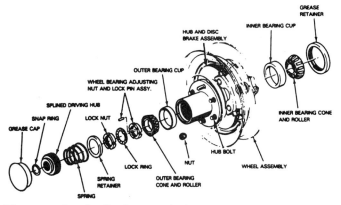

Figure 11-8 A front hub from a 4WD vehicle. Note the different style of adjusting nut lock. *(Courtesy of Ford Motor Company)*

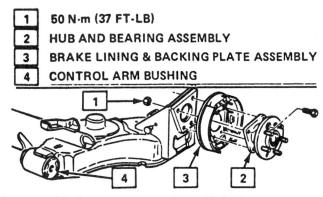

Figure 11-9 A rear suspension with brake and hub and bearing assemblies from a FWD car. This hub and bearing cannot be disassembled or serviced. It is replaced when there is a problem. *(Courtesy of General Motors Corporation, Service Technology Group)*

Most passenger cars use a style of axle and bearing that is called a **semifloating axle** (Figure 11-10). The inner end of the axle **floats** in the axle gear inside the differential. The term *floats* indicates that the axle is not directly supported by a bearing; the gear into which the axle is splined supports the axle. The axle bearing supports the vertical load on the axle but not always the side loads. Larger pickups and trucks use **full-floating axles**, and the bearing does not touch the axle at either end. The rear hubs of a full-floating axle use a large pair of tapered roller bearings that are very similar to common front wheel bearings, except much larger. A full-floating axle carries no vehicle load, only torque going to the rear tires (Figure 11-11).

Two major styles of bearings are used on semifloating axles. These styles somewhat determine how the axle is held in the axle housing. In some axles, the bearing's inner race is pressed onto the axle, and a secondary retainer is pressed onto the shaft right next to the bearing. The retainer helps ensure that the axle does not slide out of the bearing and the axle housing. The outer race of the axle bearing fits snugly into the axle housing and is held in place by a bearing retainer that is bolted to the axle housing. The brake backing plate is usually held in place by the same bolts. This arrangement is often called a **bearing-retained axle.** (See Figure 11-10.)

The other style of axle uses a C-clip to keep the axle in the housing; it is usually called a **C-clip axle.** The outer end of this axle shaft, just inboard of the wheel mounting flange, is hardened and ground smooth to serve as the inner race of a roller bearing (Figure 11-12). The outer bearing race fits snugly into the axle housing with the axle passing through it. The C-clip is placed into a groove in the inner end of the axle. This C-clip also fits into a recess in the axle gear and is locked in place when the differential pinion shaft is installed. The axle is prevented from sliding outward by the C-clip and inward by the differential pinion shaft. If this axle breaks outboard of the C-clip, the outer portion of the axle, with the tire

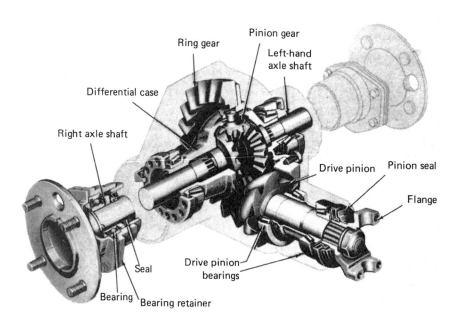

Figure 11-10 The outer end of a drive axle showing one style of axle bearing. This style is called a semifloating axle because the axle's inner end is supported by a gear. The outer end of this axle supports some of the vehicle's weight. *(Courtesy of Ford Motor Company)*

Chapter 11 ■ Wheel Bearing Theory

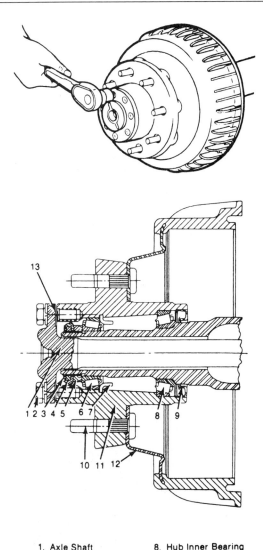

1. Axle Shaft
2. Shaft-to-Hub Bolt
3. Retainer
4. Key
5. Adjusting Nut
6. Hub Outer Bearing
7. Snap Ring
8. Hub Inner Bearing
9. Oil Seal
10. Wheel Bolt
11. Hub Assembly
12. Drum Assembly
13. R.T.V.

Figure 11-11 The outer end of a full-floating style of drive axle as used on many trucks and large pickups. Note that the drive axle (1) does not carry any vehicle loads; these are carried by the bearings (6 and 8). The bearings are adjusted and held onto the hollow spindle by the adjusting nut (5), key (4), and retainer (3). *(Courtesy of General Motors Corporation, Service Technology Group)*

and wheel and brake drum attached, can slide out of the housing, at least until the tire runs into the fender.

Both of these types of bearings are normally lubricated by either a mist of gear oil from the axle gears or by grease that was packed and sealed into the bearing during manufacture. A lip seal is often placed just inboard of the axle bearing to keep excess gear oil from passing by the bearing, out the end of the axle housing, and onto the brake shoes.

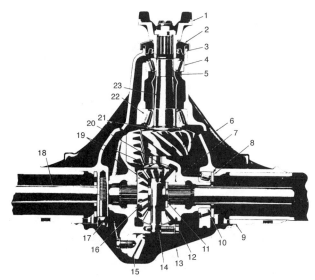

Figure 11-12 A cutaway view of a rear axle assembly showing how the C-lock (11) fits into the differential side gear (16) and is kept from coming out by the differential pinion shaft (14). The axle cannot slide outward because of the C-lock or inward because of the pinion shaft. *(Courtesy of General Motors Corporation, Service Technology Group)*

11.8 Drive Axle Wheel Bearing, Independent Suspension

The drive axle wheel bearing is found at the rear drive axle on a car with independent rear suspension or on the front end of many FWD cars. (See Figure 5–14.) Cars with independently suspended driving wheels usually transmit power to the tire through a short **stub** axle. This axle is connected to the differential by a short driveshaft or **half shaft.** The ends of these driveshafts are connected to universal joints on most RWD cars or to **constant velocity universal joints (CV joints)** on all FWD cars and some RWD cars. The CV joints are needed on FWD cars to allow sufficient turning angles and to reduce drivetrain vibrations while turning (Figure 11–13).

Independent rear suspension RWD cars usually use a pair of tapered roller bearings to support the stub axle. Like other paired, tapered roller bearings, these bearings allow the shaft to rotate while eliminating any side motions. The bearings are often mounted a few inches apart in the bearing support to give them better leverage in controlling shaft position. The spindle support is connected to the suspension members in such a way that it can control the alignment of the tire and wheel as the tire moves over the road surface. Lubrication of these bearings is by periodic disassembly, cleaning, and repacking with grease. Some manufacturers use permanently packed and sealed

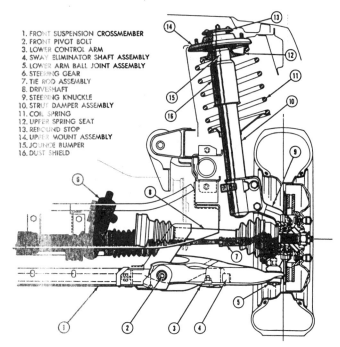

1. FRONT SUSPENSION CROSSMEMBER
2. FRONT PIVOT BOLT
3. LOWER CONTROL ARM
4. SWAY ELIMINATOR SHAFT ASSEMBLY
5. LOWER ARM BALL JOINT ASSEMBLY
6. STEERING GEAR
7. TIE ROD ASSEMBLY
8. DRIVESHAFT
9. STEERING KNUCKLE
10. STRUT DAMPER ASSEMBLY
11. COIL SPRING
12. UPPER SPRING SEAT
13. REBOUND STOP
14. UPPER MOUNT ASSEMBLY
15. JOUNCE BUMPER
16. DUST SHIELD

Figure 11-13 The front suspension of a FWD car showing the drive axle or driveshaft (8). *(Courtesy of Chrysler Corporation)*

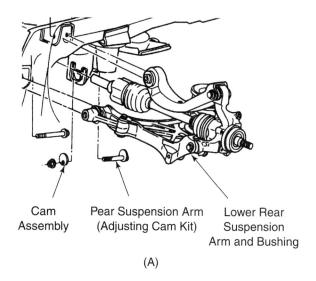

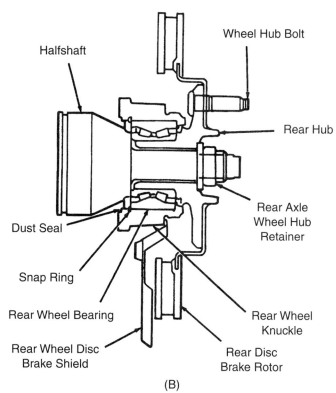

Figure 11-14 The rear suspension from a RWD IRS car (A). It uses a nonserviceable rear wheel bearing between the rear hub and the knuckle (B). *(Courtesy of Ford Motor Company)*

bearings. The end play or preload of this bearing type is usually controlled by a spacer between the bearings.

Most FWD front wheel bearings must be compact to fit in the small space provided for them. The stub axle is often the splined extension of the outer portion or housing of the CV joint. The splines of the CV joint housing pass into the splines in the front hub, and these two parts are held together by a nut at the outer end. The hub is supported by a pair of ball or tapered roller bearings that are mounted in the steering knuckle. These bearings can be packed with lubricant and sealed during manufacture or require periodic lubrication. The vehicle manufacturer's guidelines should be checked to determine the maintenance requirements for a specific car. Some manufacturers recommend that these bearings be replaced whenever the front hub is removed from the spindle. Bearing end play or preload is usually controlled by the size of the parts. As they are assembled and tightened into place, the adjustment is automatically made.

11.9 Nonserviceable Drive Axle Wheel Bearings

Nonserviceable drive axle wheel bearings are found at the front end of some FWD cars. They are very similar to the nonserviceable nondrive axle wheel bearings. The only real difference is that the drive axle bearing is hollow with a splined hole in the hub so the CV joint can attach to it. Like other FWD wheel bearings, this assembly attaches to the steering knuckle, and like other nonserviceable bearing assemblies, this unit requires no maintenance and is serviced by replacement. (See Figure 21-29.)

A nonserviceable bearing assembly is also used in some IRS. For repair, the bearing assembly is pressed in and out of the support (Figure 11-14).

11.10 Practice Diagnosis

You are working in an independent repair shop. A customer brings in a 4-year-old, compact, FWD car saying that she hears a funny kind of growling noise when traveling at about 30 mph. You suspect that this could be caused by a faulty wheel bearing. How can you confirm your suspicion using a road test and in the shop (without removing anything except the wheels)?

Terms to Know

axle bearing	dynamic (moving)	R & R
ball bearings	end play	roller bearings
bearing	floats	semifloating axle
bearing-retained axle	frictionless	static (stationary)
bushing	half shaft	stub
C-clip axle	lock or jam nut	tapered roller bearings
constant velocity (CV) universal joints	needle bearings	wheel bearing
	preload	

Review Questions

1. Statement A: Most of the rear wheel bearings, hubs, and spindles of FWD cars are similar to those on the front wheels of RWD cars.
   Statement B: Some wheel bearings cannot be adjusted.
   Which statement is correct?
   a. A only
   b. B only
   c. Both A and B
   d. Neither A nor B

2. A frictionless bearing uses _____ as the rolling element of the bearing.
   a. Balls
   b. Rollers
   c. Needles
   d. All of these

3. The purpose of the cage in a ball bearing is to:
   A. Keep the balls separated and properly spaced
   B. Hold the balls between the races
   Which option best completes the statement?
   a. A only
   b. B only
   c. Both A and B
   d. Neither A nor B

4. Statement A: A roller bearing can usually carry more side and thrust load than a ball bearing.
   Statement B: Needle bearings can carry greater thrust loads than ball bearings.
   Which statement is correct?
   a. A only
   b. B only
   c. Both A and B
   d. Neither A nor B

5. Technician A says that free movement of a hub in and out is called bearing end play.
   Technician B says that preload causes a drag on the bearings.
   Who is correct?
   a. A only
   b. B only
   c. Both A and B
   d. Neither A nor B

6. The inner race of a tapered roller bearing is called a:
   A. Cup
   B. Cone
   Which option best completes the statement?
   a. A only
   b. B only
   c. Both A and B
   d. Neither A nor B

7. Technician A says that the inner lip of a seal should always point toward the outside to keep dirt from entering under the seal lip.
   Technician B says that the sealing pressure of a lip seal is often increased by a garter spring.
   Who is correct?
   a. A only
   b. B only
   c. Both A and B
   d. Neither A nor B

8. Bearing preload ensures that the shaft or hub will not:
   A. Move in or out
   B. Rock sideways
   Which option best completes the statement?
   a. A only
   b. B only
   c. Both A and B
   d. Neither A nor B

9. The lip seal used in the front hub of a RWD car will make a dynamic seal with the:
   A. Spindle.
   B. Hub
   Which option best completes the statement?
   a. A only   c. Both A and B
   b. B only   d. Neither A nor B

10. Technician A says that the bearings used for the front wheel bearings for most RWD cars are of the tapered roller type.
    Technician B says that the spindle nut is used to adjust the bearing end play.
    Who is correct?
    a. A only   c. Both A and B
    b. B only   d. Neither A nor B

11. Nonserviceable wheel bearing assemblies can be found on the:
    A. Rear wheels of some RWD cars
    B. Front and rear wheels of some FWD cars
    Which option best completes the statement?
    a. A only   c. Both A and B
    b. B only   d. Neither A nor B

12. Most RWD cars use a rear axle of the _____ floating type.
    a. One-quarter   c. Semi
    b. Three-quarter   d. Full

13. The rear axle shaft of most RWD cars is held in place by:
    A. The axle bearing and retainer
    B. A C-shaped lock
    Which option best completes the statement?
    a. A only   c. Both A and B
    b. B only   d. Neither A nor B

14. The front wheel bearings of a FWD car fit between the:
    a. Spindle and the hub
    b. Steering knuckle and the hub
    c. Steering knuckle and the CV joint housing
    d. Spindle and the CV joint housing

15. The front wheel bearings of a FWD car are:
    A. Adjusted by the tightness of the hub nut
    B. Repacked in a manner similar to the front wheel bearings on a RWD car
    Which option best completes the statement?
    a. A only   c. Both A and B
    b. B only   d. Neither A nor B

12 Front-Wheel-Drive Driveshaft Theory

Objectives

Upon completion and review of this chapter, you should be able to:

❑ Comprehend the terms commonly used with the driveshafts on FWD cars.

❑ Understand the differences among the various CV joints.

12.1 Introduction

In the late 1970s, the evolution of the automobile progressed to FWD. This design is very efficient in terms of fuel mileage and passenger space per pound of vehicle. Compact, small-engined cars commonly use **transverse** (crosswise to the car) mounted engines with a **transaxle** (transmission and final drive with differential) at the end of the engine or right alongside of it. Two driveshafts extend from the differential in the transaxle to the front wheel hubs in the steering knuckles. These driveshafts are fitted with a **constant velocity (CV) universal joint** at each end. These driveshafts are also called half shafts or axles (Figure 12–1).

Replacement or service of the driveshafts, CV joints, or boots requires some front suspension disassembly and occasionally the disassembly of the front hubs and wheel bearings. This is when the front-end technician becomes involved with the drivetrain. Traditionally, in our world of specialization, the front-end technician had very little to do with the car's drivetrain, and the powertrain technician had very little to do with front suspensions. As mentioned earlier, FWD has blurred or mixed up more than one aspect of car repair.

12.2 Driveshaft Construction

Most FWD driveshafts consist of two CV joints connected by a small-diameter solid steel shaft. CV joints have the ability to transfer power without a speed fluctuation that would cause a noticeable vibration or steering wheel shake on corners.

CV joints are specially designed and constructed so they can operate at sharp angles and transfer power smoothly without fluctuations in speed. The output of

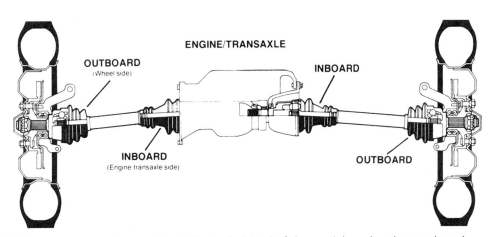

Figure 12–1 A FWD car uses two axles or driveshafts. Each driveshaft has an inboard and an outboard CV joint. *(Courtesy of Neapco Inc.)*

these joints turns at a constant velocity relative to the input. Also, a CV joint allows about a 40-degree operating angle. This large angle is necessary to provide sufficient turning angles and suspension travel. Most current outboard CV joints are of the **Rzeppa** design or variations of it. This joint usually uses six precision balls that operate in precisely sized and shaped grooves in the inner and outer races of the joint (Figure 12–2). All types of CV joints share a high degree of precision machining, and all are relatively expensive. They must be kept clean and lubricated, or they will soon fail (Figure 12–3).

Normal **cross and yoke** or **cardan universal joints (U-joints),** as used on RWD driveshafts, can transfer power and allow a change of angle. But when they run at an angle, they cause a speed fluctuation. The driven end speeds up and slows down twice in each revolution. Cross and yoke U-joints are usually used in pairs. The two joints are run at equal angles and are timed so that the speed fluctuations generated by one joint are cancelled by those generated by the other one (Figure 12–4).

The two joints on a FWD driveshaft are usually of two different types. Suspension movement causes the steering knuckle and hub to travel in an arc, relative to the car, as they move up and down. The path of this arc is determined by the geometry of the suspension. The outer end of the driveshaft also travels in an arc. The inner CV joint is the center point, and the length of the driveshaft is the radius of this arc. The arcs of the driveshaft and the hub are slightly different, so the driveshaft must change length while the tire and wheel move up and down. If it does not, there will be a binding of the driveshaft or suspension or both. The inner CV joint is usually of a **plunging** or **plunge** design. The driveshaft can move in and out of this joint while it is operating (Figure 12–5). This joint is often referred to as a **tripot, tripod, tripode,** or **tulip joint.** This style of joint uses three precision grooves in which needle bearing–equipped balls or rollers mounted on a three-legged cross operate (Figure 12–6).

The outer joint is often called a **fixed joint** because it does not allow any side motion of the driveshaft. This keeps the hinge point of the joint directly in line with the

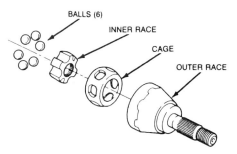

Figure 12–2 A Rzeppa-design CV joint. This is the most popular style of outboard, fixed joint. *(Courtesy of General Motors Corporation, Service Technology Group)*

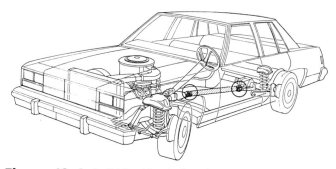

Figure 12–4 A RWD driveshaft using two universal joints (circled). These joints are indexed so the vibrations developed by one joint are canceled out by the other. *(Courtesy of Dana Corporation)*

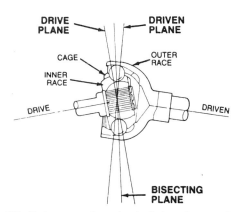

Figure 12–3 In operation, the balls in a Rzeppa joint bisect the operating planes of the driving and driven shafts. *(Courtesy of General Motors Corporation, Service Technology Group)*

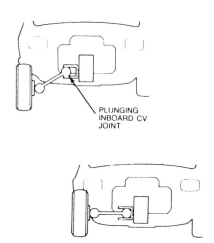

Figure 12–5 The inboard, plunge joint allows the driveshaft to change length as the suspension travels up and down. *(Courtesy of General Motors Corporation, Service Technology Group)*

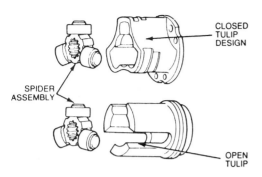
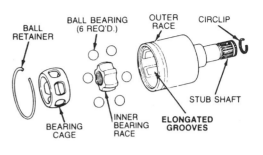

Figure 12–6 The tripot (left) and double offset (right) designs are the most popular styles of plunge joints. *(Courtesy of General Motors Corporation, Service Technology Group)*

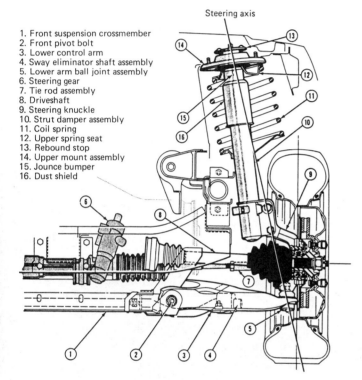

1. Front suspension crossmember
2. Front pivot bolt
3. Lower control arm
4. Sway eliminator shaft assembly
5. Lower arm ball joint assembly
6. Steering gear
7. Tie rod assembly
8. Driveshaft
9. Steering knuckle
10. Strut damper assembly
11. Coil spring
12. Upper spring seat
13. Rebound stop
14. Upper mount assembly
15. Jounce bumper
16. Dust shield

Figure 12–7 The fixed, outboard joint must stay centered with the steering axis to prevent binding during turns. *(Courtesy of Chrysler Corporation)*

steering axis. Steering or CV joint bind would occur on turns if the CV joint axis and the steering axis were not in exact alignment (Figure 12–7).

The driveshaft itself is usually a small-diameter, solid steel shaft that is splined at each end to accept the CV joints. In most FWD designs, one driveshaft, usually the left, is shorter. This is because the differential is not in the center of the car. Having one driveshaft longer than the other can cause **torque steer;** during a hard acceleration, engine torque tries to twist the two driveshafts. The longer shaft allows more twist, so torque will reach the tire on the shorter shaft first (Figure 12–8). This will cause the car to steer away from that tire. The car will steer straight when there is a steady throttle, but it will pull to one side under heavy throttle. Some cars use a larger-diameter tubular shaft on the longer side to balance out the shaft-twisting tendencies that cause torque steer. Another cure is to use equal-length driveshafts of the same size. A short, intermediate shaft is added to the long side to make up for the differences in length. Equal-length shaft installations require the use of a third U-joint, usually a simple cross and yoke type, and a support bearing (Figure 12–9).

CV joints are kept clean by a flexible, pleated, sealed **boot.** The boot also keeps the lubricant in the joint. The boot is made from a synthetic rubber and of an accordion or bellows type of construction for flexibility. Severe-duty boots are used where heat or extreme angles would cause early failure of rubber boots. These plastic boots are harder and have a slicker, shinier appearance. The boot is usually retained in place on the CV joint housing and driveshaft by metal bands. The boot is usually the first part of a joint to fail. A failing boot often shows up as a **grease spray.** Centrifugal force throws the grease out of the joint and onto the area around it. It is said that the joint will last for about 10 to 20 hours of operation after it starts spraying grease. A broken or torn boot can also let water and dirt enter the joint, which will also cause early failure.

12.3 Practice Diagnosis

You are working as an alignment technician in a front-end shop. After removing the front tire from a 5-year-old Ford Taurus, you notice a spray of grease on the strut and in the wheel well, and the CV joint boot is badly torn in a couple of places. What recommendation should you make to the customer? What is the probable condition of the joint?

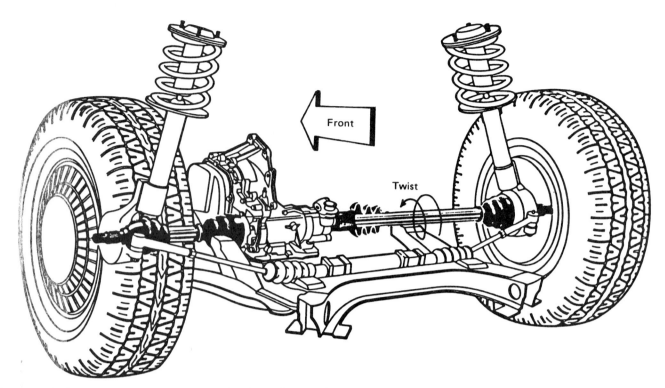

Figure 12-8 Torque steer can occur on hard acceleration because torque can twist the longer and weaker right-side driveshaft. Torque arrives at the left tire earlier than the right, and the car steers to the right. *(Courtesy of General Motors Corporation, Service Technology Group)*

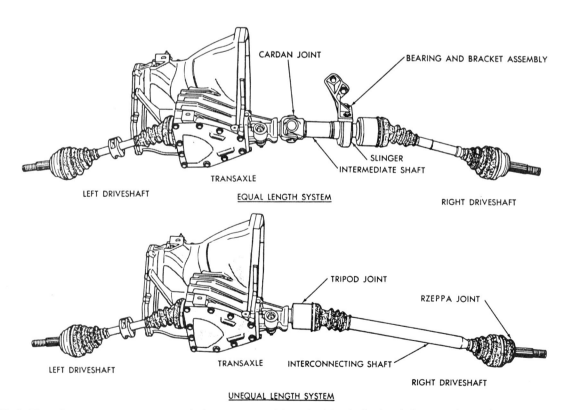

Figure 12-9 To reduce torque steer, some designs use equal-length driveshafts (top) that require an intermediate shaft and support bearing. Other designs use an oversize shaft on the long side (bottom). *(Courtesy of Chrysler Corporation)*

Terms to Know

cardan universal joints (U-joints)
constant velocity (CV) universal joint
cross and yoke joints
fixed joint
grease spray
plunge
plunging
Rzeppa
torque steer
transaxle
transverse
tripod
tripode
tripot
tulip joint

Review Questions

1. The term *CV joint* is short for:
   a. Continuous volume joint
   b. Constant volume joint
   c. Constant velocity universal joint
   d. Continuous velocity universal joint

2. Most CV joints will:
   A. Allow a large amount of angle change between the two shafts
   B. Transmit power at an angle with no speed fluctuation
   Which option best completes the statement?
   a. A only   c. Both A and B
   b. B only   d. Neither A nor B

3. Statement A: The outer CV joint is usually a plunge type.
   Statement B: The inner CV joint is usually a fixed type.
   Which statement is correct?
   a. A only   c. Both A and B
   b. B only   d. Neither A nor B

4. A Rzeppa joint is:
   A. A type of fixed joint
   B. Also used on RWD driveshafts
   Which option best completes the statement?
   a. A only   c. Both A and B
   b. B only   d. Neither A nor B

5. Many plunge joints are of the _____ type.
   a. Tripod joint   c. Tripot joint
   b. Tulip joint   d. Any of these

6. FWD designs with one long driveshaft and one short driveshaft are prone to have:
   A. Torque steer
   B. Vibrations on sharp corners
   Which option best completes the statement?
   a. A only   c. Both A and B
   b. B only   d. Neither A nor B

Front Suspension Types

Objectives

Upon completion and review of this chapter, you should be able to:

❑ Comprehend the different styles of automotive front suspensions.

❑ Comprehend the terms commonly used with front suspensions.

❑ Understand how a short-long arm suspension operates.

❑ Understand how a strut suspension operates.

❑ Understand the operating differences between RWD and FWD front suspensions.

13.1 Introduction

The front suspension of a vehicle is designed so the steering knuckle and spindle can pivot on the **steering axis** to allow steering of the vehicle. The spindle must also rise and fall, relative to the body, to allow the springs and shock absorbers to reduce bump and road shock from the vehicle's ride. The suspension system allows the springs and shock absorbers to absorb the energy of the bump, so passengers can have a smooth ride. While doing these two jobs, the suspension system must not allow loose, uncontrolled movement of the tire and wheel and must keep the alignment of the tire as correct as possible.

There are essentially five types of front suspensions used on cars, pickups, and trucks: **short-long arm (S-L A), multilinks, MacPherson strut (strut), swing axle,** and **solid axle.** Of these, only the first three are commonly used on passenger cars. All of these suspension types can be used in FWD, RWD, or 4WD vehicles.

These suspension types have several differing aspects: simplicity or complexity, size, cost, weight, alignment stability, ride and handling characteristics, and how well they adapt to automated, robot-aided construction. When a car design is developed, these aspects are carefully compared. Often, the type of suspension and drive system at one end of the car has to be balanced with the same handling characteristics at the other end. For example, a rear-engine, RWD car tends to oversteer. If a front suspension with superior tire adhesion characteristics were put on this car, it would be dangerous to drive in some conditions.

As described in Chapter 19, wheel alignment is controlled by the tilt of the spindle, the angle of the steering axis, and the angle of the steering arms. Briefly, these alignment angles are:

1. **Camber, positive or negative:** the vertical angle of the tire and wheel (viewed from the front); caused by a horizontal—downward or upward—tilt of the spindle

2. **Toe-in or toe-out:** inward or outward angle of the tires and wheels (viewed from above); caused by a forward or rearward angle of the spindles and controlled by the length of the tie-rods

3. **Caster, positive or negative:** an angle of the steering axis (viewed from the side); controlled by the position of the ball joints (S-LA), the ball joint and upper pivot damper unit (strut), or the kingpin; the steering axis is the imaginary line on which the steering knuckle pivots for turns

4. **Steering axis inclination:** an angle of the steering axis (viewed from the front); controlled by the same things as caster

5. **Toe-out on turns:** the relative position of the tires while turning; controlled by the angle of the steering arms

The first three of these angles are adjustable in many cars. The last two are not.

13.2 Short-Long Arm Suspensions

Short-long arm, or S-L A, is the typical RWD car's front suspension. It is also called an **unequal arm suspension.** It consists of two control arms (a short upper arm and a longer lower arm), a steering knuckle with spindle, and the necessary bushings and ball joints (Figure 13–1). The control arms are usually triangular, A-arm shapes with two inner pivot bushings, mounted on the car's frame or reinforced body structure. These control arms are also called **A-arms** or **wishbones.** Another type of lower control arm uses a single inner pivot bushing along with either a **trailing** or a **leading strut rod.** A trailing strut rod is also called a tension rod because it has the pivot bushing at the front and the control arm end at the rear. A leading strut rod is also called a compression rod; it has the pivot bushing at the rear. The strut rod prevents forward or rearward movement of the free (outer) end of the control arm. The outer ends of both control arms connect to the steering knuckle, which includes the spindle, through a ball joint. At one time, before the development of ball joints, a kingpin was used to connect the spindle to the spindle support. The ball joints or kingpin form the steering axis (Figure 13–2).

The length of the control arms and the mounting location of the pivot points are selected by engineers with the following factors in mind:

1. Camber change during tire and wheel movement reduces track change and tire scrub.
2. **Instant center** and **roll center** locations affect roll axis, camber change, and the resulting tire scrub during leaning motions of the car (Figure 13–3).
3. **Antidive,** a lowering of the front end during braking, is reduced by mounting the inner control arm pivots in a plane that creates a lever arm converging near the car's center of gravity (Figure 13–4).
4. Bump severity is less pronounced if the path that the tire travels when it goes over a bump is toward the rear of the car; this path is affected by the mounting location of the inner control arm pivots (Figure 13–5).
5. Camber change during body lean improves cornering ability.

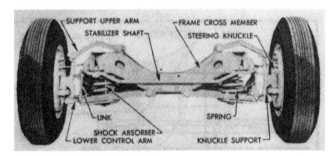

Figure 13–2 Older cars with S-L A suspensions used kingpins to connect the steering knuckle or spindle to the knuckle support; metal bushings connected the knuckle support to the control arms. *(Courtesy of Ammco Tools)*

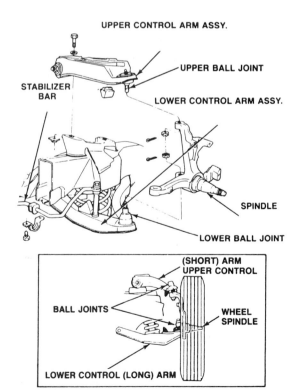

Figure 13–1 Two views of an S-L A suspension system. *(Courtesy of Ford Motor Company)*

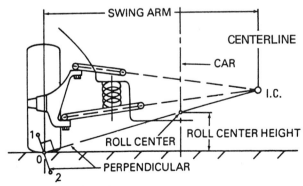

Figure 13–3 If imaginary lines are drawn along the control arm pivots, they meet at a certain point called the instant center (IC). If a line is drawn from the IC to the center of the tire, it crosses the center of the car at the roll center (RC). *(Courtesy of General Motors Corporation, Service Technology Group)*

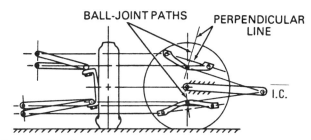

Figure 13-4 Resistance to brake dive can be generated in the front suspension. The lengthwise angle of the control arm bushings forms an imaginary lever arm. Lines drawn along the planes of the control arms show us the length and location of the end of the lever arm. *(Courtesy of General Motors Corporation, Service Technology Group)*

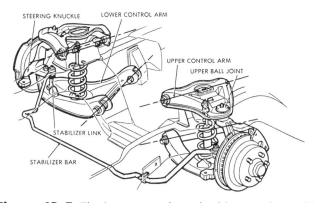

Figure 13-5 The inner control arm bushings are located in what appears to be rather odd angles, but when we consider roll center, antidive, and smooth operation over bumps, their relationship begins to make more sense. *(Courtesy of General Motors Corporation, Service Technology Group)*

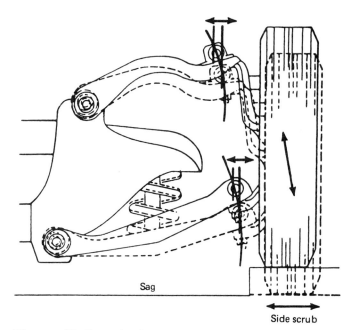

Figure 13-6 As the front suspension moves up and down, the end of the lower control arm travels in an arc; side scrub of the tire can occur if the springs sag and the lower control arm has an upward angle. *(Courtesy of Dana Corporation)*

13.2.1 Control Arm Geometry

Control arm design is matched with the size of the spring to provide optimum control arm position, tire travel over bumps, and comfortable car ride characteristics. In a static position (i.e., no bump condition), the lower control arm is horizontal or, in a new car, slightly lower at the outer end.

The ball joint at the end of this arm travels in an arc, pivoting at the inner bushing as the tire and wheel move up and down. The center point of this arc is the inner bushings; the radius of the arc is the length of the control arm. A longer control arm has a longer radius with a flatter arc. This arc is not only up and down but also slightly horizontal, depending on the midpoint and end locations of the arc (Figure 13-6). If the control arm is horizontal at the midpoint of the arc, there is minimal side motion. The more the control arm is inclined, the more side motion will occur. In a car, this side motion causes tire side scrub. If the springs sag, the lower control arms incline more and more upward at the outer end, resulting in more tire scrub and wear.

The amount of side scrub resulting from movement of the lower control arm can be reduced by the length and mounting angle of the upper control arm. The arc of travel for the upper ball joint is designed not to match that of the lower ball joint. As a consequence of the arc differences, the steering knuckle changes vertical position during wheel travel, which results in a change of the camber of the tire.

When a car goes over a bump, the tire moves upward, compressing the spring. The upward motion at the end of the control arm tends to move the tire inward, shortening the track and trying to scrub the tire. But at the same time, the upper control arm is on a more severe arc, causing the tire to move to a negative camber position (inward at the top). As the camber changes, the bottom of the tire moves outward enough to compensate for the track change; the road-contact point of the tire travels straight down the road with little or no side scrub. This is the major reason for S-L A suspensions. There are operational limits to any design. When you do a wheel alignment and lower a car's tire onto a turntable, a certain amount of side motion, which is evidence of tire scrub, can be noticed.

In the 1950s and 1960s, almost every passenger car used an S-L A front suspension. By the 1980s, most cars used strut front suspensions, mainly because of the lower cost and weight and increased space provided. S-L A or

Chapter 13 ■ Front Suspension Types

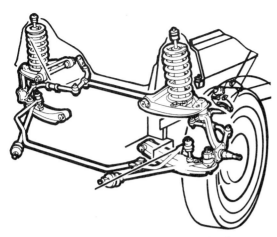

Figure 13-7 Cars without frames, or unibody cars, usually position the spring between the upper control arm and reinforced fender well. *(Courtesy of Moog Automotive Inc.)*

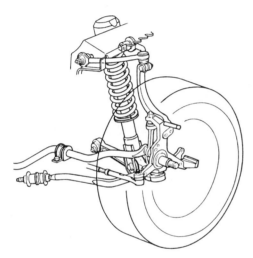

Figure 13-8 This multilink suspension uses a spring mounted over the shock absorber, like a strut. The lower end is attached to the lower control arm through a bushing. *(Courtesy of Moog Automotive Inc.)*

multilink suspensions are used on cars in which superior handling and tire wear characteristics are desirable or on vehicles that require a suspension that can support heavier loads.

13.2.2 S-L A Springs and Ball Joints

The car's weight is transferred to the front wheels through a spring. On S-L A suspensions, this is usually a coil spring; torsion bars, leaf springs, or air springs also can be used. The spring is commonly mounted between the car's frame and the lower control arm. The spring also can be mounted between the upper control arm and the fender well, which is reinforced to accept this load (Figure 13-7). Newer multilink, long-knuckle S-L A designs use a strut-mounted spring (Figure 13-8).

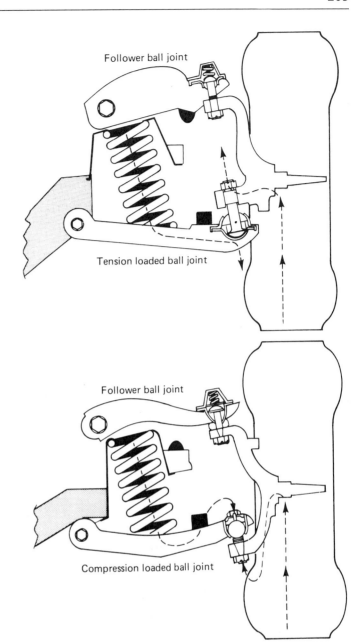

Figure 13-9 A vehicle's load travels from the frame or body, through the spring, along the control arm, through a ball joint, and on through the steering knuckle, wheel bearings, wheel, and tire, to the ground. *(Courtesy of Ammco Tools, Inc.)*

The ball joint attached to the control arm with the spring is called the **load-carrying** ball joint because it must be constructed to carry a portion of the car's weight. Since a substantial amount of load is on this joint, the joint can be constructed with running clearance. The vehicle load holds the joint tightly together, eliminating any free play (Figure 13-9).

Some load-carrying ball joints are arranged with the control arm positioned above the boss or mount on the steering knuckle, so the load squeezes the ball joint together. These are called **compression-loaded** ball joints.

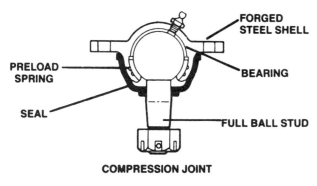

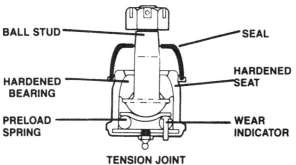

Figure 13–10 Depending on whether the control arm is above (top) or below (bottom) the steering knuckle boss, the load-carrying ball joint will be of a compression or tension design. Note the different bearing locations inside each joint. *(Courtesy of Dana Corporation)*

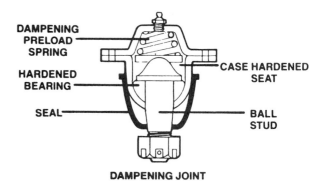

Figure 13–11 The follower ball joint uses a preload spring to keep the joint tight so it will have no play or clearance. *(Courtesy of Dana Corporation)*

Most cars position the control arm under the steering knuckle mount, so the load tries to pull the ball joint apart. These are called **tension-loaded** ball joints (Figure 13–10).

The other ball joint is called the **follower, friction-loaded, steering,** or **dampening ball joint.** The follower joint carries no vertical load; its major job is to keep the tire and wheel in alignment. This joint is built with a slight internal preload to prevent any looseness or free play; this preload is not great enough to cause steering drag. The follower and load-carrying ball joints form the steering axis, the pivot points for steering the front wheels (Figure 13–11).

Some vehicles use a type of ball joint and tie-rod end called **low friction.** These joints use a very smooth ball and a socket that is lined with a very slick polymer plastic material. These joints are normally permanently lubricated, use an improved grease seal, and have no grease fitting. Low-friction joints are most common on lighter vehicles.

13.2.3 S-L A Wear Factors

If the front springs sag too far from their original length, the geometry of the control arms will change. Static and dynamic camber will no longer be correct, and side scrub during bumps will increase. Vehicle drivability will deteriorate, and the rate of tire wear will increase.

Time and wear also tend to loosen the pivot points. The torsilastic rubber bushings commonly used at the inner pivot points tend to harden, crack, and then break apart as they get older, harder, and less resilient. Rubber bushings also tend to sag as they get old, letting the control arm change position. This position change results in a change in alignment angles. Ball joints tend to wear and cause looseness at the outer pivot points. This wear can occur fairly quickly if the boots tear or rupture. If these pivot points wear, the control arms and steering knuckle will change position, causing a change in alignment, increased tire wear, and reduced drivability. Inspection of these wear points is covered in Chapter 28.

13.3 Multilink Suspensions

Some S-L A designs have evolved so the steering knuckle has become taller, about to the top of the tire, and the spring is strut mounted over the shock absorber. This design is also called **double-wishbone, wishbone-strut, long-knuckle,** or **long-spindle S-L A** (Figure 13–12).

13.4 Strut Suspensions

A **strut suspension,** often called a **MacPherson strut,** has no upper control arm or upper ball joint. The steering knuckle connects to a spring and shock absorber assembly, which is the strut. The upper end of this assembly connects to the car body through a pivot-damper unit. A lower control arm is used; it serves the same purpose as the lower arm of an S-L A suspension (Figure 13–13). Many strut systems use a control arm with a single inner pivot, along with a strut rod. Some strut systems mount

Chapter 13 ■ Front Suspension Types

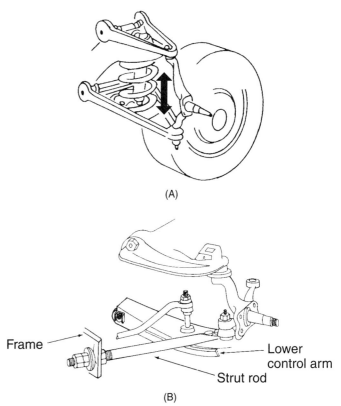

(A)

(B)

Figure 13-13 Most control arms use two inner bushings so both wheelbase and track can be maintained (A). Some control arms have a single inner bushing and use a strut rod to maintain wheelbase (B). *(Courtesy of Federal-Mogul Corporation)*

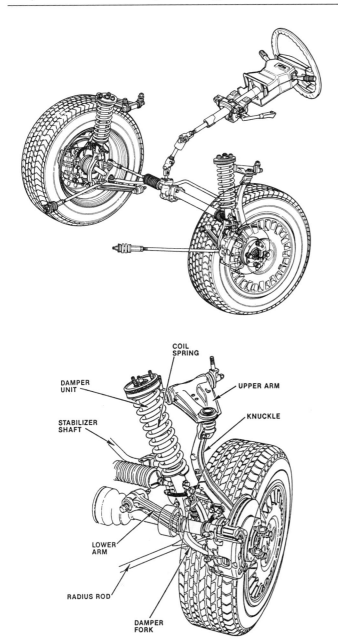

Figure 13-12 This double-wishbone suspension uses a spring and damper unit much like a strut suspension. The long curved knuckle and very angled upper control arm allow its use in areas of limited size. *(Courtesy of American Honda Motor Co., Inc.)*

spring seats at an angle or off center to try and reduce strut rod bind (Figure 13-15).

As the tire and wheel move up and down, the ball joint in the lower control arm travels in an arc similar to that of an S-L A suspension. As this occurs, the lower end of the strut must move inward and outward relative to the car; this movement changes the vertical position of the strut and therefore the camber angle. The strut's angle change is allowed by the design flexibility of the upper strut mount or pivot-damper assembly. The damper is also called an insulator (Figure 13-16).

The **upper strut mount** or **damper** serves several purposes, including the following:

1. The flexibility of the mount allows the strut angle to change to follow the travel of the lower ball joint.
2. The rubber in the mount dampens or reduces road vibrations so they are not transmitted to the body of the car.
3. A bearing is built into the mount to serve as the upper end of the steering axis.

When the front wheels are turned, the entire strut assembly, from the pivot bearing to the ball joint, pivots or

the stabilizer bar so that the end of the stabilizer bar can serve as the strut rod (Figure 13-14).

The strut is basically a coil-over shock, or a shock absorber with a coil spring mounted around it. The strut becomes shorter when the tire moves upward over a bump and longer if the tire drops into a hole. An oversize shock piston rod is required to withstand sideways bending forces; vertical loads on the tire and wheel result in sideways loads on the strut. This side load also tends to put a bind on the strut motion. Some manufacturers mount the

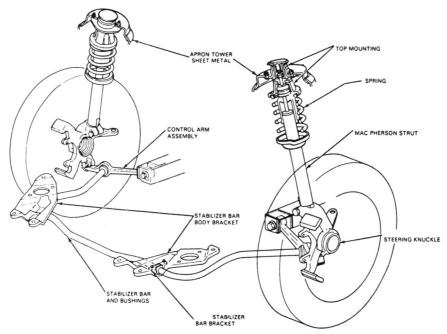

Figure 13-14 This strut suspension uses a control arm with a single inner bushing and controls the track position of the tire. The wheelbase position is controlled by the end of the stabilizer bar, which also serves as a trailing strut. *(Courtesy of Ford Motor Company)*

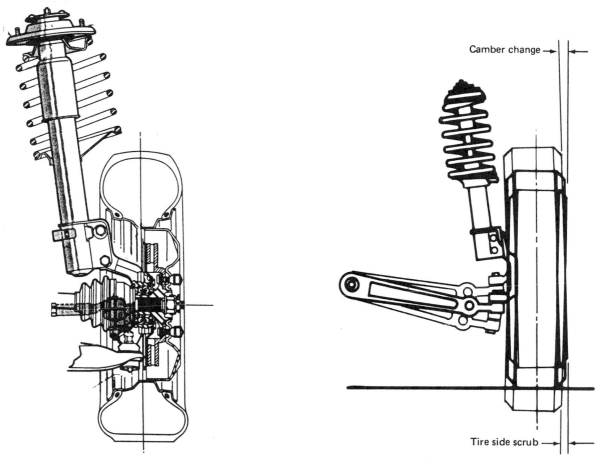

Figure 13-15 Many struts have the spring seats located off center to reduce the side load on the strut piston rod and bushing. Vehicle load tends to bend the strut sideways; the spring loads help offset this tendency. *(Courtesy of Chrysler Corporation)*

Figure 13-16 The strut must change length during suspension travel. The outer end of the control arm must travel in an arc, just like the lower control arm on an S-L A suspension.

Chapter 13 ■ Front Suspension Types

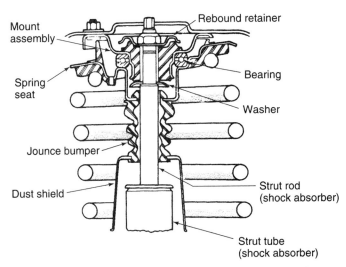

Figure 13–17 An exploded view of the upper strut mount. *(Courtesy of Chrysler Corporation)*

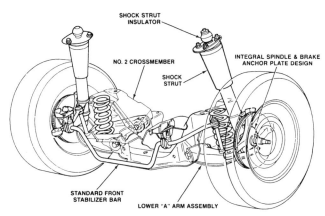

Figure 13–18 A modified strut uses a spring mounted on the lower control arm like an S-L A suspension. The strut serves as the upper end of the steering axis with the shock absorber. *(Courtesy of Ford Motor Company)*

turns. The upper mount also transfers the load of that corner of the car into the strut and spring (Figure 13–17). In most cars, the lower control arm carries no load; a friction-loaded ball joint is used.

Some struts are made so they bolt to the steering knuckle. On some of these cars, there is a provision for adjusting camber at this connection. Other struts are welded directly onto the steering knuckle. (See Figure 31–35.)

13.4.1 Modified Struts

Some cars use a **modified strut,** in which the spring is mounted on the lower control arm instead of the strut. This is also called a **single control arm** suspension (Figure 13–18). In this suspension, the strut is essentially the same as a standard strut without the spring mounts; the upper strut mounts or dampers and pivots are the same. The lower control arm and ball joint are essentially the same as the lower control arm on an S-L A suspension, and the lower ball joint is a load-carrying ball joint. A torsion bar can also be used with a strut. On these cars, the lower control arm usually becomes the lever arm for the torsion bar (Figure 13–19).

13.4.2 Advantages of Struts

In comparing a strut suspension with an S-L A suspension, the strut is simpler to build and is also lighter in weight and less expensive. Because there is no upper control arm, this suspension allows more room in the engine compartment for the transverse engine mounting plus transaxle and driveshafts of a FWD car. The under-hood area of a FWD, transverse engine car is very crowded. There is not enough room for the upper control arm of an S-L A suspension. Also, struts have only a small amount of camber change during travel over bumps.

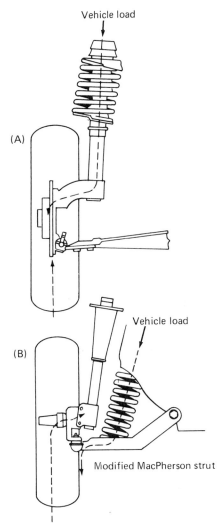

Figure 13–19 In a strut suspension, the vehicle load passes from the body, through the upper strut mount, down through the spring to the steering knuckle, and on through the wheel bearings, wheel, and tire, to the ground (A). In a modified strut, the load travels through the spring and lower control arm, like an S-L A suspension (B).

13.4.3 Strut Wear Factors

Strut suspensions are also affected by spring sag and bushing wear. In some cases, they are affected more than an S-L A type. Many strut systems do not have any, or have a very limited, provision for camber or caster adjustment to compensate for wear. Most S-L A systems are adjustable, so a readjustment can be made for slight misalignment caused by spring sag or bushing wear.

If the shock absorber wears out on a strut suspension, the strut needs to be removed for service. Depending on the strut or the availability of parts, the shock absorber is rebuilt, the shock absorber cartridge is replaced, or the entire strut is replaced. These operations are described in Chapter 31.

13.5 Solid Axles

As mentioned earlier, solid axles are not used on passenger cars because of their harsher ride and inferior handling characteristics on uneven roads. They are commonly used on trucks, 4WDs, and some pickups because of their simpler, stronger, and less expensive construction. They generally require less maintenance because of their minimum number of wear points. A **solid axle** is simply a strong, solid beam of steel (usually I shaped) with a kingpin at each end to connect to the steering knuckle. This axle is called a **monobeam** by one manufacturer (Figure 13–20).

A solid axle is usually connected to the vehicle's frame by a pair of leaf springs. The springs used are sturdy enough to position the axle and control wheelbase and caster. Camber, steering axis inclination, and track are controlled by the length and shape of the axle beam and the steering knuckles. These parts are usually durable enough to maintain proper wheel alignment for a relatively long period of time.

Solid axle suspensions have more potential design problems than independent suspensions. Because each style of independent suspension attaches each suspension unit separately to the frame and therefore each wheel moves separately from the other, the only wheel travel an engineer is concerned with is bounce travel. A solid axle is normally attached to the frame through a flexible coupling—that is, the springs. In addition to bounce travel, the axle can move in the following directions:

1. **Windup:** a twisting of the axle when the brakes are applied; windup can also occur in a drive axle during acceleration
2. **Side shake:** a sideways motion of the axle relative to the frame
3. **Yaw:** a rotation of the axle—forward on one end and rearward on the other

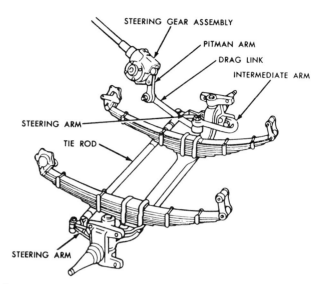

Figure 13–20 A solid axle with springs and steering linkage. *(Courtesy of Ford Motor Company)*

4. **Tramp:** a rotation of the axle—upward on one end and downward on the other; tramp is often called **shimmy**

These potential problems can be cured with additional control arms, Panhard rods, or radius arms, but these additional parts increase the complexity, cost, and number of wear points.

13.6 Swing Axles, Twin I-Beam Axles

Swing axles, also called twin I-beam axles, have been used by Ford Motor Company on its pickups, 4WDs, and light trucks. Twin I-beam axles combine some of the sturdiness and simplicity of a solid axle with some of the improved ride and handling characteristics of an independent suspension. A twin I-beam axle is a compromise between these two suspension types (Figure 13–21).

A twin I-beam axle consists of two shortened I-beams that have a kingpin boss at one end to connect to the steering knuckle, just like one end of a solid axle. Newer designs use a pair of ball joints in place of the kingpin. The inner end of these axles has a bushing boss to accept the bushing that connects the axle to the frame. This rubber bushing allows the wheel end of the axle to pivot, or swing up and down, for suspension movement. A radius arm is at the free or outer end of each axle beam to control the wheelbase. It also controls the vertical position of the kingpin and therefore the caster. The radius arm is bolted securely to the axle and connected to the frame using a rubber bushing (Figure 13–22).

Chapter 13 ■ Front Suspension Types

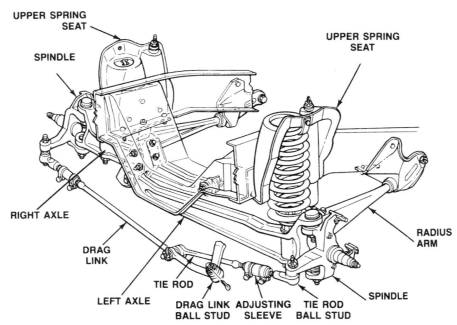

Figure 13-21 A twin I-beam front suspension. This particular swing axle design uses two ball joints to connect the steering knuckle to the axle. *(Courtesy of Ford Motor Company)*

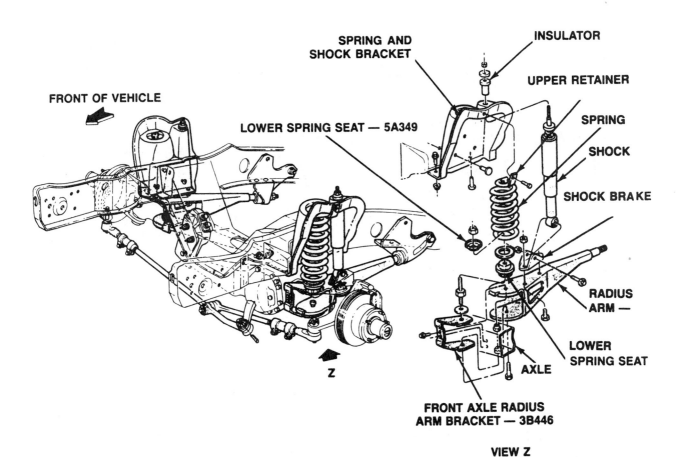

Figure 13-22 The radius arm controls the front end of the wheelbase on this twin I-beam suspension. Track is controlled by the lengths and angles of the two axles. *(Courtesy of Ford Motor Company)*

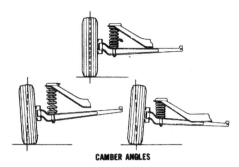

Figure 13-23 Camber and track change as a swing axle moves up and down. *(Courtesy of Ford Motor Company)*

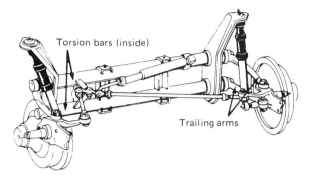

Figure 13-24 A twin trailing arm front suspension. The steering knuckle is supported by the two trailing arms. Torsion bars are enclosed in the two transverse tubes to function as the front springs. *(Courtesy of Moog Automotive Inc.)*

Camber and steering axis inclination, along with track, are controlled by the length and shape of the I-beams and the steering knuckle and by the height of the spring. As the spring changes length, the free end of the axle moves up and down relative to the pivoting end. As this occurs, camber and track change. In actual operation, the free or wheel end stays about one-half the tire diameter off the ground, while the pivoting end moves up and down along with the frame of the vehicle. Caster and wheelbase, controlled by the radius arm, also change slightly during vertical vehicle movement (Figure 13-23).

Because of the four additional pivots, there are more wear points on a twin I-beam axle than on a solid axle. Wear at these points, plus camber and track change at different spring heights, can cause the tire and wheel alignment to change and cause tire wear or drivability changes.

13.7 Miscellaneous Suspension Types

Other front suspension types have been used on cars, but these are not in current use or commonly used by any major manufacturer. The last one of these used to a large extent was the **trailing arm suspension.** The steering knuckle support is attached to one or a pair of trailing arms. During travel over bumps, the tire and wheel swing upward at the ends of the trailing arms. There is zero camber or track change. This suspension has a limitation in that vehicle lean causes an equal change in camber. This usually results in poor front-tire adhesion while cornering (Figure 13-24).

13.8 Front-Wheel Drive Axles

Most of the front-wheel suspension types described in this chapter have been adapted to FWD or 4WD by various manufacturers. The biggest change required is using a steering knuckle with a hollow spindle for the driveshaft. In some vehicles, the wheel bearings and the hub are mounted over the spindle, whereas in others, wheel bearings and a spindle that will fit into the steering knuckle are used. Both of these methods have been described in Chapters 11 and 12. Most FWD cars use a strut suspension.

On S-L A suspensions, the control arm length, angles, and mounting are essentially the same as on RWD cars. The major difference, other than the hollow spindle in the steering knuckle, is the common usage of torsion bars for springs. A coil spring would interfere with the driveshaft (Figure 13-25).

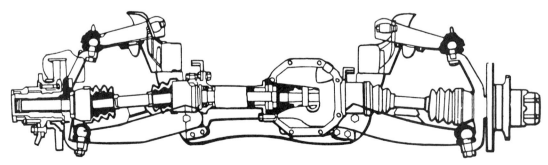

Figure 13-25 A FWD, S-L A suspension; this unit can be used with two- or four-wheel drive. *(Courtesy of Chrysler Corporation)*

Chapter 13 ■ Front Suspension Types

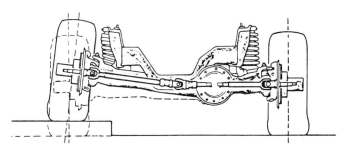

Figure 13–26 A 4WD swing axle front suspension. This particular axle system is not used on two-wheel drive vehicles. *(Courtesy of Dana Corporation)*

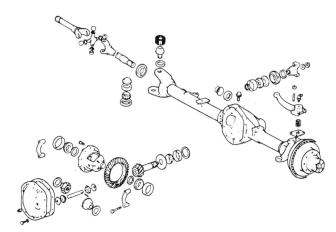

Figure 13–27 An exploded view of a FWD solid axle; this axle is common on older 4WD vehicles. *(Courtesy of Ford Motor Company)*

Front-wheel-drive solid axles and twin I-beam axles cannot use a kingpin; it would be in the way of the drive axle. A pair of ball joints are used above and below a hollow section of the axle housing, where the CV joint or universal joint is positioned. These ball joints form the steering axis. Some 4WDs use a universal joint in place of a CV joint, which usually limits the 4WD to off-pavement use only. A universal joint is less expensive than a CV joint, but it can cause vibrations or pulsations during turning maneuvers (Figure 13–26 and Figure 13–27).

13.9 Practice Diagnosis

You are working as an alignment technician in a front-end shop and encounter the following problems:

CASE 1: The customer is planning to restore his 1971 Mustang and has brought it in for an estimate to rebuild his front suspension. The suspension is like that shown in Figure 13–7. Which of the ball joints is load carrying and which one is the follower? Does the load-carrying ball joint have a tension or compression load on it?

CASE 2: You are doing a prealignment inspection on a late-model Mustang (the suspension is similar to that shown in Figure 13–18). Is the ball joint checked as a load-carrying or a follower joint?

Terms to Know

A-arms or wishbones	low friction	strut suspension
antidive	MacPherson strut (strut)	swing axle
camber (positive or negative)	modified strut	tension-loaded ball joints
caster (positive or negative)	monobeam	toe-in or toe-out
compression-loaded ball joint	multilinks	toe-out on turns
dampening ball joint	roll center	trailing arm suspension
damper	shimmy	trailing or leading strut rod
double wishbone	short-long arm (S-L A)	tramp
follower ball joint	side shake	unequal arm suspension
friction-loaded ball joint	single control arm	upper strut mount
instant center	solid axle	windup
load-carrying ball joint	steering axis	wishbone strut
long knuckle	steering axis inclination	yaw
long-spindle S-L A	steering ball joint	

Review Questions

1. A car's front suspension system allows the front wheels to:
   A. Turn for steering maneuvers
   B. Rise and fall to provide for a smooth ride
   Which option best completes the statement?
   a. A only
   b. B only
   c. Both A and B
   d. Neither A nor B

2. A control arm for a front suspension has a single pivot at the outer end and:
   A. A single pivot at the inner end
   B. Two pivots at the inner end
   Which option best completes the statement?
   a. A only
   b. B only
   c. Both A and B
   d. Neither A nor B

3. Statement A: An S-L A suspension system causes the front-end alignment to change as the wheel travels up and down.
   Statement B: The short and long arms are always parallel.
   Which statement is correct?
   a. A only
   b. B only
   c. Both A and B
   d. Neither A nor B

4. Statement A: An A-arm always has two inner pivot bushings.
   Statement B: A strut rod is never used with an A-arm.
   Which statement is correct?
   a. A only
   b. B only
   c. Both A and B
   d. Neither A nor B

5. A major advantage of the S-L A suspension design is that:
   A. The camber angle remains constant during suspension travel
   B. Very little side scrub of the tire occurs during suspension travel
   Which option best completes the statement?
   a. A only
   b. B only
   c. Both A and B
   d. Neither A nor B

6. Statement A: The load-carrying ball joint is on the lower control arm of most RWD cars with S-L A suspension.
   Statement B: The ball joint used in strut suspensions is also a load-carrying design.
   Which statement is correct?
   a. A only
   b. B only
   c. Both A and B
   d. Neither A nor B

7. The inner metal sleeve of a torsilastic bushing is:
   a. Designed to rotate on its mounting bolt
   b. Designed to allow the rubber insert to rotate on it
   c. Locked in place when the mounting bolt is tightened
   d. None of these

8. In a strut suspension, the shock absorber piston rod:
   A. Provides a turning pivot point for steering
   B. Holds the steering knuckle in alignment during suspension travel
   Which option best completes the statement?
   a. A only
   b. B only
   c. Both A and B
   d. Neither A nor B

9. Statement A: A strut suspension allows suspension travel with no change in camber angle.
   Statement B: This suspension allows suspension travel with no change in track.
   Which statement is correct?
   a. A only
   b. B only
   c. Both A and B
   d. Neither A nor B

10. The upper strut-to-fender well mount provides:
    a. A bearing for the steering axis
    b. A damper to isolate suspension vibrations
    c. Flexibility so the strut can change angle
    d. All of these

11. Statement A: A modified strut uses a load-carrying ball joint.
    Statement B: There are no spring seats on a modified strut.
    Which statement is correct?
    a. A only
    b. B only
    c. Both A and B
    d. Neither A nor B

12. A strut suspension:
    A. Is slightly heavier but less complex than an S-L A suspension
    B. Allows more room in the engine compartment than an S-L A suspension
    Which option best completes the statement?
    a. A only
    b. B only
    c. Both A and B
    d. Neither A nor B

13. Statement A: Spring sag in a strut suspension causes the camber to change.
    Statement B: Shock absorber replacement is more difficult and expensive on a strut suspension than on an S-L A suspension.
    Which statement is correct?
    a. A only
    b. B only
    c. Both A and B
    d. Neither A nor B

Chapter 13 ■ Front Suspension Types

14. Which of the following axle motions is not a potential problem on vehicles using a solid front axle?
    a. Windup that can cause an alignment change
    b. Camber change because of spring sag
    c. Tramp that can result from an imbalance or worn parts
    d. Yaw as a result of vehicle lean

15. A major advantage of a solid axle is that it is:
    A. Sturdy with few wear points
    B. Lightweight and simple
    Which option best completes the statement?
    a. A only   c. Both A and B
    b. B only   d. Neither A nor B

16. A solid axle can use _____ for the steering axis pivot.
    A. Ball joints
    B. A kingpin
    Which option best completes the statement?
    a. A only   c. Both A and B
    b. B only   d. Neither A nor B

17. Statement A: Suspension travel causes a camber change on swing axle suspensions.
    Statement B: A track change results from suspension travel on swing axles.
    Which statement is correct?
    a. A only   c. Both A and B
    b. B only   d. Neither A nor B

18. A swing axle suspension:
    A. Is classed as an independent suspension
    B. Uses a pivot bushing to connect the inner ends of the two axles
    Which option best completes the statement?
    a. A only   c. Both A and B
    b. B only   d. Neither A nor B

19. Statement A: The radius arm used with a swing axle helps control the track of the front suspension.
    Statement B: This radius arm prevents axle windup during braking.
    Which statement is correct?
    a. A only   c. Both A and B
    b. B only   d. Neither A nor B

20. Which of the following is not a front suspension type?
    a. Unequal-length control arm
    b. Semitrailing arm
    c. Swing axle
    d. Double wishbone

14 Rear Suspension Types

Objectives

Upon completion and review of this chapter, you should be able to:

❑ Comprehend the different styles of automotive rear suspensions.

❑ Comprehend the terms commonly used with rear suspensions.

❑ Understand the purpose for the various components of a rear suspension.

❑ Understand the operating differences between RWD and FWD rear suspensions.

14.1 Introduction

Rear suspensions are very similar to front suspensions in that they allow vertical tire movement. However, in most cases, they do not allow steering, and in cases in which four-wheel steering (4WS) is used, the steering is very limited. 4WS is used to improve front steering characteristics. Rear tires and wheels are normally set at or near zero camber (straight up and down) and zero toe (straight forward). Excessive camber or toe would cause tire wear and, possibly, a car that **dog tracks** and goes down the road slightly sideways.

An imaginary line drawn midway between the two rear wheels is called the **thrust line.** This thrust line should run down the middle of the car. The car's path during straight-ahead driving follows this thrust line. If the thrust line runs at an angle to the car's centerline, the rear tires will not track or follow in direct alignment with the front tires. A rear tire and wheel thrust line that points to the right of the vehicle's centerline (at the front) will cause the rear tires to steer to the right. The front tires will have to be steered to the right in order to go straight down the road, and the rear tires will leave tracks to the right side of the tracks left by the front tires (Figure 14–1).

Rear-wheel alignment has traditionally been taken care of by the strength of the solid rear axle housing. More of today's cars have independent rear suspension or lighter-weight rear axles with spindles that bolt on. A greater need has developed for rear-wheel alignment to correct tire wear or thrust problems because of these changes. Rear-wheel alignment is discussed in Chapters 19 and 34.

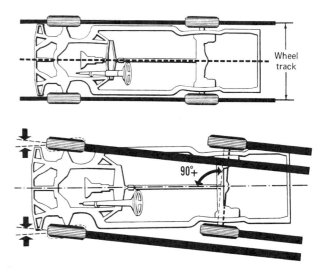

Figure 14–1 A vehicle follows a path parallel to the rear tires. If they are parallel to the car's centerline (top), the rear tires will follow the path of the front tires. If the rear tires are at an angle (bottom), the rear track will be to the side of the track of the front tires. *(Reprinted by permission of Hunter Engineering Company)*

14.2 Solid Axle, RWD Suspension

Cars, pickups, vans, and trucks have traditionally used driven, solid rear axles that are sometimes called live axles. This sturdy assembly holds the tires and wheels in alignment, transfers the vehicle load from the springs to the tires and wheels, and provides the gearing necessary

Chapter 14 ■ Rear Suspension Types

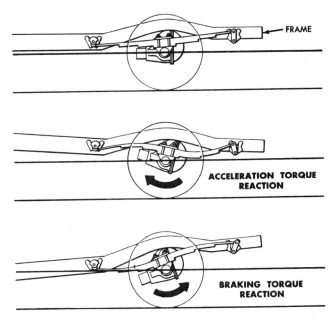

Figure 14-2 The rear axle of a RWD car tends to wind up or rotate in response to acceleration and braking torque. *(Courtesy of Ford Motor Company)*

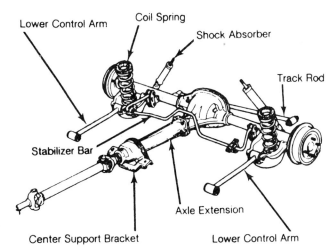

Figure 14-3 The center support bracket and the axle extension prevent axle rotation from acceleration or brake torque response. The driveshaft and axle extension are a type of torque tube driveshaft. *(Courtesy of Mitchell Repair Information Company)*

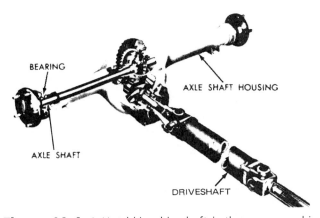

Figure 14-4 A Hotchkiss driveshaft is the common driveshaft with a universal joint at each end; it transmits torque from the transmission to the rear axle. *(Courtesy of Ford Motor Company)*

to transfer the power from the driveshaft to the tires and wheels. The bearings used to transfer the vehicle weight were described in Chapter 11.

A solid rear axle is subject to the same movements possible from a front axle: windup, side shake, yaw, and tramp. Tramp is normally a much greater problem on steerable axles than nonsteerable axles. **Axle windup** is a greater problem on live axles than dead axles. One law of physics is that for every action there is an equal and opposite reaction. In a live axle, the action of sending power to the tires and wheels and causing them to turn in a forward direction produces a reaction of the axle housing trying to rotate it in the opposite direction. If we could see it, a hard acceleration of the car produces a turning force at the axle that tries to lift the front tires and wheels. A "wheelie" by a drag race car is a dramatic example of this (Figure 14-2).

Two styles of driveshaft design have been used with live rear axles: **torque tube** and **Hotchkiss.** Torque tube driveshafts enclose the driveshaft in a steel tube that is bolted solidly onto the axle housing and connected to the car's transmission or frame member through a flexible pivot. This tube acts as a long lever, preventing axle windup. Normally, only a single universal joint is used at the forward end of the driveshaft. This style of driveshaft has seen limited use since the mid-1950s because it tends to be heavy and cumbersome (Figure 14-3).

Hotchkiss driveshafts are the most commonly used RWD driveshafts. This design usually has two universal joints, one at each end, and simply transfers power from the transmission to the rear axle. Axle windup must be controlled by other methods, usually the rear leaf springs or control arms (Figure 14-4).

One manufacturer uses a **torque arm** to control axle windup on some car models. This long stamped-steel beam bolts solidly to the axle housing and attaches to the transmission with a rubber bushing. The torque arm functions much like the tube of a torque tube driveshaft.

Rear-wheel alignment on solid axles, camber, toe, and track are controlled by the strength and length of the axle and axle housing. The rear-wheel end of the wheelbase is controlled by the rear springs or control arms (Figure 14-5).

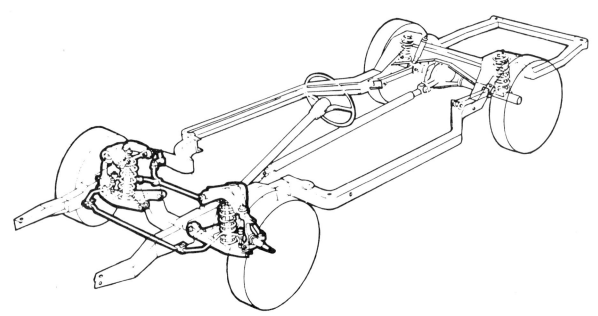

Figure 14–5 On most RWD cars, the rear axle assembly controls rear-wheel camber and toe angles; the wheelbase and the crosswise position of the axle are controlled by the rear leaf springs or suspension control members. *(Courtesy of Moog Automotive Inc.)*

14.2.1 Solid Axle, Leaf Spring Suspension

The solid axle, leaf spring suspension is the simplest form of rear suspension. At one time, it was also the most common. A pair of leaf springs attaches to the frame through a rubber bushing at the front and through rubber bushings and a shackle at the rear. The rear axle housing bolts solidly to the center of the spring with a set of U-bolts. The front portion of the spring acts as a control arm, positioning the rear axle housing and establishing the wheelbase (Figure 14–6).

The front portion of the spring is often shorter than the rear to provide better axle position control. If this portion were made longer, it would be more flexible, and axle yaw and windup would increase. Severe windup can cause **wheel hop,** a rapid vertical bouncing of the tire and wheel. Yaw, usually a result of vehicle lean, causes the rear of the car to steer right or left as the rear thrust line changes.

14.2.2 Solid Axle, Coil Spring Suspension

Because a coil spring does not have the ability to locate an axle, coil spring solid axles require more parts. A coil spring rear end usually uses two lower suspension arms (sometimes called links) to control the rear axle end of the wheelbase. It also uses one or more upper suspension arms to control axle windup and side motion. When two upper arms are used, they are usually skewed diagonally (mounted at an angle) to control side motion as well as

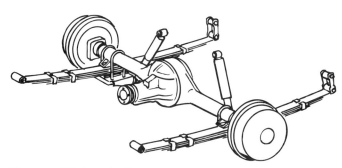

Figure 14–6 A leaf spring rear suspension consists of the springs and shock absorbers. *(Courtesy of Moog Automotive Inc.)*

windup. The two lower arms are usually in a trailing and parallel position to allow bind-free travel. Some designs use only one upper arm to control axle windup plus a **Panhard rod,** also called a **track bar,** to control side motion (Figure 14–7). A Panhard rod connects to the axle, runs across the car, and connects to the frame of the car. A flexible rubber bushing is used at each end. Sometimes a **Watt's link** is used in place of a Panhard rod. Some coil spring suspensions use two lower control arms and a Panhard rod with a torque arm (Figure 14–8).

The control arms are usually connected to the car's frame and the rear axle through rubber bushings. These bushings are very similar to those used on front ends. They offer maintenance-free service, a high degree of compliance, and the ability to dampen road vibrations.

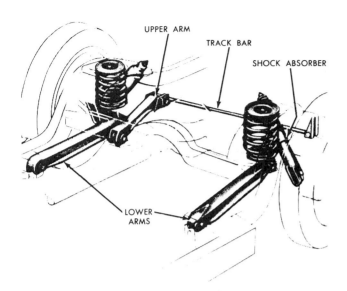

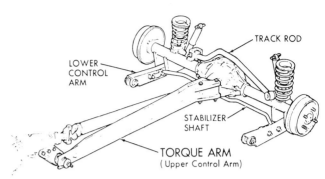

Figure 14-8 This suspension uses a torque arm to prevent axle windup. The track rod prevents side motion, the stabilizer shaft reduces lean on corners, and the lower control arms control the wheelbase. *(Courtesy of General Motors Corporation, Service Technology Group)*

Figure 14-7 This rear suspension has a single upper control arm to prevent axle rotation. Side motion is controlled by the track bar, also called a Panhard rod; the track bar connects to the axle at the left end. *(Courtesy of Ford Motor Company)*

14.3 Independent Rear Suspension

Independent rear suspension (IRS) on RWD vehicles has been used only on cars of a luxury or sporty nature. IRS is more expensive to build and has more wear points than a solid axle, but it provides better ride qualities and usually better camber and toe control of the rear tires.

One reason for an improved ride is the reduction of unsprung weight. The housing and final drive gears, which include the ring and pinion plus differential, are mounted on the frame. Their weight is carried by springs, not directly by the tires and wheels. Reducing the unsprung weight allows better spring and shock absorber control of tire and wheel movement. (See Figure 14-11.) Power is transmitted from the differential to the tires and wheels by a pair of driveshafts or half shafts, much like those on a FWD car.

14.3.1 IRS, Semitrailing Arm Suspension

In the IRS, semitrailing arm suspension design, each rear tire and wheel is mounted on a single control arm, which is usually mounted to the frame through one or a pair of rubber bushings (Figure 14-9). A single pivot can be used if an additional control arm or strut rods are used. A pure trailing arm has the pivot mounted at a 90-degree angle to the centerline of the car. The path that the tire follows as it pivots is exactly parallel to the car's centerline. If the car

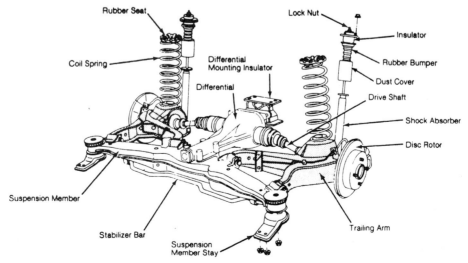

Figure 14-9 Semitrailing arms support the rear wheel bearings to control camber and toe angles, as well as control the wheelbase. Note the angles of the front pivot bushings that make these semitrailing arms. *(Courtesy of Mitchell Repair Information Company)*

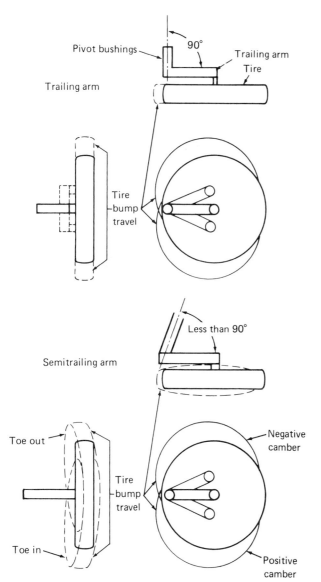

Figure 14–10 The camber and toe angles of a tire do not change as a trailing arm suspension moves through its bump travel. The tire of a semitrailing arm suspension has more toe-out and more negative camber during upward travel and more toe-in and more positive camber during downward travel.

leans, there will be an equal change in camber angle. A 5-degree lean as the car corners causes a 5-degree change (more positive on the outside tire) in camber. This reduces the amount of traction and tire adhesion (Figure 14–10).

Semitrailing arms have their pivots angled or skewed about 10 to 15 degrees from those of a pure trailing arm. With this change, tire angle will change with suspension movement. Camber will remain closer to vertical as the vehicle leans on a corner and the control arm sweeps through this different arc of travel. This is an improvement, but along with this camber change, there is a toe change that can cause a change in the rear tire and wheel thrust line. Rear-wheel steering occurs when the thrust line changes. This suspension design is relatively simple, inexpensive, and lightweight.

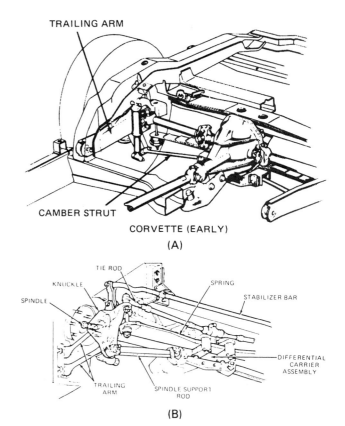

Figure 14–11 In early Corvettes (A), a single trailing arm controlled wheelbase, brake torque, and toe angle; camber angle was controlled by the axle shaft and lower strut. In newer Corvettes (B), a five-link suspension uses two trailing arms to control wheelbase and braking torque, a tie-rod to control toe angle, and a spindle support rod and axle shaft to control camber angle. *(Courtesy of General Motors Corporation, Service Technology Group)*

14.3.2 IRS, Trailing Arm Suspension

Chevrolet Corvettes use a trailing arm IRS, but additional control links have been added to provide better tire angle control. Early IRS designs used a single trailing arm on each side. Newer models use a pair of trailing links. In the early design, the trailing arm controlled the toe angle and the rear end of the wheelbase. In the newer models, the trailing links control the wheelbase, and a separate link controls toe. In both styles, the half shaft and a lateral strut rod control camber. These three or five links produce very good tire and wheel control in the three-link suspension and excellent control in the five-link suspension. A few other cars use a similar type of rear suspension (Figure 14–11).

14.3.3 IRS, Strut Suspension

Strut suspension in an IRS closely resembles the strut suspension used on front ends. RWD strut suspensions are sometimes called Chapman struts. Rear struts do not

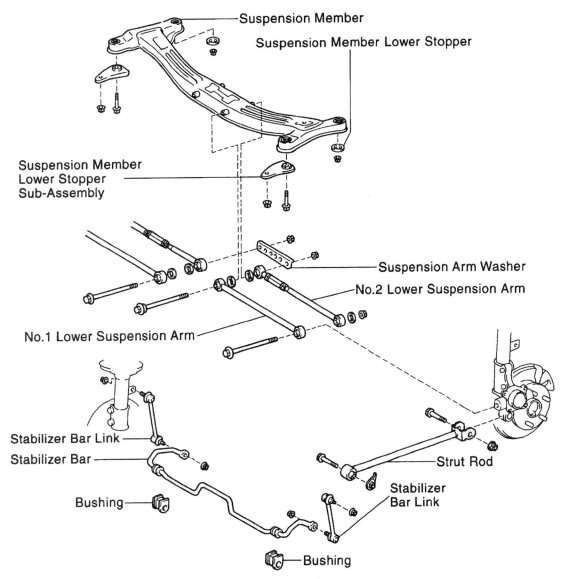

Figure 14-12 This strut suspension uses two lower suspension arms to control track and toe angle; the strut rod controls wheelbase.

use steering knuckles. The A-arms used often have a wide spread where they connect to the strut to provide good toe control. These often become H-arms, because of the two pivot bushings at each end. Many suspensions use a pair of transverse arms or links, one in front and one in back (Figure 14–12). Acceleration or braking forces try to change rear-wheel toe. Toe should be held very close to zero during acceleration and braking. One or two rubber bushings are used at this connection between the control arm and the strut. The upper mount or damper does not have a pivot bearing. It is mainly used to allow strut-angle changes and to dampen road vibrations (Figure 14–13).

Strut suspensions are relatively inexpensive, fairly lightweight, and offer fairly good tire and wheel control. However, they tend to take up space in the car's storage area or in part of the space used by the rear seat.

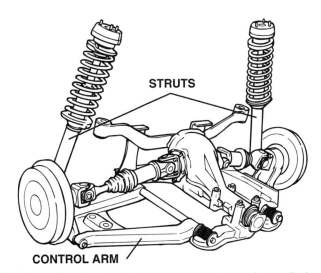

Figure 14-13 A strut rear suspension, also called a MacPherson or Chapman strut. The lower control arm and strut support the axle bearings. *(Courtesy of Moog Automotive Inc.)*

14.4 Miscellaneous RWD Axle and Suspension Types

At least three other RWD suspension types have been used on cars, but their use is very limited at this time. They include the following:

1. **Swing axle.** The axle shafts pivot from the universal joint at the inner end. A single or double trailing arm is usually used to control wheelbase and to absorb brake loads. This design has definite drawbacks in that camber and track change with vehicle height. The swing axle length is fairly short, so this camber and track change is fairly severe. Potentially dangerous suspension **jacking** can also occur. Hard cornering forces tend to tuck or fold the inside rear tire under the car, which tends to lift the car up and over that axle (Figure 14–14).
2. **Low pivot axle.** This is essentially a solid axle that bends in the middle. A single pivot is placed at the bottom of the axle housing, near the middle, and a universal joint is built into the axle shaft at this pivot point. The pivot allows independent vertical tire and wheel movement.
3. **de Doin axle.** This is a solid axle with a separate gear housing and axle shafts. This is not an IRS design; the tires and wheels are mounted on the ends of a solid axle. The rear axle gears are mounted to the frame in a manner similar to IRS cars. A de Doin axle provides the simplicity and tire angle control of a solid axle, while reducing the unsprung weight for better tire and wheel control (Figure 14–15).

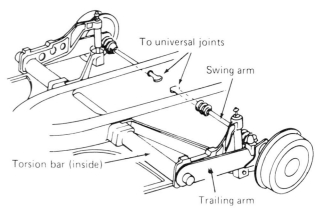

Figure 14–14 A trailing arm with swing axle rear suspension. The axles that control camber connect to universal joints at the differential assembly. Torsion bars are used for rear springs. *(Courtesy of Moog Automotive Inc.)*

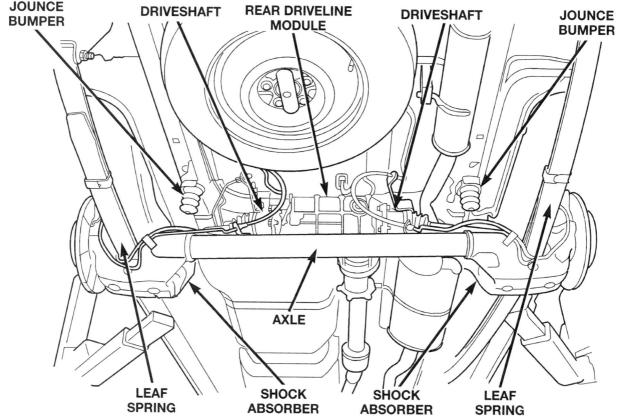

Figure 14–15 This all-wheel-drive van uses a de Doin style axle. The driveline module containing the differential is attached to the vehicle underbody. *(Courtesy of Chrysler Corporation)*

14.5 FWD Rear Axles

FWD cars use rather simple rear axles. All that is necessary is to allow the tire and wheel to move up and down, while staying in alignment. The suspension design can be a variation of any of the types previously discussed. Most manufacturers use variations of three styles: the solid axle, trailing arm, and strut.

With many FWD cars, the rear suspension carries a minor portion of the car's weight. The front tires carry most of the weight, plus the driving and steering forces. The rear tires have very limited traction because of the lower weight they carry. They also have a fairly low wear rate. Consequently, rear tires do not wear out at the same rate as front tires. The rear tires on FWD cars can wear as diagonal wipe, an odd, scalloped wear pattern, probably because of this lighter loading. Regular tire rotation is usually necessary to obtain an even wear rate.

14.5.1 FWD, Rear Solid Axle Suspension

A solid axle beam, usually of stamped steel, connects the two tires and wheels. A pair of trailing arms are bolted rigidly or welded onto the axle beam and connect to the frame through rubber bushings. The axle and trailing arms are somewhat flexible, to allow for slight twists when the car leans on corners. Side motion of the axle is controlled by a Panhard rod (Figure 14–16). Some designs use two lower links, with rubber bushings at both ends, and one or two upper links to control axle windup during braking. Both of these designs normally use coil springs. These axles can also use two leaf springs to locate the axle and provide the spring action (Figure 14–17).

Camber, toe, and track are controlled by the length and strength of the axle beam. Wheelbase is controlled by the trailing arms, links, or leaf springs. This is a simple, relatively lightweight, inexpensive design. Like other solid axles, it is not an independent suspension.

14.5.2 FWD, Rear Trailing Arm Suspension

Each rear tire and wheel is attached to a trailing arm. The trailing arm is attached to the frame through a pivot bushing at the front. Most FWD, rear trailing arm designs connect the two trailing arms to each other with a cross beam. This cross beam strengthens the trailing arms (in a crosswise direction) and tends to reduce body roll; it has to twist during body lean. Without the cross beam, the trailing arms would operate completely independently (Figure 14–18).

If we imagine a trailing arm suspension with the cross beam moved to the rear, in line with the wheel centers, we would have a solid axle with trailing arms as described in the previous section. Designers can place the cross beam at the front, rear, or anywhere in between, depending on the ride and handling characteristics they desire.

Another style of trailing arm rear suspension is used on one import car model. It is called a **double-wishbone** suspension by the manufacturer. This trailing arm pivots on a rubber bushing at the front, and it is bolted solidly to the knuckle to which the hub and wheel are attached. This knuckle is located laterally by two lower arms and one upper arm. Vehicle loads are carried by the damper unit and spring, which is fitted between the knuckle and the body. The trailing arm establishes the wheelbase and absorbs braking torque. The upper and lower arms control camber, and the two lower arms control toe. The

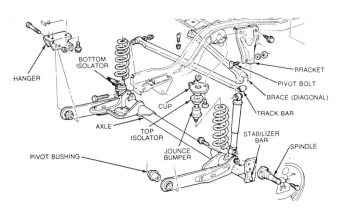

Figure 14–16 A trailing arm, solid axle rear suspension from a FWD car. The axle controls track and the camber and toe angles, the trailing arm controls wheelbase, and the track bar controls side motion. *(Courtesy of Chrysler Corporation)*

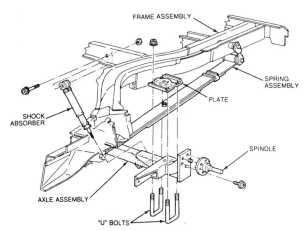

Figure 14–17 A solid axle, leaf spring suspension. The leaf springs control wheelbase and side motion. *(Courtesy of Chrysler Corporation)*

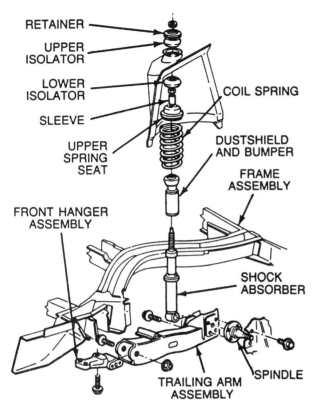

Figure 14–18 A trailing arm rear suspension. A lateral beam (not shown) connects the two trailing arms. *(Courtesy of Chrysler Corporation)*

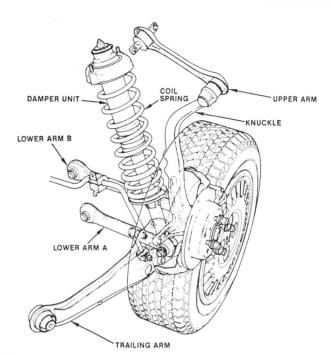

Figure 14–19 The manufacturer calls this a double-wishbone rear suspension. It combines a trailing arm with a long, curved knuckle. The top of the knuckle is located by the upper arm. The two lower arms locate the rear hub laterally and combine to give almost zero toe change during bump travel. *(Courtesy of American Honda Motor Co., Inc.)*

geometry of the knuckle and the control arms produces a controlled camber change and toe change during suspension travel (Figure 14–19).

14.5.3 FWD, Rear Strut Suspension

Nonpowered rear struts normally use a strut, control arm, and strut rod to control rear tire and wheel movement. The spring can be mounted on the strut or between the car body and the lower control arm to save mounting room. In some designs, the lower control arm is replaced by a pair of lateral links (Figure 14–20).

With strut suspensions, camber is controlled by the strut, toe and track are controlled by the control arm, and wheelbase is controlled by the strut rod or lower control arm. Strut suspensions are completely independent, lightweight, fairly simple, and fairly inexpensive to manufacture.

14.5.4 FWD, Rear Short-Long Arm Suspension

An S-L A rear suspension is used on some Ford Motor Company station wagons. This is a very compact system. Although it is more complex than the strut suspensions used on other car models, it provides a maximum of cargo space in the rear area. Variable-rate coil springs are used to combine load-carrying capacity with a comfortable ride and small space.

14.6 Rear-Wheel Steering

Several developments have been attempted to improve a car's handling and maneuverability by steering the rear wheels. Two major approaches have been used: (1) to actively steer the wheels by an input from the car's steering wheel and (2) to let road forces and suspension geometry realign the toe and camber angles. The first approach is called an active design, and the second is called a passive design. Both of these systems add more moving parts, complexity, weight, and cost to the rear suspension. Many people believe that these disadvantages outweigh any gain that rear-wheel steering offers the average passenger car.

A car with rear- or four-wheel steering has a steering gear mounted at the rear wheels, two rear tie-rods, and the rear knuckle mounted on bushings to provide a steering axis. The rear steering gear is interconnected to the front steering gear by either a mechanical shaft or hydraulic lines (Figure 14–21). Active 4WS changes the wheel alignment procedure.

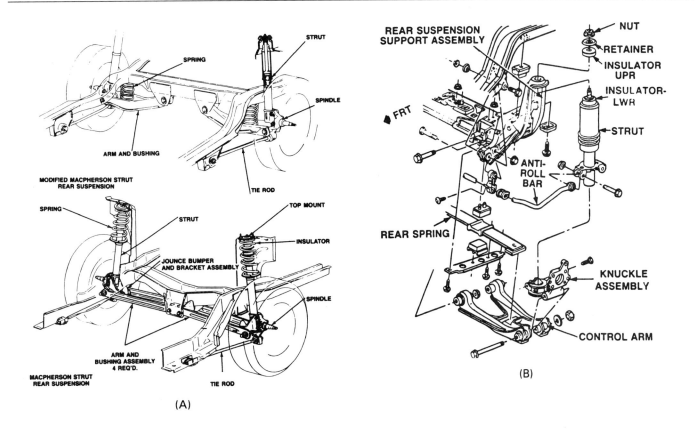

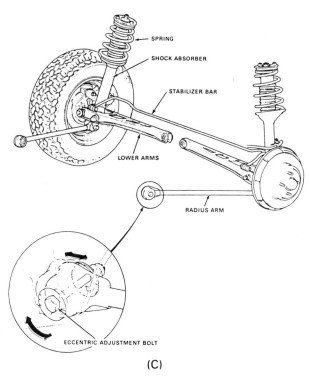

Figure 14–20 Several styles of rear strut suspensions for FWD cars. Note the type and location of the spring and the different styles of lower control arms. *(A is courtesy of Ford Motor Company; B is courtesy of General Motors Corporation, Service Technology Group; C is courtesy of American Honda Motor Co., Inc.)*

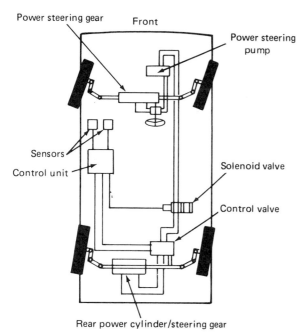

Figure 14-21 The rear wheel of this car can be steered by a hydraulically operated rear steering gear.

The second style of rear-wheel steering is called **Dynamic Tracking Suspension System** by its manufacturer. The dynamic tracking system has a toe control mechanism located in a unique two-part rear hub system. It also has a camber control mechanism located with the semitrailing suspension arm. In both of these assemblies, a combination of rubber and metal or ball-and-socket bushings is used. The rubber bushings are designed to allow a controlled amount of movement in specific directions. The metal bushings are used as pivot points. Each bushing is carefully positioned to take advantage of pressures generated by turning maneuvers, braking and acceleration forces, and vehicle lean to produce the following rear-wheel reactions:

- Braking or deceleration: toe-in
- Acceleration: toe-in
- Cornering: initial toe-out of the outer wheel that changes to toe-in at a lateral force of 0.4 to 0.5 **g**
- Cornering or bounce: positive camber during bounce or compression travel and negative camber during rebound or extension travel (Figure 14-22)

14.7 Practice Diagnosis

You are working in a brake and front-end shop. A customer has brought her car in for a front-end check and a complaint of the steering wheel being off center (low on the right side). Your road test confirms the steering wheel

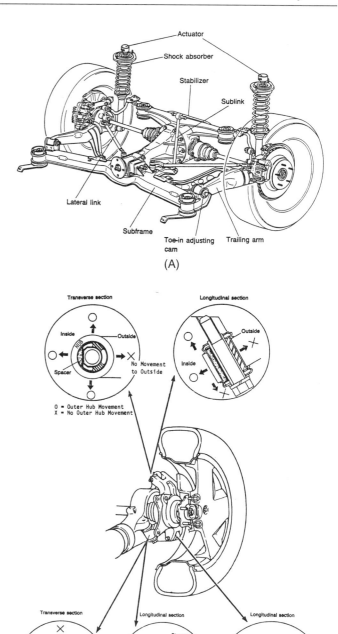

Figure 14-22 A Dynamic Tracking Suspension System (DTSS) is designed to produce carefully planned toe and camber changes during braking, acceleration, and cornering maneuvers (A). Three bushings between the inner and outer DTSS hubs allow specific motions for these changes (B). *(Courtesy of Mazda Motor Corporation)*

condition, but other than that, the car drives normally. You also notice, though, that the track is a little off, with the back of the car a little to one side. What is the probable cause of this problem? How should you check it out?

Terms to Know

axle windup
de Doin axle
dog tracks
double wishbone
Dynamic Tracking Suspension System
Hotchkiss

jacking
low pivot axle
Panhard rod
swing axle
thrust line

torque arm
torque tube
track bar
Watt's link
wheel hop

Review Questions

1. Statement A: The alignment of the rear wheels is not important because they do not do any steering.
   Statement B: Straight ahead in a car is a line that is parallel to the rear tires.
   Which statement is correct?
   a. A only
   b. B only
   c. Both A and B
   d. Neither A nor B

2. The rear axle of most RWD cars:
   A. Is a solid axle
   B. Provides driving action to the rear tires and supports the car
   Which option best completes the statement?
   a. A only
   b. B only
   c. Both A and B
   d. Neither A nor B

3. Torque tube driveshafts or torque arms are designed to prevent rear axle:
   A. Tramp
   B. Windup
   Which option best completes the statement?
   a. A only
   b. B only
   c. Both A and B
   d. Neither A nor B

4. Statement A: A solid axle with leaf spring suspension uses the leaf springs to position the axle under the car.
   Statement B: Coil spring rear suspensions also use the spring to position the axle.
   Which statement is correct?
   a. A only
   b. B only
   c. Both A and B
   d. Neither A nor B

5. A Panhard rod:
   A. Controls the fore–aft position of the axle relative to the frame
   B. Uses rubber bushings at the mounting points
   Which option best completes the statement?
   a. A only
   b. B only
   c. Both A and B
   d. Neither A nor B

6. Three or four control arms are used with coil spring rear suspensions to:
   A. Position the axle lengthwise in the chassis
   B. Prevent rear axle yaw, windup, and tramp
   Which option best completes the statement?
   a. A only
   b. B only
   c. Both A and B
   d. Neither A nor B

7. Rear axle windup in a coil spring solid axle can be controlled by:
   a. One or two upper links
   b. A single lower link
   c. A Panhard rod
   d. The upper and lower links

8. The major purpose of the two lower arms on a coil spring solid rear axle is to:
   a. Prevent side sway
   b. Prevent axle windup
   c. Control the wheelbase
   d. None of these

9. Cars that use independent rear suspension generally offer:
   A. Better traction on rough roads
   B. An improved ride quality
   Which option best completes the statement?
   a. A only
   b. B only
   c. Both A and B
   d. Neither A nor B

10. Statement A: A semitrailing arm rear suspension causes a slight camber change during suspension travel.
    Statement B: This suspension design usually uses a single inner pivot bushing on the lower control arm.
    Which statement is correct?
    a. A only
    b. B only
    c. Both A and B
    d. Neither A nor B

11. Which of the following is not a type of IRS?
    a. Trailing arm
    b. Swing arm
    c. Sliding pillar
    d. Strut

12. Statement A: A trailing arm rear suspension can use as many as five control links for each rear wheel.
    Statement B: The major purpose for the trailing arm is to absorb braking loads and wheel hop during acceleration.
    Which statement is correct?
    a. A only
    b. B only
    c. Both A and B
    d. Neither A nor B

13. Statement A: A de Doin rear axle can be used on FWD cars.
    Statement B: RWD swing axles can have problems with jacking of the rear suspension.
    Which statement is correct?
    a. A only
    b. B only
    c. Both A and B
    d. Neither A nor B

14. Statement A: FWD cars with a solid axle front suspension must use a solid axle rear suspension.
    Statement B: An independent front suspension with a solid axle rear suspension is never used on FWD cars.
    Which statement is correct?
    a. A only
    b. B only
    c. Both A and B
    d. Neither A nor B

15. The rear axle type used in many FWD cars is a:
    a. Solid axle with trailing arm
    b. Strut suspension
    c. Trailing arm
    d. Any of these

15 Springs

Objectives

Upon completion and review of this chapter, you should be able to:

- ❑ Comprehend the terms commonly used with springs.
- ❑ Understand the purpose for springs as a suspension component.
- ❑ Understand the various types of spring systems and what suspension components are required with each spring type.
- ❑ Comprehend electronically controlled air suspension systems and determine the correct diagnosis and repair procedure to correct faults with this system.

15.1 Introduction

Springs are the elastic portion of the suspension that allow vertical tire and wheel movement. They carry the weight of the car, but they also have the ability to absorb more weight. When additional load is placed on a spring because of more passengers or objects or because a tire meets a bump, the spring absorbs this load by compressing further. In reality, a spring stores energy by deflecting rather than absorbing weight. The springs are the most important suspension component that provides comfort in a car ride. Good spring action keeps the tires in contact with the road surface, adding greatly to the handling and stopping ability of the car.

When discussing springs, the term **bounce** refers to the vertical movement of the suspension, either up or down. Upward suspension travel that compresses the spring is called **jounce**. A lowering of the tire and wheel that extends the spring is called **rebound** (Figure 15–1).

To dramatize the need for springs, let us imagine a car traveling down the road at 55 mph (88.5 km/h), or 4,480 ft per minute. This vehicle is traveling at 968 in. per second (24.587 cm per second). If the road has a bump in it that is 1 in. (2.54 cm) high, the tire will have only a small part of a second to rise 1 in. and pass over this bump. Part of this rapid impact will be absorbed by the tire, deflecting the tread and part of the sidewall. Much of the bump will be absorbed by the suspension. The portion of the impact that is absorbed by the car body will lift the body slightly. Ideally, the bump is absorbed by compressing the springs. Now imagine the same bump, but on a vehicle without

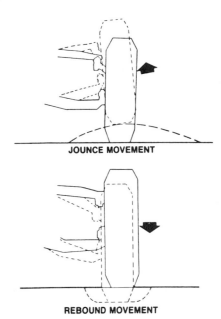

Figure 15–1 A tire and suspension have two directions for bounce travel. The upward motion is called jounce, and the downward motion is called rebound. *(Courtesy of Ford Motor Company)*

springs and, worse yet, with solid metal tires and wheels. When this vehicle meets the same bump at the same velocity, the bump will lift the whole vehicle the height of the bump. The entire vehicle will have to rise 1 in. in a fraction of a second. The vertical motion will be very fast, and the inertia involved will place a large amount of pressure on the tire when it impacts the bump. Also, the inertia of the rapidly rising vehicle will tend to cause the vehicle to keep rising after it is past the bump. The tire will leave the road, and the vehicle will probably fly for a certain distance after leaving the bump.

15.2 Sprung and Unsprung Weight

The **sprung weight** of a car is weight that is carried by the springs in the car's suspension. If the sprung weight is increased, the springs will compress further. **Unsprung weight** is not carried by the springs; it consists of those components that support or carry the springs. All the weight of the car is carried by the tires; the sprung components transfer their weight through the springs to the unsprung components and then through the tires to the ground. The tires, wheels, brake assemblies, and any other suspension parts that travel directly with the tires and wheels are unsprung weight. The frame, body, engine, transmission, and any other parts that move directly with the frame and body are sprung weight. The parts that connect the sprung weight to the unsprung weight, the suspension components, are partially sprung and partially unsprung, depending on how much movement they have during suspension travel (Figure 15–2).

In reality, the tires are springs. Some off-road vehicles rely entirely on the elastic nature of their tires for suspension travel. The spring action of the tires normally softens most of the smaller bumps that the tires meet. Those who have ridden in a car before and after switching the tires to a different type have probably noticed this difference. A change in ride quality from bias tires to belted radials is usually quite noticeable. Most discussions of car springs and sprung and unsprung weight disregard the spring action of the tires (Figure 15–3).

The springs allow the frame and body to ride relatively undisturbed while the tires and suspension members follow the bumps and holes in the road surface. The lower weight of the unsprung parts allows them to be less affected by inertia. A high sprung weight and a low unsprung weight provide improved ride quality and also improved tire traction. The springs have an easier job to do in absorbing bumps and keeping the tires in contact with the road.

15.3 Spring Rate and Frequency

Rate is the system used to measure spring strength. It is the amount of weight that is required to compress the spring a certain distance, usually 1 in. Traditionally, spring

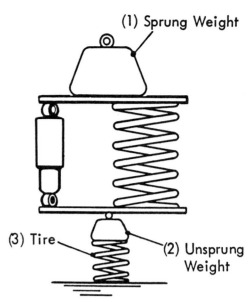

Figure 15–2 The tires and axles support the springs. They are classified as unsprung weight even though the tires are really a type of pneumatic spring. The springs support the frame, body, and other car parts that are classified as sprung weight. *(Courtesy of Tenneco Automotive)*

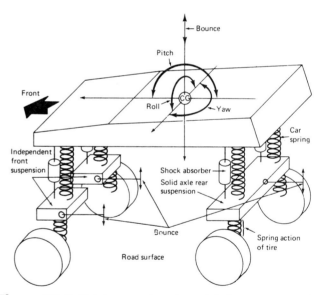

Figure 15–3 This is an engineer's model of a car's suspension system showing the various types and directions of motion.

rates have been measured in pounds per inch or Newtons (N) per millimeter. A Newton is a force equal to 0.225 lb. If it takes a force of 120 lb (54 kg or 533.76 N) to compress a certain spring 1 in. (25.4 mm), that spring would have a rate of 120 lb per inch. It would take another 120 lb of force to compress the spring one more inch, and 120 lb of force for each additional inch of compression until the spring bottomed out or used up all of the available travel (Figure 15–4).

Chapter 15 ■ Springs

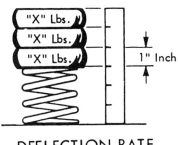

DEFLECTION RATE

Figure 15-4 Spring rate, also called deflection rate, refers to the amount of weight that it takes to compress a spring 1 in. In this example, weight X compresses the spring 1 in; each additional amount of weight X will compress the spring an additional inch. *(Courtesy of Tenneco Automotive)*

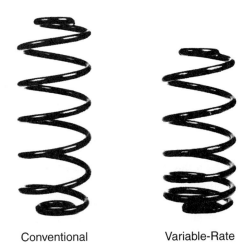

Conventional   Variable-Rate

Figure 15-5 A variable-rate spring will accept larger and larger loads as it compresses. In the example shown here, the rate of the variable spring is 240 lb for the first inch and 500 lb per inch for the third inch of compression. *(Courtesy of Sealed Power Corporation)*

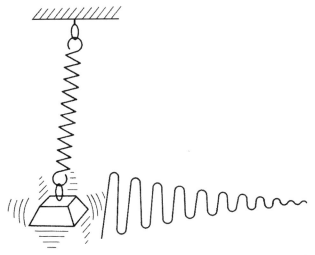

Figure 15-6 A spring will try to bounce at its natural frequency when the load is released. This frequency is the number of bounce oscillations per second. As the energy in the spring is used up, the amplitude or size of the bounce will decrease; the frequency does not change. *(Courtesy of General Motors Corporation, Service Technology Group)*

Most springs have a constant rate of resistance from the point where compression begins—the free, uncompressed length—to the point where the spring bottoms out—the coils touch each other. Springs can be made with a **variable rate;** that is, the rate changes during part of the travel. Normally, variable-rate springs increase in rate as they are compressed. Variable rate is usually accomplished by constructing the spring from material of different widths or thicknesses or by winding the spring so the coils will progressively bottom out. A variable-rate spring is more ideal in that it can be fairly soft at ride height and progressively stiffer as the suspension gets closer to the end of its travel. This would give us the softest ride possible on smooth roads yet not bottom out the suspension stops on rough roads (Figure 15-5).

Spring **frequency** is closely related to spring rate. Frequency refers to the speed of the natural oscillations of a spring. If we hung an object from a spring and let it drop, stretching the spring, we would see the object bounce up and down. Each travel period from the uppermost point to the lowest point and back to the highest point is called a cycle. The number of cycles completed in 1 second is called **cycles per second (CPS).** This is the normal spring frequency measuring system. A soft spring (low spring rate) has a fairly slow natural frequency; a strong spring bounces at a much higher rate (Figure 15-6 and Figure 15-7).

Springs of different rates can be used in a particular installation depending on the ride quality or load-carrying ability desired. Let us say a particular installation requires a spring that is 10 in. (25.4 cm) long when it is at normal ride height and has a static, installed load on it of 500 lb (227.3 kg). A spring with a free or unloaded length of 15 in. (38.1 cm) and a rate of 100 lb per inch (57 N per mm) would compress to the right length when installed. An 11-in. (free length) spring with a rate of 500 lb per inch (285 N per mm) would also compress to 10 in. long when it is installed. In comparing the two springs, the first one would give a softer ride at a slower frequency but would allow more leaning on turns and bottom out easier. The second, stronger spring could carry more load, would give a firmer and harsher ride, and would bounce at a higher frequency (Figure 15-8).

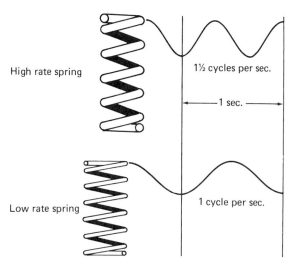

Figure 15–7 A strong, high-rate spring has a higher natural frequency than a soft, low-rate spring.

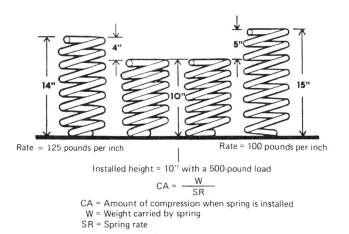

Figure 15–8 Different springs can be fitted in a particular car; they would produce different ride qualities and bounce frequencies.

15.4 Wheel Rate and Frequency

Wheel rate is a measurement of the spring's strength at the tire and wheel instead of directly at the spring. In many cases, the spring does not operate in direct relationship with the tire and wheel. The spring is compressed through the leverage of a control arm. An S-L A suspension might have the spring mounted halfway between the tire and the inner pivots on the control arm. This would produce about a 2:1 ratio between the tire and wheel motion and the spring motion. The tire would move vertically 2 in. while the spring moved 1 in. This leverage would also require a spring with twice the strength of the load it carried. A 400-lb (182-kg) vehicle load would make the spring support

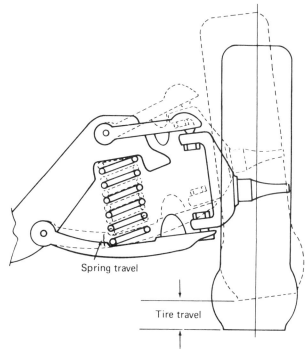

Figure 15–9 Because of the leverage of the lower control arm, the wheel rate of the spring is different from the spring rate. Note that the tire travel is about double that of the spring.

800 lb (364 kg), but this spring needs to be able to travel only half as far as the tire and wheel (Figure 15–9).

The springs in most strut suspensions operate much closer to a direct relationship with the tire and wheel. In this case, a 400-lb spring would carry about 400 lb of vehicle load, and 1 in. of tire and wheel movement would cause almost 1 in. of spring travel (Figure 15–10).

It should be noted that the track width also affects the ratio between the tire and spring lever. Installing wheels with an increased negative offset and a reduced backspacing that moves the tire outboard and increases the track width increases the tire travel relative to the spring. This change tends to reduce the wheel rate of the springs and usually causes the car to sit lower.

Suspension or **wheel frequency** varies depending on the spring's frequency and the amount of sprung weight carried by that spring. Suspension frequency is the natural speed at which the body and frame would bounce if the shock absorbers were disconnected. It is also called the **natural frequency.** Increasing the sprung weight lowers the natural frequency of the suspension. A typical suspension is designed to have a natural frequency of about 1 CPS. The natural frequency of a racing car suspension is about 2 CPS. Most drivers would find that higher suspension frequencies produce an annoying, disagreeable ride quality. A car with no suspension bounces at about 10 CPS on the tire alone; 10 CPS is the natural frequency of most passenger-car tires.

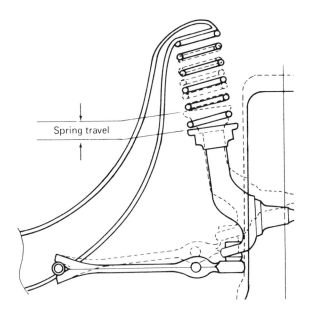

Figure 15-10 The wheel rate on a strut suspension is almost the same as the spring rate because the amount of travel is almost the same as that of the spring.

Many manufacturers design their cars so the front and rear suspensions have slightly different natural frequencies to reduce or eliminate **pitching** motions. This is called **flat ride tuning.** Pitch is an unpleasant oscillation of the car characterized by lowering at the front while rising at the rear and then rising at the front while lowering at the rear. Pitch often results when a car goes over a bump. The front tires strike the bump first, and then, a short time later, depending on the car's speed and wheelbase, the rear tires strike the same bump. Pitch begins when the rear suspension is in the jounce portion of its travel while the front suspension is in the rebound portion. It can be very disturbing when regularly spaced bumps, such as highway expansion joints, keep this pattern going.

If the vehicle has two different natural frequencies, the bouncing motions will get out of synchronization with each other (level out) after a few cycles, and pitch will soon become a simple bouncing up and down. Flat ride tuning adjusts the natural frequency of the front suspension to about 80 percent of that of the rear (Figure 15–11).

15.5 Spring Materials

Traditionally, springs have been made from steel alloy. This material, which is very strong to begin with, is heat treated to be more resilient. Springs flex through thousands of compression and expansion cycles without breaking and still retain their original shape. The most common spring failure is **sag,** a gradual reshaping of the spring that lowers the car.

Tempering a spring is a spring maker's art. It requires heating the metal to certain temperatures and then cooling the metal at a carefully controlled rate. Cooling the metal too slowly might cause annealing. An annealed spring is soft and bends or sags very easily. Cooling the spring too quickly might cause brittleness or hardening. An overly hard spring will snap because it will be too hard to bend.

Composite leaf springs are being used by a few manufacturers. These springs are made from **fiberglass-reinforced plastic (FRP)** or **graphite-reinforced plastic (GRP).** Advantages of composite springs are that they are lighter weight (about one-quarter to one-third the weight of steel, mostly unsprung) and free from corrosion and can be easily made with a variable spring rate that produces an improved ride. Composite springs are usually constructed to give a variable rate by varying the width or thickness of the spring's cross section.

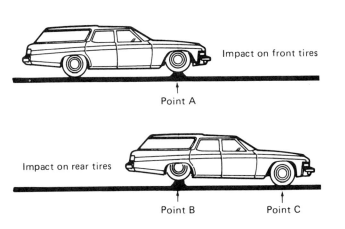
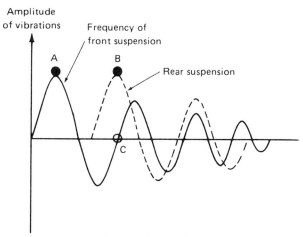

Figure 15–11 Flat ride tuning reduces vehicle pitch by using different suspension frequencies at the front and rear. The bounces of the two suspensions synchronize after a few oscillations.

They are usually more expensive than steel springs. A composite leaf spring is often mounted in a **transverse** position with one end used for the right suspension and the other end for the left suspension. A composite leaf spring is found on the rear suspension of some General Motors luxury cars and the front and rear suspensions of Corvettes (Figure 15–12).

Cars with **electronically** or **computer-controlled suspensions** use air trapped in a flexible chamber for a spring. The chamber is made partially from metal with a reinforced rubber diaphragm or reinforced rubber cylinder to form the flexible section. Air is a very good variable-rate spring; suspension travel compresses the air, which in turn increases the spring rate (Figure 15–13). The spring rate and length of air springs are easily adjustable. All that is needed is more air pressure. In most electronic systems, one or more **height sensors** note when the car is too low; this will cause the computer to either start up an onboard air compressor or open an air valve to increase the air pressure in the spring chambers. The computer will close the control valves to trap the correct amount of pressure in the chambers when the sensors note that the car is at the correct height. Air suspensions are discussed more completely later in this chapter (Figure 15–14).

Rubber is one of the better elastic materials because of its ability to store a lot of energy relative to its weight. At this time, pure rubber springs are only used for suspension travel stops, also called limiters or strikeout bumpers. As the suspension members approach the end of their upward and downward travel, they run into a **strikeout bumper** or **bump stop** instead of crashing into a metal frame or body member. The bump stops are of a conical or partially hollow shape so as to cause a **rising rate.** The further the bump stop is

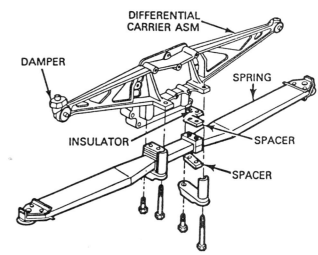

Figure 15–12 This Corvette rear spring is made from fiberglass-reinforced plastic. Note the varied width and thickness, which give the spring a variable rate. *(Courtesy of General Motors Corporation, Service Technology Group)*

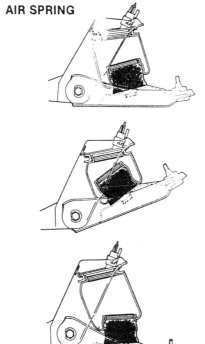

DESIGN
- Air spring is at normal trim height
- Air pressure contained in rubber membrane maintains vehicle height and acts like coil spring
- Air spring valve mounted in end cap opens to allow air to enter and exit spring
- When air is added, vehicle will rise
- When air is removed, vehicle will lower

JOUNCE
- When control arm moves upward, piston moves upward into rubber membrane
- As the arm moves upward toward jounce the rate of the air spring increases

REBOUND
- When control arm moves downward, piston extends outward from rubber membrane
- Rubber membrane unfolds from around piston to allow downward suspension movement

Figure 15–13 An air spring is usually a flexible rubber membrane that contains pressure; the trapped air compresses and expands during suspension movement. *(Courtesy of Ford Motor Company)*

Chapter 15 ■ Springs

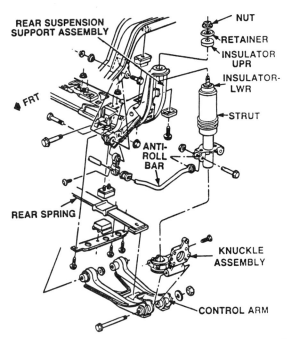

Figure 15–14 This suspension combines a transverse, fiberglass leaf spring with an air spring in the strut; the air spring allows for automatic level control. *(Courtesy of General Motors Corporation, Service Technology Group)*

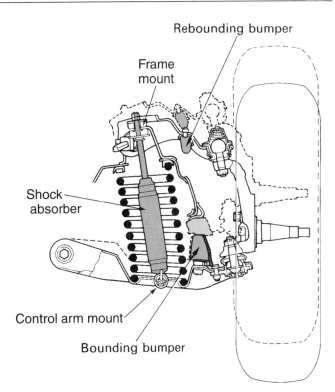

Figure 15–15 The jounce travel stop (bounding bumper) and the rebound travel stop (rebounding bumper) are common examples of rubber springs; they serve to keep the control arms from crashing into the frame during extremes of suspension travel.

compressed, the harder it becomes to compress. Occasionally, one or both of the bump stops are built into the shock absorber. In the future, rubber may be used for the major suspension springs. Rubber is lightweight and has good variable-rate characteristics. Some drawbacks of rubber springs are that rubber is affected by various chemicals, and it will change spring rate with temperature changes (Figure 15–15).

15.6 Leaf Springs

Most leaf springs serve a dual function as an axle-control member as well as a load-carrying member. They are normally of a **semielliptical** shape (i.e., shaped like a portion of an ellipse), have a main leaf with an eye at each end, and are assembled from a group of leaves of different lengths (Figure 15–16). Some cars use a single spring for both ends of the axle mounted transverse or across the car (Figure 15–17). Springs mounted in this manner are sometimes called buggy springs. A few cars have used **quarter-elliptical** springs, which look like a spring that has been cut in half. The thick end of the spring is bolted solidly to the car's frame, and the free end is attached to the axle through a bushing.

As the spring absorbs loads and deflects, the leaves change from a curved shape to a flat shape and then re-curve in the other direction. The leaves also change length as they flatten out and bend. A shackle is used with most leaf springs to allow for this change; there is also a certain amount of sliding of the ends of one leaf over the next leaf. This sliding, often called **interleaf friction,** could cause binding in the spring action (Figure 15–18). Pads of low-friction material are usually put between the leaves to reduce this friction. Interleaf friction is difficult to control on older cars and often causes a harsh ride. Instead of using a shackle, some overload spring installations on trucks use a mount that sits on top of the spring and allows the spring to slide. **Monoleaf** or **single leaf springs** are used in some installations. When only one leaf is used, it is often tapered to produce a variable rate. A monoleaf spring has no interleaf friction.

Leaf springs are made stronger by making the leaves thicker or wider or both. They are made more flexible by increasing the length of the leaves. The formula for calculating the spring rate for a leaf spring is given in Figure 15–19. Adding leaves will increase the spring rate, and if the leaves are of different lengths, the rate will be variable. The ends of the longest leaf will bend first, then the ends of the next longest leaf will bend with the main leaf, and so on. These leaves are usually bolted together using a **center bolt.** The center bolt is also used

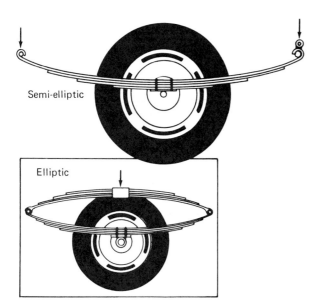

Figure 15–16 The most common curved leaf springs are of a semielliptical shape (top); some very early cars used an elliptic spring. It was usually a pair of semielliptical springs (bottom).

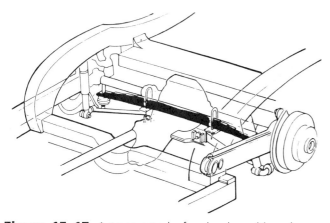

Figure 15–17 A transverse leaf spring is positioned across the car. The center is attached securely to the frame or body, and each end is connected to a right or left suspension. *(Courtesy of Dana Corporation)*

to align the axle to the spring. Metal clips are normally bent around the spring to hold the leaves in alignment with each other (Figure 15–20).

Rubber bushings are normally used in the spring eyes to allow the necessary spring motion. They also provide the compliance needed for slight side and twisting motions that occur during suspension travel and vehicle lean (Figure 15–21). The rubber bushings also dampen road vibrations from traveling into the frame and body. Some manufacturers place rubber pads between the spring and axle for this same purpose. Some vehicles use metal spring bushings, which require periodic lubrication (Figure 15–22).

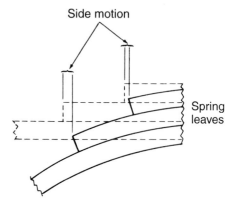

Figure 15–18 As a leaf spring flattens, the end of one leaf must slide over the leaf next to it; this generates interleaf friction, which adds to spring rate.

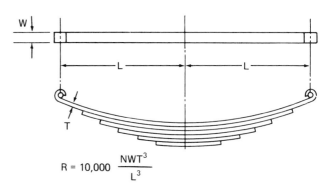

$$R = 10{,}000 \, \frac{NWT^3}{L^3}$$

R = Spring rate at point of force (pounds per inch)
N = Number of leaves
W = Width of leaf (inches)
T = Thickness of leaf (inches)
L = Length from eye to center (inches)

Figure 15–19 This formula shows the factors that affect the spring rate for a leaf spring.

Suspension designers prefer to use a spring with the flattest curve possible and a shackle that is just long enough to allow the necessary spring travel. Excessive spring curve or shackle length allows greater amounts of axle side and yaw motions, which produce roll oversteer and raise the roll center.

15.7 Coil Springs

Coil springs are made from a round steel alloy wire that is wound into the common spiral or helical shape. Since there is no interleaf friction as in leaf springs, coil springs offer a good ride quality for an extended period. The strength of the spring is basically determined by the diameter and length of the wire. An increase in wire diameter produces a stronger spring, while an increase in length makes it more flexible. The formula for calculating the rate of a coil spring is given in Figure 15–23.

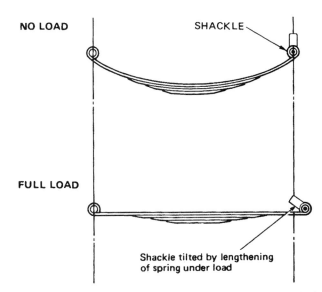

Figure 15–20 A leaf spring suspension from a light truck that uses a solid front axle. Note the action of the shackle. *(Courtesy of General Motors Corporation, Service Technology Group)*

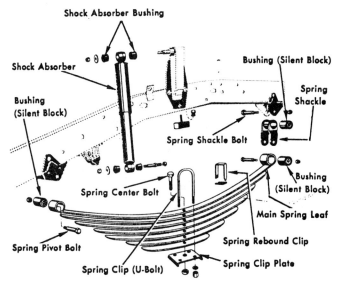

Figure 15–21 A disassembled leaf spring. Note the center bolt and spring rebound clip, which hold the leaves together, and the spring eye bushings. *(Courtesy of Mitchell Repair Information Company)*

Coil springs differ in terms of wire diameter, coil diameter, coil-free length, number of coils, coil winding direction, and the shape of the coil ends (Figure 15–24). The different coil end shapes are:

- **Tapered:** if the end of the wire is flattened
- **Tangential:** if the coil is a continuous spiral
- **Square:** if the last coil is bent to be square with the coil
- **Pigtail:** if the last coil is wound to a smaller diameter

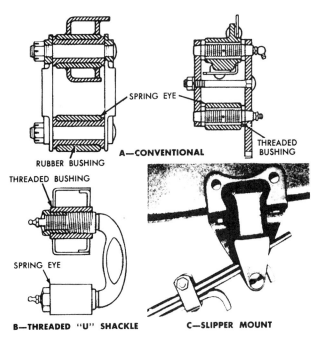

Figure 15–22 Several types of bushings are used to connect the ends of a leaf spring to the frame bracket and shackle. Threaded bushings require periodic lubrication; the slipper mount is a sliding connection. *(Courtesy of Ford Motor Company)*

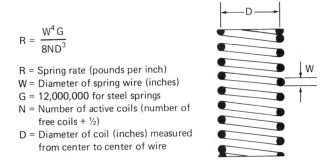

$$R = \frac{W^4 G}{8ND^3}$$

R = Spring rate (pounds per inch)
W = Diameter of spring wire (inches)
G = 12,000,000 for steel springs
N = Number of active coils (number of free coils + ½)
D = Diameter of coil (inches) measured from center to center of wire

Figure 15–23 This formula shows the factors that affect the rate of a coil spring.

These end treatments are shown in Figure 15–25. Coil springs are often wound with two different end shapes. When this occurs, the spring has a top and a bottom, and each end fits into a specific pocket. Coil springs can be wound from a tapered wire to get a variable rate, but this is more expensive. Sometimes coil springs are wound with a varying pitch between the coils. As the spring compresses, some of the coils will bottom out and the spring rate will increase.

Coil springs are normally trouble free; the common type of failure is sag. Sag can shorten the spring enough to allow the suspension to drop too far below design height for correct suspension geometry. When this occurs, tire wear or handling difficulties will result. Sag will also lower the car to the point where the suspension members are constantly hitting the bump stops (Figure 15–26).

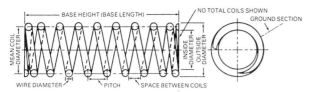

Figure 15–24 These are the commonly used coil spring dimensions. *(Courtesy of General Motors Corporation, Service Technology Group)*

Figure 15–25 Coil springs have different styles, depending on how the end of the wire is cut or bent. A square spring end is bent so the spring will stand up straight. A tangential end is simply a cut coil; it will lean if stood on end. *(Courtesy of Sealed Power Corporation)*

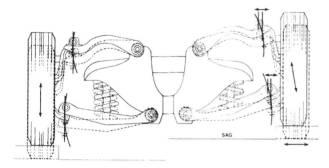

Figure 15–26 If a spring sags an excessive amount, the front suspension geometry will also change. Wheel alignment, tire wear, and vehicle handling will be affected. *(Courtesy of Dana Corporation)*

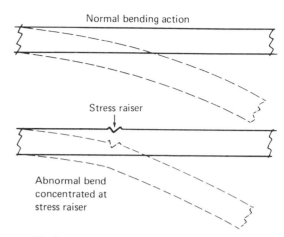

Figure 15–27 A stress raiser begins when corrosion or a nick removes metal. This portion is now weaker and will bend easier, and the increased and concentrated bending action will cause the spring to fail.

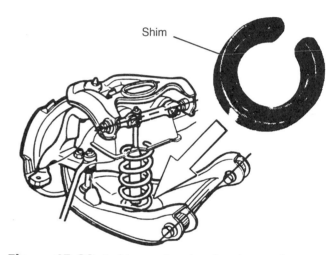

Figure 15–28 A shim can be placed under a coil spring to adjust the car to the correct trim height.

Coil spring breakage is fairly rare. If one breaks, it is usually the result of a **stress raiser.** A stress raiser is a flaw in the metal surface, sometimes just a scratch, nick, or pit caused by rust or corrosion. Many newer springs are painted or coated with epoxy to retard rust action and reduce the possibility of a stress raiser occurring. The stress raiser weakens the metal so that future bending or twisting motions tend to concentrate and bend or twist this portion of the spring more than the other areas. If it is serious enough, the increased working will cause work hardening, also called metal fatigue, which hardens the metal to the point that it will no longer bend, and the spring will break (Figure 15–27).

The correct repair for a sagged spring is to replace it. This can be expensive, and in cases of older, high-mileage cars, which are usually the ones with sagged springs, the cost can be very expensive relative to the value of the car. One less expensive cure for sagged springs is to place a shim at the end of the spring. The shims are made from rubber or soft metal, usually aluminum (Figure 15–28).

The drawbacks with using a shim are that the labor cost of installing a shim is the same as replacing the spring and that the shortening of the spring by the shim can cause coil clash. Coil clash occurs when the coils bottom and touch each other. When this occurs, the coil becomes solid, with a drastic increase in spring rate. Normally, coil clash does not occur, because the suspension will bottom out on the stops before the coil

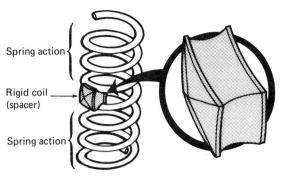

Figure 15–29 Several styles of solid spring spacers and boosters, designed to be placed between the coils, are available. These spacers are not recommended, because they place a severe bending load on the spring next to the spacer.

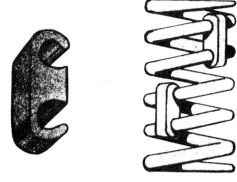

Figure 15–30 Rubber spacers can be inserted in a coil spring to restore a car to the correct trim height.

bottoms. Another cure for sagged springs is to install spacers between the coils (Figure 15–29). Solid metal spacers are not recommended because they eliminate spring action in the coil where the spacer is located, and they also make the ride quality stiffer and more harsh. The spring rate will increase because of this, and a severe bending stress is placed on the active coils right next to the spacer. Rubber spacers are less severe and can be used without serious effects (Figure 15–30). Some styles of rubber spacers have a problem of staying in place and might fall out. Spacers can usually be installed in a few minutes so they offer cost savings in terms of labor as well as parts.

15.8 Torsion Bars

A **torsion bar** is a bar of steel that is held stationary at one end and forced to twist at the other. It can be made from a single piece of metal or a group of laminations. Some cars use an L-shaped torsion bar. The rate or strength of a torsion bar can be calculated using the formula given in Figure 15–31. If you have ever placed a large amount of torque on a long extension bar while using a socket and handle to either tighten or loosen a bolt, you may have noticed a twisting of the extension bar. While you are doing this, the extension bar acts like a torsion bar. Most torsion bar suspensions anchor one end of the torsion bar solidly to the frame and connect the other end to a suspension control arm or trailing arm. The control arm becomes the lever arm for the torsion bar. L-shaped torsion bars provide their own lever arm. As the suspension allows vertical tire and wheel motion, the torsion bar twists tighter or looser. If you think about it, you will realize that the metal wire in a coil spring also twists, much like a torsion bar, as the spring compresses and extends (Figure 15–32).

An adjustable connection is usually placed at one of the ends of the torsion bar to provide an adjustment to compensate for bar sag. If the bar sags, the adjustment allows the bar to be twisted a little tighter, restoring the vehicle to correct ride height (Figure 15–33).

The long, skinny shape of a torsion bar gives the suspension engineer an alternative to the shorter, fatter coil spring in some tight mounting installations. A torsion bar can easily be used with S-L A, trailing arm, or strut FWD and RWD suspensions. Another possible advantage of a torsion bar is that a large portion of the spring load is transmitted to the fixed end of the bar. Depending on how the torsion bar is mounted, this can move some of the front suspension load from the weaker suspension area to the stronger body cowl area.

Torsion bars are fairly trouble free. With an occasional adjustment, they will normally last the life of the car. They can break, but this is rare; breaking of the mounting points can occur. As with coil springs, breakage of the mounting points is usually the result of a stress raiser. All metal springs should be handled somewhat gently to avoid scratches in the protective painted or plastic covering or in the metal, which might become stress raisers.

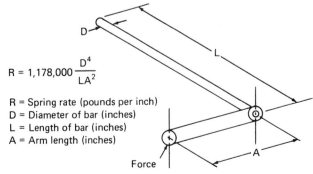

$$R = 1{,}178{,}000 \frac{D^4}{LA^2}$$

R = Spring rate (pounds per inch)
D = Diameter of bar (inches)
L = Length of bar (inches)
A = Arm length (inches)

Figure 15–31 This formula shows the factors that affect the rate of a torsion bar.

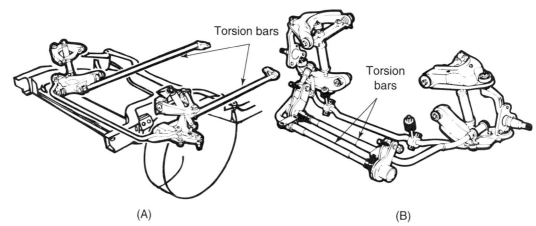

Figure 15-32 Torsion bars are usually straight, round bars of spring steel, but they can also be bent (L-shaped) so the end of the bar is also the lever. Torsion bars can also be flat, laminated, or both. *(Courtesy of Moog Automotive Inc.)*

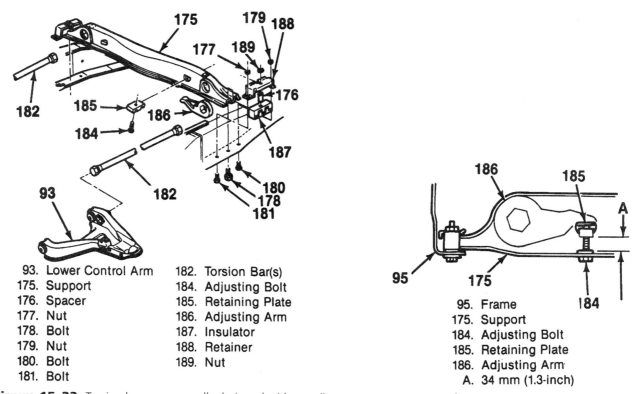

93. Lower Control Arm
175. Support
176. Spacer
177. Nut
178. Bolt
179. Nut
180. Bolt
181. Bolt
182. Torsion Bar(s)
184. Adjusting Bolt
185. Retaining Plate
186. Adjusting Arm
187. Insulator
188. Retainer
189. Nut

95. Frame
175. Support
184. Adjusting Bolt
185. Retaining Plate
186. Adjusting Arm
A. 34 mm (1.3-inch)

Figure 15-33 Torsion bars are normally designed with an adjustment to compensate for spring sag. Part 184 is the adjusting bolt. *(Courtesy of General Motors Corporation, Service Technology Group)*

If a torsion bar needs to be replaced, it should be noted that the bars on most cars are not interchangeable between the right and left sides. Many bars are pre-stressed during manufacture. The bars are usually marked or coded in some way to tell them apart, and a left-side bar should only be used on the left side of the car and vice versa.

15.9 Electronically Controlled Air Suspensions

Air suspensions are complex and expensive, but they have the ability to stay at a constant height, regardless of load. In some cases, air suspensions provide a variable height that is controlled by the driver or a **computer-controlled**

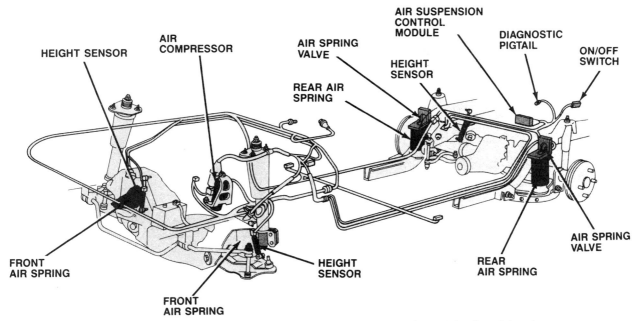

Figure 15–34 A computer-controlled air suspension system. Air springs support the car's load, and the air compressor provides the air pressure. When the height sensors sense a change in the car's height, the control module causes the air spring valves to change the pressure in the springs. *(Courtesy of Ford Motor Company)*

variable rate and height for improved handling or vehicle aerodynamics. Air suspensions use the compressibility of air for a spring. The spring rate is determined by the size of the air chamber and the amount of air pressure. If the car is too low, the air pressure in the spring chambers is increased; if the car is too high, some of the air pressure is released. The no-load spring rates are usually lower than if steel springs are used, because air suspensions have the ability to compensate for heavier loads. Metal springs of the same lower rate would bottom out if more load was added (Figure 15–34).

Depending on the manufacturer, there are several versions of electronically controlled air suspensions. Nearly all of the systems currently in use are computer controlled. The computer, also called a **control module,** monitors the height of the car through three or four **height sensors,** one for each wheel or one for each front wheel and one for the rear axle. Some systems use brake or door sensors so automatic height adjustments are canceled during stopping and while passengers are getting in or out of the car (Figure 15–35). Some systems use speed sensors so that the car can be lowered at higher speeds for aerodynamic improvement. In some systems, the spring rate is increased at the same time. Some systems also use a **g** sensor and a steering shaft velocity sensor so the spring rate can be increased during fast cornering, braking, and acceleration maneuvers (Figure 15–36).

Some systems use dual-chamber air springs or coil or leaf springs plus air springs. Some systems use single air springs only. The flexible air chamber for the air spring can be built into a strut or mounted in place of the standard coil spring. A rigid air chamber can be connected into the flexible chamber to provide a lower spring rate. All systems use an onboard **air compressor** to maintain a supply of compressed air. Some systems pressurize and bleed down the air chamber using a valve mounted between the compressor and the air chamber. Other systems

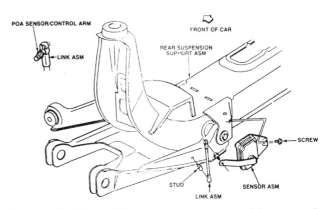

Figure 15–35 Height sensors are connected between the body and a suspension member; they send a signal to the control module if the height changes. *(Courtesy of General Motors Corporation, Service Technology Group)*

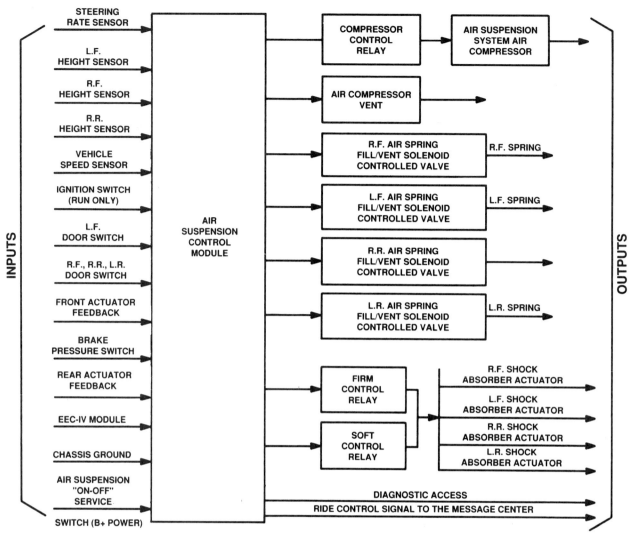

Figure 15–36 The control module circuit. The inputs to the module are to the left, and the outputs that the module controls are to the right. *(Courtesy of Ford Motor Company)*

pressurize the air chamber using the compressor and bleed down the chamber using a valve mounted on the air chamber. Some systems use a two-wheel-only system, mounted at the rear; the air chambers are built into the rear shock absorbers. Some two-wheel systems are a dealer-installed option (Figure 15–37).

The actual air chamber for the spring can take several shapes, the most common of which are somewhat tubular or diaphragm shapes. Reinforced rubber is used for the flexible membrane. Various ply materials, much like those in the sidewall of a tire, provide the flexible reinforcement. Each air chamber is connected to the air compressor or control valve through a plastic tube.

Most present-day systems use onboard air compressors that are driven by a 12-V electric motor. A **desiccant,** to remove moisture, is usually placed in the compressed airflow. Water vapor can enter the system with the incoming air. If this water vapor condenses in the system, it can cause freeze-up in cold weather or cause corrosion of the metal parts (Figure 15–38).

The height sensors electronically measure the distance between the control arm or axle and the frame. They signal the computer if this distance is longer or shorter than the design height. The computer normally ignores changes that occur over a short period of time; these are usually bumps. A height change that lasts longer than 5 or 6 seconds is responded to by changing the pressure in the air springs. Some systems wait as long as 45 seconds during certain sequences of events before responding with a height adjustment. The computer uses the height sensor information to adjust the pressure in the air chambers to keep the car at the correct ride height (Figure 15–39).

Electronically controlled air suspension systems are complex, and they have many possibilities for failure. Their operation varies from one manufacturer to another.

Chapter 15 ■ Springs

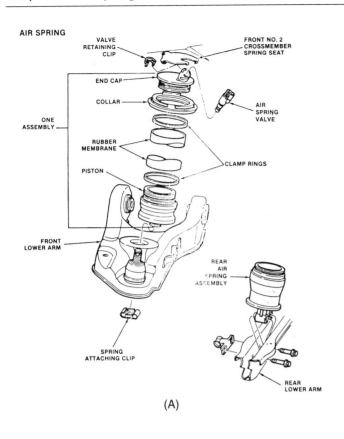

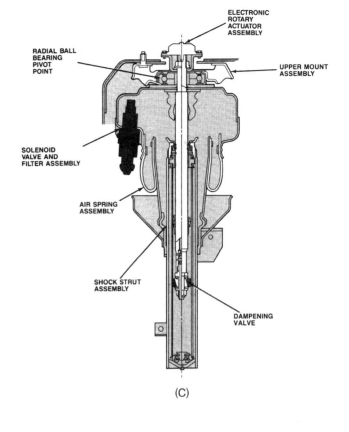

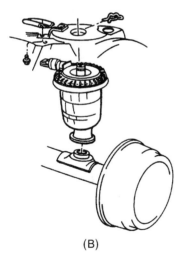

Figure 15–37 The air spring consists of a reinforced, tubular rubber membrane (A); various styles are shown here. A rear unit is shown in (B). Note that (C) also includes an adjustable shock absorber (dampening valve). *(Courtesy of Ford Motor Company)*

technician of the cause of the problem. The process of diagnosing and curing electronic air suspension problems is described in Chapter 18. This information is also available in vehicle manufacturer and technician service manuals.

Caution should be used when raising some cars with electronic air suspension. Some systems stay on all the time; they are not deactivated when the ignition is turned off. A separate switch is usually provided for that purpose. If the car is lifted on a frame contact hoist, which is recommended, the suspension will hang down, the height sensors will lengthen, and the computer will respond to a high-car condition and bleed the air pressure from the spring chambers. When the car is lowered, not only will the car be too low, but the air chambers might collapse and fold inward. This could damage and require replacement of the flexible rubber membranes. Some cars are equipped with a switch to deactivate the suspension system in these cases and also if necessary for the wheel alignment procedure.

15.10 Active Hydraulic Suspensions

The action of a tire moving upward over a bump and compressing a spring places an upward lifting action on the body of the car, which we feel as a bump. The ideal

Most systems have diagnosis features built into them so that by using the right equipment, usually very simple and inexpensive, a code that gives the nature of the problem can be read out of the computer. This code informs the

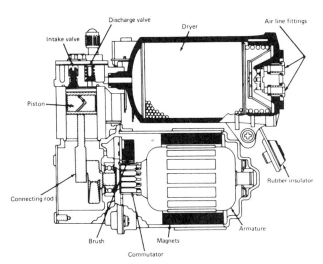

Figure 15-38 A 12-V motor drives this compressor to provide the air volume and pressure needed for an air suspension system. The drier contains a desiccant to remove moisture from the compressed air. *(Courtesy of Ford Motor Company)*

suspension system would not do this. It would let the tires rise as necessary with no resistance, but it would still support the car. The legs of a skier or motocross rider flexing to absorb bumps are examples of how a bump can be absorbed.

A fully active suspension system uses hydraulic cylinders or an actuator as the major load-carrying component, a variable-output hydraulic pump with an output up to 4 gallons per minute (15.14 liters per minute [lpm]), a series of sensors to detect each spring height as well as body motions in all directions, one or two microprocessors as system control units, a hydraulic flow control and pressure control valve for each wheel unit, one or more hydraulic accumulators, the necessary hydraulic lines to connect these parts, and a special silicone fluid that provides the proper flow characteristics under all temperature conditions (Figure 15–40). When the sensors note the pressure increase and/or upward suspension motion of the wheel as it passes over a bump, the pressure in the actuator is decreased, which greatly reduces the harshness and the upward body motion. As the wheel passes over the bump or a dip in the road, fluid is pumped back into the actuator to stop a downward body motion. One system has the ability to bring the actuator pressure from zero to full pressure in one-tenth of a second. Each actuator is controlled separately for bump control. An added advantage of active suspension is that the pressure in the actuators on one side of the car can be increased as the car goes around a corner or at one end of the car during starting or stopping. This reduces the amount of lean on turns, dive under braking, or squat during acceleration. Also, the height of the car is automatically maintained at a level position as load changes and can be made higher over bumpy roads by a manual control switch.

The major disadvantages of active suspension are its cost and complexity. It is easy to see that there are many more possibilities for problems than in a system with four springs and shock absorbers. It appears that future suspensions of this type will use conventional springs with computer-controlled shock absorbers, as described in Section 16.8. Most shock absorbers do not carry much vehicle load, so the mounting points only have to absorb spring-dampening loads.

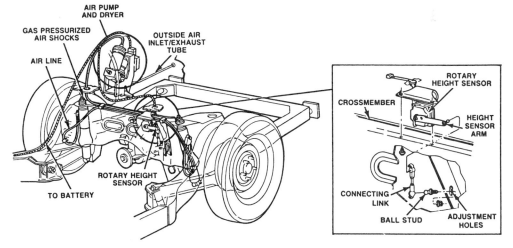

Figure 15-39 A two-wheel automatic leveling system. When weight is added to the back of this car, the air pressure in the shock absorbers is increased to raise the car back to the normal height. *(Courtesy of Ford Motor Company)*

Chapter 15 ■ Springs

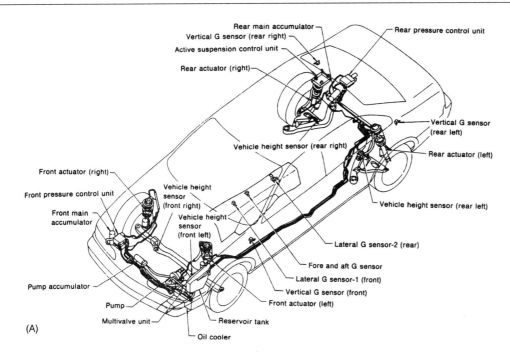

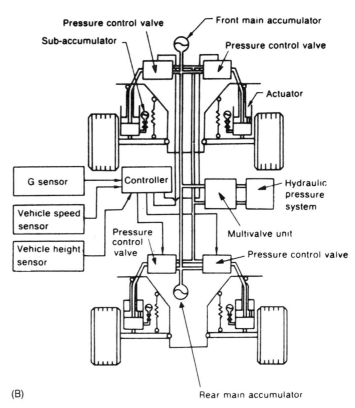

Figure 15–40 An active suspension system uses a hydraulic pump and actuators (A). The schematic shows the relationship of the various components (B). *(Courtesy of Nissan Motor Corporation)*

15.11 Aftermarket Air Suspensions

Aftermarket air suspensions are add-on systems that are sold by various aftermarket suppliers. They are two-wheel systems designed for use on the rear of the car; the air chambers are built into the shock absorbers. They are commonly called **air shocks.** The air chambers are used to supplement the standard springs already on the car. The air chambers in the shock absorbers are connected to an inflator by small plastic tubing. When the added load-carrying ability is needed, air pressure is added through the inflater. Some systems use a manual, driver-controlled, 12-V air compressor to provide the air pressure adjustment; others are inflated just like a tire (Figure 15–41).

When adding air shocks to a vehicle, it is important to ensure that the frame mounts are strong and secure. These mounting points will be carrying more of the vehicle's load.

Figure 15–41 An aftermarket air-adjustable shock absorber. Adding air pressure will allow this shock absorber to carry some of the weight of the car. *(Courtesy of Tenneco Automotive)*

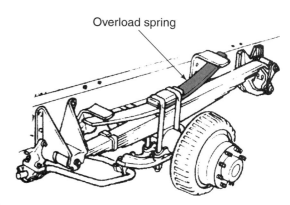

Figure 15–42 Aftermarket overload or helper springs can be added to passenger cars or pickups and light trucks. *(Courtesy of General Motors Corporation, Service Technology Group)*

15.12 Overload Springs

Add-on springs are available for many cars, pickups, and trucks. These are commonly called **overload springs.** These springs allow the vehicle to carry additional load without bottoming out the suspension.

The easiest system to install is a shock absorber with a coil spring mounted around it. This unit is commonly called a **coil-over shock.** Like air shocks, a coil-over shock places a greater load on the shock absorber mounts. Overload springs can also be in the form of a leaf spring that is attached to the leaf springs already on the car, a coil spring that is mounted between the rear axle and the frame, or a flexible air chamber that is placed inside of the coil springs already on the car or ones that can be added (Figure 15–42).

Overload springs and air shocks increase the spring rate, which tends to raise the vehicle or make the ride harsher. If you are adding overload devices to a vehicle, do not forget that the maximum load to be carried should not exceed the capacity of the tires and axle rating.

15.13 Stabilizer or Antiroll Bar

The stabilizer or antiroll bar is used with many front and rear suspensions. A stabilizer bar is not really a spring like those we have already studied; it acts like a spring only when the car leans. The ends of the stabilizer bar usually connect to the control arm for each tire or onto the outer ends of the axle (Figure 15–43). The center section of the bar connects to the frame through a pair of rubber bushings (Figure 15–44). The bushings allow the bar to rotate and even deflect vertically a small amount. When both tires on an axle meet a dip or rise in the road, the bar merely rotates in the bushings. When only one tire meets a dip or rise or when the car leans on a corner, the bar is forced to twist, much like a torsion bar (Figure 15–45).

Stabilizer bars are used primarily to reduce body roll and lean on turns. The bar adds to the roll resistance of the springs and torsilastic control arm bushings. Stabilizer bars have a drawback in that they tend to increase the spring rate of one-tire bumps and therefore remove some of the independence of independent suspensions. The formula given in Figure 15–46 can be used to calculate the strength of a stabilizer bar. The compliance of the rubber bushings in the bar end-link and center bushings tends to soften and reduce the effects of one-tire bumps.

Stabilizer bars can also be built into the control arms and trailing arms on rear suspensions. The axle beam of a FWD, solid rear axle is often designed to function as a stabilizer bar (Figure 15–47). Stabilizer bars are trouble free. The only problems that are normally encountered are the deterioration of the rubber pivot and end-link bushings.

Chapter 15 ■ Springs 247

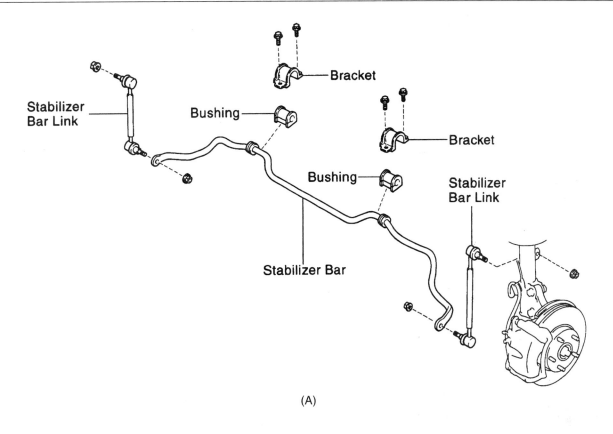

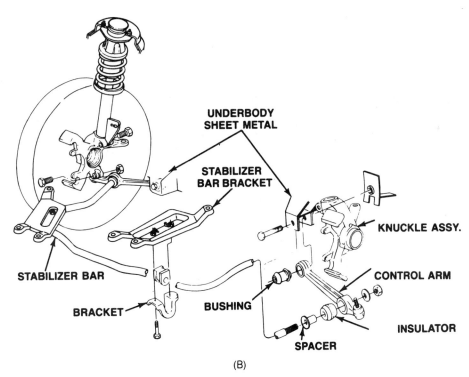

Figure 15–43 A stabilizer bar, also called an antiroll bar, is connected to the suspension arm at each side of the car. It can be connected through a link (A) or a rubber bushing or insulator (B). The end of the stabilizer bar can also act as a trailing strut, as shown in (B). *(B is courtesy of Ford Motor Company)*

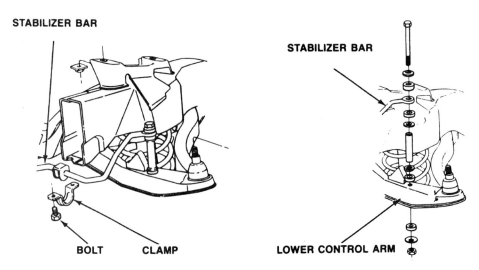

Figure 15–44 The stabilizer bar is often connected to the suspension by an end link that consists of four rubber bushings, support washers, and a bolt and nut. *(Courtesy of Ford Motor Company)*

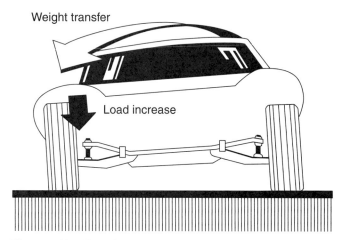

Figure 15–45 When a car goes around a corner, body lean raises the suspension on one side of the car and lowers the suspension on the other side. This action twists the stabilizer bar, and the strength of the bar reduces the amount of lean.

Changing to stronger or weaker stabilizer bars should be done with caution. Many suspension designers use this bar to adjust the oversteer and understeer characteristics of the car. Changing the strength of the bar will definitely change these important handling characteristics.

15.14 Practice Diagnosis

You are working in a front-end and wheel alignment shop and the customer has brought in a 7-year-old, midsized station wagon for a wheel alignment. His complaint is poor tire mileage. As you are doing the prealignment inspection, you notice that the rubber bump stops are broken up and the metal of the control arms that contacts the bump stops is worn clean. Have you found a cause for the tire wear? What should you do to confirm this?

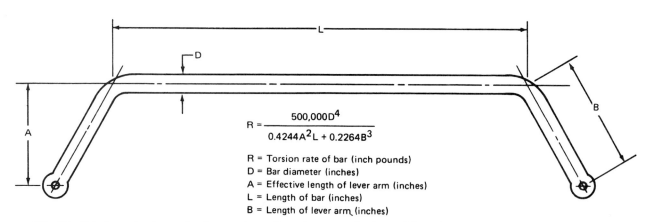

$$R = \frac{500{,}000 D^4}{0.4244 A^2 L + 0.2264 B^3}$$

R = Torsion rate of bar (inch pounds)
D = Bar diameter (inches)
A = Effective length of lever arm (inches)
L = Length of bar (inches)
B = Length of lever arm (inches)

Figure 15–46 This formula shows the factors that affect the strength of a stabilizer bar.

Chapter 15 ■ Springs

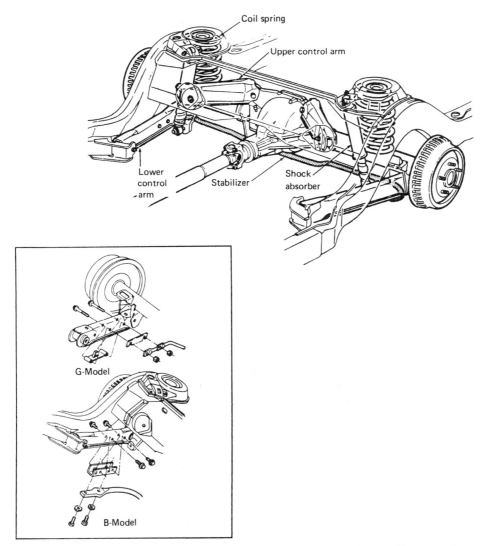

Figure 15–47 This rear-mounted bar is bolted solidly to each control arm by one of two different methods (inset). *(Courtesy of General Motors Corporation, Service Technology Group)*

Terms to Know

air compressor	electronically controlled suspensions	pigtail
air shocks	elliptical	pitching
bounce	fiberglass-reinforced plastic (FRP)	rate
bump stop	flat ride tuning	rebound
center bolt	frequency	rising rate
coil-over shock	graphite-reinforced plastic (GRP)	sag
composite leaf springs	height sensors	semielliptical
computer controlled	interleaf friction	single leaf spring
computer-controlled suspensions	jounce	sprung weight
control module	monoleaf spring	square
cycles per second (CPS)	natural frequency	stress raiser
desiccant	overload springs	strikeout bumper

suspension
tangential
tapered
torsion bar
transverse
unsprung weight
variable rate
wheel frequency
wheel rate

Review Questions

1. Which of the following terms is correct?
   A. Vertical travel of the suspension is called jounce.
   B. Downward travel of the suspension is called rebound.
   a. A only
   b. B only
   c. Both A and B
   d. Neither A nor B

2. Statement A: The tires, wheels, and brake assemblies are all unsprung weight.
   Statement B: Increasing the unsprung weight of a car places a greater load on the springs.
   Which statement is correct?
   a. A only
   b. B only
   c. Both A and B
   d. Neither A nor B

3. The car's springs:
   A. Are the flexible part of the suspension system that carries no load
   B. Provide a compressible link between the frame and the axle
   Which option best completes the statement?
   a. A only
   b. B only
   c. Both A and B
   d. Neither A nor B

4. The amount of force or weight that is required to compress a spring 1 in. is called:
   a. Spring rate
   b. Spring weight
   c. Rebound weight
   d. Any of these

5. The speed at which a spring bounces:
   A. Is measured in cycles per second
   B. Increases in proportion to the strength of the spring
   Which option best completes the statement?
   a. A only
   b. B only
   c. Both A and B
   d. Neither A nor B

6. Statement A: The spring on an S-L A suspension has a higher rate than the load at that corner of the car.
   Statement B: The spring on this suspension compresses about 1 in. for each 2 in. of vertical tire motion.
   Which statement is correct?
   a. A only
   b. B only
   c. Both A and B
   d. Neither A nor B

7. A variable-rate spring:
   A. Has a rate that gets lower as the spring is compressed
   B. Provides a softer ride as more load is placed on it
   Which option best completes the statement?
   a. A only
   b. B only
   c. Both A and B
   d. Neither A nor B

8. Which of the following is not a common spring type?
   a. Coil
   b. Multileaf
   c. Composite
   d. Torsion bar

9. Statement A: Air springs provide a softer ride because the system can compensate for high load conditions.
   Statement B: Air spring systems are not simple and inexpensive.
   Which statement is correct?
   a. A only
   b. B only
   c. Both A and B
   d. Neither A nor B

10. Which of the following spring types is the most difficult to make with a variable rate?
    a. Coil
    b. Composite leaf
    c. Leaf
    d. Torsion bar

11. The shape used for most leaf springs is the:
    a. Elliptical
    b. Semielliptical
    c. Quarter-elliptical
    d. Half-elliptical

12. Multiple leaves are used in a leaf spring to provide:
    A. Interleaf friction
    B. A variable spring rate
    Which option best completes the statement?
    a. A only
    b. B only
    c. Both A and B
    d. Neither A nor B

13. Statement A: Rubber leaf spring bushings provide enough compliance for all of the various axle motions.
    Statement B: Rubber bushings keep road vibrations from traveling from the spring to the car body.
    Which statement is correct?
    a. A only
    b. B only
    c. Both A and B
    d. Neither A nor B

14. Statement A: When a coil spring has an end that is simply cut off with no reshaping, it is called a square end.
    Statement B: A spring with a pigtail end has the last coil wound tighter.
    Which statement is correct?
    a. A only
    b. B only
    c. Both A and B
    d. Neither A nor B

15. Variable-rate coil springs are made by winding the spring:
    A. From tapered wire
    B. With tighter-spaced coils at one end
    Which option best completes the statement?
    a. A only
    b. B only
    c. Both A and B
    d. Neither A nor B

16. Coil spring rate is affected by the:
    a. Diameter of the wire
    b. Diameter of the coil
    c. Number of turns in the coil
    d. All of these

17. A torsion bar is always:
    A. Mounted so it is anchored at one end and free to turn at the other
    B. A long, straight bar of steel
    Which option best completes the statement?
    a. A only
    b. B only
    c. Both A and B
    d. Neither A nor B

18. A definite advantage of torsion bars is that they:
    a. Have a variable rate
    b. Can be adjusted to compensate for sag
    c. Offer a lower unsprung weight
    d. All of these

19. Electronically controlled suspensions are being discussed.
    Statement A: They use air springs.
    Statement B: Height sensors are used so the computer will know when the tire meets a bump.
    Which statement is correct?
    a. A only
    b. B only
    c. Both A and B
    d. Neither A nor B

20. Which of the following is not part of an average air ride system?
    a. Air compressor
    b. Coil spring
    c. Air bladder
    d. Height sensor

Math Questions

1. You are replacing the front springs on a 1985 pickup, and while looking up the specifications, you find that they have a spring rate of 500 lb per inch, a free length of 18 in., and an installed height of 16 in. How much weight is carried by each spring?

2. Your friend tells you he is going to lower the rear of his car by cutting off one coil from each spring. The stock springs have a rate of 140 lb per inch, and the coil is made from a wire 0.565 in. in diameter, with ten active coils, and an inside diameter of 5.545 in. What will happen to the spring rate if one of the coils is cut off? What is the new spring rate?

16 Shock Absorbers

Objectives

Upon completion and review of this chapter, you should be able to:

❑ Comprehend the terms commonly used with shock absorbers.

❑ Understand the purpose for shock absorbers as a suspension component.

❑ Comprehend the internal operation of the two major shock absorber designs.

16.1 Introduction

Most technicians realize that shock absorbers do not absorb shock. The term **damper,** used in many countries outside of the United States, describes much more accurately what they really do: they dampen the spring oscillations. When a spring is deflected, it absorbs energy; when it is allowed to, it will extend and release this energy. Spring inertia causes the spring to bounce too far and overextend itself. It then recompresses, but it will travel too far again. The spring will continue to bounce back and forth and oscillate at its natural frequency until the energy that was originally put into the spring is used up by spring molecular friction. This usually takes quite a few oscillations. In a car, if shock absorbers are not fitted, a bump will cause the car to bounce up and down at the natural frequency of the suspension for an uncomfortably long period of time. With shock absorbers, the suspension is allowed to oscillate through one or two diminishing cycles. The shock absorber absorbs much of the energy from the spring (Figure 16–1).

Shock absorbers also give the suspension engineer another method of tailoring the suspension movement. Shock absorbers resist suspension movement: bounce, jounce, body roll, brake dive, or acceleration squat. The added resistance of the shock absorber during the compression or extension portions of the spring cycle can do much to change the action of the suspension travel. A firm shock absorber can add to the spring rate during the compression phase.

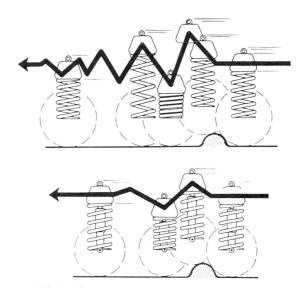

Figure 16–1 If an undampened suspension meets a bump, the sprung weight will bounce at the frequency of the suspension until the energy of the bump is used up (top). The shock absorber will limit or dampen these motions after a few oscillations (bottom). *(Courtesy of Tenneco Automotive)*

Some suspension designs use the shock absorber as the limiting member for suspension travel. **Bump stops** or **rubber cushions** can be installed on the piston rod just above (outside the body) or just below the upper main body bushing (inside the body). These bushings stop the piston rod and suspension travel. One manufacturer incorporates a hydraulic extension or rebound travel stop in some of its models. Be aware that the removal of the

Chapter 16 ■ Shock Absorbers

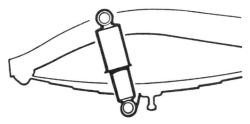

Figure 16–3 All modern shocks are of the direct-acting, telescoping type; one end is connected to the frame, and the other end is connected to the suspension. *(Courtesy of Tenneco Automotive)*

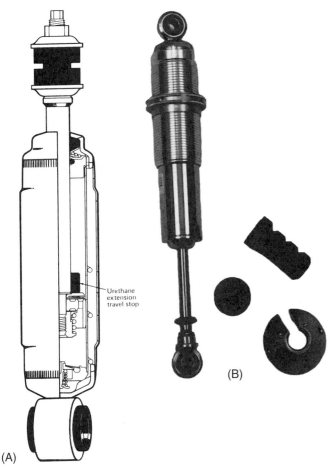

Figure 16–2 Some shocks are designed with flexible travel stops to serve as rebound (A) or jounce (B) travel stops. *(A is courtesy of Tenneco Automotive; B is courtesy of CARRERA SHOCKS, Atlanta, GA)*

shock absorber from some suspensions can cause unexpected results. It is possible for the car to be driven, but if the suspension drops from driving over a severe bump, the rear spring can fall out (Figure 16–2).

16.2 Shock Absorber Operating Principles

The shock absorber must absorb energy to reduce the spring oscillations. Laws of physics tell us that energy cannot be created or destroyed. However, energy can be converted from one form to another. Another name for a shock absorber could be energy converter. The energy that is absorbed by the shock absorber is converted into heat and then dissipated into the surrounding air. As the heat leaves to the air, much of the energy absorbed by the shock absorber goes with it. A shock absorber is really a heat machine. A hard-working shock absorber gets quite hot, sometimes reaching temperatures that cause changes in operation or even failure.

Some early styles of shock absorbers used friction to absorb the spring energy. They were constructed with one lever that connected to the axle and one lever that attached to the frame. The levers were separated by a friction material, similar to clutch or brake lining, and tightened snugly against each other. Vertical suspension movement caused a rotation and rubbing of the levers; the friction produced by the rubbing retarded the suspension and generated heat. The amount or rate of energy absorption was controlled by the tightness of the levers against each other. If they were tightened too much, they would eliminate all motion and lock up the suspension. Also, **friction shock absorbers** were very prone to wear because of the constant rubbing.

Modern shock absorbers operate **hydraulically.** They are basically oil pumps that force oil through small **openings** or **orifices.** Heat is generated when a liquid is forced through a restriction. Because the internal friction is fluid friction, hydraulic shock absorbers can operate through many cycles without wearing (Figure 16–3).

All of the shock absorbers used today are **direct acting.** One end of the shock absorber is connected to the suspension, and the other end is connected to the frame. Rubber bushings are used at these connections to offer compliance (i.e., allow for slight mounting angle changes) and to dampen vibrations. The shock absorber telescopes and changes length as the suspension moves up and down. At one time, **lever shock absorbers** were fairly common. The shock absorber body was bolted solidly to the frame, with a lever extending from the body. The lever was connected to the suspension so that suspension movement operated the lever and worked the shock absorber. On some cars, the shock absorber lever was also the upper control arm for the front suspension (Figure 16–4).

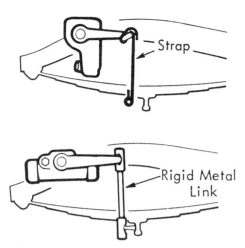

Figure 16-4 Two older, indirect action shock designs: a one-way (rebound direction only) lever shock (top) and a two-way lever shock (bottom). *(Courtesy of Tenneco Automotive)*

16.3 Shock Absorber Damping Ratios

Hydraulic shock absorbers are constructed so they offer different resistances during the two different operating directions of (1) **compression** or jounce and (2) **extension** or rebound. The differences in resistance are referred to as the **shock absorber ratio**. A friction shock absorber has a ratio of fifty-fifty. It has the same resistance on compression as on extension. A hydraulic shock absorber can be constructed with whatever ratio the ride engineer desires. The ratios are normally given with the extension control resistance printed first. Confusion occurs because the ratio numbers are usually given in the reverse order by people involved with racing. This chapter uses ratios as referred to by the engineer. A 90/10 shock absorber has 90 percent of its control ability on extension and 10 percent on compression. This shock absorber would probably compress easily and be difficult to extend. A 10/90 would compress with difficulty (90 percent of control) and extend easily (10 percent) (Figure 16-5).

The ratio is designed into a shock absorber by ride engineers with the requirement that the compression cycle allow the suspension system to rise during jounce while the tire is going over a bump and fall back down after the tire passes the bump. If the shock offers too much resistance during the compression cycle, the bump will become more severe. If too much resistance occurs during extension, it might be possible to hold the tire off the road momentarily. The shock ratio and resistance must add enough resistance to dampen the

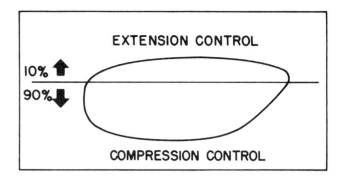

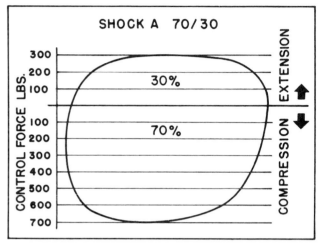

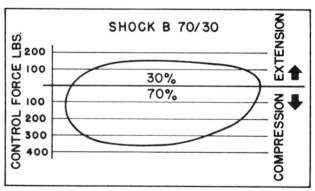

Figure 16-5 A 90/10 shock ratio tells us that 90 percent of the shock absorber's control occurs during compression (top). The curves for two 70/30 shocks are also shown (middle and bottom). Note that one of them has about twice the control force of the other. *(Courtesy of Tenneco Automotive)*

spring oscillations yet allow free enough suspension travel to give a good ride. The actual ratio of most shock absorbers is known only by the ride engineers. The term has a very limited use in the repair field. Shock specifications are seldom printed in service manuals or shock absorber catalogs. It is very difficult and seldom necessary for a technician to determine the ratio of a particular shock absorber.

16.4 Shock Absorber Damping Force

The amount of **damping** or **control force** that a shock absorber exerts is of more importance than the ratio. The ratio is just a comparison of the two control forces, compression and extension. A 50/50 shock absorber with a control force of 200 lb (91 kg) would offer a softer but bouncier ride than a 50/50 shock absorber with a control force of 400 lb (182 kg) on the same car. The amount of control force is designed into the shock absorber by ride engineers. Too much control force will give a harsh, jiggly ride with the possibility of tire hop on washboard roads. Too little control force can give a soft, bouncy ride with excessive roll on corners and excessive dive on braking. Some shock absorbers are constructed so that the control force can be adjusted for one or both directions of travel. This allows tailoring the shock absorber to a particular car and to the driver's preferences (Figure 16–6).

Shock absorber control is measured on a **shock absorber dynamometer.** This machine allows the resistance to be measured over various operating speeds. The resulting graph allows the ride engineer to see the various operating stages plus any weak spots or lags in operation (Figure 16–7).

Ride engineers design the rate of control force to change as the speed of suspension travel changes. These changes are often referred to as **stages of operation.** A car operating at slow speeds over small bumps does not require or want much shock absorber control. The same car operating at high speeds over a rough road needs a lot more control; the shock absorber will need to absorb much more energy. Shock absorber valving is sensitive to the speed at which the suspension travels (Figure 16–8).

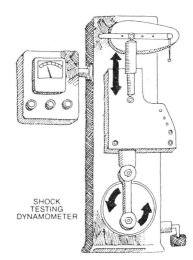

Figure 16-7 A shock absorber dynamometer. This machine is used to measure shock resistance at various speeds. *(Courtesy of Tenneco Automotive)*

Hydraulic shock absorbers have a natural tendency to increase resistance with an increase in velocity. Laws of fluid dynamics state that a liquid's resistance to flow through an orifice increases with the square of the flow velocity. In other words, as the shock absorber piston speeds up because of a high bump rate, the higher fluid flow rate that results meets increased resistance at the fluid control orifices, so the shock absorber causes a higher resistance to motion. The shock absorber resistance will increase. The second and third stages of shock absorber operation open up additional orifices or flow paths to keep the resistance from becoming too great at high speeds. These stages are controlled by spring pressure and orifice size (Figure 16–9).

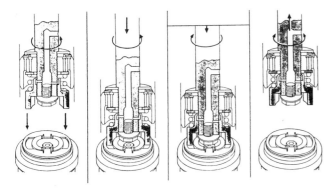

Figure 16-6 This shock uses an adjustable piston rod orifice. Rotating the adjuster nut upward on the piston rod will increase the rebound/extension resistance. Compressing the shock completely engages the adjuster nut with the foot valve, and turning the piston rod will then turn the adjuster. *(Courtesy of KONI)*

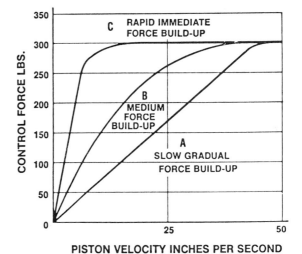

Figure 16-8 The resistance of a shock increases with the speed of the piston. It is also controlled by the internal valving. These three curves show three different stage 1 valvings. *(Courtesy of Tenneco Automotive)*

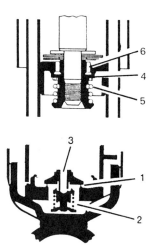

Figure 16–9 Most shocks have three or more operating stages in each direction. During compression, stage 1 (slow speed), oil passes through a small orifice (1); stage 2 (faster speed), oil flows through a spring-controlled valve (2); and stage 3 (very high speed), oil flow is controlled by the size of the valve pin orifice (3). During extension, stage 1, oil flow is controlled by a small orifice (4); stage 2, oil flow is controlled by a spring-controlled valve (5); and stage 3, oil flow is controlled by the sizes of the passages (6). *(Courtesy of Tenneco Automotive)*

Some shock absorbers have very small grooves cut into the cylinder wall at the center travel portion to soften the ride.

16.5 Double-Tube Shock Absorbers

The traditional, direct-acting shock absorber is of a **double-tube** design. The shock absorber is made with two major steel tubes. The inner pressure tube is the cylinder in which the piston operates, and the outer tube forms the outside of the reservoir. Many shock absorbers have an additional tube attached to the piston rod called a **dust** or **stone shield.** It protects the piston rod and upper seal from road damage (Figure 16–10).

16.5.1 Construction

The upper end of the piston rod connects to the car's frame. The piston, which has some of the control valves built into it, fits snugly into the pressure tube. Metal, teflon, or plastic sealing rings are often used to ensure a good seal between the piston and pressure tube. The smoothly ground, chrome-plated piston rod passes through a bushing and seal in the upper end of the pressure tube. This seal is usually a neoprene or silicone rubber of a multilip design (Figure 16–11). The bushing and seal allow the piston rod to move freely in and out

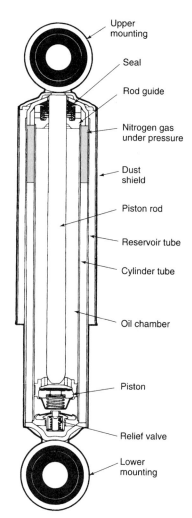

Figure 16–10 A double-tube shock. The piston travels in the inner pressure or cylinder, tube while the reservoir tube forms the outside of the reservoir. *(Courtesy of Delco Products Division of General Motors Corporation)*

of the pressure tube while keeping all of the hydraulic oil inside. A drain hole, just below the seal, allows any oil that passes through the bushing to drain back into the reservoir. This hole also allows air to bleed out of the pressure tube during operation. Some shock absorbers place a valve in this opening to offer more control (Figure 16–12).

The base mount of the shock absorber serves several functions. Besides being the mount connection to the axle or suspension arm, it holds the **base** or **foot valve** assembly that forms the bottom of the pressure cylinder and the reservoir chamber. Passenger-car shock absorber mounts are usually of a **bayonet type,** also called a **single-stud** or **eye ring type.** Rubber bushings are used with each mount type to allow a somewhat flexible mounting so the shock absorber can align itself during suspension movement and also to reduce or dampen road vibrations (Figure 16–13).

Chapter 16 ■ Shock Absorbers

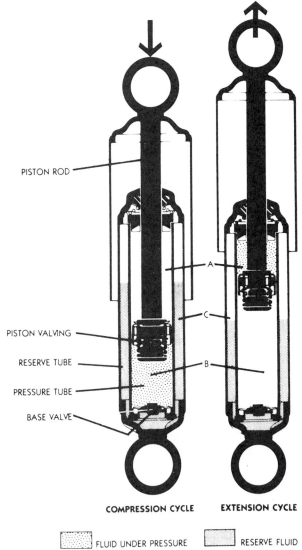

Figure 16-11 During compression, the oil between the piston and the base valve (area B) is under high pressure. During the extension cycle, the oil between the piston and upper bushing (area A) is under pressure. *(Courtesy of Tenneco Automotive)*

16.5.2 Operation

As the suspension moves up and down, the shock absorber piston and rod stroke up and down in the pressure cylinder. There are essentially three different fluid chambers in a double-tube shock absorber: (1) a low-pressure reservoir, (2) the area above the piston, and (3) the area below the piston in the pressure cylinder. One of the cylinder areas will have a very high pressure and the other a very low pressure, depending on the direction of piston travel. The pressures in the pressure cylinder are affected by the piston rod as well as the piston. The rod displaces a relatively large amount of fluid as it enters the shock absorber during compression travel. At this time, an equal volume of fluid must leave the pressure cylinder and go into the reservoir.

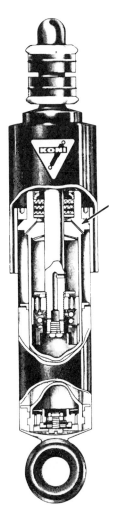

Figure 16-12 As the piston rod strokes up and down, some oil flows through the upper guide bushing. This oil can run back into the reservoir through a passage (arrow). *(Courtesy of KONI)*

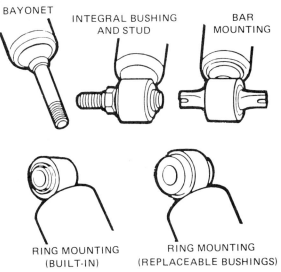

Figure 16-13 Various types of shock mount types. All of these types use a rubber bushing between the shock end and mounting point. *(Courtesy of Ford Motor Company)*

During compression, the piston and rod are moving into the pressure cylinder. The piston and rod motion are generating a high pressure below the piston and a low or negative pressure above it. Oil flow between the chambers above and below the piston are controlled by valves and orifices in the piston. Oil flow from the lower-pressure cylinder to the reservoir is controlled by valves and orifices in the base valve assembly (Figure 16–14).

These valves are usually staged. A small orifice provides the control for small bump operation. If the piston speeds are faster, the increased oil pressure that is generated will work against a coil or flat reed spring to open an additional orifice. Higher pressures will open the valve farther and farther. Maximum flow is controlled by the size of this additional orifice passing through the valve.

During compression, most of the shock absorber control is at the oil flow from the pressure tube to the reservoir. Excessive restriction at the piston valve might cause the upper cylinder pressure chamber to drop low enough to pull air in through the rod seal or the drain hole.

The amount of piston travel should be limited so it never is far enough for the piston to bump into the base valve. This can damage the base valve or piston or both, which would ruin the shock absorber. The compression travel is normally limited by external bump stops on the car's suspension or by rubber cushions between the upper shock absorber bushing and the upper shock absorber mount.

During extension, the piston and rod are moving upward in the pressure cylinder, and oil flows back from the reservoir to the pressure cylinder. There is a very high pressure in the chamber above the piston and a lower pressure below it. The flow into the lower chamber from the reservoir is almost a free flow with little restriction. Nearly all of the extension control takes place in the piston valves and orifices as the oil moves downward through them. Because of the flow from the reservoir to the pressure cylinder, a double-tube shock absorber must be in a somewhat vertical position to operate. If it were mounted upside down, air, not oil, would flow into the pressure cylinder during the extension stroke (Figure 16–15).

Operation on rough roads causes a certain amount of agitation in the reservoir oil as the body of the shock absorber is shaken up and down along with the suspension. This agitation tends to foam or aerate the oil. All double-tube shock absorbers leave a certain amount of expansion space in the

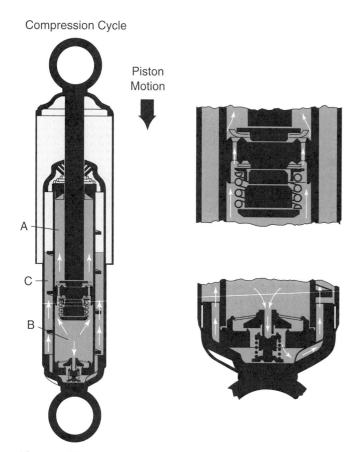

Figure 16–14 During the compression cycle, oil must flow from chamber B into chambers A and C; shock control is provided by the orifices and spring-controlled valves in the piston but mostly in the base valve. *(Courtesy of Tenneco Automotive)*

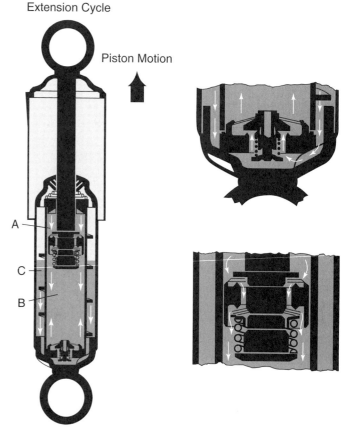

Figure 16–15 During the extension cycle, oil must flow from chambers A and C into chamber B; shock control is provided by the valve stages in the piston. *(Courtesy of Tenneco Automotive)*

reservoir to allow for heating of the oil; the most common expansion space is an air space above the oil.

Air bubbles that mix with the oil and enter the pressure cylinder interfere with shock absorber operation because air or any other gas can be compressed and also because a gas will flow through an orifice much faster than a liquid. Shock operation will be mushy. Normal shock absorber operation will eventually move this air into the upper chamber and then through the drain passage back to the reservoir, but while the air is in the pressure cylinder, a loss of shock absorber control takes place. Many shock absorber designs use a spring-shaped baffle in the reservoir or a spirally grooved reservoir tube to reduce the air or gas that can work its way down to the base valve and into the pressure tube.

Modern shock absorber designs place a gas-filled plastic bag in the reservoir and fill the reservoir completely with oil. The gas-filled bag compresses and expands during shock absorber operation or heat expansion of the oil. The plastic bag holds the gas, which is usually a nitrogen or Freon gas under a pressure of about 150 psi (1,034 kPa), captive and prevents aeration. Another design pressurizes the reservoir with a nitrogen gas charge of about 100 to 150 psi (690 to 1,034 kPa) (Figure 16–16). These are often called **gas-charged shocks.** This increased internal pressure increases the load placed on the piston rod seal. The seal has to keep in this added pressure. A shock absorber with internal pressure tends to extend itself, much like a spring. In special installations,

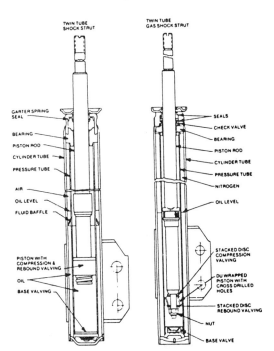

Figure 16–16 These two shocks are very similar. The one on the right has a pressurized reservoir, charged with nitrogen gas at about 100 psi (690 kPa). The gas pressure prevents oil aeration. *(Courtesy of Ford Motor Company)*

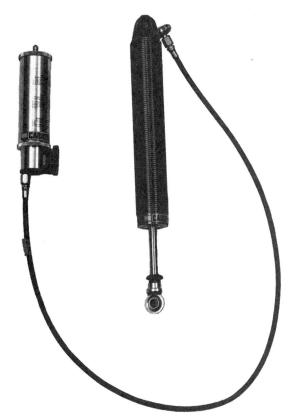

Figure 16–17 This racing shock has a remote reservoir that can be run with adjustable pressure. The remote mounting reduces vibration-caused aeration and improves cooling. Also note the adjustable spring seat. *(Courtesy of CARRERA SHOCKS, Atlanta, GA)*

the reservoir can be mounted remotely and connected to the base valve and pressure tube with tubing. A remote reservoir has a much better chance of providing cool, air-free oil (Figure 16–17).

16.6 Monotube Shock Absorber Operation

Monotube shock absorbers are also called **de Carbon, single-tube,** or **gas pressure** shock absorbers. There is only one shock absorber tube, the pressure tube, but there are two pistons, a **dividing piston** and a **working piston** (Figure 16–18). This design was developed so that gas, separated by the dividing piston, and its pressure are prevented from fluid foaming possibilities. In double-tube designs, gas can mix with oil, and as this mixture is agitated by suspension action, cavitation and fluid foaming occur. Another advantage is that the pressure tube is closer to the outside air so it will dissipate heat better and run cooler; in addition, it can be mounted with the shock body to the car frame. As the oil in a shock absorber heats up, it will thin out. This can cause reduction in the control resistance, a lag or skip in the operation, or both. A potential problem with a monotube shock

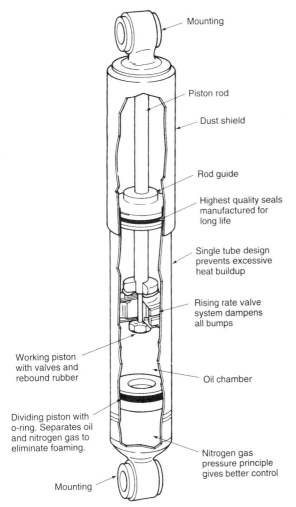

Figure 16–18 A monotube shock. Note that there is no reservoir and that there are two pistons in the pressure tube. *(Courtesy of Delco Sealed Power Corporation)*

absorber is that a bend or dent in the tube can destroy the piston-to-cylinder seal and upset or destroy shock absorber operation. The reservoir of a double-tube shock absorber adds quite a bit of protection for the pressure tube.

16.6.1 Construction

The working piston and rod of a monotube shock absorber are very similar to those of a double-tube shock absorber except that they are often upside down. A monotube shock absorber will operate with either end up. Many designers prefer to mount the body of the shock absorber on the frame to help reduce unsprung weight and agitation of the major part of the shock absorber. The pressure tube is longer than needed for piston travel and is sealed at the top where it connects to the frame mount. A free-floating, dividing piston travels in the mount end of the pressure tube. The chamber between the dividing piston and the mount end is pressurized to about 75 to 350 psi (517 to 2,400 kPa) with nitrogen gas. This gas pressure tends to move the dividing piston and the working piston toward the extended position.

16.6.2 Operation

Since a monotube shock absorber does not use a base valve, both the extension and compression control valves must be built into the working piston. The dividing piston moves up or down in the pressure tube as the piston rod moves in or out of the shock absorber. Movement of the dividing piston compensates for the oil displaced by the piston rod (Figure 16–19).

During compression, the piston rod and working piston move inward into the pressure tube, and the dividing piston is forced toward the mount end of the pressure tube. This action tends to compress the gas charge. Shock absorber control takes place as the oil pressure increases in the chamber between the two pistons and forces oil through the orifices and valves in the working piston. Monotube shock absorbers use staged valving much like that used in double-tube designs.

During extension, the piston rod and working piston move outward in the cylinder, and gas pressure moves the dividing piston toward the working piston. The oil pressure in the chamber on the rod side of the piston increases and forces oil to flow through the orifices and valves in the working piston. Shock absorber control occurs in the working piston while this oil flow takes place.

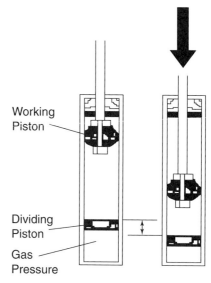

Figure 16–19 The lower dividing piston separates a high-pressure gas charge in the lower chamber from the oil in the rest of the shock. As the working piston and rod move in and out of the pressure tube, or as the oil expands or contracts, the dividing piston will move up or down to compensate.

16.7 Strut Shock Absorbers

The body of most struts is essentially a large shock absorber with an oversize piston rod and spring seats. Most OEM struts are of a double-tube design, but some are monotube. The operation and internal construction of a strut is essentially the same as that of a shock absorber (Figure 16–20).

Some older OEM struts are constructed so they can be taken apart. The upper bushing and seal are retained by a **gland nut.** Removal of the gland nut allows removal of the piston rod bushing and seal, the piston and rod, the pressure tube and base valve, and the oil. Any or all of these components can be replaced to rebuild this strut (Figure 16–21).

Normal shop practice is to remove all of these parts and replace them with a new, sealed shock absorber cartridge. Most OEM struts are welded together so they cannot be easily disassembled. When service is required, they are serviced by replacement of the strut assembly or, in some cases, by cutting the top off the strut, replacing the internal parts with a cartridge, and threading a gland nut into threads already in the outer tube. Replacement shock absorber cartridges can be either double-tube or monotube designs.

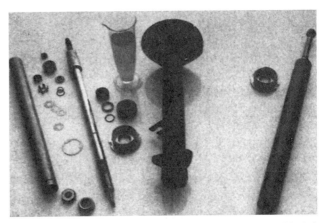

Figure 16–21 Some older struts can be disassembled for service (left). It is common practice to replace the internal parts with a cartridge (right). *(Courtesy of Tenneco Automotive)*

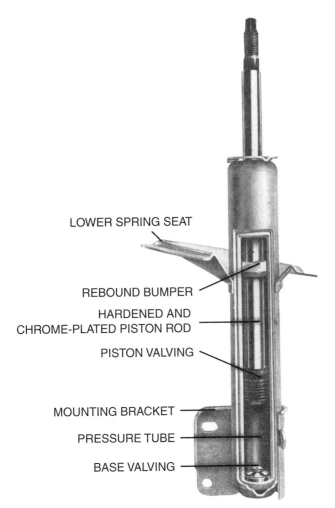

Figure 16–20 A strut. This unit is essentially an oversize shock with a spring seat and a mounting point for the steering knuckle. Note that this unit cannot be disassembled. *(Courtesy of Tenneco Automotive)*

16.8 Computer-Controlled Shock Absorbers

More manufacturers are incorporating electronic control into the shock absorbers of strut suspensions. These systems use a computer module to electronically change the dampening rate of the shock absorber to suit various driving conditions. These systems are sometimes called **semi-active** suspension. Computer-controlled shock absorbers are less expensive and more durable than active suspension systems. They are also very effective in achieving a variable ride control system; one system is said to be able to reset the shock absorbers 100 times a second.

The major parts of these systems are the four shock absorbers with their **variable valving,** an **actuator** for each shock absorber to change the valve setting, the **control module** that controls the actuators, the **sensors** that provide data needed for the control module, a switch that allows the driver to select the type of ride desired, and the wiring to connect these various parts (Figure 16–22).

The shock absorbers used with computer-controlled systems have a **variable-sized orifice** that can allow fluid to bypass the usual extension and compression control valves. At the soft setting, the orifices are at the widest opening, which gives the least amount of dampening resistance. At the very firm setting, the bypass orifices are completely closed, which gives the greatest amount of dampening resistance. A third, intermediate setting included in some systems allows some fluid to bypass and provides a dampening resistance between the other two.

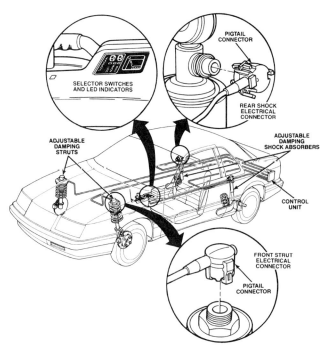

Figure 16–22 A computer-controlled ride control system. Actuators in each of the four shocks or adjustable dampening units can change the valving to firm, normal, or soft settings as determined by the driver's switch or computer programming. *(Courtesy of Chrysler Corporation)*

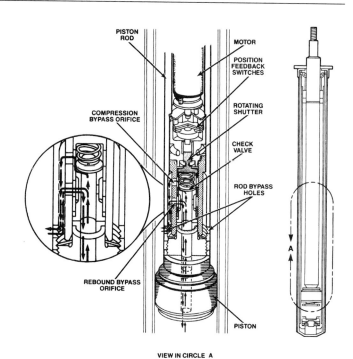

Figure 16–23 This shock has a motor inside that can turn the shutter to open or close orifices; opening the bypass orifices produces softer settings. *(Courtesy of Chrysler Corporation)*

The valve settings are changed by either rotating or changing the height of a control rod that is located in the shock absorber's piston rod (Figure 16–23).

The actuators are electronic and receive their operating signal directly from the control module. The control module is programmed to operate the actuators under various driving conditions depending on the particular system and the position of the selector switch. Some of these conditions include the following:

1. **Antidive.** During braking, a brake fluid pressure switch causes a firm to very firm setting depending on the speed of the car and the position of the selector switch.

2. **Antiroll.** During cornering, a steering wheel angle sensor, a steering shaft angular velocity sensor, or a g sensor can cause a firm or very firm setting depending on the car speed and the position of the selector switch (Figure 16–24).

3. **Antisquat.** During acceleration, the rate at which the throttle position sensor changes position can cause a firm or very firm setting depending on the car speed and selector switch setting.

4. **Speed reset.** Depending on the setting of the control switch, the setting can change to firmer settings above predetermined speeds (Figure 16–25).

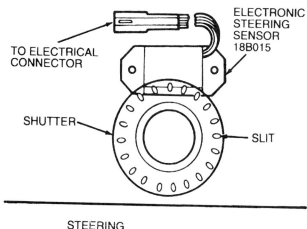

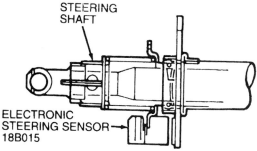

Figure 16–24 This electronic steering sensor provides an input to the control unit of how fast the steering wheel is being turned. *(Courtesy of Ford Motor Company)*

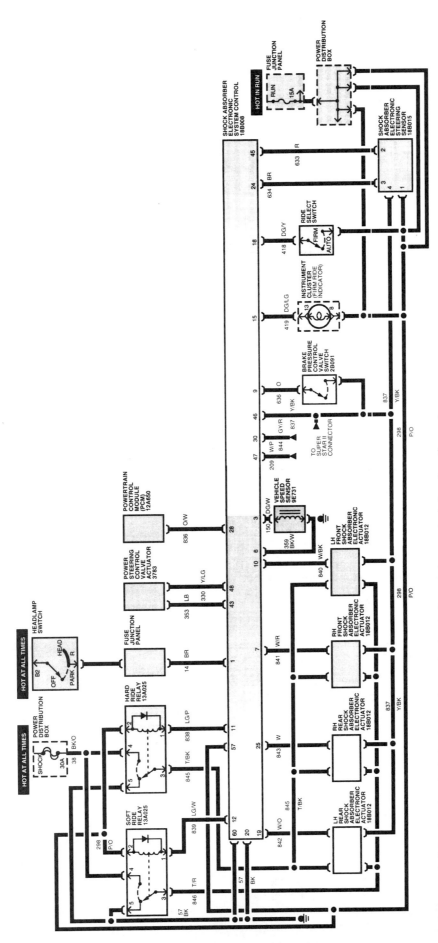

Figure 16–25 The electrical circuit for electronic shock dampening control showing the various sensors and switches, which are inputs, and the actuators, which are outputs. (*Courtesy of Ford Motor Company*)

16.9 Load-Carrying Shock Absorbers

Shock absorbers are normally not designed to carry any vehicle load except for the slight spring rate caused by the internal gas pressure of a gas-filled or monotube design. Removal or replacement of the shock absorbers does not affect the height of the vehicle. Aftermarket shock absorbers are available that can help the springs carry additional load. These are commonly used on the rear ends of vehicles that carry heavier than normal loads. They are also available for the front end on some vehicles. It should be noted that the use of load-carrying shock absorbers places more load on the frame mounts for the shock absorbers. There have been cases of mount breakage and failure because of this added load. Also, some vehicles use a height-sensitive, brake-proportioning valve. Altering the height on these vehicles will upset the bias or balance of the braking action.

Coil-over shock absorbers mount a coil spring around the shock absorber body (Figure 16–26). The upper spring seat is built into the upper shock absorber mount, and the lower spring seat is built into the lower portion of the shock absorber body. Coil-over shock absorbers use a constant-rate spring. One coil-over shock absorber design uses a spring that can act in tension as well as compression. In actual practice, this design performs much like the action of an antiroll or stabilizer bar.

Air shock absorbers connect a reinforced tubular-shaped membrane to the piston rod dust shield and to the shock absorber body. The area inside of the dust shield becomes the air chamber for an air spring (Figure 16–27). This air chamber is connected to an air pressure source by a thin plastic tube. Air shock absorbers are variable-rate air springs, depending on how much air pressure is used.

The air supply can be a tirelike Schrader valve at the rear of the car or an onboard air compressor. A small amount of air pressure must be kept in the air chambers at all times, even when there is no load. Operation with no pressure might cause the rubber membrane to be

Figure 16–26 A load-leveler or coil-over shock. A pair of these particular shocks at the rear will increase the load-carrying ability of the vehicle by as much as 1,000 lb; front units do not have as much load-carrying ability. *(Courtesy of Tenneco Automotive)*

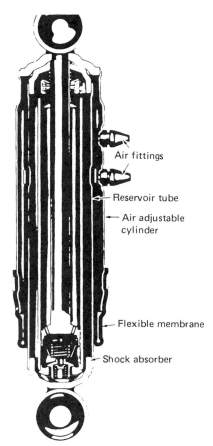

Figure 16–27 A superlift or air shock. A flexible membrane connects the reservoir tube to the outer shield to form an air cylinder. This unit has one air fitting to connect to the other shock and one fitting to go to the air supply. *(Courtesy of General Motors Corporation, Service Technology Group)*

folded and pulled inward if a rapid extension and contraction of the shock absorber occurs (Figure 16–28 and Figure 16–29).

16.10 Shock Absorber Quality

Replacement shock absorbers vary substantially in price. The cost-conscious motorist is often concerned about paying more for one product that looks just like another product that is much less expensive. Some of the difficult-to-see features of a better quality shock absorber include the following:

1. **Piston rod.** A very smooth and hard finish is required to get maximum seal life and smooth operation. An oversize diameter is sometimes used to provide a stronger shock absorber and to displace more oil during the compression stroke; the increased oil displacement provides better control capabilities.

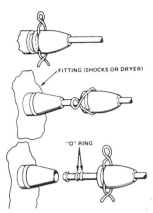

Figure 16–28 The shock air lines on some OEM units are sealed by O-rings and held in place by a quick-disconnect system. *(Courtesy of General Motors Corporation, Service Technology Group)*

2. **Upper bushing and seal.** A smooth precision bore is required for smooth operation, and a long-lasting seal is required to retain the oil for a longer operating life, especially with gas pressure designs. A shock absorber cannot function if it loses its oil.

3. **Piston and ring.** The piston must be concentric with the piston rod and with the ring and must provide a good, fluid-tight, low-drag seal with the pressure tube.

4. **Pressure tube.** The pressure tube must be strong enough to withstand the fluid pressure without deforming and must be straight and smooth to provide a good seal with the piston and ring.

5. **Valves.** The valves must be able to operate through thousands of cycles.

6. **Oil.** The oil must not break down from age or high operating temperatures.

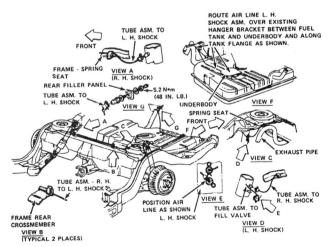

Figure 16–29 The air line routing for a superlift air shock system. Note that this system is filled from an outside air supply. *(Courtesy of General Motors Corporation, Service Technology Group)*

16.11 Shock Absorber Failures

Shock absorbers should be replaced when:

1. A piston rod or mount has broken
2. They are leaking oil
3. They are noisy or give a rough operation
4. They do not provide the correct amount of resistance

Most of these faults are easy to check. The first two can be done with a quick visual inspection. The third requires disconnecting one end of the shock absorber and manually compressing and extending the shock absorber while feeling and listening to its operation.

Checking a shock absorber for correct resistance is a seat-of-the-pants, subjective operation. Some technicians use a **bounce test**; that is, they bounce the car up and down and watch for an excessive number of oscillations, which indicates a weak shock absorber. Most technicians believe that the bounce test is almost useless because of the very slow bounce speed. All this test is really good for is to compare the first stages of two shock absorbers that are on the same axle and to check for noise. A road test over a rough road with a series of undulations is probably the best method of testing shock absorbers. Weak, worn shock absorbers allow an excessive amount of bouncing (especially after the bump), sway, or front-end dive under hard braking. Worn shock absorbers can also allow wheel hop, which results in longer stopping distances. It is best to develop a standard test area and drive it with cars that have worn shock absorbers and then cars that have good shock absorbers, so a comparison can be made. Remember that large, softly sprung cars have a much greater tendency to bounce and sway than smaller, lighter cars. On-car inspection of shock absorbers is described in more detail in Chapter 28.

16.12 Practice Diagnosis

You are working in a tire shop that also specializes in front-end repair, wheel alignment, and brake work and encounter the following problems:

CASE 1: A customer has brought in his 8-year-old Cadillac with a complaint of a noise from the left rear. Your road test confirms the noise as you go over bumps, and you also notice a loose, uneven ride quality that has a wallowing feel on turns. What is the probable cause of the noise? What should you do next?

CASE 2: The customer complaint is poor stopping ability on his 5-year-old pickup. On your road test, you confirm this: it takes a lot of pedal pressure to stop. You also notice that the back end of the empty pickup is higher than normal because of the addition of an aftermarket set of overload springs. When you pull a pair of wheels, a front rotor and a rear drum, you note there is very heavy wear on the front brake shoes but little wear on the rear brake shoes. What is wrong? How are you going to fix this vehicle?

Terms to Know

actuator	dividing piston	piston rod
air shock absorber	double-tube shock absorbers	pressure tube
antidive	dust shield	rubber cushions
antiroll	extension	semiactive suspension
antisquat	eye ring mount	sensors
base valve	foot valve	shock absorber dynamometer
bayonet mount	friction shock absorbers	shock absorber ratio
bounce test	gas-charged shocks	single-stud mount
bump stops	gas pressure shock absorbers	single-tube shock absorber
coil-over shock absorber	gland nut	speed reset
compression	hydraulically	stages of operation
control force	lever shock absorbers	stone shield
control module	monotube shock absorbers	upper bushing and seal
damper	oil	valves
damping	openings	variable-sized orifice
de Carbon shock absorber	orifices	variable valving
direct acting	piston and ring	working piston

Review Questions

1. Statement A: Shock absorbers are called dampers in many countries.
   Statement B: Shock absorbers help support the car's weight.
   Which statement is correct?
   a. A only
   b. B only
   c. Both A and B
   d. Neither A nor B

2. When driving on a rough road, a shock absorber will:
   A. Get hot
   B. Dampen excess spring oscillations
   Which option best completes the statement?
   a. A only
   b. B only
   c. Both A and B
   d. Neither A nor B

3. A shock absorber with a 70/30 ratio:
   A. Offers a different amount of resistance during compression and extension
   B. Allows the spring to travel more easily in one direction than the other
   Which option best completes the statement?
   a. A only
   b. B only
   c. Both A and B
   d. Neither A nor B

4. Statement A: Too much compression resistance in a shock can cause the tire to leave the ground as it passes over a sudden dip in the road at high speeds.
   Statement B: The amount of resistance is determined by the amount of oil in the shock absorber.
   Which statement is correct?
   a. A only
   b. B only
   c. Both A and B
   d. Neither A nor B

Chapter 16 ■ Shock Absorbers 267

5. Which of the following terms does not describe a modern shock absorber?
   a. Hydraulic
   b. Direct acting
   c. Multistage operation
   d. Friction dampening

6. Internal resistance in a shock absorber changes as the _____ changes.
   A. Size of the fluid orifice
   B. Speed of the piston
   Which option best completes the statement?
   a. A only
   b. B only
   c. Both A and B
   d. Neither A nor B

7. Multistages are used in shock absorbers to provide:
   A. An increased amount of resistance at ride height
   B. A steadily reducing amount of resistance as the suspension moves into jounce or rebound travel
   Which option best completes the statement?
   a. A only
   b. B only
   c. Both A and B
   d. Neither A nor B

8. Double-tube shock absorber operation is being discussed.
   Statement A: The amount of oil in the reservoir increases during the compression cycle.
   Statement B: There is a very high pressure between the piston and the base valve during the compression cycle.
   Which statement is correct?
   a. A only
   b. B only
   c. Both A and B
   d. Neither A nor B

9. Statement A: Oil flows from the reservoir into the pressure tube during the extension cycle of a double-tube shock absorber.
   Statement B: Aerated fluid in the reservoir causes mushy shock operation.
   Which statement is correct?
   a. A only
   b. B only
   c. Both A and B
   d. Neither A nor B

10. A double-tube shock absorber:
    A. Cannot be operated upside down
    B. Has the piston rod attached to the car body and the shock body attached to the suspension
    Which option best completes the statement?
    a. A only
    b. B only
    c. Both A and B
    d. Neither A nor B

11. Fluid aeration in the pressure tube is reduced by:
    a. Placing a coil-shaped wire in the reservoir
    b. Pressurizing the reservoir with a nitrogen gas charge
    c. Filling the reservoir completely with a plastic envelope of Freon gas and oil
    d. Any of these

12. Monotube shock absorber operation is being discussed.
    Statement A: There are two pistons in the pressure tube.
    Statement B: Monotube shock absorbers extend when disconnected.
    Which statement is correct?
    a. A only
    b. B only
    c. Both A and B
    d. Neither A nor B

13. Aeration is a lesser problem in monotube shock absorber design because:
    A. There is no air in contact with the fluid in the pressure tube
    B. Of the high pressure from the nitrogen charge
    Which option best completes the statement?
    a. A only
    b. B only
    c. Both A and B
    d. Neither A nor B

14. Monotube shock absorber operation is being discussed.
    Statement A: During the compression cycle, the dividing piston will move away from the working piston.
    Statement B: The pressure inside the pressure tube will increase during the compression cycle.
    Which statement is correct?
    a. A only
    b. B only
    c. Both A and B
    d. Neither A nor B

15. Shock absorber operation is being discussed.
    Statement A: Oil moves through the piston in a direction away from the dividing piston during the compression cycle in a monotube shock absorber.
    Statement B: The oil flow in a double-tube shock absorber is downward through the piston during the compression cycle.
    Which statement is correct?
    a. A only
    b. B only
    c. Both A and B
    d. Neither A nor B

16. The shock absorber portion of a strut:
    A. Is an oversize, standard shock absorber
    B. Can be of a monotube or double-tube design
    Which option best completes the statement?
    a. A only
    b. B only
    c. Both A and B
    d. Neither A nor B

17. Struts are being discussed.
    Statement A: The upper shock absorber piston rod bushing is retained by a gland nut in some shocks.
    Statement B: This bushing is welded in place in some struts.
    Which statement is correct?
    a. A only
    b. B only
    c. Both A and B
    d. Neither A nor B

18. Computer-controlled suspensions are being discussed.
    Statement A: Shock valving can be made softer or more firm by an electric actuator.
    Statement B: The suspension will be made more firm during braking and hard cornering maneuvers.
    Which statement is correct?
    a. A only
    b. B only
    c. Both A and B
    d. Neither A nor B

19. Fluid loss in a shock absorber:
    a. Improves the shock and strut dampening characteristics
    b. Makes the shock cooler
    c. Reduces the dampening capability
    d. None of these

20. A(n) _____ shock absorber can raise the height of the car.
    a. Coil-over
    b. Air
    c. de Carbon
    d. All of these

17 Steering Systems

Objectives

Upon completion and review of this chapter, you should be able to:

❑ Comprehend the terms commonly used with automotive steering systems.

❑ Understand the operation of both major styles of steering gears.

❑ Understand the operation of a power steering system.

❑ Understand the operation of the different types of steering linkage.

❑ Comprehend the operation of variable-assist power steering systems.

17.1 Introduction

The steering system begins at the steering wheel, where the driver decides which direction the car should go. The steering motions are transmitted down the steering shaft and column, through one or more flexible couplings, to the steering gear. Two types of steering gears are used: **conventional,** or **standard,** and **rack and pinion.** The steering gear in turn connects to the tie-rods. The steering system transfers the steering motions from the steering wheel to the steering arms (Figure 17–1).

Both types of steering gears function to change the rotating motion of the steering wheel into a reciprocating, back-and-forth motion at the tie-rods (Figure 17–2). They also provide a gear reduction to slow down the steering speed, reduce the amount of steering effort required, and make steering more precise. Automotive steering systems provide a **steering ratio** somewhere between 10:1 and 25:1. A steering ratio of 20:1 means that 20 degrees of steering wheel motion will turn the front tires 1 degree (Figure 17–3). Steering ratios usually vary depending on the weight of the car and whether manual or power steering is used. Lighter cars or ones with power steering generally use faster ratios (toward 10:1) to allow quicker steering; heavier cars with manual steering use slower steering ratios (toward 25:1) to allow easier steering. Some power steering gears provide a **variable ratio.** Variable ratios are usually slower in the center position to provide precision steering while going in a straight line and faster toward the ends to provide

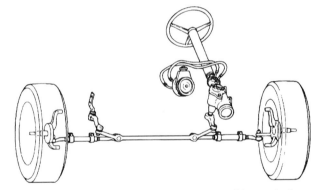

Figure 17–1 A standard steering system; this particular one uses an integral power steering gear. *(Courtesy of Moog Automotive Inc.)*

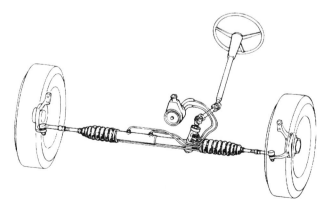

Figure 17–2 A rack-and-opinion steering system; this one has a power steering rack. *(Courtesy of Moog Automotive Inc.)*

269

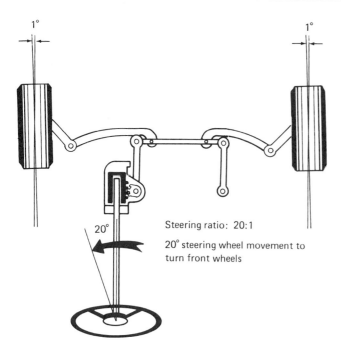

Figure 17–3 A steering ratio of 20:1 tells us that it will require 20 degrees of steering wheel rotation to produce a turn of 1 degree; a 30-degree turn will require 20 × 30, or 600, degrees of steering wheel motion.

quicker steering during cornering and parking maneuvers (Figure 17–4 and Figure 17–5).

Steering ratios appear to the driver as the number of turns it takes to move the wheel from **lock to lock,** a term referring to the number of turns the steering wheel must move from one steering stop to the other. Faster ratios are about three turns lock to lock; slower ratios are four or five turns.

Overall steering ratios are controlled by the ratio of the steering gear plus the length of the steering arms and

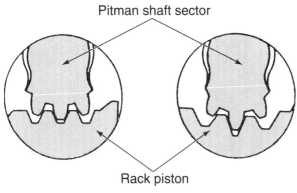

Figure 17–4 The different profile of the steering gear teeth on the right provides a variable ratio. This design provides a slower ratio for the center one-half turn for precise steering and a quicker ratio beyond that. (Courtesy of General Motors Corporation, Service Technology Group)

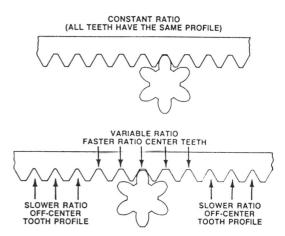

Figure 17–5 The different profile of the rack-and-pinion teeth at the bottom provides a variable ratio. This design gives a fast ratio at the center for fast, responsive highway steering and a slower ratio at the ends for easy cornering and parking. (Courtesy of Ford Motor Company)

Pitman arm. Longer steering arms or a shorter Pitman and idler arm will produce slower steering ratios. A few cars have steering arms with two sets of holes for the tie-rod ends. The holes closest to the steering axis provide the fastest steering and are normally used with lighter-weight engines or power steering. The holes closest to the ends of the steering arms provide easier but slower steering (Figure 17–6).

Steering gears are connected to the steering arms by the steering linkage. On cars with rack-and-pinion gears, the steering linkage is the two tie-rods. Several different styles of linkage are used with standard steering gears; the **parallelogram** style is the most popular (Figure 17–7). Several styles of linkage are used with trucks and pickups also; the exact style depends on the types of axle and suspension used (Figure 17–8).

Figure 17–6 This Corvette steering arm has two tie-rod mounting holes; the tie-rod on this car is in the inner, faster position.

Chapter 17 ■ Steering Systems

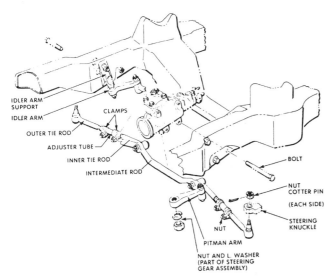

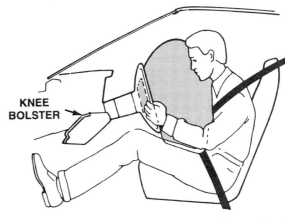

Figure 17–9 If an accident occurs, the driver-side airbag will deploy and cushion the driver's forward impact. *(Courtesy of General Motors Corporation, Service Technology Group)*

Figure 17–7 A parallelogram style of steering linkage; it is commonly used with standard steering gears and an S-L A suspension. *(Courtesy of General Motors Corporation, Service Technology Group)*

17.2 Airbag or Supplemental Restraint System

Airbag systems, also called **airbags, supplemental inflatable restraints (SIR),** and **supplemental restraint systems (SRS),** are a modern safety feature mounted in the steering wheel of most new cars; they have been required in all new cars sold in the United States since 1996. Steering wheel units are called **driver-side** airbags; many new cars also have passenger-side systems mounted in the center or passenger side of the instrument panel. Some cars also have side impact systems mounted in the side of the seat or car. The airbag is a flexible bag that is inflated instantly if the car is involved in a serious accident, and the inflating bag cushions the driver or passenger for a brief period to prevent serious head or upper body injury (Figure 17–9).

An airbag system consists of:

- One or more **airbag modules** that contain the bag and **inflator;** the inflator contains a fuselike firing device called a squib or igniter and the explosive-like propellant (sodium azide or potassium nitrate) (Figure 17–10).

- Three or more **collision sensors,** a **safing sensor** mounted under the instrument panel or with the control module, and usually a pair of **crash sensors** mounted near the front corners of the car (Figure 17–11).

- A **control** or **diagnostic module** that monitors the system to make sure it is ready; it also contains an electrical storage capacitor that can deploy the bag(s) if the car's battery is destroyed.

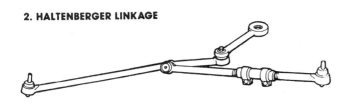

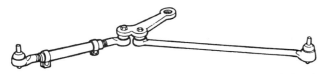

Figure 17–8 Cross-steer linkage is used with solid axles, Haltenberger linkage is used with swing axles, and center arm steer is used with several different independent suspensions. *(Courtesy of Moog Automotive Inc.)*

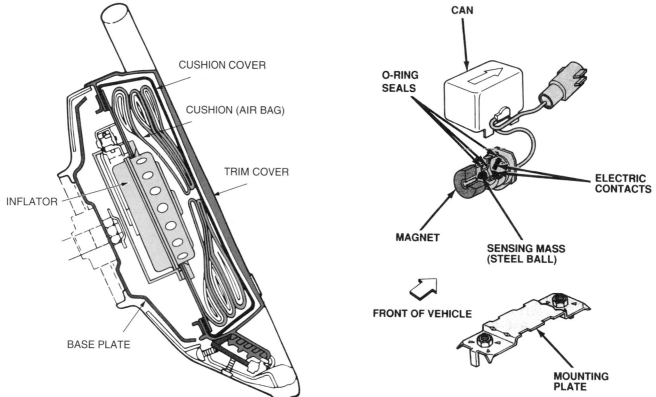

Figure 17-10 A cutaway view of an airbag assembly. (*Courtesy of General Motors Corporation, Service Technology Group*)

Figure 17-11 A crash sensor. The sensing mass will break loose and complete the circuit at the contacts if a very rapid deceleration occurs. (*Courtesy of Ford Motor Company*)

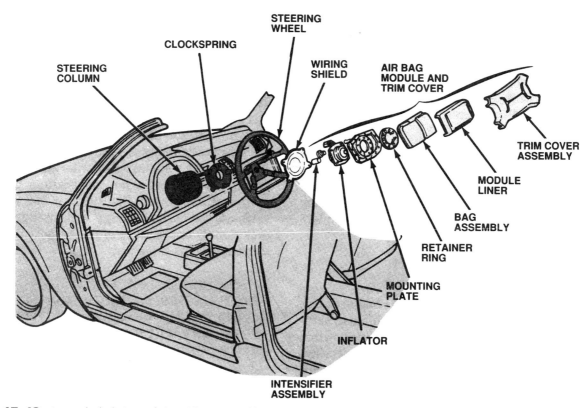

Figure 17-12 An exploded view of the airbag assembly, steering wheel, and clock spring electrical connector. (*Courtesy of Ford Motor Company*)

Chapter 17 ■ Steering Systems

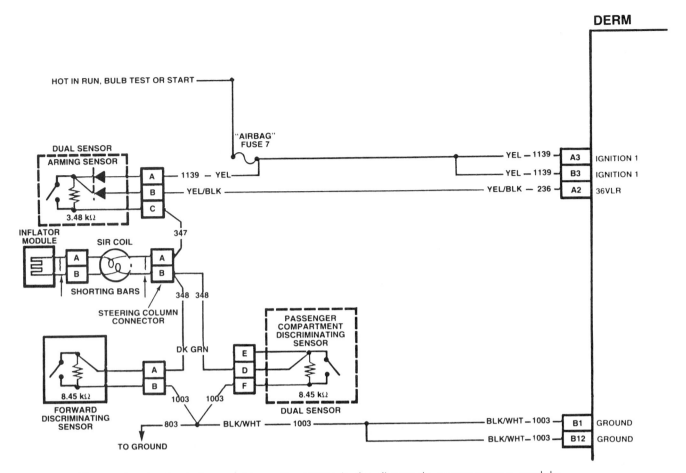

Figure 17-13 An electric circuit for an airbag: the DERM is the diagnostic energy reserve module. *(Courtesy of General Motors Corporation, Service Technology Group)*

- A connector called a **clock spring** that makes the electrical connection between the rotating steering wheel and the steering column. It is a spiral-wound (shaped like a clock spring) pair of wires that winds looser or tighter as the car is steered (Figure 17-12).
- The **warning lamp** in the instrument panel.

As shown in Figure 17-13, the airbag electrical circuit is armed, and current passes through it at all times while the car is running. The sensors contain a parallel circuit through a resistor, along with the inertia switch. The inertia switch can trigger airbag deployment. The current flow through the resistors is very weak, just enough to arm the airbag igniter and for the diagnostic module to determine proper circuit condition. The inertia switch is controlled by a metal ball (a mass) held in place by a specially fitted magnet; a very high rate of deceleration pulls the mass away from the pull of the magnet to make contact with the switch terminals. The increased current flow through the inertia switch then ignites the airbag propellant. The igniting propellant rapidly generates nitrogen to inflate the bag. The safing sensor and at least one of the crash sensors must close in order to trigger the airbag. The safing sensor is also called an **arming sensor,** and the crash sensor is also called a **discriminating sensor.** Sensor circuit designs vary; they can use strain gauges instead of moving mass and/or multiple wire connections to improve airbag triggering or diagnostics.

A yellow or orange connector is used at the steering column to connect the airbag module to the car's electrical system; disconnecting this connector will disarm or deactivate the system. If an airbag deploys, it, along with certain other parts specified by the vehicle manufacturer, must be replaced; airbag service is described in Section 33.4.1.

17.3 Steering Columns

Traditionally, **steering columns** have been used to support the steering shaft. Bearings are provided to allow free rotation of the shaft and wheel without sloppy side or end motions. As the years pass, the steering column is evolving into one of the more complex parts of the car. We are all familiar with the horn in the steering wheel and the turn indicator switch just in front of the steering wheel. We are also familiar with the ignition switch, which not only controls the starting and running of the engine but also locks the steering shaft from turning and the transmission gear selector from moving when the ignition is locked and the key is removed. Some steering column ignition switches are also interconnected to the transmission linkage through an inhibitor system that requires that the vehicle be in a certain transmission gear before the key can be removed. Many newer cars have replaced this inhibitor linkage with a simple lever or button in the column that must be moved before removing the key (Figure 17–14).

Collapsible steering columns have become standard, and they are required by federal law. These columns and steering shafts are designed to shorten under impact, a feature that has probably saved many drivers from chest and upper body injury during front-end accidents. Another safety feature, the driver-side airbag, has been added; the controls run through the column. Care should be taken when working on steering columns or steering wheels to ensure that the collapsible feature remains in proper operation; excessive force by a mechanic can sometimes cause damage. Steering columns or shafts that have collapsed should be replaced (Figure 17–15).

Several more automobile controls have moved onto the steering column, including cruise control switches, windshield wiper and washer switches, headlight dimmer switches, and hazard switches. The control levers for these switches often combine several functions; movement of the lever one way controls the operation of one system, while movement in a different direction controls a different function (Figure 17–16 and Figure 17–17).

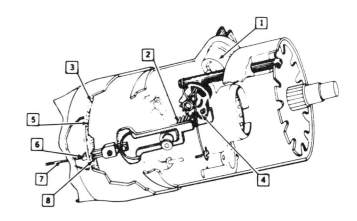

1. LOCK CYLINDER
2. RACK
3. BOWL PLATE
4. SECTOR
5. PARK POSITION
6. WEDGE SHAPE FINGER
7. ACTUATOR ROD ASSY
8. NEUTRAL POSITION

Figure 17–14 This upper steering column section contains the ignition switch and column lock. *(Courtesy of General Motors Corporation, Service Technology Group)*

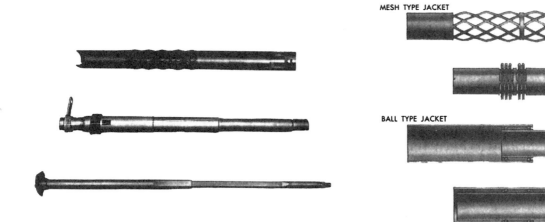

Figure 17–15 The collapsible portions (left) of a steering column are the steering shaft, the gear selector tube, and the steering column jacket. At the right are two styles of column jackets before and after collapsing. *(Courtesy of General Motors Corporation, Service Technology Group)*

Chapter 17 ■ Steering Systems

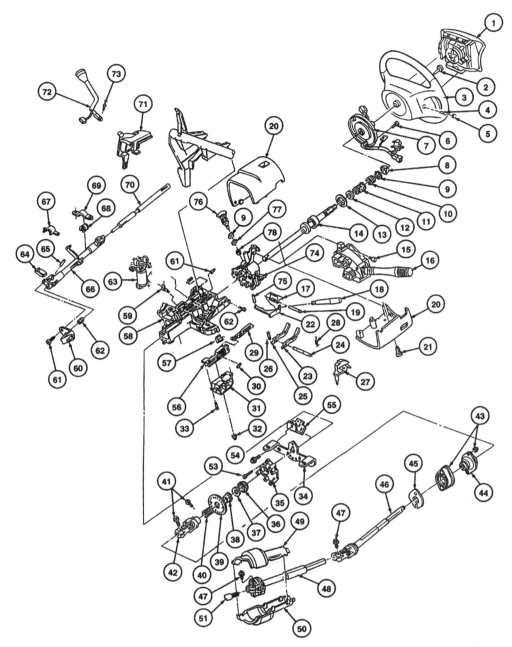

Figure 17–16 A disassembled steering column showing the internal components. *(Courtesy of Ford Motor Company)*

Some columns are adjustable for steering wheel position, steering wheel angle, or forward–backward position. These adjustments are used to place the steering wheel in a more comfortable position, a feature that most benefits a driver of nonaverage size (Figure 17–18).

Flexible couplers, and sometimes universal joints, are used to connect the steering shaft to the steering gear. These couplers allow for a slight angle difference between the two shafts and dampen road vibrations and noises from being transmitted up the shaft to the steering wheel (Figure 17–19). Universal joints are used for more severe angle changes. Most of the couplers are made from rubber, usually reinforced with fabric. They can and do fail. A safety interconnection is made between the two portions of the coupler; in case of failure, the driver will still be able to steer. Coupler failure introduces a large amount of looseness and sloppiness in the steering.

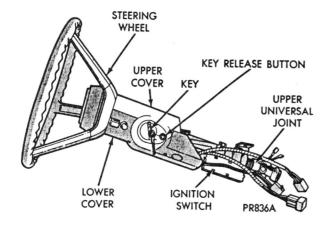

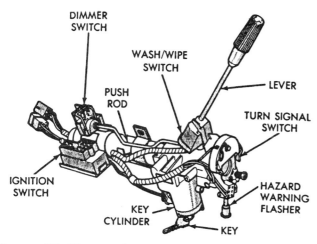

Figure 17-17 A modern steering column contains several electrical switches. *(Courtesy of Chrysler Corporation)*

Steering column service is too complex to describe in a text of this type. An adequate manufacturer or technician service manual should be consulted before attempting to disassemble or reassemble a modern steering column.

17.4 Standard Steering Gears

All steering gears must provide a change in motion, have a gear ratio, and operate as smoothly and easily as possible without excess free play. Most standard steering gears used in today's cars are of the **recirculating ball nut** type; two other design types are **worm and sector** or **worm and roller (Gemmer)** and **cam and lever (Ross)** (Figure 17-20).

Let us begin with the worm and sector steering gear because it is the easiest design to describe and understand. All steering gears use two shafts, an input connected to the steering wheel and an output connected to the linkage. The input shaft is commonly called the **steering shaft** or the **worm shaft** because it contains a worm gear; a worm gear resembles a large screw thread. The output shaft is commonly called the **Pitman shaft** because the Pitman arm connects to it, or the **sector shaft** because the gear that is made on it is a section of a gear (Figure 17-21). In some steering gears, the gear teeth on the sector have been replaced with a hardened-steel roller that has the same side shape as gear teeth; this gear set is called a **worm and roller.** The roller is attached to the Pitman shaft through a set of ball bearings to help reduce gear friction. The sector gear meshes with the worm gear; when the worm gear turns, the sector gear and shaft are forced to swing one way or the other. This motion causes the end of the Pitman arm to move one way or the other, giving the steering action. The steering shaft and the Pitman shaft are each supported by a set of bearings or bushings.

A recirculating ball nut steering gear also has a sector gear on the Pitman shaft; this gear meshes with the ball nut. The **ball nut** has an internal thread that resembles a spiral groove. It fits over the steering shaft, which also has a spiral thread groove (Figure 17-22). One or two sets of steel balls are fitted into these two grooves to form the interconnecting threads. As the shaft rotates, these balls force the ball nut to move up or down on the steering shaft, much like a nut on a bolt. The balls rotate and roll during this motion and circulate on a path provided by a set of ball guides. The balls provide an almost frictionless connection between the steering shaft and the ball nut. When the steering shaft is rotated and the ball nut moves up or down, the sector gear, which is meshed with the ball nut, is forced to rotate or swing one way or the other, producing the steering motion.

Both styles of standard steering gears allow freer motion to be transmitted through them from the steering shaft to the Pitman shaft than from the opposite direction. Steering motion is freely transmitted from the steering wheel to the Pitman arm, but road shock trying to move from the Pitman arm to the steering shaft is resisted. This action is the natural result of the type of gears used and tends to dampen road shock and provide steering that is free from kickback on rough roads.

All standard steering gears provide two adjustments to help ensure free turning without free play. These are **worm shaft bearing preload** and **gear lash,** sometimes called the **over center** adjustment. These adjustments are described in Chapter 33 (Figure 17-23).

Standard manual (nonpower) steering gears contain their own supply of lubricant; the gear housing is normally filled through a filler plug or one of the sector gear cover retaining bolts. The lubricant type varies among manufacturers; it is usually gear oil or a thin, semifluid

Chapter 17 ■ Steering Systems

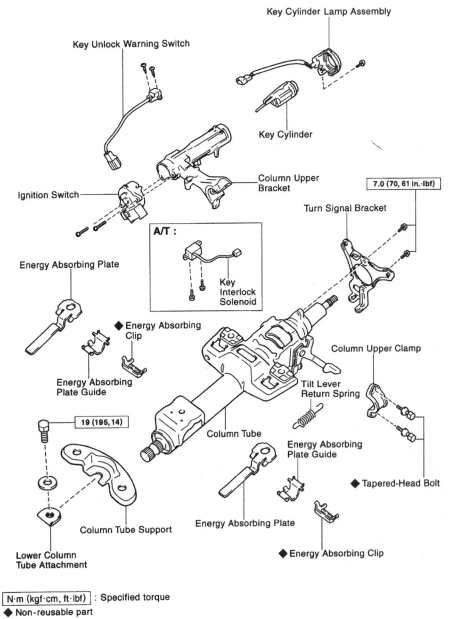

Figure 17–18 The tilt mechanism is in the column tube of this unit.

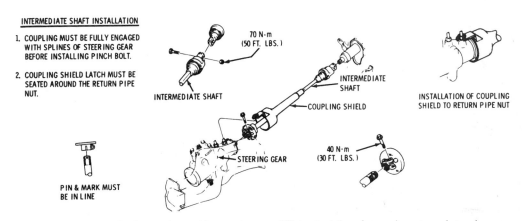

Figure 17–19 The intermediate shaft used on this car has a different style of coupler at each end. *(Courtesy of General Motors Corporation, Service Technology Group)*

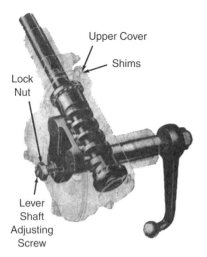

Figure 17–20 A Ross, or cam and lever, steering gear. Note the two studs on the lever (sector gear) that engage the cam (worm shaft). *(Courtesy of SPX Corporation, Aftermarket Tool and Equipment Group)*

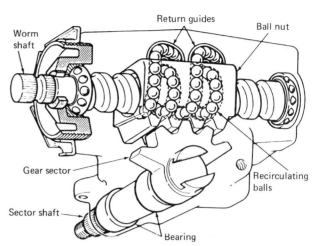

Figure 17–22 The recirculating balls run in grooves inside the ball nut and on the ball nut; as the worm shaft turns, the ball nut will be threaded up or down the shaft. *(Courtesy of Ford Motor Company)*

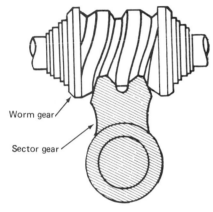

Figure 17–21 Many early steering gears were of the worm and sector type; when the worm gear was turned by the steering wheel, the sector gear turned the Pitman shaft. *(Courtesy of General Motors Corporation, Service Technology Group)*

type of grease. Seals are used at the Pitman shaft and the steering shaft to keep the lubricant in the gear housing and dirt and water out.

17.5 Rack-and-Pinion Steering Gears

Rack-and-pinion steering systems are simpler and lighter in weight than standard systems. Many people credit rack-and-pinion systems with being more responsive and giving the driver a better feel of the road; this is often due to a faster gear ratio. Rack-and-pinion steering gears

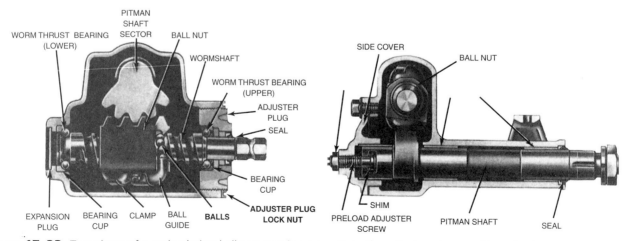

Figure 17–23 Two views of a recirculating ball nut steering gear. Note the adjuster plug used to adjust worm shaft bearing preload (left) and the preload adjuster screw used to adjust gear lash (right). *(Courtesy of General Motors Corporation, Service Technology Group)*

Chapter 17 ■ Steering Systems

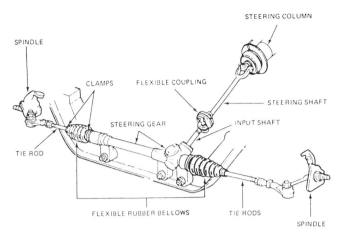

Figure 17–24 A rack-and-pinion steering gear with tie-rods and steering shaft. *(Courtesy of Ford Motor Company)*

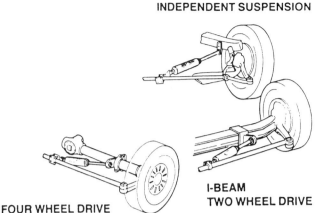

Figure 17–25 One end of a steering damper is connected to the frame or body and the other end is connected to the steering linkage. *(Courtesy of Dana Corporation)*

were traditionally fitted to smaller, lighter cars, whereas standard steering gears were fitted to larger, heavier cars, pickups, vans, and trucks (Figure 17–24).

A disadvantage with rack-and-pinion steering gears is that they transmit more road shock to the steering wheel. Most rack-and-pinion gears are mounted through **rubber bushings, grommets,** or **insulators** to help reduce this kickback; some systems use a **steering damper** to also reduce kickback. A steering damper is a unit built much like a shock absorber. The body of the damper is connected to the frame or body, and the piston rod is connected to the steering linkage. A damper adds turning resistance to the steering. The power steering piston in a power rack-and-pinion steering gear is very effective in dampening these motions (Figure 17–25).

The steering shaft in a rack-and-pinion gear set becomes the **pinion gear;** this pinion gear meshes with the teeth on the rack. When the pinion rotates, the rack is forced to move to the right or the left. The rack is supported by bushings in the gear housing, which allow it to slide sideways during turning maneuvers. **Flexible rubber bellows** are used to seal the ends of the gear housing to keep water or dirt from entering or gear lube from escaping (Figure 17–26).

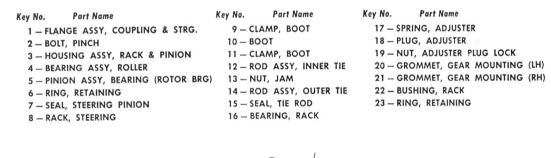

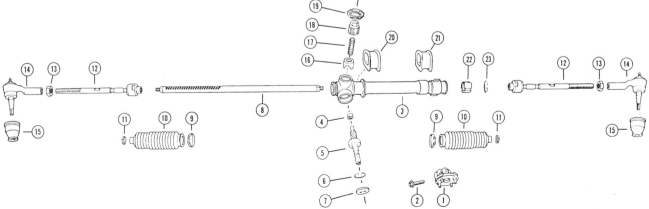

Figure 17–26 An exploded view of a rack-and-pinion steering gear with tie-rods. Note that the rack is rotated to show the gear teeth. *(Courtesy of General Motors Corporation, Service Technology Group)*

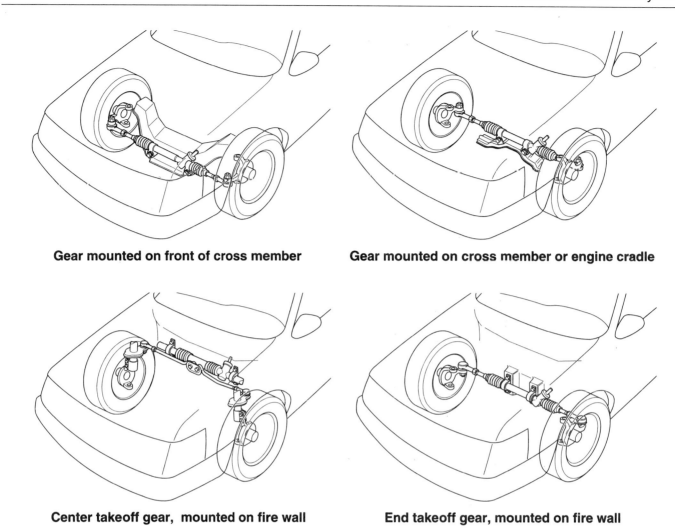

Figure 17-27 A rack-and-pinion gear can be mounted at several locations. Note how the tie-rods are connected on the center takeoff gear (lower left). *(Courtesy of Moog Automotive Inc.)*

A rack-and-pinion gear can be mounted on the front cross member, engine cradle, or front body bulkhead. Body-mounted racks can be either **end takeoff** or **center takeoff** depending on where the tie-rods connect to the rack (Figure 17-27).

Some rack-and-pinion gear sets provide a method of adjusting pinion gear bearings or the mesh clearance or lash between the rack-and-pinion gears. A rack-and-pinion gear is normally lubricated during assembly with gear oil or semifluid grease. Rack-and-pinion service is described in Chapter 33.

17.6 Power Steering

Automotive power steering units use conventional steering gears with a **hydraulic assist.** In addition to the steering gear, which is similar to those just described, there is a hydraulic pump, a control valve that is sensitive to the driver's turning effort, a pair of hoses, and a hydraulic actuator—piston and cylinder. The actuator can be mounted separately on the steering linkage, or it can be integral—built internally into either style of steering gear (Figure 17-28).

Most systems are of the integral type, with the piston being a modified recirculating ball nut or a modified rack and the cylinder being a modified gear housing. The control valve is also built into the gear housing and is constructed into the steering shaft. When the driver turns the steering wheel, the torque or turning effort on the steering shaft moves the valve. This redirects the pressurized hydraulic fluid to the proper side of the actuator to give a steering assist. The power steering fluid also serves to lubricate the gears and bearings in the steering gear (Figure 17-29).

The power for power steering comes from the engine; it is the turning effort exerted by the drive belt to the power steering pump. This pump delivers fluid under

Chapter 17 ■ Steering Systems

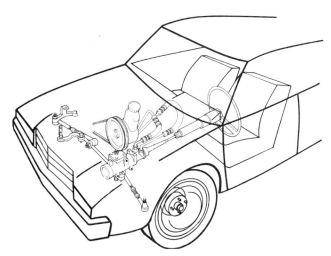

Figure 17-28 A standard power steering system. The major parts are the pump, steering gear, and hoses. *(Courtesy of Wagner Brake)*

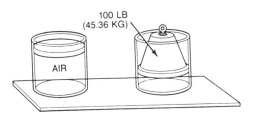

GAS CAN BE COMPRESSED

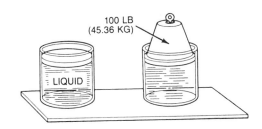

LIQUID CANNOT BE COMPRESSED

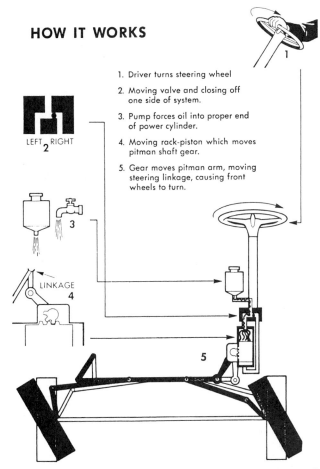

Figure 17-29 All power steering systems contain a valve (2), which is sensitive to steering wheel pressure (1); the pump (3), which provides the pressure; and the actuator or piston (4), where the pressure delivers the assist. *(Courtesy of General Motors Corporation, Service Technology Group)*

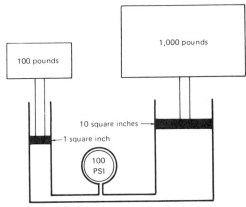

Figure 17-30 Two important principles of hydraulics are that liquids cannot be compressed and that liquids transmit pressure equally in all directions. The amount of pressure is determined by the piston size and the force on it.

pressures of about 600 to 1,300 psi (4,100 to 8,950 kPa) to the control valve in the steering gear or linkage. Two reinforced rubber hoses connect the pump to the control valve. A high-pressure hose carries the fluid to the valve, and a low-pressure hose returns the fluid to the pump reservoir.

To understand power steering, we need to have an understanding of some of the principles of hydraulics. **Hydraulics,** or **fluid power,** is based on the fact that liquids are fluid and can flow easily but cannot be compressed (squeezed into a smaller volume). Also, fluids can transmit pressure, and fluids under pressure exert force equally in all directions (Figure 17-30). Fluid pressures do not come from just the pump. In fact, the pump does not pump pressure; it pumps volume or flow. Pressure is created when this flow is restricted, and then the pump's effort generates pressure. If we connected the output hose of a pump directly into the reservoir, with no restriction, there would be very little pressure in the hose. If we took this same pump and closed the end of the hose, there would be no place for the fluid to go and a lot of pressure would be generated. The amount of pressure would be controlled by the strength of the hose and pump, the amount of power driving the pump, the ability of the drive belt to transmit this power, and possibly a relief valve. Hydraulics is a means to transfer energy from the engine to the actuator; energy is put into the fluid at the pump and removed at the actuator or piston and cylinder. The amount of power transmitted or consumed by a hydraulic system is determined by how much fluid flows and how much pressure is used. Hydraulic horsepower is equal to the flow times the pressure; this number is then adjusted by a small factor to obtain the actual horsepower amount.

Cars with small engines and power steering are often constructed with a pressure switch and computer controls that increase the engine idle speed while the power steering is used. It is possible that the load of the power steering pump would cause engine stalling. Other new cars use computer controls on the power steering system to change the amount of assist relative to vehicle speed. This is often called **variable assist.** The most assist is needed when the car is stopped or moving very slowly, and these driving conditions require the greatest turning effort from the driver. A reduced amount of assist is desired by many drivers so they can get a better feel of the road at higher speeds. Electronic controls activate a solenoid, which opens a fluid bypass above certain speeds. This bypass of fluid drops the pressure available for assist and, in turn, reduces the amount of assist (Figure 17–31).

17.6.1 Electronic Power Steering

Electronic power steering offers several advantages over the common hydraulic versions:

- It eliminates the power steering pump and hoses from the crowded engine compartment.
- It is more energy efficient in that it consumes no power while driving straight.
- It offers engine-off power assist.
- It is easily tuned to give variable assist under many driving modes.

A variety of electronic power steering systems have been developed over the past years, but most have suffered from being very expensive or causing a major power drain on the electrical system. One design mounted an electric motor around the steering rack, another mounted the motor around the pinion, and the most recent mounts the assist motor on the steering shaft (Figure 17–32).

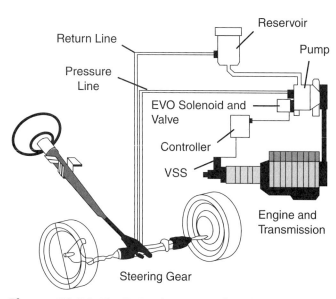

Figure 17–31 The EVO valve mounted on the power steering pump is an electronic variable orifice that allows full steering assist at low speeds. It reduces the amount of assist at higher road speeds for improved road feel.

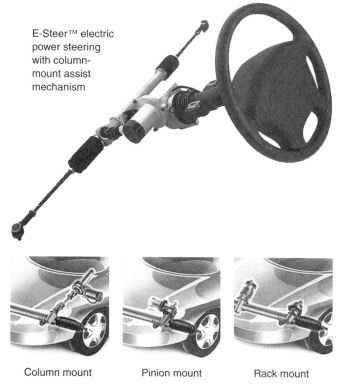

Figure 17–32 An E-Steer power steering system uses an electronically controlled assist motor mounted on the steering column or rack. *(Courtesy of Delphi Automotive Systems)*

17.6.2 Power Steering Pumps

Power steering pumps are usually **balanced vane** hydraulic pumps. They are often built into an assembly that includes the pump, a **pressure and flow control valve**, and a **reservoir;** some pumps mount the reservoir remotely to provide more room on the engine (Figure 17–33). The valve limits the amount of pressure that the pump can generate. A vane pump has a series of slots in the center rotor in which the vanes fit; these vanes can slide in or out of these slots in order to follow the contour of the pump cam or chamber ring. Some pumps use slippers or rollers instead of vanes to serve the same function (Figure 17–34). The pump cam is somewhat elliptically shaped to create two pumping chambers, one on each side of the rotor, which balance the pump. Fluid pressure in the pumping chamber tends to force the rotor off center; a balanced pump is balanced by the same pressure on each side of the center rotor.

When the pump pulley is turned, the pump shaft turns the rotor and vanes. The vanes move into the rotor on the sections of the cam that are close to the rotor and outward on the cam sections that are farther from the rotor. Outward movement of the vanes is caused by hydraulic pressure under the vanes and centrifugal force; some pumps use spring-loaded vanes or slippers. Pumping chambers are formed between two vanes and the rotor and outer cam surfaces. These chambers get bigger in the areas where the cam-to-rotor space is increasing and smaller where the space is decreasing. As the rotor chamber passes an area where a chamber is increasing in size, fluid flows into the pumping chamber through ports at the ends of the rotor; these ports are connected to the reservoir. After rotation of about one-fourth turn (90 degrees), the rotor and vanes enter a cam section that reduces the chamber size, and the fluid is forced out of the chamber through a second set of ports. These ports are connected to the control valve and pump outlet (Figure 17–35).

This style of pump is called a **positive displacement pump;** it pumps a specific volume of fluid on each revolution of the pump rotor. The displacement of a particular pump is determined by the length of the rotor, vanes, and cam ring and the shape of the cam ring. Most automotive power steering pumps can pump about 1 to 1.5 gallons (3.8 to 5.7 liters) per minute at the idle speed at low pressure. The pump is normally designed to pump a large enough volume to supply the needs of the power steering

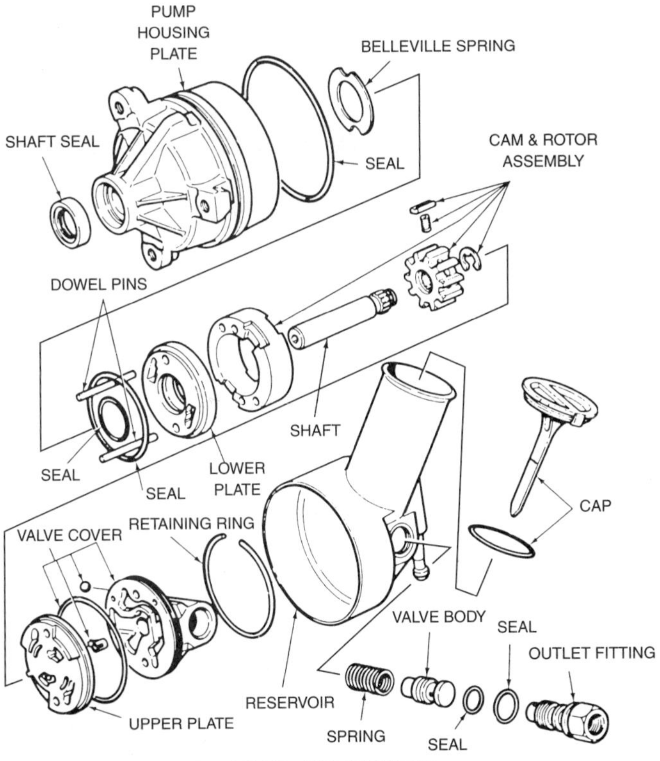

Figure 17–33 An exploded view of a power steering pump showing the internal parts. *(Courtesy of Ford Motor Company)*

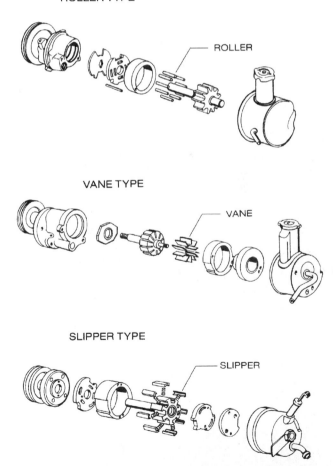

Figure 17–34 The pumping member can be a set of rollers, vanes, or slippers; the actual operation of each of these is very similar. *(Courtesy of Moog Automotive Inc.)*

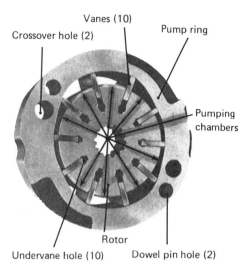

Figure 17-35 The pumping elements of a vane pump. Note the pumping chamber on each side of the rotor (top and bottom); their placement gives one complete pump cycle for each one-half turn of the rotor and balances the fluid pressure across the rotor. *(Courtesy of General Motors Corporation, Service Technology Group)*

gear with the engine idling. If the pump is too small, there is loss of hydraulic assist during faster steering maneuvers; this is called pump **catch-up.**

The reservoir of the pump is the plastic or stamped-metal portion that encloses the back and most of the body of the pump. The filler neck and cap with dipstick are usually built into the reservoir. Passages in the pump body open into the reservoir for the pump inlet. The pump outlet passage and pressure control valve pass through the reservoir surface and into the pump body. Remotely mounted reservoirs are usually mounted at the fender well and are connected to the pump inlet by a flexible hose and to the steering gear by the return line (Figure 17–36).

The control valve on most power steering pumps serves two purposes: flow control and maximum pressure control. Since pumps must be sized to provide an adequate flow at engine idle speeds and because they are positive displacement pumps, they deliver more fluid than necessary at speeds above idle. As the engine speeds up, the increased flow causes the flow control valve to open and bypass part of the flow back to the inlet port. This helps reduce the power needed to drive the pump, reduces fluid temperatures, and increases the life of the fluid. The flow control valve function is in constant use while the car is driven at road speeds; the pressure control valve is seldom used (Figure 17–37).

As we will see later, the steering gear control valve allows a free flow of fluid to return to the pump reservoir when a steering maneuver is not being performed. When there is a turning pressure on the steering wheel, the pressurized fluid is directed to one end of the actuator. Trapping the fluid coming from the pump generates the pressure needed to give us the steering assist, and the volume of the pump is used to move the actuator through its travel to produce a turn. When the actuator reaches the end of its travel, it will accept no more fluid, and additional pump output will instantly try to create excessive fluid pressures. At this time, the pressure release function of the control valve should occur. This valve will open and let part of the pump's output flow back to the pump inlet. The valve is noisy; we can usually hear it squeal or hiss when it is relieving pump pressure. Squeal along with a screeching noise is often caused by a slipping drive belt (Figure 17–38).

A turning pressure should not be kept on the steering wheel after the pressure control valve has started working; the high pressure can cause the fluid to heat and possibly damage the pump, high-pressure hose, or other

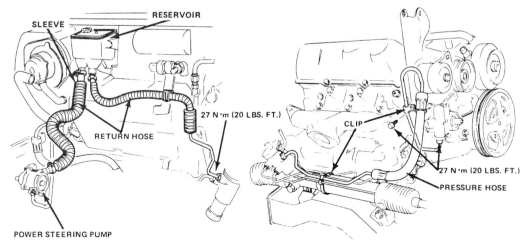

Figure 17-36 The reservoirs of some pumps are located remotely on the inner fender or rear bulkhead. *(Courtesy of General Motors Corporation, Service Technology Group)*

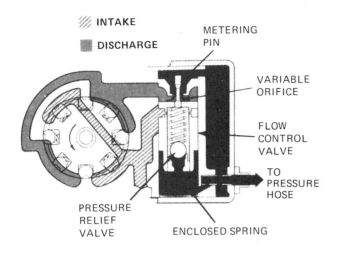

Figure 17-37 When the pump output becomes sufficient to create pressure at the variable orifice, the flow control valve moves downward and allows some of the pump's output to return to the intake cavities. *(Courtesy of Ford Motor Company)*

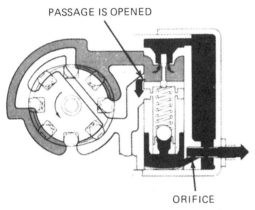

Figure 17-38 If the pump's pressure becomes too high, the pressure relief valve ball will be pushed off its seat and fluid will flow back to the pump intake cavity. *(Courtesy of Ford Motor Company)*

components of the system. Releasing the turning pressure on the steering wheel will let the control valve in the steering gear recenter itself and allow fluid to flow back to the pump. The car will turn just as sharply.

Power steering pump problem diagnosis and service operations are described in Chapter 33.

17.6.3 Steering Gear Control Valve

Most modern power steering gears use a rotary type of valve built into the steering gear to control fluid flow. This valve is normally positioned where the steering shaft enters the steering gear. This style of valve uses a small torsion bar to control the amount of assist delivered by the power steering unit.

The valve assembly consists of an inner valve spool, an outer valve sleeve, a valve housing, and the torsion bar. The input shaft of the steering gear is built into or connected to the inner valve spool and is also connected to one end of the torsion bar. The other end of the torsion bar is connected to the pinion gear of a rack-and-pinion set or the worm shaft of a standard steering gear; it is also connected to the outer valve spool. The valve spool fits into the valve housing and has passages that connect to four separate fluid ports. Sealing rings are used so it can rotate without fluid leakage. The four fluid ports are pump pressure, pump return, and each end of the actuator. The sleeve and spool are machined to give a very precise fit so as to control fluid leakage. Steering torque, being transmitted through the torsion bar, can cause the torsion bar to twist. This action will change the position of the valve spool relative to the sleeve and, in turn, change the fluid flows, producing or stopping hydraulic assist. The strength of the torsion bar controls the amount of assist and **steering feel.** A very weak torsion bar would twist easily; a lot of power assist would occur quickly and the steering would be very soft. An extremely strong torsion bar might not twist at all; we would get little or no assist and heavy steering (Figure 17-39).

In most valve assemblies, a projection of the valve spool fits into a slot in the worm shaft. The length of the slot allows enough spool-to-sleeve travel for valve operation along with a means to turn the worm shaft manually. Hard and quick steering maneuvers are done by manual force with power assist.

All power steering valves must have the four ports. The actuator ports and passages are internal on standard steering gears and external on rack-and-pinion gears and linkage booster systems. When there is no turning effort and the torsion bar is relaxed, fluid enters the valve through the inlet port and passes through the valve to the outlet port; both actuator ports are open to inlet and outlet pressure, allowing low-pressure fluid to fill both ends of the actuator (Figure 17-40). The fluid pressure, at this time, is controlled by any restriction in the steering valve or hose or anywhere between the pump outlet port and

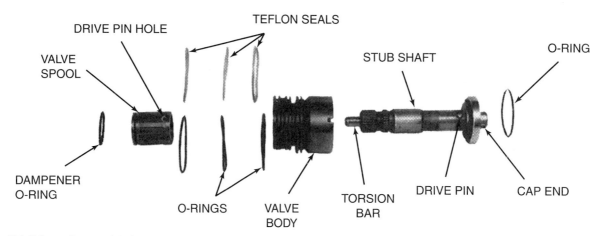

Figure 17-39 A disassembled rotary valve. The torsion bar connects the stub shaft at the left end and the cap end at the right. The valve spool turns with the stub shaft, and the valve body moves with the cap end. *(Courtesy of Ford Motor Company)*

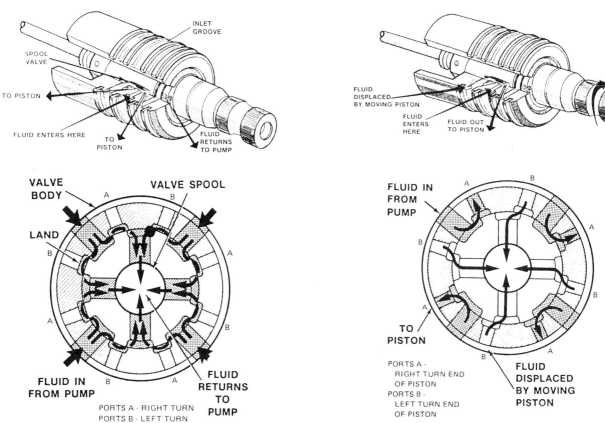

Figure 17-40 When there are no turning pressures, the torsion bar is relaxed, and the passages in the valve spool and body are aligned so the fluid flows in from the pump to both steering gear ports and to the return port back to the pump. *(Courtesy of Ford Motor Company)*

Figure 17-41 Turning pressure from a steering motion can twist the torsion bar; this action repositions the passages in the valve spool and body. Pump pressure will now be sent only to one end of the power piston; the other end will be connected to pump return. This action reverses if the steering motions are in the opposite direction. *(Courtesy of Ford Motor Company)*

the pump reservoir. When a turning effort is exerted that is strong enough to twist the torsion bar, the valve spool and sleeve become realigned. Depending on the direction of twist or turn, inlet fluid will be directed to one of the actuator ports while the other actuator port is connected to the return port. Reversing the direction of the turn would reverse these two connections. As power assist occurs, the twisting pressure on the torsion bar relaxes; movement of the actuator tends to take the pressure off the torsion bar, recenter the valve, and stop the assist. This normally occurs as the front tires reach the turning angle desired by the driver (Figure 17-41).

The valves used with linkage-type power steering and earlier standard power steering gears built by Chrysler Corporation are a sliding valve type. The operation is essentially the same as far as the fluid flows are concerned, but the mechanical action is a spool valve sliding in a bore instead of a spool rotating in a sleeve. The valve spool is centered by a pair of reaction springs when there is no steering action desired. These springs provide the feel of the amount of steering assist. An important thing to remember is that most sliding valves are adjustable for centering. If they are adjusted incorrectly, they can cause a power assist and produce turning in one direction with no steering input. Check this by raising the front tires off the ground and then starting the engine. If the steering wheel turns one direction or the other by itself, the valve needs to be adjusted. The adjustment procedure is described in the manufacturer and technician service manuals (Figure 17–42).

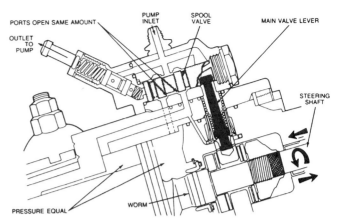

Figure 17–42 A sliding power steering valve. Steering motions can cause the steering shaft to move slightly to one side or the other in the housing; the main valve lever transfers this motion to the valve spool. Movement of the valve spool redirects the fluid flow for power assist. *(Courtesy of Chrysler Corporation)*

17.6.4 Rack-and-Pinion Power Steering Gears

Rack-and-pinion power steering gears have become very common on most modern cars. The simple design matches the simplicity of the rack-and-pinion gear set. The rack gear is built with a piston on the end opposite to the gear teeth; this piston runs in a bore in the rack housing with a sealed, sliding fit. Steel tubing is used to make the fluid connections between each end of the housing bore and the control valve (Figure 17–43).

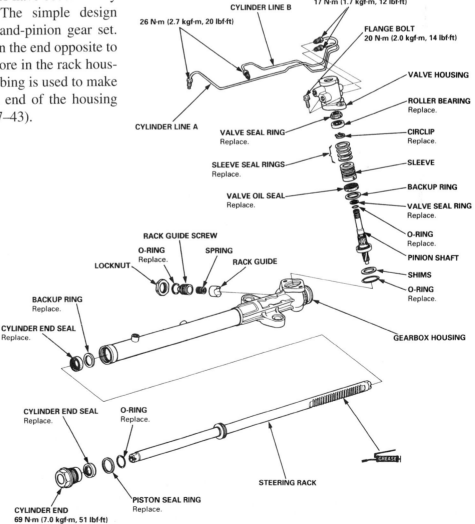

Figure 17–43 An exploded view of a rack-and-pinion power steering gear. Note the raised flange near the center of the rack; this is the power piston. *(Courtesy of American Honda Motor Co., Inc.)*

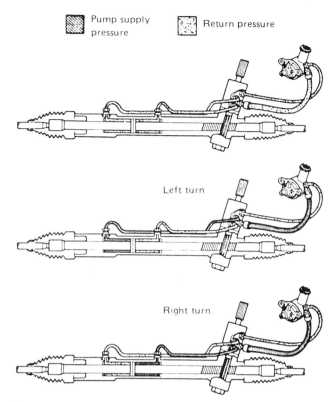

Figure 17-44 The pressure exerted on each side of the rack piston changes from a straight-ahead (top), a left-turn (center), or a right-turn (bottom) position. *(Courtesy of Ford Motor Company)*

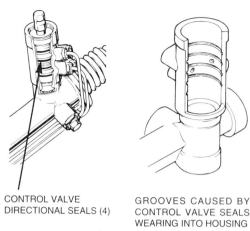

Figure 17-45 Most OEM racks use a cast-aluminum housing; the valve seals can wear into the bore, which in turn causes leakage and poor assist. *(Courtesy of Moog Automotive Inc.)*

During straight-ahead operation, the control valve allows a pressure equal to return line pressure to be exerted at each end of the piston bore. There is no hydraulic assist because there is an equal pressure at each side of the piston. A turning force on the steering wheel will move the control valve off center so that pump pressure is sent to one end of the piston bore while the other end of the bore is connected to pump return. This pressure difference on the rack piston will force the piston toward the side of lower pressure and produce the steering assist. It should be noted that once the rack reaches the end of its travel, there is no more room for additional fluid. Fluid flow will stop at this point, and the pressure will increase (Figure 17-44).

A power rack-and-pinion gear also has a small vent tube interconnecting the two end bellows. During a turn, one of the accordion bellows extends while the other contracts; air must transfer from one of the bellows to the other to allow this to happen. In a manual rack, air can move through the center of the rack, but a power rack has three separate seals that prevent air movement, making the exterior vent necessary.

Many OEM power racks are manufactured with a potential problem. The rack body is made from cast aluminum, and the valve assembly rotates in this soft material. If the fluid gets dirty, the abrasive dirt particles will cause valve bore wear, leakage at the valve bore, and a problem of poor assist, especially at low temperatures. This problem is often called morning sickness (Figure 17-45).

17.6.5 Standard Power Steering Gears (Integral)

Integral standard power steering gears use an actuator piston that is built into the ball nut, and the ball nut has a sliding, sealed fit in the gear housing. Two internal passages connect the actuator chambers to the ports of the control valve. The flows in this steering gear are similar to those in a rack-and-pinion gear (Figure 17-46).

When there is steering input and the control valve has all ports open to each other, both chambers will have the same (return) pressure, and there will be no assist. Steering input that is strong enough to twist the torsion bar and realign the valve spool and sleeve will direct pump pressure to one side of the ball nut piston and connect the chamber on the other side to the pump return. This difference in pressure across the piston will force the ball nut and sector gear in the correct direction to produce a power assist. Reversing the steering direction reverses the fluid flows and the assist direction. It should be noted that when the ball nut reaches the end of the chamber, it can move no farther, and there will be no more room for additional fluid. The fluid flow will stop at this point, and the pressure will increase (Figure 17-47).

17.6.6 Variable-Assist Power Steering

There are several styles of **variable-assist power steering (VAPS),** also called **variable-effort steering, speed-dependent steering, speed-sensitive steering, electronic**

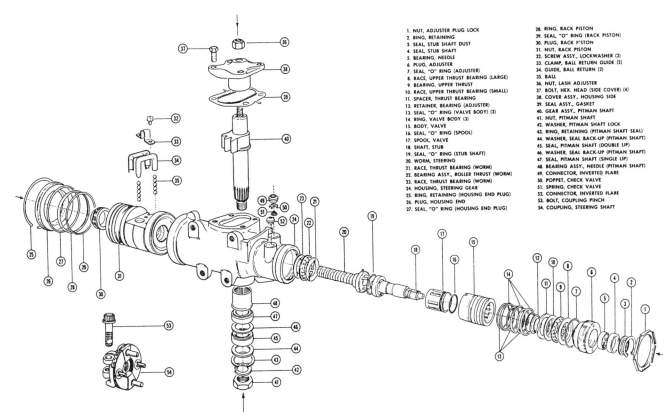

Figure 17–46 A disassembled view of an integral, standard power steering gear commonly found on RWD cars. *(Courtesy of General Motors Corporation, Service Technology Group)*

variable-orifice (EVO) steering, and **progressive power steering (PPS).** The purpose of VAPS is to reduce the amount of power assist as vehicle speed increases, to provide better road feel. As mentioned earlier, power assist is sized to provide easy steering for parking maneuvers with the engine running at low speeds. Higher speeds do not require much assist for steering, and at higher speeds, the higher engine speeds produce additional pump output (Figure 17–48). Many drivers dislike the too-soft, mushy feel of power steering at freeway speeds.

VAPS is an electronically controlled system. It uses an electronic controller, one or more sensors, and an actuator or solenoid (Figure 17–49). The controller can be a single-purpose steering control module or part of the vehicle ABS or skid control module or vehicle dynamics module. The sensors vary between systems from different manufacturers. All systems use a **vehicle speed sensor (VSS)** to determine vehicle speed; some use the **ABS wheel speed sensors (WSS).** Some VAPS systems also use a steering wheel rotation sensor and/or a driver selection switch. The steering wheel rotation sensor provides input on how fast the steering wheel is turned so that fast steering maneuvers can be performed at high speeds. The driver selection switch allows the driver to select the amount of steering assist desired (Figure 17–50).

In most systems, the VAPS actuator is an electric solenoid that operates a control valve, often called the *electronic variable orifice.* When the actuator is operated, the EVO is opened to reduce fluid pressure. The EVO and actuator solenoid can be mounted on either the pump or steering gear (Figure 17–51 and Figure 17–52).

Another style of VAPS is the General Motors **Magnasteer.** The actuator is attached to the steering gear input shaft–power steering valve input, the power steering valve output, and the gear housing (Figure 17–53). When the control module sends a current flow to the actuator, an electromagnetic force occurs that increases the amount of steering effort required to turn the steering gear input shaft.

17.6.7 Linkage Booster Power Steering

Linkage booster power steering uses a normal standard steering gear; the control valve and actuator are mounted in the steering linkage. At one time, this type of power steering was offered as a dealer-installed option. It is sometimes called a **hang-on** style. Linkage booster power steering was also factory installed on several models of cars (Figure 17–54).

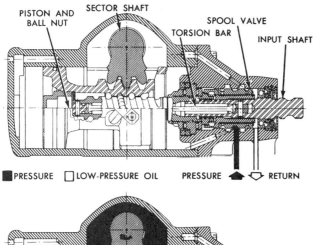

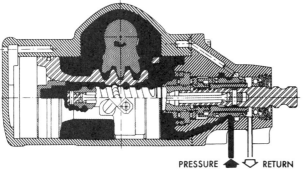

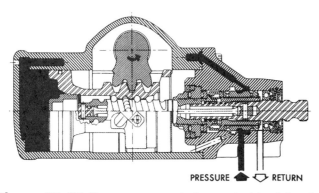

Figure 17-47 The pressure exerted on each side of the piston and ball nut changes from a straight-ahead (top), a left-turn (center), and a right-turn (bottom) position. *(Courtesy of Ford Motor Company)*

The sliding type of control valve is normally mounted at the Pitman arm end of the center steering link. The actuator is a separate hydraulic cylinder that has one end attached to the car's frame or body and the other end to the center steering link. Two high-pressure steel tubes or reinforced rubber hoses connect the ends of the actuator to the control valve.

When there is no steering input, the two reaction springs hold the control valve in the center position. Fluid pressure from the pump is directed to the pump return port and both actuator ports. When the driver turns the steering wheel, pressure between the Pitman arm and center link moves the control valve spool to an off-center position. Pump pressure is now directed to one end of the actuator, and the other end is connected to pump return.

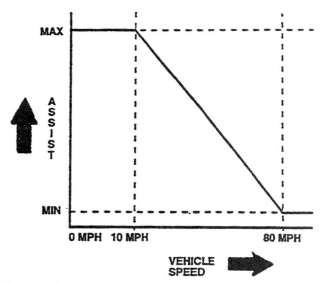

Figure 17-48 This VAPS-II system provides the maximum amount of steering assist at speeds below 10 mph. As the speed increases, the amount of steering assist decreases. *(Courtesy of Ford Motor Company)*

Power assist will occur because the pressures on both sides of the actuator piston are not equal. Like the other types of power steering, reversing the steering direction will reverse the fluid flows, power-assisted movement will remove the valve spool pressure and recenter the valve, and fluid flow will stop and pressure will increase when the actuator reaches the end of its travel.

17.6.8 Power Steering Hoses and Fluid

Two reinforced rubber hoses connect the power steering pump to the control valve. They are the pressure hose and the return hose. The return hose connections are usually made by slipping the hose onto a connector and securing it in place with a band or screw-type clamps (Figure 17–55). The pressure hose is stronger and requires pressure-tight connections. High-pressure swaged metal bands connect the hose to the steel end tubes, which connect to the pump and gear using flare or O-ring seals. These hoses come in various lengths with various-shaped end tubes to fit different installations. Some pressure hoses have built-in restrictions that resemble a reduction in hose diameter (Figure 17–56).

Some late-model cars use **hydro-boost brakes;** the power assist for brake application comes from the power steering pump. These systems require at least two additional hoses (Figure 17–57).

Some systems use a cooler for the power steering fluid; this cooler can resemble a miniature radiator or just an extra length of metal tubing with or without fins. Hydraulic fluid tends to break down and oxidize at

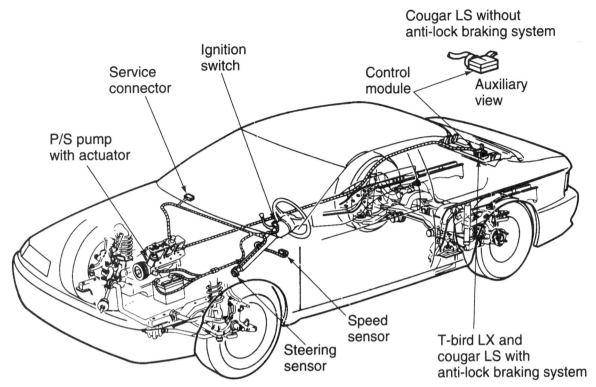

Figure 17–49 The components of this EVO system include the EVO actuator on the power steering pump and the control module in the trunk area. *(Courtesy of Ford Motor Company)*

higher temperatures. Oxidized fluid tends to turn dark brown, smell like varnish, thicken, and leave gummy, varnishlike deposits on metal surfaces. Fluid life can be lengthened if the fluid is kept relatively cool. The cooler, if used, is mounted in the return line, where it is subject to low pressures.

Traditionally, power steering systems have used automatic transmission fluid (ATF) for hydraulic fluid; the reddish color makes the fluid easy to identify. Many late-model systems are currently using power steering fluid; this fluid is dyed yellow or yellow-green. Power steering fluid has a lower viscosity (is thinner), which reduces some of the heat buildup and power loss in the steering system. The manufacturer's recommendations for fluid type should be followed when adding or changing fluid.

Power steering hose replacement and fluid changes are described in Chapter 33.

17.7 Steering Linkage

As mentioned earlier, two different styles of steering linkage are used on modern cars: a fairly simple system with rack-and-pinion gears and a more complex system with standard steering gears. The linkage transmits the side-to-side steering motions from the steering gear to the front wheels.

As the front tires raise and lower during normal suspension travel, the steering knuckle and arm travel vertically in a slight arc relative to the frame or body. The outer tie-rod end must travel in a similar arc or **bump steer** will result. Bump steer causes the car to turn right or left as the tires move up or down; this can cause potentially dangerous steering or tire wear (Figure 17–58).

Suspension designers try to match the arcs of travel of both the tie-rod end and the steering arm so they will coincide. The arc of travel of the steering arm is controlled by the length or angles of the control arms, and the arc of travel of the tie-rod end is controlled by the length of the tie-rod and height of the inner tie-rod end. Bump steer is affected by the height of the ends of the steering arms, Pitman arm, idler arm, or steering rack (Figure 17–59). Some manufacturers recommend a steering linkage parallelism check to help ensure the correct inner tie-rod end height and a minimum of bump steer effects; this check is recommended when replacing idler arms.

A type of bump steer can also occur on solid axles if the vertical arc of travel of the axle and the third arm does not match the arc of travel for the forward end of the

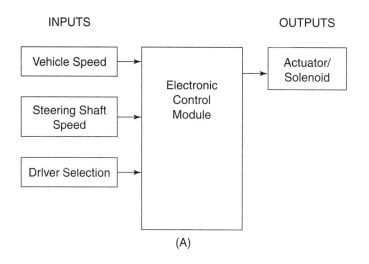
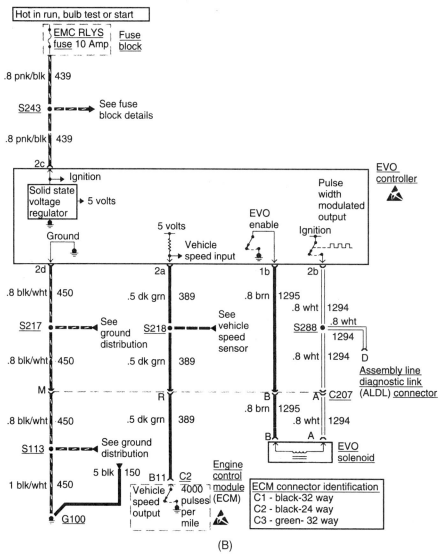

Figure 17–50 A simplified VAPS control circuit (A) showing the three inputs and one output, and a diagram of an EVO system (B) showing the components. *(B is courtesy of General Motors Corporation, Service Technology Group)*

Chapter 17 ■ Steering Systems

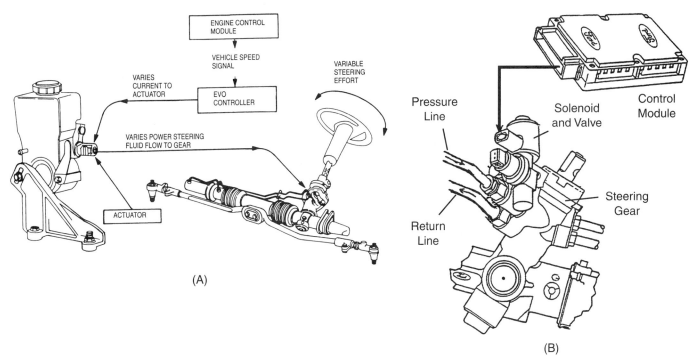

Figure 17-51 The EVO can be mounted on the power steering pump (A) or on the steering gear (B). *(A is Courtesy of General Motors Corporation, Service Technology Group; B is courtesy of Ford Motor Company)*

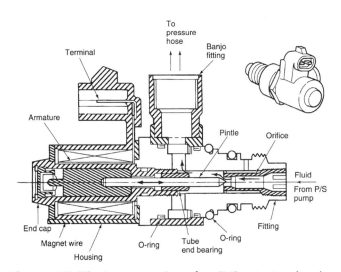

Figure 17-52 A cutaway view of an EVO actuator showing the solenoid portion at the left and the valve. *(Courtesy of Ford Motor Company)*

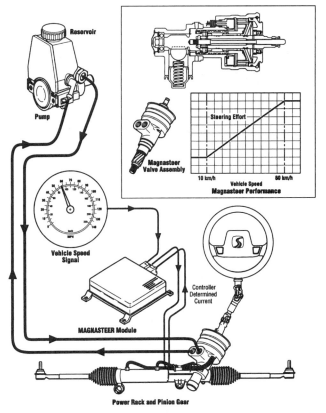

Figure 17-53 The Magnasteer actuator and valve assembly is mounted on the power steering gear. Note that steering effort is controlled by the Magnasteer module responding to vehicle speed. *(Courtesy of General Motors Corporation, Service Technology Group)*

drag link. Raising a vehicle very far above design height specifications usually aggravates this situation; a drag link should be horizontal at ride height (Figure 17-60).

17.7.1 Outer Tie-Rod Ends

Most outer tie-rod ends are a ball-and-socket joint very similar to a ball joint. These tie-rod ends use a preload spring, either metal or rubber, to maintain a slight friction

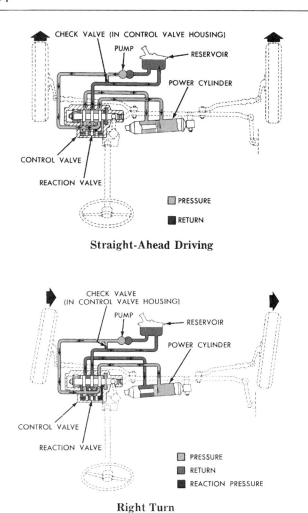

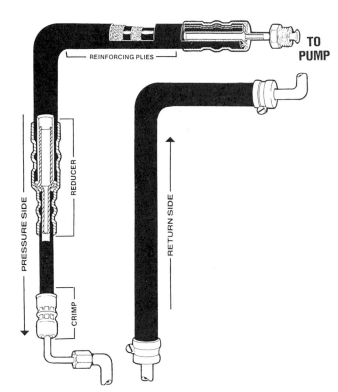

Figure 17–56 The pressure hose is reinforced with several layers of fabric and uses crimped connectors; some include a flow reducer. The return hose connections are often secured with screw or spring clamps. *(Courtesy of Moog Automotive Inc.)*

Figure 17–54 A linkage booster or nonintegral power steering system. Note that the fluid flow for a right and left turn is reversed. The control valve is at the end of the center steering link. *(Courtesy of Ford Motor Company)*

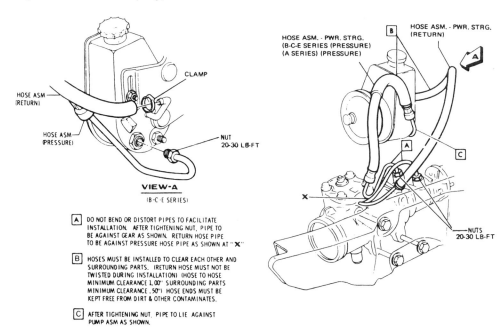

Figure 17–55 Power steering hoses and the connection procedure. *(Courtesy of General Motors Corporation, Service Technology Group)*

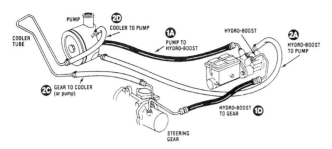

Figure 17–57 Vehicles equipped with hydro-boost braking systems require additional hoses and connections. Hoses 1A and 1D are pressure hoses; 2A, 2C, and 2D are return hoses. *(Courtesy of Moog Automotive Inc.)*

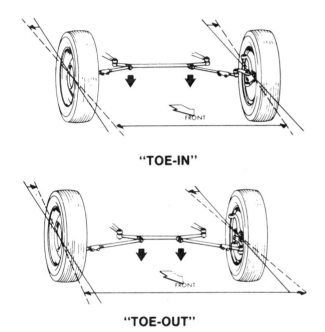

Figure 17–58 Bump steer, also called toe change, can occur as the suspension moves up and down if the steering geometry does not match the suspension geometry. *(Courtesy of McQuay Norris)*

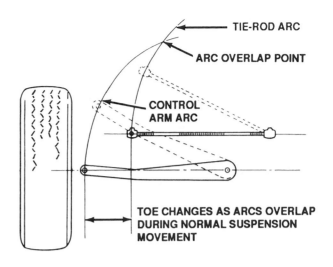

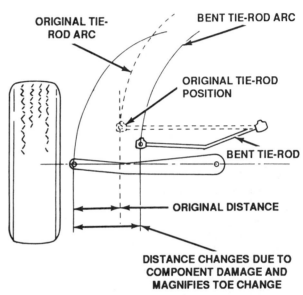

Figure 17–59 Toe change can result from a bent tie-rod or misadjusted steering components. *(Courtesy of General Motors Corporation, Service Technology Group)*

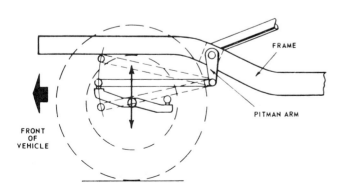

Figure 17–60 This solid axle steering system uses a drag link; turning on bumps can occur if the path of the front of the drag link and third arm does not match the path of the drag link. *(Courtesy of Ammco Tools, Inc.)*

load between the ball and socket to eliminate side play. The stud of the ball joint is a locking taper, so there will be no side play when it is installed in the steering arm. Most tie-rod ends use a flexible rubber boot to keep the lubricant in the joint and dirt and water out. A tie-rod will fail if the boot is torn or broken. At one time, most tie-rod ends had some provision for lubrication; this was either a grease or zerk fitting or a plug for installing a fitting. Many modern tie-rod ends are lubricated and sealed during manufacture; there is no provision for additional lubrication (Figure 17–61).

One manufacturer has begun using a tie-rod end with a rubber-bonded bushing. The rubber stretches or twists during steering maneuvers much like the rubber in a torsilastic control arm bushing. Like control arm bushings, this

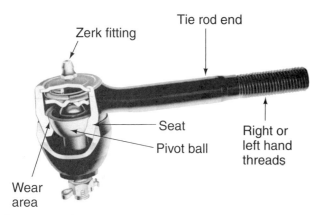

Figure 17-61 Most tie-rod ends use a steel ball stud that pivots in a metal or nylon bearing surface; a preload spring maintains zero clearance in the joint. *(Courtesy of Federal-Mogul Corporation)*

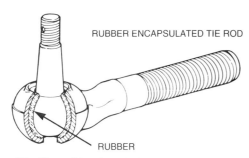

Figure 17-62 Rubber-bonded tie-rod ends use a steel ball encapsulated in rubber; the rubber must twist like a torsilastic bushing for turns. *(Courtesy of Moog Automotive Inc.)*

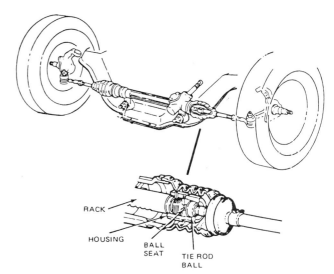

Figure 17-63 A rack-and-pinion gear inner tie-rod is a ball-shaped end of the tie-rod enclosed in a socket at the end of the rack. A flexible rubber bellows protects the joint and end of the rack. *(Courtesy of Ford Motor Company)*

tie-rod end requires no lubrication (Figure 17–62). Also, it requires a straight-ahead position as it is installed.

17.7.2 Tie-Rods, Rack-and-Pinion Steering Gear

The inner end of the tie-rods used with rack-and-pinion steering becomes the ball for the ball-and-socket joint; the socket encloses the ball and is secured onto the end of the rack. In newer designs, a sealed, permanently adjusted socket is built onto the tie-rod. Some older designs used an outer socket, which could be replaced separately from the tie-rod. The older design required an adjustment for tightness during replacement. These joints are protected by the flexible bellows or boot at the ends of the gear housing (Figure 17–63).

The threads of the inner tie-rod end socket are locked onto the end of the rack by one of these methods: a jam nut, a jam nut plus solid pin, a roll or spiral lock pin, a set screw, a staked end, or a tabbed washer. These locking methods help ensure that the tie-rod socket cannot come loose from the rack. Each of these locking methods requires a different disassembly and reassembly procedure, as described in Chapter 33.

Some FWD cars use a center takeoff design in which the tie-rod ends connect to the center portion of the rack. Replaceable rubber bushings are used between the ends of the tie-rods and the tie-rod mounting bolts. Center-mounted tie-rods use a single large bellows or boot to protect the internal gear parts.

The tie-rods themselves can be extensions of the inner tie-rod end, which threads directly into the outer tie-rod end (two-piece assembly), or use an adjuster sleeve, which threads into both the inner and outer ends (three-piece assembly). Mechanically, there is little difference between these two styles unless you are adjusting toe. The inner end of the tie-rod is turned to make the adjustment on two-piece units, while only the adjuster sleeve is turned on three-piece units. The toe adjustment is locked by a jam nut at the outer tie-rod end or a clamp at the threads of the adjuster. Toe and tie-rod adjustments are described more completely in Chapter 34.

17.7.3 Tie-Rods, Standard Steering Gears

The inner tie-rod ends used with standard steering gears are almost the same as the outer ends. The major difference is that one has right-hand threads and the other left-hand threads; the length of the threaded portion that connects to the threaded adjuster sleeve is usually much longer in the tie-rod end (Figure 17–64).

All of these tie-rods are three-piece units: an outer end, an inner end, and the adjuster sleeve, which connects the

Chapter 17 ■ Steering Systems

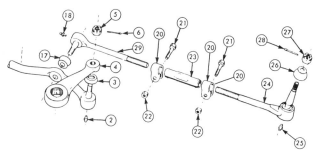

Figure 17-64 Most RWD vehicle tie-rods consist of two ends (24 and 29) connected by an adjusting tube or sleeve (23) and held together by a pair of clamps (20). *(Courtesy of General Motors Corporation, Service Technology Group)*

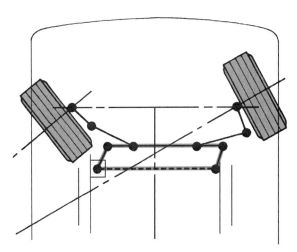

Figure 17-65 The Pitman arm, center link, and idler arm form three sides of a parallelogram; this ensures a true side-to-side movement of the inner tie-rod ends. *(Courtesy of General Motors Corporation, Service Technology Group)*

two ends. The sleeve is locked to the tie-rod ends by clamps that squeeze the slotted sleeve tightly onto the tie-rod ends.

17.7.4 Idler Arm and Center Link

Parallelogram steering linkage is used with most standard steering gears. This style uses a center steering link with the inner tie-rod ends connected to it in the correct location for bump steer. The center link is supported at one end by the Pitman arm on the steering gear, which also puts the steering motions into the steering linkage. The other end of the center link is supported by the idler arm. The idler arm is the same length and parallel to the Pitman arm and placed in exactly the same relative position on the opposite side of the car. This ensures that the center link will move straight across the car. The center link is connected to the Pitman arm and idler arm by ball-and-socket joints, which are quite similar to tie-rod end joints. These two pivots plus the center axis of the Pitman shaft and the idler arm pivot form the corner points of the parallelogram, which gives this linkage type its name. In a straight-ahead position, the idler arm, Pitman arm, and center link form three sides of a rectangle or parallelogram (Figure 17-65).

The idler arm pivots from a bracket that is bolted to the frame or body. The pivot bushing can be a rubber bushing, a threaded metal bushing, or a bearing type depending on the manufacturer. Each bushing style has different service requirements, which are described in Chapter 33 (Figure 17-66).

17.8 Drag Links

Many trucks and 4WD pickups use a solid axle and a single tie-rod to connect the two steering arms. Use of two tie-rods and parallelogram steering linkage would cause

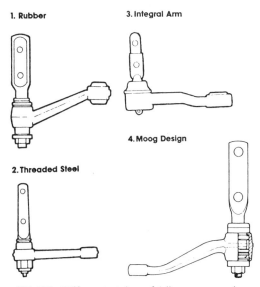

Figure 17-66 Different styles of idler arms and mounting brackets. *(Courtesy of Moog Automotive Inc.)*

bump steer. A **drag link** connects the Pitman arm on the steering gear to the tie-rod or an extension of the steering arm at the right steering knuckle, across the car.

Some trucks and pickups use a drag link to connect the Pitman arm to a **third arm** at the left steering knuckle. This design also uses a single tie-rod to connect the two steering arms. A drag link uses a ball-and-socket joint at each end connection; this unit can be similar to a normal tie-rod end or a drag link socket. The ball stud of a drag link socket is solidly connected, sometimes riveted or welded, to the Pitman arm or third arm. Drag link sockets are disassembled or loosened to allow the socket to lift off the ball stud for removal; turning the adjuster screw enlarges the socket enough so that it can be slipped

Figure 17-67 A cutaway view of a drag link showing the internal socket parts. Note that it can be removed from the ball studs by loosening the end screws. *(Courtesy of Federal-Mogul Corporation)*

off the ball. Adjustment of a drag line socket is described in Chapter 33 (Figure 17-67).

17.9 Four-Wheel Steering

Four-wheel steering (4WS), also called rear-wheel steering or all-wheel steering, provides a means to actively steer the rear wheels during turning maneuvers. Depending on the system, the rear wheels can be turned in a direction that is **in-phase** or **antiphase.**

When both the front and rear wheels steer toward the same direction, they are said to be in-phase, and this produces a kind of sideways movement of the car at low speeds. When the front and rear wheels are steered in opposite directions, this is called antiphase, counterphase, or opposite phase, and it produces a sharper, tighter turn. All 4WS cars steer the rear tires in-phase for improved handling during moderate- to high-speed turning maneuvers. This is the greatest benefit of 4WS. The amount of rear steering angle is quite small, limited to a maximum of about 5 degrees, relative to the front steering angle. Some 4WS cars steer the rear wheels in an antiphase direction at small amounts of steering wheel motion, less than 127 degrees, or at low speeds below 22 mph (35 km/h) and then change to in-phase steering at greater steering wheel motion or higher speeds (Figure 17-68).

The method used to produce 4WS varies among manufacturers. One uses a purely mechanical system with a rear steering gear driven from a shaft connected to a second pinion gear in the front steering gear. The very complex rear steering gear controls the in-phase and antiphase motions at the rear wheels. Another system uses a rear steering gear that is operated by hydraulic pressure from the front power steering circuit, with the pressure controlled by an electronic controller. The hydraulic pressure controls the amount and direction of rear steer. Other systems use a combination of mechanical, electrical or electronic, and hydraulic operation (Figure 17-69).

4WS is complex and expensive. It is designed to return the rear wheels to a straight-ahead position in case of failure. This system also requires a more refined alignment procedure to make sure the rear wheels are straight ahead when the steering is centered and that they have the correct amount of toe. At this time, 4WS seems to be too expensive for most motorists.

17.10 Practice Diagnosis

You are working in a brake and front-end shop and encounter the following problems:

CASE 1: A customer has brought in his 8-year-old, 3/4-ton Dodge pickup with a complaint of sloppy steering. A quick check of the steering wheel shows about 4 in. of free movement. What kinds of things can cause this problem? Where should you begin checking?

CASE 2: A customer has brought in her 3-year-old Oldsmobile Cutlass, complaining that it is noisy when she turns the steering wheel. As you back out of the shop, a sharp turning of the wheel causes it to make a squealing noise. What is probably causing this problem? What should your first check be?

Chapter 17 ■ Steering Systems

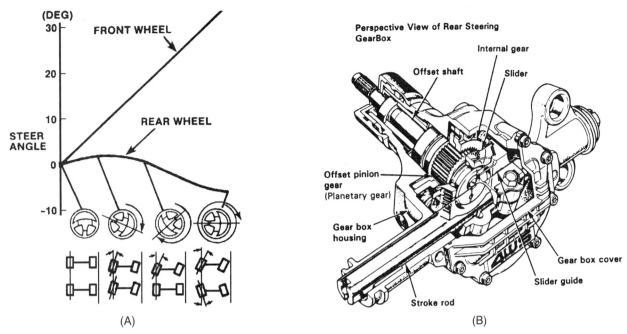

Figure 17-68 In this 4WS car, the rear wheels steer in-phase during the first portion of a turn and antiphase after that (A). This gives improved handling for lane changes and tighter turning for parking maneuvers. The steering gear for this type of system is shown at (B). *(Courtesy of American Honda Motor Co., Inc.)*

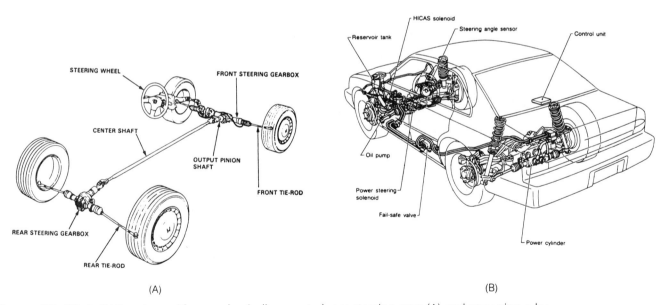

Figure 17-69 A 4WS system with a mechanically operated rear steering gear (A) and one using a hydraulically operated rear steering gear (B). *(A is courtesy of American Honda Motor Co., Inc.; B is courtesy of Nissan Motor Corporation)*

Terms to Know

airbag	flexible rubber bellows	rubber bushings
airbag modules	fluid power	safing sensor
airbag systems	gear lash	sector shaft
antiphase	grommets	speed-dependent steering
arming sensor	hang-on	speed-sensitive steering
balanced vane hydraulic pump	hydraulic assist	steering columns
ball nut	hydraulics	steering damper
bump steer	hydro-boost brakes	steering feel
cam and lever (Ross) gear	inflator	steering ratio
catch-up	in-phase	steering shaft
center takeoff	insulators	supplemental inflatable restraints (SIR)
clock spring	linkage booster power steering	supplemental restraints
collapsible steering columns	lock to lock	third arm
collision sensors	Magnasteer	variable assist
control module	over center adjustment	variable-assist power steering (VAPS)
conventional or standard steering gears	parallelogram	variable-effort steering
crash sensors	parallelogram steering linkage	variable ratio
diagnostic module	pinion gear	vehicle speed sensor (VSS)
discriminating sensor	Pitman shaft	warning lamp
drag link	positive displacement pump	wheel speed sensor (WSS)
driver-side airbag	pressure and flow control valve	worm and sector or worm and roller (Gemmer) gear
electronic variable-orifice (EVO) steering	progressive power steering (PPS)	worm shaft
end takeoff	rack-and-pinion steering gears	worm shaft bearing preload
flexible couplers	recirculating ball nut gear	
	reservoir	

Review Questions

1. Steering ratios are being discussed.
   Statement A: A 15:1 ratio will give one complete revolution of the tie-rod for each 15 turns of the steering wheel.
   Statement B: This ratio would require 150 degrees of turning the steering wheel to produce 10 degrees of turning at the front wheels.
   Which statement is correct?
   a. A only
   b. B only
   c. Both A and B
   d. Neither A nor B

2. Which of these statements describes the effects of steering ratios?
   A. Fast ratios produce quick, maneuverable steering.
   B. Slow ratios produce stable, easy steering.
   a. A only
   b. B only
   c. Both A and B
   d. Neither A nor B

3. Besides containing the steering shaft, the steering column contains or holds:
   a. The collapsible column and shaft features
   b. Several electrical switches for different circuits
   c. A locking mechanism for the steering wheel
   d. All of these

Chapter 17 ■ Steering Systems

4. The steering shaft is normally connected to the steering gear through a:
   a. Rigid, splined connector
   b. Universal joint
   c. Flexible coupler
   d. All of these

5. A worm and roller style of steering gear is also called a:
   a. Gemmer gear
   b. Ross gear
   c. Recirculating ball nut
   d. Rack and pinion

6. Steering gears are being discussed.
   Statement A: A steering shaft is also called a worm shaft.
   Statement B: The sector gear is part of the Pitman shaft.
   Which statement is correct?
   a. A only   c. Both A and B
   b. B only   d. Neither A nor B

7. The most commonly used style of standard, manual steering gear is the:
   a. Cam and lever
   b. Worm and sector
   c. Recirculating ball nut
   d. Worm and roller

8. Statement A: The ball nut moves up or down the worm shaft when the worm shaft is rotated.
   Statement B: The balls roll along grooves in the worm shaft and ball nut when the worm shaft is rotated.
   Which statement is correct?
   a. A only   c. Both A and B
   b. B only   d. Neither A nor B

9. Which of the following is true about rack-and-pinion steering gears?
   A. The rack is a long, straight gear.
   B. The pinion is a small, round, normally shaped gear.
   a. A only   c. Both A and B
   b. B only   d. Neither A nor B

10. Which of the following is true when comparing standard steering gears with rack-and-pinion steering gears?
    A. Rack-and-pinion gears are heavier and more complex.
    B. Rack-and-pinion gears do a better job of dampening road shock.
    a. A only   c. Both A and B
    b. B only   d. Neither A nor B

11. Which of the following is not true about standard power steering gears?
    a. The power assist is provided by hydraulic pressure.
    b. The control valve is mounted in the Pitman shaft.
    c. The power piston is built into the ball nut.
    d. The steering gear is lubricated by power steering fluid.

12. Fluid flow through a power steering system is being discussed.
    Statement A: The control valve in the steering gear can cause a pressure change in the high-pressure hose.
    Statement B: The maximum pressure in the high-pressure hose is controlled by the relief valve in the pump.
    Which statement is correct?
    a. A only   c. Both A and B
    b. B only   d. Neither A nor B

13. Which of the following is true about power steering hoses?
    A. The return hose has very little pressure in it.
    B. The pressure hose contains a fairly low pressure with no effort on the steering wheel and a pressure as high as the relief valve setting during turns.
    a. A only   c. Both A and B
    b. B only   d. Neither A nor B

14. Most power steering pumps are of the:
    A. Sliding or roller vane type
    B. Positive displacement type
    Which option best completes the statement?
    a. A only   c. Both A and B
    b. B only   d. Neither A nor B

15. The control valve in the pump is being discussed.
    Statement A: This valve allows greater and greater flow as the engine speeds up.
    Statement B: This valve usually hisses or squeals when it relieves pressure back to the pump inlet.
    Which statement is correct?
    a. A only   c. Both A and B
    b. B only   d. Neither A nor B

16. The power piston section of the power steering can be mounted:
    a. Outside of the steering gear
    b. On the ball nut
    c. On the rack gear
    d. Any of these

17. The outer tie-rod ends are:
    A. Usually ball-and-socket joints
    B. Preloaded with an internal device to eliminate rotating motions
    Which option best completes the statement?
    a. A only
    b. B only
    c. Both A and B
    d. Neither A nor B

18. An idler arm is:
    a. Used to support one end of the center link
    b. The same length as the Pitman arm
    c. Attached to the frame through a pivot bushing
    d. All of these

19. Inner tie-rod ends of a rack-and-pinion steering gear are being discussed.
    Statement A: They are always locked onto the end of the rack by a steel pin, jam nuts, or set screw.
    Statement B: They are enclosed by a flexible, accordion-like boot.
    Which statement is correct?
    a. A only
    b. B only
    c. Both A and B
    d. Neither A nor B

20. The tie-rods:
    A. Connect the steering arms to the Pitman arm
    B. Are arranged so that they can move up and down at the outer end as well as laterally
    Which option best completes the statement?
    a. A only
    b. B only
    c. Both A and B
    d. Neither A nor B

18 Chassis Electrical and Electronic Systems

Objectives

Upon completion and review of this chapter, you should be able to:

- ❑ Comprehend the terms commonly used with automotive electrical and electronic components and their circuits.
- ❑ Understand basic electrical troubleshooting procedures.
- ❑ Understand common electrical repair procedures.
- ❑ Perform the ASE tasks relating to electrically and electronically controlled brake, suspension, and steering system component inspection, diagnosis, and repair.

18.1 Introduction

Until recently, the front-end technician had little need for knowledge about electricity and electrical diagnostic procedures. Now the need to understand electrical systems is steadily increasing with such innovations as electronically controlled airbags, air suspensions, shock absorber systems, and variable-assist steering systems. The ability to measure voltage and resistance and interpret these measurements has become very important.

Solid-state electronics is at the heart of computer-controlled circuits. It includes transistors, diodes, and integrated microchip devices. These control and sensing devices are quite fragile relative to other automotive electrical devices.

18.2 Basic Electricity

A course in basic automotive electricity is necessary to thoroughly understand electricity and how to measure it. The description that follows is merely a brief review.

A technician is normally concerned with three measurable aspects of electricity: **volts, amperes** (or **amps**), and **ohms** (Figure 18–1). Voltage, also called electrical pressure, is the push that forces electricity to flow through a wire, conductor, or component. This flow is known as amperes, often abbreviated to amps. In a car, voltage is supplied by a source of electrical power—either the battery or the alternator. The battery supplies voltage while the engine is off, and when the engine is running, the alternator supplies voltage for the electrical systems and to recharge the battery. Amperage multiplied by voltages gives **watts,** the measurement for electrical power.

Ohms are units of electrical resistance. The amount of current flow in a circuit is determined by the resistance of the components. A large amount of resistance will stop or severely limit current flow. The symbol Ω (omega) is used to signify ohms.

18.2.1 Circuits

A circuit is a complete electrical path that allows amps to flow from the power source, through the electrical components, and back to the power source (Figure 18–2). This path is composed of wires, commonly called conductors; safety devices such as fuses, fusible links, or circuit breakers; usually one or more switches; and the electrical component(s). Some circuits have all of the components arranged so that one follows another, making the current flow pass through each of them, and this is called a **series circuit.** Other circuits divide into branches, with the current flow taking separate paths, and these are called **parallel circuits** (Figure 18–3). Many automotive circuits are a combination of series and parallel circuits. In a car, a **ground circuit** is used to conduct electricity from a component back to the source of power. The ground circuit uses the metal of the car body, frame, engine block, and so on as an electrical conductor. In

303

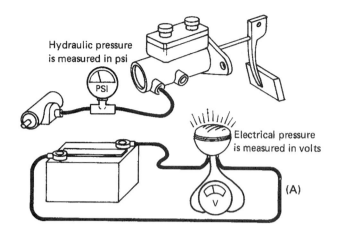

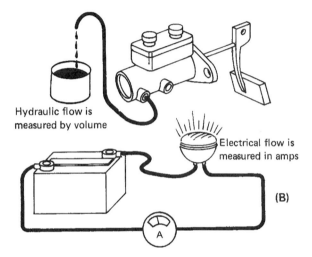

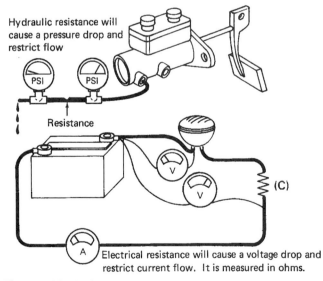

Figure 18–1 If we compare hydraulics with electricity, pressure and voltage (A), fluid flow and current flow (B), and resistance to flow (C) are very similar.

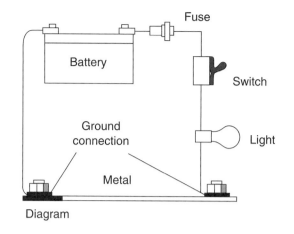

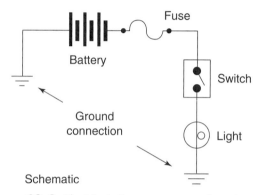

Figure 18–2 Electrical diagrams are used to show the path through a circuit. They can be drawn as a diagram (top) or a schematic (bottom), which is most common.

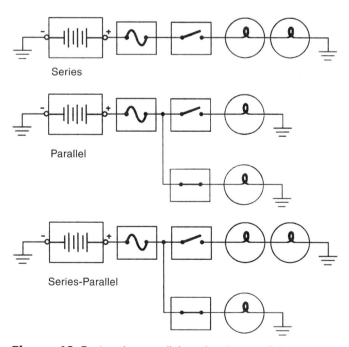

Figure 18–3 A series, parallel, and series-parallel circuit.

modern vehicles, the negative (−) terminal of the battery and alternator are connected to ground, and the positive (+) side is insulated.

Most automotive circuits have five major parts:

1. **Power source:** the battery or alternator
2. **Circuit protection:** fuse, fusible link, or circuit breaker; opens the circuit if there is excessive current
3. **Switch:** one or more open or close to control a circuit
4. **Output:** the device that does the work (e.g., light, motor, radio)
5. **Wires:** used to connect the insulated parts of the circuit

Wires provide a metal conductor that is surrounded by the insulator, which keeps the current flow contained in the wire. Wire sizes are selected with a gauge size large enough to conduct the necessary current flow with no resistance.

18.3 Basic Electronics

The basis of modern computers and automotive control modules is solid-state electronics. In solid-state electronics, nothing moves except electrons; there are no movable switches or relays. These are transistors, diodes, and integrated microchips that combine various electrical parts. Solid-state devices are quite reliable because they contain no parts that wear out. However, high current flow, voltage, and temperature and rough handling can damage them. Electrostatic discharge (ESD) can produce high voltage and ruin an electronic device (Figure 18–4). ESD is the spark we feel when we touch the door handle of some cars with certain tires or after sliding across some seat covers. It can easily be several thousand volts, and this can burn out the insulators or conductors in a microchip. Solid-state electronic devices are commonly used for sensing and control portions of ride and steering control systems.

In most automotive electronically controlled circuits, the sensors monitor different things and create an electric signal relative to that operation. This signal goes to the **electronic control module (ECM),** which is similar to a computer. The ECM can also receive signals from driver selection switches, and when the signals match the computer program, the ECM turns on or activates one or more actuators. The actuators are the outputs; they do the work or perform the duty of that system. Most actuators are electrical relays, motors, or solenoids.

18.3.1 Sensors

Sensors monitor different phases of vehicle operation and provide an electronic input to the control module (Figure 18–5). The various sensors used in the different chassis systems are:

- **Acceleration:** senses rate of vehicle acceleration; can use throttle position sensor (TPS) or mercury switches
- **Brake:** senses brake application; can use a pressure transducer or the brake pedal switch
- **Deceleration (crash and safing):** senses very rapid deceleration; can use a metal mass and electrical contacts or a pressure transducer or strain gauge
- **Door:** senses door opening
- **Height:** senses vehicle height; can use magnetic switches, slide, or Hall effect (can be either linear or rotary)
- **Steering angle:** senses speed that steering shaft is turned; uses a slotted shutter and photo cell or diode
- **Vehicle speed (VSS):** senses vehicle speed; can use magnetic induction or a magnetic switch

Some of these sensors generate their own electrical signal, and with some, such as the VSS, the frequency of the signal indicates the speed of motion. Other common sensors are like a variable resistor. When these are used, the control module sends them a low voltage (about 5 V), and their resistance determines the voltage signal returned to the control module.

Figure 18–4 This warning symbol is used to identify components and circuits that can be damaged by ESD. Special precautions such as not touching them unless you are grounded should be followed when working with such components and circuits.

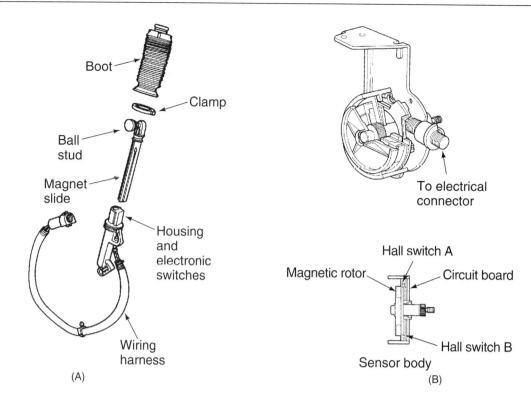
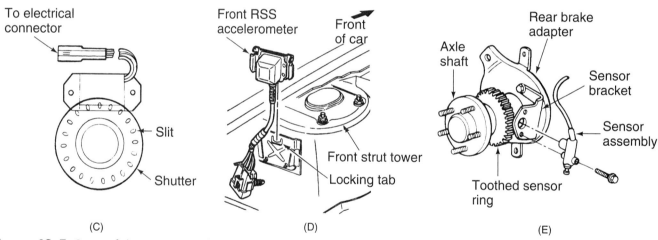

Figure 18–5 Some of the sensors used with electronically controlled circuits are a height sensor (linear) (A), rotary height sensor (B), steering wheel rotation sensor (C), accelerometer (D), and wheel speed sensor (E). *(A, B, and C are courtesy of Ford Motor Company; D is courtesy of General Motors Corporation, Service Technology Group)*

18.3.2 Control Module

The ECM is the computer that is programmed to monitor the values from the various sensors and switches, commonly called **inputs,** and turn on or off the various **outputs,** which can be called actuators (Figure 18–6). The internal parts of the control module are solid-state electronics. Most control modules use a circuit breaker or relay for a power supply. They cannot supply much electrical power for the actuators because excess current can damage the ECM. When there are high current demands, the control module will operate a relay. A relay is an electromagnetic switch that uses a low-current control circuit to operate contacts that control a much larger current flow.

Some of the more complex electronic systems share input and output information with other control modules and have the ability to operate quite a few actuators.

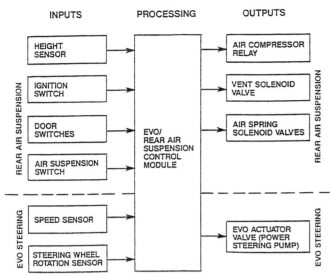

Figure 18–6 This ECM (center) controls a relay and two solenoids for the rear air suspension and the EVO power steering actuator. It uses four sensors for the suspension system and two sensors for steering system input. *(Courtesy of Ford Motor Company)*

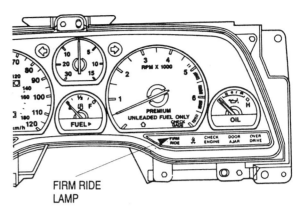

Figure 18–7 The FIRM RIDE lamp is the malfunction indicator light for this electronically controlled suspension system. *(Courtesy of Ford Motor Company)*

Some sensor information, including vehicle speed, engine temperature, and throttle position, is used by several control modules.

Many control modules are programmed to run **self-diagnosis** at start-up (each key turn cycle). If incorrect electrical values occur, a failure is indicated, and a **diagnostic trouble code (DTC),** also called an error code, is set. Failure is indicated by the instrument panel **malfunction indicator light (MIL)** (Figure 18–7). Technicians read the DTCs using special procedures to determine the nature of the problem, as described in Section 18.7.

18.3.3 Actuators

Actuators are devices controlled by the control module. Some of the actuators used in the various chassis circuits are:

- The solenoids in an ABS modulator
- The compressor relay that provides current to the compressor for the air springs
- The motors that change the settings inside the shock absorbers
- The squib in an airbag
- The solenoid that adjusts the orifice size in a VAPS system

Some multiposition actuators include a feedback circuit so the control module knows what position it is in (Figure 18–8). The indicator light on the instrument panel is an output that indicates proper or improper operation of the system.

18.3.4 Wiring and Connectors

Wire connections at the sensors, control module, and actuators must be clean and tight to prevent any changes in electrical value. The voltage and current flows through the sensor portion are very low; a change of 0.1 or 0.2 V is often enough to cause a response from the ECM. Any corrosion can produce resistance that can cause a significant voltage change and problems. Mechanical locking, weather-tight connectors are used, and many also include a waterproof conductive compound at these connections for more reliability.

18.4 Electrical Circuit Problems

Electrical problems normally fall into the following three categories: **open, shorted,** or **grounded.** Except in some solid-state units, these problems are fairly easy to check. An open circuit is a broken, incomplete circuit through which no current will flow (Figure 18–9). Open circuits are usually caused by a broken wire or a burned-out fuse or lightbulb. A loose or dirty connection can cause a high-resistance, partially open circuit. A switch opens a circuit intentionally when it is turned off. An open circuit has power source voltage in it up to the point where the circuit is open.

A grounded circuit occurs when a current-carrying wire or component touches ground, such as the bare metal of the car body. Normally, the insulation on the wire prevents a grounded circuit, but it can wear through and allow the metal conductor to touch the ground metal (Figure 18–10). A grounded circuit provides an unwanted low-resistance path for current flow, and the rate of current flow increases because of this drop in resistance. This will usually burn

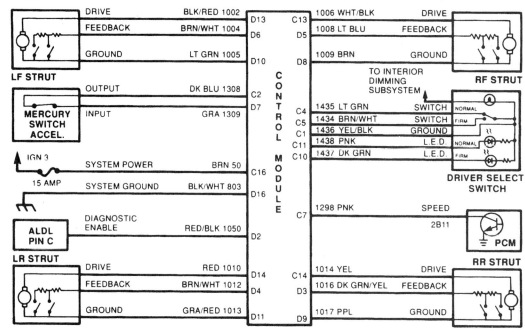

Figure 18-8 This computer command ride (CCR) system uses an actuator motor in each strut with a feedback circuit to the ECM so it will know the motor position. Also note the sensors and other outputs in this system. *(Courtesy of General Motors Corporation, Service Technology Group)*

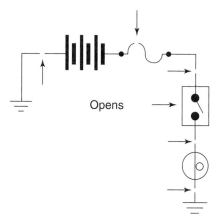

Figure 18-9 An open circuit is a break in the current path; it can occur at any of these points in a circuit.

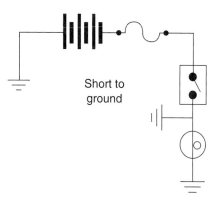

Figure 18-10 If the insulation wears off of a wire and the conductor touches ground, the current will take the easiest and shortest path to ground; this is a grounded circuit.

out a fuse, open a circuit breaker, or burn up a wire. A grounded circuit often produces an open circuit when it burns out a fuse, fusible link, or wire. A grounded circuit is also called a **short to ground.**

A shorted circuit is sometimes found in electrical units that use coils of wire. If the coils lose their insulation and make electrical contact with each other, an unwanted, shorter-than-normal electrical path is created. A short lowers the resistance of the component, which increases the current flow. It also reduces the operating efficiency of the unit (Figure 18-11). A short can occur between two wires if they lose their insulation and make contact; this can allow current to enter the wrong circuit. Some call a short a copper-to-copper connection.

18.5 Measuring Electrical Values

A technician often uses a **test light** or **volt-ohmmeter** when troubleshooting an electrical circuit. A test light is a simple, inexpensive device that can quickly indicate whether a circuit has voltage (Figure 18-12). The brightness of the light also gives an indication of the amount of voltage. A test light is a handy device for checking simple circuits, but it should not be used on electronically controlled circuits. The current flow through the light can damage the relatively fragile transistors and integrated circuits in some solid-state electronic equipment. A test light that uses a light-emitting diode (LED) can be used with

Chapter 18 ■ Chassis Electrical and Electronic Systems

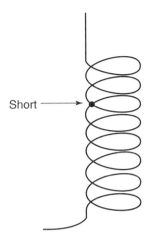

Figure 18-11 If the insulation wears between two coils, current will bypass a coil, taking the shorter path. This is a short circuit and will cause the coil to have a lower resistance and lower magnetic strength.

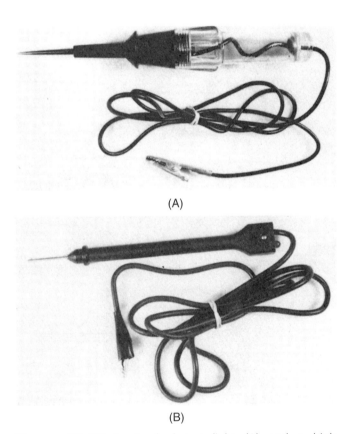

Figure 18-12 A standard test light (A) and a high-impedance test light (B). The wire clip is usually connected to ground, and if the probe is touched to voltage, the light will light up.

Figure 18-13 A digital multimeter (A) and an analog volt-ohmmeter (B). *(A is reproduced with permission from Fluke Corporation)*

these circuits. The very high resistance of the LED limits the current flow. High-impedance test lights that draw very little current are now available for checking computerized circuits. Some test lights are self-powered by an internal battery; these can be used to check **continuity** or the completeness of a component or portions of a circuit.

Volt-ohmmeters allow accurate measurements of voltage and resistance. These meters are commonly available in **analog** or **digital** form (Figure 18-13). An analog meter uses a needle that sweeps over a scale to give a reading. A digital meter displays a number that is the actual reading. Analog meters are simple and relatively inexpensive, but some of them should not be used with solid-state electronic units. Like the test light, some draw enough current from the circuit to damage the electronic components. The current draw of the meter depends on its internal resistance. If the internal resistance is at least 10 megohms (10,000,000 Ω), the meter can be safely used on solid-state circuits. A digital volt-ohmmeter, also called a **digital multimeter (DMM),** has more than 10 megohms of internal resistance. Only digital meters or high-impedance meters should be used for checking solid-state circuits.

Voltage is normally measured by connecting the negative lead of the voltmeter to a good, clean body ground and making contact between the positive lead and various connection points in the insulated portion of the circuit (Figure 18–14). The voltage in the circuit at that point will be shown on the meter. On some meters, it is important to

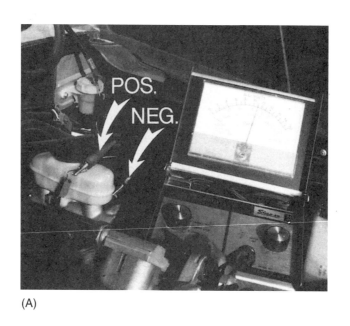

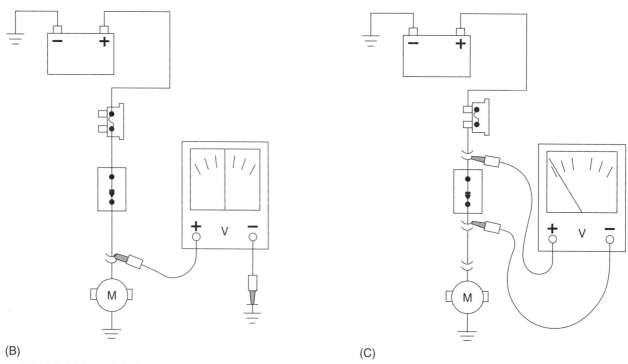

Figure 18–14 Voltage is being measured at a brake fluid level switch using a bent paper clip to make contact inside the connector (A). A schematic view of a similar check is shown in (B); connecting the meter leads to both sides of the switch checks for voltage drop across the switch (C). *(B and C are courtesy of Chrysler Corporation)*

select and set the meter to the correct voltage range before taking a reading. Some meters can be damaged if you measure a voltage higher than the meter setting. Always select a value higher than the value you expect to read; the meter can then be reset to a lower voltage scale if desired—as long as the range of the lower scale exceeds the voltage being read. Many newer meters are **self-ranging,** making voltage selection unnecessary. Also remember that voltage drops as current passes through a resistance. **Voltage drop** is the difference in voltage at the in and out terminals of a component. Most of the voltage drop is across the major actuator; voltage drop across a connector or closed switch should be less than 0.2 V.

Resistance is measured by connecting both leads of an ohmmeter to the two connections of a component or both ends of a wire (Figure 18–15). A reading indicates whether the circuit is complete (has continuity) and the level of electrical resistance. An ohmmeter is self-powered. It causes a small amount of current flow and measures the current flow to determine the amount of resistance. An ohmmeter should never be connected to a circuit that contains voltage or is connected to a battery. The added voltage from the circuit will probably damage the meter.

When using an ohmmeter that is not self-ranging, the range of the meter is selected and set to a value higher than you expect to read on the meter. Some meters have a range selection switch with ranges such as × 1, × 1,000 (1 k), × 10,000 (10 k), and so on. The reading on the meter should be multiplied by the amount of resistance. After selecting a range, the meter leads should be connected together while the meter is read. It should read zero, because there is no resistance between the leads. If not, the meter needs to be calibrated to zero by turning the calibration knob. When the leads are separated, the meter should read at the top of the scale, often marked as infinity (∞). Most digital meters read **OL** for out of limits or infinity. The leads of the meter are then connected to the terminals of the component or to the ends of a wire. A zero reading indicates a complete circuit with no resistance. A high reading indicates a large amount of resistance or a possible open circuit.

18.6 Interpreting Readings

Meter readings help only the technician who knows what the readings should be. Readings taken from different locations in a circuit or from different circuits often vary. An experienced technician is familiar with many of the fairly simple circuits such as a parking brake warning light circuit. He or she knows what the voltage or resistance should be at different points in the circuit. For a different or new and more complex circuit, such as a brake stoplight and turn indicator circuit, a wiring diagram is used.

A wiring diagram shows the electrical path through the various switches and other components in the circuit. It usually indicates the color coding of the wires, the sizes of the bulbs and fuses, and the locations of fuses, fusible links, circuit breakers, relays, and connectors in the circuit (Figure 18–16). A technician uses this

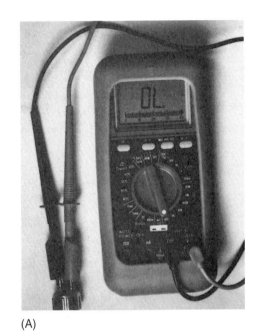

(A)

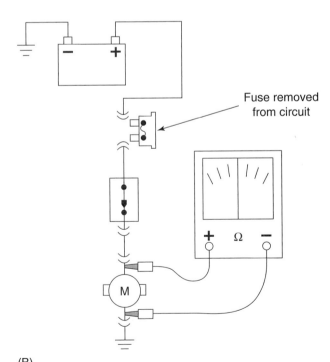

(B)

Figure 18–15 The stoplight switch has a resistance of OL (infinity), showing an open circuit (A); the resistance should drop to zero when the brakes are applied. A schematic view of a similar check is shown in (B). *(A is courtesy of SPX-OTC; B is courtesy of Chrysler Corporation)*

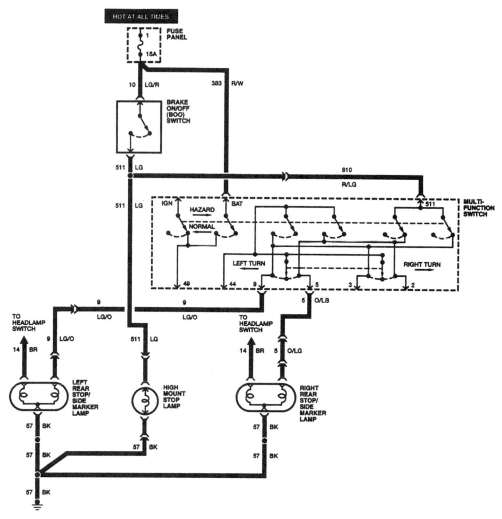

Figure 18–16 This schematic shows the stoplight circuit for a late-model vehicle. Note how the circuit from the brake on/off switch is routed through the turn indicator switch. *(Courtesy of Ford Motor Company)*

diagram to locate points where voltage or resistance checks can be made, much like a motorist using a road map, and to determine what kinds of readings should be taken at these locations.

The diagram of the rear turn and stoplight circuit shown in Figure 18–16 is for a late-model domestic vehicle that uses combination stoplights and turn lights along with a high-mounted stoplight at the center of the rear window. A system with brake lights that are separate from turn indicators is simpler. This circuit, including stoplight and turn indicator switches, can be diagnosed using a test light or a voltmeter. If they are disconnected from the circuit, the switches can be checked using an ohmmeter. The stoplight portion of this circuit should have power to it at all times. Referring to Figure 18–16, power leaves the 15-amp fuse at the lower connection and should also be present at the top (or in) connection to the brake on–off switch (stoplight switch). When the brakes are applied, electricity should flow through the brake switch and through wires 511 and 810 to the high-mounted stop lamp and the multifunction switch. Wire 511 is color-coded light green, and wire 810 is red and light green (red with a light green stripe or tracer). The multifunction switch includes the turn light switch with connection 9 to the left stop lamp and 5 to the right stop lamp; with the turn switch centered, electricity should be at both of these connections when the brake switch is on. Electricity should flow through wire 9 (light green and orange) to the left stop lamp and through wire 5 (orange and light green) to the right stop lamp. All three stop lamps are connected to ground through wire 57 (black).

In this circuit, a technician can often determine the problem by observing how it operates or fails to operate because it is divided into two parallel circuits. A total failure is probably caused by a power loss at the fuse or brake switch. A failure of both side lights can be caused by a faulty turn–hazard switch or the wires between this switch and the fuse. A partial failure of the right or left side is probably caused by a faulty turn–hazard switch or an open circuit in the wires from the switch to the rear junction of the lights. Failure of a single light is probably caused by a faulty bulb, a faulty light ground, or a faulty wire between the bulb and the junction. Knowing the voltage at a given point or the resistance between two given points will let the technician determine the exact location of the fault. At that point, correcting the problem is usually easy.

18.7 Electronically Controlled System Diagnosis

Diagnosis of electronically controlled systems begins like other diagnostic procedures by verifying the customer's complaint. You should operate the system in the shop or on a road test to see if you can duplicate the problem; this ensures that you know what the problem is so you can plan your repair procedure. This step often requires a service manual to help you find out how this system works, what parts or components it has, and where these parts are located.

Next make a visual inspection of the system to locate any faults visually. The majority of problems with electronically controlled systems are proving to be faulty electrical connections, so give these a close inspection, making sure each connector is properly latched.

If the problem was not located visually, you should refer to service and diagnosis information for the system you are working on. It is fairly easy to damage an electronically controlled system by doing the wrong thing; for example, the ECM in one system can be damaged by removing its fuse with the key on. Be sure to check for additional service bulletins that might apply to a particular system. Most manufacturers recommend that diagnosis begin with a certain procedure or cycle (Figure 18–17). This is usually the quickest and most thorough procedure to find all of the faults. Most electronic systems perform self-diagnosis on start-up and sometimes during operation. If the control module determines an electrical value that is out of normal parameters—voltage or resistance too high or too low—it will shut the system off, returning it to non–electronic control or fixed operation, set a code, and turn on the malfunction indicator light (MIL). The diagnostic trouble code (DTC) can be either hard or soft (Figure 18–18). A soft code is temporary; it is lost or erased when the key is turned off. A hard code is semipermanent; it stays in the memory until there are a certain number of key cycles or until it is erased or cleared. Clearing requires the technician to perform a certain procedure. Codes can be cleared by disconnecting the fuse or battery cable or by performing certain operations using a scan tool. Most technicians do not like to disconnect a battery cable because this also erases other electronic memories such as clock and radio station presets.

Codes can be displayed in various ways depending on the system; electronically controlled systems with code capability have a diagnostic terminal that can be manipulated, usually by connecting particular terminals (Figure 18–19). The code is then read by counting the flashes of the MIL or reading the display of a scan tool; **handheld scanners** have become a common electrical diagnostic tool (Figure 18–20). Each vehicle manufacturer uses a scan tool of its own specifications, and several aftermarket manufacturers have generic scan tools available. A wide variety of vehicle systems can be serviced by plugging in information or upgrade modules with data for certain vehicles or system types. Scan tools have the ability to read and clear codes, control certain electronic functions, monitor sensor outputs, and measure electrical values such as voltage and resistance. On these systems, the scan tool is used to read the code to determine the nature of the problem, read signals from sensors, operate the actuators, make electrical checks to locate the exact cause of the problem, and clear the DTC. The vehicle is often returned to the test conditions that set the code and then rechecked to make sure the DTC does not come back.

On vehicles without code capability, the components need to be checked individually. Some manufacturers provide **diagnostic flowcharts,** often called trouble trees, to guide the technician through the tests for a particular circuit (Figure 18–21). We also make these continuity, current draw, resistance, and voltage checks to verify the condition of any component or circuit whenever necessary. For example, the solenoid windings of one make of EVO actuator should have a resistance of 5 to 20 ohms. We can check this by connecting an ohmmeter to the two terminals; if we read between 5 and 20 ohms, the solenoid is electrically sound. If there is less than 5 ohms, the solenoid windings are shorted. If the resistance is more than 20 ohms, there is high resistance or open windings. Components with metal cases can also

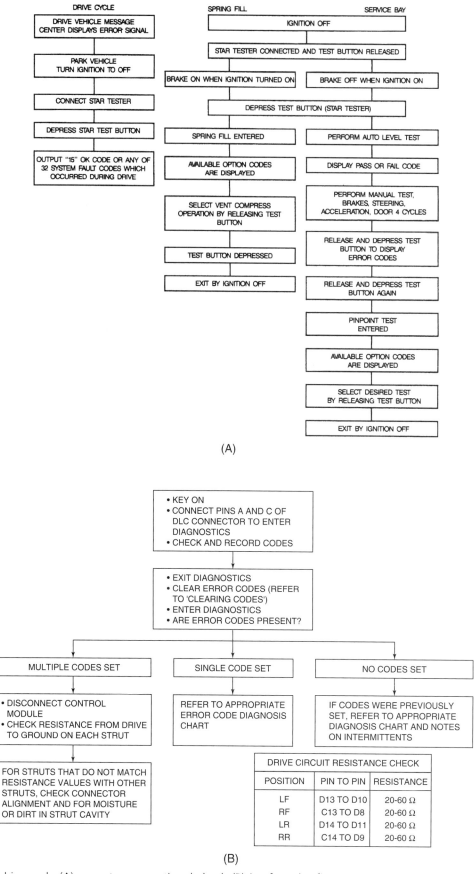

Figure 18–17 A drive cycle (A) or system operational check (B) is often the first step in diagnosing a problem in an electronically controlled system. *(A is courtesy of Ford Motor Company; B is courtesy of General Motors Corporation, Service Technology Group)*

Chapter 18 ■ Chassis Electrical and Electronic Systems

AIR SUSPENSION DIAGNOSTICS
SERVICE CODES

SERVICE CODE	PINPOINT TEST	DESCRIPTION	SERVICE PRIORITY
10		DIAGNOSTICS ENTERED, AUTO TEST IN PROGRESS	
11		VEHICLE PASSES. IF VEHICLE IS STILL LOW OR HIGH IN REAR, CHECK REAR RIDE HEIGHTS.	
12		AUTO TEST PASSED, PERFORM MANUAL INPUTS	
13		AUTO TEST FAILED, PERFORM MANUAL INPUTS	
16	*	EVO SHORT CIRCUIT	
17	*	EVO OPEN CIRCUIT	
18	*	EVO ACTUATOR VALVE	
39	A	COMPRESSOR RELAY CIRCUIT SHORTED TO BATTERY	2ND
42	B	AIR SPRING SOLENOID CIRCUIT SHORTED TO GROUND	2ND
43	C	AIR SPRING SOLENOID CIRCUIT SHORTED TO BATTERY	2ND
44	D	VENT SOLENOID CIRCUIT SHORTED TO BATTERY	2ND
45	E	AIR COMPRESSOR RELAY CIRCUIT SHORTED TO GROUND OR VENT SOLENOID CIRCUIT SHORTED TO GROUND	2ND
46	F	HEIGHT SENSOR POWER SUPPLY CIRCUIT SHORTED TO GROUND OR BATTERY	2ND
51	G	UNABLE TO DETECT LOWERING OF REAR	3RD
54	H	UNABLE TO DETECT RAISING OF REAR	3RD
68	J	HEIGHT SENSOR OUTPUT CIRCUIT SHORTED TO GROUND	2ND
70	K	REPLACE AIR SUSPENSION/EVO MODULE	1ST
71	L	OPEN HEIGHT SENSOR CIRCUIT	3RD
72	M	FOUR OPEN AND CLOSED DOOR SIGNALS NOT DETECTED	4TH
74	*	STEERING WHEEL ROTATION NOT DETECTED	
80	N	INSUFFICIENT BATTERY VOLTAGE TO RUN DIAGNOSTICS	1ST
23		FUNCTIONAL TEST, VENT REAR	
26		FUNCTIONAL TEST, COMPRESS REAR	
31		FUNCTIONAL TEST, TOGGLE COMPRESSOR	
32		FUNCTIONAL TEST, VENT SOLENOID TOGGLE	
33		FUNCTIONAL TEST, SPRING SOLENOID TOGGLE	

— AIR SUSPENSION FUNCTIONAL TESTS, NEVER PERFORM A FUNCTIONAL TEST UNLESS DIRECTED TO DO SO IN A PINPOINT TEST.

*FOR CODES 16, 17, 18 AND 74, REFER TO PINPOINT TEST STEP A1 IN SECTION 13-54. FOR ALL OTHER CODES, REFER TO SECTION 14-40 FOR THE TEST INDICATED ON THIS CHART.

Figure 18-18 Service codes or DTCs for a rear air suspension and EVO system indicate the nature of an electrical problem and what test to perform next. *(Courtesy of Ford Motor Company)*

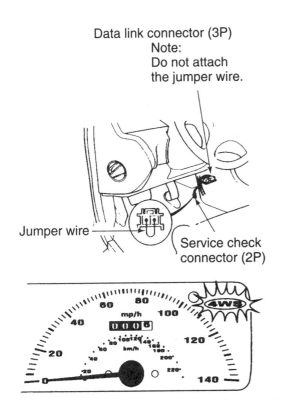

Figure 18-19 Service check is entered by inserting a jumper wire in the service check connector of this 4WS system. The codes are read at the 4WS light in the instrument panel. *(Courtesy of American Honda Motor Co., Inc.)*

be easily checked for shorts to ground; leave one of the ohmmeter leads connected to a terminal and transfer the other lead to the case mounting point. The reading should show infinite ohms or OL; a reading less than this shows a ground, indicating a faulty component (Figure 18-22). In either case, improper readings indicate a bad solenoid, and it needs to be replaced. These checks can be made directly at the component or at the wiring harness connector. Most wiring diagrams identify the connector terminals so we can determine which terminals supply power, are grounded, or connect to various components (Figure 18-23).

Most electronic components and many electrical components used in modern vehicles are not serviceable; they must be removed and replaced.

18.8 Practice Diagnosis

You are working in a shop that specializes in tire sales, wheel alignment, and brake and front-end repair and encounter the following problems:

CASE 1: While inspecting the brakes on a 3-year-old Toyota, you find that the center stoplight does not light

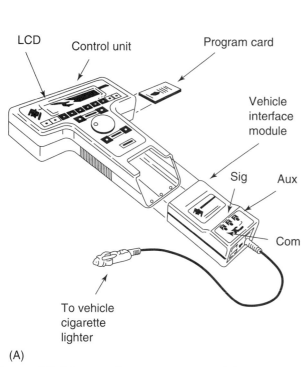 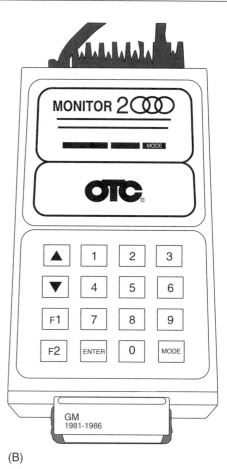

Figure 18–20 A Ford motor STAR tester (A) and an OTC Monitor 2000 (B) are both scan tools for performing diagnostic tests on electronic systems. *(A is courtesy of Ford Motor Company)*

when the pedal is depressed; the other two stoplights do come on. What is probably wrong? If you check out this possibility, and you find it is correct, what should you do next?

CASE 2: The complaint on an 8-year-old pickup is a very dim stoplight on the right side. You measure the voltage at the contacts for the bulb and find that one of them has 6 V, the other one has 0 V, and the bulb socket also has 0 V. What is probably wrong? Have you found the cause of the dim bulb? What should you do next?

CASE 3: A vehicle is 2 years old, but it is still under warranty. It is a top-of-the-line model with all of the extra equipment. The driver's complaint is that the vehicle stays in the soft ride mode, even when the firm ride selection switch is moved, and the firm ride light does not come on. What could be causing this problem? Where should you begin to find the cause?

CASE 4: A new car is being detailed for delivery to the customer, and the technician making the predelivery has noticed that the steering is harder than normal, especially while moving the car around the shop. He has asked for your help to find the cause, and you find that this car is equipped with variable-assist steering. Could a fault in the VAPS system be causing this problem? If so, how could it cause hard steering? What should you do next?

Chapter 18 ■ Chassis Electrical and Electronic Systems

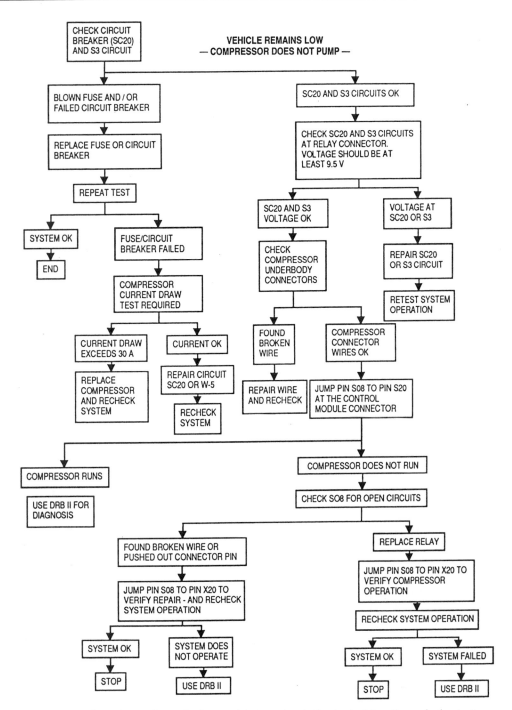

Figure 18–21 This diagnostic flowchart, also called a trouble tree, leads the technician through the test procedure of an air suspension system. *(Courtesy of Chrysler Corporation)*

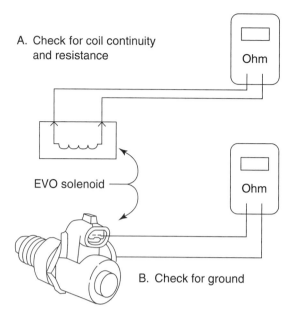

Figure 18-22 The ohmmeter checking coil continuity or resistance (A) should give a reading within the specifications. The ohmmeter checking for ground (B) should give an OL or infinity reading.

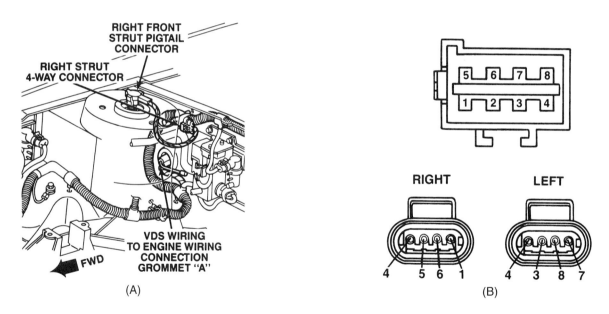

Cavity	Circuit	Gauge	Color	Function
1	Z02F	20	BK/LG*	Ground
2	S61	20	BK	Soft PFS, left front
3	—	—	—	Not Used
4	S50	20	LB	Strut Motor, right front
5	S55	20	DB/YL*	Firm PFS, left front
6	S51	20	LB/BK*	Strut Motor, left front
7	S54	20	DB	Firm PFS, right front
8	S60	20	BR	Soft PFS, right front

(C)

Figure 18-23 Electrical checks can be made of this variable-damping suspension (VDS) at various electrical connectors or at the strut actuator (A); (B) and (C) show where the circuits are in each of the connectors. *(Courtesy of Chrysler Corporation)*

Chapter 18 ■ Chassis Electrical and Electronic Systems

Terms to Know

- amperes (amps)
- amplitude
- analog
- continuity
- cycle
- diagnostic flowcharts
- diagnostic trouble code (DTC)
- digital
- digital multimeter (DMM)
- electronic control module (ECM)
- frequency
- ground circuit
- grounded
- handheld scanners
- inputs
- intensity
- malfunction indicator light (MIL)
- natural frequency
- ohms (Ω)
- OL
- open
- order
- outputs
- parallel circuit
- self-diagnosis
- self-ranging
- series circuit
- shorted
- short to ground
- test light
- voltage drop
- volt-ohmmeter
- volts
- watts

Review Questions

1. Which of the following is the unit for electrical resistance?
   a. Amps
   b. Ohms
   c. Volts
   d. Pressure

2. A test instrument that can be used on any automotive electrical circuit is the:
   a. DMM
   b. Ohmmeter
   c. Analog volt-ohmmeter
   d. Test light

3. Electronically controlled ride systems are being discussed.
   Technician A says that an ECM can be ruined if you touch the electrical connector pins.
   Technician B says that you should spray the ECM with antistatic spray before disconnecting it.
   Who is correct?
   a. A only
   b. B only
   c. Both A and B
   d. Neither A nor B

4. Technician A says that the VSS used as input for the ride control system is also used in other electronically controlled systems.
   Technician B says that this sensor measures the speed of the vehicle.
   Who is correct?
   a. A only
   b. B only
   c. Both A and B
   d. Neither A nor B

5. Technician A says that most analog volt-ohmmeters are self-ranging.
   Technician B says that meters that are not self-ranging should be set to a range above what you expect to read before taking a reading.
   Who is correct?
   a. A only
   b. B only
   c. Both A and B
   d. Neither A nor B

6. Technician A says that all power should be shut off in a circuit before checking portions of it with an ohmmeter.
   Technician B says that excessive resistance is a sign of a short circuit.
   Who is correct?
   a. A only
   b. B only
   c. Both A and B
   d. Neither A nor B

7. Technician A says that a DTC from a Ford product should indicate the same problem in a Honda.
   Technician B says you should follow the procedure for a particular make and model when diagnosing problems in an electronically controlled system.
   Who is correct?
   a. A only
   b. B only
   c. Both A and B
   d. Neither A nor B

8. Technician A says that after you repair the electronic problem, you should clear the codes, rerun the quick check or road test, and recheck for DTCs.
   Technician B says that the best way to clear the codes in all vehicles is to disconnect the battery cable.
   Who is correct?
   a. A only
   b. B only
   c. Both A and B
   d. Neither A nor B

9. The specification for a VAPS solenoid is 5 to 20 ohms, and you measure a resistance of 300 ohms. Technician A says that this solenoid has a short and should be replaced.
   Technician B says that the solenoid should be replaced because it probably has a weak internal connection.
   Who is correct?
   a. A only
   b. B only
   c. Both A and B
   d. Neither A nor B

10. You are looking for the cause of a complaint of hard steering on a car with a VAPS system.
    Technician A says that this could be an electronic problem.
    Technician B says that this could be caused by a faulty power steering pump.
    Who is correct?
    a. A only
    b. B only
    c. Both A and B
    d. Neither A nor B

19 Wheel Alignment: Front and Rear

Objectives

Upon completion and review of this chapter, you should be able to:

❑ Comprehend the terms commonly used with wheel alignments.

❑ Understand the purpose for the various front- and rear-wheel alignment angles.

❑ Understand the effect that improper alignment angles have on vehicle operation.

19.1 Introduction

For a tire to roll down the road in an efficient and controlled fashion, it must be aligned correctly. Usually, the tire is vertical or straight up and down (zero camber), and the center of the tire points straight down the road (zero toe). **Camber** and **toe** are **tire position angles.** If they are not correct, tire wear and direction control problems can result. If we did not steer the car, this is all that wheel alignment would be, but the car would travel only in a straight line (Figure 19–1).

Since we do steer the car so that we can turn corners, we also have to think about the position of the **steering axis.** The steering axis is an imaginary line that runs through the two ball joints of a car with S-L A suspension, the lower ball joint and the steering pivot of a car with strut suspension, or the kingpin of many solid or swing axles. **Caster** and **steering axis inclination (SAI)** are thought of as two different angles of the steering axis; they ensure that the front wheels are easily and safely steerable. They are actually two different planes of the steering axis, although we normally think of them as two separate and distinctly different angles. They are called **directional control angles.** Since they do not directly control the position of the tire, they cannot cause tire wear. They do affect the drivability and the handling of the vehicle (Figure 19–2).

The fifth alignment angle, **toe-out on turns,** also called **turning angle** or **turning radius,** ensures that the tires are turned to the correct angle while the vehicle is turning. Tire wear will occur during turns if this angle is not correct.

Before we begin a closer study of front-end alignment, we must remember that the rear wheels must also be aligned to the car. Camber and toe must be correct in

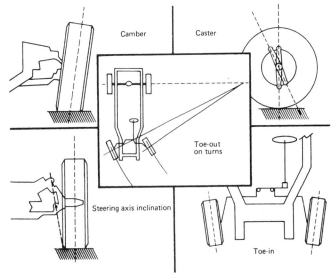

Figure 19–1 The five different front-wheel alignment angles. *(Courtesy of Chrysler Corporation)*

order to have the maximum traction for vehicle control and maximum tire life. Also, the car's direction, while traveling in a straight line, is mainly controlled by the rear wheels. The front tires turn the car, but the rear tires steer, at least in the straight-ahead direction. The straight-ahead direction of a vehicle is on a path that is parallel to the rear wheels. This is often called the **thrust line.** It is also called **track;** the rear tires should make tracks directly behind the front tires (Figure 19–3).

One additional aspect of wheel alignment is **setback.** This is the amount that one wheel on an axle is behind the other. Setback and thrust are used as **diagnostic angles;** they often indicate a problem.

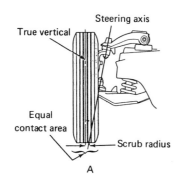

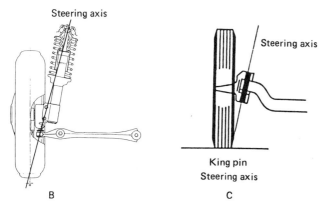

Figure 19-2 The steering axis is the point where the steering knuckle pivots on turns. It is an imaginary line through the ball joints of an S-L A suspension (A), through the ball joint and upper pivot of a strut suspension (B), and through the kingpin on a solid axle (C). *(A is courtesy of Ford Motor Company; B is courtesy of Volkswagen of America, Inc.; C is courtesy of Ammco Tools, Inc.)*

19.2 Measuring Angles

Most wheel alignment specifications are printed in degrees or parts of degrees. A degree is a single slice out of a circle in which 360 equally sized slices have been made. A degree is an international measurement; all countries use the same system. The symbol ° is sometimes used to indicate degree.

There are some differences when measuring parts of a degree. Traditionally, in the United States, a portion of a degree has been referred to as its fractional or decimal part—for example, 1/2 or 0.5 degree. Our scientists and

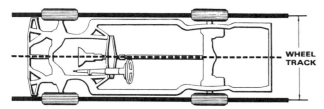

Figure 19-3 On most cars, the rear wheels do not steer so we only need to maintain correct camber, toe, and track; if the track is not correct, the car will go down the road slightly crosswise. *(Reprinted by permission of Hunter Engineering Company)*

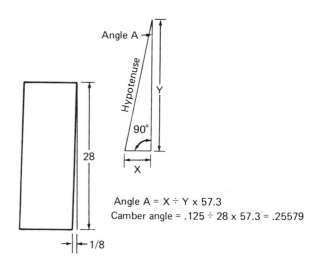

Figure 19-4 If we know the vertical distance from the ground to the top of the tire (side Y) and the amount the tire is from being vertical (side X), we can determine the camber angle (angle A) by using the given formula.

engineers and most of the rest of the world have measured portions of a degree in minutes (') and seconds ("). Much like the division of an hour, there are 60' in a degree and 60" in a minute; 1/2 or 0.5° would equal 30'. An equivalency chart is included in Appendix C of this book to help you convert fractional or decimal portions of a degree into minutes and seconds.

Those familiar with trigonometry know that if you know the length of two sides of a right triangle, it is fairly easy to compute the angle between the hypotenuse and the long side. For example, if I know that a tire 28 in. in diameter is leaning 1/8 in. from vertical, I can compute that the tire has a camber angle of 0.256, or about 1/4 degree. This is described in Figure 19-4.

Most of the alignment angles have exact specifications determined by the vehicle manufacturer (Table 19-1). Problems can result if the actual angles vary from specification.

19.3 Camber

Camber is a term used to describe the position of the tire as seen from the front or back. If the tire is exactly vertical, it has a camber angle of zero. If the top of the tire leans outward, away from the car's center, the tire has a **positive camber angle;** an inward-leaning tire has **negative camber** (Figure 19-5). Camber is directly controlled by the spindle; the centerline of the tire is at a right angle, 90 degrees from the spindle (assuming there is no looseness in the wheel bearings). A spindle that drops 1 degree at the outer end will cause 1 degree of positive camber. FWD cars do not have a front spindle,

Chapter 19 ■ Wheel Alignment: Front and Rear

Table 19–1 Alignment Angle Effects.

Angle	If	Result
Camber	Just right	Good tire wear, ride quality and directional stability
	Too positive	Tire wear at outer tread; reduced cornering ability
	Too negative	Tire wear at inner tread; excessive road shock; reduced ride quality; improved cornering ability
	Unequal	Pull toward most positive side
Rear camber	Just right	Good tire wear
	Too positive	Wear at outer tread
	Too negative	Wear at inner tread; improved cornering ability
Caster	Just right	Good directional stability and returnability
	Too positive	Excessive road shock; hard steering; possible shimmy; possible camber wear on turns
	Too negative	Wander; unstable steering
	Unequal	Pull toward least positive side
Toe	Just right	Good tire wear
	Too positive (in)	Scuffing and wear at outside shoulder
	Too negative (out)	Scuffing and wear at outside shoulder; possible wander
Rear toe		Same as front except tire wear can show as diagonal wipe
Thrust	Just right	Good tracking
	Not right	Crooked steering wheel from improper tracking
SAI	Just right	Good directional stability and returnability
	Not right	Improper camber and included angle
Turning angle	Just right	Good tire wear on turns
	Not right	Scuffing and squealing with tire scrub on turns
Setback	Just right	Normal driving
	Not right	Indicates possible cradle shift or collision, crooked steering wheel, torque steer

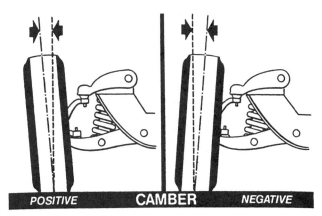

Figure 19–5 A tire with positive camber leans outward at the top; it leans inward if it has negative camber. *(Reprinted by permission of Hunter Engineering Company)*

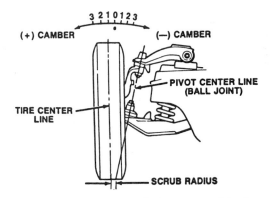

Figure 19–6 The camber angle is measured in degrees of tilt from true vertical; if the tire is straight up and down, it has 0 camber. This tire is leaning outward with 1 degree of positive camber. *(Courtesy of Ford Motor Company)*

but the hub and bearings in the steering knuckle serve the same purpose as the front spindle of a RWD car.

Many modern cars use about 1/4 to 1/2 degree of positive camber on their front wheels. This is usually just enough to cause near-to-zero camber when driving on the road (Figure 19–6).

Normally the angles of the tires at each side of a suspension or at each end of an axle are the same. If they are different, there is a **spread** or a **cross-car reading;** a camber spread or a cross camber of 1/4 degree means one wheel has 1/4 degree more camber than the other.

19.4 Camber and Scrub Radius

At one time, when tires were taller and narrower, cars had more positive camber to help reduce the scrub radius. **Scrub radius** refers to the distance between the steering axis (at ground level) and the center of the tire (Figure 19–7). A **positive scrub radius,** where the steering axis is inboard of the tire center, tends to cause the tire to turn or toe outward as the car is driven. This effect is especially common if the tire is flat or if it meets a bump or pothole in the road; a definite pull or dart will occur. The drag of the tire pushes it toward the rear of the car while the car is pushing forward on the steering axis. FWD cars usually have a **negative scrub radius;** the steering axis meets the road outboard of the tire centerline. This offsets the effect of the tires trying to toe inward as they pull the car down the road. It also produces an effect that cancels front-end pull if a front tire goes flat while the car is being driven.

A slight scrub radius helps give the front end directional stability. There is always a slight amount of compliance in the moving parts of the steering linkage, and this can cause the front tires to pivot in and out slightly. Compliance results from the slight flexibility of the ball joints and rubber bushings; it allows the tire to change position slightly. With positive scrub radius, the tire, trying to turn outward, will load the steering parts in one direction and compress the compliance to a stable, loaded position (Figure 19–8).

Besides reducing scrub radius, positive camber was said to be used to help strengthen the wooden spokes of wheels used in the past. This reason, of course, is now obsolete, and scrub radius is now controlled more by the steering axis inclination and wheel offset than by camber. There is no longer a valid reason for having very much positive camber on a rolling tire. We might set a tire to a slightly positive camber angle on the alignment rack, but it should have zero camber when it rolls down the road.

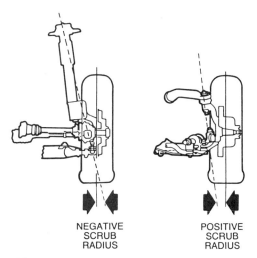

Figure 19-7 The distance between the road-contact point of the steering axis and the center of the tire is called scrub radius. If the tire center is inward of the steering axis, it is called negative scrub (left), and if it is outward, it is called positive scrub (right). *(Courtesy of Chrysler Corporation)*

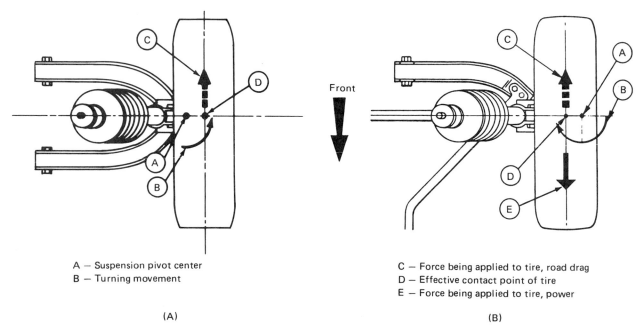

Figure 19–8 Road drag on the tire of a RWD car causes it to turn from the effects of positive scrub radius (A). The tire of a FWD car tries to turn inward from the negative scrub radius, but front-wheel driving forces offset this effect with an outward turning action (B). *(Courtesy of Ford Motor Company)*

19.5 Camber-Caused Tire Wear

The major theory with modern tires, which are lower and wider, is that the tread should be flat on the road so it has full tread contact. Camber tends to lift one side of the tire upward. Reducing the pressure between the tread and the road can reduce tire traction, causing a loss of vehicle handling and stopping ability.

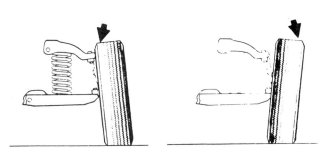

Figure 19–9 Incorrect camber angle causes a tapered wear pattern across the tire tread. *(Courtesy of Dana Corporation)*

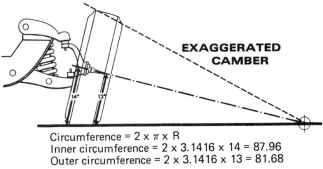

Circumference = 2 × π × R
Inner circumference = 2 × 3.1416 × 14 = 87.96
Outer circumference = 2 × 3.1416 × 13 = 81.68

Figure 19–10 In this exaggerated example, the circumference difference between the inside of the tread (88 in.) and the outside (82 in.) will cause 6 in. of tread slippage on each tire revolution. *(Reprinted by permission of Hunter Engineering Company)*

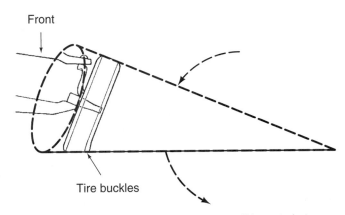

Figure 19–11 A cambered tire tries to roll in a circle just as if it were the end of a cone.

It is easy to see that if we lift part of the tread, the remaining tread portion will carry a higher load, more than its share. Consequently, incorrect camber will cause abnormally fast tire wear. Camber that is too positive will cause rapid wear on the outside of the tread, and camber that is too negative will cause rapid wear on the inner side of the tread. Wear will be greatest on one side of the tread, and the amount of wear will steadily decrease across the tread (Figure 19–9).

Some people believe that camber wear is a result of tire scrub caused by a shorter tire radius on one side of the tire than on the other. The effective tire radius is the distance between the ground and the horizontal center of the tire. The circumference of the tire, which is the distance the tire will roll down the road in one revolution, is the result of multiplying the radius (r) by 2 and then by pi (π), or 3.1416 (circumference = $2\pi r$). A tire that is cambered has two different tire radii, a different radius at each side of the tread. This causes the tire to have two different circumferences, which results in a tire trying to travel two different distances on each revolution. This, of course, is impossible, so portions of the tread will slip or scrub as the tire rolls (Figure 19–10). In any event, regardless of which theory seems most correct, incorrect camber will cause tire wear. That is why it is referred to as a tire wearing angle.

Another negative effect of incorrect camber is tire pull. A cambered tire rolls as if it were the end of a cone. A cone will not roll straight; it rolls in a circle, with the point or apex of the cone being the center of the circle. We see this effect when we are riding a bicycle or motorcycle and turn corners by leaning toward the direction we want to turn. Excessively positive camber will cause the car to pull or turn toward that side. Excessively negative camber will cause a pull toward the opposite side. A rule of camber is that the car will pull toward the most positive side (Figure 19–11).

19.6 Camber Spread and Road Crown

Roads are usually crowned or raised in the middle so that rain water will run off. Like a ball, a tire rolling across an inclined surface will tend to turn and roll downhill. This effect would cause the car to pull to the right on most crowned roads. Some manufacturers specify slightly more positive camber (about 1/4 degree) on the left wheel than on the right; again, this is called **camber spread** or **cross camber.** This spread generates a camber pull to the left; if the camber pull equals the road crown pull, the car will go straight (Figure 19–12).

Figure 19-12 If a car is driven on a hillside or high crowned road, it has a tendency to drift or turn downhill. *(Reprinted by permission of Hunter Engineering Company)*

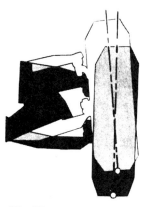

Figure 19-13 On cars with independent suspension, the tires change camber angle as they move up and down. *(Courtesy of Chrysler Corporation)*

Because camber is a tire wearing angle, many alignment technicians prefer to adjust camber to the least tire wearing angle and use a caster spread to compensate for road crown (see Section 19.16). Caster can also cause a slight pull to offset road crown, and it will not cause tire wear.

19.7 Camber Change

Vertical wheel travel on cars with independent suspension causes the camber to change. Depending on the suspension type, this can be an undesirable change, as seen on swing axles, or a carefully planned change that is designed into some S-L A suspensions. The important thing to remember is that vehicle height is important. If the car is too high or too low, the camber angle will probably be out of specification (Figure 19-13).

The procedure for measuring and adjusting camber is described in Chapter 34.

19.8 Toe-In

Toe-in, like camber, is an angle of the tire that will cause tire wear if it is incorrect. We commonly use the term *toe-in,* but we probably should use the term *toe* because many FWD cars use **toe-out** (Figure 19-14). Toe is a comparison of the distance between the front of the tires and the rear of the tires. If the two tires are parallel, these two distances are the same, and there is zero toe. If the front measurement is 1/8 in. (3.17 mm) shorter than the rear, there will be 1/8 in. (3.17 mm) of toe-in. If the rear measurement is 1/8 in. (3.17 mm) shorter, there will be 1/8 in. (3.17 mm) of toe-out. Toe is usually measured at the height of the center of the hub to ensure measuring at

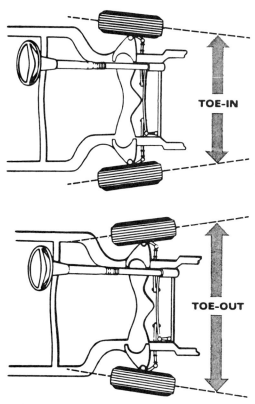

Figure 19-14 Toe-in refers to the tires being closer together at the front; if they are closer at the rear, it is called toe-out. *(Reprinted by permission of Hunter Engineering Company)*

the most forward and rearward portions of the tire. Toe is controlled by the length of the tie-rods. A longer tie-rod (mounted to the rear of the tire centerline) increases toe-in (Figure 19-15).

Traditionally, toe specifications have been printed in inches for domestic cars and millimeters for imported cars. Many modern cars are now specifying **toe angles.** This is the angle between the two tires (combined) or between a

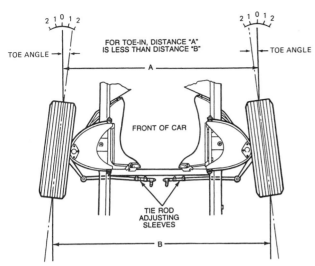

Figure 19-15 Toe can be measured as the difference in distance between the front and rear of the tires or as the angle of the tires. A positive toe angle is shown here. *(Courtesy of Chrysler Corporation)*

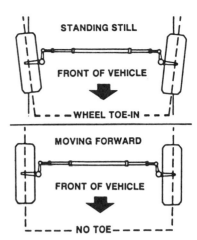

Figure 19-16 On a RWD car, road forces cause the toe to change from toe-in while standing still to zero toe while moving forward. *(Courtesy of Ford Motor Company)*

single tire and straight ahead (individual). Toe-in is considered a positive (+) angle, and toe-out is considered a negative (−) angle. A toe specification can be converted from a distance to an angle by using this formula:

$$\text{Toe Angle in Degrees} = \frac{\text{Toe Distance}}{\text{Tire Diameter}} \times 57.3$$

A conversion chart for toe distance and degrees is included in Appendix C of this book.

A small amount of toe-in, about 1/16 in. (1.58 mm), is usually used on the front tires of RWD cars. This amount of toe-in is used to offset the tendency of the front tires to toe outward as they are pushed down the road. Each tie-rod end has a small amount of compliance that allows the tie-rod to lengthen or shorten slightly under pressure; the ideal amount of toe-in exactly matches the change in length of the tie-rods so the tires will be exactly parallel, with zero toe, when the car is going down the road.

A small amount of toe-out is often used on FWD cars; the front tires of FWD cars tend to toe inward as the front tires pull the car down the road. This tendency produces zero toe if the tires are adjusted to the correct amount of toe-out.

19.9 Factors That Affect Toe

Toe is affected by the rolling resistance of the tires and the scrub radius. Tires with more road drag (lower air pressure, wider tread, bias ply) tend to toe out more than tires with less road drag (higher air pressure, narrower tread, radial ply). If the scrub radius is increased by the use of wheels with a greater negative offset than stock, the toe-out tendency will also be increased. Worn bushings, idler arms, and/or tie-rod ends also affect the amount the tires toe out on the road because of the increased compliance. These factors are usually considered when the skilled alignment technician adjusts toe (Figure 19-16).

19.10 Toe-Caused Tire Wear

If toe is not correct, the tire will have to scuff or scrub sideways as the car travels down the road. This side scuffing causes an easily recognizable wear pattern on the tread called a **featheredge** or **sawtooth**. It is most noticeable if you run your hand inward across the tread and then back out. If the roughness is felt as your hand is passed inward, excess toe-out is indicated; if the roughness is felt on the way out, excess toe-in is indicated (Figure 19-17).

Figure 19-17 Incorrect toe on bias-ply tires causes an easily recognizable, abnormal tread wear pattern. Note the feather edge toward the left; this tire was scuffing from the left toward the right. *(Courtesy of Ford Motor Company)*

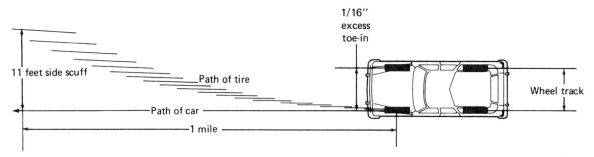

Figure 19-18 A tire that is toed 1/16 in. more or less than it should will have to scrub sideways about 11 ft during every mile it travels.

Modern belted radial tires show toe wear more as a scuffing rather than the feather edge wear of older bias tire types.

A toe error of 1/16 in. (1.58 mm) causes the tire to scuff sideways about 11 ft (3.35 m) for each mile (1.61 km) the car is driven. This is because the car is traveling one way while the tire tries to go in a different direction. The resulting side scuffing not only wears tires, it also causes a reduction in gas mileage. It takes a lot more power to push a tire sideways than it does to roll it. For example, try pushing a car forward or backward (in neutral with the brakes released); then try pushing it sideways (Figure 19-18).

Incorrect toe at the rear tires of a FWD car often appears as a scalloped, tapered diagonal wipe wear. Because of the lower weight holding the tire on the road, the tire scuffs and hops slightly as it is pulled along the road.

19.11 Toe Change

On most cars, toe changes slightly as the front wheels travel up and down over bumps. This toe change is often called **bump steer** and is caused by the steering knuckle and spindle moving in an arc as they rise and fall (Figure 19-19). The tie-rod end also moves in an arc, with its center being the inner tie-rod end. If these two arcs do not match exactly, the steering knuckle will be turned slightly, and toe will change. It is hoped that this change is very slight and only occurs for a brief period. However, a change in the position or length of the steering linkage components or suspension components can cause more severe toe change and therefore tire wear and handling problems.

Some toe gauges allow a fairly easy check to see whether toe change is occurring; toe gauges are discussed in later sections. Toe change is often measured very precisely when setting up a competition car.

Because of toe change, toe is correct only at one vehicle height—curb height. Like the other wheel alignment

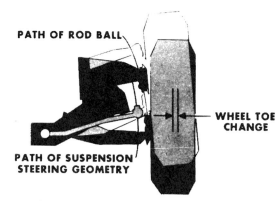

Figure 19-19 A toe change occurs during bump travel if the suspension geometry does not match the geometry of the tie-rod. *(Courtesy of Chrysler Corporation)*

angles, toe is always checked at ride height. The procedure for measuring and adjusting toe along with centering of the steering wheel is described in Chapter 34.

19.12 Caster

Caster is a directional control angle that helps the driver keep the front tires turned straight down the road and return the steering to straight ahead after turning a corner. It is an angle of the steering axis when viewed from the side; the steering axis leans toward the front or rear of the car. If the steering axis were exactly vertical when viewed from the side, there would be zero caster. If the upper end of the steering axis is toward the rear of the car, caster is **positive;** it is **negative** if the upper end of the steering axis is toward the front (Figure 19-20).

Caster can be understood more easily if we consider a furniture caster or caster on a bicycle (even though there is a slight difference in operation). A furniture caster has a vertical pivot axis that can be compared with the steering axis of a car. The wheel axle, and with it the road contact of the wheel, is positioned to one side of the pivot axis. As this distance increases, caster effect gets

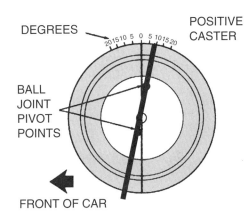

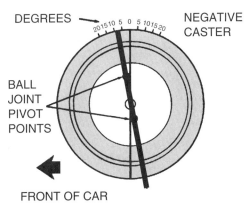

Figure 19–20 Caster is defined as a forward or rearward tilt of the steering axis. If the top is toward the rear, the caster angle is positive (top); if the top is toward the front, the angle is negative (bottom). *(Courtesy of Ford Motor Company)*

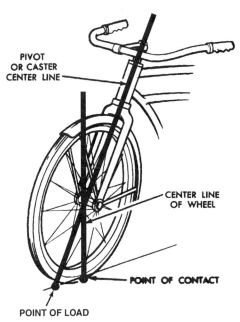

Figure 19–21 Caster effect can be demonstrated using a furniture caster or bicycle fork. In each case the pivot center (comparable to the steering axis) leads the contact point of the tire; road drag will cause the tire to follow the pivot axis. *(Courtesy of Ford Motor Company)*

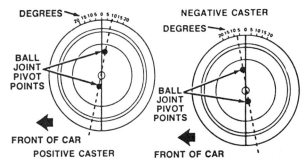

Figure 19–22 Caster is measured in degrees; if the steering axis is exactly vertical, there is 0 degrees of caster. *(Courtesy of Ford Motor Company)*

stronger. When the furniture is moved, wheel drag will cause the wheel to fall back and follow the vertical pivot. This action brings the wheel in alignment with the direction of travel. To get caster effect on a bicycle, the steering axis (front fork pivots) is inclined so the steering axis ends up ahead of the road contact of the tire. More inclination of the steering will increase the caster effect, and this will cause the vehicle to be more stable while riding straight (Figure 19–21).

If the steering axis is vertical (in a sideways plane), there is 0-degree caster. The actual amount of caster equals the amount of lean of the steering axis; 5 degrees of caster tells us the steering axis is leaning 5 degrees. Negative or positive tells us which way the steering axis is leaning. Forward at the top is negative caster, and rearward is positive caster. Positive caster angles are much more common because this is a more stable condition.

Caster is measured in degrees between the steering axis, viewed from the side, and true vertical. Most domestic cars specify a caster angle of somewhere between 0 and 3 or 4 degrees positive; some imported cars use as much as 10 degrees of caster. Unless they are specified as being negative, most wheel alignment angles are considered to be positive. A few cars use 1 or 2 degrees of negative caster (Figure 19–22).

19.13 Caster Effects

Positive caster places the ground level point of the steering axis toward the forward end of the road-contact patch of the tire. The drag of the tire on the road relative to the steering axis has the same effect that the drag of the wheel has on the pivot axis of the furniture caster. The ground level point of the steering axis is also affected by SAI, which we will study later in this chapter (Figure 19–23).

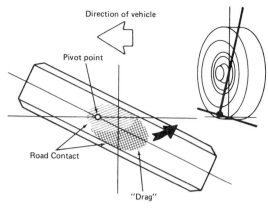

Figure 19–23 With positive caster, the road-contact point of the steering axis leads the contact patch of the tire much like that of a furniture caster. Road drag tries to keep the tire in a straight-ahead position. *(Courtesy of Ford Motor Company)*

Caster causes a tire's camber angle to change as the tires are turned to the right or left. This is often called **caster roll.** A right turn causes the camber on the right tire to become more positive and the left tire camber to become more negative (with positive caster); the camber changes are just the opposite with negative caster. This camber change is considered when selecting caster angles for competition cars raced on road circuits (Figure 19–24).

Negative caster places the ground level point of the steering axis toward the rear of the tire contact patch. Placing the steering axis behind the center of the contact makes the car easier to steer. This effect can be illustrated by comparing the effort required to steer while the car is going forward and backward. Some manufacturers specify negative caster on their cars with manual steering gears to provide easier steering. Positive caster is nearly always used on cars equipped with power steering.

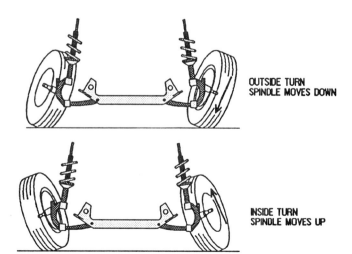

Figure 19–24 Because of the angle of the steering axis, positive caster causes a camber angle change on turns. *(Reprinted by permission of Hunter Engineering Company)*

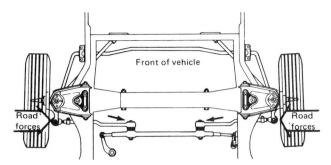

Figure 19–25 Caster places a load on the tie-rod, pushing or pulling across the car from each tire; if the loads are equal, there is no turning force. *(Courtesy of Chrysler Corporation)*

Higher caster settings can increase the amount of road shock because they align the direction of bump movement with the steering axis.

On an automobile, the caster effect of one wheel acts in opposition to the caster effect of the other wheel. These opposing forces, like those of toe, work against each other through the tie-rod. If the forces are equal, the tie-rod will be centered and the car will go straight ahead. If the forces are not equal, the car will pull or turn away from the side of the car with the most positive caster (Figure 19–25). A caster spread greater than 1/2 degree will usually cause a pulling problem.

19.14 Double Ball Joint Steering Axis

Some vehicles use separate control arms, each with an outer ball joint and inner pivot at the bottom, top, or both ends of the steering axis. This creates an instant center called a virtual steer center for the steering axis at the point where lines drawn through the center of each control arm meet. The effect is that the steering axis moves laterally and longitudinally as the car is steered, which, in turn, changes the steering axis inclination, caster, and camber along with other factors (Figure 19–26). Though more expensive, this feature, introduced on performance-oriented European cars, adds to the drivability and handling quality of the vehicle.

19.15 Factors That Affect Caster

Like toe, caster is affected by the road drag of the tire. Tires with a greater rolling resistance generate a greater caster effect than tires with less rolling resistance. An example of this can be seen by comparing the caster specifications for 1975–1980 model cars with those

Chapter 19 ■ Wheel Alignment: Front and Rear

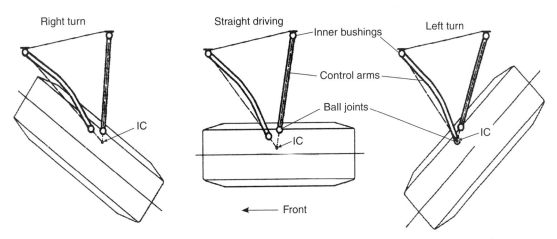

Figure 19–26 The lower control arms of a double ball joint steering assembly and the instant center (IC), which is the lower end of the steering axis. Note how the IC and the relative positions of the two ball joints move as the spindle is turned for steering.

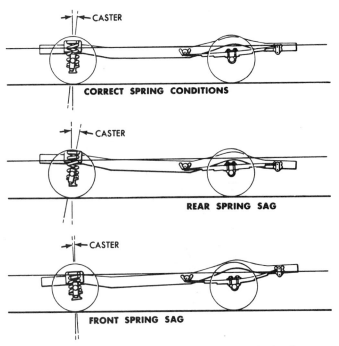

Figure 19–27 If the fore–aft angle of the vehicle changes, caster angle will change with it; this type of change often occurs from rear spring sag or when weight is added or removed. *(Courtesy of Ford Motor Company)*

made before 1970. The older cars, with less caster, used bias-ply tires while the newer ones are designed for radial tires and usually specify more caster.

Caster is also affected by the frame angle of the vehicle. Adding a heavy load to the trunk of a car usually lowers the back of the car. This lowering changes the frame or body angle and, therefore, the angle of the steering knuckle and caster. Sagging rear springs also increase caster, making it more positive. Raising the back of the vehicle or lowering the front causes caster to become more negative. The amount of caster change relative to the change in height is determined by the car's wheelbase. If you want to know exactly how much, use this formula to compute the actual amount of change: Wheelbase × 2 × π ÷ 360 = change in height that causes a 1-degree change in caster (Figure 19–27).

19.16 Caster and Road Crown

Caster is often adjusted so it is slightly more positive (1/4 to 1/2 degree) on the right wheel to offset the effects of the crown in the road. If the right wheel has the correct amount of more positive caster than the left wheel (called **caster spread** or **cross caster**), the car's caster pull (toward the left) will equal the road crown pull (toward the right), and the car will go straight down the road. Many alignment technicians prefer to use caster rather than camber to correct for road crown because caster angle does not affect tire wear (Figure 19–28).

The amount of caster spread used on a car varies with the crown of the road, the weight of the car, and the road drag of the tire. This is often a trial-and-error adjustment until experience teaches you to evaluate the vehicle and where it is driven.

19.17 Effects of Too Little or Too Much Caster

The correct amount of caster gives us a car that drives straight down the road, requires only a little or occasional correction to go straight, and steers easily. Too little caster gives us a car that **wanders**; it goes right or left and requires continuous correction to travel straight. Too much caster gives us a car that travels straight, with no

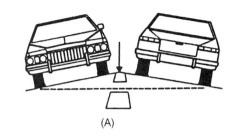
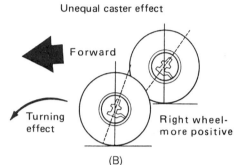

Figure 19–28 Road crown (A) often causes a car to steer toward the side of the road; more positive caster on the right tire (B) can be used to offset the effects of road crown. *(Courtesy of SPX Corporation, Aftermarket Tool and Equipment Group)*

wander, but is harder to steer. A high amount of caster can cause excessive wear at the edges of the tire tread because of caster roll if cars are driven in towns with many corners. Too much caster might even cause **tramp**, also called **shimmy**, especially in vehicles with a solid axle. Tramp is a violent and continuous turning of the steering back and forth from right to left to right, continuing on and on. Tramp places the car in a potentially dangerous situation; the driver of a shimmying car should slow down immediately. Many alignment technicians vary the amount of caster to tailor the steering to the most stable point.

19.18 Steering Axis Inclination

Like caster, **SAI** is an angle of the steering axis and is a directional control angle. It is a more powerful directional control angle than caster. The steering axis is always angled outward at the bottom so the ground level end is somewhere under the contact patch of the tire. SAI, with camber, controls scrub radius to minimize the effects that bumps and chuckholes have on steering (Figure 19–29). This angle is also called **steering axis**

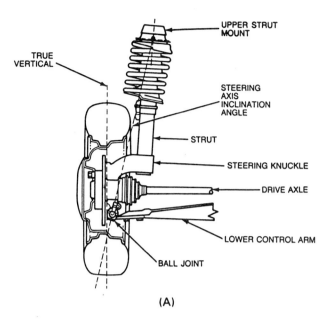
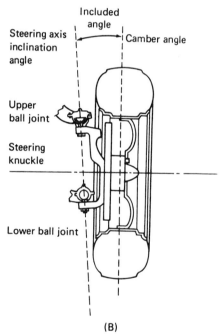

Figure 19–29 Steering axis inclination on a strut (A) and S-L A suspension (B). Note that the strut is on a FWD car so there is a negative scrub radius. *(Courtesy of Chrysler Corporation)*

angle (SAA), ball joint inclination or **angle (BJI or BJA),** or **kingpin inclination** or **angle (KPI or KPA).**

SAI works through our simplest and most reliable physical force, gravity. Because of the angle of the steering knuckle, the spindle lowers or drops (relative to the car) when the wheels are turned from straight ahead. A similar type of height change will also occur because of the caster angle (Figure 19–30). The end of the spindle cannot drop; it is supported by the tire and wheel. So the effect of SAI during a turn is to lift the car. This can be seen if you watch the front fender of a car rise and fall as

Chapter 19 ■ Wheel Alignment: Front and Rear

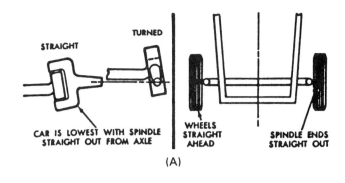

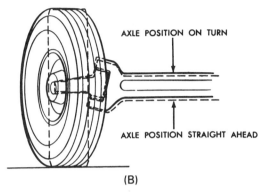

Figure 19-30 Because of SAI, a vehicle is lifted whenever the front tires are turned from straight ahead; gravity acting on the car will always try to pull the car to its lowest position, with the tires straight ahead. *(Courtesy of Ford Motor Company)*

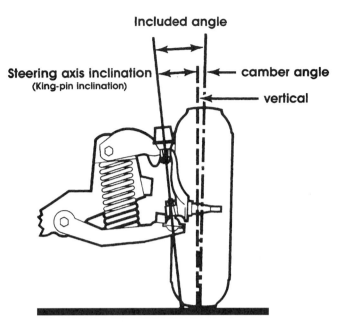

Figure 19-32 The included angle is the amount of angle between the steering axis and the centerline of the tire. *(Reprinted by permission of Hunter Engineering Company)*

the wheels are turned into a turn and back. Gravity is always trying to pull the car to the lowest position possible, which, because of SAI, is straight ahead. When you place a car's front tires on turntables, you can demonstrate SAI by turning the tires into a sharp turn and then releasing them. As shown in Figure 19-31, they will tend to turn

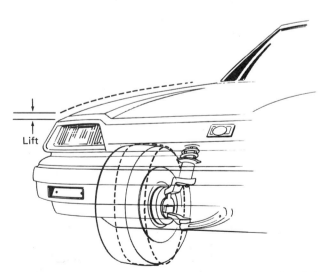

Figure 19-31 As the steering is turned, the angle of the spindle lifts the car, and the car's weight tries to turn the wheel back to straight ahead.

back toward a straight-ahead position. Most cars have an SAI angle of about 7 to 12 degrees; a few have angles as wide as 16 degrees.

19.19 Included Angle

The SAI specification can be given in degrees alone, relative to true vertical, or as part of the **included angle** (Figure 19-32). The included angle is the angle between SAI and camber; it includes camber and SAI in one angle. If camber is positive, SAI will equal the included angle minus camber; for example, a 7-degree included angle minus 1/2-degree camber = 6½-degree SAI. If the camber angle is negative, SAI will equal the included angle plus the camber angle; for example, a 7-degree included angle plus −1/2-degree camber = 7½-degree SAI (Figure 19-33). We go to the trouble of manipulating camber from the included angle because our measuring equipment measures both camber and SAI from true vertical, and different manufacturers print their specifications in different manners. It should be noted that all alignment-measuring equipment does not measure SAI. Some manufacturers publish separate camber and SAI specifications (which makes comparison easy); others publish camber and included angle specifications (making you compute the SAI angle).

SAI is not always found with the other alignment specifications because, traditionally, this angle is not always checked during a front-end alignment. A change in

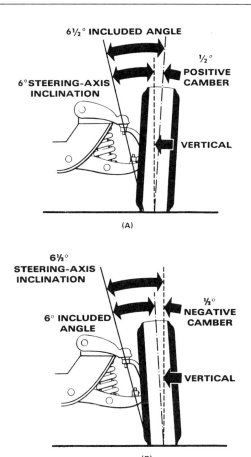

Figure 19–33 If the camber angle is positive, it is added to SAI to get the included angle (A: 6 + 1/2 = 6½); if camber is negative, it is subtracted from SAI to get the included angle (B: 6½ - 1/2 = 6). *(Reprinted by permission of Hunter Engineering Company)*

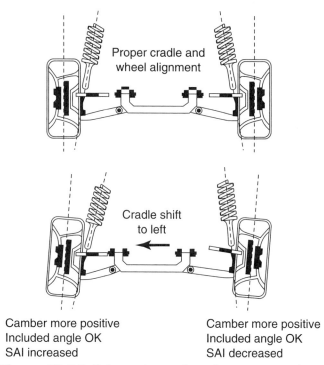

Figure 19–34 If the engine cradle or front cross member is shifted to the left, it will cause a camber change on each wheel but will not change the included angles. A problem with the strut or steering knuckle is indicated if the included angle is out of specification.

SAI, without a change in camber, would require that the motorist bend the steering knuckle or spindle without damaging any other suspension component. This is not very probable, except for with some strut suspensions, so many alignment technicians are not in the habit of checking SAI. Competent technicians check SAI when there is a problem such as wander, pull, or excess road shock that is not cured by the normal alignment steps.

SAI is a very helpful diagnostic angle when the alignment technician is trying to correct a camber problem on a car with a nonadjustable strut suspension. If camber is not adjustable, the correct way to obtain the right camber angle is to replace the parts that are worn or bent. It is sometimes difficult to determine exactly which parts are or are not bent. If SAI is at the wrong angle, one of the pivots, either the upper or lower, is in the wrong location. If the included angle is correct and the camber angle is wrong, the strut or steering knuckle is not bent, and it will not do any good to replace it. If the included angle is not correct, the strut, or at lease the spindle, is bent and should be replaced or corrected (Figure 19–34).

19.20 Toe-Out on Turns

As its name implies, toe-out on turns causes the front wheels to toe outward when a car turns a corner. This is necessary because the inner wheel has to make a sharper turn than the outer wheel; the inner wheel travels around a smaller circle. This angle is also called turning angle or turning radius. It is sometimes referred to as **Ackerman angle** because the linkage that causes toe-out on turns was developed by Rudolph Ackerman (sometime between 1818 and 1820) (Figure 19–35).

The **Ackerman linkage,** which causes toe-out on turns, consists of a pivoting steering axis with nonparallel steering arms (Figure 19–36). If these steering arms were parallel, the two front wheels would turn exactly the same amount during a turn, and one or both tires would have to scrub or scuff when turning corners. To understand toe-out on turns, we have to look at what happens when a lever moves. We can accurately think of the steering arm as a lever and the steering axis as the fulcrum. When a tie-rod moves the end of the steering arm across the car, the end of the steering arm also moves closer to or farther away from the axle. On the outside wheel, the end of the steering arm moves slightly away from the axle as it moves toward the outside of the car; it will pass through an arc where most of

Chapter 19 ■ Wheel Alignment: Front and Rear

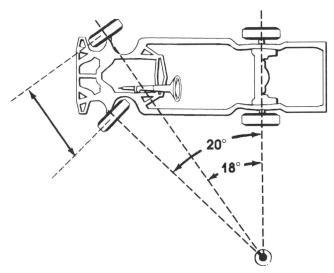

Figure 19-35 To keep from scuffing, a tire must be at a 90-degree angle to the center point of the turn; this makes it necessary for the front tires to toe out on a turn. *(Reprinted by permission of Hunter Engineering Company)*

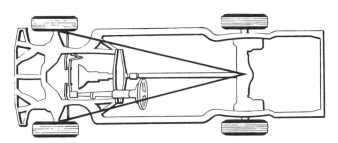

Figure 19-36 Toe-out on turns is a result of Ackerman angle, which is an angle of the steering arms pointing toward the center of the rear axle. *(Reprinted by permission of Hunter Engineering Company)*

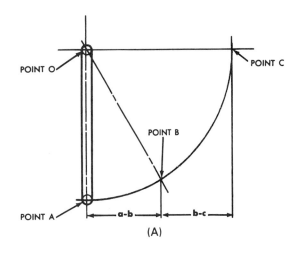

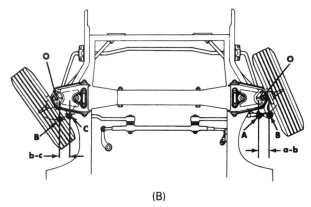

Figure 19-37 Note how distance a-b is equal to b-c, but the distance from point A to point B is much less than from point B to point C, where the lever is moving upward as well as sideways. Distances a-b and b-c compare with the sideways movement of the tie-rods, and the angles compare with the turning angles of the tires in the lower picture. *(Courtesy of Ford Motor Company)*

 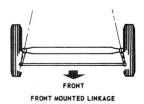

Figure 19-38 When the tie-rods are mounted behind the axle (rear steer), the steering arms angle inward; when the tie-rods are in front of the axle (front steer), they should angle outward to get the correct Ackerman angle. *(Courtesy of Ammco Tools, Inc.)*

the motion is across the car (parallel to the axle). On the other side of the car, the steering arm is passing through an arc toward the axle, as well as across the car. This added motion—toward the axle—produces a greater turn on the inside wheel (Figure 19–37).

As a car is driven in a circle, the center of a turn is almost in line with the rear axle. Each tire must be at a right angle (90 degrees) to a line from the center of the circle to the center of the tire or else the tire will scrub or scuff.

To produce these turning angles, the steering arms are mounted onto the steering knuckles so they follow a line that points approximately toward the center of the rear axle. If the steering arms are mounted in front of the axle, they must form this same angle. This is sometimes difficult because the tie-rods usually interfere with the tires or brakes, a common problem on many 4WD vehicles, causing tire scuff and wear (Figure 19–38).

The amount of turning difference in the angles of the front tires is usually about 1 or 2 degrees when one of the tires turns 20 degrees. The actual amount of toe-out on turns depends on the track width and the wheelbase of the vehicle.

Figure 19-39 This vehicle is making a sharp right turn; the faint scrub marks (arrow) behind the outer edge of the tire are a result of incorrect toe-out on turns.

19.21 Toe-Out on Turns: Problems

If the front wheels do not turn to the correct angles, one or both of them will have to scrub as they turn a corner (Figure 19-39). Sometimes this scrubbing can be heard as tire squeal as the car makes a slow turn on a very smooth, slick surface; rubber scuff marks are also evidence of this problem. This tire scrub tends to roll rubber off the edge of the tire tread, giving a wear pattern that is similar to camber wear but with a more rounded outside edge.

The procedures for measuring and adjusting (when possible) toe-out on turns are described in Chapter 34.

19.22 Setback

Setback should be considered during any discussion of wheel alignment even though it is not a normal wheel alignment angle; it is an important diagnostic angle. Most people believe that a car is made geometrically square or rectangular, with each of the four tires in perfect placement at each corner. This is true with some cars but not true with many others. Some cars have a wider track at the front or rear. Some cars, through damage, wear, or design, have one of the axles at an angle

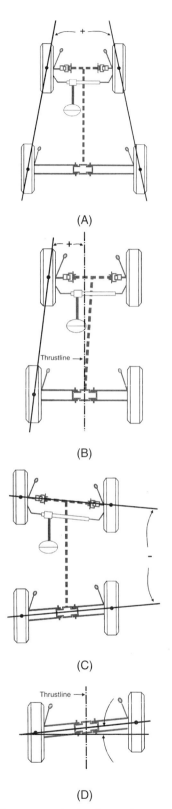

Figure 19-40 Many vehicles have the tires placed at the corners of a rectangle when viewed from above. Some vehicles are built with a track width difference (A). Some problems can be a lateral offset, with the tires at the front or rear to one side (B); a wheelbase difference, in which the wheelbase on one side is shorter than that on the other side (C); and a setback of a front or rear tire (D). *(Reprinted by permission of Hunter Engineering Company)*

Chapter 19 ■ Wheel Alignment: Front and Rear

other than 90 degrees to the car's centerline. On rear axles, this is called track or thrust and is discussed in Section 19.24. When one of the front tires is behind the other, it is called *setback*. Setback can be the result of worn suspension or accident damage; some new cars are built with as much as 1½ in. (38 mm) of setback of the right front wheel. This setback is intended to divert the car sideways during a severe accident to help protect the occupants from severe injury or reduce the effects of torque steer (Figure 19–40).

Improper setback on a driving axle can cause **torque steer:** the vehicle turns to one side during a hard acceleration. It can also cause odd tire wear or handling difficulties. Setback can also cause difficulties in adjusting toe to get a straight steering wheel; this issue is discussed more completely in Chapter 34.

19.23 Rear-Wheel Alignment

Because the rear wheels of most cars do not turn or steer, we only worry about two alignment angles: camber and toe (Figure 19–41). These are the tire wearing angles, and they can cause rear-wheel tire wear as easily as front-wheel tire wear. Those cars with four-wheel steering have such slight amounts of rear-wheel steering that caster, SAI, and toe-out on turns are not concerns.

With most RWD cars, rear-wheel alignment is controlled by the rear axle housing; its size and strength usually keep the wheels aligned without difficulty. RWD cars with IRS usually have provision for camber and toe adjustments (Figure 19–42). Most FWD cars also have methods in which rear-wheel camber or toe can be adjusted. **Four-wheel alignment** has become very common with the increased popularity of FWD cars (Figure 19–43).

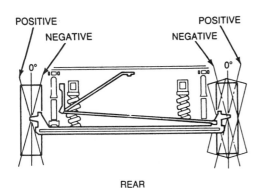

Figure 19–41 Rear-wheel camber is the same as front-wheel camber; toe is also the same. *(Courtesy of Chrysler Corporation)*

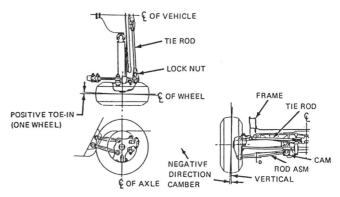

Figure 19–42 Camber on this Corvette rear suspension can be adjusted by turning a cam; toe is adjusted by changing the length of the tie-rod. *(Courtesy of General Motors Corporation, Service Technology Group)*

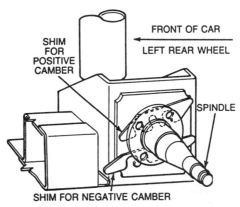

Figure 19–43 Partial-contact shims (shown) or full-contact tapered shims can be placed behind the rear spindle to adjust camber (as shown) or toe. *(Courtesy of Chrysler Corporation)*

19.24 Track or Thrust Line

In addition to camber and toe, rear-wheel alignment should be concerned with **track** or **thrust.** If the rear wheels are parallel to each other and to the centerline of the car, they follow (track) directly behind the front wheels, and the car is straight as it goes down the road. If the rear tires are parallel to each other but not parallel to the car's centerline, the car goes down the road slightly sideways, with the rear-wheel tracks to the right or left of the front-wheel tracks. The car follows a line that is parallel to the average toe of the rear wheels, called the thrust line. If this line does not run down the center of the car, as it should, the driver turns the steering wheel slightly to compensate, and the car **dog tracks** (goes slightly sideways) down the road. Some cars are adversely affected by thrust angles as small as 0.06 degree (Figure 19–44).

The procedures for measuring and adjusting (when possible) rear-wheel alignment are described in Chapter 34.

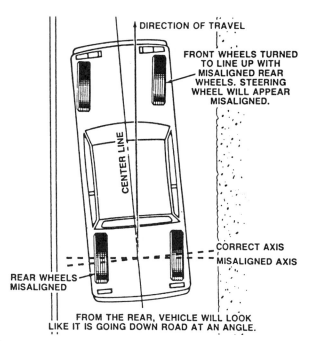

Figure 19–44 If the rear wheels are not aligned to the centerline of the car, the car will go down the road slightly sideways, or dog track; a vehicle's direction of travel is the same as the rear-wheel thrust line. *(Courtesy of Ford Motor Company)*

19.25 Practice Diagnosis

You are working in a wheel alignment shop and encounter the following problems:

CASE 1: A customer has brought in her 3-year-old Ford Mustang with a complaint of a pull to the right. You have tried cross switching the two front tires and then from the front to the back, but the pull is still there. What should you do next? If this is a wheel alignment problem, what could be causing it?

CASE 2: A 1987 Toyota Tercel was brought in for a wheel alignment with a complaint of tire wear. Both front tires show a scuffing tread wear at the outer edges, and the left rear tire shows an uneven wear across the tread. What alignment angles could be causing this wear?

CASE 3: A customer has brought in his 5-year-old Ford F150 pickup with a complaint of wandering; he has to keep his hands on the steering wheel to keep correcting the travel direction. What faults could be causing this problem?

Terms to Know

Ackerman angle
Ackerman linkage
ball joint inclination or angle (BJI or BJA)
bump steer
camber
camber angles
camber spread
caster
caster roll
caster spread
cross camber
cross-car reading
cross caster
diagnostic angles
directional control angles
dog tracks
featheredge
four-wheel alignment
included angle
kingpin inclination or angle (KPI or KPA)
negative camber
negative caster
negative scrub radius
positive camber
positive caster
positive scrub radius
sawtooth
scrub radius
setback
shimmy
spread
steering axis
steering axis angle (SAA)
steering axis inclination (SAI)
thrust
thrust line
tire position angles
toe
toe angles
toe-in
toe-out
toe-out on turns
torque steer
track
tramp
turning angle
turning radius
wanders

Review Questions

1. Statement A: Caster and camber are angles of the tire.
   Statement B: Caster and SAI are angles of the steering axis.
   Which statement is correct?
   a. A only
   b. B only
   c. Both A and B
   d. Neither A nor B

2. If the front spindle is at a slight downward angle at the outer end:
   A. Camber is negative
   B. Caster is positive
   Which option best completes the statement?
   a. A only
   b. B only
   c. Both A and B
   d. Neither A nor B

3. The major reason for using positive camber on front tires is to:
   a. Produce scrub radius
   b. Place more of the car's load on the outer wheel bearing
   c. Offset road crown effects
   d. None of these

4. Technician A says that excessively positive camber on the left front tire will cause a pull to the left.
   Technician B says that excessively positive camber will cause the tire to wear at the inner edge of the tread.
   Who is correct?
   a. A only
   b. B only
   c. Both A and B
   d. Neither A nor B

5. The distance between the center of the tires is 54½ in. at the rear and 54⅜ in. at the front. This car has:
   a. 1/16 in. of toe-in
   b. 1/8 in. of toe-out
   c. 1/8 in. of toe-in
   d. None of these

6. Technician A says that toe-in will increase if the tie-rods (rear mounted) are made longer.
   Technician B says that the tie-rod sleeves are used to adjust toe-out on turns.
   Who is correct?
   a. A only
   b. B only
   c. Both A and B
   d. Neither A nor B

7. Incorrect toe settings cause a characteristic tread wear pattern that:
   a. Has a featheredge at one side of each tread bar
   b. Is worn heavy at one side of the tread
   c. Is worn heavy at both sides of the tread
   d. Either a or b

8. The front tires of a RWD car are normally toed in slightly so:
   A. A slight amount of toe-out will occur as road forces push on the tire
   B. Camber angle will not cause a pull
   Which option best completes the statement?
   a. A only
   b. B only
   c. Both A and B
   d. Neither A nor B

9. Statement A: Caster is a directional control angle that helps keep the car going straight down the road.
   Statement B: Caster is caused by the forward or rearward tilt of the steering axis.
   Which statement is correct?
   a. A only
   b. B only
   c. Both A and B
   d. Neither A nor B

10. Technician A says that negative caster is usually used on cars with power steering.
    Technician B says that negative caster places the upper ball joint to the rear of the lower one.
    Who is correct?
    a. A only
    b. B only
    c. Both A and B
    d. Neither A nor B

11. If a lot of weight is placed in the trunk of a car, causing the back end to lower a few inches, the:
    a. Caster angle will become more negative
    b. Caster angle will become more positive
    c. Camber angle will become more positive
    d. None of these will occur

12. Technician A says that road crown pull can be offset by adjusting the left wheel's caster to be slightly more positive.
    Technician B says that this is called caster spread.
    Who is correct?
    a. A only
    b. B only
    c. Both A and B
    d. Neither A nor B

13. Too much caster can cause:
    A. Vehicle wander
    B. Hard steering
    Which option best completes the statement?
    a. A only
    b. B only
    c. Both A and B
    d. Neither A nor B

14. The angle of the steering axis, leaning inward at the top, is called:
    a. SAI
    b. BJA
    c. KPI
    d. All of these

15. Statement A: SAI causes the front wheels to seek a straight-ahead position.
    Statement B: SAI is a directional control angle.
    Which statement is correct?
    a. A only
    b. B only
    c. Both A and B
    d. Neither A nor B

16. The included angle equals the SAI angle:
    A. Plus the camber angle (if positive)
    B. Minus the camber angle (if negative)
    Which option best completes the statement?
    a. A only
    b. B only
    c. Both A and B
    d. Neither A nor B

17. The included angle is controlled by the:
    a. Position of the ball joints
    b. Angle of the spindle
    c. Shape and dimensions of the steering knuckle
    d. None of these

18. The angle created by the steering arms pointing inward at the rear is called:
    a. Ackerman angle
    b. Turning radius
    c. Toe-out on turns
    d. All of these

19. Technician A says that the front tires need to toe out on turns so they will steer more easily. Technician B says that the outside tires turn sharper than the inside tires.
    Who is correct?
    a. A only
    b. B only
    c. Both A and B
    d. Neither A nor B

20. The thrust line is:
    A. A line that is between and parallel to the rear tires
    B. The direction that the car goes while traveling straight ahead
    Which option best completes the statement?
    a. A only
    b. B only
    c. Both A and B
    d. Neither A nor B

Math Questions

1. The measurement across the car at the rear of the tire is 62 in., and it measures 61⅞ in. at the front. How much toe is there? Is this toe-in or toe-out?

2. The tire is 26 in. tall and is leaning 1/4 in. from vertical; the top is inward. What is the camber angle? Is this positive or negative camber?

3. The car you are aligning has a camber angle of +3/4 degree and an SAI angle of 7½ degrees. What is the included angle?

4. The station wagon you are aligning is 2 in. lower at the rear because of a load. The wheelbase is 115 in. What is the amount of change for the frame angle? What will this change do to the caster angle?

20 Brake System Service

Objectives

Upon completion and review of this chapter, you should be able to:

- ❏ Perform a detailed brake inspection.
- ❏ Check brake pedal operation for excessive side motion, sticking, or binding while checking the pedal for proper free travel and reserve.
- ❏ Check brake pedal feel for indications of fluid leakage or sponginess.
- ❏ Inspect a master cylinder for correct fluid level and fluid leaks.
- ❏ Inspect a master cylinder reservoir for proper fluid motions during brake application and release.
- ❏ Inspect rigid and flexible brake lines and fittings for leaks, wear, and damage.
- ❏ Remove a brake drum, inspect for lining or drum wear, and determine the cause of poor stopping, noise, pull, or other problems that can be visually located.
- ❏ Remove a brake caliper, inspect for lining or rotor wear, and determine the cause of poor stopping, noise, pull, or other problems that can be visually located.
- ❏ Remove and replace a wheel and tighten the lug nuts.
- ❏ Road test brakes to check for correct operation and identify any problems.
- ❏ Use a diagnostic chart to determine the possible cause of a brake problem.
- ❏ Have familiarity with metal surface finishes.
- ❏ Understand basic electrical troubleshooting procedures.
- ❏ Perform the ASE tasks relating to brake system inspection and problem diagnosis (see Appendix A).

20.1 Introduction

After determining that a brake component is faulty, the next step is to repair that particular component or the whole system. Brake systems can be repaired one component at a time, but most of the repairs are made as a *brake job*. Specific problems such as brake pull, pedal pulsation, or a failing power booster can often be corrected by repairing the faulty component. A good technician will not fix anything that does not need fixing. However, a car that is 5 years old or older, has been driven 50,000 mi. (80,450 km) or more, and has a lining that is close to wearing out is becoming a candidate for a brake job—a total brake overhaul. As mentioned earlier, it should be the intent of the brake technician to produce a car with *new-car brake performance*. This is the same goal as that of a technician working on an engine overhaul, a tune-up, a front-end repair and alignment, and so on. Because of the safety ramifications, the brake technician must be even more concerned about top-quality performance once the job is completed. In fact, the brake repair must not only produce a car that stops well, but one that stops well enough to meet unforeseen emergencies for a reasonable length of time. A quality brake job—like the brakes of a new car—will allow a car to stop well for the life of the new friction material.

Unfortunately, there is no simple and inexpensive way to thoroughly test a brake system, especially the hydraulic system, other than to dismantle it and make visual checks. It would be nice to be able to plug a meter into the system to measure its present and future effectiveness, but no such instrument is currently available. As we will see, the sight and other senses of the brake technician are the major means for determining whether the brake system and its parts are good or bad.

Remember that customers have different perceptions of how a car stops. We know that a vehicle is capable of stopping at a certain rate before wheel lockup, that a certain pedal pressure is required, and that the pedal feel can vary among car models. Competent technicians should become familiar with the proper braking action of the vehicle models they work on so that valid recommendations can be made to the motorist (Figure 20–1).

At this time, there is no trade standard for a brake job. However, most experts recommend that it include the following:

- Replacement of all friction material
- Careful checking, reconditioning, and/or resurfacing of friction surfaces on the rotors and drums
- Replacement of brake shoe return springs
- Cleaning, inspection, and lubrication of backing plates
- Cleaning, inspection, and replacement of certain caliper mounting hardware parts and lubrication of caliper mounting hardware
- Disassembly, cleaning, and inspection of all major hydraulic units—wheel cylinders, calipers, and master cylinder; note that some master cylinders are nonrebuildable
- Replacement of all old brake fluid and bleeding of all air from the system
- Repacking and adjusting of serviceable wheel bearings
- Inspection and/or replacement of hub and axle grease seals
- Inspection of all steel and rubber brake lines
- Adjustment of the parking brake
- Checking and adjustment, if necessary, of brake pedal free travel and the stoplight switch
- Checking of the operation of the hydraulic valves and switches

A coalition group of the Brake Manufacturers' Council (BMC) has developed a set of uniform guidelines to be followed during a brake inspection as part of the **Motorist Assurance Program (MAP).** These guidelines help ensure honest and proper recommendations to the motorist as to the condition of the braking system. They also help define the fine line that dictates whether the part *must* be replaced or *should* be replaced. According to MAP, a part must be replaced or repaired if it no longer performs its intended purpose, does not meet its design specification, or is missing. You should suggest that a part be replaced if it is close to the end of its useful life; if the motorist wants better performance, less noise, or no worry; or to comply with the vehicle manufacturer's maintenance recommendation. The MAP guidelines for brake service and repair can be obtained from one of the member companies or the Motorist Assurance Program, Washington, D.C.

Most brake service operations begin with an inspection and/or a set procedure for diagnosing a particular problem so that the best, quickest, and least expensive repairs can be performed. This chapter describes these procedures.

SAFETY TIP: Brake repair must produce proper and safe brake operation. To produce brake operation and to protect yourself from injury, the following general precautions should always be followed:

- Wear eye protection.
- Follow the repair procedure that is recommended by the vehicle manufacturer.
- Support the car in a safe and secure manner before working under it.
- Do not stir up brake dust; protect yourself so you do not breathe this dust. The use of compressed air to blow brake assemblies clean is strictly prohibited. The asbestos fibers it might contain could be harmful to you. Use proper asbestos dust control methods and equipment.
- When using an air hose for cleaning up or drying parts, be careful of the air blast and the particles it might contain.
- Use the proper tool for the job and use that tool in the correct manner.

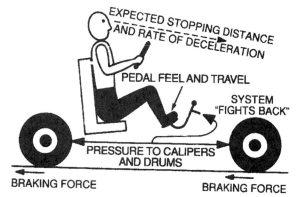

Figure 20–1 When a driver applies the brakes, a combination of feelings and perceptions indicates brake operation. If things do not feel or seem right, there may be a problem. *(Courtesy of Ford Motor Company)*

Chapter 20 ■ Brake System Service

- Do not allow a strain to be placed on a brake hose.
- Do not allow grease, oil, solvent, or brake fluid to get on the brake lining or the braking surfaces of the rotors or drums.
- Do not let oil or solvents enter the hydraulic system.
- Do not allow brake fluid to be sprayed or squirted into your face or eyes.
- Do not allow brake fluid to be spilled on painted surfaces.
- Replace all damaged, worn, or bent parts.
- Check replacement parts against the old parts to ensure exact or better-quality replacement.
- Make certain that all replacement bolts and nuts are of the same size, type, and grade as the OEM parts.
- Tighten all bolts, nuts, fittings, and bleeder screws to the correct torque and lock them in place by the correct method.
- Do not move the car after working on it until a firm brake pedal is obtained.
- Carefully road test the car after working on it to make certain that it is operating safely and correctly.

20.2 Brake Inspection

A brake inspection is performed to determine the condition of the brake system. The inspection can determine the cause of a complaint or serve as preventative maintenance to determine when and if service is necessary (Figure 20–2).

At the same time that the brakes are being inspected, the competent technician will also note the condition of the car's tires, wheel bearings, shock absorbers, wheel alignment (by noting tire wear), suspension system, and drivetrain. The average car owner has little knowledge of what is happening under his or her car. In most cases, this person is very appreciative of any information that will help ensure safe and efficient operation of the car. Most technicians feel a strong obligation to inform the motorist that a part is failing or will probably fail in the near future.

Most of the operations performed during a brake inspection are described more completely in the chapters dealing with drum, disc, hydraulic system, or parking brake service. While making a brake inspection, it is good practice to follow a set procedure and mark your findings on an inspection form or checklist. This helps ensure that no important steps are forgotten, and it also provides a record for the car owner (Figure 20–3).

To perform a brake inspection, you should:

1. Depress and release the brake pedal several times (the engine should be running on cars with power boosters). The pedal should have 1/16 to 1/8 in. (1.5 to 3 mm) of free travel before engaging the master cylinder piston. Then it should move smoothly and quietly to a firm pedal, and there should be no excessive side-to-side motion (Figure 20–4).

2. Depress the pedal heavily. On older cars, there should be no sponginess, and the pedal should stop with at least one-half of the available pedal travel left in reserve. It should be noted that many new cars have a soft, almost spongy pedal instead of the rock-hard pedal of the past and with only about 20 percent reserve (Figure 20–5).

3. Depress the pedal moderately, about 25 to 35 lb (11 to 16 kg), for about 15 seconds, making sure it does not sink to the floor. Relax the pedal pressure to a light pressure of about 5 to 10 lb (2 to 4 kg) and hold this pressure for a little longer to make certain that the pedal does not sink under light pressure. The first check is for an external leak or an internal master cylinder leak. The second step checks for an internal master cylinder leak, often called bypassing (Figure 20–6).

4. Depress and release the pedal several times under varying amounts of pressure as you watch the warning light on the dash. Also, have an assistant

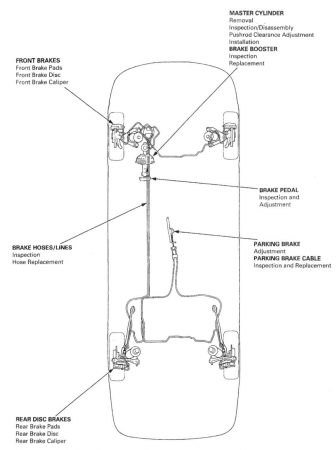

Figure 20–2 The entire brake system should be checked during an inspection. *(Courtesy of American Honda Motor Co., Inc.)*

Complete Brake Job Checklist

Customer Name _____

Address _____ Telephone Number: Work _____ Home _____

Vehicle Make _____ Year _____ Mileage _____

License Number _____

Approximate time required for service _____

Inspected by _____ Date _____

The condition of the brake system on this vehicle is as follows:

Brake System Components	Required Services	$ Estimate Parts	Labor
Disc Pads*	☐Semi-metallic ☐Conventional		
Disc Caliper*	☐Recondition ☐Replace		
Disc Hardware*	☐Replace		
Disc Rotor*	☐Recondition ☐Replace		
Grease Seals*	☐Replace		
Front Wheel Bearings*	☐Repack ☐Replace		
Wheel Cylinders*	☐Recondition ☐Replace		
Rear Brake Hardware*	☐Replace		
Brake Shoes*	☐Replace		
Brake Drums*	☐Recondition ☐Replace		
Parking Brakes	☐Adjust ☐Lubricate ☐Replace		
Power Brake Booster	☐Service ☐Replace		
Master Cylinder	☐Recondition ☐Replace		
Brake Fluid*	☐Drain old, and add new fluid		
Lines, Hoses, Combination Valve	☐Replace		
Stop Light	☐Replace Bulb ☐Replace Switch ☐Adjust ☐Replace		
Other _____			
	Total		

*Wagner's Complete Brake Job includes reconditioning or replacement of these items

McGRAW-EDISON

Figure 20–3 A checklist to use while inspecting a brake system. *(Courtesy of Wagner Brake)*

watch the stoplights. The warning light should not come on, but the stoplights should come on each time the pedal is depressed and go off each time it is released.

5. Check the brake warning light operation by cranking the engine. The light should come on as the engine is cranking. On cars equipped with ABS and/or airbags, these warning lights should also come on during cranking and remain on for a few seconds after the engine starts.

6. Apply the parking brake. The lever should not travel more than two-thirds of the available distance and should provide enough braking power to hold the car in place. Some technicians test this power by trying to drive the car with the brake applied. Check the parking brake warning light; it should come on as the parking brake is applied.

7. On cars equipped with power brakes, with the engine off, depress the brake pedal several times to exhaust the booster reserve. Hold the pedal down

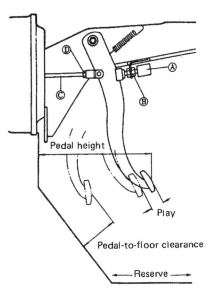

Figure 20–4 A brake pedal should have 1/16 to 1/8 in. of free play, only a small amount of side play, and a firm, hard pedal before it gets halfway to the floor. In this unit, free travel is adjusted by loosening the lock nut (D) and turning the pushrod (C). Stoplight switch (A) adjustment should be checked after adjusting free travel. *(Courtesy of Mazda Motor Corporation)*

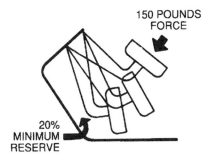

Figure 20–5 There should be a minimum of 20-percent reserve with a pedal pressure of 150 lb on this late-model vehicle. *(Courtesy of Ford Motor Company)*

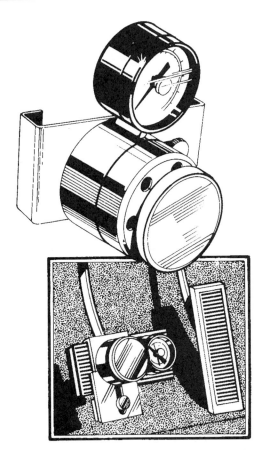

Figure 20–6 A brake pedal pressure gauge can be used to measure the force on the brake pedal. *(Courtesy of SPX Kent-Moore, Part # J-28662)*

with a light pressure and start the engine. As the engine starts, the pedal should drop slightly but noticeably. With hydraulic boosters, the pedal should drop and then rise back up (Figure 20–7). After running the engine several moments more, shut the engine off, wait 90 seconds, and depress the pedal lightly. One or more assisted brake applications should occur.

8. Check the master cylinder for external leaks at the line fittings, mounting end, or reservoir cover. Leaks at the line fittings can often be repaired by loosening and retightening the tubing nut. Fluid leaks at the booster where the master cylinder is mounted indicate internal failure of a master cylinder cup.

9. Remove the reservoir cover, note the condition of the diaphragm, and make sure the vent hole in the cover is open.

10. Check the fluid level. It should be 1/4 to 1/2 in. (6 to 13 mm) from the upper edge of the reservoir or at the level indicated by the marks on the reservoir (Figure 20–8). On cars with electrohydraulic boosters, pump the brake pedal about twenty to twenty-five times with the key off (up to forty times on some systems) to return the fluid from the accumulator to the reservoir before checking the fluid level (Figure 20–9). On master cylinders equipped with fluid level switches, make sure the wires are in good condition and have sound connections at the switch.

11. Run a clean finger around the bottom of the reservoir and check for rust, dirt, or other contamination. Clean the fluid from your finger immediately. If the fluid appears contaminated, place a sample in a clean glass jar so it can be inspected more completely. If test equipment is available, test the fluid for contamination and lowering of the boiling point (Figure 20–10).

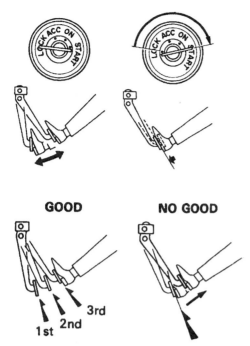

Figure 20-7 When checking a power brake system, pump the brake pedal and note the change in reserve (A). Then, with your foot on the pedal, start the engine (B); the pedal should drop slightly. After a few minutes, there should be a pedal reserve.

12. Watch the fluid in the reservoirs as the pedal is depressed. A slight spurt or swirl should occur over the compensating ports during application and also in the drum brake section of a tandem system during release. Then observe the fluid levels in both master cylinder reservoirs during several hard pedal depressions to make certain that one does not rise as the other one falls.

CAUTION *Do not lean directly over a master cylinder during brake application or release. Fluid can spurt or spray high enough to get on your face or into eyes. It is a good practice to cover the reservoir with clear plastic wrap to contain the fluid so you can observe the movement.*

NOTE: Because of the fluid bypassing through the quick-take-up valve, the fluid level in the reservoir of a quick-take-up master cylinder will normally rise as the brakes are applied and fall during their release.

13. On cars equipped with power brakes, inspect the vacuum hose and check valves, hydraulic lines, and electrical connections to make certain they are in good condition.

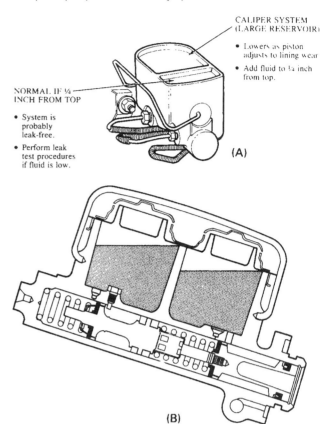

Figure 20-8 Normal brake fluid is 1/4 in. below the top of the reservoir (A); on angled master cylinders, the fluid level is almost at the top of the reservoir (B). *(A is courtesy of Ford Motor Company; B is courtesy of General Motors Corporation, Service Technology Group)*

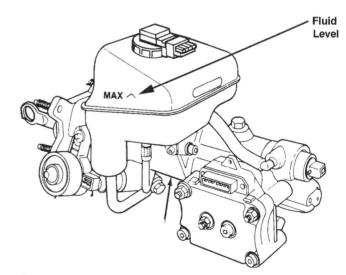

Figure 20-9 With the accumulator charged, fluid level on this plastic ABS unit reservoir should be at the maximum (MAX) level. Some units should have the accumulator discharged before you check the fluid level. *(Courtesy of Ford Motor Company)*

Chapter 20 ■ Brake System Service

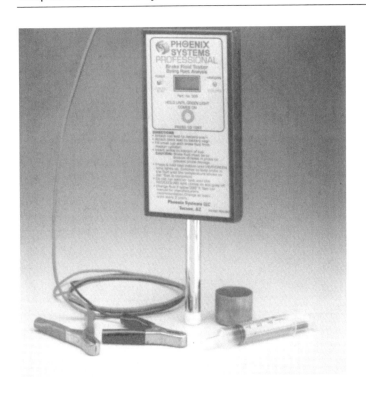

Figure 20–11 Two different types of jackstands that should be placed under a car for safe support. Never work under a vehicle supported only by a jack.

Figure 20–10 This tester checks the boiling point of the brake fluid, which provides a quick and accurate check for moisture contamination. *(Courtesy of Phoenix Systems, L.L.C.)*

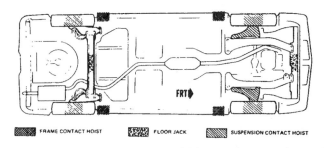

Figure 20–12 The jacking and lifting points are shown in the manufacturer or technician service manual. These points are strong enough to support the car. *(Courtesy of General Motors Corporation, Service Technology Group)*

14. Raise and securely support the car. Place a jack or the support arms of a hoist under some part of the car that is strong enough to support the weight. Some cars have a jacking pad at the bumper or on the side of the car. Other places that are commonly used are the frame, a frame cross member, a reinforced unibody box section, an axle, or a suspension arm. Lift the tire and wheel off the ground and, if using a jack, place a jackstand under some strong portion of the car (Figure 20–11).

NOTE: The rear axles of some FWD cars will bend if lifted by the center of the axle. These axles should be lifted one side at a time, with the jack placed near the spring.

NOTE: If you are unsure of where to lift the car, consult the "Lifting and Jacking" section of the car manufacturer service manual or a technician service manual (Figure 20–12).

15. Rotate and shake or rock each of the wheels as you check for excessive looseness and free rotation of the wheel or axle bearings. A rough, harsh feeling accompanied by a growling, grinding noise indicates a faulty wheel bearing (Figure 20–13).

16. Remove at least one front and one rear wheel using the following procedure. Some technicians prefer to inspect all four brake assemblies.

    a. Remove any wheel covers (often locked).
    b. Put an indexing chalk mark on the end of the lug bolt that is closest to the valve stem. This is done to ensure replacement of the wheel assembly in the same location so that no change in the balance or the runout of the tire and wheel will occur. If the valve core is positioned between two studs, mark both of them or draw a line on both the wheel and the hub (Figure 20–14).
    c. Remove the lug nuts using a six-point lug nut socket or lug wrench. If you are using an air-impact wrench, excessive speed can cause lug

Figure 20–13 The wheel bearing is quickly checked by rocking the tire as shown here. Push in at the top while pulling out at the bottom. *(Reprinted by permission of Hunter Engineering Company)*

Figure 20–14 Placing a chalk mark at the end of the wheel stud closest to the valve stem (circled) will help you replace the wheel in the correct position.

bolt or nut galling. Adjust the wrench speed or lubricate the threads to ensure that galling does not occur.

SAFETY TIP: The exhaust of an air-impact wrench can cause dust hazards by blowing asbestos fibers into the air.

    d. Remove the tire and wheel assembly.

NOTE: Steps 17 through 22 are described in more detail in Chapter 22.

Figure 20–15 Before removing a brake drum, place a chalk mark next to the marked wheel stud. This index mark helps you replace the drum in the same position.

17. Mark the drum next to the previously marked stud(s) and remove the drum (Figure 20–15).

SAFETY TIP: OSHA requirements state: "There should be no visible dust during brake inspection and repair." Removal of a brake drum can release dust and asbestos fibers. The recommended method of preventing this is to flood the brake assembly using a brake washer; rotate the drum as you thoroughly wet the inside components. Aerosol sprays and vacuum enclosures can also be used.

NOTE: On brake drums that are mounted at the front of RWD cars or at the rear of FWD cars, the drum can often be removed at the wheel bearing as a wheel, hub, and drum assembly.

NOTE: Follow the manufacturer's recommendations on drum removal if you are not sure of how to do it.

18. Check the brake lining for the amount and pattern of wear. There should be a minimum of 1/32 to 1/16 in. (0.8 to 1.5 mm) of lining at the thinnest point above the rivets on riveted shoes or above the shoe rim on bonded shoes. The wear should be even and equal across the shoe (Figure 20–16 and Figure 20–17). Also check the lining for contamination, glazing, or cracking (Figure 20–18 and Figure 20–19).

19. Check the brake springs for distortion or stretched or collapsed coils, twisted or nicked shanks, or severe discoloration (Figure 20–20).

Chapter 20 ■ Brake System Service

CHECK BRAKE SHOES

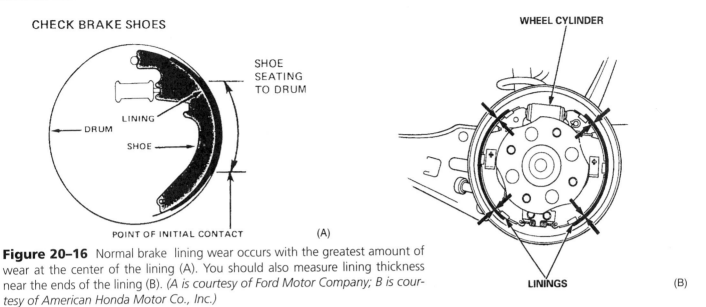

Figure 20-16 Normal brake lining wear occurs with the greatest amount of wear at the center of the lining (A). You should also measure lining thickness near the ends of the lining (B). *(A is courtesy of Ford Motor Company; B is courtesy of American Honda Motor Co., Inc.)*

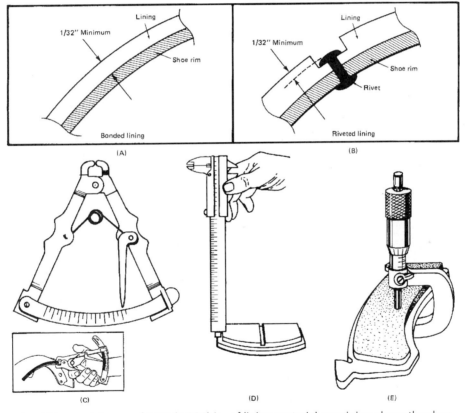

Figure 20-17 There should be a minimum of 1/32 (0.031) in. of lining material remaining above the shoe rim on bonded shoes (A) or above the rivets on riveted shoes (B). The thickness can be measured using a special gauge (C), a vernier caliper (D), or a micrometer (E). If measuring the overall width, remember to subtract the thickness of the metal backing. *(C is courtesy of KD Tools; D is courtesy of Mazda Motor Corporation)*

Check the self-adjuster and parking brake linkage for distorted or stretched parts and correct operation. Check the self-adjuster operation by prying the secondary shoe away from the anchor or pulling the cable. The adjuster mechanism should operate and turn the adjuster (Figure 20-21). Also check that the released shoe returns to its anchor.

20. Check the wheel cylinder for leakage. On units with external boots, pull a boot back to look inside. A small amount of fluid dampening is normal. Actual wetness is not.

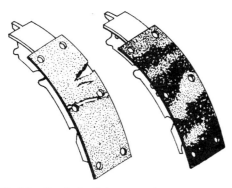

Figure 20–18 Check the lining to make sure it is not chipped, cracked, or contaminated with oil, grease, or brake fluid.

Figure 20–19 This lining is badly glazed. Note the shiny appearance, the fine cracks, and the darker color toward the center.

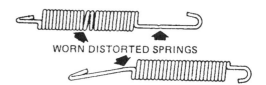

Figure 20–20 Badly bent, nicked, or stretched springs can cause dragging brakes or pull during stops. These springs should be replaced. *(Courtesy of Wagner Brake)*

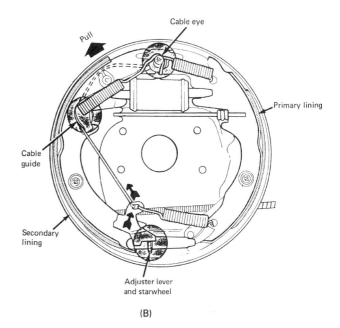

Figure 20–21 Operation of the self-adjuster on a duoservo brake can be checked by prying the secondary shoe away from the anchor (A) or pulling on the adjuster cable (B). The motion should cause the lever to turn the starwheel. *(B is courtesy of Ford Motor Company)*

21. Check the drum friction surface for cracks, unusual wear, or a worn or distorted surface. Measure the drum in several locations for size and roundness. Each measurement should be smaller than the maximum diameter indicated on the drum and within 0.010 in. (0.25 mm) of each other (Figure 20–22).

NOTE: Steps 22 through 27 are described in more detail in Chapter 23.

22. Remove the caliper in the manner recommended by the manufacturer. On some cars, it is possible to visually check the lining thickness with the caliper in place (Figure 20–23). Many technicians

Chapter 20 ■ Brake System Service

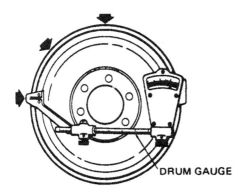

Figure 20–22 A brake drum can be measured using a drum gauge or micrometer to make sure it is not too large. Measuring at several locations helps to determine whether it is round. *(Courtesy of Wagner Brake)*

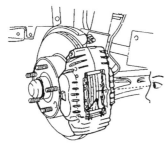

Figure 20–23 On some cars, it is possible to check lining thickness without removing the shoe. Many technicians prefer to remove the shoe for a more accurate check. *(Courtesy of Mazda Motor Corporation)*

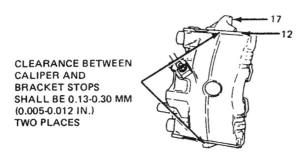

Figure 20–24 As a caliper is being removed, the clearance between the caliper and mounting bracket should be checked. *(Courtesy of General Motors Corporation, Service Technology Group)*

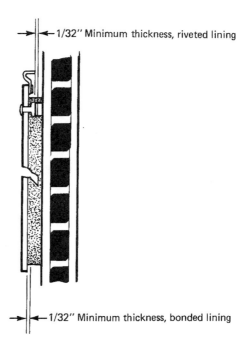

Figure 20–25 Disc brake pads should have a minimum of 1/32 (0.031) in. of lining above the pad backing on bonded shoes or above the rivet on riveted lining. The wear should be even across the lining. *(Courtesy of Wagner Brake)*

prefer to remove the caliper to allow a more complete check of the lining, rotor, and caliper. During removal of floating calipers, check the amount of clearance at the caliper-to-mount abutments (Figure 20–24).

23. Inspect the lining for wear, noting the amount and pattern. There should be a minimum of 1/32 to 1/16 in. (0.8 to 1.5 mm) of lining at the thinnest point above the rivets on riveted lining or above the backing on bonded lining (Figure 20–25). The pad wear should be even and equal from the inner edge to the outer edge, with no more than 1/16 in. (1.5 mm) more wear at one end than the other. Check the lining for cracks, contamination, or glazing.

24. Check the caliper mounting hardware for wear or distortion.

25. Check the caliper piston boot for cracks or tears and leakage. No fluid seepage is considered acceptable.

26. Check the friction surfaces of the rotor for unusual wear. Measure the rotor in several locations at the center of the friction surface. Make sure that its width is greater than the minimum width dimension indicated on the rotor (Figure 20–26). If there is a complaint of pedal pulsation or brake lockup, or if it is probable that a lining replacement will be needed, measure the amount of rotor runout and parallelism. Then compare these values with the manufacturer's specifications (Figure 20–27).

27. Replace the caliper using the manufacturer's recommended procedure, being sure to tighten the caliper mounting bolts or guide pins to the correct torque.

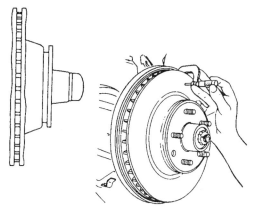

Figure 20–26 Rotor thickness is measured using a micrometer. The measurements should be taken at the center of the friction surface. Note that the rotor (left) will have various measurements indicating a parallelism problem. *(Reprinted from Mitchell Anti-Lock Brake Systems, with permission of Mitchell Repair Information, LLC)*

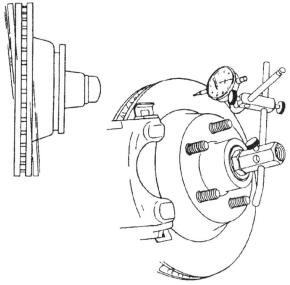

Figure 20–27 Rotor runout is measured using a dial indicator; any wobble will show up as the rotor is rotated. *(Reprinted from Mitchell Anti-Lock Brake Systems, with permission of Mitchell Repair Information, LLC)*

NOTE: When brake drag or abnormally fast pad wear is the complaint, it is a good practice to check for excessive brake drag. Do this by measuring the amount of force required to turn the rotor (with the caliper removed) using a scale as shown in Figure 20–28. Next install the caliper, apply the brakes a few times, and rotate the rotor at least ten revolutions. Then remeasure the force required to turn the rotor. Subtracting the first measurement from the second will give the amount of drag caused by the brakes. If it is excessive—more than 20 to 25 lb—the caliper or caliper mounts should be serviced.

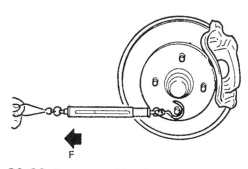

Figure 20–28 Drag caused by the caliper can be measured using a spring scale. If there seems to be too much drag, it can be measured with the caliper removed. *(Courtesy of Nissan North America, Inc.)*

28. If one or more problems were located, or there was a complaint of a specific problem and the cause has not been found, one or both of the other wheels need to be removed and the brake assemblies inspected. If a preventative maintenance inspection is being made or the cause of the complaint has been determined, replace the wheels. The following procedure should be used:

   a. Check the lug bolts and the wheel nut bosses for worn or elongated holes. Damaged wheels should be replaced. Place the wheel over the lug bolts with the valve core next to the previously marked stud(s).

   b. Snug down the lug nuts, making sure the tapered portion of the lug nut enters the tapered opening of the wheel nut bosses. While the tire and wheel are in the air, it is difficult to apply enough torque to completely tighten the nuts because the wheel will spin. However, they can and should be tightened enough to hold the wheel in the correct position.

   c. Lower the car onto the ground and immediately complete step d.

   CAUTION *Always remember step d.*

   d. Tighten the lug nuts to the correct torque using a tightening pattern that moves back and forth across (not around) the wheel (Figure 20–29). A brake drum or rotor can be distorted by overtightening the lug bolts or using the wrong order.

Chapter 20 ■ Brake System Service

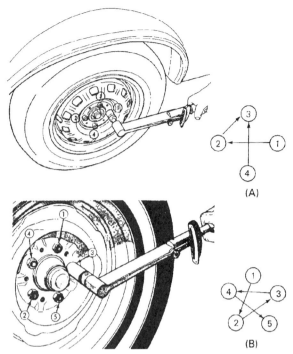

Figure 20-29 Wheel studs or lug bolts should be tightened to the correct torque using a crisscross (A) or star-shaped pattern (B). *(Courtesy of Chrysler Corporation)*

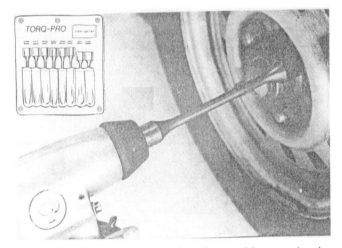

Figure 20-30 Torque sticks (inset) resemble a torsion bar with a socket built onto it; they are used with an air-impact wrench to tighten lug nuts to the correct torque.

Torque sticks used with an air-impact wrench have become a popular method for many shops to tighten lug nuts quickly (Figure 20-30).

e. Replace the wheel cover or hubcap using a rubber hammer or the hammer portion of the hubcap tool.

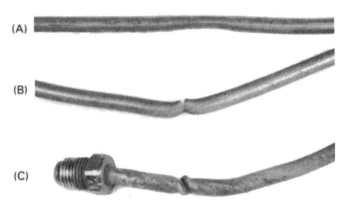

Figure 20-31 Some of the brake tube problems encountered are dents, which can cause restriction (A); kinks, which can cause leaks (B); and twists, which can cause both problems (C).

NOTE: Uneven or excessive lug nut torque can cause braking problems. As a rotor is heated and then cools through normal driving, parts of the rotor will cool at different rates because of heat transfer to the wheel. This can cause the rotor to distort, which results in runout or thickness variation.

Lug nut torque varies depending on the size of the lug bolt and the material of the wheel center section. Aluminum wheels sometimes use a slightly higher torque because of the solid center section. However, a slightly lower torque is sometimes recommended to reduce nut-to-wheel galling and compression of the wheel nut bosses. If the manufacturer's torque specifications are not available, use the following, based on the lug bolt size:

Lug Bolt Size	Torque	
3/8 × 24	35–45 ft-lb	(48–61 N-m)
7/16 × 20	55–65 ft-lb	(75–88 N-m)
1/2 × 20	75–85 ft-lb	(102–115 N-m)
M10 × 1.5	40–45 ft-lb	(54–61 N-m)
M12 × 1.5	70–80 ft-lb	(95–108 N-m)
M14 × 1.5	85–95 ft-lb	(115–129 N-m)

SAFETY TIP: It is a good practice to retighten the lug nuts after driving 10 to 20 mi. (16 to 30 km). With alloy-center wheels, it is a good idea to check lug nut tightness again after another 100 mi. (161 km).

29. Check all visible steel lines for kinks or collapsed sections that might cause fluid restriction or leaks (Figure 20-31). Check all flexible hoses for leaks, deterioration, cuts, chafing, rubbing, excessive cracks, bulges, or loose supports. Surface cracks in the outer cover are normal, but replacement

should be recommended if the inner cording is exposed (Figure 20–32). If there is a complaint of a spongy pedal, it is a good practice to check for hose swelling as a partner applies the brakes.

30. Check the parking brake cables, equalizer, and linkage. The cables should move freely in the guides and housing and not show signs of fraying.
31. Lower the car and operate the brake pedal through several slow, complete strokes until a firm pedal is obtained.
32. Road test the car on streets with little or no traffic and make several stops from speeds of 20 to 25 mph (32 to 40 km/h) at different pedal pressures. While the stop is occurring, check for pull, grab, squeal, or other unusual noises; excessive dive; or a pulsating pedal. Any faults indicate a need for further inspection to determine their cause.

CAUTION *Be sure to conduct the road test in such a way that there is no danger to yourself or other motorists and no violation of any applicable traffic regulations.*

33. Evaluate the findings of your inspection and make your recommendations on the inspection report.

20.3 Troubleshooting Brake Problems

Proper diagnosis is essential in locating the cause of a brake problem. Once the cause has been determined, it is usually fairly easy to make the repair. The repair procedures for the various systems are described in the following chapters.

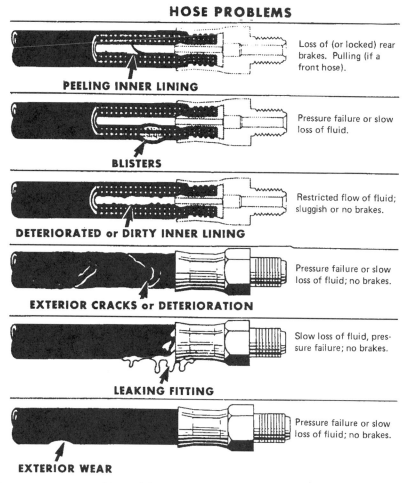

Figure 20–32 The most common causes of hose failure are shown here. *(Reprinted from Mitchell Anti-Lock Brake Systems, with permission of Mitchell Repair Information, LLC)*

When a particular brake problem shows up, an experienced brake technician often knows the cause based on his or her experience. However, beginners usually need to refer to a troubleshooting guide. These guides are published by the various vehicle manufacturers and by aftermarket brake part manufacturers. They generally list common problems that are encountered, along with the probable causes of each problem, with the most likely causes listed first. The following troubleshooting guide is divided into sections dealing with different types of problems. It is possible for a particular problem to involve two or three of these sections.

It should be noted that some problems, such as pull, chatter, and pedal pulsations, can be caused by faults in the car's suspension system, tires, or wheel bearings and that pedal pulsation during hard braking can be the result of normal ABS operation. Table 20–1 provides definitions for the common terms referring to brake problems.

SAFETY TIP: When making a decision concerning a repair, remember that lining replacement must always be done in pairs. Any change at one end of an axle must be accompanied by the same change at the other end. Rotor or drum machining must also be done in pairs. The friction surfaces of both rotors on an axle must have the same finish. Both drums on an axle must have the same finish and diameter. If one end of the axle has a different friction material or friction surface, the car will probably pull during braking; this unsafe condition must be avoided.

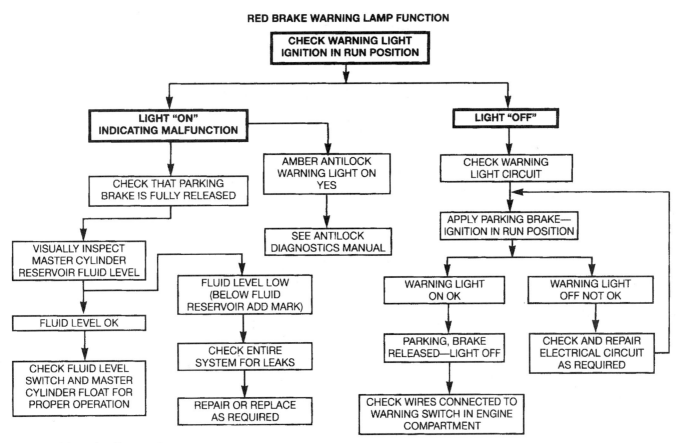

Courtesy of Chrysler Corporation

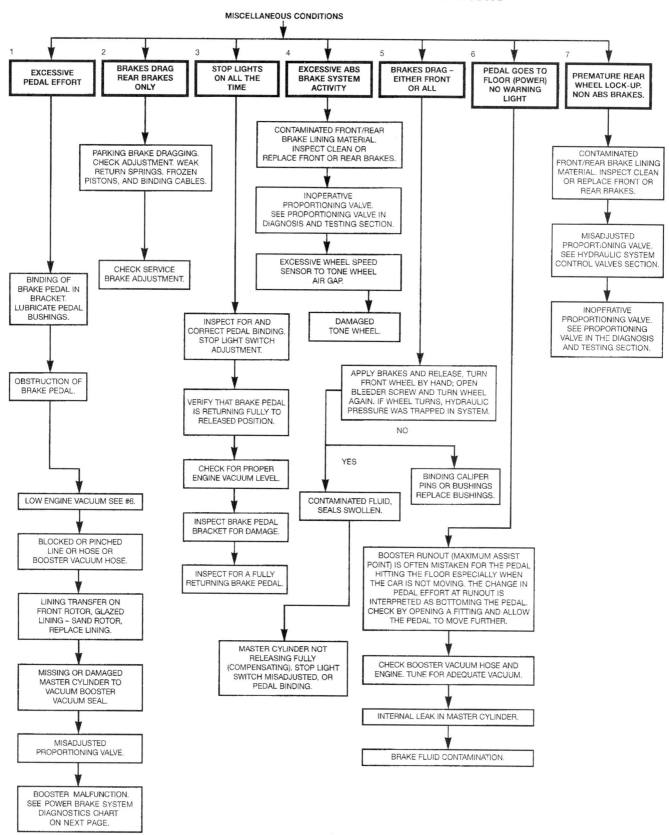

Courtesy of Chrysler Corporation

Chapter 20 ■ Brake System Service

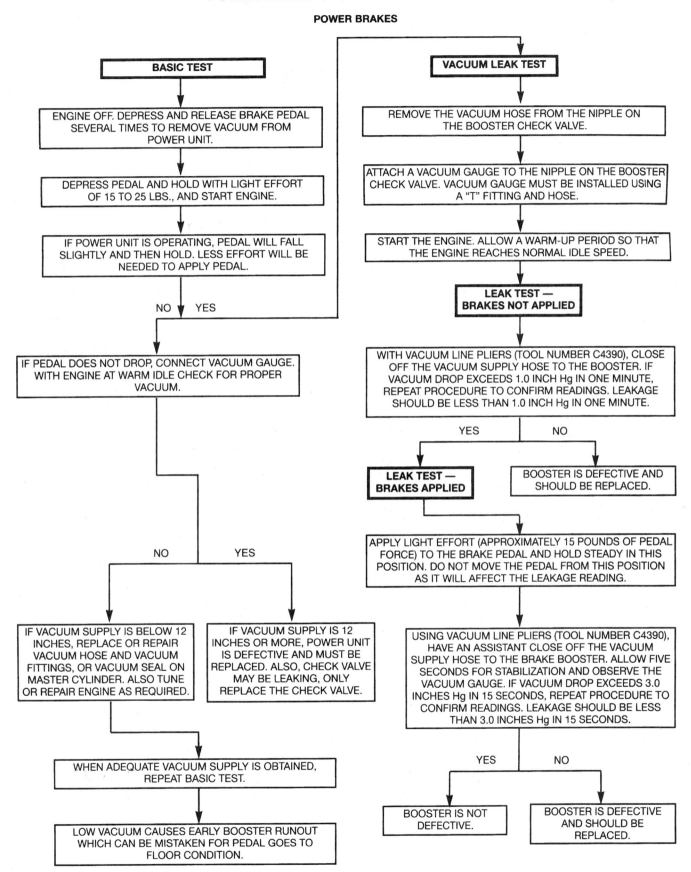

Courtesy of Chrysler Corporation

BRAKE NOISE

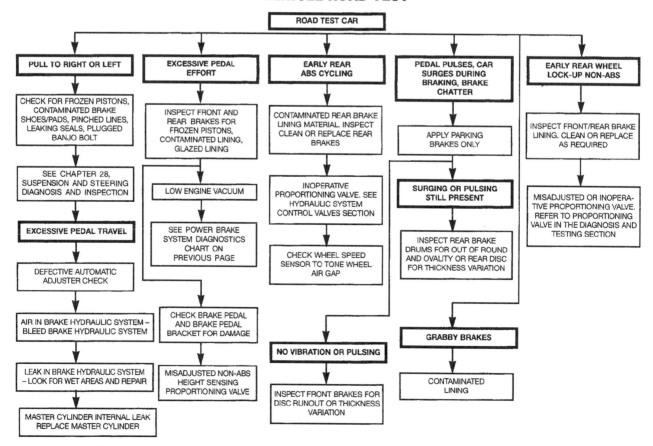

Courtesy of Chrysler Corporation

Chapter 20 ■ Brake System Service

Table 20-1 Common Brake Problems

There are a variety of brake problems, and the terms to describe them vary somewhat among technicians. The following should help you understand these terms and improve your communication with other technicians and future customers.

Symptom	What It Is
Chatter	One or more brakes do not apply smoothly but in an abrupt, apply-and-release manner.
Dive	The front of the vehicle lowers during a stop.
Drag	One or more brakes do not release completely but remain partially or completely applied; also called binding.
Fade	Braking power decreases or disappears during a stop.
Grab	One or more brakes lock up too easily.
Grind	Harsh, abrasive sound like someone grinding metal with a low-speed grinder is heard.
Hard pedal	The pedal is very firm, and it takes more than normal pressure for braking.
Knock	A harsh knocking sound is heard as the brake is applied; also called thump or clunk. This problem can be caused by worn suspension parts.
Lock	One or more brakes apply completely so the wheel does not revolve.
Low pedal	The pedal travels farther than normal.
Moan	A low-frequency noise is heard during a stop; also called groan.
Noise	A noise is heard that is different from that heard during normal braking.
No pedal	The brake pedal goes all the way down with no brake application.
Pedal squawk	A chirp or squawk is heard as the pedal is applied.
Pull	The vehicle tends to turn right or left as the brakes are applied.
Pulsating pedal	The pedal moves up and down as the car is stopping; also called pedal surge.
Sinking pedal	The pedal slowly lowers under a steady pedal pressure; also called fading pedal.
Soft pedal	The pedal goes down too easily; also called "pedal falls away."
Spongy pedal	The pedal action is the same as pushing on a spring; also called springy pedal.
Squeal	A high-pitched noise is heard during a stop; this usually can be varied by changing pedal pressure; also called squawk.
Surge	A pulsating, not smooth, stopping action is felt.

20.4 Brake Repair Recommendations

After completing a brake inspection or following a troubleshooting guide to determine the cause of a problem, it is time to make a decision about the repair. The competent technician presents the findings to the car owner, along with the various choices to be made and the cost and ramifications of each choice. The car owner always makes the final decision about what repairs are to be made. Never make a repair that is unsafe or not industry approved, and do not forget the MAP guidelines.

When there is a specific problem that can be corrected by an adjustment or replacement of one or a few parts, and the system is sound and has an adequate amount of good brake lining and a good hydraulic system, it is best to simply repair the defect. However, with older cars with a tired hydraulic system, worn lining, or both, it is often better to recommend a complete brake job. After a certain point, it is more economical to make all the necessary repairs at one time. This should also give the car owner more security. Remember that a brake job should give safe, new-car operation for several years and many thousands of miles.

20.5 Special Notes on Surface Finish

At times, a brake technician is concerned with the surface finish or texture of a metal surface. The finish refers to the flatness or roundness—also called the waviness—as well as the roughness of the metal surface. The finish of the friction surfaces of the brake drums and rotors is important because these surfaces must mate—that is, cause the new lining to wear to an exact fit. This is called **lining break-in** or **burnishing**. The surface finish of the bore of

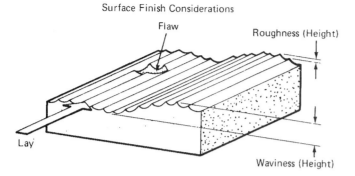

Figure 20-33 Metal surface finish is determined by a combination of roughness, waviness, and any flaws.

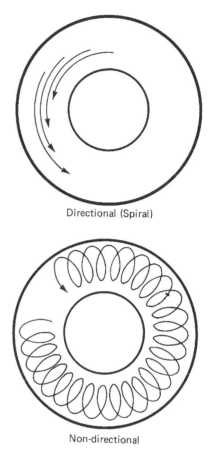

Figure 20-34 The finish (machine marks) on rotor A are in a circular, directional pattern. Rotor B has a nondirectional finish with the machine marks going in all directions.

a wheel cylinder or master cylinder is important because a rough or faulty bore can cause a poor seal or fast rubber cup wear.

The actual surface finish is determined by the following factors: roughness, the finely spaced irregularities caused by the machining process; waviness, an irregular surface height caused by unequal wear during operation or deflection during machining; scoring, a severe form of waviness; and flaws, very irregular imperfections caused by rust, impact, or defects in the metal (Figure 20-33). The pattern left by the machining operation is often called a **directional finish.** It runs in a predominantly parallel path. It is called a **nondirectional finish** if there is no regular pattern (Figure 20-34). The direction of the pattern machine marks is often called **lay** by machinists.

The surface finish is measured as the average depth of the irregularities. It is affected by the four factors involved (roughness, waviness, scoring, and flaws). A certain surface finish is often referred to as 20 to 100 rms, meaning the average scratch depth is between 20 and 100 microinches. The term *rms* refers to *root mean square,* which is a method of averaging scratch depth. A micro-inch is very small, 1/1,000,000, or 0.000001, in. (0.0254 mm).

The cylinder bore surface is fairly easy to deal with. It needs to be a very smooth cylinder shape, which is an almost perfect circle with straight sides. Normal piston operation in a hydraulic cylinder does not cause bore wear. The most common problem is corrosion, which causes pitting and other flaws. Normally, a corroded bore is not repaired, only cleaned, and if pitted excessively, the cylinder is replaced.

More problems are encountered with the surface finish of a rotor or drum because of the wear caused by contact with friction material during stops. Problems involving the surface of a drum are even worse because of its cylindrical shape. As it wears, the friction surface of a drum or rotor normally burnishes to an extremely smooth surface (Figure 20-35). A certain amount of waviness often accompanies this burnishing. Any waviness in the surface will require that the replacement lining—which is essentially flat—wear to make full contact with the drum or rotor surface. If the worn surface of the drum or rotor is too smooth, this wear during lining break-in will be very slow. There is a good possibility that the new lining might be used in a hard stop before it is broken in. An improperly fitted lining—one that is not broken in—will not develop a full braking force and will also tend to overheat in those areas that are in contact with the drum or rotor surface.

The correct surface finish for lining break-in is about 40 to 80 microinches of roughness with no waviness. A smoother finish will cause a very slow break-in. A rougher finish will probably cause faster break-in but

Chapter 20 ■ Brake System Service

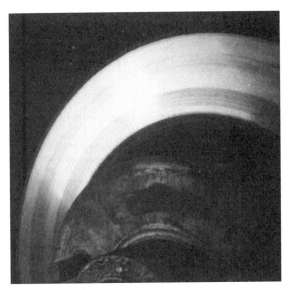

Figure 20-35 This used rotor has a very smooth, shiny surface that is common because of the burnishing of the lining.

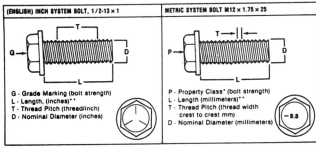

Figure 20-36 The standard dimensions used to determine bolt sizes vary between the English and the metric systems. *(Courtesy of Ford Motor Company)*

also rapid lining wear and possibly grabbiness. A nondirectional pattern is also desirable because scratches that run across the direction of rotation promote rapid lining break-in.

20.6 Fastener Security

Fasteners (commonly called bolts, nuts, or screws) used on steering and suspension parts must be strong enough to hold the parts securely in place and provide a high degree of safety. The size and strength of the fastener are designed to provide this security. They are also designed with some method of keeping them in place.

The size of a bolt or screw is determined by measuring the diameter of the threads. Traditionally, in the United States, this diameter was sized by an inch or a fraction of an inch (e.g., 1/4, 5/16, 3/8, and so forth). Today, most bolt diameters are sized by the metric system and are a certain number of millimeters in diameter. Some of the common sizes used in cars are 6, 6.3, 8, 10, 12, 14, 16, and 20 mm. When a bolt or nut is replaced, the thread pitch is also important. Thread pitch is the distance between the threads. One system of bolt sizing specifies thread pitch by the number of threads in an inch. A 1/4-20 would be a bolt that is 1/4 in. in diameter with 20 threads in 1 in. This bolt is classified as a **National Coarse (NC)** thread bolt. A 1/4-28 is a **National Fine (NF)** thread, with 28 threads in 1 in. A metric fastener specifies thread pitch by the actual pitch dimension. An 8 × 1.25 is a bolt that is 8 mm in diameter with a distance of 1.25 mm from the point of one thread to the point of the next one (Figure 20-36).

The tensile strength of an inch-sized fastener is indicated by a group of lines that are embossed onto the head of the bolt. No lines indicate grade 2 (weakest), three lines indicate grade 5, five lines indicate grade 7, and six lines indicate grade 8 (strongest). The number of lines plus two is the grade of the bolt. A metric bolt has a number embossed onto the bolt head; this number corresponds to the bolt's strength, with a larger number indicating a stronger bolt.

A nut is normally held in place on a bolt by interthread friction, friction between the bolt and nut threads, and friction between the bolt or nut head and the part. To ensure bolt or nut retention, cotter pins through the nut and bolt or lock washers have traditionally been used (Figure 20-37). The metal-cutting action of a lock washer increases the resistance to turning of a fastener in a counterclockwise direction. A class of nuts and bolts referred to as **prevailing torque** nuts or bolts are very common. These are commonly called **self-locking** nuts or bolts (Figure 20-38). This group of fasteners uses a distortion of the bolt or nut, a nylon insert in the threads of the bolt or nut, or a dry or liquid adhesive coating on the bolt or nut. Several brands of liquid lock washer are available to the technician for this same purpose. These

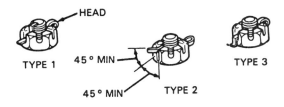

TYPES 1, 2, & 3 ARE OPTIONAL INSTALLATIONS. AFTER REACHING TORQUE REQUIRED, INSTALL COTTER PIN. IF SLOT IN NUT IS NOT ALIGNED WITH PIN HOLE, NUT MUST ALWAYS BE TIGHTENED (UP TO 1/6 TURN) FURTHER, NEVER BACK-OFF, TO INSERT COTTER PIN. INSTALL COTTER PIN TIGHTLY INTO NUT SLOT WITH HEAD OF PIN SHOULDERED AND BEND ONE OR BOTH LEGS AS SHOWN ABOVE.

Figure 20-37 Cotter pins have been used for many years to lock the nut on a bolt in critical areas. *(Courtesy of General Motors Corporation, Service Technology Group)*

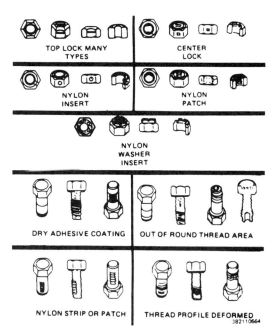

Figure 20-38 Today there are many types of prevailing torque nuts and bolts to lock the nut onto the bolt. These are commonly called self-locking nuts and bolts. *(Courtesy of General Motors Corporation, Service Technology Group)*

METRIC SIZES								
		6 & 6.3	8	10	12	14	16	20
NUTS AND ALL METAL BOLTS	N•m	0.4	0.8	1.4	2.2	3.0	4.2	7.0
	In. Lbs.	4.0	7.0	12	18	25	35	57
ADHESIVE OR NYLON COATED BOLTS	N•m	0.4	0.6	1.2	1.6	2.4	3.4	5.6
	In. Lbs.	4.0	5.0	10	14	20	28	46
INCH SIZES								
		2.50	.312	.375	.437	.500	.562	.625	.750
NUTS AND ALL METAL BOLTS	N•m	0.4	0.6	1.4	1.8	2.4	3.2	4.2	6.2
	In. Lbs.	4.0	5.0	12	15	20	27	35	51
ADHESIVE OR NYLON COATED BOLTS	N•m	0.4	0.6	1.0	1.4	1.8	2.6	3.4	5.2
	In. Lbs.	4.0	5.0	9.0	12	15	22	28	43

Figure 20-39 A prevailing torque nut or bolt is worn out if it turns too easily. If it takes less force than indicated in this chart to turn it before seating, the nut or bolt should be replaced. *(Courtesy of General Motors Corporation, Service Technology Group)*

are usually anaerobic materials—that is, they harden or set when oxygen is removed.

The use of prevailing torque fasteners by manufacturers has placed some added responsibilities on the front-end technician. If these fasteners wear, they might lose their ability to stay put and work loose. A suspension or steering part must not come loose after the car has been worked on. Reuse of a worn prevailing torque nut does not always provide the same degree of security as reuse of a nut locked by a cotter pin or lock washer.

SAFETY TIP: The following are some rules to observe for the safe reuse of prevailing torque fasteners:

1. Check the manufacturer's recommendations. Some nuts or bolts should be replaced with new ones if they are removed.
2. The nuts and bolts must be clean.
3. The nuts and bolts must be inspected for damage. Discard any that show signs of abuse, overtightening, cracks, elongation, or wear.
4. Start the nuts and bolts by hand to observe the feel; also observe the feel (amount of turning resistance) before they seat. The nut or bolt should develop as much resistance as indicated in Figure 20-39. Any prevailing torque fasteners that turn too easily should be replaced.
5. Replace all damaged or rusty fasteners with new parts of equal or greater strength.

When prevailing torque nuts are used on ball joints or tie-rod ends, it is difficult to thread the nut onto the stud. The tapered stud is kept from rotating by the locking of the tapered stud into the tapered hole. The turning resistance of the prevailing torque nut is great enough to cause the stud to rotate. Turning the nut merely rotates the stud. This job is made possible by first tightening the stud into the boss, to set the taper, using an ordinary nut of the correct size or a special nutlike tool. After the taper is locked, the nut or special tool is removed and the prevailing torque nut is installed to the correct torque (Figure 20-40).

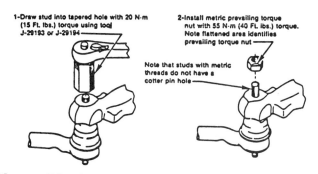

Figure 20-40 It is often difficult to install a prevailing torque nut onto a ball joint or tie-rod end stud, because the stud rotates easier than the nut does. The stud taper can be locked using a special tool or a plain nut of the correct size before installing the prevailing torque nut. *(Courtesy of General Motors Corporation, Service Technology Group)*

Chapter 20 ■ Brake System Service

20.7 Practice Diagnosis

You are working in a brake and front-end shop and encounter the following problems:

CASE 1: The complaint on the brakes for a 3-year-old Toyota is pedal pulsation with a grinding noise coming from the front. What is probably wrong? What checks should you make?

CASE 2: The complaint on a 6-year-old Dodge pickup is a lockup of the right rear brake under light to moderate pedal pressure. When you pull the right rear brake drum you find a little bit of gear oil on the brakes and some darker, crumbly brake dust. Have you found the cause of the problem? What is your recommendation for repair?

CASE 3: The complaint is a lockup of the rear brakes on another 6-year-old pickup under moderate braking. When you remove the rear brake drums, you find nothing wrong. What could be causing this problem? What should you do to locate the cause?

Terms to Know

burnishing

directional finish

fasteners

lay

lining break-in

Motorist Assurance Program (MAP)

National Course (NC)

National Fine (NF)

nondirectional finish

prevailing torque

self-locking

Review Questions

1. Two technicians are discussing brake inspection procedures.
   Technician A says that a brake pedal must have a slight amount of free movement before it starts to move the master cylinder piston.
   Technician B says that the pedal should become firm before it travels halfway to the floor board.
   Who is correct?
   a. A only
   b. B only
   c. Both A and B
   d. Neither A nor B

2. A sinking brake pedal is an indication that:
   a. There is air in the system
   b. The shoes require adjustment
   c. There is a leak in the system
   d. The lining is worn

3. Technician A says that the brake warning light should come on as the engine cranks.
   Technician B says that this light usually comes on when the brake pedal is very hard.
   Who is correct?
   a. A only
   b. B only
   c. Both A and B
   d. Neither A nor B

4. Technician A says that a low brake pedal is a definite indication that there is air in the hydraulic system.
   Technician B says that a low brake pedal can usually be corrected by bleeding the brakes.
   Who is correct?
   a. A only
   b. B only
   c. Both A and B
   d. Neither A nor B

5. Technician A says the best test for a power booster is to apply the brakes hard during a stop.
   Technician B says that a good booster will cause the brake pedal to lower if the engine is started while the brakes are being applied.
   Who is correct?
   a. A only
   b. B only
   c. Both A and B
   d. Neither A nor B

6. A small swirl or spurt in the fluid of the master cylinder reservoir as the brakes are applied indicates:
   a. Normal operation
   b. A faulty piston seal
   c. A low brake fluid level
   d. All of the above

7. Technician A says that a brake lining should be replaced when it wears to a thickness of 0.005 to 0.010 in.
   Technician B says that lining thickness is always measured from the shoe rim or backing to the outer edge of the lining.
   Who is correct?
   a. A only
   b. B only
   c. Both A and B
   d. Neither A nor B

8. During an inspection, a brake drum should be checked to make certain that:
   a. There are no cracks
   b. There is a smooth friction surface
   c. The diameter is less than the maximum specified diameter
   d. All of the above

9. A brake rotor being checked should always be:
   a. Thicker than the thickness dimension on the rotor
   b. Thinner than the maximum size dimension on the rotor
   c. Measured with a ruler
   d. All of the above

10. Technician A says that rotor runout is checked using a dial indicator.
    Technician B says that a pedal pulsation problem can be caused by nonparallel friction surfaces on the rotor.
    Who is correct?
    a. A only
    b. B only
    c. Both A and B
    d. Neither A nor B

11. Technician A says that improper tightening of the wheels can distort the shape of the rotor or drum.
    Technician B says that the lug bolts can come loose if they are not tightened correctly.
    Who is correct?
    a. A only
    b. B only
    c. Both A and B
    d. Neither A nor B

12. A brake hose should be replaced if:
    a. It leaks
    b. It has bulges or bubbles in the outer cover
    c. It has rub marks extending into the outer layer of the cord
    d. Any of the above

13. Technician A says that new brake springs should be installed if an overly stretched spring is discovered.
    Technician B says that if oil-soaked lining is found on one wheel, the lining at both ends of the axle should be replaced.
    Who is correct?
    a. A only
    b. B only
    c. Both A and B
    d. Neither A nor B

14. Technician A says that a nondirectional surface finish has no scratch marks on it.
    Technician B says that the actual surface finish is a combination of waviness, irregular flaws, and machining scratches on the metal surface.
    Who is correct?
    a. A only
    b. B only
    c. Both A and B
    d. Neither A nor B

15. Technician A says that a worn drum or rotor surface is usually too smooth to properly break in new brake lining.
    Technician B says that the correct surface finish for a rotor friction surface is about 60 microinches.
    Who is correct?
    a. A only
    b. B only
    c. Both A and B
    d. Neither A nor B

16. The strength grade of a bolt is indicated by:
    A. The number of embossed lines on the head
    B. A number embossed on the head
    Which option best completes the statement?
    a. A only
    b. B only
    c. Both A and B
    d. Neither A nor B

17. A nut can be secured onto a bolt by a(n):
    a. Lock washer
    b. Anaerobic liquid
    c. Distortion of the nut
    d. Any of these

18. Technician A says that a prevailing torque nut is worn out if it turns too easily on the bolt.
    Technician B says that lock washers should always be used with prevailing torque nuts.
    Who is correct?
    a. A only
    b. B only
    c. Both A and B
    d. Neither A nor B

Wheel Bearing Service

Objectives

Upon completion and review of this chapter, you should be able to:

- ❏ Clean, inspect, repack, reassemble, and adjust serviceable wheel bearings.
- ❏ Remove and replace nonserviceable wheel bearings.
- ❏ Remove and replace axle bearings and seals.
- ❏ Perform the ASE tasks relating to wheel bearing diagnosis, adjustment, service, and repair (see Appendix A, Section B, 3, Task 3).

21.1 Introduction

Periodic maintenance is normally required on the serviceable types of wheel bearings used on nondrive axles. The other styles of bearings are usually serviced on an as-needed basis. If the bearing is noisy, leaking grease, or loose, it should be serviced. The term **service** means different things depending on the type of bearing. It can mean removing and replacing the entire bearing assembly for nonserviceable units. On the rear axle of a RWD car, it usually means removing an axle shaft to replace a bearing or seal. On nondrive axle bearings, it usually involves disassembling the hub and bearings, cleaning and repacking the bearings, and adjusting the bearing end play or preload during reassembly. The rear axle of an independent rear suspension, RWD car usually has similar service requirements. In cases where you are not sure of the lubrication or service requirements for a particular bearing set, it is wise to check the car manufacturer or technician service manual. Diagnosis of wheel bearing problems is described in Section 28.11.

21.2 Repacking Serviceable Wheel Bearings

Many RWD front wheel bearings and FWD rear wheel bearings should be repacked at regular intervals. This operation includes disassembly, cleaning, packing with grease, reassembling, and adjusting. The interval, which varies with different car manufacturers, is about 15,000 to 20,000 mi. or greater. In some cases, bearing service is required only during a brake reline.

Whenever the wheel or axle bearings are serviced, the brake shoes and the drums or rotors should be checked for wear. The average driver has only a vague idea of what the brake shoes are or where they are located. He or she does not realize the damage that will be done to the drum or rotor shortly after the brake shoes wear out or the poor and unsafe stop that will result. This same person usually appreciates any information that you can provide that will make his or her car perform better, safer, or at a lower cost. Most technicians feel a strong responsibility to inform the car owner about things that are or might become unsafe.

SAFETY TIP: The following good service practices should be observed while servicing wheel or axle bearings:

1. Always lift the car at the correct lifting points and support it with carefully placed jack stands.
2. Never just add grease to a bearing; always clean, inspect, and repack it with new grease.
3. Never let grease or solvent get on the braking friction surfaces; always clean up any spilled grease, solvent, or greasy fingerprints.
4. Never let brake calipers or driveshafts hang by their hoses or universal joints; always support them and protect the rubber hoses and CV joint boots.
5. Do not let a tire and wheel fall and bounce during removal; keep these heavy parts under control.

6. Keep a clean, neat work area; slide the tire and wheel under the car and out of the way during service operations that require tire and wheel removal.
7. Always replace the seal with a new one; reusing an old seal is taking a chance on ruining the brakes or losing a bearing because of possible grease leakage.
8. Always replace lock washers and cotter pins; use all recommended locking devices and tighten nuts and bolts to the correct torque.
9. Remember that the service operations you are performing should last for a few years and thousands of trouble-free miles.

21.2.1 Disassembling Wheel Bearings

Disassembly is the first step when repacking serviceable wheel bearings.

To disassemble wheel bearings, you should:

1. Raise the car so the tires are off the ground.

CAUTION *Position a jack stand under a secure portion of the car and lower the car onto the jack stand or raise and support the car on a hoist.*

Figure 21–1 A simple S-shaped hook has been bent from 1/8-in. welding rod to hang the caliper.

2. If disc brakes are used on the wheel, remove the tire and wheel as described in Section 4.6. If drum brakes are used, the tire and wheel with the brake drum and hub can usually be removed as a unit.
3. On cars equipped with disc brakes, remove the caliper as described in a technician service manual and suspend it from the steering knuckle or other convenient point, using a rubber tie strap, bungee cord, or mechanic's wire (Figure 21–1).

SAFETY TIP: Do not let the caliper hang from the hose; if this hose is bent too sharply, it can be damaged and cause a brake failure.

4. Remove the grease or dust cap from the wheel hub; a dust cap remover will do this quickly and easily. Slip joint pliers can also be used for this procedure (Figure 21–2 and Figure 21–3).
5. Locate and remove the locking device for the spindle nut. The most common type is a cotter pin. Straighten the bent end using a pair of diagonal- or side-cutting pliers commonly called dikes. Grip the head of the cotter pin with the dikes and pry it out of the spindle and spindle nut (Figure 21–4).

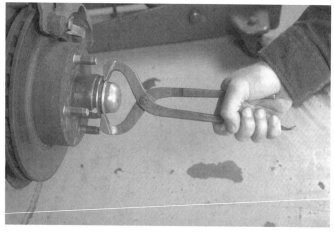

Figure 21–2 The wheel bearing dust cover can be removed using a pair of special dust cover pliers.

Another good tool for this procedure is a cotter pin puller. If a staked spindle nut is used, some manufacturers recommend merely unscrewing the nut. The staked portion will bend out of the way during removal. Other manufacturers recommend that you bend the staked portion of the nut upward, using a small, sharp chisel or using a drill and drill

Chapter 21 ■ Wheel Bearing Service

Figure 21-3 A pair of water pump or slip joint pliers can be used to remove the dust cover. *(Courtesy of Federal-Mogul Corporation)*

Figure 21-5 After the nut and washer have been removed, rock the hub to work the outer bearing out of the hub. *(Courtesy of The Timken Company)*

Figure 21-4 A cotter pin can usually be removed easily by gripping the head of the cotter pin (arrow) in the jaws of a pair of diagonal-cutting pliers (dikes) and prying the pin out.

Figure 21-6 After the seal puller has been hooked into the seal, a prying action (in this case toward the left) will lift the seal out of the hub.

    bit to drill through the staked portion of the nut. Be careful not to drill the spindle. In either case, the staked nut should be replaced.

6. Remove the spindle nut. It can usually be removed with your fingers; if not, use slip joint pliers or a wrench to unscrew it.
7. Remove the washer and outer bearing and bearing cone. Rocking the tire and wheel or hub will usually work the bearing out to where you can grasp it (Figure 21-5).
8. Slide the hub off of the spindle and pry the seal out of the back of the hub using a seal puller (Figure 21-6). Alternate methods of removing the seal are (1) to tap the inner bearing and bearing cone and seal out using a large wooden dowel and hammer (Figure 21-7) or (2) to replace the nut back on the spindle and, while sliding the hub off of the spindle, catch the bearing on the spindle nut so that the bearing and seal are jerked out of the hub. Neither of these alternate methods is recommended, because they might bend the bearing cage.

Figure 21-7 A wooden dowel or punch can be placed on the inner face of the inner bearing and tapped to remove the seal and inner bearing. *(Courtesy of Chicago Rawhide)*

21.2.2 Cleaning and Inspecting Wheel Bearings

After the spindle is disassembled, the bearings and hub should be cleaned to get rid of all of the old grease, dirt, or metal fragments it might contain to allow a thorough inspection.

SAFETY TIP: Many sources recommend wearing rubber gloves to protect your skin from the chemicals used for cleaning these parts.

To clean and inspect wheel bearings, you should:

1. Wipe off the spindle and inspect it for damage. The bearing cones must have a slip fit on the spindle. This fit should leave a marking on the spindle to show that the cone has been creeping, slowly rotating about one revolution per mile. Bearing creep keeps changing the loaded part of the cone and produces longer cone and bearing life (Figure 21-8).
2. Also check the spindle in the area where the seal runs; it should be clean and smooth.
3. Wipe all of the old grease out of the hub and wipe the bearing cups clean. Inspect the cups for pitting, spalling, or other signs of failure. If the cup is damaged, the cup and the bearing and cone should be replaced. The bearing might appear good, but it is possible to miss seeing some damage because of the cage (Figure 21-9).

NOTE: Many technicians use the part number from the old bearing or seal to ensure correct replacement.

Figure 21-8 After the spindle has been cleaned and inspected, coat it lightly with grease. Note the marks (black arrows) where the cones have been creeping on the spindle. This is good. *(Courtesy of The Timken Company)*

4. Check the cups to ensure that they are tight in the hub. A loose cup usually requires replacement of the hub.
5. If a cup is damaged, it can be pulled out of the hub using a puller or driven out using a punch and hammer (Figure 21-10). The new cup is installed using a cup driving tool and hammer (Figure 21-11). Be sure the new cup is fully seated. The hammer blows will make a different, more solid sound when the cup becomes seated.
6. Wash the bearings and cones in clean solvent and blow them dry using shop air. The direction of the air blast should always be parallel to the rollers. Do not spin the bearing with the compressed air blast. An air blast across the rollers does not really clean the bearing, and it tends to spin the bearing too fast (Figure 21-12).

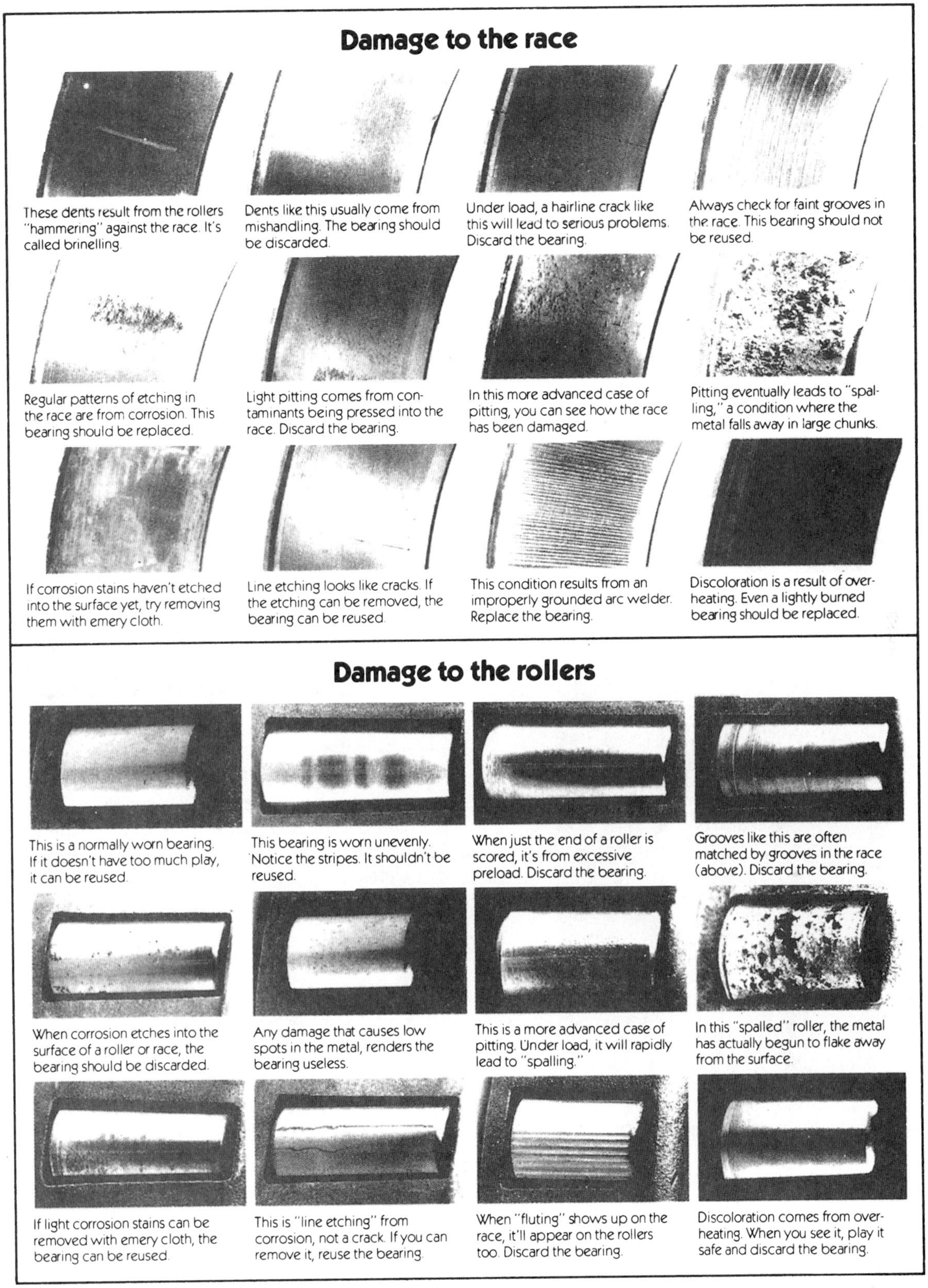

Damage to the race

These dents result from the rollers "hammering" against the race. It's called brinelling.

Dents like this usually come from mishandling. The bearing should be discarded.

Under load, a hairline crack like this will lead to serious problems. Discard the bearing.

Always check for faint grooves in the race. This bearing should not be reused.

Regular patterns of etching in the race are from corrosion. This bearing should be replaced.

Light pitting comes from contaminants being pressed into the race. Discard the bearing.

In this more advanced case of pitting, you can see how the race has been damaged.

Pitting eventually leads to "spalling," a condition where the metal falls away in large chunks.

If corrosion stains haven't etched into the surface yet, try removing them with emery cloth.

Line etching looks like cracks. If the etching can be removed, the bearing can be reused.

This condition results from an improperly grounded arc welder. Replace the bearing.

Discoloration is a result of overheating. Even a lightly burned bearing should be replaced.

Damage to the rollers

This is a normally worn bearing. If it doesn't have too much play, it can be reused.

This bearing is worn unevenly. Notice the stripes. It shouldn't be reused.

When just the end of a roller is scored, it's from excessive preload. Discard the bearing.

Grooves like this are often matched by grooves in the race (above). Discard the bearing.

When corrosion etches into the surface of a roller or race, the bearing should be discarded.

Any damage that causes low spots in the metal, renders the bearing useless.

This is a more advanced case of pitting. Under load, it will rapidly lead to "spalling."

In this "spalled" roller, the metal has actually begun to flake away from the surface.

If light corrosion stains can be removed with emery cloth, the bearing can be reused.

This is "line etching" from corrosion, not a crack. If you can remove it, reuse the bearing.

When "fluting" shows up on the race, it'll appear on the rollers too. Discard the bearing.

Discoloration comes from overheating. When you see it, play it safe and discard the bearing.

Figure 21-9 After the bearings and bearing cups have been cleaned, they should be carefully inspected for reuse. These are the commonly found causes of failure. *(Courtesy of Chicago Rawhide)*

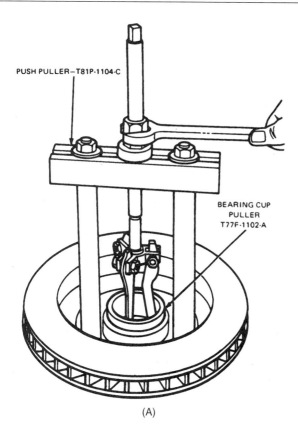

Figure 21-10 A faulty bearing cup can be removed by pulling it out with a puller (A) or driving it out with a punch (arrow) (B). *(A is courtesy of Ford Motor Company)*

CAUTION *Do not spin a bearing at high speeds. A spinning bearing might generate enough centrifugal force to explode. Also, running the bearing at high speeds without lubrication could damage the bearing. Spinning a bearing with shop air might sound neat, but it does not do any good; it is dangerous and can damage the bearing.*

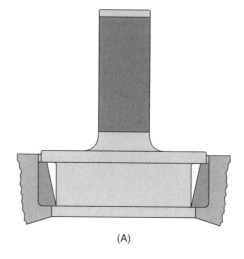

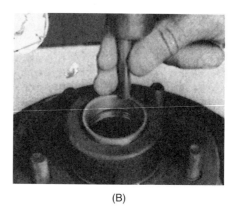

Figure 21-11 A new bearing cup can be installed quickly and easily using a bearing cup driver (A). Make sure that you hear the cup seat completely. If a cup driver is not available, a cup can be driven in place using a soft metal punch (B). Be careful to drive the cup in straight and make sure it is installed completely. *(A is courtesy of Chicago Rawhide; B is courtesy of The Timken Company)*

7. Rewash and dry the bearings until they are thoroughly clean.

8. Inspect the bearings for roller or cage damage. The commonly encountered causes of bearing failure are shown in Figure 21-9. All damaged or questionable wheel bearings and races should be replaced.

9. If the bearings, cones, cups, or spindle shows signs of misalignment, check for a bent spindle. A spindle runout gauge can be placed over the spindle and adjusted to a sliding fit on the inner bearing (Figure 21-13). Next, the dial indicator is adjusted to zero and the gauge unit is rotated around the spindle. A bent spindle is indicated if the dial indicator reading changes. Another check for a bent spindle is to measure the distance from a straightedge to the location of the outer bearing cone on each side and the top of the spindle. The measurements should be the same.

Chapter 21 ■ Wheel Bearing Service

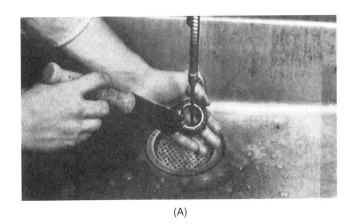

(A)

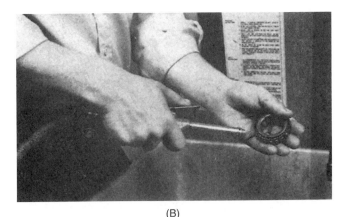

(B)

Figure 21–12 The bearing should be washed in clean solvent (A) and then dried with clean, compressed air (B). Never spin the bearing with the air blast while drying it. *(Courtesy of Chicago Rawhide)*

21.2.3 Packing and Adjusting Wheel Bearings

Wheel bearings should be packed with a suitable grease, as recommended by the manufacturer, and adjusted to the correct end play or preload during assembly. A good grade of chassis grease can usually be used to pack wheel bearings if the label on the container states that it is suitable for wheel bearings. Many technicians prefer to use a grease formulated especially for wheel bearings, NLGI classification GC. **High-temperature** or **heavy-duty** grease is recommended if the car is to be driven hard or if the brakes will get a lot of use. Packing a bearing fills the area between the cage and the cone, alongside of the rollers or balls, with grease.

A properly adjusted tapered roller wheel bearing should have 0.001 to 0.005 in. (0.03 to 0.13 mm) of end play. With this clearance, you should notice a perceptible play if you rock the tire and wheel. Ball-type wheel bearings should have zero end play; they are normally adjusted to a slight preload. Ball bearings should not have a perceptible play when the tire and wheel are rocked.

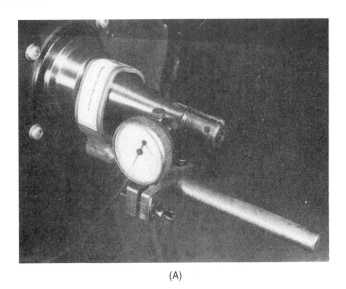

(A)

(B)

Figure 21–13 If a bent spindle is suspected, it can be checked using a spindle runout gauge (A) or a straightedge and caliper (B). The runout gauge is set and then rotated around the spindle while the technician watches the dial indicator. The straightedge is placed against the inner cone surface or brake boss while measurements are made to the outer cone surface on several sides of the spindle. If the measurements are different, the spindle is bent. *(B is courtesy of Volkswagen of America, Inc.)*

To pack and adjust wheel bearings, you should:

1. Pack the large, inner bearing first. The outer, small bearing should be packed when you are ready to install it. Some technicians prefer to pack both bearings at the same time and place the small bearing on a clean shop towel until it is needed. Packing is normally done with a bearing packer but can be done by hand. When using a bearing packer, which is much faster, follow the procedure for that particular packer (Figure 21–14).

Figure 21–14 The quickest way to pack a bearing is to use a bearing packer. Pushing downward on the packer forces grease into the area inside the cage. *(Courtesy of SPX Kent-Moore)*

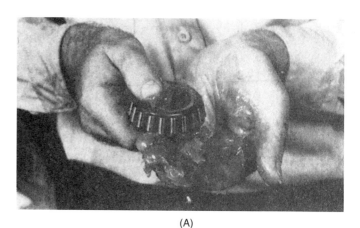

(A)

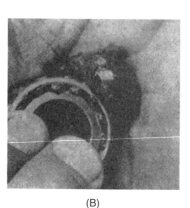

(B)

Figure 21–15 A bearing can be repacked by placing grease in the palm of your hand (A) and then repeatedly forcing the bearing into the grease until grease comes up between the rollers and cage (B). *(A is courtesy of Chicago Rawhide; B is courtesy of The Timken Company)*

When packing by hand, place a tablespoon of grease in the palm of your hand, and push the bearing, open side of the cage downward, into the grease and against the palm of your hand. Repeat this step until you see grease oozing through the upper part of the cage. Work all the way around the bearing so you fill the entire cage (Figure 21–15).

2. After packing the bearing, smear a liberal coating of grease around the outside of the cage and rollers.
3. Smear a coating of grease around the inside of both bearing cups. It is also recommended that you place a ring of grease in the hub, just inside of the cups. This ring of grease will act like a dam to prevent the grease from running out of the bearing and into the hub cavity when the grease gets hot. Smear a thin film of grease on the spindle, also being sure to include the area where the seal lip rubs (Figure 21–16).
4. Place the packed inner bearing into its cup.
5. Position the seal so the seal lip faces inward, toward the grease. Using a suitable driver, drive the seal into the hub so it is even with the end of the hub. Some technicians fill the inner area of the seal with light grease to help keep the garter spring in the proper position while the seal is driven into place. The back side of a bearing cup installer is often convenient and of the proper size. On seals without a lip, the seal is usually positioned so the part numbers face outward (Figure 21–17).
6. Wipe away any grease that might be left on the outside of the seal or on the end of the bearing cone. Any grease left on the ends of the cones or on the side of the retaining washer might cause a bearing to change adjustment. Grease that stays in these areas during adjustment can be squeezed out by side motions of the vehicle. It is said that each layer of grease might increase bearing clearance by 0.001 in. (0.025 mm) (Figure 21–18).
7. Slide the hub partially onto the spindle.
8. Repeat steps 1, 2, 3, and 4 to pack the outer bearing and slide it onto the spindle and into the hub.
9. Install the washer and spindle nut.

Chapter 21 ■ Wheel Bearing Service

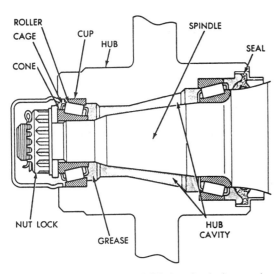

Figure 21-16 Grease should fill the shaded areas between the hub and spindle to ensure that the bearings stay lubricated. *(Courtesy of Chrysler Corporation)*

Figure 21-17 After the cup has been greased and the inner bearing installed into the cup, the area between the cone and the cup should be filled with grease, and the top edge of the cone (arrow) should be wiped clean of grease.

Figure 21-18 With the seal lip pointing inward, the new seal should be installed using a flat, properly sized seal driver. *(Courtesy of The Timken Company)*

10. Adjust the wheel bearing clearance as recommended by the car manufacturer. If the specifications are not available, the following is a standard method of adjusting tapered roller bearings:

   a. While rotating the hub by hand, tighten the spindle nut to about 15 to 20 ft-lb (20 to 27 N-m) of torque. This ensures complete seating of the bearings. This is about as tight as you can tighten the nut using slip joint pliers and normal hand pressure.

   b. Back off the spindle nut one-quarter to one-half turn.

   c. Retighten the spindle nut using your thumb and forefinger. This is equivalent to about 5 in.-lb (0.56 N-m) of torque (Figure 21-19).

11. Replace the nut lock if used and position a new cotter pin through the nut or nut lock and the spindle. Using the dikes, bend the long leg of the cotter pin outward, upward, and across the end of the spindle. The other leg can be cut off. Occasionally, if there is a static suppression spring in the dust cap, it is necessary to bend the two legs of the cotter pin sideways and around the spindle nut (Figure 21-20).

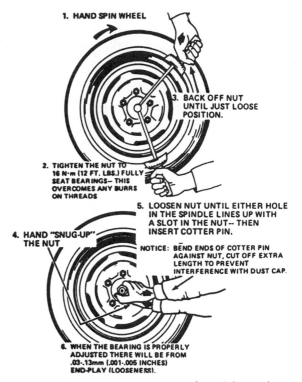

Figure 21-19 Wheel bearings are first tightened to seat the bearings and then adjusted to the correct running clearance. This is a commonly used procedure. Be sure to rotate the hub while seating the bearings. *(Courtesy of General Motors Corporation, Service Technology Group)*

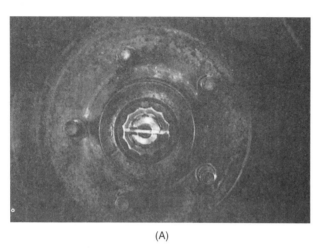
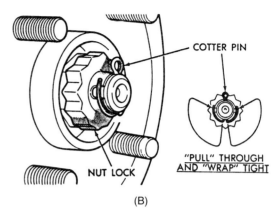

Figure 21–20 After the bearing is adjusted, the cotter pin is installed, trimmed if necessary, and bent across the end of the spindle (A) or wrapped around the spindle (B). Method B is required for clearance in some dust covers. *(A is courtesy of Federal-Mogul Corporation; B is courtesy of Chrysler Corporation)*

12. If you have doubts about the adjustment, measure the bearing or hub end play by mounting a dial indicator on the hub with the indicator stylus on the end of the spindle. Pull outward on the hub, adjust the dial indicator to zero, push inward on the hub, and read the amount of travel on the dial (Figure 21–21). Another way to measure end play is to mount the dial indicator on the steering knuckle with the stylus on the hub or rotors.
13. Replace the dust cap. Some technicians wipe a film of grease around the inside of the cap or around the lip of the cap to serve as a water seal.
14. Replace the caliper, being sure to follow the recommended procedure and tightening specifications.
15. Replace the wheel, being sure to tighten the lug bolts correctly and to proper torque specifications.

CAUTION *Make sure that all nuts and bolts are properly tightened and secure and that the brake pedal has a normal feel before moving the car.*

21.3 Repairing Nonserviceable Wheel Bearings

As stated earlier, the only repair for a faulty nonserviceable wheel bearing is to remove and replace the hub and bearing assembly. This fairly simple procedure varies slightly on different makes and models of cars. It is

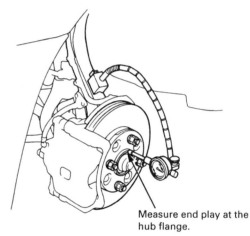

Figure 21–21 Wheel bearing adjustment can be checked using a dial indicator. A correctly adjusted bearing set will have a clearance as recommended by the vehicle manufacturer. A rule of thumb is 0.001 to 0.005 in. *(Courtesy of American Honda Motor Co., Inc.)*

wise to follow the procedure described in the manufacturer service manual.

To remove a nonserviceable hub and bearing assembly, you should:

1. Raise and support the car securely on a hoist or jack stands as previously described.
2. Remove the wheel.
3. On cars with drum brakes, remove the drum. On cars with disc brakes, remove the caliper and then remove the rotor.

CAUTION *Be sure to support the caliper to prevent damage to the brake hose.*

Chapter 21 ■ Wheel Bearing Service

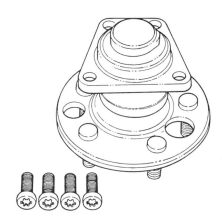

1	50 N·m (37 FT-LB)
2	HUB AND BEARING ASSEMBLY
3	BRAKE LINING & BACKING PLATE ASSEMBLY
4	CONTROL ARM BUSHING

Figure 21-22 This permanently sealed and adjusted hub and bearing assembly is removed and replaced with a new one if it becomes faulty. *(Courtesy of General Motors Corporation, Service Technology Group)*

4. Remove the bolts that secure the hub and bearing assembly to the control arm or axle. Note that a hole is provided in the mounting flange so that a socket and extension bar can be used for removal and installation (Figure 21–22).

On some rear-wheel units, you might find a shim behind the bearing assembly. As you will learn in Chapter 34, this is a tapered shim that is used to adjust camber or toe at this wheel. You should replace it in the exact position that it was found.

To install a hub and bearing, you should:

1. Make sure the recess in the control arm or axle in which the hub and bearing assembly fits is clean.
2. Install the hub and bearing assembly into position and tighten the bolts to the correct torque.
3. Install the brake drum or rotor. On cars with disc brakes, install the caliper and tighten the caliper mounting bolts to the correct torque.
4. Install the wheel, being sure to tighten the lug bolts correctly.

21.4 Repairing Drive Axle Bearings, Solid Axles

Traditionally, drive axle work has not been done by the front-end technician. It has been included in this text for use by the technician who works in a general repair shop and because the evolution of the automobile is blurring or changing the traditional areas of repair. A front end on a FWD car includes the drive axle and axle shafts. The rear end of a FWD car has wheel bearings much like those on the front end of a RWD car.

If a drive axle bearing becomes loose, rough, or noisy, the bearing must be replaced. If a grease leak develops from the end of the axle housing, the axle seal or bearing, depending on the axle design, must be replaced. To replace either the bearing or the seal, the axle must be removed from the axle housing. As previously mentioned, there are two ways of securing the axle in the housing. The inner end of the axle in both designs is supported by the axle gear in the differential unit inside of the axle housing.

A **bearing-retained axle** uses a bearing that is pressed onto the axle shaft and held in the axle housing by a retainer. A **C-lock axle** holds the axle in place using a C-shaped piece of metal that fits in a groove in the inner end of the axle. This axle slides through the axle bearing that is pressed into the end of the axle housing.

All C-lock designs use a seal that is separate from the bearing. Some bearing-retained designs use a seal incorporated into the bearing, some use a separate seal, and others use both.

Axle bearing and seal repair procedures vary on different cars. It is recommended that you always follow the procedure listed in a technician service manual.

21.4.1 Removing a Bearing-Retained Axle

To remove a bearing-retained axle, you should:

1. Raise and support the car securely on a hoist or jack stands.
2. Remove the wheel.
3. Mark the brake drum or rotor next to the previously marked stud to ensure replacement in the same location. On cars with disc brakes, remove the caliper and secure it to the axle using a rubber tie strap, bungee cord, or wire.
4. Remove the brake drum or rotor.

Figure 21–23 When removing an axle that is retained by the bearing, the nuts holding the bearing retainer and the brake backing plate onto the axle housing must be removed. *(Courtesy of Chicago Rawhide)*

Figure 21–24 This special adapter connects a slide hammer to the axle so the slide hammer can pull the axle out of the housing. *(Courtesy of SPX-OTC)*

5. Remove the nuts and bolts that secure the bearing retainer and, usually, also the backing plate to the end of the axle housing. Note the hole that is provided in the axle flange to allow the use of a socket and extension during removal and installation of these nuts or bolts (Figure 21–23).

6. Attach a slide hammer to the axle flange. Using whatever force is required, remove the axle from the housing. As you slide the axle and bearing out of the housing, support the axle so it does not drag over the seal. Also make sure the bearing passes through the backing plate bore and the backing plate stays at the end of the rear axle housing (Figure 21–24 and Figure 21–25).

21.4.2 Removing and Replacing a Bearing on an Axle

Removing and replacing a bearing on an axle requires an axle bearing removal tool, an installer set, and a lot of caution. A large amount of pressure is usually required to push an axle out of a bearing, and there is only enough room to support the bearing by the outer race.

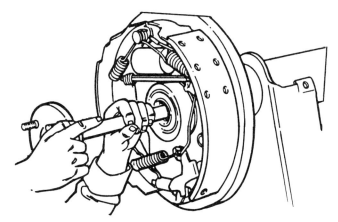

Figure 21–25 The axle should be supported as it is removed or installed into the housing to keep it from dragging across and damaging the seal. *(Courtesy of Ford Motor Company)*

SAFETY TIP: Force must be put on the axle and inner bearing race across the balls to the outer race. This force causes an outward pressure on the outer bearing race, and this outward force can cause the race to fracture and explode violently. Newer axle bearing removal tools enclose the bearing in an adapter and provide some degree of protection. If the bearing is exposed during removal operations, most technicians place a shield around the bearing during this operation to contain an explosion if it occurs (Figure 21–26).

To remove an axle bearing, you should:

1. Place the axle shaft over a large vise or anvil so it is supported under the bearing retainer ring. Using a large chisel and hammer, slice straight into the retainer in four to six places. Usually one good blow at each location is sufficient. This action should stretch the retainer so it becomes loose on the shaft. Note that the area next to the bearing is used for a seal surface on some axles; it must not be damaged. A short length of exhaust tubing or pipe can be slid over the axle to provide a shield (Figure 21–27).

2. Slide the bearing retainer plate toward the axle flange and attach the bearing removal adapter onto the bearing (Figure 21–28).

3. Place the axle and bearing remover into a press and press the axle out of the bearing.

CAUTION *After the axle is pushed a few inches, it will fall free. Be ready to catch it to prevent injury or any damage to the lug bolts (Figure 21–29).*

4. Remove the bearing retainer plate and clean the axle and retainer plate in solvent.

Chapter 21 ■ Wheel Bearing Service

Figure 21-26 Axle bearings can explode violently as they are being pressed off of the axle. A bearing separator (arrow) is positioned under the bearing to grip the bearing. An old generator or starter case can be used as a scatter shield during the removal operation.

Figure 21-27 Before pressing the axle bearing off, make about five deep chisel marks in the bearing retainer to expand it and loosen it from the shaft. Be careful not to mark or damage the seal area on the shaft next to the retainer on some axles. *(Courtesy of Ford Motor Company)*

To install an axle bearing, you should:

1. Place the bearing retainer plate over the axle next to the axle flange.
2. Place the new bearing over the axle. Note that some bearings have an inner and outer side; they should be positioned correctly.
3. Support the bearing by the inner race and press the axle completely into the bearing (Figure 21-30).
4. Place the new bearing retainer ring onto the axle, support the retainer ring on a press plate, and press the axle into the ring so the ring is solidly against the bearing.

21.4.3 Installing a Bearing-Retained Axle

Axle shaft installation is essentially the reverse of the removal process. If seal replacement is necessary on an axle that uses a seal separate from the bearing, the seal should be removed and replaced while the axle is removed. Use a procedure like that described in Section 21.4.5 to replace the seal.

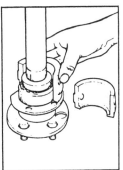
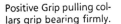
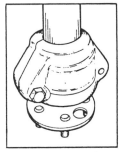

Positive Grip pulling collars grip bearing firmly.

Pulling collets completely encompass bearing during removal.

Figure 21-28 Some axle bearing puller or installer sets use a collet to enclose the bearing and attach it to the puller tube. *(Courtesy of SPX-OTC)*

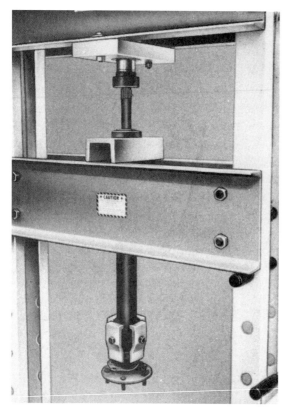

Figure 21-29 After attachment to the axle, the axle and puller assembly are placed in a press to push the axle out of the bearing. *(Courtesy of SPX-OTC)*

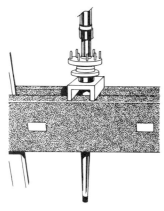

Figure 21-30 As the new bearing is pressed onto the axle, make sure that the adapters support the inner race of the bearing. A new retainer ring is pressed on after the bearing is in place. *(Courtesy of SPX-OTC)*

To install a bearing-retained axle, you should:

1. Slide the axle into the housing, being careful not to damage the axle seal, if used.
2. When the axle stops a few inches from being fully installed, grip the axle by the flange so the inner end is lifted enough to enter the axle gear in the differential. It might be necessary to rotate the axle to align the splines. Slide the axle inward so the bearing enters the housing.
3. Install the bearing retaining flange nuts and bolts and tighten them to the correct torque.
4. Replace the brake drum or rotor and caliper.
5. Replace the wheel and tighten the lug nuts to the proper torque.
6. Check the axle lubricant level and add the correct lubricant if needed.

21.4.4 Removing a C-Lock Axle

To remove a C-lock–retained axle, you should:

1. Raise and support the car securely on a hoist or jack stands.
2. Remove the wheel.
3. Mark the brake drum or rotor next to the stud that was previously marked. On cars with disc brakes, remove the caliper and secure it to the axle with a rubber tie strap, bungee cord, or wire.
4. Remove the drum or rotor.
5. Remove any dirt around the rear axle cover. Position a drain pan under the cover and remove the cover-retaining bolts. Note that most of the rear axle grease will drain out as you loosen the cover.
6. Remove the differential pinion shaft lock bolt and the pinion shaft. Note that the pinion shaft will not slide inward far enough for removal; it has to slide outward. If necessary, drive the axle shaft inward from the end opposite the lock pin hole. When the lock pin hole is past the differential case, insert a punch into the hole so the shaft can be pulled out (Figure 21-31).

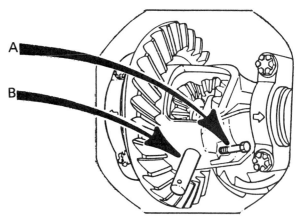

Figure 21-31 When removing an axle that is retained by a C-lock, the differential pinion shaft locking bolt (A) is removed and then the differential shaft (B) is removed. *(Courtesy of Ford Motor Company)*

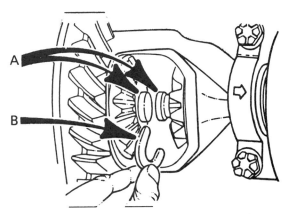

Figure 21-32 With the differential pinion shaft removed, the axles (A) can be slid inward far enough to remove the C-locks (B). *(Courtesy of Ford Motor Company)*

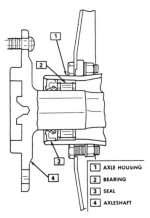

Figure 21-33 The axle serves as the inner bearing race on C-lock axles. The axle easily slides out of the bearing. If the bearing or seal surface is damaged, the axle usually needs to be replaced. *(Courtesy of General Motors Corporation, Service Technology Group)*

7. Slide the axle inward until the C-lock can be removed from the axle and remove the C-lock (Figure 21–32).

8. Slide the axle out of the housing. Be sure to support it to prevent damage to the seal.

21.4.5 Removing and Replacing a Bearing and Seal for a C-Lock Axle

The outer race of this bearing is fitted tightly into the outer end of the axle housing; the inner race is a hardened and ground portion of the axle. When an axle is removed, this portion of the axle should be inspected for damage; a rough or worn surface will require axle replacement. One aftermarket supplier provides a bearing that is positioned slightly differently to permit reuse of a worn axle. The axle seal is positioned just outboard of the bearing with the seal lip running against the axle shaft (Figure 21–33). When curing seal leakage problems, be sure to check the axle housing vent to make sure it is not plugged.

To remove a bearing or seal, you should:

1. Pry out the axle seal using a seal puller or pry bar (Figure 21–34).
2. Attach a bearing puller or hook to a slide hammer; position the puller inside the bearing and pull the bearing from the housing (Figure 21–35).

To install a new bearing, you should:

1. Lubricate the new bearing with gear lubricant.
2. Position the bearing in the housing bore; using a driving tool that contacts the entire side of the bearing, drive the bearing inward until it contacts the shoulder in the housing (Figure 21–36).
3. Lubricate the lip of the seal with gear lubricant (Figure 21–37).

Figure 21-34 The seal on C-lock axles can be easily pried out of the housing. Some bearing-retained axles use a seal farther inside of the housing, requiring the use of a puller. *(Courtesy of Chicago Rawhide)*

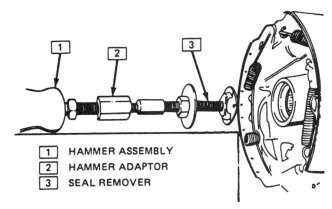

Figure 21-35 This adapter allows a slide hammer to be attached to the bearing so a faulty bearing can be pulled from the axle housing. *(Courtesy of General Motors Corporation, Service Technology Group)*

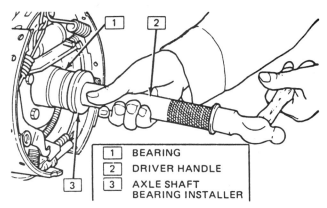

Figure 21-36 A new bearing can be driven in using a flat, correctly sized driving tool. *(Courtesy of General Motors Corporation, Service Technology Group)*

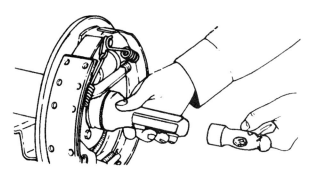

Figure 21-38 A flat, correctly sized installer should be used to drive the new seal into the housing. *(Courtesy of General Motors Corporation, Service Technology Group)*

Figure 21-37 Before a new seal is installed, it should be prelubed using the same type of grease that it is going to seal. *(Courtesy of Chicago Rawhide)*

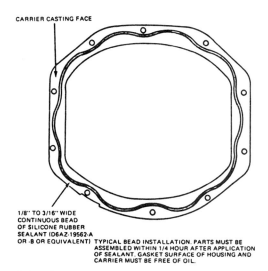

Figure 21-39 Older axle covers were sealed with gaskets. Many newer covers use a formed-in-place gasket, usually of a silicone rubber material. *(Courtesy of Ford Motor Company)*

4. If the outside edge of the seal case does not have a sealant coating, apply a thin film of **room temperature vulcanizing (RTV) silicone rubber** or nonhardening gasket sealer around the seal case.

5. Position the seal in the axle housing with the seal lip facing inward toward the lubricant. Using a driving tool that contacts the entire side of the seal case, drive the seal inward until it is flush with the edge of the axle housing (Figure 21–38).

21.4.6 Installing a C-Lock Axle

Installing a C-lock axle follows a procedure that is essentially the reverse of the removal procedure.

To install a C-lock axle, you should:

1. Lubricate the bearing and seal area of the axle.
2. Slide the axle into the axle housing and through the bearing and seal, being careful not to let the axle splines or rough axle surface drag across the seal.
3. When the axle stops a few inches from being fully installed, grip the axle flange so the inner end is lifted enough to enter the axle gear in the differential. It might be necessary to rotate the axle slightly to align the splines. Gently slide the axle inward as far as possible.
4. Slide the C-lock into the groove at the inner end of the axle.
5. Slide the axle outward so that the C-lock seats completely in the recess in the axle gear.
6. Slide the differential pinion shaft into the differential case, with the shaft hole aligned with the hole for the locking bolt.
7. Install the locking bolt and tighten it to the correct torque.

8. Install a new cover gasket and the rear axle cover and tighten the retaining bolts to the correct torque. If RTV sealant is used in place of a solid, paper gasket, thoroughly clean the gasket surfaces on the cover and the back of the axle housing and apply a bead of RTV sealant 1/16 to 1/8 in. (1.5 to 3 mm) wide around the cover or housing gasket surface, circling each of the bolt holes (Figure 21–39).
9. Add the proper gear lubricant through the filler hole to bring the gear oil to the correct level in the axle housing.
10. Replace the brake drum, with the marks aligned; replace the tire and wheel and tighten the lug nuts to the correct torque.

21.5 Repairing Serviceable FWD Front Wheel Bearings

There are several different styles of FWD front wheel bearings: ball bearings, tapered roller bearings, and sealed ball or roller bearings. Most of these do not require periodic maintenance. The major reason for servicing is to replace a damaged hub or to eliminate rough, loose, or noisy operation (Figure 21–40).

Servicing the front wheel bearings makes it necessary to remove the outer end of the axle or CV joint housing from the hub. In cases where a prevailing torque hub nut is used, a new nut should be installed during replacement. This nut can usually be removed using an air-powered, impact type of wrench, but it should be installed by hand to prevent damage to the bearings. Hand installation allows a good feel of the nut's condition. Although it is a popular trade method, using an air-impact wrench to remove and replace the hub nut is not recommended because the hammering effect of the wrench can cause brinelling of the bearings. Care should also be taken to prevent damage to the CV joint or the CV joint boot. Servicing procedures for the CV joint and boot are described in Chapter 30.

Many FWD cars with ABS have a gearlike tone wheel on the outer CV joint housing. All ABS cars have a wheel speed sensor mounted at the steering knuckle; the tone wheel is on the hub or rotor. On these cars, special care is needed to avoid damaging the tone wheel, the speed sensor, or the wires connecting the sensor to the car's body or frame.

The repair procedures for the various FWD wheel bearings vary among different makes of cars. It is highly

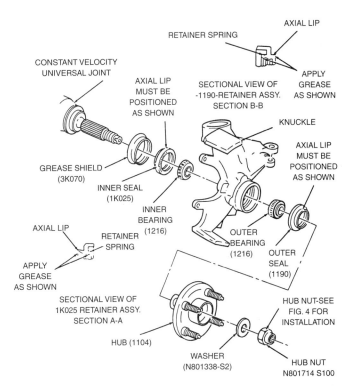

Figure 21–40 A disassembled view of the front knuckle and hub of a FWD car. This particular car uses tapered roller bearings; some cars use ball bearings. *(Courtesy of Ford Motor Company)*

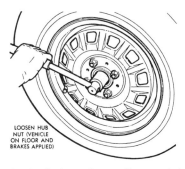

Figure 21–41 The hub nut is usually very tight; it should be loosened with the tires on the ground. This nut must be removed for access to the front wheel bearings on a FWD car. *(Courtesy of Chrysler Corporation)*

recommended that the specific repair steps for a particular car be followed while servicing the wheel bearings.

In general, to remove the front wheel bearing from a FWD car, you should:

1. With the tires on the ground, the transmission in park or first gear, and the parking brake applied, remove any locking devices and loosen the hub nut (Figure 21–41).

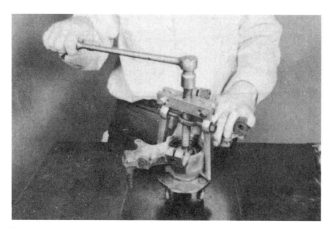

Figure 21–42 This puller is being used to pull the knuckle from the front hub. *(Courtesy of Chicago Rawhide)*

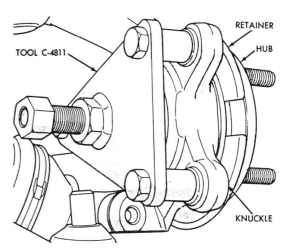

Figure 21–43 Tool C-4811 is being used to remove the hub from the knuckle. *(Courtesy of Chrysler Corporation)*

2. Raise and support the car on a hoist or jack stands.
3. If available, install a boot protector over the CV joint boots.
4. Remove the hub nut, the tire and wheel, and the brake caliper. Using a rubber tie strap, bungee cord, or wire, suspend the brake caliper from some point inside the fender.
5. Remove the hub from the steering knuckle. This operation often requires the use of a puller. The steering knuckle must be disconnected from either the lower control arm or the strut. In some cars, the tie-rod should be disconnected from the steering arm. Disconnecting these parts allows the steering knuckle freedom to move outward so the hub can be separated from the CV joint housing (Figure 21–42).
6. Remove the bearing and seal from the steering knuckle. On some cars, the bearing needs to be removed from the hub. This operation can also require the use of a special puller (Figure 21–43 and Figure 21–44).

Like the disassembly procedure, bearing installation should follow the manufacturer's procedure.

In general, to install front wheel bearings on a FWD car, you should:

1. Install the new bearing and seal into the steering knuckle. Lubricate the seal lips and the bearing, if required, with the proper amount and type of lubricant (Figure 21–45).
2. Install the hub into the steering knuckle.
3. Install the hub and steering knuckle onto the axle end with the CV joint housing.
4. Replace the hub nut and brake caliper. Tighten the hub nut and the caliper mounting bolts to the correct torque. Lock the hub nut in place as required (Figure 21–46, Figure 21–47, and Figure 21–48).
5. Replace the tie-rod, the lower end of the steering knuckle, and the steering knuckle-to-strut bolts if any of them were disconnected during disassembly. Tighten all of these nuts and bolts to the correct torque. Remove the CV joint boot protector, if used.
6. Replace the tire and wheel, and tighten the lug nuts to the correct torque.

21.6 Repairing Nonserviceable FWD Front Wheel Bearings

Servicing of this style of bearing is limited to removing and replacing the hub and bearing assembly, much like the procedure used on nonserviceable wheel bearings. This operation makes it necessary to remove the axle end with the CV joint housing from the hub. A boot protector or cover should be used to prevent damage to the CV joint boot. A damaged, cracked, or torn CV joint boot should be replaced; this operation is described in Chapter 30.

To remove a nonserviceable wheel bearing, you should:

1. With the tires on the ground, the transmission in park or first gear, and the parking brake applied, remove the locking device from the front hub nut, if used, and loosen the hub nut.
2. Raise and support the car on a hoist or jack stands.

Chapter 21 ■ Wheel Bearing Service 383

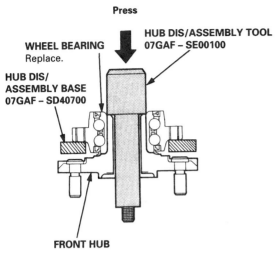

Figure 21–44 The tool is being used to press the hub out of the bearing. *(Courtesy of American Honda Motor Co., Inc.)*

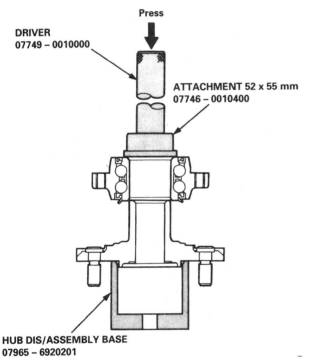

Figure 21–45 The tool is being used to press a new bearing onto the hub. Note how the attachment is putting pressure on the inner race. *(Courtesy of American Honda Motor Co., Inc.)*

Figure 21–46 The hub is being pushed back onto the driveshaft after the bearing and hub have been installed in the knuckle. *(Courtesy of Chicago Rawhide)*

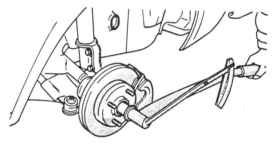

Figure 21–47 The hub nut should be tightened to the torque recommended by the vehicle manufacturer; this is somewhere between 100 and 250 ft-lb. You will probably need to install the wheel and lower the car to the ground for final tightening. *(Courtesy of Chrysler Corporation)*

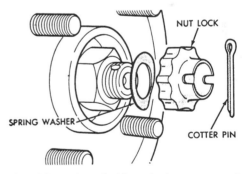

Figure 21–48 Various locking devices are used on hub nuts; if used, these should be replaced in the correct manner. *(Courtesy of Chrysler Corporation)*

3. If available, install a boot protector over the CV joint boot.
4. Remove the hub nut and the wheel.
5. Remove the brake caliper. Using a rubber tie strap, bungee cord, or wire, suspend it from some point inside of the fender and remove the rotor.
6. Remove the hub and bearing mounting bolts and the rotor splash shield (Figure 21–49).
7. Install a hub puller onto the hub flange and pull the hub and bearing assembly off of the end of the CV joint housing and out of the steering knuckle.
8. Some car models also have a steering knuckle seal that should be removed at this time.

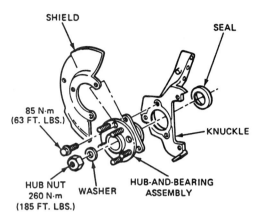

Figure 21–49 The hub and bearing assembly on this FWD car is permanently sealed, lubricated, and adjusted. It is replaced as an assembly if there are problems. *(Courtesy of General Motors Corporation, Service Technology Group)*

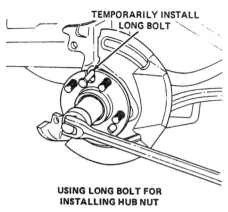

Figure 21–50 Some vehicles have a provision for installing a bolt through a notch in the hub to keep the hub from turning while the hub nut is tightened. *(Courtesy of General Motors Corporation, Service Technology Group)*

To replace a nonserviceable type of front wheel bearing, you should:

1. Clean the bore of the steering knuckle and the splined area of the CV joint housing.
2. Lubricate the lip of the new steering knuckle seal, if used, and install the seal in the steering knuckle.
3. Position the new hub and bearing assembly over the axle and into the steering knuckle. Install the hub nut and the hub and bearing assembly mounting bolts. Tighten them alternately to move the hub and bearing assembly into the correct location. Tighten the hub nut temporarily to about 70 ft-lb. (100 N-m) of torque and tighten the mounting bolts to the correct torque. A long bolt can be used in place of one of the mounting bolts to keep the hub from turning while tightening the hub nut (Figure 21–50).
4. Remove the boot protector.
5. Replace the rotor and caliper. Tighten the caliper mounting pins to the correct torque.
6. Install the wheel and tighten the lug nuts to the correct torque.
7. Lower the car to the ground and tighten the hub nut to the correct torque. Lock the hub nut in place as required.

21.7 Practice Diagnosis

You are working in a brake and front-end shop and encounter the following problems:

CASE 1: The customer's complaint is a lockup of the right rear wheel under a medium pedal pressure. The car is a mid-1980s, RWD station wagon. While on a road test, you hear a slight growling sound coming from the right rear, and the brake lockup is confirmed by a skidding right rear tire when you apply the brake. In the shop, with the rear end lifted, you hear the same growling noise coming from inside the right rear wheel; when you remove the wheel and brake drum, you find that the rear brakes are oily. Can you fix this problem? If so, what will you probably need to do?

CASE 2: The car is a midsize FWD General Motors "W" body car, and the customer's complaint is a strange noise coming from the left rear. On a road test, you hear a growling noise that seems to come from the left rear, and it becomes quieter when you make left turns. What should you do when you get this car back to the shop to confirm this problem? What is the probable cause? What will you probably need to do to repair this car?

Terms to Know

bearing-retained axle

C-lock axle

heavy-duty grease

high-temperature grease

room temperature vulcanizing (RTV) silicone rubber

service

Review Questions

1. Technician A says that faulty wheel bearings usually change noise levels as the car is turned in different directions.
   Technician B says that faulty wheel bearings usually change noise levels under different throttle and brake conditions.
   Who is correct?
   a. A only
   b. B only
   c. Both A and B
   d. Neither A nor B

2. When checking the rear axle bearings of a RWD car, in all cases there should be:
   A. No vertical play
   B. No in-and-out play
   Which option best completes the statement?
   a. A only
   b. B only
   c. Both A and B
   d. Neither A nor B

3. Technician A says that wheel bearing play is checked by gripping the tire at the top and bottom and trying to rock it.
   Technician B says that the front wheel bearing of a RWD car should have a slight amount of play when rocked.
   Who is correct?
   a. A only
   b. B only
   c. Both A and B
   d. Neither A nor B

4. A faulty wheel bearing _____ when spun during the checking process.
   A. Is noisy
   B. Operates with a roughness
   Which option best completes the statement?
   a. A only
   b. B only
   c. Both A and B
   d. Neither A nor B

5. Technician A says that the best way to remove the inner wheel bearing from a front hub is to drive the bearing and seal out with a wooden dowel.
   Technician B says that the best tool to remove a cotter pin from the spindle is either a cotter pin puller or diagonal-cutting pliers.
   Who is correct?
   a. A only
   b. B only
   c. Both A and B
   d. Neither A nor B

6. When cleaning wheel bearings, spinning the bearing with an air blast is:
   A. An effective way of cleaning the bearing
   B. Not an unsafe shop practice
   Which option best completes the statement?
   a. A only
   b. B only
   c. Both A and B
   d. Neither A nor B

7. Wheel bearings are repacked by forcing clean grease into the bearing using:
   A. Hand pressure
   B. A bearing packer
   Which option best completes the statement?
   a. A only
   b. B only
   c. Both A and B
   d. Neither A nor B

8. Technician A says that front wheel bearings of the tapered roller type should be adjusted to a slight end play.
   Technician B says that after the bearings have been seated, the spindle nut should be tightened finger tight.
   Who is correct?
   a. A only
   b. B only
   c. Both A and B
   d. Neither A nor B

9. Technician A says that you must remove the axle to repair a leaky axle seal on the rear axle of a RWD car.
   Technician B says that the rear axle housing cover must always be removed to remove an axle.
   Who is correct?
   a. A only
   b. B only
   c. Both A and B
   d. Neither A nor B

10. On C-lock axles:
    A. A faulty axle bearing can easily ruin the axle
    B. The axle bearing must be pressed off of the axle
    Which option best completes the statement?
    a. A only
    b. B only
    c. Both A and B
    d. Neither A nor B

11. As an axle shaft is slid into the housing, the:
    a. Shaft should be kept from contacting the seal
    b. Shaft splines must be aligned with the gear in the differential
    c. Bearing should be lubricated with gear oil
    d. All of these.

12. Lip seals should always be _____ before installing them.
    A. Coated with sealant
    B. Lubricated on the seal lips
    Which option best completes the statement?
    a. A only
    b. B only
    c. Both A and B
    d. Neither A nor B

13. Technician A says that it is necessary to remove the CV joint from the front hub when servicing the front wheel bearings of a FWD car.
    Technician B says these front wheel bearings should be serviced every 15,000 to 20,000 mi.
    Who is correct?
    a. A only
    b. B only
    c. Both A and B
    d. Neither A nor B

14. Technician A says that you should loosen the front hub nut of a FWD car with the tires on the ground and the parking brake applied.
    Technician B says that it is a good practice to place a boot protector over the CV joint boot when servicing the front wheel bearings of a FWD car.
    Who is correct?
    a. A only
    b. B only
    c. Both A and B
    d. Neither A nor B

15. Nonserviceable wheel bearings are repaired by:
    A. Removing and replacing the entire hub and bearing assembly
    B. Removing, replacing, and adjusting the sealed bearing unit
    Which option best completes the statement?
    a. A only
    b. B only
    c. Both A and B
    d. Neither A nor B

16. Technician A says that the front hub nut of some FWD cars is secured in place by bending the lip of the nut into a groove at the end of the threads.
    Technician B says that this nut is secured by a cotter pin.
    Who is correct?
    a. A only
    b. B only
    c. Both A and B
    d. Neither A nor B

17. The front hub nut of a FWD car must be:
    A. Tightened using an air-impact wrench
    B. Tightened enough to obtain the correct bearing preload
    Which option best completes the statement?
    a. A only
    b. B only
    c. Both A and B
    d. Neither A nor B

Math Question

1. While adjusting the wheel bearing on a Toyota pickup, you find that you need to tighten the nut to 34 N-m and have 0.5 to 1.0 mm of end play. You only have an inch-pound torque wrench, and your dial indicator reads in inches. What specifications should you use?

22 Drum Brake Service

Objectives

Upon completion and review of this chapter, you should be able to:

- ❏ Remove, clean, measure, and inspect brake drums for wear or damage and determine whether they should be machined or replaced.
- ❏ Mount a drum on a lathe and machine it according to the manufacturer's procedures and specifications.
- ❏ Clean and remove brake shoes, springs, and any related hardware and determine the necessary repair.
- ❏ Clean, inspect, and lubricate a backing plate.
- ❏ Lubricate and install brake shoes, springs, and related hardware.
- ❏ Adjust brake shoes and install brake drums.
- ❏ Perform the ASE tasks for drum brake diagnosis and repair (see Appendix A).

22.1 Introduction

A typical brake job on a drum brake assembly consists of the following: drum removal, drum machining, shoe removal, rebuilding or replacement of the wheel cylinder, thorough cleanup of the backing plate and all small parts, careful reassembly of the parts, lubrication of these parts as they are assembled, adjustment of the shoe-to-drum clearance, and adjustment of the parking brake. The parts should be carefully inspected at least three times—before disassembly, during cleanup, and during reassembly—to locate any faulty or damaged components that might affect the final job. All damaged parts should be replaced. A brake system cannot work as it should if one or more parts are faulty.

After the drum has been removed, the disassembly inspection serves three major purposes: to locate damaged parts, to determine the relationship of all of the parts, and (if there are unusual wear patterns) to determine problems involving the backing plate alignment or drum surface. A bell-mouthed, concave, or convex drum surface or a distorted backing plate is indicated by the abnormal lining wear it causes. Some newer brake designs have subtle differences in spring or linkage positioning (Figure 22–1). These differences should be noted so that mistakes will not be made during reassembly.

Competent brake technicians realize that in disassembling a unit, they become the authority on how that unit goes together. A service manual should be followed as a guide to the disassembly and reassembly steps. These

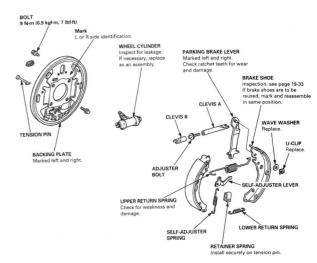

Figure 22–1 An exploded brake assembly showing various checks. *(Courtesy of American Honda Motor Co., Inc.)*

Figure 22–2 This return spring was installed upside down and was nearly worn in two from rubbing on the hub.

manuals, however, tend to be somewhat general and sometimes fail to show all the important details. As you inspect a unit to locate damaged parts, you should check the hold-down and return spring attachment positions, how and where the self-adjusters are attached, and how and where the parking brake mechanism is positioned (Figure 22–2). Determine any obvious differences that can help you, because you are the one who is going to put the unit back together. For some of the more complicated assemblies, it might be necessary to draw a simple sketch to remember the details. As a last resort with complex units, work on one side at a time. Most cars have a right-side brake that is a mirror image of the left one. If one side is left intact, it can serve as a pattern until the other one is assembled.

SAFETY TIP: The following sections describe operations on a single brake assembly. All brake operations are normally done in pairs. The same operation is done at each end of the axle to ensure even, straight braking. Do not forget the safety tips that were mentioned in Chapter 20.

22.2 Drum Removal

Several different methods are used to secure a drum onto an axle or hub. For the drive axle of RWD cars, the most common style is a floating drum. It is slid over the axle pilot, which centers it and the wheel studs (commonly called lug bolts). It is held in place with speed nuts and the wheel lug nuts (Figure 22–3). The speed nuts, also called Tinnerman nuts, are used to keep the drum on the axle during manufacture and assembly of the car. After the wheel is installed, they become unnecessary. Other drive axle drums are secured to the axle flange by a pair of bolts or cap screws (Figure 22–4). In a few cases, the center of the drum is splined for the axle and used to drive the wheel. This style of drum is usually secured to the axle by a large nut at the end of the axle (Figure 22–5).

Drums on nondrive axles—the front drum on a RWD car and the rear drum on a FWD car—use several different drum-to-hub attachment methods also. The hub and drum are normally treated as an assembly. If a faulty drum or hub requires replacement, it is usually less expensive to separate them so the faulty portion can be replaced. The most common method of securing the drum to the hub is to use shouldered wheel studs that are

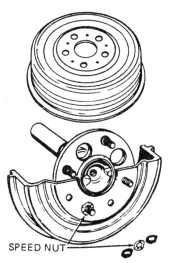

Figure 22–3 A floating brake drum is centered on the axle by the pilot hole in the center and is often held on the axle by speed nuts. *(Courtesy of Wagner Brake)*

Figure 22–4 This drum is held on the axle flange by two Phillips-head screws. Removing these screws allows drum removal. Another screw under the hole (arrow) can be turned to push the drum off if it is stuck.

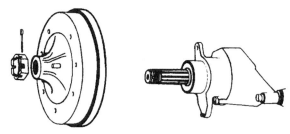

Figure 22–5 The brake drums on some cars are splined onto the axle and held in place by a nut. *(Courtesy of Bendix Brakes, by AlliedSignal)*

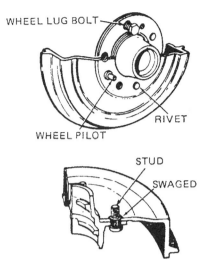

Figure 22-6 Most front drums are secured to the hub by swaging a portion of the wheel stud shank. Other front drums are riveted to the hub, and a few float on the hub like a rear drum. *(Courtesy of Wagner Brake)*

Figure 22-7 On some cars, the wheel stud is pressed through the hub and drum and then swaged. The swaging should be removed before driving or pressing the stud out. *(Courtesy of Wagner Brake)*

Figure 22-8 Occasionally a frozen drum can be loosened by squirting penetrating oil around the axle flange and wheel studs and then striking the area between the wheel studs with a punch and hammer.

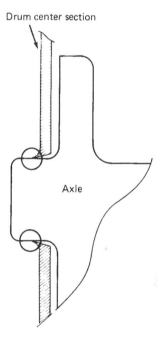

Figure 22-9 The center hole of some drums can catch and grip the axle pilot tighter as you pull the drum off.

swaged into place (Figure 22-6). **Swaging** is also called *upsetting*, *peening*, or *riveting*. The drum is normally placed over the hub, the wheel studs are pressed through the hub and drum, and the shoulder of the stud is swaged to lock it in place (Figure 22-7). Some drums are positioned on the inner side of the hub, and some are positioned on the outer side. Other methods of drum-to-hub attachment use rivets to lock them together or hold the drum in place with speed nuts, much like the floating drum on a rear axle.

Removal of a drum can be made difficult by three different problems: failure to release the parking brake, freezing of a drum onto the axle flange or pilot, and shoes that have adjusted to a worn drum surface. The solution for the first problem is to simply release the parking brake, but solving the other two problems can be more difficult. With frozen drums, apply penetrating oil around the axle pilot and wheel studs and then strike the area between and inside the wheel studs using a hammer and punch (Figure 22-8). If this does not work, use a plastic hammer and strike the outer edge of the drum with a blow directed toward the axle, repeating this procedure several times. A drum's stamped-steel center section occasionally catches and digs into the axle pilot, and the hammer blows tend to remove the interference (Figure 22-9). If this method works, be sure to check the drums for cracks that might result. If this technique does not work, apply heat on the drum face around the axle pilot in the area inside the wheel studs. This method is effective only on drums that have stamped-steel inner sections (Figure 22-10). Move the torch flame evenly around the axle flange until the center section expands and loosens its grip. A puller is also available to pull floating drums

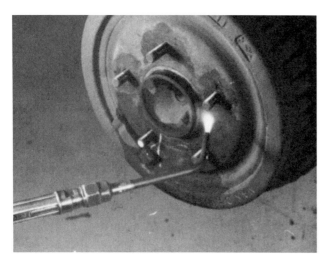

Figure 22-10 A drum with a tight pilot hole can usually be easily removed by heating the center area, inside the wheel studs.

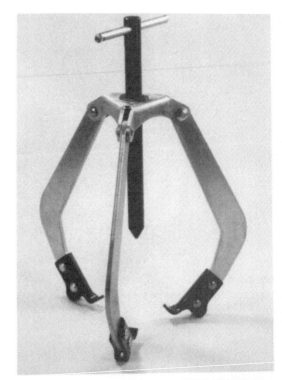

Figure 22-11 This puller is used to remove tight drums. If much force is used, be sure to check the drum for straightness after removal. *(Courtesy of KD Tools)*

off the axle (Figure 22–11). It should be noted that heat and pullers are last-resort methods of drum removal because they can damage the drum. If you use these severe drum removal methods, it is always a good practice to check the drum for straightness. Place the drum—open side down—on a flat surface and lay a straightedge across the center section. The straightedge should be parallel to the top of the surface (Figure 22–12).

The drums on some import cars and light trucks that have a cast inner section often have a pair of holes drilled and topped close to the axle flange. Machine screws or bolts can be threaded into these holes, where they will press against the axle flange and push the drum off the axle.

If the drum is loose on the axle but will not pass over the shoes, the shoes must be adjusted to a smaller size. The drum is probably grooved, and the shoes have adjusted outward into the groove (Figure 22–13). On cars with self-adjusters, this can be difficult. It is necessary to locate the adjuster screw, obtain access to the screw and adjuster lever, move the self-adjuster lever out of the way, and turn the adjuster screw in the correct direction. Most cars have an access plug positioned in the backing plate or the face of the drum that can be removed so you can reach the mechanism. This plug can be an easily removed rubber or metal plug or a stamped knockout slug. Usually you can reach the lever using a long, thin screwdriver or an ice pick and carefully push it away; it needs to move only about 1/16 in. (1.5 mm). With the lever pushed back, use a brake adjuster tool or an old screwdriver to rotate the adjuster screw (Figure 22–14). It is a good idea to plan how you will release the self-adjuster each time you work on a different brake design. As a last resort, if you cannot back off the self-adjuster, cut off the ends of the shoe hold-down pins where they extend through the backing plate. Side- or diagonal-cutting pliers (dikes) can be used for this procedure. This usually will allow you to pull the drum outward far enough to reach in and release the adjuster. The pins are relatively inexpensive, and the time you would spend fighting a drum would cost much more (Figure 22–15).

SAFETY TIP: It is recommended that the brake drum be wet down thoroughly or enclosed in the chamber of an OSHA-approved brake cleaning system during the removal steps (Figure 22–16).

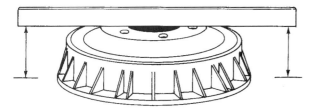

Figure 22-12 A drum is checked for straightness by placing it on a flat surface and placing a straightedge across the mounting surface. The straightedge should be parallel to the flat surface.

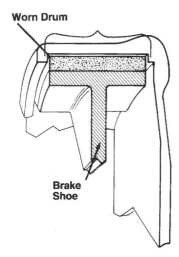

Figure 22-13 If the shoes have been adjusted to fit a badly worn drum, the drum will catch on the lining as you try to remove it.

To remove a brake drum, you should use the following procedure:

1. Raise and support the car in a secure manner.
2. Remove the wheel as described in Chapter 20.
3. Release the parking brake.
4. Prevent the release of brake dust by thoroughly wetting down the exterior of the drum using a low-pressure brake washer. Rotate the drum and try to get the solution inside of the drum to wet the interior. An aerosol spray or vacuum enclosure can also be used.
5. Determine the method of drum attachment and remove the drum.
    a. On nondrive axle drums, remove the dust cap, wheel bearing cotter pin, and adjusting nut. Then slide the drum and hub along with the wheel bearings off the spindle (Figure 22-17). This process is described in Section 20.2.
    b. If the drum is a floating drum, remove the speed nuts and slide it off the axle flange and lug bolts.

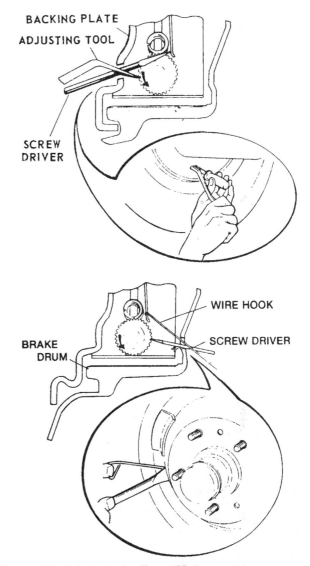

Figure 22-14 To back off a self-adjuster, it is necessary to push or pull the lever away from the starwheel so you can rotate the starwheel. *(Courtesy of General Motors Corporation, Service Technology Group)*

   c. If the drum is secured with bolts, remove the bolts and slide it off the axle. If the drum is stuck, check the drum for two holes threaded into the face. The bolts used to retain the drum or any bolts of the correct size can be threaded into these holes and used as pullers to remove the drum.
   d. If the drum is secured by a single nut on the axle, remove the nut and slide the drum off the axle. If the drum is stuck, replace the nut so it is free of the drum and even with the end of the axle; then install a brake drum puller. The nut is used to protect the threads at the end of the axle. Tighten the puller to remove the drum (Figure 22-18).

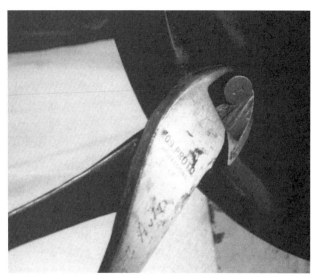

Figure 22–15 If all else fails, cut the heads off the hold-down pins. This lets you pull the drum outward and gain access to the adjuster.

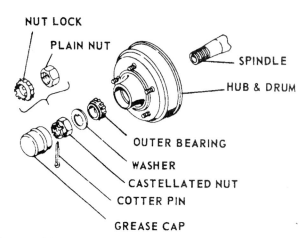

Figure 22–17 The front drums on light trucks and older RWD cars are normally retained by the front wheel bearings. Rear drums on many FWD cars are similar. *(Courtesy of General Motors Corporation, Service Technology Group)*

Figure 22–16 An enclosure has been placed over the brake assembly, and it has been connected to a HEPA filter–equipped vacuum cleaner. Any asbestos fibers released during drum removal will be safely caught in the filter. *(Courtesy of NILFISK OF AMERICA, INC.)*

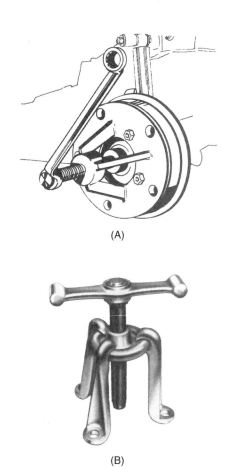

Figure 22–18 If the rear drum is splined to the axle, it is usually necessary to use a puller to remove it. Many technicians prefer to loosen the nut a few turns and leave it on the threads to protect them while pulling the drum. *(A is courtesy of Volkswagen of America, Inc.; B is courtesy of SPX-OTC)*

6. Thoroughly vacuum any dust or dirt from the inside of the drum or wash the drum in soapy water or solvent and air dry it. If the drum is washed in a petroleum-based solvent, the friction surface must be reconditioned or cleaned using denatured alcohol or a commercial brake friction surface cleaner to remove all traces of the solvent residue (Figure 22–19).

Figure 22-19 After removal, it is a good practice to thoroughly clean the drum to remove any traces of asbestos and to make inspection easier.

Figure 22-20 Support the drum loosely as shown and strike it lightly with a hammer. You should hear a bell-like sound. A dull thud indicates a cracked drum.

22.3 Drum Inspection

Most technicians check the brake drum as it is being removed. Brake drum inspection should include checks for cracks; a scored friction surface; a bell-mouthed, concave, or convex friction surface; hard spots and heat checks; an oversize diameter; and an out-of-round diameter. Faulty drums must be replaced or remachined. Drum **machining** is also called *turning, truing,* or *remachining.*

A check for a cracked drum includes a visual inspection and an auditory inspection. Look for cracks across the friction surface and at the lug bolt holes. For the auditory check, lightly support the drum by the inner hole and strike the outer edge lightly with a steel hammer. The drum should make a ringing, bell-like sound. If it makes a dull thud or the sound of plain metal striking metal, it is cracked (Figure 22-20). A cracked drum should be replaced.

A scored friction surface is caused by grit on the lining or contact with a rivet or shoe rim. A scored drum has a lining with ridges or grooves (Figure 22-21). Any scores that are deeper than 0.010 in. (0.25 mm) require that the drum be turned. An experienced technician can usually guess the depth of the grooves; the most practical way to measure them is to use a brake lathe to see how deep a cut is needed to remove them.

A **bell-mouthed drum,** one with a concave or convex friction surface, is accompanied by shoes that have

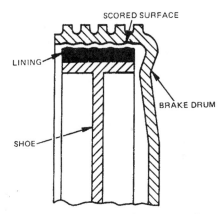

Figure 22-21 A badly scored drum. Note how the lining wears to match the scored surface. Such drums should be machined or replaced. *(Courtesy of Wagner Brake)*

uneven wear on the lining (Figure 22-22). These problems produce a drum friction surface with varying diameters. If the diameters vary more than 0.010 in., the drum should be turned.

Hard spots, also called *chill spots* and *heat checks,* are the result of very high heat conditions. Hard spots are caused by a metallurgical change from cast iron to steel; an overheated drum often shows a bluish or golden tint. Heat checks are visually evident in the worn drum surface, and excessive checking requires drum replacement (Figure 22-23). As the drum is turned, hard spots

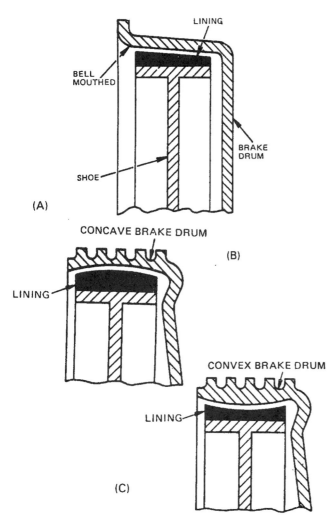

Figure 22-22 A bell-mounted drum (A), a concave drum (B), and a convex drum (C). Note that in each case the lining wears to match the drum surface. *(Courtesy of Wagner Brake)*

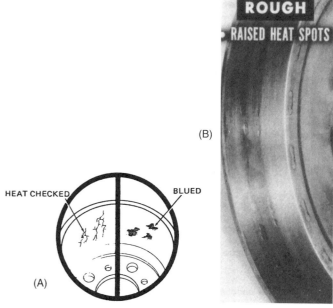

Figure 22-23 Heat checks or blued spots on the friction surface are signs of an overheated drum with hard spots (A). The hard spots will show up as the drum is turned (B). *(A is courtesy of Wagner Brake; B is courtesy of General Motors Corporation, Service Technology Group)*

will appear as raised, hard areas or islands in the drum's surface. These spots also cause a series of clicking noises when they contact the cutter bit as the drum is being turned. The raised portion of a hard spot can often be removed with a carbide cutter bit (although the raised area is often a result of drum deflection) or by grinding. Drum grinding is a slower, more expensive operation that requires a grinder attachment for the lathe (Figure 22-24). Some sources consider grinding a hard spot a temporary repair and recommend replacement of any drum with hard spots.

Drum diameter is measured using a **brake drum micrometer,** which is commonly called a **drum mike.** Measuring the diameter in three or four directions allows the drum to be checked for out-of-roundness (Figure 22-25). A drum with a diameter greater than the maximum allowable should be replaced. A drum with diameter differences greater than 0.005 in. (0.12 mm) should be turned.

Since 1971, all brake drums have been made with the maximum diameter dimension indicated on them (Figure 22-26). This is the wear or *discard diameter* of the drum, and a drum larger than this must not be used. There are three commonly used drum dimensions: the *original diameter*, the *maximum machining* or *rebore diameter* (which is usually 0.060 in. [1.5 mm] larger than the original), and the *wear diameter* (which is about 0.090 in. [2.3 mm] larger than the original). These sizes can vary; always check the manufacturer's specifications. A drum should never be turned to a size larger than the maximum machining diameter. When a drum is marked with "Max. Dia." (maximum diameter) or "Discard Dia.," this size can be considered the maximum size that the drum can be machined to. On the drums used in domestic cars before 1971, the original diameters were usually even inches or half inches, such as 9, 9½, and 10 in., and the maximum allowable diameters were 0.060 in. greater, or 9.060, 9.560, and 10.060. However, a few manufacturers used drums with original diameters that were 0.030, 0.060, and 0.090 in. larger than an 11- or 12-in. diameter. The maximum diameters for these drums are listed in service manuals.

Any drum that is too large must be replaced. When a drum is too thin, it loses its ability to absorb heat, and the operating temperatures can rise excessively. A drum that is too thin also loses its structural strength. An excessive

Chapter 22 ■ Drum Brake Service

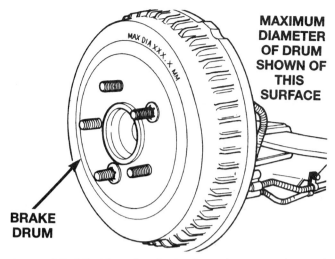

Figure 22–26 All modern brake drums have a maximum allowable dimension indicated on them. *(Courtesy of Chrysler Corporation)*

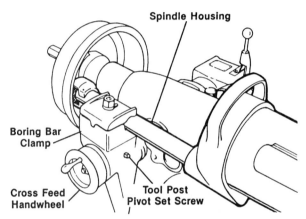

Figure 22–24 A brake drum grinder is secured to the tool post of a brake drum lathe and is used to grind hard spots so they are flush with the drum surface. *(Courtesy of Ammco Tools, Inc.)*

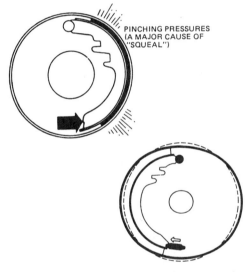

Figure 22–27 A brake drum will deflect under high brake pressures, and thinner drums will deflect more. Deflection can cause a spongy pedal or a pinching of the lining ends, which can cause squeal. *(Courtesy of Wagner Brake)*

amount of deflection can occur during braking, and excessive deflection will cause a spongy brake pedal (Figure 22–27).

To measure drum diameter, you should:

1. Adjust the drum micrometer, or mike, to the original diameter of the drum (Figure 22–28).

NOTE: A drum mike can be checked against a standard or a large outside micrometer to determine its accuracy. Inaccurate drum mikes should be recalibrated (Figure 22–29).

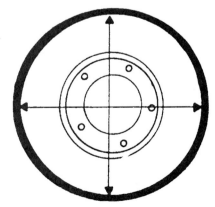

Figure 22–25 Drum diameter should be measured in two or more directions. If the measurements are different, the drum is out-of-round. *(Courtesy of Bendix Brakes, by AlliedSignal)*

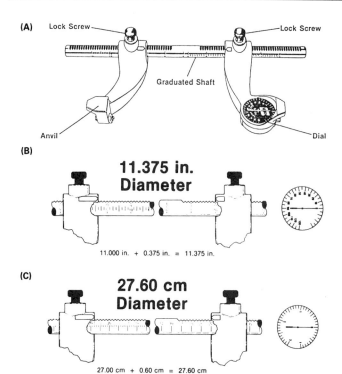

Figure 22-28 A brake drum micrometer, or mike, must be set to the drum diameter by moving the anvil and dial assemblies to the proper location on the graduated shaft. The mike at (B) has been set to 11.375 in.; the metric mike at (C) has been set to 27.60 cm, or 276 mm. *(Courtesy of Ammco Tools, Inc.)*

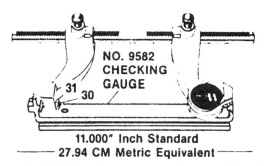

Figure 22-29 Checking gauges or large outside micrometers can be used to determine the accuracy of a drum mike. The checking standard shown has an inside width of exactly 11 in. If the reading differs, the mike should be calibrated by adjusting the set screw (30). *(Courtesy of Ammco Tools, Inc.)*

Figure 22-30 To prevent damage, a drum mike is always inserted into the drum with the dial end entering first. It is always removed with the dial end leaving last.

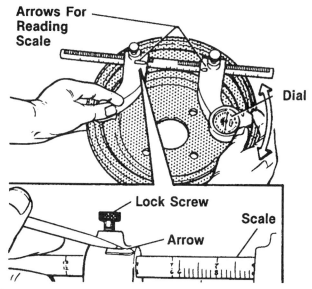

Figure 22-31 After the mike has been placed into the drum, the anvil end is held tightly against the drum while the dial end is moved to the highest reading on the dial. *(Courtesy of Ammco Tools, Inc.)*

2. Place the mike in the drum so the measuring end, next to the dial, enters first and the rigid end enters last (Figure 22-30).

3. Hold the rigid end of the mike firmly against the drum's inner surface and swing the measuring end in an arc as you watch the scale reading (Figure 22-31). Position the mike at the point of highest reading on the dial. This is the amount of drum oversize. When added to the size setting of the mike, it is the actual drum diameter. For example, if the mike is set to 10 in. and the dial reads 0.020, the drum diameter is 10.020 in. (Figure 22-32).

4. Lift the mike out of the drum so the rigid end comes out first. The mike can be damaged if the measuring end leaves first and snaps out.

5. Remeasure the drum in two or three more locations and compare the measurements. A difference in readings indicates drum out-of-roundness. A brake drum can also be checked for *radial runout*, which can cause a pulsating brake pedal and possible brake grab. Remount the drum onto the hub backwards and position a dial indicator on the

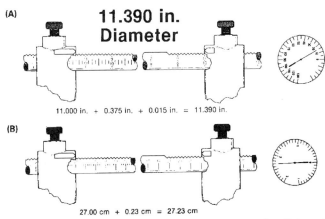

Figure 22–32 A drum mike is read by adding the dial reading to the graduated shaft setting. In (A), the size is 11.390 in. (11.00 + 0.375 [3/8] + 0.015). In (B), the size is 27.23 cm, or 272.3 mm (27.00 + 0.23). If the mike was set to the actual drum diameter, the reading on the dial would be the amount of drum oversize. *(Courtesy of Ammco Tools, Inc.)*

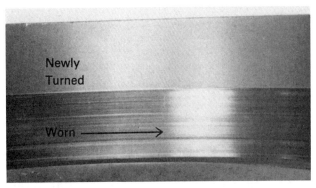

Figure 22–34 The inner portion of this drum shows light scoring, but note how the friction surface has been burnished to an extremely smooth finish. The outer portion of this drum has been turned to a smooth surface with the correct surface finish for breaking in the new lining.

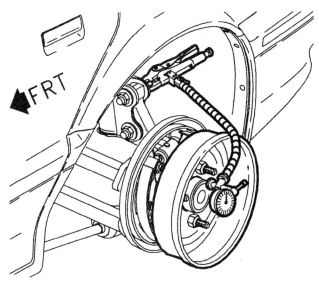

Figure 22–33 Mounting the drum onto the hub backwards allows a dial indicator to be used to measure a drum for runout or out-of-round. More than 0.011 in. (0.28 mm) of runout can cause problems. *(© Saturn Corporation, used with permission)*

drum's inner surface (Figure 22–33). Then rotate the drum while watching the dial indicator. More than 0.006 in. (0.15 mm) of runout can cause a problem. This check can be affected by hub runout. If the amount of runout is excessive, re-index the drum on the hub and repeat your check; if it is still excessive, the drum should be turned or replaced.

After the drum checks, the technician makes a decision about what to do. Any faulty drums must be replaced, and it is recommended that drums be replaced in pairs. Besides the problem of turning the new drum to match the old one at the other end of the axle, the structural differences between the new and old drums can cause stopping differences during a hard stop. Excessively worn or out-of-round drums should be turned, and drums should always be turned in pairs. The right and left drums on an axle should have the same surface finish and almost the same diameter. A size difference of 0.010 in. (2.5 mm) is allowable.

In the past, most sources recommended always turning a drum, as long as it stays within a safe diameter. The very smooth surface we see on a used drum is not always as flat as it appears. When a new lining is installed, it has to wear or be broken in to exactly match the drums' friction surface. There is only partial lining-to-drum contact until the lining wears in, and a worn, burnished drum surface is usually too smooth to wear in the shoes within a reasonable length of time. Many people want to test their brakes with a really hard stop as soon as the job is done. A hard stop with a lining that has poor contact can severely overheat those portions of the lining in contact with the drum. It is recommended that a drum have a friction surface finish of about 40 to 80 microinches. This surface is very flat, having an average scratch depth of 40 to 80 microinches. If we disregard the slight waviness on a worn drum surface, the burnished surface of an average smooth, worn drum is probably smoother than 10 microinches (Figure 22–34).

22.4 Drum Machining

A brake drum lathe is used to turn a drum. There are several different popular makes of drum lathes, and their operation is fairly similar (Figure 22–35). The drum is mounted on an **arbor,** also called a *mandrel* or

(A) (B) (C)

Figure 22-35 Three different drum and disc combination lathes with adapters. (A) is a standard lathe, (B) is a computerized version, and (C) is a dual spindle lathe. *(A and B are courtesy of Ammco Tools, Inc.; C is reprinted by permission of Hunter Engineering Company)*

spindle, which rotates the drum as a **cutter bit** or tool is moved straight across the friction surface.

When turning a drum, use the following recommendations:

- Use only sharp cutter bits that are sharpened to a slightly curved or rounded cutting point.
- Always use the correct drum mounting adapters.
- A vibration, silencer, or chatter band must always be installed snugly around the drum.
- Be aware that the surface finish of the drum will be affected by the sharpness of the cutter bit, the use of the vibration band, the depth of cut, the feed rate, the speed of the arbor, and the mechanical condition of the brake lathe.
- Turn the worst drum first and then turn the other drum to the same finished diameter.
- Front drums should be turned with the hub. Drums that are loose or easily removed should be secured to the hub using lug nuts.
- Drums with wheel bearings should be checked to make sure that the bearing races or cups are in good condition and tight in the hub. Damaged races should be replaced before turning the drum.
- Never bump the cutter bit or push it against a drum that is not revolving. The hard carbide bits will chip or break rather easily.
- The inner and outer wear ridges should be removed as an initial step.
- Always remove the smallest amount of metal possible except when matching drum diameters.
- It is a good practice to deburr the outer edge of the surface after machining with 80-grit sandpaper held lightly against the edge of the rotating drum surface in the hand or on a very small disc.
- A turned drum should be cleaned using a commercial braking surface cleaner or denatured alcohol. Wash the turned surface with a clean shop cloth and the liquid until the cloth shows no more contamination.
- After cleaning, do not touch the new friction surface. Pick up the drum by the outer diameter or by the center hole.

When a drum is turned, minute particles of metal and carbon from the cast iron will become embedded in the surface scratches. Unless they are washed out, these particles can become embedded in the new lining surface. Once there, they can cause abnormal braking, squeal, or excessive future drum wear.

NOTE: Most brake shops do not have occasion to store brake drums, but if they do, brake drums should be stored flat. A drum stored on its edge tends to go out-of-round. For this reason, new drums should always be checked for roundness.

SAFETY TIP: When machining a drum, use the caution that should always be exercised around moving machinery. Be particularly concerned about the revolving

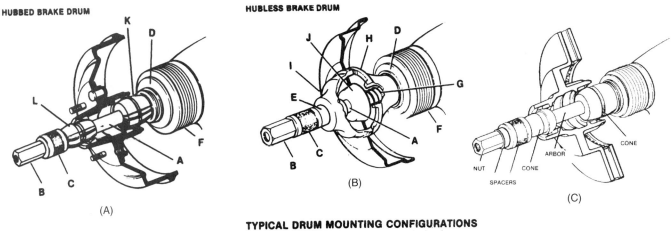

Figure 22-36 Brake drums with hubs are normally mounted on the lathe arbor as shown in (A). Floating rear drum mounting is shown in (B), and rear drum with a small center hole is mounted as shown in (C). *(Courtesy of Ammco Tools, Inc.)*

drum, the cutter bit, the revolving vibration band, and the revolving lug bolts on some drums.

The following machining procedure is very general. You should always follow the method recommended by the manufacturer of the brake drum lathe you are using. To turn a brake drum, you should:

1. Clean the drum or drum and hub. All oil and grease in the drum and hub should be washed off with a solvent or soap and water and then air dried.

2. Select the correct mounting adapters. Always make sure there is no grit or dirt between the mounting adapters and the drum, so that the drum will be centered as it is locked securely to the arbor. Make sure the cutter bit is out of the way while the drum is being mounted on the arbor.

   a. Floating rear drums are centered onto the arbor using a properly sized cone that should fit partially through the center hole. This cone is usually spring loaded to ensure a snug fit into the drum. A pair of bell-shaped adapters fit on each side of the drum's face to prevent drum runout as they secure the drum to the arbor (Figure 22-36).

   b. Drums with hubs and wheel bearings are normally centered and held in place by a pair of cones that enter partially into the wheel bearing cups. On drums that are attached to the hub with speed nuts, several lug nuts should be installed and torque tightened to ensure a tight fit. It is usually necessary to also use a few flat washers to ensure complete tightening.

NOTE: Some sources recommend attaching a wheel center to the drum with the lug bolts tightened to the correct torque. This is done to introduce any distortion that will be caused by the lug bolts so that this distortion can be machined out as the drum is turned.

3. Wrap the silencer band around the drum so it is snug, but not overly tight, and interlock the end clasp under the band. The band should be wrapped so that the free end comes over the top of the drum toward you as you are installing it from the side of the lathe with the cutter (Figure 22-37). A loose silencer band will allow the drum to vibrate and ring as it is being turned. This will leave a rough drum surface with a herringbone pattern (Figure 22-38). A loose band will also tend to loosen further and fly off. A too-tight silencer band will stress the belt and can distort the drum.

4. Select the correct spindle speed for the drum diameter. Then adjust the speed mechanism, usually a V-belt and various-sized pulleys, to produce the correct speed (Figure 22-39).

5. Adjust the position of the cutter bit tool holder so there is enough cutter or drum travel to machine the drum. Keep the extension of the tool holder or

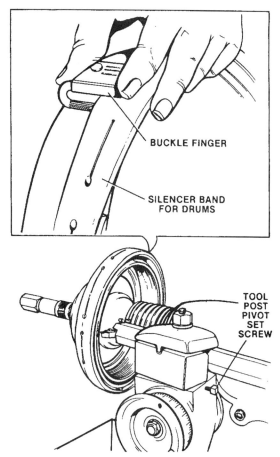

Figure 22–37 A silencer band should be wrapped snugly around the drum with the buckle coming toward you over the top. The buckle finger is then slid under the band to secure it in place. *(Courtesy of Ammco Tools, Inc.)*

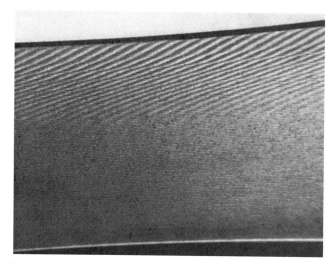

Figure 22–38 Note the odd, angled pattern on the outer portion of this drum. It resulted from the drum surface vibrating as it was being turned. This is not an acceptable surface finish.

drum as short as practical. This reduces flexing and usually produces a better drum finish.

6. Make sure the automatic feed is disengaged and then turn on the lathe. Adjust the feed handwheel to position the cutter and drum lengthwise so the cutter is midway across the friction surface. Carefully adjust the cross-feed diameter handwheel so the cutter bit makes light contact with the drum. At this point, reset the cross-feed dial to zero or note the diameter indicated on the dial (Figure 22–40).

7. Turn the cross-feed dial inward about 0.010 or 0.020 in. and turn off the machine so you can check the scratch mark (Figure 22–41). If the scratch mark made by the cutter bit extends all the way around the drum, the drum is round and mounted properly. If there is a short scratch partway around the drum, the drum is out-of-round or mounted off center. An off-center drum will force you to remove more metal than necessary to clean it up. If the drum is off center, you should:

   a. Loosen the arbor nut and rotate the drum one-half turn relative to the center mounting cone. Also, rotate each of the mounting bells one-half turn relative to the drum. While doing this, rub off any dirt or rust between the mounting bells and the drum.

   b. Retighten the arbor nut.

   c. Move the cutter sideways a small amount and repeat steps 6 and 7.

8. With the drum mounted as true as practical and with the lathe running, adjust the cross-feed handwheel back to zero so the cutter is lightly contacting the drum. Move the cross-feed handwheel so that the cutter bit is brought to the outer ridge. Continue hand feeding the handwheel slowly so that this ridge is cut away (Figure 22–42).

NOTE: Watch the scratch marks across the drum from the cutter to easily determine the position of the cutter bit.

9. Reverse the handwheel so the cutter bit moves inward across the drum surface to the inner ridge. Note any grooves in the friction surface. If they are deeper than a few thousandths of an inch, turn the cross-feed handwheel so the cutter touches the bottom of the deepest groove. Note the depth of the groove indicated on the dial (Figure 22–43). Return to a zero setting, move the cross-feed handwheel so the cutter bit is next to the inner ridge, and continue hand feeding the feed handwheel carefully to cut away the inner ridge (Figure 22–44).

Chapter 22 ■ Drum Brake Service

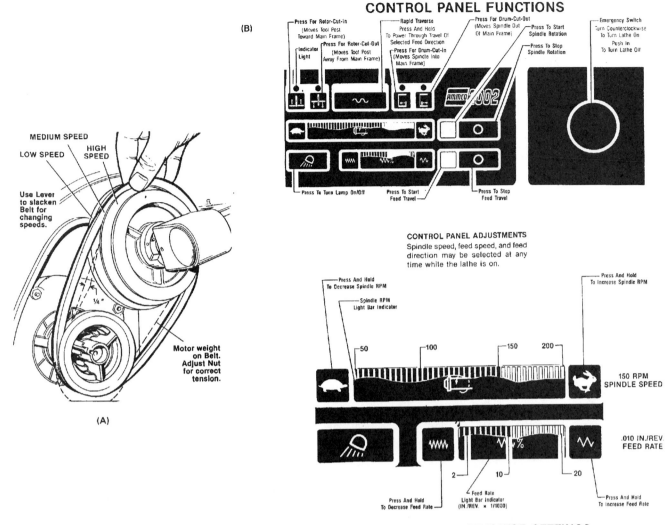

Figure 22-39 The arbor speed is adjusted by changing the drive belt position on some lathes (A) or by pressing the correct button on a computerized lathe (B). *(Courtesy of Ammco Tools, Inc.)*

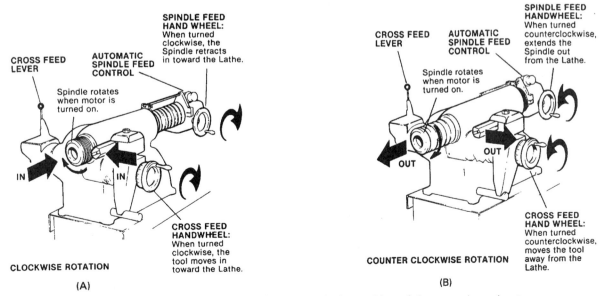

Figure 22-40 On most lathes, the spindle-feed handwheel controls the position of the cutter in and out of the drum, whereas the cross-feed handwheel controls the depth of cut. *(Courtesy of Ammco Tools, Inc.)*

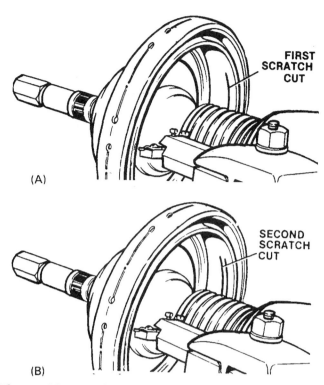

Figure 22-41 After the drum is mounted, make a scratch cut to see how true the mounting is. A short cut indicates an off-center drum (A). In this case, loosen the drum, rotate it half a turn, and make another scratch cut alongside the first one (B). If the cuts are side by side, the drum has worn off center. *(Courtesy of Ammco Tools, Inc.)*

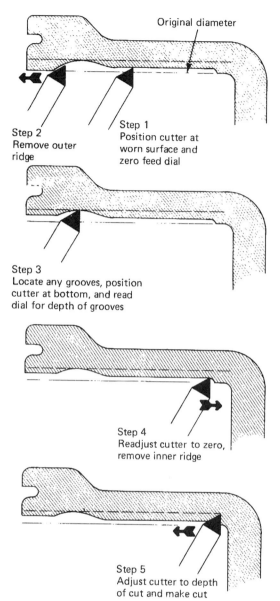

Figure 22-42 The steps in machining a drum are as follows: (1) position the cutter to the drum surface and set the feed dial to zero, (2) remove the outer ridge, (3) determine the depth of cut, (4) remove the inner ridge, and (5) make a cut across the friction surface.

10. Decide on the depth of cut that will be necessary to clean up the drum. The last finish cut should ideally be about 0.004 in. (0.1 mm) deep and at a feed rate of 0.002 to 0.006 in. per revolution. The slower rate will produce a smoother finish but take three times as long as a rough cut. Too fast or too deep a cut will leave a rough finish (Figure 22-45). A good, smooth, properly mounted drum can usually be cleaned up with one finish cut. A grooved, rough, out-of-round, or off-center drum will require one or more cuts. You should have a good idea as to the size of the cut from the trial cut you made in step 9. The maximum depth of a rough cut is determined by the strength or capacity of the lathe. A rate of 0.020 in. per revolution is usually used on a fast, roughing cut.

11. With the machine running, adjust the feed rate to the desired rate of feed and tighten the feed rate lock to secure that setting. Adjust the cross-feed handwheel to the desired depth of cut and tighten the cross-feed lock to secure that setting. Engage the automatic feed and, as the cut begins, observe the machine, cutter, and drum for any unusual noises or movement. If something appears wrong, stop the machine and recheck the settings.

12. After the cut is finished and the cutter bit moves past the outer edge of the drum, disengage the automatic feed. If you have made the final cut and the friction surface of the drum is cleaned up, shut off the lathe. If you have only made a rough pass or part of the surface was skipped, repeat steps 10 through 12.

13. Remove the silencer band from the drum and remove the drum from the arbor.

Figure 22–43 The depth of a groove in a drum can be determined by adjusting the cutter bit to the bottom of the groove and noting the reading on the cross-feed dial. Note that the outer ridge has been removed.

Figure 22–45 A drum with a ground finish (A), a finish cut (B), and a roughing cut (C) allows a comparison of the surface finishes.

Figure 22–44 This cutter has been positioned next to the inner wear ridge; it will be hand fed inward to remove this ridge before making the first cut across the drum surface.

Figure 22–46 After the drum has been turned, it should be washed to remove small metal chips and carbon particles. Use denatured alcohol, soapy water, or brake cleaner and air dry it.

14. Remeasure the drum to be sure that it is not oversized. One new brake lathe has a digital readout that shows the actual diameter as you are machining.
15. Mount the second drum from the same axle on the arbor and turn it. Be sure to use the same feed rate and cross-feed dial setting on the final finish pass.
16. Wash both newly machined drum surfaces with denatured alcohol (Figure 22–46).

22.5 Shoe Assembly: Predisassembly Cleanup

The brake assemblies should be given a preliminary cleanup before disassembly to remove the brake and road dust. This cleanup prevents this dust from being stirred up and allows a better view of the parts and their relationship.

A straight air blast *must not* be used because it will drive the dust into the air, where you or others might breathe it. Two methods are approved for this cleaning operation: vacuum and wet cleaning.

Several companies market a vacuum cleaner attachment that mounts onto the backing plate over the brake shoe assembly. These attachments include an air hose port and, in some cases, a viewing window. The attachment is connected to the backing plate and to a HEPA filter–equipped vacuum cleaner. Then the vacuum cleaner is started, and the air hose is operated. The air blast knocks the residue loose from the various brake parts, and this dust is pulled into the vacuum cleaner, where it is collected for disposal. A HEPA filter–equipped vacuum cleaner must be used (Figure 22–47). Small asbestos fibers can pass through the filter of a standard shop vacuum cleaner and be blown into the air.

The wet method uses an air-powered suction nozzle or a hand spray bottle to spray soapy water or an aerosol can of brake cleaner. A pan is placed under the brake assembly, and the soapy water or brake cleaner is sprayed over the parts (Figure 22–48). The first spraying action should be soft enough to wet the dust particles and not

Figure 22–47 This technician is using an air nozzle to knock loose all dirt and dust so these particles can be drawn into the HEPA filter on the vacuum attachment. *(Courtesy of NILFISK OF AMERICA, INC.)*

blow them into the air. Dust and asbestos particles are now trapped in the wet solution and washed down into the pan. Asbestos has a good tendency to wet and to mat when wet and can be collected by this method. The pan is often left under the backing plate so that during removal the parts can be dropped into it for soaking and further cleaning. The used cleaning solution and any cleaning rags that are contaminated with asbestos dust must be disposed of in an approved manner.

SAFETY TIP: Many commercial brake cleaners contain perchloroethylene (or Perc), which has been identified as a possible human carcinogen. It is thought to cause an increase in cancer rate. This chlorinated solvent is being studied by various air quality groups to have its usage limited or reduced.

22.6 Brake Spring Removal and Replacement

Most brake shoes are held in place on the backing plate and anchors by the return springs and the hold-down springs. Removal of these two sets of springs allows removal of the shoes. Before the shoes come loose from the backing plate, it is very important to note the relationship of the parts. Some brake assemblies almost fall apart after the springs have been removed. Also, carefully note the way that the springs are connected or hooked into the shoe and if they have a right and left, top and bottom, or in and out.

Several styles of return springs are used, and several styles of brake spring tools are available to aid in their removal and installation. Each particular style of spring requires a specially designed tool. All of these tools are called **brake spring tools** or **brake spring remover-installers.** These tools and their usage are illustrated in Figures 22–49 through 22–53. Do not try to remove these springs with ordinary pliers, except when recommended by the manufacturer. Damage to the spring or injury to yourself may result.

The tool shown in Figure 22–49 is called a **Bendix brake spring tool,** and it works on most duoservo brakes. It is available in three basic shapes, the most common being an offset form having two bends. It is also available in a straight form with a plastic handle or built into one of the legs of a pair of brake spring pliers. To remove a return spring, the tool is placed over the anchor pin with the tool's projection inside the spring eye—the curved tang at the end of the spring. The tool is rotated to move the spring away from the anchor and then leaned toward the spring to lever the spring over the end of the anchor pin. To replace a spring, the end of the tool with the slightly concave area is first placed over the anchor with the spring eye around it. Then the tool is levered to stretch the spring far enough to slide it over the anchor pin. Do not overstretch the spring, because this will weaken it. A competent technician pushes the spring over onto the anchor as soon as it is stretched far enough.

A tool, shown in Figure 22–50, has been developed to remove and replace the return springs used on some late-model General Motors cars. This tool has a wire hook to catch and pull the spring. It is hooked into the spring eye and placed over a rivet that holds the shoe guide and anchor block in place. Levering the tool will stretch the spring enough to remove it from its anchor post (Figure 22–51). The spring is replaced in a manner similar to that used with the Bendix brake spring tool.

Brake spring pliers are used on nonservo brakes of both past and present designs. This tool, shown in Figure 22–52, has two pointed jaws. One of them is pointed inward to hook into a rivet hold or into the lining, and the other is hook shaped for catching the spring eye. The tool is placed over the lining, with the hooked jaw engaging the end of the spring, and squeezed to stretch the spring enough for removal or replacement. Many technicians

Chapter 22 ■ Drum Brake Service

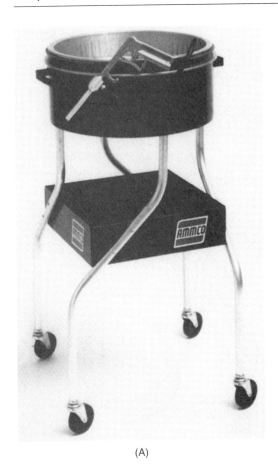

(A)

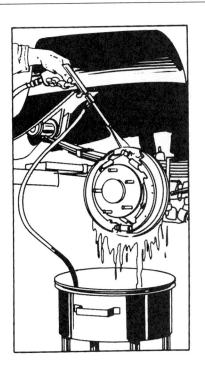

(C)

(B)

Figure 22–48 A brake assembly can be cleaned with a soapy water mist (A). The asbestos particles along with other dirt are trapped in the solution and washed into the pan. Another method is to use a household cleaner and spray bottle (B) or brake cleaner (C). *(A is courtesy of Ammco Tools, Inc.; C is courtesy of Locktite Corp.)*

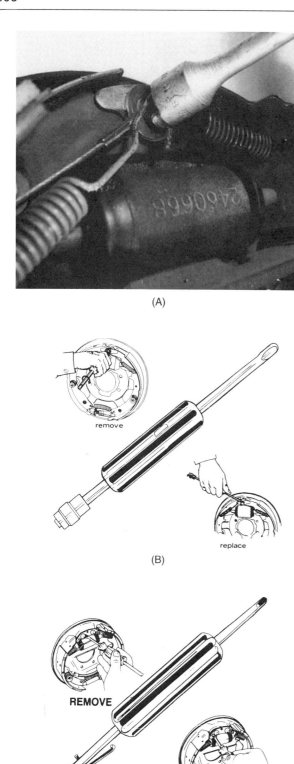

Figure 22-49 A brake spring tool is placed over the anchor pin and rotated to lift the spring over the anchor and remove it (A). Two different styles are shown in (B) and (C). *(A is courtesy of Ford Motor Company; B and C are courtesy of Lisle Corporation)*

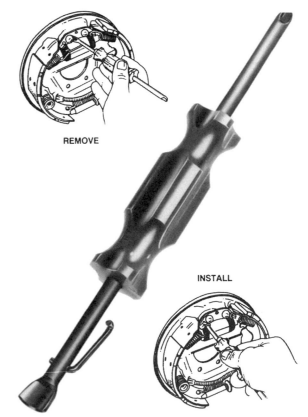

Figure 22-50 This special tool is used to remove or replace the return spring on some late-model General Motors brake assemblies. *(Courtesy of SPX Kent-Moore, Part # J-29840)*

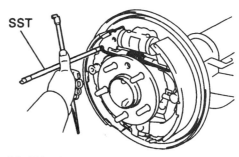

Figure 22-51 This special service tool is used to remove the return spring on this nonservo brake.

place a thin piece of plywood between the plier jaw and the brake lining to keep from gouging the lining.

Another brake spring tool from the past, shown in Figure 22-53, is sometimes called a **Lockheed brake spring tool.** This somewhat L-shaped tool has one pointed end and a washerlike eccentric attachment at the other end. To remove a spring, the tool end is inserted into the hole in the brake shoe web with the end of the spring in the notch in the eccentric. The tool is rotated so the spring is away from the shoe web and then leaned to

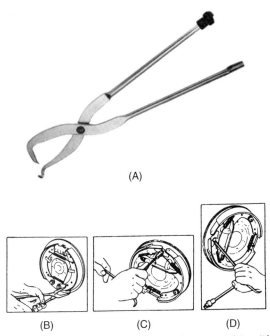

Figure 22-52 Brake spring pliers (A) are used in different ways to remove or replace return springs (B, C, and D). *(Courtesy of KD Tools)*

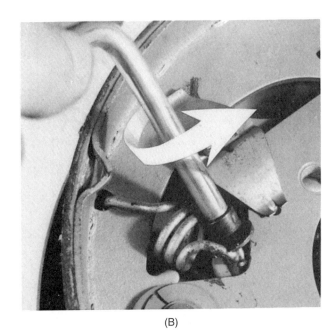

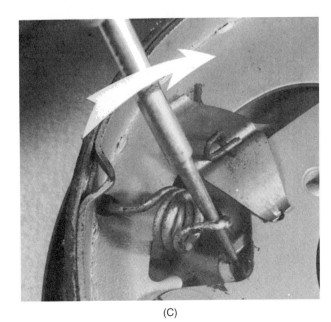

Figure 22-53 This Lockheed brake spring tool (A) is placed so the notch engages the spring (B) and is rotated to remove the spring. The pointed end is used to replace the spring (C).

lever the spring end out of the shoe. To replace the spring, the pointed end is passed through the spring eye and into the hole in the shoe web. Then the spring end is levered into the hole.

In a very few cases, the use of pliers is recommended by a manufacturer to grip the shank of the spring and stretch it to allow removal and replacement. Normally, the use of pliers is avoided because of their tendency to nick or scratch the spring. Pliers usually cannot grip stronger return springs tight enough to prevent slippage without nicking or bending the spring (Figure 22-54).

The return springs on some nonservo designs are removed by lifting the shoe over the shoe guide at the anchor or sliding it off the end of the anchor. This allows the shoe to be moved over so the spring can be unhooked. The springs are replaced by placing one shoe in position, connecting the springs, and pulling the second shoe far enough to allow it to be placed back on the anchor. With this design, the hold-down springs are removed first and installed last (Figure 22-55).

There are essentially five different styles of hold-down springs and a few different special tools to aid in their removal and installation (Figure 22-56).

The most common style of hold-down spring has a short coil, one or two concave washers with a slot in the center, and a pin, sometimes called a nail, which has a flattened, somewhat arrow-shaped end (Figure 22-57). For removal, the tubular end of the brake shoe retaining

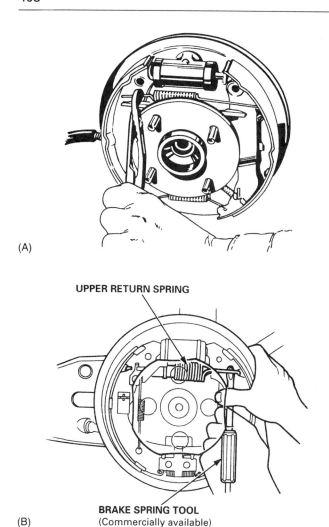

Figure 22-54 On some brakes, the manufacturer recommends using pliers to grip the return spring (A) or sliding a brake spring tool to hook onto the spring (B) so the spring can be stretched far enough for removal. *(A is courtesy of General Motors Corporation, Service Technology Group; B is courtesy of American Honda Motor Co., Inc.)*

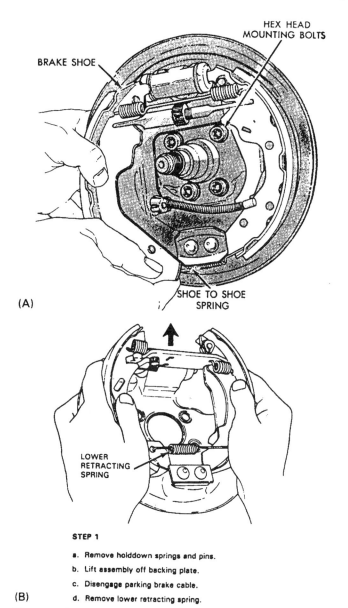

Figure 22-55 After removing the hold-down springs on some brakes, the brake shoes (with return springs still attached) are pulled off the anchor block. *(A is courtesy of Chrysler Corporation; B is courtesy of Ford Motor Company)*

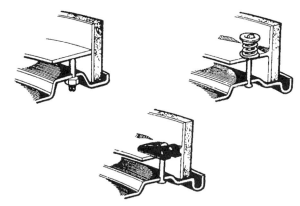

Figure 22-56 Several styles of hold-down springs are used. *(Courtesy of LucasVarity Automotive)*

spring tool is first placed over the washer. Then the washer is pushed inward, turned to align the slot with the flats of the pin, and released. Installation is the reverse of removal, but it is important to make sure that the flats on the pin align with the depressions in the washer. When working on many General Motors cars with self-adjusting duoservo brakes, note that the bottom spring cup has an extension that serves as a pivot for the self-adjuster lever.

Some hold-down springs use a flat, U-shaped spring that engages the flattened end of the hold-down pin. It can be gripped with a pair of common pliers and pushed inward to allow release or installation (Figure 22-58).

In the past, the hold-down spring used on some domestic cars was a coil spring that passed partially through the hole in the shoe web and hooked onto a tang that

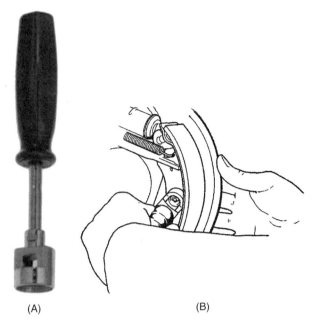

Figure 22-57 This hold-down tool (A) is used to push the retainer inward and rotate it during spring removal and installation. It is usually necessary to hold the retainer pin with one hand while the other hand uses the tool (B). *(A is courtesy of KD Tools; B is courtesy of LucasVarity Automotive)*

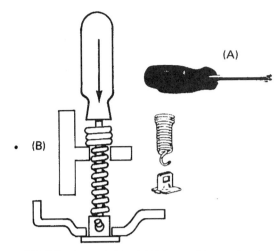

Figure 22-59 A special tool (A) is used to push a beehive-style hold-down spring inward so it can be unhooked (B). *(Courtesy of Snap-on Tools Corporation)*

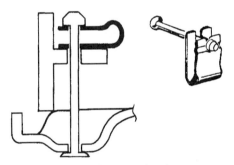

Figure 22-58 This hold-down spring is compressed while the pin is rotated to disengage it.

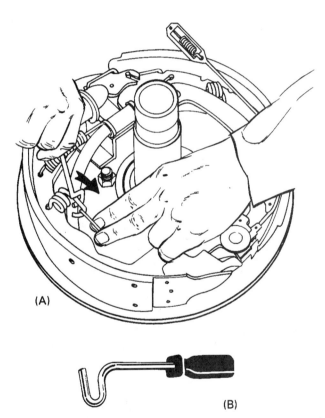

Figure 22-60 This coil-type hold-down spring must be pulled toward the brake shoe so the pin can be disconnected (A). A special tool (B) makes this easy. *(A is courtesy of Chrysler Corporation; B is courtesy of Snap-on Tools Corporation)*

passed through the backing plate. Because of its shape, it was sometimes called a *beehive spring*. It required a special tool that passed into the spring and pushed on the last coil so the spring could be stretched enough to be hooked or unhooked (Figure 22-59). If this special tool is not available, a small screwdriver that will enter but not pass through the spring will work.

Some pickups and light trucks use a mousetrap-shaped style of hold-down spring on the shoe that hooks into a curved wire extending from the backing plate. The odd-shaped tool shown in Figure 22-60 is required to bend the spring far enough to hook or unhook it.

Some brake designs use a U-shaped, flat spring that hooks into the backing plate and rests on the shoe web. The brake shoe is simply slid out from under or slid back under this spring during removal or during installation.

22.7 Brake Shoe Removal

During shoe removal operations, many technicians position a pan of soapy water or solvent under the brake assembly. As the parts are removed, they are placed in the liquid for soaking. This allows for quick and easy cleanup and reduces the amount of dust being stirred up (Figure 22–61).

When removing and replacing brake shoes, the manufacturer's recommended procedure for the particular car should be followed. The following procedure is very general. To remove a pair of brake shoes, you should:

1. Carefully check the shoes to note the types of return and hold-down springs and how they are connected to the shoes and backing plate. Also, carefully note how the adjuster mechanism and parking brake linkage are connected to the shoes, usually the secondary shoes. You should note that different colors are used on the springs, because they can help you remember the correct spring positions.
2. If you are not planning to service the wheel cylinder, install a wheel cylinder clamp to prevent the wheel cylinder pistons from popping out of their bores (Figure 22–62).
3. Remove the return springs. Also, remove any self-adjuster mechanism that has been released.

NOTE: On some duoservo brakes, the shape of the anchor causes the outer return spring to be stretched a little farther than the inner. It is a good practice to note which spring is removed first so it can be installed last.

4. Remove the hold-down springs and any self-adjuster mechanism that has been released.

NOTE: On brakes in which the shoe is pulled or slid off the anchor block to release the return spring, the hold-down springs are normally removed first.

5. If necessary, disconnect any remaining parking brake or self-adjuster linkage from the shoes. Some parking brake levers merely hook into the shoe, and others are held in place by an E-clip or a C-clip. An E-clip is removed by prying it off with a small screwdriver (Figure 22–63). A C-clip is most easily removed with pliers or a sharp punch, as shown in Figure 22–64. If it is necessary to disconnect the parking brake lever from the cable, place the jaws of a pair of side- or diagonal-cutting pliers over the cable. Then slide the pliers

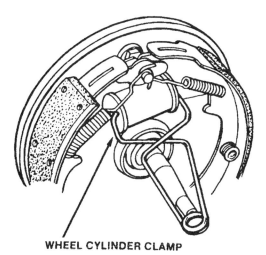

Figure 22–62 A wheel cylinder clamp can be put on a wheel cylinder to keep the pistons from coming out when the shoes are removed. *(Courtesy of Bendix Brakes, by AlliedSignal)*

Figure 22–61 The cleaning pan was left under the brake assembly and the parts were dropped into it as they were removed. Cleanup of the parts and backing plate is quick and easy.

Figure 22–63 An E-clip (left) or C-clip (right) retainer can be removed using a small screwdriver.

down the cable to compress the spring and tighten the pliers to lightly grip the cable, holding the spring compressed while you disconnect the cable from the lever (Figure 22–65).

6. If the wheel cylinder is to be serviced, remove it. This procedure is described in Chapter 24.

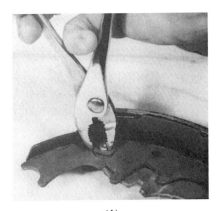

(A)

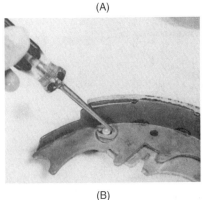

(B)

Figure 22–64 A C-clip retainer can be removed by squeezing it with a pair of pliers (top) or by tapping it out of its groove using an awl and hammer (bottom).

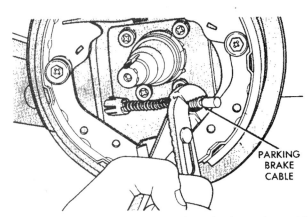

Figure 22–65 Gripping the parking brake spring with a pair of pliers allows you to hold the spring compressed so the brake cable can be removed or replaced on the lever. *(Courtesy of Chrysler Corporation)*

22.8 Component Cleaning and Inspection

After shoe removal, the backing plate and any part that is going to be reused should be cleaned and inspected. These parts should be washed in soapy water or solvent and air dried. On brake drums mounted on hubs with wheel bearings, the wheel bearings should also be cleaned and repacked. This operation is described in Chapter 21.

As mentioned earlier, many technicians normally install new return springs and hold-down springs. Several companies market a kit, often called a hardware set, that includes these springs and other small parts that are often worn or damaged or for which replacement is advisable (Figure 22–66). A shoe return spring is a critical part. If it is weak or stretched, early shoe application or drag can occur, which in turn can cause pull or rapid shoe wear (Figure 22–67). No strength or length specifications for return springs are available to the brake technician, and their low cost does not warrant spending much time to check them. Return springs are made in many different sizes and shapes. Check the replacement spring against

Figure 22–66 A Combi-Kit or hardware kit. This set includes new return and hold-down springs plus other small parts that should be replaced along with the brake shoes. *(Courtesy of Rolero-Omega, Division of Cooper Industries)*

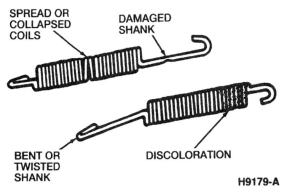

Figure 22–67 Damaged return springs should be replaced to ensure proper brake shoe application and release. *(Courtesy of Ford Motor Company)*

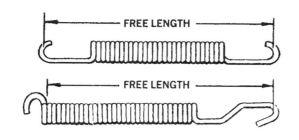
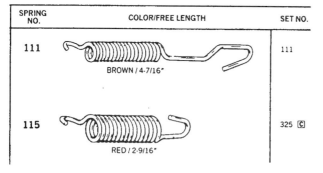

Figure 22-68 There are many shapes and sizes of return springs. Replacements should be carefully checked against the old ones to make sure they are correct. *(Courtesy of Rolero-Omega, Division of Cooper Industries)*

the old one to be sure that you have the correct spring (Figure 22-68). A popular but questionable method for testing springs is to listen to the sound as the spring is dropped on the floor. A spring that makes a ringing noise is faulty because the coils are stretched apart. Always replace the return springs in sets at both ends of the axle. The springs at each end must be the same strength to reduce the possibility of brake pull. Check any spring that is to be reused for severe discoloration, rusting, or stretched coils. Any damaged or doubtful springs must be replaced.

On cars with self-adjusters, the adjusting lever should be checked for wear. With cable-type self-adjusters, be sure to check the lever for incorrect bends or cracks in the pivot area and the cable for signs of fraying or stretching (Figure 22-69). Each style of self-adjuster has its own wear points. Any damaged parts should be replaced. For example, on 1971 to 1975 General Motors cars with leading-trailing shoes and nonservo brakes, special pliers are necessary to disassemble or shorten the strut and adjuster assembly. This strut lengthens to

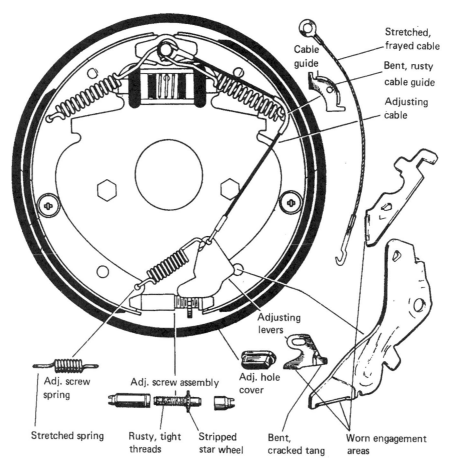

Figure 22-69 The various parts of a self-adjuster should be checked for wear or damage.

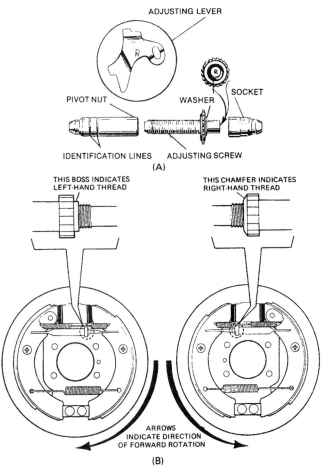

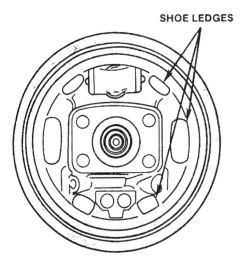

Figure 22-71 After cleaning the backing plate, it should be checked for worn shoe ledges/platforms and loose mounting bolts and anchors. *(Courtesy of Chrysler Corporation)*

Figure 22-70 The right and left adjuster mechanisms are identified by various markings. *(A is courtesy of General Motors Corporation, Service Technology Group; B is courtesy of LucasVarity Automotive)*

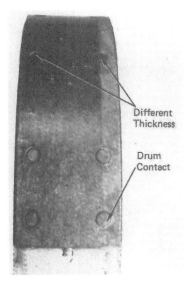

Figure 22-72 This brake shoe has worn so the right rivet was touching the drum, yet there was 1/32 in. of lining above the left rivet. This indicates that either the drum was badly tapered or the backing plate was cocked.

adjust for lining wear, and it must be shortened when new lining is installed. Failure to use the correct tool will result in broken adjuster locks.

Note that adjuster screws and self-adjuster levers are specific for the right and left sides and cannot be switched. Many adjuster screws are stamped with an *L* or an *R*, indicating the thread direction. Usually, on duo-servo brakes, the left-hand brake uses right-hand threads and vice versa (Figure 22-70). When cleaning adjuster screws, they should be unthreaded to allow thorough cleaning of all dirt or rust from the threads and also to allow proper lubrication. Any adjuster with damaged threads or teeth should be replaced.

While checking a backing plate, make sure that the platforms are clean and smooth and not grooved badly enough to catch a shoe (Figure 22-71). Small imperfections can be cleaned up with a fine file or emery cloth.

If the shoe shows a lining wear pattern indicating a cocked shoe, the backing plate should be checked to see if the platforms are worn or distorted (Figure 22-72). On brakes with an axle flange, the relative height of the platforms can be easily checked by using the following procedure: Attach a dial indicator to the axle flange, adjust the indicator stylus and dial to zero on one of the platforms, and rotate the axle so the remaining platforms can be measured (Figure 22-73). At one time, a simple tool was positioned over the spindle to allow platform height checks. This tool is shown in Figure 22-74. A badly bent backing plate or one with badly worn platforms should be replaced. When replacing a backing plate, be sure to tighten the mounting bolts to the correct torque.

Figure 22–73 The dial indicator is mounted on the axle flange to check the platform heights; if they are different, the shoe will be crooked with the drum.

Figure 22–74 This simple gauge is used to check the relative height of the brake platforms; after adjusting it to fit one of the platforms, the axle is rotated to allow checking of the others.

22.9 Component Lubrication

Drum brake components are carefully lubricated before and during assembly to reduce future wear and noise (Figure 22–75). The lubricant used must be an approved

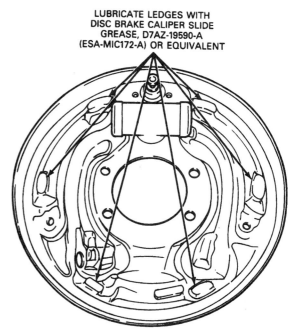

Figure 22–75 The shoe ledges/platforms should be covered with a thin film of lubricant before installing the shoes. *(Courtesy of Ford Motor Company)*

grease. It must remain where it is applied and not run onto the friction surfaces. Several of the aftermarket brake component manufacturers market a brake assembly grease. Note that this product should not be confused with hydraulic brake assembly fluid, which is used inside the hydraulic system. Some brake technicians prefer to use an antiseize compound for assembly grease because it has superior heat tolerance characteristics and an ability to retain excellent lubricating quality over a long period of time.

Normally a thin film of grease is applied to each surface where there will be a rubbing metal-to-metal contact. The smallest amount of grease should always be used to prevent the possibility of grease falling onto the friction surfaces. Grease will contaminate the lining and will probably cause grabbing. A thin layer of grease should be applied to each of the backing plate platforms. Most technicians grease the threads of the adjuster screws before threading them together. This grease seals the threads to prevent water or dirt from entering (Figure 22–76). A thin film of grease is also applied between the adjuster screw and its socket. During reassembly of the shoes, most brake technicians apply a thin film of grease at the parking brake lever pivot, at each end of the parking brake strut, at each pivot of the self-adjuster mechanism, at the point where the adjuster screw contacts the shoes, at the anchor pin or blocks, and at every metal-to-metal contact (Figure 22–77).

Chapter 22 ■ Drum Brake Service

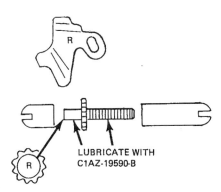

Figure 22-76 The adjusting screw should be lubricated before installation. *(Courtesy of Ford Motor Company)*

22.10 Brake Shoe Preinstallation Checks

Before installation on the backing plate, the brake shoes should be checked for proper lining attachment and proper curvature. Use care to keep the new lining clean. Clean your hands before handling the new shoes and develop the habit of handling the shoes by the web or the edges of the rim, not by the lining. Lay the shoes out on a clean bench in the position in which they will be installed and determine whether there are shoe differences (e.g., a primary and a secondary shoe or a leading and a trailing shoe). Compare the new shoes with the old shoes if there are any doubts about the correct lining length or lining or hole placement.

After determining the correct position for a particular shoe, check the lining-to-shoe attachment to be sure that the lining is positioned straight and does not hang over the edges of the shoe rim. Any overhanging lining will require that the shoe be returned to the supplier or that the excess lining be removed with a file or coarse sandpaper (Figure 22-78). If it is necessary to remove overhanging lining, be careful to remove it in an OSHA-approved manner and do not breathe any of the sanding dust.

It is a good practice to always check the lining-to-drum fit. This is sometimes called a heel and toe clearance check. The lining should almost fit the curvature of the drum to ensure quick break-in of the lining. There should be a slight clearance at the ends of the shoe, and this slight clearance should allow the shoe to rock in the drum (Figure 22-79). There should never be a clearance between the center of the shoe and the drum, because this would cause erratic, grabby brake action. Heel and toe clearance helps ensure that the ends of the lining are not pinched because of drum deflection during a hard stop. This clearance is checked by placing the shoe in the drum

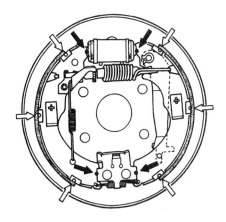

Figure 22-77 This manufacturer recommends lubricating the parts with brake cylinder grease (A), Molykote 44 MA (B), and rubber grease (C). *(Courtesy of American Honda Motor Co., Inc.)*

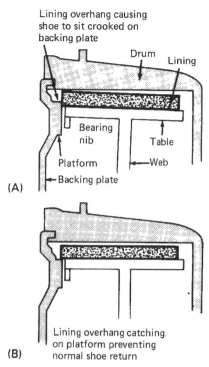

Figure 22-78 If the lining overlaps the shoe table/rim, it can move the shoe out from the platform and cause it to sit crookedly (A). This overhanging lining can also catch on the platform (B) and prevent the shoe from returning properly.

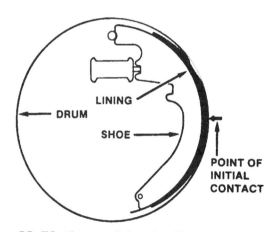

Figure 22-79 The new lining should contact the drum at the center of the lining. *(Courtesy of Bendix Brakes, by AlliedSignal)*

and squeezing the lining into the drum. Place one hand so it grips the web of the shoe and the outside of the drum. With the other hand, gauge the clearance at the ends of the lining. A 0.005-in. (0.13-mm) feeler gauge should fit between the lining and the drum at each end, and a 0.010-in. (0.25-mm) gauge should not (Figure 22-80).

In the past, brake shoes were commonly available with a standard or an oversize lining. As a drum is turned to a larger diameter, the curvature or arc of the friction

Figure 22-80 Lining clearance is checked by squeezing the shoe against the drum and trying to fit a feeler gauge at each end.

surface becomes flatter and the heel and toe clearance increases. An oversize lining was slightly thicker, and it was ground with a slightly flatter arc. The lining on brake shoes is normally ground on a radius that is one-half of the brake diameter minus 0.030 in. (0.76 mm). This is sometimes called an eccentric or cam grind. A standard lining was ground to fit drums that were standard size to 0.030 in. oversize. Oversize lining was ground to fit drums that were more than 0.030 in. oversize. Today, the lining on most relined shoes is ground to a standard size.

After finding that the shoe is the correct one, that there is no lining overhang, and that the heel and toe clearance is acceptable, some brake technicians cover the lining with one or two strips of loosely applied masking tape. This tape keeps the lining clean while the shoe is being installed on the backing plate. The tape is removed just before the drum is installed (Figure 22-81).

22.11 Regrinding Brake Shoes

In the past, most brake shops would regrind or re-arc a lining if the shoe did not have the correct heel and toe clearance. This ensured the best fit and shortest break-in period for the lining. Most shops had a brake shoe grinder as well as a brake drum lathe. Because of concerns about asbestos fibers, very few of today's shops grind brake shoes. Most shoe grinders are designed with vacuum dust-collection systems, some of which

Chapter 22 ■ Drum Brake Service

Figure 22–81 After checking the lining, some technicians cover it with masking tape to keep it clean until the drum is installed.

Figure 22–82 A brake shoe grinder is used to regrind brake lining to an arc that matches the drum. *(Courtesy of Ammco Tools, Inc.)*

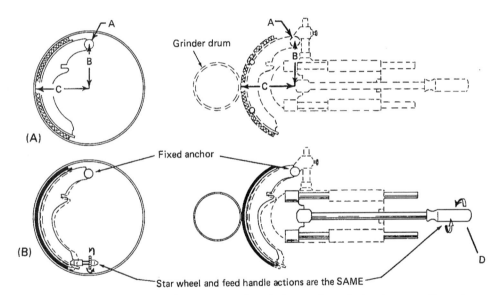

Figure 22–83 The shoe grinder positions the lining correctly, relative to the grinder drum, so it can be ground to the correct arc. *(Courtesy of Ammco Tools, Inc.)*

meet OSHA standards. As the lining is being re-arced, the dust-collection system collects the dust in a container so it can be disposed of easily and safely (Figure 22–82).

SAFETY TIP: At this time, because of the hazards connected with asbestos dust, many shops are no longer regrinding brake shoes in the field (outside manufacturing plants). However, some shoe grinders meet OSHA requirements and can be safely used. It is important that these units be maintained properly to ensure compliance with the law.

There are two styles of shoe regrinding: a standard or plain grind and a fixed-anchor grind. Standard arc grinding creates a lining that is curved to match the drum diameter minus 0.030 in. and is used for sliding-anchor or adjustable-anchor shoes. A fixed-anchor grind positions the shoe so the anchor eye is in the same position relative to the shoe clamp pivot that the brake's anchor pin is in relative to the center of the axle or spindle. A good fixed-anchor grind produces a shoe that has the correct arc and is centered on the drum when it is installed (Figure 22–83). When using this equipment, be sure to follow the directions of the manufacturer.

22.12 Brake Shoe Installation

In most cases, the assembly of the shoes on the backing plate follows a procedure that is the reverse of the disassembly procedure. It is a good practice to always follow the method recommended by the vehicle manufacturer. The backing plate and adjusting screw should be lubricated, and a slight amount of lubricant should be used on the other parts as they are installed.

To install brake shoes on a backing plate, you should:

1. Check the shoes to determine the correct placement—primary or secondary, leading or trailing—and install the wheel cylinder if it has been removed.
2. If necessary, install the parking brake lever on the secondary or trailing shoe and connect the parking brake cable to the lever.

On vehicles with manually adjusted duoservo brakes, overlap the anchor ends of the shoes as you hook the adjuster spring between the two shoes. Then place the adjuster screw in position and spread the shoes to their normal position to retain the spring and adjuster screw. Check that both the spring and the screw are in the correct position relative to the shoes and backing plate (Figure 22-84).

3. Install the hold-down spring and parts on the primary shoe. Make sure that the hold-down pin is completely in place in the detents of the washer (Figure 22-85).
4. Install the hold-down spring and parts on the secondary shoe. As this is being done, attach the self-adjuster mechanism and parking brake strut, if necessary.
5. Install the return springs. The normal order of installation on a duoservo brake is to install the primary shoe return spring first and then the secondary shoe return spring. If necessary, attach the self-adjuster mechanism as the return springs are being installed. The spring should be hooked into the shoe and then installed on the anchor or second shoe.

NOTE: On cable-type self-adjusters, make sure the cable guide stays against the shoe web and does not lift and allow the cable to go under the guide.

NOTE: Some technicians recommend using a pair of pliers to close the spring eye to the point where the spring end is parallel to the shank (Figure 22-86). This prevents the spring eye from ever opening up (Figure 22-87).

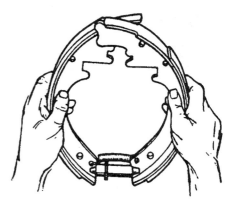

Figure 22-84 The first replacement step with manually adjusted duoservo brakes is to connect the adjuster spring and screw. *(Courtesy of Bendix Brakes, by AlliedSignal)*

Figure 22-85 When installing this style of hold-down spring, make sure the flats at the end of the pin are completely in the washer recesses.

Figure 22-86 After installing return springs, some technicians use pliers to squeeze the tang of the spring until it is parallel with the shank.

6. Inspect installation to be sure that everything is correctly installed. On most types of brakes, the shoes should be returned to the anchors, and there should be a slight clearance at the parking brake strut. Test

Chapter 22 ■ Drum Brake Service

Figure 22-87 This severely worn drum is probably the result of a broken return spring.

Figure 22-89 A brake shoe gauge speeds up the procedure to preset lining clearance before installing the drum. The first step is to adjust the gauge to the drum diameter. *(Courtesy of Ford Motor Company)*

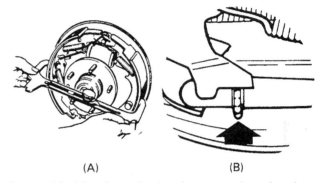

Figure 22-90 After adjusting the gauge size, place it over the lining (A) and turn the adjuster screw to bring the lining to almost the size of the gauge (B). *(Courtesy of Ford Motor Company)*

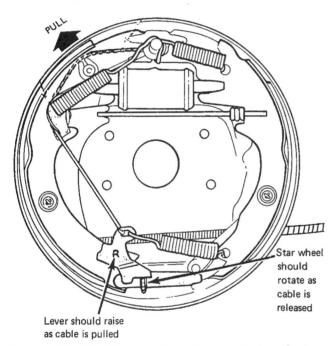

Figure 22-88 It is a good practice to check self-adjuster operation before installing the drum. *(Courtesy of Ford Motor Company)*

the self-adjuster to check its operation and then remove the masking tape, if used (Figure 22-88).

7. On cars that use adjuster screws, preadjust the brake shoe clearance. This can be done easily with a **brake shoe gauge.** First, adjust the gauge to fit inside the drum (Figure 22-89). Then adjust the shoes to a slight gauge clearance (Figure 22-90). If a gauge is not available, expand the shoes until a drag is felt when trying to slide the drum in place. Then remove the drum and turn the adjuster inward about one-half turn to increase the clearance.

8. Install the drum.

NOTE: On drums with a hub and wheel bearings, the wheel bearings should be cleaned, repacked, and adjusted. These operations are described in Chapter 21.

NOTE: On a few older cars, adjust the anchor to perform the major brake adjustment required by some manufacturers. This procedure is described in older shop manuals.

9. If the wheel cylinder has been rebuilt or replaced, bleed the air from the hydraulic system.
10. Install the wheel and tighten the lug bolts to the correct torque using the proper tightening pattern. Note that overtightening the lug bolts can distort the brake drum.

11. Complete the brake shoe adjustment.
    a. On duoservo brakes with self-adjusters, slowly and carefully back up the car while applying and releasing the brakes. You should feel the brake pedal height increase as the brakes self-adjust.
    b. On nonservo brakes with self-adjusters, slowly apply and release the parking brake. Some designs completely adjust in one application, but some require several applications. As you apply and release the service brake, you should feel an increase in the brake pedal height as the adjustment occurs.
    c. On manually adjusted duoservo brakes, turn the adjuster screw using a brake-adjusting tool or spoon as you slowly rotate the wheel. Use the tool to pry the teeth of the adjuster screw downward (tool handle upward) to expand the screw in most cases. Expand the shoes until the wheel is almost locked and then back off the adjustment until the brake is just free of drag. You should be able to hear a slight rubbing as the wheel is turned (Figure 22–91).

NOTE: After completing a brake adjustment, some sources recommend that you tap the brake drum with a hammer. It should make a ringing noise; a dull noise indicates that the shoes are in contact with the drum surface. This test is also useful to help diagnose the problem of an overheating, dragging brake. It is not a good practice to adjust the lining to the point of a definite drag.

If the brake pedal is still low after the final adjustment, remove the drums, one at a time, and use the brake shoe gauge as described in step 7 to determine whether there is excessive clearance in any of the brake assemblies. Once the faulty brake is located, a careful inspection will usually show the cause of the excess clearance. On cars with four-wheel drum brakes, save time by using the parking brake as a diagnostic tool. If the brake pedal height increases after the parking brake has been applied, there is excess clearance in one or both of the assemblies where the parking brake is located.

22.13 Completion

Several checks should always be made after completing a brake job or brake service operation:

- Be sure that all nuts and bolts are properly tightened to the manufacturer's specifications and that all locking devices are properly installed.
- Be sure that there is no incorrect rubbing or contact between parts or brake hoses.
- Be sure that the master cylinder has the correct fluid level.
- Be sure that there are no fluid leaks.
- Be sure that there is a firm, high brake pedal.
- Clean any fingerprints or grease from the vehicle.
- Perform a careful road test to make sure that the brakes operate correctly and safely.

New friction material must be properly broken in. Unless it is specifically stated otherwise by the lining manufacturer, new lining must never be "burned in." New lining can be ruined if it is overheated. Light or moderate stops should be made until the lining and drum surfaces are fully burnished and seated in. This varies with the type of driving. Some sources recommend about 250 to 2,000 mi. (400 to 3,000 km) of driving. The brake pedal will firm up and become more solid as the lining wears to exactly match the drum radius, and drum and shoe deflection will disappear. Poorly fitted lining will require an even longer, more gradual break-in period.

If a technician believes that the customer will not break in the friction material properly, the technician should make what is referred to as a 30-minute road test, which is really a break-in step. Select a road where you can make a series of stops without interfering with traffic. Drive to a speed of 30 to 40 mph and make a light to

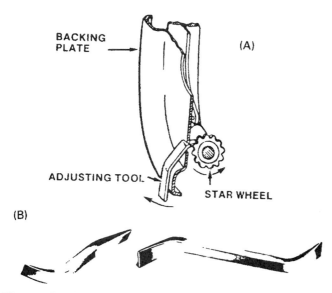

Figure 22–91 On vehicles with manual adjustment, clearance is adjusted after drum installation. Turn the starwheel to lock up the drum and then back it off until there is a very slight drag (A). An adjustment tool or spoon should be used (B). *(A is courtesy of Bendix Brakes, by AlliedSignal; B is courtesy of Snap-on Tools Corporation)*

moderate stop. Note any braking problems for later correction and accelerate slowly back up to speed. Give the brakes time to cool and repeat the previous stop. Depending on the lining and the service that was done, repeat the stops through five to ten cycles, making sure that the friction material does not overheat and the braking action is normal.

22.14 Practice Diagnosis

You are working in a brake and front-end shop and encounter the following problems:

CASE 1: A 10-year-old car has high mileage and shows a lot of abuse. It has been brought in with a complaint of noisy brakes, and on the road test, you hear metal-to-metal contact at the rear when the brakes are applied. With the car back in the shop, you remove the rear tires, but when you try to pull the drums, you find that they come out only about 1/8 in. What is happening? What should you do next? Is this going to be an inexpensive repair?

CASE 2: A car has 80,000 miles on it and is in pretty good condition. When you pull the rear drums, you find riveted lining that is worn down to the rivets on the outer edge, but there is about 1/16 in. of lining above the inner rivets. What is wrong? How can you check this?

CASE 3: A 1990 passenger car has been brought in with a complaint of overheating rear brakes. Your road test confirms a dragging condition, and when you check the brakes, you find that the right rear is very hot, the right drum makes a dull sound when you tap it with a hammer, and the left rear makes a ringing sound. What is wrong? What should you do to locate the cause of this problem? Could this be a bad proportioning valve?

Terms to Know

arbor	brake spring pliers	hard spots
bell-mouthed drum	brake spring remover-installers	Lockheed brake spring tool
Bendix brake spring tool	brake spring tools	machining
brake drum micrometer	cutter bit	swaging
brake shoe gauge	drum mike	

Review Questions

1. Two technicians are discussing a brake job. Technician A says that brake operations should be done in pairs, performing the same operation at each end of the axle.
   Technician B says that usually the only operation that is really necessary is to remove and replace the shoes.
   Who is correct?
   a. A only
   b. B only
   c. Both A and B
   d. Neither A nor B

2. The rear drums of most FWD cars are usually removed:
   a. By sliding them off the axle flange after the speed nuts have been removed
   b. Along with the hub and wheel bearings
   c. From the axle flange after removing two bolts
   d. None of the above

3. Technician A says that care should be exercised to avoid breathing the dust released while removing a drum and shoes.
   Technician B says that brake dust can be safely handled by using a HEPA filter–equipped vacuum cleaner or a soap and water spray system.
   Who is correct?
   a. A only
   b. B only
   c. Both A and B
   d. Neither A nor B

4. Technician A says that the drum must be catching on the axle flange in cases where it slides outward about 1/4 to 1/2 in. and then seizes.
   Technician B says that it is possible to back off and loosen a self-adjuster using a brake spoon and ice pick.
   Who is correct?
   a. A only
   b. B only
   c. Both A and B
   d. Neither A nor B

5. Technician A says that a hammer can be used to check a brake drum for cracks.
   Technician B says that a standard dial indicator can be used to check for an out-of-round drum.
   Who is correct?
   a. A only
   b. B only
   c. Both A and B
   d. Neither A nor B

6. There will not be good contact between the new lining and the drum's friction surface if the drum is:
   a. Scored
   b. Bell mouthed
   c. Barrel shaped
   d. Any of the above

7. Technician A says that the diameter of a brake drum is measured using a drum mike.
   Technician B says that the maximum diameter for a drum is usually indicated on the drum.
   Who is correct?
   a. A only
   b. B only
   c. Both A and B
   d. Neither A nor B

8. The diameter of a 10-in. drum has been measured at three locations, and the readings are 10.005, 10.010, and 10.015 in. The drum is:
   a. Good, and it can be safely used
   b. Oversize, and it should be replaced
   c. Out-of-round, and it should be turned
   d. None of the above

9. Two rear drums measure 9.535 and 9.505 in., and both are slightly scored. The maximum allowable size for the car is 9.560 in. The technician should:
   a. Replace both rear drums with new ones
   b. Turn both drums to 9.560 in.
   c. Turn the largest drum to the smallest possible diameter and the smaller drum to the same size
   d. Turn each drum to the smallest possible diameter

10. Technician A says that the lining can overheat if the drum is turned too thin.
    Technician B says that a spongy brake pedal can be the result if the drums are turned to too large a diameter.
    Who is correct?
    a. A only
    b. B only
    c. Both A and B
    d. Neither A nor B

11. Technician A says that a rubber belt should be wrapped around a brake drum that is being turned to deflect the metal chips.
    Technician B says that a herringbone pattern in a drum that was just turned is the result of taking too deep a cut.
    Who is correct?
    a. A only
    b. B only
    c. Both A and B
    d. Neither A nor B

12. Brake shoe return springs are normally removed using:
    a. Brake spring pliers
    b. A Bendix brake tool
    c. A pair of combination pliers
    d. Any of the above depending on the brake design

13. Before removing the brake shoes and springs, you should:
    a. Clean off the dirt and dust using an air gun
    b. Purchase new shoes and drums
    c. Clean off the dirt and dust using an OSHA-approved method
    d. Bleed the fluid out of the wheel cylinder

14. A car has a faulty rear axle seal, and the lining has been soaked with grease.
    Technician A says that the grease can be washed off using the correct friction surface cleaner.
    Technician B says that the bad axle seal and the lining on both rear brakes must be replaced.
    Who is correct?
    a. A only
    b. B only
    c. Both A and B
    d. Neither A nor B

15. Technician A says that a weak brake shoe return spring can cause a pull during a stop.
    Technician B says that a weak return spring can cause rapid lining wear.
    Who is correct?
    a. A only
    b. B only
    c. Both A and B
    d. Neither A nor B

16. Technician A says that a common shoe hold-down spring with a coil spring, pin, and washer can be removed by pushing in on the washer and turning it one-quarter turn.
    Technician B says that the hold-down spring can be used to secure the self-adjuster cable guide in place.
    Who is correct?
    a. A only
    b. B only
    c. Both A and B
    d. Neither A nor B

17. A self-adjusting, duoservo rear brake is being assembled.
    Technician A says that the shoe with the shortest lining should be placed in the position closest to the front of the car.
    Technician B says that the adjuster screw threads and the backing plate platforms should be coated with a thin film of lubricant.
    Who is correct?
    a. A only
    b. B only
    c. Both A and B
    d. Neither A nor B

18. When a brake shoe is placed in the drum so the lining is against the inner friction surface, there should be:
    a. About 0.010 in. of clearance at the center of the lining
    b. About 0.007 in. of clearance at each end of the lining
    c. A perfect fit so there is no clearance anywhere between the lining and the drum
    d. A 0.005-in. clearance between the lining and the drum at the center and each end

19. A self-adjusting duoservo brake is being assembled. Technician A says that the parking brake lever should be attached to the secondary shoe. Technician B says that the self-adjuster cable guide or actuator lever should be attached to the primary shoe.
    Who is correct?
    a. A only
    b. B only
    c. Both A and B
    d. Neither A nor B

20. After new brake lining has been installed, it is a good practice to:
    a. Burn in the lining with about ten hard stops from 50 mph
    b. Adjust the lining to a definite drag so it will wear in
    c. Make a series of slow stops from moderate speeds so as not to overheat the lining as it wears in
    d. All of the above

23 Disc Brake Service

Objectives

Upon completion and review of this chapter, you should be able to:

- ❏ Remove, clean, measure, and inspect rotors for wear or damage and determine if they should be machined or replaced.
- ❏ Mount a rotor on a lathe and machine it according to the manufacturer's procedures and specifications.
- ❏ Remove a caliper assembly and clean and inspect it for leaks and damage.
- ❏ Clean and inspect caliper mounts and slides for wear and damage and lubricate them as required.
- ❏ Remove, clean, and inspect pads and retaining hardware to determine needed repairs, adjustments, and replacements.
- ❏ Lubricate and install pads, calipers, and related hardware.
- ❏ Perform the ASE tasks for disc brake diagnosis and repair (see Appendix A).

23.1 Introduction

A typical brake job on a disc brake assembly consists of the following: caliper removal, rotor resurfacing or reconditioning, caliper rebuilding, lining replacement, caliper hardware replacement, and careful reassembly with lubrication of each part. The parts should be carefully inspected at least three times—before disassembly, during cleanup, and during reassembly—to locate any faulty or damaged components that should be replaced.

The disassembly inspection begins as soon as the wheel is removed and serves two major purposes: to locate damaged parts and to determine the relationship of the various parts. After the caliper is removed, the inspection continues on to the lining and the rotor. Some caliper designs provide visual access to check lining wear (Figure 23–1). But with this limited view, you often cannot tell whether the lining is riveted or if it is developing tapered wear.

Some newer brake designs have subtle differences in the caliper mounts and hardware items. These differences often provide opportunities for mistakes to be made during reassembly, which can change the performance of the

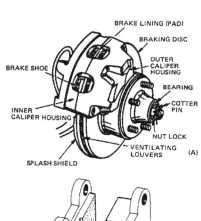
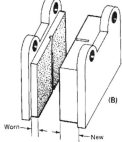

Figure 23–1 Some calipers allow the thickness of the lining to be quickly checked by looking through the openings (A). When the lining is worn to 1/32 to 1/16 in. above the rivets or pad backing (B), it should be replaced. (*A is courtesy of Wagner Brake; B is courtesy of ITT Automotive*)

system. Again, competent brake technicians realize that when they disassemble a unit, they become the authority on how the unit goes together. A service manual should be followed for guidance through the disassembly and reassembly steps, but many manuals fail to show the exact positioning of each and every small part. As in the case of drum brakes, when technicians encounter a new and usually more complicated design, they often work on one caliper at a time, saving the other to use as a pattern if needed.

When one pad wears faster than the other, a simple rule can be followed. Excessive wear of the outboard pad indicates a sticking, dragging caliper, and excessive wear of the inboard pad is caused by a sticking piston (Figure 23–2). During brake application, the amount of hydraulic force available is enough to move a partially stuck piston or caliper to apply the shoes. But since disc brakes have very weak release mechanisms, the piston or caliper will tend to stay applied and will drag. Rubber parts in the caliper and caliper hardware will harden and lose their resiliency with age and from the heat of braking. As they harden, they can no longer cushion or reposition the caliper or piston at release as they were intended to do.

In some cases, disc brake pads can simply be removed and replaced—the old ones are slid out of the caliper, and the new ones are slid in. This simple procedure is not described in this text because, in the author's opinion, this pad replacement is not a brake job. This operation cannot guarantee new-car braking performance for the life of the friction material. As with drum brakes, the entire braking assembly should be reconditioned when the linings are replaced. A thorough and fast disc brake repair has been made very easy with the increased availability of **loaded calipers** (Figure 23–3). These are rebuilt calipers that come equipped with new hardware and shoes installed in them. Simple pad replacement should be done only in instances where the brake technician knows the rest of the system is in good, safe operating condition.

SAFETY TIP: The following sections describe service operations on a single brake. But remember that brake operations are performed in pairs. The same operation is done at each end of the axle to ensure even, straight braking. Also, remember the safety points mentioned in Chapter 20.

23.2 Removing a Caliper

Before the caliper is removed, it is a good practice to make sure that the master cylinder reservoir is no more than one-half to two-thirds full. You should remove some of the fluid if necessary (Figure 23–4). One of the first steps in actual caliper removal is to retract the pistons—move them partway back into the caliper. This is necessary to get enough pad clearance to move them past the wear or rust ridge at the outer edge of the rotor. As the pistons are retracted, the fluid returning to the reservoir can cause it to overflow, resulting in spilled brake fluid. Spilled brake fluid will create a mess or damage the painted surfaces under the master cylinder. A much better way of solving this problem is to attach a small hose between the bleeder screw and a container. Then open the bleeder screw while the pistons are being retracted and allow the excess fluid to flow into the container (Figure 23–5). This is the preferred procedure on all vehicles because it does not force dirty, contaminated fluid back into the ABS hydraulic modulator. Debris being forced back into an ABS modulator can cause sticking or leaking problems of the modulator valves, and if forced into a master cylinder, the debris can cause seal wear or leaking of a quick-take-up valve.

For the most part, calipers that include a parking brake can be serviced in the same way as other calipers, except that the parking brake cable will have to be disconnected and reconnected. Another difference is that the piston must be retracted by methods other than those mentioned in the next section; these procedures are described in most service manuals. The internal hydraulic rebuilding services for calipers are described in Chapter 24.

It is a good practice to remove heavy rust deposits from the outer edge of the rotor before removing the pads or caliper. This can be done fairly easily and quickly by resting a scraper against the caliper as the rotor is turned by hand. The scraping can be followed by rubbing with coarse sandpaper if a cleaner surface is desired (Figure 23–6).

23.2.1 Retracting Caliper Pistons

There are several commonly used methods for retracting caliper pistons. On calipers with two or more pistons, note the amount of effort that is required to retract each piston. A piston that is really hard to retract is probably dragging. On fixed calipers, a large pair of slip joint pliers can grip the edge of the pad and the outside of the caliper. Slowly squeezing the pliers will retract the pads and pistons into the caliper body (Figure 23–7). Another way to retract the pistons on a fixed caliper is to use a large screwdriver or a special tool (Figure 23–8). Slide the tool between the pad and the piston or between the pad and the rotor. Then slowly and carefully push the

Figure 23–2 These two rotors were ruined because the outboard lining wore out completely. In both cases, the inboard lining and rotor surfaces are still good. This uneven wearing was caused by calipers that did not release completely and dragged.

Chapter 23 ■ Disk Brake Service   427

Figure 23-3 A loaded caliper is a rebuilt caliper with new hardware and shoes. *(Courtesy of Federal-Mogul Corporation)*

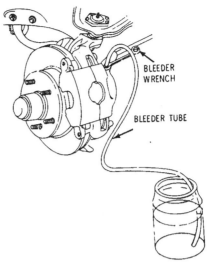

Figure 23-5 A better way to prevent reservoir overflow is to connect a bleeder hose and catch container to the bleeder screw and open the bleeder screw while the piston is being retracted. This method prevents dirty fluid and debris from being pushed into ABS modulator valves. *(Courtesy of General Motors Corporation, Service Technology Group)*

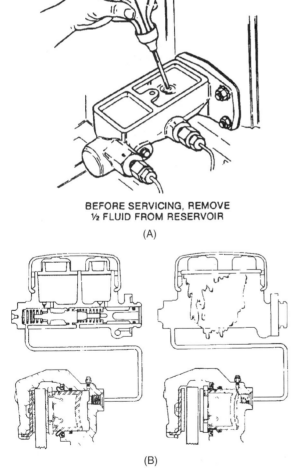

Figure 23-4 Before compressing caliper pistons, remove fluid from the reservoir (A). Fluid pushed back from the calipers will cause the reservoir to overflow (B). *(A is courtesy of Ford Motor Company)*

Figure 23-6 A heavy rust deposit at the outer rotor edge can be removed by scraping it with a chisel or old screwdriver as you turn the rotor by hand. *(Courtesy of LucasVarity Automotive)*

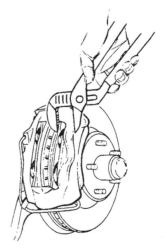

Figure 23-7 Large pliers can be used to squeeze the brake pad and force the piston to retract. *(Courtesy of General Motors Corporation, Service Technology Group)*

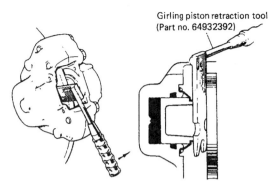

Figure 23-8 This special tool is slid between the rotor and piston and twisted to force the piston back. *(Courtesy of LucasVarity Automotive)*

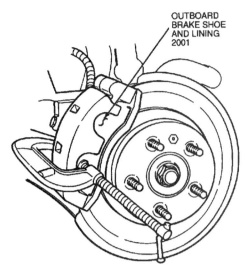

Figure 23-10 The C-clamp positioned over the caliper and outboard pad can be tightened to force the piston into its bore. *(Courtesy of Ford Motor Company)*

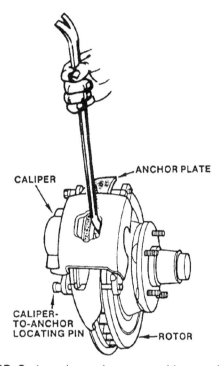

Figure 23-9 A pry bar or large screwdriver can be hooked to the rotor and pulled outward to retract the piston.

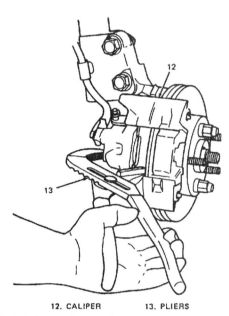

Figure 23-11 These large pliers are gripping the pad and caliper; squeezing the pliers will retract the piston. *(Courtesy of Delco Moraine Division, General Motors Corporation, Service Technology Group)*

piston back into its bore. Be careful not to scratch the rotor, piston boot, or lining. If the pads have been removed, there are several types of spreader tools that can be slid between the two pistons to retract them further.

Several different methods can be used to retract the piston in a floating caliper. A large screwdriver or pry bar can be hooked onto the cooling fins of a vented rotor or onto the edge of a solid rotor, and the caliper can be pried outward. This action will push the piston into the caliper (Figure 23-9). Another method is to place a large C-clamp over the caliper body and outboard pad. Tightening the C-clamp will force the caliper outward and retract the piston (Figure 23-10). In some cases, it is possible to position large slip joint pliers over the edge of the inner brake shoe or caliper support and the inboard side of the caliper body. Slowly squeezing the pliers will retract the piston (Figure 23-11). On calipers with an integral parking brake mechanism, the piston is retracted by rotating it with a special tool (Figure 23-12).

23.2.2 Caliper Removal

Caliper removal should follow the procedure recommended by the manufacturer, usually one of three general procedures. As a floating or sliding caliper is being

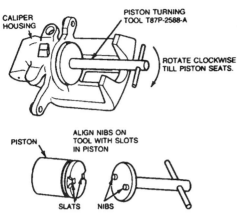

Figure 23-12 A special tool (T87P-2588-A) can be used to rotate the piston of this rear caliper to retract the piston. *(Courtesy of Ford Motor Company)*

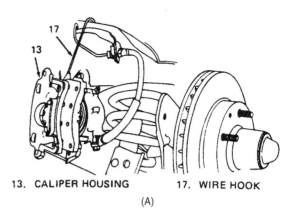

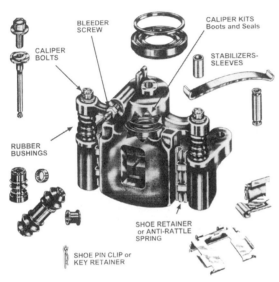

Figure 23-13 A variety of hardware is used with disc brakes. *(Courtesy of Rolero-Omega, Division of Cooper Industries)*

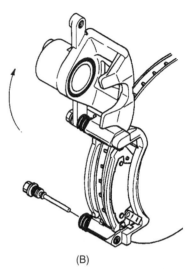

Figure 23-14 When a caliper is removed from its mounts, it should be suspended by a wire (17) and not allowed to hang on the hose (A). Some calipers can be pivoted to a vertical position and left on one of their mounting bolts (B). *(A is courtesy of General Motors Corporation, Service Technology group; B is ©Saturn Corporation, used with permission)*

removed, the relationship of the small clips and springs, often called **caliper hardware,** should be noted so that correct replacement can be made. Depending on the caliper, the hardware consists of shims, bushings, support keys, and antirattle springs. In a few cases, you will find a pair of components, such as antirattle springs or support keys, that are almost alike but not interchangeable (Figure 23-13).

The brake hose presents another problem. This hose will be damaged if it is twisted or kinked too tightly. The caliper should never be allowed to hang on this hose. A piece of wire or bungee cord should be kept handy so the caliper can be suspended from some convenient location inside the fender well after it has been removed from the mounts (Figure 23-14). Some technicians use a piece of 1/8- or 3/16-in. rod bent into an S shape about 3 or 4 in. long, just for this purpose.

If the caliper is to be removed completely, the hose will need to be disconnected. This presents a choice as to which end of the hose to remove (Figure 23-15). If the hose threads directly into the caliper, there is no choice. The other end—attached to the steel brake line at the frame—must be disconnected to prevent twisting of the hose. If the hose is secured to the caliper using a banjo fitting and a drilled bolt, there is a choice. The inlet fitting or banjo fitting bolt can be unscrewed, releasing the fitting and the hose from the caliper. If this is done, two new copper washers should be used when this fitting is replaced. In some areas, these washers are not easily available at local parts houses. Many brake technicians normally disconnect this hose at the frame end. This is usually a simple operation of unscrewing the steel tube nut from the hose end, sliding the hose retaining clip from the hose, and slipping the hose end out of the frame bracket.

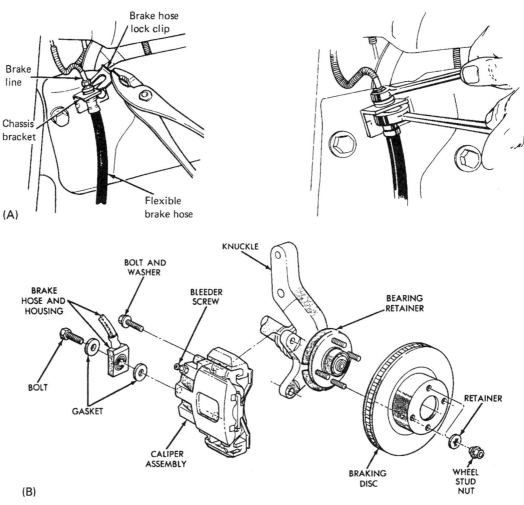

Figure 23-15 When a caliper is removed, the brake hose can be disconnected at the frame (A) or the caliper (B). The copper gaskets/washers should be replaced if the hose is disconnected from the caliper. *(A is courtesy of Brake Parts, Inc.; B is courtesy of Chrysler Corporation)*

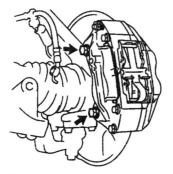

Figure 23-16 The caliper mounting bolts are removed to allow removal of a fixed caliper.

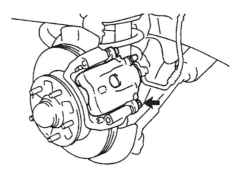

Figure 23-17 Most floating calipers are removed by removing the mounting bolts or guide pins.

Fixed calipers are normally removed by unscrewing the caliper mounting bolts at the steering knuckle or axle. Removal of these bolts allows the caliper assembly and the brake hose to be lifted off the rotor (Figure 23-16).

Some floating calipers are removed by unscrewing the **guide pins,** also called *guide bolts* or *locating pins* (Figure 23-17). Removal of these pins allows the caliper to be lifted out of the caliper mount and off the rotor. Caliper abutment clearance should be checked on floating and sliding calipers as they are removed. This check is described later in this chapter.

Several types of sliding calipers are held in place by one or two *support keys* or *guide plates*. These keys or plates are secured by a retaining screw or by cotter or

Chapter 23 ■ Disk Brake Service

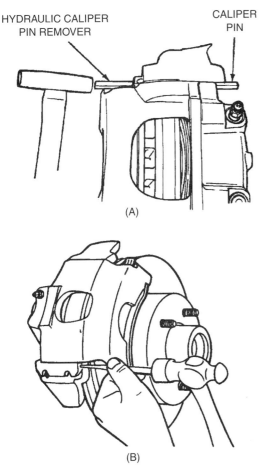

Figure 23-18 The caliper pin or support key is driven out using a special tool (A) or punch (B) so the caliper can be removed from the mounting bracket. *(Courtesy of Ford Motor Company)*

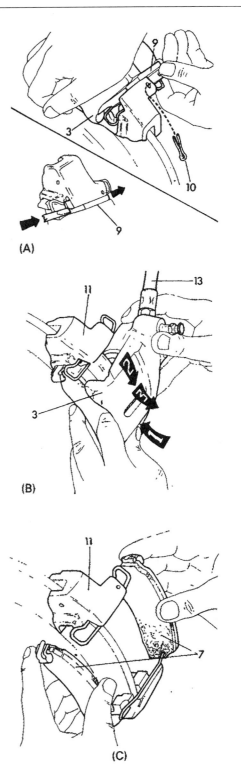

Figure 23-19 This caliper is removed (A) by the split pins (10) and guides (9), and then (B) by lifting the caliper (3) out of the bracket. This allows removal of the pads (C). *(Courtesy of LucasVarity Automotive)*

split pins. After removing the retaining screw or pins, the support key(s) can be driven out from between the caliper and the mounting bracket. Then the caliper can be lifted out of the mounting bracket and off the rotor (Figure 23-18 and Figure 23-19).

Several import calipers use a large support yoke or mounting adapter or bracket. In most cases, the caliper or support yoke is removed similarly to a floating or sliding caliper. Occasionally it is necessary to remove the mounting support to remove the rotor. A service manual should always be checked to determine the exact procedure (Figure 23-20).

23.3 Rotor Inspection

The rotor should be carefully inspected as part of the routine when doing a brake job and when trying to determine the cause of a particular complaint. Problems such as a pulsating brake pedal, pulsating or vibrating brake action, and a grabby brake are often caused by a rotor that has an excessive runout or parallelism problem. Excessive runout can also knock the pads back and cause a low brake pedal.

In the past, most sources recommended that when the pads were replaced, the rotor should be *reconditioned* or

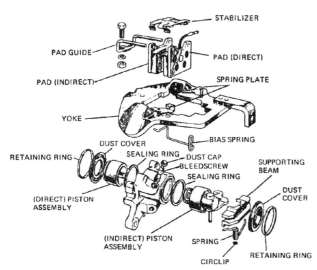

Figure 23-20 A yoke caliper like this one should be removed according to the manufacturer's procedure. *(Courtesy of LucasVarity Automotive)*

Figure 23-21 This rotor is ruined because the scoring is too deep; machining cuts to true the surface will produce a rotor that is too thin.

resurfaced. A rotor is reconditioned by **turning** it on a rotor lathe; this is also called *truing* or *remachining.* As in the case of used drums, the smooth, burnished surface of the rotor is too smooth for good lining break-in. The ideal surface finish for breaking in organic linings is about 30 to 60 microinches (0.76 to 1.5 μm). For semimetallic linings it is about 40 to 80 microinches (1 to 2 μm). Surface finishes are not normally measured in the field. They are discussed here only to give you an idea of how smooth or rough a rotor surface should be. When you install a new rotor, check the surface finish and use that as a guide.

In deciding whether a particular rotor should be resurfaced, reconditioned, or replaced, a brake technician considers the following: the type of rotor (vented or solid), the amount of scoring, the width of the rotor, the amount of thickness variation (parallelism), the amount of runout, how the rotor is mounted, the type of repair equipment available, and the manufacturer's recommendations. One manufacturer recommends that the rotors be replaced in pairs; if one needs replacement, replace both. Also, mixing rotor types is not recommended; if a car with composite rotors needs a replacement, replace that rotor with a composite type or replace both rotors with cast types. A rotor cannot be turned if the machining process will make it too thin.

Some manufacturers do not recommend turning the rotors on certain vehicles; these rotors should be replaced if there is excessive scoring or thickness variation. Runout problems can often be cured by reindexing the rotor on the hub. Some manufacturers require using an on-car lathe if turning a rotor.

One domestic manufacturer has stated that there are only three reasons that justify turning a rotor: excessive runout, excessive thickness variation, and excessive grooving (0.006 in. or deeper); even then, resurfacing can only be done if there will be sufficient thickness after turning. The manufacturer also states that if more than 0.015 in. has worn from the rotor, it should be replaced.

Begin with a visual inspection. Check both the inner and outer friction surfaces for scoring and wear. Normal wear usually produces a smooth but wavy rotor surface. Scores are abnormal wear caused by a metal or abrasive object digging into the rotor surface (Figure 23-21). If the scores or waves are deeper than 0.010 to 0.015 in. (0.25 to 0.38 mm), the rotor should be turned to produce a flat surface for breaking in the new lining. Many technicians treat the inner and outer rust ridges as scoring because they can interfere with the contact between the new lining and the rotor. Some older General Motors cars were produced with a rotor that had a deep groove machined midway around the friction surface; this groove should not be confused with scoring (Figure 23-22).

Also, check visually for rotor glaze (a highly glassy surface, which is somewhat normal); blueing; contamination from foreign materials such as oil, paint, or silicone sprays; and heat checks. The smooth, glassy surface should be roughened by resurfacing the rotor to ensure proper pad break-in. If the rotor shows severe heat checking, it should be turned or replaced (Figure 23-23). A contaminated rotor should be thoroughly cleaned by using a friction surface cleaner or by machining.

23.3.1 Rotor Measurements

Before a rotor is reused, it should be measured for size; specifically, the thickness of the friction surface is measured with an outside micrometer or caliper. This measurement is normally taken midway between the inner and

Chapter 23 ■ Disk Brake Service

Figure 23-22 The groove in this rotor was produced by the manufacturer; it is unworn.

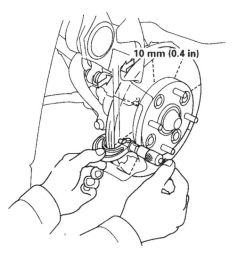

Figure 23-24 Rotor thickness is measured using a micrometer positioned in the middle of the friction surface. Measuring at eight points helps ensure that you find the thinnest spot and also checks for thickness variation. *(Courtesy of American Honda Motor Co., Inc.)*

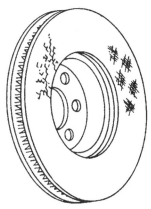

Figure 23-23 An overheated rotor has small cracks or heat checks on the rotor surface. *(Reprinted from Mitchell Anti-Lock Brake Systems, with permission of Mitchell Repair Information, LLC)*

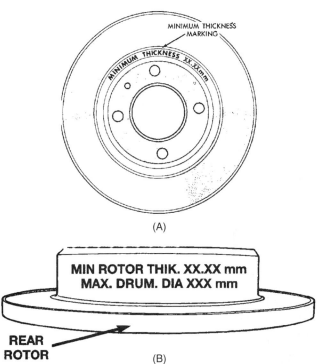

Figure 23-25 All modern rotors have their minimum thickness indicated on them (A). Rotor (B) is a rear rotor with a drum parking brake; it also bears the maximum drum diameter. *(Courtesy of Chrysler Corporation)*

outer edges of the friction surface (Figure 23-24). The minimum thickness of a rotor is specified by the manufacturer; since 1971, this dimension has been indicated on the rotor (Figure 23-25). A rotor that is too thin should not be used. As in the case of a drum, a rotor that is too thin will not have sufficient heat sink capability, and overheating of the lining will result. This is currently a potential problem with nonvented rotors used on FWD cars. Another problem is that too thin a rotor will position the caliper piston too far out of the bore.

While the rotor is being measured for thickness, it can also be checked for **parallelism.** This is sometimes called **thickness variation** because it is a variation in the thickness across the rotor's friction surface, which causes a lack of parallelism. If the two friction surfaces of a rotor are not parallel to each other, the pads must move together and apart during stops as they try to follow the changing friction surfaces. This action can be felt as a pulsation—a rising and falling—of the brake pedal or as a vibrating, grabby action of the brakes (Figure 23–26). To check parallelism, the rotor is carefully measured at eight to twelve different locations around the rotor as the technician tries to find the thickest and thinnest rotor widths. The measurements are normally taken in the middle of the rotor surface. Subtracting the thin measurement from the thick measurement will give the amount of thickness variation or parallelism (Figure 23–27). Compare these findings with the manufacturer's specifications (Figure 23–28). As a general rule, if the thickness varies more than 0.0005 to 0.001 in. (0.013 to 0.025 mm), the rotor should be turned or replaced. Some technicians prefer to check for parallelism problems using one or two dial indicators. This procedure is described later in this chapter.

Also, some manufacturers recommend checking the rotor for wedge-shaped wear while measuring the rotor thickness. Measure the rotor as close to the center of the rotor's friction surface as the micrometer frame will allow and then measure it again at the outer edge. A difference in measurements indicates wedge wear (Figure 23–29 and Figure 23–30). If it is excessive, the rotor should be turned or replaced.

Brake Specifications	
Disc thickness (new)	0.75000 in.
Disc thickness (minimum)	0.6850 in.
Disc parallelism (thickness variation, maximum)	0.0007 in.
Disc runout (maximum TIR)	0.0030 in.
Surface finish	15–80 microinches

Figure 23–28 Typical rotor specifications. *(Courtesy of General Motors Corporation, Service Technology Group)*

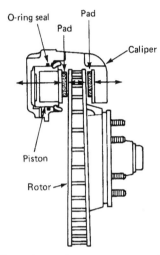

Figure 23–26 If the rotor surfaces are not parallel, the pads will move inward at the thin area and then back out when the thick area passes between them. This causes a pulsation of the brake pedal and surging brake action.

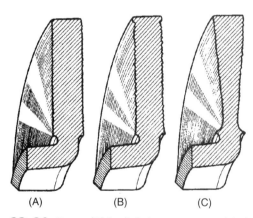

Figure 23–29 Rotor (A) is slightly worn, rotor (B) shows severe scoring and roughness, and rotor (C) is worn to a wedge shape. *(Courtesy of ITT Automotive)*

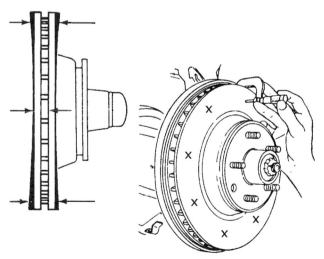

Figure 23–27 Rotor thickness variation is checked by measuring the rotor at eight to twelve places and comparing the measurements. *(Reprinted from Mitchell Anti-Lock Brake Systems, with permission of Mitchell Repair Information, LLC)*

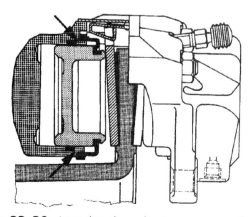

Figure 23–30 A wedge-shaped rotor can cause the piston to cock in the bore, a caliper to cock on its mounts, or a caliper to fracture. *(Courtesy of ITT Automotive)*

Rotor runout is a wobbly, side-to-side motion of the friction surfaces that occurs as the rotor revolves. It will knock or kick the pads away from the rotor as the car is driven. The result is too much pad clearance and a low brake pedal the next time the brakes are applied. It can also cause side-to-side motion of the caliper during a stop, and, if the caliper does not slide easily, it can cause a pulsating, vibrating stopping action (Figure 23–31).

Rotor runout, often called lateral runout, is measured using a dial indicator (Figure 23–32). The dial indicator is attached to the steering knuckle, spindle, caliper mounting bracket, or caliper. Then the indicator stylus is positioned in the middle of the friction surface. Be sure to mount the indicator so the stylus is at a right (90-degree) angle to the rotor surface. Rotate the rotor as you watch the indicator needle. If it moves in one direction, reverses, moves in the other direction, and reverses again, there is runout. Stop the rotor at the most downscale position of the indicator needle, adjust the indicator dial to zero, and mark this position on the rotor. Now rotate the rotor until the needle is reversing at its most upscale position, note the reading, and mark the rotor again. This is the highest, most outward part of the rotor, and the indicator reading is the amount of runout. Compare the measured amount of runout with the manufacturer's specifications. If the measured amount is greater, the rotor must be turned or replaced. As a general rule, runout should not exceed 0.005 in. (0.13 mm).

As you are measuring runout, note the location of the high and low spots. If they are on opposite sides of the rotor, the problem is truly one of runout, and the rotor is wobbling. If the indicator needle dips for only a portion of the rotor and most or part of the rotor is flat (no indicator motion), there is a hollow, concave area in the rotor's surface. This is more a problem of parallelism than of runout. Note that loose or rough wheel bearings will affect runout readings. It is recommended that you snug up the wheel bearings to zero clearance before measuring runout. Be sure to readjust the bearings to the correct clearance before the car is driven (Figure 23–33).

Tapered shims are available to remove runout at the rear rotors of older Corvettes (Figure 23–34). On these cars, as well as on some others, a distorted axle flange can produce rotor runout. This runout can be eliminated only by turning the rotors on the car or on the axle, which usually requires removal of the axle shaft from the car. This particular rotor has a new thickness of 1.250 in. (31.75 mm) and a machining limit of 1.230 in. (31.24 mm). Only 0.020 in. (0.5 mm) can be machined off in this rotor to true it. A shim, which is available in increments of 0.001 in. (0.25 mm), is placed between the rotor and the axle flange with the thick part of the shim in the correct position to reduce the amount of runout to the allowable specifications. The rotor and shim are indexed as needed to find a location where the amount of runout is acceptable.

Remember that you are really trying to determine what needs to be done about the rotor. Any rotor that exceeds the runout or parallelism limits must be turned or replaced, and any rotor that will become too thin must be replaced.

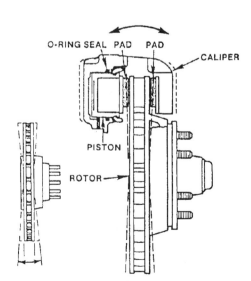

Figure 23–31 Excessive rotor runout forces the pads and caliper to move back and forth sideways when the brakes are applied. This can cause surging of the brakes, vibration, and steering wheel oscillation.

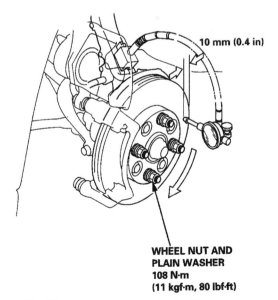

Figure 23–32 Rotor runout is checked by mounting a dial indicator on the rotor friction surface and rotating the rotor. Note the wheel nuts holding the rotor tight; on some cars the wheel bearings should be snugged up. (Courtesy of American Honda Motor Co., Inc.)

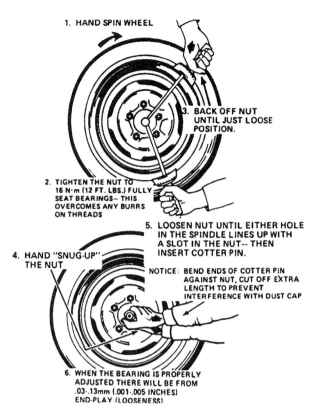

Figure 23-33 If wheel bearings are tightened for a rotor runout check, they must be readjusted before the car is driven. *(Courtesy of General Motors Corporation, Service Technology Group)*

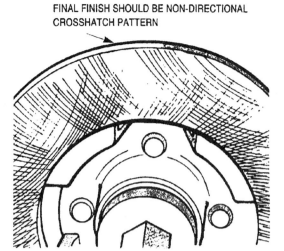

Figure 23-35 A nondirectional/crosshatch surface has a series of circular scratches that run in all directions. *(Courtesy of Chrysler Corporation)*

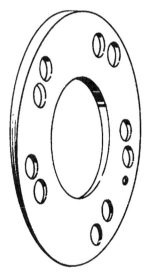

Figure 23-34 This tapered shim can be placed between the rotor and axle flange to correct runout; it is available in varying amounts of taper. *(Courtesy of Stainless Steel Brakes Corp.)*

23.4 Rotor Refinishing

A brake technician has four methods of servicing a rotor: resurfacing by hand, off-car turning, on-car turning, and replacing. When choosing a method, the brake technician must remember that both rotors on an axle must have the same surface finish.

The ideal rotor surface is a flat surface with very shallow scratches. These scratches should be about 40 to 80 microinches (0.76 to 1 μm) deep, and they should run in every direction. This is called a **nondirectional** or **swirl finish.** A rotating sander will produce a series of more desirable circular but irregular scratches running in every direction, and the overlapping scratches produce the nondirectional or swirl finish (Figure 23-35). When a lathe cuts a rotor, it produces a scratch that slowly spirals from the center to the outer edge, constantly flowing in the same circular direction (Figure 23-36).

Rotor replacement becomes mandatory if the rotor is too thin or will become too thin if turned to remove defects such as deep scoring or excessive runout. In other cases, where the rotor is worn thin but still not undersize and a new rotor is inexpensive and easily replaceable, it is a good idea to replace the rotor. If the rotor has worn a large amount (more than halfway to the discard thickness) on the original pads, it is a good idea to replace it because it will likely wear past the minimum thickness before the next set of pads needs to be replaced. A new rotor, with its greater heat sink capability, usually results in a longer lining life. In some cases, the replacement rotor can be very expensive or rotor replacement can be a difficult, time-consuming operation (Figure 23-37). In this case, the added expense does not warrant replacement, and it is usually better to refinish the old rotor.

Resurfacing a rotor by hand is not recommended by many manufacturers because of the possibility of produc-

Figure 23-36 A drum lathe produces a directional finish with all the scratches running in the same direction.

ing a different surface finish on the two rotors. Some technicians resurface a rotor to break the glaze on an otherwise good rotor while removing as little metal as possible. Some FWD cars experience very rapid lining wear after the rotors have been turned; this rapid wear can be blamed on a loss of the rotor's heat sink capability. As mentioned earlier, a car will have a brake pull toward one side if the two rotors do not have the same surface finish. This pull will probably disappear after a few stops, however, because the pads will quickly burnish out the small difference in the scratches. Resurfacing is desirable to remove glaze, old lining residue, or light rusting on true rotors, because it removes a minimum amount of metal. Resurfacing will not true up a rotor. If there is excessive runout or thickness variation, the rotor must be turned or replaced. Resurfacing a rotor by hand sanding or using a disc sander normally produces a nondirectional finish. This procedure is described later in this chapter.

A rotor is turned on a brake lathe (Figure 23-38). At one time, turning a rotor was an off-car operation only. Today, on-car lathes and turning devices are common. When an off-car lathe is used, the rotor is removed from the car and mounted on the lathe mandrel or arbor. As the rotor is rotated, a pair of cutter bits are moved straight across the friction surfaces. Some of the earlier lathes used a single cutter that required two cutting operations, one for each side of the disc. All new machines use twin cutters that cut both sides at the same time. This method is faster and produces truer friction surfaces. As the worn

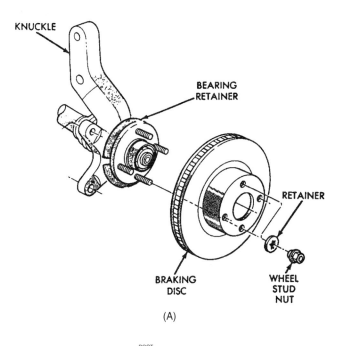

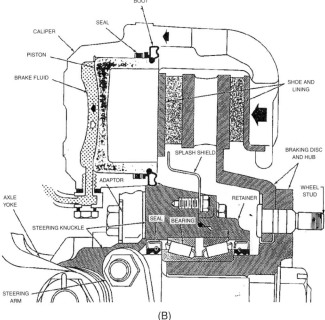

Figure 23-37 The rotor on many FWD cars (A) is held in place by the wheel and lug nuts so it is easy to remove and replace. Some rotors are held captive behind the hub flange (B); they are much harder to remove and replace. *(Courtesy of Chrysler Corporation)*

friction surface is removed by the cutters, truly parallel but thinner friction surfaces are produced.

When an on-car lathe is used, the rotor is left mounted in its normal manner, and the lathe mechanism is attached to either the caliper mount or the wheel mounting bolts. Depending on the unit, the rotor is rotated by the vehicle's engine, an electric motor that is part of the lathe unit, or a separate electric motor.

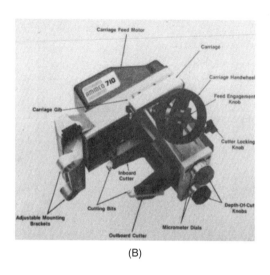
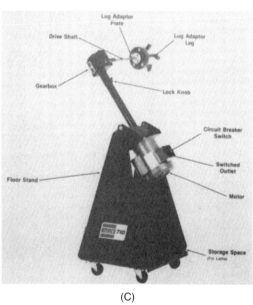
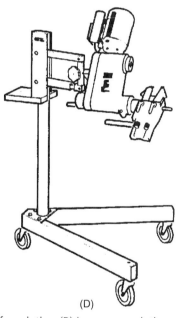

Figure 23-38 There are various types of rotor lathes. (A) is a standard off-car lathe. (B) is an on-car lathe that is mounted onto the caliper brackets. The rotor is driven by the car's engine or a rotor-driving unit (C). (D) is an on-car lathe that is mounted onto the hub and includes a drive motor. *(A, B, and C are courtesy of Ammco Tools, Inc.; D is reprinted by permission of Hunter Engineering Company)*

When a rotor is turned, the thinnest cut made should be at least 0.004 in. (0.1 mm). The outer edge of the rotor surface tends to harden from use, which could be the result of work hardening from the pads or heat treating that occurs during hard stops. With very shallow cuts, the point of the cutter bit has trouble penetrating the hard surface and will wear rapidly from the pressure and the heat generated. Shallow cuts tend to smear the rotor surface. A slightly deeper cut will use the side of the cutter to cut the harder surface skin, and the increased contact between the cutter bit and the rotor will transfer more heat and help keep the cutter bit from becoming too hot (Figure 23-39). In this way, a turned rotor will be reduced a minimum of 0.008 in. (0.2 mm)—0.004 in. on each side. If there is thickness variation, rotor runout, or mounting runout, the rotor will be even thinner because deeper cuts will be required. If the rotor is not mounted on the arbor correctly, mounting runout will occur. This runout will

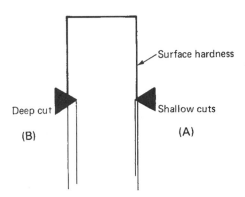

Figure 23-39 A shallow cut uses the very tip of the cutter bit (A), which places a great load on a small area. A deeper cut (B) uses more of the stronger part of the lathe bit.

Figure 23-41 This front rotor is being turned while on the car with the rotor in its original, undisturbed position. *(Reprinted by permission of Hunter Engineering Company)*

Some on-car lathes include a drive rotor to spin the rotor. Since the rotor mounting is not disturbed, removal and replacement time is saved and there are no problems with mounting runout. The rotor surfaces will be true and parallel, with no runout. One turning device uses abrasive discs instead of cutters to smooth the rotor. This device produces a nondirectional finish. Several manufacturers market an on-car lathe with a portable drive motor that allows any rotor to be turned while on the car.

When turning a rotor, use the following recommendations:

- Use only sharp cutter bits. The sharp edge at the point of the bit will wear to a flat area that will appear as a white spot as it becomes dull.
- Always use the correct rotor mounting adapters.
- A vibration silencer or antichatter device must always be installed snugly around or against the rotor.
- Be aware that the surface finish of the rotor will be affected by the sharpness of the cutter bit(s), the use of the vibration band, the depth of cut, the feed rate, the speed of the lathe arbor, and the mechanical condition of the brake lathe.
- Front rotors (RWD cars) should be turned with the hub. Rotors that are loose or easily removed should be secured with lug nuts.
- Composite rotors must be mounted using clamping adapters that simulate the clamping action of a wheel. Composite rotors have a greater tendency to vibrate, and this produces a poor finish.

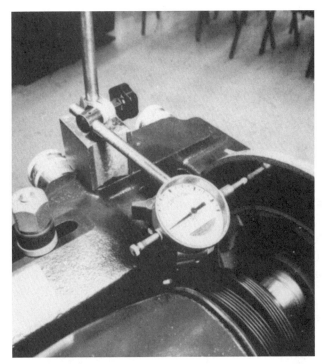

Figure 23-40 This dial indicator is set up to measure the amount of runout with the rotor mounted on the lathe. It should indicate the same high and low spots as when the rotor was on the car.

require a deeper cut on each side and also cause runout when the rotor is remounted on the car. When turning rotors that have excessive runout on the car, it is a good practice to mark the high spot of the runout. After the rotor is mounted on the lathe, it should have the same amount of runout in the same location (Figure 23-40).

On-car rotor turning is ideal for rotors that are on a drive axle—the rear axle of a RWD car or the front axle of a FWD or 4WD vehicle. The lathe is mounted on the caliper support (adapter), and the car is started and put in gear. As the engine or drive motor rotates the rotor, the cutter bits are moved past the friction surfaces (Figure 23-41).

- Rotors with wheel bearings should be checked to make certain that the bearing races are in good condition and tight in the hub. Damaged races should be replaced before turning the rotor.
- Never bump the cutter bit or push it against a rotor that is not revolving. The hard carbide bits that are commonly used chip and break easily.
- Always check and remove mounting runout before turning a rotor.
- The inner and outer ridges should be removed as a beginning step.
- Always remove the smallest amount of metal possible, but do not make too shallow of a cut. Use a minimum cut of 0.003 in. with positive rake bits or 0.004 in. with negative rake bits. Use a maximum cut of 0.010 in. with either bit.
- Always cut both sides of the rotor before removing it from the lathe.
- Rotors used with fixed calipers should have the same amount of metal machined from each friction surface.
- To achieve an acceptable final finish, hold a swirl tool with 120-grit paper or a sanding block with 150-grit paper against the rotating friction surfaces for 60 seconds using a medium pressure (Figure 23–42).
- Remember that sanding will further reduce the rotor thickness.
- It is a good practice to deburr the outer edge after machining using a file or 80-grit sandpaper—either handheld or on a small disc—placed lightly against the rotating rotor surface.
- A turned rotor should be cleaned using a commercial brake surface cleaner or denatured alcohol (Figure 23–43). Wash the turned surfaces with a clean shop cloth and the liquid until the cloth shows no more contamination.
- After cleaning, do not touch the new friction surfaces. Pick up the rotor by the outer diameter or by the center hole.

Rotor Stopping Efficiency	
New OEM rotor	100%
Standard turning procedure	60%
Standard turning procedure	
Plus sanding block for 60 seconds	110%
Plus swirl tool for 60 seconds	90%
Plus swirl tool for 20 seconds	60%
Rotor in good condition, no resurfacing	96%

Figure 23–42 A comparison of the stopping efficiency of various rotor friction surfaces. *(Reprinted by permission of Hunter Engineering Company)*

Figure 23–43 After a rotor has been turned, the friction surfaces should be flushed clean to remove all traces of carbon and small metal chips.

As in the case of a drum, when a rotor is turned, minute particles of metal and carbon from the cast iron will become embedded in the surface scratches. Unless they are washed out, these particles will become embedded in the new lining surface. Once there, they can cause abnormal braking, squeal, and/or excessive future rotor wear.

23.5 Off-Car Machining of a Rotor

SAFETY TIP: When machining a rotor, use the caution that should always be exercised around moving machinery. Be particularly concerned about the revolving drum, the cutter bit, the revolving vibration band, and the revolving lug bolts on some drums.

When a rotor is machined, follow the directions furnished by the manufacturer of the machine you are using. The description that follows is very general.

To turn a rotor, you should:

1. Clean the rotor and hub. All oil and grease in the hub or on the rotor should be washed off with solvent, and the parts should be air dried.
2. Select the correct mounting adapters. Always make sure there is no grit between the adapters and the rotor so the rotor will be exactly centered as it is mounted, with no runout. Make sure the cutter is out of the way while the rotor is being mounted.
   a. Rotors with hubs are normally centered and held in place by a pair of tapered cones that enter partway into the wheel bearing cups (Figure 23–44).
   b. Rotors without hubs are normally squeezed between a single cone passing partway through the center hole and a hubless adapter. The cone centers the rotor, and the hubless adapter eliminates lateral runout.

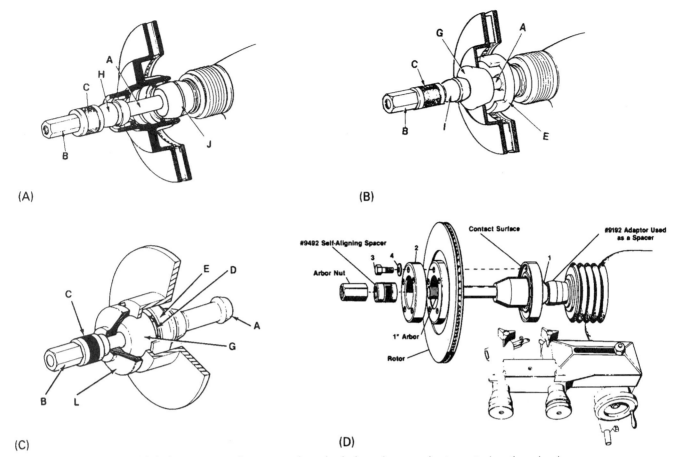

Figure 23-44 Rotors with hubs are normally mounted on the lathe using an adapter entering the wheel bearing races (A). Hubless rotors are centered using a cone (G) and kept from wobbling by one or two hubless adapters (B, C). Composite rotors require special mounting adapters to secure the center section (D). *(Courtesy of Ammco Tools, Inc.)*

  c. Composite rotors are normally clamped between two adapters that also center the rotor.

3. Mount rotors with the proper adapters and tighten the mounting nut to the correct torque. *Do not overtighten.*

NOTE: It is very easy to cause runout in a rotor as you machine it on an off-car lathe by overtightening the nut. Some sources recommend that you tighten this nut finger tight; rotation of the lathe will tighten the nut further. Prove this to yourself by doing the following: measure the runout after you turn a rotor, remove the rotor, reinstall the rotor and tighten the nut very tight, and remeasure the runout. A significant increase in the amount of runout shows the distortion created by overtightening the nut.

4. Select the correct arbor speed for the rotor diameter. Then adjust the speed mechanism, usually a V-belt and various-sized pulleys, to produce the correct speed. Too fast a speed will shorten the life of the cutter bit(s), and too slow a speed wastes time (Figure 23-45).

	Rough Cut	Finish Cut
Spindle speed		
10 in. and under	150–170 rpm	150–170 rpm
11–16 in.	100 rpm	100 rpm
17 in. and larger	60 rpm	60 rpm
Depth of cut		
(Per side)	0.005–0.010 in.	0.002 in.
Tool cross feed		
(Per revolution)	0.006–0.010 in.	0.002 in. max.
Vibration dampener	Yes	Yes
Sand rotors	No	Yes
Final finish		

Figure 23-45 Before turning a rotor, select the correct drive or spindle speed depending on the rotor diameter. Also note the other recommendations. *(Courtesy of Ammco Tools, Inc.)*

5. Adjust the position of the cutter bit holder so the cutter bits are centered or even across the rotor and there is enough travel to completely true the rotor surfaces (Figure 23–46). The extension of the tool holder should be kept as short as possible to produce the best finish.

6. Make sure the automatic feed is disengaged and turn on the lathe. Adjust the cross-feed mechanism to position the cutter(s) in the center of the friction surface. Slowly and carefully feed one of the cutter bits inward until it just begins to cut the rotor surface. It should make a shallow scratch. As you are doing this, observe the mirror image of the cutter bit in the rotor surface (Figure 23–47). You hope it will stay relatively still and not move in and out. A lot of in-and-out movement indicates runout. This can also be observed by watching the clearance between the cutter bit and the rotor.

7. Stop the lathe so you can inspect the scratch (Figure 23–48). If little runout was observed and the scratch runs almost all the way around the rotor, the rotor is mounted true and can be turned. If the scratch is short and runout was observed, the rotor is either mounted crooked or has substantial runout. To check for mounting trueness, you should:

   a. Loosen the nut securing the mounting adapters and rotor on the arbor.

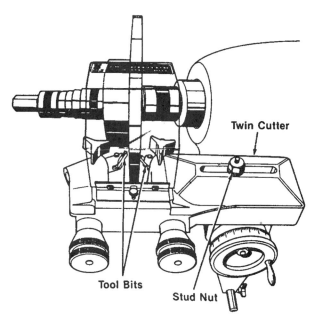

Figure 23–46 After mounting a rotor, loosen the tool holder stud nut and adjust the tool holder so the cutter bits are exactly opposite each other. *(Courtesy of Ammco Tools, Inc.)*

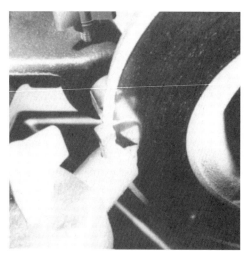

Figure 23–47 Note the mirror image of the cutter bit in the smooth rotor surface; if it moves in and out when the lathe is turned on, the rotor has runout.

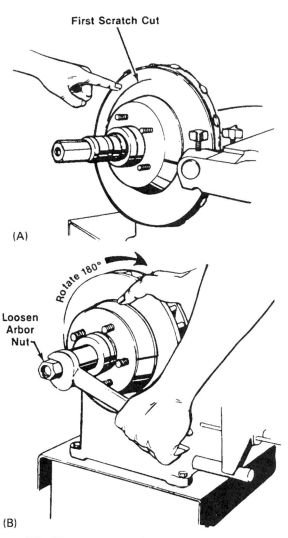

Figure 23–48 Runout can be checked by making a light scratch cut (A); a very short cut indicates too much runout. Next, rotate the rotor one-half turn on the lathe (B) and make a second scratch cut. The relative position of the second scrach cut gives a good indication of what is causing the runout. *(Courtesy of Ammco Tools, Inc.)*

b. Hold the centering cone(s) and hubless adapter (if used) stationary while you rotate the rotor one-half turn. Then retighten the arbor mounting nut.

c. Move the cross-feed handwheel one revolution and repeat steps 6 and 7.

d. If the new scratch indicates an acceptable mounting, the rotor can be turned. If the new scratch is alongside the first scratch, runout is being caused by the rotor. The rotor can be turned. If the new scratch is one-half turn from the first scratch and there is still an excessive amount of runout, there is a mounting problem. This can be caused by a bent, loose, or improperly indexed arbor; the wrong adapters; grit between the adapters and the rotor; or loose wheel bearing cones in the hub.

8. Attach the silencer or vibration device to the rotor. It is usually a band wrapped around the outer edge of the rotor or a pair of pads resting against the rotor surfaces (Figure 23–49).

9. Start the lathe and adjust the cutter bits to just contact the rotor surfaces. Adjust the cutter feed dials to zero or note the readings (Figure 23–50).

10. Make sure that the safety shield is correctly positioned over the cutters and safety glasses are worn (Figure 23–51).

11. Turn the cross-feed handwheel to bring the cutters to the outer rust ridge and slowly hand feed the machine to remove this ridge (Figure 23–52).

12. Turn the cross-feed handwheel to move the cutters inward, or toward the rotor's hub. As you do this, watch for any grooves. If you locate any, feed the cutter bit to the bottom of the groove to determine how deep a cut is necessary to remove it (Figure 23–53). Return the cutter bit to zero and continue turning the cross-feed handwheel to bring the cutters to the inner wear ridge. Hand feed the machine to remove the inner wear ridge (Figure 23–54).

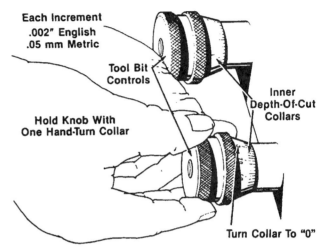

Figure 23–50 Before starting a cut, the cutter bits are adjusted to lightly touch the rotor, and the cutter feed dials should be set to zero. *(Courtesy of Ammco Tools, Inc.)*

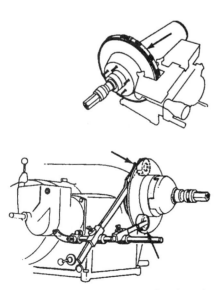

Figure 23–49 The most common vibration dampeners or silencers (arrows) are a large rubber band that wraps around the edge of the rotor or a pair of pads that drag on the friction surfaces. A vibrating rotor causes a rough surface finish. *(Courtesy of Ammco Tools, Inc.)*

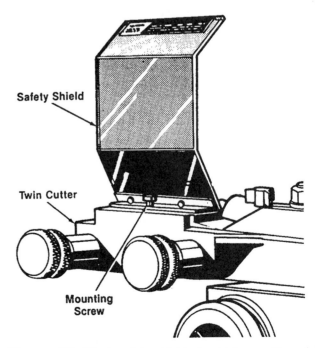

Figure 23–51 A safety shield should be positioned between you and the cutters as the rotor is machined. *(Courtesy of Ammco Tools, Inc.)*

Figure 23-52 With the cutter bits adjusted to a very light cut, they are hand fed outward to remove the outer wear or rust ridge before cutting the friction surface.

Figure 23-54 Following the cut for the outer ridges, hand feed the cutters inward to remove the inner wear ridges.

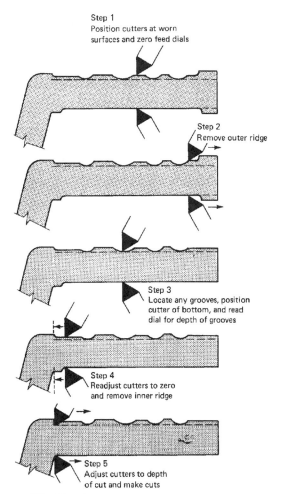

Figure 23-53 The steps used in machining a rotor are (1) set the feed dials to zero, (2) remove the outer ridges, (3) determine the depth of cuts, (4) remove the inner ridges, and (5) cut the friction surfaces.

13. Select the depth of cut. If no grooves were found, the rotor can often be turned in one finish pass. The ideal **finish cut** is between 0.004 and 0.005 in. (0.1 and 0.15 mm) deep—on each side of the rotor—with a feed rate of 0.002 in. (0.05 mm) per revolution. If the rotor requires a cut deeper than 0.006 in., it is recommended that two or more cuts be made. The maximum depth of the cut and the correct feed rate are determined by the machine being used. The depth is generally about 0.010 in. (0.25 mm) per side, with a feed rate of about 0.006 in. (0.15 mm) per revolution. This is called a **roughing cut** and should always be followed by a finish cut (Figure 23-55).

14. With the machine running, adjust the cutters to the desired depth of cut and the automatic feed to the rate of cut. Engage the automatic feed and, as the cut begins, observe the machine, cutter bits, and rotor for any unusual noise or movement. If something appears wrong, stop the machine and recheck your settings.

15. After the cut is finished and the cutter bits move past the outer edge of the rotor, disengage the automatic feed. If the cut was the finish cut and the friction surfaces are cleaned up, go to step 16. If the cut was a rough pass or part of the friction surface was skipped, repeat steps 14 and 15.

16. Remove the silencer device.

17. Many technicians hold a disc sander (with a 120-grit disc on it) against the rotating rotor for a few moments to produce a nondirectional finish. This should be repeated on both sides of both rotors using the same time interval and sander pressure on

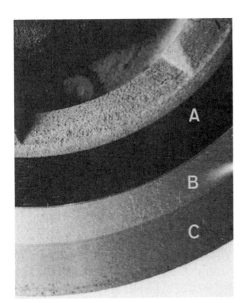

Figure 23-55 This rotor has an uncut, burnished surface (A), a section cut at a fast feed rate for a rough cut (B), and a section cut at a slow feed rate for a finish cut (C).

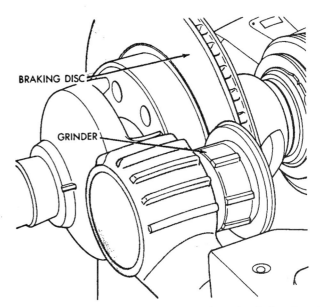

Figure 23-57 This grinder, mounted in the lathe, is adjusted to run lightly against the turning rotor to produce a nondirectional finish. *(Courtesy of Chrysler Corporation)*

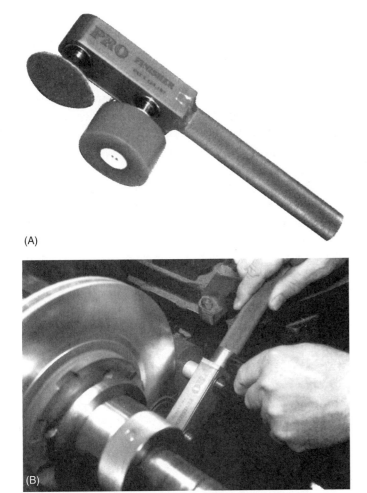

Figure 23-56 This Pro Finisher (A) is handheld and driven by the drive wheel against the outer edge of the turning rotor (B). About 30 seconds on each side produces a smooth, nondirectional finish. *(Courtesy of RSR Enterprises LTD)*

Figure 23-58 It is a good practice to stroke a file along the outer edges of the rotor to remove the sharp corners (inset). While doing so, avoid the spinning rotor and lug bolts.

each surface (Figure 23-56). A simple, handheld refinisher can be positioned on each rotor surface for about 30 seconds on each side to produce a nondirectional finish (Figure 23-57).

18. It is a good practice to stroke a smooth file along the outer edge of the rotor to deburr or chamfer the sharp corner. This will prevent shaving or chipping of the brake lining as the caliper is installed (Figure 23-58).

19. Remove the rotor from the lathe.

20. Remeasure the rotor to make sure that it is not undersize. One new lathe model includes a digital readout that displays the actual rotor width.
21. Mount the second rotor on the arbor and turn it. Be sure to use the same feed rate on the finish pass.
22. Wash the newly machined friction surfaces using denatured alcohol or brake surface cleaner.

23.6 On-Car Machining of a Rotor

Several different methods are currently used to machine rotors on the car. One of these methods uses twin cutter tools, and the rotor is turned by the vehicle's engine. Another uses twin abrasive pads, and the rotor is turned by the vehicle's engine. A third uses twin cutter tools, and the rotor is turned by an electric motor contained in the unit. This self-powered unit is different in that it is connected to the rotor using the wheel lugs, with a stabilizing link to the caliper support bracket. The other two are attached to the caliper support bracket. When using any of these units, be sure to follow the operating procedure recommended by the manufacturer.

SAFETY TIP: If performing this operation, you will be working next to a turning rotor in a confined area. Be careful, especially with vehicle-powered units. It is difficult to shut them off quickly in case of emergency.

To turn a rotor on the car, you should:

1. Raise and securely support the car.
2. Remove the wheels and the calipers.
3. Attach the lathe unit to the steering knuckle.
   a. On caliper bracket–mounted units, bolt the mounting adapters to the caliper support bracket or to the caliper mounting bosses on the steering knuckle. Then bolt the lathe assembly to the mounting adapters (Figure 23–59).
   b. On hub-mounted units, mount the adapter to the hub and adjust the adapter to compensate for any runout caused by mounting errors. Mount the lathe onto the adapter (Figure 23–60).
4. Adjust the cutter units or abrasive discs so they are centered on the rotor.
5. If applicable, attach the antivibration band around the rotor and the protection band around the exposed wheel lug bolts (Figure 23–61). On abrasive disc units, select and install the disc—fine, medium, or coarse grit—for the desired cutting rate. This rate is determined by how much material must be removed from the disc.
6. On vehicle-powered units, attach a clamp to the rotor on the other side of the car. The clamp should be positioned against the caliper abutments to prevent rotation of the rotor (Figure 23–62).
7. On vehicle-powered units, start the engine and carefully put the car in gear, first or reverse as required to turn the rotor in the correct direction. On self-powered units, turn on the switch to start the lathe.
8. On units using cutter bits, turn the cutter bit crossfeed handwheel to position the cutters midway on the rotor and turn the feed knobs so the cutters just touch the rotor. From this point, the remainder of the cutting procedure is quite similar to that with an off-car lathe (Figure 23–63). On abrasive disc units, turn the knobs to apply pressure between the

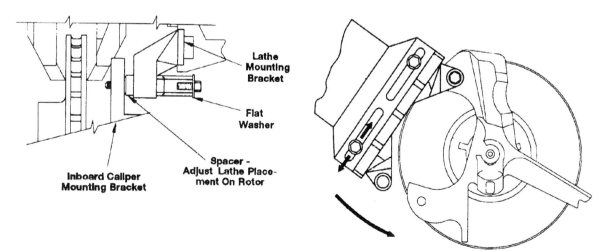

Figure 23–59 This on-car lathe is attached to the caliper mounting points by adjusting the attaching bracket and using the proper spacers. *(Courtesy of Ammco Tools, Inc.)*

Chapter 23 ■ Disk Brake Service

Figure 23–60 This on-car lathe is mounted onto the hub by first attaching an adapter and adjusting it to compensate for any runout (A). The lathe is then connected to the adapter (B). *(Reprinted by permission of Hunter Engineering Company)*

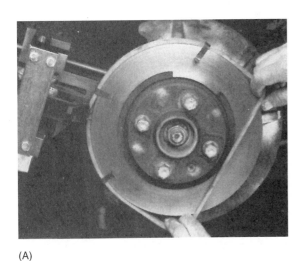
(A)

(B)

Figure 23–61 A vibration band is being positioned around the rotor with the aid of metal clips (A). The protector band placed over the lug bolts (B) helps keep the technician from being caught on them. *(Courtesy of Kwik-Way Mfg. Co.)*

abrasive pads and the rotor. Continue applying feed pressure until the rotor is cleaned up. If the operation was begun with coarse discs, stop cutting when the rotor is almost clean and switch to fine pads (Figure 23–64).

9. When the cutting or grinding operation is complete, stop the rotor and remove the machining unit.
10. Machine the second rotor in the same manner.
11. Wash the metal and carbon residue from the newly machined rotor surfaces using denatured alcohol or brake surface cleaner.

23.7 Resurfacing a Rotor

Resurfacing a rotor by hand is not recommended by all manufacturers, but it has been used successfully by many brake technicians. Its advantages are that it is quick, does not require removal of the rotor, leaves a nondirectional finish, and removes a minimum amount of metal from the rotor. This method can be used only on good rotors because it will not true a rotor. If there is excessive runout, thickness variation, or grooving, the rotor must be turned or replaced.

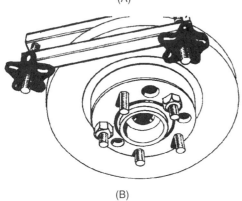

Figure 23-62 If using the car's engine to turn the rotor, a C-clamp (A) or rotor clamp (B) will prevent the opposite rotor from turning. Note the cardboard to keep the clamp jaws from marking the rotor. *(Courtesy of Can Am)*

Figure 23-63 This rotor is being machined using a Kwik-lathe. The cutter bits are fed outward by an electric motor turning the drive belt (arrow).

SAFETY TIP: With this method, you will be using a disc sander in confined quarters. You should wear eye protection and, in some cases, breathing and ear protection. Also, be careful of the rapidly rotating abrasive disc in the limited space.

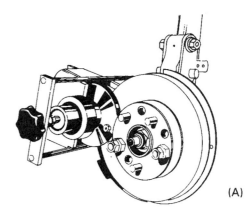

Figure 23-64 A special grinder with an abrasive pad on each side is being used to machine the rotor while the rotor is turned using the car's engine (A). A vacuum cleaner attachment (B) can be used to remove dust and debris. *(Courtesy of Can Am)*

To resurface a rotor, you should:

1. Raise and securely support the car.
2. Remove the wheels and the calipers.
3. Attach a sharp, medium-grit (120- to 160-grit) disc to a drill sanding pad or a small disc grinder; a 3- or 4-in. disc is ideal. With the drill or disc sander running and the abrasive disc held almost flat against the rotor to prevent gouges, move the disc back and forth around the rotor to obtain an even scratch pattern on the friction surfaces. Rotate the rotor as necessary to sand the entire friction surface. Try to obtain an even pattern that will remove all or nearly all of the glazed surface. Remove the rust ridges at the inner and outer edges as you work (Figure 23-65).
4. Repeat this operation on both sides of both rotors. The caliper mounting area is used for access to the inner friction surface (Figure 23-66).
5. Wash the metal and carbon grit from the newly ground friction surfaces using denatured alcohol or brake surface cleaner.

Figure 23–65 This rotor is being refinished by sanding with a drill-powered sanding disc.

Figure 23–66 This rotor was refinished with a disc sander. Note that the shiny surface has been roughened by the non-directional sanding scratches.

23.8 Rotor Replacement

If a rotor is faulty, it must be replaced. On many cars this is simply a matter of lifting the rotor off the hub and setting a new one in its place. As this is done, always make sure there is no dirt or grit between the rotor and hub, which can cause runout. Rotor runout should be checked after mounting by securing the rotor to the hub with two or three properly tightened lug nuts and washers. Measure the amount of runout using a dial indicator, as previously described (Figure 23–67).

If the runout is not within specifications, **index** it—move the rotor to one of the other positions on the hub—and recheck it. When an acceptable position is found, mark one of the lug bolts and, right next to it, the rotor. If an acceptable position cannot be found, the rotor will need to be turned, preferably on the hub.

In some cases, the rotor is bolted to the hub. In this case also, replacement is simply a matter of unbolting the old rotor and bolting on a new one. Again, be sure to check the runout of the new rotor.

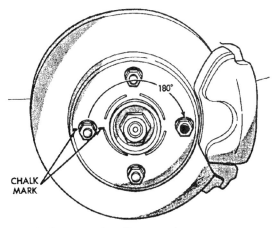

Figure 23–67 Occasionally, excessive rotor runout can be reduced to tolerable limits by reindexing the rotor. A four-bolt rotor can be turned 180 or 90 degrees in either direction. *(Courtesy of Chrysler Corporation)*

The front rotors on many RWD cars are secured to the hubs using swaged lug bolts, as in many brake drum and hub combinations. To separate the rotor from the hub, it is recommended that the first step be removal of this swaged area using a special cutter. Next, the lug bolts should be pressed out, one at a time, using a special fixture or hub anvil so the rotor or hub flange is not distorted (Figure 23–68). The new rotor is placed on the hub, and new lug bolts are pressed in place, one at a time, again using the special fixture or support anvil. Be careful that no pressing force is placed on the rotor, because it could crack or distort the rotor. The new lugs should be locked in place by using a center punch to upset the metal at the shoulder of the new lug bolts (Figure 23–69). When selecting the new lug bolts, be aware that they have several important dimensions that should be considered (Figure 23–70). New rotors might have to be turned to remove runout.

Some rotors are riveted onto the hub or axle flange. These rivets need to be removed to replace the rotor (Figure 23–71). In some cases, the rivet head can be shaved off with a sharp chisel. In many cases, it is helpful to first drill a hole slightly smaller than the rivet's diameter partially through the rivet. After cutting the rivet heads, the rest of the rivet can be driven out using a punch (Figure 23–72). The rotor can then be worked off the hub or axle. A new rotor is usually not reriveted because it will be held in place by the wheel lug bolts. It is a good practice to check the runout, as previously described in this section. As mentioned earlier, tapered shims that can be used to reduce runout are available for some cars.

The rotors on some FWD cars are mounted on the inside of the wheel hubs. On these cars, it is necessary to

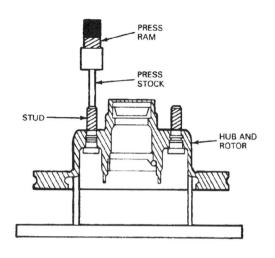

Figure 23-68 Some rotors are secured to the hub by the wheel studs that are pressed in place. If replacing a rotor, hub, or wheel stud, support the rotor properly while pressing the studs out (A) or back in (B). *(Courtesy of Ford Motor Company)*

Figure 23-69 A series of punch marks were made around this new wheel stud to lock it in place.

Figure 23-71 An older Corvette rotor is secured to the axle by a set of rivets. *(Courtesy of Stainless Steel Brakes Corp.)*

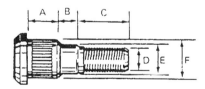

Figure 23-70 The important dimensions when selecting a replacement wheel stud are serration length (A), shoulder length (B), thread length (C), thread diameter (D), shoulder diameter (E), and serration diameter (F), along with thread pitch and direction. A replacement wheel stud should be an exact match of the original. *(Courtesy of Dorman® Products, Division of R&B Inc.)*

pull the front hubs, and this procedure disturbs the front hub bearings. The service procedure for these bearings is described in Chapter 21. Be sure to follow the manufacturer's recommendations for rotor replacement and bearing service.

23.9 Caliper Service

As a caliper is being removed, the abutment areas where the floating and sliding calipers meet the caliper support should be checked to make sure that they are clean and

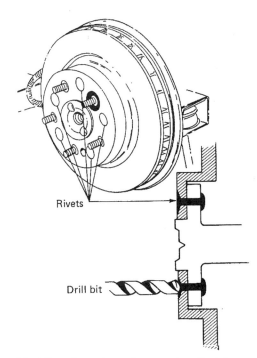

Figure 23-72 When removing a rotor that is secured to the axle or hub by rivets, drill through the head of the rivet with a drill bit about the same size as the rivet shank. Stop drilling after passing through the head and slice the rivet head off with a chisel. The remaining rivet shank can be driven out with a punch.

free of rust so the caliper will not bind during application and release. With fixed calipers, the areas where the pad ends press against the caliper should be checked. These areas must allow pad or caliper movement during application and release, but they should not allow excess clearance. A sloppy fit will cause a knock when forward and then reverse braking occurs (Figure 23-73).

The caliper abutment clearance is checked after the caliper guide pins or mounting bolts have been removed. At this time, pry the caliper up against one of the stops or abutments and measure the clearance at the other abutment (Figure 23-74). This clearance is specified by most manufacturers. If this specification cannot be located, a rule of thumb is 0.005 to 0.020 in. (0.1 to 0.5 mm). Excessively tight calipers will cause a bind and probably drag. The flats at the abutments can be finely filed to remove rust or dirt or increase the clearance. Excessive clearance will probably cause a knock. The normal solution is to replace the caliper or caliper support. On some General Motors cars, it is recommended that the caliper abutment clearance be measured at each end of the caliper with the mounting bolts in place (Figure 23-75). One caliper, used on light trucks, has five different support keys. The correct key size is selected after measuring the abutment clearance.

After the caliper has been removed, it is a good practice to remove the sleeves, bushings, insulators, antirattle springs, and other pad or caliper hardware (Figure 23-76). Some of these items are commonly replaced during pad replacement. The bushings and insulators can usually be pried out using a small screwdriver or seal pick. The new parts should be carefully worked into place so they will not be cut or torn. Water can be used as a lubricant to aid in the installation if necessary. Make sure that the new bushings and insulators are correctly seated and not twisted (Figure 23-77).

The caliper should be cleaned with a wire brush to remove any dirt, rust, or grease deposits. It is recommended that the caliper be rebuilt at this time to ensure proper caliper operation and no future leakage for a reasonable period of time. Caliper rebuilding requires removing the piston, cleaning the piston and bore, and reassembling the caliper using a new piston seal and boot. This procedure is described in Chapter 24. If the caliper is not going to be rebuilt, carefully inspect the boot to make sure it is not cracked, cut, or torn. Remember that it will have to last as long as the new lining. Also, compress the piston farther into the caliper to test its action. A sticky piston, any fluid leakage, or a faulty boot indicates that a caliper must be rebuilt.

23.10 Removing and Replacing Pads

As mentioned earlier, the pads on some fixed-caliper designs can be easily removed and replaced. This operation in its simplest and most basic form is as follows: remove the wheel, remove the pad retainer, retract the pistons, lift out the old pads, retract the pistons (farther if needed), slide in the new pads, and replace the pad retainer and wheel. However, this simple procedure is discouraged for the average car. It should be done only when you can guarantee that the rotor and caliper are in excellent operating condition. Most manufacturers and competent brake technicians recommend following a procedure similar to that described in this chapter. The extra service steps really do not take that much more time, and they give you the assurance that the job will result in new-car performance for the life of the new lining.

Some fixed-caliper designs and all floating- or sliding-caliper designs require removal of the caliper to replace the pads. The pads are retained in the caliper by slightly different methods in each design, but there are some similarities.

Pads used with fixed calipers and the inboard pad on floating and sliding calipers must both move relative to the caliper during application and release. These pads must be able to slide freely relative to the caliper

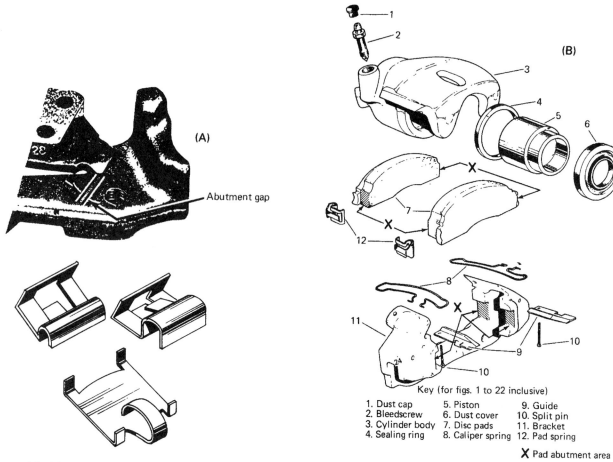

Figure 23-73 Clearance between the inboard pad and caliper mount (A) ensures the pad is free to move during application and release. An antirattle clip or caliper pad spring (B) is used to prevent rattles. *(A is courtesy of Ford Motor Company; B is courtesy of LucasVarity Automotive)*

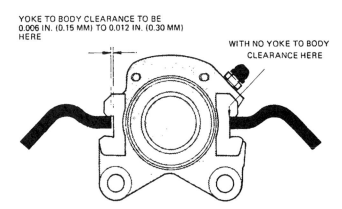

Figure 23-74 Manufacturers recommend checking abutment clearance when replacing brake pads. *(Courtesy of LucasVarity Automotive)*

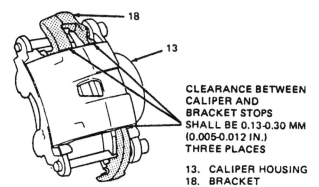

Figure 23-75 This manufacturer recommends checking abutment clearance between the caliper and the bracket stops with the guide pins installed. *(Courtesy of General Motors Corporation, Service Technology Group)*

but must not move too far or rattle while released. An antirattle clip or antirattle spring is sometimes used with such pads to prevent rattles (Figure 23-78). The inboard pad on many rear calipers has a tab that must be aligned with a piston cutout (Figure 23-79). Some designs use a retainer to attach the pad to the caliper piston (Figure 23-80). This serves two purposes: stopping rattles and pulling the pad back with the piston to provide a slight clearance between the pad and the rotor when the piston retracts.

Chapter 23 ■ Disk Brake Service

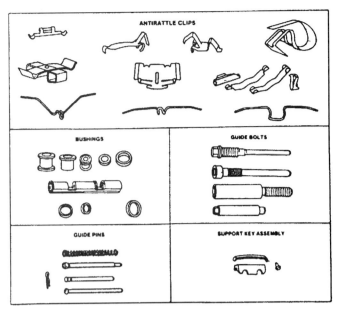

Figure 23-76 Caliper hardware varies among different brake assemblies; some of the different types of hardware are shown. *(Courtesy of Bendix Brakes, by AlliedSignal)*

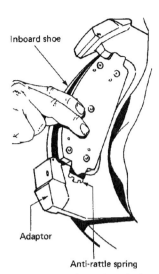

Figure 23-78 The antirattle spring is installed along with the inboard shoe/pad. *(Courtesy of Chrysler Corporation)*

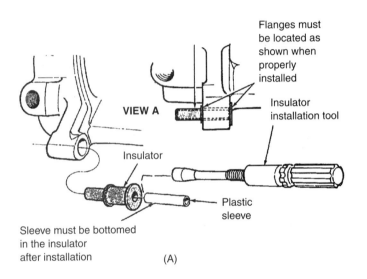

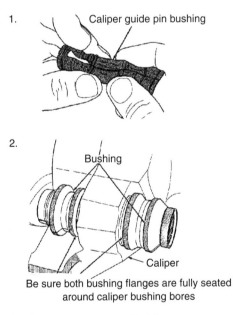

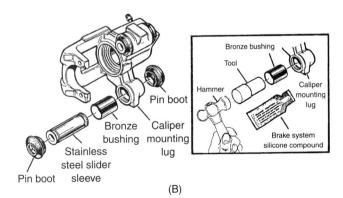

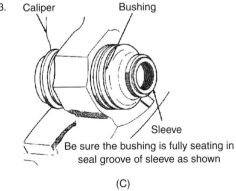

Figure 23-77 Caliper guide bushings are installed in different ways. Caliper (A) requires a special tool (which can be shop-made). (B) can be repaired using an aftermarket kit. (C) is installed in several steps. *(A is courtesy of Ford Motor Company; B is courtesy of Rolero-Omega, Division of Cooper Industries; C is courtesy of Chrysler Corporation)*

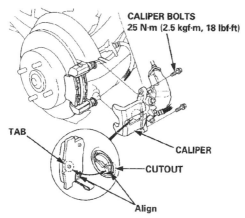

Figure 23-79 The piston on many rear calipers must be turned to index with the inboard pad. (Courtesy of General Motors Corporation, Service Technology Group)

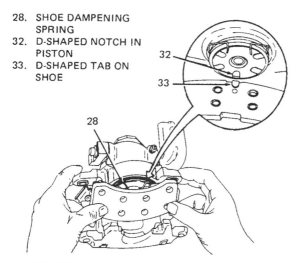

Figure 23-80 Many calipers use a retainer spring to hold the inboard pad to the piston. (Courtesy of General Motors Corporation, Service Technology Group)

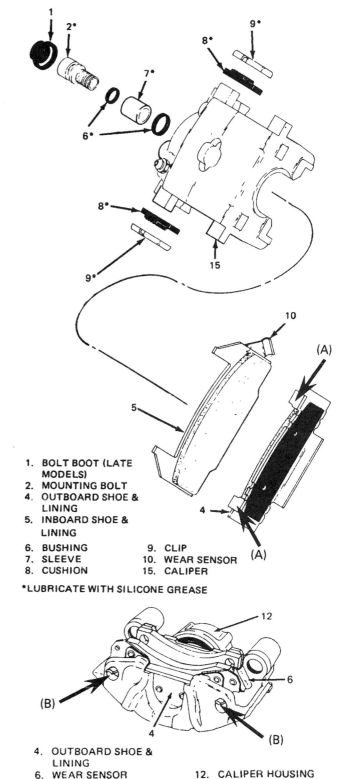

Figure 23-81 Some outboard pads have tabs that are bent to secure the pad to the caliper. (Courtesy of General Motors Corporation, Service Technology Group)

The outboard pad on a floating or sliding caliper is nearly always secured tightly to the caliper. This stops pad rattle, helps pull the pad away from the rotor during release, and helps reduce squeal. Many cases of brake squeal can be attributed to vibration between this pad and the caliper. Several methods are used to stop pad vibration. The most common involve tabs or spring clips on the pad, which fit over lugs on the caliper (Figure 23-81 and Figure 23-82). Occasionally, brake squeal can be eliminated by coating the pad backing with a special compound or by making it fit the caliper more tightly (Figure 23-83).

Some brake pad sets have four identical shoes. A few fixed-caliper sets have shoes that are almost the same except that the lining is offset on the pad. The leading edge of the pad is the end that has the largest lining gap (Figure 23-84). Many floating- and sliding-caliper lining sets have two different shoes: two inboard shoes and two outboard shoes. Many sets include wear sensors,

Chapter 23 ■ Disk Brake Service

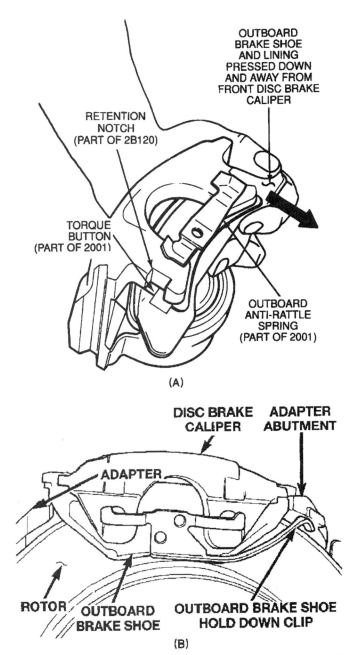

Figure 23–82 This pad is being removed; the new shoe will be slid into position with the antirattle spring in the same location and the torque buttons in the retention notches (A). The hold-down clip on the outboard shoe (B) fits over the caliper and under the adapter abutment. *(A is courtesy of Ford Motor Company; B is courtesy of Chrysler Corporation)*

Figure 23–83 A center punch was used to tighten the tabs on this outboard pad to stop a squeal.

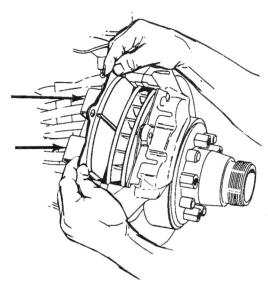

Figure 23–84 The pads on this older Corvette are slid into the gap between the caliper and rotor. Note the two clips (arrows) that are being used to keep the pistons retracted in their bores. *(Courtesy of General Motors Corporation, Service Technology Group)*

which are usually placed at the trailing edge of the outboard pad. A few pad sets have four different pads; each pad has a specific location such as inboard right hand and so on. It is a good practice to lay out the pads from a set on a clean surface and check for pad backing or lining differences to ensure that each of the pads will be positioned correctly.

It is a common trade practice to use a *noise suppressant* with disc brake pads. The most popular noise suppressants are (1) a compound that is applied to the back of the pad and (2) a plastic or fiber shim (Figure 23–85). Do not use more than one at a time because the suppressant can cause the shim to pull or creep away, reducing its effectiveness.

When replacing pads on a caliper, it is wise to follow the manufacturer's recommendations to ensure proper and complete installation.

To remove and replace brake pads in a caliper, you should:

1. Remove the inboard pad and antirattle spring from the caliper, the caliper support, or the piston.

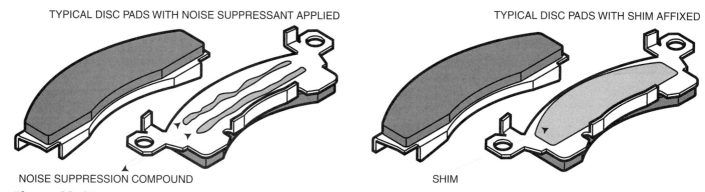

Figure 23-85 Noise-suppression compound or a pliable shim is often used on the back side of disc brake pads to prevent squeal or other noises; do not use both on the same pad. *(Courtesy of Wagner Brake)*

2. Remove the outboard pad from the caliper.
3. Check the replacement pads to make sure they are the correct pads and to determine where they will be installed.
4. Make sure the piston is bottomed in the caliper bore.
5. Install the outboard pad. It is important that this pad fit tightly against the caliper and be held securely in this position. Some pads use retainer tabs that should be bent into position after the caliper is installed and the brakes are applied (Figure 23-86).

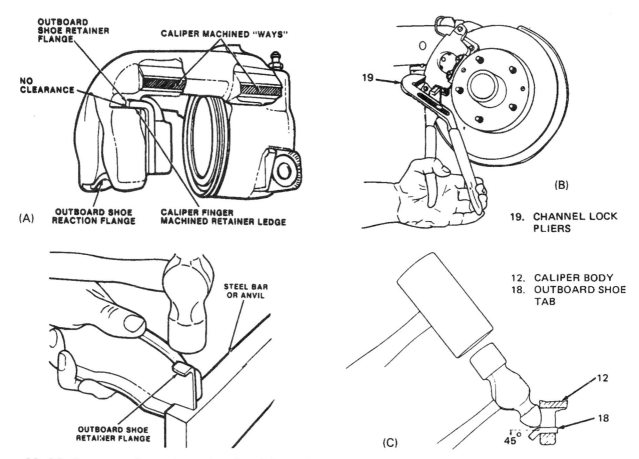

Figure 23-86 On many calipers, the outboard pad should be locked onto the caliper (A). This can be done using pliers (B) or a hammer (C). *(A is courtesy of Chrysler Corporation; B and C are courtesy of General Motors Corporation, Service Technology Group)*

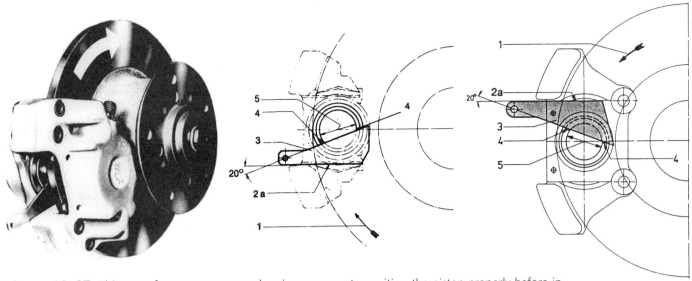

Figure 23–87 This manufacturer recommends using a gauge to position the piston properly before installing the pads. *(Courtesy of ITT Automotive)*

6. Install the inboard pad and antirattle spring or retainer clip. Make sure that this pad fits squarely against the piston.

Note that on some fixed-caliper designs, a piston or pad shim must be installed and aligned during pad replacement (Figure 23–87).

23.11 Caliper Installation

At this point, the rotor has been checked and serviced; the caliper has been cleaned and rebuilt; the caliper bushings, insulators, or other hardware parts have been replaced; and new pads have been installed. All that is left is to lubricate the unit in the proper areas and to remount it (Figure 23–88). Molylube and antiseize compound are

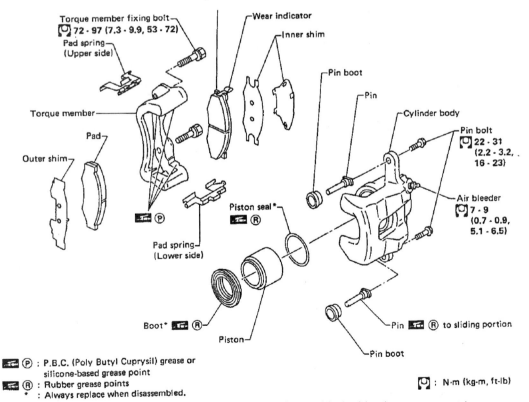

Figure 23–88 As a caliper is assembled, all of the bolts and nuts, along with the bleeder screw, must be tightened to the correct torque, and lubricant must be applied to the proper locations. *(Courtesy of Nissan. North America, Inc.)*

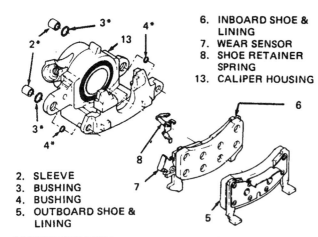

2. SLEEVE
3. BUSHING
4. BUSHING
5. OUTBOARD SHOE & LINING
6. INBOARD SHOE & LINING
7. WEAR SENSOR
8. SHOE RETAINER SPRING
13. CALIPER HOUSING

* LUBRICATE WITH SILICONE GREASE (OR EQUIVALENT)

Figure 23-89 When this caliper is installed, a film of silicone grease should be applied to the sleeves (2) and bushings (3 and 4). *(Courtesy of General Motors Corporation, Service Technology Group)*

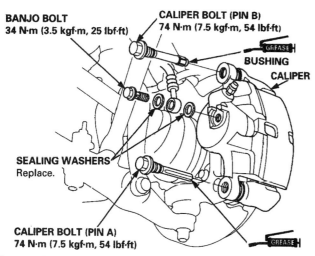

Figure 23-90 The torque ratings for the caliper bolts and banjo bolt along with the lubrication recommendation for this caliper. *(Courtesy of American Honda Motor Co., Inc.)*

popular lubricants for metal-to-metal contact areas, but only silicone-based, high-temperature lubricant should be used at rubber-to-metal contacts. Again, follow the manufacturer's installation procedure.

To replace a caliper, you should:

1. Apply the correct amount of the proper lubricant to the bushings, sleeves, or slides as required by the manufacturer. When lubricating external areas where grease might fall onto the lining or friction surfaces, use it sparingly (Figure 23-89).

2. On brakes with serviceable wheel bearings, clean, repack, and readjust the wheel bearings. These operations are described in Chapter 21.

Figure 23-91 On this caliper, the support key is driven back in place and the retaining screw tightened to the correct torque. *(Courtesy of Bendix Brakes, by AlliedSignal)*

3. Place the caliper over the rotor and lubricate the mounting bolts or guide pins or support key if required. Then install the mounting bolts or guide pins and tighten them to the correct torque (Figure 23-90 and Figure 23-91). On calipers that use support keys, drive the keys into position and replace the retaining bolt or pins (Figure 23-92).

4. Connect the brake hose and refill the master cylinder. If the caliper includes a parking brake, reconnect the parking brake cable.

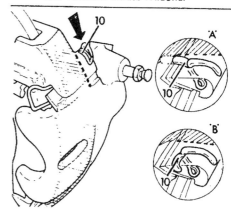

Figure 23-92 During installation, all retaining pins and clips must be properly installed to ensure that they stay in place and do not interfere with caliper operation. *(Courtesy of LucasVarity Automotive)*

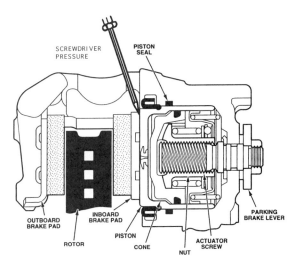

Figure 23-93 Some rear disc brakes self-adjust easily; others do not. Self-adjustment can be hastened by placing a screwdriver (as shown) to hold the inboard pad in the applied position while the brake pedal is applied and released.

5. Bleed the air out of the caliper cylinders. This operation is described in Chapter 24.
6. Apply the brakes using slow and complete pedal strokes until there is a firm, high brake pedal.
7. Finish clinching the outboard pads if necessary.
8. Install the wheel and tighten the lug bolts to the correct torque using the correct tightening pattern. Overtightening the lug bolts can distort the rotor.

On most disc brake assemblies, it is very easy to adjust the lining clearance: simply apply the brake using slow and smooth pedal applications until there is a firm pedal. With many rear brake calipers using a mechanical parking brake, the mechanical adjuster has to reset. Applying the brake pedal several times with slow, smooth, complete strokes should make the adjustment. If the shoes fail to adjust on General Motors units, hold the inboard shoe in the applied position using a screwdriver while an assistant applies and releases the brakes (Figure 23-93).

23.12 Completion

Several checks should always be made after completing a brake job or brake service operation:

- Be sure that all nuts and bolts are properly tightened and that all locking devices are properly installed.
- Be sure there is no incorrect rubbing or contact between parts or brake hoses.
- Be sure that the master cylinder has the correct fluid level.
- Be sure that there are no fluid leaks.
- Be sure that there is a firm, high brake pedal.
- Perform a careful road test to ensure that the brakes are operating safely and correctly.
- Tighten the lug nuts to the correct torque using the proper tightening pattern. Although the road test will show no problems, uneven wheel nut torque can cause later brake problems.

New friction material must be properly worn in. This is especially important if the rotors have not been turned. Unless it is specifically stated otherwise by the lining manufacturer, new lining must never be burned in. New lining can be ruined if it is overheated. Light or moderate stops should be made until the lining and rotor surfaces are fully burnished and seated in. The breaking-in period varies with the type of driving. Some sources recommend about 250 mi. (400 km) of driving.

If a technician believes that a customer will not break in the friction material properly, the technician should make what is referred to as a 30-minute road test, which is really a break-in step. Select a road where you can make a series of stops without interfering with traffic. Drive to a speed of 30 to 40 mph and make a light to moderate stop. Note any braking problems for later correction and accelerate slowly back up to speed. Give the brakes time to cool and repeat the previous stop. Depending on the lining and the service done, repeat the stops through five to ten cycles, making sure that the friction material does not overheat and the braking action is normal.

23.13 Practice Diagnosis

You are working in a brake and front-end shop and encounter the following problems:

CASE 1: You did a brake job on a 5-year-old Toyota last week. You put in new rotors and pads and rebuilt the calipers at the front and put in new lining, turned drums, and new wheel cylinders at the rear, but it has come back with a complaint of pedal pulsations. What could have gone wrong? What checks should you make?

CASE 2: A 1982 Chevrolet has a problem of a grabby right front brake. When you pull the right front wheel, you find worn but still good lining and no fluid or grease contamination. What else could be causing this problem? What should you do next?

CASE 3: A customer is complaining of brake squeal on a 5-year-old Escort, and your road test confirms a squeal that changes in level as you change the pedal pressure. What could be causing this noise? What should you do to find the cause of this problem? What methods can you use to correct it?

CASE 4: A 5-year-old car has been brought in with a complaint of a rattle. Your road test confirms a light rattle noise, and it seems to be coming from the left front. Could this be a brake problem? What might be causing it? What should you do to locate the exact cause?

Terms to Know

caliper hardware
finish cut
guide pins
index

loaded calipers
nondirectional finish
parallelism
roughing cut

swirl finish
thickness variation
turning

Review Questions

1. Two technicians are discussing disc brake service procedures.
   Technician A says that rapid wear of the outboard pad is an indication that the caliper mounts are not properly lubricated.
   Technician B says that a sticky caliper piston can cause abnormally fast wear of the inner pad.
   Who is correct?
   a. A only
   b. B only
   c. Both A and B
   d. Neither A nor B

2. Before removing a caliper, you should:
   a. Loosen the bleeder screw
   b. Remove a portion of the fluid from the master cylinder
   c. Tighten the wheel bearings
   d. All of the above

3. To make removal of the caliper from the rotor easier, you should retract the piston into the caliper using a large:
   a. C-clamp
   b. Pair of pliers
   c. Screwdriver
   d. Any of the above

4. Technician A says that too much clearance between the caliper and the mounting bracket can cause a knock as the brakes are applied.
   Technician B says that if this fit is too tight, the only solution is to replace the caliper or the mounting bracket.
   Who is correct?
   a. A only
   b. B only
   c. Both A and B
   d. Neither A nor B

5. Technician A says that as a caliper is being removed, it should be lowered down gently so it will not damage the hose as it hangs.
   Technician B says that the caliper should be removed completely or suspended from a wire.
   Who is correct?
   a. A only
   b. B only
   c. Both A and B
   d. Neither A nor B

6. A floating caliper is normally removed from a car by disconnecting the brake hose and removing the:
   a. Caliper guide pins
   b. Mounting bracket from the steering knuckle
   c. Caliper bridge bolts
   d. Brake pads

7. As you are doing a brake job, you discover a cracked caliper. You should:
   a. Weld up the crack
   b. Fill the crack with epoxy sealant
   c. Replace the caliper
   d. Just rebuild the caliper and not worry about the crack

8. Technician A says that it is a good practice to turn a rotor during lining replacement to ensure a flat surface for the new pads to run against.
   Technician B says that the smooth, shiny surface of the average worn rotor is ideal for new lining break-in.
   Who is correct?
   a. A only
   b. B only
   c. Both A and B
   d. Neither A nor B

9. The thickness of a rotor has been measured in six places; the largest measurement is 0.9345 in. and the smallest is 0.9333 in. The specifications for this rotor are nominal thickness, 0.882 in.; allowable runout, 0.003 in.; finish, 16 to 79 rms; thickness variation, 0.0005 in.
   Technician A says that this rotor should be replaced. Technician B says that the rotor should be turned before it is reused.
   Who is correct?
   a. A only
   b. B only
   c. Both A and B
   d. Neither A nor B

10. The correct tool for measuring rotor thickness is a(n):
    a. Dial indicator
    b. Outside micrometer
    c. Brake drum micrometer
    d. Ruler that shows millimeters

11. When servicing disc brakes, brake technicians normally do not check a rotor for:
    a. Runout
    b. Scoring
    c. Parallelism
    d. Bell mouth

12. Technician A says that the ideal rotor friction surface finish has a series of somewhat circular scratches and is called a nondirectional finish. Technician B says that this finish is a result of the normal cutting action of most rotor lathes.
    Who is correct?
    a. A only
    b. B only
    c. Both A and B
    d. Neither A nor B

13. You are doing a brake job on a car with one slightly scored rotor (about 0.015 in. deep); the other rotor is smooth. Both rotors measure close to the new rotor thickness. You should:
    a. Turn both rotors to the minimum allowable thickness
    b. Turn both rotors to the largest size possible that will clean up the friction surface on the scored rotor
    c. Turn each rotor the minimum amount that will clean up the surfaces on that particular rotor
    d. Replace the scored rotor with a new one and run the smooth rotor

14. A car vibrates excessively during stops, and the rotor has been found to have an excessive amount of runout. Technician A says that runout can sometimes be corrected by reindexing the rotor on the hub. Technician B says that runout can be corrected on some cars by installing a shim between the hub and the rotor.
    Who is correct?
    a. A only
    b. B only
    c. Both A and B
    d. Neither A nor B

15. Technician A says that the thickness of the rotor has no effect on brake lining temperature. Technician B says that a semimetallic brake lining is often used in OEM installations where lining temperatures are higher.
    Who is correct?
    a. A only
    b. B only
    c. Both A and B
    d. Neither A nor B

16. Technician A says that if replacement is necessary, new bearing cones should be installed before turning a rotor.
    Technician B says that it is a good practice to wash a rotor using alcohol after it has been turned.
    Who is correct?
    a. A only
    b. B only
    c. Both A and B
    d. Neither A nor B

17. When new pads are installed in a floating caliper, the outboard pad is usually:
    a. Fastened tightly to the caliper
    b. Lubricated using a high-temperature silicone grease
    c. Attached to the caliper with new antirattle clips
    d. All of the above

18. As a caliper is being replaced, be sure to:
    a. Lubricate the required contact points with the correct lubricant
    b. Tighten the mounting bolts to the correct torque
    c. Bottom the caliper piston before trying to place the caliper over the rotor
    d. All of the above

19. Technician A says that a caliper can be distorted if the lug bolts are overtightened.
    Technician B says that a rule of thumb is to tighten all lug bolts to 45 ft-lb of torque.
    Who is correct?
    a. A only
    b. B only
    c. Both A and B
    d. Neither A nor B

20. After new lining has been installed, you should:
    a. Check it out with a good, hard stop from 55 mph
    b. Make sure you have a firm brake pedal before moving the car
    c. Adjust the lining clearance by turning the starwheel with the correct brake spoon
    d. All of the above

24 Hydraulic System Service

Objectives

Upon completion and review of this chapter, you should be able to:

- ❏ Diagnose stopping problems that are caused by improper operation of a master cylinder, wheel cylinder, caliper, brake line or hose, or brake valve.
- ❏ Remove and replace a master cylinder and adjust the pedal pushrod length.
- ❏ Disassemble, clean, inspect, and measure a master cylinder bore and other internal parts to determine if the master cylinder should be rebuilt or replaced.
- ❏ Reassemble a master cylinder.
- ❏ Hone a master cylinder, wheel cylinder, or caliper bore as recommended by the manufacturer.
- ❏ Remove and replace a wheel cylinder.
- ❏ Disassemble and clean a wheel cylinder, inspect it for wear or damage, and reassemble it.
- ❏ Disassemble and clean a caliper, inspect it for wear or damage, and reassemble it.
- ❏ Remove and replace brake lines or hoses, fittings, and supports.
- ❏ Select, handle, store, and install brake fluids.
- ❏ Bleed or flush a brake hydraulic system by any of the approved methods.
- ❏ Test, inspect, and replace brake valves.
- ❏ Inspect, test, and replace brake light switches, wiring, and bulbs.
- ❏ Adjust load- or height-sensing proportioning valves.
- ❏ Reset a brake pressure differential valve.
- ❏ Test the pressure of a brake hydraulic system.
- ❏ Perform the ASE tasks for hydraulic system diagnosis, testing, and adjustment (see Appendix A).

24.1 Introduction

The hydraulic components of a brake system can be serviced on an individual, as-needed basis or as part of a brake job. A major brake job normally includes careful inspection and rebuilding of the hydraulic components. Many service shops prefer to remove and replace hydraulic components such as master cylinders and wheel cylinders rather than rebuild them. The hydraulic system is arguably the most critical portion of a brake system. A failure in this system can produce a total, or at least a substantial, loss in braking ability.

The hydraulic components begin to deteriorate as the car is being assembled. The hygroscopic nature of brake fluid causes it to absorb moisture from the atmosphere. This moisture starts the chemical reactions that produce corrosion of the metal parts of the system, especially the bottoms of the cylinder bores. Rubber parts are vulcanized to form various shapes. Before vulcanizing, rubber is very pliable. Vulcanizing, primarily a heating-under-pressure process, changes rubber to the familiar stable but elastic substance. Heat, in any form, can cause rubber parts to self-vulcanize. They will continue to harden and lose their elastic nature. As this happens, cracks will begin to appear in the surface of the rubber, indicating that hardening is occurring. A certain amount of mechanical wear also occurs on dynamic, moving seals, especially on those in the master cylinder and somewhat on those in the wheel cylinders, because these particular seals undergo a lot of movement. If a seal has to move over rough bores or debris, the critical areas at the edges of the seal will wear even faster or possibly be cut. A cut seal will leak.

After any hydraulic component is serviced, it becomes necessary to bleed all the air from the system. As mentioned earlier, air is compressible and will cause a low, spongy brake pedal. Some components, especially master cylinders, are bench bled before they are installed. **Bench bleeding** involves filling the component with fluid and removing the air before mounting it on the car. Some wheel cylinders are also bench bled because they are mounted in such a way that they cannot be completely bled when installed on the car.

SAFETY NOTE: Hydraulic system repair must produce safe brake operation. To obtain proper brake operation and to protect yourself from injury, the following general precautions should be observed:

- Wear face or eye protection.
- Do not allow brake fluid to be sprayed on or splashed in your face or eyes. Do not hold your face over the master cylinder reservoir while the pedal is being pumped.
- Do not use petroleum products such as solvent, gasoline, kerosene, and so on for cleaning hydraulic components. Only denatured alcohol or commercial hydraulic brake system cleaners should be used. Immersion in hot water can be used to remove traces of petroleum products if necessary. The part should then be dried immediately.
- Handle hydraulic components and parts with clean hands. The smallest trace of petroleum from greasy hands can damage rubber parts.
- Never soak parts, especially rubber ones, in alcohol. They should be cleaned and dried immediately.
- Be cautious when using compressed air for cleaning. The air blast sometimes contains dirt or rust particles and can drive these particles, along with other debris from the parts you are cleaning, into your skin or eyes.
- Make sure that the air you are using to clean hydraulic brake parts comes from a clean source and does not contain oil.
- Do not hone aluminum bores or use abrasive methods to clean them. Any scratch in an anodized bore can allow rapid corrosion and future failure.
- Do not leave brake fluid containers, the master cylinder reservoir included, open to the atmosphere. Close them as soon as practical.
- Never reuse brake fluid.
- Store and handle brake fluid in clean containers.
- Do not allow brake fluid to spill, especially onto painted surfaces. If it spills, wipe it up immediately and flush the area with clean water.

24.2 Working with Tubing

When removing or replacing a hydraulic brake component, it is often necessary to disconnect or connect a steel brake line. A tubing or flare-nut wrench should always be used for this purpose (Figure 24–1A). Using an open-end wrench can round out corners of the tube nut

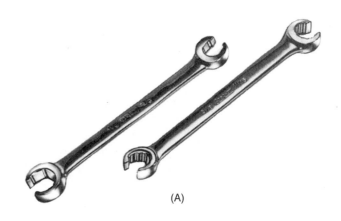

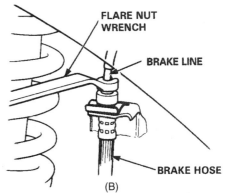

Figure 24–1 Either a six- or a twelve-point flare-nut wrench should be used when loosening or tightening a tube nut (A). The wrench design allows it to fit over the nut with the best gripping power (B). *(A is courtesy of Snap-on Tools Corporation; B is courtesy of American Honda Motor Co., Inc.)*

and make removal extremely difficult. A brake technician usually has flare-nut wrenches in both a six- and a twelve-point style in metric and fractional sizes. The greater gripping power of six-point wrench allows maximum torque for breaking the nut loose, and the twelve-point style allows working in tighter locations. It is also a good practice to use two wrenches when removing a line fitting from a union or a component that tends to turn with the fitting nut. This will often cause the line to twist. One wrench is used to prevent the twisting while the other unscrews the nut.

Many technicians make a practice of finger tightening a connection at least two revolutions before using a wrench. A tube nut that is slightly misaligned will tend to cross-thread. This can ruin the unit to which the brake line is being connected or the tube nut. Most of us are not strong enough to cross-thread a fitting with our fingers, and two revolutions of the tube nut assures us the nut is started truly straight.

Like any other fastener, nut, or bolt, a tube nut should be tightened to the correct torque. If it is too loose, the connection will probably leak under pressure. If it is too tight, the threads might strip (especially brass ones) or the tube seat or the tubing flare might be damaged. Also, over time, an overtightened connection becomes nearly impossible to take apart. A torque wrench should be used as the connection is being tightened, but at many connections this is extremely difficult if not impossible. Most brake technicians use their experience and their ability to judge correct tightness as a guide when tightening line fittings. The correct torque for tube nuts and other screwed connections is provided in Appendix E (Figure 24–2).

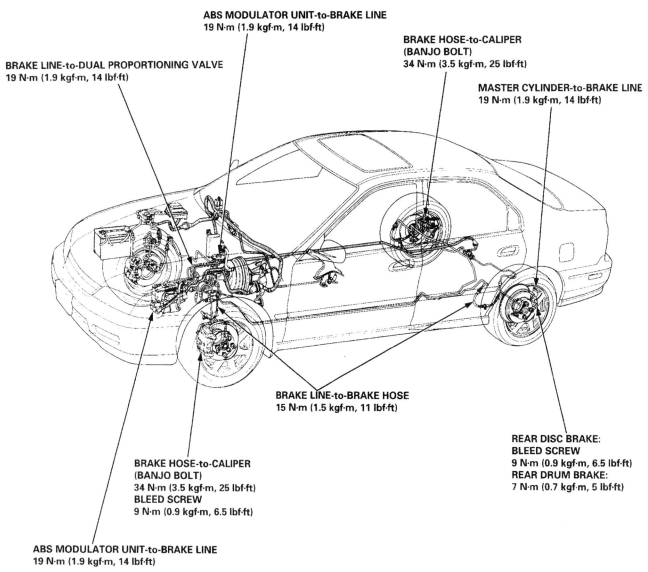

Figure 24–2 All tubing and bleed screws (bleeder valves) should be tightened to the correct torque when a component or tube is replaced. *(Courtesy of American Honda Motor Co., Inc.)*

24.2.1 Tubing Replacement

Two flare styles are used for brake tubing: double flare and ISO (Figure 24–3). Double-flare fittings use fractional sizes, and ISO fittings are metrically sized. Sometimes both are used on a vehicle (Figure 24–4).

Normally a faulty steel tube or one with a faulty tube nut should be replaced. Replacement tubes are available in various lengths with an end type and tube nut to match those on the car. The new tube should be carefully bent to the right shape using the old tube as a guide. A tubing bender should be used to prevent kinks while making sharp bends. If a tube is kinked, it will probably fracture at the point of the kink and develop a leak. If a proper tubing bender is not available, some technicians place the tubing in the groove of a pulley (alternator or water pump) and use both hands and thumbs to carefully hand bend the tubing (Figure 24–5). If a tube is too long for a particular location, it can be bent into a coil to use up some of the extra length. Be sure to lay the coil loops in a horizontal position to avoid air traps, which will cause bleeding problems later. Tubing that is too short can be joined to another section of tubing by using a union to make it longer.

If necessary, a new flare can be put on a tube to allow it to be shortened, repaired, or have a new tube nut installed. Flaring is often avoided because steel tubing is strong and difficult to flare and special flaring tools are

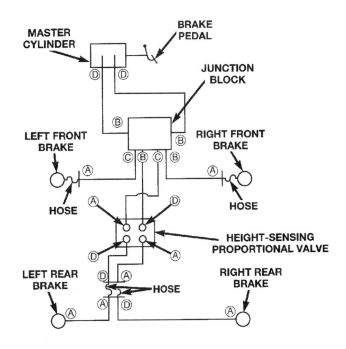

Ⓐ = 3/8 x 24 THREAD DOUBLE INVERTED FLARE
Ⓑ = M10 x 1 THREAD ISO FLARE
Ⓒ = M12 x 1 THREAD ISO FLARE
Ⓓ = 7/8 x 24 THREAD DOUBLE INVERTED FLARE

Figure 24–4 A modern vehicle might use both ISO and inverted flare fittings. Note that ISO fittings use metric threads, and inverted flare fittings use fractional-inch-sized threads. *(Courtesy of Chrysler Corporation)*

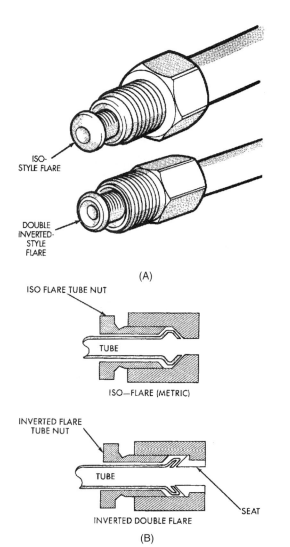

Figure 24–3 ISO and inverted flares can be identified by the shape of the flare (A). The fitting seat and tube nut also differ (B). *(Courtesy of Chrysler Corporation)*

Figure 24–5 Three different styles of tubing benders. These tools allow tubing to be bent to a short radius without kinking. *(Courtesy of Snap-on Tools Corporation)*

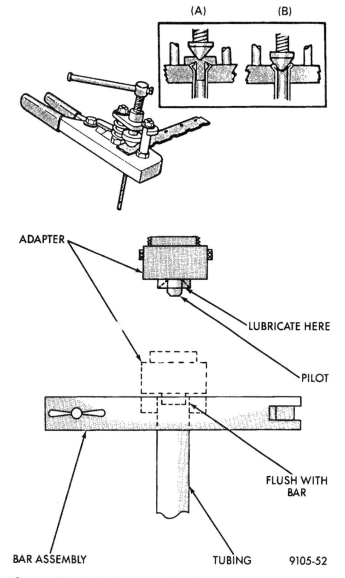

Figure 24–6 Making a double flare (top) uses a two-step process. The first step (A) begins the flare using an adapter, and the second step (B) completes the process. An ISO flare is also made using a special adapter (bottom). *(Courtesy of Chrysler Corporation)*

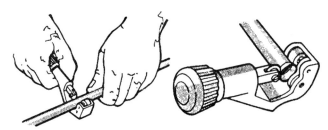

Figure 24–7 A tubing cutter is the best tool for cutting tubing because it makes a straight, neat cut with no metal chips. Note the groove in the cutter rollers so a cut can be made close to the flare (right). *(Courtesy of Chrysler Corporation)*

required. Steel tubing must be double flared when using an SAE flare (Figure 24–6). This requires an extra step and the use of a special adapter. ISO flares can also be made in the field if the special flaring tool is available. Compression fittings, sometimes used on fuel lines, must never be used on brake lines. Also remember that copper tubing must never be used for hydraulic brake lines. When repairing tubing, always use a tubing cutter to avoid any small metal chips and to ensure a straight, smooth cut (Figure 24–7).

24.3 Hydraulic Cylinder Service

A stroking rubber seal must have a straight bore of a precise size. A caliper that uses a stationary, square-cut O-ring seal around the piston has different bore requirements. It can tolerate some imperfections in the bore. Rust or corrosion deposits, dirt, and other residue must be removed from a bore used with a stroking seal. On cast-iron bores, small scratches, pits, or deposits can sometimes be removed by honing. On aluminum bores, they cannot. Most technicians will not take the gamble of rebuilding a unit with a questionable bore. A leaky wheel cylinder can easily ruin a brake lining set. A leaky master cylinder can cause a loss of braking power. Many sources recommend against rebuilding aluminum wheel cylinders or master cylinders.

NOTE: Cylinders with aluminum bores cannot be honed. They can be cleaned using a fiber or nylon brush.

A cylinder is slightly larger than the piston, but if it is too large, heel drag can occur as a result of the rubber cup being forced between the piston and the bore. Remember that heel drag can cause sticking or nibbling of the cup by the piston. Bore size is normally measured using a narrowed feeler gauge or a strip of shim stock. You can cut a feeler gauge to a narrow width, about 1/8 to 1/4 in. (3 to 6 mm), using sharp tin snips. After cutting, you should smooth any burrs with a wet stone or fine sandpaper. The gauge thickness should be as follows:

Cylinder Size	Gauge Thickness
3/4 to 1 3/16 in.	0.006 in.
19 to 30 mm	0.15 mm
1 1/4 to 1 7/16 in.	0.007 in.
32 to 37 mm	0.18 mm
Over 1 1/2 in.	0.008 in.
Over 38 mm	0.20 mm

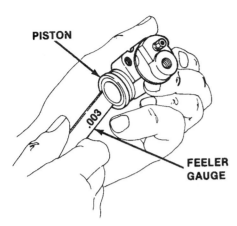

Figure 24–8 A cylinder bore size can be checked by placing a feeler gauge in the bore and then trying to slide the piston in also. If the piston enters past the feeler gauge, the bore is too large. *(Courtesy of Brake Parts, Inc.)*

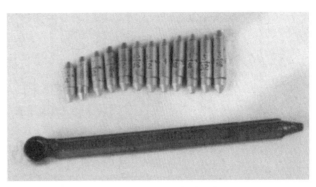

Figure 24–10 At one time a cylinder bore gauge set was available for measuring cylinders. Each gauge rod was sized to the maximum allowable for each bore. If it passed through the bore, the bore was too large.

To measure bore clearance, place the gauge strip lengthwise in the cylinder and try to slide the piston over it (Figure 24–8). If the piston will not enter past the gauge strip, the cylinder is the correct size. If the piston slides past, the cylinder is oversize and should be replaced. It should be noted that the primary face of some master cylinder pistons has an undersize land and therefore should not be used when gauging bore sizes. This particular face does not have replenishing holes, and during master cylinder release, replenishing fluid flows around this piston instead of through it.

The size of a cylinder can be gauged by using a hole gauge that is adjusted to the size of the piston plus the allowable clearance (Figure 24–9). In the past, a special gauge set was marketed for brake cylinders. A gauge was included for each common cylinder size, and each gauge was the maximum size for the cylinder. If the cylinder is the same size as the gauge or larger, it is oversize and should not be rebuilt (Figure 24–10).

To service a bore that uses a stroking seal in a master cylinder, wheel cylinder, or some calipers:

1. Wash the bore with denatured alcohol.
2. Inspect the cylinder bore under a strong light. Move the cylinder around so that the light is concentrated on the area you are checking. Then turn the cylinder so you can check the entire inner surface where the seals stroke. Pay particular attention to the bottom side of the bore (Figure 24–11).
3. If the bore is clean and smooth and has no imperfections, it can be reused. If it is not clean or imperfections are found, one of the following procedures should be used:
   a. *On aluminum bores:* Use a small fiber or nylon brush or a clean piece of cloth wrapped around a wooden dowel to swab the bore with

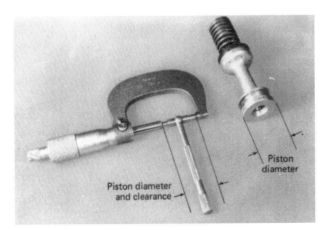

Figure 24–9 The micrometer is set to the piston diameter plus the maximum clearance, and the bore gauge is set to the micrometer. If the bore gauge is smaller than the cylinder, the cylinder is too large.

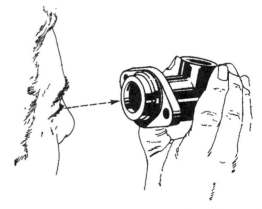

Figure 24–11 After cleaning, a bore should be carefully inspected to determine if there are any pits or corrosion. A strong light source is necessary for a thorough inspection. *(Courtesy of ITT Automotive)*

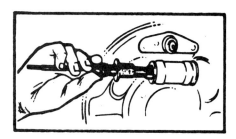

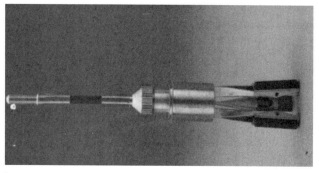

Figure 24–12 A wheel cylinder hone is normally turned using a drill. Be sure to keep the stones lubricated and clean; do not remove the hone from a cylinder while it is spinning. *(Courtesy of KD Tools)*

Figure 24–13 A carbide-tipped brush or Flex-Hone. This device is also spun by a drill in a bore to clean the cylinder. It should be lubricated during use. *(Courtesy of KD Tools)*

denatured alcohol or brake cleaner. Then air dry it and inspect it again. If the bore is clean and has no imperfections (discolorations should be disregarded), it can be reused.

b. *On cast-iron bores:* Use either crocus cloth wet with alcohol and moved in a circular direction or a brake cylinder hone (Figure 24–12). A carbide-tipped brush sold under the name Flex-Hone can also be used (Figure 24–13). Either type of hone should be wet with brake fluid or alcohol while it is used to prevent the abrasive from clogging. The hone or brush should be turned with an electric drill while moving it with slow, complete strokes through the entire cylinder. Several strokes are usually all that are necessary or allowable. Be careful not to overstroke against the bottom of the cylinder or out the end of the bore. Many technicians prefer to use a Flex-Hone because the cutting stones tend to stay cleaner and it removes debris faster.

CAUTION *A hone can fly apart if it is removed from the bore while spinning.*

4. Repeat steps 1 and 2. If the bore is still not clean, it should be replaced. Any further honing will make it oversize and unusable.

24.4 Master Cylinder Service

Normal master cylinder service includes removal, rebuilding, and replacement. Many service shops replace the master cylinder with a new rebuilt unit rather than rebuilding the one from the vehicle. If the master cylinder is faulty, a new or commercially rebuilt unit can be installed. A rebuilt master cylinder usually offers substantial price savings over a new unit, but rebuilt units should be carefully checked to ensure adequate quality control.

Normally, any master cylinder with a good bore can be rebuilt in a repair shop if a parts kit can be obtained. Many technicians prefer to rebuild the master cylinder that is on a car because they are sure that it will fit and have line fittings of the correct size. Some replacement master cylinders have outlets of a different size or at slightly different locations. Outlets of different size require step-up or step-down fittings (Figure 24–14). The tubing can usually be bent slightly to fit a different location.

When installing a new master cylinder, it is a good practice to flush the cylinder with clean brake fluid. This is done to remove any debris that might be left over from the manufacturing process or chemical coatings that were used to protect the cylinder from corrosion. To flush a master cylinder, simply fill the reservoirs and the cylinder bores about one-third full with clean brake fluid, install the reservoir cover, plug the line ports, shake the cylinder to work the fluid all around, and drain out all the fluid. Finally, bench bleed and install the cylinder in a normal manner.

24.4.1 Master Cylinder Removal

Removal and replacement procedures for most power brake (vacuum or hydraulic booster) units are very similar. Manual brake (without power assist) master cylinder removal often requires that the pushrod be disconnected

Chapter 24 ■ Hydraulic System Service

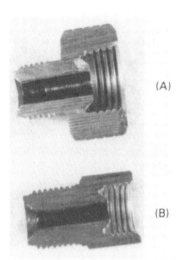

Figure 24–14 A step-down (A) and a step-up (B) adapter. These fittings allow an inverted fitting to be connected to a different-sized opening.

from the brake pedal. If you are not sure of the exact procedure, check a service manual (Figure 24–15).

To remove a master cylinder:

1. Take off the reservoir cover and remove the fluid from the reservoir.
2. Disconnect any wires connected to the reservoir or master cylinder body.
3. Disconnect the brake tubes. Depending on the system, there will be one, two, or four tubes. Be sure to use a tubing wrench for this operation.
4. Remove the nuts or bolts attaching the master cylinder to the power booster or vehicle bulkhead.
5. Slide the master cylinder off the booster or bulkhead. If it will move only a short distance but no farther, replace one of the nuts or bolts (finger tight) to support the master cylinder and then disconnect the pushrod from the brake pedal. After disconnecting the clip or pin and clip holding the pushrod to the pedal, repeat steps 4 and 5 (Figure 24–16).

Remove

1. Disconnect electrical lead lead and four hydraulic lines.
2. Remove two attaching nuts.
3. Remove master cylinder as shown.

Install

Notice: See notice at the beginning of this section.

1. Install master cylinder as shown and torque attaching nuts to 30-40 N·m (22-30 ft. lbs.)
2. Attach electrical lead and four hydraulic lines. Torque tube nuts to 13.6-20.3 N·m (120-180 in. lbs.).

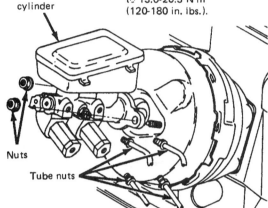

Figure 24–15 After the tubing and electrical connections have been disconnected, most master cylinders can be removed by unscrewing the two attaching nuts and lifting the master cylinder off the booster. *(Courtesy of General Motors Corporation, Service Technology Group)*

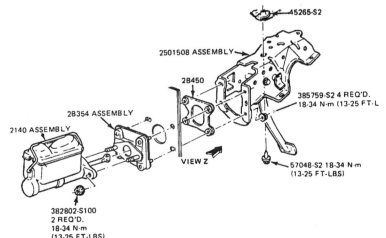
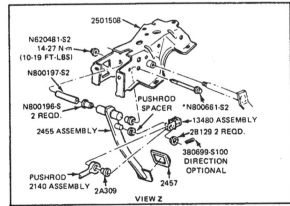

Figure 24–16 It is usually necessary to disconnect the pushrod from the brake pedal when removing a non–power brake master cylinder. *(Courtesy of Ford Motor Company)*

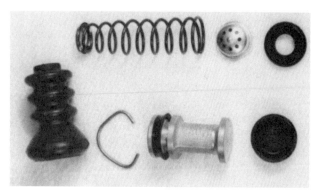

Figure 24-17 An overhaul kit for a single master cylinder. It includes (clockwise from the top) a new return spring, residual valve, grommet, primary cup, piston with secondary cup, retaining ring, and dust boot. A kit for a tandem master cylinder also includes primary and secondary seals for the secondary piston.

NOTE: With some power boosters, the pushrod will be free; test this by trying to pull it outward. If it comes out, tape the pushrod to the front of the windshield to keep it from getting lost or forgotten.

24.4.2 Rebuilding a Master Cylinder

Master cylinder rebuilding is a process of disassembling; cleaning, servicing, and checking the bore; and reassembling using new rubber parts. A master cylinder rebuilding kit is required (Figure 24-17). The kit for a tandem master cylinder usually contains a complete primary piston assembly, primary and secondary seals for the secondary piston, and a primary piston retaining ring. In some cases, it will also contain, as necessary, a new residual check valve and tube seats or reservoir grommets. Some reservoirs can be removed by simply loosening the clamp and sliding the reservoir off the master cylinder (Figure 24-18).

Master cylinder disassembly and reassembly procedures vary slightly depending on whether the reservoir is removable; whether residual check valves are used; whether it is a single, tandem, or quick-take-up unit; and whether it contains internal valves or external switches (Figure 24-19). Again, it is a good practice to follow the procedure recommended by the manufacturer.

To rebuild a master cylinder:

1. Some reservoirs have a vacuum seal that can be pried from the master cylinder assembly (Figure 24-20).
2. Remove the reservoir cover and pour out any fluid that remains. Hold the unit over a container and stroke the pistons a few times to pump any fluid out of the bore. Use a rounded wooden dowel or metal rod for a pushrod if necessary.

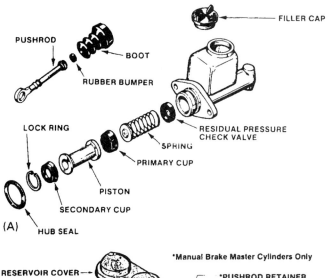

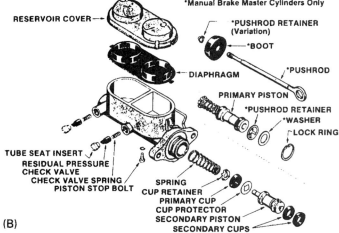

Figure 24-18 A disassembled single (A) and tandem (B) master cylinder. Their service procedures are very similar. *(Courtesy of Bendix Brakes, by AlliedSignal)*

3. To remove the reservoir on units with plastic reservoirs:
   a. Clamp the master cylinder body in a vise by gripping a master cylinder mounting ear.
   b. Remove any retaining pins or clips.
   c. Insert a large screwdriver or pry bar between the cylinder body and the reservoir and pry the reservoir off the body. Use care, because it can break (Figure 24-21). Note that a retaining pin is used on some reservoirs; this pin must be removed first.
   d. Remove the rubber grommets from the cylinder body (Figure 24-22).
4. On some tandem units, you will need to locate and remove the secondary piston stop bolt or pin (Figure 24-23). This bolt or pin enters the cylinder bore from the bottom of the reservoir or at the side or bottom on the outside of the cylinder body. Many tandem master cylinders do not use this stop bolt.

Chapter 24 ■ Hydraulic System Service

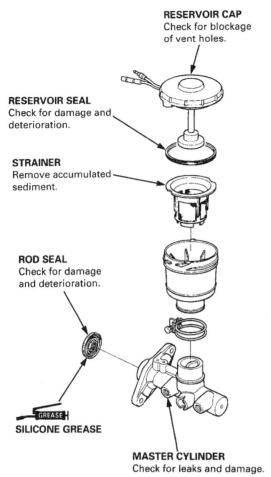

Figure 24-19 Loosening the clamp allows removal of the reservoir with its strainer, seal, and cap from the master cylinder body. Note that the fluid level switch is contained in the cap. *(Courtesy of American Honda Motor Co., Inc.)*

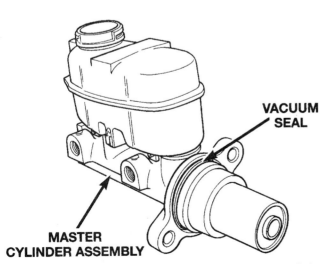

Figure 24-20 This master cylinder has a vacuum seal that can be removed by prying it off. *(Courtesy of Chrysler Corporation)*

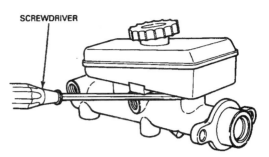

Figure 24-21 The reservoir of many composite master cylinders is removed by prying it off the cylinder body. Some use a retaining pin or pins that must be removed first. *(Courtesy of Ford Motor Company)*

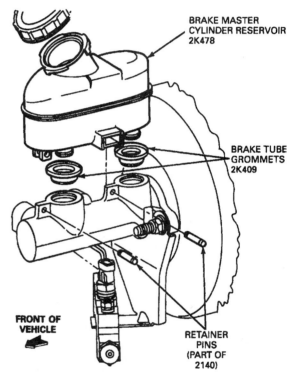

Figure 24-22 Reservoir grommets are pulled or pried out. New grommets should be lubricated as they are installed. *(Courtesy of Ford Motor Company)*

5. Clamp the cylinder body in a vise by gripping a mounting ear as described in step 3, push inward slightly on the primary piston, and remove the primary piston retaining ring. Then remove the primary piston and spring.

HELPFUL HINT: Some technicians push the primary piston inward, slide a iron wire or the shank of a small drill bit through the primary bypass port, and let the primary piston slide back slowly and catch on the wire. This keeps the spring pressure off the primary piston while the retaining ring is being removed or reinstalled (Figure 24-24).

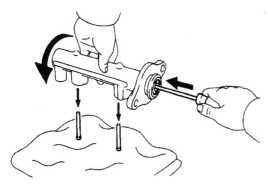

Figure 24-23 Some master cylinders use a stop pin for one or both pistons; they are removed by turning the cylinder body upside down and pushing the piston inward.

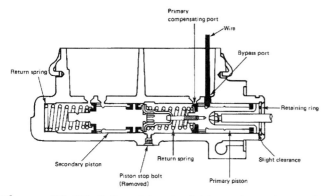

Figure 24-24 If the piston is slid inward, a wire or the shank of a small drill bit can be placed through the primary bypass port to hold the piston and return spring slightly compressed as the retaining ring is removed.

6. Slide the secondary piston out of the bore. If it is stuck, either grip it with needle-nose pliers and pull it out, slam the cylinder body onto a block of wood (bore opening down), or use air pressure.

CAUTION *If air pressure is used, be careful that the piston does not fly out. It can be contained by wrapping a shop cloth around the cylinder body, covering the bore end, and holding the cloth securely while applying air pressure to the secondary outlet port (Figure 24-25).*

7. Check the placement and direction of the seals and disassemble the secondary piston. Do not disassemble the primary piston or disturb the position of the screw unless so directed by the manufacturer.

8. If the rebuilding kit includes a replacement check valve and a tube seat, probe the outlet ports with a small wire or a straightened paper clip to determine if residual valves are used in the master

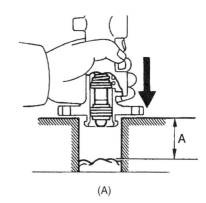

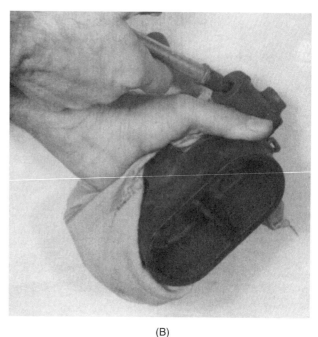

Figure 24-25 A stuck master cylinder piston can be removed by tapping the cylinder down onto two wooden blocks (A) or using air pressure to blow it out (B). Note the shop cloth to protect the piston as it drops (A) and the shop cloth to catch the piston (B).

cylinder. If the outlet port contains a residual valve, you should be able to feel the rubber valve with the wire. It will be about 1/4 in. (6.3 mm) past the tube seat. Do not remove the tube seat unless the outlet contains a check valve and you have a replacement. If there is a check valve and you have a replacement, remove the tube seat according to the following procedure:

a. Thread a #6-32 or a #8-32 self-tapping machine screw through the outlet port. It should thread in about 1/4 in.

b. Using two screwdrivers placed as shown in Figure 24-26, pry the screw upward to lift out the tube seat.

c. Remove the tube seat, check valve, and spring.

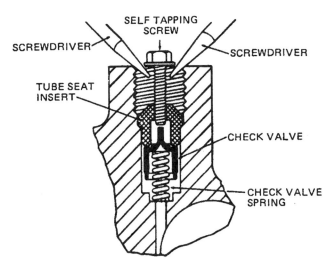

Figure 24-26 A tube seat and check valve assembly can be removed by threading a self-tapping screw into the insert and prying upward as shown here. *(Courtesy of Wagner Brake)*

NOTE: Never install a check valve in a master cylinder port if it did not originally have one.

9. Remove any other valves or switches as directed.
10. Using denatured alcohol or brake system cleaners, thoroughly clean the reservoir, cylinder body, and any other parts that will be reused. Dry these parts with compressed air, making sure that all ports and passages are clean and open. Inspect all the parts to be reused to make sure they are in good condition. If necessary, service the cylinder bore as described earlier in this chapter.

If the bore is acceptable and the other parts are in good condition, the master cylinder can be assembled using the following procedure:

1. If a check valve has been removed, place the new spring, valve, and seat in position and lightly tap the new seat downward using a flat punch. Make sure the seat remains straight as you start it into the bore. The tube seat will be completely seated as the brake tube is tightened into place.
2. Carefully install the new seals on the secondary piston in the same position as the original ones. This installation is easier if they are wet with brake assembly fluid or brake fluid (Figure 24-27).

NOTE: A dry rubber cup should never be slid into a dry bore. Assembly fluid is a rather thick fluid that is used to lubricate the rubber cups. It also helps reduce future rusting in dry areas of the bore. Silicone (preferred) or glycol brake fluid can also be used.

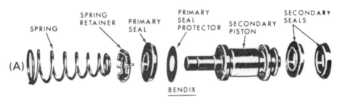
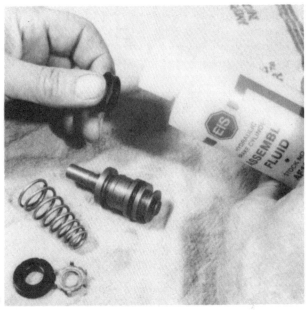

Figure 24-27 During assembly the seals must be positioned correctly on the piston (A) and lubricated with assembly fluid (B) or brake fluid as they are slid over the bands on the piston. *(A is courtesy of General Motors Corporation, Service Technology Group)*

3. Coat the new secondary seals and the cylinder bore with assembly fluid and slide the secondary piston and its spring into the bore. Make sure the lips of the cups do not catch on the edges of the bore as they enter.
4. Coat the seals on the new primary piston assembly with assembly fluid and slide it into the bore (Figure 24-28). Note that on some manual brake master cylinders, the pushrod is attached to the primary piston using a retainer inside the piston. To separate the piston, clamp the pushrod in a vise and pry upward on the piston (Figure 24-29). To install the pushrod in the new piston, insert the pushrod into the new piston and press it inward until the retainer snaps into place.
5. Push the primary piston into the bore, compressing the spring so the retaining ring can be installed. Some technicians use a holding device for the primary piston, as described in step 4 of the disassembly procedure.

6. If the master cylinder uses a secondary piston stop screw, install it. In a few cases, it is necessary to push inward on the primary piston so the secondary piston will move inward far enough to allow the screw to enter.

7. If the reservoir has been removed, wet the new grommets with assembly fluid and install them in the cylinder body. Wet the extensions of the reservoir with assembly fluid. With the cylinder body held in a vise, push the reservoir onto the body using a rocking motion. The bottom of the reservoir should contact the top of the grommets (Figure 24–30). Some technicians find it easier to place the reservoir upside down on a bench and push the cylinder body downward onto it with a rocking motion.

8. Replace all other switches or valves.

The master cylinder is now ready for bench bleeding and installation.

24.4.3 Bench Bleeding a Master Cylinder

Master cylinders have several areas that can easily trap air. This is especially true of master cylinders that are mounted at an angle. Also, step bore master cylinders with quick-take-up valves offer additional places to trap air. Bench bleeding easily removes the air from these pockets. In some cases, it is almost impossible to completely bleed a master cylinder after it is installed on the car. When bench bleeding master cylinders that have four outlet ports, plug the lower port of each section and bleed both sections using the two upper ports. Some replacement master cylinders include bleeder screws, which

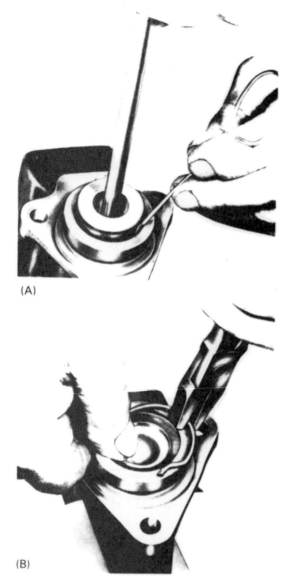

Figure 24–28 As a piston is slid into the bore, it is often necessary to coax the lip of the seal past the end of the bore with a smooth instrument (A). With the piston in place, the retaining ring can be replaced (B). *(Courtesy of ITT Automotive)*

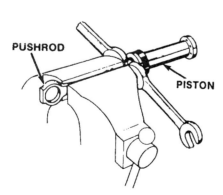

Figure 24–29 On many manual brake master cylinders, the pushrod is attached to the piston by a retainer. Some can be removed by prying with two wrenches as shown here. *(Courtesy of Bendix Brakes, by AlliedSignal)*

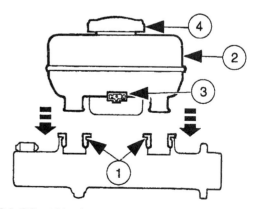

Figure 24–30 With the new grommets (1) installed and wet with assembly fluid or brake fluid, the reservoir (2) is pushed onto the body until it snaps securely into the grommets. (3) indicates the connection for the fluid level warning switch. *(Courtesy of Ford Motor Company)*

make bleeding a fairly easy, on-car operation. There are two commonly used methods of bench bleeding master cylinders.

To bench bleed a master cylinder using tubes:

1. Secure the cylinder body in a vise by gripping a mounting ear.
2. Select a tube nut or adapter of the correct size and install a tube in each outlet port. The tubes should curve up and over into the reservoir (Figure 24–31).
3. Fill the reservoir about one-half to three-quarters full with brake fluid. The fluid level should be above the ends of the tubes.
4. Push the primary piston inward using slow, complete strokes and allow the piston to return slowly. You should observe air bubbles leaving the tubes on each pumping stroke. Bleeding will go faster if you close off the tubing during the return stroke. This is done by pinching the tubing (if it is plastic or rubber) or by putting your finger over the end of a metal tube (Figure 24–32).
5. Continue step 4 until no more air bubbles are expelled on the pumping stroke. Occasionally, it will help to tilt the master cylinder bore up or down during the pumping stroke. On quick-take-up master cylinders, you need to generate 75 to 100 psi of pressure to open the quick-take-up valve and bleed it.
6. After bleeding, remove the bleeding tubes and plug the ports. Test your bleeding operation by applying pressure to your pushrod; it should not move inward. Any movement of the pushrod indicates that air is remaining in the master cylinder pressure cylinder or that there is some other fault in the master cylinder.

The master cylinder is then ready to be installed. The bleeder tubes can usually be left in place until the brake lines are attached.

To bench bleed a master cylinder using an EIS Sur-Bleed syringe:

1. Secure the master cylinder in a vise with the bore tilted upward at the pushrod end. When clamping the sides of the master cylinder, do not clamp by the cylinder bore and do not clamp the reservoir too tightly.
2. Install a plug in each of the outlet ports.
3. Fill the reservoir about half full with brake fluid.
4. Remove one of the plugs, depress the plunger of the syringe completely, press the syringe firmly against the port to make a seal, and slowly pull outward on the plunger. You should observe fluid and air entering the syringe through the cylinder port (Figure 24–33).
5. Remove the syringe and, while holding it vertically, depress the plunger until all the air is removed (Figure 24–34).
6. Place the syringe back against the outlet port and depress the plunger, pushing the fluid left in the syringe back into the cylinder. You should observe some air and fluid entering the reservoir (Figure 24–35).
7. Repeat steps 4 through 6 until there are no more air bubbles and then replace the plug in the outlet port.
8. Repeat this operation on the other cylinder port. On quick-take-up master cylinders, use the syringe to pull fluid and any trapped air through the valve.

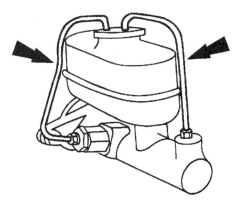

Figure 24–31 The two bleeding tubes (arrows) run from the outlet ports to below the brake fluid level in the reservoir. As the master cylinder piston is stroked, air is pumped from the cylinder through the tubes. *(Courtesy of Ford Motor Company)*

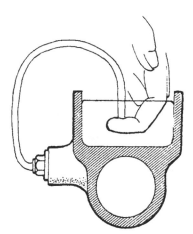

Figure 24–32 When bench bleeding using tubes, the process will often go faster if the return flow is stopped during the piston return stroke, as shown here. *(Reprinted from Mitchell Anti-Lock Brake Systems, with permission of Mitchell Repair Information, LLC)*

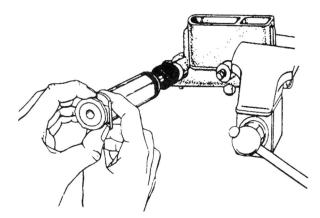

Figure 24–33 A special syringe is available for bench bleeding master cylinders. The first step is to partially fill the reservoir and then suck fluid out of the outlet port. Air will come out with the fluid. *(Reprinted from Mitchell Anti-Lock Brake Systems, with permission of Mitchell Repair Information, LLC)*

Figure 24–34 After sucking air and fluid out of the outlet port with the syringe, the air should be expelled from the syringe as shown here. *(Reprinted from Mitchell Anti-Lock Brake Systems, with permission of Mitchell Repair Information, LLC)*

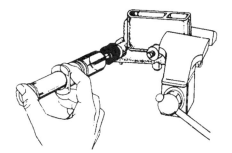

Figure 24–35 With the air removed from the syringe, the fluid is pushed back into the outlet port. The fluid flow should flush any remaining air back through the compensating port and into the reservoir. *(Reprinted from Mitchell Anti-Lock Brake Systems, with permission of Mitchell Repair Information, LLC)*

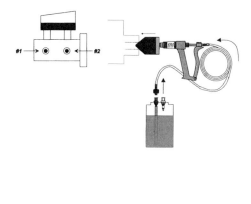

Figure 24–36 The Phoenix reverse injector can be used to pump fluid from a fluid source (right). To bleed a master cylinder, a port adapter allows it to inject fluid through the outlet ports and remove all of the air. *(Courtesy of Phoenix Systems, L.L.C.)*

The master cylinder is then ready to be installed. The plugs should be left in place until the master cylinder is mounted and the lines are being connected. You can test your bleeding operation using step 6 of the tubing bleeding method.

To bench bleed a master cylinder using a Phoenix reverse injector:

1. Secure the master cylinder in a vise and install plugs for the outlet ports if they are not already plugged.
2. Remove the plug from the secondary outlet port.
3. Inject fluid into the secondary outlet port until no more bubbles are seen in the fluid stream entering the front reservoir (Figure 24–36). Or, on master cylinders with four outlet ports, inject fluid into the lower outlet port until fluid drips from the upper port. Then plug the lower port and inject fluid into the upper outlet port until the fluid stream entering the front reservoir is free of air bubbles.
4. Replace the plug in the secondary outlet.
5. Repeat steps 2 through 4 on the primary section.
6. Test your operation by trying to push the primary piston inward; it should not move.

24.4.4 Master Cylinder Replacement

Master cylinder replacement is essentially the reverse of removal. If the only repair of the hydraulic system was the master cylinder, a careful bleeding of the lines as they are being connected will usually save bleeding the whole system. If other parts of the system were worked on, they will need to be bled also.

Chapter 24 ■ Hydraulic System Service

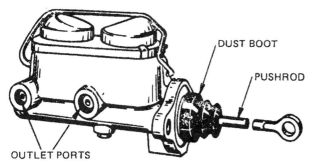

Figure 24-37 Many manual brake master cylinders use a rubber dust boot to keep the end of the cylinder bore clean. *(Courtesy of Wagner Brake)*

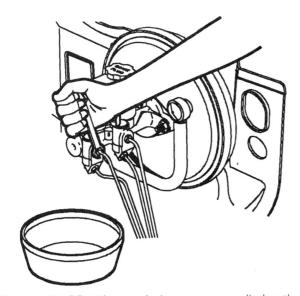

Figure 24-38 When replacing a master cylinder, the line connections can be bled by having someone lightly apply the brake pedal as you tighten the tube nut. After fluid leaks out and the air bubbles stop, tighten the nut before the pedal is lifted. *(©Saturn Corporation, used with permission)*

If the old master cylinder used a boot or hub seal, a new one should be installed as the master cylinder is being replaced (Figure 24-37).

To replace a master cylinder:

1. Place the master cylinder in position on the booster or bulkhead, replace the mounting bolts or nuts, and tighten them to the correct torque. *On manual brake cars,* reconnect the pushrod to the brake pedal as necessary.
2. Remove the plugs or bleeder tubes from the outlet ports as you connect the brake lines. Do not tighten the lines yet. Place a shop cloth under each line fitting to catch any fluid that may leak out.
3. Fill the reservoir about three-fourths full with brake fluid.
4. Have an assistant slowly push the brake pedal as you observe the connections at the outlet ports. They will probably be leaking some fluid with air bubbles. Continue the pedal strokes until only fluid with no air bubbles leaves the connection. At this point, tighten the connection while the pedal is being pushed downward (Figure 24-38).
5. Fill the reservoir to the correct level and replace the cover.
6. Reconnect any wires that were disconnected.
7. Check the brake pedal free travel and adjust it if necessary. There should be 1/16 to 1/8 in. (1.6 to 3.1 mm) of free travel before the pushrod engages the primary piston in the master cylinder.

24.5 Wheel Cylinder Service

Wheel cylinder service includes removing, rebuilding, and replacing rubber parts. If the wheel cylinder is faulty, a new one should be installed. Wheel cylinders are normally serviced during lining replacement. The installation of new lining disturbs the position of the wheel cylinder pistons and cups and moves them inward over the rust, corrosion, or other debris in the center of the cylinder. An otherwise sound-appearing wheel cylinder often starts leaking soon after a lining replacement, and the fluid leaking onto the new lining will ruin it (Figure 24-39).

Most service shops install a new wheel cylinder rather than rebuild the old one. If installing a new wheel cylinder, it is a good practice to flush the cylinder with clean brake fluid in a manner similar to that recommended in Section 24.4. Flushing the new cylinder ensures that small metal particles remaining from the machining process will not damage the hydraulic system.

A wheel cylinder with a good bore can usually be rebuilt. In many cases this can be done with the cylinder body fastened to the backing plate. In some cases, the wheel cylinder is mounted on a backing plate with piston stops. These stops prevent the pistons from sliding out of the end of the cylinder bore (Figure 24-40). If the mounting prevents disassembly of the unit, the wheel cylinder must be removed from the backing plate, either entirely or at least an inch or so. Some sources recommend always moving the wheel cylinder to a bench for service, because you will be able to clean and inspect it more completely.

24.5.1 Wheel Cylinder Removal

The following three methods are commonly used to attach a wheel cylinder to the backing plate: nuts or bolts, a spring lock retainer, and a U-shaped lock plate with

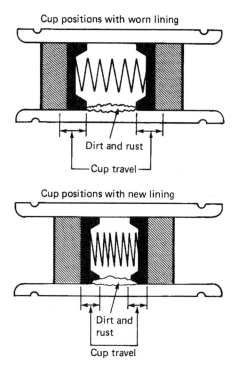

Figure 24-39 As brake lining wears, the piston and cups often move outward in the bore, and dirt and rust deposits form in the bottom. When new lining is installed, the cups are positioned back toward the center, and leaks often result because the cups are pushed over the dirt and rust.

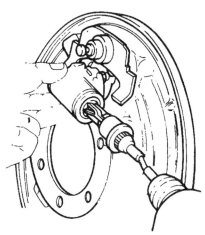

Figure 24-40 Many wheel cylinders can be rebuilt while mounted on the backing plate. This unit uses piston stops, so the mounting bolts were removed for piston and boot removal and access to the ends of the bore. *(Courtesy of Brake Parts, Inc.)*

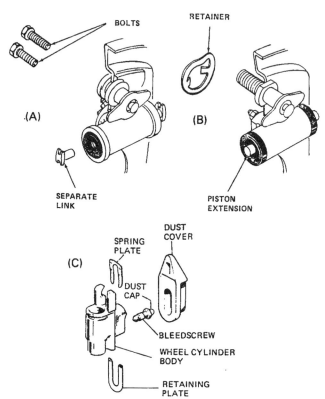

Figure 24-41 The most common methods of attaching wheel cylinders are bolts (A), a ringlike retainer (B), and a U-shaped retainer plate (C). *(A and B are courtesy of General Motors Corporation, Service Technology Group; C is courtesy of LucasVarity Automotive)*

shims (Figure 24-41). On a few older cars, the anchor pin passed through a boss on the wheel cylinder and was threaded into the steering knuckle. The hydraulic tube or hose is also attached to the cylinder body. Wheel cylinder removal is normally done during a lining replacement. On a single-fault repair of a leaky wheel cylinder, it is usually necessary to remove and replace the contaminated lining.

To remove a wheel cylinder:

1. Remove the brake shoes as described in Chapter 22. If you are servicing a faulty wheel cylinder and the lining is still good, sometimes it is possible to merely remove the return springs and slide the shoes out of the way.

2. Disconnect the brake line from the wheel cylinder. Be sure to use a flare wrench and not to bend the metal tubing any more than necessary. Bending the brake line will make its replacement more difficult (Figure 24-42).

3. Disconnect the wheel cylinder mounting.
   a. *If bolts are used,* remove the bolts or nuts.
   b. *If a spring lock retainer is used,* remove the retainer using a pair of awls or a special tool. Insert the points of the awls or tool into the recess slots between the retainer tabs and the wheel cylinder pilot. Then bend both tabs outward and over the pilot at the same time (Figure 24-43 and Figure 24-44).

Chapter 24 ■ Hydraulic System Service

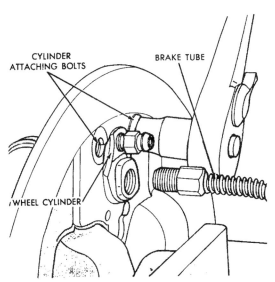

Figure 24–42 After removing the brake shoes, a wheel cylinder can be removed by unscrewing the attaching bolts. *(Courtesy of Chrysler Corporation)*

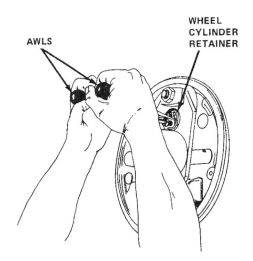

Figure 24–44 A wheel cylinder retaining ring can be removed using a pair of awls to pry the retainer tabs outward. *(Courtesy of General Motors Corporation, Service Technology Group)*

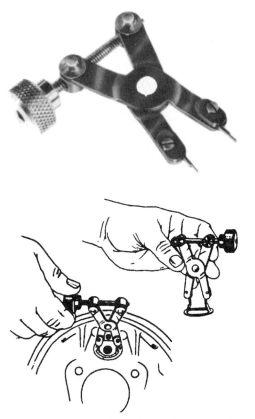

Figure 24–43 This special tool is used to expand the retaining ring so it can be removed or replaced. *(Courtesy of SPX Kent-Moore, Part # J-29839)*

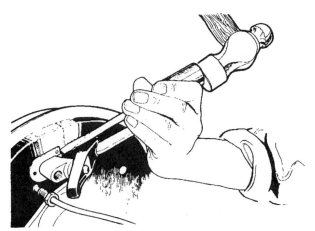

Figure 24–45 This U-shaped retainer is slid out of its groove using a punch or screwdriver and hammer. *(Courtesy of LucasVarity Automotive)*

HELPFUL HINT: When a service operation that requires the removal of a wheel cylinder, caliper, or line is done, the fluid drip from the open line can be stopped by partially applying the brake. Prop the brake pedal to the point where the primary piston cup moves past the compensating port; this will block the fluid flow from the reservoir (Figure 24–46).

24.5.2 Reconditioning a Wheel Cylinder

Wheel cylinder rebuilding is normally a process of disassembling; cleaning, servicing, and checking the bore; and reassembling using new rubber parts. Some technicians believe that a wheel cylinder is not worth rebuilding, especially if it has a stuck or broken bleeder screw or a

   c. *If a lock plate is used,* tap out the retainer plate using a screwdriver and hammer. Be sure to note the relationship and position of the lock tab and any shims used (Figure 24–45).

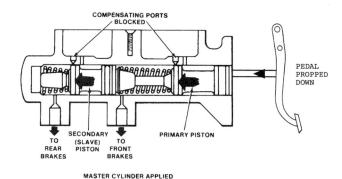

Figure 24–46 Propping the brake pedal downward positions the primary cups to shut off flow from the compensating ports and reservoir. This stops flow toward the master cylinder or drips from a disconnected line.

Figure 24–48 An exploded view of a single- and a double-piston wheel cylinder. Note the position of the rubber cups relative to the pistons. *(Courtesy of Wagner Brake)*

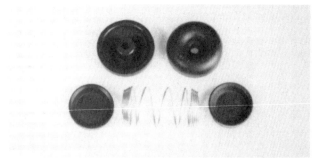

Figure 24–47 A wheel cylinder rebuilding kit includes new cups, boots, and a spring. Note that the spring is shaped to place an outward pressure on the lips of the cups.

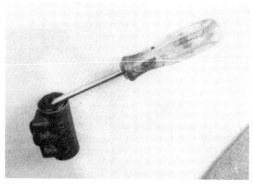

Figure 24–49 This boot, which uses an internal retainer, is removed by inserting a screwdriver under the edge of the boot and twisting the screwdriver to pry it upward.

stuck piston. Others believe that it is easy to check out a wheel cylinder and then go ahead and rebuild it if it is good enough. A wheel cylinder rebuilding kit is required. This kit normally contains new cups and boots and sometimes a new spring or cup expanders (Figure 24–47). Some larger shops purchase cups and boots in bulk rather than stocking individual kits.

To rebuild a wheel cylinder:

1. Loosen the bleeder screw. If it snaps off, discard the wheel cylinder and replace it with a new one. The procedure to loosen a stuck bleeder screw is described later in this chapter.
2. Remove the brake shoe links (Figure 24–48).
3. *If external boots are used,* pull them off. *If internal boots are used,* insert a screwdriver through the center opening in the boot to the edge of the boot and wheel cylinder and pry the boot loose. Be careful not to damage the cylinder bore (Figure 24–49).
4. Slide the pistons out of the bore. If one of them is stuck, insert a wooden dowel through the bore and tap it out. If both pistons are stuck, wrap a shop cloth around the cylinder body so it covers both ends of the bore. Then grip the cloth tightly around the cylinder body and use air pressure through the cylinder port to blow the pistons loose (Figure 24–50).
5. Remove the piston cups and spring.
6. Thoroughly clean the cylinder body, piston, spring, bleeder screw, and any other small parts that are to be reused. Air dry the parts, making sure that all the ports and passages are clean and open. Inspect the parts to make certain they are in good condition. If necessary, service the bore as described in Section 24.3. Some technicians have successfully rebuilt wheel cylinders (duoservo brakes) that have small pits in the center of the bore, away from the area where the piston cup strokes.

If the bore is acceptable and the other parts are in good condition, the wheel cylinder can be assembled using the following procedure:

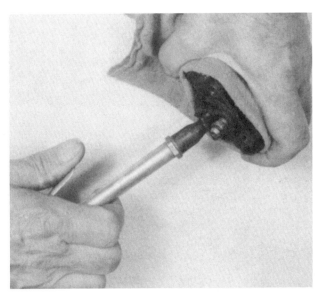

Figure 24–50 If the pistons are stuck in a wheel cylinder, they can be blown loose. Note the shop cloth wrapped around the wheel cylinder to prevent the cups from flying out.

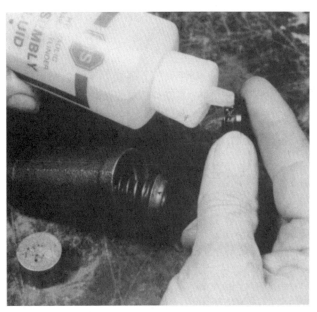

Figure 24–52 As a wheel cylinder is assembled, the bore, pistons, and cups should be lubricated with assembly fluid or brake fluid.

1. Wet the new cups and bore with assembly fluid or brake fluid and slide the cups and spring into the cylinder bore. Be sure that you:
    a. Position the cups so their lips are toward each other.
    b. Do not let the lips of the cups catch on the edge of the cylinder.
    c. Do not push the cups past the ports in the center of the bore; the lips of the cups might be damaged because they tend to catch in the ports.
    d. Do not push the cups inward in the cylinder so far that they cover the fluid inlet or bleeder ports (Figure 24–51).
    e. Fit the expander in position if required.
2. Wet the piston with assembly fluid and slide it into position, making sure the flat side of the piston is next to the flat side of the cup (Figure 24–52).
3. Replace the boots.
    a. *If they are external boots,* slide them into place, making sure they fit snugly into the grooves on the outside of the cylinder ends (Figure 24–53).

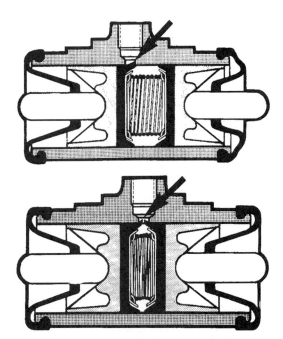

Figure 24–51 If the cups are not in the right positions, they will leak or the lips of the cups can be damaged. *(Courtesy of ITT Automotive)*

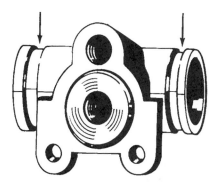

Figure 24–53 The boot retaining grooves (arrows) must be clean to ensure a tight seal between the dust boots and cylinder body. *(Courtesy of ITT Automotive)*

b. *If they are internal boots,* press them into place by hand. They should work completely into position.

4. Install the bleeder screw finger tight.
5. Install the brake shoe links if required.

24.5.3 Wheel Cylinder Replacement

Wheel cylinder replacement is essentially the reverse of the removal procedure. Many technicians prefer to connect the brake line first. This allows freedom to move the cylinder slightly to align it with the tube. Some wheel cylinders are in very tight locations, and a little misalignment makes it extremely difficult to connect this line.

Some wheel cylinders are difficult to bleed while mounted on the backing plate. These cylinders should be bench bled or filled with fluid before they are installed. A vertically mounted wheel cylinder and one with the bleeder screw mounted in the line fitting are examples of wheel cylinders that are difficult to bleed. Any portion of the wheel cylinder above the bleeder valve seat cannot be bled. To bench bleed a wheel cylinder, simply position it with the port in the uppermost position and fill it with fluid. Next, place a plastic or rubber plug or cap over the port until the brake tube or fitting is ready to be attached (Figure 24–54).

To replace a wheel cylinder:

1. Place the wheel cylinder in position on the backing plate and thread the tube nut or brake hose into the wheel cylinder. Finger tighten. Note that on a wheel cylinder using a spring lock retainer, the retaining ring cannot be installed using a socket with the brake line in place.
2. Replace the mounting attachment.
   a. *If bolts are used,* replace the nuts or bolts and tighten them to the correct torque (Figure 24–55).
   b. *If a spring lock retainer is used,* it is recommended that a new retaining ring be used. Two methods can be used to replace this retainer. The simplest is to use the special tool shown in Figure 24–43 to expand the retainer tabs enough for them to be placed in position. Then remove the tool. Another method is to wedge a block of wood between the axle flange and the wheel cylinder to hold the wheel cylinder in position. Then press in on the retainer ring until the tabs snap into place. A 1⅛-in., twelve-point socket is the right size to use as a pushing tool (Figure 24–56).
   c. *If a lock plate is used,* lubricate the wheel cylinder, backing plate, and shims as required. Then

 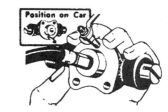

Figure 24–54 Some wheel cylinders are mounted with the bleeder screw in a position other than the top, and they must be bled before mounting on the backing plate. Vertical wheel cylinders can be bled by sliding a thin feeler gauge past the cup to allow the air to leak out. *(Reprinted from Mitchell Anti-Lock Brake Systems, with permission of Mitchell Repair Information, LLC)*

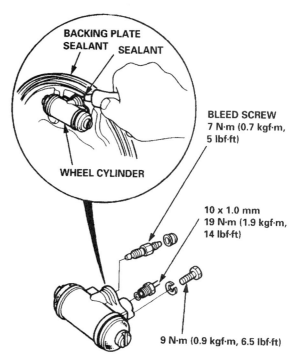

Figure 24–55 Some manufacturers recommend using a sealant between the wheel cylinder and backing plate as the wheel cylinder is replaced to help keep water and dirt out of the brake assembly. *(Courtesy of American Honda Motor Co., Inc.)*

place the wheel cylinder in position and slide the shims and lock plate into place.

NOTE: It is possible for a wheel cylinder with a spring lock retainer to rotate in the backing plate so the ends of the pistons will pass the brake shoes; this movement can allow the wheel cylinder pistons to pop out of the bore. Make sure that the wheel cylinder is fully seated in the backing plate and the retainer is fully seated in the wheel cylinder cavity.

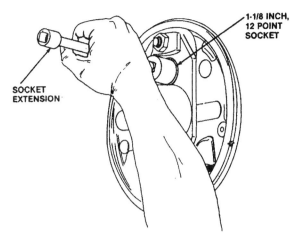

Figure 24–56 A wheel cylinder retaining ring can be installed by driving it into place as shown here. The wheel cylinder must be blocked in place as the ring is installed. *(Courtesy of General Motors Corporation, Service Technology Group)*

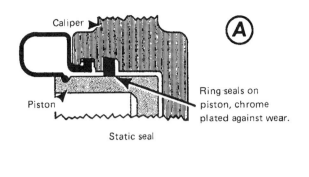

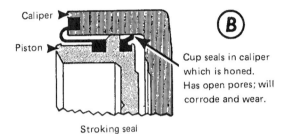

Figure 24–57 Most calipers use a static seal—a square O-ring seal in the caliper body (A). Some older calipers use a stroking seal—a lip seal on the piston (B). A stroking seal demands a good, straight, and smooth bore. *(Reprinted from Mitchell Anti-Lock Brake Systems, with permission of Mitchell Repair Information, LLC)*

3. Tighten the brake line to the correct torque.
4. Bleed the air from the brake line and wheel cylinder.

24.6 Caliper Service

Normal caliper service includes removing, rebuilding, and replacing. Caliper removal and replacement were described in Chapter 23. If the caliper is faulty, a new or commercially rebuilt unit can be installed. Rebuilt units offer substantial cost savings over new units. The use of loaded calipers has become a popular repair method. Loaded calipers are rebuilt calipers that are loaded with new hardware and brake shoes. The technician simply removes the old calipers and worn parts and replaces them with the loaded/rebuilt unit. Calipers that use stroking seals are often rebuilt using stainless steel cylinder sleeves to prevent future bore corrosion (Figure 24–57).

When installing a new or rebuilt caliper, it is a good practice to flush the inside of the caliper with clean brake fluid. Simply slosh brake fluid around the inside of the unit in a manner similar to that described in Section 24.4 to remove any debris that might contaminate the hydraulic system.

Normally, a caliper with a good bore can be rebuilt in a repair shop if a caliper kit can be obtained. Many technicians prefer to rebuild the caliper that is on a car because it is a relatively quick and easy operation and an exact replacement is ensured. Remember that the bleeder screw positions on the right- and left-side calipers are different; they are mirror images of each other.

Caliper rebuilding during lining replacement is strongly recommended. Most calipers use a square-cut O-ring seal. You should remember that this O-ring not only seals in hydraulic pressure but also returns the piston to a released position. Time and heat tend to harden this rubber seal, and when it loses its ability to return the piston, lining drag will result. Time and heat also cause the boot rubber to crack and break. Cracking allows bore corrosion, which will also cause piston sticking and drag. Remember that a brake job should last quite a few years.

24.6.1 Reconditioning a Caliper

Caliper reconditioning is normally a process of disassembling; cleaning, servicing, and checking the cylinder bore and piston; and reassembling using new rubber parts. A caliper rebuilding kit, which usually includes a new boot and piston seal, is required.

Many newer calipers use phenolic pistons. These pistons should be treated gently because they can crack, chip, or break. Some cracks and chips are acceptable as long as they do not extend completely across the piston face, are not in the area of the piston seal, and do not enter the dust boot groove (Figure 24–58). A phenolic piston will swell if it contacts a petroleum product, grease, oil, or solvent. If contact occurs, wash the piston in denatured alcohol and wipe it dry with a clean paper towel.

After removal of the pistons, some fixed calipers, mostly domestic, can be split in half. Removal of the

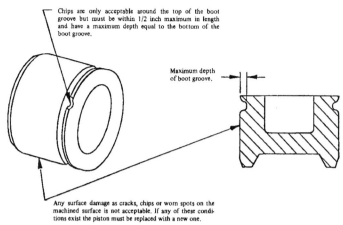

Figure 24-58 A phenolic piston with small chips at the end is acceptable for reuse, but any damage in the sealing area requires replacement. *(Courtesy of Brake Parts, Inc.)*

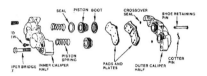

Figure 24-59 When servicing domestic fixed calipers, it is usually recommended that they be completely disassembled. *(Courtesy of Bendix Brakes, by AlliedSignal)*

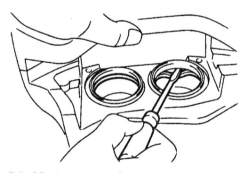

Figure 24-60 Some manufacturers recommend rebuilding a fixed caliper while the two caliper halves are still together.

bridge bolts will separate the two halves, giving better access to the pistons and cylinders for service, cleaning, and inspection (Figure 24-59). Some manufacturers, mostly foreign, do not recommend separating the caliper halves (Figure 24-60).

Removing the pistons is sometimes a difficult operation in caliper rebuilding. There are several off-car methods of piston removal. They are easier if the piston is partially removed while still on the car. When removing the caliper, first remove the brake pads and place the caliper back over the rotor. Next, pump the brake pedal. The hydraulic pressure will move the pistons outward in

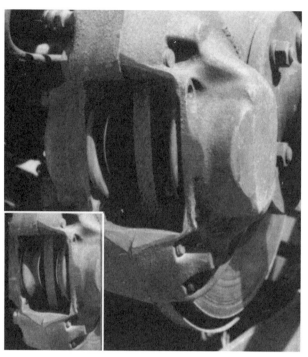

Figure 24-61 Caliper pistons can be moved outward in their bores by removing the pads and applying the brake pedal (inset). This operation must be done before removing the caliper.

their bores (Figure 24-61). Moving each of the pistons about 1/2 in. or so is usually sufficient for easier removal later. When the caliper is off the car, the pistons are commonly removed using air pressure. Some manufacturers do not recommend this method for phenolic pistons. The on-car method just described is recommended instead. A phenolic piston that is really stuck can be removed by breaking it into several pieces using a hammer and chisel. When doing this, be sure to wear eye protection. Other off-car methods that can be used with metal pistons employ several types of piston pullers. These tools all grip the piston so that it can be pulled out (Figure 24-62). For really tight pistons, some shops adapt a grease or zerk fitting to the caliper port and, using a grease gun, force the piston out. If this method is used, be sure to thoroughly remove all traces of grease.

The description that follows is very general. The procedure recommended by the manufacturer should be followed. To rebuild a caliper:

1. Remove the caliper as described in Chapter 23.
2. Use a wire brush to remove dirt and grease from the outside of the caliper. Light grease deposits and oil can be washed off with denatured alcohol.
3. Loosen the bleeder screw and drain the old fluid out of the caliper. If the bleeder screw breaks off, the caliper can be saved by installing a replacement from a bleeder screw kit. This requires that the broken bleeder screw be drilled out and

Chapter 24 ■ Hydraulic System Service

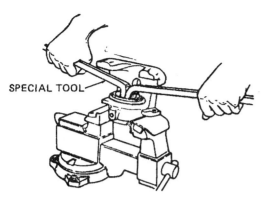

Figure 24–62 Several styles of caliper piston removal tools are available to help pull a piston out of the bore. *(Courtesy of Wagner Brake)*

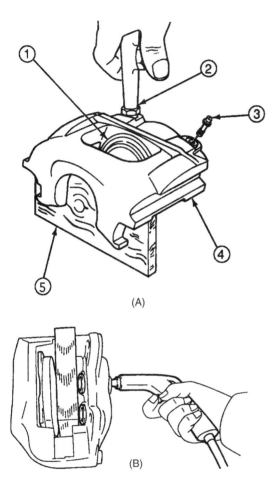

Figure 24–63 An air gun (2) can be used to help remove the piston (1); the wood block (5) keeps the piston from traveling too far (A). This procedure works on fixed calipers with multiple pistons (B). Note the wood block and brake pad to limit piston travel. *(A is courtesy of Ford Motor Company)*

threads be cut into the caliper so the bleeder screw fitting can be installed. Each kit contains directions for installation. If a bleeder screw is so tight that you feel breakage is probable, follow the steps given in Section 24.7 to loosen it.

4. Place a wood block (a one-by-four or two-by-four, about 6 in. long) in the caliper opening so the piston cannot fall or fly out of the bore. Then carefully apply air pressure to the caliper port (Figure 24–63). It is recommended that the air pressure be limited to about 30 psi (207 kPa). Try to move the piston slowly out of the bore in a controlled fashion. Note that the bleeder screw should be closed.

CAUTION *Do not place your fingers in the way of the piston or try to catch it.*

NOTE: On multipiston calipers the loosest piston will move out first, leaving the tightest piston to be removed by another method. Observe the pistons. After one of them moves a little way, wedge it in place or hold it with a clamp so the other one is forced to move.

5. Remove the boot. If it is retained by a metal ring, lift the ring off or use a screwdriver to pry the boot off the caliper (Figure 24–64).

6. Remove the piston seal or O-ring using a sharpened wooden dowel or plastic rod to pry it out of the groove in the caliper or piston (Figure 24–65). If a metal device is used, be careful not to damage the bore or O-ring groove.

7. Remove the bleeder screw.

8. Thoroughly wash the caliper, piston, and bleeder screw. Dry these parts with compressed air, making sure that all the ports and passages are clean and open. Then inspect them for damage. If necessary, service the cylinder bore on calipers with stroking seals using the procedure described in

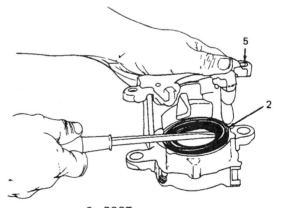

2. BOOT
5. CALIPER HOUSING

Figure 24–64 To remove a boot held in place by a metal retainer ring, slide a screwdriver under it and twist the screwdriver. *(Courtesy of General Motors Corporation, Service Technology Group)*

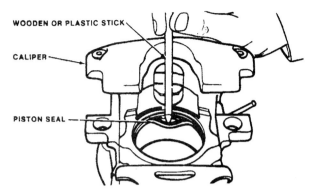

Figure 24–65 With the piston out, the O-ring can be removed from its groove using a pointed wooden or plastic stick. *(Courtesy of Chrysler Corporation)*

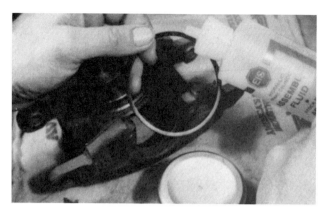

Figure 24–67 The O-ring seal and piston should be lubricated with assembly fluid or brake fluid as they are being installed.

Section 24.3. For calipers with an O-ring seal, the primary sealing surface is on the piston, and it must be clean and smooth. A rough or otherwise damaged piston must be replaced. Also, the O-ring and boot grooves in the caliper must be clean. Remove any rust or dirt from these grooves (Figure 24–66).

If the caliper bore and piston are acceptable, the caliper can be reassembled using the following procedure:

1. *If an O-ring seal is used,* place the new O-ring in its groove, making sure that it is properly seated and not twisted. Lubricate the caliper bore, O-ring, and piston with assembly fluid or brake fluid and slide the piston partially into the bore (Figure 24–67).

If a stroking seal is used, lubricate the piston and seal with assembly fluid or brake fluid and work the seal into its groove in the piston, making sure it is properly seated. Lubricate the cylinder bore and carefully slide the piston and spring, if required, into the bore. Be careful that the seal lip does not catch on the edge of the bore (Figure 24–68). A seal installer will facilitate this job. A 0.005-in. (0.10-mm) feeler gauge with smooth edges can also be used to guide the seal lip into the bore.

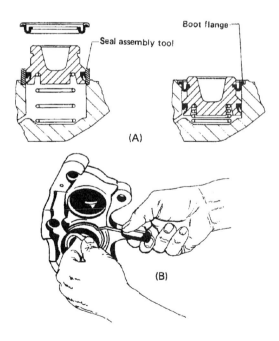

Figure 24–68 The lip of a stroking seal will tend to catch on the edge of the cylinder bore and be damaged. A seal assembly tool (A) or a smooth seal installer (B) can be used to guide the seal into the bore. *(A is courtesy of General Motors Corporation, Service Technology Group; B is courtesy of Brake Parts, Inc.)*

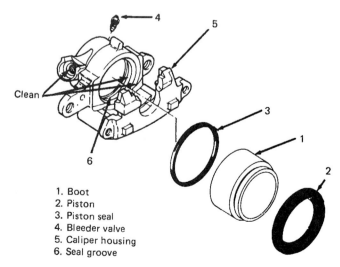

1. Boot
2. Piston
3. Piston seal
4. Bleeder valve
5. Caliper housing
6. Seal groove

Figure 24–66 As a caliper is cleaned, be sure that the O-ring groove, boot retainer groove, and passages into the caliper and to the bleeder screw are clean. *(Courtesy of General Motors Corporation, Service Technology Group)*

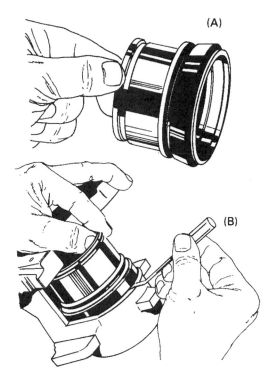

Figure 24-69 The lower edge of the boot shown here must be fitted into its groove before the piston is installed. The recommended way of doing this is to slide the boot over the piston (A), work the boot into the groove (B), and then slide the piston into the bore. *(Courtesy of Brake Parts, Inc.)*

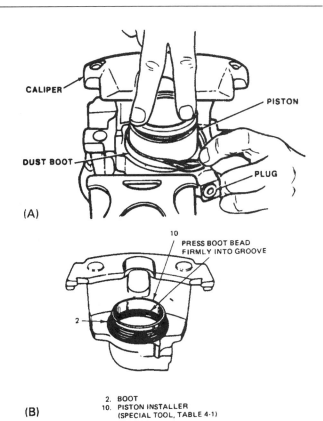

Figure 24-70 Two alternate methods of installing a boot into a Kelsey-Hayes caliper are positioning the boot into the groove and stretching it outward as the piston is slid through it (A) or using a piston installer to keep the boot expanded as the piston is slid into place (B). *(A is courtesy of Chrysler Corporation; B is courtesy of General Motors Corporation, Service Technology Group)*

2. Install the boot.

   a. *If the outer edge of the boot is secured by the bore,* the boot should be installed on the piston and worked into its groove before the piston is slid into the bore. As the piston enters the bore, the boot will be locked in place (Figure 24-69). An alternate method of installing this type of boot is to place it into its bore groove, insert a special piston installer, and slide the piston through the installer (Figure 24-70).

   b. *If the outer edge of the boot is retained by a metal ring molded into the boot,* with the piston partially installed in the bore, slide the boot over the piston. Then position the boot in the counterbore in the caliper and use a boot installer and hammer to seat it (Figure 24-71).

   c. *If the outer edge of the boot is secured by a separate metal ring,* with the piston partially in the bore, slide the boot over the piston. Then position the boot in the groove in the caliper and work the retaining ring into position.

   Note that some boots are locked into the caliper before the piston is installed. Note also that the boot on Delco Moraine fixed-caliper pistons should be sealed to the piston and caliper using thin beads of silicone sealant. This will prevent moisture from entering and causing bore corrosion (Figure 24-72).

3. Make sure the boot seats properly and retract the piston completely into the bore. This can be done by rocking it inward using both thumbs or by pushing it in with a hammer handle (Figure 24-73 and Figure 24-74).

4. For fixed calipers, if the caliper halves were separated, place new O-ring seals between them if required. Then install and tighten the bridge bolts to the correct torque (Figure 24-75).

5. Install and finger tighten the bleeder screw.

6. Install the lining in the caliper and the caliper on the car as described in Chapter 23.

7. Bleed the brake line and caliper.

A caliper test bench is used by some rebuilders and service shops to pressurize a rebuilt caliper and check for leaks and proper piston movement (Figure 24-76).

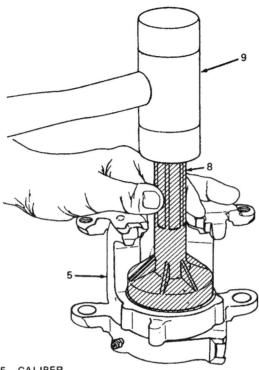

5. CALIPER
8. BOOT SEATING TOOL
9. PLASTIC MALLET

Figure 24–71 Boots that have metal support rings built into them should be installed in the caliper using the properly sized seating tool. *(Courtesy of General Motors Corporation, Service Technology Group)*

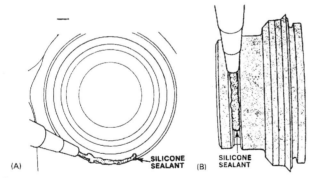

Figure 24–72 It is recommended that the boot of a fixed caliper with stroking seals be sealed to the caliper (A) and piston (B); this prevents moisture from entering the bore and causing corrosion. *(Courtesy of Bendix Brakes, by AlliedSignal)*

24.6.2 Reconditioning a Caliper with a Mechanical Parking Brake

Currently, the following two basic caliper designs are used on domestic cars with a mechanical parking brake mechanism: Delco Moraine (used on General Motors cars) (Figure 24–77) and Kelsey-Hayes (used on Ford products) (Figure 24–78). Several different designs are used on import cars. The hydraulic portion of these calipers uses a square-cut O-ring seal and a boot that are

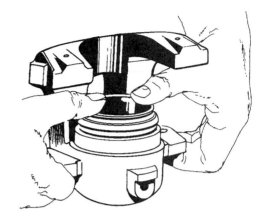

Figure 24–73 Once the piston enters the O-ring, it can usually be worked to the bottom of the bore using both thumbs, as shown here. *(Courtesy of Brake Parts, Inc.)*

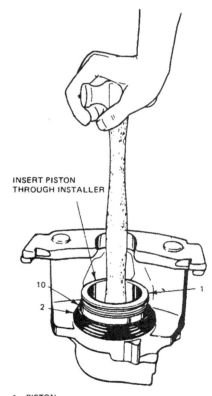

1. PISTON
2. BOOT
10. PISTON INSTALLER

Figure 24–74 If the piston is too tight to be worked in using your thumbs, a hammer handle can be used to tap it into place. *(Courtesy of General Motors Corporation, Service Technology Group)*

essentially the same as those used on the standard caliper; a mechanical mechanism has been added. The additional mechanism requires different disassembly, checking, and reassembly procedures, and the various calipers differ in the procedure required. On cars that use rear disc brakes with drum-type parking brakes, the caliper is serviced in

Chapter 24 ■ Hydraulic System Service   489

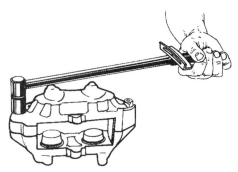

Figure 24–75 If a fixed caliper has been disassembled, new O-rings should be installed and the bridge bolts tightened to the correct torque. *(Courtesy of Brake Parts, Inc.)*

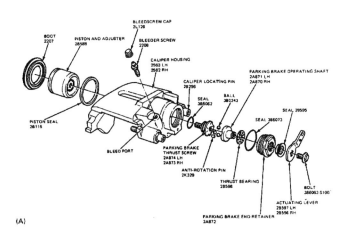

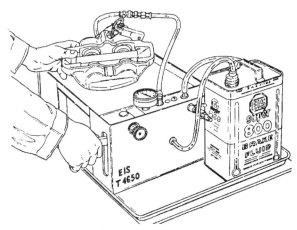

Figure 24–76 A caliper test bench is used to apply pressure to a caliper to make sure there are no leaks or a sticky piston. *(Reprinted from Mitchell Anti-Lock Brake Systems, with permission of Mitchell Repair Information, LLC)*

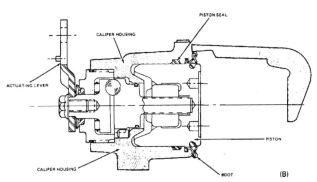

Figure 24–78 An exploded view (A) and cutaway view (B) of a Kelsey-Hayes rear caliper. Note that the parking brake mechanism is removed from the rear of the caliper. *(Courtesy of Ford Motor Company)*

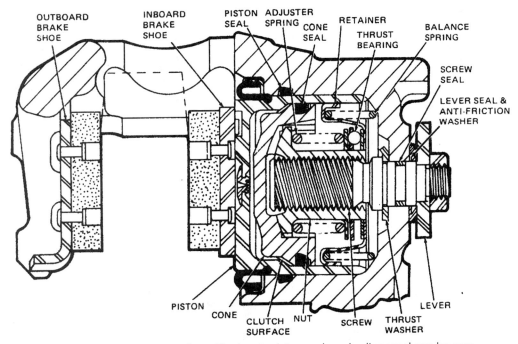

Figure 24–77 A cutaway Delco Moraine rear caliper. The boot, piston seal, and caliper seal can be serviced. *(Courtesy of General Motors Corporation, Service Technology Group)*

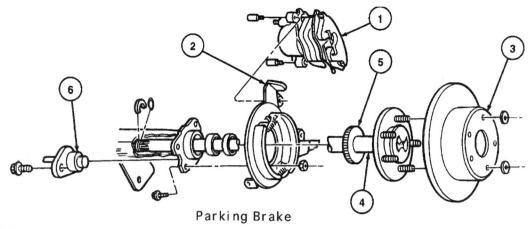

Figure 24–79 This rear disc brake assembly uses a caliper (1) that is similar to a front caliper. The drum-type parking brake uses the inner portion of the rotor (3). *(Courtesy of Ford Motor Company)*

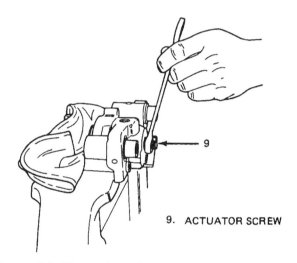

Figure 24–80 On this caliper, using a wrench to turn the actuator screw in the application direction will move the piston out of the caliper bore. *(Courtesy of General Motors Corporation, Service Technology Group)*

the same manner as a front caliper; the service of this parking brake is described in Section 26.2 (Figure 24–79).

The manufacturer's overhaul procedure should be followed when rebuilding these calipers. In general, to rebuild a caliper that includes a mechanical parking brake mechanism:

1. Remove the caliper from the car.
2. Clean the outside of the caliper using a wire brush. Grease and oil can be washed off using denatured alcohol or brake cleaner.
3. Loosen the bleeder screw and drain the old fluid.
4. Remove the piston.

On Delco Moraine units, place a wooden block or folded shop cloth in the caliper opening to protect the piston and then remove the nut and parking brake lever.

Attach a wrench to the hex of the actuator screw and rotate the screw to move the piston out of the bore. The actuator screw is turned clockwise on right-side calipers and counterclockwise on left-side calipers (Figure 24–80).

On Kelsey-Hayes units, use the following procedure:

1. Remove the parking brake lever.
2. Unscrew the retainer and lift it off with the operating shaft and thrust bearing.
3. Remove the three metal balls and, using a magnet, take out the antirotation pin. If the pin catches, tightly rotate the thrust screw (Figure 24–81).

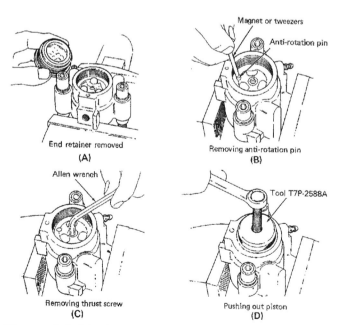

Figure 24–81 The parking brake mechanism is removed from this caliper by unscrewing the end retainer (A), lifting out the balls and antirotation pin (B), and unscrewing the thrust screw (C). The caliper piston can then be pushed out using the special tool (D). *(Courtesy of Ford Motor Company)*

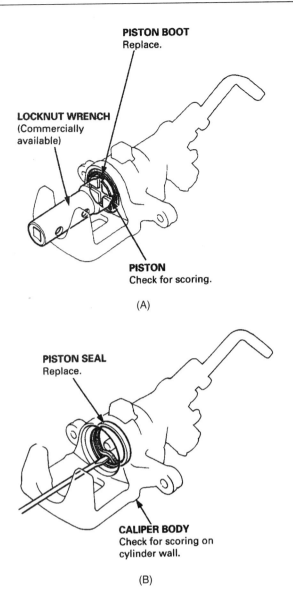

Figure 24-82 The locknut wrench is used to unscrew the piston and remove it from this rear caliper (A). With the piston removed, the piston seal can be removed using a sharp wood or plastic tool (B). *(Courtesy of American Honda Motor Co., Inc.)*

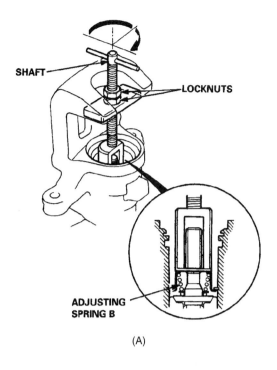

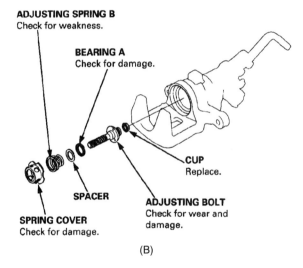

Figure 24-83 This special tool is used to compress the adjuster spring so an internal circlip or retaining ring can be removed (A). With the circlip removed, the adjusting spring and bolt can be removed for cleaning and inspection (B). *(Courtesy of American Honda Motor Co., Inc.)*

4. Remove the thrust screw by unscrewing it.
5. Install the special tool and push the piston out of the caliper.

On other units, follow the procedure given by the manufacturer; it will be similar to the following:

1. Remove the piston by rotating it using a special tool and remove the piston seal (Figure 24-82).
2. Using the special tool, compress the adjuster spring and remove the retaining ring or circlip (Figure 24-83).
3. Remove the adjusting bolt and spring parts.
4. Remove the sleeve piston and O-ring seal (Figure 24-84).
5. Remove the lever and cam assembly with seal (Figure 24-85).
6. Remove the boot, O-ring, and bleeder screw in the same manner as for a standard caliper.

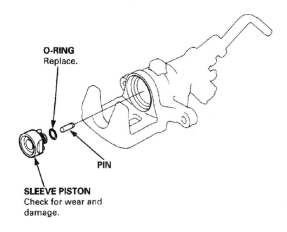

Figure 24–84 With the adjustor bolt removed, the next step is to remove the sleeve piston and pin for cleaning, inspection, and replacement of the O-ring. *(Courtesy of American Honda Motor Co., Inc.)*

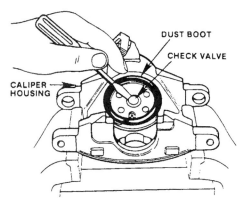

Figure 24–86 When rebuilding a Delco Moraine rear caliper, you should remove the plastic two-way check valve and inspect the inner area. If the piston is usable, a new plastic check valve should be installed. *(Courtesy of Bendix Brakes, by AlliedSignal)*

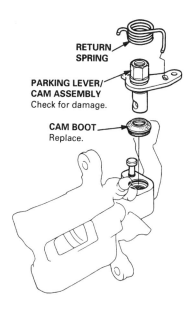

Figure 24–85 The final disassembly step on this caliper is to remove the parking lever/cam assembly. *(Courtesy of American Honda Motor Co., Inc.)*

7. Wash the caliper and internal parts with denatured alcohol, air dry them, and inspect them as you would a standard caliper, with the following additional steps:

    a. *On Delco Moraine units,* remove the rubber check valve from the center of the piston face and inspect the inner area. If there are signs of brake fluid, moisture, pitting, or corrosion, the piston should be replaced. Lubricate a new check valve with brake fluid and install it in the piston (Figure 24–86).

    b. *On Kelsey-Hayes units,* thread the thrust screw into the piston, hold the piston firmly, and pull upward on the thrust screw. It should move upward about 1/4 in. (6.3 mm), and the adjuster nut sleeve in the piston should rotate. As the adjuster screw is released, the adjuster nut should not turn. If the adjuster does not work correctly, the piston should be replaced (Figure 24–87).

If the piston and caliper are acceptable, the caliper can be assembled using the following procedure:

1. Install the O-ring in the caliper bore and lubricate the piston and cylinder with assembly fluid or brake fluid. The piston and boot should be installed in the same manner as they would be on a standard caliper, with the following additions:

    a. *On Delco Moraine units,* install a new actuator shaft seal and lubricate the actuator shaft and seal with assembly fluid. Then position the actuator shaft, balance spring, and thrust bearing on the piston and start the piston in the bore (Figure 24–88). An installer tool can be used to push the piston to the bottom of the bore. As the piston enters the bore, the actuator screw should move through the back of the caliper (Figure 24–89). Seat the boot in position in the caliper bore recess using the correct boot-installing tool (Figure 24–90).

    b. *On Kelsey-Hayes units,* with the boot correctly positioned, start the piston in the bore. Install a new O-ring seal on the thrust screw and wet the seal with assembly fluid. Then thread the thrust screw into the piston until the piston is bottomed in the bore and the thrust screw is bottomed in

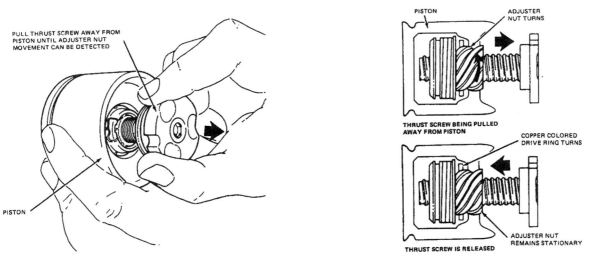

Figure 24-87 The adjuster of a Kelsey-Hayes piston is checked by pulling the thrust screw. The adjuster nut should rotate as the piston is lifted but remain stationary as the adjuster screw moves inward. *(Courtesy of Ford Motor Company)*

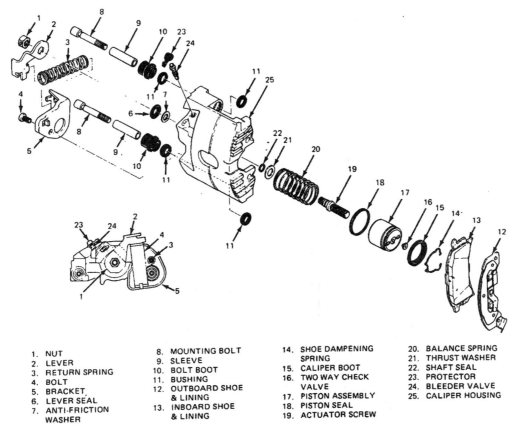

1. NUT
2. LEVER
3. RETURN SPRING
4. BOLT
5. BRACKET
6. LEVER SEAL
7. ANTI-FRICTION WASHER
8. MOUNTING BOLT
9. SLEEVE
10. BOLT BOOT
11. BUSHING
12. OUTBOARD SHOE & LINING
13. INBOARD SHOE & LINING
14. SHOE DAMPENING SPRING
15. CALIPER BOOT
16. TWO WAY CHECK VALVE
17. PISTON ASSEMBLY
18. PISTON SEAL
19. ACTUATOR SCREW
20. BALANCE SPRING
21. THRUST WASHER
22. SHAFT SEAL
23. PROTECTOR
24. BLEEDER VALVE
25. CALIPER HOUSING

Figure 24-88 A disassembled view of a Delco Moraine rear caliper. *(Courtesy of General Motors Corporation, Service Technology Group)*

its recess (Figure 24-91). Index the thrust screw notch and install the antirotation pin. Lubricate the three steel balls with silicone grease and position them. Place the operating shaft in position and lubricate the thrust bearing with silicone grease. Install a new O-ring seal on the end retainer and tighten it to the correct torque.

It is a good practice to plug the inlet port and fill the inner cavity with brake fluid before installing the thrust screw (Figure 24-92).

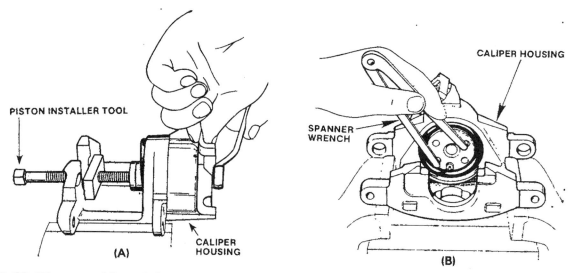

Figure 24–89 When assembling a Delco Moraine rear caliper, a piston installer can be used to push the piston through the O-ring seal and into the bore (A). After installation, the piston should be rotated so that it is correctly aligned with the inboard pad (B). *(Courtesy of Bendix Brakes, by AlliedSignal)*

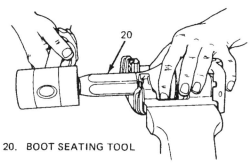

Figure 24–90 The boot of a Delco Moraine rear caliper is seated using a special tool (20), the same way as on a front caliper. *(Courtesy of General Motors Corporation, Service Technology Group)*

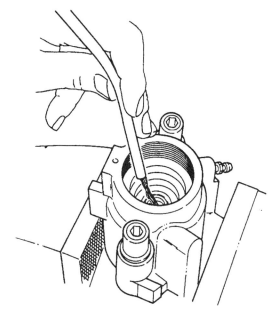

Figure 24–92 After installing the piston, it is a good practice to fill the piston chamber with brake fluid before replacing the thrust screw and parking brake mechanism. *(Courtesy of Ford Motor Company)*

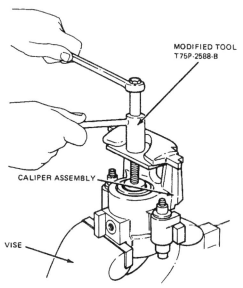

Figure 24–91 The piston of a Kelsey-Hayes caliper is being pushed into the bore using this special tool. *(Courtesy of Ford Motor Company)*

2. Install the parking brake lever in the correct position and tighten the retainer nut or bolt to the correct torque.
3. Install the caliper on the car in the same manner as for a standard caliper, with the additional step of connecting and adjusting the parking brake cable. This procedure is described in Chapter 26.
4. Bleed the brake line and caliper.

On other units, with the unit disassembled, clean and air dry the parts and inspect them for wear, corrosion, or

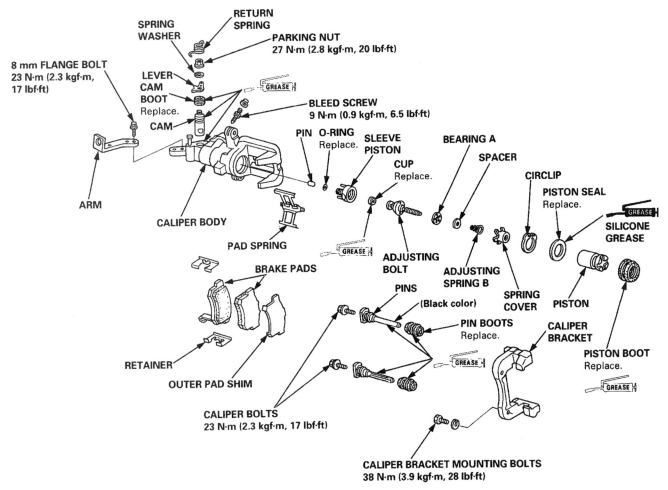

Figure 24–93 A disassembled view of a Nissan-type rear caliper. Note the locations where silicone or rubber grease should be used. *(Courtesy of American Honda Motor Co., Inc.)*

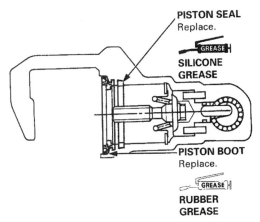

Figure 24–94 A cutaway view of an assembled Nissan-type caliper. *(Courtesy of American Honda Motor Co., Inc.)*

other damage (Figure 24–93). Lubricate the various parts using silicone grease or rubber grease and reassemble the unit following the manufacturer's recommendations (Figure 24–94).

24.7 Bleeding Brakes

The process of removing air from the hydraulic system is called **bleeding**. As mentioned earlier, air is compressible, and any air in the system will be compressed during brake pedal application and cause a spongy pedal.

Air is much lighter than brake fluid and has a natural tendency to rise above it. Anyone who has filled a bottle or glass with water has seen air rise above water. To facilitate bleeding, bleeder screws (actually valves) are placed in strategic locations at the top of caliper, wheel cylinder, and some master cylinder bores (Figure 24–95). Loosening these screws allows air and fluid flow through the valve. The flow will be outward as long as the internal pressure is great enough to cause fluid flow. Bleeding can also take place at any line connection by loosening the connection.

Some systems almost bleed themselves, by gravity. Imagine a system with the master cylinder mounted in the conventional location and with a brake tube running

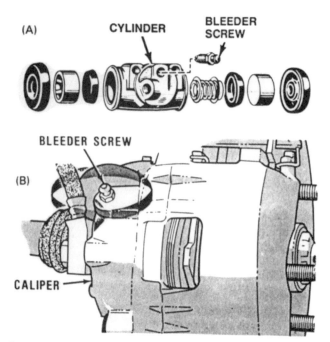

Figure 24-95 The bleeder screw is positioned in the uppermost location in a wheel cylinder (A) or caliper (B). *(A is courtesy of Ford Motor Company; B is courtesy of Chrysler Corporation)*

straight in a downward direction to each caliper and wheel cylinder. As fluid is poured into the master cylinder reservoir, it runs downward through the compensating ports and through the master cylinder bore. Then it flows out the outlet ports and down through the tubing to the caliper and wheel cylinder bores. As this happens, the air in the calipers and wheel cylinder rises upward on a reverse path to the reservoir (Figure 24–96). Because of

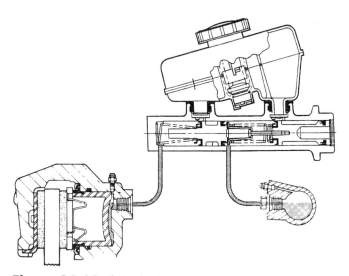

Figure 24-96 If the brake components were arranged as shown here, brake fluid could run downward through the outlet ports from the reservoir, through the brake lines, and into the wheel cylinders and calipers. The air—except for that trapped at the top of the wheel cylinder or caliper—would move upward to the master cylinder reservoir.

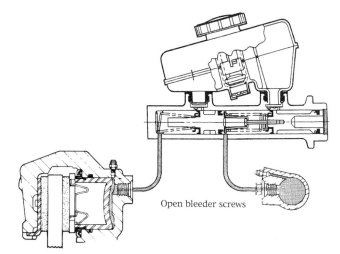

Figure 24-97 If the bleeder screws of the wheel cylinder and caliper are opened, the air can escape, along with some brake fluid.

the small compensating port opening and the two opposite flows, this process is slow but reliable. Given enough time, if enough fluid is poured into the reservoir—about a pint (0.47 liter)—the system will fill with fluid to the level of the line port in each caliper and wheel cylinder. If the bleeder valves are opened, the rest of the air in the cylinders will escape, and the cylinders will fill with fluid (Figure 24–97).

The system just described cannot exist because the brake tubing has to be routed around many obstacles on the way to the calipers and wheel cylinders. As the tubing follows this sometimes tortuous path, it often makes several up-and-down bends. These bends trap air bubbles. These bends, along with any restrictions in control valves, can make bleeding difficult. The air bubbles must be flushed out of the bends and down to the calipers and wheel cylinders or back into the master cylinder, where they can be removed (Figure 24–98).

Bleeding can also be made difficult if fluid agitation or motion breaks the air bubbles into foam and still more difficult if molecular attraction causes small air bubbles to collect in tiny spaces such as between the caliper bore and piston. Even more difficult to bleed is the interior of a caliper with a parking brake mechanism inside. Each added component presents another surface for air bubbles to cling to or become trapped in. Occasionally it helps to tap the caliper firmly with a plastic hammer to dislodge these bubbles (Figure 24–99).

The metering valve can also present a bleeding problem when using pressure bleeders. Remember that this valve closes at about 10 to 15 psi (69 to 103 kPa) and reopens at about 100 to 150 psi (689 to 1,034 kPa). Most pressure bleeders operate in the pressure range that closes

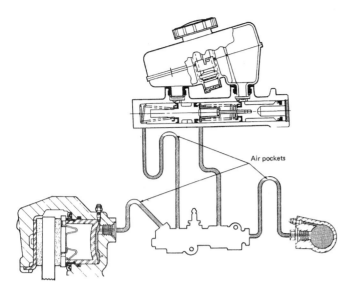

Figure 24–98 Bleeding some systems is complicated when the brake lines go upward and then downward; each upward section becomes an air trap.

the valve, and the valve will shut off the flow to the front brakes. If the valve closes, special tools are available for pulling on the stem of a pull-type valve or pushing on a push-type valve to hold the valve in an open position (Figure 24–100). Push-type valves are normally covered by a rubber boot, whereas pull-type valves have a metal stem extending through the rubber seal. Do not use a clamp or locking pliers for this purpose because they can damage the valve.

There are several different commonly used brake-bleeding methods, including gravity, manual, pressure, and vacuum bleeding. A technician often uses one or two of these methods. Occasionally, several are combined and used at the same time.

Bleeder screws often seize in a closed position because an overtightened screw seizes onto the seat or rust and corrosion build up at the threads. When you try to open a bleeder screw do not use excessive force or it will snap off; remember that the screw is small and hollow. Some technicians insert the shank of a small drill bit (the largest possible) into the bleeder screw; if the drill bit locks up as you try to loosen the bleeder screw, the screw is ready to break. Some methods used to free a stuck bleeder screw are:

- Apply brake fluid to the threads and let it soak in.
- Tap the end and side of the screw using a punch and small hammer to try to break it loose.
- Heat the screw to smoking hot and touch a candle to the hot screw; the heat will expand the screw and the contraction as it cools might loosen it. In addition, the melted candle wax will be drawn into the threads and lubricate them.

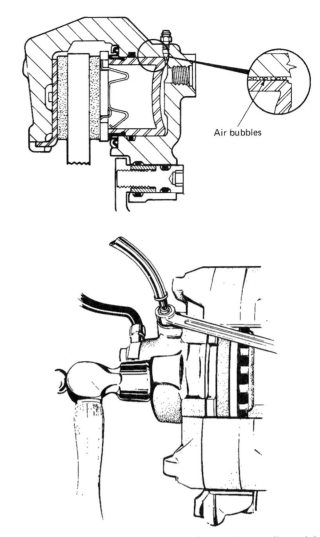

Figure 24–99 Bleeding some calipers is complicated because small air bubbles tend to cling in the tiny space between the caliper bore and piston. Tapping the caliper (with a plastic hammer) often dislodges these bubbles so they will move to the bleeder screw.

When tightening a bleeder screw, remember the torque specifications and that bleeder screws are hollow. They twist apart fairly easily. Bleeder screw torque specifications are provided by most manufacturers and are also provided in Appendix E of this book. It is a good practice to cap a bleeder screw so it will stay clean inside. A plugged bleeder screw can be cleaned using a small drill bit turned by hand, working the drill through both parts of the passage (Figure 24–101). A broken or damaged bleeder screw should be replaced (Figure 24–102). A six-point box wrench should be used to loosen bleeder screws to prevent the possibility of rounding off the corners of the screws. A box wrench is also handy in that it will hang on the bleeder screw and stay in place during the bleeding operation. Some cars require special bleeder wrenches bent in configurations necessary to reach past obstructions near the bleeder screw (Figure 24–103).

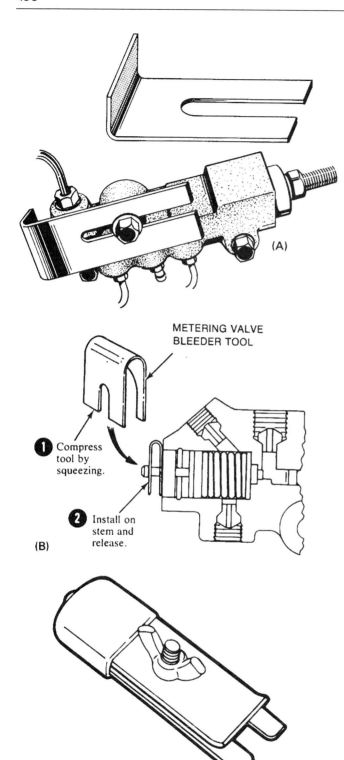

Figure 24–100 The pressure from a pressure bleeder can close the metering valve. Special tool (A) is used to compress the stem of a push-type valve, tool (B) is used to open a pull-type valve, and tool (C) depresses the valve in some RWAL systems. *(A is courtesy of SPX Kent-Moore, Part # J-23709; B is courtesy of Ford Motor Company; C is courtesy of SPX-OTC)*

Figure 24–101 A small drill bit can be turned by hand to clean the passages of a plugged bleeder screw.

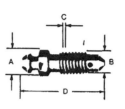

A = Hex Diameter
B = Thread Diameter
C = Thread Pitch
D = Overall Length

No.	A	B	C	D
3201	10mm	10mm	1.5mm	28mm
3202	3/8	3/8	24	1-33/64
3203	3/8	3/8	24	1-7/64
3204	8mm	8mm	1.25mm	24mm
3205	5/16	5/16	24	1-1/32
3206	1/4	1/4	28	15/16
3207	7/16	7/16	20	1-9/32
13209	10mm	10mm	1.25mm	33mm
13210	10mm	10mm	1mm	33mm
13211	7mm	7mm	1mm	30mm

Figure 24–102 Replacement bleeder screws are available. Note the various sizes and dimensions. *(Courtesy of Rolero-Omega, Division of Cooper Industries)*

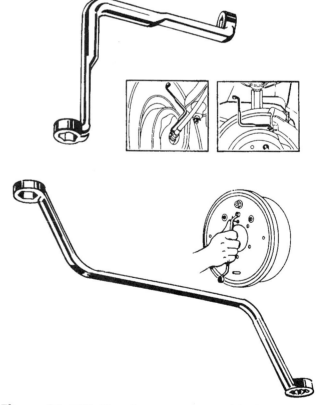

Figure 24–103 The placement of some bleeder screws requires a special wrench to get past the obstructions. *(Courtesy of KD Tools)*

When bleeding brakes, always bleed until the fluid runs clear, making sure that you have removed all of the old, contaminated fluid and all of the air bubbles.

Fluid that is bled from a system must be discarded in the proper manner or recycled.

24.7.1 Bleeding Sequence

It is recommended that a set sequence be followed when bleeding a system. This sequence helps save time and ensures that the whole system is bled. The best bleeding sequence for any vehicle is the one recommended by the manufacturer, especially on vehicles equipped with ABS.

The usual sequence is to bleed the components in the following order:

1. Master cylinder at the bleeder screw, if so equipped, or by loosening the lines at the outlet ports
2. Combination valve if equipped with a bleeder screw
3. Wheel cylinders and calipers in succession beginning with the longest brake line and ending with the shortest brake line; on most cars this sequence is right rear, left rear, right front, left front (Figure 24–104)

With diagonal split systems, this sequence is changed to bleed the secondary master cylinder circuit first, beginning with the longer brake line and then the shorter brake line, followed by the longer line of the primary section and the shorter line of the primary section. On many cars this sequence is right rear, left front, left rear, right front (Figure 24–105).

If a caliper has two bleeder screws, bleed the inboard section first and then the outboard section. If a drum brake has two wheel cylinders, bleed the lower one first, followed by the upper one. Most vehicles with ABS require a special bleeding procedure. Some are bled by manual methods, some by pressure bleeding, and some require the use of a scan tool to activate the pump or solenoids. Be sure to check the manufacturer's procedure.

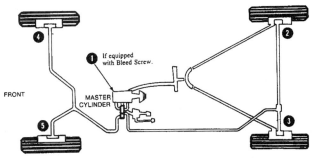

Figure 24–104 The sequence normally recommended for bleeding the brakes of a tandem split system. *(Courtesy of Ford Motor Company)*

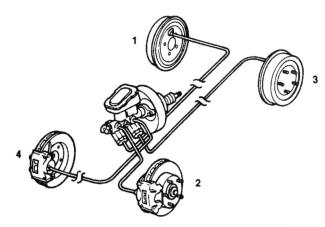

Figure 24–105 The sequence normally recommended for bleeding a diagonal split system. *(Courtesy of Bendix Brakes, by AlliedSignal)*

24.7.2 Gravity Bleeding

The **gravity bleeding** method simply lets the fluid run down into the calipers and wheel cylinders; if done correctly, it can save a substantial amount of time. Like other methods, it often becomes a step in performing a brake job. A possible drawback with gravity bleeding is that it does not always remove all the air. It sometimes must be followed with another bleeding method, but it is too simple a process to skip.

An experienced technician arranges the sequence of a brake job so the complete hydraulic system is assembled before the brake shoes are assembled or the caliper replaced. As soon as the last brake line or hydraulic component is assembled and connected, fluid is poured into the master cylinder reservoir. Now, while the individual disc or drum brake assemblies are being put together, they can be bled. Open the bleeder screw as you begin to install the shoes, hardware, or other items on a backing plate or caliper and keep an eye on the bleeder screw. Normally, before you have completed the assembly, the bleeder screw will start to drip fluid; close the bleeder screw. This indicates that the caliper or wheel cylinder is full of fluid. Be sure to wipe up the spilled fluid.

Add fluid to the master cylinder reservoir and repeat this operation as you assemble each of the other brake units. When the last unit has been assembled and the drums installed, finish filling the master cylinder reservoir and test the brake pedal. If it is spongy, use one of the other bleeding methods to force the remaining air out of the pockets in the system.

24.7.3 Manual Bleeding

Manual bleeding uses the master cylinder and brake pedal as a pump to cause fluid flow through an open bleeder screw (Figure 24–106). This fluid flow should

Figure 24-106 When manually bleeding a system, the pedal is pushed slowly downward while the bleeder screw is open.

flush air out of any pocket or trap. Manual bleeding should be done as smoothly as possible to avoid turbulence in the fluid, which could cause foaming. Foamy fluid contains tiny air bubbles that are very hard to bleed out. Excessively fast pumping of the pedal tends to cause foaming, and if the master cylinder has not been replaced or rebuilt, the longer-than-normal piston travel can damage the rubber cups as they pass over debris or corrosion in the cylinder bore.

Manual bleeding is done with the aid of a bleeder hose and a bottle, which serve several major purposes. When the end of the hose is immersed in fluid, air bubbles are produced that are easily seen and air is prevented from flowing back into the bleeder screw. It also contains waste fluid and helps prevent a mess. A short section, about 1 ft (30 cm), of 3/16- or 1/4-in. (4.7- or 6.3-mm) rubber or plastic tubing may be used as a bleeder hose. Choose a size that will slide onto the nipple of the bleeder screw and stay in place. Clear plastic hose is preferred because it allows you to watch the air bubbles and fluid condition. Any small, half-pint or pint glass or plastic bottle can be used. Plastic is preferred because glass breaks easily (Figure 24-107).

A recent innovation is a mechanical device that can stroke the brake pedal by remote control. This unit allows the technician to be at the wheel or brake assembly and push or release the brake pedal. It can be used for diagnostic checks as well as for bleeding operations.

To manually bleed a system:

1. Fill the master cylinder reservoir at the start and enough times during the operation to keep it at least half full.
2. Instruct your helper to keep a moderate, steady pressure on the pedal, to push with a slow and steady motion, to inform you when it reaches the floor, and to release it slowly when told to.
3. With pressure on the brake pedal and the bleeder hose connected to the first bleeder screw in the sequence, open the bleeder screw and observe the flow from the hose. When you are told that the pedal has reached the floor or the flow stops, close

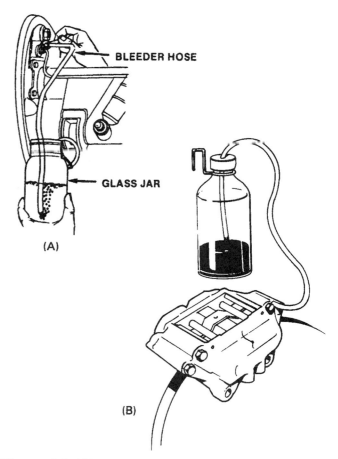

Figure 24-107 A bleeder hose is placed on the bleeder screw, with the free end in a jar partially filled with brake fluid (A). This allows the waste fluid to be collected, allows easy observation of any air bubbles, and prevents reentry of air into the system. If a hanger is used and enough room is available, it is even better to suspend the catch container above the bleeder screw (B). *(A is courtesy of Bendix Brakes, by AlliedSignal; B is courtesy of ITT Automotive)*

the bleeder screw and tell your helper to release the pedal. Some sources recommend a 15-second wait between the time the pedal is released and the time it is reapplied.

Note that each time the bleeder hose is disconnected, the fluid will run out and the hose will fill with air. Bubbles normally appear at the start of each bleeding step as this air is bled out of the hose.

4. Repeat step 3 until the fluid flow from the bleeder hose is clear and without bubbles. With the pedal held downward, tighten the bleeder screw to the correct torque and check the fluid level in the reservoir. Repeat this operation on the next brake in the sequence.
5. After bleeding the last brake assembly, fill the reservoir and check the pedal feel.

Chapter 24 ■ Hydraulic System Service

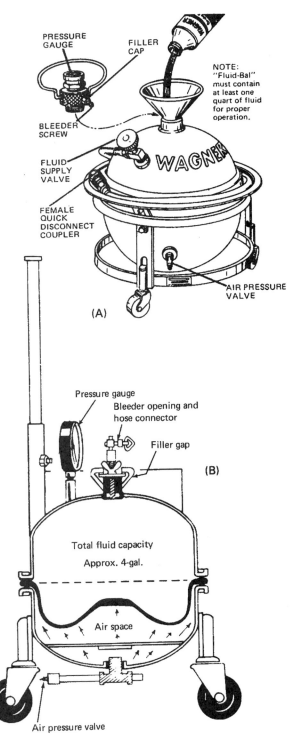

Figure 24-108 The top portion of a pressure bleeder is filled with fluid (A). Note that the filler cap contains a bleeder screw so the air can be removed. Also note that the upper and lower sections are separated by a rubber diaphragm so the air pressure will not contaminate the fluid (B). *(A is courtesy of Wagner Brake; B is courtesy of Branick Industries, Inc.)*

24.7.4 Pressure Bleeding

Pressure bleeding normally uses a pressurized tank of brake fluid to cause fluid flow through the bleeder screws (Figure 24-108). This tank should have a fluid chamber that is separate from the air chamber to prevent the fluid from being contaminated by the compressed air. The air chamber is normally pressurized to a working pressure of 10 to 15 psi (69 to 103 kPa). If using silicone fluid, the pressure should be dropped to about 5 psi (35 kPa). Higher pressures might introduce turbulence in the fluid, which could cause foaming. Pressure bleeding has the advantage that only one person is required and that the fluid level in the reservoir is continuously maintained.

Another recent innovation is a brake fluid injector. This unit is a hand-operated pump that can pump 10 to 20 milliliters (0.6 to 1.2 cu. in.) per stroke, depending on the model, at pressures up to 125 psi (862 kPa) into the brake system. Besides bleeding in the normal direction, this unit can be used to **reverse bleed** or **back bleed** the system by forcing fluid into a bleeder screw and out the master cylinder. The unit can also **cross bleed** parts of the system by forcing fluid in one bleeder screw and out another (Figure 24-109). These styles of bleeding allow the technician more flexibility in bleeding air that might become trapped in odd locations; they are described in the following sections.

A brake fluid injector can be used for pressure bleeding. Using a port adapter, this unit is used to pump fluid into the master cylinder compensating port, through the circuit, and out the bleeder valve (Figure 24-110).

Adapters are required to connect the pressure bleeder unit to the master cylinder, and a pressure-tight connection must be made to prevent fluid leakage. The adapter is normally clamped to the top of a cast-iron reservoir. Special adapters are required to fit into plastic reservoirs, because these reservoirs are not strong enough to withstand the clamping pressure (Figure 24-111).

To pressure bleed a system:

1. Fill the master cylinder reservoir about half full.
2. Attach the correct adapter to the master cylinder (Figure 24-112).
3. Check the air pressure in the pressure bleeder unit and adjust it if necessary.
4. Connect the hose between the bleeder unit and the adapter and open the fluid supply valve. Check for leaks at the adapter or in the brake system.
5. Attach a bleeder hose and bottle to the first bleeder screw in the sequence. Open the bleeder screw, observe the fluid flow, and close the bleeder screw when the flow is clear and free of bubbles.
6. Repeat step 5 on the last brake in the sequence.
7. After bleeding the last brake in the sequence, shut off the fluid supply valve. Disconnect the hose, remove the adapter, and wipe up any spilled fluid.

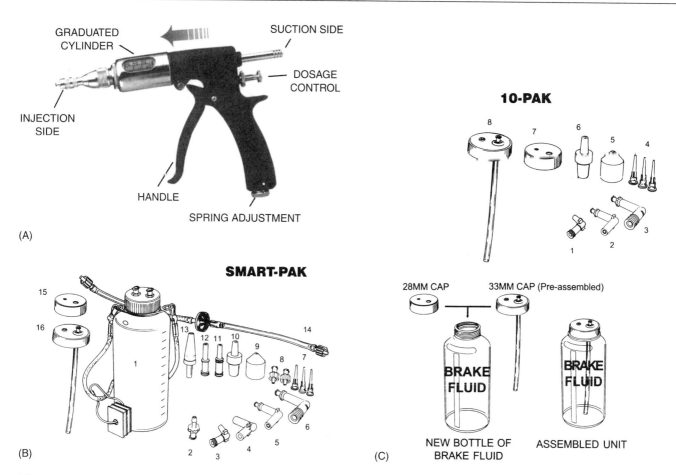

Figure 24–109 The Phoenix Injector system consists of a hand-operated injector (A), a fluid bottle, and connector tubing and fittings (B and C). The injector can be used to create a fluid pressure or vacuum depending on how the tubing is connected. *(Courtesy of Phoenix Systems, L.L.C.)*

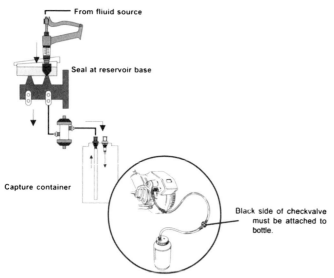

Figure 24–110 With the use of a port adapter, the Phoenix Injector can be used to pressure bleed portions of a brake system. *(Courtesy of Phoenix Systems, L.L.C.)*

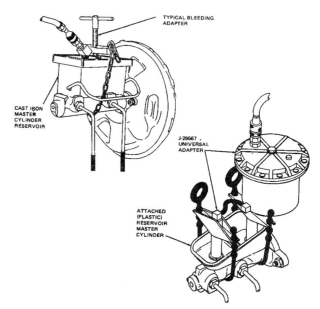

Figure 24–111 An adapter is required to attach the pressure bleeder to the master cylinder reservoir. This adapter can clamp onto the top of a metal reservoir, but it must connect to the fluid recesses of a plastic reservoir. *(Courtesy of General Motors Corporation, Service Technology Group)*

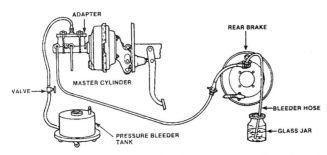

Figure 24–112 Pressure bleeding a system. Opening the valve (at left) allows fluid under bleeder tank pressure to flow through the system to the open bleeder screw, pushing any air and fluid into the catch container. *(Courtesy of Bendix Brakes, by AlliedSignal)*

8. Fill the reservoir, if necessary, and check the pedal feel.
9. Some pressure bleeders include a system pressure gauge and allow you to shut off the pressure supply. If your unit is so equipped, turn off the pressure, note the pressure in the system, and recheck the pressure after a few minutes. A loss of pressure indicates a leak.
10. Discard the contaminated fluid in the proper manner.

24.7.5 Vacuum Bleeding

Vacuum bleeding uses a pump to pull fluid and air out of the bleeder screw (Figure 24–113). A hand- or air-powered vacuum pump can be used for this procedure. Vacuum bleeding is a rather simple and effective operation, but several guidelines should be observed. As in manual bleeding, the master cylinder reservoir will empty and let air enter the system. Also, the bleeder screw threads are not airtight. Placing a vacuum on the bleeder screw will pull air past the threads, as well as air and fluid through the screw. Air can also be pulled past the cups in a wheel cylinder. This air does no harm because it is usually removed as it enters. Bubbles will nearly always be present in the fluid flowing from the bleeder screw.

To vacuum bleed a system:

1. Fill the master cylinder reservoirs at the start and enough times during the operation to keep them at least one-fourth full. Note that some systems have an attachment that will maintain the correct reservoir fluid level during bleeding operations (Figure 24–114).
2. Attach the bleeder unit to the first bleeder screw in the sequence. Open the bleeder screw, operate the bleeder pump, and observe the flow. After fluid flow begins and the bubble rate shows a significant drop, close the bleeder screw and stop the vacuum pump.
3. Repeat step 2 on the remaining brake units, being sure to check the reservoir fluid level after bleeding each brake.
4. After bleeding the last brake, fill the reservoir and check the pedal feel.
5. Discard the contaminated fluid in the proper manner.

By reversing the connectors to the fluid injector, it can also be used for vacuum bleeding (Figure 24–115).

24.7.6 Reverse-Flow Bleeding

Some brake systems have the lines or brake valves arranged so there are air traps close to the master cylinder. Also, because air bubbles naturally move upward through fluid, it makes more sense to bleed from the bottom up than to bleed from the top down, as in other bleeding methods. In reverse-flow or back bleeding, fluid is forced into the bleeder valve, through the circuit, and out the master cylinder reservoir.

To reverse bleed a system:

1. Remove enough fluid from the master cylinder reservoirs for three to ten injection strokes. Do not allow the reservoir to overfill.
2. Open the bleeder valve and attach the injector to the bleeder valve, pumping fluid to the end of the adapter to expel all air from the hose (Figure 24–116).
3. Gently depress the handle of the injector to pump fluid into the bleeder valve in slow and steady strokes. Overly fast strokes can loosen debris and move it into the brake valves. Continue pumping for three to ten strokes.
4. Remove the injector from the bleeder valve and allow a small amount of fluid and any air to escape from the bleeder. Then tighten the bleeder valve.
5. Repeat steps 2 to 4 on the remaining brake units.
6. Discard the contaminated fluid in the proper manner.

24.7.7 Changing Brake Fluid

To ensure a maximum brake fluid boiling point and to reduce interior system corrosion, a system that uses DOT 3 or 4 brake fluid should have the fluid changed every year or every other year. As mentioned earlier, this is not commonly done, although it is a relatively simple operation.

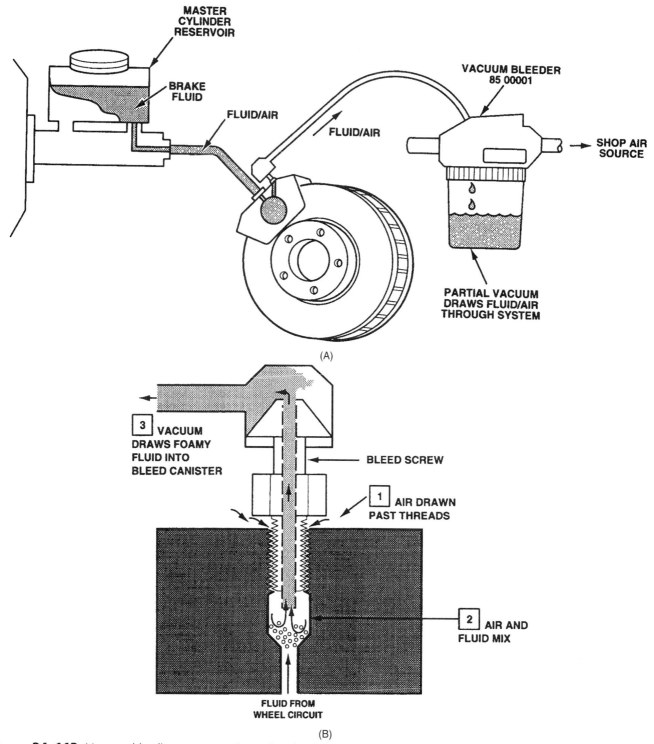

Figure 24–113 Vacuum bleeding a system. Operating the vacuum bleeder pulls air and fluid from the open bleeder screw (A); the fluid is coming from the master cylinder reservoir. A vacuum bleeder normally removes foamy fluid because air is drawn past the threads of the bleeder screw (B). *(Courtesy of General Motors Corporation, Service Technology Group)*

With today's cars and driving practices, changing the brake fluid every other year can be highly recommended for two major reasons: safety and economics. Many drivers of FWD cars in heavy-traffic situations have brake fluid that is close to the boiling point, and old, contaminated fluid has a lower-than-normal boiling point. Quick-take-up master cylinders are expensive, costing more than several hundred dollars, and ABS hydraulic modulators are very expensive, with replacement costs of some units well over a thousand dollars. Old, contaminated brake fluid can easily cause improper operation and ruin these parts.

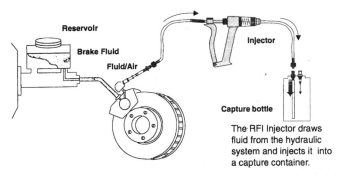

Figure 24–115 Using the proper adapters allows the fluid injector to vacuum bleed a system. *(Courtesy of Phoenix Systems, L.L.C.)*

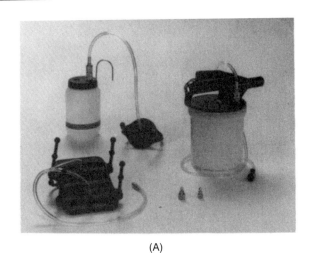

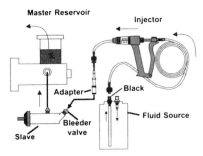

Figure 24–116 In reverse-flow bleeding, the injector is pumping fluid into the bleeder valve, which forces fluid and air upward and out of the master cylinder. *(Courtesy of Phoenix Systems, L.L.C.)*

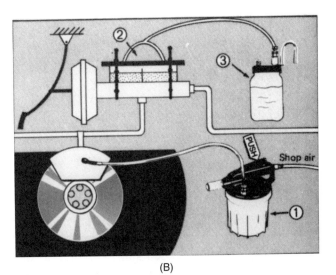

Figure 24–114 This vacuum-bleeding system includes a refiller for the reservoir (A). When using this system, the reservoir is filled and the cover (2) is installed and connected to the refiller (3). The bleeder (1) is connected to the bleeder screw, which is loosened, and the bleeder is operated to produce the bleeding operation (B). *(A is courtesy of Vacula Automotive Products)*

Several styles of brake fluid testers are available. Some units test a fluid sample from the master cylinder reservoir for water contamination; others measure its boiling point. (See Figure 20–10.) One tester uses a paper strip that is dipped into the fluid. These methods provide a clear indication that the fluid condition may cause a brake loss or is causing corrosion in the system.

Changing the brake fluid, also called **flushing** a system, is the same as bleeding a brake, except that the major purpose is to remove all of the old fluid, not just the air. Most brake technicians do this when they bleed the brakes after a major brake overhaul. Most manufacturers recommend a thorough bleeding—until clean, new fluid leaves the bleeder screw—during each major brake repair to ensure that the system is filled with new, clean fluid.

Any of the previously described brake-bleeding techniques can be used for this operation. Pressure bleeding is preferred by many technicians because it is quickest and maintains a supply of new fluid.

In the past, it was common to flush the dirty fluid out of a system using a flushing fluid or denatured alcohol. This is currently considered a poor practice because any leftover flushing fluid will lower the boiling point of the new brake fluid. Never flush a system with alcohol; use only clean brake fluid. Alcohol is a good cleaning agent when a unit is disassembled because it can be completely dried off the parts.

If the system is being changed over to silicone fluid, the bleeding operation should be done smoothly and slowly. The slightly higher viscosity of silicone fluid causes it to trap air bubbles, especially if it is mixed with glycol fluid. When making a fluid changeover, it is important to remove all of the old glycol fluid. Silicone fluid is lighter than glycol fluid, so the glycol fluid will stay in the bottom of the cylinders and cannot be bled out. The best time to make a fluid changeover is during a major overhaul. The calipers, wheel cylinders, and master cylinder should be rebuilt, cleaned, and emptied of old fluid.

To change brake fluid:

1. Using a syringe, remove the old fluid from the reservoir and wipe out the reservoir with a clean shop cloth to remove any contaminants and residue.
2. Connect a pressure bleeder to the reservoir and bleed the system as described in Section 24.7.4.
   Or, fill the reservoir with new fluid and bleed the system using one of the other methods described.
3. Continue the bleeding operation at each bleeder screw until clean, fresh, new fluid flows from each one.

With ABS cars, changing fluid is more difficult and time consuming. An alternate method of removing most of the contaminated fluid is to cross bleed portions of the system. The usual cross-bleeding circuits are both front brakes and rear brakes of a tandem split system and the two front–rear brake circuits of a diagonal split system. While cross bleeding, the brake pedal can be depressed to stop any flow to the master cylinder and ABS modulator valves.

To cross bleed a system:

1. Depress the brake pedal about 1/2 to 1½ in. and install a pedal depressor to hold it in this position.
2. Install a catch container at one end of the circuit to be bled and open this bleeder valve.

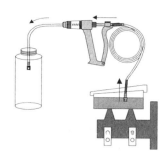

Figure 24–118 The fluid injector can be used to remove old, contaminated fluid from the master cylinder reservoir. *(Courtesy of Phoenix Systems, L.L.C.)*

3. Inject fluid through the bleeder valve at the other end of the circuit until fluid entering the catch container runs clear (Figure 24–117).
4. Remove the old fluid from the master cylinder using the fluid injector, a syringe, or a vacuum pump (Figure 24–118).
5. Refill the master cylinder reservoirs with new fluid and discard the contaminated fluid in the proper manner.

24.8 Diagnosing Hydraulic System Problems

Occasionally a problem occurs in a hydraulic brake system that cannot be corrected by the service steps previously described. Several service operations can be used to help you pinpoint the cause of the problem. The following procedures deal with these common problems.

24.8.1 Diagnosing a Spongy Brake Pedal

A spongy brake pedal is usually caused by air in the system. It can also be caused by a brake hose that is expanding under pressure or a drum that is being deflected under pressure. These two possibilities can be visually checked while an assistant applies and releases pressure on the brake pedal.

CAUTION *Operating the brakes with the reservoir cover removed will allow fluid to spray upward, sometimes well above the master cylinder. Never have your face over or near the reservoir at this time unless protected by a face shield or fluid barrier (Figure 24–119).*

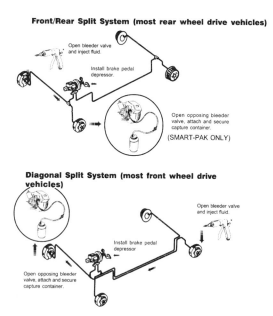

Figure 24–117 In cross bleeding, fluid is injected into the bleeder valve at one wheel while the old fluid and any air bubbles are being removed at the bleeder valve of the other wheel. *(Courtesy of Phoenix Systems, L.L.C.)*

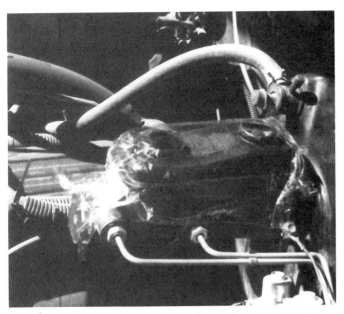

Figure 24–119 Clear plastic (food wrap) can be placed over an open master cylinder reservoir to catch any fluid sprays.

A diagnostic approach to checking for air is to have an assistant rapidly pump the brake pedal about twenty times and then keep it depressed while you remove the master cylinder reservoir cover. Release the pedal and watch the reservoir for abnormally large swirls. If there is a larger-than-normal swirl, air is probably trapped in that section of the system, and it should be rebled (Figure 24–120). It is a good idea to cover the reservoir with a clear plastic film before releasing the brake pedal to contain the fluid swirl.

If the system has been bled several times and you are sure all the air has been bled out, but the pedal is still spongy, isolate the source of the problem by closing off parts of the system. Plug up the ports of the master cylinder one at a time. If the pedal becomes firm, the problem is in the section that has just been closed off. If the pedal is still low or spongy with both ports plugged, the master cylinder is faulty. The outlet ports can be plugged using a coupler and a plug threaded onto the tube nut. Or a

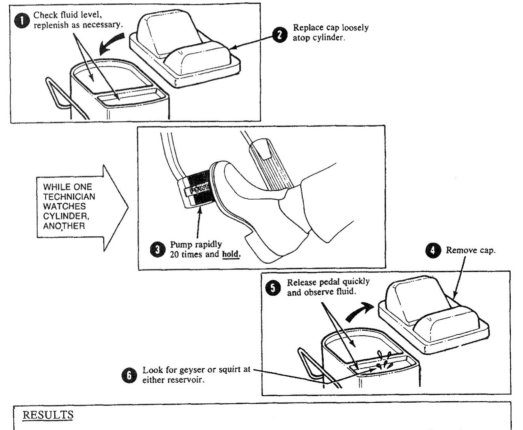

Figure 24–120 An air-entrapment test. This quick check is used to determine which portion of the brake system has air trapped in it. *(Courtesy of Ford Motor Company)*

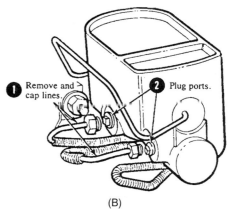

Figure 24–121 A group of inverted flare plugs of different sizes (A). These plugs can be used to isolate sections of a hydraulic system to help locate the cause of a low or spongy pedal (B). *(B is courtesy of Ford Motor Company)*

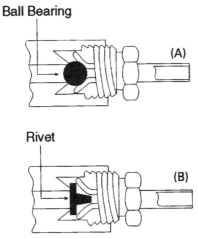

Figure 24–122 A ball bearing (A) or a rivet (B) can be used as a line plug. Be careful not to overtighten it and damage the flare seat or wedge the ball into the brake line.

connection can be plugged by placing a steel ball between the tube flare and the tube seat (Figure 24–121). When using a steel ball, be careful not to damage the flare or the seat. Choose a ball large enough so that it will not become wedged in the tube (Figure 24–122). A properly sized copper rivet can be used in place of the ball. As a plug is installed, it must be bled. Loosely tighten the plug and have an assistant apply the brake pedal. Then tighten the plug as soon as the air bubbles stop and there is a clear fluid leak.

After the problem has been isolated to a particular section of the system, the exact location can be determined by moving the plug to the various connections in that section.

24.8.2 Diagnosing a Sinking or Bypassing Brake Pedal

A sinking or bypassing brake pedal is one that sinks or falls to the floor under pressure. This problem is usually caused by an internal leak in the master cylinder or an external fluid leak from the system (Figure 24–123). Bypassing is a fluid leak from one fluid section to the other, past the secondary cups on the secondary piston in the master cylinder. Although an external fluid leak is more common, it is easier to check for bypassing. Have an assistant watch the levels of the fluid in both reservoirs as the pedal sinks. On plain master cylinders, the level in both reservoirs should drop slightly during normal brake application. With quick-take-up master cylinders, the level in the primary reservoir will rise when the quick-take-up valve opens. Bypassing will cause a fluid transfer that lowers the level in one reservoir while raising the level in the other.

An external leak is located by looking for a leak in the system. The probable location of the leak will be a loose line fitting, a faulty hose, a faulty wheel cylinder, or a leaky caliper. As you look for the leak, follow each of the lines until fluid is located. It should be noted that the rear brake lines on some cars pass through the passenger compartment. In these cases, it is necessary to remove the rear seat to see all the lines.

If an external leak cannot be located, the probable cause of the sinking pedal is bypassing a leak past the primary cup of the primary piston or secondary cup of the secondary piston in the master cylinder. This condition often causes a pedal to sink under light pressure but not under very heavy pressure. Sometimes it causes fluid turbulence in the reservoir as the pedal is sinking. If you are

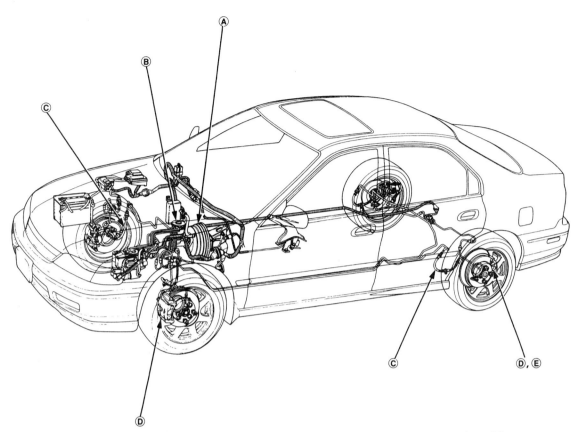

Figure 24–123 The probable locations of an external fluid leak are the master cylinder connections (B), hoses (C), calipers (D), and wheel cylinders (E). *(Courtesy of American Honda Motor Co., Inc.)*

still not sure whether a master cylinder is at fault, plug the outlets as previously described. If the pedal sinks with the ports tightly plugged, the master cylinder is clearly defective.

24.8.3 Diagnosing a Brake That Will Not Apply

If a brake unit will not apply, have an assistant exert a moderate pressure on the brake pedal while you open the bleeder screw. Be careful, because it should spurt fluid. If it does, the problem is being caused by a stuck wheel cylinder or caliper piston. If fluid does not spurt from the bleeder screw, a line is restricted. Check for a kinked or compressed line, an internally collapsed hose, or a plugged control valve. Loosen the line connections closer to the master cylinder as an assistant maintains pedal pressure to locate the cause of the restriction.

24.8.4 Diagnosing a Brake That Will Not Release

If a brake drags, apply and release the brakes and then open the bleeder screw at the dragging brake. In most modern systems, fluid should drip out. In older systems using residual valves, there should be a small spurt of fluid as the residual pressure is released. If fluid spurts out in a greater volume than normal and the brake releases, there is a restriction in the line that prevents fluid release back to the master cylinder. If there is a normal fluid release and the brake still drags, a stuck wheel cylinder or caliper piston is probably the cause.

24.8.5 Diagnosing a Rear- or Front-Wheel Lockup Problem

Occasionally a proportioning valve fails to split the front-to-rear pressure properly, which can result in insufficient or excessive rear brake pressure. On cars with diagonal braking systems, lockup of a single rear wheel can be caused by a malfunction in one of the proportioning valves. This problem can be diagnosed using a pair of pressure gauges (Figure 24–124).

A gauge set can be assembled in the shop. Obtain two 0–1,000 or 0–2,000 psi (0–10,000 kPa) gauges. These gauges should have a small dial, about 2 in. in diameter, and a side-mounted port. The port will normally have a 1/8-in. national pipe thread (NPT) male thread. Adapters can be purchased, or made in the shop from bleeder screws that will thread into the wheel cylinder or caliper using the bleeder screw port (Figure 24–125). The gauge can also be installed in a line connection using a tee that

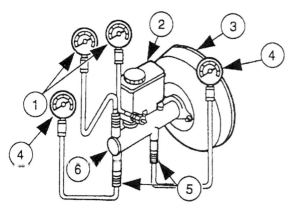

Figure 24–124 Gauges can be connected to the master cylinder (2 and 6) to determine pressure at the front brake lines (1) and rear brake lines (4). A difference between the front and rear pressure shows the operation of the differential or proportioning valves (5). *(Courtesy of Ford Motor Company)*

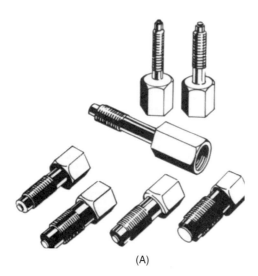

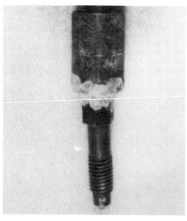

Figure 24–125 Commercial (A) or shop-made (B) adapters can be used to connect a pressure gauge to a caliper or wheel cylinder. The shop-made adapter is a bleeder screw that has a hole drilled through the end and a 1/8-in. NPT pipe coupler brazed onto it. *(A is courtesy of ITT Automotive)*

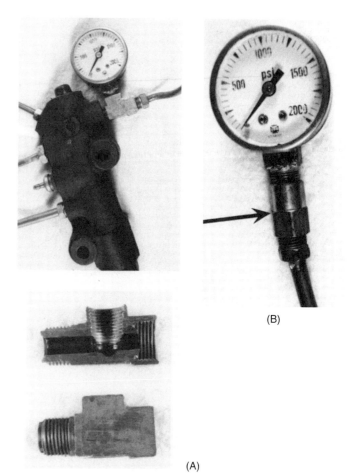

Figure 24–126 A pressure gauge can be installed in a tee placed between a component and the tubing by using a tee fitting (A) or onto the end of a brake line by using a female adapter (arrow in B).

has one 1/8-in. NPT female port and the same male and female ports as the line flare nut (Figure 24–126). The gauge can also be installed directly on tubing or into a fitting by using a connector fitting.

To use these pressure gauges, connect one to the bleeder port of the front caliper and one to the rear wheel cylinder or to a tee fitting installed on each side of the proportioning valve. Be sure to bleed each connection as it is being made (Figure 24–127). Next have an assistant apply the brake pedal as you monitor the gauge readings. The gauge pressures should rise equally until about 300 to 700 psi (2,068 to 4,826 kPa). From that point on, the pressure in the front brakes should rise much faster than that in the rear brakes. In some systems, the metering valve will cause a lag in the front pressure increase of between 10 and 100 psi (69 and 690 kPa). Occasionally, the exact pressures are given in the manufacturer's specifications. A defective proportioning valve should be replaced. On cars with diagonal braking systems, install a gauge in each rear brake circuit and apply the brakes with a firm pedal. Each of the rear brakes should have the same pressure. A pressure difference indicates a faulty valve.

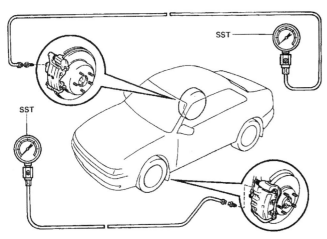

Figure 24–127 A proportioning valve can be tested by installing a gauge at the front and rear brakes. A difference in pressure with a hard brake application shows proportioning valve operation.

24.8.6 Height-Sensing Proportioning Valve Adjustment

Pickups, vans, and some cars use a height-sensing proportioning valve to control rear brake application pressure relative to vehicle height and load. This valve is also called a *pressure control valve* or a *regulated proportioning valve*.

The linkage (the mounting position) of this valve is adjustable to allow proper valve-to-axle positioning and ensure that the valve will produce the correct control pressure for a particular load. The actual adjustment varies in different car models. The service manual should be consulted to determine the correct procedure for checking and adjusting this valve (Figure 24–128).

24.9 Testing a Warning Light

When completing a brake job, it is a good practice to test the operation of the warning light system to be certain of its future operation if the brakes fail. It is common for the switch piston to stick on center after years of remaining stationary in this position (Figure 24–129).

To check the operation of a warning light switch:

1. First verify that the warning light operates as you crank the engine. Then test the warning lightbulb circuit by turning on the ignition, disconnecting the lead from the warning light switch, and connecting a jumper wire from this lead to ground (Figure 24–130). As the jumper is being connected, the warning light should come on. If it does not, replace the bulb or repair the circuit. Reattach the lead to the switch.

2. Check the fluid level in the master cylinder reservoir to make sure it is at least half full.

3. Connect a bleeder hose and bottle to a bleeder screw.

4. With the ignition on, open the bleeder screw while an assistant applies pressure on the brake pedal. The light should come on when there is light-to-moderate pressure on the pedal. If it does not, the warning switch or valve is faulty and should be replaced. The bleeder screw should be closed before the pedal is released.

In some cases, after a system failure has been repaired, the switch piston will stick in an off-center position. This will keep the warning light on even though the system is in good condition. Also, some late 1960s and early 1970s cars have valves constructed so they cannot recenter themselves.

To center a warning light switch or valve:

1. Attach a bleeder hose and bottle to a bleeder screw in the hydraulic circuit opposite the one that has failed or was bled last.

2. Turn on the ignition, open the bleeder screw, and have an assistant push slowly on the brake pedal while watching the warning light. At the instant that the light goes out, close the bleeder screw. If the light flickers and comes back on or does not go out at all, repeat this operation at a bleeder screw on the other hydraulic section.

3. Fill the master cylinder reservoir to the correct level.

24.10 Brake Light Switch Adjustment

On many cars the brake light switch position is adjustable. An improperly adjusted switch might fail to turn off the stoplights if it is too loose or cause brake drag if it is too tight. This switch is usually closed as the pedal is depressed to apply the brakes and opened to break the circuit as the pedal is released (Figure 24–131).

On some cars, the brake light switch is a double switch. When the brake is applied, this switch deactivates the cruise control at the same time that the brake lights are switched on (Figure 24–132). On some cars, a separate cruise control switch is used with the brake control switch.

On some cars with automatic transmissions, the brake light switch is used to disconnect the torque convertor clutch during deceleration. An improperly adjusted switch will cause improper torque convertor clutch operation.

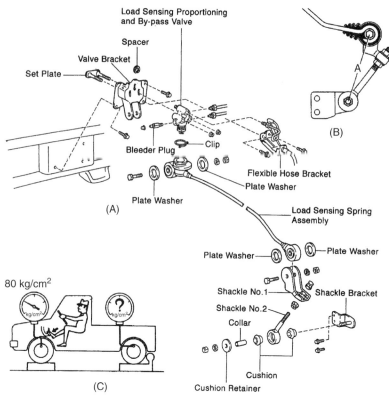

Figure 24-128 A load-sensing proportioning valve assembly (A). Its adjustment has a specification for shackle number 2 (B), and it can be tested using two pressure gauges. Added weight to the rear of this pickup should change the valve setting and the rear pressure (C).

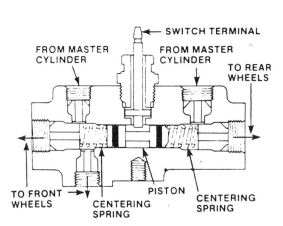

Figure 24-129 This pressure differential switch uses centering springs to keep the piston centered. A loss in pressure in the front or rear system will move the piston off center and contact the switch. *(Courtesy of Bendix Brakes, by AlliedSignal)*

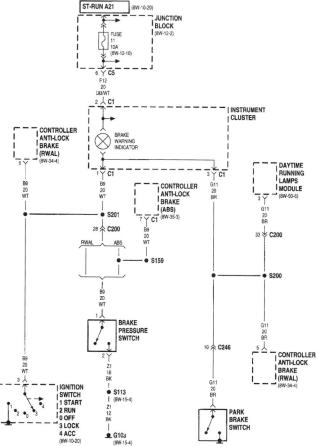

Figure 24-130 A brake warning light circuit. The brake warning indicator should light when the ignition switch is turned to start; it can also be turned on by the brake pressure switch or the parking brake switch. It is integrated into the ABS and daytime running lights on some models. *(Courtesy of Chrysler Corporation)*

Chapter 24 ■ Hydraulic System Service

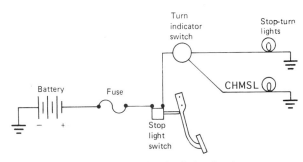

Figure 24-131 A simple brake light circuit.

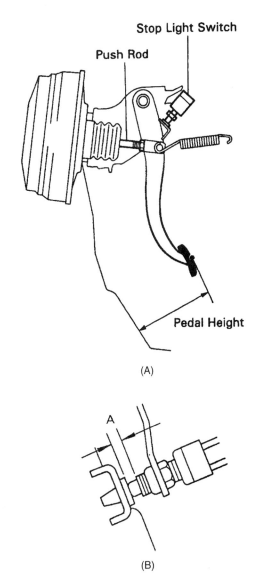

Figure 24-133 Many stoplight switches are mounted on a bracket near the brake pedal (A). This switch has a specified clearance, A, between the pedal and end of the switch body (B).

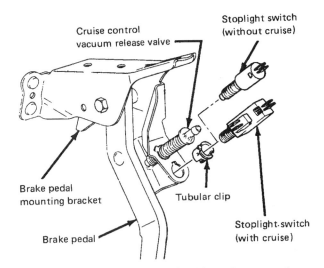

Figure 24-132 Cars equipped with cruise control use a brake switch to turn off the cruise control when the brakes are applied. Many cars use one switch for both brake lights and cruise control; if two switches are used, they should both be adjusted to operate at the right time. *(Courtesy of General Motors Corporation, Service Technology Group)*

Brake light switch adjustment varies in different cars. Many cars have an adjustable bracket or an adjustable switch position on the bracket. Some cars have a switch position that is automatically adjusted on the bracket as the pedal is released. A service manual should be checked, especially on cars with cruise control and/or automatic transmissions, to determine the correct adjustment method when working on a particular car (Figure 24-133). After any brake light switch adjustment, it is important to check the amount of brake pedal free play to make certain that the adjustment has not reduced the amount of free play, which will result in brake drag.

24.11 Practice Diagnosis

You are working in a brake and front-end shop and encounter the following problems:

CASE 1: A 10-year-old car was brought in with a complaint of brake drag; the right front brake does not release completely. This problem was easy to confirm because the wheel was very hot to touch. Could this be a hydraulic system problem? If so, how can you determine this? After that check, what should you do next to locate the exact cause?

CASE 2: The delivery van is driven through hilly areas, and the driver is complaining of occasional brake pedal fade. It gets bad enough that he can barely stop, and then he has to let the van sit for a while for the pedal to come back. What is happening? What should you do to find the cause? What can you do that would probably help this vehicle?

CASE 3: A car is 15 years old and has 140,000 mi. on the odometer. The complaint is a very low brake pedal. On checking the master cylinder, you find one of the reservoirs empty and the other overflowing. What is probably wrong? What should you do to fix it? What should you do or at least recommend before returning this car to the customer (besides collecting the money for the repair)?

CASE 4: A 5-year-old car is brought in with a complaint of a grabby right rear brake. The customer says he doesn't know what it can be because his neighbor helped him replace the brake shoes about 1,000 mi. ago. When you pull the brake drum you find fairly new lining that is wet with brake fluid from a wheel cylinder leak. What went wrong? What should you do to correct this problem?

Terms to Know

back bleed	flushing	pressure bleeding
bench bleeding	gravity bleeding	reverse bleed
bleeding	manual bleeding	vacuum bleeding
cross bleed		

Review Questions

1. Two technicians are discussing brake service.
   Technician A says that air in the hydraulic system will cause a low but firm brake pedal.
   Technician B says that air is removed by bleeding the brakes.
   Who is correct?
   a. A only   c. Both A and B
   b. B only   d. Neither A nor B

2. During service, a master cylinder can be cleaned using:
   a. Gasoline   c. Denatured alcohol
   b. Solvent   d. Any of the above

3. Technician A says that using an open-end wrench to loosen a tube nut can ruin the nut.
   Technician B says that during replacement, it is always a good idea to thread the nut into the connection several turns with your fingers.
   Who is correct?
   a. A only   c. Both A and B
   b. B only   d. Neither A nor B

4. Technician A says that when you are rebuilding an aluminum wheel cylinder or master cylinder you should be very careful as you hone out the bore if it is corroded.
   Technician B says that the rubber cups can be damaged if a cylinder bore is too large.
   Who is correct?
   a. A only
   b. B only
   c. Both A and B
   d. Neither A nor B

5. The bore size of a hydraulic brake cylinder is usually checked using a(n):
   a. Piston from the cylinder and a feeler gauge
   b. Inside micrometer
   c. Special plug gauge
   d. Feeler gauge and a rubber cup

6. If a secondary piston stop bolt is used in a master cylinder, it will enter the bore from:
   a. The bottom
   b. Inside the reservoir
   c. The side
   d. Any of the above

7. Technician A says that removal of a stuck master cylinder piston requires the use of a special puller.
   Technician B says that a residual check valve can be removed from a tandem master cylinder with the aid of a self-tapping screw.
   Who is correct?
   a. A only
   b. B only
   c. Both A and B
   d. Neither A nor B

8. A plastic reservoir is removed from the master cylinder body by:
   a. Removing the hold-down bolts
   b. Removing the retainer pin and prying it off
   c. Using air pressure to carefully blow it off
   d. Any of the above depending on the master cylinder

9. Technician A says that the lips of all the sealing cups in a master cylinder face away from the pushrod.
   Technician B says that the rubber cups should be lubricated as a master cylinder is being assembled.
   Who is correct?
   a. A only
   b. B only
   c. Both A and B
   d. Neither A nor B

10. Technician A says that it is a wise practice to always bench bleed a master cylinder before installing it.
    Technician B says that brake pedal free travel should always be checked after installing a master cylinder.
    Who is correct?
    a. A only
    b. B only
    c. Both A and B
    d. Neither A nor B

11. A wheel cylinder can be rebuilt while it is mounted on the backing plate unless there are:
    a. Piston stops
    b. Deep pits in the cylinder bore
    c. Raised areas on the backing plate that prevent dust boot removal
    d. Any of the above

12. The first step in rebuilding a wheel cylinder should be to:
    a. Remove it from the backing plate and hone the bore
    b. Loosen the bleeder screw
    c. Remove the boots, cups, and pistons
    d. Any of the above

13. Technician A says that a steel caliper piston must be replaced if the chrome plating is flaking or coming loose.
    Technician B says that a cracked phenolic piston must never be reused.
    Who is correct?
    a. A only
    b. B only
    c. Both A and B
    d. Neither A nor B

14. A caliper piston can be removed from a caliper bore with the aid of:
    a. Shop air pressure
    b. The car's hydraulic brake pressure
    c. A special puller
    d. Any of the above

15. Technician A says that a major difference between the right and left calipers is the location of the bleeder screw.
    Technician B says that a caliper with a bleeder screw broken off in it must be replaced.
    Who is correct?
    a. A only
    b. B only
    c. Both A and B
    d. Neither A nor B

16. Technician A says that after a brake job is completed, the brakes should be bled to remove all the air from the hydraulic fluid.
    Technician B says that all the old brake fluid should be bled out of the system.
    Who is correct?
    a. A only
    b. B only
    c. Both A and B
    d. Neither A nor B

17. During brake-bleeding operations, fluid flow is generated by:
    a. Pressure from a pressure bleeder
    b. Movement of the brake pedal
    c. A vacuum at the bleeder screw
    d. Any of the above

18. Technician A says that the normal bleeding sequence of most cars built in the 1970s is right rear, left front, left rear, right front.
    Technician B says that this is the correct sequence for some cars with a diagonal split hydraulic system.
    Who is correct?
    a. A only
    b. B only
    c. Both A and B
    d. Neither A nor B

19. Technician A says that the master cylinder can be damaged by improper pedal pumping during manual bleeding operations.
    Technician B says that a bleeder adapter should not be clamped onto the top of a master cylinder with a plastic reservoir.
    Who is correct?
    a. A only
    b. B only
    c. Both A and B
    d. Neither A nor B

20. Technician A says that the only way to solve a spongy brake problem is to continue bleeding until the pedal becomes firm.
    Technician B says that it is normal on all modern systems for a small spurt of fluid to emit from a caliper bleeder screw as it is being loosened, after the brake pedal has been applied and released.
    Who is correct?
    a. A only
    b. B only
    c. Both A and B
    d. Neither A nor B

25 Power Booster Service

Objectives

Upon completion and review of this chapter, you should be able to:

- ❑ Test power booster operation using the brake pedal.
- ❑ Check the vacuum supply using a vacuum gauge.
- ❑ Examine a vacuum booster and check valve for leaks and proper operation.
- ❑ Check a hydro-boost system for leaks and proper operation.
- ❑ Check a Powermaster system for leaks and proper operation.
- ❑ Remove and replace a power booster.
- ❑ Disassemble and reassemble a power booster.
- ❑ Perform the ASE tasks for a power assist unit diagnosis and repair (see Appendix A).

25.1 Introduction

In most cases, power booster service consists of inspection and replacement of the vacuum hose or check valve and removal and replacement of the booster. A boost—vacuum, hydraulic (hydro-boost), or electrohydraulic (Powermaster)—can be dismantled and repaired. In the average repair shop, many technicians prefer to install a new or rebuilt replacement unit because of the special tools, skills, and parts that are required when rebuilding a unit.

Most booster service operations are undertaken because of customer complaints concerning poor or unsafe brake operation. By following an inspection procedure or a troubleshooting guide (both given in Chapter 20) or by referring to past experience, the technician will be led to the probability of faulty booster operation. In some cases booster faults are easy to see. For example, power steering fluid leaking from a hydro-boost unit is an obvious indication that repair or replacement is necessary. For other problems, such as a faulty control valve, other checks should be made to confirm that the booster is faulty (Figure 25–1 and Figure 25–2).

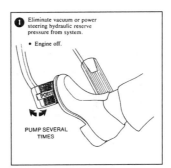

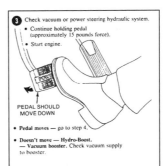
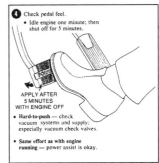

Figure 25–1 There are two basic tests for power system operation. Steps (1) and (2) ensure the booster operates properly, and steps (3) and (4) ensure it has enough reserve to function with the engine off. *(Courtesy of Ford Motor Company)*

517

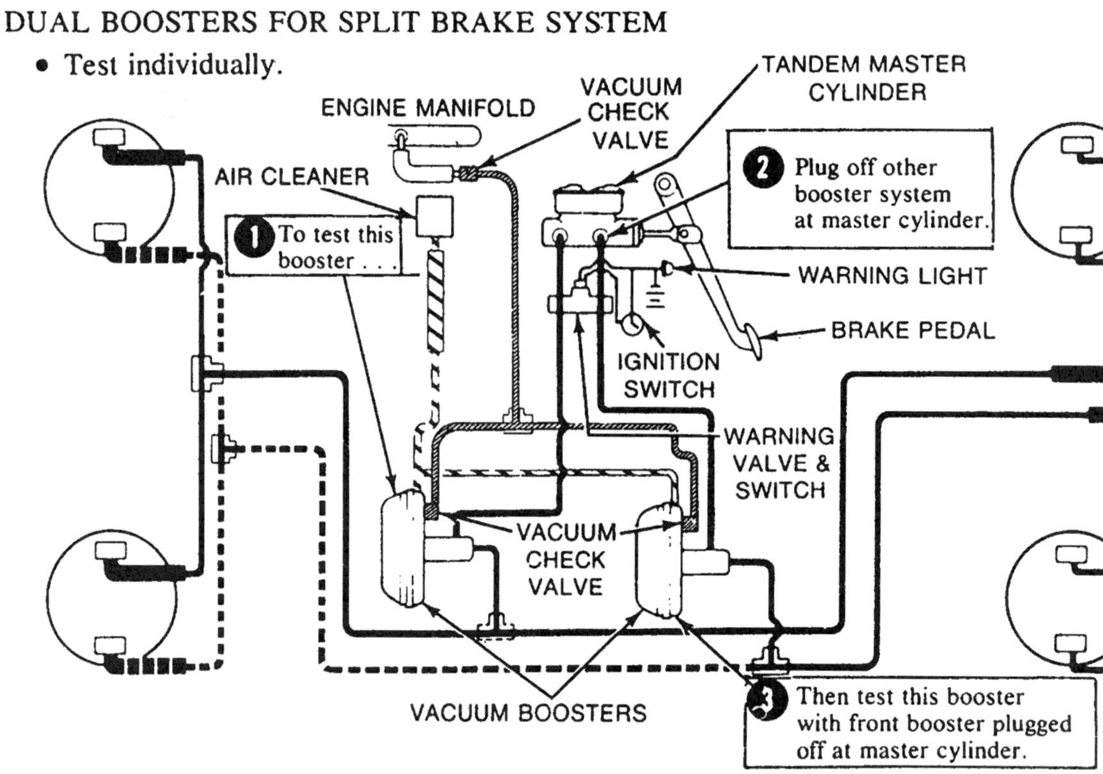

Figure 25-2 A remote booster is tested in the same manner as an integral booster. Some light-truck systems use dual boosters, and these units are checked separately. *(Courtesy of Ford Motor Company)*

Occasionally a faulty booster can cause brake drag. Brake drag causes rapid lining wear, especially with disc brakes. A quick check for booster-caused drag is as follows: Raise both front wheels and (with the engine off) operate the brake pedal enough times to use up the vacuum or hydraulic pressure reserve. Rotate the front wheels to check them for drag. Next start the engine and recheck for drag. An increase in the amount of drag indicates that the booster is trying to apply the brakes. Check to make sure there is free travel at the brake pedal. If there is no free travel, adjust it; if there is free travel, the booster should be replaced or rebuilt.

On Powermaster units, the booster and master cylinder sections are serviced as a single unit.

25.2 Vacuum Supply Tests

A vacuum booster cannot develop full power without a good, strong supply of vacuum. This vacuum can be checked using a vacuum gauge connected to the intake manifold as close to the booster connection as possible (Figure 25-3). With the engine running at a fast idle, there should be about 17 to 20 in. Hg of vacuum, with a minimum of 14 in. Hg of vacuum. This reading will change as the throttle is opened and closed. With the vacuum supply clamped off, a vacuum loss greater than 1 in. Hg in 15 seconds indicates a faulty check valve and internal booster leak. Some technicians prefer to take vacuum readings while the car is being driven at cruising speeds. If the vacuum supply readings are low, check for loose or missing manifold vacuum hoses or an engine that needs a tune-up or needs to be rebuilt.

The booster vacuum supply hose should be inspected visually for cracks or breaks. It can be checked for restrictions by disconnecting it from the booster with the engine running. A substantial flow of air into the hose should occur, and a vacuum leak this large will cause most engines to stall. A faulty hose should be replaced. Also check the inside of the hose to see if it is wet. Signs of brake fluid indicate a faulty secondary cup (primary piston) in the master cylinder. If the hose is wet from engine oil, a faulty booster vacuum check valve is indicated. A faulty check valve can allow crankcase fumes—passing through the positive crankcase ventilation (PCV) valve—to enter the booster after engine shutoff. These fumes can ruin the rubber valve seals and diaphragm in the booster.

The vacuum check valve can be checked by prying it out of the grommet in the booster and trying to blow through it (Figure 25-4). You should be able to blow through it in the booster-toward-manifold direction but

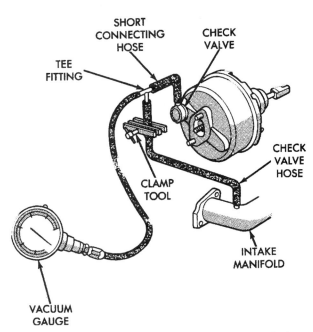

Figure 25–3 A vacuum gauge can be used to check the vacuum supply to the booster; by adding the clamp, you can check whether it holds the vacuum. A drop of more than 1 in. Hg in 15 seconds indicates a leak. *(Courtesy of Chrysler Corporation)*

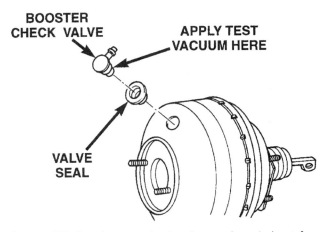

Figure 25–4 A booster check valve can be pried out for replacement of the valve or seal/grommet or to check the valve. Wet the valve with water to make removal or installation easier. *(Courtesy of Chrysler Corporation)*

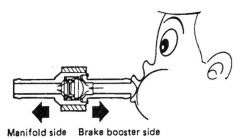

Figure 25–5 A check valve can be tested by trying to blow through it in both directions. It should allow flow only toward the manifold. *(Courtesy of Nissan North America, Inc.)*

Figure 25–6 This simple tool to test for a booster leak is made by fastening two pieces of 3/8-in. tubing to the lid of a small container. Note that one of the tubes extends almost to the bottom of the container.

not in the opposite direction (Figure 25–5). A faulty check valve or grommet should be replaced. It is a good practice to wet the grommet and check valve with water for lubricant as they are being installed.

Internal air leaks in the booster can prevent proper booster operation. A booster with internal leaks should be replaced or rebuilt. An **internal leakage check,** also called an airtightness test, is made by running the engine for 1 or 2 minutes and then shutting it off. The brake is then applied with normal pedal pressure and held for at least 30 seconds. If the pedal remains steady, the booster is probably good. If the pedal slowly rises, the booster has an internal leak.

Internal booster leaks can also be checked by fabricating a simple tool, as shown in Figure 25–6. Choose a glass jar that has a tight-fitting lid and attach two 3/8-in. (9.5-mm) OD tubes to the lid. One tube should be long enough to reach almost to the bottom of the jar, and the other should end just inside the lid. Fill the jar about half full with water and connect the short tube to the booster supply hose from the intake manifold. Then use a short piece of hose to connect the long tube to the booster inlet. Start the engine and watch for air bubbles in the jar as you operate the brake pedal a few times. Airflow from the booster to the manifold will cause air bubbles to pass through the water in the jar. As the brake pedal is released, there should be a large flow of bubbles, but after a moment this flow should stop. With the brake pedal released or applied, there should be no flow. A constant flow of bubbles with the brake applied indicates a leaky piston diaphragm. A constant flow with the brakes released indicates a leaky control valve (Figure 25–7).

A vacuum booster can also be checked for internal leakage with a vacuum pump. Remove the vacuum hose from the check valve and connect the vacuum pump directly to the check valve or inlet fitting (Figure 25–8). With the pedal released, you should be able to draw a 17- to 20-in. Hg vacuum, and this vacuum reading should hold steady for several minutes. A leak in the control valve or booster chamber is indicated if the vacuum drops. Next, apply the brakes with moderate pressure on the pedal. An immediate drop in the reading should occur as the pedal moves. Draw the vacuum back to 17 to 20 in. Hg and observe the reading to make sure it does not drop more than 2 in. Hg in the next 30 seconds. A leaky diaphragm, control valve, or vacuum chamber is indicated if the reading does drop.

25.3 Booster Replacement

A faulty booster can be removed for rebuilding or replacement with a new or rebuilt unit. Booster replacement can be a relatively easy operation depending on the location of the bolts and nuts securing the booster to the car's bulkhead (Figure 25–9).

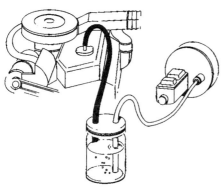

Figure 25–7 The leak tester is partially filled with water and connected between the booster and intake manifold; then the engine is started. A constant stream of air bubbles with the pedal released indicates a leaky valve in the booster, and a stream of bubbles with the pedal applied indicates a leaky booster diaphragm.

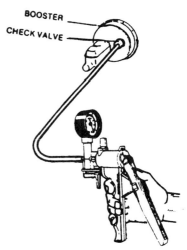

Figure 25–8 If a vacuum pump is connected to a booster, it should be possible to draw a 17- to 20-in. Hg vacuum, and this vacuum should hold steady. Repeating this check with the pedal applied should give the same readings. *(Courtesy of Mityvac/Prism)*

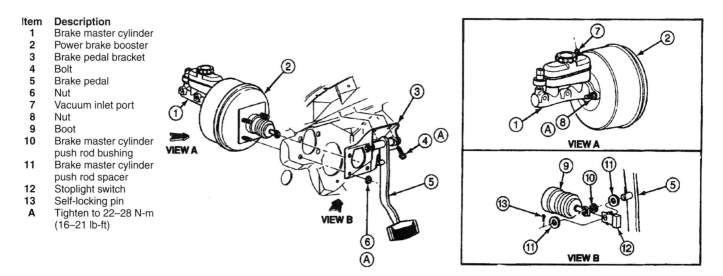

Item	Description
1	Brake master cylinder
2	Power brake booster
3	Brake pedal bracket
4	Bolt
5	Brake pedal
6	Nut
7	Vacuum inlet port
8	Nut
9	Boot
10	Brake master cylinder push rod bushing
11	Brake master cylinder push rod spacer
12	Stoplight switch
13	Self-locking pin
A	Tighten to 22–28 N-m (16–21 lb-ft)

Figure 25–9 A vacuum booster can be removed by disconnecting the pushrod (B) and the mounting nuts (6). Usually the master cylinder is disconnected (A) and moved aside or removed. *(Courtesy of Ford Motor Company)*

NOTE: On some vehicles, the brake pedal will drop and activate the brake lights when the booster is removed, and the extended operation of the center brake light can overheat the rear window or operation of all of the lights can discharge a battery. Check the light operation after booster removal; if the lights are lit, prop up the pedal to turn them off.

To remove a power booster:

1. On hydro-boost and Powermaster units, first remove the pressure in the accumulator. With the engine off, apply and release the brake several times until you feel no change in the hardness of the pedal. On Powermaster units, ten to twenty pedal applications are recommended.
2. Disconnect the master cylinder from the booster. On many cars, the brake lines are long and flexible enough to allow the master cylinder to be moved away from the booster without disconnecting them. If not, remove the master cylinder as described in Chapter 24.

NOTE: On some boosters, the master cylinder pushrod will come free. Test this by trying to pull it out; if you can pull it out, tape the pushrod onto the front of the windshield to keep it from becoming lost or forgotten.

On Powermaster units, the brake lines must be disconnected because the booster and the master cylinder are combined (Figure 25–10).

3. Disconnect the booster power supply:
    a. On a vacuum booster, disconnect the vacuum line.
    b. On a hydro-boost unit, disconnect the power steering pump and steering gear fluid lines (Figure 25–11).
    c. On a Powermaster unit, disconnect the electric connectors.
4. Disconnect the pushrod from the brake pedal.
5. Remove the nuts or bolts securing the booster to the bulkhead and remove the booster.

Booster installation is essentially the reverse of the removal procedure. During installation, check the master cylinder **pushrod adjustment.** A pushrod that is too long can position the primary piston too deep in the cylinder bore. The primary cup will then cover the compensating port and cause brake drag. A short pushrod can cause a low brake pedal. Three different procedures can be used to check this adjustment: the gauge method, the air method, and the fluid swirl method. Many boosters use an adjustable pushrod that can be lengthened or shortened as necessary (Figure 25–12).

The gauge method is the quickest. A gauge can be easily shop-made from cardboard or thin sheet metal if the dimensions are available. It is normally a two-step, "go–no go" gauge. It is placed in position with the pushrod touching the short step of the gauge but not the long one (Figure 25–13).

For the fluid swirl method, after the booster and master cylinder are mounted, have an assistant apply the

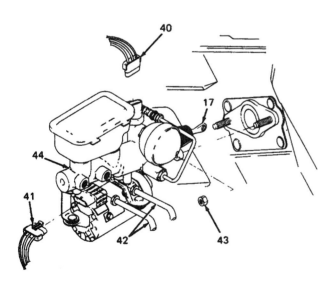

17. PUSHROD
40. ELECTRICAL CONNECTOR
41. ELECTRICAL CONNECTOR
42. BRAKE PIPE
43. NUT
44. POWERMASTER UNIT

Figure 25–10 A Powermaster unit can be removed after disconnecting the pushrod, mounting nuts, and electrical and hydraulic connections. *(Courtesy of General Motors Corporation, Service Technology Group)*

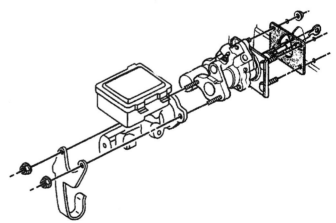

Figure 25–11 A hydro-boost unit is removed by disconnecting the master cylinder and pushrod and removing the mounting nuts. *(Courtesy of General Motors Corporation, Service Technology Group)*

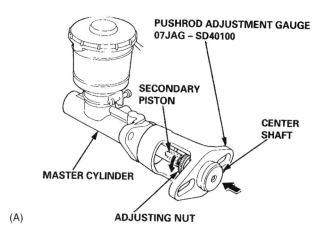

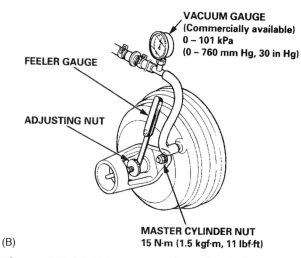

Figure 25-12 This gauge is adjusted to fit the master cylinder piston (A) and then placed onto the booster to check the pushrod adjustment (B). Note that the engine is running during step (B). *(Courtesy of Ford Motor Company)*

brake pedal as you observe the fluid in the primary reservoir section. The reservoir should be about half full of fluid, and a clean plastic film should be placed over the reservoir to contain the fluid. A slight swirl should be seen in the fluid as the brakes are applied and also during release. No swirl on application indicates that the pushrod is too long. A larger-than-normal swirl on application indicates that the pushrod is too short. If you suspect the pushrod is too long, loosen the master cylinder mounting bolts about 1/4 in. and repeat this test. If a normal swirl is now present, the pushrod is too long.

With the air method, the master cylinder is mounted on the booster, and the brake line is disconnected from the primary outlet port. Then clean, low-pressure, compressed air is blown into the outlet port. If the air passes through the compensating port and into the reservoir, the pushrod is not too long. Note that this method requires bleeding of the primary section of the master cylinder.

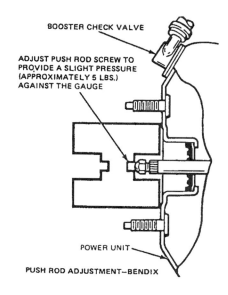

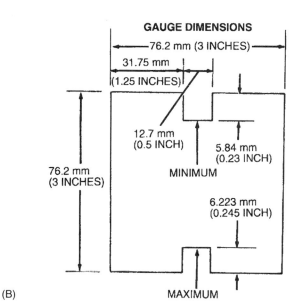

Figure 25-13 The booster pushrod can be checked using a gauge (A) to see if it is the correct length. Some manufacturers provide dimensions for making a gauge (B). *(Courtesy of American Honda Motor Co., Inc.)*

To replace a booster:

1. Place the booster in position, install the nuts or bolts securing it, and tighten them to the correct torque.

NOTE: A Powermaster unit must be bench bled in the master cylinder section before installation.

2. Reconnect the pushrod to the brake pedal.

3. Remount the master cylinder on the booster and tighten the nuts or bolts to the correct torque.

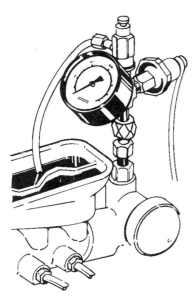

Figure 25-14 A pressure gauge has been installed on this Powermaster unit to measure pump output pressure. Abnormal pressure can indicate a faulty pump, switch, accumulator, or booster. *(Courtesy of General Motors Corporation, Service Technology Group)*

4. Connect the booster's power supply. *On vacuum boosters,* reconnect the vacuum hose. *On hydroboost units,* reconnect the lines to the power steering pump and gear, tighten them to the correct torque, and bleed the air out of the lines. To bleed the lines:

   a. Fill the power steering pump reservoir.
   b. Crank the engine for several seconds but do not start it.
   c. Check the power steering reservoir and refill it as necessary.
   d. Start the engine and slowly turn the steering wheel from stop to stop twice. Do not hold the steering wheel against the stops.
   e. Stop the engine and apply and release the brake pedal several times to bleed the pressure out of the accumulator.
   f. Check the reservoir and refill it if necessary.

   On Powermaster units, reconnect the electric connector, fill all three reservoir sections with brake fluid to the correct level, and turn on the ignition. The pump should run and shut off within 20 seconds. If it is still running after 20 seconds, shut off the ignition and refer to the diagnosis chart for the solution to this problem. Powermaster pumping pressures are tested using a pressure gauge, as shown in Figure 25–14.

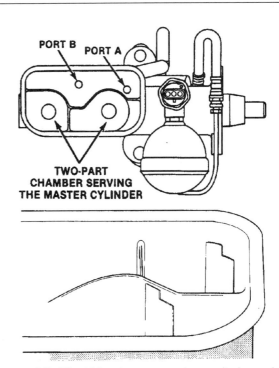

Figure 25-15 With the accumulator discharged, the Powermaster fluid level should be at the step in the reservoir baffle. *(Courtesy of General Motors Corporation, Service Technology Group)*

5. Check the master cylinder reservoir and refill it if necessary. Start the engine and check the booster operation. Note that if the master cylinder has been completely removed, it will be necessary to bleed the brake hydraulic system.

Note that on Powermaster units, the pump section of the reservoir will have the correct fluid level only when the accumulator is discharged. Filling the reservoir after the accumulator is charged will cause an overflow and spillage when the accumulator discharges (Figure 25–15).

25.4 Booster Service

As mentioned earlier, boosters can be rebuilt in the repair shop if the proper tools and equipment, service information, and replacement parts are available. Most service shops prefer to install a new or rebuilt unit. In most cases, the special tooling required is rather minor, and the larger tools can be shop-made. Repair information is available in some factory service manuals, technician service manuals, and repair manuals published by some of the major aftermarket brake part manufacturers. These manuals provide a step-by-step procedure for making the necessary repairs and adjustments. Repair parts can be obtained from some new-car dealerships and parts houses.

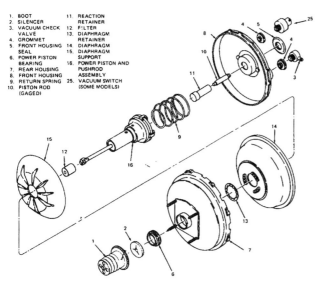

Figure 25–16 An exploded view of a vacuum booster. Note the tabs on the front housing (8) and the sockets in the rear housing (7) with which the tabs interlock. *(Courtesy of General Motors Corporation, Service Technology Group)*

25.4.1 Vacuum Booster Repair

Most vacuum boosters are held together by interlocking tabs between the front and rear housings, and they can be separated by twisting one housing relative to the other (Figure 25–16). Separation is made difficult because of the drag of the rubber diaphragm that is clamped between the two sections. Another potential problem is caused by the return spring, which tries to spread the two housing sections apart. Repair of a booster includes diaphragm replacement, control valve cleaning or replacement, and air cleaner or silencer cleaning or replacement. During reassembly, the parts should be properly lubricated and the pushrod adjustment checked.

The following method is very general. Be sure to follow the exact procedure recommended by the manufacturer. To service a vacuum booster:

1. Remove the vacuum check valve and scribe a mark on the front and rear housing so you can properly align them during reassembly.

2. Place the booster in the disassembly fixture with the booster mounting studs entering the holes in the fixture and the long bar or wrench over the master cylinder mounting studs.

   Adjust the fixture to apply a slight downward pressure and twist the upper housing to unlock the two housings. Adjust the fixture to allow the return spring to extend and separate the parts of the booster (Figure 25–17).

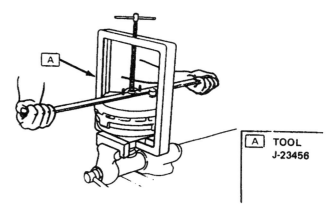

Figure 25–17 This booster is attached to a fixture while the tool is used to rotate the front housing and unlock the tabs. The clamp portion keeps the internal spring compressed during this operation. *(Courtesy of General Motors Corporation, Service Technology Group)*

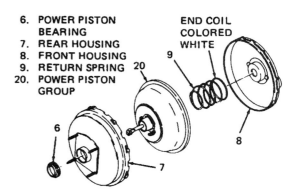

Figure 25–18 A partially disassembled vacuum booster. *(Courtesy of General Motors Corporation, Service Technology Group)*

3. Remove the power piston bearing, return spring, and piston group from the housing (Figure 25–18).

4. Disassemble the piston and valve assembly to remove the diaphragm (Figure 25–19).

5. Disassemble the valve assembly (Figure 25–20).

6. Clean the various parts with alcohol and air dry them. Rust deposits can be cleaned from the inside of the housings with fine sandpaper. Inspect the parts for damage. There should be no cuts or tears in any of the rubber components.

7. Lubricate the valve parts as required and reassemble them.

8. Lubricate the components as required and reassemble the piston and valve assembly.

Chapter 25 ■ Power Booster Service

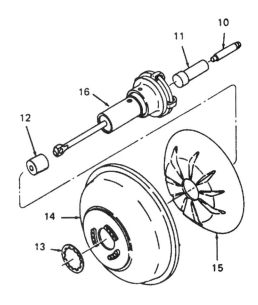

10. PISTON ROD
11. REACTION RETAINER
12. FILTER
13. DIAPHRAGM RETAINER
14. DIAPHRAGM
15. DIAPHRAGM SUPPORT
16. POWER PISTON AND PUSHROD ASSEMBLY

Figure 25–19 This power piston group has been disassembled, allowing diaphragm replacement. *(Courtesy of General Motors Corporation, Service Technology Group)*

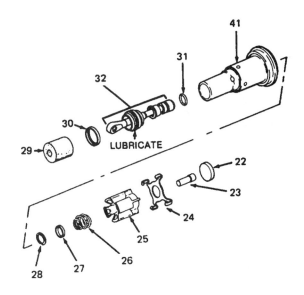

22. REACTION DISC
23. REACTION PISTON
24. REACTION BODY RETAINER
25. REACTION BODY
26. AIR VALVE SPRING
27. REACTION BUMPER
28. RETAINING RING
29. FILTER
30. RETAINER
31. O-RING
32. AIR VALVE PUSHROD ASSEMBLY
41. POWER PISTON

Figure 25–20 A disassembled vacuum booster valve assembly. *(Courtesy of General Motors Corporation, Service Technology Group)*

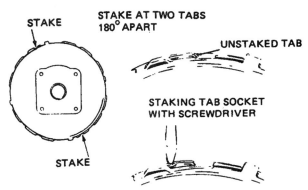

Figure 25–21 After a vacuum booster is reassembled, two tab sockets should be restaked to lock the chambers together. *(Courtesy of General Motors Corporation, Service Technology Group)*

9. Place the piston assembly and return spring in the housings, aligning the scribed marks. Then place them in the assembly fixture and adjust the fixture to hold the two halves together.

10. Twist the housings so the tabs interlock and stake two housing tabs to lock the assembly (Figure 25–21).

11. Check the pushrod adjustment.

25.4.2 Hydro-boost Repair

Normal hydro-boost service includes disassembly, cleaning and inspection, and replacement of the inner seals (Figure 25–22). It is recommended that a special seal installer be used when installing the seals on the input rod. Before disassembling a hydro-boost unit, make sure that the accumulator pressure has been released. The following method is very general. Be sure to follow the exact procedure recommended by the manufacturer (Figure 25–23).

To service a hydro-boost unit:

1. Secure the unit in a vise.

2. Remove the spool valve retainer, plug, and spool valve spring. On some units, remove the spool valve sleeve (Figure 25–24).

3. Remove the power piston spring retainer if present. *On hydro-boost II units,* remove the power piston spring.

4. Remove the return line fitting.

5. *On hydro-boost I units,* remove the output rod retainer, spring, and output rod.

6. Remove the housing-to-cover bolts and separate the parts of the unit, being careful not to drop the spool valve.

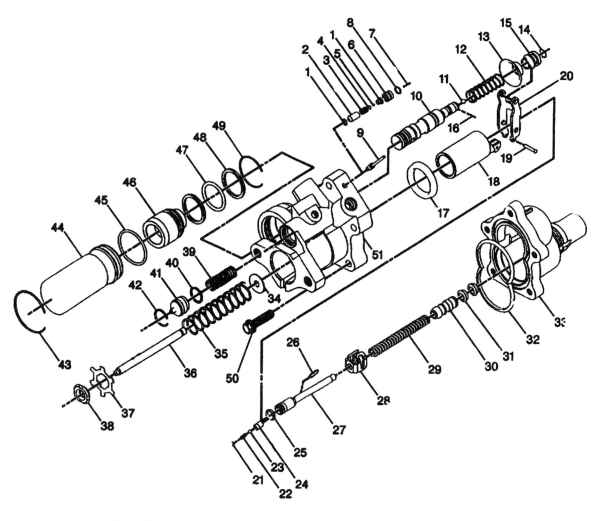

1. Seat, Check Valve
2. Body, Check Valve
3. Spring, Check Valve Relief
4. Washer, Check Valve
5. Ball, Check
6. Insert, Body
7. Plunger
8. O-ring
9. Plug, Housing
10. Valve, Spool
11. Ball, Spool Valve Check
12. Spring, Sleeve
13. Actuator
14. Ring, External Retainer
15. Sleeve, Spool
16. Pin, Spool Valve
17. Seal, Piston
18. Piston
19. Pin, Input Lever
20. Lever, Input
21. Spring, Retainer
22. Spring, Relief Valve
23. Ball, Relief Valve Check
24. Seat, Relief Valve
25. Ring, Input Rod
26. Rod and Plunger
27. Rod, Input
28. Bracket, Input Rod
29. Spring, Input Rod
30. End, Input Rod
31. Seals, Input Rod
32. Seal, Housing/Cover
33. Cover, Booster
34. Retainer, Output Push Rod
35. Spring, Piston Return
36. Push Rod, Output
37. Spring, Piston Retainer
38. Spring, Baffle Retainer
39. Spring, Spool
40. O-ring, Spool Plug
41. Plug, Spool
42. Ring, Spool Plug Retaining
43. Ring, Accumulator Retainer
44. Accumulator
45. Ring, Accumulator
46. Piston, Accumulator
47. O-ring, Accumulator Piston
48. Rings, Accumulator Piston
49. Ring, Retainer
50. Bolts, Housing/Cover
51. Housing, Booster

Figure 25–22 An exploded view of a hydro-boost assembly. *(Courtesy of General Motors Corporation, Service Technology Group)*

Chapter 25 ■ Power Booster Service

A. INPUT SEAL LEAK — Fluid leakage from housing cover end of booster near reaction bore. Replace input assembly kit.

B. POWER PISTON/ACCUMULATOR SEAL LEAK — Fluid leakage from vent at front of unit near master cylinder. Replace power piston/accumulator seal kit.

C. HOUSING — Fluid leakage between the housing and housing cover. Replace housing seal kit.

D. SPOOL VALVE SEAL — Fluid leakage near plug area. Replace spool plug seal kit.

E. RETURN PORT FITTING SEAL — Replace "O" Ring seal.

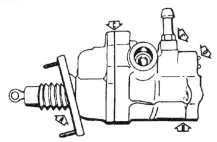

Figure 25–23 Areas to check for fluid leaks on a hydro-boost unit. *(Courtesy of General Motors Corporation, Service Technology Group)*

7. Remove the housing-to-cover seal, detach the spool valve, and remove the input rod and seal.

   a. *On hydro-boost II units,* remove the power piston and accumulator assembly.

   b. *On hydro-boost I units,* install a suitable clamp, remove the accumulator retainer, and loosen the clamp to remove the accumulator (Figure 25–25).

CAUTION *Accumulators contain a strong spring or pressurized gas and can cause injury.*

8. Clean all the parts using denatured alcohol or power steering fluid. Do not use automatic transmission fluid. Inspect the parts for damage, paying special attention to the spool valve and bore. If you locate any scratches deep enough to feel with a fingernail, the unit should be replaced. During reassembly, all the parts should be lubricated with power steering fluid.

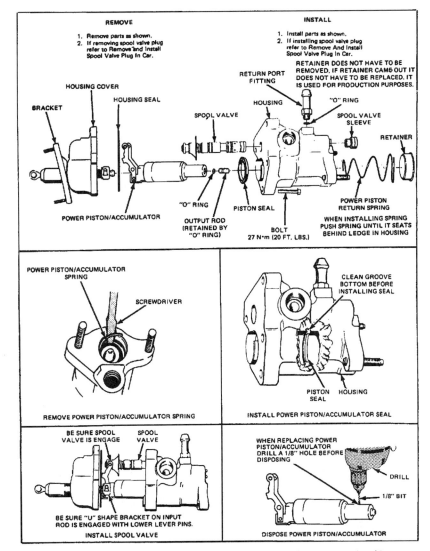

Figure 25–24 The procedure for disassembling and reassembling a hydro-boost II unit. *(Courtesy of General Motors Corporation, Service Technology Group)*

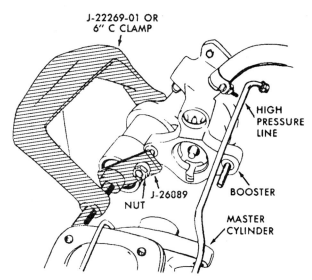

Figure 25–25 This clamp is placed over the accumulator to hold it compressed while removing the retainer. *(Courtesy of General Motors Corporation, Service Technology Group)*

9. Install the return line fitting and O-ring and tighten them to the correct torque.
10. *On hydro-boost I units,* install the accumulator, O-ring, and accumulator seal. Then check the valve assembly and the dump valve assembly.
11. Install the input rod (with new seals) into the cover (Figure 25–26).
12. Install the power piston seal, housing seal, and power piston with the spool valve attached in the housing.
13. Install the housing-to-cover bolts and tighten them to the correct torque.
14. Install the spool valve spring, spool valve, and retainer.

25.4.3 Powermaster Repair

Powermaster service includes replacement of the pressure switch, accumulator, pump assembly, or return hose and overhaul of the unit. A unit overhaul requires

Figure 25–26 The tapered tool (left) is being used to guide the input rod seals into the proper position. *(Courtesy of Bendix Brakes, by AlliedSignal)*

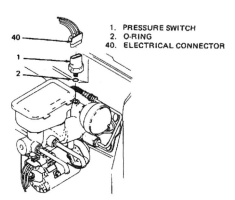

Figure 25–27 The pressure switch can be removed and replaced on a Powermaster unit with the unit still in the car. *(Courtesy of General Motors Corporation, Service Technology Group)*

removal of the assembly from the car. The other repairs can be made with the Powermaster unit on the car. Before any of these operations are performed, the accumulator must be depressurized by applying and releasing the brake pedal at least ten times. Do not turn on the ignition because the pump will run and recharge the accumulator.

The on-car operations are essentially removal and replacement of the components. It is very important that new O-rings or grommets be used where necessary and that they be tightened to the correct torque during reassembly (Figure 25–27, Figure 25–28, and Figure 25–29).

A unit overhaul requires disassembly, cleaning, inspection, and reassembly. The procedure is very similar to the overhauling of a master cylinder with additional components. It is very important to note the relationship and position of these additional parts (Figure 25–30). The procedure recommended by the manufacturer should be followed.

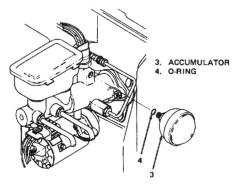

Figure 25–28 A Powermaster accumulator can be removed and replaced. *(Courtesy of General Motors Corporation, Service Technology Group)*

Chapter 25 ■ Power Booster Service

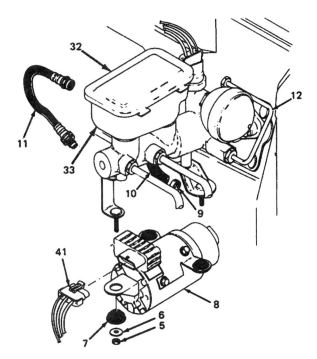

5. NUT
6. WASHER
7. GROMMET
8. E-H PUMP
9. HOSE CLAMP
10. SUMP HOSE
11. PRESSURE HOSE ASSEMBLY
12. TUBE AND NUT ASSEMBLY
32. RESERVOIR COVER AND DIAPHRAGM
33. RESERVOIR
41. ELECTRICAL CONNECTOR

Figure 25-29 The pump and pressure hose can be removed and replaced while the unit is mounted in the car. *(Courtesy of General Motors Corporation, Service Technology Group)*

25.5 Practice Diagnosis

You are working in a brake and front-end shop and encounter the following problems:

CASE 1: A customer is at a meeting in your city; he lives in another town. He is having a problem of brake drag on his 1994 Camry, and he mentions that he had to replace the vacuum brake booster the day before he left his home. You lift the car on a rack to check for drag, and you find that the wheels turn freely until the engine is started. What is probably wrong? What will you need to do to fix this problem?

CASE 2: While inspecting the brake system of a 5-year-old Dodge Caravan, you find the brake pedal very hard. When you start the engine, the brake pedal drops, indicating a good booster operation; however, when you check the pedal a minute after turning the engine off, you find the pedal rock hard again. What is wrong? What should you do next?

CASE 3: A customer complains of a very hard brake pedal on his 10-year-old Chevrolet pickup. Your road test confirms the problem: you have to push very hard on the pedal of this hydro-boost–equipped vehicle. What could be wrong? What should you do next?

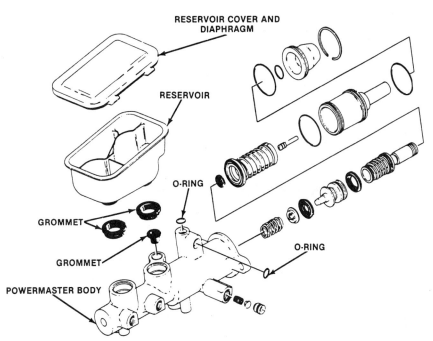

Figure 25-30 An exploded view of a Powermaster unit showing the relationship of the internal parts. *(Courtesy of General Motors Corporation, Service Technology Group)*

Terms to Know

internal leakage check

pushrod adjustment

Review Questions

1. The standard check for power booster operation is to pump the brake pedal until the reserve is exhausted and then start the engine with the brake applied. As the engine starts, the:
   a. Brake pedal should rise slightly
   b. Brake pedal should fall a noticeable amount
   c. Brake warning light should come on and then off
   d. Booster should make a clicking sound

2. Two technicians are discussing power booster checks.
   Technician A says that you should be able to blow through a vacuum check valve in one direction only.
   Technician B says that a faulty check valve can let oil fumes into the booster, which can cause damage.
   Who is correct?
   a. A only   c. Both A and B
   b. B only   d. Neither A nor B

3. Technician A says that an airtight booster will provide an assisted brake application after the engine has been shut off.
   Technician B says that the engine should speed up slightly as the brakes are applied.
   Who is correct?
   a. A only   c. Both A and B
   b. B only   d. Neither A nor B

4. If the output pushrod on a power booster is too long, the result can be:
   a. An overly sensitive brake application
   b. A very low brake pedal
   c. Brake drag, especially after a few stops
   d. All of the above

5. Air is bled from the hydraulic section of a hydro-boost unit by:
   a. Starting the engine and slowly turning the steering wheel from stop to stop and then applying the brake pedal with the engine off
   b. Starting the engine and opening the bleeder screws on the booster
   c. Pumping the pedal slowly until it gets hard
   d. Applying the brake pedal while the engine is started

6. Technician A says that a vacuum booster can be disassembled by turning the master cylinder end of the booster while the other housing is held stationary.
   Technician B says that you should be careful while doing this because fluid can spray out.
   Who is correct?
   a. A only   c. Both A and B
   b. B only   d. Neither A nor B

7. A faulty hydro-boost unit will:
   a. Leak
   b. Not provide an assist
   c. Cause chatter on application
   d. Any of the above

8. Technician A says that a charged accumulator can spray fluid all over as a hydro-boost unit is being disassembled.
   Technician B says that the spring in the accumulator of some units can cause injury as the booster is being disassembled.
   Who is correct?
   a. A only   c. Both A and B
   b. B only   d. Neither A nor B

9. Technician A says that the master cylinder and the booster sections of a Powermaster unit are completely separate.
   Technician B says that the fluid level of all three reservoir sections of a Powermaster unit should be checked with the engine running.
   Who is correct?
   a. A only   c. Both A and B
   b. B only   d. Neither A nor B

26 Parking Brake Service

Objectives

Upon completion and review of this chapter, you should be able to:

- ❏ Check a parking brake system; inspect the cables and parts for wear, rusting, or corrosion; clean or replace parts as necessary; and lubricate the assembly as needed.
- ❏ Adjust the parking brake assembly and check for correct operation.
- ❏ Test the parking brake indicator light, switch, and wiring and adjust the switch as necessary.
- ❏ Perform the ASE tasks relating to parking brakes (see Appendix A).

26.1 Introduction

Normal parking brake service includes inspection and adjustment of the parking brake cable and an occasional adjustment of the warning light switch (Figure 26–1). Sometimes this service includes replacement of one or more faulty cables (the front control cable, the rear application cable, or both), adjustment or replacement of the lining or shoes on a separate parking brake assembly, or replacement of a faulty control assembly (Figure 26–2).

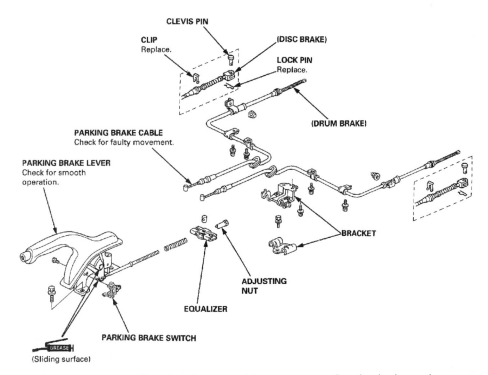

Figure 26–1 A parking brake system with a hand-operated lever; areas to be checked are shown. *(Courtesy of American Honda Motor Co., Inc.)*

531

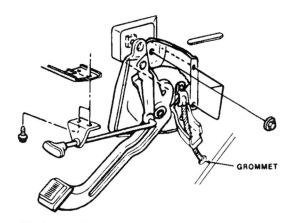

Figure 26–2 A pedal-operated parking brake control assembly. Foot pressure on the pedal pulls the cable to apply the brakes, and pulling the control handle releases the ratchet mechanism to release them. *(Courtesy of Chrysler Corporation)*

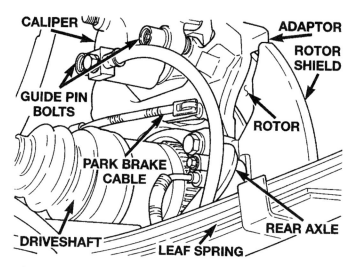

Figure 26–4 This four-wheel disc system uses a drum brake for the parking brake. Note the parking brake cable separate from the caliper. *(Courtesy of Chrysler Corporation)*

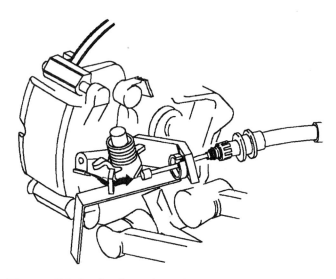

Figure 26–3 This four-wheel disc brake system uses a mechanical operation of the caliper for the parking brake. Note that the cable is being connected as the caliper is replaced. *(© Saturn Corporation, used with permission)*

Most parking brakes are mounted on the rear wheels and fall into one of three categories: drum brakes with an integral parking brake mechanism, disc brakes with an integral parking brake mechanism (Figure 26–3), or disc brakes with a separate parking brake mechanism (Figure 26–4). In the past, some cars were equipped with drum brakes with a separate parking brake mechanism mounted at the transmission output shaft. Many trucks use a separate auxiliary parking brake.

In most cases, a parking brake uses the shoes, lining, and drum or rotor of the service brake, and these portions are repaired as a brake job is done. This procedure was described in Chapters 22 and 23. In a few cases, a parking brake has its own set of shoes, and their servicing is a separate operation. Because a parking brake normally encounters only static friction, there is very little wear. Lining replacement or even adjustment of separate parking brake assemblies is seldom done. The actual replacement of shoes on these units follows a procedure very similar to that for a drum brake assembly. This operation is described in the service manual for each particular car (Figure 26–5).

26.2 Cable Adjustment

A parking brake is properly adjusted if it applies completely in less than one-half the travel distance of the lever (about five clicks) and releases completely with no drag when the lever is released (Figure 26–6). Usually an adjustment is made by lengthening or shortening the cable attachment at the equalizer assembly. The major purpose of an adjustment is to compensate for cable stretch or wear of the linkage at the various contact points. Lining clearance is adjusted by the normal operation of most modern brake assemblies.

Some vehicles are equipped with an automatic self-adjusting mechanism to maintain correct adjustment of the parking brake cable. This mechanism exerts a pull on the parking brake cable while it is released; a ratchet mechanism allows normal operation when the parking brake is applied. The automatic adjuster mechanism must be locked up with a pin when the brake control or cables are disconnected (Figure 26–7).

The service brake lining should have the correct clearance when the cable adjustment is made. In most cases, clearance can be checked by applying the service brake. A high brake pedal indicates correct lining clearance. If the

Chapter 26 ■ Parking Brake Service

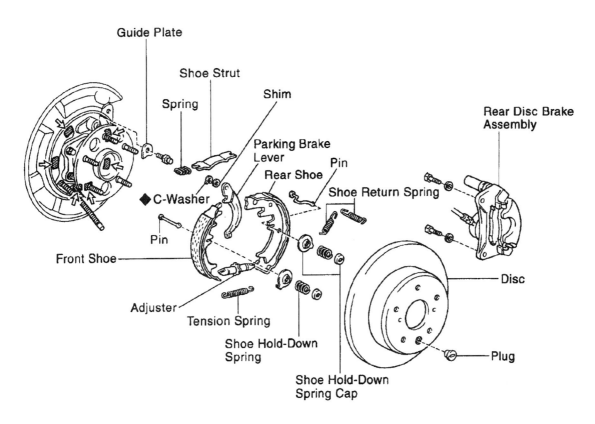

Figure 26–5 An exploded view of a drum parking brake–disc service brake combination. Parking brake shoe clearance is adjusted manually.

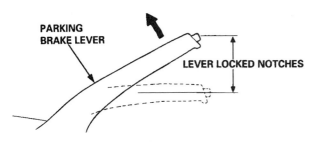

Figure 26–6 The parking brake lever should become locked when it travels four to eight notches (rear drum brakes) or seven to eleven notches (rear disc brakes). *(Courtesy of American Honda Motor Co., Inc.)*

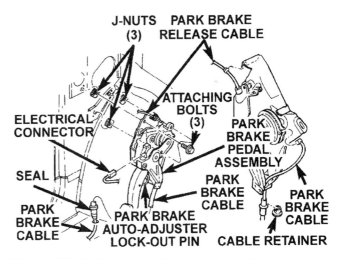

Figure 26–7 The park brake auto-adjuster lock-out pin was installed in this parking brake pedal assembly to hold the cable tensioning spring from unwinding. This should be done to allow safe disconnection of a parking brake cable or removal of the pedal mechanism. *(Courtesy of Chrysler Corporation)*

pedal is low on a car with rear-wheel drum brakes, the brake shoe adjustment should be checked as described in Chapter 22. If the car has rear-wheel disc brakes with an integral parking brake mechanism, apply and release the parking brake about five times, using a firm application pressure, and then recheck the brake pedal height. If it has increased, continue applying and releasing the parking brake until there is no further improvement. If the pedal height does not increase, have an assistant tap sharply on the caliper with a hammer while the brakes are being applied with a firm pressure on the pedal. If the pedal height still does not improve, check the self-adjuster mechanism in the rear calipers as described in Chapter 23.

If the car has rear-wheel disc brakes with a separate parking brake, the brake shoe clearance should be adjusted before changing the cable length. This operation is very similar to adjusting the brake shoe clearance on a non-self-adjusting duoservo brake and is described in various service manuals (Figure 26–8).

On some cars, a front cable moves an **intermediate lever;** it is also called an **equalizer lever** or **ratio bar.** This front cable should be adjusted first if it is adjustable. A front cable adjustment is also done to compensate for cable stretch and linkage wear (Figure 26–9). It ensures complete movement of the intermediate lever without excessive play.

To adjust parking brake linkage:

1. Inspect the cable to make sure that it can move freely in the guides and housings, has no excess wear at equalizer or other contact points, and shows no signs of broken strands (Figure 26–10).

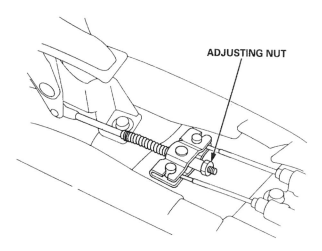

Figure 26–9 Turning this adjusting nut tightens the cable and reduces lever travel. *(Courtesy of American Honda Motor Co., Inc.)*

Replace any defective parts and lubricate the cable as necessary. On disc brake calipers with an integral parking brake mechanism, check for proper positioning of the parking brake lever at the caliper, the cable and housing, and the lever return spring if required.

2. Apply and release the service brake several times using a firm pedal pressure.
3. Apply and release the parking brake several times using a firm pressure on the lever.
4. Apply the parking brake lever to the first notch.
5. Adjust the cable length at the equalizer to remove all slack (Figure 26–11). On some cars, this adjustment should be made at the parking brake lever. A slight side pull on the cable should just start to cause shoe drag when the wheel is turned.
6. Release the parking brake lever and rotate the wheel. There should be no sign of drag caused by the parking brake linkage (Figure 26–12).
7. If a second jam or check nut is used on the adjuster, make sure that it is correctly tightened.

26.3 Cable Replacement

Occasionally it is necessary to replace a parking brake cable. Essentially this is a simple job of removing and replacing the cable and housing. On some cars, a single cable runs from one rear brake, through the equalizer, to the other rear brake. On other cars, separate cables are used that are joined by a connector or attached individually to the equalizer or control lever assembly. Cable replacement is easier in the second case (Figure 26–13).

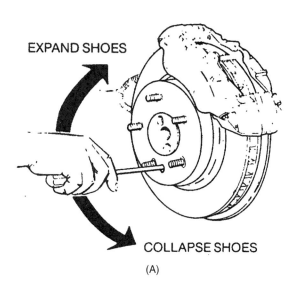
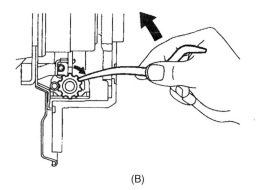

Figure 26–8 A screwdriver is being used to turn the adjuster nut to adjust the lining clearance on an older Corvette parking brake (A); a brake spoon is used to adjust the lining clearance on a Camry (B). *(A is courtesy of Stainless Steel Brakes Corp.)*

Chapter 26 ■ Parking Brake Service

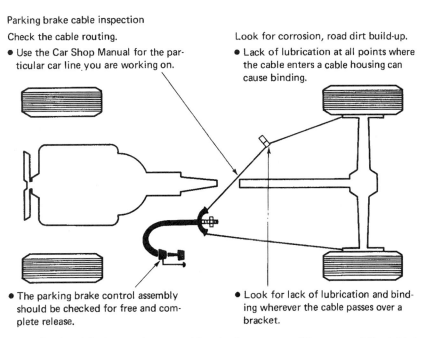

Figure 26-10 The items to check while inspecting a parking brake system. *(Courtesy of Ford Motor Company)*

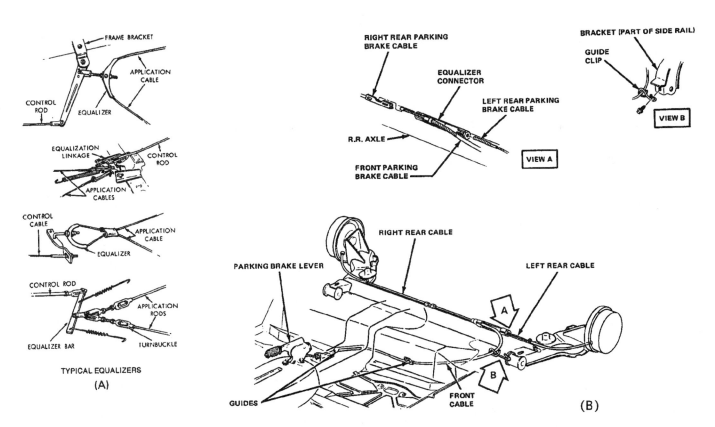

Figure 26-11 In many cars, the equalizer is at the connection between the front control cable or rod and the rear application cable (A). Some cars connect the control cable to one rear brake and the housing to the other one (B). *(A is reprinted from Mitchell Anti-Lock Brake Systems, with permission of Mitchell Repair Information, LLC; B is courtesy of General Motors Corporation, Service Technology Group)*

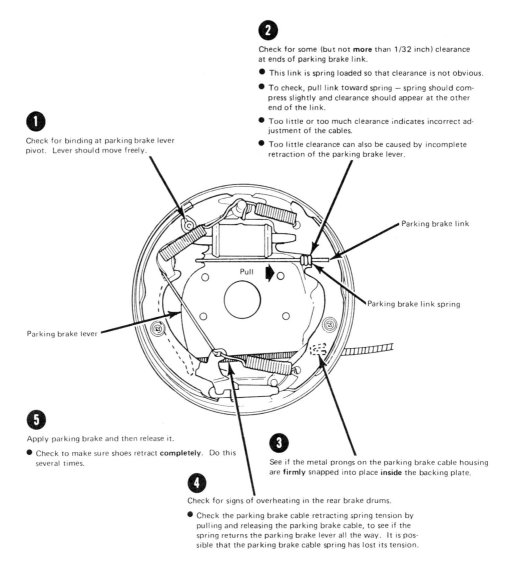

Figure 26-12 The inspection procedure for parking brake drag problems. *(Courtesy of Ford Motor Company)*

To remove a parking brake cable:

1. Raise and support the car in a secure manner.
2. Disconnect the cable from the equalizer.
3. Remove any clips that secure the cable housing to frame or body brackets.
4. *On cars with disc brakes with an integral parking brake,* disconnect the cable from the caliper at the lever (Figure 26-14).

On cars with drum brakes, follow this procedure:

a. Remove the wheel and brake drum.

b. Disengage the cable end from the parking brake lever. Side- or diagonal-cutting pliers can be used to compress the spring and grip the cable to hold the spring in a compressed position during this operation (Figure 26-15).

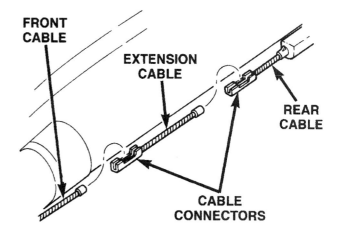

Figure 26-13 The cable connector allows cables to be connected or disconnected for replacement. *(Courtesy of Chrysler Corporation)*

Chapter 26 ■ Parking Brake Service

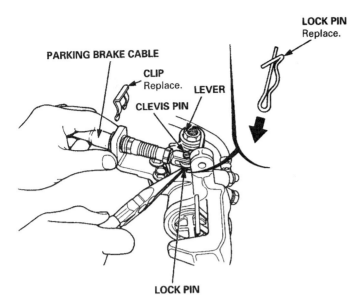

Figure 26-14 Parking brake cable housings are often attached to the cable bracket by a U-shaped clip; this cable is attached to the caliper lever by a lock pin. *(Courtesy of American Honda Motor Co., Inc.)*

Figure 26-15 A pair of side- or diagonal-cutting pliers over the application cable. Holding the parking brake lever and moving the pliers and spring to the right compress the spring; squeezing the pliers holds the spring compressed while the lever is disconnected.

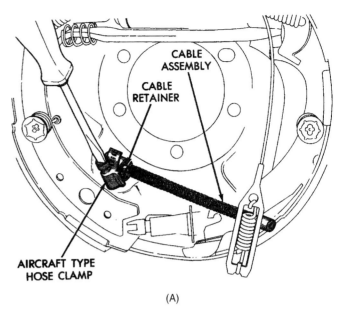

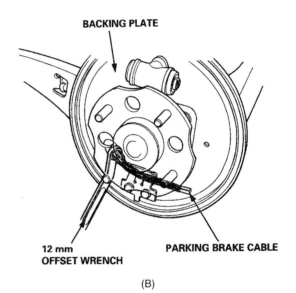

Figure 26-16 The prongs must be compressed using a hose clamp (A) or box-end wrench (B) so the parking brake cable can be removed from the backing plate. *(A is courtesy of Chrysler Corporation; B is courtesy of American Honda Motor Co., Inc.)*

    c. Depress the cable retainer prongs using a small hose clamp and slide the cable and housing through the backing plate (Figure 26–16).

To replace a brake cable:

1. *On cars with drum brakes,* follow this procedure:
   a. Push the cable and housing into the backing plate, making sure that the prongs on the retainer are completely spread and locked in place (Figure 26–17).
   b. Slide the cable end outward or compress the spring. Then connect the cable end to the parking brake lever.
   c. Install the brake drum and wheel.

   On cars with disc brakes, connect the cable end to the caliper lever.

2. Route the cable and housing through the brackets or guides and connect the cable to the equalizer.

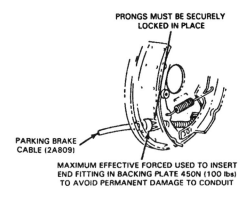

Figure 26–17 When installing a cable housing into a backing plate, make sure the prongs expand to lock it in place. *(Courtesy of Ford Motor Company)*

3. Replace any clips that were removed.
4. Adjust the parking brake cable.
5. Check the application and release of the parking brake.

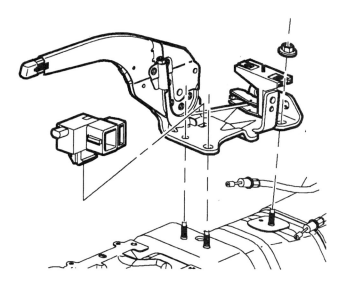

Figure 26–18 The warning light switch is usually mounted next to the parking brake lever; this one is removed by squeezing the retaining tabs. (© *Saturn Corporation, used with permission*)

26.4 Parking Brake Warning Light Service

On most cars, the parking brake warning light circuit is controlled by an additional switch and is in a parallel circuit with the brake warning light switch on the pressure differential valve. Both these switches use the same warning lightbulb. In some cars, the parking brake warning light uses a separate bulb and switch. The switch for the parking brake warning light is mounted at the control lever. It is closed when the parking brake is applied and opened when the lever is released.

NOTE: Remember that parking brake drag can cause enough heat to boil the fluid in the wheel cylinders; in a diagonal split system, this can cause total brake failure. Be sure that the warning light comes on when the parking brake is applied.

System faults usually result from a defective bulb, wire connection, or switch or from an incorrect switch adjustment. In many cases, switch adjustment is controlled by the mounting bracket, and the bracket must be bent to change the adjustment (Figure 26–18).

When troubleshooting circuit electrical problems, it is advisable to check a wiring diagram to determine how the wires are connected in the circuit (Figure 26–19). In most cases, the closing of the switch completes the ground circuit for the warning lightbulb. In such cases, a trouble light or voltmeter can be used to make certain that there is voltage at the switch. If there

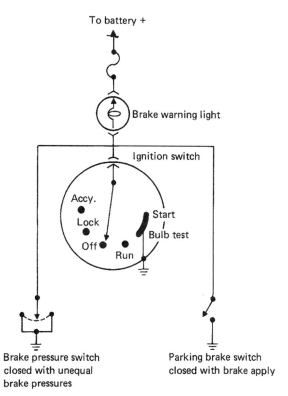

Figure 26–19 A typical parking brake warning light circuit. Note that the switch is closed to make a path to ground and turn on the light when the brake lever is applied; the brake pressure switch is in a parallel circuit.

is no voltage, the circuit is open. There is a break in the wiring, a bad connection, or a burned-out bulb or fuse. If there is voltage at the switch, there is probably a faulty switch or switch adjustment or a faulty ground connection.

26.5 Practice Diagnosis

You are working in a brake and front-end shop and encounter the following problems:

CASE 1: A customer has brought in her 10-year-old 4×4 Nissan pickup (V6 engine and automatic transmission) with a problem of an overheating right rear brake. The wheel is hot and smells of very hot lining. After pulling the drum, you find the lining glazed, with light cracks down the center; the shoe metal is slightly blued, and there is no clearance at the parking brake strut. What went wrong? What should you check next? What will you probably have to do to fix this problem?

CASE 2: A Corvette owner has brought in his car to have the parking brake adjusted; the brake handle goes full travel without meeting resistance. What is the probable cause of the parking brake getting this far out of adjustment? What check should you make as you adjust the cable? What other check should you recommend to the customer?

CASE 3: While inspecting the brakes on a 7-year-old Olds Cutlass, you notice that the brake light does not come on as you apply the parking brake. When you crank the engine, the light does come on. What is probably wrong? What should you do to check this?

Terms to Know

equalizer lever

intermediate lever

ratio bar

Review Questions

1. Two technicians are discussing parking brake service.
   Technician A says that a parking brake can use the same lining and friction surfaces as the service brake.
   Technician B says that a parking brake must be applied by hydraulic pressure.
   Who is correct?
   a. A only
   b. B only
   c. Both A and B
   d. Neither A nor B

2. The parking brake lever on a certain car travels almost its full distance when applied.
   Technician A says that the service brake lining might need adjustment.
   Technician B says that the parking brake cable needs adjustment.
   Who is correct?
   a. A only
   b. B only
   c. Both A and B
   d. Neither A nor B

3. A domestic car with four-wheel disc brakes has an excessive amount of travel at the parking brake lever during application. To correct this problem, you should:
   a. Loosen the rear cable at the equalizer
   b. Tap the caliper body with a hammer as the brakes are applied
   c. Use the park gear in the automatic transmission
   d. None of the above will correct the problem

4. Parking brake cable adjustments are normally made at the point where the:
   a. Rear cable is attached to the control cable
   b. Rear cable is attached to the parking brake lever
   c. Front cable is attached to the parking brake pedal
   d. Strut is attached to the application cable

5. Technician A says that a parking brake must be capable of stopping a car in 45 ft from a speed of 20 mph.
   Technician B says that a parking brake must be able to lock up the two wheels to the limit of traction.
   Who is correct?
   a. A only
   b. B only
   c. Both A and B
   d. Neither A nor B

6. Technician A says that a parking brake should apply completely in less than one-half of the available travel distance of the lever.
Technician B says that a correctly adjusted parking brake cable will cause a slight amount of lining drag when released.
Who is correct?
   a. A only
   b. B only
   c. Both A and B
   d. Neither A nor B

7. On most cars, looseness of the rear cable is adjusted at the:
   a. Equalizer
   b. Control assembly
   c. Brake shoes
   d. Backing plate

8. On some cars with four-wheel disc brakes, in order to completely adjust the parking brake it is sometimes necessary to adjust the:
   a. Slack in the rear cables
   b. Lining-to-drum clearance at the brake shoes
   c. Lining-to-rotor clearance at the brake shoes
   d. Both a and b

9. The parking brake warning light on a certain car fails to work.
Technician A says that this could be caused by a faulty switch adjustment at the application lever.
Technician B says that you can sometimes determine the cause of the problem by cranking the engine.
Who is correct?
   a. A only
   b. B only
   c. Both A and B
   d. Neither A nor B

10. A parking brake cable should be replaced if it:
    a. Begins to fray
    b. Is frozen in the housing
    c. Has a broken end button
    d. Any of the above

27 Antilock Braking System Service

Objectives

Upon completion and review of this chapter, you should be able to:

- ❏ Check the operation of an antilock braking system.
- ❏ Diagnose improper antilock braking system operation.
- ❏ Check the operation of the system sensors, actuators, control module, connectors, and wiring.
- ❏ Remove, replace, and adjust faulty components in an antilock braking system.
- ❏ Perform the ASE tasks relating to antilock braking system diagnosis and repair (see Appendix A).

27.1 Introduction

Normal antilock braking system (ABS) service includes diagnosing, locating, and correcting problems that might occur while driving or stopping the car. Since the ABS components serve as controlling devices that prevent wheel lockup, there are essentially no wearing points that require periodic service or maintenance. When troubleshooting brake problems, remember that an ABS is a controlling system only (Figure 27–1). Problems such as drag, pull, a low

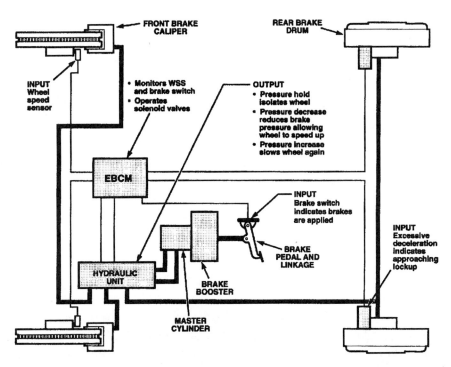

Figure 27–1 ABS adds the monitoring and control features to the base brake system. *(Courtesy of General Motors Corporation, Service Technology Group)*

pedal, and unusual noises are probably caused by defects in the basic brake hydraulic system or wheel assemblies. A thorough inspection, like that described in Chapter 20, and proper service usually locate and correct these faults.

When diagnosing ABS problems, remember that tire diameter and tread condition are important because they affect the rotating speed of wheels. If the motorist has changed the tire size significantly, this can fool the electronic brake control module (EBCM) into thinking there is a fault in the system. This is an easy item to check because the original tire size is usually on a sticker in the glove compartment or on the door jamb.

ABS uses either a vacuum or an electrohydraulic power booster, and some of the electrohydraulic systems use the hydraulic pressure from the pump to operate the rear brakes. With these systems, pump or motor failure results in failure of the rear brakes. The hydraulic modulator on these units is usually integral with the master cylinder. On most systems, a pair of hydraulic lines connect the master cylinder to the **hydraulic modulator,** which is also called a *hydraulic control unit, ABS actuator,* or *modulator unit.* Most modulator assemblies are not serviceable internally, even though they can be disassembled (Figure 27–2). Some manufacturers allow partial disassembly to replace subcomponents and provide repair parts; these procedures are described in their service manuals. Replacement ABS components and parts are available from new-car dealerships; aftermarket suppliers also have new and rebuilt parts available.

Because ABS is essentially an electronic control system with speed sensor inputs and electric solenoids or motors as outputs, most problems are electrical in nature. The other major parts are in the hydraulic modulator. Faulty electrical connectors are the most commonly encountered problem.

Some guidelines should be observed when testing computer controls. It is important not to touch the connector pins of the EBCM. An electrostatic charge can be generated in your body as you move around a car (Figure 27–3), especially over the interior carpet and seat covers. If this high-voltage charge is discharged into the EBCM, it can damage it. Some technicians connect a ground wire between themselves and the body of the car or spray the carpet and seat covers with an antistatic laundry spray to help prevent this type of damage. As mentioned earlier, it is also important not to use test devices, which could cause increased current flow through the computer system circuits. If electric arc welding is done on the car, the EBCM should be disconnected to prevent possible damage from a voltage surge. Also, if the car is to be painted and the paint cured in an oven at high temperatures, the EBCM can be damaged by exposure to the high temperatures.

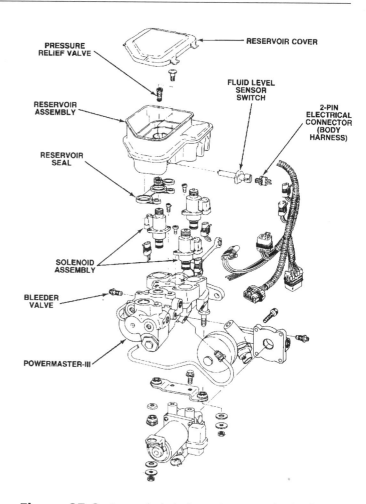

Figure 27–2 An exploded view of an ABS hydraulic unit containing the master cylinder, booster, and solenoid valves. *(Courtesy of General Motors Corporation, Service Technology Group)*

Extra care should be exercised when repairing ABS components to prevent unusual or unexpected occurrences. Imagine what could happen if the wheel speed sensor wire on a RWAL system came loose during a stop with a loaded pickup or van. The EBCM would think the rear wheels were locked up and would then reduce the rear braking pressure. This reduction in braking power could lead to tragic consequences.

ABS units are quite complex, and there are some definite differences in the operation of the various systems. Many systems have variations, such as traction control. In a 10-year period, one domestic manufacturer used eleven different systems, and another one had eight different systems. And some of these systems had variations between different vehicle models. Some systems are unique to one or two car models, and others, such as RWAL, are very similar on several makes of pickups. An

Chapter 27 ■ Antilock Braking System Service

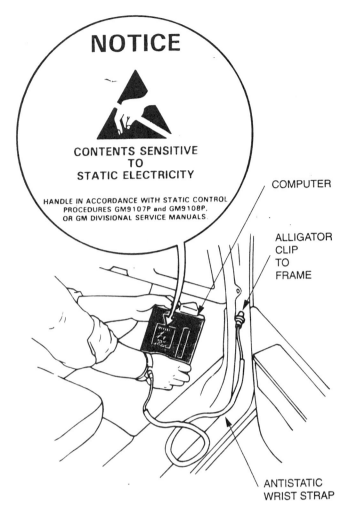

Figure 27–3 This warning symbol is used to identify components or circuits that can be damaged by electrostatic discharge (ESD). Special precautions, such as not touching the components unless you are grounded, should be followed when you are working with them. (*Courtesy of General Motors Corporation, Service Technology Group*)

adequate service manual must be followed while diagnosing or repairing the faults in these systems. One new-car manufacturer has an eighty-page section of the service manual for one car model devoted to ABS problem diagnosis and service. It is also a good idea to familiarize yourself with the normal system operation. These units self-check each time the car is started. In some systems, if you apply the brake with a firm pedal during starts, you should feel a bump on the pedal and hear clicking or chunking sounds as the EBCM checks out the various components. Another design produces the brief whining noise of electric pump operation after the car reaches a certain speed, about 4 mph. On a road test, if the brake is applied with a firm-to-hard pedal, one or several chirps should be heard from the rear tires as they

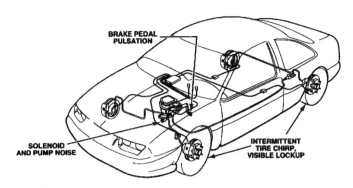

Figure 27–4 These conditions are normal in ABS-equipped vehicles, although they might appear to be problems. (*Courtesy of General Motors Corporation, Service Technology Group*)

try to lock up. At this time, pulsations or bumps on the brake pedal are a normal sign as the ABS goes to work (Figure 27–4). While doing this test, be sure that there is no traffic following closely behind.

The following typically encountered problems can be attributed to faulty ABS components: poor tracking during hard stops that involve ABS operation, a spongy brake pedal, and improper warning light operation. A spongy pedal is corrected by bleeding the system using the correct procedure for that particular system. Some ABS units require special bleeding procedures, and these are described in service manuals.

A problem that is becoming more common as ABS units are getting older is leaking modulator valves. Modulator valve leaks are caused by debris between the valve and seat. The source of this debris has been attributed to dirty brake fluid, formation of copper crystals from chemical activity in the fluid, crud pushed up from a caliper as the piston is depressed, and material left in the system from manufacturing processes. It is highly recommended that the brake fluid be changed every 2 or 3 years to reduce the likelihood of this problem.

When working on an ABS, a service manual for that particular car should be followed and these precautions should be observed:

- Make sure the system is depressurized before opening any hydraulic circuit. Depressurization is done by pumping the brake pedal an adequate number of times as instructed by the manufacturer of that system.
- Follow the manufacturer's directions on the use of the proper equipment when bleeding a system.
- Use only specially designed brake hoses and lines.
- Use only the recommended brake fluid; do not use silicone brake fluid in an ABS.

- Make sure the ignition is turned off before disconnecting or reconnecting any ABS electrical connectors to prevent damage to the EBCM.
- Do not touch any of the connectors to an EBCM with your fingers or with a meter probe unless directed to by a service manual; then only do so while following the directions given in the manual.
- Disconnect the EBCM and any other onboard computers while using an electric welder on the car.
- If installing a transmitting device such as a telephone or citizens band (CB) radio, make sure the electrical connections and antenna do not interfere with the ABS.
- Do not hammer or tap on speed sensors or sensor rings; they can be demagnetized, which can affect their signal accuracy.
- Use only anticorrosion coatings on speed sensors; do not contaminate them with grease.
- Always check the sensor air gap when either a sensor or sensor ring has been replaced on a rotor, axle, or CV joint.
- Tighten wheel lug nuts to the correct torque; overtightening can distort a rotor or drum and affect speed sensor signals.
- If replacing the tires, do not mix the sizes; usually the diameters of all four tires must be equal and the same as the original tires.
- Do not subject the EBCM to excessive heat.

27.2 ABS Warning Light Operation

ABS systems use two brake warning lights, a **red brake warning light** and an **amber antilock warning light.** The function of the red brake warning light is similar to that in vehicles without ABS. The amber warning light warns of problems in the hydraulic and electrical portions of the ABS controls.

On some systems, the amber light glows whenever the brake fluid pressure or reservoir level is low. On all systems, the amber warning light comes on when the EBCM receives or sends out signals that are not in the correct electrical range. If this light is on while the car is being driven, it indicates that the ABS has shut itself off and that the car's brakes will operate using non-ABS braking. In normal operation, this light should come on for 3 to 6 seconds when the ignition is turned on, during the time that the ignition switch is in the crank or start position, and for 3 to 6 seconds after the car starts and is running. With some integral systems, if the accumulator is completely discharged, the light may stay on longer as the pump charges the accumulator.

On some systems, the red warning light comes on whenever the hydraulic pump motor is running. If the pump motor runs longer than 3 minutes on some systems, this light will flash on and off. The red warning light on most systems also comes on when the pump or accumulator pressure is low, the brake fluid level is low, the parking brake is applied, and the ignition is turned to start. The pump motor in the Teves unit, like the pump motor in the Powermaster unit, should not run longer than 30 seconds at a time. If it does, there is a problem, and the excessively long operation can cause the motor to overheat and burn out. Figure 27–5 illustrates how the ABS warning light is connected into the circuits.

Sometimes inexperienced technicians blame base brake system problems on ABS. If you are unsure of the cause of a particular problem, you can remove the ABS fuse and road test the vehicle. With the fuse pulled, ABS will turn off and the vehicle will revert to base brakes. If the problem disappears, it was ABS related; if the problem remains, it is in the base brakes.

27.3 Warning Light Sequence Test

A technician uses the improper operation of the warning lights or a diagnostic trouble code from the EBCM to determine the correct test procedure for detecting the cause of a particular problem. The use of self-diagnosis codes is described in the next section. A light sequence test is the second step in locating the cause of a particular problem. The first step is a visual inspection of the system to note any faults that can be seen, such as a loose wire connector, low fluid level, or fluid leak. Thus far, the majority of ABS problems have been caused by loose or faulty wire connections.

To perform a light sequence test, you should observe the operation of the two warning lights as you proceed through the following six steps:

1. With the ignition off for at least 15 seconds, turn the ignition to the run position. If the lights come on for 30 seconds or less, repeat this step.
2. Turn the ignition to start and start the engine.

Chapter 27 ■ Antilock Braking System Service

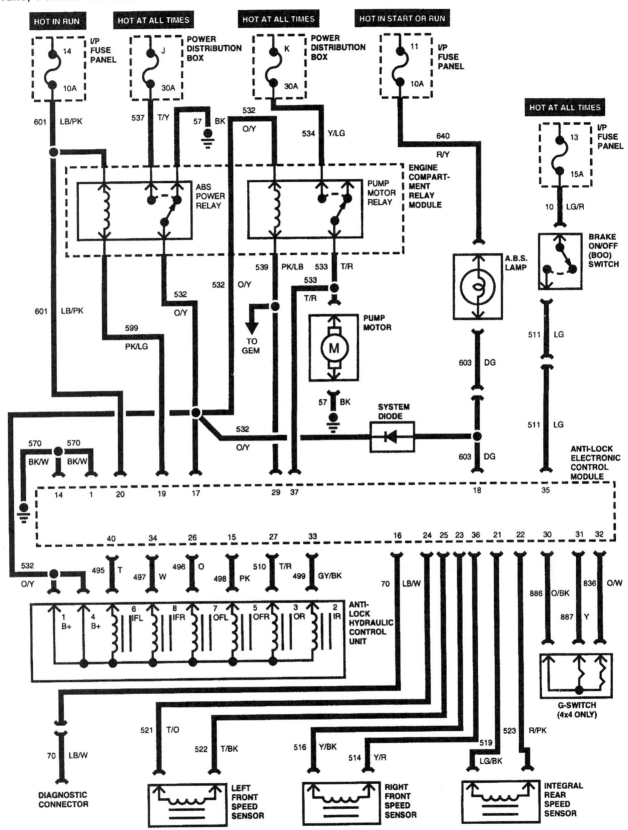

Figure 27-5 An ABS wiring diagram. Note the use of numbers to identify a wire or wire connection and the color coding to help locate that wire. *(Courtesy of Ford Motor Company)*

3. As soon as the engine starts, release the switch to the run position.
4. Drive the car at a minimum speed for a short distance.
5. Brake the car to a stop.
6. Place the transmission selector in park and let the engine idle for a few seconds.

During this time, the lights should be on or off as indicated in Figure 27–6.

Normal light operation indicates that the electrical values in both the hydraulic and the electrical portions of the brake system are normal. On integral systems using electrohydraulic boosters, the values include the hydraulic pressure in the brake lines and in the booster. On all systems, problems in the electrical circuits or components set trouble codes. If there are problems with these systems, they will probably show up in one of the following ways:

- Normal light operation along with a stopping problem (e.g., pull or drag). Check the operation of the ABS control valves and the brake assemblies.
- Pump motor runs longer than 1 minute. Check for a low fluid level, low pump pressure, or defective pump switch.
- Constant amber light and normal red light operation. Check all circuits to or from the EBCM (e.g., sensors, solenoids, pump pressure, power relays, fuses, and ground) for proper electrical values and trouble codes.
- Normal amber light operation in steps 1 and 2, constant light in steps 3 through 7, and normal red light operation. Check for an improper electrical value in all wheel speed sensor circuits.
- Normal amber light operation in steps 1 through 3, constant light in steps 4 through 6, and normal red light operation. Check for a problem in the wheel sensors (e.g., wrong sensor air gap, defective wheel bearing, or damaged sensor ring).
- Normal amber light in steps 1 through 3, intermittent light in steps 4 and 5, and normal red light operation. Check for loose electrical connectors in all circuits or a problem in the fluid level or pump pressure circuit.
- Constant amber and red light operation. Check for a problem in the pump or pump pressure circuit.
- Normal amber light and constant red light operation. Check for a low fluid level or a problem in the parking brake switch, fluid level switch, or pump pressure circuit.
- No amber or red light operation. Check for problems in the bulb circuits.

If these types of problems occur, the technician should proceed to self-diagnosis to measure the resistance or voltages of the particular circuit involved to locate the faulty circuit and the problem within that circuit. Improper sensor clearance or excessive runout, end play, or damage to the sensor ring usually causes problems that will appear when the vehicle is driven (Figure 27–7).

27.4 ABS Problem Codes and Self-Diagnosis

Most ABS systems store a **diagnostic trouble code (DTC),** also called a trouble or error code, in the memory of the EBCM. This code indicates the nature of the problem that occurred. When the EBCM determines a fault and turns the warning light on, it also stores the code in its memory (Figure 27–8).

The code can be a *soft* code, which is temporary, or a *hard* code, which is more permanent. A soft code is removed from the memory when the ignition key is cycled (i.e., turned off and on). If the problem is intermittent or temporary and goes away, the code will disappear when the key is cycled. If you road test a car, park it, and turn off the ignition, you will erase any soft codes from the road test. A hard code stays in the memory for a certain number of ignition cycles or until cleared, depending on the system. Some systems can display the number of times a problem has occurred or the number of ignition cycles since the problem has occurred.

The codes are displayed in various ways. On some vehicles with digital instrument panels, the code is displayed as the actual code number. On some other vehicles, the code is read by counting the number of flashes of the brake warning light. Many vehicles require that a test light, voltmeter, dedicated tester, or scanner be connected to the special diagnostic terminal. A dedicated tester is a unit developed by the vehicle manufacturer especially for that particular make or model of vehicle or ABS. When using a test light or analog voltmeter, the code is read by counting the number of flashes of the test light or sweeps of the meter needle.

Handheld scan tools have become popular in diagnosing faults in engine electronic control systems. These units can communicate with the EBCM and are very useful in reading ABS trouble codes. In some cases, they can

ABS Light Operation

Vehicle Status

	Step 1	Step 2	Step 3	Step 4	Step 5	Step 6
	Engine off, ignition on	Engine cranking	Engine running	Driving	Stopping	Stopped, engine running
Light status						
Red[a]		On	Off	Off	Off	Off
Amber	On 3–6 seconds	On	On 3–6 seconds	Off	Off	Off

[a] The light might come on for 30 seconds or less.

Figure 27–6 The amber and red brake warning lights should go on as the engine cranks, and the amber light should stay on for a brief period after the engine starts. Problems are indicated if either of them lights up at the wrong time. *(Courtesy of General Motors Corporation, Service Technology Group)*

Symptom	Action
Poor tracking during antilock braking	Do test 9, the wheel valve functional tests.
Spongy brake pedal	Bleed brakes. Check mounting of hydraulic unit. Check condition of calipers and rotors.
Pump motor runs longer than 1 minute, antilock light normal, flashing brake light	Check the brake fluid level. Do test 10, the hydraulic system pressure test. Do tests 2 and 5, the pressure switches test and the pump relay test. If the brake light flashes but the pump motor turns off normally in less than 1 minute, do test 7, the timer–flasher module test.
Antilock light on solid, brake light normal	Do tests 1, 2, 3, 4, 5, and 6. Start with test 1, the pin-out box test.
Antilock lights come on while moving, brake light normal	Measure the wheel speed sensor resistances and voltage in test 1, the pin-out box test. If a speed sensor voltage is missing or low, check the gap at the toothed wheel. Also check wheel bearing end play and runout.
Antilock light on solid, brake light on solid	Check brake fluid level and do test 3, the fluid level switches test. Do tests 5 and 6, the pump motor test and the pump relay test. Do test 2, the pressure switches test. If condition persists, do the hydraulic system pressure test.
Intermittent antilock light, brake light normal	Do tests 2 and 3, the pressure switches test and the fluid level switches test. Check the connectors and contacts at the wheel speed sensors.
Antilock light normal, brake light on solid	Check circuit 33 (TAN/WHT), the parking brake switch, and the ignition switch for a short to ground. Disconnect the timer–flasher module. If light goes out, replace the module. Do tests 2 and 3, the pressure switches test and the fluid level test.
No antilock light during start-up, brake light normal	Check bulb. Check for battery voltage at terminal 27 of the pin-out box, test 1. Ignition on. Repair open in circuit 852 (GRY/WHT) if no voltage. Check diode, terminals 27 and 3 for breakout box.
Antilock and brake lights on while braking	Do test 10, the hydraulic system test, to check the accumulator precharge pressure. If the pump motor runs more than a few seconds after the car sits overnight, check for external leaks. If none, replace the hydraulic unit.

Figure 27–7 A symptom and action chart can be used to locate the cause of a problem. The tests are described in the service manual for the particular vehicle. *(Courtesy of General Motors Corporation, Service Technology Group)*

be used on a road test to display live, real-time operation of the system, by displaying values such as battery voltage, wheel speed sensor output, and solenoid operation (Figure 27–9). Most dedicated testers are designed to plug directly into the diagnostic terminal. An aftermarket scan tool usually requires an adapter to make this connection (Figure 27–10). Some of the scan tools have the ability to indicate the proper or improper operation of the various switches and solenoids, the voltage at the control module and ignition switch, and the speed indicated at each of the wheel speed sensors, as well as the trouble codes.

Reading a code usually requires a special operation. In some cars, this is as simple as turning the ignition on and holding the brake pedal down for 5 seconds or longer. The procedure for one manufacturer is to connect a special shorting tool to the service check connector, turn on the key, and record the frequency and timing of light blinks from the ABS warning light. The light flashes indicate the codes, as shown in Figure 27–11. A table given in the service manual is then consulted to interpret the codes.

	None	Blink Codes	Bi-directional Scan Tool	Driver Information Center
Bosch 2U/2S	*	X	X	
Bosch III		X		X
Delco III			X	
Delco VI			X	
Kelsey-Hayes RWAL		X	X	
Kelsey-Hayes 4WAL		X	X	
Teves Mark II	*	*		*
Teves Mark IV			X	

* Depending on model year and application

Figure 27–8 Diagnostic codes are read using different methods depending on the ABS type. *(Courtesy of General Motors Corporation, Service Technology Group)*

Most vehicles require that the diagnostic terminal be connected to ground or that two terminals be connected together (Figure 27–12). Reading a code can also be accomplished using a scan tool or dedicated tester. Some systems will have as many as fifty or sixty different codes.

A chart like the one in Figure 27–13 is used to interpret the code to determine what the fault is and what further tests and repairs are necessary. It should be remembered that codes indicate faulty electrical values, which in some cases can be caused by improper mechanical operation.

Technicians normally perform tests such as those in the next section to determine the exact cause of the problem. They then repair the problem, clear the codes, and, finally, road test the vehicle to make sure the problem is corrected and the problem codes do not come back. Clearing a code to erase it from the memory also requires a special operation. As mentioned earlier, a soft code is cleared by simply turning off the ignition. In some cars, driving the car above a certain low speed or removing the diagnostic connector clears all codes. In other cars, a dedicated tester or scan tool is used to clear codes.

As a last resort, codes can be cleared by removing the power fuse to the control module or disconnecting the battery cable, but doing so erases all of the other electrical memories in the car. Some vehicles with electronic engines and transmissions have to go through a computer relearn process if the battery or that particular control module is disconnected. Drivability problems such as rough idle, hesitation or stumble, or odd shifting patterns can be the result. It will take driving through a certain number of cycles or miles for the computer to relearn the car's operation and for things to return to normal.

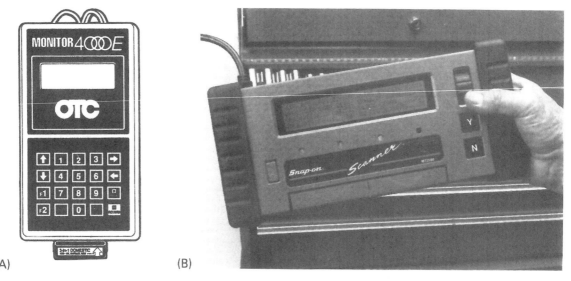

Figure 27–9 Handheld scanners can be used on some systems to read the problem codes, read ABS component operation on some systems, and clear codes on some systems. *(A is courtesy of Snap-on Tools Company, Copyright Owner; B is courtesy of SPX-OTC)*

Chapter 27 ■ Antilock Braking System Service

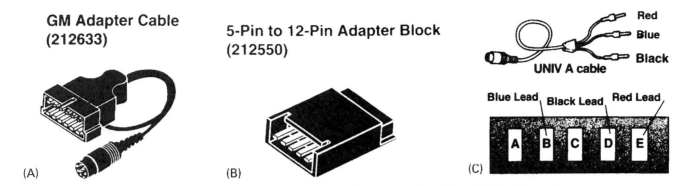

Figure 27–10 Adapter cables are used to connect an aftermarket scan tool to the vehicle's diagnostic connector. *(Courtesy of SPX-OTC)*

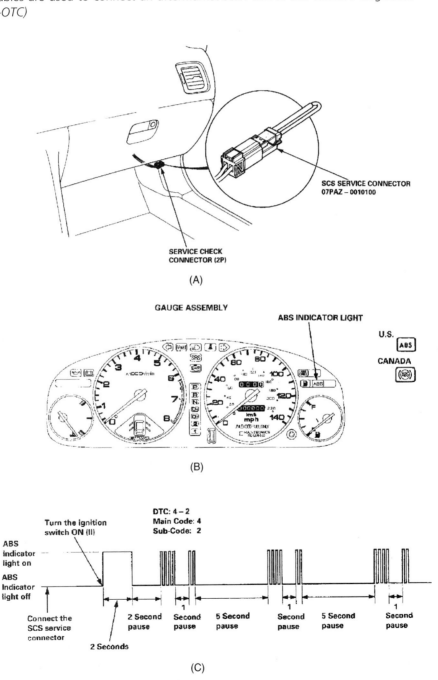

Figure 27–11 When the SCS service connector tool is connected to the service check connector (A) and the key is turned on, the ABS light will blink out any codes stored (B). The timing and length of the light blinks tells the code; a code 4-2 is shown here (C). *(Courtesy of American Honda Motor Co., Inc.)*

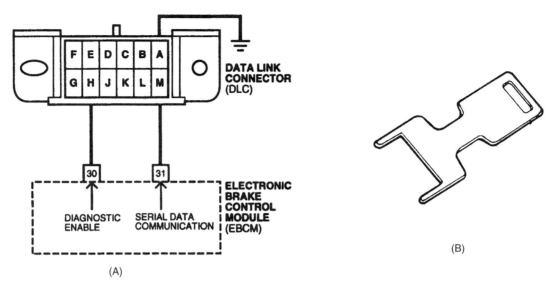

Figure 27-12 A diagnostic terminal (data link connector [DLC] or assembly line data link [ALDL]) is manipulated to obtain diagnostic codes (A). A shorting tool (B) can be used to make the necessary connection. *(A is courtesy of General Motors Corporation, Service Technology Group; B is courtesy of SPX-OTC)*

ABS Diagnostic Codes

Code	Description
1	LF valve problem
2	RF valve problem
3	RR valve problem
4	LR valve problem
5	LF wheel speed sensor low output
6	RF wheel speed sensor low output
7	RR wheel speed sensor low output
8	LR wheel speed sensor low output
9	LF/RR diagonal wheel speed signal error
10	RF/LR diagonal wheel speed signal error
11	Replenishing valve problem
12	Valve relay error
13	Improper pressure switch signal
14	Improper travel switches sequence
15	Improper brake switch signal
16	EBCM errors

Figure 27-13 This chart identifies sixteen diagnostic codes. For example, code 2 indicates that a problem exists in the electrical portion of the right front control valve. Some vehicles have up to fifty-six codes. *(Courtesy of General Motors Corporation, Service Technology Group)*

27.5 Electrical Tests

To make the system electrical checks more convenient, easier, and quicker, a pin-out or breakout box is connected to the system. This unit is installed by removing the connector from the EBCM and attaching it to the pin-out box (Figure 27-14). The pin-out box provides a set of measuring pins to which the leads of a digital volt-ohmmeter can be easily connected to measure the resistance or voltage of the various wheel speed sensors, hydraulic modulators, and power relays in the ABS (Figure 27-15).

All electrical checks should begin at source voltage by making sure that the battery and alternator are good and can supply the proper voltage (Figure 27-16).

Next, the circuit where a problem is indicated is checked; let us say it is in the left front wheel sensor. For example, on a particular car, pins 5 and 23 might lead to the left wheel sensor, and this sensor should have 800 to 1,400 ohms of resistance. One ohmmeter lead is connected to pin 5, and the other to pin 23. A reading between 800 and 1,400 ohms indicates a good sensor and a good sensor circuit. A reading less than 800 ohms indicates a shorted sensor, and a reading over 1,400 ohms indicates an open sensor or wire or a dirty or loose connection. If the sensor resistance is correct, the meter is then switched to AC volts and the leads are left connected to pins 5 and 23 while the left front wheel is rotated by hand. A fluctuating reading between 0.05 and 0.7 V indicates a good sensor, sensor ring, and air gap. No reading or a reading less than 0.05 V indicates a fault in the sensor ring or air gap.

If the electrical value—resistance or voltage—of a component is wrong when measured at the pin-out box, the technician then measures the value directly at the component. If the component has the wrong value, it should be replaced. If the measurement is correct at the component

Chapter 27 ■ Antilock Braking System Service

(A)

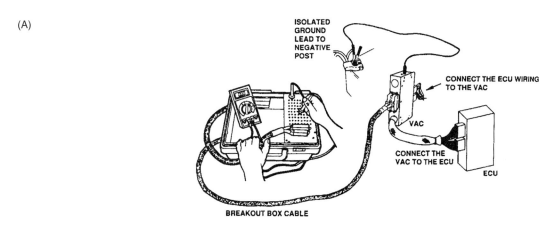

(B)

ANTI-LOCK QUICK CHECK SHEET USING ROTUNDA EEC-IV 60-PIN BREAKOUT BOX 014-00322

Item to be Tested	Ignition Mode	Measure Between Pin Numbers	Tester Scale/Range	Specification	Pinpoint Test
Battery Power	OFF or ON	1 and 13	dc Volts	10 V min.	B
		2 and 14	dc Volts	10 V min.	B
Ignition Feed	OFF	15 and 13	dc Volts	0 V	B
	ON	15 and 13	dc Volts	10 V min.	B
LF Sensor Resistance	OFF	5 and 6	kOhms	0.8-1.4 k Ohms	C
RF Sensor Resistance	OFF	7 and 8	kOhms	0.8-1.4 k Ohms	D
LR Sensor Resistance	OFF	9 and 10	kOhms	0.8-1.4 k Ohms	F
RR Sensor Resistance	OFF	11 and 12	kOhms	0.8-1.4 k Ohms	E
Sensor Continuity To Ground					
LF	OFF	5 and 13	Continuity	No continuity	C
RF	OFF	6 and 13	Continuity	No continuity	D
LR	OFF	7 and 13	Continuity	No continuity	F
RR	OFF	8 and 13	Continuity	No continuity	E
Sensor voltage: Rotate wheel @ one revolution per second					
LF	OFF	5 and 6	ac mVolts	100-3500 mV	C
RF	OFF	7 and 8	ac mVolts	100-3500 mV	D
LR	OFF	9 and 10	ac mVolts	100-3500 mV	F
RR	OFF	11 and 12	ac mVolts	100-3500 mV	E
Diagnostic Link	ON	28 and 13	dc Volts	10 V min.	A
ABS Warning Indicator	OFF	21 and 13	dc Volts	0 V	G
	ON	21 and 13	dc Volts	10 V min.	G
Ground Continuity	OFF	13, 14 and 22	Continuity	< 5 ohms	A

Figure 27–14 A pin-out or breakout box (A) is connected into the wiring harness at the EBCM and provides a terminal for testing each of the circuits for continuity, resistance, or voltage. Quick checks for the most common problems are shown in (B). *(A is courtesy of SPX-OTC; B is courtesy of Ford Motor Company)*

but wrong at the pin-out box, faulty wiring is indicated. For example, if the resistance of the left front wheel sensor measures greater than 1,400 ohms at the pin-out box but less than 1,400 ohms at the sensor connector, the sensor is good, but there is a bad connection somewhere between the sensor and the pin-out box (Figure 27–17).

Service manuals usually provide a detailed procedure to locate the cause of the problem. Figure 27–18 illustrates the circuits between the control module and the four wheel speed sensors from the service manual of a late-model car; the figure shows the wire/circuit numbers, wire color coding, and all the connectors. The diagnostic chart for code 32, left front wheel circuit fault, is shown in Figure 27–19. Note how this chart takes a technician through a series of diagnostic checks, using the circuit diagram, until the problem is located.

A dedicated ABS tester is available for some car models. It is attached to either a diagnostic connection or, like the pin-out box, the EBCM connector (Figure 27–20). The tester allows the technician to make electrical checks

Cavity	Circuit/Color	Function
1	B7 WT	RF wheel speed sensor (+)
3	B116 GY	Pump/motor relay control
4	G84 LB/BK	Traction control warning lamp
5	D1 VT/BR	C2D bus (+)
6	B6 WT/DB	RF wheel speed sensor (−)
7	A20 RD/DG	Fused battery feed
8	B28 VT/WT	Rotation sensor (−)
9	B27 RD/YL	Traction control switch sense
10	B30 RD/WT	Brake pedal travel sensor feed
11	Z1 BK	Body ground
12	Z1 BK	Body ground
13	B120 BR/WT	Switched battery feed
15	B9 RD	LF wheel speed sensor (+)
16	G19 LG/OR	ABS amber warning lamp control
17	B21 DG/WT	Low brake fluid switch #2 feed
18	G9 GY/BK	Red brake warning lamp control-CAB
19	B3 LG/DB	LR wheel speed sensor (−)
20	G83 GY/BK	Traction control function lamp
21	B29 YL/WT	Rotation sensor (+)
22	L50 WT/TN	Stop lamp switch (to lamps)
24	Z1 BK	Body ground
25	B120 BR/WT	Switched battery feed
27	D2 WT/BK	C2D bus (−)
28	B4 LG	LR wheel speed sensor (+)
29	B2 YL	RR wheel speed sensor (+)
30	B8 RD/DB	LF wheel speed sensor (−)
31	B1 YL/DB	RR wheel speed sensor (−)
32	B58 OR/BK	ABS main relay control
33	F20 WT	Ignition feed
35	B20 DB/WT	Low brake fluid level switch #2 return
36	B31 PK	Brake pedal travel sensor return
37	B120 BR/WT	Switched battery feed

Figure 27-15 The terminals of the thirty-seven-pin connector are identified. Note that it is easy to become confused when trying to locate a particular terminal. *(Courtesy of Chrysler Corporation)*

for various portions of the ABS circuits, monitor pump operation, obtain output readings from the wheel speed sensors, and operate the pressure control valves.

27.5.1 Speed Sensor Mechanical Problems

Each wheel speed sensor generates an electrical signal from the speed of that wheel; that signal must be regular, with a smooth transition as the speed and frequency increase and decrease with the speed of the car. The speeds of the different sensors must match, with only a slight variation allowed between them. Something as simple as mismatched tire size can cause speed sensor problems.

Other mechanical sensor problems can be caused by a broken or chipped tooth on a sensor ring or a ring that is either not round or is mounted so that runout occurs. Loose or faulty wheel bearings are more commonly encountered problems. Anything that can change sensor

Chapter 27 ■ Antilock Braking System Service

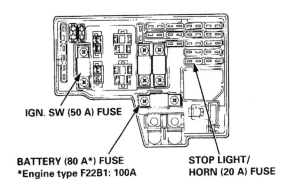

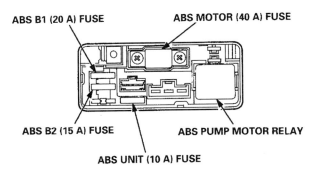

Figure 27-16 The power source for most ABS units is the vehicle's power center/fuse-relay box. *(Courtesy of American Honda Motor Co., Inc.)*

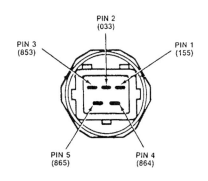

Resistance Pressure Switches—Ignition Off (System Discharged)

Pins	Resistance 200-Ohm Range
3 and 5	Infinite ohms
1 and 4	0 ohms
1 and 2	0 ohms

Figure 27-17 A pressure switch connector (top). An ohmmeter connected between pins 3 and 5 should read infinite resistance and, when connected between pins 1 and 2 or 4, should read 0 ohms. *(Courtesy of General Motors Corporation, Service Technology Group)*

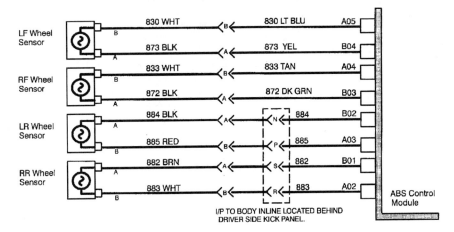

Figure 27-18 The circuit for the wheel speed sensors identifies the wire color and number coding and the numbers for the various wire connections. It can be used along with the diagnostic chart for the same vehicle. *(©Saturn Corporation, used with permission)*

gap during rotation can cause the EBCM to shut the system down, turn on the warning light, and set a problem code. Remember that the center core of a sensor is a magnet, and it will pick up metal debris that can affect its operation. Sensors that are mounted into an axle or transmission can attract enough iron particles to cause an erratic signal.

When speed sensor problems that do not appear to be electrical are encountered, the tire sizes, wheel bearings, sensor rings, and sensors should be carefully inspected. Check the sensor gap using a nonmagnetic feeler gauge between the sensor pole and the toothed ring (Figure 27-21). If a steel feeler gauge is used, the magnetic pull will increase the drag and cause a false reading. Also

DIAGNOSTIC CHART   CODE 32

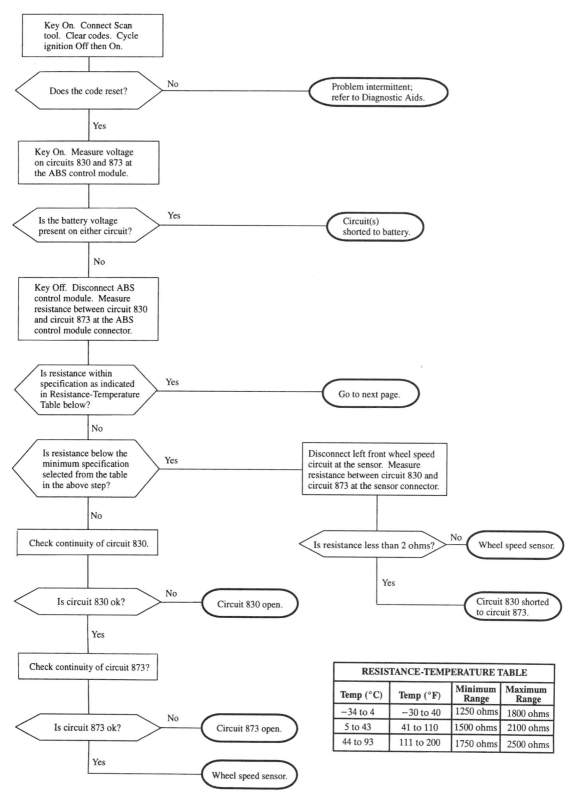

Figure 27–19 The diagnostic procedure to locate the problem causing code 32 (left wheel speed circuit fault) on this car. The circuit numbers are shown in Figure 27–18. (©*Saturn Corporation, used with permission*)

DIAGNOSTIC CHART CODE 32

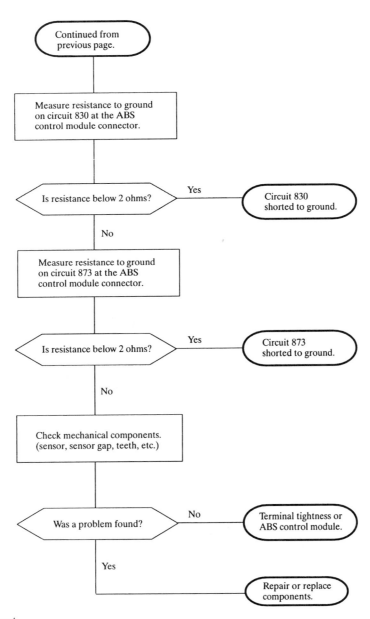

Figure 27-19 *continued*

rotate the wheel and ring and check the gap in several locations around the ring. It should be the same in each location. When sensors are mounted down into a part so you cannot measure the gap directly, measure a depth down to a tooth and then subtract the height of the sensor from this dimension (Figure 27–22).

Sensor electrical output can be checked using an oscilloscope or lab scope to get a better look at its value. This unit displays the electrical signal on a screen; a wheel speed sensor should produce a smooth sine wave (Figure 27–23 and Figure 27–24). An irregular signal indicates the type of problem.

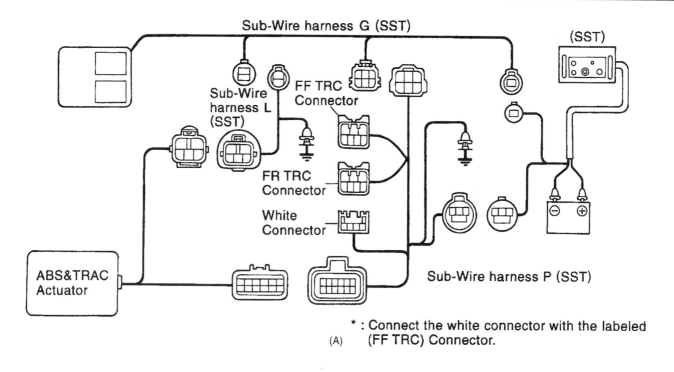

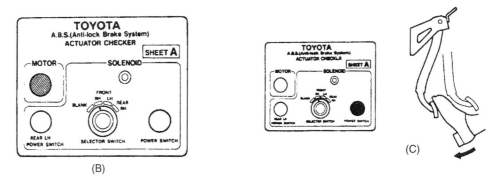

Figure 27-20 Most vehicle manufacturers have developed a tester for their ABS systems; it is connected into the diagnostic connectors (A) and (B) and is used to test various components (C).

27.6 Hydraulic Pressure Checks

Some ABS problems can be caused by poor hydraulic pump output or sticky or leaking modulator valves. Pump output can be checked by attaching a pressure gauge (Figure 27-25). A pressure gauge allows you to check the amount of pressure that the pump develops, the settings of the switch or switches to start and stop the pump, the cycle time between start and stop, and the leak rate after the pump stops. In most systems, after the pump runs, builds pressure, and shuts off, the pressure should hold steady. The pump should not restart unless the brakes are applied to drop the pressure.

Some diagnostic scan tools allow the operator to cycle the modulator valves. With these units, the vehicle can be raised on a hoist, and, with the brakes applied, the modulator valves can be operated while the wheels are checked to determine if they are locked up or free to rotate. A good modulator operates correctly in each of its circuits.

A sinking brake pedal is normally diagnosed by blocking off the suspected portion of the system using plugs, as described in Chapter 24.

Most modulator valves are not serviceable and can only be replaced (Figure 27-26). If replacement becomes necessary, special bleeding procedures, as described later in this chapter, should be followed. Most experts recommend periodic brake fluid replacement to help prevent dirt and corrosion that might cause problems in these expensive units.

Chapter 27 ■ Antilock Braking System Service

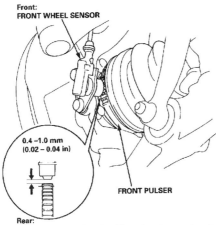

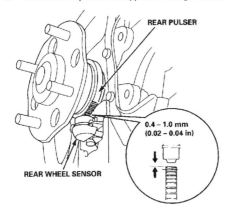

Figure 27-21 Sensor air gap is the clearance between the sensor pole and the teeth of the reluctor/sensor/pulser ring. *(Courtesy of American Honda Motor Co., Inc.)*

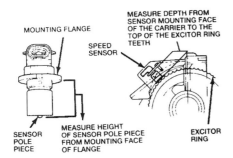

Figure 27-22 This sensor is mounted into the rear axle, so the gap cannot be measured. The amount of air gap can be determined by measuring the height of the sensor and the depth to the teeth of the sensor ring. *(Courtesy of Ford Motor Company)*

27.7 Repair Operations

All the mechanical and electrical ABS components are serviceable to some degree. Some units allow replacement of faulty modulator components such as the motor pack, pump, or accumulator (Figure 27-27). When replacing any of these components, follow the manufac-

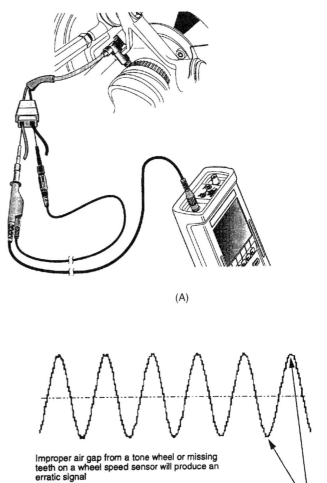

Figure 27-23 This Fluke 98 lab scope is connected into a wheel speed sensor (A). It will display a waveform of the sensor output in which the time value is shown horizontally and the electrical voltage value is shown vertically (B). *(Reproduced with permission from Fluke Corporation)*

turer's procedure to prevent damage to any of the components or parts and to ensure a reliable and safe repair (Figure 27-28). Replacement is the normal service procedure for the electrical components. As mentioned earlier, the manufacturer's instructions should be followed when making replacements. Service operations include removal and replacement of the hydraulic control unit, hydraulic accumulator, hydraulic pump motor, reservoir, EBCM, individual wheel sensors, sensor rings, or electric relays and switches. Some manufacturers require replacement of an entire wire harness rather than repair of a wire or connector fault to ensure the integrity of that electrical circuit.

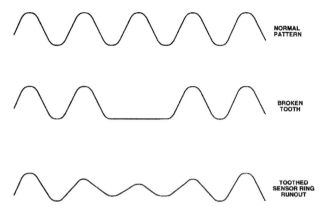

Figure 27-24 If the output of a wheel speed sensor is displayed on an oscilloscope, it should appear as a smooth sine wave (top); a broken tooth or an uneven sine wave indicates problems. *(Courtesy of General Motors Corporation, Service Technology Group)*

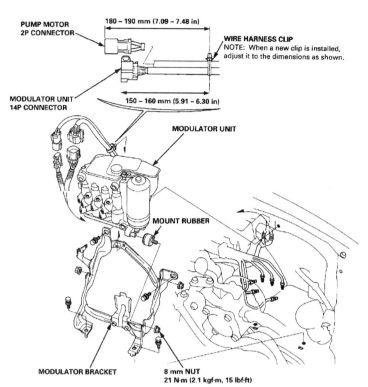

Figure 27-26 In most cases, a faulty modulator is serviced by replacing the entire unit. *(Courtesy of American Honda Motor Co., Inc.)*

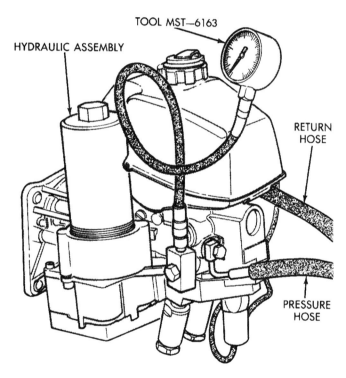

Figure 27-25 Tool MST-6163 is a pressure gauge, used to check the pump, switch, accumulator, and seals in this hydraulic assembly. *(Courtesy of Chrysler Corporation)*

A faulty sensor ring can usually be replaced by pressing the old ring off of the CV joint, rotor, or axle. The new ring should be carefully pressed into place so it is not damaged and so it is in the correct position (Figure 27-29). Most sensors are held in place by one or two bolts and can be easily replaced; be sure to route the wires in the same manner and location as the original (Figure 27-30). When replacing sensors or sensor rings, it is necessary to adjust the air gap. On some sensors, the air gap is adjusted by installing the sensor so the paper spacer on the sensor touches a tooth of the sensor ring. A new sensor includes a paper spacer; when a used sensor is reinstalled, a new paper spacer of the correct thickness can be glued onto the sensor. The paper spacer is designed to wear off as the car is driven. When paper spacers are not used, sensor gap is checked using nonmagnetic feel gauges as previously described.

The operation of the hydraulic pump, accumulator, and control switches can be checked by measuring the pressure. Special adapters are usually required to connect a gauge to the system. A particular sequence is usually required to determine what the correct pressure should be under various operating conditions. Switch problems are indicated by a failure to turn the pump on or off at the correct pressure. Pump problems are indicated by a failure to reach high enough pressure or too fast a leak-down rate. A leak-down rate that is too fast can also be caused by a leaking valve. A faulty accumulator is indicated by a very fast pressure increase when the pump runs and a very fast pressure decrease when the brakes are operated with the pump off.

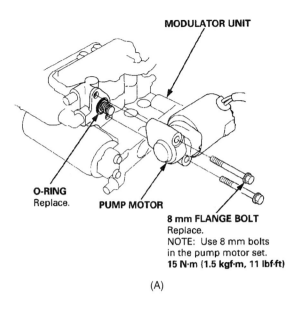

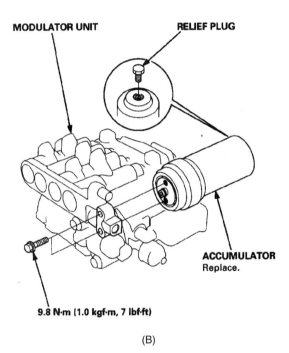

Figure 27-27 On this particular modulator assembly, a faulty pump and motor (A) or accumulator (B) can be removed and replaced. *(Courtesy of American Honda Motor Co., Inc.)*

When making pressure checks or replacing the hydraulic control unit, accumulator, pump, or pressure switch, it is important to discharge the pressure from the accumulator before beginning the service operations. It is recommended that you apply and release the brake pedal twenty-five to thirty times with the key off. A definite increase in pedal resistance should be noted as the pressure is bled off.

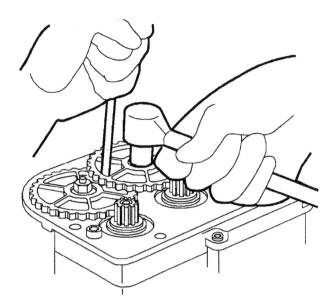

Figure 27-28 Modulator drive gears are a serviceable portion of this modulator assembly. *(©Saturn Corporation, used with permission)*

27.7.1 Bleeding ABS

When service work on the hydraulic components is completed, a special bleeding process for that particular system is usually required. This procedure is described in the manufacturer service manual and in various technician service manuals from aftermarket component manufacturers.

Some systems can be bled manually or by pressure. Some of these systems include a bleeder screw at the hydraulic modulator or use a bleeding procedure such as loosening a line fitting (Figure 27-31). Some systems use standard tools and procedures but require that the key be off or on while purging air from certain locations. Some systems use flow from the electric booster pump to bleed certain portions of the system. Other systems require the use of a particular ABS tester to cycle solenoid valves for a thorough bleeding operation. These systems can usually be bled using a more lengthy procedure without the special equipment.

Remember that in an ABS that uses an electric pump and accumulator, air can become trapped in the accumulator. This air will not affect system operation until there is an ABS stop, and then the air can be forced into the system, producing a low and spongy brake pedal.

When bleeding ABS, it is best to check the service manual for the correct procedure.

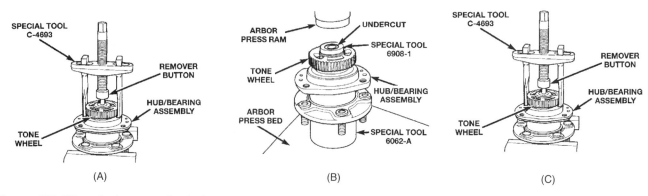

Figure 27-29 A faulty tone wheel/reluctor can be pulled or pressed off the hub (A). A new unit can be pressed into place; sometimes special press adapter tools are required (B). This new tone wheel should be flush with the end of the hub (C). *(Courtesy of Chrysler Corporation)*

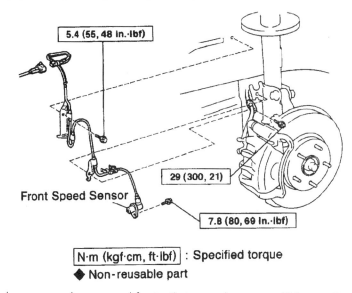

Figure 27-30 A wheel speed sensor can be removed for testing or replacement. *(©Saturn Corporation, used with permission)*

27.8 Completion

At the completion of ABS repair, you should check to ensure that all of the diagnostic trouble codes have been cleared. Next, take the vehicle for a test drive to make sure that all of your repairs and adjustments are working properly. After the road test, check that the red brake light and amber ABS light are working properly and that no additional trouble codes have been set.

27.9 Practice Diagnosis

You are working in a new-car dealership and encounter the following problems:

CASE 1: A customer purchased a new car last month and has brought it back in with a complaint that the ABS light stays on. He mentions that he noticed it come on as he was driving away from a stoplight, and it did not go out after that. What may be wrong with this system? Where should you start checking for problems?

CASE 2: A customer has brought in her car with a complaint of a pulsating brake pedal. She bought her new car in the summer and had no problems until a cold, icy winter morning when she was in a hurry to get to work. On your road test, you notice normal brake and warning light operation. What should you do next? What was probably the cause of the complaint?

CASE 3: A 3-year-old car had a torn CV joint boot that caused the CV joint to fail. You replaced the CV joint and road tested the car to make sure everything was right, and the ABS light came on when you got to about 10 mph. What went wrong? What checks should you make?

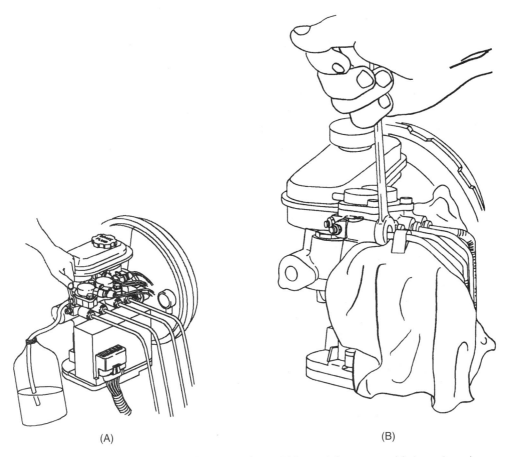

Figure 27-31 Some ABS units require special bleeding procedures. This modulator assembly is equipped with a bleeder screw (A); line connections are bled by loosening the line fitting (B). *(Courtesy of General Motors Corporation, Service Technology Group)*

Terms to Know

amber antilock warning light

diagnostic trouble code (DTC)

hydraulic modulator

red brake warning light

Review Questions

1. Two technicians are discussing antilock braking system service.
   Technician A says that the EBCM can be ruined if you touch the electrical connector pins.
   Technician B says that the EBCM should be sprayed with antistatic spray before disconnecting it.
   Who is correct?
   a. A only
   b. B only
   c. Both A and B
   d. Neither A nor B

2. A car with ABS is started while the brake pedal is depressed, and a bump is felt in the pedal as the engine starts. This bump is caused by:
   a. The normal ABS start-up operation
   b. A faulty control module
   c. A faulty control valve assembly
   d. All of the above

3. The red brake warning light in an ABS-equipped car should light:
   a. During engine start-up
   b. If the brake fluid level is low
   c. When the parking brake is applied
   d. All of the above

4. In most cars, the amber warning light warns of a failure in the:
   a. Brake hydraulic system
   b. EBCM or one of its circuits
   c. Mechanical portion of a control valve assembly
   d. All of the above

5. Technician A says that a problem in one of the wheel sensors will cause the amber warning light to come on as the car is being driven.
   Technician B says that an improper sensor air gap or faulty wheel bearing can cause this problem.
   Who is correct?
   a. A only
   b. B only
   c. Both A and B
   d. Neither A nor B

6. Technician A says that a wheel sensor should produce a fluctuating DC voltage of about 9.5 V as the wheel is turned.
   Technician B says that the resistance of a wheel sensor can be measured at the EBCM connector or at the connector closest to the sensor.
   Who is correct?
   a. A only
   b. B only
   c. Both A and B
   d. Neither A nor B

7. A thorough check of an ABS electrical circuit requires a:
   a. Pin-out box
   b. Ohmmeter
   c. AC voltmeter
   d. All of the above

28 Suspension and Steering System Diagnosis and Inspection

Objectives

Upon completion and review of this chapter, you should be able to:

❑ Understand the commonly used term *NVH*.

❑ Have a basic understanding of how NVH problems occur.

❑ Follow the procedure used to diagnose suspension and steering system problems.

❑ Understand how vibration frequency can be used to locate the problem source.

❑ Follow a diagnostic trouble tree.

❑ Determine the cause of an unusual tire wear pattern and recommend the needed repair.

❑ Determine the cause of a tire-related vibration and recommend the needed repair.

❑ Measure tire, wheel, and hub runout in a radial and lateral direction.

❑ Determine whether a steering pull is caused by a tire-related problem and, if so, correct this problem.

❑ Determine the cause of wheel or axle bearing problems.

❑ Measure vehicle ride height and determine whether repairs are needed.

❑ Determine the cause of noises, excessive body sway, or uneven ride heights and recommend the needed repairs.

❑ Inspect springs, shock absorbers, and stabilizer bars for wear or damage.

❑ Inspect and measure ball joints and kingpins for excessive clearance, faulty seals, or binding.

❑ Check control arm and strut rod bushings for wear or excessive clearance.

❑ Check struts and strut mounts for wear or damage.

❑ Check conventional steering gears and steering linkage systems for wear, excessive clearance, or binding.

❑ Check rack-and-pinion steering gears and steering linkage for wear, excessive clearance, or binding.

❑ Check power steering systems for abnormal noises, leakage, or improper operation.

❑ Determine what repairs are needed to correct any suspension and steering faults located during this inspection procedure.

❑ Perform ASE tasks relating to suspension and steering system diagnosis and inspection (see Appendix A).

28.1 Introduction

Most chassis service operations result as a cure for a problem, and most mechanical problems can be placed into three categories:

- NVH (noise, vibration, and harshness): these are unpleasant feelings as we drive the car
- Handling problems: the car does not drive as it should; an example is a pull to one side
- Excessive or unusual tire wear

When an experienced technician encounters problems of these types, he or she draws from experience and training to begin a diagnostic procedure to find the cause. Once the cause is located, it is usually fairly easy to repair it. As a student, you still have a lot to learn, and this should occur as you complete this text and your class instruction. This chapter should help you understand diagnostic procedure.

28.2 NVH

Noise, vibration, and harshness are unpleasant motions or noise as we drive a car. With modern cars, we expect a smooth, soft ride, and with the sound systems turned off, some drivers expect to hear only a mild hum. We certainly do not want anything to interfere with our favorite music or a conversation with another passenger. Of these, vibration is the most commonly encountered problem.

28.2.1 Noise

Noise is an **audible** sound coming from something that moves, such as the rattle of a loose or broken part, the clunk of a part that moves within the range of a worn bushing, or the constant rumble of a harsh tire tread or growl of a bad bearing (Figure 28–1). The type of noise often tells us what to look for, and the location of the noise is usually the source of the problem. A constant noise from a tire tread or bearing often varies in **intensity** or **frequency** as vehicle speed changes. Noise intensity is simply how loud it is; frequency is the tone, from a low bass moan to a high-pitched squeal. Intensity is measured in decibels, and frequency is measured in **hertz (Hz)**. The type of noise is often hard to describe because certain terms mean different things to different people. Table 28–1 helps relate common terms to common sounds.

Loud noises can cause hearing loss. OSHA places work area limits of 100 decibels for a maximum of 2 hours, 95 decibels for a maximum of 4 hours, and 90 decibels for a maximum of 8 hours to reduce chances of hearing injury for workers.

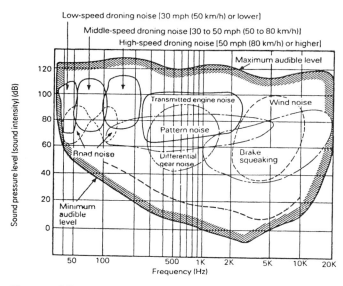

Figure 28–1 Noise loudness is measured in decibels (dB), and frequency is measured in hertz (Hz). Various vehicle noises and sounds are shown here. *(Courtesy of Chrysler Corporation)*

As an aid to diagnosing a noise problem, remember that tire noise changes with the road surface; if the offending noise changes as we pass from an asphalt surface onto a concrete surface, it is probably coming from a tire. Also, wheel and axle bearing noise changes with load. On a road test, you can vary bearing load by swerving the vehicle or lightly applying the brakes; if either of these actions changes the noise, it is probably caused by a faulty axle or wheel bearing.

28.2.2 Vibration

Vibration is a **tactile** motion coming from something that moves; it is usually considered a back-and-forth oscillation. Tactile means it is perceptible by touch or feel. Vibration can be felt or seen as a rhythmic motion that can also be heard (Figure 28–2). The sound frequency will match the vibration frequency in that a slow vibration will set up a moan or bass tone and a very rapid vibration will cause a squeal.

An unbalanced tire causes an outward motion at the heaviest spot on the tire that is not counterbalanced at the opposite side (Figure 28–3). Each time the tire revolves, an oscillation **cycle** will occur, and as the tire speeds up, a **frequency** of X rpm (revolutions per minute) or X hertz (cycles per second) occurs. The frequency will match the rotational speed. If the **amplitude** or strength of vibration gets strong enough, it will pass through the suspension and body of the car to be felt by the driver (Figure 28–4).

To be annoying, three things must occur:

- There must be a source or something to cause the vibration.

Chapter 28 ■ Suspension and Steering System Diagnosis and Inspection

Table 28–1 Noise and Vibration Problems.

Feels like

Buzz: High-frequency vibration (50 to 100 Hz), like holding an electric razor

Roughness: Higher-frequency vibration (20 to 50 Hz), like holding a jigsaw

Shake: Low-frequency vibration (20 to 95 Hz), like driving with an out-of-round or unbalanced tire

Shimmy: See shake

Shudder: See shake

Tingling: Highest-frequency vibration, causes pins-and-needles sensation, like when your foot "goes to sleep"

Tramp: See shake; also called high-speed shimmy

Waddle: See shake

Wobble: See shake

Sounds like

Boom: A cycling, rhythmic sound, like distant thunder or a bass drum

Buzz: A distant buzz saw or door buzzer

Click: The operation of a retractable ballpoint pen or a camera shutter

Clunk: A heavy metallic sound or a heavy door closing

Drone: Low-frequency (60 to 120 Hz), constant noise, like a bumblebee, blowing air across an empty soda bottle, or a bagpipe's background tone

Grind: A garbage disposal or a grinder grinding metal

Grunt: A short, low-frequency, raspy sound, like someone trying to clear his or her throat

Hiss: Letting the air out of a tire

Howl: Midrange frequency (120 to 300 Hz), like wind howling

Knock: Two hammers hitting each other or someone knocking at your door

Moan: See drone

Ping: Like marbles inside of a tin can

Pop: Popping popcorn or a Champagne cork being removed

Rattle: A baby's rattle or rocks moving inside of a tin can

Roar: The sound of a busy freeway or waterfall

Rumble: A bowling ball rolling down the alley or distant thunder

Screech: A tire's noise being driven into a parking garage or screaming people on an amusement park ride

Spit: Water drops in a hot frying pan

Snap: Breaking a pencil

Squeak: A short-duration noise, like a tennis shoe on a wooden court

Squeal: Longer duration than squeak, like a door hinge that needs oil

Tap: Bumping a knife and fork together

Thud: Dropping a bowling ball

Whir: An electric fan running

Whine: High-pitched (300 to 500 Hz) noise, like a mosquito or high-speed dentist's drill

Note: Many terms are used to describe problem vibrations and sounds; this listing compares the more commonly used terms with things you might be familiar with.

- There must be a transfer path to go from the source to the passengers.
- There must be a responder for the passengers to see, feel, or hear the vibration (Figure 28–5).

The source is often a tire or wheel with excessive runout or unbalance that causes the vibration; the source is the primary concern of the technician. The transfer path is the axle or wheel bearings, suspension parts and mounts,

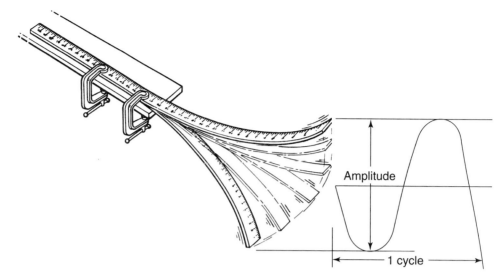

Figure 28–2 A vibration cycle is one upward and one downward motion to return to the starting point; the height of this cycle is called amplitude.

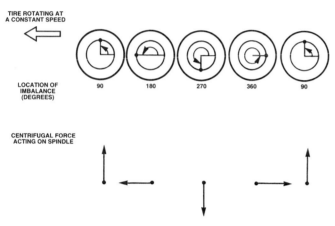

Figure 28–3 This tire with a high spot will cause one vibration cycle per revolution; it is called a first-order vibration. *(Courtesy of General Motors Corporation, Service Technology Group)*

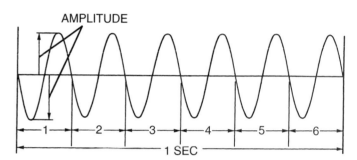

Figure 28–4 If a vibration goes through six cycles in 1 second, it has a frequency of 6 cycles per second (6 CPS), or 6 Hz. *(Courtesy of General Motors Corporation, Service Technology Group)*

body, and steering column. Rubber mounting points are designed to break this path to stop vibration transfer and are a primary concern of the design engineer (Figure 28–6). The responder is where we notice the vibration, and this also is a concern for the design engineer.

Another factor concerning vibration is **natural frequency** and **resonance.** Many items try to vibrate at a certain frequency that is natural for that particular item. An example is the different strings on a guitar; each one is tuned to a different sound frequency. Exciting an item at the same speed as its natural frequency causes a resonance, which intensifies the effect of the vibration (Figure 28–7). This becomes important to us if we realize that when the vibration cycle of a tire reaches the natural frequency of the suspension, the vibration going through this transfer path is greatly intensified. With many cars,

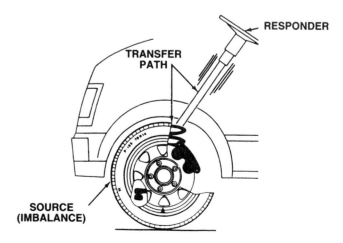

Figure 28–5 A vibration needs a source or cause (an unbalanced tire), a transfer path (the suspension and steering column), and a responder where it is felt or heard (the steering wheel); without any one of these, it would not be annoying. *(Courtesy of General Motors Corporation, Service Technology Group)*

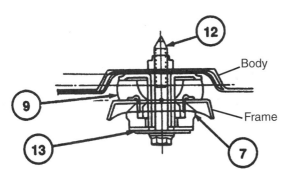

Figure 28–6 Rubber insulators (7 and 9) are used where the body (top) attaches the frame (center), and the bolt (12) passes through the rubber so there is no metal-to-metal contact. Any vibrations will be dampened by the rubber. *(Courtesy of Ford Motor Company)*

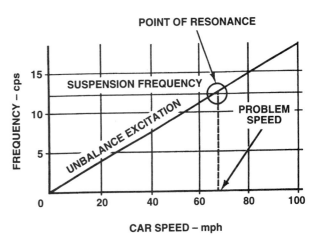

Figure 27–8 This chart shows an unbalance excitation matching the frequency of the suspension at about 67 mph; there could be a noticeable vibration in the car at that speed. *(Courtesy of General Motors Corporation, Service Technology Group)*

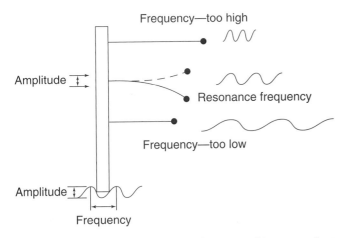

Figure 28–7 If an item is excited by an outside source (bottom) at a frequency that matches its natural frequency, it will resonate. Items with a higher or lower frequency will not be affected.

Figure 28–9 The engine speed was increased until the vibration frequency from a slight unbalance of the engine matched the natural frequency of the antenna; it demonstrates resonance frequency.

intensified vibration occurs at about 50 to 55 mph. In many cases, a technician can locate the source of a vibration from the vibration frequency relative to the vehicle or engine speed (Figure 28–8).

With many cars, you can see harmonic resonance by simply increasing the engine speed while in neutral and watching the antenna. When the frequency of the vibrations coming from the engine matches the natural frequency of the antenna, the antenna will begin vibrating (Figure 28–9).

A related factor is called **beating** or **phasing.** Two different and slightly unbalanced parts that operate at different speeds, such as a driveshaft and wheel, can add to or cancel out the vibration of the other. When the two vibrations are in phase, the combined force is very noticeable; when they are out of phase, the vibration becomes less noticeable or goes away. In most cases the vibration cycles in and out (Figure 28–10).

Another vibration characteristic important to the technician who uses a **reed tachometer** or **vibration analyzer** to measure vibration frequency is **order,** or how often the vibration occurs per revolution of the source. Most vibrations are first order and create one disturbance per revolution. Let us say we have a tire with one heavy spot (a first order) that is revolving at 600 rpm; this

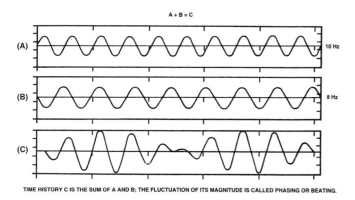

Figure 28-10 Vibration (A) has a frequency of 10 Hz, and vibration (B) has a frequency of 8 Hz. (C) is the vehicle's response to these vibrations; at times the amplitude is very high, and at other times it almost disappears. *(Courtesy of General Motors Corporation, Service Technology Group)*

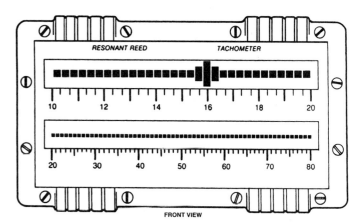

Figure 28-12 The face of a reed tachometer. The reed above 16 Hz is vibrating, which indicates the vibration frequency of where it has been placed. *(Courtesy of General Motors Corporation, Service Technology Group)*

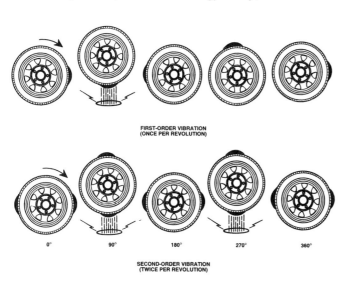

Figure 28-11 Most tire and wheel problems are first-order vibrations; they will cause one vibration cycle per revolution. Some can cause two vibration cycles per revolution; they generate a second-order vibration. *(Courtesy of General Motors Corporation, Service Technology Group)*

would create a disturbance of 600 vibrations per minute, or 10 Hz (10 cycles per second) (Figure 28-11). If this same tire also has a flat spot, it would create two disturbances per revolution. This second-order vibration would have 1,200 disturbances per minute, with a frequency of 20 Hz. The frequency of vibration will increase at a multiple of the speed (2 × speed for second order). A third-order vibration would cause three disturbances per revolution and create a frequency that is 3 × speed.

A reed tachometer is a mechanical device that contains groups of reeds that are tuned to different frequencies. We can determine what the vibration frequency is by seeing which reed is vibrating at the greatest amplitude (Figure 28-12). Once the frequency is known, we can find the source if we know what is revolving at that

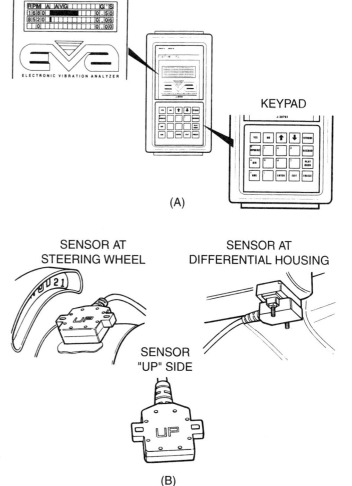

Figure 28-13 An electronic vibration analyzer (EVA) showing the display screen and key pad (A); it requires a sensor (B) that is attached to the vibration transfer path or responder. *(Courtesy of General Motors Corporation, Service Technology Group)*

speed. It is usually necessary to know the tire diameters and gear ratios to determine the number of revolutions per minute of the tires and driveshaft. An electrical vibration analyzer (EVA) is also used to measure vibration frequency (Figure 28–13). It is a handheld device similar to a scan tool, and it uses a sensor attached to various locations to pick up the disturbance. The sensor is attached to a probable source, the vehicle is driven through the problem speed, and the frequency of any disturbance at that source is displayed on the sensor.

28.2.3 Harshness

Harshness is closely related to noise and vibration and is much harder to explain. Instead of the rhythmic motion of a vibration, harshness is just an unpleasant motion; it is an audible and tactile high-frequency vibration, above 20 Hz. Like the other two, harshness comes from something that moves.

28.3 NVH Diagnostic Procedure

With an experienced technician, the procedure often varies depending on the problem type and the vehicle. The first step, often without even thinking about it, is to decide if the problem is NVH alignment or some other type. For example, let us imagine a vibration at the right front of a fairly new car with only 5,000 mi. This problem is most likely caused by the right front tire, and the most likely cause is NVH, either a runout or unbalance condition. If we find and cure one of these conditions, we probably would not need to check anything else. For another example, let us imagine a 7-year-old car with 95,000 miles on it that shows fairly heavy usage. The problem is a pull to the right under braking. This car probably has badly worn suspension and steering components, sagged springs, worn tires, worn brakes, and, because of the worn and tired suspension, poor alignment. This problem could be caused by unbalance of the front brakes or an alignment change that is occurring because of brake action and worn suspension bushings. It could also be caused by worn or mismatched tires. In either case, a wise technician realizes his or her job is to (1) cure a problem and (2) make sure nothing is going to fail under the car for a reasonable period after it leaves the shop. Most technicians start this diagnostic procedure by giving the car a road test followed by a thorough under-car inspection. This inspection is twofold: to locate the problem cause and to provide a record of the under-car condition for the car owner. Remember that most drivers do not know what is under their cars and rely on their mechanics to keep them safe.

28.3.1 Road Test

A technician uses a road test to confirm the nature of the problem and also to try to determine the cause. Most technicians have a preferred test route that provides various bumps and road surfaces and is away from heavy traffic. During a road test, the vehicle is driven in a manner to try to simulate the particular problem. In other words, if you are checking for a noisy shock, you want a rough road; if you are checking a pull condition or vibration problem, you want a smooth road. Finding a good test route that is close to the shop can be difficult in many metropolitan areas.

While conducting a road test, you should observe the following guidelines:

- Make sure the vehicle is safe to operate by quickly checking the tires, brake pedal operation, and steering wheel feel.
- Observe all pertinent traffic laws.
- Use the vehicle's seat and shoulder belt.
- Drive the vehicle in a normal manner so you will not cause any additional wear or damage.
- Note the operation of the ABS and airbag warning lights if the vehicle is so equipped.

During the road test, the technician will try to accomplish the following:

- Confirm the customer's complaint.
- Pin down the problem area, front or back and right or left side.
- Determine the nature of the problem (i.e., noise, vibration, shake, or pull).
- Determine whether the problem is vehicle speed sensitive and at what speed it is most noticeable, whether it is torque sensitive, and whether it is engine speed sensitive or gear ratio sensitive.
- Determine whether it is a tire, suspension, steering alignment, or brake problem.

Note any unusual or unsafe condition to be brought to the attention of the owner.

28.3.2 Diagnostic Charts and Trouble Trees

Diagnostic charts are available to help you locate the causes of common problems. Many problems can have more than one cause, so it becomes the technician's job

Chart 28–1 Suspension and Steering Problem Diagnosis.

Condition	Possible Cause	Action	Refer to Chapter
Vibration	Wheel balance	Check balance	28
	Tire or wheel runout	Check runout	28
	Wheel or axle bearings	Check bearings	28
	Brake rotor or drum balance	Check balance	28
	Driveline balance[a]	Check balance	28
	Damaged driveshaft or U-joints	Inspect driveshaft	28
	Engine balance[b]	Check balance	28
Excessive noise	Rough tires	Check tire condition	28
	Loose or broken shock mounts	Check mounts	28
	Damaged or worn suspension mounting insulators	Check insulators	28
Improper tire wear	Incorrect air pressure	Check inflation	28
	Excessive toe	Check alignment	34
	Excessive camber	Check alignment	34
Rough ride	Damaged shocks	Check shocks	28
Vehicle leans or is too high or low	Broken or sagged springs	Check ride height	28
Pulls or drifts right or left	Unequal tire pressure	Check pressure	28
	Mismatched tires	Check tire sizes and types	28
	Brake drag	Check for free turning	28
	Steering gear or linkage worn or damaged	Check steering	28
	Vehicle leans sideways or lengthwise	Check ride height	28
	Tight or worn suspension parts	Check ball joints or strut mount	28
	Incorrect toe settings	Check alignment	34
	Incorrect caster	Check alignment	34
	Excessive camber or caster spread	Check alignment	34

[a]Drive line balance is vehicle speed sensitive and sometimes torque sensitive. This means the vibration is controlled by the speed and also can be affected by the throttle.
[b]Engine unbalance is engine speed sensitive and will occur at different vehicle speeds depending on the gear used, and it will go away while the vehicle is coasting in neutral.

to check out and eliminate some of them, usually working from the easiest or least expensive to the most difficult to check or the most expensive to repair. This procedure often follows what is commonly called a **trouble tree**.

For example, if we check Chart 28–1 for vibration, we find seven probable causes. The next step would be to check these causes until we locate the problem. In this text, tire vibration problems are described in Section 28.8, and this section includes further vibration problem checks and troubleshooting charts. A diagnosis chart or trouble tree for locating the cause of pull is shown in Figure 28–14. Following this chart shows the need to check tire pressure, vehicle trim height, several tire switching operations, and adjustment of the alignment angles.

28.3.3 Under-Car Inspection

Another important step in locating a problem cause is a thorough under-car inspection. This inspection is normally done to determine if a car is in good condition, to locate the faulty part causing a problem, or as a preliminary check to a wheel alignment. Under-car inspection is described completely later in this chapter.

28.3.4 Matching Component rpm to Vibration Frequency

When the vibration source is not easily found, a clue to its cause can be found by calculating the speed of the things that cause vibrations and comparing them with the vibration frequency. The major causes of vibrations are the tires and wheels and the driveshaft. Some manufacturers

Chart 28-1 Continued

Condition	Possible Cause	Action	Refer to Chapter
Pulls during braking	Excessive caster spread	Check alignment	34
	Worn suspension bushings	Check suspension	28
	Brake problems	Check brakes	20
Wanders or requires constant steering correction	Insufficient caster	Check alignment	34
	Improper toe	Check alignment	34
	Improperly adjusted steering gear	Check adjustment	33
Torque steer	Incorrect tire pressure	Check inflation	29
	Loose or worn engine or cradle mounts	Check mounting	
	Improper driveshaft angles		
Steering wheel off center	Mismatched tires	Check tires	29
	Incorrect alignment	Check alignment	34
	Steering system problem	Center steering	33
Hard steering	Excessive caster	Check alignment	34
	Binding steering gear	Check steering gear	33
	Power steering problem	Check power steering	33
	Binding ball joint or linkage	Check steering system	28
Excess steering play	Worn steering linkage	Check steering system	28
	Improper steering gear adjustment	Check adjustments	33
	Worn rack-and-pinion mounts	Check mount bushings	28
Improper tracking	Rear axle damage	Check axle and mounting	34
	Misaligned rear axle	Check alignment	34

provide charts that make it very easy to determine wheel or driveshaft rpm, but if these charts are not available, it is fairly easy to calculate the rpm.

From the formula given in Chapter 9, we can compute tire circumference, $2 \times ER \times \pi = C$. Let us say the tire has an effective radius (center of tire to ground) of 12½ in.; $2 \times 12½ = 25$, $25 \times 3.1416 (\pi) = 78.54$. This tire will roll 78.54 in. (or $78.54 \div 12 = 6.545$ ft) each revolution. There are 5,280 ft in a mile, so dividing that by 6.545 gives us 806.7 revolutions per mile. Now, if the car is going 50 mph, the tire will be turning $806.7 \times 50 = 40,336.13$ revolutions per hour, or $40,336.13 \div 3,600 = 11.2$ revolutions per second (there are 3,600 seconds in an hour). At this speed, a first-order vibration from an unbalance or runout problem would cause a vibration with a frequency of 11 Hz. A second-order vibration would be $2 \times 11 = 22$ Hz, and a third-order vibration would be $3 \times 11 = 33$ Hz.

Driveshaft speed in a RWD vehicle is slower than tire speed by the same amount as the axle gear ratio. If this vehicle has an axle ratio of 3.09:1, we can divide 11.2 by 3.09 and get a speed of 3.6 revolutions per second. A first-order driveshaft vibration (balance problem) would have a frequency of 3.6 Hz. At 50 mph, a second-order vibration (improper angles or phasing) would have a frequency of $3.6 \times 2 = 7.2$ Hz.

Now let us say the customer complaint was a vibration at 50 mph, and we were to measure a vibration frequency of 20 Hz at that speed. From our calculations, we know that this is very close to the speed of a second-order tire or wheel problem. By concentrating our checks on the tires and wheels, we should be able to locate the cause.

At first glance, the calculations seem tedious and too much work. By using a calculator, they took less than 5 minutes, and the results point directly toward the source of the problem. Some of us have seen mechanics spend hours balancing and rebalancing tires or replacing various parts trying to find the cause of a vibration.

28.4 Tire Wear Inspection

As a tire wears out, the wear should occur evenly across the tread. When a tire is worn to a point where there is less than 2/32 in. (1/16 in., 1.587 mm) in two adjacent

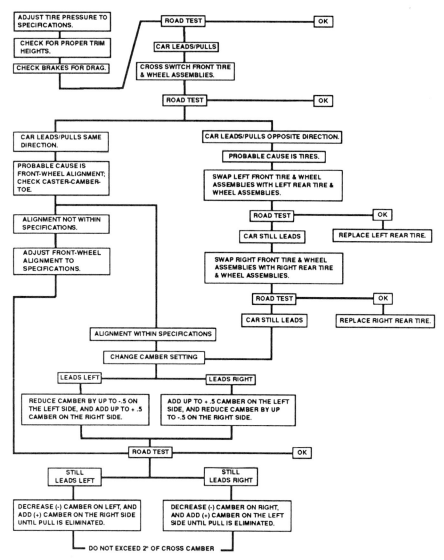

Figure 28-14 This is a procedure to locate the cause of a lead or pull condition. We begin at the top and work our way downward until the cause is located and cured. *(Courtesy of General Motors Corporation, Service Technology Group)*

tread grooves, the tire is legally worn out in many states. Because of the loss in the ability of water to run out through the tread, a worn-out tire will give poor driving control, stopping control, and stopping ability if operated on wet pavement. Besides being unsafe, the probability of flats and blowouts also increases as the tread wears thinner. A thinner tread becomes more vulnerable to cuts and penetration from road objects as the tread rubber wears off. Tire wear is checked visually by looking at the tread, by feeling the tread, and by using a tread depth gauge. Some tire shops rub a tire crayon on a spinning tire to make any defects show up better (Figure 28–15).

Tread depth can be quickly checked by observing the wear bars, also called tread wear indicators. These are raised areas, 1/16 in. (1.6 mm) high, running across the bottom of the tread grooves at six places around the tire (Figure 28–16). Another easy way to check tread depth is to place a U.S. penny in the tread groove, as shown in Figure 28–17. The distance from the edge of the penny to the top of Abraham Lincoln's head is a little greater than 1/16 in. A more accurate method of checking tread wear is to use a tread depth gauge (Figure 28–18). This gauge allows a careful check of each tread groove and will show normal or abnormal wear much better than an unaided visual inspection. A regular check with a tread depth gauge can give an early warning if inflation pressures or alignment angles are not correct.

Normal tire wear occurs evenly across the tread. If a tire wears faster on one side or location than another, something is wrong (Figure 28–19). Depending on the

(A)

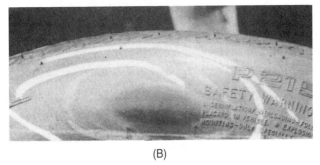

(B)

Figure 28-15 A tire crayon was rubbed on a tire (A) while it was spinning on a balancer; the darker areas show cupping and other unusual wear patterns. The bulge of a tire (B) is a sign of internal separation. *(B is courtesy of B.F. Goodrich. © The B.F. Goodrich Company)*

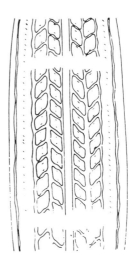

Figure 28-16 Wear bars are raised strips across the bottoms of the tread grooves. When they appear, the tire is worn out. *(Courtesy of General Motors Corporation, Service Technology Group)*

Figure 28-17 A penny placed in the tread groove with Lincoln's head upside down. If you can see the top of his head, the tire is nearly worn out.

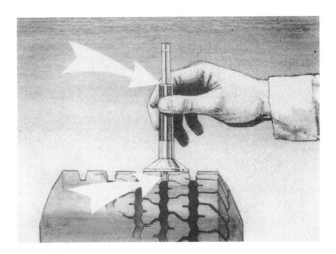

Figure 28-18 A tread depth gauge. The measuring stem (lower arrow) should touch the bottom of the tread groove; the depth of the groove is read on the scale at the top arrow. *(Courtesy of B.F. Goodrich. © The B.F. Goodrich Company)*

location of the wear, a competent technician can tell if the tire has been inflated incorrectly, driven severely, aligned wrong, or mounted on a suspension with worn parts (Figure 28-20).

Incorrect tire pressures cause a tire tread to wear abnormally fast at either the center, if the pressure is too high, or at the edges, if it is too low (Figure 28-21). A properly inflated tire has a relatively even road contact pressure across the tread. Excessive pressure tends to

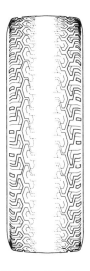 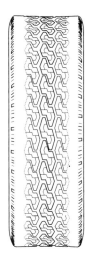 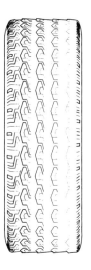 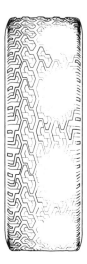 

Center Rib Wear
CAUSES:
TIRE AIR PRESSURE ABOVE MFG. SPECIFICATION, BELT SEPARATION.

Outer Edge Wear
CAUSES:
AIR PRESSURE BELOW MFG. RECOMMENDATIONS, HIGH SPEED CORNERING, HIGH CASTER AFFECTING WEAR DURING CORNERING. EXCESSIVE POSITIVE OR NEGATIVE CAMBER.

Camber/Toe Wear
CAUSES:
MISALIGNMENT, LOOSE PARTS, INCORRECT RIDE HEIGHT.

Cupped Wear
CAUSES:
LOOSE OR WORN PARTS, IMPROPER WHEEL BALANCE.

Diagonal Wear
CAUSE:
INCORRECT REAR TOE ADJUSTMENT, WEAR WILL VARY DEPENDING ON TIRE.

Figure 28–19 Commonly encountered abnormal tire wear patterns and their probable causes. *(Courtesy of Moog Automotive Inc.)*

CONDITION	RAPID WEAR AT SHOULDERS	RAPID WEAR AT CENTER	CRACKED TREADS	WEAR ON ONE SIDE	FEATHERED EDGE	BALD SPOTS	SCALLOPED WEAR
EFFECT							
CAUSE	UNDER-INFLATION OR LACK OF ROTATION	OVER-INFLATION OR LACK OF ROTATION	UNDER-INFLATION OR EXCESSIVE SPEED*	EXCESSIVE CAMBER	INCORRECT TOE	UNBALANCED WHEEL OR TIRE DEFECT*	LACK OF ROTATION OF TIRES OR WORN OR OUT-OF-ALIGNMENT SUSPENSION.
CORRECTION	ADJUST PRESSURE TO SPECIFICATIONS WHEN TIRES ARE COOL ROTATE TIRES			ADJUST CAMBER TO SPECIFICATIONS	ADJUST TOE-IN TO SPECIFICATIONS	DYNAMIC OR STATIC BALANCE WHEELS	ROTATE TIRES AND INSPECT SUSPENSION SEE GROUP 2

*HAVE TIRE INSPECTED FOR FURTHER USE.

Figure 28–20 Another set of tread wear patterns with their causes and corrections. *(Courtesy of Chrysler Corporation)*

Figure 28–21 Overinflation will cause excessive wear at the center of the tread (left), underinflation will cause a tire to wear fastest at the sides of the tread (middle), and correct inflation will give an even tread-to-road contact and wear. *(Courtesy of Moog Automotive)*

push the center of the tire harder onto the road, whereas too little pressure does not push hard enough. Correct tire pressure depends on the size of the tire and the weight of the car. The recommended pressure is usually printed in the owner's manual and on a decal that is attached to the car (Figure 28–22). A belted tire often tends to hide or reduce the effects of incorrect pressure.

Checking and maintaining the correct pressure is the most important thing a motorist can do to obtain maximum

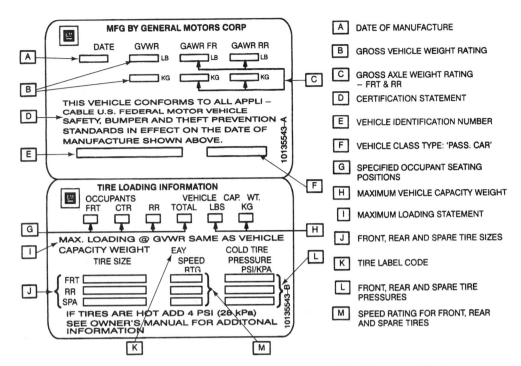

Figure 28–22 All cars, pickups, vans, and light trucks have a decal or placard like this to provide the correct tire inflation pressure for that particular vehicle. *(Courtesy of General Motors Corporation, Service Technology Group)*

tire life and safety. When checking tire pressure, it is important to use an accurate gauge and to observe the following recommendations:

1. Check tire inflation once a month and before taking a major trip.
2. Check inflation when a change in weather temperature occurs. A temperature change of 10°F (6°C) causes a tire pressure change of about 1 psi.
3. Check the pressure when the tires are cold or after driving only a short distance, 1 mi. (1.6 km) or less.
4. Never reduce the pressure from a hot tire. The pressure of a hot tire should be greater than normal.
5. Make sure that all tire valves have a valve cap to keep dirt and moisture out of the valve.
6. Most tire manufacturers and tire service technicians recommend running a tire at the maximum inflation pressure printed on the sidewall for the best tire service and performance.

A bias-belted tire often shows rapid wear on the tread rib that is next to the outside on each side of the tire. This unusual tread wear is normal for this type of tire. It occurs because the centrifugal force acting on the edges of the belt tends to increase the road pressure on these two tread bars.

Figure 28–23 Incorrect camber angles will cause a tapered tire wear that begins at one of the tread shoulders. *(Courtesy of SPX Corporation, Aftermarket Tool and Equipment Group)*

Narrower, steel-belted radial tires mounted on front wheels often wear the shoulder areas and leave high tread in the center. This wear pattern will appear to be underinflation wear, but it can occur even if the tires are operated at maximum inflation pressures. This wear pattern may be caused by a fault in the tire's internal construction (pyramid belts) or by the tire rolling under during turns.

A tire that is cambered excessively tends to have faster tread wear on the side of the tire that it is leaning toward. The tread wear will be greatest at the outer groove, with less wear at the groove next to it, and still less wear at each additional groove across the tread. Also, there will be a fairly sharp corner between the worn tread and the sidewall (Figure 28–23). This sharp corner indicates an alignment problem that is covered more thoroughly in Chapters 19 and 34.

A tire that is toed in or toed out excessively will be forced to scuff sideways as it rolls down the road. With radial tires, a scuffing wear will occur at the outer shoulder if the tires have excess toe-in. Excess toe-out will produce scuffing wear at the inner shoulders (Figure 28–24).

Bias-ply tires tend to wear in a featheredge or sawtooth pattern that is easier felt than seen. Run your hand across the tread from the outside to the inside and then back to the outside. Toe-in wear is most noticeable as you pass your hand toward the outside; toe-out wear will cause a featheredge pattern in the opposite direction, with the sharp edge toward the outside. These patterns indicate an alignment problem.

A type of wear pattern similar to positive camber wear, or wear on the outside edge, can be caused by hard cornering. This wear tends to cause a rounded shoulder, whereas camber wear causes a sharper corner at the edge of the tread. When a tire is driven hard into a corner, the tread will tend to roll, causing a heavy load on the shoulder and sometimes even on the sidewall. This wear pattern can also be caused by an incorrect turning radius or toe-out on turns.

Spotty wear that occurs in one or more spots around the tire is usually caused by improper balancing, worn suspension parts, or worn shock absorbers (Figure 28–25). This wear pattern, often called **cupping,** is easier felt than seen, at least in the early stages. Pass your hand around the tire and feel for spotty, worn areas. Incorrect balance causes excessive tread pressure in one or more locations because the heavy spots of the tire spin with the tire and push the tread harder into the road. Worn suspension parts allow the tire to change camber or toe angle as the tire spins. These intermittent changes in alignment cause intermittent camber or toe wear in spots of the tread. Occasionally, this type of tire wear can also be seen on properly balanced tires mounted on a good suspension system. In these cases, it may be caused by internal tire construction in the belt area. Cupping is also occasionally seen on the rear tires of FWD cars. In these cases, it may be started by a severe brake application, causing skidding on the lightly loaded tread. One worn spot on a tire can cause the tire to hop, which, in turn, will cause additional flat spots. Flat spots produce a tire wear pattern called diagonal wipe (Figure 28–26).

Figure 28–25 These two tires show spotty tire wear that was caused by loose suspension parts in combination with wrong alignment settings; the loose parts allowed the tires to move or dance around, producing a scrubbing action. *(Courtesy of Moog Automotive Inc.)*

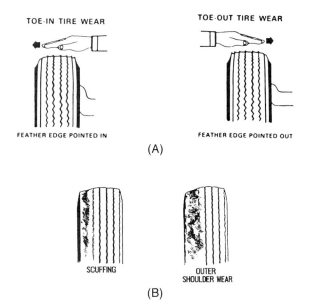

Figure 28–24 Incorrect toe will cause a featheredge wear pattern across the tire tread of a bias-ply tire that can be felt as you slide your hand across it (A). With radial tires (B), toe wear will show more as scuffing or outer shoulder wear. *(A is courtesy of SPX Corporation, Aftermarket Tool and Equipment Group; B is courtesy of Hunter Engineering Company)*

Figure 28–26 Diagonal wipe is a rear tire wear pattern caused by excessive toe. *(Reprinted by permission of Hunter Engineering Company)*

28.5 Tire-Related Problem Diagnosis

Faulty tires or wheels can cause several different driving problems that usually fall into the categories of pull and vibration. Car **pull** is a tendency of the car to try to turn right or left by itself. Pull, sometimes called **lead,** is normally caused by a fault or a misalignment of one or both front tires.

Car vibrations are caused by one or more parts that spin or turn rapidly. Annoying vibrations can be seen, heard, or felt. Vibrations occur when the spinning objects have runout or are unbalanced. Tire and wheel problems cause more vibration complaints than anything else on the car. Runout occurs when an object spins off center or wobbles. The intensity of the vibration usually depends on the amount of unbalance and the speed of the spinning object; in some cases, however, it depends on the harmonic frequency of the things affected by it. A harmonic vibration occurs when the frequency of the vibration force of an unbalanced or runout condition matches the frequency of the suspension. The harmonic frequency of the car's suspension is usually most affected by tire unbalance or runout in wheel speed frequencies equivalent to 40 to 60 mph (64.4 to 96.6 km/h).

28.6 Tire Pull

Conicity is a term that describes a tire's tendency to pull. It is caused by a tire being built with a belt that is off center or by a belt that is slightly shorter on one side than the other. It can also be caused by the beads not being on the same plane. Anyone who has tried to roll a cone knows that it will not roll straight. It rolls in a circle, with the center of the circle being the point of the cone. A tire with a conicity problem will also try to roll in a circular fashion, just like the cone. Tire conicity problems usually show up when a new tire is first mounted on a car, but occasionally a tire can develop conicity as it wears (Figure 28–27).

A tire pull problem usually shows up when the tires are changed or rotated. It is fairly easy to determine if a pull is caused by a tire problem. Merely switch the two front tires; if the pull direction changes, the pull is caused by one or both of the tires. Determining which of the two tires is the culprit is almost as easy. Switch one of the front tires with a rear tire or one that you know is good. Again, if the pull changes, you have found the problem tire. If it is new, a tire with a conicity problem can be returned to the dealer, or it can be run on the rear with no problems. It is possible to run this tire on the front with an adjustment in the amount of caster spread (see Chapters 19 and 34) if the car has adjustable caster angles, but another caster adjustment will be required when that tire is changed or moved (Chart 28–2).

Torque steer can be caused by a tire problem. This pull occurs only during acceleration or deceleration; there is no pull when the car is driven at even speeds. Torque steer is caused by one of the two tires on the drive

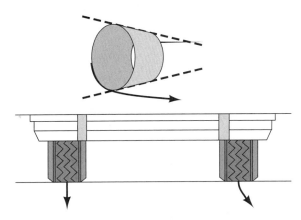

Figure 28–27 Tire conicity will cause the tire to turn in the direction of the cone's apex and to pull or turn the steering mechanism in that direction. *(Reprinted by permission of Hunter Engineering Company)*

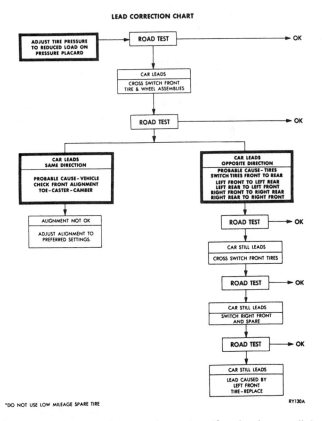

Chart 28–2 Procedure to determine if a lead or pull is caused by a tire or wheel alignment and, if tire caused, which is the problem. *(Courtesy of Chrysler Corporation)*

axle. Some tire types are more prone to torque steer if the tire pressures are unequal. The term *torque steer* is usually used to describe a phenomenon encountered with FWD axles, which is described in Chapter 12.

28.7 Radial Tire Waddle

If the belt is placed in a crooked position as the tire is built, **radial tire waddle** will occur. The belt tends to go straight down the road, which can force the tire and wheel to waddle back and forth. Waddle usually occurs at slow speeds, about 5 to 30 mph (8 to 50 km/h). Most noticeable at the rear, it feels like someone is slowly shaking the back of the car back and forth, in time with the tire rotation. If this tire is at the front, it can cause the front end to appear as though it is moving back and forth. A tire that causes waddle will often also cause rough operation at 50 to 70 mph (80 to 110 km/h). Roughness or harshness is a problem that can be felt, heard, or both, much like vibration, except that the roughness is irregular.

A tire with a tread or ply separation can cause problems similar to radial waddle. At low speeds, it can feel the same as radial tire waddle but becomes a vibration at higher speeds. A separated tire can be located by careful inspection for bulges or soft spots.

28.8 Vibrations

A **vibration** is a regular motion that can be felt, seen, or heard. In most cases, tire-caused vibrations are speed sensitive; that is, they come and go and change intensity depending on the speed of the car. The suspension system is able to absorb the tire motions until their speed matches the resonant frequency range of the suspension. Knowing the speed at which the vibration occurs will often give the technician a clue as to the nature of the vibration. The possible tire-, wheel-, and suspension-related vibrations and the probable speeds at which they occur are as follows:

Tire wear (spotty or uneven)	Felt at 30 to 70 mph
Radial tire runout	Felt at 20 to 70 mph
Lateral tire runout	Felt at 60 to 70 mph
Tire balance	Heard and felt at 30 to 60 mph
Wheel bearing	Heard at 0 to 70 mph Felt at 50 to 60 mph

With the proper equipment, a technician can measure the speed of the vibration in cycles per second or hertz. Then, by determining the speed of the tire (by comparing the tire size and gear ratio to the vehicle speed) or the speed of the drivetrain, he or she can determine the cause of the vibration.

The procedure for eliminating a tire-caused vibration is usually to check and cure, if necessary, tire radial and lateral runout, check and cure, if necessary, faulty wheel bearings, and then rebalance the tire (Chart 28–3).

Most tire shops use a procedure similar to the following to cure a tire vibration:

1. Make a preliminary check for tire defects and mount the tire and wheel on an off-car balancer. With the tire mounted, make a more thorough check for flat spots, separation, and other defects.
2. A quick check for lateral and radial runout is made by watching the edges of the tire tread while the tire is spinning on the balancer. If runout appears excessive, a more accurate check will be made. These checks are described in Sections 28.9.1 and 28.9.2.
3. The balance of the tire and wheel is checked and corrected, if necessary.
4. The tires are replaced on the car, and a test drive is made to check the results.
5. If a problem still exists, the tire and wheel are spin balanced on the car. While doing this, the operator should watch for bad wheel bearings, a bent hub or axle flange, faulty motor mounts or driveline, as well as any other defect that might cause a vibration (Chart 28–4).

28.9 Tire and Wheel Runout

Runout is a commonly used term to describe something that does not spin evenly or true. A tire with **lateral runout** wobbles from side to side; a tire with **radial runout** hops or wobbles up and down (Figure 28–28). A tire with runout cannot run true; it will try to move the suspension up and down or in and out, which will result in a vibration. Excessive runout is easily seen by lifting the tire off the ground and spinning it. Watch the position of the tread relative to some stationary object. If the tread appears to move up and down, it has radial runout; if it appears to move sideways, it has lateral runout. The amount of runout is determined by measuring it with a runout gauge or a dial indicator. If a standard dial indicator is used, it should have a wide tip to ensure that the tip will not get caught in the tread grooves.

Chapter 28 ■ Suspension and Steering System Diagnosis and Inspection

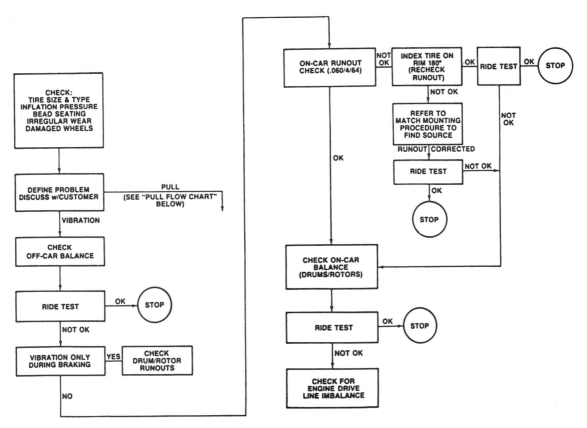

Chart 28–3 Cause and correction procedures to cure a vibration. Note that many of them are tire problems. *(Courtesy of Chrysler Corporation)*

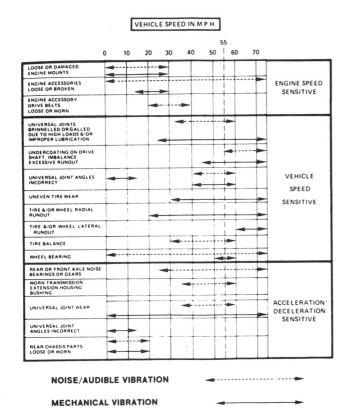

Chart 28–4 Most vehicle vibrations occur at certain speed and load conditions, giving the technician a valuable clue as to what is causing them. *(Courtesy of Dana Corporation)*

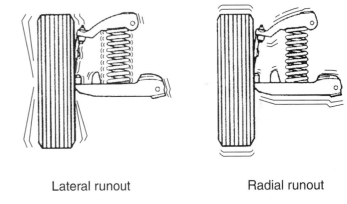

Lateral runout   Radial runout

Figure 28–28 Lateral tire runout can cause a sideways tire shake, whereas radial runout can cause a hop. *(Courtesy of Ford Motor Company)*

28.9.1 Radial Tire and Wheel Runout

Radial runout of the tire tread can be caused by an out-of-round or nonconcentric tire, an out-of-round or nonconcentric wheel, a nonconcentric hub bolt pattern, or a tire with a variation in sidewall stiffness. Tire flat spotting affects radial runout. If a tire has been sitting for a while with the car's weight on it, the car should be driven to warm up the tires and round out any flat spots before checking runout (Figure 28–29).

Figure 28–29 A runout gauge being used to measure radial tire and wheel runout. *(Courtesy of Hickok Incorporated)*

Tire stiffness variation, or a tire that has strong or weak portions of the sidewall or belt, can only be checked with the tire and wheel still mounted on the car and carrying the weight of the car. The device used to make this check, called **loaded radial runout,** is a tire problem detector (TPD). A TPD is a motor-driven roll that rotates the tire and measures any up-and-down motions of the car that occur because of the tire's rotation. This device is no longer on the market in a simple form; an on-car tire truer uses a TPD sensor. A tire with a stiffness variation problem, such as one with radial runout, can sometimes be saved by index mounting the stiff or high spot of the tire on the low spot of the wheel. Also, it can usually be saved by truing the tire using an on-car tire truer (Figure 28–30).

Unloaded runout is commonly checked in garages and tire shops. The technician lifts the tire off the ground, places an indicator at the center of the tread, and rotates the tire while watching the indicator movement. If you are checking a tire with an aggressive tread, a strip of tape can be placed around the tread to provide a smoother checking surface. The amount of radial runout that a car will tolerate varies depending on the size and type of tire, the weight of the car, and the tolerance or sensitivity of the driver. Radial runout tolerances are usually listed in the manufacturer's manual. A rule of thumb is that a very

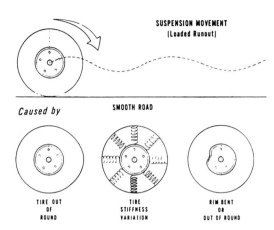

Figure 28–30 The most common causes of loaded radial runout, which in turn can cause vertical suspension movement. *(Courtesy of General Motors Corporation, Service Technology Group)*

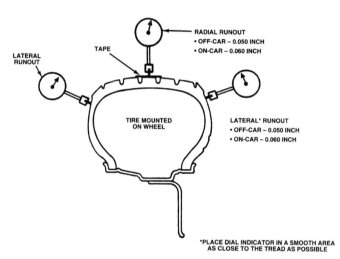

Figure 28–31 Radial and lateral tire runout is checked by positioning a dial indicator on the tire as shown. Note the tape to provide an even surface on tires with a rough tread pattern. *(Courtesy of General Motors Corporation, Service Technology Group)*

sensitive driver might notice and complain about 0.030-in. (0.75-mm) runout. Most drivers will notice and find objectionable 0.060-in. (1.5-mm) runout, and everyone will probably complain about runout greater than 0.090 in. (2.3 mm) (Figure 28–31).

If excessive runout is found at the tire tread, the high spot should be marked, and then the wheel should also be checked for runout. To measure the wheel's radial runout, the indicator is mounted at the inner side of the wheel's bead seat, and the wheel is rotated while the technician watches for indicator movement. A more accurate method of measuring wheel runout is to remove the tire and check for runout at the tire side of the bead seats. The manufacturer's manual should be checked to determine the allowable runout. The normally accepted limit for

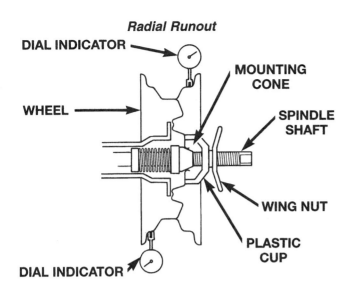

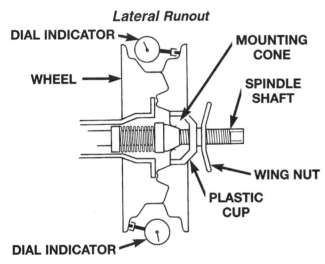

Figure 28-32 Wheel runout is checked by putting the wheel back on the hub or on the spindle of a wheel balancer, positioning a dial indicator as shown, and rotating the wheel. *(Courtesy of Chrysler Corporation)*

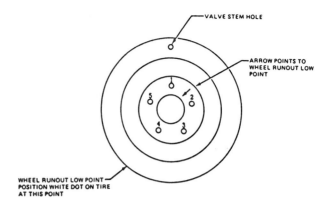

Figure 28-33 One of the systems used for match mounting a tire and wheel. Match mounting reduces runout to produce a smoother running assembly. *(Courtesy of General Motors Corporation, Service Technology Group)*

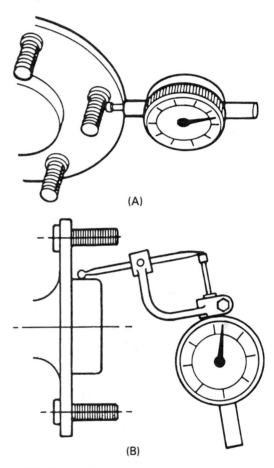

Figure 28-34 Radial tire and wheel runout can be caused by runout of the wheel stud circle (A) or hub wheel pilot (B). Runout can be checked using a dial indicator as shown here. *(Courtesy of Ford Motor Company)*

wheel radial runout is 0.030 to 0.040 in. (0.75 to 1.00 mm), which varies slightly depending on the type of wheel. A faulty tire is indicated if the wheel runout is acceptable and the tire runout is not (Figure 28-32).

Occasionally, tire runout can be reduced to acceptable standards by **tire matching**—that is, matching the tire to the wheel. It is also called match mounting, indexing, or vectoring. Tire matching is done on many OEM tire and wheel mountings. Tire stiffness is measured, and the high (or low) spot is marked on the tire. Wheel runout is measured, and the wheel is marked. The tire is then placed on the wheel with the marks in the proper relationship, so the high or stiff spot of the tire is located over the low spot of the wheel. Tire and wheel marking systems vary among manufacturers; at this time, there is no universal system (Figure 28-33).

Excessive wheel radial runout can be caused by lug bolt circle runout.

To measure lug bolt runout, you should:

1. Remove the wheel and mount a dial indicator so the indicator stem is on one of the lug bolts (Figure 28-34).

CAUTION *Be sure the car is properly supported on a hoist or jack stands.*

2. Rotate the hub slightly to get the highest possible reading and adjust the indicator to read zero.
3. Carefully pull back the indicator stem. Do not let the indicator body move as you rotate the hub to bring the next lug bolt into position with the indicator stem.
4. Rotate the hub slightly to get the highest possible reading. Compare this reading with zero; it is the amount of runout between these two lug bolts.
5. Repeat steps 3 and 4 on the remaining lug bolts. The lug bolt circle runout should not exceed 0.015 in. (38 mm).

NOTE: Some wheels are centered using the pilot of the hub or axle; in these cases it is important that the runout of the pilot be checked.

Lug bolt circle runout is corrected by replacing the hub or axle. The effect of lug bolt circle runout can sometimes be reduced by positioning the low spot of the wheel to coincide with the high lug bolt. Usually the quickest way to do this is to measure the tire runout, relocate the wheel one lug bolt different, and remeasure the runout. Continue doing this in the other two or three wheel positions until an acceptable position is found.

When straight-shank lug nuts are used on alloy wheels, wheel runout can result from the clearance between the wheel holes and the shanks of the nuts. Some of this runout can be reduced by snugging the wheel onto the hub or axle mounting flange using two standard tapered lug nuts to center the wheel to the lug bolts. Then two straight-shank nuts should be installed and partially tightened and the tapered nuts replaced with straight-shank nuts. The wheel should then be tightened in place.

28.9.2 Lateral Tire and Wheel Runout

Lateral tire runout can be caused by a faulty tire, a bent or poorly machined wheel, or a bent hub or axle flange. Lateral tire runout is measured by placing an indicator at the side of the tire on a smooth section of rubber that is close to the shoulder of the tread. Next, the tire is rotated while watching the indicator movement. Limits for lateral runout are published by various manufacturers. If they are not available, use the same rule of thumb as for radial runout—0.030 in. (0.75 mm), 0.060 in. (1.5 mm), and 0.090 in. (2.3 mm). If excessive lateral runout is found at the tire, mark the tire and check the lateral wheel runout.

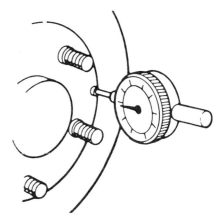

Figure 28–35 Lateral tire runout can be caused by a bent wheel mounting flange; it is checked using a dial indicator mounted as shown here. *(Courtesy of Ford Motor Company)*

Lateral wheel runout is measured by placing the indicator on the side of the bead flange and rotating the wheel while watching the indicator movement. Acceptable wheel runout is less than allowed at the tire, depending on how much closer the checking point is to the center of the hub; 0.010 in. of wobble, 5 in. out from the hub, would become 0.020 in. of wobble if it could be measured 10 in. out from the hub. A faulty tire is indicated if the wheel runout is acceptable while the tire runout is not (Figure 28–35).

If wheel runout is excessive, the hub or axle flange should be checked for runout. Mount the indicator on the side of the hub flange and rotate the hub while watching for indicator movement. About 0.010 in. (0.25 mm) or less runout is acceptable. Excessive flange runout is corrected by replacing the hub or axle. The effect of excessive flange runout can sometimes be reduced by relocating the wheel on the flange using the same procedure that was described for lug bolt runout. Remount the wheel in the different positions, being careful to use the correct amount of torque and the correct pattern to tighten the lug bolts. Measure the wheel runout at each of the possible positions until an acceptable position is found. If an acceptable position cannot be found, the hub or axle will have to be replaced.

28.10 Tire Balance

Tire and wheel balancing is the process of adding weights to the wheel rim to counterbalance heavy spots in the tire and wheel assembly and, sometimes, the brake rotor or drum. If not counterbalanced, these heavy spots are acted on by centrifugal force when the tire is spinning. Centrifugal force can cause the tire to hop up and

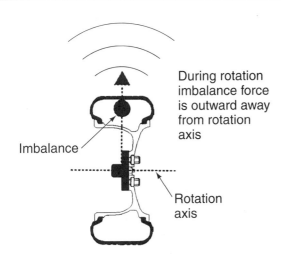

Figure 28-36 When a tire is spun, differing amounts of centrifugal force can cause a tire to go out of balance if there are heavy portions on different radii such as the tire tread, wheel rim, or brake drum. This is called kinetic unbalance. *(Courtesy of John Bean Company)*

down (or at least try to) or wiggle sideways while it is rolling. Balance problems are placed into two general classifications, **static** and **dynamic.**

Static is a term that normally refers to a stationary object. For a tire, static balance refers to the distribution of weight around the wheel. Any heavy spots on the tire should be counterbalanced by an equally heavy weight on the other side of the wheel.

Static balance probably got its name because the original balancers checked balance with the tire stationary on a bubble balancer or turning slowly, by gravity, on a free-turning arbor. A statically balanced tire will probably spin true about 75 percent of the time, but in some cases it will still try to hop or wiggle. Several styles of spinning static balancers are available to correct these problems. They are called **kinetic** or **single-plane** (radial) balancers. Kinetic balancing is recommended over static balancing because a certain weight on the rim can counterbalance a certain weight on the tread while the tire is static; however, when the wheel is spinning, the different radius of each point will generate differing amounts of centrifugal force, which would unbalance the tire (Figure 28-36).

Dynamic is a term that normally refers to a moving object or thing. A tire that is unbalanced dynamically has a heavy spot on the inside or the outside. Centrifugal forces cause the heavy spot to try to move to the tire centerline when the tire spins. If the heavy spot moves toward the centerline when it is in a forward position, and then also in a rearward position, it causes the tire to wiggle back and forth, or shimmy. Dynamic unbalance is noticeable on front, steerable wheels because they are the only ones that will allow this sideways movement. Dynamic balance can only be checked by spinning the tire and wheel. A dynamic balancer is sometimes called a **two-plane** balancer because it can balance a tire in the radial plane as well as the lateral plane. Most spin balancers can be dynamic or kinetic or both dynamic and kinetic.

28.11 Wheel Bearing Problems: Diagnostic Procedure

Faulty or excessively loose wheel or axle bearings can appear to the driver of the car as noise, road wander, wheel shake, play in the steering, cuppy tire wear, or a low brake pedal on disc brakes. If some or all of these problems are encountered, a systematic procedure should be followed to determine if loose or faulty wheel or axle bearings are at fault and which particular ones are faulty.

The first step in diagnosing faults such as these is to check the tires for proper inflation and wear. After adjusting the tire pressure to normal, if necessary, and ensuring that the tires are in a safe condition, road test the car. When road testing, first drive at varying speeds until you find the conditions that make the problem show up. When noise is the complaint, try to determine from which corner of the car the noise is coming. Then, if traffic and road conditions permit, make slow-to-moderate left and right turns. Faulty wheel or axle bearings usually change noise level as the vehicle load on them changes because of the weight transfer. During the road test, apply the service and parking brakes lightly. A reduction in noise level as the brake shoes apply pressure on the drums and rotors indicates a faulty wheel bearing. Use caution while doing this test. It is possible for the parking brake to lock up the rear wheels and cause a skid. Apply the brakes gently and be ready to release them quickly if necessary (Chart 28-5).

If the road test confirms a problem, raise the car on a hoist and check for loose bearings. Try pushing the tire and wheel straight up and down, pushing straight in and out, and rocking it sideways. Depending on the bearing type, some of these motions are not permitted and some of them are. A vertical motion is not permitted in any wheel or axle bearing type. More than barely perceptible in-and-out motion is permitted only on C-lock axle RWD rear axle bearings. On these axles, end play up to 0.010 in. (0.26 mm) is permitted on most axles; some axles are allowed up to 0.030 in. (0.79 mm). Tapered roller wheel and axle bearing sets (i.e., a pair of bearings) should have about 0.001 to 0.005 in. (0.03 to 0.13 mm) of end play. Nonserviceable wheel and axle bearings are allowed up to 0.005 in. (0.13 mm) of end play.

When rocking a tire and wheel to check for bearing looseness, grip the tire at the top and bottom. Push inward

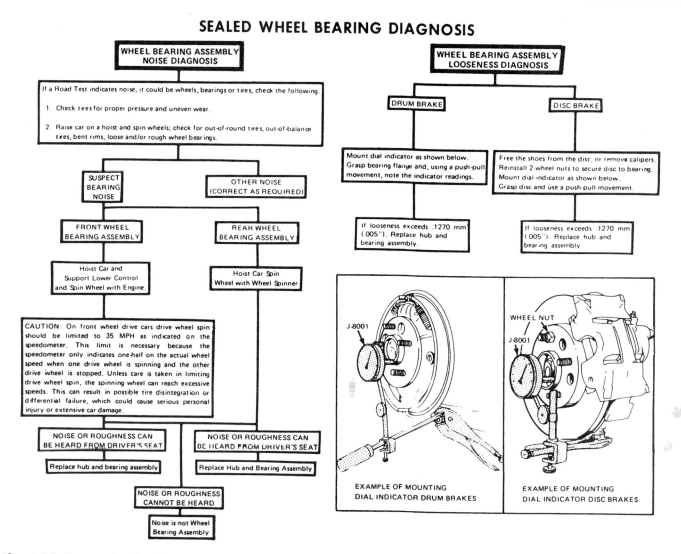

Chart 28-5 Procedure for diagnosing problems with sealed, nonserviceable wheel bearings. The same procedure is used for serviceable bearings except for the repair steps. (Courtesy of General Motors Corporation, Service Technology Group)

with one hand while pulling outward with the other. Then, reverse these motions. A perceptible motion is acceptable on tapered roller wheel bearings and axle bearings on independent suspensions. More than barely perceptible rocking motion is not permitted on ball-type wheel or rear drive axle bearings (Figure 28-37).

Next, spin a nondrive tire and wheel by hand or by a wheel balancer spinner motor. Spin a drive axle tire and wheel with the engine. As the tire spins, listen for a harsh, grating sound. You can often confirm the sound origin and the bearing roughness by gently placing your fingertips on the steering knuckle or axle housing, as close to the bearing as possible. A rough wheel or axle bearing will cause a rough, irregular feel.

The proper repair method can be determined after the faulty bearing has been located and the type of bearing has been identified. Loose bearings on some types can be adjusted; others require replacement. Rough bearings of some types require replacement of the whole assembly; on others, only the faulty bearing part needs replacement.

SAFETY TIP: Always use extreme caution when moving or working around a spinning tire and wheel. When spinning a tire with the engine, limit the tire speed to about 55 mph (88 km/h). Do not forget that the differential gears can cause one of the tires to run at twice the speed shown on the speedometer, and excessive speed can cause tire explosion. Higher speeds are not necessary.

Figure 28–37 When checking for loose bearings, push in at the top of the tire while pulling outward at the bottom; then reverse these pressures while feeling for excessive play. A small amount of play is normal *(Reprinted by permission of Hunter Engineering Company)*

28.12 Suspension and Steering System Inspection

The purpose of an inspection is to determine the cause for the vehicle owner's complaint and to determine what steps will be needed to cure that complaint. It is a good practice to note any other parts that show signs of failing in the near future so the customer can be aware of them. The suspension should operate for many miles and a year or so until the next time it is inspected; the average motorist does not check suspension components very frequently (Figure 28–38).

SAFETY TIP: A suspension, steering, or brake failure can place the car and its passengers in a highly dangerous situation. While making an inspection, this fact is in the forefront of the front-end technician's mind. Any item that might fail in the near future and cause an accident is noted and brought to the car owner's attention.

Sometimes an inspection will determine that a simple adjustment or realignment is all that is needed to correct the situation; often a worn bushing or ball joint will show up. Worn parts must be replaced before an

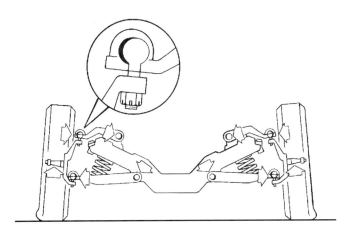

Figure 28–38 As the suspension parts wear, they allow the alignment of the front tires to change and permit uncontrolled, sloppy tire movement. *(Courtesy of McQuay Norris)*

alignment can be done. It does no good to do a wheel alignment if the suspension parts are sloppy. In most cases on an older car, when a realignment is necessary, that need is probably caused by worn parts or sagged springs. Remember that the rear wheels also have a suspension system and that their parts also wear out. Rear suspension bushings and pivots are checked in the same manner as those at the front.

As an inspection is being performed, it is a good practice to follow a set procedure to ensure that portions of the suspension and steering systems are not skipped or forgotten. When checking a modern car, another good practice is to note the instrument panel lights as you start the engine. They will indicate if the car has ABS or an airbag and if these systems are operating properly. A suspension and steering system inspection should include checks of the:

1. Steering wheel for excessive steering looseness or binding
2. Tires for correct inflation
3. Tires for wear pattern to give an indication of incorrect alignment, balance, or worn parts, and also for physical defects that might cause failure
4. Vehicle for correct height and attitude
5. Vehicle for optional springs, shock absorbers, or overload devices that might change the ride quality or alignment
6. Tire spinning (by hand) for tire runout and wheel bearing condition
7. Tire and wheel shake (top and bottom) for wheel bearing looseness

8. Tire and wheel shake (side to side) for steering component looseness
9. Ball joints for excessive looseness, boot condition, and binding
10. Control arm bushings for wear or deterioration
11. Strut rod bushings for wear or deterioration
12. Stabilizer bar bushings and end links for wear or deterioration
13. Springs for loose or broken parts
14. Shock absorbers or strut for leakage, loose or broken mounts, or broken parts
15. ABS sensors, sensor wires, and tone rings for damage (Figure 28–39)
16. Tie-rod ends for looseness or torn boots
17. Steering gear, center link, and idler arm for loose pivot points or loose mounting bolts
OR Steering rack and tie-rods for loose mounting bushings, loose inner tie-rod ends, or torn boots or bellows
18. Power steering pump, hoses, and gear for leaks
19. Power steering pump drive belt for wear and adjustment
20. Power steering pump for correct fluid level and fluid condition

OR Manual steering gear for correct lubricant level

If the owner complained of vibration, the tires should be checked for excessive runout and spun up to speed to check for an unbalanced condition. Loose, worn, or misaligned suspension or steering parts do not normally cause vibrations; vibrations originate from a part that spins or rotates.

Some of the suspension checks are made with the tires on the ground; most of them are made with the vehicle raised on a frame contact hoist to provide good access to the tires and wheels, suspension components, and steering components. As indicated in the steps, some of the inspection procedures have already been covered in earlier chapters. This chapter describes the procedure for checking the remaining components.

Many technicians follow an inspection checklist such as the one illustrated in Figure 28–40. The checklist helps ensure that none of the checks is missed or forgotten, and it allows a more professional discussion with the car owner.

CAUTION *Serious problems can result from the improper replacement of a tire and wheel. The proper procedure for this operation is described in Section 20.2.*

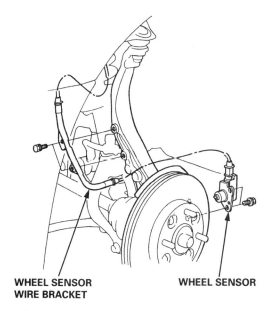

Figure 28–39 Many modern cars with ABS have a wheel speed sensor mounted on the front suspension and a tone wheel or reluctor ring on the front hub or CV joint. *(Courtesy of American Honda Motor Co., Inc.)*

28.13 Spring and Shock Absorber Inspection

An inspection of the springs and the shock absorbers often begins with a customer complaint of noise, tire wear, low vehicle (one end, one side, or all over), excessive vehicle leaning on turns, or front-end dive under braking. Any of these complaints might indicate weak or broken springs or shock absorbers. The best test for weak shock absorbers is a road test followed by a visual inspection as described here. Begin your inspection by parking the car on a smooth, level surface so you can perform a **bounce test** and a **height check**. A **computerized suspension analyzer** has been developed that can give a quick, 2-minute inspection and provide a printed readout of the adhesion and damping ability of each suspension. The readout gives the overall ability and the balance between the left and right sides (Figure 28–41).

28.13.1 Bounce Test

The bounce test is a simple and quick test that should give an indication of the condition of the suspension system.

To perform a bounce test, you should:

1. Grip one end of the bumper and alternately pull upward and push downward several times until you get that corner of the car bouncing up and

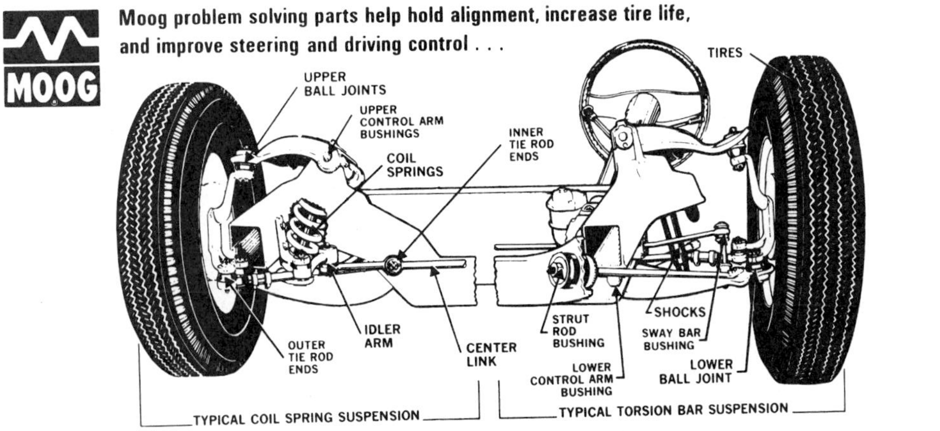

YOUR CAR DESERVES A

WHEEL TO WHEEL SECURITY CHECK

Owner______ Date______ Phone______

Make______ Model______ Year______

Mileage______ License Number______ Engine Size______

PARTS DESCRIPTION	OK	COMMENTS	PARTS	LABOR
SPRINGS				
CONTROL ARM BUSHINGS				
POWER STEERING				
LOWER BALL JOINT				
UPPER BALL JOINT				
WHEEL BRGS.				
BALANCE				
TIRES				
TIE ROD ENDS				
IDLER ARM				
PITMAN ARM				
CENTER LINK				
SWAY BAR BUSHINGS				
STRUT ROD BUSHINGS				
SHOCK ABSORBERS				
ALIGNMENT				
REMOVED PARTS REQUESTED	YES / NO	SUB TOTAL		
		TOTAL		

CAR HEIGHT	Left	Right	Specs.
FRONT			
REAR			
BALL JOINT READINGS			
Load Carrier	Left	Right	Specs.
AXIAL			
RADIAL			
ALIGNMENT			
	Left	Right	Specs.
CAMBER			
CASTER			
TOE			

REMARKS:

Inspector

FORM 2000C • Litho in U S A

Figure 28–40 Many technicians follow a checklist like this to ensure that they do not skip any checks and also to give the car owner a record of what was found during the inspection. *(Courtesy of Moog Automotive Inc.)*

down as far as you can. While the car is bouncing up and down, listen for any unusual noises that might indicate worn or broken parts.

2. With the car at the upper end of a bounce, release the bumper and watch the remaining oscillations until they stop. Two or more oscillations indicate the possibility of worn shock absorbers or, less likely, worn front-end bushings (Figure 28–42).

3. Repeat steps 1 and 2 at the other end of the bumper and compare the bouncing action of the two sides of the car. They should be the same; a difference indicates a weak shock absorber or worn suspension bushings.

4. Repeat steps 1, 2, and 3 at the other end of the car. Do not compare the number of bounces of the front with the rear; they are often different.

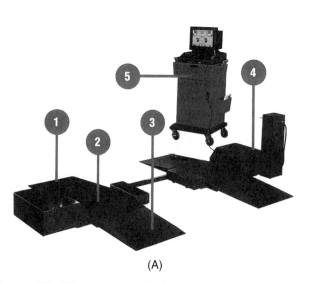
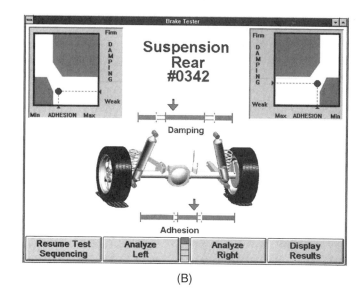

(A) (B)

Figure 28–41 A computerized suspension analyzer (A). The vehicle is positioned onto the ramps (2), and the device bounces the vehicle and measures the weight change during the bounce cycles. It displays the response (B), comparing the right and left sides, and prints out a comparison of the side-to-side balance and effectiveness at both the front and the rear. *(Reprinted by permission of Hunter Engineering Company)*

A car with no more than one or two bounce oscillations after releasing the bumper at each corner of the car and with smooth, quiet operation probably has good springs and shock absorbers, if the height and ride quality are good. Unusual or excessive noises, a differing number of bounce oscillations at each side of the car, or excessive bouncing indicates a need to follow up the bounce test with one of the remaining tests.

28.13.2 Suspension Ride Height Check

A suspension ride height check is a simple and quick way to determine if the car is too low; weak, sagging springs let the suspension height drop. Excessive spring sag can cause excessive bottoming of the suspension (suspension members striking the bump stops severely) or tire wear and handling difficulties as the suspension system loses its proper geometry.

Ride height is a term meaning the standard no-load height of the vehicle; it is also called **trim height, curb height,** and **chassis height.** Ride height specifications are published by most vehicle manufacturers and are also available in technician service manuals and specialized manuals available from some aftermarket manufacturers (Figure 28–43).

Ride height specifications are not published by all vehicle manufacturers; in these cases, it is possible to look at the position of the lower control arm and the condition of the bump stops to get an indication of whether the vehicle is too low. At ride height, the lower control arm of most independent suspensions should be level or slightly lower at the ball joint than at the inner pivot; sagging

Figure 28–42 A shock absorber bounce test is manually performed by bouncing each corner of the car; with the car bouncing as much as possible, it is released and the bouncing action is observed. If it bounces more than one and a half oscillations, the shock is probably weak. *(Courtesy of SPX Corporation, Aftermarket Tool and Equipment Group)*

springs will cause the pivoting or frame end of the control arm to be lower. If the bump stops show a lot of activity or damage, the springs have probably sagged (Figure 28–44).

The measuring locations for ride height checks vary among manufacturers. Sometimes they are easy-to-measure places like the bottom of the bumper to the ground, the top of the fender opening to the ground, or the bottom of the rocker panel to the ground. If any of these measurements are used, remember that they are affected by the diameter and inflation pressure of the tire and wheel; low tires cause low height readings. Other measuring points can be the distance between the top of the axle and the bottom of the frame, between the con-

Chapter 28 ■ Suspension and Steering System Diagnosis and Inspection

MODEL	YEAR	BODY STYLE	FRONT Dim. Measurement Point	FRONT Specifications	REAR # Dim. Measurement Point	REAR # Specifications
CALAIS, DeVILLE	71-72	All (Exc. ALC)	"A"	3⅞ — 4⅝	"B"	6⅛ — 6⅞
		ALC	"A"	3⅞ — 4⅝	"B"	4⅞ — 5⅝
	73-74	All (Exc. ALC)	"A"	3⅞ — 4⅝	"B"	5⅜ — 6⅛
		ALC	"A"	3⅞ — 4⅝	"B"	4⅞ — 5⅝
	1975	All (Exc. ALC)	"A"	3⅛ — 3⅞	"B"	5¾ — 6½
		ALC	"A"	3⅛ — 3⅞	"B"	4⅞ — 5⅝
	1976	All (Exc. ALC)	"A"	3⅞ — 4⅝	"B"	5⅜ — 6⅛
		ALC	"A"	3⅞ — 4⅝	"B"	4⅞ — 5⅝
	77-79	Std.	"D" minus "E"	2⅛ — 2⅞	"B"	5¾ — 6½
		ALC	"D" minus "E"	2⅛ — 2⅞	"B"	5⅜ — 6⅛

*With Radial Tires, Add ⅜" to Front and ½" to Rear.
#With ALC and ELC depressurized.

Figure 28–43 Suspension height specifications are used to determine if the springs are sagged and the vehicle is too low. The commonly used measuring points are indicated in Figure 28–45. *(Courtesy of Moog Automotive Inc.)*

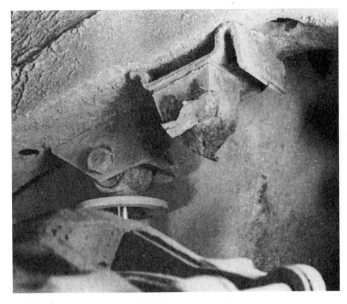

Figure 28–44 This bump stop or strikeout bumper is showing severe damage; the springs are probably sagged on this vehicle, causing a low ride height.

trol arm and the bump stop, or the difference in ground clearance between the ball joint and the inner control arm bushing (Figure 28–45). Measuring this second group is more difficult, but they are not affected by tire diameter. Ride height measurements are usually minimum heights; a car with new springs is normally somewhat higher than specifications. An alternate method with a **locator tool** has been developed by Moog, an aftermarket suspension component manufacturer. The locator snaps onto the wheel and provides a measuring point from the center of the wheel (Figure 28–46).

To measure ride height, you should:

1. Park the car on a smooth, level surface; the ramps of a wheel alignment rack are ideal because they are level and allow easy access to the suspension members.
2. Check for unusual amounts of weight that might be in the trunk or backseat of the car. They should be removed or allowances made for any added weight; ride height specifications are given for unloaded cars.
3. Check the tire pressure and correct it, if necessary. Note whether the tires are of stock size; if not, allowances must be made in the checking dimensions.
4. Obtain the ride height specifications and the locations of the measuring points.
5. Measure the distances at each measuring point and compare them with the specifications. Sagged springs are indicated if the measured distances are shorter or lower than the specifications.
6. Compare the left and right measurements; they should be almost equal.

When one side of the car sags more than the other, it is necessary to determine whether the lean is caused by a weak front spring, a weak rear spring, or both; either will cause this problem (Figure 28–47). The solution is simple; merely lift one end of the car and see if the lean disappears. Carefully position a jack so it is exactly centered under a frame cross member at one end of the car and lift that end of the car until the tires are off the ground.

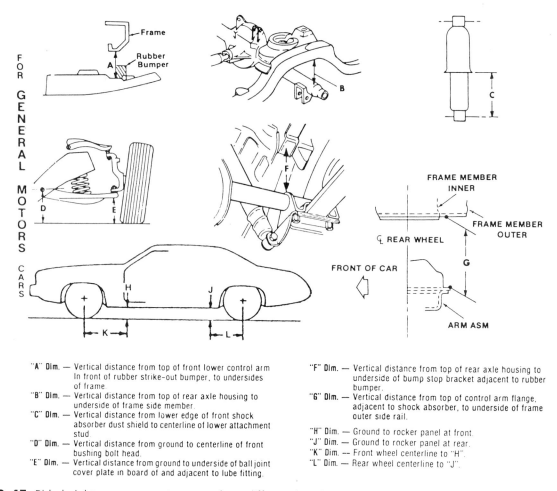

Figure 28–45 Ride height measurements are made at different locations as determined by the vehicle manufacturer. *(Courtesy of Moog Automotive Inc.)*

Observe or remeasure the ride height; if the front end was lifted and a lean still exists, one of the rear springs is weak. If the lean disappears, a front spring is weak (Figure 28–48).

28.13.3 Visual Inspection

Visual inspection is an important check in cases of noises; to make this check, you should:

1. Lift the vehicle on a frame contact hoist or a jack so the suspension will drop to allow a better view of the components.

2. Carefully inspect the springs and shock absorbers for these conditions:

   a. Excessive contact with the bump stops
   b. Loose, worn, or broken shock absorber mounts
   c. Worn, rough, or broken shock absorber piston rod
   d. Badly dented shock absorber body
   e. Leaky shock absorbers (disregard moist spots but condemn the shock if oil is dripping or running down the body of the shock absorber); also look for reddish rust dust
   f. Shiny, worn metal spots on the springs, shock absorber, or nearby portions of the frame or body
   g. Improper positioning of the ends of coil springs
   h. Loose spring or shackle mounting bolts
   i. Worn spring or shackle bushings
   j. Broken torsion bar adjusters, mounting brackets, or insulators

3. If any corner of the car was noisy during the bounce test, extra care should be taken while inspecting that suspension. If the cause of the noise cannot be seen, disconnect one end of the shock absorber and stroke it through its compression and extension travel to feel its internal condition.

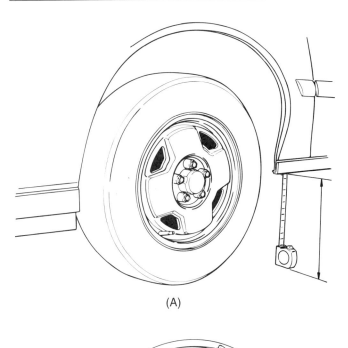

(A)

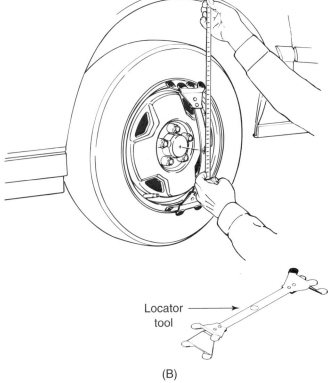

Locator tool

(B)

Figure 28–46 Ride height measurements are made with the vehicle on a flat, level surface and often are made between the rocker panel and ground (A). Make sure the measuring tape is straight and add in the amount for the base. The locator tool (B) is clipped onto the wheel and provides a quick means of measuring between the wheel center and the fender well. *(Courtesy of Moog Automotive Inc.)*

Figure 28–47 This vehicle is probably leaning because of a weak right front or rear spring; lifting the car in the exact center at the front or rear will usually show if it is the front or rear spring that has sagged. *(Courtesy of McQuay Norris)*

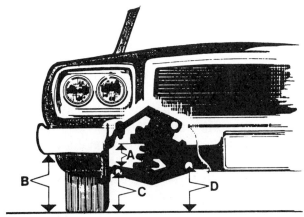

Side-to-side height comparisons.

MAXIMUM ALLOWABLE DIFFERENCE

A	B	C	D
1/4"	3/8"	3/4"	3/4"

Figure 28–48 Car lean is excessive if the difference between the right and left side is greater than the dimensions given here. Note that all of these measuring points except A are affected by tire size. *(Courtesy of McQuay Norris)*

For this check, the bottom end of the shock absorber is usually disconnected because it is easier to access. After it is disconnected, pull down and push up on the shock absorber body as far and as fast as you can. You should hear only a faint swish of the oil through the valves, and you should feel a steady and even amount of resistance on each stroke. Hops or skips in the resistance or excessive noise indicates a faulty shock absorber. If the resistance feels high or low, compare it with that of the other shock absorber on the same axle.

28.14 Ball Joint Checks

As mentioned earlier, ball joints form the steering axis for S-L A suspension. The steering axis for a strut suspension is the lower ball joint and the upper strut mount. These ball joints also form the pivot between the control arm and the steering knuckle to allow suspension travel. To perform these tasks, the ball joint must allow free movement without free clearance, which would allow sloppy tire and wheel or steering motions. Ball joint checks are to ensure that this free motion is within the wear tolerances set by the manufacturer. They also ensure that the ball joint is lubricated and that the boot needed to retain this lubricant is in good condition.

Ball joint boots are checked visually; the area behind the boot and ball joint where you cannot see can be checked by running your finger around the boot and feeling for problems. Look or feel for grease outside of the boot, which indicates breaks or tears. If the boot is torn, the ball joint will probably fail, if it has not already, and should be replaced (Figure 28–49). While checking the boot, squeeze it to ensure that there is grease inside of the boot. An empty boot indicates a need for lubrication.

Lubrication requirements for a ball joint vary among manufacturers. The lubrication intervals for modern joints are rather long; one manufacturer, for example, requires lubrication every 3 years or 30,000 mi. (48,000 km). Long intervals such as this make it easy for the average motorist to forget about lubricating ball joints completely. Low-friction ball joints are permanently sealed and require no further lubrication.

To lubricate a ball joint, you should:

1. Clean off all dirt or grease that might have accumulated around the grease/zerk fitting and boot. If a small plug is threaded into the ball joint instead of a grease fitting, remove the plug and install a grease fitting or use a grease gun attachment that threads directly into the joint. After lubricating the ball joint, some technicians leave the grease fitting in place; other technicians prefer to replace the plug for a more positive seal (Figure 28–50).

2. Using a grease gun, pump enough of the correct type of grease into the joint until the boot begins to balloon or grease begins to flow from the bleed area of the boot.

It is a good practice while checking the boot to also check the visual condition of the ball joint and the control arm for cracks or breaks in the metal, which indicate a probable failure. These cracks often show up as reddish-colored, loose rust streaks or dirt-free areas in otherwise

Figure 28–49 A cut or torn ball joint boot will let the grease escape and allow dirt and water to enter the joint; if the joint is not already worn out, it will soon fail. *(Courtesy of McQuay Norris)*

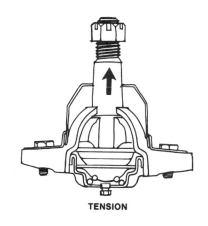

TENSION

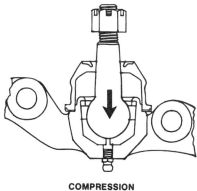

COMPRESSION

Figure 28–50 Most ball joints have either a plug or grease fitting in the housing so the joint can be greased. *(Courtesy of McQuay Norris)*

dirty parts. A sudden separation of a ball joint from a control arm can have catastrophic results (Figure 28–51).

A complaint of **hard steering** can be caused by a tight ball joint. This is not a common complaint, but it does occur. Place your fingers on the ball joint as you turn the steering knuckle; noisy operation or a rough feeling indicates a **dry** or **tight joint.** It should be checked further. Disconnect the tie-rod end from the steering arm and ro-

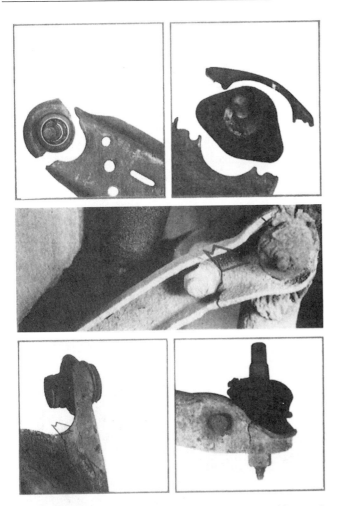

Figure 28–51 When lubricating ball joints or making an inspection, keep an eye out for cracks, which indicate probable failure of the component; cracks often show up as clean or rusty brown streaks in otherwise dirty areas. *(Courtesy of Federal-Mogul Corporation)*

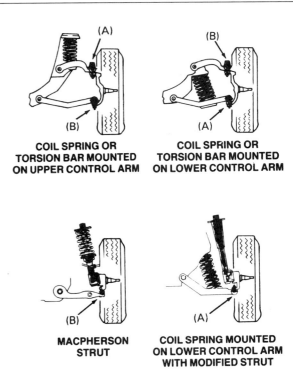

Figure 28–52 Depending on the suspension type and where the spring is located, a ball joint is either (A) a load-carrying or (B) a friction-loaded type. *(Courtesy of Dana Corporation)*

tate the steering knuckle by itself. The steering knuckle should rotate freely and smoothly, requiring only a slight pressure at the steering arm.

The most commonly encountered problem with ball joints is excessive clearance. The amount of clearance allowed in a ball joint varies depending on where the ball joint is used and the manufacturer. Generally speaking, a friction-loaded or follower style of ball joint (used on the control arm without the spring on an S-L A or strut suspension) should have no clearance or free play (Figure 28–52). A load-carrying ball joint is often manufactured with an operating clearance; the load on the joint removes this clearance and keeps the joint tight. When the load is taken off this style of joint, it usually feels loose because the clearance is now perceptible. If the clearance is less than the manufacturer's limits, the joint is still good (Figure 28–53). Some states require that the car owner be informed of the allowable clearance for that joint and how much clearance was measured in the worn ball joint during the inspection. Ball joint replacement cannot be sold without this information.

Four different checking methods are used for ball joints depending on their style and location: wear-indicating ball joints, load-carrying ball joints on lower control arms, load-carrying ball joints on upper control arms, and friction-loaded ball joints. Specifications for the allowable ball joint clearance and the exact checking method for a particular ball joint can be found in the manufacturer's service manual, a technician service manual, and specialized service manuals published by aftermarket suspension component manufacturers. A rule of thumb followed by many technicians when specifications are not available is that a load-carrying ball joint with 0.060 in. (1.5 mm) of vertical clearance or less with no visible damage should still be usable.

28.14.1 Checking a Wear Indicator Ball Joint for Excess Clearance

Wear indicator ball joints are designed to easily show if the ball joint has excessive clearance; they are used as load-carrying ball joints on S-L A and modified strut suspensions. They can often be identified by inspecting the lower face of the ball joint. Wear indicator joints have an opening in the lower cover through which the grease fitting or a metal boss projects. The exact checking procedure varies

Model	Year	Vertical Movement	Model	Year	Vertical Movement
CHEVROLET (Cont'd)			OLDSMOBILE (Cont'd)		
Camaro	67-69*	.060"	Toronado	66-76	.125"
	70-73	.020"**	Omega	73-74	.0625"
	74-76	Wear Indicator♦		75-76	Wear Indicator♦
Nova (Chevy II)	62-67	See Table II	Starfire	75-76	Wear Indicator♦
	68-70*	.060"	PLYMOUTH	57-67	.050"
	71-74	.0625"		68-72	.070"
	75-76	Wear Indicator♦	Valiant, Barracuda	60-67	.050"
Vega,	71-74	.0625"		68-76	.070"
Vega, Monza	75-76	Wear Indicator♦	Volare	1976	.020"**
Chevette	1976	Wear Indicator♦	Fury	1973	.070"
CHRYSLER	57-64	.050"		74-76	.020"**
	65-73	.070"	Satellite	73-74	.020"**
	74-76	.020"**	PONTIAC		
Cordoba	75-76	.020"**	Catalina, Bonneville, Grandville, etc.	58-64	.060"
COLT/ARROW	71-76	.020"**		65-70	.050"
CRICKET	71-72	.020"**		71-72	.020"**
DODGE	57-67	.050"		73-76	Wear Indicator♦
	68-72	.070"	Tempest	61-63*	.093"
Dart	60-67	.050"		1964*	.060"
	68-76	.070"	LeMans (Tempest)	65-69	.050"
Challenger	70-73	.070"	Grand Prix, LeMans	70-72	.0625"
Aspen	1976	.020"**			

Figure 28-53 Clearance specifications are printed for non-wear-indicating, load-carrying ball joints; the ball joint is considered good if the internal clearance is less than the specification. *(Courtesy of Federal-Mogul Corporation)*

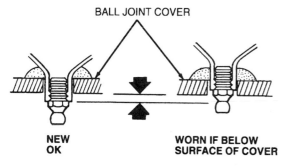

Figure 28-54 As a wear indicator ball joint wears, the checking surface moves upward and into the joint. *(Courtesy of Ford Motor Company)*

Figure 28-55 You can check a ball joint with a screwdriver; if the screwdriver blade catches on the checking surface, the joint is good. *(Courtesy of McQuay Norris)*

slightly among manufacturers; the method described here is a general approach (Figure 28-54).

To check a wear indicator ball joint for excessive clearance, you should:

1. Park the car on a level surface that allows access to the lower control arms and ball joints. The ramps of an alignment rack are ideal. The weight of the car should remain on the tires.

2. Wipe off any grease or dirt on the checking surface or the lower face of the ball joint.

3. On some styles, slide a plain flat screwdriver or other flat, metal object about 1/4 to 1/2 in. (6 to 12 mm) wide across the bottom surface of the ball joint; it should bump into the checking surface. If the checking surface has moved up into the ball joint, the ball joint is excessively loose and should be replaced (Figure 28-55).

OR On some styles, grip the grease fitting with your fingers and try to rotate it; if the grease fitting can be rotated, the ball joint is excessively loose and should be replaced (Figure 28-56).

28.14.2 Checking a Load-Carrying Ball Joint on a Lower Control Arm for Excessive Clearance

When the vehicle load passes from the spring and through the lower control arm to the steering knuckle, the lower ball joint is the load-carrying ball joint. This is true in cases where either a torsion bar, coil spring, or air spring is attached to the lower control arm. This load squeezes a compression-loaded ball joint tightly between the control arm and the steering knuckle or tries to pull a tension-loaded joint apart. The ball joint must be unloaded (vehicle and spring load removed) to measure the

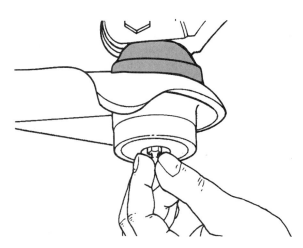

Figure 28-56 If the grease fitting can be easily rotated on some ball joints, the joint is worn out and should be replaced. *(Courtesy of Chrysler Corporation)*

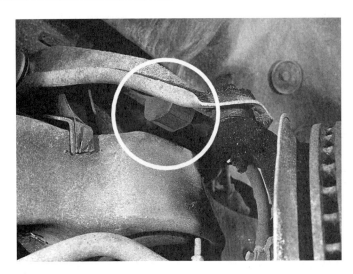

Figure 28-58 As the load is removed from the lower joint, a gap will be evident between the extension bump stop and the frame. *(Courtesy of Moog Automotive Inc.)*

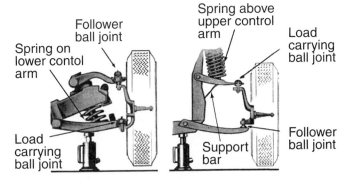

Figure 28-57 When checking the clearance of a load-carrying ball joint, the load of the vehicle must be removed from the ball joint using a jack placed in the correct location depending on whether the load-carrying joint is the lower or upper one. Note the support wedge that is used when the load-carrying joint is on the upper arm (right). *(Courtesy of Federal-Mogul Corporation)*

amount of clearance in the ball joint. This is usually accomplished by lifting the car by the lower control arm so the spring is compressed. If the car was lifted by the frame, the spring would push the lower control arm downward until the rebound/extension bump stop contacted the upper control arm. In this position the spring pressure holds both ball joints tightly. The ball joints appear to have zero clearance (Figure 28-57).

Wheel bearing clearance can be confused with ball joint clearance while shaking a tire. This problem can be eliminated by tightening the wheel bearing to eliminate its clearance or applying the brakes to lock the rotor to the steering knuckle (use a brake pedal jack for this). Ball joint clearance is usually measured using a dial indicator; it can also be measured using vernier or dial calipers. Some states require that the amount of clearance be measured by devices capable of measuring to a thousandth of an inch so the customer can be accurately informed as to the actual amount of clearance in the worn joints.

To check a non-wear-indicating, load-carrying ball joint on a lower control arm, you should:

1. Place a jack or the lifting pads of a suspension contact hoist under the lower control arm. The jack or lifting pad should be positioned as close to the ball joint as possible to ensure that the spring is compressed and the ball joint is unloaded.

2. Raise the vehicle so the tire is off the ground and check the rebound/extension bump stop to ensure that it is not under a load. This ensures that the ball joints are unloaded (Figure 28-58).

3. Position a dial indicator so that it will read the **axial/vertical** motion of the ball joint (Figure 28-59).

OR Measure the distance from the lower face of the ball joint to the end of the ball joint stud using a vernier or dial caliper (Figure 28-60).

4. Raise the tire, wheel, and steering knuckle as far as you can without lifting the car; many technicians place a lever under the tire or steering arm to make this easier. Note the amount of travel on the dial indicator; this is the amount of clearance in the ball joint (Figure 28-61).

OR Remeasure the length of the ball joint with the calipers and subtract the first measurement from the second. The result is also the amount of clearance.

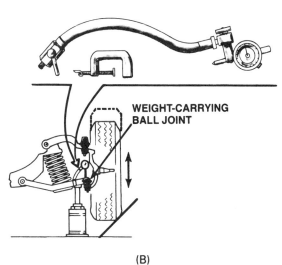

Figure 28-59 Ball joint vertical (axial) clearance can be checked by mounting a dial indicator as shown. *(A is courtesy of Moog Automotive Inc.; B is courtesy of Dana Corporation)*

5. Compare the amount of clearance measured with the specifications. An excessively worn ball joint is indicated if the measured clearance exceeds the specifications. It should be replaced.

6. Some manufacturers prefer to check ball joint clearance in a **horizonal/radial** direction. This is done by mounting the dial indicator in a horizontal rather than a vertical position and moving the tire, wheel, and steering knuckle in and out rather than up and down while making the check. If the radial clearance exceeds the specifications, the ball joint should be replaced (Figure 28-62).

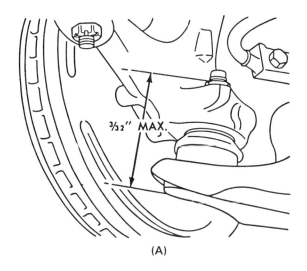

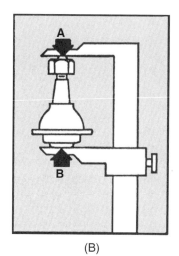

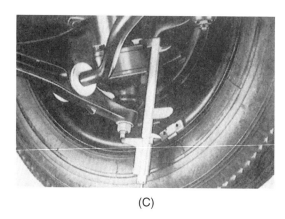

Figure 28-60 Ball joint vertical (axial) clearance can be checked using calipers to measure the overall distance (A) with the load on the joint and again with the load removed from the joint (B). A measurement is being made in (C). *(A is courtesy of General Motors Corporation, Service Technology Group; B is courtesy of Dana Corporation; C is courtesy of Volkswagen of America, Inc.)*

Figure 28–61 The tire and wheel can be lifted using a bar while watching the dial indicator measure vertical clearance. *(Courtesy of Moog Automotive Inc.)*

Figure 28–63 When checking a load-carrying ball joint on the upper control arm, a prop is positioned under the control arm to hold it in its normal position as the car is lifted by the frame. *(Courtesy of Moog Automotive Inc.)*

Figure 28–62 When a technician checks ball joint horizontal (radial) clearance, the dial indicator is mounted in a horizontal position. *(Courtesy of Moog Automotive Inc.)*

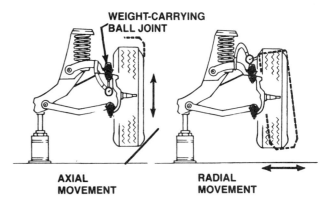

Figure 28–64 Upper load-carrying ball joints are checked for axial or radial clearance as shown here. *(Courtesy of Dana Corporation)*

28.14.3 Checking a Load-Carrying Ball Joint on an Upper Control Arm

Checking a load-carrying ball joint on an upper control arm is essentially the same operation as described in Section 28.14.1 except for the procedure used to unload the ball joint. If the car is lifted by the lower control arm, vehicle load transmitted from the spring to the upper control arm will keep the ball joint tight between the steering knuckle and the upper control arm; both ball joints will have zero clearance. A prop or brace should be placed between the upper control arm and the frame to hold the control arm in a near-normal position, and the vehicle should be lifted by the frame. The prop will transfer the spring load to the frame, unloading the ball joint while the control arm is held in a near-normal position. Both sides of the car should be braced and checked at the same time; if you try to check one side at a time, the stabilizer bar will twist and load the ball joints, making them appear tight.

To check a load-carrying ball joint on an upper control arm, you should:

1. Position the car on a flat surface so there is access to the upper control arm.
2. Place a prop/support bracket between the upper control arm and the frame on each side of the car so that when the car is lifted the control arm will be held upward (Figure 28–63).
3. Lift the car so the tire is off the ground.
4. Follow steps 3 through 6 of Section 28.14.2 (Figure 28–64).

28.14.4 Checking a Friction-Loaded or Follower Ball Joint

Friction-loaded or follower ball joints are manufactured with no clearance and in most cases have a slight preload. This joint is mounted on the control arm without a direct connection to a spring or torsion bar. Two common methods are used to check a follower ball joint: free play and preload.

To check a follower ball joint for **free play**, you should:

1. Raise the vehicle in such a way as to unload the load-carrying ball joint (described in Sections 28.14.2 and 28.14.3); on strut suspensions, lift the car by the frame.
2. Grasp the tire at the top or bottom, next to the ball joint being checked, and push in and out on the tire.

Watch for any motion between the control arm and the steering knuckle that would indicate clearance. If you have trouble pushing on the tire while watching the ball joint, place your hand so your fingers span the gap between the steering knuckle and the control arm while the tire is being shaken. If you can feel or see any perceptible clearance in the ball joint, it has excessive clearance and should be replaced (Figure 28–65 and Figure 28–66).

To measure the **preload** of a follower ball joint, you should:

1. Raise and support the car on a hoist or jack stands.
2. Disconnect the ball joint stud from the steering knuckle; this operation is described in Chapter 32.
3. Replace the nut onto the ball joint stud and run it to the end of the threads or use a jam nut so the stud is forced to turn with the nut.
4. Use a low-reading torque wrench to measure the amount of torque required to rotate the stud under the internal ball joint resistance. Many manufacturers publish a torque specification for this check; a rule of thumb, if specifications are not available, is 2 to 8 ft-lb or 24 to 96 in.-lb (3 to 11 N-m). If the stud rotates too hard or too easily, the ball joint should be replaced (Figure 28–67).

28.14.5 Checking Ball Joint Clearance on an I-Beam or 4WD Solid Axle

I-beam and 4WD solid axles are used on some pickups and 4WD vehicles. The ball joints are used in pairs, with both of them carrying some of the vehicle load. There are two methods of checking these ball joints depending on

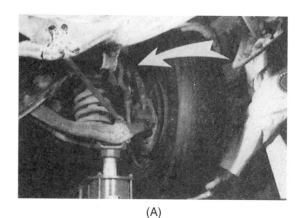

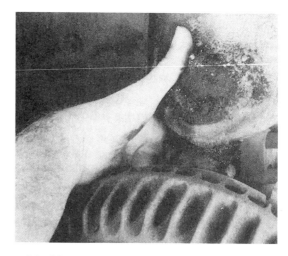

Figure 28–65 When checking a non-load-carrying/friction-loaded ball joint, shake the tire in and out (A) while watching for movement between the control arm and steering knuckle (B). *(A is courtesy of Ford Motor Company; B is courtesy of SPX Corporation, Aftermarket Tool and Equipment Group)*

Figure 28–66 As the tire is shaken in and out, there should be no apparent movement in an unloaded ball joint. *(Courtesy of SPX Corporation, Aftermarket Tool and Equipment Group)*

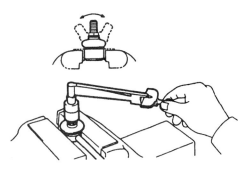

Figure 28-67 The preload of a friction-loaded ball joint can be measured using a low-reading torque wrench attached to a nut on the stud. Swing the stud back and forth to loosen the joint before taking the measurements.

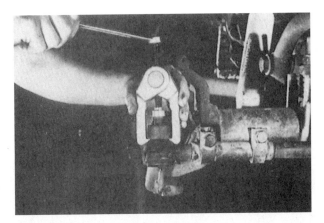

Figure 28-68 The tie-rod needs to be disconnected before measuring the turning effort on this 4WD axle; it can be separated from the steering arm using a special puller. (Courtesy of Moog Automotive Inc.)

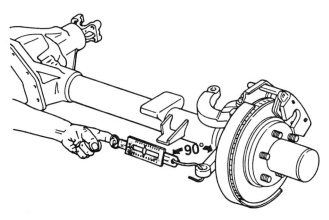

Figure 28-69 Turning effort is measured using a spring scale attached to the steering arm as shown. Note the scale to read the amount of pull required to turn the steering knuckle. (Courtesy of Dana Corporation)

Figure 28-70 Some 4WD axles have a provision for adjusting the turning effort; the adjusting sleeve is turned inward to increase it. (Courtesy of Dana Corporation)

the specifications given by the vehicle manufacturer, **turning torque** and **side clearance**.

To measure the turning torque, you should:

1. Raise and support the vehicle on a hoist or jack stands.
2. Disconnect the tie-rod end from the steering arm; this operation is described in Chapter 33 (Figure 28–68).
3. Attach a spring scale to the tie-rod boss in the steering arm or a torque wrench onto a ball joint stud nut and measure the force needed to turn the steering knuckle. Excessive ball joint clearance is indicated if the steering knuckle turns too easily (Figure 28–69).
4. Some manufacturers use an adjustment sleeve at the ball joint to provide a means of adjusting the turning effort, provided the ball joint is still good (Figure 28–70).

To check ball joint side clearance, you should:

1. Raise and support the vehicle on a hoist or jack stands.
2. Push the bottom of the tire in and out while watching for movement between the steering knuckle and the axle in the area next to the lower ball joint (Figure 28–71 and Figure 28–72).
3. Repeat step 2, pushing the top of the tire in and out while watching for side clearance at the upper ball joint.
4. If the amount of side motion exceeds the manufacturer's specifications, the ball joint should be replaced; if specifications are not available, a rule of thumb is 0.031 or 1/32 in. (0.8 mm).

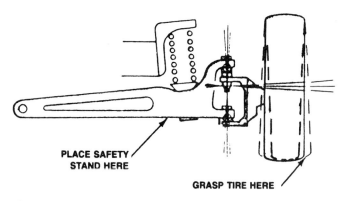

Figure 28-71 The ball joints on a twin I-beam axle are checked in the same way as a friction-loaded ball joint; shake the tire in and out while looking for any visible side motion between the axle and steering knuckle. *(Courtesy of Ford Motor Company)*

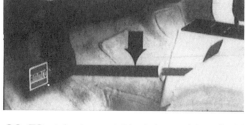

Figure 28-73 A brake pedal jack (arrow) installed between the seat and brake pedal will hold the brakes in an applied position; brake application locks the brake rotor or drum to the steering knuckle to eliminate play between them. *(Courtesy of Ammco Tools, Inc.)*

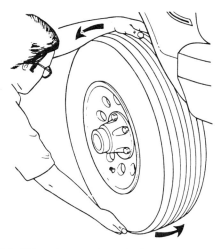

Figure 28-72 When making ball joint checks, grasp the top and bottom of the tire, push inward with one hand while pulling outward with the other, and then reverse these pressures. *(Courtesy of Ford Motor Company)*

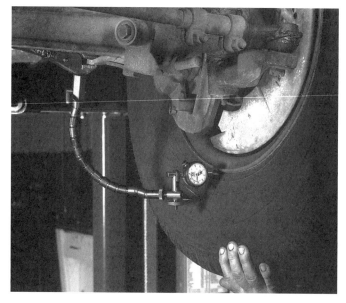

Figure 28-74 Kingpin clearance is measured by mounting a dial indicator as shown and moving the bottom of the tire in and out while watching the amount of travel on the dial indicator. *(Courtesy of Moog Automotive Inc.)*

28.14.6 Checking Kingpin Clearance

Although kingpins are not ball joints, they have been included with these checks because a kingpin performs a similar function and the checking method is similar. Kingpins are used on solid axles and some twin I-beam axles; they are normally checked by measuring the **side shake** of the tire.

To check a kingpin for excessive clearance, you should:

1. Raise and support the vehicle on a hoist or jack stands.
2. Eliminate wheel bearing clearance by installing a brake pedal jack to apply the brakes or by tightening the spindle nut (Figure 28-73).
3. Position a dial indicator at the lower part of the tire with the dial indicator stylus in a horizontal position (Figure 28-74).
4. Push in and out on the tire while observing the dial indicator readings. Worn kingpins are indicated if there is more than:

    a. 1/4 or 0.250 in. (6.35 mm) of side motion on 16-in. or smaller wheels

    b. 3/8 or 0.375 in. (9.5 mm) for 17- to 18-in. wheels

    c. 1/2 or 0.500 in. (12.7 mm) for wheels larger than 18 in.

28.15 Control Arm Bushing Checks

As previously mentioned, most control arm bushings are made from rubber and are of a torsilastic style. The outer metal sleeve is pressed into the control arm, and the tubelike inner sleeve is clamped tightly into the frame mount. As the control arm pivots up and down, the rubber portion twists and stretches to allow for suspension travel. This action transfers a certain amount of energy into the rubber bushing material, which generates heat. Heat from any source tends to harden rubber; as the rubber bushings become harder and stiffer, they tend to crack, break, and then disintegrate. The life of rubber bushings is determined somewhat by their temperature; rough, bumpy roads will cause more suspension action, more heat, and a shorter life. This bushing should not be lubricated; besides doing no good, a petroleum product will attack and shorten the life of the rubber.

Worn suspension bushings allow the control arm to move inward and outward or forward and backward as well as up and down. This results in an alignment change of the tires, which, in turn, will cause tire wear and handling difficulties. This looseness often causes suspension noises, usually "clunks," when driving over rough roads or when the brakes are applied. Faulty rubber control arm bushings can usually be seen during a visual inspection. In locations where the bushings are difficult to see, faulty bushings are identified by excessive control arm motion through either an in-and-out or a sideways direction.

To check rubber control arm bushings, you should:

1. If possible, check the upper control arm bushings from under the hood. Use a light so you can get a good look at the rubber parts of the bushing. Ignore small, light cracks as long as the rubber is still solid and resilient. Look for heavy cracks, rubber material breaking out, or rubber distortion, which allows the control arm to change position. The pivot bolt should be centered in the bushing. Bushings that are distorted, breaking up, or getting ready to break up should be replaced (Figure 28–75).
2. Raise and support the car on a hoist or jack stands.
3. Visually check the bushings on the lower control arm, looking for the same sort of problems. Also, check the sides of the control arm and the frame metal next to it for signs of metal contact, which indicate bushing failure.
4. Swing the tire rapidly back and forth while forcing it to bump at the steering stops; also, force the tire in and out. While doing this, watch the control arm for any motions that would indicate bushing failure.

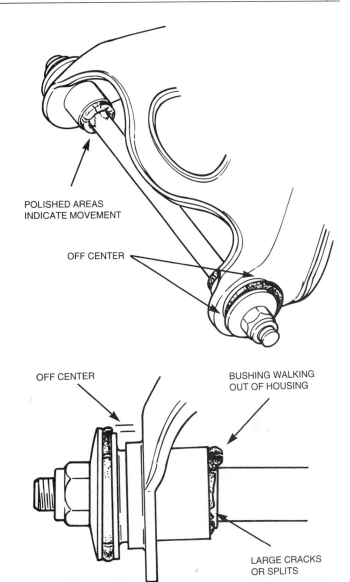

Figure 28–75 These faults indicate control arm bushings that need replacement. *(Courtesy of Moog Automotive Inc.)*

NOTE: The bellows at the end of the rack can be distorted if the checking motions are made too quickly on some cars with manual rack-and-pinion steering gears.

5. On single lower control arms, try prying the inner end of the control arm sideways using a pry bar or large screwdriver. A slight motion is acceptable; larger motions indicate weak bushings.

Metal control arm bushing faults are more difficult to see. These bushings consist of a pair of large nuts that are locked into the control arm; the inner control arm shaft is bolted to the frame and is threaded on the ends to accept the large nutlike bushings. Suspension action turns the bushings on the shaft much like a nut turning on a bolt. These bushings must be lubricated to prevent wear, and they must be sealed to prevent dirt or water from entering

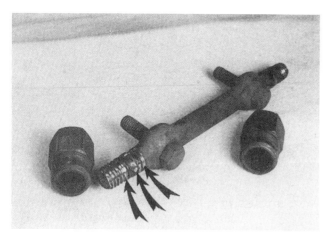

Figure 28–76 A badly worn metal control arm bushing and shaft. Note the worn, shiny areas on the control arm shaft (arrows).

them. When this bushing runs out of lubricant, it will begin to wear and squeak; noise is usually the first indication of metal bushing failure (Figure 28–76).

To check metal control arm bushings, you should:

1. Bounce the suspension while listening for squeaks or other bushing-related abnormal noises. If possible, place your finger lightly on the bushing while bouncing the front end; a noisy bushing will often have a rough, harsh feel. Noisy bushings can sometimes be cured by greasing them, but if they have squeaked for very long, they are probably worn and should be replaced.

2. Raise and support the car on a hoist or jack stands.

3. Swing the tire back and forth rapidly, so the turning stops strike rather hard, and watch the control arm bushings. A very slight amount of side motion is acceptable, but a definite motion or jumping of the control arm on the shaft indicates a faulty bushing.

28.16 Strut Rod Bushing Checks

Strut rod bushings are rubber bushings that are compressed tightly against each side of an opening in the frame bracket. If they become weak, the outer end of the lower control arm will have an excessive amount of travel in a forward and backward direction.

Strut rod bushing failure is often indicated by a "thump" or "clunk" as the brakes are applied. These bushings are checked visually.

To check strut rod bushings, you should:

1. Raise and support the car on a hoist or jack stands.

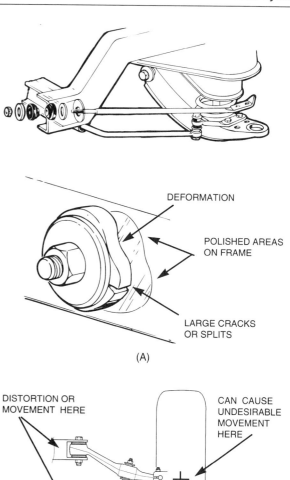

Figure 28–77 Faulty strut rod bushings (A) allow the outer end of the lower control arm to move forward or rearward (B). *(Courtesy of Moog Automotive Inc.)*

2. Grip the bushing end of the strut rod and shake it up and down; any free play indicates a faulty bushing.

3. Inspect the bushing for hard cracks, rubber breaking, and severe distortion of the rubber; also, check for signs of contact between the metal backup washer and the bushing bracket. Any of these indicate a faulty bushing (Figure 28–77).

28.17 Strut Checks

Occasionally, the tire on a strut suspension shows excessive camber wear, which indicates that the strut body or strut piston rod might be bent. Several checks can be made to determine if this has occurred.

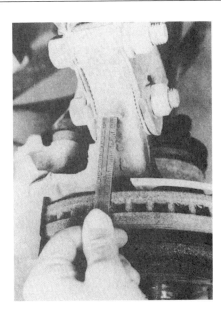

Figure 28-78 A ruler is being used to measure the rotor-to-strut distance, checking for a bent spindle or strut; this distance should be the same on the other side of the car.

One check is to compare the included angle or the camber angle and steering axis inclination angles with the specifications. This procedure requires wheel alignment equipment and is described in Chapter 34.

A quick check for a bent strut body is to measure the distance between the strut body and the brake rotor on both sides of the car and compare the measurements. If they differ by more than a few thousandths of an inch, one of the struts is bent (Figure 28-78).

To check for a bent piston rod, loosen the piston rod at the upper mount and then rotate the piston rod while you watch for side motion at the top of the tire or strut body. A good strut rod should rotate evenly with no side motion of the tire or strut body.

28.18 Strut Damper or Insulator Checks

A faulty strut damper assembly can cause noises; it will also allow the upper end of the strut to sag to an improper position, changing the alignment of the strut as well as the tire and wheel (Figure 28-79).

To check a strut damper, you should:

1. With the car's weight on the tires, note the relative position of the upper strut mount to the fender opening.
2. Raise the car by the frame and note any change in the position of the mount assembly. A slight downward movement is normal. Any free motion is unacceptable. The manufacturer's specifications should be checked to determine the allowable clearance for a particular car. One manufacturer allows up to 1/2 in. (12.7 mm).
3. With the tires off the ground, reach around the tire and grip the strut's spring as close to the upper mount as possible. Alternately push inward and pull outward on the strut and spring while you watch the upper end of the strut piston rod. A slight movement is normal; there should be no free movement. One manufacturer allows up to 3/16 in. (4.8 mm) of side motion (Figure 28-80).

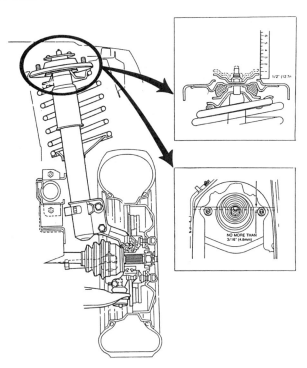

Figure 28-79 A faulty strut damper allows excessive movement of the strut mount as the car is lifted and the load is removed from the tires; excessive play as the strut is pushed in and out is another indication of failure.

If there is excessive damper motion or if a visual inspection reveals separation of the rubber portions, the damper portion of the strut mount should be replaced.

28.19 Steering Linkage Checks

The checks for steering linkage wear are commonly done at the same time as the suspension checks. It is easier for us to study the suspension system separately from the steering system, but the average front-end technician does not separate them when it comes to repairing them or making a wheel alignment. These two systems are usually serviced together.

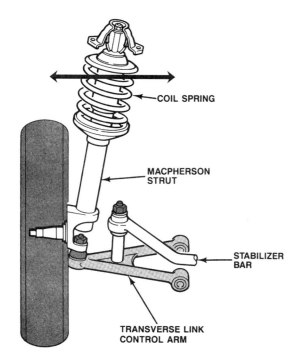

Figure 28–80 Excessive side play indicates a faulty strut or strut mount.

There are two major styles of steering systems: standard steering gears with parallelogram linkage and rack-and-pinion steering gears and linkage. Each system has some different components that must be checked. The most common steering system fault is looseness; any clearance in any portion of the steering linkage will cause loose, sloppy steering or the tires to change toe angle as they go down the road.

The various portions of the steering system can be checked individually, as described in the following sections, or together by performing a **"dry park"** test. Dry park is the steering action some people make while parking a car—turning the steering wheel while the car is standing still. This places a much higher load on the steering components than turning with the tires rolling.

To perform a dry park test, you should:

1. Park the car on a level surface where there is access to the steering linkage. An alignment rack is ideal because the car's weight should remain on the tires; if an alignment rack is used, the turntables should stay locked so they do not rotate.

2. Have a helper turn the steering wheel in a back-and-forth manner, keeping a constant, alternating motion through any free play.

3. Observe the entire steering system for signs of play, being sure to check for:

    a. Loose or worn steering shaft couplers

    b. Loose steering gear mounts and bushings

    c. Loose steering gear, internal

    d. Worn Pitman-arm-to-center-link connection

    e. Worn idler-arm-to-center-link connection

    f. Worn idler arm bushing

    g. Worn outer and inner tie-rod ends.

A rotational or reciprocating motion of these parts is normal; free or sloppy motion indicates worn parts.

28.19.1 Tie-Rod End Checks

Tie-rod ends are ball-and-socket joints that are preloaded to have zero clearance. Do not forget that the tie-rod ends must allow a pivoting and swiveling action between the tie-rod and the steering arm. Most sources recommend checking tie-rod ends with the load on the tires so they will be in their normal position.

Tie-rod ends must be lubricated to prevent wear, and they must be sealed to keep dirt and water out. If the sealing boot fails, the tie-rod end will soon fail. Traditionally, tie-rod ends have had either a grease/zerk fitting or a plug for the purpose of installing a grease fitting so grease could be added to the joint. Many newer tie-rod ends are lubricated and sealed during manufacture or use a rubber bushing. These joints require no lubrication; they have no provision for adding grease.

Tie-rod ends that use rubber bushings (rubber bonded) should have no free play at all; resistance should be felt in the joint in a rotary as well as in an up-and-down or a sideways direction.

Worn tie-rod ends allow loose, sloppy steering action with steering wheel free play; they can also sometimes cause a **pop** or **snap** noise while turning.

To check tie-rod ends, you should:

1. Raise and support the car on a hoist or jack stands.

2. Grip the tie-rod next to the end and firmly push up and pull down while watching for movement. Any free play is excessive. About 1/16 to 1/8 in. (1.5 to 3 mm) of motion, resisted by a spring, is allowable; travel greater than this is excessive and indicates a faulty tie-rod end (Figure 28–81).

OR One manufacturer recommends measuring the distance from the side of the tie-rod end to the end of the stud using vernier or dial calipers. This distance should be measured with the tie-rod end pushed together and then pulled apart; the difference between these two measurements is the amount of clearance in the tie-rod end. The manufacturer has a specification for the maximum allowable clearance.

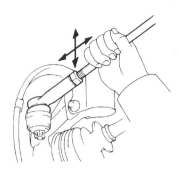

Figure 28–81 A quick check of a tie-rod end is made by pushing and pulling on the tie-rod in the direction of the stud; a good tie-rod will have no visible looseness.

3. Grip both front tires and firmly try to pull them together and then push them apart. Feel and watch for free play while doing so; free play indicates a faulty pivot point (Figure 28–82).

4. Rotate the tie-rod; it should turn smoothly with no catches or binding for about 10 to 30 degrees.

5. Inspect the boots for breaks and tears to ensure they are in good condition; a tie-rod end with a bad boot should be replaced.

28.19.2 Parallelogram Steering Linkage and Idler Arm Checks

Parallelogram steering linkage, which is used with standard steering gears, uses a center link that is supported on one end by the Pitman arm from the steering gear and on the other end by an idler arm. The idler arm pivots from a bracket that is bolted to the frame. The idler arm is the same length and in the same relative position as the Pitman arm. The center link travels sideways across the car during turning maneuvers, and the two tie-rods connect to the center link through a tie-rod end. There is also a pivot joint, similar to a tie-rod end, at each connection between the center link and the idler or Pitman arm.

Wear checks for the center link pivot points are made in the same manner as tie-rod end checks. Worn Pitman arm and center link joints will often show up when the tires are pulled together and then spread apart. At the same time, a worn idler arm bushing will let the center link end of the idler arm raise and lower. Worn idler arm or Pitman arm bushings will also cause loose, sloppy steering (Figure 28–83).

To check an idler arm bushing, you should:

1. Raise and support the car on a hoist or jack stands.
2. Grip the center link near the end of the idler arm and firmly push up and pull down while watching the vertical motion (Figure 28–84). Movement of 1/8 in. (3 mm) or so is normal. The manufacturer's

Figure 28–82 To check the tie-rods along with other steering linkage parts, push the tires apart and pull them together while you watch and feel for play or free movement. Any free play indicates a fault. *(Courtesy of Moog Automotive Inc.)*

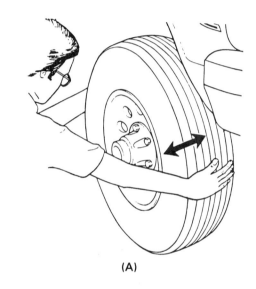

Figure 28–83 Steering linkage wear can be measured by lifting one tire off the ground and rocking the tire sideways (A). The amount of wear can be measured using a dial indicator as shown in (B). *(A is courtesy of Ford Motor Company; B is courtesy of General Motors Corporation, Service Technology Group)*

specifications should be used to determine the allowance for each car; some manufacturers allow up to 1/4 in. (6 mm). Greater movement indicates a faulty idler arm (Figure 28–85).

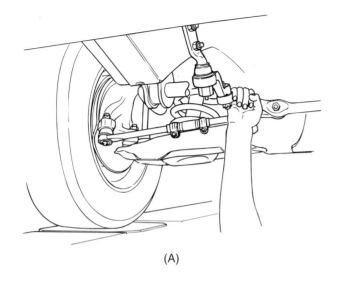

(A)

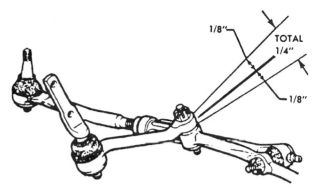

Figure 28-85 This manufacturer allows a total movement of 1/4 in. at the end of the idler arm; it should be replaced if there is more play. *(Courtesy of General Motors Corporation, Service Technology Group)*

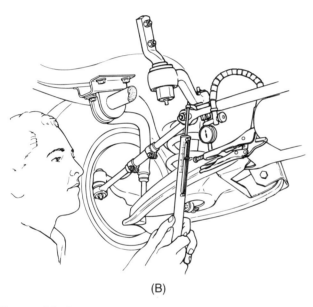

(B)

Figure 28-84 An idler arm is checked by pushing up and pulling down on the center link close to the idler arm (A); some manufacturers recommend using a scale and dial indicator to be more precise (B). *(Courtesy of Moog Automotive Inc.)*

3. Inspect the frame area surrounding the idler arm mounting bolts for cracks and other signs of breakage. Cracks or breaks should be repaired by someone competent in frame repair.

28.19.3 Rack-and-Pinion Steering Linkage Checks

Rack-and-pinion steering systems are much simpler than the standard style. A tie-rod connects to each end of the steering rack, which attaches to the frame through rubber bushings. The inner tie-rod end, where the tie-rod connects to the rack, is a ball-and-socket joint that is enclosed in the flexible, rubber bellows or boot.

All outer tie-rod ends are checked in the same manner as described in Section 28.19.1. The inner tie-rod ends are a little more difficult to check because they are enclosed in the boot.

To check inner tie-rod ends, bellows, and rack bushings, you should:

1. Raise and support the car on a hoist or jack stands.

2. Turn the steering to one of the stops and carefully check the bellows, which has been stretched out, for excessive cracks in the rubber, tears, breaks, or distorted pleats. Turn the steering to the other stop and inspect the remaining bellows. Collapsed bellows can often be straightened by loosening the clamps and turning one end of the bellows. Torn or broken bellows should be replaced. Power steering fluid leaks from a bellows indicate a faulty steering gear.

3. Squeeze the bellows so you can feel the inner tie-rod end socket and tie-rod and push and pull on the side of the tire. If you can feel any free play between the tie-rod and the socket, the inner tie-rod end should be replaced (Figure 28-86). If you have difficulty feeling the inner joint, loosen the large bellows clamp and slide the bellows aside so you can see the joint.

OR Some manufacturers recommend disconnecting the outer tie-rod end from the steering arm and measuring the amount of force needed to swing the tie-rod back and forth. If the tie-rod swings too easily, the inner end should be replaced (Figure 28-87).

4. Carefully inspect the rack mounting bushings for signs of wear, deterioration, or excessive rack movement. Faulty or oil-soaked bushings should be replaced (Figure 28-88).

Chapter 28 ■ Suspension and Steering System Diagnosis and Inspection

Figure 28-86 The inner tie-rod socket is checked by squeezing the bellows so you can feel the socket and tie-rod while you push and pull on the tire; if you feel any movement between them, the tie-rod end should be replaced. The bellows should also be checked to ensure there are no cracks or splits. *(Courtesy of Moog Automotive Inc.)*

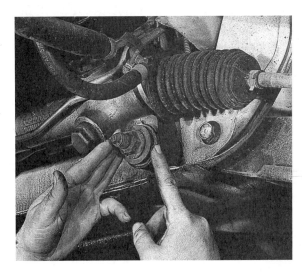

Figure 28-88 The rack mounting bushings should be checked for wear and deterioration; faulty bushings should be replaced. *(Courtesy of Moog Automotive Inc.)*

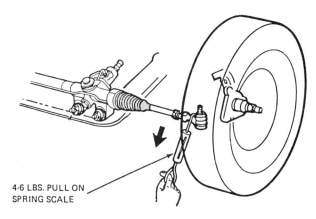

Figure 28-87 The inner tie-rod socket preload is measured by checking the articulation effort with a spring scale as shown; if the tie-rod swings too easily, the tie-rod socket should be adjusted (some models) or the tie-rod end should be replaced. *(Courtesy of Ford Motor Company)*

Figure 28-89 Check the steering shaft couplers and U-joints for looseness and wear; worn units should be replaced. Also rotate the coupler and steering shaft and try to move the shaft in and out of the gear; excessive play indicates the need for steering gear service. *(Courtesy of Moog Automotive Inc.)*

28.20 Steering Gear Checks

Steering gear looseness can be checked at two separate points: from the steering wheel at the beginning of the inspection procedure and from under the car during the procedure. The first point gives an indication of whether free play is present; the second check helps pinpoint the exact location of the free play. Steering gear clearance checks should be made with the tires in a straight-ahead position. A standard/conventional steering gear normally has a slight play in the turning positions.

Rock the steering wheel back and forth using a very light pressure and feel for free movement. The engine should be running on cars equipped with power steering. No clearance is acceptable; a very slight amount of play is tolerable. If it is difficult to feel any free play, watch the left front tire—a perceptible turning motion should be seen as soon as the steering wheel is turned.

On most cars, it is possible to reach up from under the car and turn the steering shaft while watching the linkage. Again, no clearance is acceptable, and a very slight clearance can be tolerated. If clearance is felt, observe the various parts to determine where the free play occurs; that part should be adjusted or repaired. On standard steering gears, you should also grip the Pitman arm and try to move it up and down as well as sideways in a steering direction; again, only a barely perceptible amount of play can be tolerated (Figure 28-89).

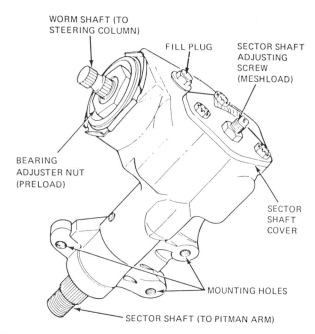

Figure 28–90 The lubricant level of a manual steering gear should be even with the bottom of the fill plug opening; if there is no fill plug, check the level at the uppermost sector shaft cover bolt. *(Courtesy of Ford Motor Company)*

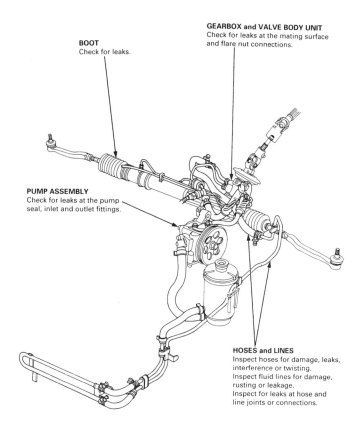

Figure 28–91 Most power steering leaks occur at these locations. *(Courtesy of American Honda Motor Co., Inc.)*

Also from underneath the car, check the security and tightness of the steering gear mounting bolts and the condition of the steering coupler. Any faults should be noted or repaired.

Conventional steering gears of the manual type should be checked to ensure they have an adequate supply of lubricant. Lubricant level is checked at the filler plug at the top surface of the gear housing or by removing one of the sector gear cover retaining bolts. The correct level is normally even with the bottom of the opening (Figure 28–90).

Adjustments and service of steering gears are described in Chapter 33.

28.21 Power Steering Checks

These additional checks should be made on cars that are equipped with power steering: hose leaks and condition, pump and gear leaks, drive belt condition and tension, and fluid level and condition.

Check both power steering hoses for leaks, signs of deterioration, excessive cracking, sponginess, and rubbing or chafing, which might lead to failure. Faulty hoses should be replaced. Hose replacement is discussed in Chapter 33.

From underneath the car, check for signs of power steering fluid leakage (Figure 28–91). Look for the reddish (if ATF is used) or yellow-green (if power steering fluid is used) colored fluid. If fluid is found, follow the leak upward and forward to the faulty component. Also, run your hand along the hoses, pump, or gear as you feel for wetness to determine the exact location of the leak (Figure 28–92).

Newer cars can use a quick-connect style of pressure hose fitting that does not lock the line in place. It has two possible leak points and two different seals that can cause leakage. The tube nut on this fitting should be tightened to the proper torque specification (Figure 28–93).

Inspect the pump drive belt for cracks, frayed fabric, very shiny or glazed sides, or other signs of deterioration; doubtful belts should be replaced. Any belt that is losing material should be replaced (Figure 28–94). Some experts recommend replacing all the drive belts on an engine every 4 years.

Potential belt failure is not easily recognized on newer belts, and it is much less expensive to replace a belt too early than too late. The belt tension should also be checked and readjusted, if necessary (Figure 28–95 and Figure 28–96).

Power steering fluid level is checked by removing the filler cap and checking the dipstick. Low fluid level can cause erratic steering motions as well as humming or buzzing noises during turning maneuvers (Figure 28–97).

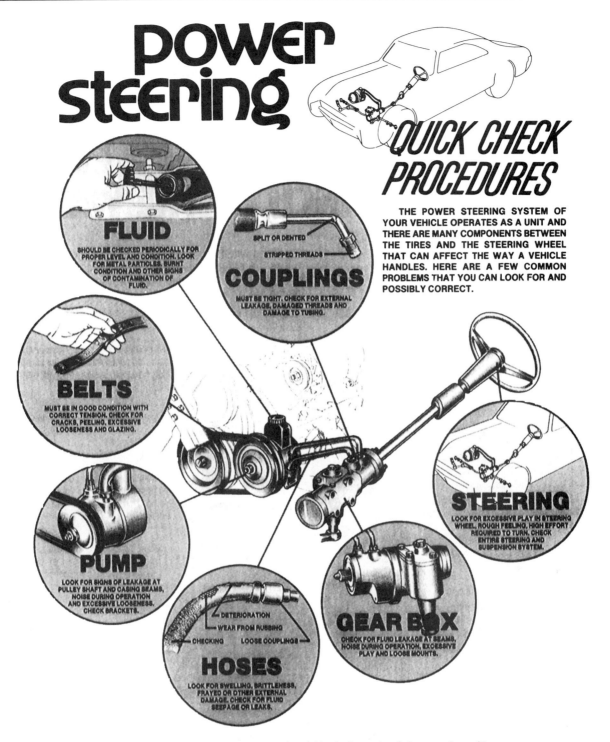

Figure 28–92 An inspection of a power steering system should include each of these points. *(Courtesy of Moog Automotive Inc.)*

Power steering fluid level should be checked with the fluid hot (above the temperature where you can comfortably hold the end of the dipstick); cold fluid will check lower than normal (Figure 28–98). While checking the fluid level, inspect the condition of the fluid. Dirty fluid that has lost its bright, new appearance or that has begun to smell like varnish should be changed and replaced with new fluid. Place a drop of fluid on white, absorbent paper. The liquid will be soaked into the paper, leaving any solid particles on the surface (Figure 28–99). Excess solid material indicates contaminated fluid and possibly a worn steering gear.

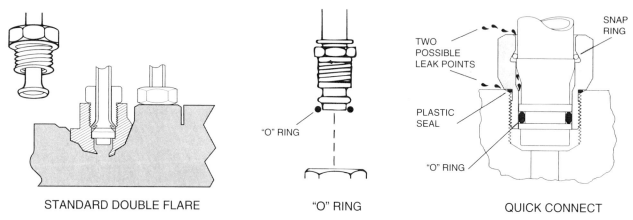

Figure 28–93 The ends of the power steering pressure hose use different types of sealing methods. Note that the quick-connect style has two possible leak points, and the hose end does not become tight when the fitting is tightened. *(Courtesy of Moog Automotive Inc.)*

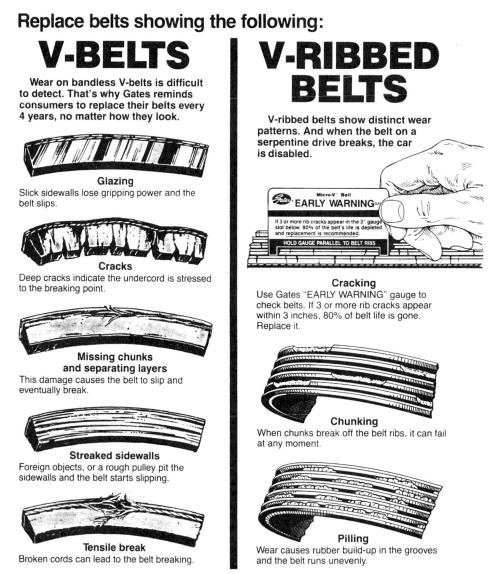

Figure 28–94 It is wise to replace drive belts if they show these indications of failure. *(Courtesy of The Gates Rubber Company)*

ADJUSTING POWER STEERING BELT

1 For a new belt, or one that has run less than 15 minutes, tension should be **120 to 150 lbs.**

2 For a belt that has run longer than 15 minutes, tension should be **90-120 lbs.**

3 Always check tension with a gauge, such as T63L-8620-A

4 TYPICAL SLIDING ADJUSTMENT
Rotate nut to adjust tension after loosening bolts in slotted holes.

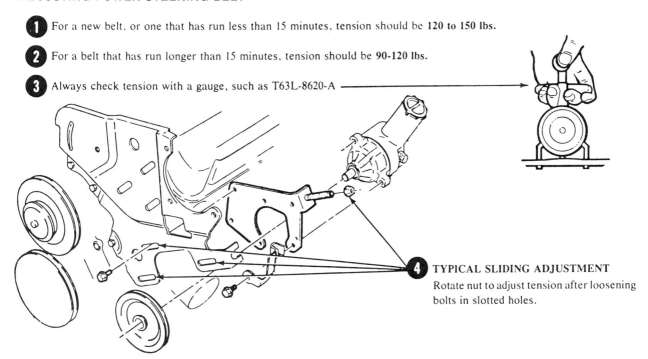

Figure 28-95 Belt tension should be measured using a tension gauge placed over the belt; the power steering pump bracket and adjustment procedure are also shown. *(Courtesy of Ford Motor Company)*

TENSION IN LBS. (KG.) USING STRAND TENSION GAUGE

Application	New Belt	Used Belt
American Motors		
"V" Belts	140 (64)	100 (45)
Serpentine	190 (86)	150 (68)
Chrysler Corp.	120 (54)	70 (32)
Ford Motor Co.		
1.6L	50-90 (23-40)	40-60 (18-27)
2.3L HSC	150-190 (68-86)	140-160 (64-73)
Diesels	120-160 (54-73)	110-130 (50-59)
2.3 OHC, V6, V8	120-160 (54-74)	90-120 (41-54)
Ribbed V-Belts		
4 Rib	110-150 (50-68)	100-130 (45-59)
5 Rib	130-170 (59-77)	120-150 (54-68)
6 Rib		
Fixed	140-180 (64-82)	130-160 (59-73)
Tensioner	85-140 (39-64)	80-140 (36-64)
General Motors		
1.6L & 2.0L	145 (66)	70 (32)
1.8L & 2.5L	165 (75)	100 (45)
V6 Gas	135 (61)	65-80 (29-36)
V8 Gas		
Exc. Corvette	150 (68)	80 (36)
Corvette	130 (59)	
All Diesels	170 (77)	90 (41)

Figure 28-96 The recommended tension for power steering drive belts is printed in many service manuals. *(Courtesy of Mitchell Repair Information Company)*

28.22 Completion

After completing the inspection, the technician has the information needed to make a decision about what to do for the car. If no worn parts are found and there are no handling or driving problems, nothing needs to be done. If no worn parts are found but there are abnormal tire or driving problems, a wheel alignment should be performed. Wheel alignment is described in Chapter 34. If worn parts are located, they should be replaced or adjusted.

28.23 Practice Diagnosis

You are working in a brake and front-end shop and encounter the following problems:

CASE 1: A good customer purchased a set of tires last summer that now have 9,670 miles on them, and he has just come back with a complaint of a vibration between speeds of 50 and 55 mph. Your road test of the car confirms the problem condition, and you also feel a slight sideways

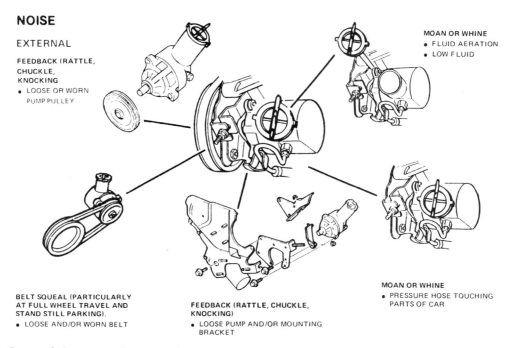

Figure 28-97 Some of the causes for unusual power steering pump noise. (*Courtesy of Ford Motor Company*)

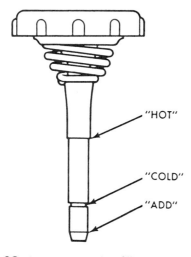

Figure 28-98 A power steering filler cap and dipstick; note the markings for three different fluid levels. (*Courtesy of General Motors Corporation, Service Technology Group*)

Figure 28-99 A drop of power steering fluid has been placed on white paper; it is clean fluid. The contaminants from dirty fluid are not present.

oscillation at the steering wheel during the periods of vibration. What procedure should you follow to locate the cause of this problem? What is the probable cause?

CASE 2: While doing a prealignment inspection on a 1985 Chevrolet Caprice, you find a torn boot on the lower ball joint, right side. You check the clearance on this ball joint and find the grease fitting boss almost even with the surface of the ball joint. What should you check next? What is your recommendation to the customer?

CASE 3: The customer's complaint is a clunking noise when the brakes are applied on her 1988 Nissan Sentra. What is the best way of locating the cause of this noise?

CASE 4: The customer has brought in his 1982 Dodge Colt with a problem of sloppy steering. When you check it, you find about 3 in. of free travel at the steering wheel rim. What is the best way of finding the source of this play?

CASE 5: A local building contractor has brought in his work pickup with this complaint: "When I apply the brakes, I get kind of a shudder and then it pulls the steering wheel to a right turn." This pickup is a 3/4-ton rig with a utility box that looks pretty well loaded. The tires are fairly well worn but still have some usable tread on them. The speedometer shows 93,560 mi. What is the probable cause of this problem? What procedure should you use to confirm this?

Chapter 28 ■ Suspension and Steering System Diagnosis and Inspection

You have changed jobs, and now you are working at a new-car dealership, specializing in the steering, suspension, and alignment area.

CASE 6: The customer took delivery of a compact pickup last week and has brought it back in with a complaint that it tries to turn to the left. Your road test confirms this. What procedure will you follow to locate and cure this problem?

CASE 7: The car is a midsize, FWD General Motors "W" body car, and the customer's complaint is a strange noise coming from the left rear. On a road test, you hear a growling noise that seems to come from the left rear, and it becomes quieter when you make left turns. What should you do when you get this car back to the shop to confirm this problem? What is the probable cause? What will you probably need to do to repair this car?

Terms to Know

amplitude	horizontal/radial movement	side shake
audible	intensity	single-plane balancer
axial/vertical movement	kinetic balancer	snap
beating	lateral runout	static
bounce check	lead	tactile
chassis height	loaded radial runout	tire and wheel balancing
computerized suspension analyzer	locator tool	tire matching
conicity	natural frequency	tire stiffness variation
cupping	order	torque steer
curb height	phasing	trouble trees
dry or tight joint	pop	turning torque
dry park test	preload	two-plane balancer
dynamic	pull	unloaded runout
free play	radial runout	vibration
frequency	radial tire waddle	vibration analyzer
hard steering	reed tachometer	wear indicator
height check	resonance	
hertz (Hz)	side clearance	

Review Questions

1. Technician A says that the loudness of a noise is measured in hertz.
   Technician B says that very loud noises can damage your hearing.
   Who is correct?
   a. A only
   b. B only
   c. Both A and B
   d. Neither A nor B

2. Technician A says that a squealing noise is at a higher frequency than a booming noise.
   Technician B says that the frequency of a squealing noise is less than 20 Hz.
   Who is correct?
   a. A only
   b. B only
   c. Both A and B
   d. Neither A nor B

3. Technician A says that a vibration needs a path to get from its cause to the driver.
   Technician B says that the steering column can be one of the paths.
   Who is correct?
   a. A only
   b. B only
   c. Both A and B
   d. Neither A nor B

4. Technician A says that the vibration frequency measured in the car will often match the speed of rotation for some part of the car.
   Technician B says that the speed can be twice the vibration frequency if it is a second-order vibration.
   Who is correct?
   a. A only
   b. B only
   c. Both A and B
   d. Neither A nor B

5. Technician A says that an unbalanced tire rotating at 45 rpm will set up a vibration of 45 Hz.
   Technician B says that if this tire had a second-order vibration, the frequency would be 90 Hz.
   Who is correct?
   a. A only
   b. B only
   c. Both A and B
   d. Neither A nor B

6. While driving at a steady speed, the car's vibration comes and goes away at regular intervals of a few seconds.
   Technician A says that this vibration is coming from two different sources.
   Technician B says that one of the sources is turning faster than the other.
   Who is correct?
   a. A only
   b. B only
   c. Both A and B
   d. Neither A nor B

7. The car has a vibration of a frequency of 24 Hz.
   Technician A says that it is coming from something that is rotating at a speed of 24 revolutions per second.
   Technician B says that the rotational speed could be 8 or 12 revolutions per second.
   Who is correct?
   a. A only
   b. B only
   c. Both A and B
   d. Neither A nor B

8. Technician A says that one of the early steps in solving a car's problem should be to take the car out for a road test.
   Technician B says that one of the first things to do on the road test is to locate and confirm the cause of the complaint.
   Who is correct?
   a. A only
   b. B only
   c. Both A and B
   d. Neither A nor B

9. Technician A says that a torque-sensitive vibration occurs during acceleration but goes away while coasting.
   Technician B says that these problems are usually caused by something in the drivetrain.
   Who is correct?
   a. A only
   b. B only
   c. Both A and B
   d. Neither A nor B

10. Technician A says that trouble trees will often guide you to the cause of a vehicle's problem.
    Technician B says that you can often skip to the middle of a trouble tree because they are very basic.
    Who is correct?
    a. A only
    b. B only
    c. Both A and B
    d. Neither A nor B

11. A tire is worn heavily on the inside of the tread with a fairly sharp corner at the sidewall; this wear was probably caused by too much:
    a. Negative caster
    b. Negative camber
    c. Toe-in
    d. Positive camber

12. Technician A says that too much air pressure will cause a tire to wear more rapidly in the center.
    Technician B says that this wear pattern can also be caused by improper alignment.
    Who is correct?
    a. A only
    b. B only
    c. Both A and B
    d. Neither A nor B

13. The air pressure in a tire should be checked:
    A. Any time during the car's operation
    B. At regular intervals, about once a month
    Which option best completes the statement?
    a. A only
    b. B only
    c. Both A and B
    d. Neither A nor B

14. Faulty construction of a radial tire can cause a:
    a. Side pull
    b. Back-and-forth wobble
    c. Hop up and down
    d. Any of these

15. Technician A says that a side-to-side steering wheel shake is probably caused by radial tire runout.
    Technician B says that this condition can be cured with a static balance of the tire.
    Who is correct?
    a. A only
    b. B only
    c. Both A and B
    d. Neither A nor B

16. Radial tire runout will cause the tread of the tire to move _____ while the tire is rotated.
    A. In and out
    B. Sideways
    Which option best completes the statement?
    a. A only
    b. B only
    c. Both A and B
    d. Neither A nor B

17. Technician A says that lateral tire runout can sometimes be improved by remounting the wheel in a different position on the hub.
    Technician B says that it can usually be improved by remounting the tire on the wheel.
    Who is correct?
    a. A only
    b. B only
    c. Both A and B
    d. Neither A nor B

18. Technician A says that a faulty wheel bearing will usually change noise levels as the car is turned in different directions.
    Technician B says that faulty wheel bearings will usually change noise levels under different throttle and brake conditions.
    Who is correct?
    a. A only
    b. B only
    c. Both A and B
    d. Neither A nor B

19. Technician A says that wheel bearing play is checked by gripping the tire at the top and bottom and trying to rock it.
    Technician B says that the front wheel bearing of a RWD car should have a slight amount of play when rocked.
    Who is correct?
    a. A only
    b. B only
    c. Both A and B
    d. Neither A nor B

20. A faulty wheel bearing will _____ when spun during the checking process.
    A. Be noisy
    B. Operate with roughness
    Which option best completes the statement?
    a. A only
    b. B only
    c. Both A and B
    d. Neither A nor B

21. Technician A says that a badly worn ball joint can allow the camber angle of the front tires to become more negative.
    Technician B says that a badly worn ball joint might come apart.
    Who is correct?
    a. A only
    b. B only
    c. Both A and B
    d. Neither A nor B

22. A wear indicator ball joint should be checked with the joint:
    a. Unloaded by a jack under the control arm
    b. Unloaded with a jack under the frame
    c. Loaded and the car's weight on the tires
    d. Loaded and the tires in the air

23. The ball joint boot:
    A. Is still all right if it has a few tiny tears in it
    B. Should be filled partially with grease
    Which option best completes the statement?
    a. A only
    b. B only
    c. Both A and B
    d. Neither A nor B

24. A load-carrying ball joint mounted in the lower control arm is being checked.
    Technician A says that the jack should be placed directly under the spring to unload the joint.
    Technician B says that the allowable radial clearance for all ball joints is 0.250 in.
    Who is correct?
    a. A only
    b. B only
    c. Both A and B
    d. Neither A nor B

25. Ball joint clearance:
    A. Is usually measured using a dial indicator
    B. Can be measured with a caliper
    Which option best completes the statement?
    a. A only
    b. B only
    c. Both A and B
    d. Neither A nor B

26. A friction-loaded ball joint is being inspected.
    Technician A says that there should be no perceptible play in the joint as the steering knuckle is being shaken.
    Technician B says that if the joint were disconnected, it would take 0 to 5 in.-lb of torque to rotate the ball stud of a good joint.
    Who is correct?
    a. A only
    b. B only
    c. Both A and B
    d. Neither A nor B

27. Kingpin bushing wear is checked by lifting the axle and measuring the:
    A. Side motion at the tire tread
    B. Vertical motion of the steering knuckle
    Which option best completes the statement?
    a. A only
    b. B only
    c. Both A and B
    d. Neither A nor B

28. A faulty rubber control arm bushing will:
    a. Have rubber portions breaking out of it
    b. Have a center sleeve sagged off center
    c. Show signs of metal contact between the control arm and the frame
    d. Any of these

29. Technician A says that a squeaky metal control arm bushing is most likely bad and should be replaced. Technician B says that with metal bushings, a side motion of up to 0.125 in. is acceptable.
    Who is correct?
    a. A only
    b. B only
    c. Both A and B
    d. Neither A nor B

30. A motorist is complaining about a loud, clunking noise when the brakes are applied.
    Technician A says that this noise can be caused by bad control arm bushings.
    Technician B says that this noise can be caused by bad strut rod bushings.
    Who is correct?
    a. A only
    b. B only
    c. Both A and B
    d. Neither A nor B

31. Which of the following is not an indicator of a worn front end?
    a. Front-end vibration at 50 mph
    b. Scalloped tire wear
    c. Sloppy steering with some degree of wander
    d. Popping and clunking from the front end during different driving maneuvers

32. A faulty strut rod bushing will show signs of:
    A. Rubber breakup
    B. Metal-to-metal contact between the backup washer and the frame
    Which option best completes the statement?
    a. A only
    b. B only
    c. Both A and B
    d. Neither A nor B

33. A strut damper is being discussed.
    Technician A says that you can check it by trying to shake the top of the strut in and out while the tires are in the air.
    Technician B says that there should be no vertical motion at the top of the strut while lifting the tires off the ground.
    Who is correct?
    a. A only
    b. B only
    c. Both A and B
    d. Neither A nor B

34. Idler arm wear checks are being discussed.
    Technician A says that a worn idler arm will allow excessive vertical motion of the right end of the center link.
    Technician B says that a worn idler arm will usually show up when the backs of the tires are pushed apart and pulled together.
    Who is correct?
    a. A only
    b. B only
    c. Both A and B
    d. Neither A nor B

35. An outer tie-rod end can be checked by watching for play in the joint as it is:
    a. Squeezed together with a pair of pump pliers
    b. Squeezed together using dust cap pliers
    c. Pushed and pulled by hand
    d. Pried apart using a screwdriver

36. Rack-and-pinion steering checks are being discussed.
    Technician A says that the steering should be turned outward to extend the bellows as it is being checked for cracks and breaks.
    Technician B says that the inner tie-rod end can be checked by gripping the end of the rack and the end of the tie-rod through the bellows and feeling for play as the tire is moved back and forth.
    Who is correct?
    a. A only
    b. B only
    c. Both A and B
    d. Neither A nor B

37. Checks of a manual rack-and-pinion steering gear normally do not include:
    a. Lubricant level
    b. Mounting bushing condition
    c. Amount of internal play
    d. Amount of internal drag

38. While inspecting a power steering system, you find a hard, glazed, cracking pump drive belt; it is probably caused by a leaky:
    A. Steering control valve
    B. Pump seal
    Which option best completes the statement?
    a. A only
    b. B only
    c. Both A and B
    d. Neither A nor B

39. Power steering problems are being discussed.
    Technician A says that a steering wheel that "kicks" in one direction as the engine is started is probably caused by excessive pump pressure.
    Technician B says that a buzzing or humming noise as the wheels are turned can be caused by air in the system.
    Who is correct?
    a. A only
    b. B only
    c. Both A and B
    d. Neither A nor B

40. Technician A says that a dry park test is made by turning the steering wheel back and forth with the tires on the ground.
    Technician B says that free movement of the steering wheel while making this check indicates wear or looseness in the steering gear or linkage.
    Who is correct?
    a. A only
    b. B only
    c. Both A and B
    d. Neither A nor B

Math Questions

1. The tire has an effective radius of 13 in. What speed (in rpm) will it be turning at 60 mph? What speed in revolutions per second? What first-order vibration frequency would an unbalance condition cause? What second-order vibration frequency would it cause?
2. You are checking ball joint clearance using a vernier caliper. With the ball joint loaded, the distance from the top of the ball joint to the end of the stud is 3.685 in. (Figure 28–60C). With a jack under the control arm and the ball joint unloaded, the distance is 3.723 in. How much clearance does this ball joint have? Using the specification in Figure 28–60A, is this ball joint good or bad?
3. You are checking a strut damper, and your first measurement is 7/8 in. with the tires on the ground (Figure 28–79, top right). With the tires off the ground, this measurement is 7/16 in. How much movement does this damper have? Is it good or bad?

29 Tire and Wheel Service

Objectives

Upon completion and review of this chapter, you should be able to:

- ❑ Correctly adjust the tire pressure for a particular car.
- ❑ Rotate a set of tires using any of the recommended rotation patterns.
- ❑ Statically and dynamically balance a tire and wheel, either on or off the car.
- ❑ Perform the ASE tasks relating to wheel and tire diagnosis and repair (see Appendix A, Section D, tasks 1–7).

29.1 Introduction

When a tire problem occurs and the motorist brings the car into a shop for repairs, that shop and the tire service technicians concerned take on an obligation to determine the cause of the problem and to repair it. The repair should be such that the car can be operated in a normal and safe manner for a reasonable length of time. The repairs should be made following methods that are approved by the car manufacturer, the RMA, and the repair industry. Also, the repairs should be made in such a way that the cost to the motorist will be as low as possible yet ensure a reasonable wage for the tire service technician and profit for the shop.

The service procedures given in this text are typical for the industry. Exact methods might vary depending on the particular make or model of car being worked on or the particular type of equipment available for use in the repair. In all cases, a repair operation should begin with a base of knowledge on the part of the technician, be followed by a diagnostic procedure to determine the cause of the problem and the type of repair needed, and then end with a planned, careful, and thorough repair procedure. Good workmanship will ensure lasting repairs.

When diagnosing a problem, the technician uses knowledge and experience to arrive at the cause of the problem. The first step is to understand what the problem is. As much as possible or necessary, the motorist should be asked to describe the problem, how he or she knows that it is there, when it began, and under what speeds and driving conditions it occurred. If possible and if necessary, the technician should verify that the problem exists by road testing the car. This procedure is described in Chapter 28. The causes of many problems often become very obvious when they are actually seen, heard, or felt by a good technician. For example, nearly every vibration problem is caused by something that moves, spins, rotates, or rolls. If a noise or problem is checked under different driving conditions, the technician can come closer to the cause of a vibration problem. An experienced technician will compare the rate of vibration to the speeds at which the various parts rotate to help determine the cause of the vibration. The cause of some problems can be more difficult to determine, but problem diagnosis flowcharts such as those shown in Chapter 28 are valuable guides.

The procedure to correct tire-related driving problems is described in the following sections. The appearance of the tires is a good indicator of the condition of the suspension system and the alignment.

29.2 Removing and Replacing a Tire and Wheel

Removing and replacing a tire and wheel on a car is a common service operation. Most tire service technicians perform this task so often that they appear very casual and nonchalant while doing so. Through experience, a good technician has developed work habits that are carefully

Chapter 29 ■ Tire and Wheel Service

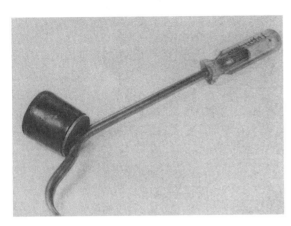

Figure 29–1 The lower end of this tool is used to pry off hubcaps and wheel covers, and the rubber hammer portion is used to drive them back on.

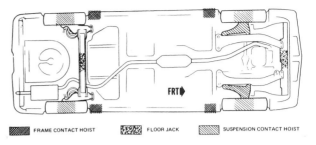

Figure 29–3 The jacking and lifting points, which are usually shown in service manuals, are places that are strong enough to support the car. *(Courtesy of General Motors Corporation, Service Technology Group)*

followed, even though he or she appears casual while doing them. Hopefully, you will not have to have a wheel come loose from a car that you have worked on to show you the effect of poor workmanship.

To remove a tire and wheel assembly, you should:

1. Remove any wheel covers, which are often locked (Figure 29–1).

CAUTION Set the parking brake. If the car is on an unlevel surface, block the wheels so the car will not roll and fall off the jack or stand.

2. Chalk mark the end of the lug bolt or the wheel stud or bolt closest to the valve stem. Such markings allow the wheel assembly to be replaced in the same location to ensure that no change in the balance or the runout of the tire and wheel will occur. If the valve stem is positioned between two studs, mark both of them or draw a line on both the wheel and the hub. This step is unnecessary during tire replacement or rotation (Figure 29–2).

3. Loosen the lug nuts using a six-point lug nut socket or lug wrench.

4. Place a jack under an appropriate part of the car that is strong enough to support the weight (Figure 29–3). Some cars have a jacking pad at the bumper or side of the car. Other places that are commonly used are the frame or subframe, a frame cross member, a reinforced unibody box section, an axle, or most suspension arms. Lift the tire and wheel off the ground and place a jack stand under some strong portion of the car (Figure 29–4).

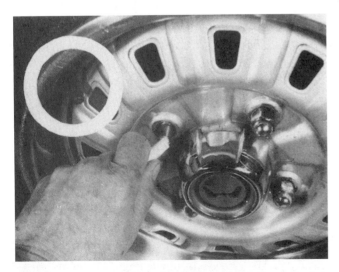

Figure 29–2 Placement of a chalk mark on the end of the wheel stud closest to the valve stem will help in replacing the wheel in the same position it was before removal.

Figure 29–4 Two different styles of jack stands that can be placed under the car for safe support. Never work under a car that is supported only by a jack.

NOTE: The rear axle of some FWD vehicles will bend if the jack is placed in the center of the axle. The axle should be lifted, one side at a time, with the jack placed near the spring.

NOTE: If you are unsure where to lift the car, consult the lifting and jacking section of the car manufacturer service manual or a technician service manual.

5. Finish removing the lug nuts. If you are using an air-impact wrench, excessive speed can cause lug bolt or nut galling. Adjust the wrench speed or lubricate the threads to ensure that galling does not occur.
6. Remove the tire and wheel assembly.

To install a tire and wheel assembly, you should:

1. Check the wheel nut bosses for worn or elongated holes. Damaged wheels should be replaced. Place the wheel over the lug bolts with the valve core next to the previously marked stud or studs.
2. Snug down the lug nuts, making sure the tapered portion of the lug nut enters the tapered opening of the wheel nut bosses. While the tire and wheel are in the air, it is difficult to apply enough torque to completely tighten the nuts because the wheel will spin. However, they can and should be tightened enough to hold the wheel in the correct position.

CAUTION *Lower the car onto the ground and immediately complete step 3. Do not allow yourself to forget it.*

3. Tighten the lug nuts to the correct torque using a tightening pattern that moves back and forth across, not around, the wheel (Figure 29–5). Lug nut torque varies depending on the size of the lug bolt and the wheel center section material.

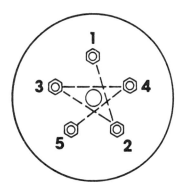

Figure 29–5 Lug bolts should be tightened in a star-shaped or crisscross pattern. *(Courtesy of Moog Automotive Inc.)*

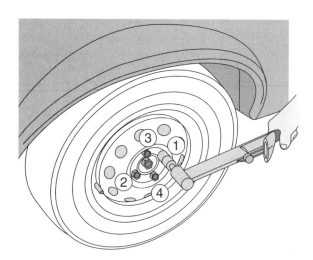

Figure 29–6 A torque wrench is used to ensure that the nuts are tightened properly.

Aluminum wheels sometimes use a slightly higher torque because of the solid center section (Figure 29–6 and Figure 29–7). However, a slightly lower torque is sometimes recommended to reduce nut-to-wheel galling and compression of the wheel nut bosses. If the manufacturer's torque specifications are not available, use the following torques, based on the lug bolt size:

3/8-24	—	35–45 ft-lb	(48–61 N-m)
7/16-20	—	55–65 ft-lb	(75–88 N-m)
1/2-20	—	75–85 ft-lb	(102–115 N-m)
M10 × 1.5	—	40–45 ft-lb	(54–61 N-m)
M12 × 1.5	—	70–80 ft-lb	(95–108 N-m)
M14 × 1.5	—	85–95 ft-lb	(115–129 N-m)

SAFETY TIP: It is a good practice to retighten the lug nuts after driving 10 to 20 mi. (16 to 30 km). With alloy center wheels, it is a good idea to check lug nut tightness again after another 100 mi. (161 km).

4. Replace the wheel cover or hubcap using a rubber hammer or the hammer portion of the hubcap tool.

29.3 Lug Bolt and Stud Replacement

If a few of the threads on a lug bolt are damaged, the threads can be **chased** to repair them, but if the threads are too badly damaged or the bolt is broken, the bolt should be replaced. A thread chaser, also called a chasing tool or die nut, must be of the same thread size and pitch

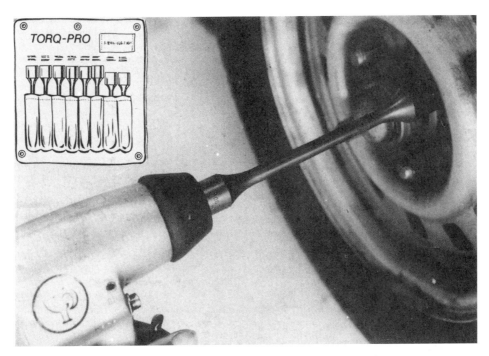

Figure 29–7 Torque sticks (inset) resemble a torsion bar with a socket built into it, and the diameter of the stick acts like a torsion bar to prevent overtightening a lug nut when used with an impact wrench. *(Courtesy of Pro-Cut Int'l., West Lebanon, NH USA)*

as the lug bolt. One style of thread chaser resembles a nut and is threaded onto and then off of the lug bolt, cutting away any metal that is in the wrong place. Another style of thread chaser is split into two pieces, and the two halves are placed onto the bolt (the threads next to the rotor or drum are usually in the best shape) and then threaded off to make the thread repair (Figure 29–8).

The procedure used to remove a lug bolt depends on where the bolt is used and how tightly it is retained. When the rotor or drum is normally removed with the hub, the lug bolt is often upset or riveted to secure the rotor or drum to the hub and secure the lug bolt in place. This is also called swaging. The term **upset** refers to the metal that is deformed or moved when parts are riveted together. This case requires a little work to replace the lug bolt. In other cases, where the rotor or drum can be easily removed separate from the hub, the lug bolt can usually be tapped loose with a hammer and punch. In either case, if force is needed to get the bolt loose, care should be taken to prevent bending of the hub flange, brake rotor, or drum.

To remove a lug bolt from a hub or axle flange that is separate from the rotor or drum, you should:

1. Tap on the outer end of the lug bolt using a punch and hammer. If the bolt loosens, proceed to step 3. If the bolt does not come loose under a reasonable hammer blow, proceed to step 2 (Figure 29–9).

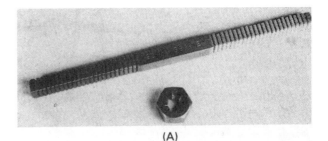

(A)

(B)

Figure 29–8 A thread file and thread chaser or die nut (A) can be used to repair the threads on a wheel stud, but they are rather slow. A wheel stud thread chaser (B) is faster and often does a better job; it is placed over the stud and then unscrewed to dress up the threads.

2. Position a pressing tool such as a universal joint press or a suspension bushing press across the hub or axle flange and lug bolt. Tighten the press to push the bolt inward until the bolt breaks loose and can be moved inward and out of the hub flange (Figure 29–10).

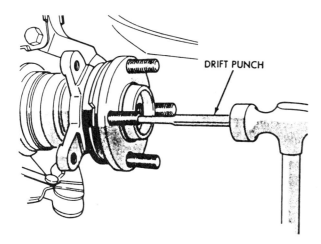

Figure 29-9 A damaged lug bolt can often be removed using a hammer and punch if it is not swaged in place. *(Courtesy of Chrysler Corporation)*

Figure 29-11 If the wheel stud is upset or swaged on the outer side, the swaged shoulder can be removed with a special cutting tool. *(Courtesy of Century Wheel, Division of Precision Switching)*

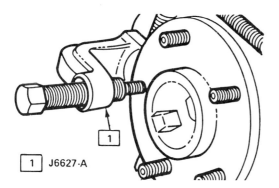

Figure 29-10 A damaged wheel stud is removed using a pressing tool (J6627-A). *(Courtesy of General Motors Corporation, Service Technology Group)*

Figure 29-12 A brake drum or rotor can be placed over this anvil, which will support the hub as the wheel stud is pressed or driven out.

3. If necessary, turn the hub or axle flange to a point where there is sufficient room and slide the broken lug bolt out of the flange.

To remove a lug bolt from a rotor and hub or a drum and hub combination, you should:

1. If the bolts are upset and you have the right size of cutting tool, cut the upset metal from the lug bolt (Figure 29-11).
2. Place a support, often called an anvil, under the hub so the hub is well supported yet there is room for the bolt to pass by (Figure 29-12).

OR Position a lug bolt press over the hub and rotor or drum and the damaged lug bolt.

3. Position either setup in the bed of a press and press the old lug bolt out of the hub (Figure 29-13).

OR Strike the old lug bolt or the arbor with a hammer to drive the old bolt out of the hub.

The replacement lug bolt should have the same diameter and length of thread and the same diameter and length of serrations as the one removed (Figure 29-14). In most cases, the replacement bolt can be installed by sliding the new bolt into position, placing several flat washers over the bolt threads to help protect the nut and to help ensure a straight installation, and then tightening the lug nut onto the bolt (Figure 29-15). The nut should be positioned so its flat face is against the washers and tightened to pull the wheel stud into place. The bolt should then pull into place. If necessary, the lug bolt can be pressed into the hub and rotor or drum assembly with either a press and hub support tool or a lug bolt press using a procedure just the opposite of the removal procedure. Whatever method is used, do not place a pressing load across the brake rotor or drum because the rotor and drum can crack or become distorted.

Chapter 29 ■ Tire and Wheel Service

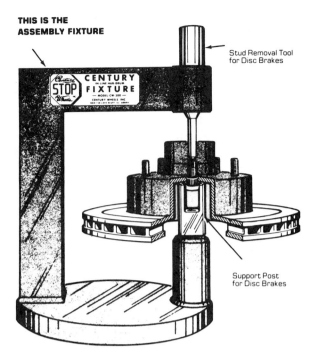

Figure 29–13 This fixture is set up for removal of a wheel stud. Note how the hub is supported and that there is no pressure on the brake rotor. *(Courtesy of Century Wheel, Division of Precision Switching)*

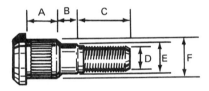

Figure 29–14 The important dimensions when selecting a replacement wheel stud are (A) serration length, (B) shoulder length, (C) thread length, (D) thread diameter, (E) shoulder diameter, and (F) serration diameter plus the thread pitch and direction. Replacement wheel studs should match the original ones. *(Courtesy of Dorman® Products, Division of R&B Inc.)*

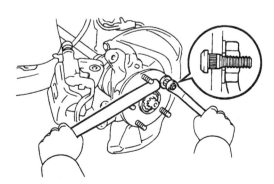

Figure 29–15 A new wheel stud can be pulled into position using a few flat washers and the lug nut.

29.4 Tire Rotation

Tires should be rotated to different positions on the car periodically to even out the wear through a set of tires. Turning motions, hard acceleration, and hard braking tend to increase tire wear. Tire wear normally occurs faster at some tire positions. For example, a high-horsepower RWD car with an impatient driver will usually wear out the right rear tire first. A FWD car will tend to wear out the front tires first, especially if the driver is aggressive. For convenience and more stable driving, most motorists replace the tires as a set of four, with the replacement time determined by the first tire to wear out. The ideal situation is for all four tires to wear out completely at the same time. Tires are rotated so that each one spends part of its life in the highest wearing position and the rest of its life in the lower wearing positions.

Tires are normally rotated when an uneven tread wear is found, or about every 10,000 mi. (16,100 km). It is a good practice to rotate the tires more frequently if the car is subjected to severe driving conditions. The recommended rotation patterns are shown in Figure 29–16. At one time, it was recommended that a radial tire never be cross-rotated so it would run in the opposite direction. This recommendation has now been dropped by the RMA and several major tire manufacturers. You can use the pattern that best suits the vehicle's needs.

Some of the rotation patterns are four-tire patterns. A standard type of spare tire can be included. If so, the

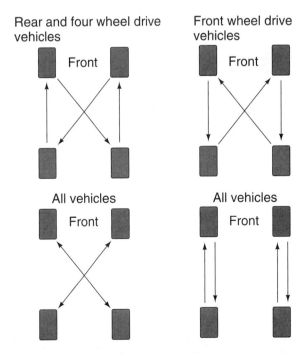

Figure 29–16 Two different tire rotation patterns are recommended depending on whether you are working on a FWD or RWD vehicle. *(Courtesy of Rubber Manufacturers' Association)*

spare is often placed on the right rear, and the tire that would have gone to the right rear is put in the trunk to become the new spare. Temporary-use spares should not be included in any rotation pattern, and unidirectional tires can only be rotated front to back and back to front.

29.5 Removing and Replacing a Tire on a Wheel

Tire replacement is normally done at a tire shop or a store that specializes in tire sales and service. Although the operation can be done with hand tools, a tire changing machine is usually used to make the operation of removing and replacing a tire on a wheel quick and easy. There are several makes and styles of tire changers, and space does not permit covering the operation of the individual machines. The operation of most tire changers is similar.

CAUTION *Although the possibility is remote, there has been a case of tire explosion and injury after a technician created a spark while repairing a tire inflated using a tire sealant–inflator canister. These canisters, designed for roadside fix of flat tires, contain an extremely flammable gas.*

CAUTION *Tire changers operate very quickly and powerfully. They present several possibilities for injury. It is important to follow the operating directions for each particular machine to avoid the possibility of injury to yourself, the tire, or the wheel.*

CAUTION *Run-flat tires, often marked EMT or ZP, have an air pressure sensor/transmitter mounted inside the tires. Care must be exercised to avoid damaging this unit when demounting and mounting the tire on the wheel.*

CAUTION *A few upscale vehicles are equipped with AH2 wheels as standard equipment. These wheels, which are designed to hold the tire bead in place if the tire loses air pressure, require special operations when mounting or demounting the tire.*

Most new cars use **match mounted** tires and wheels to reduce tire runout to a minimum. Match mounting places the highest or stiffest part of the tire next to the lowest portion of the wheel rim. With new tires, the

Figure 29–17 A crayon is being used to place an index mark next to the valve stem; this will ensure the tire is mounted in the same location.

match mark, commonly called a locator dot, is often a sticker or dye mark; with wheels, it can be a sticker on the rim, a paint mark in the drop well, or the tire valve hole (on most aluminum wheels). When you are removing a tire to repair either the tire or the wheel, it is a common practice to mark the valve location onto the tire with chalk or a tire crayon; doing so allows you to reindex the tire and wheel when you put them back together. When you install a new tire onto a wheel, check them both for match marks and align the marks before inflating the tire. If you cannot locate any other marks on the wheel, use the valve hole (Figure 29–17).

NOTE: Many aluminum wheels are painted with a clear-coat paint to protect the aluminum from corrosion. Do not scratch this clear coating.

To remove a tire from a wheel, you should:

1. Locate the side of the wheel with the shortest distance between the bead flange and the drop well and clamp the wheel into the tire changer with this side up (Figure 29–18 and Figure 29–19).
2. Remove the valve from the valve stem to ensure complete tire deflation. Also, remove all balance weights from the upper bead flange. If this tire is going to return to the same wheel after repair, mark the location of the weights and valve stem on the tire sidewall.

Chapter 29 ■ Tire and Wheel Service

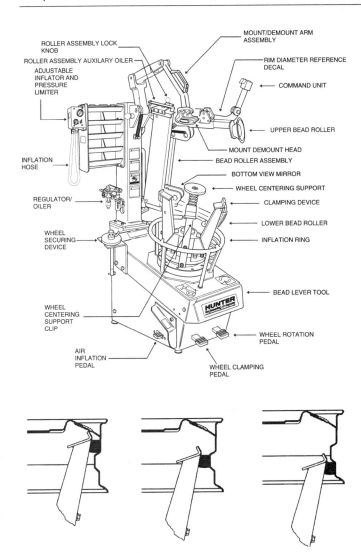

3. Position the bead breaker next to the bead flange and operate the bead breaker (Figure 29–20). Most tire changers will break loose both the upper and lower beads from the bead flange at the same time. The beads need to be loosened so they can be moved into the drop well during tire removal. It is sometimes necessary to rotate the wheel assembly and break the bead at several locations to loosen it.

NOTE: Care should be exercised when breaking the bead on alloy wheels. A strong bead breaker can crack or break the wheel flange if it catches it.

4. Lubricate the upper tire bead and wheel flange with tire or rubber lube. This lubricant is specially formulated for this purpose; it is often a mixture of vegetable soap and water (Figure 29–21).

5. Insert the removal end of the tire tool between the upper tire bead and the wheel flange and pry the bead up and over the flange. It is usually necessary to push the bead down into the well on the opposite side of the tire as the bead is being pried upward (Figure 29–22 and Figure 29–23).

6. Operate the tire changer to rotate the removal tool around the wheel, removing the top bead. If the tire has a tube in it, remove the tube at this time.

7. Repeat steps 4, 5, and 6 to remove the lower bead and the tire over the upper bead flange (Figure 29–24).

Installing a tire onto a wheel usually follows this procedure, which is just the reverse of removal:

1. Clamp the wheel into the tire changer with the narrowest bead seat upward (Figure 29–25).

2. Lubricate the wheel flanges, bead seats, and tire beads with tire lube (Figure 29–26).

3. Slide the lower tire bead partially over the wheel flange. If a tube is to be installed, place the tube in the tire and inflate it with just enough pressure to hold its shape. Install the valve cap to hold the pressure in the tube.

4. Position the tire installing tool so it hooks between the wheel flange and the tire bead (Figure 29–27). If you can locate the index mark on a new tire, usually a colored mark or sticker, and the index mark on the wheel, align these two marks so they match.

5. Operate the tire changer to rotate the installing tool around the wheel, sliding the tire bead over the wheel flange. When about one-half to two-thirds of the bead enters the wheel, make sure that

Figure 29–18 With modern tire changers, the tire and wheel are placed over the wheel clamps and the wheel clamps are expanded to lock the wheel. The clamping mechanism rotates the wheel through the tire demounting and mounting operations. *(Reprinted by permission of Hunter Engineering Company)*

Figure 29–19 After positioning the tire onto this older style of tire changer (narrow ledge upward), the hold-down cone is installed to fasten it securely in place. *(Courtesy of Rubber Manufacturers' Association)*

(A)

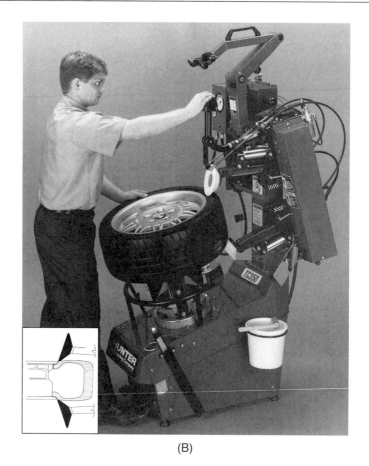
(B)

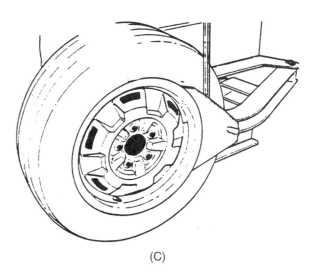
(C)

Figure 29–20 The bead breaker (old style) is positioned at the rim (right next to it on many changers) and operated to break the bead loose from the rim (A). Both beads need to be loosened. This new-style machine uses two rollers (inset) that push inward on the tire as it is rotated (B). Many new-style machines use a bead breaker mounted on the side of the machine (C). *(A is courtesy of Rubber Manufacturers' Association; B is reprinted by permission of Hunter Engineering Company; C is courtesy of Accu Industries)*

the bead is also entering the drop well. If a tube is being used, remove the valve cap, work the valve stem through the valve hole, and install a valve stem fishing tool onto the stem to prevent it from being lost.

CAUTION *Be careful not to catch your fingers or hands between the tire and rim or between the bead installing tool and the rim while mounting the tire.*

6. Repeat steps 3, 4, and 5 to install the upper bead.
7. Inflate the tire to seat the beads using a maximum pressure of 40 psi (276 kPa). Tubeless, belted tires usually require an inflation device to rapidly move a volume of air between the bead and bead flange to get enough air into the tire quickly enough to seat the beads. A bead expander can be used on nonbelted tires to force the beads up to the bead flanges for easy inflation (Figure 29–28). In either case, watch the tire beads during inflation; they

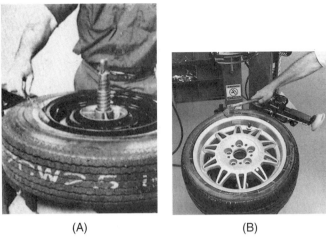

Figure 29–21 An approved tire lubricant should be placed on the tire bead and inside the flange (A). With some tires and wheels, it is a good practice to lubricate before breaking the bead (B). *(A is courtesy of Rubber Manufacturers' Association; B is reprinted by permission of Hunter Engineering Company)*

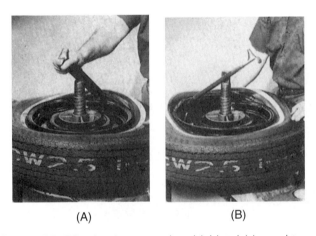

Figure 29–22 The tire removal tool (old style) is used to pry the bead over the wheel flange; at the same time, the bead at the opposite side of the tire must go into the wheel well (A). The changer is then operated to move the tire removal tool around the rim, lifting the upper bead past the flange (B). *(Courtesy of Rubber Manufacturers' Association)*

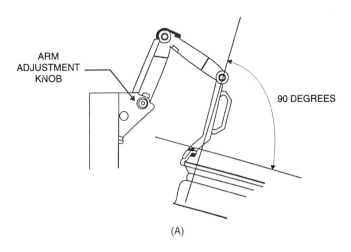

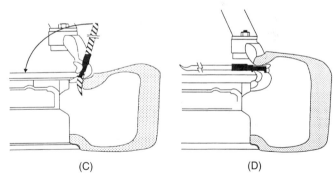

Figure 29–23 With new-style machines, the mount–demount arm is positioned next to the rim (A), the bead lever tool is used to lift the bead over the rim and onto the mount–demount head (B and C), and the wheel is rotated to lift the bead from the rim (D). *(Reprinted by permission of Hunter Engineering Company)*

should creep upward and snap into position as the tire is inflated. If a tube is used, make sure the tube is not caught between the bead and wheel flange during inflation.

CAUTION Do not position yourself over or in front of a tire that is being inflated; tires and wheels can separate violently during inflation, causing severe injury (Figure 29–29).

CAUTION If the beads do not seat with 40 psi (276 kPa) or less, break the bead down and clean and relubricate the beads and bead seats.

Figure 29–24 After lubricating the lower bead and upper bead flange, repeat the removal steps to lift the lower bead off the rim. *(Courtesy of Rubber Manufacturers' Association)*

Figure 29–25 The wheel should be cleaned and placed on the changer with the narrow bead flange upward. Both bead seats should be lubricated. *(Courtesy of Rubber Manufacturers' Association)*

Figure 29–26 The sides and base of both bead seats should be lubricated as the tire is placed over the wheel. *(Courtesy of Rubber Manufacturers' Association)*

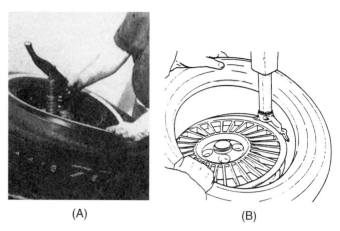

Figure 29–27 With the bead in the center well, the top bead is slid over the flange by rotating the tire tool around the rim (A). With new-style machines, the bead is positioned next to the mount–demount head and the wheel is rotated (B). *(A is courtesy of Rubber Manufacturers' Association; B is courtesy of Accu Industries)*

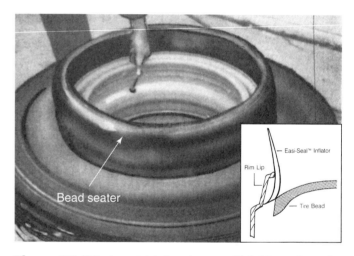

Figure 29–28 A portable bead seater. Air is blown from the base ring into the tire through the gap between the bead and wheel flange. A similar unit is built into most tire changers. *(Courtesy of Hardline International, Inc.)*

8. After inflation, check the positioning of the tire-to-wheel flange to ensure proper bead seating (Figure 29–30).

29.6 Removing and Replacing a Tubeless Tire Valve

A weathered, brittle rubber, snap-in type of tubeless tire valve should be replaced when a tire is replaced to ensure that it will last the life of the new tire. With the tire off the wheel, valve stem replacement is an easy matter.

Chapter 29 ■ Tire and Wheel Service

Figure 29-29 *Do not stand over the tire while inflating it.* Use an extension to allow you to stay clear of the tire in case it comes apart. Inflate the tire slowly until the bead slides or pops up to the wheel flanges. *Warning:* If more than 40 psi (275 kPa) are required to seat the bead, deflate the tire and relubricate the bead and seats. *(Courtesy of Rubber Manufacturers' Association)*

Figure 29-31 After the valve core is lubricated with tire lube, it is pulled into place from the inside of the tire.

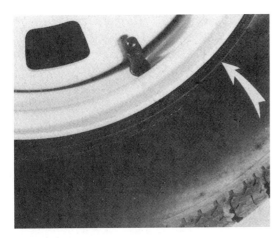

Figure 29-30 After the tire bead is seated, check the concentric ring (arrow) next to the wheel flange to ensure the bead is properly seated; if the ring and wheel flange are uneven, the bead is not fully seated.

An old valve is usually removed by either cutting the inner portion off with a knife or pulling it apart using a valve installer tool. Thread the installing tool onto the old valve and try pulling it out of the rim. It will either break apart or pull out. Some tire shops have a special tool that allows installation of a rubber valve stem from the outside of the wheel, which makes removal of the tire from the wheel unnecessary. A metal valve is removed by removing the retaining nut and pushing the old valve stem into the wheel well.

To install a new rubber, snap-in valve, you should:

1. Position the new valve inside of the wheel, pass the valve installer through the wheel boss, and thread the tool onto the valve (Figure 29-31).

2. Lubricate the valve and the wheel hole with tire lube.

3. Pull the new valve into position using a prying motion of the tool. Rubber from the valve should creep outward on both sides of the rim.

To install a metal, clamp-in type of valve core, merely lubricate the sealing area with tire lube, place the new valve stem in position, and tighten the retaining nut.

29.7 Repairing a Tire Leak

Although a punctured tire can be repaired using a plug from the outside, it is not a good practice. A plug installed from the outside is considered a temporary repair because it is not always safe or permanent. Inside tire repair offers two real advantages: (1) the use of a better patching method and (2) the ability to inspect the inside of the tire to make sure that it is safe. Failure to seal the tire's inner liner might allow pressurized air to enter the cord body; the air might work around the cords and cause a separation between the ply layers or between the plies and the rubber layers.

SAFETY TIP: All good tire service technicians feel responsible for ensuring the safe operation of every car they repair. The only way to ensure that a tire to be repaired is sound is to check the inside of it. The person who repairs an unsafe tire and sends it back on the road can be held liable for damages if that tire fails (Figure 29-32).

The first step in repairing a leaky tire is to locate the leak. Location is usually determined by visually inspecting

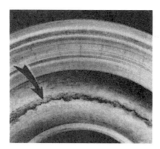

Figure 29-32 This tire damage was not visible from the outside of the tire and was caused by driving a short distance while severely underinflated. It is a dangerous, nonrepairable condition. *(Courtesy of Rubber Manufacturers' Association)*

the tread area, looking for the puncturing object. If nothing is found, check for a bead leak by laying the tire down, running water around the tire bead, and looking for the telltale bubbles.

If the leak still has not been located, immerse the tire in a test tank. Again, you will be looking for the telltale stream of bubbles coming from the leak. Tire repair shops immerse the tire as the first step in looking for a leak; it is important to locate all of the leaks before any repairs are made. Also, bubbles at the wheel flange or at a weld indicate a possible crack. Cracked wheels are unsafe and should be replaced. Once the leak has been located, its location can be marked using a tire crayon, thus allowing you to find the leak after the tire has been broken down (Figure 29-33).

Figure 29-33 A tire should be immersed in water before dismounting to show all of the leaks in the tire and any cracks or breaks in the wheel. *(Courtesy of Rubber Manufacturers' Association)*

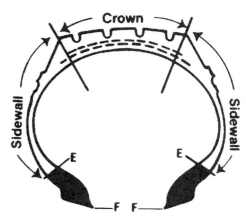

Figure 29-34 The repairable areas of a radial tire. The tire must be inspected internally, the hole must be filled, and the inner liner must be sealed for a proper repair. *(Courtesy of Patch Rubber Company—A Myers Industries Company)*

If the leak is at the valve or valve core, the defective valve stem or core will need to be replaced. If the bead is leaking, the tire bead will need to be broken loose from the bead seat so the wheel bead seat and flange and the tire bead can be cleaned or checked for cracks or breaks. If the leak is in the tread portion, it can be patched. Sidewall damage to a radial tire is normally not considered to be repairable. This is because the puncturing object might have cut enough cords in the body plies to severely weaken the tire, and the flexing of the sidewall might work the patch loose. However, a tire shop with the proper vulcanizing equipment can often repair sidewalls and shoulder areas if the damage is not greater than 1/4 in. (0.63 cm) in size (Figure 29-34).

Two methods are used to repair a hole in the tread. If the hole is big enough to cause a void, it is recommended that both be used. A plug can be inserted through the hole from the inside or outside. This plug not only seals the air leak but also keeps water and dirt from entering into the belt plies. A patch on the inside of the tire is the best method of sealing a leak. Many shops prefer patches because the plug-installing tool tends to enlarge the hole or damage the ply materials.

To plug and patch a leak, you should:

1. Spray an area several inches in diameter around the puncture on the inside of the tire with a pre-buffing cleaner. Let it soak in for about 15 seconds and scrape or wipe off the cleaner plus any silicones, waxes, or other contaminants from the inner liner (Figure 29-35).

2. Probe the hole with a probe or rasp to clean out the hole (Figure 29-36).

3. Coat the inside of the hole and the plug, also called a filler or insert, with vulcanizing cement and push or pull the plug into the hole (Figure 29-37).

Chapter 29 ■ Tire and Wheel Service

Figure 29-35 The first step in repairing a tire is to clean the area around the injury with a prebuff cleaner and a scraper. *(Courtesy of Patch Rubber Company—A Myers Industries Company)*

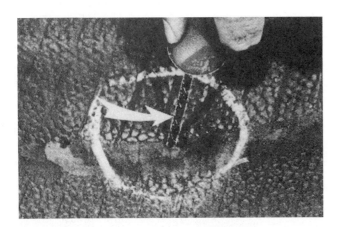

Figure 29-36 The hole or injury through the tire should be cleaned using a tapered carbide cutter at a slow speed. Be sure to follow the angle of penetration and wear eye protection. *(Courtesy of Patch Rubber Company—A Myers Industries Company)*

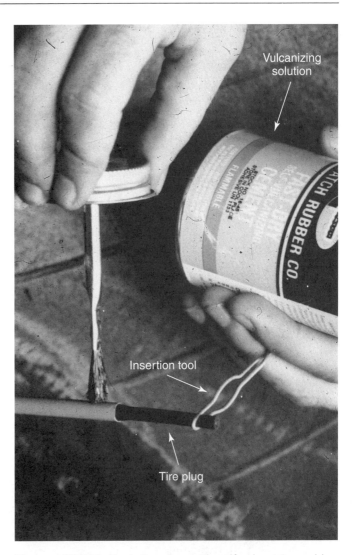

Figure 29-37 Coat the plug with self-vulcanizing rubber cement; while the cement is still wet, pull the plug through the hole. *(Courtesy of Patch Rubber Company—A Myers Industries Company)*

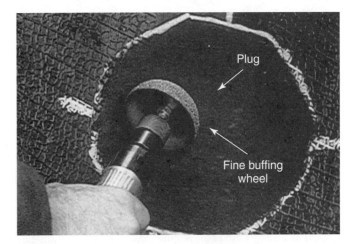

Figure 29-38 With the plug inserted and trimmed to length, the area for the patch is buffed. Be careful not to buff through the inner liner. *(Courtesy of Patch Rubber Company—A Myers Industries Company)*

4. Trim off the inner and outer ends of the plug with a knife.
5. Buff the inner end of the plug and the area around it using a powered buffer. Properly buffed rubber should have a dull, velvety appearance (Figure 29-38).
6. Coat the buffed area with vulcanizing cement (Figure 29-39).

Figure 29–39 A coating of self-vulcanizing rubber cement is applied to the buffed area and allowed to dry completely. *(Courtesy of Patch Rubber Company—A Myers Industries Company)*

7. Allow the vulcanizing cement to dry thoroughly, about 10 to 20 minutes, depending on temperature and humidity. Remove the protective backing from the patch, center the patch over the hole, and stitch the patch securely in place. Stitching is done with a special, narrow roller. You should begin at the center of the patch and work to the sides. Stitching should work any air bubbles out and press the patch firmly into contact with the tire (Figure 29–40). Remove the protective film from the patch after stitching it in place.

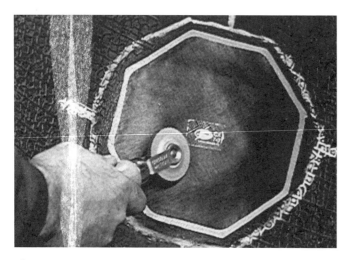

Figure 29–40 Remove the protective backing from the patch, apply the patch to the inner liner, and stitch the patch firmly into place. Begin at the center and work outward to remove any air bubbles. Note that this patch is positioned relative to the beads. *(Courtesy of Patch Rubber Company—A Myers Industries Company)*

Other methods have been developed to repair tubes and heavier truck and equipment tires. Instructions for these methods are available from manufacturers of tire repair materials.

29.8 Tire Truing

Tire truing is a process of cutting rubber off of a tire to make it round. Tires can be trued either on or off the car, depending on the type of equipment used.

Off-car truing involves mounting the tire and wheel in a machine that rotates the tire while cutting off the high portions of the tread. A motor-driven blade or rasp-type cutter is used to remove the excess rubber, although there is seldom any excess rubber on a tire. This type of tire truing is not recommended by some because it reduces tire life and does not always make a permanent repair (Figure 29–41).

On-car tire truing involves using a machine that measures tire-loaded radial runout and cuts rubber from the edges of the tread in relationship to the runout. This machine uses a pair of rasp-type cutters, one on each side of the tread. This machine is fairly expensive, but it has a reputation of being an effective tire repair method. These machines should be operated following the procedure recommended by the manufacturer. This operation is becoming less common with automotive tires.

Figure 29–41 An off-car tire truer. The motor turns a cutter blade (inset) that removes the high spots from the tread as the tire is rotated. *(Courtesy of Amermac, Inc., Ellaville, GA)*

Figure 29–42 A tire siper is used to cut sipes across the tread of the tire in some tire shops. The sipes increase traction in wet or icy conditions. *(Courtesy of Saf-Tee Siping & Grooving, Inc.)*

29.9 Tire Siping

Tire siping is a process of cutting sipes in the tread to improve wet traction; these sipes allow increased flexibility of the tread blocks. The machine makes a series of slices across the tread as the tire is rotated (Figure 29–42). This service is provided by larger tire stores.

29.10 Balancing Operations

Balancing a tire usually requires the placement of a number of weights, occasionally only one but usually two or four, on the wheel rim. Two weights or two pairs of weights are generally used to statically balance a tire and wheel, one on the inside and one on the outside (Figure 29–43). The amount of weight added should equal the amount of unbalance. If the unbalance is very slight, only one weight can be used; however, its placement on only one side of the rim could upset dynamic balance. A balancer is used to tell the operator where to place the weight and give some indication of how much weight to use (Figure 29–44).

Two equal-sized weights are normally used to correct dynamic unbalance. These two weights must generate a force equal to the amount of unbalance, and these two weights must be placed 180 degrees apart on the rim, with one weight on the inside of the wheel and the other on the outside (Figure 29–45).

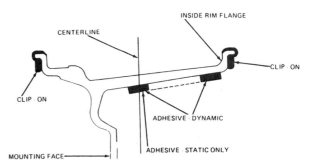

Figure 29–43 Clip-on weights are secured to the rim flange. Adhesive weights can be placed at the rim centerline or any flat surface inside the rim; the closer they are to the rim, the more effect they will have on dynamic balance. *(Courtesy of General Motors Corporation, Service Technology Group)*

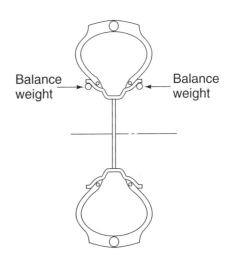

Figure 29–44 Weights to correct static or kinetic unbalance are placed on the rim directly opposite the heavy spot. If the amount of weight to be added is more than an ounce or so, it is usually split in half and added to each side of the rim.

It is common and often less confusing to use four weights, two for static and two for dynamic balance. Computer balance machines help an operator reshuffle these weights so placement of the static weights will reduce or eliminate the need for dynamic weights. An experienced tire and wheel balance technician never allows two weights to end up 180 degrees apart on the same side of the wheel; in this position, they are balancing each other out. A computer balancer, either off or on the car, normally uses only one or two weights, one on the inside and one on the outside of the wheel. Occasionally, only a single weight is needed. The computer has the ability to measure the various forces and quickly make the necessary calculations so the placement of two weights will balance both radial and lateral forces.

SAFETY TIP: Always use caution when spinning tires and wheels for balancing purposes (or for other reasons) because there are several possibilities for injury. Getting

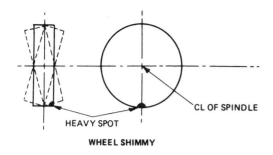

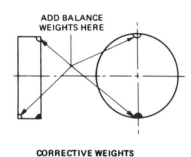

Figure 29–45 Weights to correct dynamic unbalance are always placed in two locations, at the inside and outside of the rim; these two weights are usually equal. *(Courtesy of General Motors Corporation, Service Technology Group)*

clothing or various body parts caught in the spinning tire and wheel or balancer and getting hit by a flying weight or debris from the tire and wheel are probably the most common causes of injury. Avoid them.

29.10.1 Wheel Weights

Wheel balance is achieved by attaching weight onto the wheel rim. Weights are available in various sizes, generally in 1/4-oz. increments. There are several styles of clip-on weights for steel and aluminum rims. Weights are also available in a stick-on style that is often used on alloy rims. These weights come in premolded sizes or in sticks that can be cut to the weight needed. Caution should be exercised when working with alloy wheels. They are easily scratched, and some weight materials can quickly damage the wheel. For example, a weight that contains zinc will cause rapid corrosion on a magnesium wheel (Figure 29–46).

Clip-on weights are installed and removed using a wheel weight tool or hammer. This tool is designed to both remove weights quickly and easily and drive them into place. Some weight tools are also made to trim the size of the weight (Figure 29–47).

29.10.2 Off-Car and On-Car Balancing

Wheel balancing can be done with the tire and wheel off the car and mounted on a balancer or with the tire and wheel mounted in their normal position and the balancer

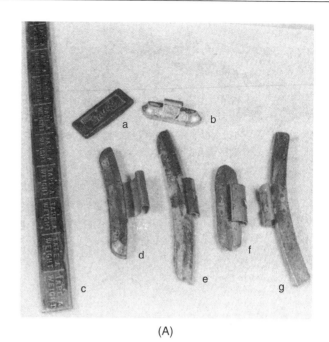

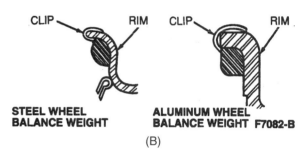

Figure 29–46 Some of the various types of wheel weights are shown in (A). Note the standard clip-on weights of two different sizes (b and e), a weight with an extra long clip to fit under some wheel covers (d), and two glue-on or stick-on weights (a and c), the longer of which is to be cut to fit under the right weight. At right center is a pickup or truck weight with an overly wide clip (f). At the right is a weight with an alloy clip for use on alloy wheels (g). The major types of clip-on weights are shown in (B). *(B is courtesy of Ford Motor Company)*

attached to the car. Off-car balancing is ideal when a new tire is mounted. The wheel is already off the car, saving the removal step. Off-car balancing machines are often simpler and less expensive because the machine is only concerned with the tire and wheel, and the tire and wheel are in a definite position relative to the machine sensors. All modern, spin-type, off-car balancers are two-plane kinetic and dynamic balancers. A possible disadvantage with off-car balancers is that the wheel commonly mounts to the machine using the wheel's center hole. A runout and balance error can occur if the wheel's center hole and the lug bolt bosses are not concentric.

On-car balancing, often called **spin balancing,** will probably produce the best balancing results because the tires and wheels are mounted in their running position

Chapter 29 ■ Tire and Wheel Service

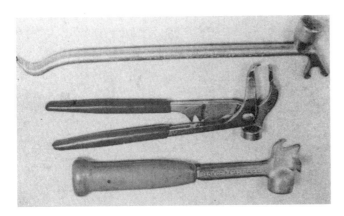

Figure 29–47 Three different wheel weight tools. They each provide a hammer to drive weights onto the rim and a way to pry weights off; the center tool can also trim weights smaller and tighten the spring clip.

Figure 29–48 A spinner motor is positioned to check tire balance. It can spin nondrive wheels fast enough to cause an unbalance condition to show up. *(Reprinted by permission of Hunter Engineering Company)*

and any slight runout errors are compensated for. Also, the brake rotor or drum is balanced with the tire, compensating for another possible unbalance. Possible disadvantages of on-car balancing are that this balance is disrupted if the tire and wheel are moved to another location on the car or rotated on the hub or axle. Also, the balancer is more complex and often more expensive because it has to be portable and adaptable to different cars. On-car balancing is used mainly to correct problems that off-car balancing cannot solve. On-car balancing can be either one or two plane and either kinetic or kinetic and dynamic. Nonsteerable wheels cannot be dynamically balanced on the car. A balancer cannot sense dynamic unbalance if the tire is not free to move in that direction.

On-car balancing requires something to turn or spin the wheels. Spinner motors are used to spin nonpowered wheels: the front wheels of a RWD car or the rear wheels of a FWD car. A spinner uses an electric motor that drives the tire through a machined metal drum that is pressed against the tread or shoulder of the tire. Spinners are available in different horsepower ratings. A spinner that is too small will not be powerful enough to spin large and heavy wheels at full speed without overheating. Spinners are also affected by brake drag. Sometimes it is necessary to loosen the shoes to allow sufficient speed and coast time. A brake is provided on spinners to allow the tire to be stopped quickly when desired. Spinners are often used to spin a tire and wheel up to speed to see if that particular tire and wheel are causing a vibration or to check a wheel bearing (Figure 29–48).

Drive wheels—the front wheels of a FWD car or the rear wheels of a RWD car—are normally spun by the car after starting the engine and putting the transmission into gear. In most cases, only the wheel being checked is lifted off the ground. On cars with limited-slip differentials, both drive wheels must be off the ground. Use of the car's engine to drive the wheels ensures that the differential gears are properly lubricated during the balancing operation. Severe differential wear has occurred on some cars through the use of a spinner motor on one tire of a drive axle.

SAFETY TIP: Following are some cautions to observe when using the engine to spin the drive wheel:

- Blocks should be placed in front of one or more of the tires remaining on the ground (Figure 29–49).
- The power should be applied gradually to make sure the car does not move forward.
- No one should be permitted in front of a car during this operation.

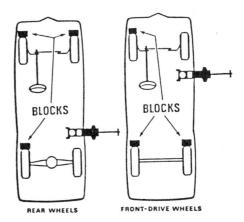

Figure 29–49 Drive wheels are spun using the car's engine; a jack is used to lift the tire to be checked off the ground while blocks are placed in front of the remaining tires. Both drive tires must be lifted if the car has a limited-slip differential. *(Reprinted by permission of Hunter Engineering Company)*

- When only one wheel is off the ground, the differential gears will cause a 2:1 speed increase of the spinning wheel; its speed will be double that read on the speedometer. Excessive tire speed can cause the tire to explode; it is recommended that the speed be limited to 40 mph (65 km/h).
- The wheels on some FWD cars hang down far enough to place severe operating angles on the axle shafts and universal joints. They should either be operated for only a short time at these angles or blocked partially up to limit the severity of these angles.

The procedure to spin balance the drive wheels on a car with a limited-slip differential is as follows:

1. Lift both drive wheels off the ground.
2. Remove the tire and wheel not being balanced and replace two lug nuts to retain the brake drum.
3. Spin the first tire and wheel to be balanced by starting the engine and placing the transmission in gear; in this case, the speedometer will be reading tire speed. Balance the first tire and wheel.
4. Replace the unbalanced tire and wheel and balance it. There is no need to remove the previously balanced tire and wheel because it is balanced.

29.10.3 Static Balancing, Bubble Balancer

A bubble balancer is an off-car, single-plane, static balancer. At one time, bubble balancers were the most common type of balancers. They are simple, reliable, inexpensive, fairly fast, and fairly accurate (Figure 29–50).

To balance a tire using a bubble balancer, the tire is placed on the balancer, and the operator watches the center bubble while trial weights are placed on the rim. The size and location of the weights are adjusted until the bubble becomes centered. Then these weights are attached to the rim (Figure 29–51).

29.10.4 Static and Kinetic Balancing, Mechanical Balancer

Static and kinetic balancing with a mechanical balancer uses a single-plane, on-car spin balancer that attaches to the wheel. Internal weights can be adjusted while the tire and wheel and the balancer unit are spinning. After these weights have been adjusted so that the assembly spins true with no vibration, the assembly is stopped. The amount and location of weight to add are indicated by the balancer unit (Figure 29–52).

To spin balance a tire and wheel using a mechanical balancer, the operator uses an adapter to attach the balancer to

Figure 29–50 A bubble balancer. Note the three adjustable feet (A) that are used to level the device and the control handle (B) that is moved downward to allow the balancer head to pivot.

Figure 29–51 Note the center bubble (arrow). This wheel is out of balance. Weight should be added to the rim directly opposite the bubble.

the wheel. The wheel is then spun up to a speed at which the vibration occurs, and then balancer knobs are used to adjust the amount and location of the internal weights to eliminate the vibration (Figure 29–53). The wheel is then stopped, the amount and location of weight are read from the balancer head, and this weight is attached to the wheel (Figure 29–54).

CAUTION *Extreme caution should be exercised if using this type of balancer, because you will be in very close proximity to a rapidly moving object that can catch clothing, hair, or other body parts.*

Chapter 29 ■ Tire and Wheel Service

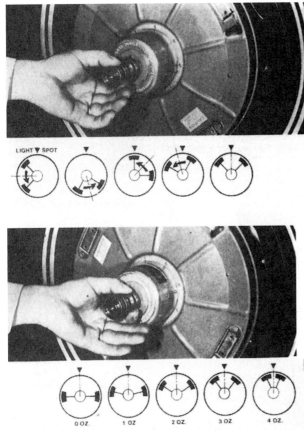

Figure 29–52 A mechanical balancer head has two internal weights. The location of the weights is moved by gripping one of the green knobs; the amount of weight is changed by gripping one of the red knobs. Both are done while the wheel is spinning. *(Reprinted by permission of Hunter Engineering Company)*

Figure 29–53 The balancer head is tuned by gripping one of the four knobs while watching a vibrating part of the car, often the bumper. When the vibration disappears, the wheel is stopped and the amount of unbalance is read on the balancer head. Note that the operator's hand is positioned so his thumb points in the direction of rotation. *(Reprinted by permission of Hunter Engineering Company)*

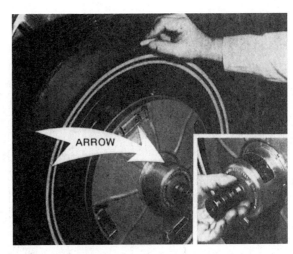

Figure 29–54 The amount of weight to add is indicated in the balancer window (inset), and the weight should be added to the rim in line with the "add weight here" arrow. *(Reprinted by permission of Hunter Engineering Company)*

29.10.5 Dynamic and Kinetic Balancing, Strobe Light Balancer

A **strobe light balancer**, sometimes called an **electronic balancer**, is an on-car, two-plane spin balancer. These units sense tire and wheel unbalance by means of a pickup, also called a transducer unit, that attaches to the suspension using a magnet. Vibrations of the suspension cause the strobe light to flash in time with the vibration; the flashes of the strobe light can cause the spinning tire and wheel to appear stopped. By noting the position of the wheel, the operator can tell where to add weights. By noting the reading on an indicator dial, the operator can tell the approximate amount of weight to add. The pickup unit is placed under the suspension in a vertical position while doing a kinetic balance (Figure 29–55). It is placed alongside of the suspension in a horizontal position while dynamic balancing.

To balance a tire and wheel using a strobe light, the operator places the pickup probe vertically under a suspension part, spins the tire up to speed, and uses the light to freeze the tire's position. The tire is then stopped and turned to this same position, and a weight is then added to the top of the wheel. The tire is then respun to check the job, and, if necessary, the amount and location of the weight are adjusted (Figure 29–56). Dynamic balancing is a similar operation with the pickup probe moved to a horizontal position (Figure 29–57). Correction of dynamic unbalance requires that two weights be used (Figure 29–58).

Figure 29–55 This pickup probe is placed under a front suspension in a kinetic balance position. Note the strobe light positioned so it will flash onto the tire. The flashing light freezes the tire's motion to show where weight needs to be added. *(Reprinted by permission of Hunter Engineering Company)*

29.10.6 Dynamic and Kinetic Balancing, Computer Balancing

Most **computer balancers** are off-car balancers. They are two-plane balancers that balance in both planes at the same time. They have the reputation of being fast, accurate, and easy to use. Most computer balancers can be forced to give a kinetic-only balance by entering an extremely narrow wheel width; this approach is rarely recommended. After the operator mounts the wheel, enters information about the tire and wheel, and starts the procedure, the balancer goes through a short spin cycle,

Figure 29–57 This pickup probe is positioned for a dynamic balance; it will sense sideways motions. *(Reprinted by permission of Hunter Engineering Company)*

stops the rotation, and then gives a readout of how much and where to add weight on each side of the wheel (Figure 29–59).

To balance a tire using a computer balancer, you should:

1. Place the wheel on the mounting cone and hand tighten the hub nut to secure it in place. Remove any heavy dirt deposits and all the old weights from the wheel (Figure 29–60).
2. Measure the machine-to-rim distance and enter it into the machine (Figure 29–61 and Figure 29–62).
3. Using the special calipers, measure the rim width and enter it into the machine (Figure 29–63).

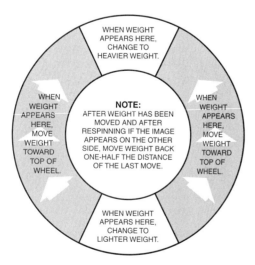

Figure 29–56 After a weight is added, the tire should be respun. If an unbalance still exists, this chart indicates the adjustment needed. *(Reprinted by permission of Hunter Engineering Company)*

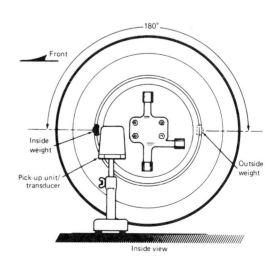

Figure 29–58 Dynamic weights are added in pairs, one right next to the pickup at the inside of the wheel and one to the outside, one-half turn away.

Chapter 29 ■ Tire and Wheel Service

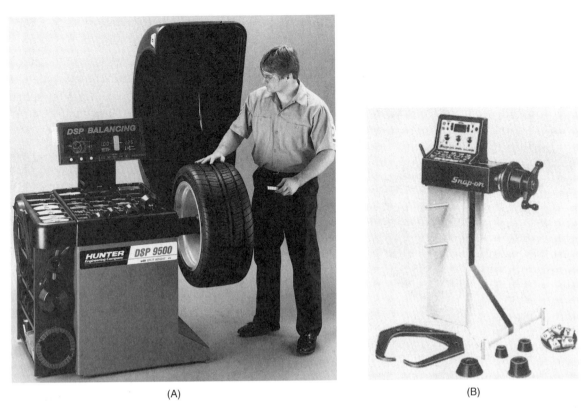

Figure 29–59 Off-car computer wheel balancers. *(A is reprinted by permission of Hunter Engineering Company; B is courtesy of Snap-on Tools Corporation)*

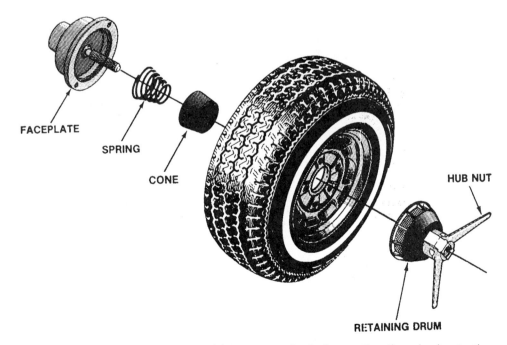

Figure 29–60 Most computer balancers use a quick-mount method of mounting the wheel onto the balancer. Most balancers center the wheel using a cone in the center hole, as shown here. *(Courtesy of Snap-on Diagnostics)*

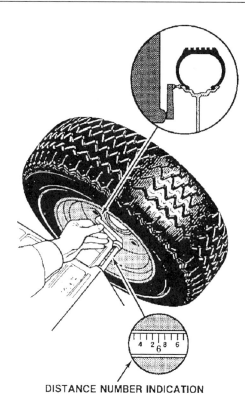

Figure 29-61 Gauging the wheel-to-balancer location. This is one of three dimensions that need to be programmed into the balancer computer. *(Courtesy of Snap-on Diagnostics)*

4. Note the wheel diameter and enter it into the machine.

5. Lower the protective cover or safety hood and press the start button. The machine will now spin the tire for a few seconds, stop spinning after a preset time, and display the amount of unbalance for each side of the tire.

6. Raise the protective cover and rotate the tire slowly by hand until the inner or upper (depending on the machine type) indicator light comes on. Attach the displayed amount of weight to the inside or upper edge of the wheel rim in line with the weight location mark on the machine. Continue rotating the tire until the second indicator light comes on and attach an amount of weight equal to the display to the outside or lower side of the wheel rim (Figure 29-64).

7. Verify the balance job by repeating step 5. The weight displays should show zero or ready for weight amounts. If they do not, adjust the amount or location of the weights as indicated.

An on-car computer balancer combines much of the operating simplicity of a computer balancer (i.e., digital readout of weight amount and location) with the ability to

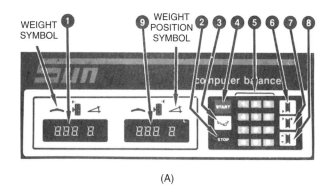

(A)

(B)

Figure 29-62 The control panels for two different computer balancers. On control panel (A), 1 is the weight readout for the inside of the wheel, 2 is the stop button, 3 is to select a special mode for styled wheels, 4 is the start button, 5 is the keyboard to input dimensions, 6 selects rim distance, 7 selects rim width, 8 selects rim diameter, and 9 is the readout for the outside weight. On control panel (B), the amount of unbalance will be a readout at the upper and lower planes where "rdy" appears. *(A is courtesy of Snap-on Diagnostics; B is reprinted by permission of Hunter Engineering Company)*

Chapter 29 ■ Tire and Wheel Service

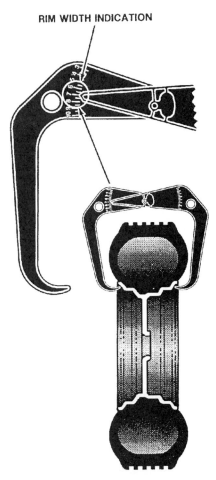

Figure 29-63 Gauging the rim using a rim caliper. *(Courtesy of Snap-on Diagnostics)*

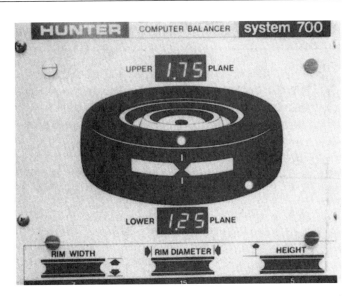

Figure 29-64 This control panel indicates that 1.75 oz of weight should be added to the upper edge of the rim with the wheel in this position. The wheel will be turned until the lower center light comes on, and then 1.25 oz of weight will be added to the lower rim. *(Reprinted by permission of Hunter Engineering Company)*

29.11 Practice Diagnosis

You are working at a new-car dealership and encounter the following problems:

CASE 1: The customer complaint is a slight vibration that occurs at 55 mph. This is a new, mid-sized, FWD car with only 137 mi. on it. Your road test confirms the vibration and the speed, and it seems to be coming from the left rear. What is probably causing the vibration? What procedure should you follow to locate the cause?

CASE 2: While checking a vibration complaint, you find a tire that has 0.095 in. of runout at a position in line with the valve stem. When you check the wheel, you find 0.025 in. of runout at a position also next to the valve stem. You decide to check the wheel lugs, and you find 0.005 in. of runout at a position opposite to the valve stem. Can you fix this problem? If so, how?

correct for wheel mounting runout errors and brake rotor or drum unbalance. A major problem in developing an on-car balancer of this type is in the construction of portable pickup units or an unbalance sensor that can give the computer an accurate signal. The car is placed on special support stands that have the sensors mounted in them. Wheel location is determined by the machine in reference to strips of reflective tape that are placed on the tire. This tape reflects an infrared light beam that is sent from the balancer or remote stand.

Terms to Know

chased
computer balancers
electronic balancer
lateral tire runout

match mounted
spin balancing
strobe light balancer

tire siping
tire truing
upset

Review Questions

1. Technician A says that it is a good idea to mark a lug bolt before removing a wheel so the wheel can be replaced in the same position.
   Technician B says that relocating a wheel on a hub can change runout and balance conditions.
   Who is correct?
   a. A only
   b. B only
   c. Both A and B
   d. Neither A nor B

2. When a wheel is replaced, the lug bolts should be torque tightened to ensure that the:
   A. Wheel does not come loose
   B. Brake rotor does not become distorted
   Which option best completes the statement?
   a. A only
   b. B only
   c. Both A and B
   d. Neither A nor B

3. Technician A says that all replacement lug bolts are the same except for the thread size.
   Technician B says that a lug bolt can often be installed using a few flat washers and the lug nut.
   Who is correct?
   a. A only
   b. B only
   c. Both A and B
   d. Neither A nor B

4. Uneven tire wear can show the need to:
   a. Improve tire pressure checking habits
   b. Get a wheel alignment
   c. Repair some worn parts
   d. All of the above

5. Technician A says that a run-flat tire driven after it had lost its air pressure should be demounted and inspected for internal damage before repairing.
   Technician B says that the air pressure sensor inside the tire can be ruined while demounting the tire.
   Who is correct?
   a. A only
   b. B only
   c. Both A and B
   d. Neither A nor B

6. When setting a tire bead, never exceed a pressure of:
   a. 10 psi
   b. 20 psi
   c. 40 psi
   d. 60 psi

7. A tire is removed from a wheel by:
   A. Pushing the bead into the drop center at one side of the wheel
   B. Prying the bead over the wheel flange
   Which option best completes the statement?
   a. A only
   b. B only
   c. Both A and B
   d. Neither A nor B

8. When removing and replacing a tire on a wheel, the tire bead and the bead seats should be:
   A. Clean
   B. Lubricated with motor oil
   Which option best completes the statement?
   a. A only
   b. B only
   c. Both A and B
   d. Neither A nor B

9. Technician A says that a leaky tire can be safely sealed with a plug inserted through the tread.
   Technician B says that it is a good practice to always inspect the inside of the tire before repairing it.
   Who is correct?
   a. A only
   b. B only
   c. Both A and B
   d. Neither A nor B

10. The best period for tire rotation is about _____ rotations through the life of the tires.
    a. one
    b. two
    c. four
    d. six

11. The recommended wheel weight for aluminum wheels is one that:
    a. Contains zinc
    b. Has an A cut into the clip
    c. Has a painted clip
    d. Uses an extra long clip

12. Technician A says that a static unbalance condition can cause a bounce or tramp condition of a tire.
    Technician B says that a dynamic unbalance will cause the same problem.
    Who is correct?
    a. A only
    b. B only
    c. Both A and B
    d. Neither A nor B

13. Technician A says that steering wheel shake is caused by either tire runout or a static unbalance condition.
    Technician B says that static unbalance can cause spotty tire wear.
    Who is correct?
    a. A only
    b. B only
    c. Both A and B
    d. Neither A nor B

14. On-car tire truing can often cure:
    A. Radial tire pull
    B. Loaded radial runout
    Which option best completes the statement?
    a. A only
    b. B only
    c. Both A and B
    d. Neither A nor B

15. Technician A says that all of the weights can be placed on the inner side of the wheel when statically balancing a tire.
    Technician B says that some clip-on weights can damage aluminum alloy wheels.
    Who is correct?
    a. A only
    b. B only
    c. Both A and B
    d. Neither A nor B

16. Technician A says that all on-car balancing systems dynamically balance the tire and wheel.
    Technician B says that a dynamic balance should cure a steering wheel side shake condition.
    Who is correct?
    a. A only
    b. B only
    c. Both A and B
    d. Neither A nor B

17. When spin balancing the drive wheels and using the car's engine to spin the tires, use caution because:
    A. Spinning the tire too fast can cause it to explode
    B. The car can move forward as power is applied to the drive wheels
    Which option best completes the statement?
    a. A only
    b. B only
    c. Both A and B
    d. Neither A nor B

18. When kinetically balancing a tire using a strobe light balancer, the:
    A. Pickup probe is placed in a vertical position
    B. Weights are added at the 12 o'clock position on the tire
    Which option best completes the statement?
    a. A only
    b. B only
    c. Both A and B
    d. Neither A nor B

19. A tire is being dynamically balanced using a strobe light balancer, and an unbalance condition equal to 6 oz is located with the pickup probe at the 3 o'clock position. Where should the weight be added to correct this problem?
    A. 3 oz on the inside at 3 o'clock and 3 oz on the outside at 3 o'clock
    B. 3 oz on the inside at 3 o'clock and 3 oz on the outside at 9 o'clock
    a. A only
    b. B only
    c. Both A and B
    d. Neither A nor B

20. The advantage with computer balancers is that they are:
    A. Accurate and sensitive
    B. Fast
    Which option best completes the statement?
    a. A only
    b. B only
    c. Both A and B
    d. Neither A nor B

Math Question

1. When you check the specification for tightening the lug nut on an import car from Japan, you find a specification of 69–78 N-m. Your only torque wrench reads in foot-pounds. Can you tighten these lug nuts properly? (Hint: See Appendix B.) If so, what torque will you use?

30 Front-Wheel-Drive Driveshaft Service

Objectives

Upon completion and review of this chapter, you should be able to:

❏ Determine the cause of a driveshaft or CV joint problem on FWD cars.

❏ Remove and replace a driveshaft on a FWD car.

❏ Remove and replace a CV joint or boot on FWD driveshafts.

❏ Rebuild a CV joint.

30.1 Introduction

CV joint service often begins with a complaint of excessive vibration or noise or as a result of seeing grease sprayed around the inside of the tire and wheel well.

30.2 CV Joint Problem Diagnosis

CV joint boots are easily checked by raising the car on a hoist and visually inspecting them. First, check for grease spray in the area around the boot. Turn the front of the wheel outward and rotate the tire and axle to get a good look at the condition of the outboard boot. Pull the accordion pleats apart while rotating the tire and axle to get a good look into the creases and folds of the inboard boot. A cut, torn, or cracked boot should be replaced.

A faulty CV joint can sometimes be felt as loose or rough operation. The first indication of possible failure is often noisy operation on turns. Most CV joint problems can be found on a road test. Listen to the operation of the joints as you drive the car slowly in right and left circles. The noise from a faulty CV joint will increase on a turn, especially if it is the inner joint. Then drive the car in a straight line while alternately applying and releasing the throttle. Listen for clunking noises during the power change. Apply the throttle in such a way that the front end bounces up and down. A faulty inner plunge joint can often be made to react during a power change while the front end is bobbing up and down.

One or more faulty joints is indicated if any of the following conditions shows up:

- A clicking, clacking, or chattering noise on turns (outboard joint)
- A clunking noise during power change (plunge joint)
- A vibration or shudder during acceleration (either joint)
- A vibration at highway speeds (either joint)

If any of these faults shows up and if it is believed that the CV joints are faulty, the driveshaft should be removed so you can physically feel the operation of the CV joint and clean and inspect the joints. Cleaning the joint requires repacking it with new grease and installing a new boot.

The service that can be done on a driveshaft varies. Some driveshafts cannot be disassembled. In those cases, the only repair that can be made is to remove and replace the entire driveshaft. It is possible, though, to install a replacement boot of the split type. This procedure is described in the next section. On driveshafts that can be disassembled, the boot can be replaced with either an OEM or split type. CV joint replacement and CV joint rebuilding are the various service operations that can be done.

30.3 Split CV Joint Boots

Quick-replacement, split-type CV joint boots are available from aftermarket suppliers. They are made with interlocking, zip-lock–type edges. An early design used an adhesive to seal the joint. A recently introduced version called the

Quick-Replacement Outer CV Boot, or CV-Easy boot, has a sealing system that uses a sealant as well as springs molded into the rubber (Figure 30–1). The quick-replacement boot has been approved by at least seven car manufacturers. These boots are said to be grease- and watertight, and they can be installed on a driveshaft that is still mounted in the car. Removal of the driveshaft for boot installation might be an advantage, but it is not necessary.

To install a split boot, cut off the clamps and the old boot. Then slip the new boot over the shaft, apply the sealant or adhesive to the lock groove, lock the boot into itself, position the boot over the CV joint, and install the new clamps. Quick-replacement outer CV boots do not need any cure time, and the vehicle can be driven immediately after the installation is completed. With the older design, it is absolutely necessary to keep the seam clean before and after applying the adhesive; a little grease or dirt can ruin the seal. Also, the shaft and boot have to remain stationary until the adhesive cures, at least an hour.

Some technicians recommend caution in the use of split boots because it is very difficult to clean out any water, dirt, or grit that might have entered through the old, torn boot. Be sure to check a grease sample for grit. It is difficult to clean and inspect the internal portions of the CV joint for damage while the joint is assembled, especially while it is still on the car. It is a waste of time and money to replace the boot on a dirty or worn CV joint.

Another device, called a cone, provides a means of installing the boot from the end of the driveshaft over the CV joint. Boot replacement using a cone is described in Section 30.6.1.

30.4 Driveshaft Removal

A FWD driveshaft is held in place by the outer splines, passing through the hub, that are secured by the hub nut and by a connection of the inner end to the transaxle. The driveshaft floats between these two points. On most domestic and some *imported cars* (made outside of the United States), the inner end is splined into the differential side or axle gear. The inner end is held in place by a **spring clip,** which is a weak internal circlip or snap ring that will release under pressure (Figure 30–2). On some cars, the splined, inner end is retained in the differential gear by a conventional snap ring. Using snap ring pliers, this ring must be removed or installed from the inside of the differential. At least one manufacturer spring loads the inner joint. The internal spring provides a constant pressure to keep the splined end of the driveshaft in the differential. Some imported cars use bolts or a spring clip to connect the inner CV joint to a flange extending from the differential (Figure 30–3).

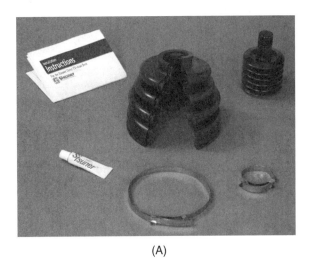

(A)

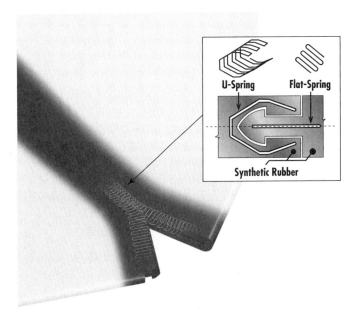

(B)

Figure 30–1 The quick-replacement outer boot can be slid over a CV joint from the side. The boot kit items (A) include the boots with seam sealant, clamps, and grease. The seam is sealed by unique springs molded into the rubber (B). *(Courtesy of Sisuner International, Inc., 1-888-290-3335)*

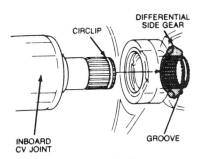

Figure 30–2 A common method of securing the inboard joint into the transaxle is a circlip on the joint splines that snaps into a groove in the differential side gear. *(Courtesy of General Motors Corporation, Service Technology Group)*

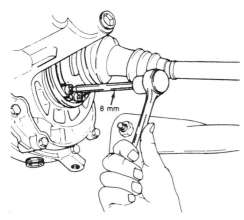

Figure 30-3 Some inboard joints are bolted onto a flange, which is part of the shaft that extends out of the transaxle differential. *(Courtesy of General Motors Corporation, Service Technology Group)*

Care should be taken during removal of the driveshaft so the boot does not get torn or damaged and the inner joint does not get overly compressed or extended. When the outer end of the shaft is removed, the shaft should be supported so it does not hang on the plunge joint. When the shaft is removed it should be carried, handled, and stored in a horizontal position (Figure 30-4). Driveshafts vary; it is a wise practice to follow the procedure listed in the service manual when servicing them.

To remove a FWD driveshaft, you should:

1. Loosen the hub nut. It is usually loosened with the tires on the ground, the transmission in park or first gear, and the parking brake set. Staked hub nuts are usually removed by merely unthreading them; new nuts should be used for replacement. As mentioned earlier, an air-impact wrench is a fast and easy method of removing and replacing this nut, but the hammering effect can brinell and damage the wheel bearing.

2. Raise and support the vehicle on a hoist or jack stands.

3. Remove the hub nut and the wheel.

4. Remove the tie-rod end from the steering arm. This operation usually requires the use of a special puller, which is described in Chapter 33.

5. Disconnect the steering knuckle from the lower control arm. On most cars, this is simply a matter of removing the ball joint stud pinch bolt and prying the control arm downward until the ball joint stud is out of the steering knuckle boss (Figure 30-5). Cars that have a tapered ball joint stud with a nut at the end of the stud require the use of a special puller (Figure 30-6). Usually, it is necessary to disconnect the stabilizer bar end links so the control arm can be moved downward far enough to move the ball joint stud from the steering knuckle boss.

Figure 30-5 Many cars use a pinch bolt to lock the ball joint stud into the steering knuckle. After the stud has been removed, the control arm can be pried away from the steering knuckle. *(Courtesy of Moog Automotive Inc.)*

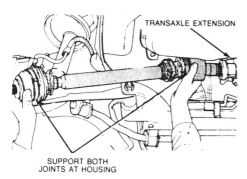

Figure 30-4 When a driveshaft is removed from a car, it should be kept in a horizontal position as it is carried or handled. *(Courtesy of General Motors Corporation, Service Technology Group)*

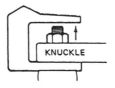 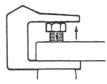

Figure 30-6 Some cars use a tapered ball joint stud that locks into the steering knuckle. A special tool is required to break the taper loose and allow separation of the control arm from the steering knuckle. *(Courtesy of General Motors Corporation, Service Technology Group)*

Chapter 30 ■ Front-Wheel-Drive Driveshaft Service

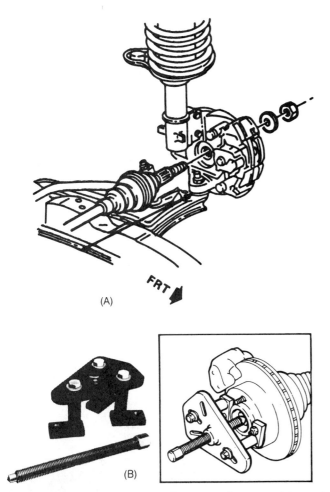

Figure 30-7 The CV joint spline must be moved inward through the hub (A). A puller is often attached to the wheel studs so it can push the CV joint inward (B). *(B is courtesy of Branick Industries, Inc.)*

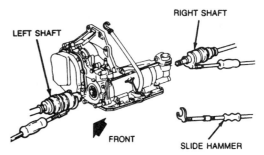

Figure 30-8 A special tool can be attached to a slide hammer to pull the inboard joint from the transaxle. *(Courtesy of General Motors Corporation, Service Technology Group)*

NOTE: Some manufacturers recommend reversing steps 6 and 7. Also, driveshafts that are attached to the differential by other methods require slightly different disassembly procedures. Always check a service manual if you are not sure of the operation.

NOTE: Many cars equipped with ABS have a tone ring (a toothed gear around the CV joint) and a wheel speed sensor mounted in the steering knuckle. Care should be exercised not to disturb or damage either of these parts.

NOTE: On some cars, if both driveshafts are removed at the same time, the differential gears can rotate inside of the differential in such a way as to make it impossible to reinstall the inner joint. It is highly recommended that these gears be held in alignment with a special tool.

6. It is often necessary to install a puller onto two of the wheel studs and tighten the puller screw to push the end of the CV joint housing into the hub. Occasionally, a tap on the end of the CV joint housing from a hammer and brass punch is sufficient. Pull outward on the steering knuckle to help remove it from the CV joint housing splines (Figure 30-7).

7. Pull the driveshaft out of the differential and out of the car. On some cars, an attachment can be installed onto a slide hammer to help pop the inner joint out of the differential (Figure 30-8). On other cars, a pry bar can be inserted between the inner joint housing and the differential case to pry the inner joint out of the differential.

 *Some driveshafts pop out rather suddenly. Be prepared to catch them.*

30.5 Driveshaft Disassembly

The driveshaft must be partially disassembled to replace a boot using the standard OEM type, to replace a CV joint, or to rebuild a CV joint. Two methods are commonly used to secure the fixed CV joint to the shaft. One method uses a visible snap ring; the other method uses a hidden spring clip or snap ring. Looking at the inner race of the CV joint will tell you which method is used (Figure 30-9). Newer Chrysler Corporation vehicles use a plastic CV joint retainer. This ring is spread wider, using snap ring pliers or a special retaining ring expander, to remove the joint (Figure 30-10).

To disassemble a FWD driveshaft, you should:

1. Mark the position of the boot on the driveshaft and cut the boot clamps and the boot so they can be removed. Discard them. You should always use a new boot and clamps when assembling a driveshaft (Figure 30-11 and Figure 30-12).

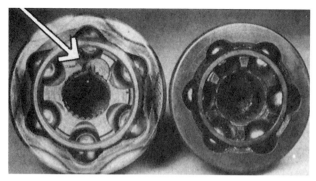

Figure 30–9 The fixed joint is secured onto the driveshaft by either a visible snap ring (left) or circlip at the end of the driveshaft where it cannot be seen. *(Courtesy of Moog Automotive Inc.)*

Figure 30–11 Before removing a boot, a scratch mark or reference should be made on the driveshaft so the new boot can be positioned correctly. *(Courtesy of Moog Automotive Inc.)*

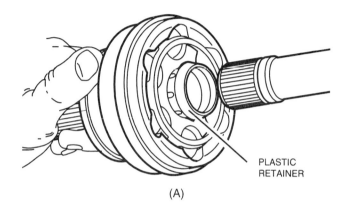

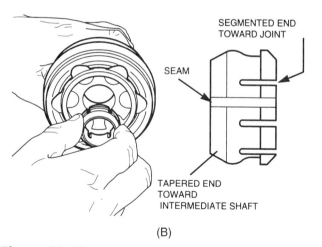

Figure 30–10 A plastic retainer holds some CV joints onto the shaft (A); when replacing the CV joint, position the new retainer correctly (B). *(Courtesy of Moog Automotive)*

Figure 30–12 A boot is removed by prying or cutting off the old clamps (top) and then cutting the old boot (bottom). *(Courtesy of Moog Automotive Inc.)*

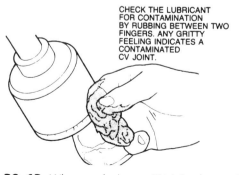

Figure 30–13 When replacing a CV joint boot, the lubricant should be checked. If grit is felt, the joint should be disassembled, cleaned, and inspected or replaced. *(Courtesy of General Motors Corporation, Service Technology Group)*

2. After removal of the boot, clean your fingers and rub a sample of the grease from inside of the joint between your fingers. Dirt or metal particles will cause a gritty feeling and are a sign of a worn joint. A CV joint with metal or dirt particles in the grease will probably need to be replaced (Figure 30–13).

Chapter 30 ■ Front-Wheel-Drive Driveshaft Service

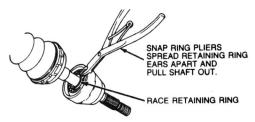

Figure 30–14 If a visible snap ring or plastic retainer is used to secure the joint to the driveshaft, spread the ring or retainer using snap ring pliers or a special tool and slide the joint off of the shaft. *(Courtesy of General Motors Corporation, Service Technology Group)*

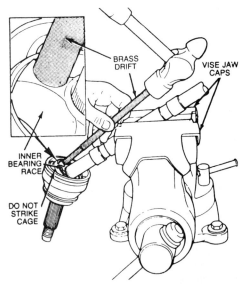

Figure 30–15 If a snap ring is not visible, a brass drift punch is used to force the joint off the shaft. A sharp, hard blow is usually required to pop the circlip loose. *(Courtesy of General Motors Corporation, Service Technology Group)*

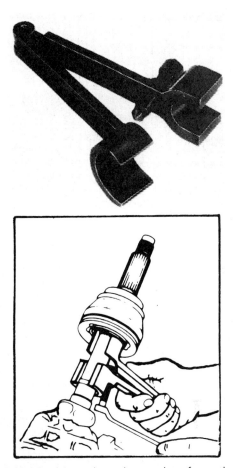

Figure 30–16 This tool can be used to force the inboard and outboard CV joint off the driveshaft for some cars. *(Courtesy of SPX-OTC)*

3. Clean the side of the inner race and look for the snap ring and recess to determine which style of retainer is used. If a snap ring is visible, expand the snap ring and slide the CV joint off the end of the shaft (Figure 30–14). If the hidden spring clip is used, clamp the shaft in a vise and tap the inner race of the CV joint sharply with a brass drift punch and hammer (Figure 30–15). The joint should pop loose and slide off the end of the shaft. Some manufacturers recommend replacement of the spring clip. If you replace this clip, do not expand it any farther than necessary to install it. A tool is available to separate the CV joint from the driveshaft on some car models (Figure 30–16).

4. Repeat steps 1 and 2 on the remaining joint. On many inner plunge joints, the outer race or tulip will slide off the joint once the boot has been removed. The tripod and balls are held in place by a visible snap ring.

30.6 Servicing CV Joints

Service parts for CV joints are available as boot kits that include a boot, grease packet, and clamps. All of the CV joint parts, including the outer race or housing, inner race, ball set, and cage, are available separately. Complete CV joint assemblies and complete driveshaft assemblies are also available. If the joint is noisy or worn or operates roughly, the old joint should be rebuilt or a new joint installed.

30.6.1 Boot and Grease Replacement

Boot replacement is usually a matter of cutting off the old boot, removing the CV joint from the shaft, sliding the new boot onto the shaft, replacing the joint onto the shaft, and clamping the new boot in the proper location.

This job becomes a little more difficult because the joint must have a supply of clean grease before the new boot is installed. After the joint has been removed from the shaft, it is fairly easy to thoroughly clean it using a solvent, denatured alcohol, or brake cleaner and a brush.

Figure 30–17 The procedure to install a CV joint boot using the cone is to lubricate the cone and boot with wax, turn the boot inside out, and place the boot onto the cone. Next, place the CV joint into the cone and with one quick motion, pull the boot into place on the driveshaft. *(Courtesy of Cosmos International, Inc.)*

After cleaning, it is important to thoroughly dry the joint and remove all traces of solvent from the inside. Compressed air can be used to blow out all of the solvent. Some shops use a hair dryer to heat the joint and help evaporate all traces of solvent. After the joint is clean and dry, check all of the balls and race grooves for wear or damage.

A packet of the correct amount of grease comes with most boot kits. This grease should be worked into the joint so it is thoroughly distributed. Normally, as much grease as possible is packed into the joint, and any remaining grease is spread around the inside of the boot. If the grease packet is not available, it is best to purchase CV joint grease from the vehicle manufacturer and use the exact type and amount of grease specified by that manufacturer.

Using a cone allows boot replacement with the CV joint still attached to the driveshaft, which should save some time. The supplier of the cone also provides the Uni-Fit boot, which is designed to fit CV joints of various sizes. To install a Uni-Fit boot using a cone, you should:

1. Thoroughly clean the CV joint and the driveshaft to prevent dirt from entering the CV joint during boot installation. Check the cone to make sure there are no cuts or nicks. Warm the boot up to 78°F (25°C). Spray the entire surface of the cone with a wax-type spray (Figure 30–17).
2. Spray the entire outer surface of the boot with a wax-type spray.
3. Turn the boot inside out. Inner boots can be slid over the CV joint right side out; they should have the inside coated with wax.
4. Place the boot over the waxed cone.
5. Attach the cone to a 1/2-in. bolt and the stand.
6. Place the CV joint inside the cone.
7. Quickly and forcefully pull the boot upward and over the cone. Roll the boot right side out.

To attach the Uni-Fit boot, you should:

1. Pack the CV joint with grease and stretch the large boot end over the CV joint until a tight fit is formed (Figure 30–18).
2. Release any trapped air using a blunt screwdriver.
3. Install the clamp at the large end in the original location, using a banding tool for the proper tension.
4. With the boot at its natural length, install the clamp at the small end.
5. If the boot is at the location shown in step 5 of Figure 30–18, you are finished. No cutting is necessary.
6. If the boot extends too far past the clamp, as shown in step 6 of Figure 30–18, the excess should be cut off.
7. Use a sharp knife to cut off the excess material.
8. Remove the excess boot material.

30.6.2 Disassembling an Outer, Fixed-Type CV Joint

If a joint has a worn or damaged cage, balls, inner race, or outer race, it can be disassembled for replacement of these parts. Often, it is less expensive to replace the entire joint assembly because a fractured or pitted ball is usually accompanied by fractured or pitted inner and outer races. If only one of the two parts is damaged, it will probably be less expensive to rebuild the joint.

Removing the inner race and cage from some CV joints is much like a puzzle. Parts have to be positioned a certain way to allow disassembly or reassembly. Look for a difference in the widths of the lands between the ball grooves of the outer race or a pair of longer windows in the cage as clues for the disassembly or reassembly procedure. The differences you find are there specifically for this purpose (Figure 30–19).

To disassemble a fixed CV joint of the Rzeppa type, you should:

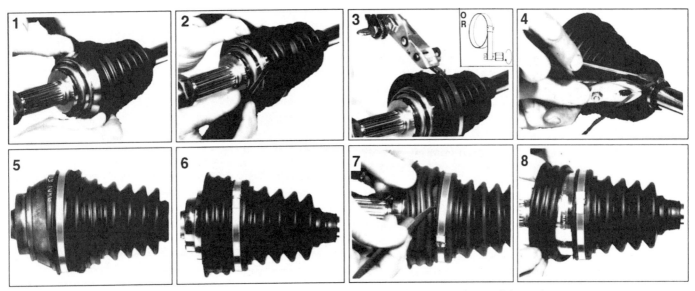

Figure 30–18 When installing a Uni-Fit boot, slide the large end of the boot over the CV joint, release any trapped air, and install the large clamp; next position and clamp the small end and then trim off the excess material from the large end of the boot. *(Courtesy of Cosmos International, Inc.)*

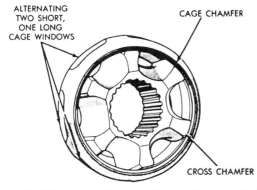

Figure 30–19 Before disassembling a Rzeppa joint, check the parts for differences in the inner race, cage, and outer race. *(Courtesy of Chrysler Corporation)*

1. Using a brass punch or special tool, push or tap inward on one side of the inner race, causing the cage and inner race to tilt. After they have tilted far enough, lift the exposed ball from the cage (Figure 30–20).
2. Repeat step 1 on the remaining balls until all of them have been removed.
3. Pivot the cage and inner race 90 degrees so the windows of the cage are aligned with the ball race lands of the outer race. Observe and mark the cage, if necessary, so you know which side goes inward. Withdraw or swing the cage and inner race up and out of the CV joint housing. (See Figure 30–23).

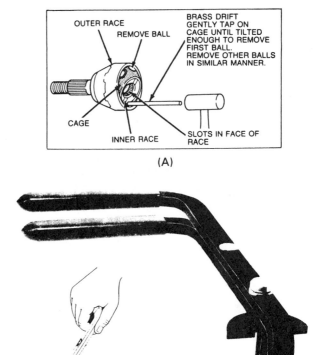

Figure 30–20 Begin the disassembly of a Rzeppa joint by tilting the cage and inner race enough to allow removal of the balls. This can be done using a brass drift (A) or a special tool (B). *(A is courtesy of General Motors Corporation, Service Technology Group; B is courtesy of Lisle Corporation)*

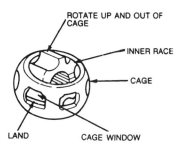

Figure 30–21 The inner race is removed from the cage by turning it to this position and then pivoting it up and out of the cage. *(Courtesy of General Motors Corporation, Service Technology Group)*

4. Observe and mark the inner race, if necessary, so you know which side goes inward. Pivot the inner race 90 degrees in the cage so one of the lands aligns with a window—the longer ones if there are any—and swing the race out of the cage (Figure 30–21).
5. Clean the parts in solvent, dry them with compressed air, and inspect them for damage (Figure 30–22).

SAFETY TIP: Many sources recommend wearing rubber gloves to protect your skin from the chemicals used for cleaning these parts.

30.6.3 Assembling an Outer, Fixed-Type CV Joint

Reassembly of a CV joint is normally a reversal of the disassembly procedure. On most joints, it is important that the parts end up in the same relative position as they were before disassembly. The inside of all the parts should face toward the inside.

To reassemble a fixed CV joint of the Rzeppa type, you should:

1. Place the inner race into the cage, align one of the race lands with a cage window, and swing the race into the cage. Pivot the inner race 90 degrees so the race and cage are parallel and check the two parts for correct positioning.
2. Place the inner race and cage into the outer race so that one of the cage windows is aligned with a race land. Swing the cage and inner race inward and pivot them 90 degrees. All three of the parts should be parallel and in the correct position (Figure 30–23).

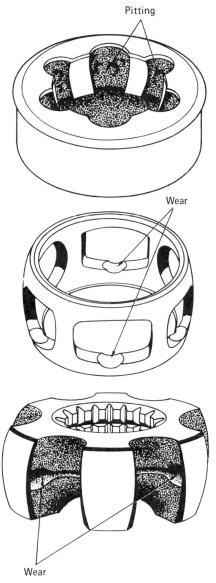

Figure 30–22 After the CV joint has been disassembled and cleaned, the ball grooves and cage should be inspected for pits, wear, and cracks. If any of these faults are found, the joint or the individual parts should be replaced. *(Courtesy of General Motors Corporation, Service Technology Group)*

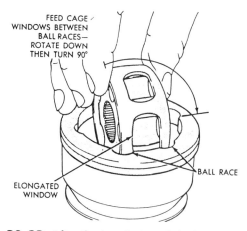

Figure 30–23 After the installation of the inner race into the cage, the second assembly step is to replace the cage and inner race into the outer race. *(Courtesy of Chrysler Corporation)*

Chapter 30 ■ Front-Wheel-Drive Driveshaft Service

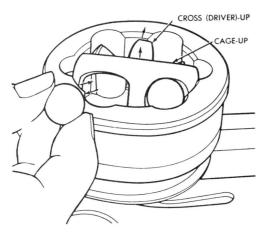

Figure 30-24 Assembly of a Rzeppa joint is completed as the balls are replaced, one at a time. *(Courtesy of Chrysler Corporation)*

3. Tilt the inner race and cage so a ball can be inserted into one of the cage windows and into an inner and outer race groove. Using a brass punch and hammer, tap the inner race so the ball enters the outer race or use the special tool to tilt the inner race so another cage window is exposed (Figure 30-24).

4. Repeat step 3 until all of the balls have been inserted.

30.6.4 Disassembling an Inner CV Joint

Some inner joints are disassembled in a manner very similar to that used for a Rzeppa-type joint. A tripod joint is much simpler. Most technicians consider a tripod type of joint much easier to work on.

To disassemble an inner joint of the tripod type, you should:

1. Mark the shaft at the edge of the boot, remove the boot clamps, and cut off the old boot (Figure 30-25 and Figure 30-26).

Figure 30-25 Before removing the boot from a plunge joint, mark the shaft to ensure replacement of the boot in the proper place. *(Courtesy of Moog Automotive Inc.)*

Figure 30-26 After sliding the boot back, place match marks on the shaft, joint, and outer race to ensure proper reassembly.

2. Slide the outer race or tulip off the driveshaft. On some joints, it will be necessary to remove a retaining clip (Figure 30-27).

3. Remove the retaining clip from the end of the driveshaft. Note which way the tripod is positioned and slide the tripod and rollers off the shaft (Figure 30-28).

4. Clean the parts in solvent, dry them with compressed air, and inspect them for damage.

30.6.5 Assembling an Inner CV Joint

Assembling a tripod joint follows a procedure that is essentially the reverse of the disassembly process.

To assemble an inner joint of the tripod type, you should:

1. Slide the new boot and clamps onto the driveshaft.

2. Slide the tripod and rollers onto the driveshaft with the correct side inward. Secure it in position with the snap ring (Figure 30-29).

3. Distribute the proper amount of grease over the tripod and rollers and in the tulip.

Figure 30-27 With the boot removed, the outer race or tulip can be slid off most driveshafts. *(Courtesy of Moog Automotive Inc.)*

Figure 30–28 Removal of this snap ring allows removal of the tripod or spider assembly from the shaft. *(Courtesy of Moog Automotive Inc.)*

Figure 30–29 The tripod assembly can be tapped into position using a brass drift or socket and hammer. *(Courtesy of Moog Automotive Inc.)*

4. Slide the outer race or tulip over the tripod and rollers and install the retainer, if used (Figure 30–30).
5. Install the boot and clamps as described in the next section.

30.7 Installing CV Joint Boots

New boots are installed as the CV joints are replaced onto the driveshaft. Normally, the boot is slid onto the shaft, the joint is installed and packed with grease, and the boot is then clamped in place. It is important that the boot be installed without twists or unusual wrinkles and

Figure 30–30 Spread the correct amount of the proper grease through the tulip and around the tripod before replacing the tulip housing. *(Courtesy of Moog Automotive Inc.)*

that the clamps make a tight seal between the boot and the driveshaft or CV joint housing. Most cars use a different boot on the inner and outer joint.

To install a CV joint boot, you should:

1. Slide the small end of the boot with the clamp in place onto the shaft (Figure 30–31).
2. On fixed outer joints, distribute the proper amount of grease through the joint and position the joint onto the driveshaft splines. Tap the joint onto the shaft until you hear or feel the spring clip snap into place. Pull outward on the joint to ensure that it is locked in place. On inner joints, refer to steps 2, 3, and 4 of Section 30.6.5 (Figure 30–32).
3. Line up the small end of the boot with the marked location on the shaft and tighten the small clamp (Figure 30–33).
4. If required, spread any remaining grease into the boot and slide the large end of the boot into position on the CV joint housing. Check to make sure that the boot is not twisted or collapsed. A collapsed boot can usually be cured by **burping the boot,** or sliding a small screwdriver under the lip

Figure 30–31 The new boot and clamp should be slid onto the shaft before replacing a fixed joint. *(Courtesy of Moog Automotive Inc.)*

Chapter 30 ■ Front-Wheel-Drive Driveshaft Service

Figure 30–32 The fixed joint is tapped onto the shaft until you hear or feel the circlip slip into position. *(Courtesy of Moog Automotive Inc.)*

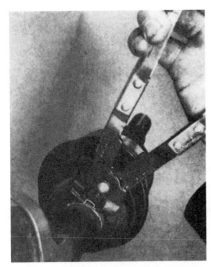

Figure 30–33 With the joint in position, the new boot is aligned with the scratch mark and clamped in place using special pliers. *(Courtesy of Moog Automotive Inc.)*

Figure 30–34 If the boot is twisted or collapsed (top), it should be straightened out and burped or vented (bottom). *(Courtesy of Moog Automotive Inc.)*

of the boot and letting trapped air pressure equalize between the inside and the outside of the boot (Figure 30–34).

5. Tighten the large clamp (Figure 30–35).

NOTE: Some joints use a boot retainer ring in place of a clamp; this ring is pressed onto the CV joint housing along with the boot (Figure 30–36 and Figure 30–37).

30.8 R & R ABS Tone Ring

Many cars equipped with ABS use a tone/reluctor ring that is pressed onto the outboard CV joint housing. If it is damaged, the tone ring can be removed and replaced with a new one. In most cases, the old ring can be driven off using a hammer and punch, preferably of brass (Figure 30–38).

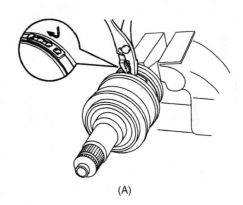

(A)

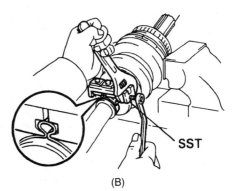

(B)

Figure 30–35 After positioning this clamp on the large end of the boot (A), it is secured using the special tool. The clamp is nearly closed when tightened properly (B).

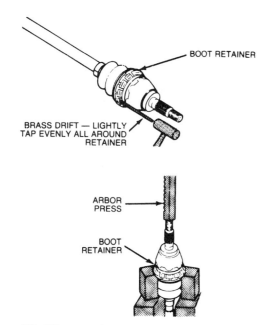

Figure 30–36 A retainer is used to secure the large end of the boot on General Motors cars. This retainer can be removed by tapping it with a punch and installed by using a press and two blocks. It must be kept straight and even during installation. *(Courtesy of General Motors Corporation, Service Technology Group)*

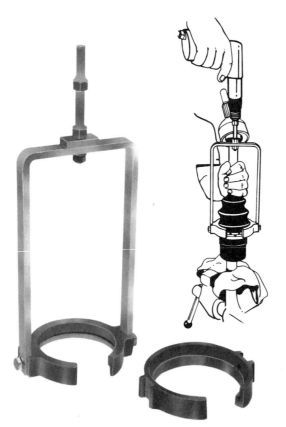

Figure 30–37 This special tool is used with an air hammer or an ordinary hammer to remove or install boot retainer rings. *(Courtesy of Lisle Corporation)*

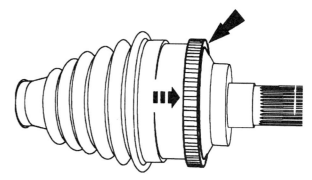

Figure 30–38 A damaged ABS tone ring (upper arrow) can be removed by tapping it in the direction of the left arrow with a brass drift. *(Courtesy of Ford Motor Company)*

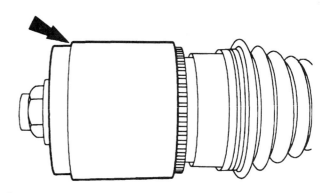

Figure 30–39 This tool (arrow) is being used to install a new ABS tone ring onto the CV joint. *(Courtesy of Ford Motor Company)*

Before installing the new tone ring, make sure that it has the same inside and outside diameters and the same number of teeth. In some cases, the new ring can be placed on press adapters and then the CV joint housing is pressed into it; in other cases, special tools are available to press the new ring into position (Figure 30–39).

30.9 Installing a FWD Driveshaft

Installing a FWD driveshaft is another operation that is the reverse of removal. One additional check to make is of the wear or deflector ring on many outer joints, on which a wheel bearing seal fits. If the ring or seal is damaged, it should be replaced. It can usually be tapped off using a punch and hammer to allow replacements to be installed. Care should be taken during installation of these parts to ensure that they are not bent or damaged. Be sure to follow the procedure listed in the service manual (Figure 30–40).

Chapter 30 ■ Front-Wheel-Drive Driveshaft Service

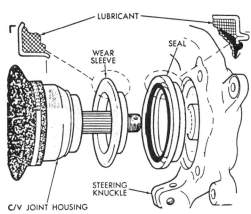

Figure 30–40 A wear sleeve is used on some outboard CV joints. This wear sleeve runs in a seal used to protect the front hub bearing. If either is damaged, it should be replaced. *(Courtesy of Chrysler Corporation)*

To install a FWD driveshaft, you should:

1. Slide the inner end of the driveshaft into the differential until you feel the spring clip snap in place; test its connection by trying to pull the CV joint back out. It should be locked (Figure 30–41).

OR Connect the inner CV joint to the differential flange and install the locking ring or retaining bolts. If bolts are used, tighten them to the correct torque.

2. Slide the hub and steering knuckle over the outer end of the driveshaft (Figure 30–42).

3. Install the washer and hub nut, using new ones if required, onto the CV joint housing. Tighten the nut to pull the hub into position on the CV joint housing. Pretighten the nut to about 50 to 75 ft-lb (70 to 100 N-m) of torque. If necessary, tap the hub onto the driveshaft splines so the nut can be started on the threads, using a soft hammer to strike the wheel mounting face on the hub.

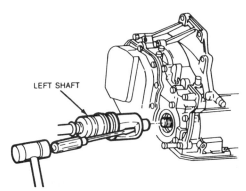

Figure 30–41 A circlip-retained axle is slid into the differential. If necessary, a punch or screwdriver can be used to tap it inward. It should go in until the circlip is engaged. *(Courtesy of General Motors Corporation, Service Technology Group)*

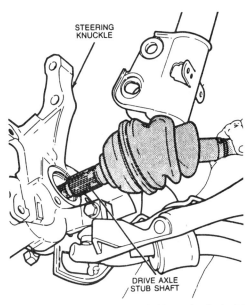

Figure 30–42 The fixed joint is slid into the hub. Make sure the seal or wear sleeve is not damaged. *(Courtesy of General Motors Corporation, Service Technology Group)*

4. Fit the ball joint stud into the steering knuckle, install the pinch bolt and nut (some manufacturers require new ones) or ball joint stud nut, and tighten the nut to the correct torque. If a prevailing torque nut is used, make sure it offers sufficient resistance. If a cotter pin is used, install and secure it (Figure 30–43 and Figure 30–44).

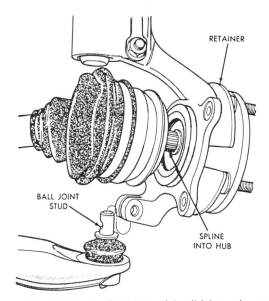

Figure 30–43 The ball joint stud is slid into the steering knuckle boss and locked in place by tightening the clamp bolt to the correct torque. If the ball joint stud has a single notch (like the one shown) for the bolt, be sure to align the notch with the bolt hole. *(Courtesy of Chrysler Corporation)*

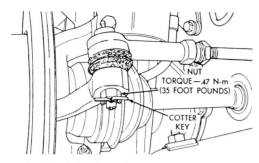

Figure 30–44 If the tie-rod was disconnected, it should be replaced and the nut tightened to the correct torque and secured with a cotter pin. *(Courtesy of Chrysler Corporation)*

5. Install the wheel and tighten the lug nuts to the correct torque.
6. Lower the car to the ground, set the parking brake, and tighten the hub nut to the correct torque. Lock the hub nut as required. If staking is required, a staking tool can be made by regrinding the point of a chisel. Stake the nut by driving a portion of the nut into the groove.

SAFETY TIP: Make sure that all nuts and bolts are properly tightened and secured and that the brake pedal has a normal feel before moving the car.

30.10 Practice Diagnosis

You are working in a brake and front-end shop and encounter the following problems:

CASE 1: The customer's complaint is a vibration in the front end of his Honda Civic that only occurs as he accelerates from a stop. Your road test confirms this vibration, and it seems to be coming from the right side. What is the probable cause of this vibration? What should you do to verify this assumption?

CASE 2: The customer has brought in his Chevrolet Corsica for an alignment. As you are doing a prealignment inspection of the front end, you find a grease spray from a torn boot on the outer right side CV joint. What should you do? What are the different choices for repairing this joint?

Terms to Know

boot
burping the boot

grease spray
imported cars

spring clip

Review Questions

1. Technician A says that most CV joints are provided with a special lube fitting so they can be lubricated. Technician B says that all FWD CV joints are protected by an accordion boot.
   Who is correct?
   a. A only
   b. B only
   c. Both A and B
   d. Neither A nor B

2. Technician A says that a fixed joint that is failing often makes a clunking noise during a load change. Technician B says that a faulty inner joint often makes a clicking noise during sharp turns.
   Who is correct?
   a. A only
   b. B only
   c. Both A and B
   d. Neither A nor B

3. Technician A says that a faulty CV joint boot can be replaced with the driveshaft still in the car. Technician B says that the joint should be cleaned and repacked with wheel bearing grease before installing a new boot.
   Who is correct?
   a. A only
   b. B only
   c. Both A and B
   d. Neither A nor B

4. Technician A says that the ball joint or strut must be disconnected from the steering knuckle to remove a FWD driveshaft. Technician B says that many FWD driveshafts are secured into the transaxle by a spring clip.
   Who is correct?
   a. A only
   b. B only
   c. Both A and B
   d. Neither A nor B

5. Technician A says that it is possible to pull the new boot over the top of the CV joint if you use a special tool.
   Technician B says that this process lets you install a new boot with the joint still on the shaft.
   Who is correct?
   a. A only
   b. B only
   c. Both A and B
   d. Neither A nor B

6. Technician A says that many fixed CV joints can be removed from the driveshaft by driving outward on the inner race.
   Technician B says that some fixed CV joints are secured to the driveshaft by a set screw.
   Who is correct?
   a. A only
   b. B only
   c. Both A and B
   d. Neither A nor B

7. When replacing a boot on a CV joint, it is a good practice to:
   A. Feel the old grease for signs of grit
   B. Clean and inspect any dirty joints
   Which option best completes the statement?
   a. A only
   b. B only
   c. Both A and B
   d. Neither A nor B

8. Technician A says that it is necessary to disassemble the fixed CV joint to replace the boot.
   Technician B says that a plunge joint is easier to disassemble than a fixed joint.
   Who is correct?
   a. A only
   b. B only
   c. Both A and B
   d. Neither A nor B

9. CV joint boots are secured in place using:
   A. RTV and mechanics wire
   B. A metal retainer ring
   Which option best completes the statement?
   a. A only
   b. B only
   c. Both A and B
   d. Neither A nor B

10. When replacing a FWD driveshaft, it is usually necessary to:
    A. Drive the inner joint into the transaxle using a special driving tool
    B. Lower the tire onto the ground before final tightening of the hub nut
    Which option best completes the statement?
    a. A only
    b. B only
    c. Both A and B
    d. Neither A nor B

Math Question

1. You have just replaced the outer CV joint on one driveshaft and the boot on the other driveshaft of a Ford Taurus. The parts charge is $22.96 for the boot and $151.79 for the CV joint. The flat rate is 1 hour for R & R of each driveshaft, 0.3 to R & R a boot, and 0.3 to R & R a CV joint. Your shop labor rate is $45 per hour. What is the cost for this job?

31 Spring and Shock Absorber Service

Objectives

Upon completion and review of this chapter, you should be able to:

❑ Remove and replace shock absorbers.

❑ Remove and replace a shock absorber cartridge in a front or rear strut.

❑ Remove and replace front or rear springs and torsion bars.

❑ Adjust a torsion bar suspension to the correct ride height.

❑ Perform the ASE tasks relating to spring, strut, and shock absorber diagnosis, adjustment, service, and repair (see Appendix A, Section B, 1, tasks 1 and 6–11; Section B, 2, tasks 1, 2, 4, and 5; Section B, 3, tasks 1 and 2; and Section C, task 2).

31.1 Introduction

Shock absorbers are usually replaced on an as-needed basis by many technicians and do-it-yourselfers; springs, especially the front ones, are commonly serviced as part of a front-end rebuilding operation. Most shock absorbers can usually be easily removed and replaced using common hand tools and limited mechanical knowledge. Spring replacement often requires special tools and equipment, and it can be very dangerous if the proper procedure is not followed or the normal cautions are not observed. Springs store energy, and they can release this energy very quickly.

Shock absorber replacement is much more complex on strut suspensions. It is no longer a driveway or shade tree operation. Special tools and a certain amount of knowledge are required.

Front spring or strut service is often done in connection with rebuilding a front end because the operation usually requires a partial disassembly of the front suspension. The ball joint stud is often disconnected when removing the spring from an S-L A suspension; as this is done, it is fairly easy to R&R the ball joint. The spring and the other strut parts are removed as part of the operation to service the shock absorber portion of a strut assembly; it is usually just as easy to install a new spring at this time as it would be to replace the old one.

Spring replacement is usually followed up with a wheel alignment. When the control arm or strut is removed to allow replacement of a spring on an S-L A or strut, the adjustable portions of the suspension are disconnected and reconnected; there is a good possibility that the alignment angles will be changed. Also, the change in height caused by the new springs changes the attitude of the vehicle and the relationship of the front suspension components, which changes some of the alignment angles (Figure 31–1).

Some of the steps needed to complete the operations given in this chapter, mainly the separation and connection of the ball joint stud on S-L A suspensions, are described in Chapter 32.

31.2 Removing and Replacing Shock Absorbers

R & R shock absorbers is usually a rather simple job requiring the removal of only two or three nuts or bolts. Two types of problems sometimes occur to make shock absorber replacement difficult or possibly hazardous: removal of the nut on single-stud mounts and the possibility of a sudden dropping of the suspension as the shock absorber is removed.

SAFETY TIP: Before beginning any job, it is wise to look over the various parts to see how they are arranged and to develop a plan for the disassembly and reassembly procedures. It is a good idea to do this after the vehicle has been raised and before removing the shock absorbers. Be careful to note what devices are used for the extension or down travel suspension stops. Many coil spring rear axles use the shock absorber as the limiter for axle down travel; when the shock absorber is removed, the axle can drop several inches until it is stopped by the brake hose or other components under the axle. This sudden movement will probably damage the brake hose and other vehicle parts and may injure you. The axle should be supported by stands or some other lifting device before removing the shock absorber.

The nut on single-stud shock absorber mounts tends to rust and seize in place. Its removal is made more difficult because the stud and piston rod often rotate during removal attempts. The polished piston rod should never be gripped with plain or locking pliers. The smooth surface will be damaged, which in turn will ruin the upper shock absorber bushing and seal, and the polished surface is too smooth to grip tightly. There are several approved ways of removing this nut; begin by applying penetrating oil to the threads. Next try one of these methods:

1. Often a fast-operating air wrench will spin the nut off.
2. The shock absorber stud has two flats on the end that can be held using a special wrench or socket (Figure 31-2).
3. If the nut is rusted in place, the shock stud can be broken off by placing a deep-well socket with extension bar over the nut or thread or a special tool made just for this purpose onto the exposed threads of the stud. Bend the shock stud back and forth until it breaks off; this, of course, ruins the shock absorber (Figure 31-3).

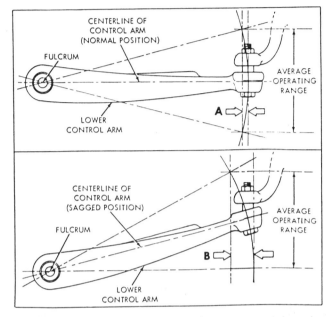

Figure 31-1 Correct front suspension geometry is based on a horizontal lower control arm at ride height; a minimum amount of side scrub (A) occurs if this arm is horizontal. Sagging springs cause a change in the operating range that increases the amount of side scrub (B) and tire wear. *(Courtesy of Moog Automotive Inc.)*

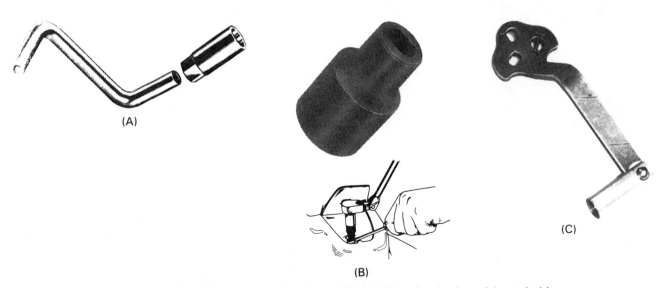

Figure 31-2 This odd-shaped tool (A) has sockets in each end to fit the flats of a shock stud; it can hold the stud from turning while the nut is unscrewed. The socket (B) is used to turn the shock stud while a wrench holds the nut. Shock wrench C has openings to fit three different shock stud shapes and can also be used to pull a shock stud into place. *(B is courtesy of Lisle Corporation)*

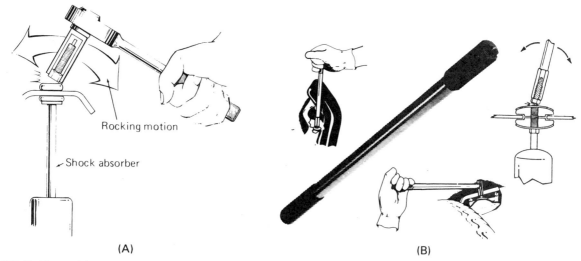

Figure 31–3 The quickest way to remove a stud mount is usually to break it; this can be done using a ratchet and deep-well socket (A) or special tool (B). *(B is courtesy of Lisle Corporation)*

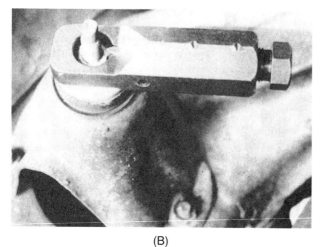

Figure 31–4 A nut-splitting tool (A) can be placed over the stud nut (B) and tightened to break the nut open.

4. A **nut breaker** can be placed over the nut and tightened to break or snap the nut open for easy removal (Figure 31–4).
5. An air-impact hammer and chisel can be used to cut down each side of the nut until the nut snaps or is cut into two parts (Figure 31–5).

To remove shock absorbers, you should:

1. Raise and support the car securely on a hoist or jack stands.

NOTE: On cars whose upper front shock absorber mounts can be reached from under the hood, it is often faster to disconnect the upper shock absorber mount before raising the car.

2. Disconnect the upper shock absorber mount, if it is not already done.

SAFETY TIP: While removing the nut on stud-type mounts, observe the rubber bushing; if it stays under a compression load while the nut is being removed, the suspension will probably drop when the nut leaves the threads. In this case, it is best to lift the suspension slightly or lower the body to remove this load.

3. Disconnect the lower mount and remove the shock absorber.

Sometimes it is necessary to purge the air out of the pressure tube of a double-tube shock absorber; this is also called bleeding a shock absorber. If the shock absorber was stored on its side, air will enter the pressure tube. Air is compressible and also will pass through the shock valving much faster and easier than oil. A shock with air in the pressure tube offers very little resistance for part or all of a stroke. When mounted on a car, a shock purges itself after enough compression and extension cycles.

Figure 31-5 An air hammer and chisel can be used to split a nut from a shock stud. *(Courtesy of SUPERIOR PNEUMATIC & MFG., INC., Cleveland, OH 44140)*

To purge the air from the pressure tube of a double-tube shock absorber, you should:

1. Compress the shock while holding it in an upside-down position.
2. Place the shock in an upright position and extend the piston rod.
3. Repeat steps 1 and 2 several more times or until an even resistance is felt for the entire stroke. A shock absorber that does not develop a full stroke resistance is faulty.

To replace a shock absorber, you should:

1. Check the bushings and other pieces of shock absorber mounting hardware to make sure they are in good condition and assemble these parts in the proper relationship (Figure 31-6).
2. Place the shock absorber with bushings in position and connect the upper mount. On single-stud mounts, tighten the nut to compress the rubber bushing, but not too much. The correct amount of tightness will squeeze the rubber bushing to about the same diameter as the metal washer. If the nut has not tightened securely against a metal shoulder or sleeve at this point, or if a self-locking, prevailing torque nut is not used, a jam or pal nut should be used to lock the retaining nut in position.

3. Place the lower bushings in position and connect the lower shock absorber mount. Tighten the retaining nuts or bolts to the correct torque.

Shock absorber service on strut suspensions is described in Section 30.6.

31.3 Removing and Replacing Coil Springs

Before beginning any major repair, you should check the procedure in a service manual. The manual will give you the quickest procedure, any safety hazards, and any special tools or steps required.

In some styles of rear suspensions that use coil springs, the spring can easily be removed once the shock absorber is disconnected. In fact, on some cars, the spring will almost fall out; operating a car that has rear coil springs with a shock absorber disconnected is not recommended for this reason. On these cars, spring replacement is a matter of disconnecting the shock absorber, lowering the rear axle, or raising the car enough to lift out the old spring; checking to make sure the spring pad insulators, if used, are in good condition; setting the new spring in place in the correct position; and replacing the shock absorbers. When the axle is raised to reinstall the shock absorber, the spring will be locked in place.

When installing new springs, be sure to get the correct replacement. Many cars use one of several different spring rates depending on the engine and the accessories. One manufacturer even uses a different spring on the right and left side on one model of car. The best method of getting the exact replacement is to use the part number, which is often noted on a paper or plastic tag fastened to the old spring. Also note that some springs have a top and a bottom; they should be installed right side up. If the car has a slight lean, a shim can be added to the spring on the low side so the car will sit level (Figure 31-7).

Springs are normally replaced in pairs; it is impossible to buy a new spring that will exactly match the rate and length of a used spring. Replacement of only one spring usually causes leaning of the vehicle or an odd ride quality.

Coil spring replacement on strut suspensions is described in Section 30.6.

31.3.1. Removing and Replacing a Coil Spring, S-L A Suspension

To remove a coil spring from an S-L A suspension, you should:

1. Raise and support the car securely on a hoist or jack stands.

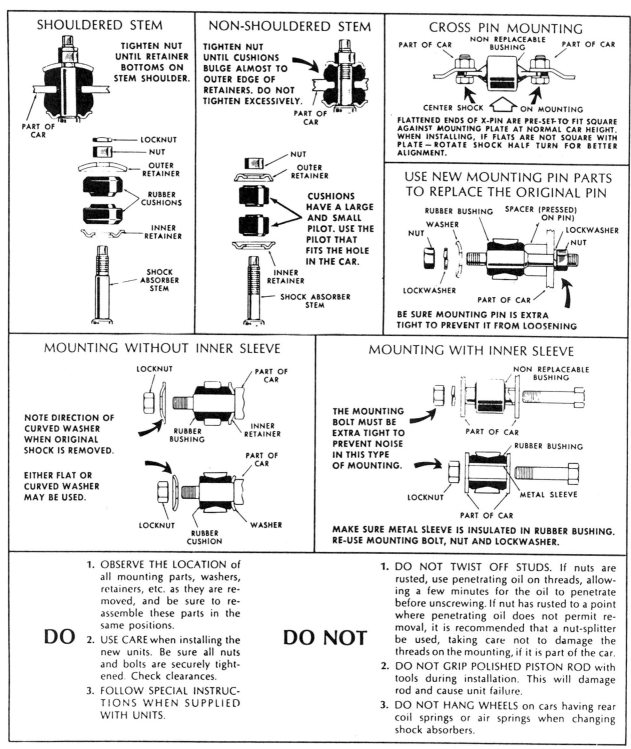

Figure 31–6 Different installation procedures are required for different types of shock mounts. *(Courtesy of Tenneco Automotive)*

2. Remove the wheel, the stabilizer bar end links, and the shock absorber.
3. Break the taper of the lower ball joint; this operation is described in Section 32.2. This step is not necessary if you plan to disconnect the lower control arm at the inner pivots.

SAFETY TIP: Removing a spring from an S-L A suspension requires a careful procedure and the use of special tools; this spring is very powerful and can cause injury. Depending on the equipment available, there are several ways of correctly performing this job; they begin in a similar fashion. Some sources recommend installing a chain loosely

Figure 31-7 Original springs often have a tag that indicates the part number and sometimes the top or bottom. The tag number helps ensure the exact replacement. This spring set tag indicates the driver-side spring. *(Courtesy of Moog Automotive Inc.)*

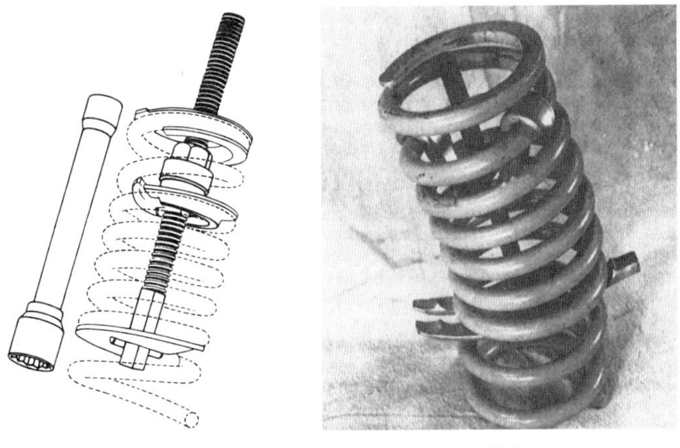

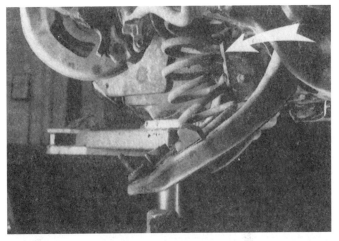

(A)   (B)

Figure 31-8 Two commonly used types of spring compressors; they are used to reduce the pressure on the control arm for safe and easy spring replacement. *(A is courtesy of Moog Automotive; B is courtesy of McQuay Norris)*

through the spring and frame or control arm to restrain the spring in case it comes loose. It is wise to follow the procedure recommended by the vehicle manufacturer and use the special spring tools in the manner recommended by their manufacturer.

If a spring compressor is used:

1. Install the spring compressor and tighten it as far as practical or until there is clearance at one end of the spring (Figure 31-8).

CAUTION *Observe the rebound bump stop; when you begin, it will be compressed from the spring pressure. It will have very little load on it when the spring has been compressed sufficiently.*

2. Disconnect the lower ball joint stud or remove the inner control arm pivot bolts and swing the lower control arm downward, releasing the compressed spring (Figure 31-9). Some sources recommend disconnecting the control arm pivot bolts at the inner bushings and pivoting the control arm downward at the ball joint; if you use this method, use care because the control arm can twist, slip, and allow the spring to escape. A jack adapter is available to make this job more secure (Figure 31-10).
3. Remove the compressed spring. If force is necessary, use a pry bar; be careful not to let your hand get trapped around a coil spring.

Figure 31-9 This spring is being removed by disconnecting the ball joint and lowering the control arm with the aid of a jack. Note the spring clip (arrow) to keep the spring partially compressed. *(Courtesy of Moog Automotive Inc.)*

If spring clips are used:

1. Select the shortest clips that can be used that will capture as many coils as possible and position the clips so they are toward the center of the car. It is possible to use a single clip, but most technicians recommend using two clips at a time for safety reasons (Figure 31-11).

HELPFUL HINT: Lifting the opposite rear corner of the car will place more load on the front corner; this will compress the spring further, usually allowing more spring coils to be captured by the spring clip.

(A)

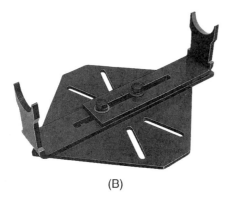

(B)

Figure 31-10 This spring is being removed by disconnecting the inner pivot bushings (A). The control arm is supported by a jack and special attachment (B). *(Courtesy of SPX Kent-Moore)*

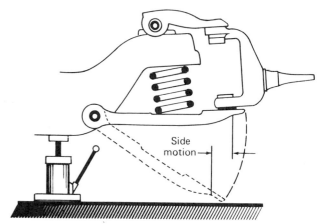

Figure 31-12 As a control arm is lowered, the free end travels in an arc. Note that the free end travels sideways as well as downward. If a jack is used under the control arm, it must be able to move inward.

Figure 31-13 A hammer has been placed between the frame and upper control arm to hold the steering knuckle out of the way while the spring is removed and replaced. *(Courtesy of Moog Automotive Inc.)*

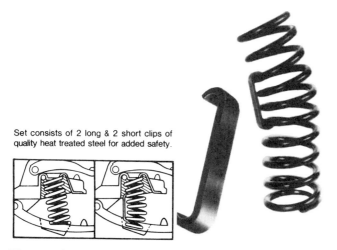

Figure 31-11 Spring clips (normally used in pairs) are placed on the spring before removal. Note the curve that the spring takes as it extends during removal. *(Courtesy of Moog Automotive Inc.)*

2. Position a jack under the control arm at the outermost position possible.

CAUTION *The jack must be able to travel inward a few inches as the control arm swings downward; the control arm might slip off the jack or the jack might tip if the jack cannot move inward (Figure 31-12).*

3. Disconnect the lower ball joint stud and carefully lower the jack to allow the control arm to swing downward. If possible, lift the upper control arm and block it in the upper position by placing a hammer or block at the bump stop (Figure 31-13).

4. After the control arm has swung downward, remove the spring; a pry bar is often needed to work the spring out, and using one is safer than placing your fingers between the spring and the control arm.

While the spring is out, check the upper and lower spring mounts to determine how the spring is positioned. Note that many spring mounts have positioning locations for the ends of the coil as well as the sideways and lengthwise locations (Figure 31-14).

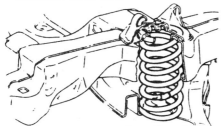

Figure 31-14 There is usually a pocket for each end of the spring, and the spring has a top and a bottom. *(Courtesy of General Motors Corporation, Service Technology Group)*

To prepare a spring for installation using a spring compressor, you should:

1. Measure the length of the old compressed spring, note the location of the spring compressor and compressor plates, and remove the spring compressor (Figure 31-15).
2. Position the spring compressor on the new spring in the same manner and in the same position as on the old spring and compress the new spring to the same length as the old one during removal. Failure to correctly position the compressor and plates

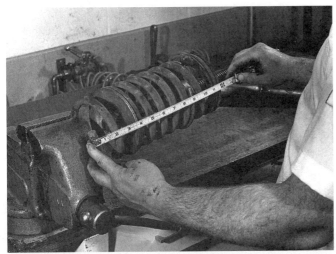

Figure 31-15 The new spring should be compressed to the same length as the old one was when it was removed. *(Courtesy of Moog Automotive Inc.)*

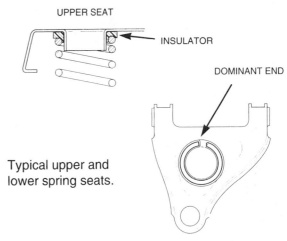

Figure 31-16 In most installations, an insulator goes between the top of the spring and the frame, and the bottom is the dominant end for positioning. *(Courtesy of Moog Automotive Inc.)*

might make them difficult or impossible to remove after the spring has been installed.

To prepare a spring for installation using spring clips you should:

1. Note the position of the clips on the old spring.
2. Install a spring compressor and compress the spring enough to allow removal of the clips.
3. Install the spring compressor on the new spring and compress the spring enough to allow installation of the clips in the same position as they were on the old spring during removal.
4. Remove the spring compressor.

To install a coil spring, you should:

1. Place the compressed spring and insulator, if used, in the pocket and position the spring in the correct relationship with the seats (Figure 31-16).
2. Swing the control arm upward, position the ball joint stud in the steering knuckle, and thread the retaining nut onto the ball joint stud (Figure 31-17). Tighten the ball joint stud retaining nut; the correct procedure is described in Section 32.7.

    If the control arm was disconnected at the inner pivots, swing the control arm upward, align the holes, and install the bolts and nuts; a punch can be used to help align the holes in the control arm bushing with those in the frame or body (Figure 31-18). Final tightening of these bolts should be made with the vehicle weight on the tires.

3. Remove the spring compressor and reinstall the shock absorber, stabilizer bar end links, and wheel.

Figure 31–17 If the ball joint stud was disconnected for spring removal, the spindle hole should be cleaned and inspected before reassembly; also turn the stud so the cotter pin can be easily put into the hole. *(Courtesy of Moog Automotive Inc.)*

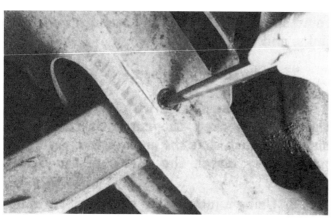

Figure 31–18 If the inner bushings were disconnected for spring removal, a punch can be used to pry the control arm into alignment.

If spring clips are used:

1. Place the spring, with clips in place, in the spring pocket; the curve of the spring should be such that the ends of the spring approximately match the angles of the spring seats.
2. Using a jack, swing the control arm upward until the ball joint stud enters the steering knuckle boss and the retaining nut can be placed onto the ball joint stud. As the control arm swings upward, check to make sure that the ends of the spring are in the correct position in the spring seats.
3. Tighten the ball joint stud retaining nut to the correct tightness, as described in Section 32.7.
4. Remove the spring clips and reinstall the shock absorber, stabilizer bar end links, and wheel.

When the coil spring is mounted above the upper control arm, a spring compressor is used to compress the spring enough to allow the spring to be lifted out of the spring pocket. If you are unsure of how to do this procedure, consult a manufacturer or technician service manual (Figure 31–19).

31.4 Leaf Spring Service

Currently, leaf springs are in limited use in passenger cars; only a small amount of service work is commonly done on them. Leaf springs are normally used at the rear of pickups, vans, utility vehicles, and light trucks. The service operations occasionally done are shimming to correct for side-to-side variations in ride height or to correct for high or low pinion shaft angles, R & R of a spring, R & R of the spring and/or shackle bushings, and R & R of the spring center bolt. Straight shims can be placed between the spring and axle to raise one side and adjust for body lean (Figure 31–20). Tapered shims are used to adjust the pinion shaft angle; this is similar to the caster adjustment procedure described in Section 34.8.5.

To remove a leaf spring, you should:

1. Raise and support the car securely on a hoist or jack stands.
2. Remove the wheel.
3. Place a support under the axle and remove the U-bolts between the axle and spring mounting.

SAFETY TIP: Be sure that both the car and the axle are safely supported.

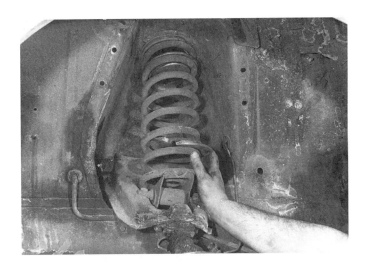

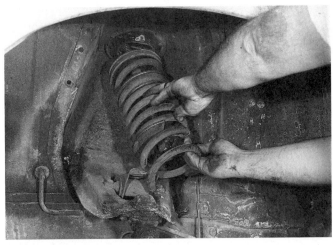

Figure 31–19 When removing a spring mounted on the upper control arm, use of a spring compressor often allows spring removal with the control arm still in place. *(Courtesy of Moog Automotive Inc.)*

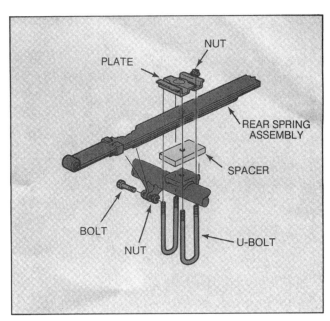

Figure 31–20 The flat spacer is being used to lift this side of the vehicle a distance equal to the spacer thickness. *(Courtesy of Ford Motor Company)*

4. Depending on its position in relation to the spring, raise or lower the axle so there is clearance between the axle and the spring.
5. Remove the front pivot bolt and allow the spring to hang down.
6. Remove the shackle bolts, the shackle, and the spring.

To replace a leaf spring, you should:

1. Assemble the shackle to connect the spring to the frame or body mounts; start the nuts onto the shackle bolts enough to hold the parts together.
2. Swing the spring up into position, place the pivot bolt in position, and thread the nut onto the bolt a few turns.
3. Bring the axle next to the spring, making sure that the axle is centered with the spring center bolt.

Install and tighten the U-bolts between the axle and spring mounting to the correct torque.

4. Tighten the pivot and shackle bolts to the correct torque.
5. Install the wheel.

Leaf spring bushings are of two basic types: a pair of straight or shouldered rubber bushings that are slid into the spring eye from each side or a one-piece metal-backed rubber bushing. The metal-backed bushings are pressed into the spring eye and can be replaced in a manner very similar to that of a control arm bushing. Control arm bushing replacement is described in Chapter 32. Also, like a control arm bushing, the bolt for this bushing should be tightened with the vehicle weight on the tires and the spring and bushing in a normal position. The two-piece rubber bushings are simply slid into the spring eye; normally no special tools are required. The operation of installing rubber bushings can be made easier by using tire mounting rubber lube (Figure 31–21).

Removing and replacing a spring center bolt is merely a matter of taking the old bolt out and putting in the new one, but the curvature of some spring leaves can cause them to fly apart as the center bolt nut is removed. If possible, a C-clamp should be placed over the spring before removing the center bolt to hold the leaves together while the center bolt is changed. Normally, a center bolt that is too long is installed to make it easier to pull the various leaves into position; after the bolt is tightened, the excess length is cut off (Figure 31–22).

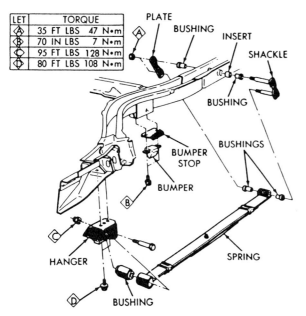

Figure 31-21 Faulty spring bushings can be removed and replaced with new ones. *(Courtesy of Chrysler Corporation)*

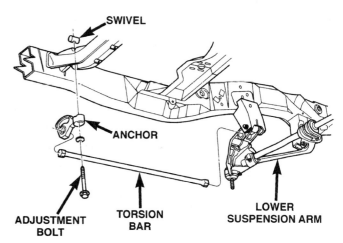

Figure 31-23 Turning the torsion bar adjustment bolt changes the vehicle's ride height. *(Courtesy of Chrysler Corporation)*

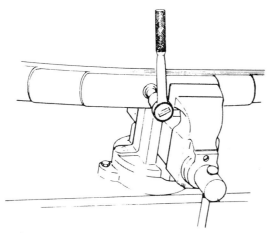

Figure 31-22 When removing and replacing a spring center bolt, it is a good practice to use a vise or clamp to hold the spring together. *(Courtesy of Chrysler Corporation)*

31.5 Torsion Bar Service

Normal torsion bar service consists of adjusting the ride height and removing and replacing the torsion bar. The exact procedure to adjust or remove and replace a torsion bar varies with the manufacturer of the car. A manufacturer or technician service manual should be followed when performing these operations on a specific car model; the procedure described here is very basic and general. When adjusting or replacing a torsion bar, it is a good practice to check the mounting areas and insulators for cracks or breakage.

To adjust a set of torsion bars, you should:

1. Park the car on a smooth, level surface; an alignment rack is ideal because it allows easy access to the measuring and adjusting points, and the side movement from the turntables will remove any binding that might result from tire scrub while the height adjustment is being made or when the car is lowered.

2. Locate the adjustment at the end of each torsion bar and lubricate the bolt threads (Figure 31-23).

3. Obtain the height specifications and measure the front ride height. If the measurements are not within limits to each other or the specifications, an adjustment is necessary. Gauges are available for some car models to make the height checks easier and more accurate (Figure 31-24).

4. Turn the adjusting screws as required to bring the heights into specification limits; it is a good practice to bounce the suspension to normalize the height after making an adjustment and before taking a new measurement. The adjustment is complete when the heights are correct.

To remove a torsion bar, you should:

1. Raise and support the car securely on a hoist or jack stands.

2. Remove the wheel, the shock absorber, and the stabilizer bar end links.

3. Turn the torsion bar adjuster screw all the way out to release as much of the torsion bar pressure as possible. It is a good practice to count and record the number of turns released so that

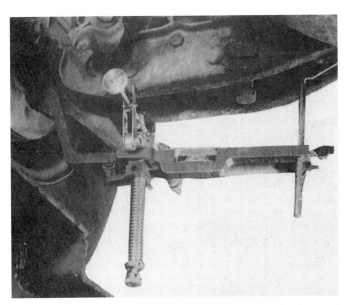

Figure 31–24 This torsion bar height gauge is clamped onto the control arm with one end in contact with the ball joint and then leveled. The measuring stem is at the right side. *(Courtesy of McQuay Norris)*

during the replacement a quick, rough adjustment can be made by retightening the adjuster screw the same number of turns.

4. On some cars, the bar can be removed by removing the mounting bolts and then the bar. On many cars, the torsion bar is slid forward or backward after removing a locating clip so the bar leaves the hex-shaped opening or splines at one end; then the bar is slid the other way and out of the second opening (Figure 31–25).

To replace a torsion bar, you should:

1. Slide the torsion bar into position and replace the retaining clip or place the bar into position and replace the mounting bolts. Tighten the bolts to the correct torque.
2. Retighten the adjusting bolt the same number of turns that it was loosened.
3. Replace the stabilizer bar end links, the shock absorber, and the wheel.
4. Adjust the torsion bars to bring the car to the correct height.

31.6 Strut Service

Normal strut service includes rebuilding or replacing the shock absorber portion, removing and replacing the spring, or removing and replacing the upper mount or damper assembly. With experience and the proper equipment, some of these operations can be performed in the fender opening with the lower end of the strut still connected. Doing so saves time and difficulty in disconnecting the lower end of the strut. In the average metropolitan garage or repair shop with high labor rates, time-saving steps become very important. Many technicians prefer to remove the strut assembly from the car for better access to the parts, and this is highly recommended. Many struts do not require much effort to remove them entirely, and the access and ease of working on the strut off the car helps ensure a better job (Figure 31–26).

The procedure to remove a strut varies depending on how the strut is attached at the bottom to the steering knuckle or control arm. Some manufacturers connect the strut to the steering knuckle using a pair of bolts or a telescoping connection and a pinch bolt. Some manufacturers weld the steering knuckle solidly to the bottom of the strut. On some, the steering arm is bolted to the bottom of the strut along with the ball joint boss; on others, the steering arm is forged as part of the steering knuckle, and the ball joint boss is bolted to the bottom of the strut. Some newer cars locate the steering arm on the upper part of the strut, just under the spring seat. Some cars require the lower ball joint, the tie-rod end, and the brake caliper to be disconnected. Currently, most FWD cars allow the separation of the strut from the steering knuckle, saving the time needed to disconnect the front hub and CV joint. More parts made into a strut assembly mean more parts need to be disconnected for strut removal (Figure 31–27).

After the spring has been removed, some struts can be easily dismantled for internal service. Depending on the availability of parts and the condition of the old parts, the shock absorber portion of these struts can be rebuilt with new parts and new oil or have a cartridge installed. It is a common practice to remove all of the old parts and oil and install a replacement shock absorber cartridge. Struts that are welded together require a replacement strut assembly. This style usually is one that bolts to the steering knuckle, so it really is not much more expensive than the cartridge alone. Some of the welded struts can be cut open using a tubing or pipe cutter to allow removal of the old shock absorber parts and the installation of a new cartridge. Internal threads are already in the upper portion of the outer tube for installation of a gland nut. Be careful if cutting a strut open, because some struts are charged with nitrogen gas under high pressure; disassembly of these struts can cause injury.

SAFETY TIP: Care should be exercised when compressing a strut spring to protect the spring as well as yourself. Many spring compressors have padded jaws to

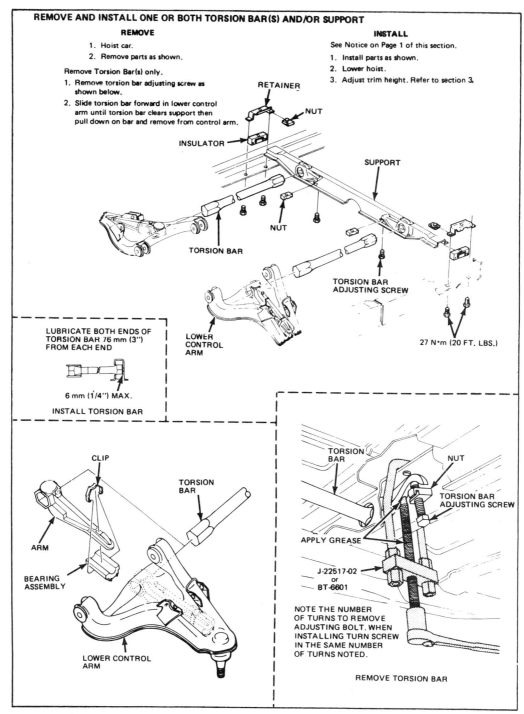

Figure 31-25 The procedure to use when removing and replacing a torsion bar on some General Motors cars. *(Courtesy of General Motors Corporation, Service Technology Group)*

protect the painted or epoxy coating on the springs. This coating is used to retard rust action, which can etch or pit the spring. Rust pitting can cause a stress raiser, which can lead to spring breakage. Spring breakage causes a sudden lowering of that corner of the car, and there have been cases in which the broken end of the spring spiraled off the strut and into the tire. The spring can cut the tire, which could cause a loss of control of the vehicle. Another potential hazard is caused by the normal tendency of a coil spring to twist as it is compressed or extended, and this tendency can cause it to slip loose. Many compressors have extra hooks or safety straps to contain the spring in case of slippage; they should be used if available. Also, many newer springs are fairly long and require a compressor with sufficient travel to allow the spring to completely extend without the compressor coming apart.

Chapter 31 ■ Spring and Shock Absorber Service

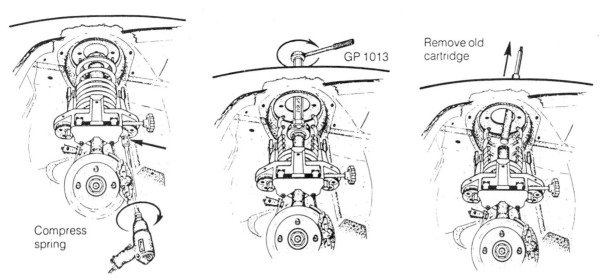

Figure 31-26 This special tool allows the strut cartridge to be removed and replaced through the upper mount.

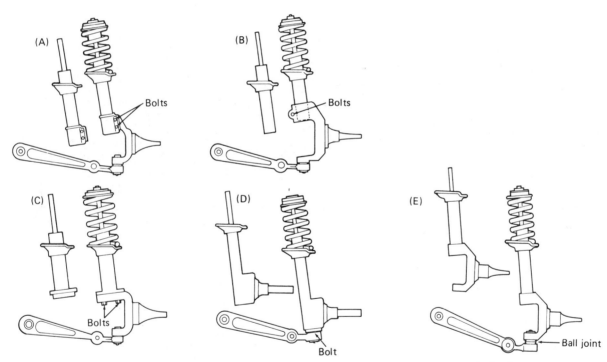

Figure 31-27 There are at least five ways to connect the strut to the steering knuckle or lower ball joint. (A) uses a pair of bolts, (B) uses a sliding connection that is locked by a pinch bolt, (C) uses two or three bolts, in (D) the strut is combined with the steering knuckle and they are bolted to the steering arm or ball joint boss, and (E) connects directly to the ball joint.

SAFETY TIP: The spring and shock assembly from a multilink suspension appears much like that of a strut suspension, but because the bottom end is connected to the control arm instead of the steering knuckle, the spring is often substantially stronger. Many strut spring compressors are *not* designed for the stronger springs and might fail under load. When changing a multilink spring or shock absorber, make sure that the spring compressor is capable of withstanding the heavier load.

31.6.1 Special Strut Service Points

Some struts require special service procedures or precautions. Some cars use spring seats that are positioned off center with the ends of the spring at an angle to the strut. The lower spring seat is welded to the strut body, but the upper seat is held in place by the upper mount and can be turned. If this seat is not positioned correctly, the spring will be positioned improperly. If the hole for the shock rod and mount is not in the center of the spring seat, the hole is usually located toward the center of the car, away from the tire. This will properly align the spring with the strut so the spring is positioned closest to the tire. It is a good practice to mark the upper mount and spring seat before disassembly to ensure the proper location during reassembly (Figure 31–28).

General Motors cars use springs that are coated with epoxy to retard rust. It is recommended that spring compressors used on these springs have jaws that are padded and free of any sharp edges that might cut the spring coating.

Most FWD Ford Motor Company cars use a connection between the strut and steering knuckle similar to that in part B of Figure 31–27; the bottom of the strut fits into the steering knuckle and is locked in place by a pinch bolt. It is recommended that the spring be partially compressed to reduce the length of the strut before removing the strut from the steering knuckle. Failure to compress the spring will require that the

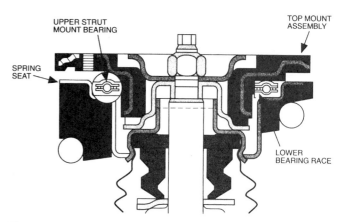

Figure 31–29 Some cars use a mount bushing that requires periodic lubrication; a dry bushing can cause hard steering, poor return, noise, or memory steer. *(Courtesy of Dana Corporation)*

steering knuckle be pried downward an excessive amount to allow separation. This increased travel can apply an excessive amount of force on the CV joints, possibly damaging them. An indication that this damage has occurred is a steering wheel shake of side-to-side rotation under acceleration.

Some Chrysler Corporation "K" cars use a sintered metal bushing for the upper strut mount steering pivot. This bushing requires periodic lubrication to prevent the car from developing poor returnability, or **memory steer.** Memory steer causes a pull in the last direction that the car was turned. To lubricate this bushing, you need to compress the spring, remove the top mount, and remove the bushing. Then a coating of grease can be put on the bushing and the bushing can be replaced. This style of bushing needs lubing every 15,000 to 20,000 mi. (24,000 to 32,000 km); it can be replaced with a mount assembly that uses a ball bearing, as used on many "K" cars. It is a good practice to check this bushing or bearing in all struts on which you work (Figure 31–29).

Some import cars use an anaerobic locking compound on the strut body gland nut to help ensure the retention of the nut. This compound makes the removal of the nut extremely difficult. It softens at temperatures of 200 to 275°F (93 to 135°C). Nut removal can be made easier by heating the top of the strut body and gland nut to temperatures slightly above the boiling point of water using a propane or oxyacetylene torch.

31.6.2 Strut Removal and Replacement

Strut removal and replacement procedures vary among different cars. It is wise to follow the manufacturer's recommended procedure.

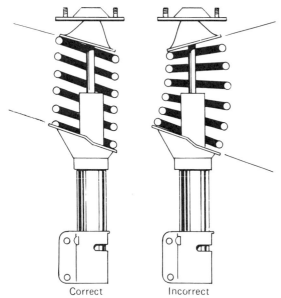

Note: Mark on the upper spring seat must line up with mark on top spring coil. Otherwise, the spring will bow.

Figure 31–28 The upper spring seat for some struts is mounted off center or at an angle; it is a good practice to mark the upper mount and spring seat to ensure replacement in the correct position.

To remove a strut assembly, you should:

1. Mark one of the strut upper mounting bolts so the upper mount can be replaced in the same position, loosen the nut on one of the mounting bolts so it is several turns from coming off, and remove the nuts from the remaining mounting bolts (Figure 31–30 and Figure 31–31).

2. If you plan to disassemble the strut, loosen the center retaining (piston rod) nut a couple of turns.

CAUTION *Do **not** remove the nut at this time. Many technicians skip this step until the strut is off the car and a spring compressor has been installed.*

Figure 31–30 Before removing the upper mounting bolts, mark one of them (the innermost or front one) with a punch, chalk, or felt pen so the mount can be replaced in the same position. *(Courtesy of Arvin Replacement Products Group—Gabriel Ride Control Products)*

3. Raise and support the car securely on a hoist or jack stands.

4. Remove the wheel, stabilizer bar end links, and strut rod if necessary.

5. Disconnect any brake hose supports that attach to the strut. On cars on which the steering arm is part of the strut, disconnect the tie-rod from the steering arm.

6. As required by the strut design:

   a. Remove the nuts and bolts between the strut and the steering knuckle (Figure 31–32).

   b. Remove the pinch bolt between the strut and the steering knuckle (Figure 31–33).

   c. Remove the brake caliper from the steering knuckle and suspend it from some point inside of the fender and remove the bolts between the steering arm or ball joint boss and strut assembly.

OR

   d. Remove the tie-rod end from the steering arm, remove the brake caliper from the steering knuckle and support it from some point inside of the fender, and remove the bolts between the ball joint boss and the strut assembly.

NOTE: Strut removal is usually easier if a spring compressor can be installed to compress and shorten the spring. This requires a small, compact compressor.

7. If necessary, pry downward on the lower control arm to separate it from the strut.

8. Remove the remaining nut from the upper mount and remove the strut assembly from the car.

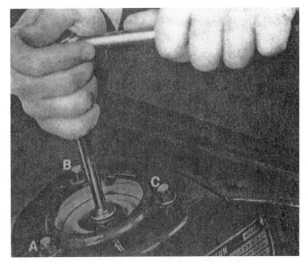

Figure 31–31 Some mechanics prefer to loosen the piston rod nut before removing the mount; do *not* remove the nut at this time. The mounting bolts A, B, and C should be removed, but one of them is often left finger tight until the lower end is disconnected. *(Courtesy of McQuay Norris)*

To install a strut assembly, you should:

1. Slide the strut into position with the previously marked mounting bolt in the correct hole and finger tighten the nut.

2. If necessary, pry the control arm downward to allow the lower end of the strut to be placed in the correct position.

3. Reattach the lower end of the strut at those locations that were disconnected during removal; tighten all of these nuts and bolts to the correct torque.

4. If necessary, replace the brake caliper and brake hose clamps that were removed; tighten all of these nuts and bolts to the correct torque.

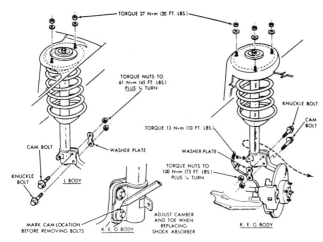

Figure 31-32 Some struts can be removed from the steering knuckle by removing two or three bolts. Note the index marks being placed on the cam location (center). *(Courtesy of Chrysler Corporation)*

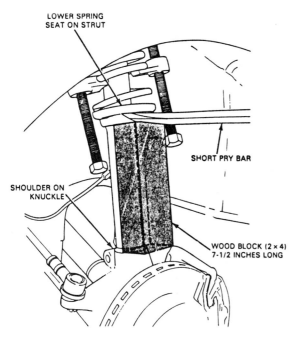

Figure 31-33 A spring compressor, block, and pry bar help when the strut must be slid upward and out of the steering knuckle. *(Courtesy of Ford Motor Company)*

5. Replace the stabilizer bar end links and/or strut rod and tighten the nuts and bolts to the correct torque.
6. Replace the wheel, correctly tightening the lug nuts, and lower the car onto the ground.
7. Replace the remaining nuts on the upper mounting bolts and tighten them to the correct torque.

On General Motors "W" body cars, the access hole in the inner fender panel is large enough to service the strut cartridge with the strut assembly mounted on the car (Figure 31-34). Using the proper tools, the strut mount cover, strut mount, jounce bumper, and strut cartridge can be removed and replaced through the access hole (Figure 31-35).

31.6.3 Strut Spring Removal and Replacement

To remove the spring from a strut assembly, you should:

1. Check the spring to determine how it fits in the mounts and secure the strut assembly in a bench holding fixture or strut vise (Figure 31-36).
2. Following the directions of the tool manufacturer, install a spring compressor and compress the spring until there is clearance between the spring and one of the spring mounts.

CAUTION *Do **not** overcompress the spring; there is no need to compress it to the point where the coils touch each other.*

3. Remove the center retaining (piston rod) nut; an air wrench can spin the nut off to make this a quick and easy operation. Many strut piston rods have wrench flats or an internal hex opening so the piston rod can be held from turning if necessary (Figure 31-37).
4. Remove the upper strut mount. The steering pivot bearing should be checked at this time to ensure its easy, smooth, and free but not sloppy rotation. Also, check the rubber damper portion to ensure that there is no excessive cracking or breakage (Figure 31-38).
5. Remove the spring and compressor.

To replace a spring onto a strut assembly, you should:

1. Secure the strut in a strut vise.
2. Place the compressed spring and spring compressor onto the strut and turn the spring until it is correctly positioned on the lower spring seat.
3. Pull the piston rod upward to the completely extended position.
4. Install the piston rod dust shield, if used, and the upper mount onto the piston rod. Align the upper spring mount with the top end of the spring and, in some cases, align the flat on the piston rod with the flat in the upper mount (Figure 31-39). A tool is available to hold the piston rod in the extended position, if necessary. If the tool is not available, a large rubber band

Chapter 31 ■ Spring and Shock Absorber Service

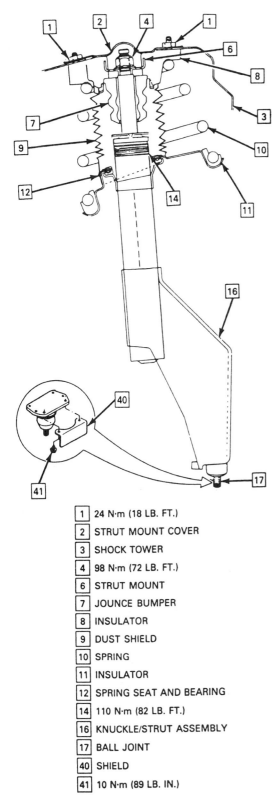

1	24 N·m (18 LB. FT.)
2	STRUT MOUNT COVER
3	SHOCK TOWER
4	98 N·m (72 LB. FT.)
6	STRUT MOUNT
7	JOUNCE BUMPER
8	INSULATOR
9	DUST SHIELD
10	SPRING
11	INSULATOR
12	SPRING SEAT AND BEARING
14	110 N·m (82 LB. FT.)
16	KNUCKLE/STRUT ASSEMBLY
17	BALL JOINT
40	SHIELD
41	10 N·m (89 LB. IN.)

Figure 31–34 When this strut mount cover (2) is removed, there is enough room on this car to remove the strut cartridge shaft nut (4), the mount bushing, and the strut cartridge with the strut in the vehicle. *(Courtesy of General Motors Corporation, Service Technology Group)*

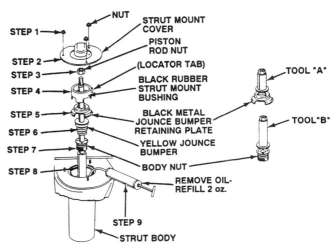

Figure 31–35 The procedure used to remove and replace the strut cartridge from the strut shown in Figure 31–34. *(Courtesy of Arvin Replacement Products Group—Gabriel Ride Control Products)*

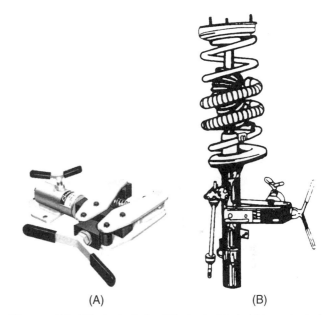

Figure 31–36 A strut vise (A). A strut is held in a strut vise so it can be worked on (B). *(A is courtesy of Branick Industries, Inc.; B is courtesy of Arvin Replacement Products Group—Gabriel Ride Control Products)*

or a short section of heater hose (sliced to allow removal) can be placed over the piston rod to help hold it in an extended position (Figure 31–40).

5. Install the center retaining nut and tighten it to the correct torque. Some manufacturers recommend that the strut be installed in the car and turned to a straight-ahead position before final tightening of this nut. Many sources recommend that a new nut be used (Figure 31–41).

6. Remove the spring compressor, making sure that the spring contacts the seats correctly.

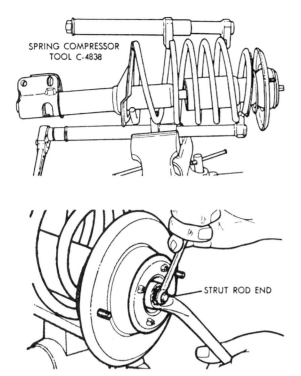

Figure 31–37 There are several styles of spring compressors for strut springs. With the spring compressed, the piston rod nut can be safely removed; if needed, a wrench can be used to hold the strut rod from turning. *(Courtesy of Chrysler Corporation)*

31.6.4 Strut Cartridge Installation

To remove the shock absorber portion of a strut, you should:

1. Secure the strut in a strut vise and remove the spring as described in Section 30.6.3.
2. Remove the large gland nut from the top of the strut tube; special wrenches are available to fit the various nut shapes (Figure 31–42). On all-welded struts, if a replacement cartridge and gland nut are available, use a pipe cutter to cut completely through the reservoir tube at the marked location (Figure 31–43).
3. Remove the O-ring that was under the gland nut and pull upward on the piston rod to dislodge the upper shock absorber bushing. If necessary, use the piston rod and piston like a slide hammer to remove this bushing. Be ready to catch some shock absorber oil as the piston comes out.
4. Remove the piston rod and piston, pressure tube, and other internal shock absorber along with all of the old oil from the strut.

This text does not describe the procedure used to rebuild the shock absorber; consult the vehicle manufacturer service manual for that information.

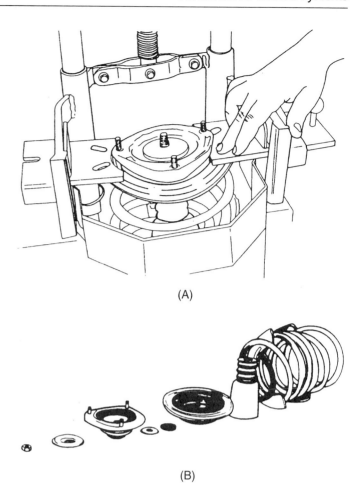

Figure 31–38 With the spring compressed and piston rod nut removed (A), the upper mount can be lifted off the strut and the spring can be removed from the compressor (B). *(Courtesy of Arvin Replacement Products Group—Gabriel Ride Control Products)*

To install a shock absorber cartridge in a strut, you should:

1. Slide the shock absorber cartridge into the strut housing, adding any internal spacers required by the cartridge manufacturer. Many sources recommend pouring a couple of spoonfuls of lightweight oil into the housing before installing the cartridge; the oil helps transfer heat from the cartridge to the outer housing (Figure 31–44). If the strut was one that had to be cut open, smooth the cut portion of the reservoir tube before installing the cartridge to ensure that the gland nut will thread into it (Figure 31–45).
2. Install a new gland nut, make sure that it has at least three threads of engagement, and tighten it to the correct torque. The cartridge should be locked tightly in the housing; there is often a slight gap between the shoulder of the gland nut and the top of the strut housing (Figure 31–46).

Chapter 31 ■ Spring and Shock Absorber Service

Figure 31-39 With the spring on its lower seat, the dust shield, if used, and the upper mount are replaced over the strut rod. Be sure to align the spring and upper seat. *(Courtesy of Moog Automotive Inc.)*

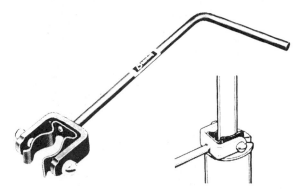

Figure 31-40 This tool can be used to hold the strut rod in an extended position during replacement of the upper mount and rod nut. *(Courtesy of Branick Industries, Inc.)*

When you dispose of a gas-charged shock absorber or strut cartridge, it is recommended that you drill a hole into the reservoir to relieve the gas pressure. This should prevent any future injury to anyone who might come into contact with it. Simply drill a small hole through the side of the shock body, but use caution, wear safety glasses or a face shield, and stay to the side as much as possible in case fluid or metal shavings are sprayed outward (Figure 31-47).

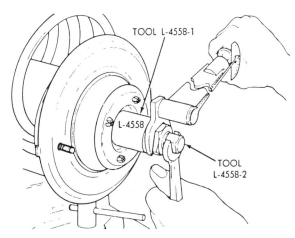

Figure 31-41 The strut rod nut must be tightened to the correct torque; some manufacturers prefer that this step be done with the strut mounted in the car. *(Courtesy of Chrysler Corporation)*

Figure 31-42 There are many different sizes and shapes of strut body gland nuts; a special wrench is recommended for their removal and replacement. *(Courtesy of Branick Industries, Inc.)*

31.7 Stabilizer Bar Service

The common stabilizer bar service operations are the removal and replacement of the end links and bushings and the center bushings (Figure 31-48).

To remove and replace the end links and bushings, you should:

1. Raise and support the car on a hoist or jack stands.
2. Note the direction in which the bolt is positioned and remove the nut, bolt, and bushings from the end link.
3. Install the new link bolt, bushings, washers, spacer (if used), and nut; the new bolt should be positioned in the same direction as the old one.
4. Tighten the nut until it seats at the end of the bolt threads; if included with the new parts, install the jam or pal nut (Figure 31-49).

To remove and replace the center bushings, you should:

1. Raise and support the car on a hoist or jack stands.
2. Remove the bolts securing the bushing brackets to the frame or body and remove the brackets and bushings.

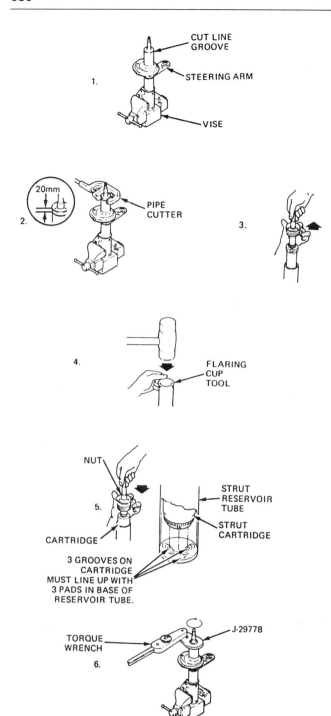

Figure 31-43 On some General Motors struts, the top of the strut body can be cut off using a pipe cutter (2) and the internal parts removed (3). The top of the strut is then deburred (4) to allow the installation of a shock cartridge (5 and 6). *(Courtesy of General Motors Corporation, Service Technology Group)*

Figure 31-44 A small amount of motor oil should be poured into the strut housing to help cool the new cartridge; the oil should almost fill the housing with the new cartridge installed. *(Courtesy of Moog Automotive Inc.)*

Figure 31-45 The new cartridge is slid into the housing. *(Courtesy of Moog Automotive Inc.)*

Chapter 31 ■ Spring and Shock Absorber Service 681

Figure 31-46 With the new cartridge in place, the gland nut should be tightened to the correct torque using the proper spanner wrench. *(Courtesy of McQuay Norris)*

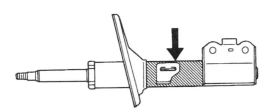

Figure 31-47 A small hole should be drilled into a pressurized shock to let the gas escape before disposing of it; drill anywhere in the shaded area, but watch out for a spray as the drill bit enters the reservoir.

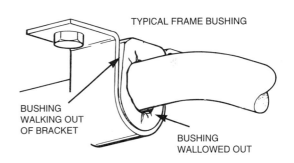

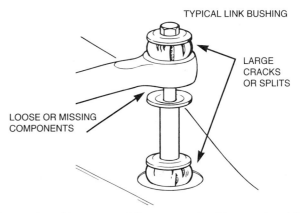

Figure 31-48 Worn stabilizer mount bushings and end-link bushings. *(Courtesy of Moog Automotive Inc.)*

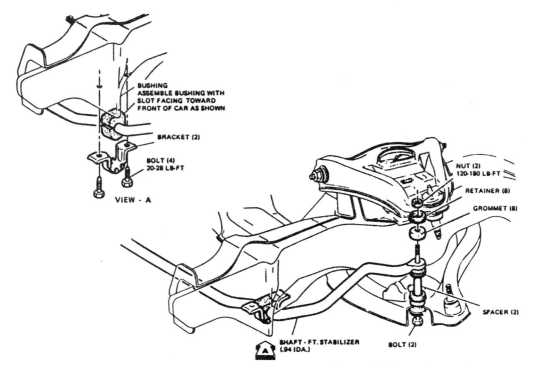

Figure 31-49 The stabilizer mount and end-link bushings. *(Courtesy of General Motors Corporation, Service Technology Group)*

3. Install the new bushings and brackets and install the retaining bolts finger tight.
4. Lower the car to the ground and tighten the retaining bolts to the correct torque.

31.8 Completion

Several checks should always be made after completing any spring or shock absorber service operation:

- Ensure proper tightening of all nuts and bolts, with all locking devices properly installed.
- Ensure that there is no incorrect rubbing or contact between parts or brake hoses.
- Ensure proper feel of the brake pedal.
- Check and readjust the wheel alignment if there is a possibility that the replacement parts have changed the alignment.
- Perform a careful road test to ensure correct and safe vehicle operation.

31.9 Practice Diagnosis

You are working in a brake and front-end shop and encounter the following problems:

CASE 1: A customer has brought in her 5-year-old Lincoln Town Car for a wheel alignment because it has not been checked for more than a year. She also mentions an occasional noise, especially when she goes over bumps. On the road test, you notice a thump as the suspension bottoms out when you go over a set of railroad tracks. When you have the car on the alignment rack and are doing a preinspection, you see that the bump stops and suspension arms are worn. What should you do next? What is the probable cause of this problem?

CASE 2: A 1984 Camry has been brought in with a complaint of noise from the front end and bad tire wear. A check of the tires shows heavy wear on the shoulders and edges of the tread, and the tire pressure seems okay. Your road test confirms an irregular clanking noise as you go over a bumpy section of your test route. The ride quality also seems loose. You do a bounce test with the car on the alignment rack and find that the suspension is loose and the noise seems to be coming from the right strut. What is probably causing this problem? What should you do next?

Terms to Know

bounce check
chassis height
computerized suspension analyzer
curb height
height check
locator tool
memory steer
nut breaker
ride height
trim height

Review Questions

1. Shock absorber tests are being discussed. Technician A says that a bounce test is a simple, quick, and very accurate test of the shock absorber's overall condition.
   Technician B says that a road test is the best way to test the second and third stages of shock absorber operation.
   Who is correct?
   a. A only
   b. B only
   c. Both A and B
   d. Neither A nor B

2. Shock absorbers should be visually inspected for:
   a. Leakage
   b. Broken piston rods
   c. Broken or damaged mounts
   d. All of these

3. Which of the following does not indicate worn shocks?
   a. Excessive dive during braking
   b. More than two cycles after being bounced
   c. Cuppy tire wear
   d. Tire wear at the center of the tread

4. Suspension ride height checks are important to see if the car has:
   a. Sagged springs
   b. Broken springs
   c. Weak shock absorbers
   d. Worn bump stops

5. Suspension ride height checks are important because if the ride height is too low, the:
   A. Front end geometry will be wrong
   B. Car will bottom out too easily
   Which option best completes the statement?
   a. A only
   b. B only
   c. Both A and B
   d. Neither A nor B

6. Ride height checks are being discussed. Technician A says that the tire size and inflation pressure can change the height and should be considered when comparing the measurements with the specifications. Technician B says that these checks should be made on a flat, level surface. Who is correct?
   a. A only
   b. B only
   c. Both A and B
   d. Neither A nor B

7. A car is leaning noticeably to the left. This lean is probably caused by a weak left:
   A. Front spring
   B. Rear spring
   Which option best completes the statement?
   a. A only
   b. B only
   c. Both A and B
   d. Neither A nor B

8. A car is a little too low at the front and rear. This is probably caused by weak:
   A. Shock absorbers
   B. Springs
   Which option best completes the statement?
   a. A only
   b. B only
   c. Both A and B
   d. Neither A nor B

9. If the shock absorber nut is rusted in place on a bayonet-type shock mount, the shock can be removed by:
   a. Breaking the shock mounting stud
   b. Cutting the nut off with a nut breaker
   c. Breaking the nut in half with an air hammer and chisel
   d. Any of these

10. Shock absorber replacement is being discussed. Technician A says that the rear axle can drop when the shock absorbers are disconnected. Technician B says that the shock absorber mounting bushings should be lubricated with motor oil during installation. Who is correct?
    a. A only
    b. B only
    c. Both A and B
    d. Neither A nor B

11. The car has 25,000 mi. on it, and the right front shock absorber is leaking. The:
    A. Right front shock absorber should be replaced.
    B. Left front shock absorber should be replaced.
    Which option best completes the statement?
    a. A only
    b. B only
    c. Both A and B
    d. Neither A nor B

12. The car has 50,000 mi. on it, and the right front spring has sagged out of specification. The:
    A. Right front spring should be replaced
    B. Left front spring should be replaced
    Which option best completes the statement?
    a. A only
    b. B only
    c. Both A and B
    d. Neither A nor B

13. In order to replace rear coil springs, it is necessary to remove the rear:
    A. Shock absorbers
    B. Control arms
    Which option best completes the statement?
    a. A only
    b. B only
    c. Both A and B
    d. Neither A nor B

14. Two technicians are discussing the replacement of a front spring that is mounted on the lower control arm of an S-L A suspension. Technician A says that it is a good practice to install a spring compressor before disconnecting the ball joint or inner pivot bolts. Technician B says that the control arm can swing downward after the ball joint has been disconnected. Who is correct?
    a. A only
    b. B only
    c. Both A and B
    d. Neither A nor B

15. A car equipped with torsion bars sits low at the front end. The torsion bars should be:
    A. Adjusted
    B. Replaced
    Which option best completes the statement?
    a. A only
    b. B only
    c. Both A and B
    d. Neither A nor B

16. New front springs are being installed in a car with S-L A suspension. The:
    A. Ball joint stud, inner control arm pivot bolts, stabilizer bar end links, and shock absorber bolts should be tightened to the correct torque before lowering the car to the ground
    B. Front-end alignment should be checked and adjusted

    Which option best completes the statement?
    a. A only
    b. B only
    c. Both A and B
    d. Neither A nor B

17. Rear-mounted leaf springs are being discussed. Technician A says that a broken center bolt is probably caused by loose U-bolts between the axle and the spring.
    Technician B says that the head of the center bolt is used to align the spring to the axle.
    Who is correct?
    a. A only
    b. B only
    c. Both A and B
    d. Neither A nor B

18. Which of the following is important when compressing the spring for removal from a strut?
    a. The compressor should have padded jaws
    b. The compressor must allow enough travel to let the spring extend fully
    c. The compressor must be able to contain the spring as it twists
    d. All of these

19. A strut with a worn-out shock absorber is being discussed.
    Technician A says that the one repair method is to disassemble the strut and rebuild it with new components.
    Technician B says that you can install a new shock absorber cartridge in most struts.
    Who is correct?
    a. A only
    b. B only
    c. Both A and B
    d. Neither A nor B

20. A small amount of oil should be poured into the strut housing as the new cartridge is being installed. The purpose of the oil is to:
    a. Lubricate the cartridge
    b. Keep the cartridge from rusting
    c. Help keep the cartridge cooler
    d. All of these

Suspension Component Service

Objectives

Upon completion and review of this chapter, you should be able to:

- ❑ Remove and replace upper and lower ball joints.
- ❑ Remove and replace upper and lower control arms.
- ❑ Remove and replace rubber and metal control arm bushings.
- ❑ Remove and replace strut rod bushings.
- ❑ Remove and replace kingpins.
- ❑ Perform the ASE tasks relating to suspension component problem diagnosis, adjustment, service, and repair (see Appendix A, Section B).

32.1 Introduction

After it is determined that a suspension component is faulty, the next operation is to remove and replace that component. This chapter concentrates on the replacement of ball joints, kingpins, and suspension bushings. Service of the components of the steering system is described in Chapter 33. As in other portions of this text, the repair methods given are very general. A service manual should be consulted for operations specific to certain car models.

The operations described in this chapter can be done separately or together in various combinations. One school of thought prefers to rebuild the front end, replacing all bushings and ball joints at one time; another school of thought prefers to replace all bushings and any suspicious ball joints at one time; and a third school of thought prefers to replace a part only when that part absolutely needs replacing. There are pros and cons with each of these approaches. A few of the major considerations when making a decision are the mileage and value of the car, how long the car will be kept by the current owner, how often the front-end or other under-car parts are checked, how often the owner is willing to be without the car, and how secure the owner needs to be about the car's condition. With front ends, there can be real labor savings when several operations are completed at one time; for example, once the ball joint is disconnected for replacement, the spring becomes very easy to replace. A properly rebuilt front end should give many thousands of miles and many years of safe, smooth, trouble-free, new-car-suspension driving with a minimum of tire wear.

SAFETY TIP: Front-end work can be hazardous; there are heavy parts and strong springs to work with. To prevent injury to yourself and to produce a safely operating car, the following practices should be observed:

- Wear eye protection.
- Properly and securely support the car when working under it.
- Follow the repair procedure recommended by the vehicle manufacturer.
- Determine what the result will be before removing each bolt or nut.
- Do not allow a part to fall or a spring to expand in an uncontrolled fashion. Use the proper tool for the job and use that tool in the correct manner.
- Install spring compressors to keep a spring compressed to a safe length.
- Do not allow a strain to be placed on brake hoses.
- Do not let grease, oil, solvent, or brake fluid get on the brake lining or the braking surfaces of the rotor or drum.
- When using an air hammer, be careful; they operate rather violently and very noisily. Face and ear protection should be worn.
- Replace all damaged, worn, and bent parts.

- Check replacement parts against the old parts to ensure an exact or better quality replacement.
- Do not heat suspension and steering parts.
- Use replacement bolts of the same size, type, and grade as the originals.
- Tighten all bolts and nuts to the correct torque and lock them in place by the correct method.
- Tighten parts with rubber bushings while the car is on the ground and the steering is straight ahead.
- Check and adjust the wheel alignment whenever there is a possibility that the replacement parts have changed the alignment angles.
- Make sure the brake pedal operates with a normal feel before moving the car.
- Carefully road test the car to ensure that it will operate in a safe and correct manner.

After replacing front-end, suspension, or steering system parts, a wheel alignment should be done to make sure that the tires are in correct alignment.

32.2 Taper Breaking

One of the difficult operations encountered during front-suspension disassembly is breaking loose the tapers on ball joints or tie-rod ends. The stud of a ball joint is locked into the steering knuckle, and the stud of a tie-rod end is locked into the steering arm (Figure 32–1). If these parts are not locked tightly together, the strong side loads they encounter will cause looseness. Not only will any loose-

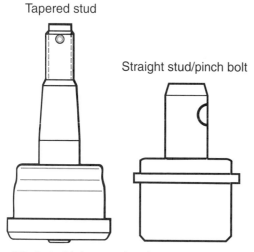

Figure 32–1 Ball joints used with S-L A suspensions have a tapered stud that locks into the steering knuckle (left); many ball joints used with strut suspensions have a straight stud that is held in place by a pinch bolt. *(Courtesy of Moog Automotive Inc.)*

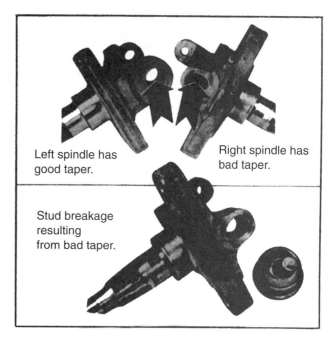

Figure 32–2 This ball joint stud (bottom right) was not tightened correctly; as a result, it worked loose, elongated the hole in the steering knuckle, and then broke the stud. *(Courtesy of Federal-Mogul Corporation)*

ness in these parts cause loose, sloppy steering and suspension action, but the movement will cause wear, which increases to the point of eventual failure (Figure 32–2).

To eliminate play and wear, these parts are connected using locking tapers. The taper used on the stud and hole is about 2 in. per foot (51 cm per 305 mm), or about 9.5 degrees. This taper can vary among car models or even, in a few cases, redesigned parts for the same car model. When the tapered stud is inserted into the tapered hole, the two tapers will interfere and connect; at that point, they should lock together over the entire length of the taper. Even after the nut is loosened, the tapers will stay locked together. Locking or Morse tapers are also used with some machine tools to lock drill bits in drill presses and centers or drill chucks in lathe tail stocks (Figure 32–3).

Several methods are commonly used to break these tapers loose to allow parts replacement: shock, separator tool, and pressure from a puller or spreader. **Shock**, usually from one or two hammers or an air hammer, is fast and often effective; it is usually the first method to try. **Separator tools**, commonly called **pickle forks** because of their shape, are fast and generally effective, but they usually destroy the ball joint or tie-rod end boot and may destroy the joints themselves. Separator tools should only be used when the ball joint or tie-rod end is going to be replaced with a new one. **Pullers** and **spreaders** are special tools for ball joints or tie-rod ends; they are fast and

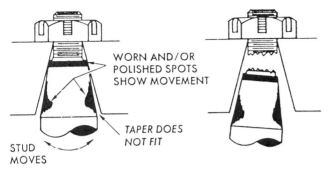

Figure 32–3 If the stud taper does not fit and lock into the steering knuckle, movement of the stud can cause breakage. Always check for worn or polished spots on the ball joint stud and in the boss, which indicate that this problem has occurred. *(Courtesy of Moog Automotive Inc.)*

effective. They also break the tapers in a professional way so as not to damage any of the parts (Figure 32–4).

When breaking the taper on ball joint studs, it is a good practice to let the spring help you or keep it from hindering you. If the ball joint is tension loaded, the spring on an S-L A suspension is trying to separate the ball joint stud; on compression-loaded ball joints, it is trying to keep the stud and the steering knuckle together. Before breaking the stud taper on tension-loaded ball joints, the car should be raised and supported by the frame, and the ball joint stud nut should be loosened a few turns. Loosen it just enough to let the taper break loose. With compression-loaded ball joints, a jack should be placed under the lower control arm or a spring

Figure 32–4 The stud taper of tension-loaded ball joints can usually be broken loose using a taper breaker (tool J-8806); it places an outward pressure on the studs. Note that the nuts on the studs to be loosened must be unscrewed several turns. *(Courtesy of General Motors Corporation, Service Technology Group)*

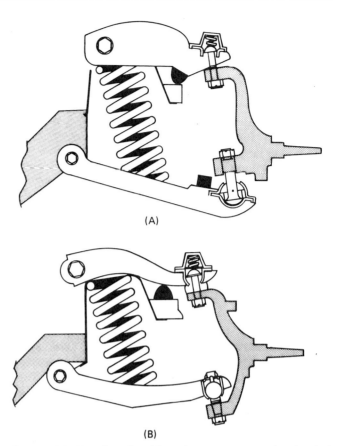

Figure 32–5 When breaking the taper on tension-loaded ball joints (A), spring pressure will aid you. When breaking the taper on compression-loaded ball joints (B), spring pressure holds the tapers together.

compressor should be installed to remove the spring pressure from the ball joint stud (Figure 32–5).

If both ball joints on one side are to be serviced, it is often faster if both tapers are broken at the same time before disconnecting either one. If the follower ball joint is tension loaded, it is easier to break the taper while the spring pressure is helping you, as described earlier.

32.2.1 Breaking a Taper Using Shock

Once the nut has been loosened, a locking taper can often be broken loose by shock or vibration. With a quick and firm blow, strike the side of the ball joint boss in the steering knuckle or steering arm, being careful not to bend or damage the part being struck or any of the surrounding parts. When breaking loose a tie-rod end, a backup hammer is necessary to prevent bending of the steering arm and to increase the amount of shock (Figure 32–6). Front-end technicians often use an air hammer with a hammerlike punch for this operation (Figure 32–7). Many technicians do not use shock from a hammer to break tapers loose.

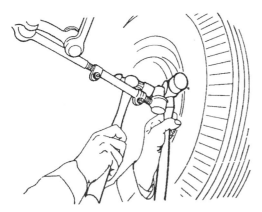

Figure 32–6 The taper of a tie-rod end and sometimes a ball joint stud can often be broken loose by a sharp, quick blow from a hammer. When striking the side of a steering arm, use a second hammer for a backup to prevent damage. *(Courtesy of General Motors Corporation, Service Technology Group)*

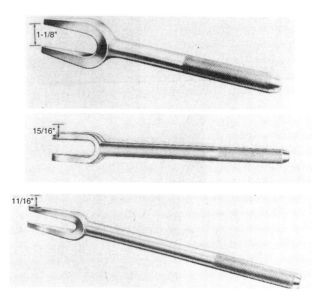

Figure 32–8 Three different separators or wedges, commonly called pickle forks; the different-sized openings allow them to be used on Pitman arms (top), ball joints (center), and tie-rod ends (bottom). Note that their use will destroy some joints and seals. *(Courtesy of SPX-OTC)*

Figure 32–7 After loosening the stud nut a few turns, the stud taper can often be broken loose by vibrating the steering knuckle boss using an air hammer and punch.

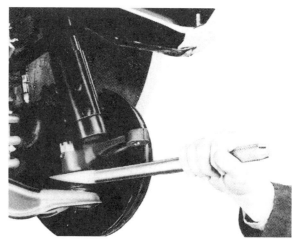

Figure 32–9 A ball joint separator being used to break loose the ball joint stud taper on this modified strut suspension; the right end of the tool will be struck with a hammer. Note that the stud nut has been loosened a few turns and left on the stud. *(Courtesy of SPX-OTC)*

32.2.2 Breaking a Taper Using a Separator Tool

Separator tools or pickle forks are available with different-sized openings designed for tie-rod end, ball joint, and steering arm removal (Figure 32–8). They are available in larger, longer sizes to be struck with standard hammers or in a shorter version as an attachment for an air hammer (Figure 32–9). The wedgelike separator is driven into the gap between the ball joint and the steering knuckle or between the tie-rod end and the steering arm. The wedging action of the separator plus the shock and vibration of the hammer blows, along with a prying action from the technician, are fairly effective in separating these parts. However, this action cuts and destroys the sealing boots and also tends to pull the stud and ball out of the socket of the joint; it can destroy the joint (Figure 32–10).

32.2.3 Breaking a Taper Using Pressure

Pressure applied by pulling or spreading tools often called **taper breakers** is a fast, quiet, neat, and professional way to separate tapers (Figure 32–11). The tools, which are designed for this special purpose, are placed in position between the ball joint studs and tightened to produce the

Chapter 32 ■ Suspension Component Service

Figure 32-10 Separator tools are also available for use in an air hammer; this one is breaking loose a tie-rod end. *(Courtesy of SUPERIOR PNEUMATIC MFG., INC., Cleveland, OH 44140)*

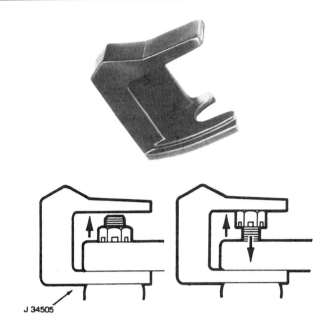

Figure 32-12 This tie-rod stud remover is used by removing the nut, turning it upside down, and then threading it against the tool to exert a downward pressure on the stud. *(Courtesy of SPX Kent-Moore)*

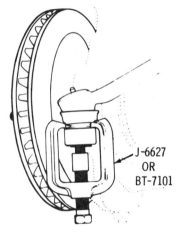

Figure 32-11 Tool J-6627 is being used to break loose the tie-rod end stud taper. *(Courtesy of General Motors Corporation, Service Technology Group)*

necessary outward force to break loose the tapers. Sometimes a slight blow on the side of the stud boss is used after the puller or spreader is tightened. Taper breakers cannot be used with compression-loaded ball joints.

Tie-rod end pullers often resemble two-jaw pullers; some designs use a single fixed jaw. The pressure from the puller screws pushes the tie-rod stud into and through the steering arm. Most ball joint taper breakers resemble a long nut and bolt; one end is positioned at each ball joint stud. When the parts of the taper breaker are threaded apart, a spreading force is exerted on the ends of the ball joint studs. Taper breakers for some FWD cars use the ball joint nut to apply pressure onto the ball joint stud (Figure 32-12).

32.2.4 Removing and Replacing Ball Joint Studs Locked by Pinch Bolts

Most cars with strut suspensions do not use locking tapers on their ball joint studs; the stud is locked in place by a **clamp** or **pinch bolt** in the steering knuckle boss. The steering knuckle boss is slotted so the pinch bolt can clamp it tightly onto the ball joint stud. The pinch bolt also passes through a notch in the stud to help ensure that the stud cannot come out of the boss, even if the pinch bolt loosens. This type of connection is also used to secure the strut to the steering knuckle on some cars.

To separate this style of ball joint stud from the steering knuckle, you simply remove the pinch bolt and slide the ball joint stud out of the steering knuckle boss. Occasionally it is necessary to place a screwdriver into the slot in the steering knuckle and pry it open to loosen its grip on the stud. To install it, you slide the stud into the boss, making sure that the notch in the stud is aligned with the pinch bolt hole, slide the pinch bolt into position, and tighten the nut to the correct torque (Figure 32-13). It should be noted that some manufacturers recommend that this bolt not be reused; it should be replaced with a new bolt and nut.

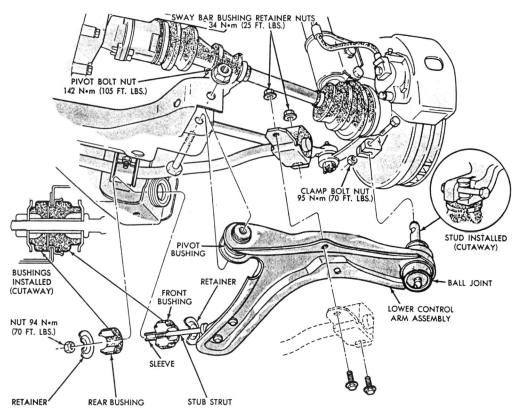

Figure 32-13 A pinch/clamp bolt locks this ball joint stud into the steering knuckle. Note that this stud has a single notch that must be properly aligned to install the clamp bolt. *(Courtesy of Chrysler Corporation)*

32.3 Ball Joint Replacement

Ball joint replacement is necessary when the clearance of a vehicle-loaded ball joint exceeds specifications, a follower-type joint shows clearance, or either style of joint has a cut or torn boot or is too tight. If one ball joint of a set shows excessive wear, many technicians prefer to replace the entire set of ball joints, using the theory that all of the ball joints have operated for the same length of time under the same operating conditions. The remaining ball joints may fail relatively quickly anyway. It is less expensive to replace all of the ball joints at one time than to replace one at a time. Also, a more complete repair should give the car owner more security for a longer period.

Ball joints can be replaced with the control arm attached to the car; many technicians prefer this approach because the arm is held securely during replacement operations and the time needed to remove and replace the control arm is saved. Ball joints can also be replaced with the control arm off the car; many technicians prefer this approach because they have total access to the control arm and ball joint. Whether the control arm should be removed depends on the skill and preference of the technician, the type of equipment available, and what other jobs are to be done. In either case, it is necessary to disconnect the ball joint stud from the steering knuckle. Removal and replacement of ball joints are described in Section 32.5.

In a few cases, it is necessary to replace the control arm with the ball joint; the ball joint is not available from the car manufacturer by itself. In some cases, individual parts are available from aftermarket suppliers; OEM parts come as assemblies.

Some manufacturers have replacement ball joint and tie-rod end boots available so that a torn boot can be replaced rather than changing the whole joint. After removing the old retaining ring and boot, lubricate the new boot and joint as shown in Figure 32-14; the boot clip guide tool is used to position the new retaining ring properly.

32.4 Removing a Control Arm

This section describes the procedure to remove a lower control arm on an S-L A suspension with a spring on the lower arm. A similar procedure is used to remove an upper arm from this same suspension or a lower arm from a strut suspension; there is no need to compress or remove

Chapter 32 ■ Suspension Component Service

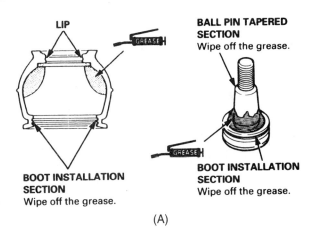

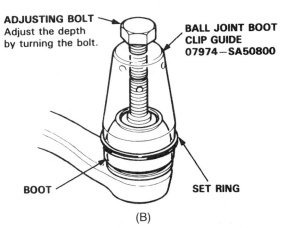

Figure 32-14 Grease should be added to the replacement boot and the ball joint stud (A) before installing the new boot; the special tool is used to install the set/retaining ring at the correct location (B). *(Courtesy of American Honda Motor Co., Inc.)*

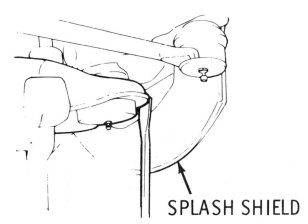

Figure 32-15 On some cars, the lower control arm needs to be guided past the brake splash shield as it is lowered; a screwdriver or small prybar can be used. *(Courtesy of General Motors Corporation, Service Technology Group)*

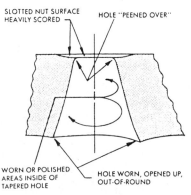

Figure 32-16 With the ball joint stud removed, check the hole in the steering knuckle boss for problem signs. *(Courtesy of Moog Automotive Inc.)*

the spring for these other styles of control arms. A similar procedure to that described here is used on an upper control arm with a spring mounted on it.

To remove a lower control arm, you should:

1. Raise and support the car on a hoist or jack stands.
2. Remove the wheel.
3. Remove the shock absorber and the stabilizer bar end link.
4. If the car has a tension-loaded lower ball joint, loosen the ball joint stud nut several turns and, using a spreader tool, separator, and/or shock, break loose the ball joint stud taper.
5. Install a spring compressor and compress the spring enough to remove the spring load from the control arm. You can be sure the spring load is removed when the spring becomes free to move around or when the control arm can be moved upward. If the car has a compression-loaded ball joint, break loose the ball joint stud taper at this time.
6. Disconnect the ball joint stud and separate the steering knuckle boss from the ball joint stud. It is often necessary to pry between the lower control arm and the brake splash shield, rotor, or steering knuckle to keep the end of the control arm from catching on the splash shield (Figure 32-15). Inspect the tapered portion of the stud for any bright, worn metal, which would indicate a worn steering knuckle boss (Figure 32-16). The upper control arm, with the steering knuckle attached, can be raised and a block can be inserted between the control arm and the frame or body bracket to hold them up and out of the way (Figure 32-17).
7. The lower control arm can be pivoted downward to remove the spring. If desired, the lower ball joint can be replaced. The lower control arm can now be removed by removing the inner bushing/pivot bolts.

Figure 32-17 A hammer has been placed under the upper control arm to hold the steering knuckle and brake assembly out of the way. *(Courtesy of Moog Automotive Inc.)*

The upper control arm can also be removed at this time by disconnecting the ball joint stud and the inner pivot bolts. If this is done before replacing the lower control arm, it will be necessary to support the steering knuckle with the brake rotor or drum and caliper or backing plate, or disconnect the brake hose and completely remove the steering knuckle and brake assembly. Many front-end technicians prefer to service the parts on the lower control arm, reattach the lower control arm to the steering knuckle and frame, tie the control arm in place (if necessary to keep it from falling and stressing the brake hose), and then service the parts on the upper control arm. The steering knuckle and brake assembly can be removed from the car, but the tie-rod end will have to be disassembled from the steering arm and the brake hose from the caliper or wheel cylinder. If the brake hose is disconnected, it will be necessary to bleed the air from that brake after reassembly.

Many General Motors cars with S-L A suspension mount the upper control arm on a pivot shaft, which is bolted to a bracket on the frame; the pivot shaft is positioned inward of the bracket. This control arm must be slid inward and off the bolts to remove it. The steering shaft or exhaust manifold often prevents enough inward movement to allow the shaft to be slid off the mounting bolts. Use a socket, extension bar, and long socket handle or air-impact wrench to turn the mounting bolts; this will loosen them. The bolts should work their way out of the frame bracket. It helps to pry outward on them as they are rotated. A special tool is available to press these bolts out of the frame bracket. With the bolts out of the way, the control arm can be lifted out. For replacement, the bolts are merely driven back in place after the control arm is repositioned. New bolts will be required if their serrations become too badly worn; if they do not fit tightly, the bolts may turn when replacing shims during a wheel alignment (Figure 32–18).

Figure 32-18 The steering shaft interferes with the removal of this control arm; if the control arm mounting bolts (arrows) are removed, the control arm can be easily removed.

32.5 Removing and Replacing a Ball Joint

In most cases, ball joints and bushings are available as individual parts and can be replaced as separate items. Ball joints are usually secured into the control arm by one of several methods: pressed into place, riveted or bolted into place, or threaded into place (Figure 32–19). If the OEM ball joint is riveted into place, the replacement ball joint will be secured by bolts. The ball joint usually enters and leaves the control arm from the side opposite to the stud; in this position, normal operating pressures tend to seat the ball joint tighter rather than try to pull it out of the control arm.

Whenever a ball joint is removed, the end of the control arm and the opening for the ball joint should be inspected for cracks or other signs of damage, which would indicate a faulty control arm (Figure 32–20).

32.5.1 Removing and Replacing a Pressed-In Ball Joint

There are several methods of removing and replacing pressed-in ball joints. Most shops purchase a specialized kit for this operation. Ball joint service kits usually include a series of pressing tools of different diameters to push on the ball joint and a series of various-sized

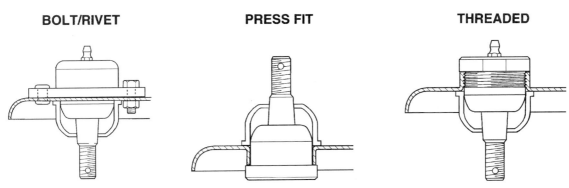

Figure 32–19 A press fit (center) is the most common way of attaching the ball joint to the control arm. *(Courtesy of Moog Automotive Inc.)*

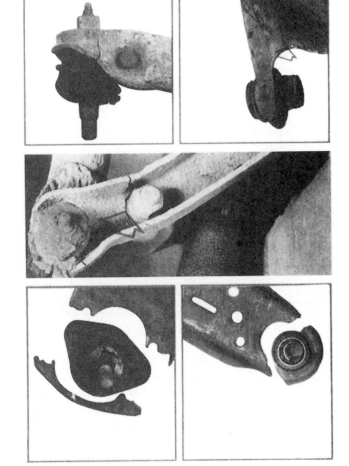

Figure 32–20 The control arm should be checked for cracks before replacing the ball joint. *(Courtesy of Federal-Mogul Corporation)*

To remove a pressed-in ball joint, you should:

1. Clean the control arm in the area around the ball joint and check for spot welds (Figure 32–21).
2. Select the smallest support or receiver adapter tube that will fit over the ball joint support boss (Figure 32–22).
3. Select the size of pressing adapter that fits the stud side of the ball joint best; this adapter must be able to pass through the control arm.
4. Assemble the ball joint press with the two adapters over the control arm and press the ball joint out of the control arm (Figure 32–23).

To install a pressed-in ball joint, you should:

1. **SAFETY TIP:** Inspect the control arm to make sure there are no cracks and the opening is not distorted; a damaged control arm should be replaced (Figure 32–24).
2. Position the new ball joint in the control arm and select a pressing adapter that fits the ball joint and receiver adapter for the control arm (Figure 32–25).
3. Place the ball joint press, support adapter, and pressing adapter over the control arm and press the ball joint completely into the control arm.
4. Some press-fit ball joints use a retaining ring to secure the joint in the control arm (Figure 32–26).

32.5.2 Removing and Replacing a Riveted- or Bolted-In Ball Joint

Many ball joints are riveted in place at the factory; the replacement ball joint is bolted in place.

To remove a riveted-in ball joint, you should:

1. Clean the control arm in the area around the ball joint.

support tools to fit around the ball joint to protect the control arm from distortion. Pressure to push the old joint out or the new joint in comes from either a hydraulic jack or a screw thread. Do not drive ball joints out or in using a hammer; there is too much danger in distorting the control arm or the new ball joint.

Figure 32–21 Some ball joints have a spot weld holding them in place (top); this must be carefully cut away before removing the ball joint (bottom). The replacement joint will use a snap ring to secure it. *(Courtesy of Moog Automotive Inc.)*

2. Working from the ball joint side of the control arm, use a sharp chisel and an air hammer to slice off the head of the rivet next to the ball joint. Be careful not to damage the control arm. If you have difficulty cutting the rivet, drill a hole, about three-fourths of the diameter of the rivet, down through the center of the rivet. The hole should extend into the depth of the ball joint flange. After drilling, the rivet head should be easy to cut (Figure 32–27).
3. Use a punch to drive the body of the rivet out of the ball joint and the control arm.
4. Remove any bolts or nuts that may be used in addition to the rivets and remove the ball joint.

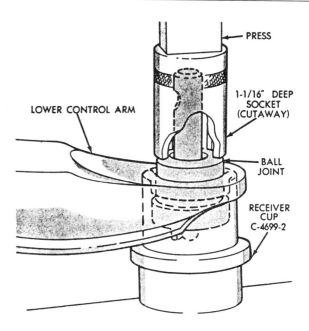

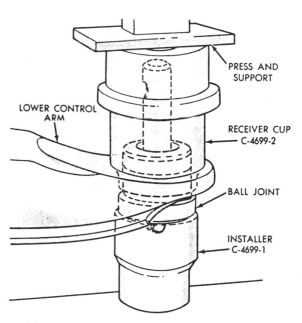

Figure 32–22 A pressed-in ball joint being removed (top) and installed (bottom). Note the placement of the pressing and receiving tools for each operation. *(Courtesy of Chrysler Corporation)*

To install a riveted-in ball joint, you should:

1. **SAFETY TIP:** Inspect the control arm to make sure there are no cracks and the opening is not distorted; a damaged control arm should be replaced.
2. Place the new ball joint in position and assemble the bolts, nuts, washers, and any retainers, following the directions provided with the ball joint (Figure 32–28).
3. Tighten the retaining bolts to the correct torque (Figure 32–29).

Chapter 32 ■ Suspension Component Service

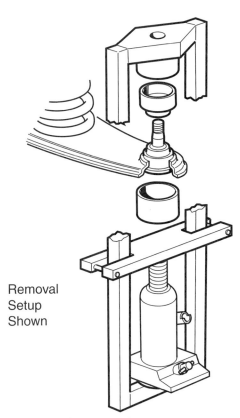

Figure 32-23 This press tool is being set up to press the ball joint downward and out of the control arm. *(Courtesy of Moog Automotive Inc.)*

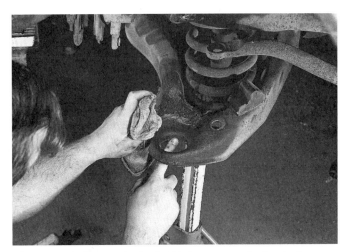

Figure 32-24 With the ball joint removed, check the control arm for damage. *(Courtesy of Moog Automotive Inc.)*

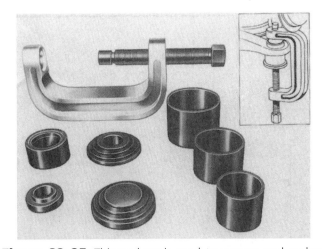

Figure 32-25 This tool set is used to remove and replace ball joints. Note the various sizes of adapters and receiver tubes. *(Courtesy of SPX-OTC)*

Figure 32-26 The snap ring being installed will ensure that the ball joint stays in the control arm. *(Courtesy of Moog Automotive Inc.)*

32.5.3 Removing and Replacing a Threaded Ball Joint

A threaded ball joint can be identified by the wrench flats on the body of the ball joint. Several different sizes of special ball joint sockets are available to fit these flats.

To remove and replace a threaded ball joint, you should:

1. Clean the control arm in the area around the ball joint.

2. Secure the control arm and, using the correct socket and rather long socket handle for leverage, carefully unscrew the ball joint from the control arm. Penetrating oil can help ensure removal without damage to the control arm.

3. **SAFETY TIP:** Inspect the control arm for cracks and damage to the ball joint threads.

4. When installing the new ball joint, make sure it does not cross-thread, and thread it into the control arm. Tighten the ball joint to the correct torque; if the correct torque cannot be obtained, the control arm must be replaced (Figure 32-30).

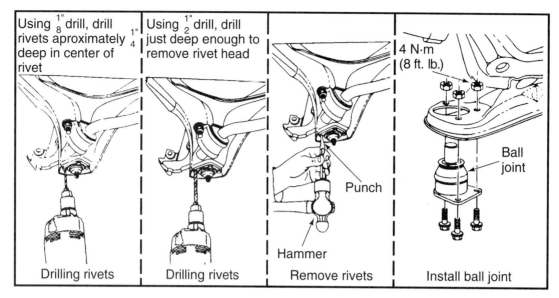

Figure 32-27 A hole drilled through the rivet head will make it much easier to remove. Note that the smaller hole makes the use of the larger drill easier and ensures that the hole stays centered. *(Courtesy of General Motors Corporation, Service Technology Group)*

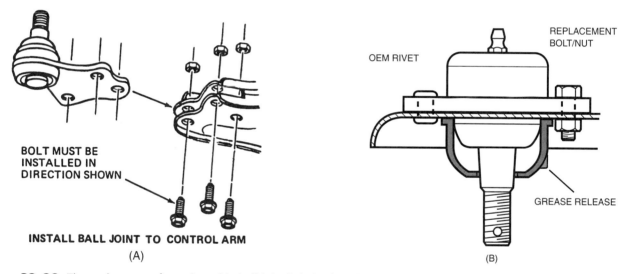

Figure 32-28 The replacement for a riveted-in ball joint is bolted in place with bolts and nuts properly positioned (A). The boot for some joints has a grease relief hole, which must be positioned away from the brakes (B). *(A is courtesy of General Motors Corporation, Service Technology Group; B is courtesy of Moog Automotive Inc.)*

32.6 Control Arm Bushing Replacement

Most domestic cars use rubber control arm bushings; at one time, some used metal bushings. Most rubber control arm bushings used in domestic cars are of a three-piece (inner sleeve, rubber insert, and outer sleeve), torsilastic type; some import and a few domestic cars use a one- or two-piece (no outer sleeve) type. The outer sleeve of the three-piece type makes it relatively easy to install.

With torsilastic bushings, the control arm pivot bolts squeeze the serrations of the inner sleeve of the bushing against the shoulder of the control arm shaft or frame or body bracket; the inner sleeve becomes locked, and the control arm is preloaded in this position. The rubber in the bushings must now twist if the control arm is raised or lowered. These bolts should be tightened in the normal ride height position to prevent overstressing the rubber in the bushings. If the pivot bolts were tightened with the control arm in the rebound position, the bushings would be twisted at ride height and twisted an excessive amount

Chapter 32 ■ Suspension Component Service

Figure 32-29 The ball joint retaining bolts must be tightened to the correct torque. *(Courtesy of Moog Automotive Inc.)*

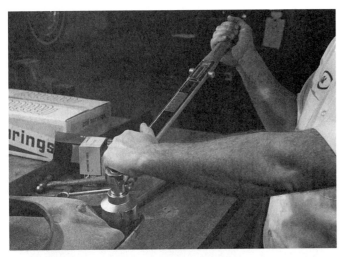

Figure 32-30 A threaded ball joint is removed and replaced using a special socket; it must be tightened to the correct torque during installation. *(Courtesy of Moog Automotive Inc.)*

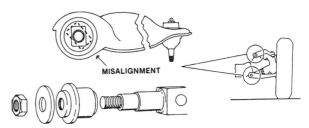

Figure 32-31 When torsilastic bushings are installed, the retaining bolts and nuts must be tightened with the control arm in the correct position and the suspension at ride height. If the bushings are tightened with the control arms hanging down, they will be severely stressed. *(Courtesy of Moog Automotive Inc.)*

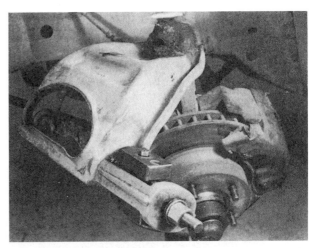

Figure 32-32 This bushing is being pulled out of the control arm with the control arm still attached at the ball joint.

at jounce. These bolts are usually left slightly loose until the front end is assembled and the car's weight is on the tires; then they are tightened to the correct torque. In cases where it is very difficult to tighten these bolts after assembly, the position of the control arm shaft to the control arm is carefully noted and marked during disassembly before the retaining bolts are loosened; then, after bushing replacement, the shaft is put back in this exact position, and the bolts are tightened (Figure 32-31).

Special bushing pullers and installers allow replacement of the bushings with the control arm still in the car without disconnecting the ball joint. In some cases, there is adequate room to remove and replace the bushings under the hood. If under-hood space is restricted, the control arm can be pivoted on the ball joint so the bushing end of the control arm is in the fender well to gain working space (Figure 32-32).

In some cases, the inner sleeve of the bushing rusts onto the control arm shaft, making removal of the bushing very difficult. It is sometimes necessary to remove the bushing in pieces; removal of the outer sleeve and rubber insert allows access to a slit in the inner sleeve. Place the control arm shaft over a vise so you can strike the sleeve with a hammer or spread the slit open with a chisel. Doing so will allow the sleeve to be easily slid off of the control arm shaft. The rust-covered surfaces of the control arm shaft can be cleaned by wire brushing to ease the installation of the new bushings (Figure 32-33).

After the bushings have been removed, the control arm shaft is free to fall out of the control arm. Most control arm shafts are symmetrical, so their position is not important, but it is a good practice to mark the position of the shaft before bushing removal to ensure its installation in the correct position.

Some General Motors vehicles have a spot weld on one or both of the retaining nuts on the control arm shaft

Figure 32–33 These bushings were removed using an air hammer and chisel; note the chisel marks. The left one came out easily; the others required more time and effort.

to prevent the nuts from unthreading and falling off. When replacing the control arm bushings on these cars, it is necessary to cut the control arm shaft in half and remove the two halves of the shaft with the bushings. During bushing replacement, a new control arm shaft will be required along with the new bushings. If you encounter one of these models that is not spot welded and the nuts can be removed, the old shaft can be reused. It is a good practice to install a self-locking nut, nut locking compound, or lock washer to ensure retention of the nut.

32.6.1 Removing Rubber Bushings Using an Air Hammer and Chisel

Removing control arm bushings with an air hammer and chisel is normally fairly fast. With experience, it can be done with the control arm on the car, but it is best to remove the control arm and secure the arm in a vise. The vibration of the air hammer probably has as much effect in removing the bushing as the force of the hammer blows.

SAFETY TIP: Care should be taken not to damage the control arm with the chisel; personal caution should also be exercised because of the noise and violent manner of operation.

To remove a control arm bushing using an air hammer, you should:

1. Secure the control arm in a vise.
2. Clean the control arm in the area around the bushing.
3. Using an air hammer and chisel against the bushing flange or biting into the body of the bushing, force the bushing out of the control arm (Figure 32–34). On control arms where the bushing passes through two surfaces of the control arm, it is a good practice to vibrate the control arm right next

Figure 32–34 The air hammer and chisel will vibrate the bushing out of the control arm. *(Courtesy of Moog Automotive Inc.)*

to the bushing (with the air hammer and chisel or punch) and loosen each contact surface between the bushing and the control arm before starting the bushing removal.

32.6.2 Removing Rubber Bushings Using a Puller Tool

Pullers have the advantage that they can usually be used with the control arm on the car. The ball joint can be left attached to the steering knuckle, and the control arm can be pivoted to an area where there is access to the bushing with a puller. Most puller sets also include adapters to press the new bushing into the control arm. For bushings that do not have an inner shaft, it is possible to make a puller installer from two pieces of tubing, a length of threaded rod, two large washers, and two nuts. Puller use is quiet and normally will not damage the control arm.

To remove a rubber bushing using a puller, you should:

1. Clean the area of the control arm around the bushing.
2. Select a sleeve large enough for the bushing to be pulled into (Figure 32–35).
3. Select an adapter that fits over the control arm shaft that will push against the end of the bushing.
4. Position the two adapters and the puller over the control arm and bushing and tighten the puller to force the bushing out of the control arm (Figure 32–36).

OR Some bushing pullers are designed to be used with a press or special puller bolts; these units are often made specifically for one brand or model of car (Figure 32–37).

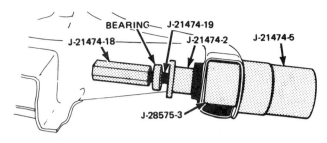

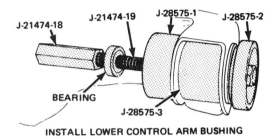

Figure 32-35 Tightening the nut (top) will cause the adapter to push the bushing out of the control arm and into the receiver adapter; part J-28575-3 keeps the control arm from collapsing. A new bushing can be installed using the setup at the bottom. *(Courtesy of General Motors Corporation, Service Technology Group)*

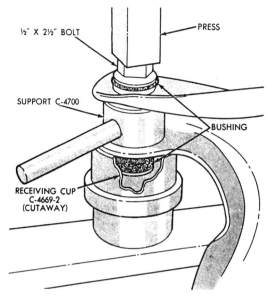

Figure 32-37 The control arm is in a press with the adapters positioned to remove the bushing. *(Courtesy of Chrysler Corporation)*

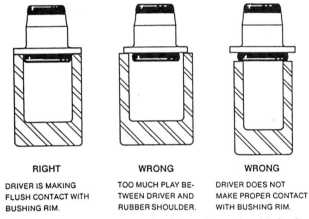

Figure 32-38 The proper bushing driver should fit flush against the rim of the bushing (left). The center driver is too big, and the right-hand one is too small. *(Courtesy of Moog Automotive Inc.)*

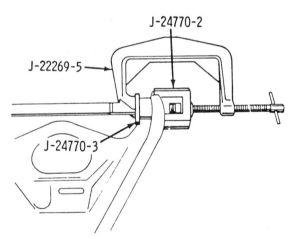

Figure 32-36 This puller is set up to remove a control arm bushing. *(Courtesy of General Motors Corporation, Service Technology Group)*

To install a rubber bushing using a driver, you should:

1. Inspect the control arm to make sure it is not damaged.
2. Select a bushing driver that fits just around the rubber portion so it seats solidly onto the bushing flange (Figure 32-38).
3. Position the control arm with the bushing bore over the jaws of a vise; adjust the vise opening so it is slightly wider than the bushing.
4. If necessary, place the control arm shaft in place. Start the bushing into the control arm by hand and drive it into position using a hammer and bushing driver (Figure 32-39).

32.6.3 Installing a Rubber Bushing Using a Driver

Using a driver to install bushings is normally quite fast, and **bushing drivers,** often called **knockers,** are fairly inexpensive. Care should be taken that the bushing is driven in straight to ensure that the control arm is not damaged.

Figure 32-39 Installing a control arm bushing using a hammer and bushing driver. Note how the control arm is supported with the vise. *(Courtesy of Moog Automotive Inc.)*

32.6.4 Installing a Rubber Bushing Using a Pressing Tool

Like the removal step, bushing installation can be done on the car, thus saving the time and trouble needed to remove and replace the ball joint stud. Pressing the bushing into place is fairly fast and usually does not damage the control arm.

To press a rubber bushing into a control arm, you should:

1. Inspect the control arm to make sure it is not damaged.
2. Select a pushing tool that fits solidly against the flange of the bushing (Figure 32-40).
3. Select a support adapter that fits around the control arm opening.
4. If necessary, place the control arm shaft in position and start the bushings into the control arm by hand.
5. Assemble the bushing press and the two adapters over the control arm and press the bushing completely into the control arm (Figure 32-41).

32.6.5 Removing and Replacing Rubber Bushings without Outer Sleeves

Rubber bushings without outer sleeves are used on some import cars as well as a few domestic cars; they are removed and replaced in a manner similar to that used for the metal-sleeved bushings. The tight fit into the control arm causes them to work like a sleeved torsilastic bushing;

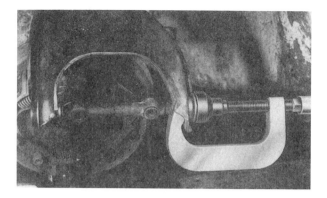

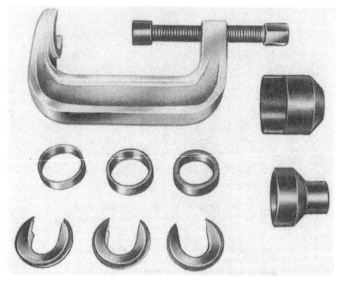

Figure 32-40 This tool is installing a new bushing (top); note the various adapters for different-sized bushings (bottom). *(Courtesy of SPX-OTC)*

this tight fit can cause difficulty on installation. Bushing removal and installation tools are available to make this job fast and fairly easy. The bushing and opening should be lubricated with tire rubber lube or a vegetable-based cooking oil before installing; do not use petroleum, motor oil, or brake fluid as a lubricant, because they will cause deterioration of the rubber (Figure 32-42).

32.6.6 Removing and Replacing Metal Bushings

Metal bushings are removed by unscrewing them from the control arm; the control arm shaft is removed after removal of the bushings (Figure 32-43). The pivot shaft is normally replaced with a new one as the bushings are replaced; a worn shaft must be replaced during bushing replacement. A kit with rubber bushings and pivot shaft to fit them is available from aftermarket suppliers for those who prefer the advantages of the rubber bushing.

Chapter 32 ■ Suspension Component Service

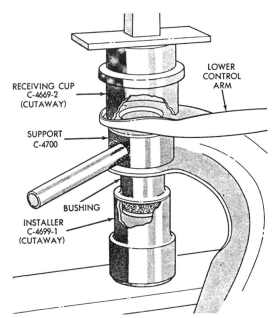

Figure 32–41 The control arm is being pressed onto this bushing. Note the relationship of the adapters and the support to keep the control arm from collapsing. *(Courtesy of Chrysler Corporation)*

During replacement, the shaft is placed in the control arm in the correct position, and the new bushings with seals are threaded into the control arm and onto the shaft (Figure 32–44). The bushings should be tightened alternately to ensure that the shaft is centered in the control arm, and they should be tightened to the correct torque. These bushings must be lubricated; in some cases, it is necessary to grease them before installing the control arm in the car because there is not enough room to grease them after installation.

32.7 Installing a Control Arm

This section describes the procedure to replace a lower control arm and spring on an S-L A suspension. A similar procedure is used on other control arms.

To install a lower control arm with spring on an S-L A suspension, you should:

1. Place the control arm in position in the frame or body brackets and install the inner pivot bolts. Replace the nuts and tighten them finger tight.
2. Turn the ball joint stud to align the cotter pin hole lengthwise to the car so the cotter pin will be easy to install (Figure 32–45). Clean and inspect the ball joint stud hole in the steering knuckle boss (Figure 32–46).

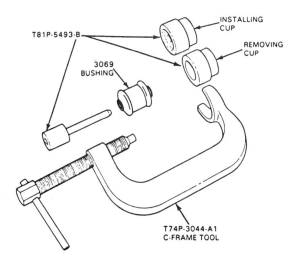

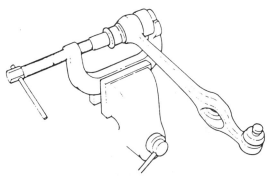

Figure 32–42 The control arm pivot bushing set (top) is used to remove and replace bushings that do not have a metal outer sleeve. A bushing installation is shown at the bottom; note that the bushing should be lubricated with vegetable oil or tire lube before it is installed. *(Courtesy of Ford Motor Company)*

Figure 32–43 Metal control arm bushings are unscrewed from the control arm to remove them. *(Courtesy of Moog Automotive Inc.)*

Figure 32–44 Installation of metal control arm bushings is the reverse procedure of removal; the control arm shaft should be centered in the control arm. *(Courtesy of Moog Automotive Inc.)*

3. Replace the compressed spring and align it to the correct position, as described in Section 31.3. Swing the control arm upward so the ball joint stud enters the steering knuckle boss and thread the nut onto the ball joint stud.

4. Tighten the ball joint stud nut to the correct torque and install the cotter pin (Figure 32–47). If a prevailing torque nut is used, it is sometimes difficult to tighten the nut because the stud tends to rotate. In this case, it is usually necessary to either pretighten the stud using a plain nut, use a stud holder, or tap the control arm upward to try to seat the stud taper (Figure 32–48).

5. Remove the spring compressor, making sure the spring seats correctly.

6. Install the shock absorber and stabilizer bar end links. Install the tires and wheels and lower the car onto the tires.

7. Tighten the inner pivot bolts to the correct torque.

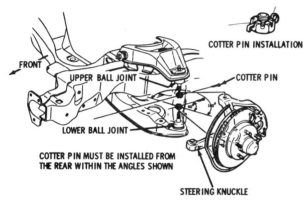

Figure 32–45 It is best to position the ball joint stud so the cotter pin can be easily installed before installing the ball joint stud into the steering knuckle. *(Courtesy of General Motors Corporation, Service Technology Group)*

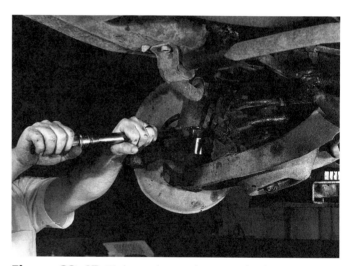

Figure 32–47 Tighten the stud nut to the lowest torque specification so it can be tightened further if necessary to align the cotter pin hole. Never just back off the nut to align the hole. *(Courtesy of Moog Automotive Inc.)*

Figure 32–46 The tapered hole in the steering knuckle should be cleaned by running a shop cloth through it before installing the ball joint stud. *(Courtesy of Moog Automotive Inc.)*

Figure 32–48 This simple tool is used to hold the ball joint stud from turning when self-locking nuts are used. Another method is to lock the tapers using a plain nut and then install the self-locking nut. *(Courtesy of Moog Automotive Inc.)*

SAFETY TIP: When ball joint studs are replaced, there are a few cautions to observe:

- Do not overtighten the retaining nut. After the tapers meet, there is not much effective bolt length to accept any further stretching; overtightening can cause enough stress to weaken the threaded end of the stud.
- Do not back off the nut to align the cotter pin slots. Backing off or loosening of the nut can cause a slight clearance between the nut and the stud boss; this can allow the tapers to come loose, which will cause play and wear.
- Make sure that the stud and the hole are clean to ensure a good locking action of the tapers. Some technicians run a shop towel in and out of the stud hole to ensure its cleanliness.
- Tighten the nut first to the minimum torque specification, and then, if necessary, continue tightening to the nearest cotter pin slot (Figure 32–49).
- On cars that use self-locking nuts, first use a plain nut of the correct thread to lock the stud taper and then install the self-locking nut to the correct torque.
- Tighten the bolts that secure the rubber pivot bushings when the suspension and the bushing are in the normal ride height position. Rubber bushings in the steering linkage are tightened with the steering in the straight-ahead position.

32.8 Removing and Replacing Strut Rod Bushings

Most strut rod bushings can be replaced while the control arm is off during a bushing or ball joint replacement or as a separate operation with the control arm connected to the steering arm and frame. On many cars, the strut rod is connected to the control arm with one or two bolts and can be easily removed after these bolts and the nut at the strut rod bushing are removed. On some cars, the strut rod passes through the lower control arm and is retained by a nut; strut rod bushing replacement on these cars is more involved. The lower control arm will need to be removed or moved away from the strut rod to make strut rod removal possible.

To replace strut rod bushings, you should:

1. Remove the strut rod bushing nut; if two nuts are used at the bushings, leave the inner nut in position so the caster adjustment will not be disturbed. Also, remove the belts that connect the strut rod and control arm (Figure 32–50 and Figure 32–51).
2. Lift the strut rod from the control arm and slide it out of the frame bracket (Figure 32–52). Remove the old bushings and washers.
3. Following the directions for the bushing set, assemble the new inner bushing and retaining washer. Slide the strut rod into the bracket, assemble the remaining bushing parts, and install the retaining nut (Figure 32–53).
4. Hold the free end of the strut rod in its normal ride height position, tighten the retaining nut to snug, and align the bushing parts.

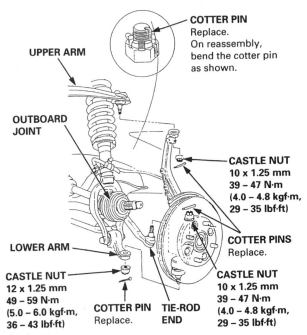

Figure 32–49 The tightening torque specifications and cotter pin installation procedure for this multilink suspension. *(Courtesy of American Honda Motor Co., Inc.)*

Figure 32–50 Strut rod removal begins by removing the nut from the strut rod end. *(Courtesy of Moog Automotive Inc.)*

Figure 32–51 This strut rod is held to the lower control arm by two bolts and nuts. *(Courtesy of Moog Automotive Inc.)*

Figure 32–53 The new bushings should be correctly positioned as the strut rod is slid into the frame bracket. *(Courtesy of Moog Automotive Inc.)*

Figure 32–52 The strut rod being removed from the frame bracket. *(Courtesy of Moog Automotive Inc.)*

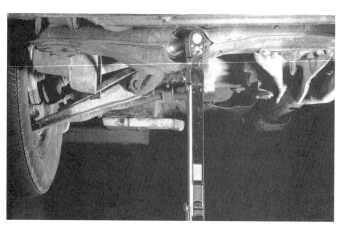

Figure 32–54 The strut rod bushing nut and other retaining bolts and nuts should be tightened to the correct torque. *(Courtesy of Moog Automotive Inc.)*

5. Install the bolts and nuts between the strut rod and control arm and tighten all nuts and bolts to the correct torque (Figure 32–54).

32.9 Removing and Replacing Kingpins

Most trucks and pickups using a solid axle or twin I-beam axles use a reversed Elliott kingpin design, which locks the kingpin in the end of the axle and has a pair of kingpin bushings in the steering knuckle. Two types of bushing materials are used: **bimetal** (bronze bushing material in a steel shell) and **synthetic** (plastic or nylon). Synthetic bushings tend to reduce steering effort and reduce road shock; another advantage is that they can be installed without machining. However, they tend to wear faster, especially under dusty conditions. Bimetal bushings tend to resist contaminants better and usually last longer than synthetic bushings under adverse conditions. Metal bushings usually have to be machined (reamed or honed) after installation to bring them to the correct size (Figure 32–55).

To remove and replace kingpins and bushings, you should:

1. Raise and support the vehicle on a hoist or jack stands.
2. Remove the wheel and the brake caliper and rotor or brake drum and backing plate. Using a wire, hang the caliper or backing plate inside the fender.
3. Remove the tie-rod end from the steering arm; this operation is described in Chapter 33.
4. Remove the retaining nut and drive the kingpin lock bolt out of the axle.

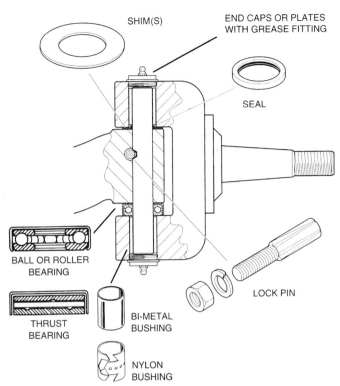

Figure 32-55 A kingpin set includes a new kingpin, shims, seal (if used), lock pin, bushings, and thrust bearing. *(Courtesy of Moog Automotive Inc.)*

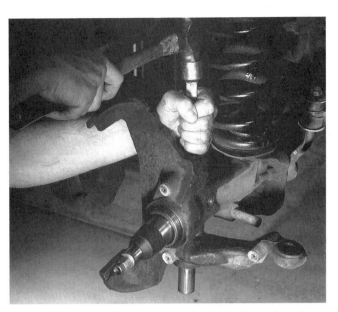

Figure 32-56 After removing the lock pin and end caps, the old kingpin is driven out of the axle and steering knuckle. *(Courtesy of Moog Automotive Inc.)*

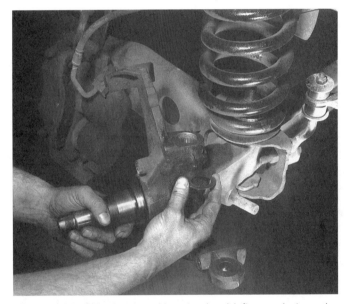

Figure 32-57 The new kingpin should fit snugly into the axle. *(Courtesy of Moog Automotive Inc.)*

5. Remove the kingpin end caps. If they have wrench flats, they can be unscrewed. If they are soft plugs/metal discs, drive a punch into them to loosen them and pry them out using a screwdriver.
6. Using a punch and hammer, drive the kingpin out of the axle and steering knuckle (Figure 32-56).
7. Remove the steering knuckle, bearing, and shims; save the shims for possible reinstallation (Figure 32-57).
8. Drive the old bushings out of the steering knuckle using the correct size of bushing driver and a hammer (Figure 32-58).
9. If using synthetic bushings, slide the new bushings into the steering knuckle bosses.

OR If using bimetal bushings, carefully drive or press the new bushings into the steering knuckle, making sure the grease holes in the bushings line up with the zerk fitting holes in the steering knuckle bosses. A correctly sized bushing driver must be used to ensure that the bushings are not damaged during installation (Figure 32-59). The bushing bores should be honed (preferred) or reamed to size the bores to the kingpin diameter and ensure that the two bores are in exact alignment. Honing or reaming of the bushing bores can be done by most automotive machine shops (Figure 32-60).

10. If used, place the sealing O-rings in the recesses in the steering knuckle bores, lubricate the O-rings and bushings, place the steering knuckle (correct side up) over the axle, pack the thrust bearing with grease, and place the bearing between the bottom of the axle and the lower steering knuckle bore (Figure 32-61).

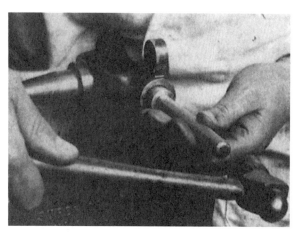

Figure 32–58 The old bushings are driven out of the steering knuckle using a hammer and bushing driver. *(Courtesy of McQuay Norris)*

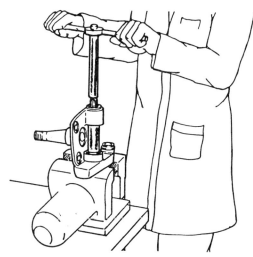

Figure 32–60 Metal kingpin bushings are normally reamed or honed to the correct size; it is important to keep the two bushing bores in straight alignment. *(Courtesy of Dana Corporation)*

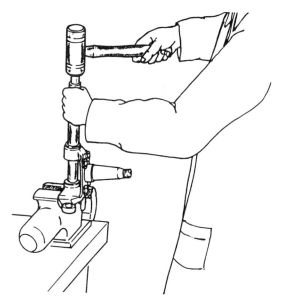

Figure 32–59 New metal bushings are pressed or driven into the steering knuckle using a properly sized driver; nylon bushings usually slide in. *(Courtesy of Dana Corporation)*

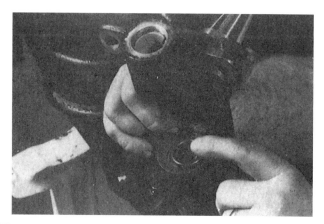

Figure 32–61 Some kingpins use an O-ring seal; it is installed and lubricated before replacing the steering knuckle. *(Courtesy of Moog Automotive Inc.)*

11. Slide a new set or the original shims into the gap between the top of the axle and the upper steering knuckle boss. Now, try to slide a 0.005-in. (13-mm) feeler gauge alongside of the shims; if it enters, larger shims should be installed so the gap is 0.005 in. or less (Figure 32–62).
12. Slide the new kingpin into the steering knuckle and axle bores, making sure that the notch or flat in the kingpin aligns with the lock pin hole.
13. Install the lock pin and tighten the nut to the correct torque (Figure 32–63).
14. Install the dust caps and lubricate the kingpin and bushings using a grease gun (Figure 32–64).
15. Replace the brake assemblies and the tires and wheels.

32.10 Rear-Suspension Service

The wearing points of rear suspensions, in most cases, are the rubber bushings, which are almost the same as those already discussed. If these bushings become excessively worn, rear-wheel misalignment or noises will result. Rear-wheel control arm bushing service is done following a procedure that is very similar to front-suspension bushing replacement; many of the same tools are used.

Rear-suspension service procedures are described in manufacturer and technician shop manuals, if there is doubt as to the exact procedure for these operations.

Chapter 32 ■ Suspension Component Service

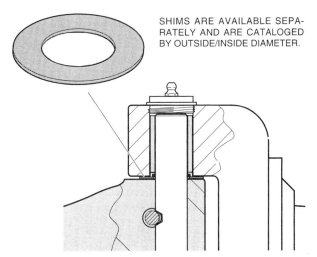

Figure 32-62 Shims are added to the top of the axle to reduce the gap size to the specification. *(Courtesy of Moog Automotive Inc.)*

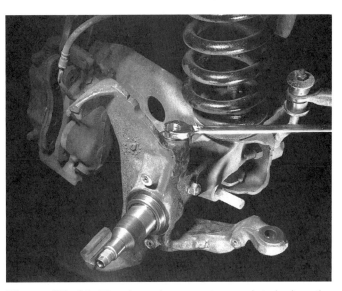

Figure 32-64 Kingpin replacement is completed when the end caps are installed and the parts are given a final greasing. *(Courtesy of Moog Automotive Inc.)*

- Check and readjust the wheel alignment if there is a possibility that the replacement parts have changed the alignment.
- Perform a careful road test to ensure correct and safe vehicle operation.

Figure 32-63 The lock pin is tapped into place, and the nut should be tightened to the correct torque. *(Courtesy of McQuay Norris)*

32.12 Practice Diagnosis

You are working in a brake and front-end shop and encounter the following problems:

CASE 1: A customer has brought in his 1988 Camaro for a wheel alignment. The car exterior and interior appear to be in very good condition; there are 88,000 mi. on the odometer. Your road test does not show any faults except for a little excess noise and kind of mushy operation. What should you do next?

CASE 2: A customer has brought in his Chevrolet Astro van for new ball joints and bushings. The vehicle shows heavy use and has 45,000 mi. on the odometer, but this could be 145,000 or 245,000 mi. When you disconnect the lower right ball joint, you find a bent stud and slightly egged-out hole for it. What will you need to do to repair this vehicle? What probably caused this problem?

32.11 Completion

Several checks should always be made after completing any suspension system service operation:

- Ensure proper tightening of all nuts and bolts with all locking devices properly installed.
- Ensure there is no incorrect rubbing or contact between parts or brake hoses.
- Ensure proper feel of the brake pedal.

Terms to Know

bimetal bushing	pickle forks	shock
bushing drivers	pinch bolt	spreader
clamp	puller	synthetic bushing
knockers	separator tool	taper breakers

Review Questions

1. A ball joint stud taper can be broken loose using a:
   a. Sharp hammer blow against the stud boss
   b. Stud separator
   c. Pickle fork
   d. Any of these

2. Technician A says that the pressure from the car's spring can help you break the stud taper of a compression-loaded ball joint.
   Technician B says that in all cases, the spring should be compressed before breaking the ball joint studs loose.
   Who is correct?
   a. A only   c. Both A and B
   b. B only   d. Neither A nor B

3. If a ball joint has a torn boot, the:
   a. Hole in the boot should be glued shut
   b. Boot should be replaced
   c. Ball joint should be replaced
   d. Any of these

4. A pickle fork is a quick way of breaking ball joint studs loose, but it:
   A. Will ruin the boot
   B. Can distort the ball joint socket
   Which option best completes the statement?
   a. A only   c. Both A and B
   b. B only   d. Neither A nor B

5. Technician A says that many ball joints are secured to the control arm by a press fit.
   Technician B says that some ball joints are threaded into the control arm.
   Who is correct?
   a. A only   c. Both A and B
   b. B only   d. Neither A nor B

6. Rivets used to secure the ball joint to the control arm can be removed using a:
   A. Cutting torch
   B. Drill and chisel
   Which option best completes the statement?
   a. A only   c. Both A and B
   b. B only   d. Neither A nor B

7. A torsilastic control arm bushing's:
   A. Inner sleeve is locked to the control arm shaft when the mounting bolt is tightened
   B. Rubber portion should be lubricated with a special grease
   Which option best completes the statement?
   a. A only   c. Both A and B
   b. B only   d. Neither A nor B

8. Technician A says that metal control arm bushings must be pressed into the control arm an equal amount on each side.
   Technician B says that metal control arm bushings must be lubricated during or after installation.
   Who is correct?
   a. A only   c. Both A and B
   b. B only   d. Neither A nor B

9. Technician A says that the installing tool for rubber control arm bushings should push against the outer sleeve.
   Technician B says that the bolt retaining rubber bushings should be tightened only with the car at correct ride height.
   Who is correct?
   a. A only   c. Both A and B
   b. B only   d. Neither A nor B

10. Common rubber control arm bushings are pressed in place:
    A. By a force applied on the rubber portion
    B. Until they seat solidly in the control arm
    Which option best completes the statement?
    a. A only   c. Both A and B
    b. B only   d. Neither A nor B

11. Metal control arm bushings are installed:
    A. By threading them into the control arm and onto the control arm shaft
    B. An equal amount on each bushing
    Which option best completes the statement?
    a. A only   c. Both A and B
    b. B only   d. Neither A nor B

12. Sleeveless rubber bushings can be installed more easily if they are lubricated with:
    a. Lithium grease   c. Vegetable oil
    b. Brake fluid   d. Any of these

13. Technician A says that a worn ball stud hole can be caused by a mechanic installing the stud into a dirty hole.
    Technician B says that a broken ball joint stud can be the result of a mechanic backing off the nut to align the cotter pin.
    Who is correct?
    a. A only   c. Both A and B
    b. B only   d. Neither A nor B

14. Technician A says that the ball joint stud can be held from turning with a special tool when installing a self-locking nut.
    Technician B says that the stud can be locked in place with a standard nut before installing the self-locking nut.
    Who is correct?
    a. A only   c. Both A and B
    b. B only   d. Neither A nor B

15. When removing or replacing the ball joint stud on a car with S-L A suspension, a spring compressor should be installed to make the job:
    A. Easier
    B. Safer
    Which option best completes the statement?
    a. A only   c. Both A and B
    b. B only   d. Neither A nor B

16. When replacing strut rod bushings, the:
    A. Control arm must always be removed
    B. Retaining nut should be tightened with the strut rod in the normal ride height position
    Which option best completes the statement?
    a. A only   c. Both A and B
    b. B only   d. Neither A nor B

17. Technician A says that ball joint studs that are locked in place using a pinch bolt are more difficult to remove than the other style.
    Technician B says that during replacement, the notch in the ball joint stud must be carefully aligned with the pinch bolt.
    Who is correct?
    a. A only   c. Both A and B
    b. B only   d. Neither A nor B

18. Technician A says that bimetal kingpin bushings are presized and do not require machining after installation.
    Technician B says that shims should be added if the gap between the axle and the thrust bearing is greater than 0.005 in.
    Who is correct?
    a. A only   c. Both A and B
    b. B only   d. Neither A nor B

19. Plastic or nylon kingpin bushings should be:
    A. Lubricated after installation
    B. Reamed to the correct size after installation
    Which option best completes the statement?
    a. A only   c. Both A and B
    b. B only   d. Neither A nor B

20. After new ball joints or control arm bushings have been installed, the:
    A. Alignment should be checked and adjusted if necessary
    B. Car should be road tested
    Which option best completes the statement?
    a. A only   c. Both A and B
    b. B only   d. Neither A nor B

Math Questions

1. While replacing the ball joint and reinstalling the control arm on a Nissan, you find a torque specification of 59–71 N-m. Your only torque wrench reads in foot-pounds. What torque should you use?

2. You have rebuilt the front end on a 1987 Dodge Caravan. You have replaced these parts: upper control arm bushings at $12.00 per side, lower control arm bushings at $8.00 per side, both upper ball joints at $44.00 per pair, both lower ball joints at $47.00 per pair, and both shock absorbers at $17.50 each. The flat rate manual gives these times for the operations: R & R all shaft bushings, 2.4; R & R ball joints, 1.6; adjust torsion bars, 0.6; and align wheels, 1.4. Your shop has a flat rate of $55.00 per hour. What is the cost for the parts? The labor? The total cost?

33 Steering System Service

Objectives

Upon completion and review of this chapter, you should be able to:

- ❏ Remove and replace tie-rods and tie-rod ends on conventional and rack-and-pinion steering systems.
- ❏ Remove and replace idler arms, center links, and Pitman arms.
- ❏ Remove and replace conventional and rack-and-pinion steering gears and mounting bushings.
- ❏ Remove and replace steering wheels, steering columns, and steering shaft couplers (including on airbag-equipped vehicles).
- ❏ Make the necessary adjustments on a conventional or rack-and-pinion steering gear.
- ❏ Overhaul a conventional or rack-and-pinion steering gear.
- ❏ Repair fluid leaks and replace the fluid in a power steering system.
- ❏ Test power steering system pressures and determine whether they are correct.
- ❏ Adjust power steering control valves.
- ❏ Overhaul a power steering pump.
- ❏ Perform the ASE tasks relating to steering system diagnosis, adjustment, service, and repair (see Appendix A, Section A).

33.1 Introduction

After it is determined that a steering system component is faulty, the next step is to repair, adjust, or remove and replace that component. This chapter focuses on the replacement, adjustment, and repair of tie-rod ends, idler arms, manual and power steering gears, and power steering hoses, pumps, and belts.

Steering system service should be done with great care so that there is no possibility of failure after the work is done. Failure of the steering system can have devastating effects on the operation of the car. Safety first must be the attitude of a professional technician while working on steering systems (Figure 33—1).

SAFETY TIP: The precautions listed in the Safety Tips in Sections 32.1 and 32.7 for suspension system repair also apply to steering system repair operations. A driver-side airbag has a serious effect on steering wheel, gear, and column service. Be sure to read and follow the manufacturer's service procedure.

Most manual steering gears can be easily disassembled, inspected, reassembled, and adjusted in a short period of time, and the service operations involved require only a few, if any, special tools. Many technicians have successfully rebuilt manual steering gears. Power steering gears can also be rebuilt in garages or shops. Special tools are usually required to perform some of the service operations, and special equipment is needed to test the effectiveness of the repair. Removal and replacement of a power rack-and-pinion steering gear on some late-model FWD cars can be a difficult and time-consuming process. It is normally considered too difficult and time-consuming to

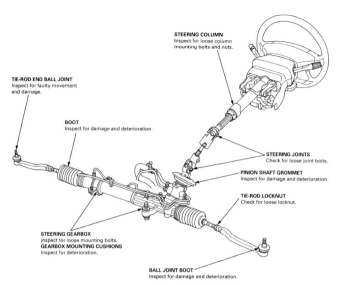

Figure 33–1 A steering system inspection includes checking each of these parts. *(Courtesy of American Honda Motor Co., Inc.)*

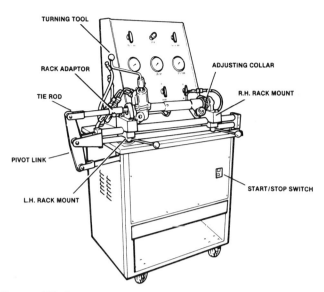

Figure 33–2 A rebuilt power rack-and-pinion gear is checked using an analyzer like the one shown; it checks for leaks as well as overall performance. *(Courtesy of Branick Industries, Inc.)*

install a unit in the car for testing with the possibility that it will have to be removed for further work. Major companies are rebuilding or remanufacturing power rack-and-pinion steering gears to new or better-than-new condition. These units are thoroughly tested to ensure proper, leak-free operation. In the case of power steering units, many technicians install a rebuilt unit to complete the job quicker with fewer potential problems (Figure 33—2).

Faulty steering linkage components are removed and replaced with new parts. Other steering linkage services are lubrication of the pivot points (if required) and adjustment of the tie-rods to correct toe. Toe and other front-end adjustments are described in Chapter 34.

The repair methods described in this chapter are presented in a general manner, so it is highly recommended that you check a service manual when doing specific service operations on most car models.

33.2 Steering Linkage Replacement

Like ball joints, most steering linkage components are connected using locking tapers. Tie-rod ends and center link connections at the Pitman arm and idler arm require taper breaking methods much like ball joint taper breaking. Hammer blows, separators or pickle forks, and pullers are the common tools for breaking loose tapers; of these, pullers are the most professional and will cause the least amount of damage to the parts.

Pickle forks should be used only on joints that are to be replaced.

After replacement of any part of the steering linkage, toe must be checked and readjusted, if necessary, to prevent tire wear or handling difficulties.

33.2.1 Removing and Replacing an Outer Tie-Rod End

Tie-rod ends can be replaced individually or together as part of a replacement tie-rod. Toe should always be checked and readjusted, if necessary, after a tie-rod end replacement. When replacing tie-rods or tie-rod ends, time can be saved by preadjusting the tie-rod length before installing the tie-rod studs. Measurements of tie-rod length are made before tie-rod disassembly, and the new parts are adjusted to the same length during replacement (Figure 33—3). It is also possible, and sometimes quicker, to count the number of turns used to remove the tie-rod end and then turn the new end the same amount during installation. When using this second method, check to ensure that the threaded portions of the old and new tie-rod ends are the same length.

To remove an outer tie-rod end, you should:

1. Raise and support the car on a hoist or jack stands.
2. Loosen the clamp bolt at the tie-rod adjuster sleeve or the jam nut at the tie-rod end.
3. Remove the cotter pin and nut from the tie-rod stud.
4. Install a tie-rod end puller and tighten it to break the taper so the stud can be removed from the steering arm (Figure 33—4).

Figure 33–3 Before replacing a tie-rod end, it is a good practice to measure the length of the tie-rod; the measurement allows you to make a quick and fairly accurate presetting of the toe adjustment. *(Courtesy of Moog Automotive Inc.)*

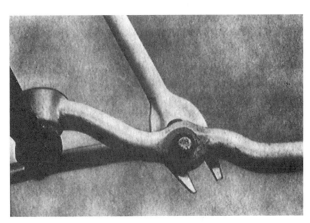

Figure 33–5 Driving the tie-rod wedge/pickle fork between the center link and tie-rod end will break loose the tie-rod stud taper. Pickle forks should be used only on joints that will be replaced. *(Courtesy of McQuay Norris)*

Figure 33–6 A sharp hammer blow will often break the tie-rod stud taper loose. Note that the pry bar is providing a separating force. *(Courtesy of SPX Corporation, Aftermarket Tool and Equipment Group)*

OR Hold the head of a hammer securely against one side of the steering arm boss while striking the other side of the boss with a sharp, quick blow from another hammer. When the taper is broken loose, lift the stud out of the boss (Figure 33—6).

NOTE: The latter method is not approved by many technicians because it can damage the steering arm.

5. Measure the distance from some point on the tie-rod end, usually the center of the stud while it is in a centered position, to some point on the tie-rod, usually the adjuster sleeve, or count the number of turns as you remove the tie-rod end from the tie-rod.

To replace a tie-rod end, you should:

1. Thread the new tie-rod end into the tie-rod sleeve or tie-rod the same number of turns used for removal or until the tie-rod length is correct. On three-piece tie-rods, each tie-rod end should have about the same amount of thread contact with the adjusting sleeve (Figure 33—7).

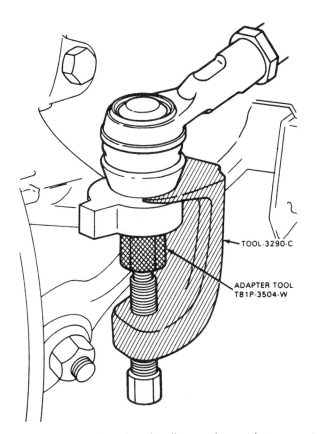

Figure 33–4 A tie-rod end puller can be used to separate the tie-rod from the steering arm or center link quickly and easily. *(Courtesy of Ford Motor Company)*

OR Drive a separator tool/pickle fork between the steering arm and the tie-rod end to break the taper and allow stud removal (Figure 33—5). Note that this tool will tear the boot and may damage the socket.

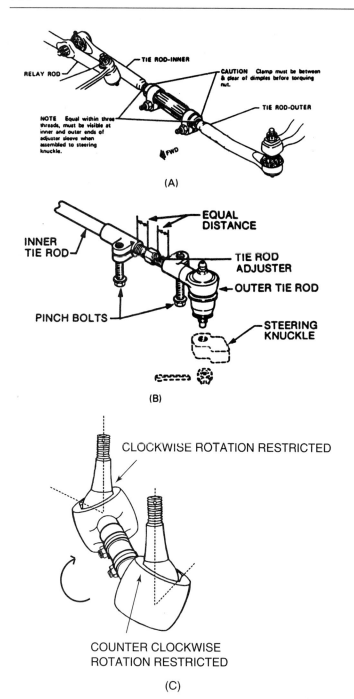

Figure 33-8 Before replacing the tie-rod stud, make sure the tapered hole is clean and not damaged. *(Courtesy of Moog Automotive Inc.)*

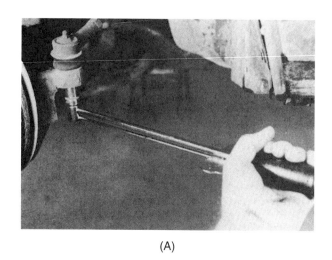

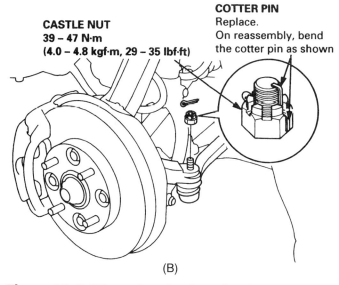

Figure 33-7 Each tie-rod end should be threaded into the adjustment sleeve an equal amount (A and B). Before tightening the clamps/pinch bolts, make sure the tie-rod end bodies, not just the studs as shown in (C), are aligned. *(A and B are courtesy of General Motors Corporation, Service Technology Group; C is courtesy of Moog Automotive Inc.)*

2. Inspect and clean the tie-rod stud hole in the steering arm (Figure 33—8).
3. Position the dust boot onto the tie-rod end, if separate, and install the tie-rod stud into the steering boss. Replace the nut on the tie-rod stud, tighten it to the correct torque, and install a new cotter pin (Figure 33—9).

Figure 33-9 When a tie-rod end is replaced, the nut should be tightened to the correct torque (A) and the cotter pin, if used, should be replaced and bent properly (B). *(B is courtesy of American Honda Motor Co., Inc.)*

Chapter 33 ■ Steering System Service

NOTE: If the tie-rod end is of the rubber bonded type, the front wheels must be in a straight-ahead position when the stud is tightened into the steering arm. A memory steer or pull will develop if the studs are installed in a wheels-turned position; this will also cause an overstretching of the bushing material when a turn is made in the opposite direction.

4. Check and adjust toe, center both the tie-rod ends to their studs, and tighten the clamp or jam nut on the tie-rod. This step is described in more detail in Chapter 34.

33.2.2 Removing and Replacing an Inner Tie-Rod End, Standard Steering Gear

Removing and replacing an inner tie-rod end on a vehicle with standard steering is nearly identical to the procedure just described. The only real difference is that the inner end is removed from the center link, whereas the outer end is removed from the steering arm when it is replaced.

If both tie-rod ends are removed from the adjuster sleeve at the same time, care should be taken to ensure that both ends are threaded into the sleeve the same number of turns and the sleeve is centered on the two ends. The concern is to ensure secure connections between the adjuster sleeve and the two tie-rod ends.

33.2.3 Removing and Replacing an Inner Tie-Rod End, Rack-and-Pinion Steering

The procedure for removing and replacing an inner tie-rod end on a vehicle with rack-and-pinion steering varies slightly depending on the method used to lock the inner tie-rod end to the steering rack gear. Currently, there are at least eight different methods used to make sure that the inner tie-rod socket does not come loose from the rack (Figure 33—10). Most tie-rods connect to the ends of the rack; one style connects to the center. In most cases, it is possible to unlock the tie-rod end socket with the steering gear mounted in the car; it is then possible to remove and

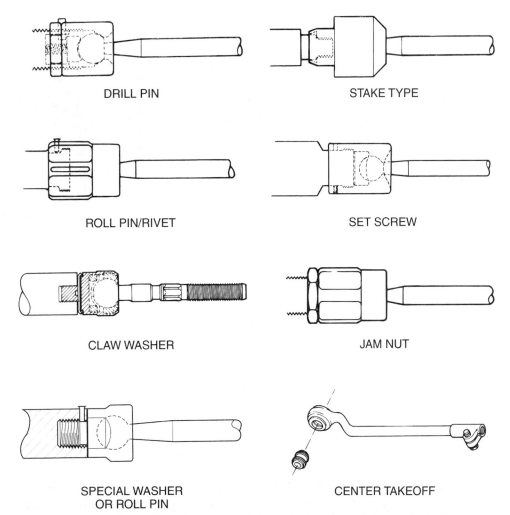

Figure 33-10 The inner tie-rod end is locked to the ends of the rack using any of the methods shown here. The lower right tie-rod is from a center takeoff rack. *(Courtesy of Moog Automotive Inc.)*

replace the inner tie-rod end without removing the steering gear. If it is not possible to remove the lock pin, it will be necessary to remove the steering gear to change the tie-rod end.

On most cars, the end-mounted, inner tie-rod socket is a complete assembly that is adjusted to the correct preload. On some cars, the tie-rod end is several parts. A preload spring and bushing is part of the socket; the tie-rod and tie-rod end housing form the rest of the socket. It is necessary to adjust the preload tension on the tie-rod end because this second style is being replaced onto the rack.

When the tie-rod end sockets are being unscrewed from the rack gear (after removal of the locking device), the turning force on the socket and rack gear can damage the teeth on the pinion or rack gear; the rack should be held from turning by placing an adjustable open-end wrench over the tooth portion of the rack. The contact area between the wrench jaws and the rack teeth is large enough to keep the rack from turning without damage.

The boot/bellows must be removed when replacing the inner tie-rod ends. It is normal for some gear oil to be in the boots of manual steering gears. On power steering units, a large amount of power steering fluid in the boots indicates faulty seals inside the steering gear. New boots and clamps should be installed during tie-rod replacement.

To remove an inner tie-rod end on a rack-and-pinion steering gear, you should:

1. Raise and support the car on a hoist or jack stands.

2. Remove the outer tie-rod end from the steering arm.

3. Remove both bellows clamps and slide the bellows down the tie-rod. Note that on power steering units, a vent/breather tube is connected into each bellows (Figure 33—11).

4. Check for the locking method used on the tie-rod socket. If it is a plain jam nut, a swaged connector, or an accessible roll pin, solid pin, or set screw, it should be possible to replace the tie-rod end while it is on the car. If a solid pin, roll pin, or set screw is in an inaccessible location, the steering gear should be removed. Removal of a rack-and-pinion steering gear is described later in this chapter. If removal of the steering gear is necessary, it should be mounted securely in a holding fixture or vise during the repair steps (Figure 33—12).

5. If a plain jam nut is used, place one wrench on the tie-rod socket and a second wrench on the jam nut and loosen the jam nut. Next, place an adjustable wrench over the rack teeth, adjust the jaws to fit tightly onto the gear teeth, and unscrew the tie-rod socket with the tie-rod from the rack.

Figure 33–11 Boot retainers are usually cut using side-cutting pliers or dikes; the boot can then be slid or cut off. *(Courtesy of General Motors Corporation, Service Technology Group)*

Figure 33–12 When working on a steering gear out of the car, it should be secured in a manner that will not damage the gear or housing; this gear is held in a special rack-and-pinion gear and strut vise. *(Courtesy of Moog Automotive Inc.)*

OR If a swaged end is used, place an adjustable wrench over the teeth on the rack, tighten the wrench jaws onto the rack, and, using a second wrench, unscrew the tie-rod socket from the rack. The swaged portion of the socket will bend outward as it is unscrewed (Figure 33—13).

OR If a roll pin is used, install a special roll pin puller into the roll pin and remove it from the tie-rod socket. A roll pin can also be removed by using a screw extractor/easy-out tool to rotate the roll pin until it works its way out of the hole. After the roll pin is removed, hold the rack gear from turning with one wrench and unscrew the tie-rod socket with a second wrench (Figure 33—14).

Chapter 33 ■ Steering System Service   717

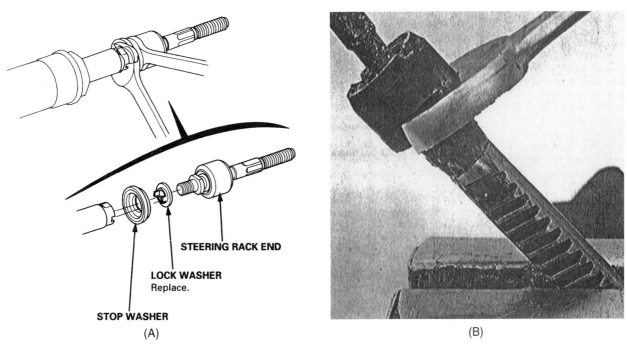

Figure 33-13 When the tie-rod end is unscrewed from the rack, the rack should be held from turning using a wrench (A) or a vise with soft jaws (B). *(A is courtesy of American Honda Motor Co., Inc.; B is courtesy of General Motors Corporation, Service Technology Group)*

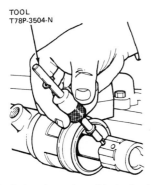

Figure 33-14 If the tie-rod end is locked by a roll pin, the roll pin can be removed using a special puller tool. It threads into the roll pin, and the slide hammer is used for the outward force. *(Courtesy of Ford Motor Company)*

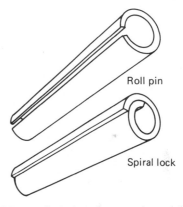

Figure 33-15 A roll pin is soft enough to drill out if necessary; a spiral pin is too hard to drill out.

NOTE: There are two styles of roll pins. A normal roll pin is a tubular-shaped strip of metal that is rolled into almost one revolution; it is soft enough to be drilled out if necessary. A spiral lock pin is also tube shaped, but it has several revolutions of hardened steel. A spiral lock pin cannot be drilled out; it is too hard (Figure 33—15).

OR If a solid pin is used, install a drill guide so it is centered over the roll pin, set the drill depth limiter so the drill depth will just touch the rack, and drill the old lock pin out. Because the tie-rod socket and jam nut are going to be replaced, it is possible to simply use an oversize drill bit to drill out the pin; be very careful not to drill any farther into the rack than absolutely necessary. After the pin is removed, place one wrench on the tie-rod socket and one wrench on the jam nut and loosen the jam nut. Next, grip the rack teeth with an adjustable wrench and unscrew the tie-rod socket with a second wrench (Figure 33—16).

OR If a set screw is used, remove the set screw with an Allen wrench, grip the rack teeth with an adjustable wrench, install a special wrench on the tie-rod socket, and unscrew the tie-rod socket from the rack (Figure 33—17).

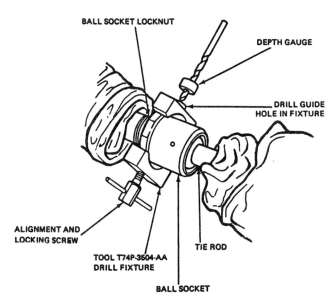

Figure 33–16 A solid locking pin is removed using a drill; this fixture guides the drill bit. *(Courtesy of Ford Motor Company)*

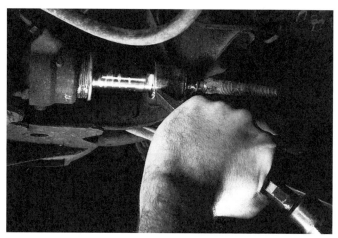

Figure 33–18 A small chisel is being used to bend the washer away from tie-rod end flats. *(Courtesy of Moog Automotive Inc.)*

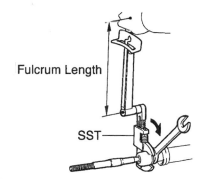

Figure 33–19 A new tie-rod end should be tightened to the correct torque. The length of a special service tool (SST) affects the amount of torque.

Figure 33–17 A locking set screw is simply removed. *(Courtesy of General Motors Corporation, Service Technology Group)*

NOTE: I have tried to describe each style of locking device used on domestic cars; however, some import cars use other methods to secure the tie-rod socket onto the end of the rack. When in doubt, check the service manual for that particular car.

To replace an inner tie-rod end on a rack-and-pinion steering gear, you should:

1. Thread the new tie-rod end socket with tie-rod onto the rack and, while holding the rack from turning with an adjustable wrench, tighten the tie-rod socket to the correct torque (Figure 33—19).

OR In a few cases, the replacement tie-rod end will require adjustment. A spring scale is used to measure the force required to swing/articulate the tie-rod in the socket; tightening the tie-rod end will increase the tightness of the joint and the amount of force needed to swing the tie-rod. The tie-rod socket should be tightened enough to obtain the correct articulation force (Figure 33—20).

OR If a tabbed washer is used, use a hammer and small chisel to bend the washer out of the locking groove (Figure 33—18).

SAFETY TIP: The rack should be checked for other lock pin drillings; each end of the rack can have a maximum of two lock pin holes. The steering gear should be replaced if one of the inner tie-rod ends has already been replaced.

Chapter 33 ■ Steering System Service

Figure 33-20 If the tie-rod end comes in several pieces (see Figure 33-24A), it must be adjusted to the correct amount of socket preload while being tightened onto the rack; socket preload is measured using a spring scale. *(Courtesy of Ford Motor Company)*

2. Lock the tie-rod end in place. If a jam nut is used, use two wrenches and tighten the jam nut against the tie-rod socket.

OR Stake or swage the new inner tie-rod socket onto the flats at the ends of the rack; a special tool is available to crimp the socket if the steering gear is on the car (Figure 33—21). If the unit is off the car, rest the tie-rod socket over a vise and strike the socket using a hammer and punch to crimp the socket onto the flats. Check for proper swaging by trying to slide a 0.010-in. (0.25-mm) feeler gauge under the staked portion; if it does not enter, the stakes/crimps are adequate (Figure 33—22).

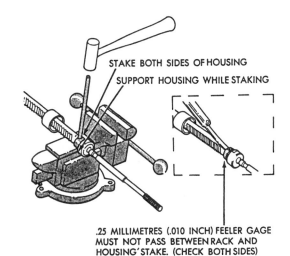

Figure 33-22 A tie-rod end can be swaged/staked by solidly supporting it and striking the socket using a punch and hammer. The swaged portion should make tight contact, as shown. *(Courtesy of Chrysler Corporation)*

OR Drive a new roll pin into the tie-rod socket (Figure 33—23).

OR Drill a new hole into the junction of the socket and the tight jam nut; the hole should be the correct depth so the pin will just enter into the rack. A drill guide is available to ensure a properly drilled hole. Drive the new lock pin into the hole and punch mark around the pin to lock it in place (Figure 33—24).

Figure 33-21 The tool at the top allows removal of a swaged inner tie-rod end with the steering gear mounted in the car; the bottom tool is used to crimp/swage the new tie-rod end as it is replaced. *(Courtesy of Moog Automotive Inc.)*

Figure 33-23 After the tie-rod end has been tightened to the correct torque, a spiral pin is driven into the hole in the socket. *(Courtesy of Ford Motor Company)*

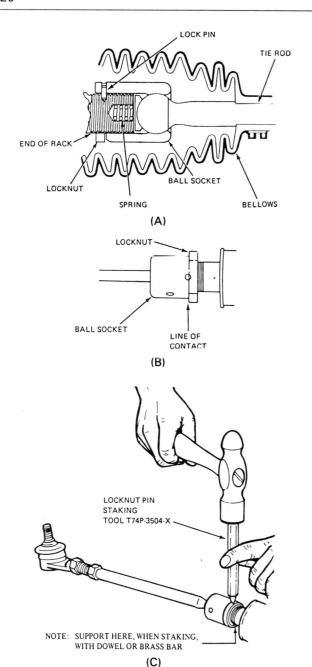

Figure 33-24 If the tie-rod socket is locked by a solid pin, a hole must be drilled to the correct depth in the rack (A); a new lock pin is driven into the hole (B) and then staked in place using tool T74P-3504-X or a center punch (C). *(Courtesy of Ford Motor Company)*

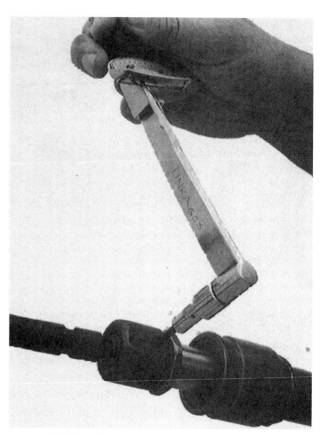

Figure 33-25 If the tie-rod socket is locked with a set screw, it should be tightened to the correct torque. *(Courtesy of General Motors Corporation, Service Technology Group)*

OR Tighten the set screw to the correct torque (Figure 33—25).

OR Bend the washer into its locking groove (Figure 33—26).

3. Slide the new bellows/boot into position and tighten the large clamp. The small clamp will be tightened after the toe check. A special crimping tool is often required to secure the clamps (Figure 33—27).

4. Replace the outer tie-rod end and check and adjust toe, if necessary. After adjusting toe, tighten the small clamp on the bellows.

33.2.4 Removing and Replacing Inner Tie-Rod End Pivot Bushings, Rack-and-Pinion Steering Gear with Center-Mounted Tie-Rods

A center-mounted tie-rod uses a bushing instead of a ball-and-socket joint; it is currently used on some late-model FWD cars. The tie-rods are secured to the steering gear by a pair of bolts that pass through a slot in the gear housing and are threaded into the rack gear (Figure 33—28).

If both tie-rods are removed, care should be taken to hold the boot and other parts in alignment. It is recommended that you remove the first tie-rod and replace the bolt before removing the second bolt and tie-rod (Figure 33—29).

To remove and replace an inner tie-rod bushing, you should:

1. Raise and support the car on a hoist or jack stands.
2. Remove the outer tie-rod end from the steering arm.

3. Bend the tabs of the lock plate away from the bolt head and remove the inner tie-rod bolt. Remove the tie-rod.
4. Using the correct tool, remove the old bushing from the tie-rod end (Figure 33—30).

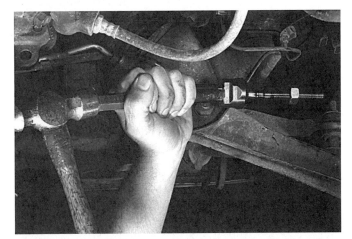

Figure 33–26 If the tie-rod end is locked with a claw washer, bend the edges of the washer over the flats of the socket. *(Courtesy of Moog Automotive Inc.)*

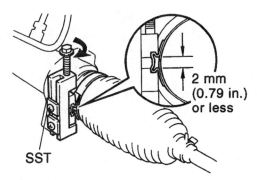

Figure 33–27 With the tie-rod end tightened and locked in place, slide the new boot into position and crimp the retainer as required to lock it in place.

5. Coat the new bushing with a light film of grease and, using the special tool, press the bushing into the tie-rod end.
6. Make sure the center housing cover washers are fitted into the rack-and-pinion boot, place the tie-rod end in position, and install the inner tie-rod bolt.
7. Tighten the tie-rod bolt to the correct torque and bend the flats of the lock plate against the bolt head to secure the bolt in position.
8. Replace the outer tie-rod end.

33.2.5 Removing and Replacing an Idler Arm

Several styles of idler arms and brackets are used with parallelogram steering. The most common is an assembly in which the idler arm is integral with the mounting bracket. A worn idler arm is usually replaced as an idler arm and bracket assembly; in a few cases, only the idler arm or idler arm bushing is replaced.

When rubber idler arm bushings are used, the idler arm should be tightened onto the bracket with the tires in a straight-ahead position. As with rubber tie-rod ends, tightening the bushings in a turned position can cause a steering pull in the direction that the arm was when it was tightened and early failure of the bushing.

To remove and replace an idler arm, you should:

1. Raise and support the car on a hoist or jack stands.

SAFETY TIP: Some car models have changed the length of their idler arms during model changeover, and in some cases, the two different arms are almost interchangeable. If an idler arm is too short, it can go over center and lock the steering at the extreme position of a turn. *Always check the new idler arm against the old one to ensure an exact replacement.*

2. Remove the cotter pin and nut from the idler arm or center link stud. Install a puller to break the stud

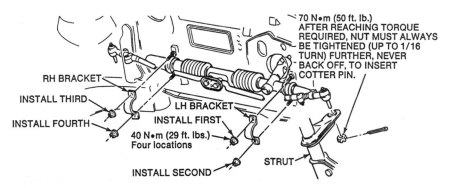

Figure 33–28 A center takeoff rack, tie-rods, and steering arms. *(Courtesy of General Motors Corporation, Service Technology Group)*

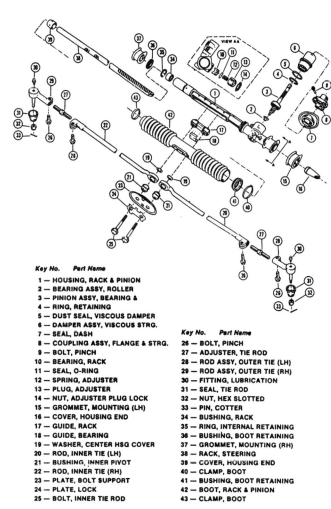

Key No. **Part Name**
1 — HOUSING, RACK & PINION
2 — BEARING ASSY, ROLLER
3 — PINION ASSY, BEARING &
4 — RING, RETAINING
5 — DUST SEAL, VISCOUS DAMPER
6 — DAMPER ASSY, VISCOUS STRG.
7 — SEAL, DASH
8 — COUPLING ASSY, FLANGE & STRG.
9 — BOLT, PINCH
10 — BEARING, RACK
11 — SEAL, O-RING
12 — SPRING, ADJUSTER
13 — PLUG, ADJUSTER
14 — NUT, ADJUSTER PLUG LOCK
15 — GROMMET, MOUNTING (LH)
16 — COVER, HOUSING END
17 — GUIDE, RACK
18 — GUIDE, BEARING
19 — WASHER, CENTER HSG COVER
20 — ROD, INNER TIE (LH)
21 — BUSHING, INNER PIVOT
22 — ROD, INNER TIE (RH)
23 — PLATE, BOLT SUPPORT
24 — PLATE, LOCK
25 — BOLT, INNER TIE ROD

Key No. **Part Name**
26 — BOLT, PINCH
27 — ADJUSTER, TIE ROD
28 — ROD ASSY, OUTER TIE (LH)
29 — ROD ASSY, OUTER TIE (RH)
30 — FITTING, LUBRICATION
31 — SEAL, TIE ROD
32 — NUT, HEX SLOTTED
33 — PIN, COTTER
34 — BUSHING, RACK
35 — RING, INTERNAL RETAINING
36 — BUSHING, BOOT RETAINING
37 — GROMMET, MOUNTING (RH)
38 — RACK, STEERING
39 — COVER, HOUSING END
40 — CLAMP, BOOT
41 — BUSHING, BOOT RETAINING
42 — BOOT, RACK & PINION
43 — CLAMP, BOOT

Figure 33–29 An exploded view of a center takeoff unit. Note that the guide (17) can slip out of position if both inner tie-rod bolts (25) are removed at the same time. *(Courtesy of General Motors Corporation, Service Technology Group)*

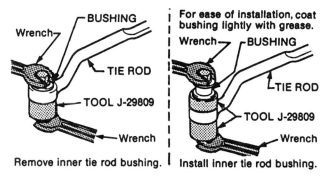

Figure 33–30 The rubber bushing can be removed and replaced from the inner end of a center takeoff tie-rod using the method shown here. *(Courtesy of General Motors Corporation, Service Technology Group)*

taper and separate the idler arm from the center link (Figure 33—31).

3. Remove the bolts securing the idler arm bracket to the frame and remove the idler arm.

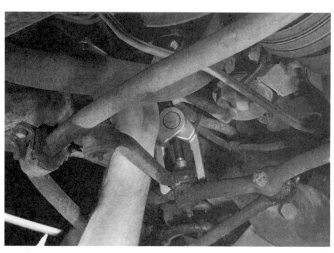

Figure 33–31 Removing an idler arm begins by removing the nut and breaking the stud taper at the end. *(Courtesy of Moog Automotive Inc.)*

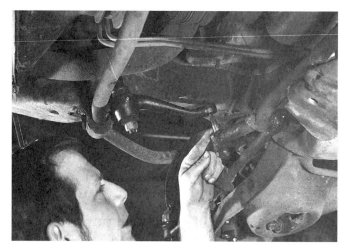

Figure 33–32 Be sure the tapers in the hole and stud are clean and undamaged before connecting them. *(Courtesy of Moog Automotive Inc.)*

4. Clean and inspect the frame mounting location and the stud taper and mounting hole to make sure they are in good condition (Figure 33—32).

5. Place the new idler arm bracket in position, install the mounting bolts, and tighten the bolts and nuts to the correct torque.

NOTE: Some idler arms have adjustable slots at the mounting bolts. On these cars, steering linkage parallelism should be adjusted. With the steering exactly centered, measure the distance from the center of the inner tie-rod stud to the center of the upper control arm mounting/pivot bolt on each side of the car. Slide the idler arm up or down so these two measurements are the same or within 0.060 in. (1.5 mm) of each other (Figure 33—33).

Chapter 33 ■ Steering System Service

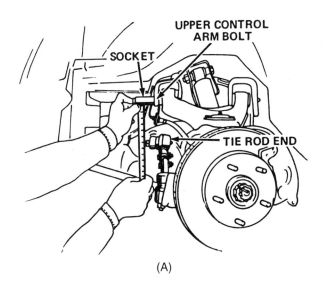

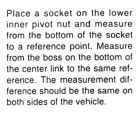

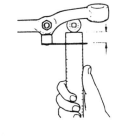

Place a socket on the lower inner pivot nut and measure from the bottom of the socket to a reference point. Measure from the boss on the bottom of the center link to the same reference. The measurement difference should be the same on both sides of the vehicle.

(B)

Figure 33–33 Idler arms with slotted holes or threaded bushings can alter steering parallelism, and improper installation can affect bump steer. Steering parallelism can be checked by comparing the side-to-side motion of the center link (A and B). *(A is courtesy of General Motors Corporation, Service Technology Group; B is courtesy of Moog Automotive Inc.)*

NOTE: Some idler arms use a threaded-style bushing; the idler arm can be unthreaded from the bracket. When installing this style of idler arm, the arm should be threaded as far as possible onto the bracket, but with enough room to rotate for normal operation. Thread the idler arm completely onto the bracket until it stops and then back it off one-half to one turn. Some idler arm assemblies have a setting dimension for this purpose (Figure 33—34).

NOTE: A simple check for steering linkage parallelism is to perform a suspension bounce test as described in Section 24.2.1 with the steering wheel unlocked. The steering wheel should not rotate when the suspension is bounced straight up and down.

6. Connect the idler arm to the center link, install the nut to the correct torque, and install the cotter pin (Figure 33—35).

NOTE: With the steering gear centered, the idler arm should point either straight forward or straight backward.

7. Check toe and adjust, if necessary.

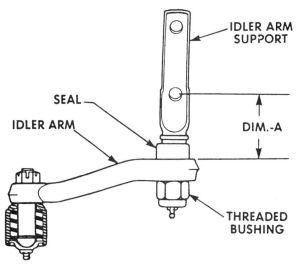

DIVISION	BODY	DIMENSION A MILLIMETRES	INCHES
Buick	A-B-C-E	59.5 ± 1.6	2 11/32 ± 1/16
Olds	A-B-C	59.5 ± 1.6	2 11/32 ± 1/16
Pontiac	F-X	75.4 ± 1.6	2 31/32 ± 1/16
	A-B	59.5 ± 1.6	2 11/32 ± 1/16
Cadillac	C-D	59.5 ± 1.6	2 11/32 ± 1/16
	K	75.4 ± 1.6	2 31/32 ± 1/16

Figure 33–34 When installing an idler arm with a threaded bushing, dimension A should be adjusted to the correct specification. *(Courtesy of General Motors Corporation, Service Technology Group)*

33.2.6 Removing and Replacing a Center Link

Two styles of center links are commonly used with parallelogram steering linkage: those with wearing sockets and those without. Center links without wearing sockets seldom require any service; the wear points are in the idler arm and Pitman arm. If the sockets are in the center link and they wear out, the center link will need to be replaced.

To remove and replace a center link, you should:

1. Raise and support the car on a hoist or jack stands.
2. Remove the cotter pins and retaining nuts from the studs of the inner tie-rod ends and from the idler arm and Pitman arm to the center link joints.
3. Using a puller, break the taper of the four studs and remove the center link.
4. Clean and inspect the four studs and holes.
5. Place the center link in position, install the idler arm and Pitman arm studs, and install the two nuts.

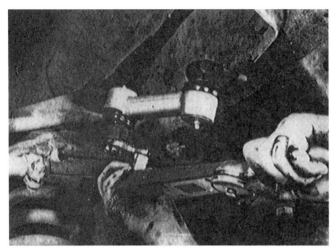

Figure 33–35 The bolts securing the idler arm to the frame should be tightened to the correct torque. *(Courtesy of McQuay Norris)*

6. Place the inner tie-rod end studs in place, install the two nuts, tighten all four nuts to the correct torque, and install new cotter pins.
7. Check toe and adjust, if necessary.

33.2.7 Removing and Replacing a Pitman Arm

The Pitman arm used with nonwearing center links has a socket in it that can wear. Pitman arms are secured to the Pitman shaft with spined and tapered connections; a puller is required to separate them. One or more **missing** or **master splines** are usually used at this connection to ensure that the arm is replaced onto the Pitman shaft in the proper position. It is still a good practice to mark the Pitman-arm-to-shaft position to ensure correct mounting (Figure 33—36).

To remove and replace a Pitman arm, you should:

1. Raise and support the car on a hoist or jack stands.
2. Remove the cotter pin and nut from the Pitman arm or center link stud. Break the taper on this stud, preferably with a puller, and separate the Pitman arm from the center link.
3. Mark the Pitman-arm-to-shaft position and loosen the retaining nut several turns or until it is even with the end of the threads.

NOTE: The nut should be left on the Pitman shaft to help protect the threads and to prevent the Pitman arm from popping completely off.

4. Install a puller and tighten it to break loose the taper between the Pitman arm and shaft. Remove the puller, the retaining nut, and the Pitman arm (Figure 33—37).
5. Clean and inspect the splines on the Pitman shaft, the splined hole in the Pitman arm, and the hole and stud at the center link connection.
6. Connect the Pitman arm to the center link and place the front tires in a straight-ahead position.
7. Align the marks between the Pitman arm and shaft or turn the steering wheel to the center-most position of the steering gear with the steering wheel centered and slide the Pitman arm onto the Pitman shaft. At this point the Pitman arm should be positioned either straight to the front or straight to the rear.
8. Install the retaining nut and lock washer on the Pitman shaft and tighten the nut to the correct torque.
9. Tighten the Pitman arm or center link stud retaining nut to the correct torque and install a new cotter pin.

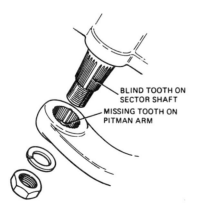

Figure 33–36 The blind tooth/master spline ensures that the Pitman arm is replaced in the correct position. *(Courtesy of Ford Motor Company)*

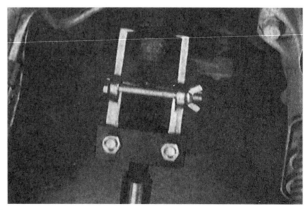

Figure 33–37 A Pitman arm puller is being used to remove a Pitman arm. Some technicians prefer to leave the loosened nut on the threads to help protect the end of the Pitman shaft. *(Courtesy of McQuay Norris)*

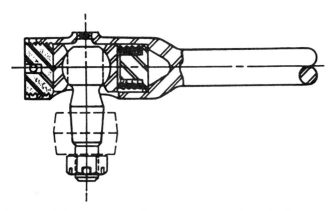

Figure 33–38 Unscrewing the screw on the left allows the drag link to be lifted off the ball stud. *(Courtesy of Federal-Mogul Corporation)*

33.2.8 Removing and Replacing an Adjustable Drag Link Socket

Drag links are used on some pickups, vans, 4WDs, and trucks to connect the Pitman arm to the third arm on the left steering knuckle. Most drag link sockets resemble a tie-rod end; some are disassembled for removal and need to be adjusted after reassembly (Figure 33—38).

To disconnect an adjustable drag link socket, you should:

1. Raise and support the vehicle on a hoist or jack stands and remove the left tire and wheel.
2. Remove the cotter pin from the drag link and unscrew the adjuster plug enough to allow the drag link socket to be lifted off the ball stud.

To connect a drag link socket, you should:

1. Clean and inspect the ball stud to ensure that it is not worn out-of-round.
2. Place the drag link socket over the ball stud and turn the adjuster plug to tighten the socket against the ball stud.
3. Turn the steering to the right and left turn stops to check for binding; loosen the adjuster screw, if necessary.
4. Install a new cotter pin and lubricate the drag link socket.

33.3 Steering Gear Service

Most standard manual steering gears are relatively easy to service. Normal service operations are worm shaft bearing preload and sector gear lash adjustments and steering gear rebuilding. Rebuilding steps include disassembly, cleaning and inspection, and reassembly. Service procedures for rack-and-pinion steering gears are similar to those for standard units; however, the adjustment methods on different makes of gear units are not as standardized.

NOTE: Remember that anytime the steering wheel is removed or a portion of the steering column is disconnected, care must be taken to ensure that the airbag clock spring remains properly indexed. This procedure is described in Section 33.4.

Service of standard power steering gears follows the same general procedures as that for manual gears, with these additions: a control valve with seals, outer and (sometimes) inner piston seals, and high-pressure seals at the shafts and at any other openings in the gear housing. Additional care must be taken to ensure that the gear housing and internal pressure areas are completely sealed during assembly to prevent leaks. Seal kits for most popular steering gears are available from the vehicle manufacturer and aftermarket sources.

The **sector gear lash** and sometimes **worm shaft bearing preload** adjustments can usually be made in the car on most standard steering gears. When making such adjustments, one end of the Pitman arm must be disconnected to allow a better feel of the steering gear. Gear adjustments are measured by the small amount of turning effort it takes to rotate the worm shaft. Some technicians prefer to remove, disassemble, and inspect the internal parts of a steering gear before adjustment. They believe that if the gears and bearings have worn enough to require an adjustment, they might be worn far enough to cause a failure. They also point out that in the case of many manual steering gears, after the Pitman arm has been disconnected, only the shaft coupler and a few bolts hold the steering gear in the car; steering gear removal is fairly quick from this point. If the steering gear adjustments are made with the unit in the car, care should be taken to feel for any suspicious rough or erratic turning motions that might indicate faulty bearings or worn gears.

In a few cases, rack-and-pinion steering gears can be adjusted in the car; in many cases, they must be removed. Adjustment procedures for rack-and-pinion steering gears vary too much to be described in a text of this type; the manufacturer or technician service manual should be consulted for the exact procedure.

33.3.1 Steering Gear Removal and Replacement, Standard Steering Gear

The removal and replacement for power steering gears is similar to that used for a manual gear with the addition of the two fluid lines. When removing and replacing a

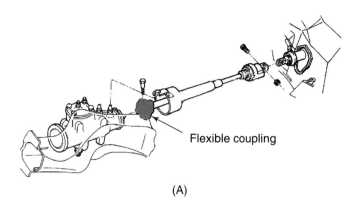

(A)

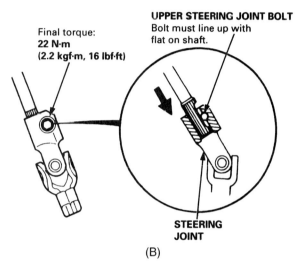

(B)

Figure 33-39 Many steering couplers (A) and U-joints (B) use a flat or master spline so they can be replaced in only one position; you should place index marks on the others to ensure correct alignment. *(A is courtesy of General Motors Corporation, Service Technology Group; B is courtesy of American Honda Motor Co., Inc.)*

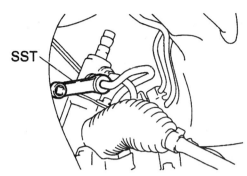

Figure 33-40 When removing a power steering gear, the fluid lines are disconnected and capped to reduce fluid loss and contamination.

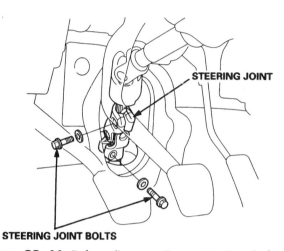

Figure 33-41 Before disconnecting a steering shaft coupling, check for a flat or master spline; if there is none, place index marks on the joint and shaft to ensure correct alignment. Note that on some FWD vehicles, the steering shaft is disconnected from inside the vehicle. *(Courtesy of American Honda Motor Co., Inc.)*

power steering gear, check the condition of the fluid and hoses. Dirty, contaminated fluid should be changed; faulty hoses or ones with internal deterioration should also be changed. As a power steering unit is replaced, the fluid level should be adjusted, any air should be bled out, and the fluid level should be rechecked. Filling and bleeding procedures are described later in this chapter.

The steering coupler or universal joint should be marked before disassembly to ensure replacement in the same position; steering wheel position will be changed if the coupler is replaced improperly. Some coupler and steering shaft splines have a flat or missing master spline to ensure proper alignment. The condition of the coupler should always be checked; damaged couplers must be replaced or rebuilt (Figure 33—39).

To remove a standard steering gear, you should:

1. If the car is equipped with power steering, disconnect and cap the two fluid lines to reduce fluid loss and contamination (Figure 33—40).

2. Mark and disconnect the steering shaft coupler. Usually the coupler cannot be slid off the shaft until the gear is removed from the car (Figure 33—41).

3. Raise and support the vehicle on a hoist or jack stands.

4. Mark the Pitman-arm-to-shaft position and remove the Pitman arm from the shaft.

5. Remove the bolts and nuts securing the gear housing to the frame and remove the steering gear from the car. The steering shaft coupler should be slid off the shaft as the steering gear is removed (Figure 33—42).

To replace a standard steering gear, you should:

1. Place the gear housing in the proper position in the frame and replace the mounting bolts. It is often necessary to slide the coupler onto the steering shaft at this time. Tighten the mounting bolts to the correct torque.

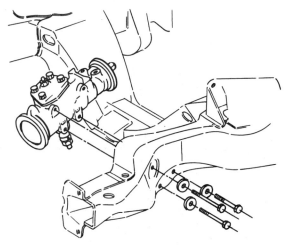

Figure 33–42 This standard steering gear is secured to the frame using three bolts. *(Courtesy of General Motors Corporation, Service Technology Group)*

2. Align the Pitman arm to the shaft and replace the Pitman arm. Tighten the retaining nut to the correct torque.

3. Align the coupler to the steering shaft and slide it onto the steering shaft. Tighten the pinch bolt to the correct torque. Note that aligning and installing the coupler onto the steering shaft are often done during step 1.

4. On power steering units, reconnect the two fluid lines, fill the system with fluid, and bleed the air from the system. After road testing, check the system for possible leaks and correct the fluid level, if necessary.

33.3.2 Steering Gear Removal and Replacement, Rack-and-Pinion Steering Gear

Removal and replacement of a rack-and-pinion steering gear follows a procedure similar to that used with a standard gear. Access to some rack-and-pinion units is very difficult, especially on some FWD cars with the rack mounted on the engine compartment bulkhead. You should follow the specific procedure outlined in the manufacturer or technician service manual. The description given here is very general.

To remove a rack-and-pinion steering gear, you should:

1. Raise and support the car on a hoist or jack stands and remove the front tires and wheels. On some cars, it is easier to do steps 2 and 3 from under the hood before raising the car.

2. If the car is equipped with power steering, disconnect the two fluid lines from the power steering pump and cap the lines to reduce fluid loss (Figure 33—43).

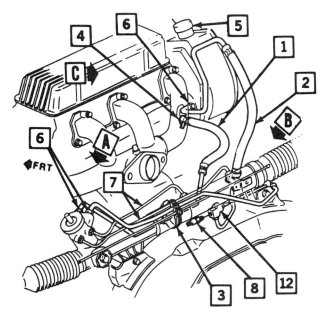

Figure 33–43 When removing a power rack-and-pinion steering gear, the pressure (1) and return (2) hoses should be disconnected and plugged to reduce fluid loss and contamination. *(Courtesy of Moog Automotive Inc.)*

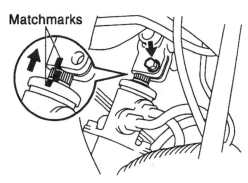

Figure 33–44 Index or match marks are placed on the steering shaft and coupling to ensure correct positioning during replacement.

3. Mark and disconnect the steering shaft coupler; on some cars, it is easier to do so after step 5 (Figure 33—44).

4. Disconnect both outer tie-rod ends (Figure 33—45).

5. Remove the bolts and nuts securing the gear housing to the frame and remove the steering gear from the car (Figure 33—46).

To replace a rack-and-pinion steering gear, you should:

1. Check the mounting bushings in the gear housing and, if necessary, replace any faulty, deteriorated bushings (Figure 33—47).

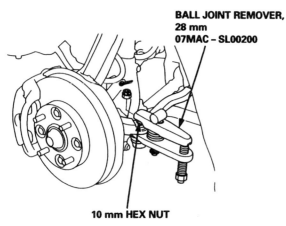

Figure 33–45 The tie-rods are normally disconnected from the steering arms during removal of a rack-and-pinion steering gear. *(Courtesy of American Honda Motor Co., Inc.)*

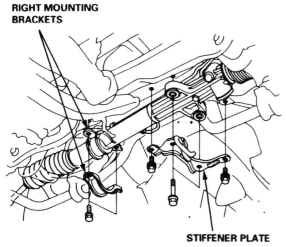

Figure 33–46 With the hoses, steering shaft, and tie-rods disconnected, remove the mounting bolts, brackets, and plates to remove the rack. Be sure to check the condition of the mount bushings. *(Courtesy of American Honda Motor Co., Inc.)*

2. Slide the steering gear into position, install the mounting bolts and nuts, and tighten them to the correct torque. On many cars, it is necessary to do steps 2 and 3 at the same time.
3. Align the coupler and the steering shaft and replace the coupler. Tighten the pinch bolt to the correct torque.
4. Reconnect the outer tie-rod ends.
5. On power steering units, reconnect the two fluid lines, fill the system with fluid, and bleed the air from the system. After road testing, check the system for any possible fluid leaks and correct the fluid level, if necessary.
6. Check and adjust toe, if necessary.

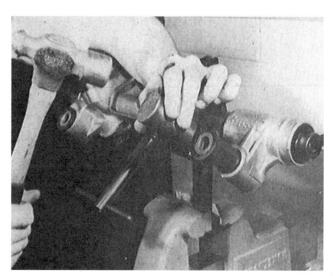

Figure 33–47 Faulty rack mount bushings are removed and replaced with new ones. *(Courtesy of Moog Automotive Inc.)*

33.3.3 Steering Gear Adjustments, Standard Steering Gear

There are normally three adjustments to be made on standard steering gears. Two of these (worm shaft bearing preload and sector gear lash) are made from the outside of the gear housing, and one (**sector gear or shaft end play**) can only be checked with the sector gear removed from the unit (Figure 33—48).

The worm shaft bearing preload adjustment provides for easy, play-free turning of the steering shaft; it is also called **worm bearing preload.** The bearings are adjusted so there is a light pressure or preload on them at all times; this pressure is just enough to eliminate bearing and worm shaft end play. Steering shaft preload can be measured with a torque wrench and socket on the worm shaft or on the steering wheel retaining nut; it can also be measured with a spring scale, measuring the pull at the rim of the steering wheel (Figure 33—49 and Figure 33—50).

The sector gear lash adjustment is also called the **over center adjustment.** This adjustment ensures that there is no lash between the worm gear (worm and sector gear set) or ball nut (recirculating ball nut gear set) and the sector gear when the steering is straight ahead. There is actually a preload, because gear lash at this time would cause sloppy steering and wander. As the steering is turned from the center position, the teeth of the sector gear swing away from the worm gear or ball nut, and lash usually occurs between the two gears. Gear lash during a turn is not critical, because the wheel alignment angles are trying to push the tires back to a straight-ahead position; this self-aligning force from the tires keeps pressure on the gears to eliminate lash. Sector gear lash, actually preload at center position, is also measured with a torque wrench (Figure 33—51).

Chapter 33 ■ Steering System Service

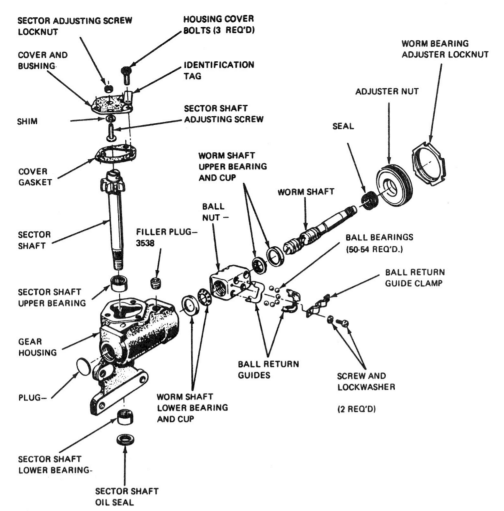

Figure 33-48 An exploded view of a recirculating ball nut steering gear; these parts are normally disassembled during a steering gear overhaul. *(Courtesy of Ford Motor Company)*

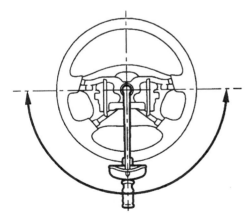

Figure 33-49 Steering gear preload/turning effort can be measured using a torque wrench attached to the steering wheel retaining nut. With power steering vehicles, the engine should be running.

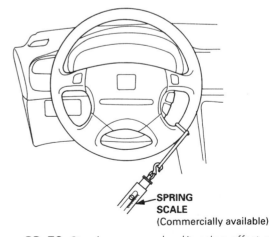

Figure 33-50 Steering gear preload/turning effort can be measured by attaching a spring scale to the rim of the steering wheel; pulling on the scale will show how much "rim pull" it takes to turn the steering shaft. *(Courtesy of American Honda Motor Co., Inc.)*

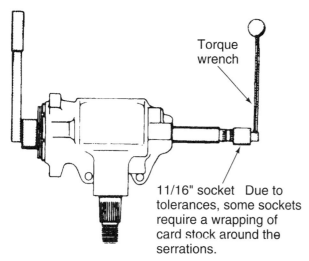

Figure 33–51 A torque wrench and socket can be attached directly to the steering shaft to measure preload. Note: Removing the sector gear allows a worm shaft adjustment with no interference. *(Courtesy of General Motors Corporation, Service Technology Group)*

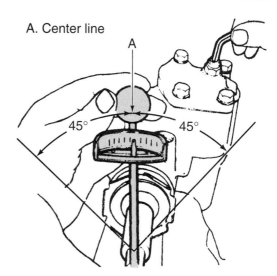

Figure 33–52 When adjusting sector gear preload/gear lash, the steering shaft preload must be measured in a 90-degree portion of the center of the shaft travel; turning the adjuster screw (with an Allen wrench) inward will increase the preload. *(Courtesy of General Motors Corporation, Service Technology Group)*

Sector gear end play is important because the sector gear tends to drop downward into the worm gear or ball nut as the steering is turned off center. If it drops too far, the gears might bind and wedge when the steering is turned back to center.

These adjustments are normally made in the following order when rebuilding a steering gear: (1) sector gear end play as the sector gear is removed or installed, (2) worm shaft bearing preload as the worm shaft and bearings are being installed into the gear housing, and (3) sector gear lash after the gear cover has been replaced (Figure 33—52). If these adjustments are made with the gear in the car, the Pitman arm must be disconnected. The turning feel of the gear set should also be checked carefully to ensure that the internal gears and bearings are in good shape. Check a service manual for the adjustment procedure and the specifications for a particular car.

Measuring Sector Shaft End Play.
To measure sector gear end play, you should:

1. Remove the sector gear/shaft and adjuster screw from the cover by turning the sector gear adjusting screw all the way inward.
2. Try to slide a 0.002-in. (0.05-mm) feeler gauge between the head of the adjuster screw and the bottom of the slot; it should not enter (Figure 33—53). If it does not enter and the screw rotates freely, no change is required. If the clearance is greater than 0.002 in., a thicker shim is required. A thinner shim is required if the screw does not rotate freely.

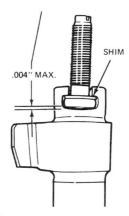

Figure 33–53 If the clearance between the sector gear and the adjusting screw is excessive, the gear might drop too far and cause binding. *(Courtesy of Ford Motor Company)*

Adjusting Worm Shaft Bearing Preload.
To measure and adjust worm shaft bearing preload, you should:

1. Select the smallest twelve-point socket that will slide over the splines of the worm shaft, and with a low-reading inch-pound torque wrench, measure the torque required to turn the worm shaft. Note that if the steering gear is assembled, the sector gear lash adjuster should be turned completely out or up to eliminate any possible drag between the two gears. If an excessive amount of slippage occurs between the socket and the shaft splines, a tighter fit can be obtained by wrapping a strip of cardboard around the shaft and forcing the socket over the shaft and cardboard.

2. Obtain the worm shaft bearing preload specification and compare it with your readings. If specifications are not available, a rule of thumb is 5 in.-lb (0.5 N-m) of torque.

NOTE: As a general rule, the average technician can turn a worm shaft with a thumb and forefinger and notice a definite turning resistance.

3. If an adjustment is required, loosen the lock nut for the adjuster plug and rotate the adjuster plug inward to increase preload or outward to reduce it. When the effort to turn the worm shaft matches the specifications, tighten the adjuster lock nut and double-check the preload to ensure that it has not changed.

Adjusting Sector Gear Lash. To measure and adjust sector gear lash, you should:

1. Rotate the steering shaft one direction until it stops. Next, rotate it the opposite way, counting the number of turns it takes to reach the other stop; divide this distance by two to determine the center of travel for the steering gear.
2. Turn the shaft one turn from either end stop and, using a low-reading inch-pound torque wrench, measure the steering shaft preload. Record this measurement.
3. Turn the steering shaft to the center-most position and remeasure the steering shaft preload. The difference between these two measurements is the sector gear lash measurement.
4. Obtain the sector gear lash specifications and compare them with your measurements. If specifications are not available, a rule of thumb is 10 in.-lb (1 N-m) total or 5 in.-lb (0.5 N-m) greater than the amount of steering shaft bearing preload.

NOTE: As a general rule, the average technician must use both thumbs and forefingers to exert 10 in.-lb of torque and rotate the steering shaft as it passes through the center of travel.

5. If an adjustment is required, loosen the adjuster lock nut and turn the adjuster screw inward to increase the turning effort or outward to decrease it. Remember that the turning effort should only be measured at the center-most position; if you check it, you will notice that the turning effort drops off when the steering shaft is turned about one-half turn off center.
6. When the lash is adjusted, tighten the adjuster lock to the correct torque.

33.3.4 Steering Gear Overhaul, Standard Steering Gear

If the steering shaft turns with a rough or erratic motion or there is free play that cannot be adjusted out, the steering gear should be replaced or overhauled. Most of the internal steering gear parts are available from the OEM manufacturer or aftermarket sources. Many steering gears have a metal identification tag under one of the bolts or a number stamped into the gear housing or cover. This identification number should be used when ordering replacement parts. The procedure and specifications recommended by the manufacturer should be followed while doing the operations.

To disassemble a standard steering gear, you should:

1. Clamp a mounting ear of the gear housing in the jaws of a vise.
2. Remove the bolts securing the sector gear cover to the gear housing and remove the gear cover with the sector gear/Pitman shaft from the gear housing. The Pitman arm must be off the Pitman shaft (Figure 33—54).
3. Loosen the lock nut and thread the sector gear adjustment screw into and through the gear cover to allow removal of the sector gear from the cover. Measure the sector shaft end play.
4. Loosen the lock nut and remove the worm shaft bearing adjuster plug. The worm shaft, with ball nut and worm shaft bearings, is removed along with the adjuster plug (Figure 33—55).

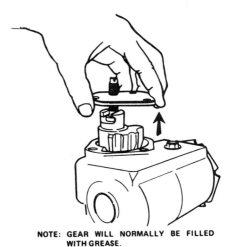

NOTE: GEAR WILL NORMALLY BE FILLED WITH GREASE.

Figure 33–54 A standard steering gear is disassembled by removing the sector cover retaining bolts and lifting the cover and gear out of the housing; turning the adjuster screw all the way inward removes the gear from the cover. *(Courtesy of Ford Motor Company)*

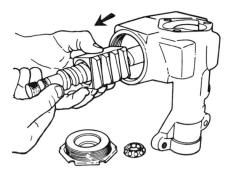

Figure 33-55 After removing the worm shaft bearing adjuster, the worm shaft and ball nut assembly can be removed from the housing. Note the angle of the ball nut teeth as the assembly is removed. *(Courtesy of Ford Motor Company)*

5. Remove the ball guides from the ball nut and remove the balls to allow separation of the ball nut from the worm shaft. It is recommended that you do this operation over a shop rag or carpet section to help prevent loss of the balls (Figure 33—56).

6. Clean the parts in solvent and dry them with air to allow inspection. Inspection of the parts should include:

   a. Worm shaft and Pitman shaft bearings
   b. Worm shaft and Pitman shaft bearing surfaces
   c. Worm shaft and Pitman shaft seals (should be replaced)
   d. Ball nut and steering shaft grooves

All the damaged parts should be replaced.

To assemble a standard steering gear:

1. As the steering gear is assembled, the parts should be lubricated with the same lubricant that will be used in the unit, usually gear oil, light grease, power steering fluid, or ATF. Place the worm shaft into the ball nut, making sure that the ball nut is in the correct position (note the teeth on the ball nut). Hold the steering shaft centered in the ball nut and insert the balls, one at a time, into the openings of the ball grooves. It is recommended that the balls be installed at only one end of each groove. Push the balls inward with a heavy wire or small punch or rotate the shaft slightly to help the balls enter. When the grooves are almost full (push a ball in one end, and the ball at the other end should lift), divide the remaining balls into two groups. Some ball guides have loading slots; in this case, place the ball guides in position and drop the remaining balls into the slots in the guides. If the ball guides do not have loading slots, use petroleum jelly to stick a group of balls into each of the ball

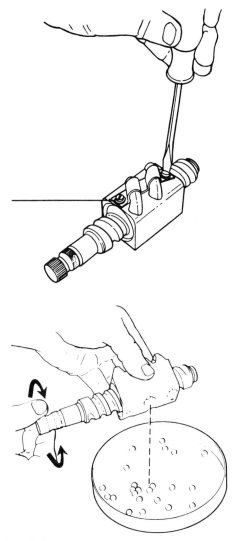

Figure 33-56 To remove the ball nut from the worm shaft, first remove the clamp for the ball return guides and then remove the return guides. Turn the assembly upside down over a container or shop cloth and rotate the shaft to work all of the balls out of the grooves. *(Courtesy of Ford Motor Company)*

guides and place the ball guides, with balls, into the ball nut. Secure the ball guides to the ball nut (Figure 33—57).

2. Test the fit between the ball nut and the worm shaft by holding the ball nut so the worm shaft is vertical; gravity should cause the worm shaft to rotate smoothly downward. Reverse the nut position, and, again, the steering shaft should rotate downward.

3. Place the worm shaft with one of the bearings into the gear housing, making sure the bearing and races are completely seated and the ball nut is positioned correctly. Position the remaining bearing and install the adjuster plug. Adjust the worm shaft bearing preload.

Figure 33–57 When assembling a ball nut and worm shaft, first center the shaft in the ball nut and, working from one end in each circuit, fill both circuits with balls. Use a punch to help work the balls into the grooves (top). When the circuits are almost full, drop the remaining balls into the ball guides (center) or load them into the guides and use petroleum jelly to hold them (bottom). *(Courtesy of General Motors Corporation, Service Technology Group)*

4. Connect the sector gear to the gear cover by threading the adjuster screw through the gear cover until the gear is almost against the cover.

5. Place the cover gasket in position and place the sector gear/shaft into the housing so the two gears are meshed correctly. Install and tighten the cover-retaining bolts to the correct torque.

6. With the steering shaft in the center-most position, thread the sector gear adjusting screw into the housing until you feel the sector gear contact the ball nut; from this point, complete the sector gear lash adjustment.

7. After installation of the steering gear in the car, fill the housing with the correct lubricant; this is usually a gear oil or a semifluid gear grease.

33.3.5 Steering Gear Adjustments, Rack-and-Pinion Steering Gear

As mentioned earlier, service on rack-and-pinion steering gears varies among manufacturers. Some manual units are nonserviceable and nonadjustable; if they become faulty, they should be replaced with a new or rebuilt unit. Most rack-and-pinion steering gears can be serviced, and the overhaul procedure of the various units is similar. It is described in the next section. Adjustment of the different units varies; most units have a rack support bearing adjustment; some have a pinion bearing adjustment. A manufacturer or technician service manual should always be followed when adjusting rack-and-pinion steering gears. These adjustments are usually made with the unit off the car.

Most modern rack-and-pinion steering gears use an adjustable rack support opposite the pinion gear; the purpose is to hold the two gears together to eliminate gear lash. The **rack support** is also called a **support yoke, rack bearing, rack plunger, friction plunger,** or **pressure pad** by the different manufacturers. A spring is usually used in the rack support to maintain a constant pressure. A common method of adjusting the rack support is to thread the adjuster plug inward to a specific torque or until it bottoms and then back it off a specified amount. The adjustment is then checked by measuring the rotating torque of the pinion shaft and comparing it with specifications; a torque wrench and socket or spring scale and adapter are used to measure the rotating torque.

A few makes of rack-and-pinion steering gears require an adjustment of the pinion bearings. When an adjustment is required, there will be a shim pack or threaded adjuster at the bearing retainer (Figure 33—58).

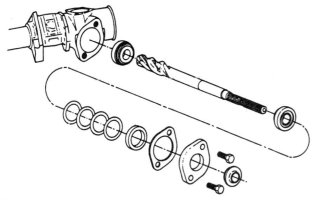

Figure 33–58 Some rack-and-pinion steering gears have adjustable pinion shaft bearings; removing shims on this one will decrease bearing preload. *(Courtesy of Ford Motor Company)*

33.3.6 Steering Gear Overhaul, Rack-and-Pinion Steering Gear

Most rack-and-pinion steering gears can be overhauled rather easily; the serviceable portions are pinion bearings and seals, the rack support, and a rack housing bushing. Lubrication of these units is usually accomplished by packing the parts with grease during assembly or by pouring the correct amount of gear oil into the housing before installing the second bellows. When overhauling a particular unit, check the service manual for the exact repair, adjustment, and lubrication procedure (Figure 33—59).

To disassemble a manual rack-and-pinion steering gear, you should:

1. Support the unit in a rack-and-pinion holding fixture or a standard shop vise by gripping a mounting ear (Figure 33—60).
2. Remove the bellows and tie-rod ends from the ends of the rack gear and housing. On some units,

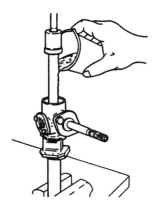

Figure 33–59 Some manual gear sets are lubricated by pouring a small amount of gear oil into the housing; make sure the boot is installed at the other end before pouring in the oil. *(Courtesy of Ford Motor Company)*

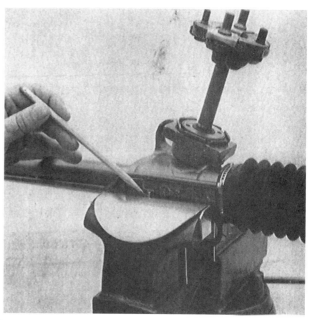

Figure 33–60 The rack housing should be supported before beginning disassembly; this particular gear housing is strong enough to clamp in a vise. *(Courtesy of General Motors Corporation, Service Technology Group)*

the tie-rod end at the end opposite the gear teeth can be left in place.

3. Remove the retaining bolts or locking ring for the pinion bearing and remove the pinion retainer, the pinion shaft and gear, and the bearing and seal (Figure 33—61).
4. Remove the rack support retainer and the rack support assembly (Figure 33—62).
5. Slide the rack gear out of the housing.
6. Wash the parts in solvent and dry them using shop air. Inspection of the parts should include:
   a. Pinion bearings and seals
   b. Rack-and-pinion gear teeth
   c. Rack support
   d. Rack housing bushing

All the damaged parts, seals, and rack bellows should be replaced.

To assemble a manual rack-and-pinion steering gear, you should:

1. Slide the rack gear into the gear housing with the rack teeth positioned to the pinion gear location. Do not forget to lubricate the parts, if required.
2. Install the pinion gear and shaft assembly and tighten the retaining bolts to the correct torque. A pinion bearing adjustment is required on some cars.

Chapter 33 ■ Steering System Service

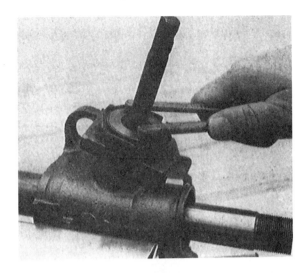

Figure 33–61 Disassembly of a rack-and-pinion steering gear often begins with the removal of the pinion shaft bearings and pinion gear. *(Courtesy of General Motors Corporation, Service Technology Group)*

3. Install and adjust the rack support assembly.
4. Check the pinion shaft turning effort and compare it with specifications. If specifications are not available, pinion shaft rotation should be smooth, not harsh or erratic, and require about 10 in.-lb (1 N·m) of torque. If the effort is too high or too low, readjust the rack support; when it is correct, tighten the adjuster lock to the correct torque.

SAFETY TIP: A rack gear tends to wear more in the center than at the ends; after making an adjustment, check the gear travel from lock to lock to ensure that there is no binding.

5. Replace the tie-rod ends and bellows; do not tighten the small bellows clamp until toe is checked and adjusted.

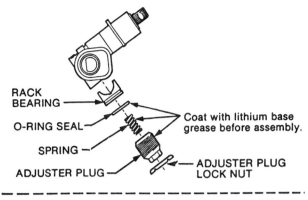

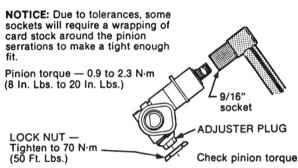

Figure 33–62 With the rack support and pinion gear removed, the rack can be slid out of the gear housing. Note the method of adjusting the rack bearing during reassembly. *(Courtesy of General Motors Corporation, Service Technology Group)*

33.4 Steering Wheel and Column Service

Until recently, steering wheel service was a matter of R & R of a steering wheel; now we have driver-side airbags with some potentially serious aspects. Steering column service can include checking the mounting points and/or replacing the column or any of the switches. The service procedure in a service manual for that particular car should be followed (Figure 33–63).

33.4.1 Airbag Service

When servicing an airbag-equipped vehicle, be aware that electrical current flow through the igniter portion of an airbag will cause it to deploy, and we do not want to cause accidental deployment. This makes a sound much like a shotgun blast (can move objects, break windshields, and the like), spreads a powdery substance, and requires a rather expensive new airbag. If you are doing an operation that you believe might accidentally cause deployment, you should deactivate the circuit. It is best to follow the manufacturer's recommendations for this procedure.

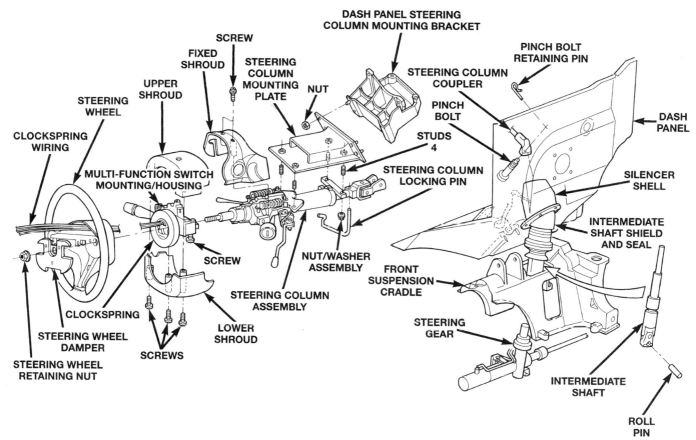

Figure 33-63 The steering column and shaft from a car or van are rather complex; a service manual should be followed when doing repairs. *(Courtesy of Chrysler Corporation)*

To disable an airbag, you should:

Disconnect the negative (−) battery cable or airbag circuit fuse and wait for the system capacitor to discharge. It should be noted that disconnecting the battery cable will cause all of the vehicle's electronic memory circuits to be lost.

OR Disconnect the yellow or orange connector at the steering column (Figure 33—64).

Other cautions to observe while working with an airbag-equipped steering wheel or column are the following:

- Position the steering wheel or column so the airbag is facing away from you and others whenever possible.
- Never place the steering wheel or column so the airbag contacts another surface.
- Never short a 12-V current supply through the airbag module unless you want to cause deployment.
- If making electrical checks on an airbag module, use only a digital meter to prevent excess current flow through the module.
- If you are restoring a vehicle to service after an airbag deployment, remember that the new airbag must deploy if involved in a future accident. The manufacturer's service procedures must be followed.
- If cleaning up the powdery residue from a deployment, the powder is mostly talc, corn starch, or sodium bicarbonate (baking soda), but it might also contain sodium hydroxide, which can irritate the skin, eyes, nose, and throat. Avoid breathing this dust. Gloves, safety glasses, and a dust mask should be worn during cleanup operations.
- The steering column is severely loaded during a crash, which deploys the airbag. Inspect the column, its mounts, the steering wheel, and all of the related parts for damage.
- All the sensors must be mounted correctly, and their mounting points must not be damaged.
- Never use an impact gun or hammer on or near a crash sensor.

A driver-side airbag is normally secured to the steering wheel, which is attached to the steering column in a normal manner (Figure 33—65). An electrical connector called a clock spring is used to complete the electrical circuit between the airbag igniter and the steering column. The clock spring is a continuous, flat conductor that is wound

Chapter 33 ■ Steering System Service   737

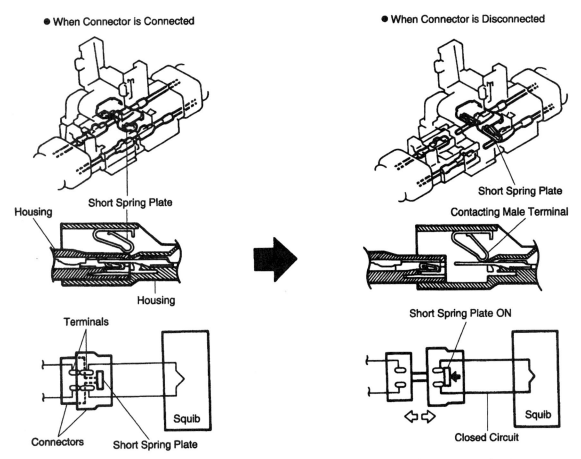

Figure 33–64 In most systems, disconnecting the airbag connector allows the short spring plate/shorting bar to short out the squib circuit. This shorting should prevent an undesired deployment; too much current through the squib will deploy the airbag.

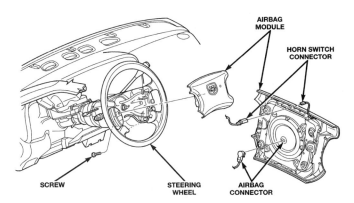

Figure 33–65 This airbag module is attached to the steering wheel by four screws and two wires; the clock spring is under the steering wheel. *(Courtesy of Chrysler Corporation)*

to allow the necessary rotation of the steering wheel. When a steering wheel, gear, or column is replaced, the wheel, clock spring, and steering must be properly centered or indexed to each other to prevent clock spring failure as the steering wheel is turned. Most manufacturers provide a centering procedure after steering wheel removal (Figure 33–66). An alternate procedure is to center the steering

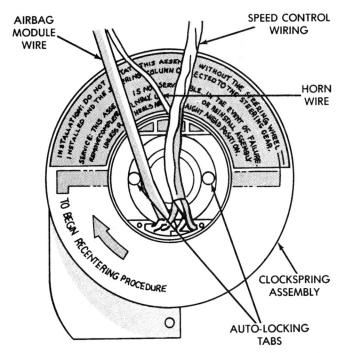

Figure 33–66 This clock spring assembly has the centering procedure printed on it. *(Courtesy of Chrysler Corporation)*

Figure 33-67 Securing the seatbelt after running it through the steering wheel keeps the steering wheel centered while allowing some steering shaft motion to align parts or gain access to clamping bolts.

before removing the wheel, connector, or gear and then to keep everything centered (as much as practical) during the repair procedure (Figure 33—67).

You should also make absolutely sure that the system is completely functional when your work is completed. Follow all of the manufacturer's service procedures and check to make sure the system is functioning properly when done. To quickly check a system in which you have made minor repairs to the mechanical portions, you should:

1. After installing all components and reconnecting all of the electrical connections, make sure no one is in the vehicle and reconnect the battery.
2. From as far away from the airbag(s) as possible, turn the ignition switch from off to run.
3. Observe the airbag light in the instrument panel; it should come on for 4 to 8 seconds and then go out. If the light does not come on or it stays on, there is a problem. In this case, the manufacturer's diagnosis and repair procedure should be followed. If the steering wheel has been removed and replaced or the steering column has been disconnected and reconnected, it is a good practice to turn the steering wheel from lock to lock while watching for the airbag warning light to come on.

The vehicle manufacturer provides complete diagnosis procedures for airbag circuit faults. Some systems set fault codes that can be read using the right electrical tester, and these codes indicate the cause of the problem (Figure 33—68). Electrical values and resistance are also provided for each component in the circuit. Electrical circuit problem diagnosis is described more completely in Chapter 18.

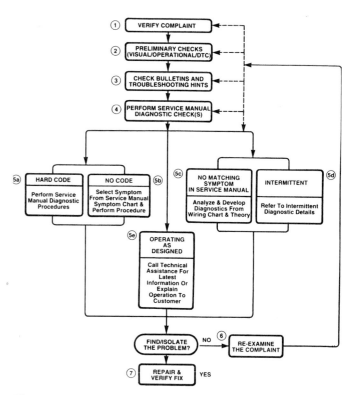

Figure 33-68 A strategy-based diagnostic flowchart for locating the cause of an airbag problem. *(Courtesy of General Motors Corporation, Service Technology Group)*

A quick check of some airbag circuits can be made using the voltage drop (as measured by a digital voltmeter) at the airbag connector. Remember that the circuit is live whenever the ignition is on; the collision sensors include a resistor, which causes a voltage drop, the airbag igniter, which also causes a slight voltage drop, and the airbag safing sensor, which causes a third voltage drop. The voltage at the airbag connector should be about half that of battery voltage, with the entering voltage slightly higher than the exiting voltage. This is determined by the exact amount of resistance in each part of the circuit. Too high or too low voltage indicates a faulty, open, grounded, or shorted circuit.

33.4.2 Removing and Replacing a Steering Wheel

Most steering wheels are indexed to the steering shaft by a mark on the steering wheel hub and another mark on the end of the shaft. Look for these marks, often at the top of the hub when the steering wheel is in the centered position. It is a good practice to disconnect the wire at the horns or horn relay before beginning the removal procedure.

To remove a steering wheel, you should:

1. Remove the horn ring to provide access to the steering wheel retaining nut. The ring is often retained

Chapter 33 ■ Steering System Service

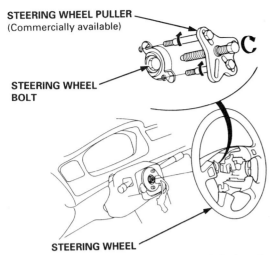

Figure 33–69 Most steering wheels are drilled and tapped at the hub so a puller can be attached for pulling the wheel off of the shaft. *(Courtesy of American Honda Motor Co., Inc.)*

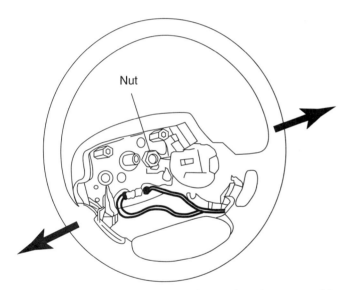

Figure 33–70 A steering wheel can often be removed by gripping the rim at each side (arrows) and then pulling in an alternating manner; first, though, tighten the nut slightly to break the splines loose and then remove it.

by two or three screws entering from the back side or a spring clip.

2. Locate the steering wheel index marks; if you cannot find any, use a scratch awl or scribe to make your own.
3. Loosen and remove the steering wheel retaining nut.
4. Install the bolts for a puller into the threaded holes in the steering wheel hub and tighten the puller to pull the steering wheel off the shaft (Figure 33–69).

OR Tighten the steering wheel retaining nut one-fourth to one-half turn to unseat the splines and then remove the nut. Grip the steering wheel with one hand at each side or at the top and bottom and pull firmly on the wheel rim in an alternating fashion, first on one side, then on the other. The steering wheel should creep off the shaft. If it does not, rotate the steering wheel one-fourth turn and repeat the alternating pulling action (Figure 33—70).

To install a steering wheel, you should:

1. Slide the steering wheel onto the splines of the steering shaft, making sure the index marks are aligned.
2. Install the retaining nut and tighten it to the correct torque.
3. Install the horn ring and reconnect the wire for the horns, if necessary.

33.5 Power Steering Service Operations

Normal power steering maintenance operations include fluid level and condition checks, drive belt tension and condition checks, and inspection of the rest of the system to ensure that there are no leaks and the pump mounts are good. In addition, there are also diagnostic, problem-solving operations when something goes wrong with the system and repair operations when the cause of the problem is determined.

The most common complaints with power steering systems are hard steering, loose or rough steering, noise, and leaks. Some of these problems, such as leaks, are often easy to diagnose; some are more difficult and require a systematic troubleshooting procedure. For example, hard steering can be caused by a number of things; the most probable faults are a faulty pump or steering gear. Most experienced front-end technicians still follow a diagnostic chart such as the one in Table 33—1 to help them locate the cause of the problem and ensure that they do not overlook or forget some possibilities. Remember that some of these problems can be caused by components outside of the steering gear and pump; they might be caused by steering linkage, suspension components, or the electronic controls for variable-assist steering systems.

33.5.1 Power Steering System Checking and Problem Diagnosis

Power steering problem diagnosis begins with a thorough inspection of the system. It is a good practice to not only cure the immediate problem but also make sure the system will operate trouble free for a reasonable period.

Table 33–1 The Most Probable Causes of Poor Power Steering System Performance.

Symptom	Possible Causes
Hard steering	Low power steering fluid level
	Low tire pressure
	Slipping drive belt or pulley
	Binding in steering linkage
	Restricted hose
	Low pump output pressure
	Overly tight steering gear adjustments
	Sticking flow control valve in pump
	Internal leak in steering gear
	Binding in steering column
Hard steering parked	All of the above
	Low engine idle speed
Poor steering wheel return	Low tire pressure
	Binding in steering linkage
	Tight steering gear
	Incorrect wheel alignment
	Binding in steering column
Hard steering in one direction	Defective seals in gear valve assembly
	Defective seal in rack housing
Excessive free play	Worn or loose wheel bearings
	Worn steering linkage
	Improper gear adjustments
	Worn gear
Pulling to one direction	Unequal tire pressures
	Incorrect wheel alignment
	Worn control valve in gear
	Improperly adjusted gear control valve
	Brake drag or pull
Noise	Low fluid level
	Air in hydraulic system
	Restricted hose
	Defective pump mount
	Improper pump control valve operation
	Slipping drive belt

Begin at the pump and then, depending on the nature of the problem, check out the entire system.

To check out a power steering system, you should:

1. Check the condition of the drive belt; if it is worn, badly cracked or frayed, or has very shiny or glazed sides, the belt should be replaced. These problems are described in Section 26.10.

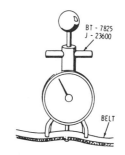

	5/16 WIDE	3/8 WIDE	15/32 WIDE
NEW BELT	350 N Max. 80 Lbs. Max.	620 N Max. 140 Lbs. Max.	750 N Max. 165 Lbs. Max.
USED BELT	200 N Min. 50 Lbs. Min.	300 N Min. 70 Lbs. Min.	400 N Min. 90 Lbs. Min.
USED COGGED BELT		250 N Min. 60 Lbs. Min.	

Figure 33–71 Power steering belt tension should be checked using a belt tension gauge; a tension that differs from the specifications indicates the need to readjust the belt. *(Courtesy of General Motors Corporation, Service Technology Group)*

Some sources recommend replacement of this belt every 4 years.

2. Check the tension of the belt; if it is too loose, readjust it. The best way to measure belt tension is with a tension gauge, comparing the reading with the specification. Most power steering belts use a tension of about 70 lb (32 kg). If a tension gauge is not available, push downward at the center of the belt with a pressure of about 5 lb (2.25 kg); the belt should deflect or move about 1/16 in. (1.5 mm) if it is new or 1/8 in. (3 mm) if it is used (Figure 33—71).

3. Readjust the drive belt, if necessary, by loosening the pivot and adjusting bolts and swinging the pump on the pivot bolt. Many pump brackets and mounts have provisions for attaching a wrench or socket handle to aid in moving the pump; do not pry on the pump reservoir (Figure 33—72).

4. Before checking the fluid level, start the engine and turn the steering almost to the right and left stops several times. Shut off the engine and remove the dipstick. Remember that the full level for hot fluid is about 170 F (77 C). A dipstick with an end that is too hot to hold with your fingers indicates the oil is hot. If necessary, add fluid and repeat this step (Figure 33—73).

5. Smell the fluid and place some on a piece of clean white paper. If the fluid has a burned or varnishlike odor, has a dull, very dark, or brown appearance, or

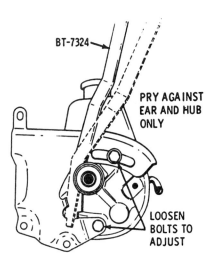

Figure 33-72 This power steering bracket is constructed so a pry bar can be used to pry outward for adjustment. Some brackets have wrench flats or square openings to allow a wrench or socket handle to be used for this purpose. Never pry against the pump reservoir. *(Courtesy of General Motors Corporation, Service Technology Group)*

Figure 33-73 The power steering fluid level is usually checked using the dipstick attached to the cap.

contains particles of hose or other debris, it should be changed. A grayish color usually indicates aluminum particles that are caused by rack housing valve bore wear. Also check for fluid aeration, which causes foamy, milky fluid. The procedure to change the fluid and bleed air out is described in Section 33.5.2.

6. Check the system for leaks at the pump, hoses, and gear (Figure 33—74). A leak in the high-pressure areas can be made to show up better by starting the engine and holding the steering wheel briefly against the right or left turn stops. This will raise the pump output pressure to the setting of the relief valve.

SAFETY TIP: *Do not hold the steering wheel against the stops longer than 4 or 5 seconds; fluid aeration and overheating or pump damage can occur.*

7. Check the hoses for heavy rubbing or abrasion, excessive rubber cracking, excessive swelling, soft, spongy areas, damaged metal ends, or other signs of deterioration.

8. Replace any damaged hoses. After replacement, fill the system with fluid, bleed the air out, and check for leaks. New O-rings should be used on hose ends that use O-ring seals. If leaks occur at the hose connections sealed by a double-flare fitting, loosen the tube nut, rotate the tubing slightly to reseat it with the brass seat, and retighten the tube nut to the correct torque (Figure 33—75).

9. If the complaint is hard steering and if the problem has not been corrected at this point, check the pump and system operating pressures. Power steering pressure gauge sets are available for testing purposes. The pressure checking procedure is described in Section 33.5.3 (Figure 33—76).

33.5.2 Power Steering Fluid Change and Air Bleeding

Contaminated power steering fluid should be changed. The valve body portion of an aluminum rack-and-pinion steering gear is susceptible to wear, especially from contaminated fluid. During a fluid change or when fluid, hoses, or any other part of the system is replaced, air can enter the system. Air causes a moaning or groaning noise or an erratic steering feel. Normally, air in the system is caught by the fluid flow and brought to the reservoir, where it will escape; occasionally air is trapped inside the power steering gear or pump. A commercial flushing unit is available to make this job easier and faster.

Be sure to use the fluid type specified by the vehicle manufacturer. This can be ATF or one of the various types of power steering fluid. The fluid requirement is often imprinted on the dipstick.

To change power steering fluid, you should:

1. Remove the return hose from the pump and place it in a container. Be ready for a fluid loss from the hose and reservoir as the hose is removed (Figure 33—77).

2. After the fluid has drained from the reservoir opening, cap the opening.

3. Refill the reservoir, start the engine, observe the flow from the hose, and be ready to add additional new fluid into the reservoir. New fluid should be added at the same rate that it is pumped out; the reservoir *should not* be allowed to become empty.

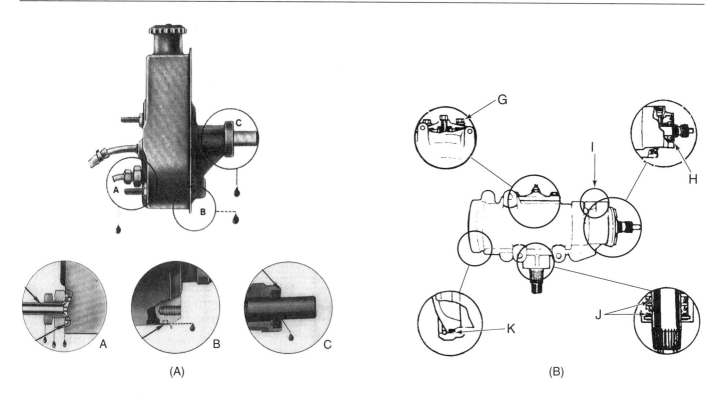
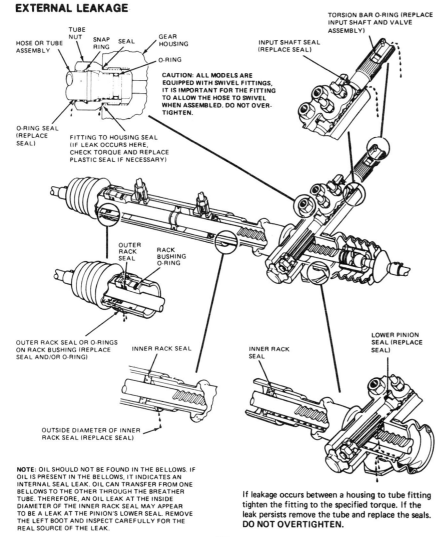

Figure 33-74 The normally encountered power steering leakage areas of a pump (A), standard steering gear (B), and rack-and-pinion gear (C). (A and B are courtesy of General Motors Corporation, Service Technology Group; C is courtesy of Ford Motor Company)

Chapter 33 ■ Steering System Service

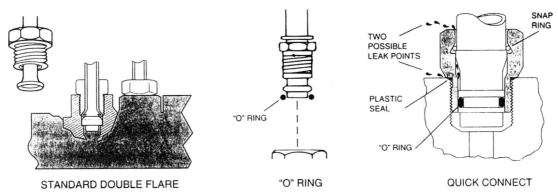

Figure 33–75 Power steering hoses seal at the flare against the tube seat, an O-ring, or a quick-connect type of hose connection that uses an O-ring and a plastic seal. *(Courtesy of Moog Automotive Inc.)*

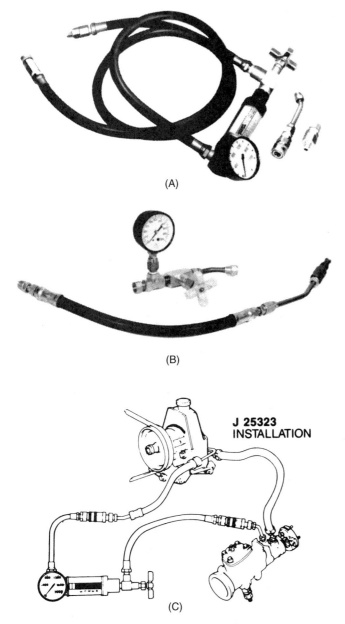

Figure 33–76 (A) and (B) are two different styles of power steering testers; unit (A) can measure flow in addition to pressure. Either unit is connected in the pressure line between the pump and gear (C). *(Courtesy of SPX Kent-Moore)*

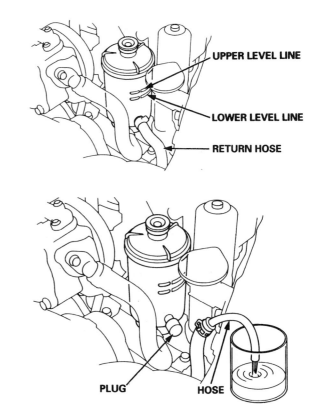

Figure 33–77 To drain the fluid from this system, remove the return hose from the reservoir, cap the reservoir fitting, and, if necessary, attach a short hose so the fluid will drain into a container. *(Courtesy of American Honda Motor Co., Inc.)*

NOTE: It is better to change fluid by cranking the engine rather than running it. Doing so requires a good battery, disconnecting the ignition coil, and cranking the engine in 10-second bursts to prevent overheating the starter.

4. When all of the old fluid is gone and the fluid flow appears new, stop the engine, reconnect the return hose, and fill the reservoir to the correct level with new fluid.

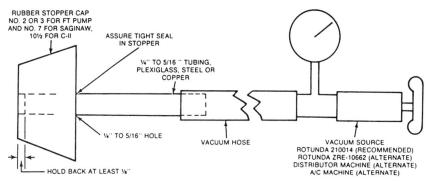

Figure 33–78 This device was constructed so a vacuum pump could be attached to the power steering pump reservoir. The vacuum pump helps remove air bubbles, which can cause moan or whine. *(Courtesy of Ford Motor Company)*

To bleed air from a system, you should:

1. Start the engine and slowly turn the steering wheel to the right and left turn stops; do not hold the steering wheel at the stops. Some technicians prefer to lift the front tires partially off the ground for this step to remove the steering load and strain from the steering linkage and gear and to reduce scrubbing of the tires.

2. If the car is equipped with a hydro-boost power brake unit, depress the brake pedal several times while the steering wheel is being turned.

3. Check the fluid condition; if it is excessively foamy or aerated, let the system sit with the engine off for several minutes to allow the foam to dissipate. In extreme cases, it should sit for several hours.

4. One manufacturer recommends using a vacuum pump to remove air bubbles in extreme cases where the moaning noises persist. A stopper is connected to the pump reservoir filler neck, and the vacuum pump is connected to the stopper. A vacuum of 15 in. Hg (51 kPa) is then applied to the pump reservoir while the engine is running at idle speeds for 2 or 3 minutes. The system is then stopped, the fluid level is checked, and the fluid is added, if necessary. Next a 15-in. Hg vacuum is reapplied, and the system is operated at idle speed for about 5 minutes while the steering wheel is slowly turned to the right and left turn stops (Figure 33—78).

33.5.3 Pressure Testing a Power Steering System

Power steering system pressure gauge sets usually include a 0- to 2,000-psi (0- to 13,790-kPa) pressure gauge, a shutoff valve, and a series of adapters to connect to the fluid ports and hose ends. Some testers include a gauge for measuring the amount of fluid flow. Pressure specifications are available for most systems in manufacturer or technician service manuals. These specifications are for fluid temperatures of about 150 to 170 F (65 to 77 C); if the system being tested is colder, warm the fluid by operating it a few minutes with the valve on the tester partially closed to provide a pressure of about 350 psi (2,400 kPa).

To test the pressure in a power steering system, you should:

1. Connect the test unit into the pressure line between the pump and the steering gear. Usually the most accessible end of the pressure hose is disconnected, the tester is connected to the port where the hose was, and the hose is connected to the tester (Figure 33—79).

2. With the tester valve open, start the engine and let it run at idle speed. Observe the system pressure; with the valve open, it should be about 30 to 80 psi (200 to 550 kPa). If the pressure is above 200 psi (1,380 kPa), check for restrictions in the hoses or steering gear valve. Many systems do not specify a pressure for this part of the test.

3. Close the tester valve completely, observe the pressure, and reopen the valve. *Do not* keep the valve closed longer than 4 or 5 seconds. The pressure should raise to within 50 psi (345 kPa) of the specifications; specified pressures are normally between 600 and 1,300 psi (4,100 and 8,950 kPa). If the pressure is low, a faulty pump or pump relief valve is indicated. A faulty relief valve is indicated if the pressure is too high.

4. Repeat step 3 two more times and compare the readings. A faulty flow control valve is indicated if the three pressures are not within 50 psi (345 kPa) of each other.

5. If the readings in steps 3 and 4 indicate problems, remove and clean the flow control valve; if the fluid contains debris, change the fluid. Repeat steps 3 and 4.

Chapter 33 ■ Steering System Service

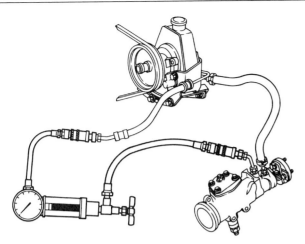

Figure 33-79 Four pressure checks can be made using a flow- or pressure-type analyzer. Test 1 is at idle speed with the gate valve fully open and flow and pressure are read. Test 2 is at idle speed with the valve closed enough to generate 750 psi and flow is read. Test 3 is at idle speed with the valve closed and opened three times and the pressure read each time. Test 4 is at 1,500 rpm with the valve fully open and the flow is read. Flows and pressures for each of these tests are compared with specifications to determine if there is a problem with the relief valve, steering gear, or pump. *(Courtesy of Ford Motor Company)*

6. Open the valve completely and observe the pressure while you turn the steering wheel to each stop. The pressure while the steering wheel is at the stop should be at the same maximum pressure as in step 3 or 4; if it is lower, the steering gear has an internal leak.

If the system pressures are correct with the steering wheel turned to each stop, the pump, the flow and pressure control valve, and the internal seals of the steering gear and valve are good.

33.5.4 Power Steering Pump Service

A variety of power steering pumps are made by the various manufacturers; these units are similar in design. The commonly performed power steering pump service operations are control valve cleaning, leak repair, and pump rebuilding. Most of these service operations are done with the pump off the car.

Cleaning a Control Valve. On some pump designs and installations, the control valve can be removed and replaced with the pump mounted on the engine. Most pumps mount the valve under the fluid outlet fitting where the high-pressure line connects.

To remove a control valve assembly, you should:

1. Remove the high-pressure hose from the pump.
2. Remove the outlet fitting from the pump.

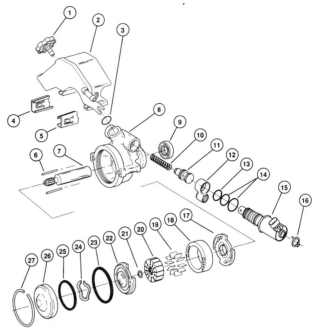

(A)

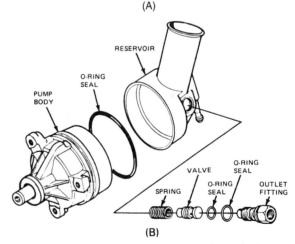

(B)

Figure 33-80 The control/relief valve (A and B) is usually under the pump's outlet fitting. Note that pump (A) has an additional valve (15) with an electronic variable orifice. *(A is courtesy of General Motors Corporation, Service Technology Group; B is courtesy of Ford Motor Company)*

3. Use a narrow magnet to lift the control valve out of the bore and also the control valve spring (Figure 33—80).

4. Some control valves can be disassembled for cleaning. Carefully clamp the valve in a soft-jawed vise and unscrew the hex-headed relief ball seat. Remove the shims and other internal valve

parts and clean them. The shims are used to adjust pump relief pressure. After cleaning, reassemble the control valve assembly (Figure 33—81).

To replace a control valve assembly, you should:

1. Slide the spring for the control valve into the bore.
2. Lubricate the control valve with power steering fluid or ATF and slide it into the bore. Make sure the valve is not cocked by pushing inward on the valve; you should be able to feel it move against the spring pressure.
3. Replace the outlet fitting and tighten it to the correct torque. Replace the hose.

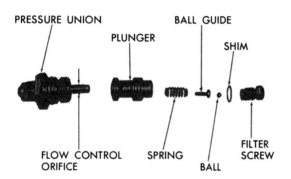

Figure 33—81 Removing the filter screw from this valve plunger allows the pressure relief ball, guide, and spring to be removed for cleaning; a thicker shim will lower pump relief pressure. *(Courtesy of General Motors Corporation, Service Technology Group)*

Removing and Replacing Pump Seals. The most commonly encountered pump leak locations are at the hose connections, at the reservoir to pump body connection, around the reservoir-retaining bolts, at the filler cap, and at the front shaft seal. Seal and O-ring kits are available from the vehicle manufacturer and aftermarket suppliers.

To repair power steering pump leaks, you should:

1. Disconnect the hoses and remove the pump from the engine.
2. Remove the front pulley. A special alternator—power steering pump pulley remover is available to do a quick and neat job; a two- or three-jaw puller can also be used (Figure 33—82).
3. Clean the dirt and rust from the driveshaft. Using a sharp chisel or punch, cut a hole in the old seal and pry it out of the pump housing (Figure 33—83).
4. Lubricate the lip of the new seal with power steering fluid or ATF and, using a driver that engages the whole face of the seal cartridge, drive the seal into the housing bore (Figure 33—84).
5. Remove the reservoir-retaining bolts and the outlet fitting. Tap and rock the reservoir to loosen it, and work the reservoir off the pump body. Discard the old O-ring seals (Figure 33—85 and Figure 33—86).

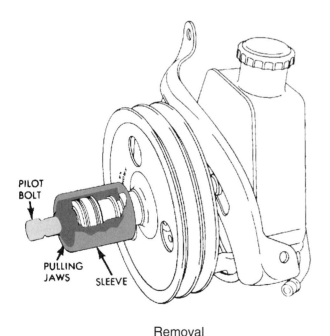
Removal

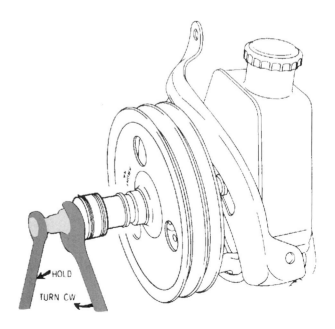
Installation

Figure 33—82 This puller slides into a groove in the pulley and puts pressure on the end of the shaft to make it easier to pull the pulley off (left). A different tool is used to push the pulley back onto the shaft (right). *(Courtesy of General Motors Corporation, Service Technology Group)*

Chapter 33 ■ Steering System Service

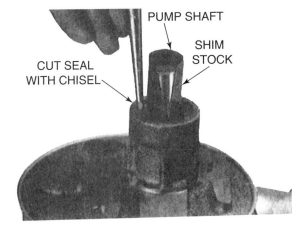

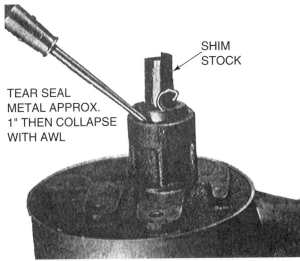

Figure 33–83 To remove the pump shaft seal, first cut into the seal body with a small chisel, collapse the seal body slightly, and then pry it out; shim stock is being used to protect the shaft. *(Courtesy of General Motors Corporation, Service Technology Group)*

Figure 33–84 To install a seal, first clean the shaft with a shop cloth (top) and, if necessary, crocus cloth; next drive the seal in using a driver tool that fits flat against the entire portion of the seal body (bottom). *(Courtesy of General Motors Corporation, Service Technology Group)*

6. Place the new O-rings in position; if necessary, stick them in place using petroleum jelly. Note that there should be an O-ring around each opening into the reservoir.

7. Place a film of petroleum jelly on the large O-ring around the pump body and reservoir sealing surface and slide the reservoir over the pump body, making sure the bolt holes are aligned.

8. Install the retaining bolts and outlet fitting and tighten them to the correct torque.

9. Install the drive pulley using an installing tool attached to the pump shaft; some pumps can be damaged by pressing the pulley directly onto the shaft. Most pumps will be damaged if the pulley is driven on using a hammer.

Power Steering Pump Overhaul. Pump overhaul should be done in a clean work area; it is a process of disassembly, inspection, replacement of worn parts, and

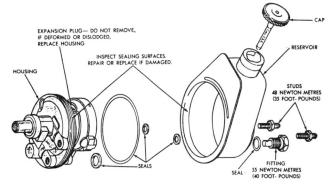

Figure 33–85 Pump reservoir leaks are cured by removing the reservoir, making sure the sealing surfaces are in good condition, installing new seal rings, and replacing the fitting and bolts to the correct torque. *(Courtesy of Chrysler Corporation)*

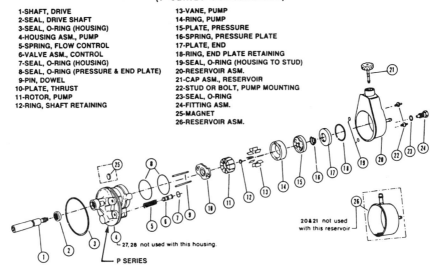

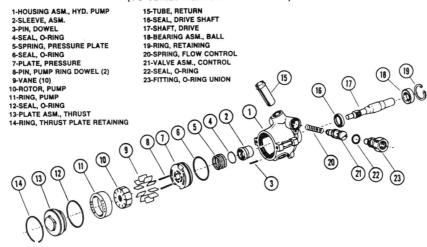

Figure 33–86 An exploded view of two different power steering pumps. Note that the internal parts are essentially the same. *(Courtesy of General Motors Corporation, Service Technology Group)*

reassembly. After assembly, the pressure plate must fit tightly against the cam ring so the pump can work. On some pumps, the drag of the pressure plate O-ring is greater than the strength of the pressure plate spring. It is a good practice to lightly press the pressure plate into position. Disassembly and reassembly procedures vary slightly among different makes of pumps; it is recommended that you follow a procedure for the specific pump you are working on as outlined in the manufacturer or technician service manual.

To overhaul a power steering pump, you should:

1. Remove the pump pulley and reservoir.

2. Remove the end plate retaining ring and tap the pump body to remove the end plate.

3. Carefully tap inward on the front of the driveshaft to remove the pump assembly, pressure plate, thrust plate, cam ring, rotor, shaft, alignment dowels, and vanes, rollers, or slippers. Note the relationship of the various parts as they come apart; some of them have slight differences on each side (Figure 33–87).

4. Most of the pump wear will occur on the inside surfaces of the cam ring, the vanes, and the shaft bushing; inspect them for damage.

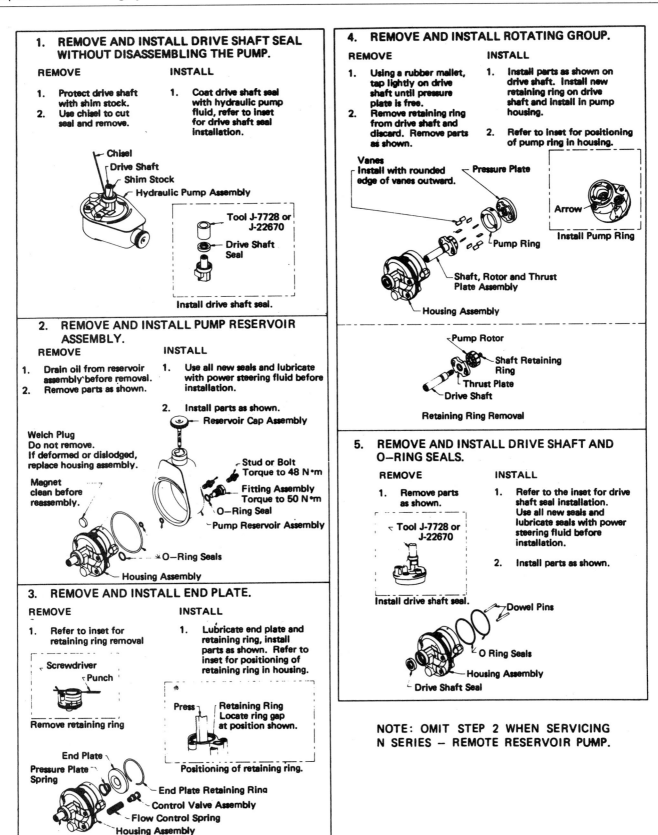

Figure 33-87 The procedure to overhaul this power steering pump. *(Courtesy of General Motors Corporation, Service Technology Group)*

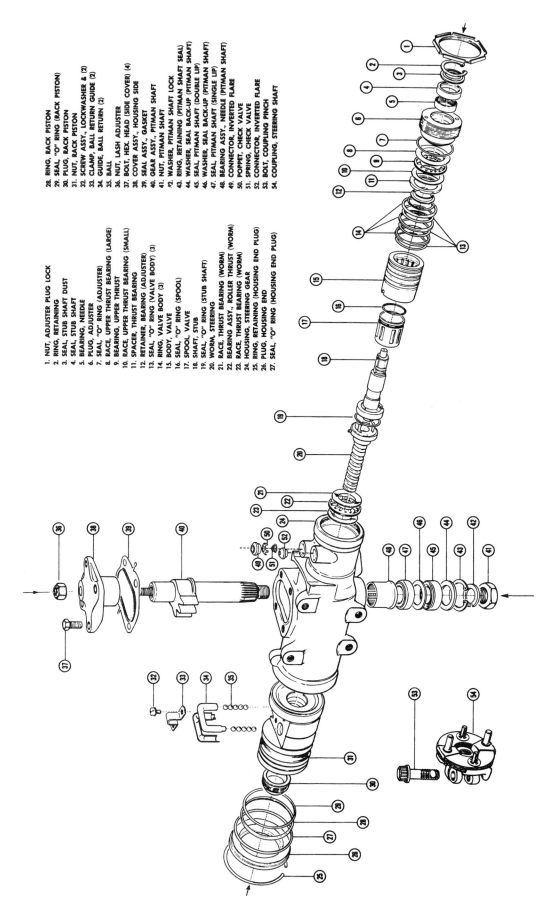

Figure 33-88 An exploded view of a standard power steering gear; the gear is normally disassembled to this point during an overhaul. (*Courtesy of General Motors Corporation, Service Technology Group*)

5. Remove the old O-rings and replace them with new ones. Apply a film of petroleum jelly on each O-ring.

6. Install the pump assembly parts into the pump body, making sure the parts are correctly aligned. Make sure the pressure plate is completely seated.

7. Pour a small amount of power steering fluid or ATF into the pump assembly, place the pressure plate spring in position, and install the end plate and retaining ring.

8. Replace the pump pulley and reservoir.

33.5.5 Power Steering Gear Service

The most commonly performed power steering gear service operations are leak repair and gear assembly overhaul. Some seals (the outer steering shaft seal and Pitman shaft seal) are relatively easy to remove and replace; steering gear disassembly is necessary to replace the remaining seals.

Disassembly, inspection, reassembly, and adjustment of a standard type of gear require a procedure similar to that used for a manual gear. This operation is done in many dealerships, garages, and repair shops. A few special tools are required when working on some gear sets (Figure 33—88).

Disassembly, inspection, reassembly, and adjustment of a power rack-and-pinion gear are also similar to those of a manual rack-and-pinion unit. However, quite a few additional steps are required because of the internal and external fluid seals and lines. Additional special tools are also required. Many technicians have found it advantageous to replace a faulty power rack-and-pinion unit; it saves them time and provides the customer with a professionally rebuilt or remanufactured unit. Most rebuilders install a steel sleeve in the housing valve bore to eliminate this wear problem area. It is often less expensive than a garage repair because of the shop time that is saved. Many rebuilt units come complete with tie-rods and bellows or, in a short rack version, without tie-rods and bellows (Figure 33—89).

Except for Pitman shaft seal replacement, power steering gear service is too varied to try to describe in a text of this type. Manufacturer and technician service manuals provide a thorough explanation of repair procedures for specific units. It is a good practice to follow these procedures if you decide to rebuild a power steering gear.

33.5.6 Pitman Shaft Seal Replacement

Pitman shaft seal replacement is an in-car operation used to replace a faulty seal. Some vehicles use a double seal at this location.

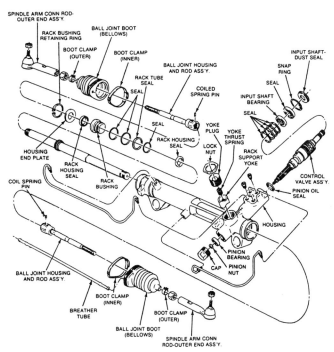

Figure 33–89 An exploded view of a power rack-and-pinion steering gear disassembled for an overhaul. (*Courtesy of Ford Motor Company*)

To replace a Pitman shaft seal, you should:

1. Raise and support the vehicle on a hoist or jack stands.
2. Mark the Pitman-arm-to-shaft position and remove the Pitman arm from the shaft.
3. Clean the exposed end of the Pitman shaft and housing.
4. Remove the seal-retaining snap ring and backup washer.
5. Start the engine and turn the steering wheel completely to the right or left position, as needed, to force out the seals and washers. Stop the engine.
6. Inspect the housing for scratches or burrs and the Pitman shaft seal area for damage. Slight shaft damage can be cleaned using crocus cloth.

To replace this seal, you should:

1. Coat the lip of the seal with petroleum jelly and install the seal, backup washer, and retaining ring.
2. Refill the power steering reservoir, start the engine, and check the new seal installation for leaks (turn the steering to both locks so full pressure is on the seal). Stop the engine.
3. Center the steering and replace the Pitman arm and nut. Tighten the nut to the correct torque.

33.6 Completion

Several checks should always be made after completing any steering system service operation:

1. Ensure proper tightening of all nuts and bolts, with all locking devices properly installed.
2. Ensure that there is no incorrect rubbing or contact between parts or brake hoses.
3. Ensure that there are no fluid leaks.
4. Ensure an even and smooth movement of the steering from one lock to the other.
5. Ensure proper feel of the brake pedal.
6. Check and readjust the wheel alignment if there is a possibility that the replacement parts have changed the alignment.
7. Perform a careful road test to ensure correct and safe vehicle operation.

33.7 Practice Diagnosis

You are working in a brake and front-end shop and encounter the following problems:

CASE 1: A customer has brought in his 1976 Dodge B-series van for a wheel alignment, with a complaint of sloppy steering. This problem is easily confirmed as you rock the steering wheel back and forth several inches. While doing a prealignment inspection, you find that the end of the idler arm jumps upward about an inch when you spread the tires, and one tie-rod end shows free play. What is your recommendation for this vehicle?

CASE 2: A customer has brought in a 5-year-old Chevrolet Cavalier with a complaint of hard steering, especially on cold mornings. What should you do to locate the cause of this problem?

Terms to Know

friction plunger
master splines
missing splines
over center adjustment
pressure pad

rack bearing
rack plunger
rack support
sector gear lash

sector gear/shaft end play
steering linkage parallelism
support yoke
worm shaft bearing preload

Review Questions

1. Technician A says that the tapered stud of a tie-rod end can be loosened using tools and methods similar to those used on ball joints.
   Technician B says that a pickle fork can damage the boot and the socket of the tie-rod end.
   Who is correct?
   a. A only
   b. B only
   c. Both A and B
   d. Neither A nor B

2. Before disassembly, the _____ of a tie-rod should be measured to allow preadjustment of the toe.
   A. Diameter
   B. Length
   Which option best completes the statement?
   a. A only
   b. B only
   c. Both A and B
   d. Neither A nor B

3. Technician A says that idler arms and tie-rod ends that use rubber bushings can be tightened in place with the steering turned to any position.
   Technician B says that these joints should be lubricated with chassis grease during installation.
   Who is correct?
   a. A only
   b. B only
   c. Both A and B
   d. Neither A nor B

4. When threading inner tie-rod ends off or on a rack gear, the rack can be held from turning with:
   A. An adjustable wrench tightened over the rack teeth
   B. A pipe wrench
   Which option best completes the statement?
   a. A only
   b. B only
   c. Either A or B
   d. Neither A nor B

5. An inner tie-rod end can be locked onto the rack gear by a:
   a. Steel pin
   b. Set screw
   c. Swaged metal collar
   d. Any of these

6. When replacing a multipiece inner tie-rod end socket, the socket should be:
   A. Adjusted to obtain the correct articulation effort
   B. Locked onto the rack gear
   Which option best completes the statement?
   a. A only   c. Both A and B
   b. B only   d. Neither A nor B

7. Technician A says that some cars mount the inner tie-rod ends to the center of the rack gear.
   Technician B says that rubber bushings are used on these tie-rod ends.
   Who is correct?
   a. A only   c. Both A and B
   b. B only   d. Neither A nor B

8. When idler arms with _____ are replaced, the installation height or length should be adjusted.
   A. Slotted mounting holes
   B. Threaded bushings
   Which option best completes the statement?
   a. A only   c. Both A and B
   b. B only   d. Neither A nor B

9. When replacing idler arms,
   A. The frame brackets and stud hole should be checked for damage
   B. All bolts and nuts should be tightened to the correct torque
   Which option best completes the statement?
   a. A only   c. Both A and B
   b. B only   d. Neither A nor B

10. Which of the following is *not* one of the steps in removing a steering gear?
    a. Disconnect shaft coupler
    b. Disconnect idler arm
    c. Disconnect Pitman arm
    d. Remove mounting bolts

11. Steering gear adjustments are being discussed.
    Technician A says that steering shaft preload and sector gear lash can be adjusted with the steering gear either on or off the car.
    Technician B says that sector gear end play can only be adjusted with the steering gear assembled.
    Who is correct?
    a. A only   c. Both A and B
    b. B only   d. Neither A nor B

12. A properly adjusted steering gear will allow steering shaft rotation with a slight:
    A. Drag
    B. End play
    Which option best completes the statement?
    a. A only   c. Either A or B
    b. B only   d. Neither A nor B

13. Technician A says that a standard steering gear should turn with the same amount of preload from one lock to the other.
    Technician B says that this gear should have a slight amount of lash between the ball nut and sector gear all the way through the gear travel.
    Who is correct?
    a. A only   c. Either A or B
    b. B only   d. Neither A nor B

14. Sector gear end play is adjusted by:
    A. Turning an adjustment screw
    B. Changing a shim
    Which option best completes the statement?
    a. A only   c. Either A or B
    b. B only   d. Neither A nor B

15. A properly adjusted manual rack-and-pinion steering gear will rotate:
    A. From one lock to the other with the same amount of preload
    B. Smoothly from one lock to the other
    Which option best completes the statement?
    a. A only   c. Either A or B
    b. B only   d. Neither A nor B

16. Technician A says that tightening the rack support bushing will increase the amount of preload in a rack-and-pinion steering gear.
    Technician B says that this gear is lubricated through an oil level plug at the top of the gear housing.
    Who is correct?
    a. A only   c. Both A and B
    b. B only   d. Neither A nor B

17. Power steering drive belt tension should be checked using:
    A. The belt deflection method
    B. A belt tension gauge
    Which option best completes the statement?
    a. A only   c. Either A or B
    b. B only   d. Neither A nor B

18. Normal power steering service includes:
    a. Checking the fluid level and condition
    b. Checking the belt tension and condition
    c. Inspecting and curing leaks
    d. All of these

19. A hard steering complaint is being discussed. Technician A says that the cause can be determined by using a special pressure gauge set. Technician B says that this gauge should be installed between the pump outlet port and the steering gear inlet port. Who is correct?
    a. A only
    b. B only
    c. Both A and B
    d. Neither A nor B

20. Technician A says that a power steering system should have about 500 to 700 psi at idle speed. Technician B says that pump pressure should increase when the valve on the tester is closed. Who is correct?
    a. A only
    b. B only
    c. Both A and B
    d. Neither A nor B

Math Question

1. You are adjusting the steering gear from a RWD vehicle. After adjusting the worm shaft preload to 6 in.-lb of torque, you find the specification for the over center adjustment is 4 to 8 in.-lb in addition to the worm shaft preload but not to exceed 18 in.-lb. How much torque should it take to turn the worm shaft through center after your adjustments are complete?

34 Wheel Alignment: Measuring and Adjusting

Objectives

Upon completion and review of this chapter, you should be able to:

❑ Determine the cause of abnormal vehicle handling such as pulling, wandering, hard steering, or poor return and recommend the needed repairs.

❑ Measure camber, caster, SAI, toe-out on turns, and toe on the front wheels of vehicles.

❑ Measure camber and toe on the rear wheels of vehicles.

❑ Adjust front and rear alignment angles that are adjustable on a given vehicle.

❑ Check a rear axle assembly for bending or misalignment.

❑ Adjust a steering wheel to a centered position.

❑ Perform the ASE tasks relating to wheel alignment diagnosis, adjustment, and repair (see Appendix A, Section C).

34.1 Introduction

Of the five different alignment angles, **SAI** is never adjustable, **toe** is always adjustable, **caster** and **camber** are often adjustable, and **toe-out on turns** can sometimes be adjusted. Three of these—camber, toe, and toe-out on turns—are angles of the tire; two of them—caster and SAI—are angles of the steering axis. The three that are angles of the tire will cause tire wear if they are not correct and are called **tire wearing** angles. Caster and SAI are referred to as **directional control** angles. All five of these angles can affect vehicle handling under various circumstances. Because the rear tires of most vehicles do not turn for steering, only two of these angles, camber and toe, apply to them.

Although they are not traditional alignment angles, the modern alignment technician is also concerned with **thrust line** and **setback.** These two items are important for diagnosis and to help us locate the cause of problems.

Performing a wheel alignment normally includes measuring these angles, comparing the measurements with the specifications for the particular vehicle, and then adjusting any of the angles that do not meet the specifications. The procedure for measuring most of the angles is fairly easy, especially with some of the more modern, computerized equipment. The procedure for correcting alignment angles can be fairly easy or extremely difficult depending on whether a means of adjustment is provided by the manufacturer. At one time, most cars provided for adjusting caster, camber, and toe; with many modern cars, there is a provision for adjusting toe only.

Manufacturers are building many cars so that caster, camber, SAI, and toe-out on turns angles are correct when the car leaves the assembly line. With modern assembly methods using precise fixtures, it is faster and less expensive to build a car with the front tires in correct alignment. Adjustable parts are more expensive to build, and adjustments take time. On cars with nonadjustable alignment angles, if any of these angles become incorrect at a later date, something must be wrong (bent, worn, sagged, or improperly positioned); in this case, the alignment is corrected by repairing the problem.

All alignment checks are made with the car on a level surface because most measuring instruments or gauges measure the angles relative to true vertical (plumb) or true horizontal (level). The vehicle should also be at the correct curb height; a lower or higher than normal height will change some of the alignment angles. It is also important

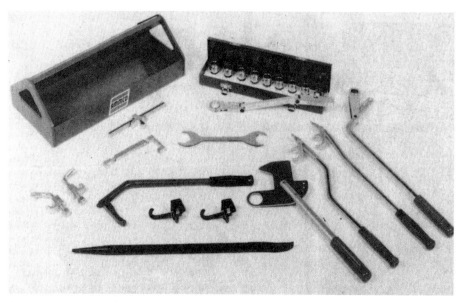

Figure 34-1 Sometimes a special set of alignment tools are used to make the adjustments easier and faster; in some cases, a particular vehicle cannot be aligned without them. (*Courtesy of Ammco Tools, Inc.*)

that the tires, at least on each axle, be the same size and have the correct inflation pressure.

In all cases, the suspension and steering systems must be in good shape before performing a wheel alignment. For example, it does little good to measure and adjust toe if a worn tie-rod end or idler arm allows the toe to change as the car goes down the road; tire wear or handling difficulties will still occur after the alignment is completed.

While doing an alignment, special wrenches and tools are available from various sources to make adjustments easier and faster. The mounting or adjusting bolts on some cars are very difficult to access and to turn using ordinary wrenches, especially if the car has optional accessories. In some cases, it is difficult to see or touch these bolts, much less turn them. The special wrenches are often bent to a configuration that will fit one or two particular bolts on one particular model of car. The special tools are designed to speed up an adjustment or make that adjustment more accurate. Special tools can be of a general nature, such as a tie-rod adjusting sleeve wrench, or of a very specific nature for one particular model of car (Figure 34-1).

34.2 Measuring Alignment Angles

Because some alignment angles change each time the tires are turned from a straight-ahead position and also change with vertical suspension travel, care should always be taken that the tires are pointed straight forward and the car is at the correct ride height. Some alignment angle checks require turning the tires a certain number of degrees from straight ahead; the tires are placed on turntables with degree scales to allow these measurements to be made. If these precautions are not followed, there will be an error in the measurements and corresponding adjustments.

SAFETY TIP: All wheel alignments should result in a vehicle that will drive and handle in a safe and efficient manner. To ensure safe and efficient handling, it is necessary to observe these precautions:

- The car must be given a prealignment inspection to ensure there are no worn or unsafe parts.
- The alignment angles should be adjusted to the specifications recommended by the vehicle manufacturer using the recommended methods.
- Safety glasses should be worn.
- All nuts and bolts should be properly tightened to the correct torque specification.
- The car should be road tested to ensure proper drivability.

Most modern shops use electronic gauges, often called **alignment machines** or **systems.** They attach to the wheel, and they measure the angles electronically and display the readings on computer-like screens. Older units project a light beam onto a graduated screen or display readings on analog meters. They usually measure camber, caster, toe, and SAI. Most of the newer computerized machines electronically compensate for mounting error or wheel runout. Most are **four-wheel systems** that can measure the location and angle of all four wheels at the

Chapter 34 ■ Wheel Alignment: Measuring and Adjusting

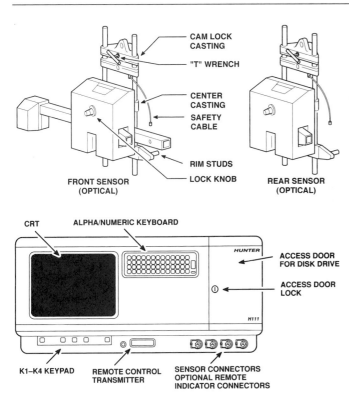

Figure 34–2 This four-wheel system consists of two front sensors, two rear sensors, a console, and a remote control transmitter. Some of the options include a printer. *Note:* Front and rear sensors (toe line) are not shown. (*Reprinted by permission of Hunter Engineering Company*)

same time, and some give a printed readout comparing the vehicle measurements with the specifications (Figure 34–2). Electronic gauges tend to be more accurate, more sensitive, more fragile, and much more expensive. These gauges are quicker and easier to use than magnetic gauges. One manufacturer claims the ability to measure front-wheel camber, caster, and toe and rear-wheel camber and toe in 5½ minutes from beginning to attach the wheel clamps to getting a readout of the measurements.

Electronic alignment systems use three basic ways of displaying alignment measurements: a light beam, a group of meters (**analog type**), and a digital readout on a computer screen; in addition, most systems will print out the readings on a sheet of paper. All alignment systems give two-wheel readouts; some measure and display the angles of all four wheels. Most systems use wheel rim mounting for the **sensing** or **projector units;** these units determine the position of the tire so the measurements can be taken (Figure 34–3). They usually must be **compensated** so they are perfectly aligned with the tire to eliminate mounting runout. Some systems require a manual runout compensation of the sending units. Newer computerized systems have electronic compensation; all the operator needs to do is push one or two switches while rotating the tire.

Alignment systems sometimes take a little longer to mount and set up on a car than magnetic gauges, but they measure and read out more angles and the display is much larger so they can be read accurately from a distance, even from under the car. Modern systems provide the vehicle specifications; a readout of all angles plus setback and thrust; coaching or prompting in setting the equipment up and making measurements; a detailed inspection report; a detailed printout of the alignment settings; and, on some units, illustrated directions of where and how to make the measurements (Figure 34–4). With a little experience, a wheel alignment can be done much faster and more accurately with these systems than with portable equipment.

A recent development is a wheel-less alignment system. This system replaces the wheels with a set of stands that have been adjusted to the same radius and offset as the vehicle's wheels. A digital camber–caster gauge and toe measuring system are built into the front stands, and, on the four-wheel system, a thrust measurement system is built into the rear stands (Figure 34–5). This system offers very good access to the suspension for making adjustments and is less expensive than the electronic alignment system. The procedure for making the measurements is very similar to that of the other types of measuring systems.

Each style of alignment system has some means of checking and **calibrating** itself to maintain accuracy. Computerized systems run their own self-checks.

Alignment systems are used in conjunction with **alignment racks.** Racks provide a convenient, level surface so the front tires (usually) are on turntables and all four tires are at the same, level height while the measurements are being made. They also provide access to the suspension and steering components so that checks or adjustments can be made quickly and easily. Most racks have built-in air-powered jacks so the car or suspension parts can be easily raised for service.

The following different styles of racks are in common use:

1. **Pit racks,** also called **half** or **stub racks.** A pit is constructed in the floor of the shop with an island or peninsula to support the turntables; the pit provides access to the suspension and steering parts. The car is driven straight over the pit with the wheels to be aligned over the turntables.

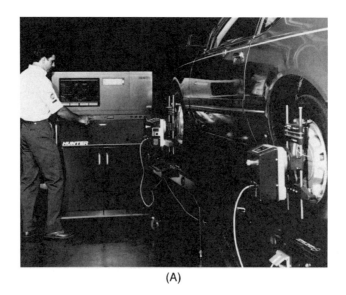

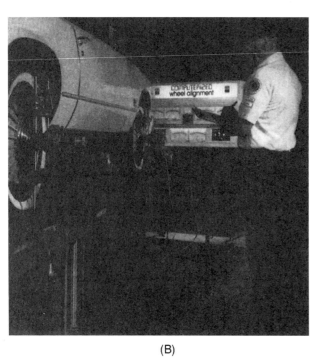

Figure 34-3 Three different styles of alignment systems: system (A) is a computerized four-wheel system with a digital readout, system (B) uses an analog meter readout, and system (C) uses a light projection readout. (*Reprinted by permission of Hunter Engineering Company*)

2. **Fixed/power rack.** Two metal runways with the turntables built into the front end are secured to the shop floor. The rear of these runways can be lowered so the car can be driven on and then raised to a level position. The height of the runways provides a convenient working height and access to the suspension and steering parts.

3. **Hoist rack.** Two metal runways with turntables are attached to a standard, single-post shop hoist. Raising the car on the hoist provides access for repairs or adjustments. The rack is lowered onto four stands (built into the rack) to provide a level surface while the measurements are made (Figure 34-6).

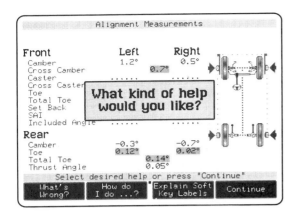

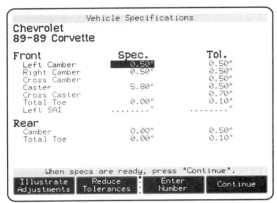
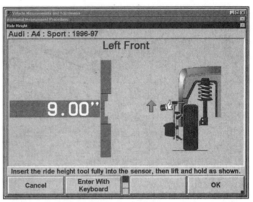
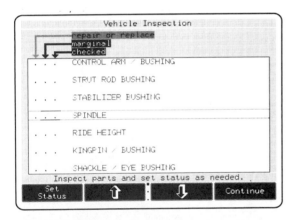
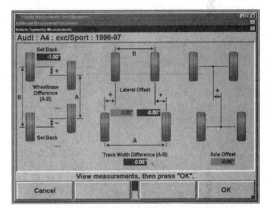

Linear Distances

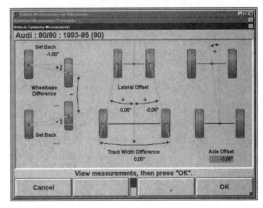

Angles

Figure 34–4 Besides measuring the alignment angles, modern four-wheel alignment systems provide help in using the equipment, alignment specifications, a neatly printed record for the vehicle inspection, videos of how to do certain operations, and a means of measuring vehicle height, setback, and offset. Some systems also provide illustrated procedures for making an adjustment on a particular vehicle. (*Reprinted by permission of Hunter Engineering Company*)

Figure 34–5 A wheel-less alignment system replaces the wheels with a set of alignment fixtures adjusted to the proper height and wheel offset. (*Courtesy of Norkar Technologies, Inc.*)

(A)

(C)

(B)

Figure 34–6 Three different styles of alignment racks: Rack (A) is attached to a single-post shop hoist; it allows raising the car for repairs and lowering it onto supports while the alignment is being made. Rack (B) is a power rack; the rails are lowered so a car can be driven on and raised to provide a level alignment surface. Rack (C) is a pit rack with the turntables set at floor level; the vehicle is driven straight onto the alignment stands. (*Reprinted by permission of Hunter Engineering Company*)

Chapter 34 ■ Wheel Alignment: Measuring and Adjusting

Figure 34–7 A tire being lowered onto a turntable. At this time the tire should be centered to the turntable and pointed straight ahead; the turntable scale should be at zero.

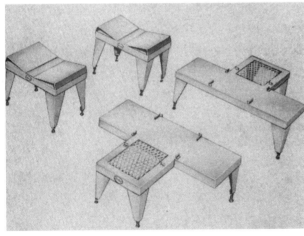

Figure 34–8 A set of floor stands. The turntables are placed on the larger pair of stands, and these two stands are placed under the front tires. The rear tires are placed onto the other pair of stands. This raises the car to a better working height and provides access under the car for adjustments. (*Courtesy of Snap-on Tools Corporation*)

Portable caster–camber gauges that attach to the front hubs using a strong magnet are commonly available. These **magnetic gauges** use a simple bubble level with a scale so the camber angle can be read. The gauge is usually used along with a pair of portable **turntables** or **slip plates,** which ensure that the tires can move or scuff sideways when they are lowered and that the suspension moves to the correct ride height (Figure 34–7). The turntables can be placed directly on the floor, but some shops use two **portable stands** for the turntables and two more stands for the rear tires to provide a level surface for the car while the measurements are being made. The stands also allow access to the suspension and steering components when the adjustments are made (Figure 34–8).

When the center of the wheel does not allow attachment of the gauge or there is damage to the end of the hub, adapters can be used to attach the gauge onto the wheel rim (Figure 34–9). Bubble level gauges are also available in a handheld type; to read camber or caster, this gauge unit is simply held against the side of the wheel (Figure 34–10).

34.3 Wheel Alignment Sequence

The three basic alignment procedures determine the sequence. They are:

- Two-wheel centerline, which assumes that the rear wheels are correctly aligned
- Two-wheel thrust line, which compensates for a slight rear-wheel misalignment (Figure 34–11)
- Four-wheel, which aligns rear wheels first

Both two-wheel procedures adjust the front wheels only.

Most technicians follow a particular sequence of operations while doing a wheel alignment. Part of the sequence is personal preference, part is not. Toe is always the last angle checked and adjusted because it is affected by the others. For example, a more positive camber adjustment will move the top and also the center of the tire outward; this movement will produce less toe-in on cars with rear-mounted tie-rods. When using modern computerized equipment, the measurement sequence is programmed into the machine.

Camber and caster are always measured first and adjusted at the same time or in an order that is determined by the technician's preference or the car. For example, on cars with S-L A suspension and a sliding adjustment, they are made together. On cars with shim or eccentric adjustments, they can be made in any order.

Toe-out on turns is measured along with caster because this is the easiest and quickest time to do it. Measurement of SAI is often skipped unless there is a problem with the car. These two angles are normally not adjusted.

Figure 34-9 A wheel clamp allows the mounting of a magnetic gauge to the wheel instead of the hub; this tool is usually more accurate because compensation can be made for mounting errors. (*Courtesy of Ammco Tools, Inc.*)

Figure 34-10 This caster-camber gauge is handheld and placed against the wheel rim while making measurements. (*Courtesy of Ammco Tools, Inc.*)

34.4 Measuring Camber

Because it is compared with true vertical, camber could be measured with a carpenter's level or a plumb bob. However, the bulge on the lower side of the tire would get in the way and affect the measurement. Each gauge

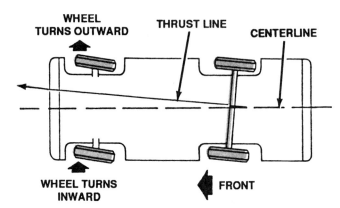

Figure 34-11 A centerline alignment aligns the front tires to the geometric centerline of the vehicle. A thrust line alignment (shown) adjusts front-wheel toe to the thrust line to compensate for rear-wheel-position errors. (*Courtesy of General Motors Corporation, Service Technology Group*)

measures camber from true vertical. The greatest error in measuring camber probably results from the attachment of the gauge to the measuring surface, either the hub or the wheel rim. If these surfaces are not true or straight, the measurement will be wrong and the resulting adjustment will also be wrong. All wheel rim mount systems should be adjusted to compensate for rim runout or mounting errors. If a hub mount gauge is used, the face of the hub should be checked to ensure that it is true.

34.4.1 Measuring Camber Using an Alignment System

Currently, there are several different types of alignment systems. The specific operation of each system varies among different models and manufacturers and is too varied to describe in a text of this type. The following description is very general.

To measure camber using an alignment system, you should:

1. Drive the car onto the alignment rack and position it with the tires over the turntables.
2. Attach the sensors/projectors securely onto the wheel (Figure 34-12).
3. Raise the tire or the turntable and, using the correct procedure, compensate for any sensor or projector mounting errors (Figure 34-13).
4. With the tires in a straight-ahead position, lower them onto the turntables, making sure that the tires are centered on the turntables and the turntables are positioned at zero.
5. Bounce the car up and down to ensure that the suspension is settled to ride height.
6. Read the camber display (Figure 34-14).

Chapter 34 ■ Wheel Alignment: Measuring and Adjusting   763

Figure 34-12 The sensor head of a wheel alignment system is placed in position at the wheel bead flange and is being locked in place. *(Reprinted by permission of Hunter Engineering Company)*

Figure 34-13 One of the runout compensating screws is being turned to correct mounting errors. Readings are taken with the wheel turned to three different positions, and the compensating screws adjust to the average position. Newer computer units self-compensate in a quick and easy step. *(Reprinted by permission of Hunter Engineering Company)*

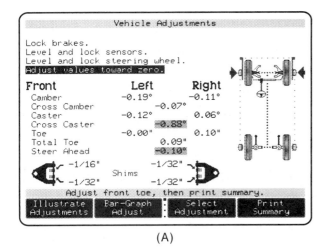

(A)

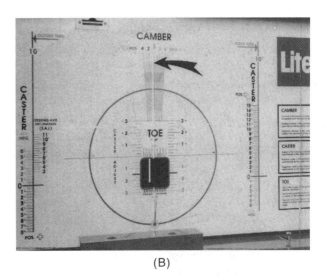

(B)

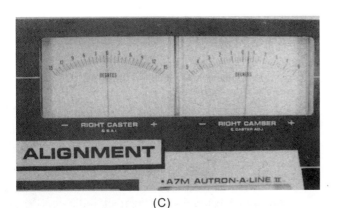

(C)

Figure 34-14 The camber readings from three types of alignment systems: The screen on the computer unit (A) is displaying several angles in a digital format plus the shim change needed for the adjustment. The light projection unit (B) is reading +1/2 degree (at arrow). The analog meter unit (C) is also reading +1/2 degree. *(Reprinted by permission of Hunter Engineering Company)*

NOTE: Many alignment systems with their live displays allow a quick check for correct ride height. Note the static camber and toe readings. Then push down on the front bumper and note any change in alignment. A slight change is normal. Next lift the front bumper the same amount and again note the change. If the changes that occurred were small and went the same direction from the static height, the ride height is probably good. If the pushed-down height went one way and the pulled-up height went the other direction, the height is probably wrong, usually too low. Remember that suspension systems must be designed to allow brief alignment changes during suspension travel, and they should come back to the proper settings at ride height.

NOTE: If you believe that spring sag has caused the alignment angles to be off, jack up the vehicle body to the proper ride height and recheck the alignment. If the angles come into specifications, you know the problem can be cured by installing new springs.

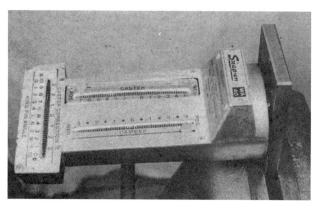

Figure 34–15 The accuracy of a magnetic caster–camber gauge can be checked by placing the gauge on each side of a flat metal plate. The camber readings should both be zero, or if one side gives a positive reading, the other side should be equally negative. If not, the gauge should be calibrated. (*Courtesy of Snap-on Tools Corporation*)

34.4.2 Measuring Camber Using a Magnetic Gauge

Many small shops prefer **portable alignment gauges** because they are less expensive and are relatively durable. They also take only a small amount of skill and are fairly fast to use.

Magnetic gauges can be checked for accuracy by one of two different and fairly easy methods. If a vertical iron surface is available—some door frames, for example—attach the gauge to the metal surface; if the surface is truly vertical, the camber gauge should read zero. The gauging surface can be checked with a carpenter's level if there is any question. If a flat piece of metal that has two smooth and parallel sides is available, clamp the metal in a vise and measure the camber on each side of the metal. If both sides read zero or if one side shows a negative reading that is equal to a positive reading on the other side, the gauge is accurate. Most magnetic gauges have a provision for calibrating or correcting the camber scale (Figure 34–15).

To measure camber using a magnetic gauge, you should:

1. Park the car on a level surface.
2. Remove the hubcap and the front hub dust cap. Clean the vertical machined surface at the end of the hub. Check this surface for any bumps or burrs that might prevent the gauge from seating completely.

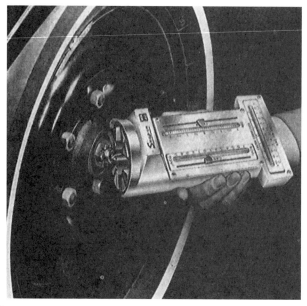

Figure 34–16 After the hub surface is cleaned, the magnetic gauge is centered and then attached to the hub. (*Courtesy of Snap-on Tools Corporation*)

3. Raise the tire off the ground, grip the tire at the top and bottom, and rock the tire to check for loose wheel bearings. Readjust the wheel bearings, if necessary.
4. Center the magnetic gauge with the spindle and carefully, because of the strong magnet, attach the gauge to the hub (Figure 34–16).
5. With the tire still in the air, cross-level (level in a horizontal or front-to-back direction) the gauge and read the amount of camber indicated on the

Chapter 34 ■ Wheel Alignment: Measuring and Adjusting

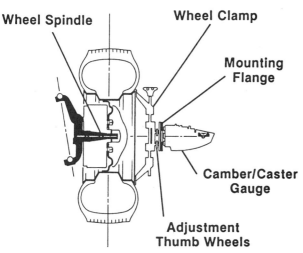

Figure 34-17 A wheel clamp attaches to the outside or inside edge of the wheel flange and provides a mounting flange for a magnetic caster–camber gauge. Note the adjustment/compensation thumb wheels to correct mounting flange runout. (*Courtesy of Ammco Tools, Inc.*)

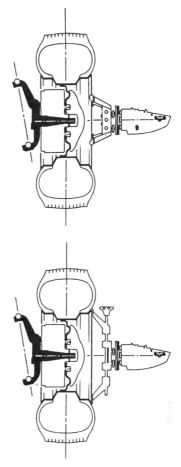

Figure 34-18 Runout of the wheel clamp is corrected by measuring camber with the wheel rotated to three different positions; the adjustment thumb wheels are then turned so the gauge reads the average reading. After compensating for runout, all three readings should be the same. (*Courtesy of Ammco Tools, Inc.*)

gauge (this reading is only to check for possible errors). Rotate the tire one-half turn or 180 degrees, rotate the gauge one-half turn, reread the gauge, and compare the two readings; if they are different, a defect in the gauge mounting is causing an error in one or both readings. If there is an error or the gauge cannot be mounted directly onto the hub, a rim mount adapter should be used or the hub surface should be trued. A rim mount adapter is attached to the wheel rim and has a provision for calibrating or compensating for errors. Calibration procedures vary among different manufacturers.

To use a rim mount adapter, you should:

1. Attach the rim mount adapter securely to the wheel rim and attach the magnetic gauge to the adapter (Figure 34-17).
2. With the adapter straight up and down, cross-level the gauge and read the camber, hold the gauge still while you rotate the tire one-half turn, and reread the gauge.

NOTE: Some rim mount adapters have three compensating screws. When using one of these units, a camber reading should be taken every one-third turn, or 120 degrees. A reading should be taken with each compensating screw straight up (Figure 34-18).

3. If the two (or three) readings are not the same, determine the average reading and adjust the compensating screw(s) until the camber gauge shows the average reading. For example, if the readings are +1/4 degree and +3/4 degree, the compensating screw would be adjusted to a reading of +1/2 degree.

NOTE: The compensating screw must be in a vertical position during this adjustment.

4. Repeat step 2 and, if necessary, step 3 until the readings taken in step 2 are the same.
5. Position a pair of turntables so they are centered under the road-contact patch of the tire. If caster or toe measurements are also going to be made, the front wheels should be straight ahead, with the amount of toe split equally at each wheel, and the turntable scales should be at zero.

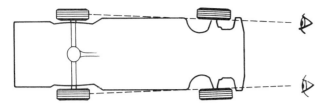

Figure 34–19 Splitting the toe can be done by sighting down the edges of both front tires. The tires are turned until the line of sight meets each rear tire at the same location (relative to the center of the car).

Figure 34–20 While doing alignment checks with the turntables placed on the floor, the rear tires should be placed on blocks to keep the car level.

Figure 34–21 The camber gauge is read by locating the center of the bubble on the camber scale and reading the scale at that point. This gauge is reading about −1/4 degree.

To **split the toe** you should:

1. Sight past the inner edge of the right front tire and note where your line of sight meets the right rear tire.
2. Sight past the inner edge of the left front tire and note where your line of sight meets the left rear tire (Figure 34–19).

NOTE: On cars with toe-out, it is necessary to sight along the outside of the tires.

3. If the lines of sight do not meet the rear tires in the same place, strike the appropriate sidewall of one of the front tires so the tires are turned slightly to the right or left.
4. Repeat steps 1, 2, and 3 until the sight lines are equal.

NOTE: If caster is going to be measured, the rear tires should be placed on blocks that are equal to the height of the turntables; the car must be level in a front-to-back as well as a side-to-side direction (Figure 34–20).

5. Pull the locking pins from the turntables and lower the car onto the turntables. Bounce the end of the car up and down; you should be able to see the turntables and the tires move sideways during the bouncing action.
6. Cross-level the magnetic gauge and read the camber scale on the gauge. The amount of camber is indicated where the center of the bubble aligns with the gauge scale (Figure 34–21 and Figure 34–22).

34.5 Camber Specifications

All car manufacturers publish wheel alignment specifications that include camber. A typical specification would be 0 to 3/4 degree with a maximum variation or spread between the two tires of 1/8 degree. This means that a camber measurement of 0 or positive 3/4 degree or anywhere in between would be correct, as long as both readings are within 1/8 degree of each other. All specifications are assumed to be positive unless otherwise noted. Another way of publishing a specification is 1/2 ± 1/4 degree. With this specification, +1/2 degree − 1/4 = +1/4 degree and +1/2 + 1/4 = +3/4 degree; anywhere in between +1/4 degree and +3/4 degree is acceptable. In this case, +1/2 degree would be the most ideal setting (Figure 34–23).

Specifications for domestic cars are usually given in degrees or fractions of a degree (2 or 1/4 degree). Decimals are occasionally used for portions of a degree (e.g., 0.25 instead of 1/4 degree). Imported car specifications are also given in degrees, but angles less than 1 degree are measured in minutes and sometimes even seconds (1/4 degree becomes 15').

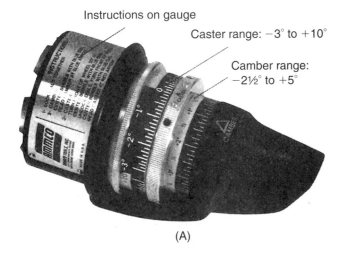

(A)

(B)

Figure 34–22 The camber gauge shows about +1 degree (A). It is read by turning the camber scale (B) until the bubble is centered and then reading the scale at the arrow. (*Courtesy of Ammco Tools, Inc.*)

Some alignment technicians believe that the most ideal camber angle for decreased tire wear and better drivability is for the front wheels of a RWD car to be 1/4 to 1/2 degree positive; they will try to make their final settings (within range of the specifications) to be within this 1/4-degree range.

34.6 Adjusting Camber

If the camber readings fall within the specifications and no camber wear is present on the tire, there is no need for adjustment. If the measurements are outside of the specification range or they are within the range and the tire is showing camber wear, camber should be adjusted. The camber range on some cars is quite wide, sometimes as much as 2 degrees, and some cars show tire wear when they are as little as 1/4 to 1/2 degree from ideal. It is very possible for a tire to be at the positive end of the camber range and within specifications and yet show positive camber wear on the tire. In such cases the angle should be adjusted to a more negative position (Figure 34–24).

The amount of camber change that will occur at the tire while an adjustment is being made varies from car to car depending on the height of the steering axis (between the two steering pivots). Because we are working with degrees, the amount of change adjustment becomes an arc of a circle, with the radius of the circle being the length of the steering axis. For example, if an adjustment is being made on a car with S-L A suspension and there is a span on the steering knuckle of 14 in. (35.6 cm) between the ball joints, we will be working with a radius of 14 in. A sideways movement of 1/8 in. (3 mm) will cause a camber change of 0.5 or 1/2 degree (Figure 34–25). The circumference of a circle with a radius of 14 in. is almost 88 in. (223 cm); if we divide 88 by 360, the result is 0.24 in. (6.1 mm). If an arc of 0.24 in. is equal to 1 degree, then an arc of 0.125 in. (1/8 in.) would equal about 0.5 degree. If we were adjusting a car equipped with a strut and there was a span of 36 in. (91 cm) between the ball joint and the upper strut mount, the radius would be 36 in., the circumference would be 226 in. (574 cm), a degree would become a span

SPECIFICATIONS

FRONT WHEEL ALIGNMENT	Acceptable Alignment Range	Preferred Setting
CAMBER—All Models	−0.2° to +0.8° (−1/4° to +3/4°)	+0.3° (+5/16°)
TOE—All Models		
Specified in Inches	7/32" OUT to 1/8" IN	1/16" OUT ±1/16"
Specified in Degrees	.4° OUT to .2° IN	0.1° OUT ±.1°

Figure 34–23 Wheel alignment specifications. Note that this manufacturer prints its specifications in both decimals and fractions and that the readings are given in an acceptable and a preferred setting. (*Courtesy of Chrysler Corporation*)

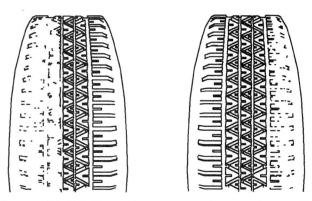

Figure 34–24 These tires are showing possible camber wear. If the outside of these tires was to the right, the camber on the left tire was too negative and the right tire was too positive. (© *Saturn Corporation, used with permission*)

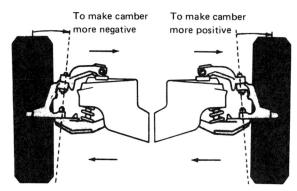

Figure 34–26 Camber is adjusted to a more positive (+) position by moving the top of the tire outward; this is done by moving the upper ball joint outward or the lower ball joint inward. Camber is made more negative (−) by adjusting in the opposite direction. (*Courtesy of Ford Motor Company*)

of 0.63 in. (1.6 cm), and the 1/8-in. adjustment would now cause a camber change of only 0.2 degree.

Many cars provide for adjusting camber. These adjustments are made by moving the top or bottom of the steering knuckle and tire inward or outward (Figure 34–26). The usual adjustment methods are to (1) add or remove shims between the control arm mounts and the frame, (2) turn eccentrics where the control arm attaches to the frame or ball joint or where the strut connects to the steering knuckle, or (3) slide the control arm pivot shaft or the top of the strut inward or outward or the connections between the strut and steering knuckle along slotted mounting holes (Figure 34–27).

Some front suspensions are manufactured with no provision for adjustment. In these cases, if camber goes out of specification, something is worn or bent. It is now the responsibility of the alignment technician to determine whether a weak spring, worn bushings, or bent suspension arms, spindle, and/or frame is at fault. A suspicion of a bent steering knuckle or strut can be confirmed by comparing the camber reading with the steering axis inclination specification or included angle specification.

Remember that the included angle is the angle between the camber and SAI angles. If the steering knuckle or strut is bent so the spindle changes position, the included angle will change. For example, let us imagine a car that comes in with a camber reading of −3/4 degree, a camber specification range of 0 to +1 degree, and an included angle specification of 9 degrees. There is no specification for SAI, but an included angle of 9 degrees minus

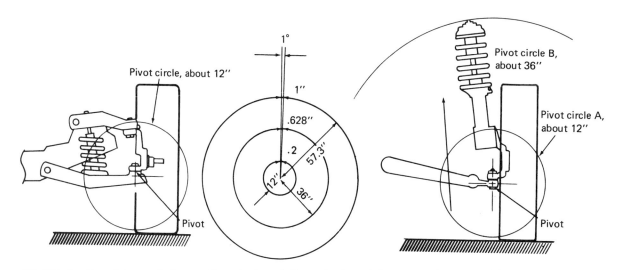

Figure 34–25 As the distance between the pivot and the adjustment increases, larger and larger changes must be made to get the same amount of angle change. A 1-degree change would require a 1-in. shim if the adjustment distance was 57.3 in., a 0.628-in. shim at 36 in., or a 0.2-in. shim at 12 in.

Chapter 34 ■ Wheel Alignment: Measuring and Adjusting

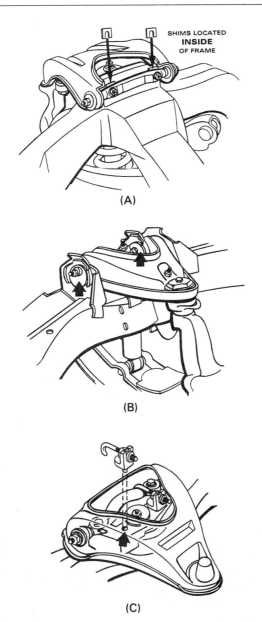

Figure 34-27 S-L A suspensions are usually adjusted by adding or removing shims (A), turning one or two eccentric cams (B), or sliding the control arm shaft inward or outward along slotted holes in the frame (C). (*Reprinted by permission of Hunter Engineering Company*)

a camber angle of +1/2 degree (ideal) gives us an SAI angle of 8½ degrees. If you were to measure the SAI on this car and it measured 8½ degrees or close to it, you would know that the strut was bent and should be replaced. If the SAI measured about 10 degrees, you would know that the included angle was almost correct (9 + 3/4), so the steering knuckle or strut is not bent; however, damage to the car has moved the upper strut mount or the lower control arm mounting locations, which has changed the SAI angle and, in turn, the camber angle (Figure 34-28). If a front-end technician were to bend the strut to correct the camber (in the second example), camber might end up correct, but the car would probably have handling difficulties because of the incorrect SAI angle. Remember that a quick and simple check for a bent strut is to compare the distances between the brake rotor and the strut on each side of the car. The faulty parts should be corrected or replaced.

Camber adjustments will always change the track width slightly and therefore will also change toe; it is always a good practice to check toe after adjusting camber or caster.

Bending tools are available from aftermarket suppliers to bend axles or steering knuckles into alignment. Bending often saves the technician time and the customer money, but it should be used with discretion because the strain placed on the parts could cause later damage or failure of the parts. Since it does not correct the cause of the misalignment, many technicians believe that bending only hides a problem instead of curing it. *No car manufacturer recommends the bending of a suspension component.*

An opportunity for adjustment is provided by mounting of the lower control arms on removable subframes. There is often room at the bolt holes to loosen the mounting bolts so the subframe and control arms can be shifted to a better position.

34.6.1 Adjusting Camber, Shims

Shim adjustments are the favorite of many alignment technicians. Shims are fairly easy to change, and it is easy to make very small or large adjustments and to keep track of how much adjustment was made. The shims can be located in pairs at the upper or lower control arm or in a single group at the lower, single control arm. The upper control arm shaft can be mounted inboard or outboard of the frame mounting bracket. Adding shims on a control arm shaft that is inboard of the mounting bracket will cause a more negative camber change; if the control arm shaft was outboard, a more positive change would occur. If the shims are used in pairs, two equal-sized shims are added or removed, front and back, for camber adjustments; a caster change will occur if the number of shims changed at one end is different from the number changed at the other end (Figure 34-29).

Shims are available in various thicknesses between 1/64 and 3/16 in. (0.4 and 4.7 mm) (Figure 34-30). Most technicians use the largest shims they can in order to reduce the total number of shims. It is a good practice to use a maximum of five shims at any one location; the chance of control arm loosening and shim loss increases as the number of shims increases (Figure 34-31). Camber and caster are usually adjusted at the same time, using the same shims. Shim-style caster adjustments are described in Section 34.8.1.

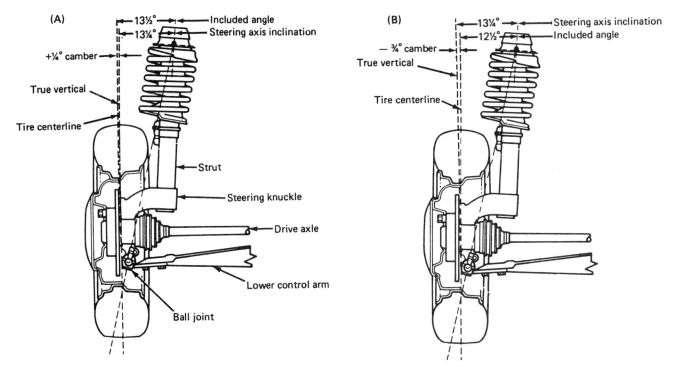

Figure 34–28 This car (A) has a camber angle of +1/4 degree, an SAI angle of 13¼ degrees, and an included angle of 13½ degrees (1/4 + 13¼ = 13½). The strut (B) with −3/4-degree camber and 13¼-degree SAI has an included angle of 12½ degrees. The included angle of (B) is out of specification, and this indicates a bent strut.

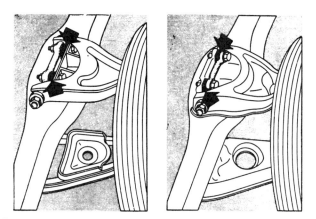

Figure 34–29 Increasing the shim pack at the left will make camber more negative, whereas increasing the shim pack at the right will make camber more positive. (*Courtesy of Snap-on Tools Corporation*)

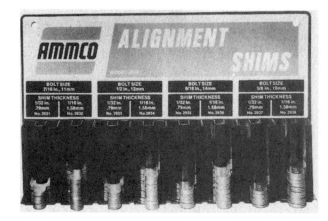

Figure 34–30 Alignment shims are available in different widths to fit different bolt sizes. They are commonly available in 1/32- and 1/16-in. thicknesses; 1/64- and 1/8-in. thicknesses are also available. (*Courtesy of Ammco Tools, Inc.*)

To adjust camber using shims, you should:

1. Measure camber, record the readings, and determine how much change is needed and in what direction the change should be made.

2. Locate the shims and determine whether shims will need to be added or removed.

3. Loosen the control arm mounting bolts, pry the control arm away from the bracket, and change the shims by the desired amount (Figure 34–32).

4. Snug down the control arm mounting bolts and remeasure camber.

5. If the camber angle is correct, tighten the mounting bolts to the correct torque. If the camber angle is still wrong, repeat steps 3 and 4. At this time, it should be fairly easy to determine how much of an additional shim change is needed by noting how much of a change occurred during the last adjustment.

Chapter 34 ■ Wheel Alignment: Measuring and Adjusting

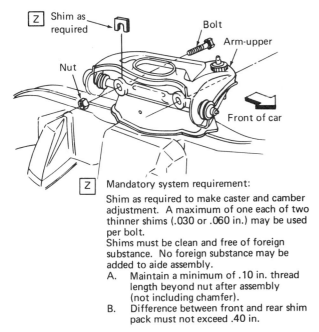

Figure 34–31 Some of the practices that should be followed when changing shims to ensure a good job. (*Courtesy of General Motors Corporation, Service Technology Group*)

34.6.2 Adjusting Camber, Eccentric Cams

Eccentric cams look like off-center washers that are attached to the ends of the inner control arm mounting bolt. The frame brackets for these cam bolts have slotted openings through which the bolts are positioned and a pair of raised shoulders that the eccentric cams act against. Turning the cam bolt will cause the bolt and the control arm bushing to move sideways, in and out, in the frame bracket. When the nut is tightened on the cam bolt, the bolt and the control arm are locked in that position (Figure 34–33).

Eccentric cams are used at the upper or lower control arms on S-L A suspensions, on some strut suspensions where the strut is connected to the steering knuckle, and on some IRS suspensions. The eccentric cams allow very fine or rather large adjustments. Like shims, eccentric cams can be used in pairs on control arms or singly on single bushing control arms. Also like shims, when cams are used in pairs, the cams, front and back, should be turned equally when adjusting camber; turning only one cam will change caster as well as camber.

Occasionally, on very rusty cars, cam bolts rust in place and become very hard to turn. Penetrating-type oils or solvents help loosen the cam bolts. They should be applied early in the alignment sequence, during the pre-alignment inspection, so they will have a chance to soak in by the time the adjustment is made. Sometimes it helps to use an air hammer with a punch against the head of the bolt to vibrate the bolt loose or vibrate the penetrating oil

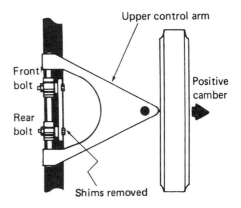

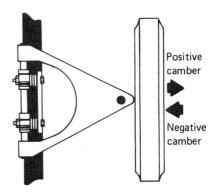

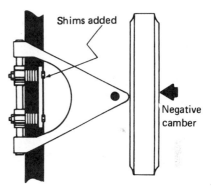

Figure 34–32 When adjusting camber, an equal change should be made at the front and rear of the shim pack.

in and around the bolt. Damaged or broken cams or cam bolts should be replaced with new ones.

To adjust camber using eccentric cams, you should:

1. Measure the camber, record the readings, and determine the amount of camber change desired.

2. Locate the cams and determine which way they need to be turned. It is a good practice for beginners to mark the position of the cam in case it becomes necessary to return to the starting point.

3. Loosen the retaining nut on the cam bolt and, using a wrench on the head of the bolt, turn the cam bolt enough to move the control arm the desired

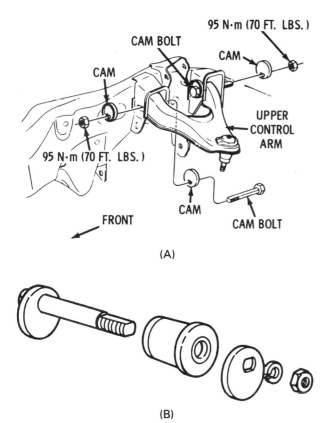

Figure 34–33 This upper control arm (A) is equipped with eccentric cams to adjust caster and camber. Note the frame brackets on each side of the cam for the cam to push against and the slotted hole for the cam bolt. A replacement cam bolt set plus a control arm bushing is shown in (B). (*A is courtesy of General Motors Corporation, Service Technology Group; B is courtesy of Moog Automotive Inc.*)

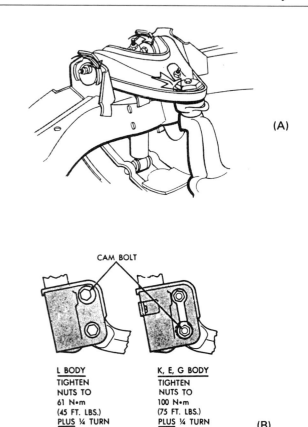

Figure 34–34 Turning each eccentric cam the same direction an equal amount will change camber without changing caster (A). Some struts use a single eccentric cam where the strut attaches to the steering knuckle (B). (*A is reprinted by permission of Hunter Engineering Company; B is courtesy of Chrysler Corporation*)

amount. If a pair of cam bolts are used, turn the other cam bolt an equal amount in the same direction (Figure 34–34).

4. Remeasure camber and, if necessary, repeat step 3. It should be fairly easy at this time to determine how far each cam needs to be turned to complete the adjustment.

5. When camber is correct, tighten the retaining nut to the correct torque.

34.6.3 Adjusting Camber, Sliding Adjustment

Sliding adjustments are used at the upper control arm inner shaft on some S-L A suspensions, at the connection between the strut and steering knuckle on some strut suspensions, and at the mount between the strut and fender well on some cars. A slotted opening is provided in the frame bracket to allow the loosened control arm shaft mounting bolts to slide to different positions across the top of the bracket. The bottom of the control arm shaft is serrated to help lock it in position on the bracket when the mounting bolts are tightened.

Many technicians have difficulty making fine adjustments with sliding adjusters; when they loosen the mounting bolts to allow an adjustment, the control arm slides, moves too far, and loses the starting point. The mounting bolts should be loosened just enough to allow the adjustment and no more. Also, tools are available to help gradually move the control arm the desired amount and to hold it in that position (Figure 34–35). Some technicians use a jack under the frame to help them; the control arm tends to move inward when the car's weight is on the tires and outward when the frame is lifted. Some technicians prefer to move one end of the shaft at one time in the following sequence: barely loosen the rear bolt, loosen the front bolt, adjust the front end of the shaft, snug the front bolt, loosen the rear bolt, and then adjust the rear end of the shaft.

Chapter 34 ■ Wheel Alignment: Measuring and Adjusting

To adjust camber using sliding adjustments, you should:

1. Measure the camber, record the readings, and determine what camber changes are needed.
2. Locate the adjustment location and determine which way the adjustment should be made. It is a good practice for the beginner to mark the location of the sliding parts in case it is necessary to return to the starting point.
3. Loosen the mounting bolt(s), attach the correct tool, and slide the control arm shaft or strut mount bracket the desired amount.
4. Snug down the mounting bolt(s) and remeasure camber. Repeat step 3, if necessary.
5. Tighten the mounting bolt(s) to the correct torque.

34.6.4 Adjusting Camber, Struts

The procedure for adjusting camber on a strut is to move the top of the strut outward or inward to make camber more positive or negative. Some strut suspensions are built with a camber adjustment; most are not. However, most can be adjusted. The exact procedure can be designed by the vehicle manufacturer or the aftermarket provider. Struts that bolt to the steering knuckle can be adjusted at this connection. This can be a cam adjustment (Figure 34–34) or a sliding adjustment (Figure 34–36). The upper mounts of some struts have enough room to allow the position to be adjusted (Figure 34–37). On some, you must drill out a spot weld or pop rivet. On some cars, the manufacturer provides a template or plan to show the location to file the mounting holes to provide for adjustment (Figure 34–38). Some strut mounts are made with the top of the strut rod off center to the mounting bolts; rotating the mount will adjust camber and/or caster (Figure 34–39).

Sometimes, the engine cradle can be shifted sideways to adjust camber. This movement will reduce the camber toward negative on one side and increase the camber on the other side. On some cars, you can also shift the cradle forward or backward on one or both sides to adjust caster. A cradle shift is accomplished by loosening the cradle mounting bolts and prying the cradle to the new position (Figure 34–40).

Aftermarket sources market special kits to provide a camber adjustment on some of the more popular imported cars with nonadjustable strut suspensions. These kits vary depending on the car model. On some cars, they are an offset inner control arm pivot bolt. On others, they are a way to convert a fixed inner control arm pivot bolt

(A)

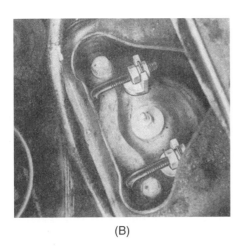

(B)

(C)

Figure 34–35 Several special tools have been developed to help make sliding adjustments easier and faster. (*Courtesy of SPX-OTC*)

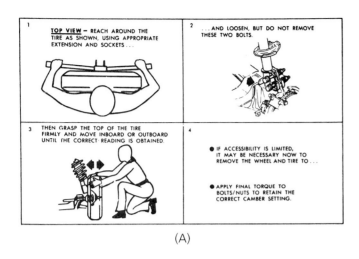

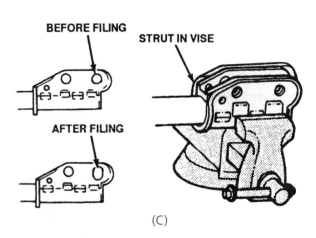

Figure 34-36 On some struts, camber can be adjusted by loosening the bolts between the strut and knuckle and pulling the tire to the correct position (A). Tool J 29862 is positioned on the strut attachment bolts and is used to pull the knuckle so camber will be correct (B). Some models require elongating the holes in the strut to allow enough adjustment (C). (*Courtesy of General Motors Corporation, Service Technology Group*)

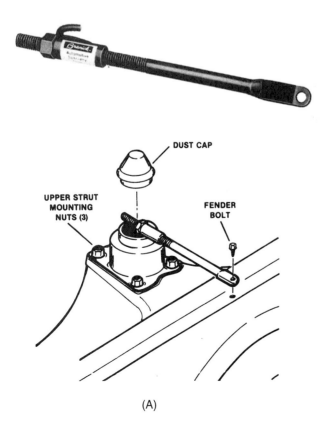

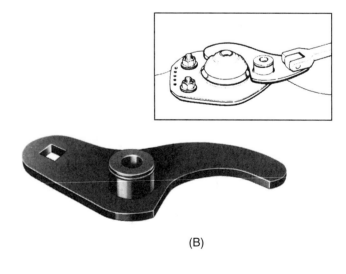

Figure 34-37 A special tool can be used to pull the top of the strut on some Camaro and Firebird models to make camber more positive (A). Another special tool is used to change the strut position on some Thunderbird models (B). (*Courtesy of Branick Industries, Inc.*)

Chapter 34 ■ Wheel Alignment: Measuring and Adjusting

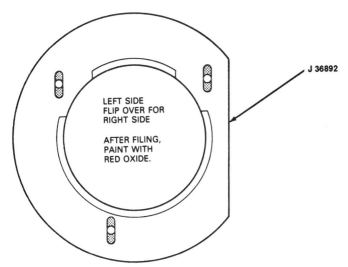

Figure 34-38 This template is used to mark the strut mounting holes. Filing the shaded areas will provide room to adjust the strut position and adjust camber. (*Courtesy of General Motors Corporation, Service Technology Group*)

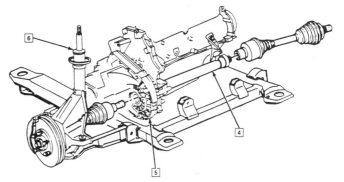

Figure 34-40 Many modern FWD cars use a subframe, commonly called a cradle, to support the engine and transaxle (5). Note that the lower control arms and struts (6) attach to the cradle, so a change in cradle position will also change the alignment. Some RWD cars also have the lower control arms attached to a subframe or cross member. (*Courtesy of General Motors Corporation, Service Technology Group*)

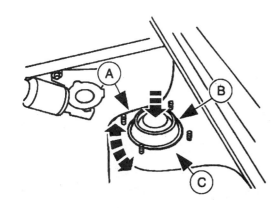

Direction Indicator	Difference From Standard Position	
	Camber Angle	Caster Angle
A	+ 14 Minutes	+ 14 Minutes
B	+ 29 Minutes	0 Degrees
C	+ 14 minutes	- 14 Minutes

Figure 34-39 This strut piston rod is off center to the mounting bolts; rotating it to a different position will change camber and/or caster. (*Courtesy of Ford Motor Company*)

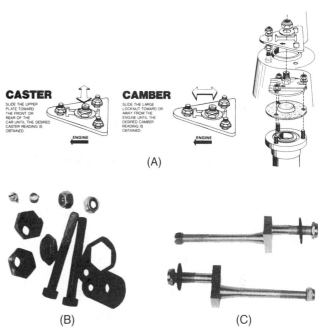

Figure 34-41 Aftermarket kits can be used to replace the upper strut mount (A) or lower control arm inner pivots (B and C) to provide an adjustment. Modification of the mounting points is often required. (*A is courtesy of Moog Automotive Inc.; B and C are courtesy of Arn-Wood Company*)

to one with an eccentric bushing; use of this style of kit usually requires some modification to the control arm mounting bracket (Figure 34-41).

34.6.5 Adjusting Camber, Miscellaneous Styles

Additional adjustment methods can be used to adjust camber on other vehicles. Most described here are built into the car by the manufacturer; a few are aftermarket modifications that provide an adjustment method on nonadjustable suspensions or provide additional adjustment on some suspensions that have a limited adjustment range. This text does not completely describe the adjustment procedure for each of these types; the procedure is usually described in manufacturer or technician service manuals.

Model year 1961 to 1976 Cadillacs used an eccentric bushing between the ball joint stud and the steering knuckle. Camber is adjusted by loosening the ball joint stud nut and rotating the bushing.

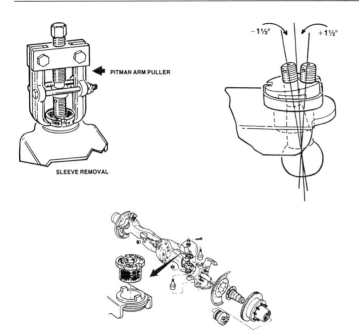

Figure 34-42 An eccentric bushing is used on some 4WD vehicles to connect the upper ball joint to the axle; this bushing can be replaced with one that has a different angle or rotated to change the camber setting. (*Courtesy of Dana Corporation*)

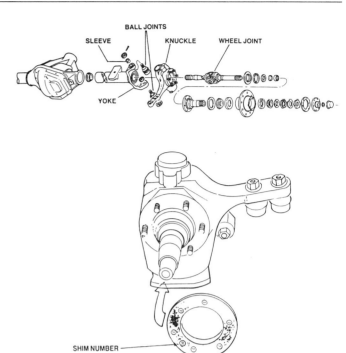

Figure 34-43 Special tapered shims are available to change camber on some 4WD vehicles; this shim, available in tapers of different degrees, is placed between the steering knuckle and spindle. (*Courtesy of Dana Corporation*)

Late-model Ford 4WDs and pickups (both 2WD and 4WD) use an eccentric sleeve between the axle and the upper ball joint stud. A similar aftermarket bushing is available for other makes of 4WDs. The sleeve can be replaced with one of a greater or lesser amount of offset, from 1/8 or 1/4 degree to 1½ or 2 degrees depending on the exact part. The bushing can also be rotated to reduce its effect or change caster, in some cases (Figure 34-42).

Tapered shims are available from aftermarket sources to fit between the spindle and the steering knuckle on 4WD axles. They come in various tapers between 1/8 and 1½ degrees. The tire and wheel, hub and wheel bearings, and spindle must be removed for the shim to be installed or changed. These shims should be used with discretion on some vehicles because they will cause a misalignment between the brake rotor and the caliper; whether they cause misalignment depends on how and where the caliper is mounted and how large a shim is required (Figure 34-43).

Solid beam axles must be bent to change camber; special fixtures or racks are required for this procedure. Securing the outer end of the axle while jacking up just inboard will bend the axle so camber is more positive; jacking up the outer end with the axle secured just inboard will cause the camber to become more negative (Figure 34-44).

Many older cars with S-L A suspensions that used kingpins provided an eccentric sleeve or bushing threaded onto the outer pivot for the upper control arm. Turning this sleeve adjusted the camber (Figure 34-45).

Figure 34-44 The fixtures on this truck axle are set up to bend the axle in such a way as to make camber more positive. (*Courtesy of Bee Line Co.*)

Aftermarket kits are also available to provide additional camber adjustment for some domestic cars. Some of these models tend to sag at the front cross member and cause camber to become too negative. The car will come in with camber excessively negative and

Chapter 34 ■ Wheel Alignment: Measuring and Adjusting

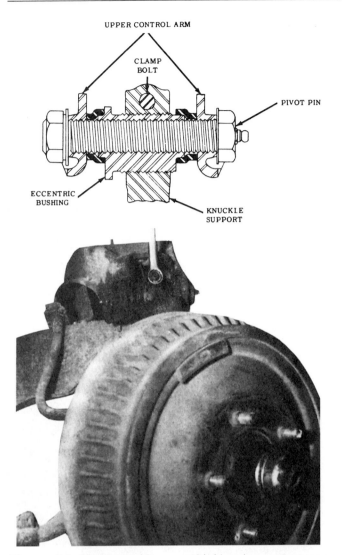

Figure 34–45 Many older cars with kingpins use an eccentric bushing to connect the top of the spindle support to the upper control arm; turning this bushing allows adjustment of caster and camber. (*Courtesy of Ammco Tools, Inc.*)

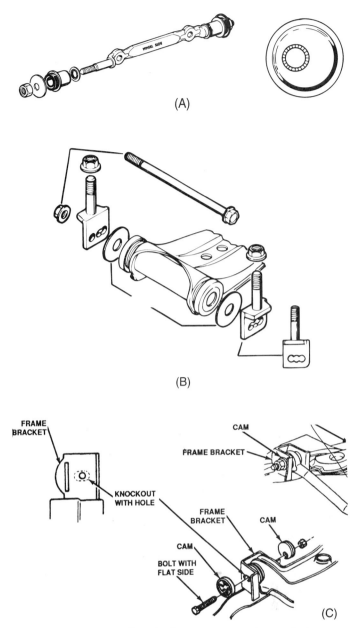

Figure 34–46 Offset bushings or control arm shafts provide more adjustment (A). A kit can be installed on some cars to allow the control arm to be adjusted (B). Another kit can be installed on some pickups (C). (Note that the frame comes with knockouts so the round hole can be changed to a slot.) (*A and B are courtesy of Moog Automotive; C is courtesy of General Motors Corporation, Service Technology Group*)

the adjustment as far as possible in the positive direction; the adjustment is used up. An increased amount of adjustment can be obtained by installing an offset inner control arm shaft, offset control arm bushings, or ball joints with slotted mounting holes; the different styles depend on the exact car model and the aftermarket supplier (Figure 34–46).

34.7 Measuring Caster

If a steering knuckle were machined smoothly so a side of it was parallel to the steering axis, caster angle could be measured with a level and protractor. However, it is not made that way, and it is usually too difficult to get clear access into the steering knuckle to measure it in that manner (Figure 34–47). Because caster inclines the steering knuckle, camber will change when the wheels are turned to the right or to the left. Positive caster will cause the camber to go more negative on the outer wheel and more positive on the inner wheel as a car goes around a corner. If you measured the camber change as a wheel was turned from a 20-degree right turn to a 20-degree left turn and divided this change by 2, you would have the caster angle. All caster gauges measure the vertical wheel position

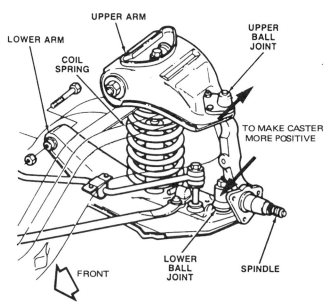

Figure 34–47 Because caster is a forward–backward tilt of the steering knuckle, it is adjusted by moving either the top or the bottom of the steering knuckle forward or backward. (*Courtesy of Ford Motor Company*)

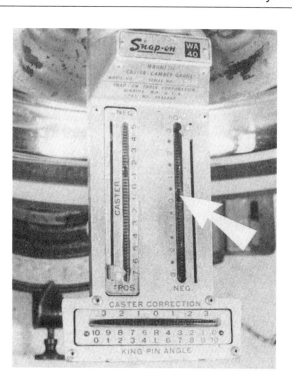

Figure 34–48 A magnetic caster–camber gauge at a straight-ahead (top) and 20-degree turn position (bottom). Note how the camber bubble moves from +1/4 to +1¾ degrees.

(camber) at two different turning angles and convert that change into a caster reading (Figure 34–48).

When caster is measured, the front tires are placed on turntables that have degree scales built into them; they allow the wheels to be turned the specified amount. Most camber gauges have a caster scale built into them. Caster is also measured with alignment systems. After the units have been set up to measure camber, a few additional steps are used to get the caster readings. These steps vary from one brand of equipment to another, but they generally follow the same routine.

Remember that caster readings vary with vehicle loading. Heavy objects in the trunk will compress the rear springs; this will cause caster to become more positive. The amount of caster change depends on the wheelbase of the car and the change in height (Figure 34–49).

Caster specifications, like camber specifications, are given in degrees and fractions of a degree for most domestic cars and in degrees and minutes for most imported cars. Caster specifications for many trucks and pickups are given with a relative frame angle; the angle of the frame is measured, and, using this angle, the correct caster angle can be determined (Figure 34–50).

A rearward cradle shift can cause a caster change to occur. The condition shown in Figure 34–51 will cause the right side caster to be too negative, and the caster spread will probably cause a pull to the right. It will also cause more setback on the right side. This cradle shift could be the result of an accident or improper installation of the cradle after an engine or transmission repair.

34.7.1 Measuring Caster Using an Alignment System

Alignment systems use various methods of measuring caster. As with magnetic gauges, measuring caster is usually the next step after reading camber. Some machines display the directions to follow while measuring

Chapter 34 ■ Wheel Alignment: Measuring and Adjusting   779

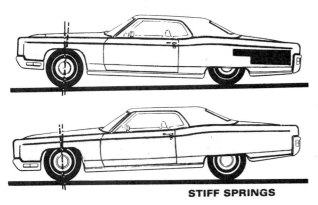

Figure 34–49 Any height change at the back or front of a vehicle that changes the attitude will also cause a change in caster angle. (*Reprinted by permission of Hunter Engineering Company*)

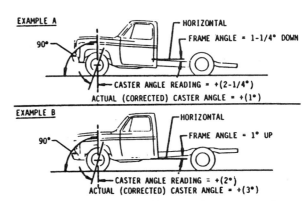

Figure 34–50 Caster specifications for trucks and vans are often given with a reference to frame angle so compensation can be made for load conditions. The case illustrated here will cause a pull. (*Courtesy of General Motors Corporation, Service Technology Group*)

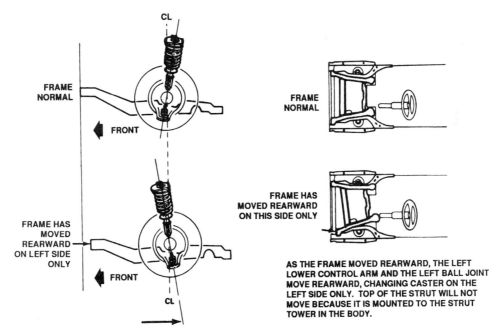

Figure 34–51 If a cradle gets moved toward the rear, caster will change to a more negative (−) position. Note that this movement will also set back the tire. (*Courtesy of General Motors Corporation, Service Technology Group*)

caster. In general, to measure caster, the unit is set up to measure camber, the wheels are turned to the right or left a certain amount, a button is pushed or the steering is held in this position for a designated time to perform the zeroing step, the wheels are turned in the opposite direction a certain amount, a second button is pushed or the steering is held for a short wait period, and the caster reading is displayed (Figure 34–52). Turning the wheels in the two directions is called making a **caster swing**.

Some modern systems can give a readout as either the actual readings or a bar graph, as shown in Figure 34–53.

Many technicians find the bar graph easier to use; as the adjustments are correctly made, the indicator moves to the center of the bar (Figure 34–53).

34.7.2 Measuring Caster Using a Magnetic Gauge

Caster is usually measured right after camber in the alignment sequence, using the same gauge and turntables. Magnetic gauges have a caster scale built into them that is separate from the camber scale; the caster scale has a means to reset it. This scale is set to zero with the front

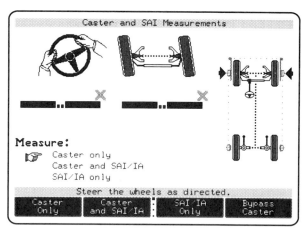

Figure 34-52 On-screen directions are being displayed to measure caster. A caster-only selection has been made, but SAI could also be measured along with caster. The operator needs to turn the steering using the horizontal bar graphs as a guide. *(Reprinted by permission of Hunter Engineering Company)*

wheels turned to a certain position determined by the gauge manufacturer; the usual position is a 20-degree turn. Turning the wheel the wrong direction at the beginning will reverse the readings; positive will become negative and vice versa. After zeroing the scale, the wheel is turned to a second specified position, and the scale is read.

To measure caster using a magnetic gauge, you should:

1. Perform steps 1 through 6 of Section 34.4.2.
2. Turn the front of the wheel inward or outward (as required by the gauge manufacturer) the correct amount, usually to a 20-degree left or right turn; the exact instructions are often printed on the gauge.
3. Cross-level the gauge and zero the caster bubble. Many technicians prefer to use the end of the bubble, rather than the center, for zeroing and reading (Figure 34–54).

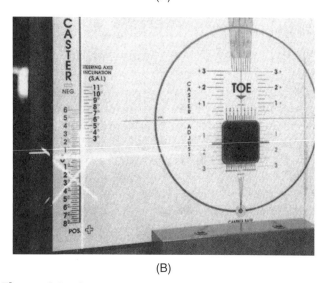

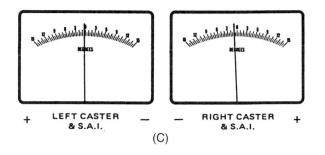

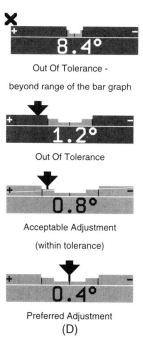

Figure 34-53 The readout on the computer unit is in the "adjust" mode and displaying both the caster and camber measurements (A). The arrow positions to the bar graphs are indicating good settings. The light projection unit is showing +2½ degree of caster (B). The analog meter unit is showing about +1/4 degree on the left wheel and +1 degree on the right (C). A bar graph indicates alignment of the wheel by the position of the indicator and the color of the bar (D). *(Reprinted by permission of Hunter Engineering Company)*

Figure 34–54 Measuring caster requires two steps. The first step is to turn the tire in or out a specified amount as read on the turntable scale and set the caster scale to zero (left). The second step is to swing the tire a specified amount and read the caster scale (right). (*Courtesy of Snap-on Tools Corporation*)

4. Turn the wheel the required amount in the opposite direction (usually enough to produce a 40-degree turn), cross-level the gauge, and read the caster scale. This is the amount of caster for that wheel.

34.8 Adjusting Caster

Caster is adjusted by moving either the top or the bottom of the steering axis forward or backward. This is usually done by changing the length of the strut rod or repositioning the upper control arm (sometimes the lower) so the ball joint moves forward or backward. Moving the upper control arm in this manner is done by moving one end of the inner pivot toward the center of the car while moving the other end toward the outside of the car. This can be done by moving a shim from one end of the pivot shaft to the other, by turning one eccentric inward and one outward, or by sliding one end of the inner pivot shaft inward and the other end outward. Equal and opposite movement on each end of the pivot shaft will cause the caster to change with no change in camber. The top of a strut is moved forward or backward for this same adjustment.

Many technicians prefer to adjust camber first and then caster. When a front-end technician really understands what is happening, both adjustments are made at the same time. For example, removing or adding a shim at one end of the control arm will change both camber and caster. In cases where it is impossible to get both adjustments exactly right, most technicians place a greater importance on camber; if it is off as little as 1/8 or 1/4 degree, tire wear

Figure 34–55 Note the caster-correction scale at the end of this gauge (arrow); it shares the level vial with the kingpin angle scale. This scale can be set to show the amount of change that occurs while caster is being adjusted. (*Courtesy of Snap-on Tools Corporation*)

will result. Caster can be off more than this, and as long as the spread is not too great, drivability problems will not occur. Incorrect caster will not cause tire wear.

As caster is being adjusted, it is rather tedious to keep remeasuring to see if you have completed the adjustment. For this reason, many caster gauges have a caster-correction scale built into them (Figure 34–55). A

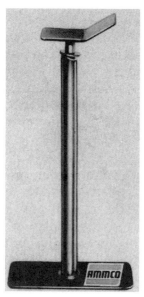

Figure 34-56 A brake pedal jack. This tool is wedged between the brake pedal and car seat to lock the brake rotor or drum to the steering knuckle; this causes the tire to rotate with the steering knuckle so the caster-correction scale can be used. (*Courtesy of Ammco Tools, Inc.*)

brake pedal jack is used to apply the brakes and lock the wheel to the steering knuckle (Figure 34–56). It is always a good practice to use a brake pedal jack while measuring caster to help keep the tire from walking off the turntable. With a brake pedal jack installed, as caster is adjusted and the steering knuckle angle is changed, the caster-correction scale will show this change through the slight rotation of the wheel. A caster change of 2 degrees will cause a wheel rotation of 2 degrees. The alignment technician can now measure caster, determine the amount of desired change, and then adjust caster until the correct change occurs on the caster-correction scale.

34.8.1 Adjusting Caster, Shims

Caster adjustment with shims is fairly easy as long as there are enough shims to work with. The adjustment is made by moving a shim from one of the shim packs to the other, from front to back or vice versa. This operation will cock the control arm slightly and cause the ball joint to move forward or backward. As the upper ball joint moves forward, caster becomes more negative; caster becomes more positive if the ball joint is moved toward the rear.

To adjust caster using shims, you should:

1. Measure caster, record the readings, and determine the change needed to meet the specifications and the desired road crown spread.

2. Adjust the caster-correction scale on the gauge so the amount of adjustment can be noted. There are several ways this can be done. Probably the simplest is to set the caster-correction scale to the existing reading, and then, when the correction scale reaches the desired reading, adjust the caster (Figure 34–57).

3. Install a brake pedal jack to lock the wheel and brake rotor or drum to the steering knuckle.

4. Locate the shims and determine which way a shim should be moved.

5. Loosen the control arm mounting bolts; then remove a shim from one of the shim packs and install it in the other.

6. Snug down the mounting bolts and read the caster-correction scale.

7. After caster has changed the desired amount, tighten the mounting bolts to the correct torque. If the wrong amount of caster correction has occurred, repeat steps 5 and 6 until it is correct. If necessary, caster or camber can be remeasured.

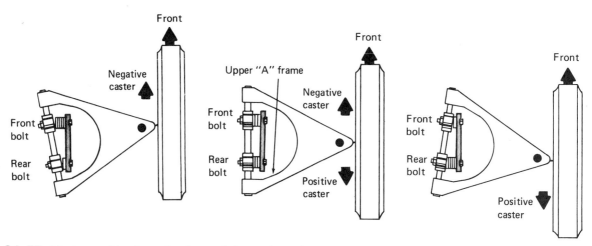

Figure 34-57 Moving a shim from the front of the pack to the rear or vice versa will cause a caster change without changing camber.

34.8.2 Adjusting Caster, Eccentric Cams

Caster adjustment using eccentric cams follows a pattern similar to a shim-type adjustment. Moving one end of a control arm outward while moving the other end inward will move the ball joint forward or backward but not inward or outward.

To adjust caster using eccentric cams, you should:

1. Perform steps 1, 2, and 3 of Section 34.8.1.
2. Locate the eccentric cams and determine which way they should be turned.
3. Loosen the retaining nuts on the cam bolts and turn one of them outward and the other one inward the same amount. Note the marks on the cams or the amount that the wrench moves to ensure that both cams turn the same amount (Figure 34–58).
4. Read the caster-correction scale; if it shows the desired amount of adjustment, tighten the retaining nuts to the correct torque. If the wrong amount of correction has occurred, repeat step 3 until it is correct. If necessary, caster and camber can be remeasured.

34.8.3 Adjusting Caster, Sliding Adjustment

Caster adjustment using a sliding adjustment method follows a pattern similar to shim and eccentric cam adjustments. It can be a little more difficult to perform because of the difficulty in determining how far the control arm has been moved. Some of the special tools help greatly by allowing control arm movement in carefully controlled amounts.

To adjust caster using a sliding adjustment, you should:

1. Perform steps 1, 2, and 3 of Section 34.8.1.
2. Locate the adjustment and determine which way the control arm should be moved.
3. Loosen the control arm mounting bolts, attach the proper alignment tool, if available, and carefully move one end of the control arm inward and the other end outward the same amount (Figure 34–59).
4. Snug down the mounting bolts and read the caster-correction scale.
5. If caster has changed the desired amount, tighten the control arm mounting bolts to the correct torque. If the wrong amount of correction has occurred, repeat steps 3 and 4 until it is correct. If necessary, caster and camber can be remeasured.

34.8.4 Adjusting Caster, Adjustable Strut Rod

Many cars with S-L A or strut suspension that use a single lower control arm use a strut rod that has an adjustable end mounting. Two nuts, one on each side of the strut rod bushings, are used to change the length of the strut rod. Making a trailing strut longer will move the lower ball joint toward the rear, which will make caster more negative.

To adjust caster using an adjustable strut rod, you should:

1. Perform steps 1, 2, and 3 of Section 34.8.1.
2. Locate the adjustment and determine whether to lengthen or shorten the strut rod.

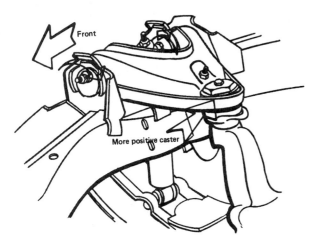

Figure 34–58 If one eccentric cam is inward while the other is rotated outward the same amount, caster will change without changing camber. (*Reprinted by permission of Hunter Engineering Company*)

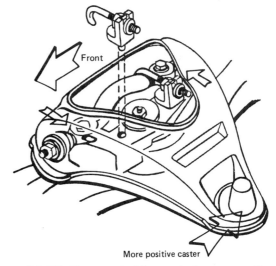

Figure 34–59 If one end of the control arm mounting shaft is slid inward while the other end is slid outward the same amount, caster will change without changing camber. (*Reprinted by permission of Hunter Engineering Company*)

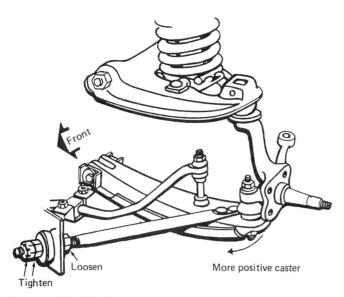

Figure 34–60 If this strut rod is made shorter by loosening the inner nut and then tightening the outer nut, caster will become more positive. (*Reprinted by permission of Hunter Engineering Company*)

3. Loosen one of the strut rod adjusting nuts and tighten the other one.
4. Read the caster-correction scale; if it has changed the desired amount, tighten the adjusting nuts to the correct torque. If the wrong amount of correction has occurred, repeat step 3 until it is correct. If necessary, caster can be remeasured; the camber angle should not have been disturbed (Figure 34–60).

34.8.5 Adjusting Caster, Miscellaneous Styles

A few additional methods of adjusting caster are of limited use. For trucks and pickups with solid beam axles and leaf springs, you can adjust caster by using **tapered shims.** These shims, which are available in several widths, lengths, and various tapers between 1/2 and 6 degrees, are placed between the axle and the spring; when the bolts are tightened, the axle will tilt an amount equal to the shims (Figure 34–61).

Caster spread on a solid axle is changed by twisting the axle; special jacking fixtures are required for this operation. Twisting the axle will increase or decrease the caster angle on only one end of the axle (Figure 34–62).

Special aftermarket eccentric bushings are available to adjust the caster on early twin I-beam axles. The radius arm bosses must be bored out to accept this bushing. Caster can also be adjusted on these axles by bending the radius arm (Figure 34–63).

Aftermarket kits are available for some imported cars with nonadjustable caster. One kit converts a fixed-length strut rod into an adjustable one (Figure 34–64).

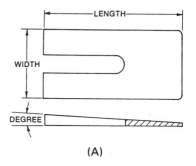

(A)

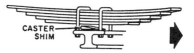

When thick part of shim is placed toward the rear, caster will increase. (Positive)

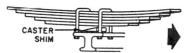

When thick part of shim is placed toward the front, caster will decrease. (Negative)

(B)

Figure 34–61 Tapered shims are available in different widths and various degrees of taper; placing a shim between the solid axle and the spring will change caster. (*A is courtesy of SPECIALTY PRODUCTS COMPANY; B is courtesy of Moog Automotive Inc.*)

Figure 34–62 These attachments are set up to twist this truck axle; operating the jack will cause caster to become more positive on the right wheel. (*Courtesy of Bee Line Co.*)

34.9 Measuring SAI

SAI, like caster, could be measured with a level and protractor if there were an accurate place from which to take the measurements, but this place does not exist. So SAI, like caster, is measured by attaching a gauge to the wheel

Figure 34–63 Eccentric bushings can be installed in the radius arm of this twin I-beam axle to provide a caster adjustment; the hole in the radius arm must be drilled oversize. (*Courtesy of Arn-Wood Company*)

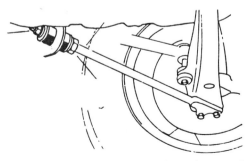

Figure 34–64 This kit is designed to change the fixed-length strut rod on an imported car to an adjustable one to make caster adjustment possible. (*Courtesy of SPECIALTY PRODUCTS COMPANY*)

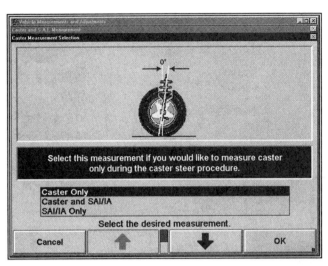

Figure 34–65 Some alignment systems display SAI along with camber and the included angle. This system provides several choices. (*Reprinted by permission of Hunter Engineering Company*)

assembly, turning the wheel through a turn, and noting the change in gauge position. SAI scales are found on modern electronic systems and on some magnetic, bubble-type caster–camber gauges (Figure 34–65). On many magnetic gauges, this is the same scale that is used for caster correction. SAI scales are found on some but not all caster–camber gauges of both bubble and magnetic types and on modern electronic alignment machines.

Specifications are not always published for SAI, but it can be easily determined if the included angle is available. Remember that SAI equals the included angle minus the amount of positive camber; the amount of negative camber plus the included angle equals SAI.

If no specifications are available, you should compare the right- and left-side readings. If they are the same, they are probably correct. The SAI and included angles on a car should be within 1/2 degree of each other.

Many magnetic gauges and some alignment systems include an SAI gauge in them. The procedure used to measure SAI is similar to that of a caster measurement (Figure 34–66).

To measure SAI using a magnetic gauge, you should:

1. Perform steps 1 through 6 of Section 34.4.2.
2. Turn the front of the wheel the correct number of degrees, usually 20, in the correct left or right direction, as specified by the gauge manufacturer.
3. Zero the SAI bubble by rotating the gauge unit or turning the adjusting screw.
4. Turn the front of the wheel in the opposite direction the required amount and read the SAI scale. If the reading goes off the end of the scale, repeat steps 2 and 3, turn the wheel to straight ahead, read the

Figure 34–66 Measuring SAI is a two-step process much like measuring caster. Here the wheel is turned, and the bubble (arrow) is set to zero. Next the wheel will be swung the required amount, and the scale will be read. (*Courtesy of Snap-on Tools Corporation*)

scale, then reset the scale to zero, turn the wheel the rest of the way to the specified degree setting, reread the scale, and add the two readings together.

34.10 Adjusting SAI

SAI is considered a nonadjustable angle because it is determined by the construction of the steering knuckle or strut and cannot be changed without bending the spindle. Bending stresses might cause fracture and later failure of the spindle or steering knuckle or strut. The possibility of a broken steering knuckle or spindle presents a very high risk. If SAI is wrong and camber is correct, the included angle is wrong; either the spindle, the steering knuckle, or the strut must be bent or damaged. A competent alignment technician will locate and replace that bent part.

There is some argument as to whether a camber adjustment really adjusts camber or SAI and whether the steering knuckle really controls SAI or camber. These points are presented here as they are treated by the majority of professionals and manufacturers in the suspension and alignment industry. For most practical purposes, it makes sense to treat these views as being correct.

34.11 Measuring Toe-Out on Turns

Specifications for toe-out on turns are usually printed with one of the wheels at 20 degrees and a specified

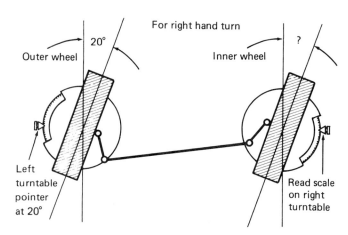

Figure 34–67 Toe-out on turns is measured by turning one of the tires to 20 degrees and then reading the amount of turn on the other turntable. The next step is to repeat this operation in the other direction.

number of degrees for the other wheel. If the specified angle is larger than 20 degrees, it will be for the inner wheel; if it is smaller than 20 degrees, it will be for the outer wheel. If specifications are not available, one side of the car can be checked relative to the other side. If the readings are different, something is wrong. If the readings are the same, toe-out on turns is probably correct. Some manufacturers allow a spread of 1.5 degrees.

Toe-out on turns is very easy to measure while caster measurements are being made. After one of the tires (depending on the specifications) has been set to a 20-degree position on the turntable, the technician can easily read the amount of turn from the scale on the other turntable. This step is then repeated as caster is measured on the second tire (Figure 34–67).

34.12 Adjusting Toe-Out on Turns

On most cars, toe-out on turns is nonadjustable; if it is not correct, one or both steering arms is bent and should be replaced. A quick check for a bent steering arm is to measure the distance from the brake rotor to the steering arm on each side of the car and compare the two measurements (Figure 34–68). They should be very close to equal. In some cases, the steering arm is bolted onto the steering knuckle, and replacing it is easy. In others, the arm is cast as part of the steering knuckle, and the whole knuckle assembly has to be replaced.

Some steering arms are bolted onto the steering knuckle in such a way that a slight error in toe-out on turns can be corrected by shimming between the arm and the steering knuckle. The amount of shim should be limited to about 1/16 in. (1.5 mm) to ensure a secure tightening of the

Chapter 34 ■ Wheel Alignment: Measuring and Adjusting

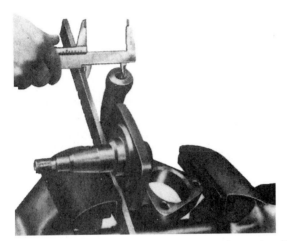

Figure 34-68 A check is being made to determine if the steering arm is bent; if this dimension is not available, it should be the same as on the other side of the car. (*Courtesy of Volkswagen of America, Inc.*)

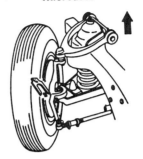

Figure 34-69 When the steering arm is bolted to the steering knuckle, toe-out on turns can be adjusted by shimming the steering arm. A shim at the rear will increase the difference between the turning angles; a shim at the front will reduce it. (*Courtesy of SPX Corporation, Aftermarket Tool and Equipment Group*)

steering arm. The rear end of the steering arm should be moved toward the center of the car (increasing toe-out on turns) if there is not enough difference measured between the two angles or moved toward the outside of the car (decreasing toe-out on turns) if there is too much difference (Figure 34-69).

34.13 Measuring Toe

Because toe is a comparison of two distances, it can be measured with a ruler or tape measure. Measure the two distances and subtract one from the other. Front- and rear-wheel toe measurements are made in the same manner. Tires and wheels, however, are not perfectly true, and the tread, where the measurements should be taken, does not lend itself to accurate measuring. Also, you would have a great deal of difficulty running a tape measure from the back of one tire to the other. Incorrect toe will cause tire wear (Figure 34-70).

To provide an accurate measuring point, a line can be scribed around the tire by pushing a sharp object against the tread while rotating the tire. **Tire scribes** are available to make this operation easy. Toe measurements are made from the scribed line on one tire to the line on the other tire. A trammel bar is a simple tool with two extended pointers to measure these distances accurately.

Another method of measuring toe is with a **light beam**. This device projects a beam of light at a 90-degree (right) angle to the wheel. The amount of toe is read where the light beam meets the scale on the projector that is attached to the other wheel. This device is actually measuring toe angle, but the gauge is usually calibrated

Figure 34-70 Tire wear caused by incorrect toe. If we were looking rearward at the top of the right (driver-side) tire, the wear on the upper tire would have been caused by excessive toe-out and that on the lower tire by excessive toe-in. (*Courtesy of Snap-on Tools Corporation*)

in fractions of an inch. This style of measurement is affected by the width of the car; a wider car would measure as having more toe than a narrower car. This discrepancy, however, is usually very slight. An advantage to this style of measuring system is that it measures **individual toe**. Individual toe readings allow easier and faster centering of the steering wheel. Individual toe readings on the rear wheels of IRS cars also give a quick indication of the rear-wheel thrust line.

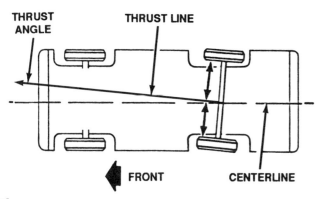

Figure 34-71 If you were to adjust the toe on this vehicle to the centerline of the car, it would drive with the steering wheel low on the right side. It will be necessary to turn slightly to the right to compensate for the thrust angle of the rear axle. (*Courtesy of General Motors Corporation, Service Technology Group*)

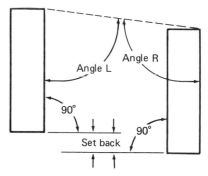

Figure 34-72 If toe is measured using equipment that actually measures the angles, setback of one of the tires will cause an error in the instrument readings. In this case there is zero toe, but the instruments will read toe-out on the left tire and toe-in on the right tire.

Modern alignment systems measure toe individually: some by inches or millimeters, some by degrees, and many by both measuring systems. The more modern equipment notes the position of the rear wheels so that toe and steering wheel position can be set relative to the rear-wheel thrustline or front-tire setback. If the rear-wheel thrust line does not run down the center of the car, the rear wheels will steer to the right or left, and the driver will have to turn the steering off center, toward the side where the thrust line is, in order to go straight down the road. This positions the steering wheel off center and is a visual annoyance to the driver (Figure 34-71). The ideal cure for this problem is to align the rear wheels to correct the thrust line. On some cars this adjustment is fairly easy and relatively inexpensive; on others, it can be very expensive. The inexpensive cure is to readjust the tie-rod lengths to center the steering wheel; this procedure is described in Section 34.16.3.

Setback also affects toe settings and steering wheel position if the equipment measures toe angles. For example, if both tires were pointed straight ahead, zero toe angle, and the right tire was set back, the angle at the right tire would be greater than 90 degrees and the angle of the left tire would be less than 90 degrees (Figure 34-72). This difference would cause the equipment to read toe-in on the right tire and toe-out on the left tire. Some toe gauges can be adjusted to compensate for setback; some computerized equipment makes this compensation automatically.

The front tires must be in a straight-ahead position when measuring toe to prevent the effect of toe-out on turns from changing the measurements. Also, the car should be at the correct ride height; the geometry of some suspension and steering systems causes a toe change, often called bump steer, when the suspension moves up and down.

34.13.1 Measuring Toe Using an Alignment System

Alignment systems use various means of measuring toe. Depending on the system, a light beam is projected from one wheel to the other (in a manner similar to the portable units) or from the wheel units to a projection screen. Other systems connect a string between the two wheel units. The display for toe readings is usually a continuous live reading for each wheel and is often calibrated for inches, millimeters, and degrees. Newer computer systems with digital readouts can be programmed to give readings in inches, millimeters, or angles. When using toe angles, toe-in is considered positive toe, and the readings will have either no sign or a plus (+) sign; toe-out is considered negative toe and should have a minus (−) sign (Figure 34-73).

Once an alignment system has been installed on a car so camber can be read, it is usually reading toe at the same time. It is sometimes necessary to connect a toe string or turn a switch.

34.13.2 Measuring Toe Using a Trammel Bar

A **trammel bar** is simple, trouble free, and inexpensive; it is merely a metal bar with a pair of stands and a pair of points. A scale is often built into one of the points to provide a means of measuring the amount of toe. Some technicians prefer to do the measuring with a scale or rule rather than slide the pointer. A trammel bar is always used in conjunction with a tire scribe to obtain an accurate measuring point; if used correctly, this is an extremely accurate measuring method.

To measure toe using a trammel bar, you should:

1. Park the car on a smooth, level surface, preferably an alignment rack, and position the steering wheel in a straight-ahead position.

Chapter 34 ■ Wheel Alignment: Measuring and Adjusting

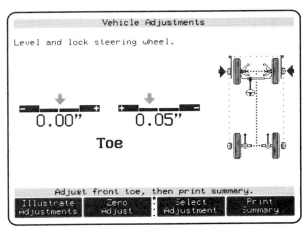

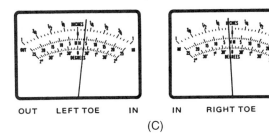

Figure 34–73 The readout on the computer unit is in the "adjust" mode and displaying the toe measurements (A). The left wheel is perfect; the right wheel is almost perfect. The light projection unit is showing between 3/16 and 1/4 in. (25 and 30 minutes) of toe-in (B). The analog meter unit is reading toe-in for both wheels (C). (*Reprinted by permission of Hunter Engineering Company*)

2. Raise the tires, place a tire scribe so the pointer presses against the most even portion of the tire tread, and rotate the tire so a fine line is scratched around the tire. If the line does not show up very well, mark a line around the tire using chalk or a tire crayon and scribe a scratch line in the marked portion (Figure 34–74).

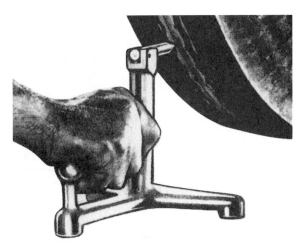

Figure 34–74 When using a trammel bar to measure toe, the first step is to scribe a line around the tire; the tire scribe is pushed against the tire while the tire is rotated. (*Courtesy of Snap-on Tools Corporation*)

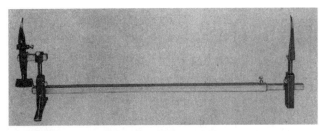

Figure 34–75 A trammel bar. The pointers of this tool are set to the width of the scribed lines at the back of the tires; the tool is then moved to the front of the tires to compare the pointer width to the scribed lines. (*Courtesy of Ammco Tools, Inc.*)

3. Position a pair of turntables or slip plates under the tires, pull the locking pins, and, with the tires in a straight-ahead position, lower the tires onto the turntables.
4. Bounce the car up and down to help the suspension settle to ride height.
5. Adjust the points so they are at hub height (even with the center of the hub) and slide the trammel bar under the car so the points are at the rear of the tires (Figure 34–75).
6. Adjust the width of both pointers so they are aligned with the scribed lines, and, if you are planning to use the measuring scale on the bar, adjust it to zero (Figure 34–76).
7. Carefully, so you do not disturb the points, slide the trammel bar out from under the car and reposition it so the pointers are at the front of the tires.
8. Slide the trammel bar so the pointer without the measuring scale is aligned with its scribed line. The other pointer will probably not be aligned.

Figure 34–76 The left pointer of this trammel bar has been adjusted to align with the scribed line at the back of this tire; the right pointer is also aligned to its scribed line. Note that the pointer is at hub height.

Figure 34–77 After setting the pointers at the rear of the tires, the trammel bar pointers are placed at the front to measure the amount of toe.

If a toe adjustment is made, it should be remembered that adjusting toe moves the rear and the front of the tire at the same time. When remeasuring toe with a trammel bar, it is necessary to repeat steps 5, 6, 7, and 8.

34.13.3 Measuring Toe Using a Light Gauge

Light gauges are more complex, fragile, and expensive than a trammel bar, but they are usually faster to use because they give continuous individual readings. They consist of a pair of projectors that attach to each front wheel. Each unit has a light source, a means of focusing and calibrating the light beam, and a target or scale for the other light beam to shine onto (Figure 34–78).

It is a good practice to check the accuracy of a light gauge periodically and recalibrate it, if necessary, especially if it has been dropped or bumped. A calibration fixture that has machined ends of the correct size to accept the gauge units is available. When the units are attached, they should give a toe reading of zero (Figure 34–79).

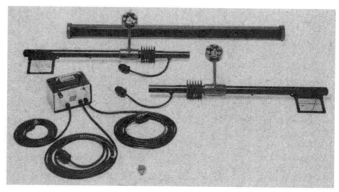

Figure 34–78 A light projection or optical toe gauge set. This unit has a projector with a scale that attaches to each front hub; the straight bar at the top is a calibration fixture. *(Courtesy of Ammco Tools, Inc.)*

9. Measure the distance between the second pointer and the scribed line; this is the amount of toe. Measurements can be taken by repositioning the pointer and reading the scale or by holding a ruler or tape in line with the pointer and scribed line (Figure 34–77).

Figure 34–79 The gauge units are positioned on the calibration fixture (arrow) to check their accuracy; while on this fixture, the gauges should read zero toe. *(Courtesy of Ammco Tools, Inc.)*

Chapter 34 ■ Wheel Alignment: Measuring and Adjusting

Figure 34–80 The light projection units are attached to each front hub or wheel clamps on the wheels; toe can be read and adjusted while watching the light beams on the scales. (*Courtesy of Ammco Tools, Inc.*)

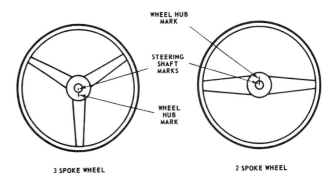

Figure 34–81 When removing a steering wheel, make sure there are index marks that can be used to replace it in the same position. A three-spoke wheel is normally centered with the middle spoke downward, and a two-spoke wheel with the spokes horizontal. (*Courtesy of Ammco Tools, Inc.*)

To measure toe using a portable light gauge, you should:

1. Place the car on a smooth, level surface, preferably an alignment rack, and turn the steering wheel to a straight-ahead position.
2. Remove the wheel covers or hubcaps and the dust cap from the hub.
3. Attach the light gauge units to the vertical machined surface of the front hubs. Wheel rim mounting adapters can be used on rear wheels or front wheels that do not allow mounting the units directly onto the hub. If a wheel rim mounting adapter is used, it should be of the type with three compensating screws so it has 360 degrees of parallelism, not just up and down compensation.
4. Level the gauge units, focus the light beam, and raise or lower the light beam as necessary to get a sharp image on the scales.
5. Read the center point of the light image on each scale. This is the amount of toe for each wheel; the readings must be added together to obtain the total amount of toe (Figure 34–80).

34.14 Toe and Steering Wheel Position

Equipment that measures individual toe is very convenient when an alignment technician is adjusting the position of the steering wheel. The steering wheel is indexed so that it is in the correct position when the steering gear is in the center of its travel (straight-ahead position) (Figure 34–81). In this position, the amount of toe at each wheel should be equal and the Pitman and idler arms should be pointed straight forward or rearward

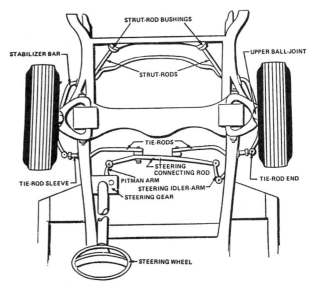

Figure 34–82 When a steering wheel is centered, both tires should be straight ahead, with the toe split equally. Note that if one tie-rod is made longer and the other one shorter, either the steering wheel or the front tires must turn. (*Reprinted by permission of Hunter Engineering Company*)

(Figure 34–82). If they are not equal, the wheels are turned slightly, and the car will travel in a circle. What really happens is that the driver will turn the steering wheel slightly so there will be equal toe when the car is going down the road, but the steering wheel will be off center. Having the steering wheel off center is an annoying visual problem, but it also moves the steering gear off center. An off-center steering gear can have excessive clearance, which can allow vehicle wander.

The procedure to center a steering wheel is simply to shorten one tie-rod and lengthen the other the same amount so as not to disturb toe. Centering a steering wheel is usually part of toe adjustment because both operations are performed by turning the tie-rod sleeves or tie-rods. A final centering adjustment is sometimes done after the wheel alignments and if the road test, which should always follow a wheel alignment, shows an off-center steering wheel. Figure 34–83 illustrates the procedure to center a steering wheel; the procedure is described more completely in Section 34.16.3.

Occasionally a steering wheel can only be centered by reindexing it on the steering shaft. The fineness of this adjustment is determined by the number of connecting splines. For example, if there are 24 evenly spaced splines, the steering wheel would turn 360 degrees ÷ 24 = 15 degrees when moved from one spline to the next.

STEERING WHEEL POSITION	TOE CORRECT (No change)	TO INCREASE TOE-IN	TO INCREASE TOE-OUT
	No adjustment needed	Make one-half of adjustment on each tie rod sleeve	Make one-half of adjustment on each tie rod sleeve
	Adjust tie rod sleeves equally in opposite directions	Adjust right tie rod sleeve	Adjust left tie rod sleeve
	Adjust tie rod sleeves equally in opposite directions	Adjust left tie rod sleeve	Adjust right tie rod sleeve

Figure 34–83 When adjusting toe, follow these recommendations to ensure a centered steering wheel and the correct toe. (*Courtesy of Snap-on Tools Corporation*)

34.15 Toe Specifications

Toe specifications have traditionally been given in fractions of an inch for domestic cars and in millimeters for imported cars. This variance often required the technician to convert one to the other depending on how the measuring scale on the equipment was made. Conversion of a dimension is fairly easy. Multiply inches by 25.4 to get millimeters; multiply millimeters by 0.04 to get inches. A chart is provided in Appendix B to make this an even easier task (Figure 34–84).

Several manufacturers are now printing their toe specifications in degrees per wheel; this is the angle that a toed wheel will take from straight ahead when viewed from above or below. Some of these manufacturers publish their specifications in both toe angle and distance dimensions. Conversion from one dimension to the other is difficult because tire diameter affects the conversion. A 28-in. (71.12-cm) tire has a smaller angle at 1/16-in. (1.58-mm) toe-in than a 22-in. (55.88-mm) tire would. The chart in Appendix B will help you make this conversion.

34.16 Adjusting Toe

Every car provides for adjusting toe; this is the only wheel alignment angle that can be adjusted on all cars. Toe is adjusted by changing the length of the tie-rods. Most cars with independent suspension have two adjustable tie-rods; vehicles with solid axles have only one.

SPECIFICATIONS

			(1) SPECIFICATIONS FOR DIAGNOSIS FOR WARRANTY REPAIRS OR CUSTOMER PAID SERVICE	(2) SPECIFICATIONS FOR PERIODIC MOTOR VEHICLE INSPECTION	(3) SPECIFICATIONS FOR RESETTING ALIGNMENT
B SERIES	CASTER		+2° TO +4°	+1° TO +5°	+3° ± 0.5°
	CAMBER		0° TO +1.6°	-0.7° TO +2.3°	+0.8° ± 0.5°
	TOE-IN	INCHES (TOTAL)	1/16" TO +1/4"	-3/16" TO +9/16"	+1/8" ± 1/16"
		DEG (PER WHEEL)	.05° TO +.25°	-.15° TO +.55°	+.15 ± .05°
G SERIES	CASTER	MAN. STR.	0° TO +2°	-1° TO +3°	+1° ± 0.5°
		POW. STR.	+2° TO +4°	+1° TO +5°	+3° ± 0.5°
	CAMBER		-0.3° TO +1.3°	-1.0° TO +2.0°	+0.5° ± 0.5°
	TOE-IN	INCHES (TOTAL)	1/16" TO +1/4"	-3/16" TO +9/16"	+1/8" ± 1/16"
		DEG (PER WHEEL)	.05° TO +.25	-.15° TO +.55°	+.15 ± .05°
ALL SERIES	CROSS CASTER		NO MORE THAN 1° SIDE TO SIDE VARIATION		NO MORE THAN 1/2° SIDE TO SIDE VARIATION
	CROSS CAMBER		NO MORE THAN 1° SIDE TO SIDE VARIATION		NO MORE THAN 1/2° SIDE TO SIDE VARIATION

Figure 34–84 Toe specifications are included in these wheel alignment specifications. Note that they are given in both inches and degrees per wheel. When resetting toe, the ideal setting would be +1/8 or +0.15 degree. (*Courtesy of General Motors Corporation, Service Technology Group*)

The tie-rods can be located either in front or in back of the tire centerline, with the rear mounting being the most common. With the tie-rods mounted at the rear, a longer tie-rod will produce more toe-in, and a shorter tie-rod will produce more toe-out (Figure 34–85).

Most tie-rods used with parallelogram steering linkage are made in three parts: two ends and a center sleeve. One tie-rod end and one end of the sleeve have right-hand threads, whereas the other side has left-hand threads. The sleeve is now like a turnbuckle; turning it one way makes the tie-rod longer, and turning it the other direction makes it shorter (Figure 34–86). Many cars with rack-and-pinion steering gears use a two-piece tie-rod. The inner tie-rod end is built partially onto the tie-rod and partially onto the end of the rack, and the tie-rod is threaded into the outer tie-rod end. Toe adjustment is made by threading the tie-rod into or out of the outer end (Figure 34–87).

Toe-in should be the last adjustment made during a wheel alignment. Other adjustments will change the position of the steering knuckle and therefore change toe. Toe should always be checked and readjusted, if necessary, each time caster or camber is adjusted.

Many RWD cars use a toe specification of 1/16 to 3/16 in. (1.5 to 4.7 mm); some cars have a range between the minimum and maximum setting as great as 1/4 in. (6.3 mm). Tire wear often results if the toe is adjusted to the minimum or maximum specification. Normally, the range is split and toe is adjusted to the middle setting. A car with a toe specification of 1/16 to 3/16 in. would have

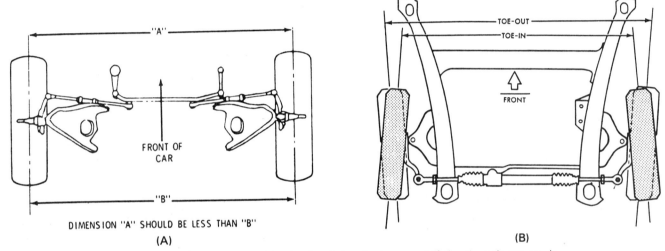

Figure 34–85 When the tie-rods are mounted in the front, toe is decreased if the tie-rods are made longer (A); if the tie-rods are mounted in the rear, toe-in is increased (B). (*Courtesy of General Motors Corporation, Service Technology Group*)

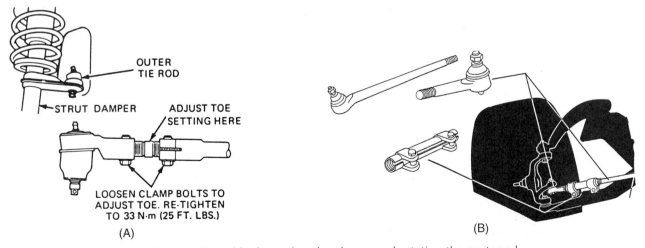

Figure 34–86 Three-piece tie-rods are adjusted by loosening the clamps and rotating the center adjuster (A and B). (*A is Courtesy of General Motors Corporation, Service Technology Group; B is courtesy of Moog Automotive Inc.*)

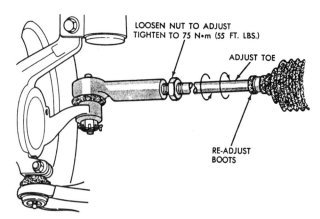

Figure 34-87 A two-piece tie-rod is adjusted by loosening the lock nut and boot clamp and then rotating the tie-rod. (*Courtesy of Chrysler Corporation*)

a preferred setting of 1/8 in. Check the tire for signs of toe wear when trying to determine the ideal setting for a particular car. For example, if a car with the given specification shows slight toe-out wear with 1/8-in. toe, the ideal setting for this particular car would be 1/16 in.

Special wrenches are available for turning tie-rod sleeves. Many amateurs are tempted to use a pipe wrench or slip joint pliers to turn the tie-rod sleeves, but this is a bad practice. The jaws of these tools not only mar the surface of the sleeve, but they also tend to collapse the sleeve, making it harder to turn. A tie-rod wrench grips the sleeve by the slot in the sleeve, and the pressure to turn the sleeve tends to open the sleeve, which makes it easier to turn. Tie-rod wrenches are available in a large or small size and a variety of shapes to fit into tight locations (Figure 34-88).

Frozen, rusty tie-rod sleeves can often be loosened by spraying penetrating oil on the threads and in the slots in the sleeve and vibrating the sleeve using an air hammer and a special wide punch (Figure 34-89).

When retightening the clamps on the tie-rod sleeves while completing a toe adjustment, two points are important: tie-rod end bind and clamp location. The tie-rod must be free to rotate within the limits of the ends; if the ends are incorrectly aligned, the amount of rotation will be limited. The easy way to align the tie-rod ends is to rotate both of them toward the front or rear of the car as far as they will go and then tighten the clamps. After both clamps are tightened, the tie-rod ends can be rotated back to center (Figure 34-90). The clamps are usually positioned between two raised areas at the ends of the sleeves, and the open portion should not be in direct alignment with the slot in the sleeve. The clamps themselves should be positioned so they will not touch or strike any part of the frame or body during steering or suspension motions. After the

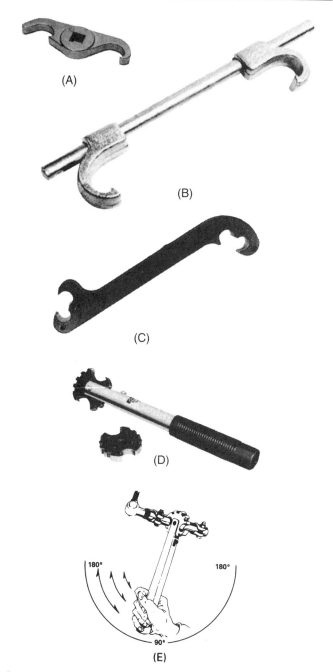

Figure 34-88 A variety of tools designed to turn the tie-rod adjuster sleeve. Note that they grip the sleeve using the slot. (*A is courtesy of SPECIALTY PRODUCTS COMPANY; D and E are courtesy of Branick Industries, Inc.*)

clamps have been tightened to the correct torque, there must be a slight gap in the open portion of the clamps and the sleeve must be locked securely onto the tie-rods (Figure 34-91).

When adjusting toe on vehicles with rubber-bonded tie-rod ends, it is recommended that if you change the toe more than 1/16 in., you should break the taper of the tie-rod end stud and allow it to recenter itself. A pull might result if you do not break the taper.

Chapter 34 ■ Wheel Alignment: Measuring and Adjusting   795

Figure 34-89 If the adjuster is stuck or rusted in place, it can be loosened by using penetrating oil and then vibrating it with an air hammer and flat punch. (*Courtesy of SUPERIOR PNEUMATIC & MFG., INC., Cleveland, OH 44140*)

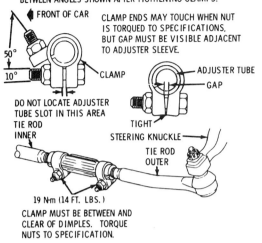

Figure 34-91 These guidelines should be followed as the clamps are tightened to ensure that the sleeves are locked securely to the tie-rods. (*Courtesy of General Motors Corporation, Service Technology Group*)

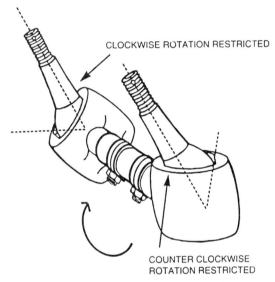

Figure 34-90 When the tie-rod sleeve clamps are tightened, both tie-rods should be centered in their housings; misalignment will cause binding during steering and/or suspension motions. (*Courtesy of Moog Automotive Inc.*)

34.16.1 Adjusting Toe, Standard Steering

Depending on the measuring system (total or individual measurements), toe is adjusted in one of two ways. Total toe is adjusted by turning each tie-rod sleeve the same amount, unless a change of steering wheel position is desired. Individual toe is adjusted by turning the tie-rod sleeve for one side until the tire on that side has the correct toe. Then the tie-rod sleeve for the second side is turned enough to adjust the second side. Toe specifications are usually for the total amount of toe; individual tire settings are one-half of the total.

Steering wheel or center link holders are available to hold the steering linkage and gear steady while individual toe adjustments are being made. When adjusting toe on cars with power steering, it is a good practice to start

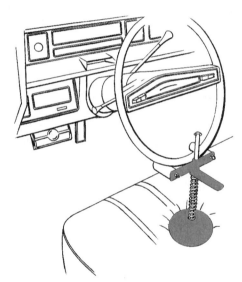

Figure 34-92 A steering wheel holder has been installed to keep the steering wheel from turning during a toe adjustment; this holder is normally used with a toe gauge that gives a live or continuous reading. (*Reprinted by permission of Hunter Engineering Company*)

the engine and rock the steering wheel slightly in a straight-ahead position to ensure that the steering gear is centered before installing the steering wheel holder; then shut off the engine (Figure 34-92).

To adjust toe on a standard steering linkage, you should:

1. Park the car on a smooth, level surface, preferably on an alignment rack, with the tires on turntables. Center and lock the steering if toe is to be adjusted individually.

2. Measure and record the amount of toe; compare the readings with the specifications and determine what change is necessary.

3. Locate the tie-rod adjusting sleeves and determine whether they need to be lengthened or shortened. Clean off the threads at the ends of the sleeves and determine which direction the sleeves need to be rotated.

4. Loosen the tie-rod sleeve clamps; special wrenches are available to make this a one-hand task (Figure 34–93).

5. Engage the tie-rod wrench in the slot in the adjuster sleeve and rotate the sleeve (Figure 34–94). When using a trammel bar, it is necessary to adjust both sleeves, remeasure toe, and then readjust both sleeves, if needed. When using equipment that gives a continuous readout, observe the toe readings while turning the adjusting sleeves. Stop adjusting when toe is correct.

6. Rotate the adjusting sleeve clamps until they are in a good position, rotate the tie-rod ends in the same direction to align them, and tighten the clamp bolts to the correct torque.

7. Road test the car to ensure that the steering wheel is centered; if necessary, center the steering wheel as described in Section 34.16.3.

34.16.2 Adjusting Toe, Rack-and-Pinion Steering

If adjusting sleeves are used on the tie-rods, toe adjustment on rack-and-pinion steering is no different than the procedure just described. When two-piece tie-rods are used, a slightly different procedure is used; the tie-rod is rotated to make the adjustment. When the tie-rod is rotated, the bellows/boot clamp must be loosened, and care must be taken to ensure that the boot is not twisted. A jam nut is provided to lock the adjustment.

To adjust toe on rack-and-pinion steering, you should:

1. Follow steps 1, 2, and 3 of Section 34.16.1.
2. Loosen the tie-rod jam or lock nuts and the clamps between the bellows/boot and the tie-rod (Figure 34–95).

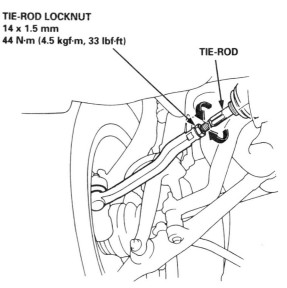

Figure 34–95 Threading the tie-rod out of the tie-rod end makes the tie-rod longer and on a rear-mounted tie-rod will increase toe-in. (*Courtesy of American Honda Motor Co., Inc.*)

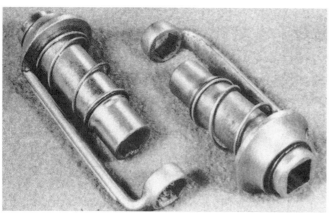

Figure 34–93 These wrenches combine a box wrench with a deep-well socket to make loosening and tightening a tie-rod sleeve clamp bolt a one-hand operation.

Figure 34–94 Rotating a tie-rod sleeve one direction makes the tie-rod longer; rotating it the opposite direction makes the tie-rod shorter. (*Courtesy of Ford Motor Company*)

Chapter 34 ■ Wheel Alignment: Measuring and Adjusting

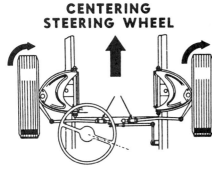

Figure 34–96 The tie-rods can be readjusted to center the steering wheel. (*Courtesy of SPX Corporation, Aftermarket Tool and Equipment Group*)

3. Rotate the tie-rods until toe is correct. Make sure the tie-rod turns inside of the boot and the boot does not get distorted.
4. When toe is correct, tighten the jam nut and boot clamps and road test the car.

34.16.3 Adjusting Toe to Center a Steering Wheel

After making an alignment adjustment, it is always a good practice to road test the car to ensure proper steering and a centered steering wheel. If the steering wheel is off center, one of the tie-rods needs to be shortened and the other lengthened the same amount. Which tie-rod to shorten or lengthen depends on which side of the steering wheel is low and whether the tie-rods are mounted in front or in back of the tire centerline. This is sometimes called a **clear vision** adjustment; the spokes of an off-center steering wheel can block the view of the dash instruments.

If the steering wheel is low on the left side, the tie-rod sleeves should be adjusted to move the tires toward a left turn. If the tie-rods are mounted behind the tire centerline, this would mean shortening the left tie-rod and lengthening the right one. The opposite would apply if the tie-rods were in front of the tire centerline or if the steering wheel was low on the right side (Figure 34–96).

34.17 Rear-Wheel Alignment

At one time, rear-wheel alignments were seldom done. Occasionally, a car with IRS showed rear tire wear or had handling difficulties that indicated a need for a rear-wheel alignment. Today, rear-wheel alignment is

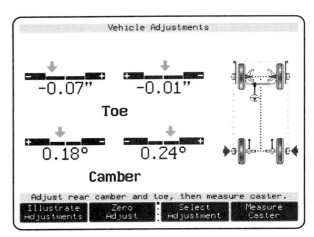

Figure 34–97 The screen of this four-wheel system is displaying the camber and toe readings for the rear wheels. Toe on the right front wheel is nearly perfect; the other three are within tolerance. (*Reprinted by permission of Hunter Engineering Company*)

becoming common, and the newer alignment systems are designed for use on four wheels (Figure 34–97).

Portable camber and toe gauges and turntables can be used on rear wheels in the same way as on the front wheels; wheel rim mounting adapters are required to attach magnetic gauges to the wheels. Some two-wheel alignment systems and racks require that the car be driven onto the rack backward; others use long enough connecting leads for the wheel sensors to reach the rear wheels. A slip plate is required if the rear tires are raised off the rack; some racks provide rear-tire slip plates. Slip plates allow the rear tires to slip sideways but not rotate like turntables. Side slippage is especially important to allow cars with IRS to return to a normal position as the rear tires are lowered onto the rack. A four-wheel system combines a rack with front turntables and rear slip plates with a measuring system using four wheel sensors and controls that allow measurement of four wheels. Alignment of some cars with 4WS requires the use of turntables at the rear and special procedures to measure the actual turning amount.

34.18 Measuring Rear-Wheel Camber

Rear-wheel camber and toe are easily measured with a four-wheel electronic system. It is simply a matter of attaching the rear measuring heads, compensating for mounting error, and lowering the tires onto slip plates. If using magnetic gauges, it is usually necessary to use rim mount adapters (Figure 34–98).

Rear-wheel camber specifications are published for most FWD cars and RWD cars with IRS (Figure 34–99).

Figure 34–98 Camber angle is being measured on this rear tire. Note the wheel clamp being used with the magnetic gauge.

If no specifications are available for cars with solid rear axles, you can safely assume that the camber should be zero (0 degrees).

Another method can be used to check for rear-wheel camber on a RWD solid axle when you suspect a bent axle housing. Scribe a line around the two tires using a tire scribe or mark a spot on the inside of the sidewall on each tire. Measure the distance between the marks or the scribed lines at the very front, the very bottom, the very back, and, if possible, the very top. If you are using the marked sidewall, it will be necessary to rotate the tires between measurements. Compare the measurements; if they are all the same, the housing is true. If the front and back are different, there is toe. If the bottom differs from

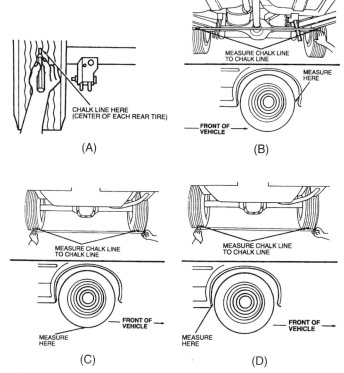

Figure 34–100 This is an accurate way to check for a bent rear axle housing. After marking of both tires (A), they are rotated to three different positions. All three measurements should be the same (B–D). (*Courtesy of Ford Motor Company*)

the top or the front and back, there is camber. The axle housing should be straightened or replaced if there is camber or toe (Figure 34–100).

34.19 Measuring Rear-Wheel Toe

Like camber, rear-wheel toe is easily measured when using four-wheel systems. Rear-wheel toe is measured the same way as front-wheel toe except that it is normally done from the rear of the car when using portable magnetic gauges.

REAR WHEEL ALIGNMENT		Acceptable Alignment Range	Preferred Setting
CAMBER			
L Body		$-1.25°$ to $-.25°$ ($-1\text{-}1/4°$ to $-1/4°$)	$-.75° \pm .5°$ (1/2°)
K,E,G Bodies		$-1.0°$ to $0°$ ($-1°$ to $0°$)	$-.5° \pm .5°$ (1/2°)
L/Z 28		$-1.1°$ to $-.1°$ ($-1\text{-}1/8°$ to $-1/8°$)	$-.6° \pm .5°$ (1/2°)
TOE*			
L Body & L/Z 28	Specified in Inches	5/32" OUT to 11/32" IN	3/32" IN
	Specified in Degrees	0.3° OUT to 0.7° IN	0.2° IN
K,E,G Bodies	Specified in Inches	3/16" OUT to 3/16" IN	0" ± 1/8"
	Specified in Degrees	.38° OUT to .38° IN	0° ± .25°
*TOE OUT when backed on alignment rack is TOE IN when driving.			

Figure 34–99 Rear-wheel alignment specifications. Note the preferred (ideal) setting and the acceptable range. (*Courtesy of Chrysler Corporation*)

This approach allows more working room for the gauges, but it also reverses the readings. Toe-in readings become toe-out and vice versa. Toe can be read using any of the different styles of toe gauges.

34.20 Measuring Rear-Wheel Track or Thrust

Rear-wheel track/thrust can be measured using different methods. **Track gauges** are relatively accurate and inexpensive, but they are becoming obsolete because they are slow and require two people. This gauge has three pointers that are adjusted so they just touch the front and rear of the rim next to the wheel flange on the rear wheel and the rear of the rim on the front wheel on one side of the car. The gauge is then moved to the other side of the car to compare the wheel positions; pointers that do not match indicate an error in wheel position (Figure 34–101).

A more popular method of checking thrust is done on four-wheel alignment systems. Measurements are made of the rear-wheel positions, and they are compared with the front-wheel positions by the machine. This system allows setting front-wheel alignment angles to compensate for any misalignment of the rear wheels (Figure 34–102).

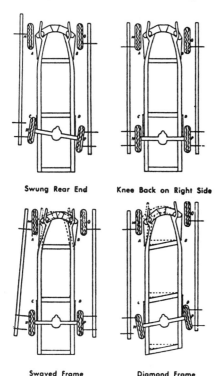

Figure 34–101 A track gauge has three pointers that are adjusted to the wheels on one side of the car; the gauge should fit the wheel positions on the other side also. (*Courtesy of SPX Corporation, Aftermarket Tool and Equipment Group*)

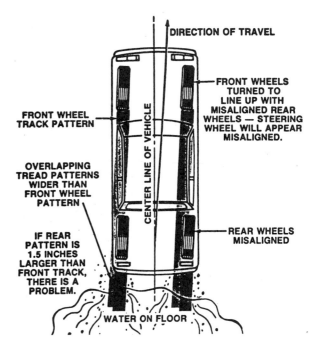

Figure 34–102 A quick check for dog tracking is to drive through a water spot and check the patterns left by the wet tires; ideally, the rear tracks will be on top of the front tire tracks. (*Courtesy of Ford Motor Company*)

As rear-wheel toe is measured using toe gauges that give individual readings, a thrust error is indicated if the two readings are not the same. For example, 1/8 in. (3 mm) of toe-in on the right rear tire and 1/8 in. (3 mm) of toe-out on the left will give zero total toe and a car that goes down the road with the rear wheels to the left of the front wheels. If each tire has the same amount of toe-in or toe-out, the thrust line will be straight down the center of the car (Figure 34–103).

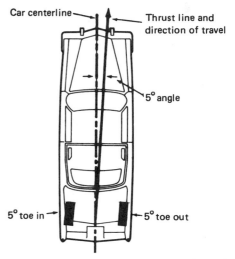

Figure 34–103 If the left rear tire has toe-in and the right rear tire has toe-out, the rear-wheel thrust line will be to the right, the car will dog track with the rear tires to the right, and the steering wheel will be low on the right side because a slight right turn will be needed to go straight ahead.

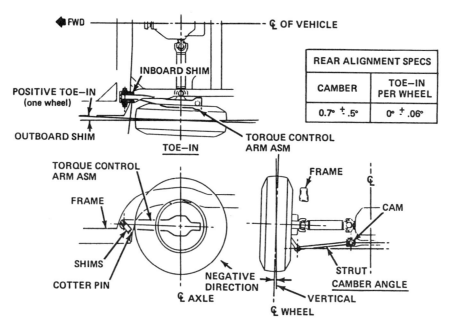

Figure 34-104 The rear-wheel alignment specifications and adjustment locations for an older Corvette. Camber is adjusted using the cam at the inner end of the strut; toe is adjusted by shims at the front of the torque control (trailing) arm assembly. (*Courtesy of General Motors Corporation, Service Technology Group*)

34.21 Adjusting Rear-Wheel Camber and Toe

Most IRS cars have some provision for adjusting rear-wheel camber and toe; the exact method varies depending on the car make and model. Older Corvettes, for example, have an eccentric cam on the lower strut mount to adjust camber and a set of shims at the front of the trailing arm to adjust toe. Newer Corvettes still use an eccentric cam for camber adjustment and a tie-rod for toe adjustment (Figure 34–104).

Many FWD cars use a rear-wheel spindle that bolts to the rear axle or spindle support; shims are available to install between the spindle and the axle (Figure 34–105). Depending on the placement of the shim, camber or toe is changed. The shims are available in a tapered, full-contact style or a smaller design that fits under only two bolts. Full-contact shims come in various angles of taper or thickness to provide various amounts of adjustment. Depending on how the shim is positioned, camber and toe are adjusted using the same shim. To adjust rear-wheel camber or toe using a full-contact shim, you should:

1. Measure the camber and toe.
2. Compare your readings with the specifications and determine how far each should be changed and in what direction.
3. Refer to the chart for the shim you are using and determine what shim (size and angle) you need and how it should be positioned. Some four-wheel alignment systems do this for you (Figure 34–106).
4. Remove the spindle and install the shim, making sure that the thick portion is indexed properly.

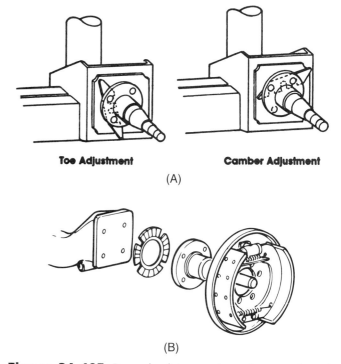

Figure 34-105 Rear-wheel toe and camber can be adjusted by using partial shims (A) or tapered full-contact shims (B). (*Courtesy of Moog Automotive Inc.*)

Chapter 34 ■ Wheel Alignment: Measuring and Adjusting

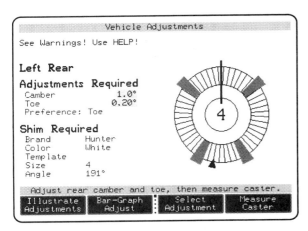

Figure 34–106 The screen on this four-wheel alignment system is displaying camber and toe readings for the left rear wheel and that an adjustment is needed. It is also showing the shim that is needed to make the correction and how it should be positioned. *(Reprinted by permission of Hunter Engineering Company)*

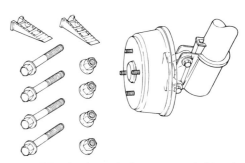

Figure 34–107 This kit includes a tapered shim that can be pushed between the strut and knuckle to adjust camber to be more positive (+). *(Courtesy of Moog Automotive Inc.)*

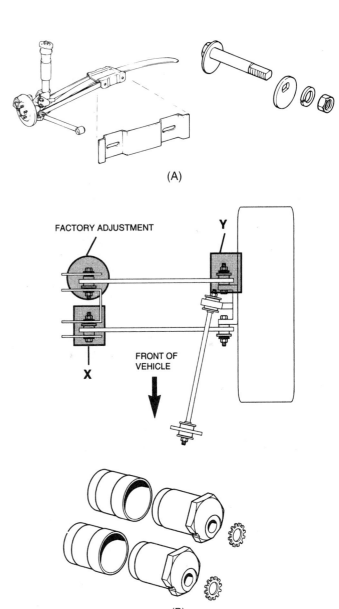

Figure 34–108 A kit includes a guide plate and cams to allow adjustment of the front or rear arms inward or outward (A). Another similar kit contains eccentric bushings (B) that can be used along with the original cams to provide more adjustment. *(Courtesy of Moog Automotive Inc.)*

Another style of shim is tapered; this shim can be pushed in between the strut and hub to make camber more positive (Figure 34–107). Other suspensions require offset bushings or cam kits to correct rear-wheel toe (Figure 34–108).

A **camber-correction gauge** is a handy tool when making adjustments where it is necessary to remove a wheel and disturb the measuring instrument. This gauge is attached to the hub, spindle, or backing plate and adjusted to the camber reading. After making a camber adjustment, the amount of change that actually took place can be read on the gauge. If the proper amount of change did not occur, another adjustment can be made before replacing the wheel and instrumentation for the final check (Figure 34–109).

34.22 Frame and Body Alignment

Occasionally, the frame or body of a car appears to be out of alignment. Frame or body alignment is normally done by a body shop that is large enough to have the equipment and the skill needed for this operation. Frame and body alignment is checked in various ways.

Frame and body dimensions are available from most manufacturers; these dimensions provide the distances between various **datums** or **measuring points** under the car (Figure 34–110). These distances are measured for a particular car and compared with the specified dimensions; if they differ, the frame or body is out of alignment. Possible frame twists are checked by attaching a set of special **frame gauges** to the car and looking along

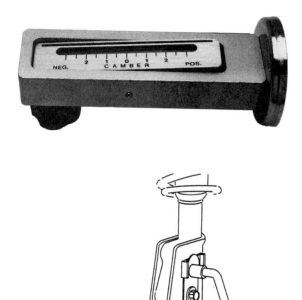

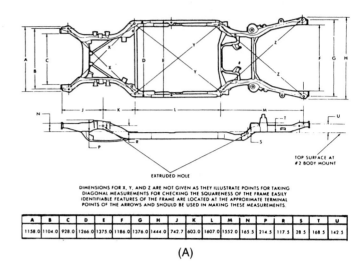

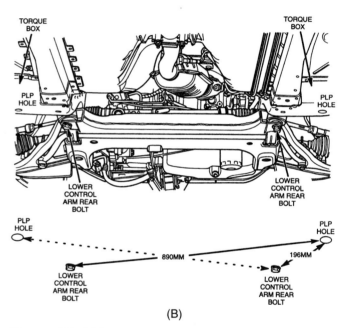

Figure 34-109 A camber-correction gauge. This gauge is attached to the brake rotor or suspension using a magnet and is adjustable so it can be set at the camber reading. As camber is adjusted, the amount of change that occurs can be read on the gauge scale. (*Courtesy of SPECIALTY PRODUCTS COMPANY*)

Figure 34-110 Frame/body dimensions are available for use in checking a frame or body for straightness (A); most measuring points are drilled or stamped holes. If dimensions are not available, measure the diagonals of different sections using groups of holes (PLP, principle locating points) as shown here (B). The diagonals should be equal. (*A is courtesy of General Motors Corporation, Service Technology Group; B is courtesy of Chrysler Corporation*)

the gauges to locate any twists or bends. Some of these gauges are large enough to fit the upper strut mounts on strut-equipped cars. Some shops use a special **jig,** with which the car is compared; if it does not fit the jig, the frame or body is bent (Figure 34-111).

34.23 Road Testing and Troubleshooting

When trying to resolve a driving or handling complaint or when completing a wheel alignment, the competent front-end technician gives the car a road test. The car is generally driven in a particular pattern while the driver checks for various possible problems.

The road test often begins by driving the car over a **sideslip** or **scuff tester;** this device gives a quick indication of whether the toe is incorrect or some other problems are causing tire scuff (Figure 34-112). On a road test, the car is driven at slow, medium, and high speeds over straight roads with some turning maneuvers so the technician can check for loose or hard steering, pull, wander, poor returnability, shimmy, or vibrations. A guide such as Table 19-1 can be used to help diagnose the cause of any problems. You should be able to recognize most of these problems.

Chapter 34 ■ Wheel Alignment: Measuring and Adjusting

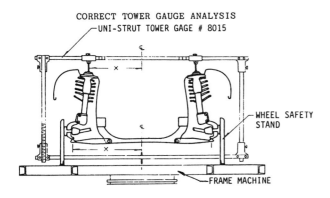

Figure 34–112 A car driving over a computerized sideslip meter or scuff gauge. Shops use this type of gauge as a quick check to determine if an alignment is needed. (*Reprinted by permission of Hunter Engineering Company*)

Figure 34–111 A frame/body gauge set. The pieces are attached to the car at various measuring points to check for bends or twists. Note how one gauge is used for checking strut towers and lower control arm locations. (*Courtesy of Arn-Wood Company*)

34.24 Practice Diagnosis

You are working in a wheel alignment shop and encounter the following problems:

CASE 1: A customer has brought in his 1989 Plymouth Voyager van for an alignment with a complaint that it drives strange: sometimes it pulls to the right, and sometimes it pulls to the left. On your road test, you confirm this problem but also notice that it pulls to the left after you make a left turn and to the right after you make a right turn. What do you call this problem? Can you cure it with an alignment? If so, what is the faulty angle? What should you do to locate the cause of this problem?

CASE 2: A customer has brought in her 1990 Ford Taurus with the problem that it pulls to the right. She also mentions that she had some front-end work done recently while visiting her sister in another city, and part of the work involved replacing a tie-rod end. Your road test confirms the pulling condition. With the car up on the rack, you see the new parts, and it appears that the work was done properly. What should you do to locate the cause of this problem? Is it an alignment problem?

CASE 3: A customer has brought in her 1992 Honda Accord for an alignment with a complaint of a pull to the right. Your road test confirms this problem, and your pre-alignment inspection shows nothing really wrong with the front end. Your alignment readings are shown in the following table. Can you cure this problem with a wheel alignment? If so, how? What else should you do to locate the cause of this problem?

Angle, Front	Reading	Specification	Angle, Rear	Reading	Specification
Caster, right	+2½	3			
Caster, left	+3½				
Camber, right	−1/2	0	Camber, right	−1	−1/2
Camber, left	+1/2		Camber, left	−1/2	
Toe	−1/16 in.	0	Toe	0	+3/32 in.
SAI, right	7	$6^{27}/_{32}$			
SAI, left	7½				

Terms to Know

alignment machines (or systems)	frame gauges	slip plates
alignment racks	half racks	split the toe
analog type	hoist rack	steering axis inclination (SAI)
calibrating	individual toe	stub racks
camber	light beam	tapered shims
camber-correction gauge	magnetic gauges	thrust line
caster	measuring points	tire scribes
caster swing	pit racks	tire wearing angles
clear vision	portable alignment gauges	toe
compensated	portable stands	toe-out on turns
datums	projector units	track gauges
directional control angles	sensing units	trammel bar
fixed or power rack	setback	turntables
four-wheel systems	sideslip or scuff tester	

Review Questions

1. Of the five wheel alignment angles:
   A. Caster and camber are the tire wearing angles
   B. Toe-in and SAI are directional control angles
   Which option best completes the statement?
   a. A only
   b. B only
   c. Both A and B
   d. Neither A nor B

2. Which of the following angles normally is not adjusted?
   a. Caster
   b. Camber
   c. SAI
   d. Toe

3. Technician A says that the car must be placed on a level surface and be at the correct height when measuring alignment angles.
   Technician B says that the front wheels must be straight ahead when measuring camber and toe.
   Who is correct?
   a. A only
   b. B only
   c. Both A and B
   d. Neither A nor B

4. Caster–camber gauges are available that will magnetically attach to:
   A. Wheel rims
   B. Hub faces
   Which option best complete the statement?
   a. A only
   b. B only
   c. Both A and B
   d. Neither A nor B

5. Technician A says that adapters are available to attach magnetic gauges to wheel rims, but they are not very accurate.
   Technician B says that caster is checked with the front wheels straight ahead.
   Who is correct?
   a. A only
   b. B only
   c. Both A and B
   d. Neither A nor B

6. Depending on the car, camber is not adjusted by:
   a. Adding or removing shims at the control arm mounting
   b. Bending the spindle
   c. Turning one or two eccentric cams at the control arm mounting
   d. Sliding the control arm mounting along a slotted hole

7. Camber is made more positive by moving the upper ball joint or top of the strut:
   A. Inward
   B. Rearward
   Which option best completes the statement?
   a. A only
   b. B only
   c. Both A and B
   d. Neither A nor B

Chapter 34 ■ Wheel Alignment: Measuring and Adjusting

8. Technician A says that it is a good practice to use a maximum of five shims at any single location. Technician B says that it is impossible to adjust camber on a solid or twin I-beam axle.
   Who is correct?
   a. A only
   b. B only
   c. Both A and B
   d. Neither A nor B

9. Technician A says that the alignment cannot be adjusted on cars with strut suspension. Technician B says that the rear wheels on most FWD cars have alignment adjustments.
   Who is correct?
   a. A only
   b. B only
   c. Both A and B
   d. Neither A nor B

10. Both front tires have a wear pattern that can be felt as your hand is slid outward over the tread but not inward. This pattern tells us that the front tires have too much:
    a. Negative camber
    b. Positive camber
    c. Toe-out
    d. Toe-in

11. When a trammel bar is used to measure toe, the measurements are taken at:
    a. The inner edge of the wheel rim
    b. The inner edge of the tire tread
    c. A scribed line on the tire tread
    d. Any of these

12. Technician A says that toe-in is increased by making the tie-rods (front mounted) longer. Technician B says that toe is always the last angle to measure and adjust when doing a wheel alignment.
    Who is correct?
    a. A only
    b. B only
    c. Both A and B
    d. Neither A nor B

13. Lengthening the left tie-rod and shortening the right one an equal amount will:
    A. Increase toe-in
    B. Change the steering wheel position
    Which option best completes the statement?
    a. A only
    b. B only
    c. Both A and B
    d. Neither A nor B

14. Technician A says that both tie-rod ends should be rotated in the same direction before tightening the clamps on the tie-rod sleeve. Technician B says that you need to make sure the rack-and-pinion gear boots do not get damaged while making toe adjustments.
    Who is correct?
    a. A only
    b. B only
    c. Both A and B
    d. Neither A nor B

15. To measure caster, the tires must be:
    A. Turned to one position where the gauge is set and then turned to a second position where the gauge is read
    B. Centered on a pair of turntables
    Which option best completes the statement?
    a. A only
    b. B only
    c. Both A and B
    d. Neither A nor B

16. Technician A says that caster is made more positive by moving the front end of the upper control arm pivot shaft outward and the rear end inward. Technician B says that caster is made more positive by making a trailing strut rod shorter.
    Who is correct?
    a. A only
    b. B only
    c. Both A and B
    d. Neither A nor B

17. Technician A says that road crown is compensated for by adjusting caster to a 1/4- to 1/2-degree spread. Technician B says that road crown is compensated for by adjusting the right wheel to a slightly more positive caster angle.
    Who is correct?
    a. A only
    b. B only
    c. Both A and B
    d. Neither A nor B

18. SAI:
    a. Cannot be measured
    b. Can be adjusted similarly to caster
    c. Cannot be adjusted
    d. None of these

19. The steering wheel is off center (down on the right side) while driving straight down the road. Technician A says that there is setback, or the rear-wheel thrustline is off center, and the tie-rods need to be adjusted to center the wheel. Technician B says that the right-wheel caster is too positive.
    Who is correct?
    a. A only
    b. B only
    c. Both A and B
    d. Neither A nor B

20. A car equipped with a manual rack-and-pinion steering gear has a shimmy. Technician A says that it could be caused by worn or loose tie-rod ends. Technician B says that it could be caused by faulty rack mounting bushings.
    Who is correct?
    a. A only
    b. B only
    c. Both A and B
    d. Neither A nor B

Math Questions

1. While doing a four-wheel alignment of a car, you find the left rear to have -0.75 degree of camber and the specification is 5/32 degree. Rear-wheel toe is okay. What degree shim do you need to correct this camber? On checking, you find that these shims are available in 1/4-degree increments. What size shim do you need?

2. The caster readings of the car you are aligning are $+0.7$ degree right and -0.3 degree left; the camber readings are $+0.3$ degree right and 0 degrees left. What are the caster and camber spreads? Would they be considered acceptable?

APPENDIX A: ASE Certification

Many automotive technicians have taken tests to become certified by ASE, the National Institute for Automotive Service Excellence. These tests are voluntary in that the technician decides on his or her own which tests to take. Certification has become a status symbol that indicates highly qualified technicians, and the shops and garages that employ these technicians can display the ASE signs.

One of the eight automotive tests is A4, Suspension and Steering. The medium and heavy truck category also has eight tests, one of which is T5, Suspension and Steering. The school bus category has six tests, one of which is S5, Suspension and Steering. The A4 test has forty questions, which are divided into these content areas:

A.	Steering Systems Diagnosis & Repair	10
	1. Steering Columns & Manual Steering Gears (3)	
	2. Power-Assisted Steering Units (4)	
	3. Steering Linkage (3)	
B.	Suspension Systems Diagnosis & Repair	13
	1. Front Suspensions (6)	
	2. Rear Suspensions (5)	
	3. Miscellaneous Service (2)	
C.	Wheel Alignment Diagnosis, Adjustment, & Repair	12
D.	Wheel & Tire Diagnosis & Repair	<u>5</u>
		40

One of the eight automotive tests is A5, Brakes. The medium and heavy truck category also has eight tests, one of which is T4, Brakes. The school bus category has six tests, one of which is S4, Brakes. The A5 test has fifty-five questions that are divided into these content areas:

A.	Hydraulic System Diagnosis and Repair	14
	1. Master Cylinders (3)	
	2. Fluids, Lines, and Hoses (3)	
	3. Valves and Switches (4)	
	4. Bleeding, Flushing, and Leak Testing (4)	
B.	Drum Brake Diagnosis and Repair	6
C.	Disc Brake Diagnosis and Repair	13
D.	Power Assist Units Diagnosis and Repair	4
E.	Miscellaneous Diagnosis and Repair	7
F.	Antilock Brake System Diagnosis and Repair	<u>11</u>
		55

These content areas are further divided into a group of tasks to aid the technician in preparing for the certification test. As you look over the task list, you can compare it with the table of contents for this text and locate the information for each task. If you are preparing to take a certification test, you can get an up-to-date task list and information concerning the test by calling ASE at 703-713-3800.

Suspension and Steering Task List

A. Steering Systems Diagnosis and Repair

1. Steering Columns and Manual Steering Gears

Task 1. Diagnose steering column noises, looseness, and binding problems (including tilt mechanisms); determine needed repairs.

Task 2. Diagnose manual steering gear (non–rack-and-pinion type) noises, binding, uneven turning effort, looseness, hard steering, and lubricant leakage problems; determine needed repairs.

Task 3. Diagnose rack-and-pinion steering gear noises, vibration, looseness, and hard steering problems; determine needed repairs.

Task 4. Inspect and replace steering shaft U-joint(s), flexible coupling(s), collapsible columns, and steering wheels (includes steering wheels with airbags and/or other steering wheel–mounted controls and components).

Task 5. Remove and replace manual steering gear (non–rack-and-pinion type) (includes vehicles equipped with airbags and/or other steering wheel–mounted controls and components).

Task 6. Adjust manual steering gear (non–rack-and-pinion type) worm bearing preload and sector lash.

Task 7. Remove and replace rack-and-pinion steering gear (includes vehicles equipped with airbags and/or other steering wheel–mounted controls and components).

Task 8. Adjust rack-and-pinion steering gear.

Task 9. Inspect and replace rack-and-pinion steering gear inner tie-rod ends (sockets) and bellows or boots.

Task 10. Inspect and replace rack-and-pinion steering gear mounting bushings and brackets.

2. Power-Assisted Steering Units

Task 1. Diagnose power steering gear (non–rack-and-pinion type) noises, binding, uneven turning effort, looseness, hard steering, and fluid leakage problems; determine needed repairs.

Task 2. Diagnose power rack-and-pinion steering gear noises, vibration, looseness, hard steering, and fluid leakage problems; determine needed repairs.

Task 3. Inspect power steering fluid level and condition; adjust level in accordance with vehicle manufacturers' recommendations.

Task 4. Inspect, adjust tension and alignment, and replace power steering pump belt(s).

Task 5. Remove and replace power steering pump; inspect pump mounts.

Task 6. Inspect and replace power steering pump seals and gaskets.

Task 7. Inspect and replace power steering pump pulley.

Task 8. Perform power steering system pressure and flow tests; determine needed repairs.

Task 9. Inspect and replace power steering hoses, fittings, and O-rings.

Task 10. Remove and replace power steering gear (non–rack-and-pinion type) (includes vehicles equipped with airbags and/or other steering wheel–mounted controls and components).

Task 11. Remove and replace power rack-and-pinion steering gear; inspect and replace mounting bushings and brackets (includes vehicles equipped with airbags and/or other steering wheel–mounted controls and components).

Task 12. Adjust power steering gear (non–rack-and-pinion type) worm bearing preload and sector lash.

Task 13. Inspect and replace power steering gear (non–rack-and-pinion type) seals and gaskets.

Task 14. Adjust power rack-and-pinion steering gear.

Task 15. Inspect and replace power rack-and-pinion steering gear inner tie-rod ends (sockets), seals, gaskets, and bellows boots.

Task 16. Diagnose, inspect, adjust, repair, or replace components of electronically controlled steering systems.

Task 17. Flush, fill, and bleed power steering system.

Task 18. Diagnose, inspect, repair, or replace components of variable-assist steering systems.

3. Steering Linkage

Task 1. Inspect and adjust (where applicable) steering linkage geometry including parallelism and vehicle ride height.

Task 2. Inspect and replace Pitman area.

Task 3. Inspect and replace relay rod (center link/drag link/intermediate rod).

Task 4. Inspect, adjust (where applicable), and replace idler arm and mountings.

Task 5. Inspect, replace, and adjust tie-rods, tie-rod sleeves, clamps, and tie-rod ends (sockets).

Task 6. Inspect and replace steering linkage damper.

B. Suspension Systems Diagnosis and Repair

1. Front Suspensions

Task 1. Diagnose front suspension system noises, body sway/roll, and ride height problems; determine needed repairs.

Task 2. Inspect and replace upper and lower control arms, bushings, shafts, and rebound bumpers.

Task 3. Inspect, adjust, and replace strut rods/radius arm (compression/tension) and bushings.

Task 4. Inspect and replace upper and lower ball joints (with or without wear indicators).

Task 5. Inspect and replace steering knuckle/spindle assemblies and steering arms.

Task 6. Inspect and replace front suspension system coil springs and spring insulators (silencers).

Task 7. Inspect and replace front suspension system leaf spring(s), leaf spring insulators (silencers), shackles, brackets, bushings, and mounts.

Task 8. Inspect, replace, and adjust front suspension system torsion bars; inspect mounts.

Task 9. Inspect and replace stabilizer bar (sway bar) bushings, brackets, and links.

Task 10. Inspect and replace MacPherson strut cartridge or assembly.

Task 11. Inspect and replace MacPherson strut upper bearing and mount.

2. Rear Suspensions

Task 1. Diagnose suspension system noises, body sway, and ride height problems; determine needed repairs.

Task 2. Inspect and replace rear suspension system coil springs and spring insulators (silencers).

Task 3. Inspect and replace rear suspension system transverse links (track bars), control arms, stabilizer bars (sway bars), bushings, and mounts.

Task 4. Inspect and replace rear suspension system leaf spring(s), leaf spring insulators (silencers), shackles, brackets, bushings, and mounts.

Task 5. Inspect and replace rear MacPherson strut cartridge or assembly and upper mount assembly.

Task 6. Inspect rear axle assembly for bending, warpage, and misalignment.

Task 7. Inspect and replace rear ball joints and tie-rod assemblies.

3. Miscellaneous Service

Task 1. Inspect and replace shock absorbers.

Task 2. Inspect and replace air shock absorbers, lines, and fittings.

Task 3. Diagnose and service front and/or rear wheel bearings.

Task 4. Diagnose, inspect, adjust, repair, or replace components of electronically controlled suspension systems including primary and supplemental air suspension systems.

C. Wheel Alignment Diagnosis, Adjustment, and Repair

Task 1. Diagnose vehicle wander, drift, pull, hard steering, bump steer, memory steer, torque steer, and steering return problems; determine needed repairs.

Task 2. Measure vehicle ride height; determine needed repairs.

Task 3. Check and adjust front- and rear-wheel camber on suspension systems with a camber adjustment.

Task 4. Check front- and rear-wheel camber on nonadjustable suspension systems; determine needed repairs.

Task 5. Check and adjust caster on suspension systems with a caster adjustment.

Task 6. Check caster on nonadjustable suspension systems; determine needed repairs.

Task 7. Check and adjust front-wheel toe.

Task 8. Center steering wheel.

Task 9. Check toe-out on turns (turning radius or angle); determine needed repairs.

Task 10. Check SAI/KPI (steering axis inclination/kingpin inclination); determine needed repairs.

Task 11. Check included angle; determine needed repairs.

Task 12. Check rear-wheel toe; determine needed repairs or adjustments.

Task 13. Check rear-wheel thrust angle; determine needed repairs or adjustments.

Task 14. Check for front-wheel setback; determine needed repairs or adjustments.

Task 15. Check front cradle (subframe) alignment; determine needed repairs or adjustments.

D. Wheel and Tire Diagnosis and Repair

Task 1. Diagnose tire wear patterns; determine needed repair.

Task 2. Inspect tires; check and adjust air pressure.

Task 3. Diagnose wheel/tire vibration, shimmy, and noise problems; determine needed repairs.

Task 4. Rotate tires/wheels according to manufacturers' recommendations.

Task 5. Measure wheel, tire, axle flange, and hub runout (radial and lateral); determine needed repairs.

Task 6. Diagnose tire pull (lead) problems; determine corrective actions.

Task 7. Balance wheel and tire assembly (static and/or dynamic).

Brakes Task List

A. Hydraulic System Diagnosis and Repair

1. Master Cylinders (Non-ABS)

Task 1. Diagnose poor stopping or dragging caused by problems in the master cylinder; determine needed repairs.

Task 2. Diagnose poor stopping, dragging, high or low pedal, or hard pedal caused by problems in the step bore master cylinder and internal valves (e.g., volume control devices, quick-take-up valve, fast-fill valve, pressure-regulating valve); determine needed repairs.

Task 3. Measure and adjust pedal pushrod length.

Task 4. Check master cylinder for defects by depressing brake pedal; determine needed repairs.

Task 5. Diagnose cause of master cylinder external fluid leakage.

Task 6. Remove master cylinder from vehicle; install master cylinder; test operation of hydraulic system.

Task 7. Bench bleed (check for function and remove air) all non-ABS master cylinders.

2. Fluids, Lines, and Hoses

Task 1. Diagnose poor stopping, pulling, or dragging caused by problems in the brake fluid, lines, and hoses; determine needed repairs.

Task 2. Inspect brake linings and fittings for leaks, dents, kinks, rust, cracks, or wear; tighten loose fittings and supports.

Task 3. Inspect flexible brake hoses for leaks, kinks, cracks, bulging, or wear; tighten loose fittings and supports.

Task 4. Fabricate and/or replace brake lines (double-flare and ISO types), hoses, fittings, and supports.

Task 5. Select, handle, store, and install brake fluids (includes silicone fluids).

Task 6. Inspect brake lines and hoses for proper routing.

3. Valves and Switches (Non-ABS)

Task 1. Diagnose poor stopping, pulling, or dragging caused by problems in the hydraulic system valves; determine needed repairs.

Task 2. Inspect, test, and replace metering (hold-off), proportioning, pressure differential, and combination valves.

Task 3. Inspect, test, replace, and adjust load- or height-sensing-type proportioning valve(s).

Task 4. Inspect, test, and replace brake warning light system switch and wiring.

4. Bleeding, Flushing, and Leak Testing (Non-ABS Systems)

Task 1. Bleed (manual, pressure, vacuum, or surge) and/or flush hydraulic system.

Task 2. Pressure-test brake hydraulic system.

B. Drum Brake Diagnosis and Repair

Task 1. Diagnose poor stopping, pulling, or dragging caused by drum brake hydraulic problems; determine needed repairs.

Task 2. Diagnose poor stopping, noise, pulling, grabbing, dragging, or pedal pulsation caused by drum brake mechanical problems; determine needed repairs.

Task 3. Remove, clean, inspect, and measure brake drums; follow manufacturers' recommendations in determining need to machine or replace.

Task 4. Machine brake drum according to manufacturers' procedures and specifications.

Task 5. Using proper safety procedures, remove, clean, and inspect brake shoes/linings, springs, pins, self-adjusters, levers, clips, brake backing (support) plates, and other related brake hardware; determine needed repairs.

Task 6. Lubricate brake shoe support pads on backing (support) plate, adjuster/self-adjuster mechanisms, and other brake hardware.

Task 7. Install brake shoes and related hardware.

Task 8. Preadjust brake shoes and parking brake before reinstalling brake drums or drum/hub assemblies and wheel bearings.

Task 9. Reinstall wheel, torque lug nuts, and make final check and adjustments.

C. Disc Brake Diagnosis and Repair

Task 1. Diagnose poor stopping, pulling, or dragging caused by disc brake hydraulic problems; determine needed repairs.

Task 2. Diagnose poor stopping, noise, pulling, grabbing, dragging, or pedal pulsation caused by disc brake mechanical problems; determine needed repairs.

Task 3. Retract integral parking brake piston(s) according to manufacturers' recommendations.

Task 4. Remove caliper assembly from mountings; clean and inspect for leaks and damage to caliper housing.

Task 5. Clean and inspect caliper mountings and slides for wear and damage.

Task 6. Remove, clean, and inspect pads and retaining hardware; determine needed repairs, adjustments, and replacements.

Task 7. Disassemble and clean caliper assembly; inspect parts for wear, rust, scoring, and damage; replace all seals, boots, and any damaged or worn parts.

Task 8. Reassemble caliper.

Task 9. Clean and inspect rotor; measure rotor with a dial indicator and micrometer; follow manufacturers' recommendations in determining need to machine or replace.

Task 10. Remove and replace rotor.

Task 11. Machine rotor, using on-car or off-car method, according to manufacturers' procedures and specifications.

Task 12. Install pads, calipers, and related attaching hardware; bleed system.

Task 13. Adjust calipers with integrated parking brakes according to manufacturers' recommendations.

Task 14. Fill master cylinder to proper level with recommended fluid; inspect caliper for leaks.

Task 15. Reinstall wheel and torque lug nuts and make final check and adjustments.

D. Power Assist Units Diagnosis and Repair

Task 1. Test pedal free travel with and without engine running to check power booster operation.

Task 2. Check vacuum supply (manifold or auxiliary pump) to vacuum-type power booster.

Task 3. Inspect the vacuum-type power booster unit for vacuum leaks and proper operation; inspect the check valve for proper operation; repair, adjust, or replace parts as necessary.

Task 4. Inspect and test hydro-boost system and accumulator for leaks and proper operation; repair, adjust, or replace parts as necessary.

E. Miscellaneous (Wheel Bearings, Parking Brakes, Electrical, etc.) Diagnosis and Repair

Task 1. Diagnose wheel bearing noises, wheel shimmy, and vibration problems; determine needed repairs.

Task 2. Remove, clean, inspect, repack wheel bearings or replace wheel bearings and races; replace seals; adjust wheel bearings according to manufacturers' specifications.

Task 3. Check parking brake system; inspect cables and parts for wear, rusting, and corrosion; clean or replace parts as necessary; lubricate assembly.

Task 4. Adjust parking brake assembly; check operation.

Task 5. Test service and parking brake indicator and warning light(s), switch(es), and wiring.

Task 6. Test, adjust, repair or replace brake stoplight switch, lamps, and related circuits.

F. Antilock Brake System (ABS) Diagnosis and Repair

Task 1. Follow accepted service and safety precautions during inspection, testing, and servicing of ABS hydraulic, electrical, and mechanical components.

Task 2. Diagnose poor stopping, wheel lockup, pedal feel and travel, pedal pulsation, and noise problems caused by the ABS; determine needed repairs.

Task 3. Observe ABS warning light(s) at start-up and during road test; determine if further diagnosis is needed.

Task 4. Diagnose ABS electronic control(s) and components using self-diagnosis and/or recommended test equipment; determine needed repairs.

Task 5. Depressurize integral (high-pressure) components of the ABS following manufacturers' recommended safety procedures.

Task 6. Fill the ABS master cylinder with recommended fluid to proper level following manufacturers' procedures; inspect system for leaks.

Task 7. Bleed the ABS hydraulic circuits following manufacturers' procedures.

Task 8. Perform a fluid pressure (hydraulic boost) diagnosis on the integral (high-pressure) ABS; determine needed repairs.

Task 9. Remove and install ABS components following manufacturers' procedures and specifications; observe proper placement of components and routing of wiring harness.

Task 10. Diagnose, service, test, and adjust ABS speed sensors and circuits following manufacturers' recommended procedures (includes voltage output, resistance, shorts to voltage/grounds, and frequency data).

Task 11. Diagnose ABS braking problems caused by vehicle modifications (tire size, curb height, final drive ratio, etc.) and other vehicle mechanical and electrical/electronic modifications (communication, security, and radio, etc.).

Task 12. Repair wiring harness and connectors following manufacturers' procedures.

Source: Courtesy of the National Institute for Automotive Service Excellence. For an up-to-date task list, call ASE at 703-713-3800 and request an Automobile Preparation Guide.

APPENDIX B: English–Metric Conversion

The following conversion factors can help you convert a dimension from one measuring system to another. Simply multiply the dimension you have by the factor to get the dimension you want.

Unit	Multiply	By	To Get
Length	inch	25.4	millimeter (mm)
	foot	0.305	meter (m)
	yard	0.914	meter
	mile	1.609	kilometer (km)
	millimeter	0.04	inch
	centimeter	0.4	inch
	meter	3.28	feet
	kilometer	0.62	mile
Area	inch^2	645.2	millimeter^2 (mm^2)
	foot^2	0.093	meter^2 (m^2)
	millimeter^2	0.0016	inch^2
	centimeter^2	0.16	inch^2
Volume	inch^3	16,387	millimeter^3 (mm^3)
	quart	0.164	liter (l)
	gallon	3.785	liter
	millimeter^3	0.000061	inch^3
	liter	1.06	quart
	liter	0.26	gallon
Weight	ounce	28.4	gram (g)
	pound	0.45	kilogram (kg)
	ton	907.18	kilogram
	gram	0.035	ounce
	kilogram	2.2	pound
Force	kilogram	9.807	newton (N)
	ounce	0.278	newton
	pound	4.448	newton
Pressure	inches of water (H$_2$O)	0.2488	kilopascals (kPa)
	pounds/inch^2	6.895	kilopascals
	kilopascals	0.145	pounds/inch^2
	kilopascals	0.296	inches of mercury (Hg)
Power	horsepower	0.746	kilowatt (kw)
	kilowatts	1.34	horsepower

Torque	inch-pound	0.113	newton-meter (N-m)
	foot-pound	1.356	newton-meter
	newton-meter	8.857	inch-pound
	newton-meter	0.737	foot-pound
Speed	miles/hour	1.609	kilometer/hour (km/h)
	kilometers/hour	0.621	miles/hour
Acceleration and deceleration	feet/second^2	0.345	meter/second^2
	inch/second^2	0.025	meter/second^2
	meter/second^2	3.28	feet/second^2
Fuel economy	miles/gallon	0.425	kilometer/liter (km/l)
	kilometers/liter	2.35	miles/gallon
Temperature	Fahrenheit, degree	0.556 (°F − 32)	Celsius, degree (°C)
	Celsius, degree	1.8 (°C + 32)	Fahrenheit, degree (°F)

APPENDIX C
Distance and Angular Equivalents

The following table provides you with the length and angular equivalencies you might need to change dimensions from one measuring system to another.

Inch (Fractional)	Inch (Decimal)	Metric (mm)	Degrees (Decimal)	Degrees (Fractional)	Degrees (Minutes)
1/32	0.0312	0.793	0.0625	1/16	3.75
1/16	0.0625	1.587	0.125	1/8	7.5
3/32	0.0937	2.381	0.1875	3/16	11.25
1/8	0.125	3.175	0.25	1/4	15
5/32	0.1562	3.968	0.3125	5/16	18.75
3/16	0.1875	4.762	0.375	3/8	22.5
7/32	0.2187	5.556	0.4375	7/16	26.25
1/4	0.250	6.35	0.5	1/2	30
9/32	0.2812	7.143	0.56625	9/16	33.75
5/16	0.3125	7.937	0.625	5/8	37.5
11/32	0.343	8.7317	0.6875	11/16	41.25
3/8	0.375	9.525	0.75	3/4	45
13/32	0.4062	10.318	0.8125	13/16	48.75
7/16	0.4375	11.112	0.875	7/8	52.5
15/32	0.4687	11.906	0.9375	15/16	56.25
1/2	0.500	12.7	1.0	1	60

APPENDIX D

Bolt Torque Tightening Chart

Torque tightening values for a bolt vary depending on the diameter of the bolt, the grade of the bolt material, the pitch of the bolt thread, whether the threads are lubricated and the type of lubricant used, and the material into which the bolt is threaded. Tightening a bolt too much might stretch the bolt to the yield point, which may cause the bolt to break or cause stripping of the threads of the bolt or nut. Tightening a bolt to too low a torque value might allow the bolt to come loose while the vehicle is in use.

If the tightening torque for a particular bolt cannot be located, the values in the following table can be used as a guide.

Grade:

	SAE	1 and 2	5	8		
	Metric	5	8	10	12	

Size (Diameter) U.S.	Metric	SAE 1&2 / Metric 5	SAE 5 / Metric 8	SAE 8 / Metric 10	Metric 12
	6	5	9	11	13
1/4		5	7	10	
5/16		9	14	22	
	8	12	21	26	32
3/8		15	25	37	
	10	23	40	50	60
7/16		24	40	60	
	12	40	70	87	105
1/2		37	60	90	
	14	65	110	135	160

Note: All torque values are given in foot-pounds and for clean, lubricated bolts. The values given are for steel-to-steel threads using motor oil for a lubricant. To convert these values to inch-pounds, multiply them by 12. To convert them to newton-meters, multiply them by 1.356.

APPENDIX E: Torque Tightening Chart for Line Connections and Bleeder Screws

Torque tightening values for a tube nut, banjo bolt, or bleeder screw vary depending on the diameter of the bolt, whether the threads are lubricated, whether a sealing washer is used, and the material into which the tube nut or bleeder screw is threaded. Tightening a connection too much can stretch the parts to the yield point, which may cause the bleeder screw or banjo bolt to break or cause stripping of the threads of the bolt, nut, or component to which it is being connected. Tightening a bolt to too low a torque value might allow a leak or cause the bolt to come loose while the vehicle is in use. Dry threads should be lubricated with brake fluid.

It is recommended that the torque values provided by the vehicle manufacturer always be used. If the tightening torque for a particular screw or bolt cannot be located, the values in the following tables can be used as a guide.

Bleeder Screws

Screw Size	Torque Inch-Pounds	Torque Newton-Meters
1/4 in.	65	7.3
7 mm	70	7.9
5/16 in.	80	9
8 mm	80	9
3/8 in.	80	9
10 mm	80	9

Banjo Bolts

Bolt Size	Torque Foot-Pounds	Torque Newton-Meters
10 mm	25	34
7/16 in.	30	40

Tube Nuts

Nut Size	Torque Foot-Pounds	Torque Newton-Meters
6 m	2	3
7 mm	3	4
8 mm	6	8
3/8 in.	15	20
10 mm	11	15
7/16 in.	15	20
1/2 in.	15	20
9/16 in.	15	20

Note: All torque values are given for clean, steel-to-steel threads lubricated with brake fluid. To convert these values from foot-pounds to inch-pounds, multiply them by 12. To convert these values from foot-pounds to newton-meters, multiply them by 1.356.

APPENDIX F

Shoe Size Chart

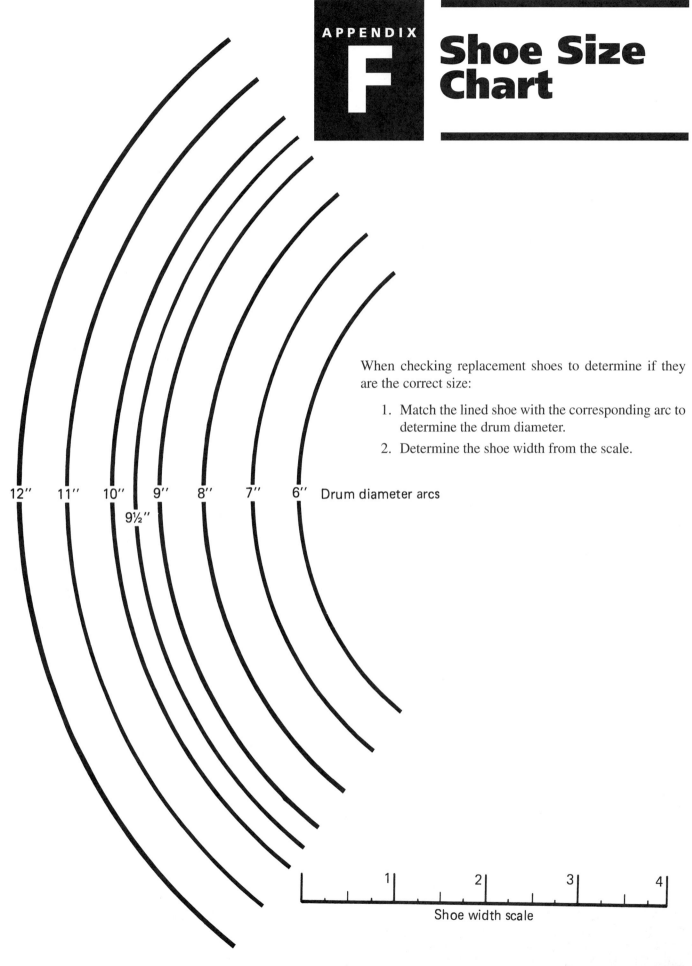

When checking replacement shoes to determine if they are the correct size:

1. Match the lined shoe with the corresponding arc to determine the drum diameter.
2. Determine the shoe width from the scale.

12" 11" 10" 9" 8" 7" 6"   Drum diameter arcs
           9½"

Shoe width scale

Glossary

A-arm: *See* Control Arm.

ABS: *See* Antilock Braking System.

Accumulator: A chamber used to store pressurized hydraulic fluid.

Ackerman Angle: The angle of the two steering arms; it produces toe-out on turns.

Active Suspension: A suspension system that uses electronic controls and hydraulic actuators in place of springs.

Adapter: A bracket used to attach the caliper to the spindle, steering knuckle, or axle.

Adjustable Anchor: An anchor pin mounted on the backing plate so that it can be adjusted up or down.

Adjusting Cam: A bolt that passes through the backing plate and is used to adjust brake lining clearance.

Aftermarket: A replacement part that is produced and sold by a company other than the original manufacturer.

Air Brakes: A brake system common on large trucks and buses that uses air pressure to activate the brakes.

Air Chamber: The component in an air brake system that converts air pressure into brake shoe application force.

Air Suspension: A system in which air-filled, elastic springs are used in place of metallic springs.

Alignment: The process of adjusting the position of the tires and steering axis to bring them to a specified, predetermined position.

Amber Antilock Warning Light: An amber-colored light mounted in the instrument panel to warn the driver of a problem in the ABS.

Ampere: The unit used to measure electrical current flow rate.

Anchor: The part of a drum brake assembly that prevents the shoe from rotating with the drum.

Anchor Pin: The steel pin attached to a backing plate that is used to prevent the brake shoes from turning with the drum.

Anodize: To use an electrochemical process to harden the surface of aluminum and increase its ability to resist corrosion.

Antidive Suspension: A suspension design to reduce the amount of dive during braking.

Antilock Braking System (ABS): A system that prevents lockup of the tires under hard brake applications.

Antirattle Springs, Clips, or Washers: Parts that prevent brake pads, shoes, or shoe parts from rattling while they are released.

Aramid: A synthetic fabric used for brake linings. It is sold under the brand name of Kevlar or Flexten.

Arcing Shoes: A procedure of regrinding the lining on brake shoes to the correct arc to match the arc of the drum.

Articulation: Vertical motion of an axle relative to the frame.

Asbestos: A noncombustible, fibrous substance used for brake linings. Caution should be used when working around asbestos because it is a health hazard.

Aspect Ratio: The relationship between the width and height of a tire.

Atmospheric Pressure: The pressure on the surface of the Earth created by the weight of the air in the atmosphere. This pressure is 14.7 psi at sea level.

Automatic Adjuster: A mechanism that will automatically adjust lining clearance when necessary.

Automotive Service Excellence (ASE): A group that promotes excellence in the automotive service industry through the voluntary testing and certification of competent technicians.

Axial: A direction that is parallel to the rotating axis.

Axis: A line or point marking the center of rotation of an object.

Axle: A cross support for a vehicle that is designed to carry the weight of the car.

Axle Bearing: The bearing used to allow drive wheels to turn freely.

Backing Plate: The mounting plate on which the components of a drum brake are assembled.

Balance, Chassis: A ride condition that gives a level, flat, front-to-rear flowing sensation without pitch.

Balance, Tire: A condition in which the tire can spin without causing a vibration of the suspension or car.

Ball Bearing: A type of bearing that uses a series of steel balls to reduce friction.

Ball Joint Angle (BJA): See Steering Axis Inclination (SAI).

Banjo: A type of hydraulic fitting in which the end of the hose has a ringlike opening.

Barrel-Shaped Drum: See Concave Drum.

Bearing: A device that allows rotation or linear motion with a minimum of friction. It usually uses a series of balls or rollers so there is a rotating motion of the internal parts.

Bearing-Retained Axle: A drive axle held in the housing by a C-shaped lock at the inner end.

Bell-Mouthed Drum: A brake drum that has been deformed so that the open end has a larger diameter than the closed end.

Bellows: A flexible, accordion-like seal used where angular or lateral motions require a large degree of movement.

Bench Bleeding: A procedure used to remove all of the air from the hydraulic portion of a component before mounting that component on the car. Master cylinders are normally bench bled before installation.

Bleeder Screw: A small, hollow screw that is loosened to "open" it, so air and fluid can flow out of a hydraulic component.

Bleeding: A procedure that removes all of the air from a hydraulic system.

Bonded Lining: A brake lining that is attached to a shoe using an oven-cured adhesive.

BOO: Brake on–off switch. *See also* Brake Light Switch.

Boot: *See* Bellows *or* Dust Boot.

Bore: The diameter and walls of a cylinder.

Boss: An area, usually enlarged, in which a bolt or fastener is to be installed.

Bottoming: A noise and jolt created when the compression cycle of the suspension ends at the bump stops.

Bounce: Straight-line motions of the sprung mass of a car in a vertical direction.

Bound Up: A rubber-mounted, normally isolated component that is stressed so it transmits NVH. *See* Neutralize.

Brake: A mechanism that converts energy to slow or stop a vehicle.

Brake Balance: The relative amount of stopping force generated by the different axles of a vehicle.

Brake Band: A flexible metal band lined with friction material that wraps around the outside of a brake drum.

Brake Block: A short lining segment that is curved and drilled so it can be bolted or riveted onto a shoe.

Brake Booster: A device that uses engine vacuum, power steering pump pressure, or electricity to reduce the amount of brake pedal pressure necessary to stop a car.

Brake Disc: *See* Brake Rotor.

Brake Drum: The rotating portion of a drum brake assembly against which the shoes push.

Brake Dynamometer: A device used to measure braking power.

Brake Fluid: A special fluid used to transmit application pressure from the master cylinder to the pistons in the wheel cylinders and calipers.

Brake Hose: A reinforced, flexible hose used to transfer hydraulic pressure from the rigidly mounted steel lines to a caliper or axle that moves or rotates.

Brake Light Switch: A mechanically or hydraulically operated switch used to turn the stoplights on when the brakes are applied.

Brake Line: A rigid steel tube that conducts hydraulic pressure to the various parts of the brake system.

Brake Lining: A special friction material attached to the brake shoes and pads. It withstands high temperatures and pressures.

Brake Pad: The lining and backing assembly used in disc brake units that is pressed against the rotor to cause braking action.

Brake Pedal: The foot-operated lever against which the driver pushes to cause brake application.

Brake Rotor: The rotating portion of a disc brake assembly against which the shoes press.

Brake Shoe: The curved metal part of the brake to which the lining is attached and that is pushed against the inside of a drum to cause braking action.

Brake Spoon: A tool, somewhat resembling a screwdriver, used to adjust brake shoe clearance.

Brake Torque: The stopping power of a brake assembly.

Brake Warning Light: A light mounted on the dash that lights up when there is a pressure loss in one of the hydraulic circuits.

Bridge Bolts: The bolts used to secure the two halves of a fixed caliper together.

Bubble Balancer: A device used to statically balance tires; it uses a bubble level as an indicator.

Build Cycle: In ABS, the action of increasing braking pressure.

Bump Steer: A steering action caused by a bounce motion of the suspension.

Bump Stop: An elastic member that increases the spring rate near the end of the compression and extension travel to reduce the effects of bottoming and/or topping.

Burnishing: The process in which the lining and the friction surface of the drum or rotor wear to conform to each other.

Bushing: A device that allows rotation or linear motion. It usually uses a sliding motion of the internal parts.

Bypass Port: The port that allows flow from the master cylinder reservoir to the system during brake release.

Caliper: The C-shaped housing of a disc brake assembly that fits over the disc and holds the pads and pistons.

Caliper Hardware: *See* Hardware.

Caliper Mount: The mounting point for a caliper.

Camber: A vertical angle of the tire seen when viewing the tire from the front or rear; used primarily to reduce tire wear.

Camber Spread: The difference between the camber angles of the front or rear wheels.

Caster: An angle of the steering axis in which the top of the steering axis leans forward or rearward; used primarily for directional control.

Caster Spread: The difference between the caster angles of the two front wheels.

Center of Gravity (CG): The balance point of a car.

Centrifugal Force: A force acting on a turning body that pushes the body outward.

Changeover: *See* Split Point.

Chassis: The portions of a car that remain after the body has been removed. It includes suspension and steering systems.

Check Valve: A valve that allows fluid or gas flow in only one direction.

Circuit: A path for electrical flow.

C-Lock Axle: A drive axle held in the housing by the axle bearing.

Coefficient of Friction: The ratio between the amount of force required to slide an object over another and the amount of force holding the objects together. The coefficient of friction of typical brake lining is 0.35 to 0.45.

Combination Valve: A brake warning light valve in combination with a proportioning and/or metering valve.

Compensating Port: One or more passages in the master cylinder that are open between the cylinder bore and the reservoir when the master cylinder is in the released position.

Compliance: The ability to yield elastically to change position.

Composite: The result of a manufacturing method for brake drums and rotors that combines a cast-iron friction surface with a stamped-steel center section.

Computer Balancer: A tire balancer that uses a computer to indicate the location and amount of weight needed to kinetically and dynamically balance a tire.

Computer-Controlled Suspension: A suspension system that uses a computer to change the shock absorber settings and/or air spring pressure to suit various driving conditions.

Computerized Plate Tester: A device using movable plates under each wheel to measure brake power.

Concave Drum: A brake drum that is deformed so that the diameter is larger in the center than at the ends.

Control Arm: A suspension member used to determine the position of a steering knuckle or axle, usually in a lateral direction.

Control Cable: A cable used to transmit application force for a parking brake.

Controller: A device used to control the current flow that applies electric brakes.

Convex Drum: A brake drum that is deformed so that the diameter is smaller in the center than at the ends.

Cornering Force: The tractional force in a lateral direction that is generated by a tire.

Corrosion: The deterioration of metal by chemical or electrochemical action.

CPS: Cycles per second.

Cross Camber: *See* Camber Spread.

Cross Caster: *See* Caster Spread.

Cup: (1) The common hydraulic piston seal used in wheel cylinders and master cylinders. (2) The outer bearing race.

Cup Expanders: A washerlike device or spring shape that places an outward pressure on the lip of a cup.

Curb Height: The distance from the ground to a specified point of the car to determine if the car is at the correct height. It is usually measured with the car at the correct curb weight.

Curb Weight: The weight of an empty vehicle that has a full supply of fuel, oil, and water.

CV Joint (Constant Velocity Joint): A universal joint that will deliver power at an angle with no change in velocity between the input and output.

Damped, Dampened: A force or action opposing a vibrating motion to reduce the amount of vibration.

Decay Cycle: In ABS, the action of reducing braking pressure.

Decelerate: To reduce speed or slow down.

Deceleration Rate: The rate of the reduction in speed.

Decelerometer: A device used to measure deceleration rate.

Deenergized: In a drum brake, the state of brake shoes when shoe application pressure is reduced by the rotation of the drum. Trailing shoes are deenergized.

Deflection: A movement that changes a shape or position in reaction to an outside force.

Denatured Alcohol: Ethyl alcohol that contains methyl alcohol and is used to clean a hydraulic brake system. It should not be consumed by humans.

Department of Transportation (DOT): A government agency that sets standards for brakes as well as other automotive units.

Diagnostic Trouble Code (DTC): A code from an electronic device self-test that indicates the nature of a problem.

Diagonal Split System: A hydraulic system that is split so one front wheel and the opposite rear wheel are in each part.

Dial Indicator: A precision instrument used to measure linear movement.

Diaphragm: A flat, flexible rubber membrane used to separate two areas in a chamber.

Digital: An electrical signal that has only two possible values, on or off.

Directional Finish: A machined pattern in which the scratches form a parallel pattern.

Directional Stability: Ability of a car to travel in a straight line with a minimum of correction from the driver.

Disc: *See* Brake Rotor.

Disc Brake: A brake design that generates stopping power from friction caused by two pads pressing against the sides of a rotating disc.

Dive: A pitching motion of the sprung mass of a car downward at the front that usually occurs during braking.

Double Flare: A type of flare, used at the ends of a brake line, in which the end of the tubing is doubled over.

Drift: *See* Wander, Weave.

Drivetrain: All the power-transmitting components from the engine to the drive wheels.

Drum: *See* Brake Drum.

Drum Brake: A brake design that generates stopping power from friction caused by shoes pushing against the cylindrical portion of the drum.

Dual Master Cylinder: A master cylinder with two pistons and pressure chambers that is used with split brake systems.

Duoservo: A drum brake design that generates servo action in both forward and reverse directions.

Dust Boot: A flexible rubber part that keeps dirt and foreign material from entering the bore of a wheel cylinder or caliper.

Dynamic Balance: A balancing of the lateral, centrifugal forces of a spinning tire and wheel.

Dynamic Seal: A seal that allows movement between it and the object that it seals.

Dynamic Suspension: *See* Active Suspension.

Eccentric: Off center; two circles having different centers.

Edge Brand: *See* Edge Code.

Edge Code: A series of letters and numbers on the edge of the brake lining that identifies the manufacturer, specific lining, and coefficient of friction.

Electronic Brake Control Module (EBCM): A computer-like unit that operates the hydraulic brake module when needed to prevent wheel lockup.

Electronic Controller: *See* Electronic Brake Control Module (EBCM).

Electronic Control Module (ECM): *See* Electronic Brake Control Module (EBCM).

Emergency Brake: *See* Parking Brake.

End Play: The distance a shaft or hub can move sideways.

Energized: In a drum brake, the state of brake shoes when shoe application pressure is increased by the rotation of the drum. Primary and leading shoes are energized during forward rotation.

Equalizer Lever: A device used with parking brake cables to ensure equal pull on each cable.

Fade: A loss in braking power usually caused by excessive heat.

Featheredge: An abnormal tire wear condition in which each tire rib wears in a tapered, angled fashion.

Finish Cut: The final cut made while turning a drum or rotor; the cut is usually at a shallow depth and slow feed rate to obtain a smooth surface finish.

Fixed Anchor: An anchor pin that is riveted or welded to the backing plate; it is not adjustable.

Fixed Caliper: A caliper design mounted securely to the steering knuckle, spindle, or axle; it uses one or two pairs of pistons to apply the pads on each side of the disc.

Flat Spots (Tires): Irregular tire tread wear patterns that result from tire-locked skids.

Float: (1) A slow, low-frequency movement of the car that produces a sensation of continuous front-to-rear, vertical movement of the suspension. (2) A driving mode during cruise in which engine speed matches vehicle speed so driving and coasting forces are not on the drivetrain.

Floating Caliper: A caliper design that uses a single piston to apply both brake pads; the caliper floats sideways on pins or bolts to apply the outboard pad.

Floating Valve: A valve in which the outer bore and inner spool move.

Flushing: The operation used to replace the brake fluid in a system.

Flutter: A forced oscillation of the steerable wheels about their axes occurring at speeds of 50 to 88 mph.

FMVSS: Federal Motor Vehicle Safety Standard.

Force: Physical power that will cause a movement.

Forward Shoe: *See* Leading Shoe.

Four-Wheel Drive (4WD): A system in which all four wheels of a vehicle are driven.

Four-Wheel Steering (4WS): A system in which the rear wheels are steered along with the front wheels.

Frequency: The speed at which an action occurs.

Friction: The resistance as one surface slides over another.

Friction-Loaded Joint: A flexible joint in which internal springs provide a friction to remove any free play from the joint.

Friction Material: *See* Lining.

Front-End Geometry: The angular relationship of the front end and steering parts in their various positions of operation.

Front-Wheel Drive (FWD): A vehicle that drives from the front wheels.

G Force (g): A measurement of the amount of acceleration, braking, or cornering force that a car can generate.

Gravity Bleeding: A method of removing air from a wheel cylinder or caliper by using the force of gravity to cause a fluid flow.

Ground Circuit: An electrical path that uses the vehicle body or frame to return the current to the source.

Grounded: An electrical connection to ground.

Guide Pins or Bolts: The pins that control the sideways motion of a floating caliper.

Handling: The relative ability of a vehicle to maneuver through turns and go where the driver wants it to go.

Hard Spots: Areas on the surface of a brake drum or disc that are harder than the metal around them.

Hardware: Clips, springs, guides, and other parts used to ensure proper action of the brake shoes or calipers.

Harmonic Vibration: An increase in vibration amplitude that occurs when the speed of the vibrating force matches the natural frequency of a vibrating object such as a spring or suspension system.

Harshness: (1) Felt and/or heard vibrations that are caused by interaction between the tire and the road surface. It can be caused by tire or road irregularities. (2) A harder-than-normal behavior of a component.

Heat Sink: A mass with the ability to absorb heat.

Heel: The end of the brake shoe where it pivots, opposite the end that is applied.

Heel Drag: A condition in which the edges of a cup get caught by a piston.

Height-Sensing Proportioning Valve: A proportioning valve that changes rear brake pressure relative to vehicle height and load.

Hertz (Hz): Measured in cycles per second.

Hold Cycle: In ABS, the action of maintaining a steady braking pressure.

Hold-Down Spring: A spring used to hold a brake shoe against the backing plate.

Hone: A set of fine abrasive stones used to true and refinish a cylinder.

Hop: Vertical oscillations of a tire that can be caused by radial runout and/or static or kinetic unbalance.

Hub: A rotating unit, mounted on bearings, that supports the wheel and brake drum or disc.

Hydraulic Brake System: A braking system that uses hydraulics to transfer the motion and pressure from the driver's foot to the brake shoes and pads.

Hydraulic Modulator: *See* Modulator Valves.

Hydraulic Pressure: The pressure within a hydraulic system measured in the amount of force on a unit of area, commonly in pounds per square inch (psi).

Hydraulics: A system that uses fluid under pressure to transfer motion and pressure.

Hydro-boost: The brand name for a brake booster that uses the hydraulic pressure from the power steering pump as a source of power.

Hygroscopic: The ability to readily absorb moisture.

Imbalance: Out of balance; a rotating component with more weight on one side than the other; causes a shake or vibration.

Inboard: Toward the center of the vehicle.

Inches of Mercury (In. Hg): A system for measuring vacuum.

Included Angle: The angle between the camber angle and steering axis inclination.

Index: To align a part, such as a rotor or wheel, to marks applied before removal; to realign parts to the best possible position.

Inertia: The physical tendency of a body at rest to remain at rest and a body in motion to remain in motion and travel in a straight line.

Integral: Made together in one unit.

Integral Parking Brake: A caliper with a mechanism that allows mechanical parking brake application.

Internal Leakage Check: A test for an air leak through a vacuum booster.

International Standards Organization (ISO) Flare: The shape of the end of the brake tubing designed by the ISO.

Inverted Flare: The type of flare fitting commonly used on brake hydraulic tubing.

Isolate: To separate from the influence of other components.

Jackknifing: An undesirable result of lockup of the brakes of a trailer, or of the rear axle of a tow vehicle, that causes the trailer to skid forward into the rear corner of the tow vehicle.

Jacobs Brake: A device, used with diesel engines, that can open the exhaust valve during part of the compression stroke to increase the amount of braking force created by the engine during deceleration.

Jounce: A bounce motion during which the tire travels upward, relative to the car, compressing the spring and shock absorber.

Kick: An undampened reaction of the suspension that causes a jolting recoil sensation.

Kinetic Balance: A balancing of the radial, centrifugal forces of a spinning tire and wheel.

Kinetic Energy: The energy contained in an object that is in motion.

King Bolt, Kingpin: A sturdy steel shaft used to connect the steering knuckle to an axle. It provides the pivot axis.

Kingpin Inclination (KPI): *See* Steering Axis Inclination (SAI).

Kingpin Offset: *See* Scrub Radius.

Knock Back: An undesirable result of excessive rotor runout that increases pad clearance and produces a lower brake pedal.

Lateral: A direction that is to the side.

Lateral Runout: A distortion of a brake disc that causes it to wobble from side to side as it rotates.

Lead: *See* Pull.

Leading Shoe: A shoe that is ahead of the axle, or a shoe that is mounted so the drum rotation will cause it to push harder against the drum.

Leading–Trailing Shoe: *See* Nonservo.

Lining: The actual friction material that is attached to the brake shoe or pad.

Lining Brake-In: The process used to help the lining mate with the drums and rotors.

Linkage: A system of levers and rods used to transmit motion or force.

Loaded Caliper: A reconditioned caliper that comes complete with new parts, shoes, and hardware.

Load Range: A system of measuring and labeling the carrying capacity of the tires.

Lockup: The point at which braking power overcomes traction and the wheel stops turning; skidding will normally occur.

Loose: *See* Oversteer.

Low-Drag Caliper: A caliper design that retracts the piston farther so it will have more clearance between the disc and the lining and less friction drag in the released position.

Lurch: Straight-line motions of the sprung mass of a car in a horizontal direction.

Machining: In servicing brakes, the procedure of mounting a drum or rotor on a lathe and recutting the friction surfaces to remove imperfections.

MacPherson Strut: *See* Strut.

Major Brake Adjustment: An adjustment of the anchor to position the curvature of the shoe to the drum.

Manual Bleeding: A bleeding procedure that requires two people; one pumps the master cylinder while the other performs the bleeding operation.

Glossary

Master Cylinder: The source of hydraulic pressure in a brake system. Operated by the brake pedal, it supplies fluid pressure to the calipers and wheel cylinders.

Maximum Diameter: The largest size allowable for a brake drum or a cylinder bore.

Metering Valve: A hydraulic valve used in some systems to cause a slight delay in operation of the front disc brakes.

Minimum Thickness: The thinnest allowable width for a brake rotor.

Modulation: In ABS, the process of turning a valve on and off at a rapid rate to maintain proper braking action.

Modulator: *See* Modulator Valves.

Modulator Valves: The ABS valves that control the power of the individual brakes during an ABS stop.

Motorist Assurance Program (MAP): Guidelines for automotive service, including brake systems.

Mounting Hardware: Parts used to properly connect a caliper to its mount.

National Highway Traffic Safety Administration (NHTSA): A government body that establishes regulations for motor vehicles.

National Institute of Automotive Service Excellence (NIASE): *See* Automotive Service Excellence (ASE).

Neutralize: To adjust a bound-up component to an unstressed position.

Neutral Steer: A cornering situation in which the slip angles of the front and rear tires are equal. No steering correction is required.

NIOSH: National Institute of Occupational Safety and Health.

Nonasbestos Organic (NAO): A brake lining compound that does not include asbestos.

Nondirectional Finish: A machined surface in which the scratches do not form any noticeable pattern.

Nonservo: A brake shoe design that applies the two shoes independently of each other.

NVH: Noise, vibration, harshness.

Occupational Safety and Health Administration (OSHA): A government body that regulates working conditions as they relate to personal safety.

Ohm: The unit used to measure electrical resistance.

Onboard Diagnostic (OBD): A system that monitors all computer input and output signals to determine any faults that might occur.

Open: An electrical circuit that is broken.

Original Equipment (OE): The parts that were originally used on the vehicle.

Original Equipment Manufacturer (OEM): The company that made the parts that were originally used on a car.

Oscillation: A back-and-forth, repeating motion.

Outboard: Toward the outside of the vehicle.

Out-of-Round: A brake drum that has two different diameters; it is not round.

Oversteer: A tendency of the car to turn more sharply than it should for the amount of steering input. A steering correction toward straight ahead or the opposite direction is required.

Pad: The brake shoe for a disc brake. It consists of the lining and the backing to support it.

Pad Wear Indicator: A device that warns the driver when the lining is almost worn out.

Panhard Rod: A device that connects the axle to the frame; it controls sideways motions of the frame or body.

Parallelism: A variation in the thickness of the friction surfaces of a rotor.

Parking Brake: A mechanically applied brake system designed to hold a car stationary.

Pascal's Law: Physical law governing the action of fluids.

Phenolic: A plastic material used to make caliper pistons.

Pitch: Rotary motions of the sprung mass of a car around the transverse axis. The front end will rise while the rear lowers and vice versa.

Pitting: Surface irregularities that result from rust or corrosion.

Play: Free movement of a part allowed by internal clearances.

Plow: *See* Understeer.

Pounds per Square Inch (psi): A measurement of fluid pressure.

Power Assist: *See* Power Brakes.

Power Booster: *See* Brake Booster.

Power Brakes: A brake system that uses a power booster.

Preload: A slight thrust load, less than zero lash, adjusted into a bearing set to eliminate all axial motion.

Pressure: A unit of force applied on a given area.

Pressure Bleeder: A container of brake fluid that is pressurized. It will supply brake fluid under pressure to the master cylinder for bleeding purposes.

Pressure Bleeding: A procedure of bleeding the brakes using a pressure bleeder to force fluid through the system.

Pressure Differential Switch: A switch, operated by a pressure difference in the two hydraulic circuits, that turns on the brake warning light to indicate a faulty hydraulic circuit to the driver.

Pressure Differential Valve: A valve that senses pressure in each part of a split system and operates the pressure differential switch if one side loses pressure.

Primary Cup: In a master cylinder, the cup that pumps fluid to the system.

Primary Piston: In a tandem master cylinder, the piston that is operated by the pushrod.

Primary Shoe: The forward shoe; in a duoservo brake, the primary shoe is energized by the rotation of the drum and applies pressure on the secondary shoe.

Proportioning Valve: A hydraulic valve used to improve brake balance by reducing the pressure applied to the rear brakes.

Pull: A tendency for a car to steer toward one side.

Pushrod Adjustment: An adjustment to ensure the proper length of the pushrod between the vacuum booster and the master cylinder.

Quick-Take-Up Master Cylinder: A dual master cylinder design that uses a stepped primary piston to displace more fluid and take up the clearance of low-drag calipers.

RABS: Rear antilock braking system.

Radial: An outward direction that is at a right angle to the center. Radial is in the plane of rotation.

Radial Runout: A variation of the tire's radius that causes a vertical oscillation or bounce of the tire.

Radius Arm: A type of control arm that attaches to an axle at one end and pivots at the other end. It is often mounted in a lengthwise direction.

Reaction Time: The amount of time that it takes a driver to recognize a problem and apply the brakes.

Rear-Wheel Drive (RWD): A vehicle in which the rear wheels are driven.

Rebound: A bounce motion during which the tire travels downward relative to the car and the spring and shock absorber extend.

Red Brake Warning Light: *See* Brake Warning Light.

Regenerative Braking: Braking that uses an electrical or hydraulic system to remove and store the kinetic energy of a vehicle to reduce its speed; the energy removed is later used to propel the vehicle.

Reservoir: The portion of a master cylinder in which fluid is stored.

Residual Pressure Check Valve: A hydraulic check valve used in some master cylinders to maintain a slight pressure in the lines while the brakes are released.

Return Spring: The spring used to return a brake shoe to a released position. Also called pullback spring or retracting spring.

Reverse Shoe: *See* Trailing Shoe.

Revolutions per Mile: The number of revolutions a tire must make to travel 1 mi.

Ride Height: *See* Curb Height.

Riveted Lining: A lining attached to the steel backing with rivets.

Road Crown: The raised portion in the center of a road to promote water runoff.

Road Shock: A harsh force transmitted from the tires through the suspension or steering linkage.

Road Test: Operation of the vehicle under conditions that are designed to recreate a particular problem.

Roll: A rotary motion of the sprung mass of a car around the longitudinal (lengthwise) axis that results in body roll.

Roller Bearing: A type of bearing that uses a series of steel rollers to reduce friction.

Roller Burnishing: The process used to produce a very smooth cylinder bore.

Roll Rate: The amount of resistance generated by the suspension components that resist roll.

Roll Steer: A steering motion of the front or rear tires that occurs as a result of the car's body rotating on its roll axis.

Room Temperature Vulcanizing (RTV): A material used for formed-in-place gaskets.

Rotor: The flat, rotating disc against which the brake pads press in a disc brake assembly.

Roughing Cut: The first cut made while turning a drum or rotor; usually made fairly deep and at a fast feed rate.

Roughness: A heard or felt vibration generated by a rolling tire on a smooth road surface that produces the sensation of rolling on a coarse or irregular road surface.

Rubber Manufacturers Association (RMA): A group that sets standards in the tire industry.

Runout: An out-of-round or wobble motion. *See also* Lateral Runout.

Scoring: Irregular grooves that are worn into the friction surface of a rotor or drum.

Glossary

Scrub Radius: The distance between the center of the tire and the steering axis when measured at the road surface.

Seal: A device used to prevent foreign material or water from entering an area; a grease seal also retains grease or oil inside an area.

Secondary Cup: On a master cylinder piston, the cup that is used to prevent fluid from running out the end of the bore or from running between the primary and secondary sections.

Secondary Piston: In a tandem master cylinder, the piston that is applied by hydraulic pressure from the primary piston.

Secondary Shoe: The rearward shoe in a duoservo brake; the secondary shoe is applied by the energized primary shoe.

Self-Adjusting: A drum brake mechanism that automatically adjusts the lining clearance when necessary.

Semimetallic: A nonasbestos lining compound that uses powdered metal or strands of steel wool.

Sensor: A device that senses a value and converts it into an electrical signal.

Service Brake: The primary brake system used to stop a vehicle under normal driving conditions.

Servo Brake: A drum brake design that uses the rotational force generated by one shoe to apply the other one.

Shake: A resonance set up in the suspension that produces a nervous ride quality.

Shim: A spacer used to adjust the distance or angle of an item.

Shimmy: A series of rotary (turning) oscillations of the front wheels around the steering axis.

Shock Absorber: A device, usually hydraulic, used to dampen or reduce the amount of spring oscillations after a bump.

Shoe: On drum brakes, the lining with its curved, steel backing.

Shoe Anchor: *See* Anchor.

Shoe Return Spring: *See* Return Spring.

Short Circuit: An undesired connection between a circuit and any other point.

Shorted: *See* Short Circuit.

Short-Long Arm (S-L A) Suspension: A suspension system that uses a relatively short upper control arm and a longer lower control arm.

Skid: The act of a tire sliding without rotating; skid usually results from a failure of the tire to grip the roadway with enough force to overcome the stopping or cornering loads placed on it. A skid often results in a loss of vehicle control.

Skid Control: *See* Antilock Braking System (ABS).

Sleeves: Tubular items on which a caliper slides sideways.

Sliding Caliper: A floating caliper that moves sideways on machined ways or keys.

Slip Angle: The angular difference between the wheel's centerline and the actual direction of travel.

Slope: The rate of difference between front and rear brake pressures above the split point.

Society of Automotive Engineers (SAE): A group that establishes standards in the automotive and petroleum industries.

Solid Rotor: A nonventilated rotor.

Solid State: A term that describes an electronic device with no moving parts that performs a variety of operations.

Split Point: The pressure at which the proportioning valve begins to reduce the pressure increase to the rear brakes.

Spongy Pedal: A condition in which the pedal feels springy under pressure instead of having the normal solid feel.

Spring: A flexible suspension member that allows bounce travel of the suspension.

Spring Rate: The change of load on a spring per unit of deflection.

Sprung Weight: The total weight of the portions of the car that are carried by the springs.

Squat: A pitching motion of the sprung mass downward at the rear that often occurs during acceleration.

Stabilizer Bar: A suspension member used to reduce body lean during cornering.

Standing Wave: A deformation of the tire sidewall and tread that results when the tire velocity exceeds a critical point.

Starwheel: A coglike attachment of the adjuster screw that allows it to be rotated by a tool or lever.

Static: Stationary, not moving.

Static Balance: A balancing of a tire and wheel so it will remain stationary and not rotate or lean because of unequal weights.

Static Seal: *See* Stationary Seal.

Stationary Seal: A seal that does not move, such as the O-ring seal in many calipers; the piston moves within the seal.

Steady State: The conditions of a car's body and suspension that exist while the body and suspension are not changing roll, pitch, or yaw attitude.

Steering Axis: The line around which the front tires turn when a car turns a corner.

Steering Axis Inclination (SAI): An angle of the steering axis in which the top of the steering axis leans inward.

Steering Geometry: The angular relationship of the various steering and suspension components during their different positions of travel.

Steering Knuckle: The front suspension component that attaches the front tires and wheels to the steering axis and steering linkage.

Stoplight Switch: *See* Brake Light Switch.

Stroking Seal: A hydraulic seal that moves with the piston.

Strut: A suspension system type that utilizes the shock absorber as the upper tire position locating member.

Strut Rod: A suspension member that is used to brace the control arm to keep it from moving forward or backward.

Swaging: A method of deforming the metal to permanently lock an item in place.

Sway Bar: *See* Stabilizer Bar.

Swept Area: The total amount of area on the drums and rotors on which the brake lining rubs.

Swirl Finish: *See* Nondirectional Finish.

Tandem Diaphragm: A power booster that uses two diaphragms.

Tandem Hydraulic System: A hydraulic brake system that is split between front and rear.

Tandem Master Cylinder: A master cylinder that uses two pistons, one mounted behind the other.

Tapered Roller Bearing: A type of bearing that uses a series of tapered steel rollers to reduce friction. This bearing type is used in pairs to control end play.

Tee: A hydraulic fitting used to connect three lines.

Thermodynamics: The physical principles that govern mechanical actions and heat energy.

Thickness Variation: *See* Parallelism.

Thump: A periodic vibration and/or sound generated by a tire producing a pounding sensation that occurs in time with the tire rotation.

TIR: Total indicator runout.

Tire Contact Area, Tire Print: The amount of tire tread that is in contact with the road surface. Also called the footprint.

Tire Deflection: Bending of the tire sidewalls.

Tire Force Variation: A condition in which portions of the tire deflect a different amount, resulting in a vibration that causes loaded runout when the tire rotates.

Toe: (1) An angle of a tire, relative to straight ahead, if viewed from above. (2) The end of a brake shoe that contacts the drum first.

Toe Angle: The actual amount that the tire differs from pointing straight ahead.

Toe-In: A condition in which both tires of an axle are positioned so they are closer together at the front than at the rear.

Toe-Out: A condition in which both tires of an axle are positioned so they are closer together at the rear than at the front.

Toe-Out on Turns: A condition in which the front tires toe outward during a turning action of the car.

Topping: A noise and jolt when the extension cycle of the suspension travel ends at the travel stop.

Torque: A twisting or turning force.

Torsilastic Bushing: A bushing that allows motion through the elastic nature of rubber.

Torsion: A rotating motion that causes a twisting action.

Torsion Bar: A spring that allows suspension motion by twisting.

Track: The center-to-center distance between the two tires on an axle.

Tracking: The degree to which the rear tires follow behind the front tires.

Traction: The frictional grip between the tires and the road surface.

Trailing Shoe: A shoe that is behind the axle; in a nonservo brake, it is the deenergized shoe.

Tramp: A rotary motion of a solid axle upward at one end and downward at the other end that reverses in a series of oscillations.

Transient State: The conditions that result in a car while the body and suspension are changing position.

Transverse: A direction that goes across the car.

Tread Width: The width from outside edge to outside edge of the two tires on an axle.

Trim Height: *See* Curb Height.

Turning: The process of using a special lathe to machine the friction surface of a drum or rotor so it is like new.

Two–Leading Shoe: A drum brake design using two leading shoes.

Two-Plane Balance: Radial and lateral balance; also called kinetic and dynamic balance.

Undamped, Undampened: A system in which there are no forces or actions that oppose vibrating motions.

Understeer: A tendency of the car that requires a greater steering motion to make a given turn; the car turns less than it should.

Unequal-Length Control Arms: See Short-Long Arm (S-L A) Suspension.

Union: A hydraulic fitting used to connect two lines.

Uniservo: A drum brake design that generates servo action in one direction only.

Unsprung Weight: The total weight of the portions of the car that support the springs.

Vacuum Bleeding: A bleeding method that pulls fluid and air from the bleeder valve using vacuum.

Vacuum Booster: A brake booster that uses engine vacuum as a power source.

Vehicle Speed Sensor (VSS): A sensor that provides vehicle speed information.

Vented Rotor: A rotor that has internal air passages between the friction surfaces.

Vibration: A periodic motion or oscillation of an item that often causes an annoying motion or sound.

Viscosity: The thickness or body of a fluid as it relates to the ability to flow.

Volt: The unit used to measure electrical force or pressure.

Wander, Weave: The tendency of a car not to follow a straight line; it requires continuous correction from the driver.

Warning Light: A light that lights up to warn the driver of a braking problem.

Watts: Electrical power determined by multiplying volts by amps.

Watt's Link: A suspension member consisting of two rods and a pivoting bell crank that is used to keep the body from moving sideways relative to the axle.

Wear Sensor: See Pad Wear Indicator.

Weight Transfer: The amount of weight that moves laterally across the car because of cornering forces or lengthwise because of braking or acceleration forces.

Wheelbase: The center-to-center distance between the front and rear tires.

Wheel Bearing: The bearing used to allow the non–drive wheels to turn freely.

Wheel Cylinder: A component in drum brakes that converts hydraulic pressure into a mechanical force to push the shoes against the drum.

Wheel Hop: A rapid vertical oscillation of the tires resulting from a loss of traction control.

Wheel Offset: The lateral distance between the centerline of a wheel and the inner side of the wheel mounting flange.

Wheel Slip: Tire rotational speed relative to vehicle speed; 100 percent slip indicates wheel lockup.

Wheel Speed Sensor: An ABS sensor that measures the speed of a wheel.

Yaw: The rotary motion of the sprung mass of a car around a vertical axis that is encountered in a spin.

Zerk Fitting: A brand of grease fitting.

Index

A-arms, 16, 17, 221
  control, 203
ABS:
  bleeding, 559
  electrical circuit for, 143
  hydraulic pressure checks on, 556
  repair operations on, 557–559
  speed sensor mechanical problems, 552–555
ABS actuator, 542
ABS brake fluid, 154
ABS systems, 148–154
ABS wheel speed sensors (WSS), 289
Acceleration sensor, 305
Accumulator, 129
Ackerman angle, 334
Ackerman linkage, 334
Active hydraulic suspensions, 243–245
Actuator, 144, 306, 307
Actuator, shock absorbers, 261
Adaptor for floating calipers, 65
Add-on antilock brake systems, 139–140
Adjustable anchors, 54
Adjustable strut rod, adjusting caster with, 783–784
Adjustable-anchor brake shoes, 417
Adjusting wheel bearings, 371–374
Aftermarket air suspensions, 246
Aftermarket wheels, 181–182
AH2 standard wheel design, 179
Air chamber for spring, 242
Air compressor in springs, 241
Air method of pushrod adjustment, 521, 523
Air shock, 246
Air shock absorbers, 264
Air springs, 205
Air suspension, aftermarket, 246
Air suspension, electronically controlled, 240–243
Air suspension system, 20
Airbag:
  accidental deployment of, 735
  cautions to observe, 736
  disabling, 736
Airbag modules, 271
Airbag service, 735–738
Airbag systems, 271–273
Airtightness test, 519
Alignment angles, measuring, 756–761
Alignment machines, 756
Alignment measures, methods of displaying, 757
Alignment racks, 757
Alignment system, 756
  measuring toe with, 788
All season tire, 162
Aluminum bores, 90, 91, 467–468
Amber antilock warning light, 544
Ambient temperature, 33
Amperes, 303
Amphibole type of asbestos, 37
Amplitude of vibration, 564
Analog volt-ohmmeters, 309
Anchor, shoe, 53, 54
Anchor pin, 53, 54
Anchor plate for floating calipers, 65
Anchors, drum brake, 3
Angle (BJA), 332
Angle (BJI), 332
Angle (KPA), 332
Angle (KPI), 332
Angles, measuring, 322
Annealing, 233
Anodized bores, 91
Anodizing, 91
Anti-skid braking, 137
Antidive, 203, 262
Antidive suspensions, 44
Antilock braking system (ABS), 2, 8, 44
  common problems with, 543
  completing repair, 460
  dedicated tester, 551–552
  diagnosing problems in, 542–543
  electrical tests of, 550–555
  problem codes and self-diagnosis, 546–549
  service, 541–560
  service manual for, 542–543
  theory, 137–155
  warning light operation, 544
Antiphase steering, 298
Antirattle clip, 71
Antirattle spring, 71
Antiroll, 262
Antiroll bar, 246–248, 264
Antiseize compound, 457
Antispin regulation, 154
Antisquat, 262
Antisqueal hardware, 70
Application cables, 77
Apportioning valve, 112–113
Aramid, 161
Arbor, 397
Arming sensor, airbag, 273
Asbestos:
  amphibole type of, 37
  brake lining, 34
  chrysotile family of, 37
  concerns about, 36–37
Asbestosis, 36
Aspect ratio, 163
  of tires, 164
Atmospheric pressure, 123
Atmospheric suspended booster, 124
Audible sensor, 70
Audible sound, 564
Automatic adjuster at parking brake lever, 79
Automatic slip regulation (ASR), 154
Automatic traction control, 154
Automatic transmission fluid (ATF), 291
Automotive brake tubing, common sizes for, 105
Auxiliary drum parking brake, 81
Axial/vertical motion of ball joint, measuring, 595
Axle, 10
Axle bearing, 187
  faulty, 583
  loose, 583
  removing and replacing, 376–377
  rough, 584
Axle bearing noise, 564

831

Axle shaft installation, 377
Axle windup, 218
Axle yaw, 218
Axles, bending tools for, 769

Back bleed, 501, 503
Backing plate, 3
Backing plate of brakes, 52–53
Backside spacing, 180
Backspacing, 180
Balance problems in tires, 583
Balanced vane hydraulic pumps, 283
Balancer, tire, 633
Balancing operations, tire, 633–641
Ball bearing, 187, 188
Ball joint, 18, 203
    breaking loose tapers on, 686–689
    checking boots, 592
    checking clearance on I-beam or 4WD solid axle, 598–599
    checking friction-loaded or follower, 598
    checking load-carrying, on lower control arm, 594–596
    checking load-carrying, on upper control arm, 597
    checking non-wear-indicating, load-carrying, 595
    checks, 592–600
    lubricating, 592
    removing and replacing, 692–695
    removing and replacing bolted-in, 693–694
    removing and replacing pressed-in, 692–693
    removing and replacing riveted-in, 693–694
    removing and replacing threaded, 695
    S-L A, 205
Ball joint, excessive clearance in, 593
    checking load-carrying ball joint on lower control arm for, 594–596
    checking wear indicator ball joint for, 593–594
Ball joint clearance on I-beam or 4WD solid axle, checking, 598–599
Ball joint inclination, 332
Ball joint replacement, 690
Ball joint studs locked by pinch bolts, removing and replacing, 689
Ball nut steering gear, 276
Banjo fittings, 108
Base valve assembly, 256
Basic electricity, 303–305
Basic electronics, 305–307
Bayonet type shock absorber, 256

Bead flanges, 177
Bead lock rim, 179
Bead seats, 177
Bead wires in tires, 158
Beam axle, 12
Bearing cups, inspecting, 368
Bearing end play, 189–190
Bearing on an axle, removing and replacing, 376–377
Bearing parts, 187–189
Bearing preload, 190
Bearing-retained axle, 192, 375
    installing, 377–378
    removing, 375–376
Bearings, 187–189
    inspecting, 368–370
Beating, 567
Beehive spring, 409
Bell-mouthed drum, 393
Belts, tire, 161
Bench bleeding, 463
Bench bleeding a master cylinder, 474–476
    using EIS Sur-Bleed syringe, 475
    using tubes, 475
Bending tools, 769
Bendix brake spring tool, 404
Bendix Hydrovac unity, 125
Bias belted tires, 161
Bias ply, 160
Bias-belted tire, wear on, 575
Bimetal bushing material, 704
Binders in brake lining, 34
Bleeder screws, 101, 105
Bleeding ABS, 559
Bleeding brakes, 495–506
Bleeding of brakes, gravity, 499
    manual, 499–500
    pressure, 501–503
    reverse-flow, 503
    sequence, 499
    vacuum, 503
Bleeding shock absorber, 662
Blueing of rotor, 432
Body and frame alignment, 801–802
Body of car, 1
Body plies, 158
Boiling point, 32
Boiling point of brake fluid, 115
Bolt, sizing of, 361
Bolt patterns, 178
Bolted-in ball joint, removing and replacing, 693–694
Bonded lining on brakes, 52
Booster, power brake, 122
Booster replacement, 520–523
Booster service, 523–528

Booster service operations, 517
Boot, CV joint, 199
Boot and grease replacement, 649–650
Boot kits for CV joints, 649
Bootlegger's turn, 43
Bore clearance, measuring, 467
Bosch-designed modulator, 145, 148, 152
Bounce, 229
Bounce test, 265, 586–588, 590
Bounce travel, 210
Brake adjuster, styles of, 56
Brake adjustment, major, 54
Brake assembly grease, 414
Brake caliper, 5
Brake component cleaning, 411–413
Brake component inspection, 411–413
Brake drag, 518
Brake drum, 48, 58–61
Brake drum assemblies, 54
Brake drum cooling, 60
Brake drum friction surface, inspecting, 350
Brake drum inspection, 393–397
Brake drum lathe, 397, 416
Brake drum lining-to-shoe fit, 415
Brake drum machining, 393, 397–403
    procedure for, 399–403
Brake drum micrometer, 394, 395–396
Brake drum removal, 388–392
Brake dynamometer, 40, 41
Brake failure warning light, 108
Brake fluid, 6, 115–119
    as booster fluid, 133
    changing, 503–506
Brake fluid injector, 501
Brake fluid testers, 505
Brake hose, removing caliper and, 429
    types of, 108
Brake hydraulic systems, 6–7
    tandem, 6, 7
Brake inspection, 343–354
    steps in performing, 343–354
Brake job, 2, 341, 342
    break-in step after, 420
    checks for completing, 420–421
    completing, 459
    on drum brake assembly, 387
Brake lathe, 437
Brake light switch, 113–114
    adjustment, 511–513
Brake lining, 2, 9, 33, 34–36
    attachment, 52
    friction, 33–37
    inspecting, 348, 350–351

Index

Brake lining-to-shoe attachment, 415
Brake lockup, 2, 8
   skids and, 43
Brake Manufacturers' Council (BMC), guidelines of, 342
Brake pads, 69–70,
   removing and replacing, 451–457
Brake pedal, 89
Brake pedal, bypassing, diagnosing, 508–509
   remote control, 500
   sinking, diagnosing, 508–509
   spongy, diagnosing, 506–508
Brake pedal travel sensor, 144
Brake power, 38–41
   measuring, 40
Brake pull, 412
Brake repair recommendations, 359
Brake sensor, 305
Brake shoe, 2–3, 51–52
   deenergizing, 3
   disc, 5
   energizing, 3
   regrinding, 416–417
   types of, 49
Brake shoe gauge, 419
Brake shoe grinder, 416
Brake shoe installation, 418–420
Brake shoe preinstallation checks, 415–416
Brake shoe removal, 410–411
Brake shoe return springs, 54
Brake shoe toe, 51
Brake spoon, 56
Brake spring, 54, 55
   inspecting, 348–349
   removal and replacement, 404–409
   tools, 404
Brake spring pliers, 404–405
Brake spring remover-installers, 404
Brake stop lights, 108–110
Brake surface finish, 359–360
Brake system, 1–2
   antilock (ABS), 8
   components of, 2–8
   federal and state inspections of, 2
   federal laws pertaining to, 27
   repair, 9–10
Brake system service, 341–362
Brake that will not apply, diagnosing, 509
Brake that will not release, diagnosing, 509
Brake torque, 38
Brake warning light, 7
   inspecting, 344
Brake-guard ABS, 139

Brakes:
   bleeding, 495–506
   disc, 5–6
   drum, 3–5
   parking, 8
   power, 7–8
   troubleshooting problems, 354–359
   wear of, 9
Braking efficiency, 29
Breakout box, 550
Bridge bolts, 105
British thermal units (BTU), 32
Bubble flare, 106
Bubble level gauges, 761
Bubble wheel balancer, 636
Buggy springs, 235
Bump steer, 22, 291, 328, 788
Bump stops, 234, 252
Burnishing, 359, 360
Burping the boot, 654–655
Bushing, 10, 187
Bushing drivers, 699
Bushing pullers, 697
Bushings for guide pins, 67, 71
Butyl liner, 169
Bypass port, 89
Bypassing brake pedal, diagnosing, 508–509

C-clip axle, 192
C-lock axle, 375
   installing, 380–381
   moving and replacing bearing and seal for, 379–380
   removing, 378–379
Cable adjustment, parking brake, 532–534
   lining clearance, 532–533
Cable-style self-adjuster brakes, 56, 57
Cables, parking brake, 77–78
Calibrating alignment system, 757
Caliper, 5, 64–68, 424–425
   basics, 101
   fixed, 5, 64
   floating, 5
   reassembling, 486–487
   rebuilding of, 483, 484–487
   reconditioning, 483–488
   removing, 425, 428–431
   service, 483–494
   sliding, 5
Caliper body, and bleeder screw, 105
Caliper boot, 103–105
Caliper brakes, 63
Caliper hardware, 429
Caliper installation, 457–459

Caliper mount, 65
Caliper mounting hardware, 70–72
   inspecting, 351
Caliper piston, 67, 88, 101–103
   retracting, 425–431
Caliper piston boot, inspecting, 351
Caliper positioning hardware, 70
Caliper service, 450–451
Caliper with mechanical parking brake, reconditioning, 488–495
Caliper-type parking brakes, 81
Calories (c), 32
Cam and lever (Ross) steering gear, 276
Cam grind, 416
Camber, 321, 322–323
   adjusting rear-wheel, 800–801
   and scrub radius, 324
   change, 13
   in rear wheel alignment, 337
   measuring rear-wheel, 797–798
   positive or negative, 202
   shims, adjusting, 769–770
   zero, 10–11
Camber, adjusting, 755, 767–777
   eccentric cams, 771–772
   miscellaneous styles, 775–777
   struts, 773–775
   tapered shims for, 776
   using sliding adjustment, 772–773
Camber, measuring, 762–766
   using alignment system, 762–764
   using magnetic gauge, 764–766
Camber and tire wear, 576
Camber change, 326
Camber gauge, 778
Camber specifications, 766–767
Camber spread, 323, 325–326
Camber-caused tire wear, 325
Camber-correction gauge, 801
Cap base, 159
Carbide cutter bit, 394
Carcass splines, 158
Cardan universal joints (U-joints), 198
Cast wheels, 184
Cast-iron bores, 468
Cast-iron reservoir, 501
Caster, 321, 328–329
   adjusting with shims, 769–770
   and road crown, 331
   effects of too little, 331–332
   effects of too much, 331–332
   factors that affect, 330–331
Caster, adjusting, 755, 781–784
   miscellaneous styles of, 784
   using adjustable strut rod, 783–784
   using eccentric cams, 783

Caster, adjusting (*Cont.*):
  using shims, 782
  using sliding adjustment, 783
Caster, measuring, 777–781
  using alignment system, 778–779
  using magnetic gauge, 779–781
Caster, positive or negative, 202
Caster effects, 329–330
Caster roll, 330
Caster specifications, 778
Caster spread, 331, 784
Caster swing, 779
Caster-camber gauges, portable, 761
Celsius (C), 32
Center bolt in leaf spring, 235
Center link, 297
  removing and replacing, 723–724
Center of gravity (CG) of car, 41
Center section of wheel, 177
Center steering link, 297
Center takeoff racks, 280
Center-mounted tie rod, 720
Centering springs, 110
Centrifugal force in tires, 582–583
Changeover point, 112, 113
Chapman strut, 14, 220
Chased threads, 620
Chasing tool, 620
Chassis components, 1, 11
  general purpose of, 1
Chassis electrical, 303–315
Chassis electronic systems, 303–315
Chassis height, 588
Chatter band, 398
Check valve, faulty, 518, 519
Chill spots on brake drum, 393
Chrysotile family of asbestos, 37
Circuit breakers, 303
Circuit protection, 305
Circuits, 303–305
Clamp to lock tapers, 689
Clear vision adjustment, 797
Clock spring, airbag, 273
Closed ABS system, 140
Codes, clearing, 313
Coefficient of friction, 33, 34, 167
Coil clash, 238
Coil spring, 19, 80, 205, 236
Coil spring, installing, 667
  using spring clips, 668
Coil springs, removing and replacing, 663–668
  from S–L A suspension, 663–665
  with spring clips, 665–666
  with spring compressor, 665
Coil spring suspension, solid axle, 218
Coil-over shock, 246

Coil-over shock absorbers, 264
Cold-cap process, 173
Collapsible steering columns, 274
Collision sensors, 271
Combination valve, 110
  hydraulic brake, 7
Compact spare tires, 170, 171
Compatibility of brake fluid, 115
Compensating port, 89
Compensation, 757
Competitively priced brake lining, 35
Composite leaf springs, 233
Composite wheels, 178
Compression of shock absorbers, 254
Compression-loaded ball joints, 205
Computer balance for tires, 633
Computer balancer, 638
  on-car, 640
Computer-controlled air suspensions, 240–241
Computer-controlled shock absorbers, 261–262
Computer-controlled suspensions, 234
Computerized plate tester, 40
Computerized suspension analyzer, 586
Conductors, 303
Conduit, cable, 78
Conicity, 577
Connectors, 307
Constant velocity (CV) universal joint, 193, 197–199
Contamination of rotor, 432
Continuity check of circuit, 309
Control arm, 10, 13, 16–17
  installing, 701–703
  removing, 690–692
Control arm bushing checks, 601–602
Control arm bushing replacement, 696–701
Control arm geometry, 204
Control cable, 77
Control force of shock absorber, 255
Control module, 306
  air suspension, 241
  airbag, 271
Control module of shock absorbers, 261
Control valve, leaky, 520
Control valve, steering gear, 285–287
Control valve assembly on power steering pump,
  removing, 745–746
  replacing, 746
Control valve on power steering pump, cleaning, 745–746
Controller antilock brake (CAB), 142
Conventional standard steering gear, 22
Conventional steering system, 269

Copper tubing, 107–108
Cord ply, 159
Cotton sidewall plies, 158
Cracked drum, check for, 393
Cradle shift, 778
Crash sensor, 271, 305
Criss-cross system, 96
Cross and yoke universal joints (U–joints), 198
Cross bleed, 501, 506
Cross camber, 325
Cross caster, 331
Cross-car reading, 323
Cup, rubber, 91
Cup expanders, 92
Cup seal, 102–103
Cupping, 576
Curb height, 588
Curing agents in brake lining, 34
Custom capping, 174
Cutter bit, 398
CV joint, inner:
  assembling, 653–654
  disassembling, 653
CV joint, outer, fixed-type:
  assembling, 652–653
  disassembling, 650–652
CV joint, servicing, 649–654
CV joint boots, installing, 654–655
CV joint grease, 650
CV joint problem diagnosis, 644
CV–Easy boot, 645
Cycles, spring, 231
Cycles per second (CPS) of spring, 231
Cylinder bore:
  in master cylinder, 89
  requirements of, 90–91
Cylinder size, gauging, 467

Dampening ball joint, 206
Damper, 252
Damping force of shock absorber, 255–256
Damping ratios of shock absorbers, 254
Dash lamp switch, 110–111
Datums, 801
de Carbon shock absorber, 259
de Doin axle, 222
Deceleration, 28–29
  measuring rates of, 29–31
Deceleration sensor, 305
Decelerometer, 29
Decibels, 564
Dedicated ABS tester, 551–552
Dedicated tester for EBCM, 546–547
  clearing codes for, 548

Index

Deenergized brake shoe, 49
Deenergizing brake shoes, 3
Delco ABS–VI modulator, 145
Delco Moraine caliper design, 488
   rebuilding, 490, 492
Delco/Delphi ABS–VI hydraulic
   modulator assembly, 153
Department of Transportation
   (DOT), 115
Desiccant, 242
Diagnostic angles, 321
Diagnostic charts for NVH, 569–570
Diagnostic flowcharts, 313
Diagnostic module, airbag, 271
Diagnostic trouble code (DTC),
   307, 313
   in EBCM, 546
Diagonal split brake system, 6–7
Diagonal split hydraulic ABS
   system, 138
Diagonal split system, 96
Diagonal wipe wear pattern, 576
Dial calipers, 595
Dial indicator, measuring ball joint
   clearance with, 595
Diaphragm, leaky, 520
Die nut, 620
Digital multimeter (DMM), 310
Digital volt-ohmmeters, 309
Direct acting shock absorbers, 253
Direct torque backing plate, 52
Directional control angles, 321
   adjusting, 755
Directional finish, 360
Directional tread, 162
Disc, 63, 72
   wheel, 177
Disc brake, 5–6
   lining, 69–70
   theory, 63–74
Disc brake calipers, 97
Disc brake service, 424–459
Disc parking brake, 81–83
Disc-balancing valve, 111–112
Discard diameter of drum, 394
Discriminating sensor, airbag, 273
Dive, 41
Dividing piston, 259
Dog tracking, 216, 337
Door sensor, 305
DOT number on tires, 163
Double ball joint steering axis, 330
Double flare tubing, 106, 465
Double-tube shock absorbers, 256–259
   purging, 663
Double-wishbone suspension, 15, 206, 223–224

Drag link, 23, 293, 297–298
Drag link socket, adjustable, removing
   and replacing, 725
Drive axle bearings, solid axles,
   repairing, 375–381
Drive axle wheel bearings:
   independent suspension, 187,
   193–194
   solid axle, 187, 191–193
Driver-side airbags, 271
Driveshaft construction, 197–199
Driveshaft disassembly, 647–649
Driveshaft removal, 645–647
Driveshaft rpm, calculating, 571
Drop center rims, 178
Drum, brake, 2
   wear of, 9
Drum brake, 2–5
   theory, 48–61
Drum brake component
   lubrication, 414
Drum brake service, 387–421
Drum diameter, measuring, 395
Drum grinding, 394
Drum inspection, 393–397
Drum machining, 393, 397–403
   procedure for, 399–403
Drum mike, 394, 395–396
Drum removal, 388–392
Drum-to-hub attachment methods,
   388–389
Dry ball joint, 592
Dry park test of steering system, 604
Dual compounding of rubber, 159
Dual master cylinder, 98
Dual master cylinder construction,
   92–94
Dual power booster, 123
Dual solenoid control valve, 149
Dual-chamber air springs, 241
Dual-diameter bore master cylinder, 97
Duoservo brake, 4, 49, 50
   adjusters on, 57, 58
   heel of, 51
   toe of, 51
Duplex brake, 50
Dust boot, 100–101
Dust shield, 256
Dynamic and kinetic balancing:
   computer balancing, 638–641
   strobe light balancer, 637
Dynamic balance of tire, 633
Dynamic brake tester, 40
Dynamic seal, 102, 190
Dynamic tire balance problems, 583
Dynamic Tracking Suspension
   System, 226

Easy-out tool, 716
Eccentric cams, 771
   adjusting caster with, 783
Eccentric cams for adjusting camber,
   771–772
Eccentric grind, 416
Economy brake lining, 35
Edge brand on brake lining, 36
Edge code on brake lining, 36
EIS Sur-Bleed syringe, 475
Electric motor-operated hydraulic
   booster, 8
Electrical ABS components, repair of,
   557–559
Electrical circuit problems, 307–308
Electrical tests on ABS, 550–555
   service manuals and, 551
Electrical values, measuring, 308–311
Electrical vibration analyzer (EVA), 569
Electricity, basic 303–305
Electrohydraulic boost, 517
Electrohydraulic booster
   construction, 133
Electrohydraulic booster operation, 133
Electrohydraulic booster theory,
   131–132
Electrohydraulic boosters, 122–123
Electrohydraulic power booster in
   ABS, 542
Electronic alignment systems, 756–757
Electronic balancer, 637
Electronic brake control module
   (EBCM), 142–144, 146, 149, 154,
   542, 543, 544
   diagnostic trouble code in, 544, 546
   hard codes in, 546
   soft codes in, 546
Electronic control module (ECM), 139,
   142, 305
Electronic controller, 142–144
Electronic modulator, 144–146
Electronic power steering, 282
Electronic variable-orifice (EVC), 289
   steering, 288–289
Electronic wheel speed sensors, 139,
   140–142
Electronically controlled air
   suspensions, 240–243
Electronically controlled modulator
   valves, 139
Electronically controlled
   suspensions, 234
Electronically controlled system
   diagnosis, 313–315
Electronics, basic, 305–307
Electrostatic discharge (ESD), 305
Elliott king pin design, reversed, 704

Elliott style king pin, 19
Emergency brake, 8, 77
End takeoff racks, 280
Energized brake shoe, 49
Energizing brake shoes, 3
Energy, 31
Energy converter, 253
Energy of motion, 31
Engine cradle, 11
Equalizer, 78
Equalizer lever, 534
Equilibrium reflux boiling point (ERBP), 116
Error code in EBCM, 546
    clearing, 548
Exciter ring, 140
Expander washers, 99
Extended mobility tires (EMT), 172
Extension of shock absorbers, 254
External brake fluid leak, 508
External contracting brake, 48
External dust boot, 100–101
Extra load tires, 166
Eye ring shock absorber, 256

Fahrenheit (F), 32
Failure warning light, 111
Fast-fill master cylinder, 97
Fastener security, 361–362
Fasteners, 361
Feather edge wear pattern, 327, 576
Federal Motor Vehicle Safety Standards (FMVSS), 27
Fiberglass-reinforced plastic (FRP), 233
Filler, tire, 159
Fillers in brake lining, 34
Finish cut, 444
Fins, vented rotor, 73
Fixed anchor, 53
Fixed calipers, 5, 64, 101,
    removing, 430
    retracting piston in, 425–428
Fixed joint, 198
Fixed-anchor brake shoe grind, 417
Fixed-caliper pistons, 64–65
Fixed-caliper units, on real-wheel disc brakes, 74
Fixed/power rack, 758
Flanges, 177
Flare-nut wrench, 463–464
Flat ride tuning, 233
Flexible couplers, steering column, 275
Flexible hoses, 105
Flexible rubber bellows, 279
Floating axle, 192

Floating calipers, 5, 65–69, 101
    in rear-wheel disc brakes, 74
    removing, 430
    retracting piston in, 428
Floating drum, 388, 391
Floating valve, 126
Fluid level, 144
    checking, 345
Fluid level switches, 115
Fluid power, 85, 281–282
Fluid pressures, computing, 86
Fluid swirl method of pushrod adjustment, 521, 522–523
Flushing brake fluid, 505
Folding spare tire, 170, 171
Follower ball joint, 206
    checking, 598
    measuring preload of, 598
Foot valve assembly, 256
Forward brake shoe, 3, 49
Four-circuit brake, 139
Four-wheel alignment, 337, 761
Four-wheel alignment systems, 756–757
Four-wheel antilock (4WAL), 138
Four-wheel disc brakes, 74
Four-wheel steering (4WS), 298
Frame and body alignment, 801–802
Frame gauges, 801
Frame of car, 1, 11
Free play in ball joint, 598
Freezing point of brake fluid, 115
Frequency, spring, 231
Frequency of noise, 564
Frequency of vibration, 564
    matching component rpm to, 570–571
Friction, 33
    coefficient of, 33, 34
Friction and heat energy, 31–32
Friction material, 2–3, 33, 34
Friction Materials Standards Institute (FMSI), 51
Friction modifiers in brake lining, 34
Friction plunger, 733
Friction shock absorbers, 253
Friction-loaded ball joint, 206
Frictionless bearings, 187
Frictionloaded ball joint, checking, 598
Front rear split, 96
Front suspension types, 202–213
Front wheel bearings, repairing nonserviceable FWD, 382–384
Front-wheel drive axles, 212–213
Front-wheel-drive (FWD) cars:
    brakes and, 6, 7
    center of gravity in, 41

    swing axles and, 13
    torsion bars in, 20
Front-wheel-drive driveshaft:
    service, 644–658
    theory, 197–199
Front-wheel lockup, diagnosing, 509–510
Full-contact shims, 800
Full-floating axles, 192
Fuses, 303
Fusible links, 303
FWD driveshaft:
    disassembling, 647–649
    installing, 656–658
FWD front wheel bearings:
    repairing nonserviceable, 382–384
    types of, 381
FWD rear axles, 223–224
FWD, rear short-long arm suspension, 224
FWD, rear solid axle suspension, 223
FWD, rear strut suspension, 224
FWD, rear trailing arm suspension, 223–224

g (gravity) in deceleration, 28
Gas pressure shock absorber, 259
Gas-charged shocks, 259
Gauge method of pushrod adjustment, 521
Gear lash, 276, 728
    adjusting sector, 731
Gland nut, 261
Glycol brake fluids, 116
Go-no go gauge, 521
Graphite-reinforced plastic (GRP), 233
Gravity bleeding of brakes, 499
Grease and boot replacement, 649–650
Grease spray in boot, 199
Grommets with rack-and-pinion steering gears, 279
Grooving of rotor, excessive, 432
Ground circuit, 303
Grounded circuit, 307
Guarded tubing, 107
Guide bolts, 430
Guide bolts for calipers, 67
Guide pins, 430
Guide pins for calipers, 67
Guide plates, 430
Gum, 159

H–arms, 221
Half axle, 197
Half racks, 757

Index

Half shaft, 193, 197
Hammer blows, 712
Handheld scanners, 313
Hang-on power steering, 289
Hard brake lining, 36
Hard code in EBCM, 546
    clearing, 548
Hard spots on brake drum, 393
Hard steering, 592
Hardware set, 411
Harmonic resonance, 567
Harshness, 569
Heat checking of rotor, 432
Heat checks on brake drum, 393
Heat energy, 32–33
Heat sink, 59
Heavy-duty brake lining, 35
Heavy-duty grease, 371
Heel and toe clearance check, 415, 416
Heel drag, 100
Heel of brake shoe, 51
Height check, 586
Height sensor, 234, 305
    air suspension, 241
Height-sensing proportioning valve, 113
    adjustment, 511
Hertz (Hz), 564
High-efficiency particulate air (HEPA) filter, 37
High-temperature grease, 371
Hoist rack, 758
Hold-down springs, 54, 407–409
    replacing, 411
Hold-off valve, 111–112
Horizontal, radial direction clearance of ball joint, 596
Hoses, 105–108
    inspecting, 353–354
Hot-cap process, 173
Hotchkiss driveshaft, 217
Housing, cable, 78
Hub and wheel bearing, installing, 375
Hub centric wheels, 183
Hydraulic assist steering, 280
Hydraulic boost, 517
Hydraulic booster, 122
    construction, 130
    operation, 130–131
    theory, 129–130
Hydraulic brake system, 85
Hydraulic control unit, 144, 542
Hydraulic control valves and switches, 108–115
Hydraulic cylinder service, 466–468
Hydraulic modulator, 542
Hydraulic pressure checks on ABS, 556
Hydraulic system, brake, 6–7, 85–88

    deterioration of, 9
    diagonally split, 96
    pressure in, 86
    tandem, 6, 7
Hydraulic system mineral oil (HSMO), 116
Hydraulic system problems, diagnosing, 506–511
    spongy brake pedal, 506–508
Hydraulic system service, 462–513
Hydraulically operated shock absorbers, 253
Hydraulics, 85, 281–282
    failure of, 94–95
    pistons, 86–87
    principles of, 85–88
Hydro-boost, 517
    removing, 521
Hydro-boost brakes, 290
Hydro-boost by Bendix, 129
Hydro-boost units, 7–8
    reconnecting hose on, 523
    repair of, 525–528
Hydroplaning, 162
Hygroscopic brake fluids, 116

Idler arm, removing and replacing, 721–723
Idler arm and bracket assembly, 721
Idler arm bushing, 721
    checking, 605–606
Idler arm with steering linkage, 297
Ignition switch, 274
In-phase steering, 298
Inboard rotor pad, 65, 67
Inboard shoe, 5
Inches of mercury (in. Hg), 123
Included angle, 333–334
Independent rear suspension (IRS), 219–221
Independent suspensions, 12, 16
Indexing, 581
Indexing runout, 449
Individual toe, 787
Inertia, 41
Inflator, airbag, 271
Inner control valve, 133
Inner CV joint:
    assembling, 653–654
    disassembling, 653
Inner plunge joint, faulty, 644
Inner tie-rod end:
    rack-and-pinion steering, removing and replacing, 715–720
    standard steering gear, removing and replacing, 715

Inner tie-rod end pivot bushings, rack-and-pinion steering gear, center-mounted tie-rods, removing and replacing, 720–721
Inorganic brake lining, 34
Inputs, control module, 306
Installing tire and wheel assembly, 620
Installing tire onto wheel, 625–626
Instant center locations, 203
Insulators for guide pins, 67, 71
Insulators with rack-and-pinion steering gears, 279
Intake port, 89
Integral antilock brake systems, 139–140
Integral booster, 124
Integral drum parking brake, 79
Integral molded brake lining, 52
Integral power steering gears, 288
Intensity of heat, 32
Intensity of noise, 564
Interleaf friction, 235
Intermediate lever, for front cable, 534
Internal booster leaks, 519–520
Internal dust boot, 100–101
Internal expanding brake, 48
Internal leakage check, 519
International Standards Organization (ISO) flare, 106–107
Interthread friction, 361
Inverted flare nut, 106, 107
IRS, semitrailing arm suspension, 219–220
IRS, strut suspension, 220–221
IRS, trailing arm suspension, 220
ISO tubing, 465
Isolation-dump valve, 149

Jacket, rubber tubing, 108
Jacking, 222
Jackknifing, 43
Jam nut, 191
Jig, 802
Jounce, 229

Kelsey-Hayes caliper design, 488
    rebuilding, 490–491, 492–494
Kelsey-Hayes parking brake calipers, 82
Kinetic energy, 31
Kinetic tire balancers, 583
King bolt, 18–19
King pin, 18–19, 203
    Elliott style, 19
    removing and replacing, 704–706
    reversed Elliott style, 19

Kingpin clearance, checking, 600
Kingpin inclination, 332
Kingpin offset, 181
Knockers, 699
Knockoff wheels, 185

L-shaped torsion bar, 239
Lateral runout, 73, 435, 578
    tire and wheel, 582
Lateral tire runout, 582
Latex, 159
Lathe-cut o-ring, 101–102
Lead, car, 577
Lead screw disc parking brake, 81
Leading brake shoe, 3, 49
Leading strut rod, 203
Leading to trailing shoe, 50
Leading-brake shoe, 49
Leaf spring, 19, 205, 210, 235–236
    removing, 668–669
    replacing, 669
    service, 668–669
Leaf spring bushings, types of, 669
Leaf spring suspension, solid axle, 218
Ledges on backing plates, 53
Lever, parking brake, 79–80
Lever and starwheel adjuster, 57
Lever shock absorbers, 253
Lever-style self-adjuster brakes, 56–57
Light beam for measuring toe, 787
Light gauge, measuring toe using, 790–791
Limiters, 234
Lining attachment, brake, 52
Lining break-in, 359, 360
Linkage booster, 124
Linkage booster power steering, 289–290
Linkage-type power steering, 287
Links, 218
Lip seal, 102–103, 190, 193
Live axles, 216
Load index, 167
Load range system, 166
Load ratings for tires, 165–166
Load-carrying ball joint, 205
    checking, 594–596
Load-carrying shock absorbers, 264
Loaded calipers, 425
Loaded radial runout, 580
Locating pins, 430
Locator dot, 624
Locator tool, 589
Lock nut, 191
Lock to lock, 270
Lockheed brake spring tool, 405–406

Locking methods for tie rods, 296
Locking tapers, 686
Locking tapers with steering linkage components, 712
Long-knuckle suspension, 206
Long-spindle S–L A suspension, 15, 206
Low brake pedal, 56, 70
Low friction ball joint, 206
Low friction tie-rod end, 206
Low pivot axle, 222
Low-brake-fluid warning lights, 115
Lubricating ability of brake fluid, 115
Lucas-Girling master cylinders, 98
Lug bolt, removing, 621–622
Lug bolt and stud replacement, 620–622
Lug bolt circles, 178
    runout, 582
Lug bolt runout, measuring, 581–582
Lug nuts, 182–184

Machining, brake drum, 393
MacPherson strut, 206
MacPherson strut front suspension, 202
MacPherson strut suspension, 14
Magnasteer, 289
Magnetic gauge for measuring SAI, 785–786
Magnetic gauges for alignment, 761, 764
Magnetic induction coil, 141
Major brake adjustment, 54
Malfunction indicator light (MIL), 307, 313
Mandrel, 397
Manual bleeding of brakes, 499–500
Manual rack, 288
Manual rack-and-pinion steering gear, assembling, 734–735
Master cylinder, 6, 89
    basic operation, 89–90
    basics and components, 89
    bench bleeding, 474–476
    construction, 90–92
    inspecting, 345
    miscellaneous designs, 98
    pistons, 91
    pushrod adjustment, 521
    rebuilding, 470–474
    removing, 468–470
    replacing, 476–477
    service, 468–477
    used with ABS, 94
Master splines, 724
Match mounted tires and wheels, 624
Match mounting, 581

Maximum machining diameter of drum, 394
Measuring alignment angles, 756–761
Measuring angles, 322
Measuring points, 801
Mechanical ABS, 139
Mechanical ABS components, repair of, 557–559
Melting point, 32
Memory steer, 674
Metal, clamp-in valve core, installing, 629
Metal control arm bushings:
    checking, 601–602
    removing and replacing, 700–701
Metal fatigue in springs, 238
Meter readings, interpreting, 311–313
Metering valve, 7, 110, 111–112
    on brakes, 496
Missing splines, 724
Modified struts, 209
Modulator, 144
Modulator unit, 542
Mold bonded brake lining, 52
Molylube, 457
Monobeam axle, 210
Monoleaf springs, 235
Monotube shock absorber operation, 259–260
Morning sickness, 288
Morse tapers, 686
Motorist Assurance Program (MAP), guidelines of, 342
Mounting hardware, 70
Mud traction ability, 162
Multilink front suspension, 202
Multilink suspension, 15, 206

National Coarse (NC) thread bolt, 361
National Fine (NF) thread bolt, 361
National Highway Safety Administration (NHSA), 166
National Highway Traffic Safety Administration (NHTSA), 115, 166
    braking systems and, 27
Natural frequency, 232
Natural frequency of vibration, 566
Needle bearings, 189
Negative camber, 202
Negative camber angle, 322
Negative caster, 202, 328–329, 330
Negative offset, 180
Negative scrub radius, 7, 96, 324
New-car brake performance, 341
Nitrogen gas-charged accumulator, 129

Index

No-flat tires, 171–172
Noise, 564
   visual inspection for, 590–591
Noise suppressant with disc brake pads, 455
Nonasbestos organic (NAO) brake lining, 34, 35
Noncorrosive brake fluid, 115
Nondirectional finish, 360
   on rotor, 436
Nonenergized brakes, disc brakes as, 5
Nonindependent suspensions, 12
Nonintegral antilock brake systems, 139–140
Nonradial tubes, 168
Nonserviceable drive axle wheel bearings, 187, 194
Nonserviceable nondrive axle wheel bearings, 187, 191
Nonserviceable wheel bearings, repairing, 374–375
Nonservo brakes, 4, 49, 50
   adjuster on, 56, 57
   disc brakes as, 5
   rear, 50–51
   styles of, 4
Nonwearing center links, 724
Nut breaker, 662
NVH (noise, vibration, and harshness), 564–569
   diagnostic charts and trouble trees for, 569–570
   diagnostic procedure, 569–571
   matching component rpm to vibration frequency, 570–571
   road test for, 569
Nylon sidewall plies, 158

O-ring, 101–102
Occupational Safety and Health Administration (OSHA):
   asbestos and, 37
Off-car rotor machining, 440–446
Off-car rotor turning, 436
Off-car wheel balancing, 634–636
Offset wheel dimension, 180
Ohms, 303
Oil in shock absorbers, 265
On-car rotor machining, 446–447
On-car rotor turning, 436, 437–440
   recommendations for, 439–440
On-car wheel balancing, 634–636
Open ABS system, 140
Open circuit, 307
Order of vibration, 567
Orifices in shock absorbers, 253

Original diameter of drum, 394
Original equipment (OE) brake lining, 34
Original equipment manufacturer (OEM) tires, 170
Oscillation cycle, 564
Outboard rotor pad, 65, 67
Outboard shoe, 5
Outer control valve, 133
Outer fixed-type CV joint:
   assembling, 652–653
   disassembling, 650–652
Outer tie-rod end, 293–296
   removing, 712–713
   removing and replacing, 712–715
   replacing, 713–715
Output:
   circuit, 305
   control module, 306
Output pistons, 99
Over center adjustment, 276, 728
Overload springs, 246
Overrelease, 102
Oversize brake lining, 416

P–metric system of tire size, 164, 166
Packing wheel bearings, 371–374
Pad antirattle hardware, 70
Pad knock back, 73
Pad mounting hardware, 70–72
Pad positioning hardware, 70
Pad retention hardware, 70
Pad wear indicators, 70
Pads, brake, 2, 69–70
   disc brake, 63
Pads, removing and replacing on caliper, 451–457
Panhard rod, 218, 223
Parallel circuit, 303
Parallelism in rotor, 73, 431, 433, 435
   checking, 434
Parallelogram steering, 721
Parallelogram steering gears, 270
Parallelogram steering linkage, 22, 297, 723
Parallelogram steering linkage and idler arm checks, 605
Parking brake, 8, 77
   disc, 81–83
   inspecting, 344, 354
   lever, 79–80
   pedal height, 533–534
   service, 531–538
   theory, 77–83
   warning light, 79
   warning light service, 538

Parking brake cable:
   adjustment, 532–534
   adjustment of front, 534
   automatic self-adjusting mechanism, 532
   removing, 536–537
   replacement, 534–538
   replacing, 537–538
Parking brake linkage, adjustment, 534
Pascal's law, 86
Pedal pulsation, 73, 74
Peening, 389
Phasing, 567
Phenolic pistons, 483
Phenolic resin pistons, 102
Phoenix reverse injector, bench bleeding with, 476
Pickle forks, 712, 713
   breaking a taper using, 686, 688
Pickup, 637
Pigtail coil springs, 237
Pin-out box, 550–551
Pinch bolt to lock tapers, 689
Pinion bearings and seals, 734
Pinion gear, 279
Pintle-type pistons, 103
Piston:
   fixed-caliper, 65
   for brake system, 89
   hydraulic, 86–87
   in master cylinder, 89, 91
   primary, 93
   secondary, 93
Piston and ring in shock absorber, 265
Piston boot, 103
Piston return springs, 94
Piston rod, 264
Pit racks, 757
Pitching motions, 233
Pitman arm, 22, 270, 605, 607
   removing and replacing, 724
Pitman shaft, 22, 276
   seal replacement, 751
Pivot bushing, 13
Plain brake shoe grind, 417
Plastic reservoirs, 501
Platforms on backing plates, 53
Plies, tire, 158
Plugging a tire, 629, 630–632
Plunge design CV joint, 198
Plunging CV joint, 198
Plus 1, 2, and 3 wheel conversion, 181
Ply arrangement, 160
Ply materials, 158–160
Polyester sidewall plies, 158
Polyglycol-based brake fluid, 115
Portable alignment gauges, 764

Portable stands for turntables, 761
Positive camber, 202
Positive camber angle, 322, 324
Positive caster, 202, 328–329, 330
Positive crankcase ventilation (PCV) valve, 518
Positive displacement pump, 283
Positive offset, 180
Positive scrub radius, 96, 324
Pounds per square inch (PSI), 86
Power assist, 122
Power booster, removing, 521
Power booster service, 517–528
Power brake, 7–8
    theory, 122–133
Power rack, 288
Power source, circuit, 305
Power steering, 280–291
    air bleeding, 744
    electronic, 282
    fluid change, 741–743
    hydraulic assist, 280
Power steering checks, 608–609
Power steering fluid, 290–291
Power steering fluid leakage, 608
    checking for, 608–609
Power steering gear service, 751
Power steering hoses, 290–291
Power steering pump, 283–285
Power steering pump overhaul, 747–751
Power steering pump seals,
    removing and replacing, 746–747
    repairing leaks, 746–747
Power steering pump service, 745–751
Power steering service operations, 739–751
Power steering system:
    checking and problem diagnosis, 739–741
    pressure gauge sets, 744
    pressure specifications for, 744
    pressure testing, 744–745
Powermaster, 131, 517
    removing, 521
Powermaster units:
    pump motor in, 544
    reconnecting electric connector on, 523
    repair of, 528
Precured recapping, 173
Precured recaps, 173
Preload, bearing 190
Preload of follower ball joint, 598
Premium quality brake lining, 35
Pressed-in ball joint:
    kits for, 692
    removing and replacing, 692–693

Pressure, breaking a taper using, 688–689
Pressure and flow control valve, 283
Pressure bleeder, 111
Pressure bleeding of brakes, 501–503
Pressure control valve, 112–113, 511
Pressure differential valve and switch, 7, 110–111
Pressure multiplier, 124
Pressure pad, 733
Pressure ratio valve, 112–113
Pressure testing power steering system, 744–745
Pressure tube in shock absorber, 265
Pressure-reducing valve, 112–113
Pressure-regulating valve, 112–113
Prevailing torque nuts or bolts, 361
Primary brake shoe, 4
Primary cup, 89, 90, 97
Primary piston, 93, 97
Primary shoe, 49–50
Progressive power steering (PPS), 289
Projector units for alignment, 757
Proportioning valve, 7, 96–97, 110, 112–113, 509
Proportioning valve adjustment, height-sensing, 511
Pull, car, 577
Pull-type valve stem, 111–112
Pullback springs, 54
Pullers, 686, 712
Pump catch-up, 284
Pump pressure, 144
Pumps, power steering, 283–285
Push-type valve stem, 111–112
Pushrod, 89
    installing, 473
Pushrod adjustment, 521
Pyramid belts, 575
Pyramid construction of plies, 161

Quantity of heat, 32
Quarter-elliptical springs, 235
Quick-change wheels, 185
Quick-Replacement Outer CV Boot, 645
Quick-take-up master cylinder, 97–98

Rack, 279
Rack and pinion steering system, 269
Rack bearing, 733
Rack gear, 716
Rack housing bushing, 734
Rack plunger, 733

Rack support, 733, 734
Rack-and-pinion gears, 270, 281
Rack-and-pinion power steering gears, 278–280, 287–288
    adjusting toe on, 796–797
    tie-rods and, 296
Rack-and-pinion steering gear, 22
    assembling a manual, 734–735
Rack-and-pinion steering gear adjustments, 733
Rack-and-pinion steering gear overhaul, 734–735
Rack-and-pinion steering gear removal and replacement, 727–728
Rack-and-pinion steering gears and linkage, checking, 604
Rack-and-pinion steering linkage checks, 606
Radial ply, 160
Radial runout, 396–397, 578
    tire and wheel, 579–582
    tolerances, 580
Radial tire balancers, 583
Radial tire waddle, 578
Radial tubes, 168–169
Radius rod, 13
Rate, spring, 230–231
Ratio, shock absorber, 254
Ratio bar, 534
Rayon sidewall plies, 158
Re-arcing brake lining, 416
Reaction disc, 133
Reaction group, 133
Reaction piston, 133
Reaction time for braking, 44
Rear antilock braking system (RABS) valve, 149
Rear caliper pistons, 74
Rear spacing, 180
Rear suspension types, 216–226
Rear wheel alignment, 337
Rear-suspension service, 706
Rear-wheel alignment, 797
Rear-wheel antilock (RWAL) system, 138, 149
Rear-wheel camber, measuring, 797–798
Rear-wheel camber and toe, adjusting, 800–801
Rear-wheel disc brakes, 74
Rear-wheel drum brakes, integral parking brake with, 80
Rear-wheel floating calipers, 74
Rear-wheel lockup, diagnosing, 509–510
Rear-wheel steering, 224–226
Rear-wheel toe, measuring, 798–799

Index

Rear-wheel track or thrust, measuring, 799
Rear-wheel-drive (RWD) vehicles, 12
   center of gravity in, 41
Rebore diameter of drum, 394
Rebound, 229
Recapping tires, 173
Recirculating ball nut steering gear, 276
Recirculation pump, 146–148
Reconditioned rotor, 431–432
Red brake warning light, 544
Reed tachometer, 567, 568
Regrinding brake shoes, 416–417
Regrooved tires, 173
Regulated proportioning valve, 511
Reluctor, 140
Remachining, brake drum 393
Remachining rotor, 432
Repacking serviceable wheel bearings, 365–374
Repair and replacement of ABS tone ring, 655–656
Repair and replacement of bearing assembly, 187
Repair operations on ABS, 557–559
Replacement tire selection, 170
Replacement tubing, 107
Replacing rotor, 436
Replenishing port, 89
Reservoir, 91
   in master cylinder, 89
   power steering pump, 283
Residual pressure, 92
Residual pressure check valves, 92
Resistance, measuring, 31
Resonance of vibration, 566
Resurfaced rotor, 432
Resurfacing a rotor, 447–448
   by hand, 436–437
Retracting springs, 54
Retreading tires, 173
Retreads, 173
Return springs, 404, 407
   brake shoe, 54
   replacing, 411
Reverse bleed, 501, 503
Reverse brake shoe, 49
Reverse-flow bleeding of brakes, 503
Reversed Elliott king pin, 19
Ride height, 588
Ride height specifications, 588
Rim, wheel, 177
Rim mount adapter, using, 765
Rising rate bump stop, 234
Riveted lining on brakes, 52
Riveted-in ball joint, removing and replacing, 693–694

Riveting, 389
Road crown, 325–326
   and caster, 331
Road test for NVH, 569
Road testing, after wheel alignment, 802
Roll center locations, 203
Roller bearing, 187, 189
Roller burnishing, 91
Room-temperature vulcanizing (RTV) silicone rubber, 380
Root mean square (rms), 360
Rotary action disc parking brake, 81
Rotating ratchetlike adjuster, 57
Rotating tires, 623–624
Rotor, 72–74
   brake, 2
   disc brake, 63
   inspecting, 351
   resurfacing, 447–448
   wear of, 9
   wedge wear of, 434
Rotor glaze, 432
Rotor inspection, 431–435
Rotor machining:
   off-car, 440–446
   on-car, 446–447
Rotor measurements, 432–435
Rotor refinishing, 436–440
Rotor replacement, 436
Rotor runout, 73, 431, 432, 435, 437, 438, 439
   and rotor replacement, 449
   measuring, 435
Rotor turning:
   off-car, 436
   on-car, 436, 437–440
   recommendations for, 439–440
Roughing cut, 444
Rubber, snap-in valve, installing, 629
Rubber bushings, 17
   with rack-and-pinion steering gears, 279
Rubber control arm bushing, 696
   checking, 601
   installing using driver, 699
   installing using pressing tool, 700
   removing using air hammer and chisel, 698
   removing using puller tool, 698
   without outer sleeves, removing and replacing, 700
Rubber cup, 91
Rubber cushions, 252
Rubber idler arm bushings, 721
Rubber latex, 159
Rubber Manufacturers' Association (RMA), 166, 178

Rubber springs, 234
Run-flat tires, 171–172
Runout, 127, 578
Runout of rotor, 431, 432, 435
RWAL, 542
RWD axle and suspension types, miscellaneous, 222
RWD suspension, solid axle, 216–218
Rzeppa CV joint:
   assembling, 652–653
   disassembling, 650–652
Rzeppa outboard CV joint design, 198

S–L A ball joints, 205–206
S–L A springs, 205–206
S–L A wear factors, 206
Safety bead, 177
Safing sensor, 271, 305
Sag:
   coil spring, 237
   spring, 233
SAI, adjusting, 755, 786
SAI, measuring, 784–786
   using magnetic gauge, 785–786
SAI scales, 785
Sawtooth wear pattern, 327, 576
Scan tools, handheld, for EBCM, 546–547
   clearing codes for, 548
Schrader valve, 168, 169, 264
Scored friction surface on brake drum, 393
Scoring of rotor, 432
Screw, sizing of, 361
Screw extractor, 716
Scrub, 13, 204
Scrub radius, 96, 181, 324
   and camber, 324
Scuff tester, 802
Sealing rings, 101
Seals, 190
Secondary brake shoe, 4
Secondary cup, 89, 97
Secondary piston, 93
Secondary shoe, 49–50
Sector gear end play, 730
Sector gear lash, 725, 728, 730
   adjusting, 731
Sector gear lash adjustments, 725
Sector shaft, 276
Sector shaft end play, measuring, 730
Self-adjusting brakes, 56
   caliper-type, 81, 82
Self-diagnosis of control modules, 307
Self-locking nuts or bolts, 361
Self-operating cams, 57

Self-ranging meters, 311
Self-sealing tires, 171–172
SEMA Foundation, Inc. (SFI), aftermarket wheels and, 182
Semi-active suspension, 261
Semielliptical leaf spring, 235
Semifloating axle, 192
Semimetallic brake lining, 34, 35
Semitrailing arm IRS, 219–220
Semitrailing arm suspension, 16
Sensing units for alignment, 757
Sensor, 141
Sensor ring, 140
Sensors, electronic, 305
Sensors in ABS, replacing, 558
Sensors in shock absorbers, 261
Separator tool, 712, 713
   breaking a taper using, 686, 688
Series circuit, 303
Service, types of, 365
Serviceable nondrive axle wheel bearings, 187, 191
Servicing CV joints, 649–654
Servo action, 4, 49–51
Servo brakes, 49–50, 4
   adjuster on, 56
   return springs on, 54
Setback, 321, 336–337, 755, 788
Shackle, spring, 235
Shim adjustments, 769–770
Shimmy, 210, 332
Shims, adjusting casters with, 782
Shock, using to break tapers, 686, 687
Shock absorber, 1, 10, 12, 14, 19, 252–265
   damping ratios of, 254
   double-tube, 256–259
   failures in 265
   load-carrying, 264
   operating principles, 253
   purging double-tube, 663
   quality of, 264–265
   removing, 660–663
   replacing, 663
   stages of operation of, 255
Shock absorber cartridge, installing in strut, 678–679
Shock absorber damping force, 255–256
Shock absorber dynamometer, 255
Shock absorber failures, 265
Shock absorber inspection, 586–591
Shock absorber portion of strut, removing, 678
Shock absorber quality, 264–265
Shock absorber ratio, 254

Shock absorber service, 660–682
   completing, 682
Shoe, disc brake, 63
Shoe adjusters, 56–58
Shoe anchors, 53–54
Shoe assembly, predisassembly cleanup, 403–404
Shoe energization, 49–51
Shoe return spring, replacing, 411
Shoe rim, 51
Shoe web, 51
Short to ground, 308
Short-long arm (S–L A) front suspension, 202, 203–206
Short-long arm (S–L A) suspension, 13, 14
   torsion bars with, 20
Shorted circuit, 307
Side clearance, checking, 599
Side shake, 210, 217
Side shake of tire, measuring, 600
Sideslip, 802
Sidewall plies, 158
Sidewall rubber, 159
Silencer band, 398, 399, 402
Silica-based tread rubber, 161
Silicone brake fluid, 505
Silicone fluid, 154
Silicone-based brake fluid, 115–116, 117
Simplex brake, 50
Single air springs, 241
Single control arm suspension, 209
Single leaf springs, 235
Single modulator, 149
Single-plane, on-car spin balancer, 636
Single-plane tire balancers, 583
Single-stud shock absorber, 256
   removing nut from, 661
Sinking brake pedal, diagnosing, 508–509
Sipes, tire, 633
Skids, 43–44
Sleeves for guide pins, 67
Sliding adjustment for camber, 772–773
Sliding calipers, 5, 68
Sliding caster adjustment, 783
Sliding valve, 287
Sliding-anchor brake shoes, 417
Slip plates, 797
   for magnetic gauges, 761
Slope, 112–113
Snow traction ability, 162
Society of Automotive Engineers (SAE): braking systems, 27, 36
Soft brake lining, 36

Soft code in EBCM, 546
   clearing, 548
Solid axle, 12, 210, 223
   coil spring suspension, 218
   front suspension, 202
   leaf spring suspension, 218
   RWD suspension, 216–218
Solid rotors, 72
Solid suspension, 10
Solid-state electronics, 305
Spacers, spring coil, 239
Spare tires, 170–171
Specialty Equipment Manufacturers' Association (SEMA), aftermarket wheels, 182
Specifications, camber, 766–767
Speed nuts, 388
Speed ratings of tires, 167–168
Speed reset, 262
Speed sensor, electronically controlled, 139
Speed sensor mechanical problems on ABS, 552–555
Speed-dependent steering, 288
Speed-sensitive steering, 288
Spider, wheel, 177
Spin balancing, wheel, 634–635
   on car with limited-slip differential, 636
Spin tire balancer, 583
Spindle, 1, 10, 14, 398
   inspecting, 368, 370
Spindle runout gauge, 370
Spinner, 635
Split boot, installing, 645
Split CV joint boots, 644–645
Split point, 112
Split rims, 178
Spongy brake pedal, diagnosing, 506–508
Spool valve, 130
Spread, camber, 323
Spreaders, 686
Spring, 1, 10, 12, 19, 229–248
   S–L A, 205
Spring, preparing for installation:
   using spring clips, 667
   using spring compressor, 667
Spring center bolt, removing and replacing, 669
Spring clip, 645
Spring frequency, 231
Spring inspection, 411–412, 586–591
Spring materials, 233–235
Spring oscillations, 19
Spring rate, 230–231
Spring service, 660–682

Index

completing, 682
Spring types, 19–22
    air suspension, 20
    coil, 19–20
    leaf, 19
    torsion bars, 20
Spring washer, in parking brake lever, 80
Sprung weight, 230
Square coil springs, 237
Square-cut o-ring, 101–102
Squeal by brakes, 64
    and pad, 69
Squirm, tread, 160
Stabilizer bar, 246–248, 264
Stabilizer bar service, 679–682
    remove and replace center bushings, 679–682
    remove and replace end links and bushings, 679
Stacked construction of plies, 161
Stages of operation of shock absorber, 255
Standard brake lining, 416
Standard brake shoe grind, 417
Standard load tires, 166
Standard power steering gears (integral), 288
Standard steering gear, 276–278, 291
    assembling, 732–733
    disassembling, 731–732
    removal and replacement, 725–727
    tie rods and, 296
Standard steering gears with parallelogram linkage, 1
    checking, 604
Standard steering linkage, adjusting toe on, 795–796
Standard steering system, 269
Standing wave phenomenon, 167
State tire balance problems, 583
Static and kinetic balancing, mechanical balancer, 636
Static balance of tire, 633
Static seal, 102, 190
Static wheel balancing, 636
Steel brake line, 463–464
Steel lines, inspecting, 353
Steel tubing, 105
Steel-belted radial tires, wear on, 575
Steering angle sensor, 305
Steering axis, 22, 202, 321
Steering axis angle (SAA), 332
Steering axis inclination (SAI), 202, 321, 332–333, 334
Steering axis offset, 181
Steering ball joint, 206

Steering column, 273–276
Steering column service, 735–739
Steering coupler, 726
Steering damper with rack-and-pinion steering gears, 279
Steering feel, 285
Steering gear, 1, 11, 22, 270
    standard, 276–278
Steering gear, standard:
    assembling, 732–733
    disassembling, 731–732
Steering gear adjustments:
    rack-and-pinion steering gear, 733
    standard steering gear, 728–731
Steering gear checks, 607–608
Steering gear control valve, 285–287
Steering gear overhaul:
    rack-and-pinion steering gear, 734–735
    standard steering gear, 731–733
Steering gear rebuilding, 725
Steering gear removal and replacement:
    rack-and-pinion steering, 727–728
    standard steering gear, 725–727
Steering gear service, 725–735
Steering knuckle, 14, 22
    ball joints and, 18
    bending tools for, 769
    bent, 768–769
    extended, 15
Steering linkage checks, 603–606
Steering linkage replacement, 712–725
Steering linkage, 22–23, 270, 291–297
    ball joints and, 18
Steering ratio, 269
Steering shaft, 22, 276
Steering system, 10–11, 22, 269–298
Steering system diagnosis and inspection, 563–611
    completing, 611
Steering system inspection, 585–586
    checklist for, 585–586
Steering system service, 711–752
    completing, 752
Steering torque, 285
Steering wheel, 11
    adjusting toe to center, 797
    installing, 739
    removing and replacing, 738–739
Steering wheel free play, 604
Steering wheel position and toe, 791–792
Steering wheel service, 735–739
Step bore master cylinder, 97, 98
Step bore wheel cylinder, 99
Stitching a tire patch, 632
Stone shield, 256

Stop Control System (SCS), 139
Stoplight switches, 113–114, 144
Stopping sequence, 44
Stowaway spare tires, 170, 171
Stress raiser, 238
Strikeout bumpers, 234
Strobe light balancer, 637
Stroking seal, 102
Strut, 10, 223
    advantages of, 209
    bent, 768–769
    dismantling, 671
    removing, 671
Strut assembly:
    installing, 675–676
    removing, 675
Strut body, bent, 602–603
Strut cartridge installation, 678–679
Strut checks, 602–603
Strut damper or insulator checks, 603
Strut piston rod, bent, 602–603
Strut rod, 16
Strut rod bind, 207
Strut rod bushing checks, 602
Strut rod bushings, removing and replacing, 703–704
Strut service, 671–679
    normal, 671
    special points, 674
Strut shock absorbers, 261
Strut spring:
    removing, 676
    replacing, 676–677
Strut suspension, 14, 206–210
    ball joints and, 18
Strut suspension IRS, 220–221
Strut wear factors, 210
Struts for adjusting camber, 773–775
Stub axle, 193
Stub racks, 757
Stud and lug bolt replacement, 620–622
Subframe, 11
Supplemental inflatable restraints (SIR), 271
Supplemental restraint system (SRS), 271–273
Support keys, 430–431
Support yoke, 733
Support, shoe, 53, 54
Suspension arms or axles, 12
Suspension bushings, 17
Suspension component service, 685–707
    competing, 707
Suspension diagnosis and inspection, 563–611
    completing, 611

Suspension frequency, 232
Suspension inspection, 585–586
   checklist for, 585–586
Suspension ride height check, 588–590
Suspension system, 10–11
Suspension travel stops, 234
Suspension types, miscellaneous, 212
Suspensions, electronically controlled air, 240–243
Swaging, 389, 621
Swept area, 39
Swing axle, 210–212, 222
Swing axle front suspension, 202
Swing axle suspension, 12–13
Swirl finish on rotor, 436
Switch, circuit, 305
Synthetic bushing material, 704
System diagnosis, electronically controlled, 313–315
System effectiveness switch, 110–111

Tactile motion, 564
Tactile sensor, 70
Tandem ABS system, 138
Tandem booster, 124
Tandem diaphragms, 127
Tandem hydraulic systems, 6, 7
Tandem master cylinder, 92
   dual, 93
   operation of, 94–95
Tandem split, 96
Tangential coil springs, 237
Taper, breaking a, 686–689
   using pickle fork, 686, 688
   using pressure, 688–689
   using separator tools, 686, 688
   using shock, 686, 687
Taper breakers, 688–689
Tapered coil springs, 237
Tapered roller bearing, 187, 189
Tapered shims:
   for adjusting camber, 776, 801
   for adjusting caster, 784
Telescoping rod and strut assembly, 57
Temperature grades of tires, 167
Temporary spare tires, 170, 171
Temporary-use spare tires, 170
Tension rod, 203
Tension-loaded ball joints, 206
Test light, 308
Teves unit, pump motor in, 544
Teves-designed modulator, 145, 148
Thickness variation in rotor, 73–74, 433
   excessive, 432
Third arm, 297

Thread chaser, 620
Threaded ball joint, removing and replacing, 695
Three-circuit brake, 139
Thrust, 337
   in rear-wheel alignment, 337
Thrust line, 216, 321, 755
Tie-rod, 11, 22, 23
Tie-rod bushing, inner, removing and replacing, 720–721
Tie-rod end, breaking loose tapers on, 686–689
Tie-rod end checks, 604–605
Tie-rod end sockets, 716
Tie-rod wrench, 794
Tie-rods, rack-and-pinion steering gear, 296
Tie-rods, standard steering gears, 296
Tight ball joint, 592
Tinnerman nuts, 388
Tire and wheel, removing and replacing, 618–620
Tire balance, 582–583
Tire balancing, 582
Tire balancing operations, 633–641
Tire construction, 158–163
Tire flat spotting, 579
Tire friction, 37–38
Tire leak:
   locating, 630
   repairing, 629–632
Tire load ratings, 165–166
Tire lockup, 2
Tire matching, 581
Tire noise, 564
Tire on wheel, removing and replacing, 624–627
Tire position angles, 321
Tire pressure, 573–575
   checking, 575
   wear and, 573–574
Tire pressure monitor systems, 172
Tire problem detector (TPD), 580
Tire pull, 577–578
Tire quality grading, 166–167
Tire rotation, 623–624
Tire runout, 578–582
Tire scribe, 787
Tire scrub, 204
Tire selection, replacement, 170
Tire sidewall information, 163
Tire siping, 633
Tire sizing, 163–164
Tire, skidding of, 37, 38
Tire stiffness variation, 580, 581
Tire theory, 157–174
Tire truing, 632

Tire wear:
   camber-caused, 325
   inspection, 571–576
   toe-caused, 327–328
Tire wearing angles, adjusting, 755
Tire width, 163
Tire-related problem diagnosis, 577
Toe, 321, 326, 755
   factors that affect, 327
   in rear wheel alignment, 337
   tire wear caused by, 327–328
Toe, adjusting, 792–797
   rack-and-pinion steering, 796–797
   rear-wheel, 800–801
   standard steering, 795–796
   to center steering wheel, 797
Toe, measuring, 787–791
   using a trammel bar, 788–790
   using alignment system, 788
   using light gauge, 790–791
   with light beam, 787
Toe, measuring rear-wheel, 798–799
Toe and steering wheel position, 791–792
Toe angles, 326–327
Toe change, 328
Toe specifications, 792
Toe-in, 10, 202, 326–327
Toe-out, 10, 202, 326
Toe-out on turns, 202, 321, 334–335
   adjusting, 755, 786–787
   measuring, 786
   problems with, 336
Toed-in tire, 576
Toed-out tire, 576
Tone wheel, 140
Tone/reluctor ring on CV joint, 655
Toothed ring, 140
Toothed rotor, 140
Torque arm, 217
Torque buttons, 69
Torque converter clutch operation, 511
Torque steer, 199, 337, 577–578
Torque tube driveshaft, 217
Torque wrench, 464
Torsilastic bushing, 17, 696
Torsion bar, 20, 205, 239–240
   adjusting set of, 670
   removing, 670–671
   replacing, 240, 671
   service, 670–671
Track, 321, 337
   change, 13
   in rear-wheel alignment, 337
   of tires, 10
Track bar, 218
Track gauges, 799

Index

Traction, 38, 157
Traction control system (TCS), 154
Traction grades of tires, 167
Trailing arm, 223
Trailing arm IRS, 220
Trailing arm suspension, 16, 212
Trailing brake shoe, 3, 49
Trailing strut rod, 203
Trammel bar, 787, 788
    measuring toe with, 788–790
Tramp, 210, 217, 332
Transaxle, 197
Transducer, 637
Transverse mounted engines, 197
Transverse mounted leaf spring, 234
Tread, 161–163
    repairing hole in, 630–632
Tread depth check, 571
Tread depth gauge, 572
Tread plies, 161
Tread squirm, 160
Tread wear indicators, 572
Trim height, 588
Tripod joint, 198
    assembling, 653–654
    disassembling, 653
Tripode joint, 198
Tripot joint, 198
Trouble code in EBCM, 546
    clearing, 548
Trouble trees, 313
Trouble trees for NVH, 569–570
Troubleshooting, 802
Troubleshooting brake problems, 354–359
Truing, brake drum, 393
Truing, tire, 632
Truing rotor, 432
Tube nut, 463–464
Tube tires, 168–170
Tubeless tire, 168–170
Tubeless tire valve, removing and replacing, 628–629
Tubeless tire valve stems, 169–170
Tubing, 105–108
    replacement 465–466
    working with, 463–466
Tubing bender, 465
Tubing wrench, 463
Tulip joint, 198
Turning, break drum, 393
Turning angle, 321
Turning radius, 321
Turning rotor, 432
Turning torque, checking, 599
Turntables for magnetic gauges, 761
Twin I-beam axles, 210–212

Two-leading shoe, 50
Two-plane tire balancer, 583
Two-trailing shoe, 50
Two-wheel centerline alignment, 761
Two-wheel thrust line alignment, 761

Under tread, 159
Under-car inspection for NVH, 570
Unequal arm suspension, 13, 203
Uni-Fit boot, installing, 650
Unibody construction, 11
Uniform Tire Quality Grading Labeling (UTQGL), 166–167
Uniservo brake, 50
Universal joint, 13, 726
    steering column, 275
Unloaded runout, 580
Unsprung weight, 230
Upper bushing and seal in shock absorber, 265
Upper strut damper, 207
Upper strut mount, 207
Upset, 621
Upsetting, 389

Vacuum, 123
Vacuum bleeding of brakes, 503
Vacuum boost, 517, 520
Vacuum booster, 7
Vacuum booster construction, 125–127
Vacuum booster in ABS, 542
Vacuum booster operation, 128–129
Vacuum boosters, 122
    reconnecting hose on, 523
    repairing, 524–525
    theory, 123–125
Vacuum chamber, leaky, 520
Vacuum check valve, 518–519
Vacuum pump, 520
Vacuum storage chamber, 124
Vacuum supply tests, 518–520
Vacuum suspended booster, 124
Valve stem, 169, 170
Valve stem replacement, 628–629
Valves in shock absorbers, 265
Vane pump, 283
Variable assist, 282
Variable rate springs, 231
Variable ratio steering gear, 269
Variable valving in shock absorbers, 261
Variable-assist power steering (VAPS), 288–289
Variable-effort steering, 288

Variable-sized orifice in shock absorbers, 261
Vectoring, 581
Vehicle speed sensor (VSS), 289, 305
Vented rotors, 72
Vernier calipers, 595
Vibration, 564–569
    amplitude of, 564
    check for, 586
    curing tire-caused, 578
    frequency of, 564
    major causes of, 570–571
    order of, 567
    tire and wheel problems and, 577
    tire-caused, 578
Vibration analyzer, 567
Vibration band, 398
Viscosity of brake fluid, 115
Visual inspection for noise, 590–591
Visual inspection of shock absorbers, 590–591
Visual inspection of springs, 590–591
Visual sensors, 70
Volt-ohmmeter, 308
Voltage drop, 311
Volts, 303
Vulcanizing rubber, 161, 463

Waddle, radial tire, 578
Wander, 331
Warning lamp, airbag, 273
Warning light, parking brake, 79
Warning light, testing, 511
Warning light on ABS:
    operation, 544
    problems in electrical circuits or components, 546
    self-diagnosis of, 546
    sequence test, 544–546
    trouble codes for, 546
Warning light service, parking brake, 538
Warning light switch, 110–111
Water tolerance of brake fluid, 115
Watts, 303
Watt's link, 218
Wear bars, 572
Wear diameter of drum, 394
Wear indicator ball joints, checking, 593–594
Wearing sockets, 723
Wedge wear of rotor, 434
Weight transfer in brakes, 41–43
Wheel:
    construction of, 177–179
    parts of, 177

Wheel alignment, 321–337
  measuring and adjusting, 755–802
Wheel alignment sequence, 761
Wheel and tire, removing and replacing, 618–620
Wheel attachment, 182–185
Wheel balancing, 582
Wheel bearing, 187
  cleaning and inspecting, 368–370
  disassembling, 366–367
  faulty, 583
  inspecting, 347
  loose, 583
  noise, 564
  packing and adjusting, 371–374
  problems, diagnostic procedure for, 583–584
  repairing nonserviceable, 374–375
  repairing nonserviceable FWD front, 382–384
  rough, 584
  service, 365–384
  theory, 187–194
Wheel cylinder:
  basics, 99
  construction, 99–101
  inspecting, 349
  rebuilding, 477, 480–482
  reconditioning, 479–482
  removing, 477–479
  replacement of, 482–483
  service, 477–482
Wheel cylinder cup, 100
Wheel cylinder pistons, 100
Wheel dimensions, 179
Wheel frequency, 232
Wheel hop, 218
Wheel lockup, 137
Wheel lug bolt and nut, 182–183
Wheel offset, 180
Wheel rate, 232
Wheel rpm, calculating, 571
Wheel runout, 578–582
Wheel size, 163, 179
Wheel speed sensors, 140–142, 144
Wheel spin, 154–155
Wheel theory, 177–185
Wheel weights, 634
Wheel-less alignment system, 757
Wheelbase, 10
Wheelie, 217
Wheels, 1
  aftermarket, 181–182
Windup, 210, 217
Wiping lip, 190
Wires, circuit, 305
Wiring, 307
Wiring diagram, 311–312
Wishbone control arms, 203
Wishbone-strut suspension, 15, 206
Wishbones, 17
Work, 31
Work hardening in springs, 238
Working piston, 259
Worm and roller (Gemmer) steering gear, 276
Worm and sector steering gear, 276
Worm bearing preload, 728
Worm gear, 276
Worm shaft, 276
Worm shaft bearing preload, 276, 725, 730
  adjusting, 730–731

Yaw, 210, 217
Yoke, 68

Zero camber, 10–11, 321, 325
Zero pressure (ZP) tires, 172
Zero toe, 326